PROCEEDINGS OF THE
SIXTH BERKELEY SYMPOSIUM

Volume I

PROCEEDINGS *of the* SIXTH BERKELEY SYMPOSIUM ON MATHEMATICAL STATISTICS AND PROBABILITY

Held at the Statistical Laboratory
University of California
June 21–July 18, 1970

with the support of
University of California
National Science Foundation
Air Force Office of Scientific Research
Army Research Office
Office of Naval Research

VOLUME I

THEORY OF STATISTICS

EDITED BY LUCIEN M. LE CAM,
JERZY NEYMAN, AND ELIZABETH L. SCOTT

UNIVERSITY OF CALIFORNIA PRESS
BERKELEY AND LOS ANGELES
1972

UNIVERSITY OF CALIFORNIA PRESS
BERKELEY AND LOS ANGELES
CALIFORNIA

CAMBRIDGE UNIVERSITY PRESS
LONDON, ENGLAND

ISBN: 0-520-01964-4

LIBRARY OF CONGRESS CATALOG CARD NUMBER: 49-8189

PRINTED IN THE UNITED STATES OF AMERICA

CONTENTS OF PROCEEDINGS
VOLUMES I, II, AND III

Volume I—Theory of Statistics

General Theory

Sequential Analysis

Asymptotic Theory

Nonparametric Procedures

Regression Analysis

Central Limit Theorem

Volume III — Probability Theory

Passage Problems

Markov Processes — Potential Theory

Markov Processes — Trajectories — Functionals

Point Processes, Branching Processes

Information and Control

PREFACE

BERKELEY SYMPOSIA ON MATHEMATICAL STATISTICS AND PROBABILITY have been held at five year intervals since 1945. The Sixth Berkeley Symposium was divided into four sessions. The first took place from June 21 to July 18, 1970. It covered mostly topics in statistical theory and in theoretical and applied probability. The second session was held from April 9 to April 12, 1971 on the special subject of evolution with emphasis on studies of evolution conducted at the molecular level. The third session held in June 1971 was devoted to problems of biology and health. A fourth session on pollution was held in July 1971.

The first three volumes of the Proceedings cover papers presented in June and July, 1970, as well as papers which were sent to us at that time, but could not be presented in person by their authors. The first volume is entirely devoted to statistics. The second and third are devoted to contributions in probability. Allocation of the papers to the three volumes was made in a manner which we hope is fairly rational, but with an unavoidable amount of arbitrariness and randomness. In the event of doubt, a general index should help the prospective reader locate the desired contribution.

The Berkeley Symposia differ substantially from most other scientific meetings in that they are intended to provide an extended period of contact between participants from all countries in the world. In addition, an effort is made to promote cross contacts between scholars whose fields of specialization cover a broad spectrum from pure probability to applied statistics. However, these fields have expanded so rapidly in the past decades that it is no longer possible to touch upon every domain in a few weeks only. Since time limits the number of invited lectures, the selection of speakers is becoming rapidly an impossible task. We could only sample the abundance of available talent. For this selection, as well as for several other important matters, we were privileged to have the assistance of an advisory committee consisting of Professors Z. W. Birnbaum and L. Schmetterer, representatives of the Institute of Mathematical Statistics, and of Professor Steven Orey, delegate of the American Mathematical Society. The visible success of our gathering is in no small measure attributable to the help we have received from this committee and other scientific friends.

A conference which extends over six weeks with participants from various parts of the world entails expenses. In this respect we feel fortunate that in spite of the general shortage of funds, the University of California and the Federal Agencies found it possible to support our enterprise. We are grateful for the allocation of funds from the Russell S. Springer Memorial Foundation, the National Science Foundation, the Office of Naval Research, the Army Research Office, the Air Force Office of Scientific Research, and the National Institutes of Health, which contributed particularly to the sessions on evolution and on problems of biology and health. In addition the pollution session received support from the Atomic Energy Commission.

The organization of the meetings fell under the responsibility of the undersigned with the able help of the staff of the Statistical Laboratory and of the Department of Statistics. For assistance with travel arrangements and various organizational matters, special thanks are due to Mrs. Barbara Gaugl.

The end of the actual meeting signals the end of a very exciting period, but not the end of our task, since the editing and publishing of over 3,000 pages of typewritten material still requires an expenditure of time and effort.

In this respect we are indebted to Dr. Morris Friedman for translations of Russian manuscripts. We are particularly grateful to Dr. Amiel Feinstein and Mrs. Margaret Stein who not only translated such manuscripts but acted as editors, checking the references and even verifying the accuracy of mathematical results.

The actual editing and marking of manuscripts was not easy since we attempted to follow a uniform style. We benefitted from the talent and skill of Mrs. Virginia Thompson who also assumed responsibility for organizing and supervising the assistant editors, Miss Carol Conti, Mrs. Margaret Darland, and Miss Jean Kettler. We are extremely grateful to all the editors for the knowledge and patience they have devoted to these manuscripts.

In the actual publication of the material the University of California Press maintained their tradition of excellence. The typesetting was performed by the staff of Oliver Burridge Filmsetting Ltd., in Sussex, England.

The meetings of the Sixth Symposium were saddened by the absence of two of our long time friends and regular participants, William Feller and Alfréd Rényi. Professors J. L. Doob and Mark Kac were kind enough to write a short appreciation of Feller. For a similar appreciation of Rényi, we are indebted to Professor L. Schmetterer. The texts appear at the beginning of the second volume.

L.LC.   J.N.   E.L.S.

CONTENTS

xi

Multivariate Analysis

UPPER AND LOWER RISKS
AND MINIMAX PROCEDURES

R. J. BERAN

University of California, Berkeley

The essential goal of R. A. Fisher's fiducial argument was to make posterior inferences about unknown parameters without resorting to a prior distribution. Over the past decade, there have been two major attempts at developing a statistical theory that would accomplish this convincingly. One of these efforts has been described in a series of publications by Fraser, the other in papers by Dempster. From the early work [4], [11], [12], [13], which was tied to a fiducial viewpoint, both authors developed statistical theories that were distinct from the fiducial argument, yet achieved the goal of non-Bayesian posterior inference [5], [6], [7], [8], [14], [15], [16].

Despite technical and other differences, the main ideas underlying this later work by Dempster and by Fraser appear to be similar. Fraser's papers, analyzing statistical models that possess a special kind of structure, arrive at "structural probability" distributions for the unknown parameters. Dempster's papers, dealing with less specialized models, derive "upper and lower probabilities" on the parameter space. Disregarding some technicalities, these upper and lower probabilities reduce to structural probabilities for the models considered by Fraser.

To this extent, upper and lower probabilities are a generalization of structural probabilities. However, there appear to be differences in interpretation. Fraser has given a frequency interpretation to structural probabilities in [11], [12] (but not in later work); this interpretation depends upon the special form of the statistical models in his theory, and does not apply to Dempster's theory. Dempster has provided no simple interpretation for upper and lower probabilities; he suggested in [7] that his theory might be "an acceptable idealization of intuitive inferential 'appreciations'." More recently, he has embedded his theory within a generalized Bayesian framework [9], [10]. The justification for the latter is unclear at present (see the discussion to [9]).

Lacking in both the Dempster and Fraser theories are systematic methods for dealing with estimation and hypothesis testing problems (or suitable analogues of such). A method of constructing tests was described by Fraser in [16], but no performance criteria were established. Dempster [5] defined upper and lower risks but did not pursue their application; the statistical meaning of these risks

This research was partially supported by National Science Foundation Grant GP–15283.

1

is not evident under his interpretation of upper and lower probabilities. Since even simple models suggest a variety of natural estimates and tests, some theory seems necessary as a guide to choice of procedure.

The results presented in this paper proceed in several directions. A statistical interpretation for upper and lower probabilities and risks is described in Section 2; this rationale leads naturally to a minimax criterion for statistical procedures and, in principle, to an alternative to standard decision theory. The desirability of such an alternative stems from well-known awkward features of standard decision theory, such as the possibility that a test of low size and high power may make a decision which is contradicted by the data (see Hacking [17]). A heuristic account of these ideas in a less general context has previously been given by the author in [1].

Section 3 of the paper develops basic mathematical properties of upper and lower probabilities and risks in the light of Choquet's [3] theory of capacities. The results include extensions of properties given by Dempster in [6].

In Section 4, convenient conditions are established for the existence of minimax procedures (as defined in Section 2). An example in a nonparametric setting follows.

2. Statistical background

An experiment is performed, resulting in observation x. It is known that the observed x was generated from a parameter value t and a realized random variable e by the mapping

$$(2.1) \qquad x = \xi(e, t).$$

Moreover, t lies in a parameter space T, x lies in an observation space X, and e is realized according to a probability measure P on an elementary space E. Both P and the mapping ξ are known. The problem is to draw inferences concerning t from x and the model.

The following formal assumptions are made: X is a Borel subset of a metric space and is endowed with the σ-algebra $\mathscr{X}$ of all Borel sets. T and E are complete separable metric spaces, endowed with σ-algebras $\mathscr{T}$ and $\mathscr{E}$, respectively. $\mathscr{T}$ consists of all Borel sets in T. P is defined on the Borel sets in E and $\mathscr{E}$ is the completion with respect to P of the σ-algebra of these Borel sets; thus $\mathscr{E}$ contains all analytic sets. The function $\xi: E \times T \to X$ is Borel measurable.

Formally, performing the experiment described above amounts to realizing, through physical operations, a specific triple $(x, t, e) \in X \times T \times E$. Before the experiment is carried out (or the outcome x is noted), the following prospective assertions can be made about the triple to be realized: the chance that $e \in B$, $B \in \mathscr{E}$, is $P(B)$; t is an unspecified element of T; the observable x is related to t and e through (2.1).

Once the experiment has been performed and x has been observed, the particular triple (x, t, e) that was realized can be described more precisely. If

$$(2.2) \qquad T_x(e) = \{t \in T : x = \xi(e, t)\},$$

it is evident that the e realized in the experiment must lie in

$$(2.3) \qquad\qquad E_x = \{e \in E : T_x(e) \neq \varnothing\},$$

and whatever that e is, the realized t must belong to the corresponding $T_x(e)$.

Since $E_x = \text{proj}_E[\xi^{-1}(x)]$, E_x is analytic under the assumptions and so lies in $\mathscr{E}$. Let $P[B|E_x]$ denote the conditional probability defined by

$$(2.4) \qquad\qquad P[B|E_x] = \frac{P[B \cap E_x]}{P[E_x]}, \qquad\qquad B \in \mathscr{E},$$

provided $P[E_x] > 0$. If $P[E_x] = 0$, it may still be possible to condition upon a suitable statistic. In any event, a modification of ξ so as to include round off error incurred in observing x will generally result in $P[E_x] > 0$.

Thus, after the experiment has been performed and x has been observed, the following prospective statements can be made about the realized triple (x, t, e): x is as observed; $e \in X_x$ and the chance that $e \in B$, $B \in \mathscr{E}$, is $P[B|E_x]$; whatever e is, t is an unspecified element of the corresponding set $T_x(e)$; relation (2.1) is necessarily satisfied. This collection of assertions about the triple (x, t, e) will be called the *posterior model* $\mathscr{M}_x$ for the experiment. Both Dempster and Fraser have previously considered reductions of this type, though not in terms of experimental triples.

Since the realized experiment (x, t, e) is described more precisely by the posterior model $\mathscr{M}_x$ than by the original model, it is proposed to evaluate statistical procedures of interest by their average behavior over a hypothetical sequence of independent experiments, each of which is generated under the assumptions of $\mathscr{M}_x$. The aim is to measure how well a statistical procedure performs when applied to hypothetical experimental triples that are as similar as can be arranged to the actual triple (x, t, e).

Let D denote a space of decisions and let $\ell : T \times D \to R^+$ be a nonnegative loss function. Let $\mathscr{R}^+$ denote the σ-algebra of all Borel sets in R^+, and assume that for every $d \in D$, $\ell(\cdot, d)$ is a measurable mapping of $(T, \mathscr{T})$ into $(R^+, \mathscr{R}^+)$. Suppose $d \in D$ is a specific decision whose consequences are to be evaluated relative to the posterior model $\mathscr{M}_x$ under the loss function ℓ.

Let $\{(x, t_i, e_i), i = 1, 2, \cdots\}$ be a sequence of independent hypothetical experiments generated under the posterior model; in other words, $e_1, e_2, \cdots$ are independent random variables, each distributed according to $P[\cdot|E_x]$, t_i is selected arbitrarily from $T_x(e_i)$, x is the observed data. For each i, the equation $x = \xi(e_i, t_i)$ will necessarily be satisfied.

Let the general notation $\text{prop}_n(\pi_i)$ denote the proportion of true propositions among the propositions $\{\pi_i, \pi_2, \cdots, \pi_n\}$. The average loss incurred over the first n hypothetical experiment as a result of taking decision d is $n^{-1} \sum_{i=1}^n \ell(t_i, d)$. Since $\ell \geqq 0$,

$$(2.5) \qquad \frac{1}{n} \sum_{i=1}^n \ell(t_i, d) = \int_0^\infty \text{prop}_n[\ell(t_i, d) > z] \, dz.$$

If $A(z, d) = \{t \in T : \ell(t, d) > z\}$, then $A(z, d) \in \mathcal{T}$ for every $z \in R^+$ and $d \in D$, and $\{\ell(t_i, d) > z\} = \{t_i \in A(z, d)\}$. Therefore,

$$(2.6) \qquad \int_0^\infty \text{prop}_n[T_x(e_i) \subset A(z, d), T_x(e_i) \neq \varnothing] \, dz.$$

$$\leqq \frac{1}{n} \sum_{i=1}^n \ell(t_i, d)$$

$$\leqq \int_0^\infty \text{prop}_n[T_x(e_i) \cap A(z, d) \neq \varnothing] \, dz.$$

Now,

$$(2.7) \qquad \int_0^\infty \text{prop}_n[T_x(e_i) \cap A(z, d) \neq \varnothing] \, dz = \frac{1}{n} \sum_{i=1}^n \int_0^\infty I_{C_i}(z) \, dz,$$

where $C_i = \{z \in R^+ : T_x(e_i) \cap A(z, d) \neq \varnothing\}$ and $I_{C_i}(z)$ is the indicator of C_i. Moreover,

$$(2.8) \qquad \int_0^\infty I_{C_i}(z) \, dz = \sup_{t \in T_x(e_i)} \ell(t, d),$$

and for every $d \in D$, the function on the right of (2.8) is a measurable mapping of $(E, \mathcal{E})$ into $(R^+, \mathcal{R}^+)$.

By Fubini's theorem,

$$(2.9) \qquad E \int_0^\infty I_{C_i}(z) \, dz = \int_0^\infty v_x[A(z, d)] \, dz,$$

where for $A \in \mathcal{T}$,

$$(2.10) \qquad v_x(A) = P[e : T_x(e) \cap A \neq \varnothing | E_x]$$

and the expectation is with respect to $P[\cdot | E_x]$.

Since $\{e : T_x(e) \cap A \neq \phi\} = \text{proj}_E[\xi^{-1}(x) \cap E \times A]$, this set is analytic, belongs to $\mathcal{E}$, and therefore $v_x(A)$ is defined. The strong law of large numbers, applied to (2.7), shows that as $n \to \infty$, the upper bound in (2.6) converges with probability one to

$$(2.11) \qquad s_x(\ell, d) = \int_0^\infty v_x[A(z, d)] \, dz.$$

A dual argument shows that the lower bound in (2.6) converges with probability one, as $n \to \infty$, to

$$(2.12) \qquad r_x(\ell, d) = \int_0^\infty u_x[A(z, d)] \, dz,$$

where for $A \in \mathcal{T}$,

$$(2.13) \qquad u_x(A) = P[e : T_x(e) \subset A, T_x(e) \neq \varnothing | E_x].$$

Thus, the lower risk $r_x(\ell, d)$ and the upper risk $s_x(\ell, d)$ measure the smallest and largest long run average loss that could be incurred as a consequence of decision d. The evaluation is made under the posterior model $\mathcal{M}_x$. The relative desirability of various decisions $d \in D$ may be assessed by reference to the corresponding risks. More generally, a decision procedure $\delta: X \to D$ may be compared with other decision procedures by studying the risks as functions on X.

For $\ell \geq 0, s_x(\ell, d)$ and $r_x(\ell, d)$ are equivalent to the upper and lower expectations defined by Dempster in [5], [6]; v_x and u_x are the corresponding upper and lower probabilities defined on $(T, \mathcal{T})$. A frequency interpretation for u_x, v_x is obtained by specializing ℓ in the foregoing; see [1] for the result.

The frequency interpretation for r_x and s_x suggests the following simple optimality criterion.

DEFINITION 2.1. *A decision $d \in D$ is minimax under loss function ℓ and observations x if $s_x(\ell, d) \leq s_x(\ell, d')$ for every $d' \in D$.*

This definition differs slightly from an earlier one given in [1]. An extension of the definition to decision procedures is

DEFINITION 2.2. *A decision procedure $\delta: X \to D$ is minimax under loss function ℓ if $s_x(\ell, \delta(x)) \leq s_x(\ell, \delta'(x))$ for every $x \in X$ and every $\delta': X \to D$.*

Finding a minimax decision procedure amounts to finding a minimax decision for each $x \in X$. The existence of minimax decisions is discussed in Section 4.

3. Formal properties

Several basic theorems about u_x, v_x, r_x, s_x are proved in this section. Some of the results have been obtained for finite T by Dempster [6]. Further related results, in different contexts, may also be found in Choquet [3], Huber [18], and Strassen [19]. For notational convenience, the subscript x is dropped throughout the rest of this paper.

Let ϕ be a real valued set function defined on $\mathcal{T}$. For $B, A_1, A_2, \cdots, A_p$ in $\mathcal{T}$, let

$$(3.1) \qquad \Delta_p = \phi(B) - \sum \phi(B \cup A_i) + \sum \phi(B \cup A_i \cup A_j)$$
$$- \cdots + (-1)^p \phi(B \cup A_1 \cup \cdots \cup A_p),$$

and let

$$(3.2) \qquad \nabla_p = \phi(B) - \sum \phi(B \cap A_i) + \sum \phi(B \cap A_i \cap A_j)$$
$$- \cdots + (-1)^p \phi(B \cap A_1 \cap \cdots \cap A_p).$$

The sums in (3.1) and (3.2) are taken over all possible distinct combinations of indices, excluding combinations that repeat indices. Following Choquet [3], we say that ϕ is alternating of order p if $\Delta_p \leq 0$ for arbitrary $B, A_1, \cdots, A_p \in \mathcal{T}$ and is monotone of order p if $\nabla_p \geq 0$ for arbitrary $B, A_1, \cdots, A_p \in \mathcal{T}$.

PROPOSITION 3.1. *The set function v is alternating of all orders. The set function u is monotone of all orders.*

PROOF (essentially due to Choquet). The probability $P[\cdot|E_x]$ is monotone and alternating of all orders. Now

$$(3.3) \qquad\qquad v(A) = P[\psi(A)|E_x], \qquad\qquad A \in \mathscr{T},$$

where $\psi(A) = \text{proj}_E[\xi^{-1}(x) \cap E \times A]$. If $A_1, A_2 \in \mathscr{T}$,

$$(3.4) \qquad\qquad \psi(A_1 \cup A_2) = \psi(A_1) \cup \psi(A_2),$$

therefore v is alternating of all orders. The complete monotonicity of u then follows by property (c) of Proposition 3.2.

PROPOSITION 3.2. *The set functions u and v defined on $\mathscr{T}$ have the following properties:*

    (a) $u(\varnothing) = v(\varnothing) = 0$;

    (b) $u(T) = v(T) = 1$;

    (c) $u(A) + v(\mathscr{C}A) = 1$;

    (d) $u(A) \leqq v(A)$;

    (e) *if $A \subset B$, $u(A) \leqq u(B)$ and $v(A) \leqq v(B)$;*

    (f) $u(A \cup B) + u(A \cap B) \geqq u(A) + u(B)$,
         $v(A \cup B) + v(A \cap B) \leqq v(A) + v(B)$;

    (g) *if $A_n \downarrow A$, $u(A_n) \downarrow u(A)$, while if $A_n \uparrow A$, $v(A_n) \uparrow v(A)$.*

PROOF. Properties (a), (b), (c), (d) are immediate from the definitions of u and v. Property (e) is equivalent to $\nabla_1 \geqq 0$ for u and $\Delta_1 \leqq 0$ for v, while property (f) is implied by $\nabla_2 \geqq 0$ for u and $\Delta_2 \leqq 0$ for v. These inequalities were established in Proposition 3.1. Finally, from (3.3), $v(A_n) = P[\psi(A_n)|E_x]$. If $A_n \uparrow A$,

$$(3.5) \qquad\qquad \psi(A_n) \uparrow \bigcup_1^\infty \psi(A_n) = \psi\left(\bigcup_1^\infty A_n\right) = \psi(A).$$

This implies the second half of (g). The first half now follows from (c).

REMARK 3.1. The counterpart of (g) with u and v interchanged does not hold in general.

REMARK 3.2 Properties (c), (d) and the first property in each of (a), (b), (e), (f), (g) imply the remaining properties. All further propositions proved in this section are consequences of Proposition 3.2 alone.

PROPOSITION 3.3. *The following inequalities hold on $\mathscr{T}$. If $A \cap B = \phi$, then*

    (a) $u(A) + u(B) \leqq u(A \cup B) \leqq u(A) + v(B)$,

    (b) $u(A) + v(B) \leqq v(A \cup B) \leqq v(A) + v(B)$.

PROOF. The lower bound in (a) and the upper bound in (b) follow from (f) of Proposition 3.2. Since $A \cap B = \varnothing$, $B \subset \mathscr{C}A$. Therefore,

$$(3.6) \qquad v(B) + v(\mathscr{C}A \cap \mathscr{C}B) = v(\mathscr{C}A \cap B) + v(\mathscr{C}A \cap \mathscr{C}B) \geqq v(\mathscr{C}A),$$

which is equivalent to

$$(3.7) \qquad v(B) + 1 - u(A \cup B) \geqq 1 - u(A).$$

The upper bound in (a) holds, consequently. A dual argument establishes the lower bound in (b).

REMARK 3.3. The upper bounds in (a) and (b) are valid without the condition $A \cap B = \emptyset$.

PROPOSITION 3.4. *The following inequalities hold on* $\mathscr{T}$:

(a) $u(A \cup B) + u(A \cap B) \leq u(A) + v(B)$,

(b) $v(A \cup B) + v(A \cap B) \geqq u(A) + v(B)$.

PROOF. Since $A \cup B = B \cup (A{-}B)$ and since $A = (A{-}B) \cup (A \cap B)$, Proposition 3.3 shows that

$$(3.8) \qquad v(A \cup B) \geqq u(A{-}B) + v(B)$$

and

$$(3.9) \qquad u(A) \leqq u(A{-}B) + v(A \cap B).$$

These two inequalities imply (b). Inequality (a) is proved by taking complements in (b).

PROPOSITION 3.5. *If* A, B, $C \in \mathscr{T}$ *and* $B \subset A$, *then*

(a) $v(A \cup C) - v(A) \leq v(B \cup C) - v(B)$,

(b) $u(A \cap C) - u(A) \leq u(B \cap C) - u(B)$.

PROOF (Choquet). If $X = B \cup C$ and $Y = A$, then $X \cup Y = A \cup C$ and $X \cap Y = B \cup (A \cap C) \supset B$. Therefore,

$$(3.10) \quad v(A \cup C) + v(B) \leq v(X \cup Y) + v(X \cap Y) \leq v(B \cup C) + v(A)$$

by Proposition 3.2. This establishes (a). Inequality (b) is derived by taking complements in (a).

PROPOSITION 3.6. *If the* $\{A_i\}$ *and* $\{B_i\}$ *belong to* $\mathscr{T}$, *and* $B_i \subset A_i$, *then*

(a) $v(\bigcup_1^\infty A_i) - v(\bigcup_1^\infty B_i) \leq \Sigma_1^\infty [v(A_i) - v(B_i)]$,

(b) $u(\bigcap_1^\infty A_i) - u(\bigcap_1^\infty B_i) \leq \Sigma_1^\infty [u(A_i) - u(B_i)]$.

PROOF. The result is established by Choquet [3] for finite unions and intersections (through induction and Proposition 3.5). Taking limits and using (g) of Proposition 3.2 completes the proof.

PROPOSITION 3.7. *For every sequence* $\{A_n\}$ *in* $\mathscr{T}$,

(a) $v(\lim_n \inf A_n) \leqq \lim_n \inf v(A_n)$,

(b) $u(\lim_n \sup A_n) \geqq \lim_n \sup u(A_n)$.

PROOF. Since $A_m \supset \inf_{n \geqq m} A_n$ and since $\inf_{n \geqq m} A_n \uparrow \lim_n \inf A_n$ as $m \to \infty$,

$$(3.11) \qquad \liminf_m v(A_m) \geq \lim_{m \to \infty} v(\inf_{n \geq m} A_n) = v(\liminf_n A_n),$$

using (e) and (g) of Proposition 3.2. A dual argument proves (b).

REMARK 3.4.   The roles of u and v cannot, in general, be interchanged in Proposition 3.7.

PROPOSITION 3.8.   *Let $\mathscr{S} = \{A \in \mathscr{T} : u(A) = v(A)\}$. Then $\mathscr{S}$ is a σ-algebra and $u = v$ is a probability measure on $(T, \mathscr{S})$.*

PROOF.   Clearly $\phi, T \in \mathscr{S}$. If $A \in \mathscr{S}$, then $\mathscr{C}A \in \mathscr{S}$; indeed $u(A) = v(A)$ implies $v(\mathscr{C}A) = u(\mathscr{C}A)$ by (c) of Proposition 3.2. If the $\{A_i\} \in \mathscr{S}$ and are disjoint, then $\bigcup_1^\infty A_i \in \mathscr{S}$; indeed by Proposition 3.3, for any integer $n > 1$,

$$(3.12) \qquad u\left(\bigcup_1^n A_i\right) \geq \sum_{i=1}^n u(A_i) = \sum_{i=1}^n v(A_i) \geq v\left(\bigcup_1^n A_i\right).$$

Therefore, $u(\bigcup_1^n A_i) = v(\bigcup_1^n A_i)$. Moreover, by Propositions 3.7 and 3.2,

$$(3.13) \qquad u\left(\bigcup_1^\infty A_i\right) = u(\lim_n \bigcup_1^n A_i) \geq \limsup_n u\left(\bigcup_1^n A_i\right)$$

$$= \limsup_n v\left(\bigcup_1^n A_i\right) = v\left(\bigcup_1^\infty A_i\right),$$

so that $u(\bigcup_1^\infty A_i) = v(\bigcup_1^\infty A_i)$. The fact that $u = v$ is a probability on $(T, \mathscr{S})$ follows from Propositions 3.2 and 3.3.

REMARK 3.5.   This theorem links upper and lower probabilities to structural probabilities. For Fraser's models, $\mathscr{S}$ contains all Borel sets. In general, however, $\mathscr{S}$ may be trivial.

Let $\mathscr{C}$ denote the vector lattice of measurable functions mapping $(T, \mathscr{T})$ into $(R, \mathscr{R})$, the real line endowed with the σ-algebra of all Borel sets. Let $\mathscr{C}^+ = \{f \in \mathscr{C} : f \geq 0\}$. The following definitions are abstracted from the upper and lower risks of Section 2.

DEFINITION 3.1.   *If $f \in \mathscr{C}^+$, the upper integral $s(f)$ and the lower integral $r(f)$ are defined as*

$$s(f) = \int_0^\infty v[f(t) > z]\, dz,$$

$$(3.14)$$

$$r(f) = \int_0^\infty u[f(t) > z]\, dz.$$

To extend the definitions to $f \in \mathscr{C}$, let $f^+ = f \vee 0$ and $f^- = -f \vee 0$, so that $f = f^+ - f^-$.

DEFINITION 3.2.   *If $f \in \mathscr{C}$, the upper integral $s(f)$ and the lower integral $r(f)$ are defined as*

$$s(f) = s(f^+) - r(f^-),$$

$$(3.15)$$

$$r(f) = r(f^+) - s(f^-),$$

excluding the indeterminate case $\infty - \infty$.

Definition 3.2 can also be motivated by a frequency interpreration.

PROPOSITION 3.9. *The following assertions hold for functions in $\mathscr{C}$:*

(a) *if $r(f)$ and $s(f)$ both exist, then $r(f) \leqq s(f)$;*

(b) $s(a + bf) = \begin{cases} a + bs(f) \text{ if } b \geqq 0 \text{ and } s(f) \text{ exists} \\ a + br(f) \text{ if } b \leqq 0 \text{ and } r(f) \text{ exists}; \end{cases}$

(c) $r(a + bf) = \begin{cases} a + br(f) \text{ if } b \geqq 0 \text{ and } r(f) \text{ exists} \\ a + bs(f) \text{ if } b \leqq 0 \text{ and } s(f) \text{ exists}; \end{cases}$

(d) *if $s(f), s(g)$ both exist and $f \leqq g$, then $s(f) \leqq s(g)$;*

(e) *if $r(f), r(g)$ both exist and $f \leqq g$, then $r(f) \leqq r(g)$;*

(f) *if the $\{s(f_n)\}$ all exist and $r(f_n^-) < \infty$ for at least one n, then $f_n \uparrow f$ implies that $s(f)$ exists and $s(f_n) \uparrow s(f)$;*

(g) *if the $\{r(f_n)\}$ all exist and $r(f_n^+) < \infty$ for at least one n, then $f_n \downarrow f$ implies that $r(f)$ exists and $r(f_n) \downarrow r(f)$.*

PROOF. Assertion (a) holds for $f \in \mathscr{C}^+$ by (d) of Proposition 3.2, and hence as stated. If $f \in \mathscr{C}^+$, $a \geqq 0$, and $b \geqq 0$, a change of variable in Definition 3.1 shows that

$$s(a + bf) = a + bs(f),$$

(3.16)

$$r(a + bf) = a + br(f).$$

Therefore, if $a \geqq 0, b \geqq 0, f \in \mathscr{C}$, and $s(f)$ exists,

(3.17) $$s(a + bf) = s(a + bf^+) - r(bf^-)$$
$$= a + bs(f^+) - br(f^-) = a + bs(f).$$

The other cases in assertions (b) and (c) are proved similarly.

For $f, g \in \mathscr{C}^+$, assertions (d) and (e) are immediate from (e) of Proposition 3.2. If $f, g \in \mathscr{C}, f \leqq g$, then $f^+ \leqq g^+, f^- \geqq g^-$, and assertions (d) and (e) follow as stated.

To prove (f) and (g), note that if the $\{f_n\} \in \mathscr{C}^+$ and $f_n \uparrow f \in \mathscr{C}^+$, then for any $z \in R^+$,

(3.18) $$\{f_n(t) > z\} \uparrow \bigcup_1^\infty \{f_n(t) > z\} = \{f(t) > z\},$$

consequently $s(f_n) \uparrow s(f)$ by (g) of Proposition 3.2. Similarly, if $\{f_n\} \in \mathscr{C}^+$ and $f_n \downarrow f \in \mathscr{C}^+$, then $r(f_n) \downarrow r(f)$. Now suppose the $\{f_n\} \in \mathscr{C}$ and $f_n \uparrow f$. By the foregoing, $s(f_n^+) \uparrow s(f^+)$ and $r(f_n^-) \downarrow r(f^-)$. Since $r(f_n^-) < \infty$ for at least one n, $r(f) < \infty$; therefore $s(f)$ exists and (f) follows. Assertion (g) is proved analogously.

REMARK 3.6. In general, the roles of r and s cannot be interchanged in (f) and (g) above.

PROPOSITION 3.10.   Let $f, g \in \mathscr{C}$.

(a) If either $s(f^+) + s(g^+) < \infty$ or $s(f^-) + r(g^-) < \infty$, then $r(f) + s(g)$ $\leq s(f \vee g) + s(f \wedge g) \leq s(f) + s(g)$.

(b) If either $s(f^-) + s(g^-) < \infty$ or $r(f^+) + s(g^+) < \infty$, then $r(f) + r(g)$ $\leq r(f \vee g) + r(f \wedge g) \leq r(f) + s(g)$.

PROOF.   Let $f, g \in \mathscr{C}^+$. Since for any $z \in R^+$,

$$(3.19) \quad \begin{aligned} \{f(t) \vee g(t) > z\} &= \{f(t) > z\} \cup \{g(t) > z\}, \\ \{f(t) \wedge g(t) > z\} &\doteq \{f(t) > z\} \cap \{g(t) > z\}, \end{aligned}$$

inequalities (a) and (b) follow from Propositions 3.2 and 3.4. If $f, g \in \mathscr{C}$, then

$$(3.20) \quad \begin{aligned} (f \vee g)^+ &= f^+ \vee g^+, \quad (f \wedge g)^+ = f^+ \wedge g^+, \\ (f \vee g)^- &= f^- \vee g^-, \quad (f \wedge g)^- = f^- \wedge g^-, \end{aligned}$$

and the proposition follows from the results on $\mathscr{C}^+$.

PROPOSITION 3.11.   Let $\{f_n\}$ be a sequence of functions in $\mathscr{C}$.

(a) If $g \in \mathscr{C}, r(g^-) < \infty$, and $f_n \geq g$ for all n, then $s(\lim_n \inf f_n) \leq \lim_n \inf s(f_n)$.

(b) If $g \in \mathscr{C}, r(g^+) < \infty$, and $f_n \leq g$ for all n, then $r(\lim_n \sup f_n) \geq \lim_n \sup r(f_n)$.

PROOF.   In (a), since $\inf_{m \geq n} f_m \geq g$, $r([\inf_{m \geq n} f_m]^-) \leq r(g^-) < \infty$ for all n, and therefore $s(\inf_{m \geq n} f_m)$ exists for all n. Similarly, $r(f_m^-) < \infty$ for all m, so that $s(f_m)$ exists for all m. By Proposition 3.9, $s(\lim_n \inf f_n)$ exists and as $n \to \infty$,

$$(3.21) \quad \inf_{m \geq n} s(f_m) \geq s(\inf_{m \geq n} f_m) \uparrow s(\lim_n \inf f_n),$$

which proves (a). A dual argument establishes (b).

Let $\{A_n : A_n \in \mathscr{T}, n = 0, \pm 1, \pm 2, \cdots\}$ be a countable partition of T, and let $\mathscr{A}$ denote the σ-algebra generated by this partition. If $B \in \mathscr{A}, B = \cup_I A_i$, where I is countable. Define a set function q on $\mathscr{A}$ as follows:

$$(3.22) \quad q(A_j) = v\left(\bigcup_{i=j}^{\infty} A_i\right) - v\left(\bigcup_{i=j+1}^{\infty} A_i\right), \qquad j = 0, \pm 1, \cdots.$$

More generally, if $B \in \mathscr{A}, B = \cup_I A_i$, define $q(B)$ by

$$(3.23) \quad q(B) = \sum_I q(A_i).$$

LEMMA 3.1.   If $\lim_{n \to \infty} v(\cup_{i=n}^{\infty} A_i) = 0$, then for every $B \in \mathscr{A}, u(B) \leq q(B) \leq v(B)$, and q is a probability measure on $\mathscr{A}$.

PROOF.   To verify that q is a probability on $\mathscr{A}$, note first that q is countably additive by definition. Since v is monotone, $q(A_j) \geq 0$ for all j and hence $q(B) \geq 0$ for $B \in \mathscr{A}$. Also

$$(3.24) \qquad q(T) = \sum_{i=-\infty}^{\infty} q(A_i) = \lim_{m, n \to \infty} \sum_{i=-m}^{n} q(A_i)$$

$$= \lim_{m, n \to \infty} \left[v\left(\bigcup_{i=-m}^{\infty} A_i \right) - v\left(\bigcup_{i=n+1}^{\infty} A_i \right) \right] = 1,$$

by Proposition 3.2 and the hypothesis of the lemma.

From Proposition 3.3, applied to (3.22),

$$(3.25) \qquad q(A_j) \leqq v(A_j), \qquad j = 0, \pm 1, \pm 2, \cdots.$$

Let $C = \cup_J A_j$, with J a finite set of natural numbers, and suppose that $q(C) \leqq v(C)$. If k is a natural number smaller than any element of J,

$$(3.26) \qquad q(A_k \cup C) = q(A_k) + q(C)$$

$$\leqq v\left(\bigcup_{i=k}^{\infty} A_i \right) - v\left(\bigcup_{i=k+1}^{\infty} A_i \right) + v(C)$$

$$\leqq v(A_k \cup C),$$

the last inequality coming from (f) of Proposition 3.2. Starting with (3.25) and applying (3.26) a finite number of times shows that for any finite J,

$$(3.27) \qquad q(\bigcup_J A_j) \leqq v(\bigcup_J A_j).$$

Taking limits establishes the inequality $q(B) \leqq v(B)$ for any $B \in \mathcal{A}$. Finally, if $B \in \mathcal{A}$, then $\mathcal{C}B \in \mathcal{A}$ and $q(\mathcal{C}B) \leqq v(\mathcal{C}B)$, hence by Proposition 3.2, $u(B) \leqq q(B)$.

A function $f \in \mathcal{C}$ is elementary if it can be represented in the form

$$(3.28) \qquad f(t) = \sum_{j=-\infty}^{\infty} a I_{A_j}(t),$$

where $\{A_n : A_n \in \mathcal{T}, n = 0, \pm 1, \pm 2, \cdots\}$ is a partition of T when repetitions are excluded, $a_0 = 0$, and $a_{j+1} - a_j \geqq \delta > 0$ for each j and some δ. If all but a finite number of the $\{A_n\}$ equal $\varnothing$, then f is a simple function.

LEMMA 3.2. *If* $f \in \mathcal{C}$ *is elementary, with representation* (3.28), *and if* $|s(f)| < \infty$, *then*

(a) $\lim_{n \to \infty} a_n v(\bigcup_{i=n}^{\infty} A_i) = 0, \lim_{n \to -\infty} a_n u(\bigcup_{i=n}^{-\infty} A_i) = 0,$

(b) $s(f) = \Sigma_{j=-\infty}^{\infty} a_j q(A_j).$

PROOF. Under the hypotheses of the lemma,

$$(3.29) \qquad f^+(t) = \sum_{j=1}^{\infty} a_j I_{A_j}(t), \qquad f^-(t) = \sum_{j=-1}^{-\infty} -a_j I_{A_j}(t),$$

and $s(f^+) < \infty, r(f^-) < \infty$. From Definition 3.1,

$$(3.30) \qquad s(f^+) = \sum_{j=1}^{\infty} (a_j - a_{j-1}) v\left(\bigcup_{i=j}^{\infty} A_i \right)$$

$$= \lim_{n \to \infty} \left[\sum_{j=1}^{n} a_j q(A_j) + a_n v\left(\bigcup_{i=n+1}^{\infty} A_i \right) \right].$$

Therefore, since $s(f^+) < \infty$,

$$(3.31) \qquad \lim_{n \to \infty} (a_n - a_{n-1})\, v\left(\bigcup_{i=n}^{\infty} A_i\right) = 0,$$

and since $a_n v(\bigcup_{i=n+1}^{\infty} A_i) \geqq 0$,

$$(3.32) \qquad \lim_{n \to \infty} \sum_{j=1}^{n} a_j q(A_j) \leqq s(f^+) < \infty.$$

Moreover, because of (3.31), $\lim_{n \to \infty} v(\bigcup_{i=n}^{\infty} A_i) = 0$ and

$$(3.33) \qquad a_n v\left(\bigcup_{i=n}^{\infty} A_i\right) = a_n \sum_{j=n}^{\infty} q(A_j) \leqq \sum_{j=n}^{\infty} a_j q(A_j).$$

From these relations follow the first part of (a) and

$$(3.34) \qquad s(f^+) = \sum_{j=1}^{\infty} a_j q(A_j).$$

A similar argument on $r(f^-)$ establishes the other part of (a) and

$$(3.35) \quad r(f^-) = -\sum_{j=-1}^{-\infty} a_j\left[u\left(\bigcup_{i=j}^{-\infty} A_i\right) - u\left(\bigcup_{i=j-1}^{-\infty} A_i\right)\right] = -\sum_{j=-1}^{-\infty} a_j q(A_j),$$

the second equality coming from Proposition 3.2. Finally, (b) is a consequence of (3.34) and (3.35).

REMARK 3.7.   From Lemma 3.1, the set function q appearing in Lemma 3.2 is a probability. Thus, (b) represents $s(f)$ as an expectation.

PROPOSITION 3.12.   *Let $f, g \in \mathscr{C}$ be such that $f + g$ is defined.*

(a) *If either $s(f^+) + s(g^+) < \infty$ or $s(f^-) + r(g^-) < \infty$, then $r(f) + s(g) \leqq s(f + g) \leqq s(f) + s(g)$.*

(b) *If either $s(f^-) + s(g^-) < \infty$ or $r(f^+) + s(g^+) < \infty$, then $r(f) + r(g) \leqq r(f + g) \leqq r(f) + s(g)$.*

PROOF.   (i) Let $f, g \in \mathscr{C}$ be elementary functions to which the hypotheses of (a) apply. Then $s(f), s(g)$ and $r(f)$ exist. Assume that $|s(f + g)| < \infty$. The sum $f + g$ may be represented in the form (3.28). If $e(\cdot)$ denotes expectation with respect to q, then by Remark 3.7 and the preceding lemmas,

$$(3.36) \qquad s(f + g) = e(f + g) = e(f) + e(g) \leqq s(f) + s(g).$$

(ii) If $f, g \in \mathscr{C}$, each may be approximated from below by a monotone increasing sequence of elementary functions. Under the hypotheses of (a) and if $|s(f + g)| < \infty$, the result of (i) applies to approximating elementary functions. Taking monotone limits establishes

$$(3.37) \qquad s(f + g) \leqq s(f) + s(g).$$

Special cases. If $s(f + g) = -\infty$ and the hypotheses of (a) hold, then (3.37) is trivial. If $s(f + g) = \infty$, then $s(f^+ + g^+) \geq s[(f + g)^+] = \infty$. Since f^+, $g^+ \in \mathscr{C}^+$, each may be approximated from below by a monotone increasing sequence of simple functions, each of which is in $\mathscr{C}^+$ and is bounded. The result in (i) for elementary functions applies; taking monotone limits shows that $s(f^+) + s(g^+) \geq s(f^+ + g^+) = \infty$. Thus, one of $s(f)$, $s(g)$ is ∞ and (3.37) is valid.

In summary, therefore, if $s(f + g)$ exists and the hypotheses of (a) hold, then (3.37) is valid. Under the same assumptions,

$$(3.38) \qquad s(g) = s(f + g - f) \leq s(f + g) + s(-f),$$

which, by Proposition 3.9, is equivalent to the left inequality in (a).

(iii) Suppose $f, g \in \mathscr{C}$ and the hypotheses of (b) hold, ensuring that $r(f)$, $r(g)$, $s(g)$ exist. Assume also that $r(f + g)$ exists. Since $r(f + g) = -s(-f - g)$, the inequalities of (b) follow from (ii).

(iv) To complete the proof, it is necessary to show that $s(f + g)$, $r(f + g)$ exist under the hypotheses of (a) and (b), respectively. Since $s(f^+ + g^+)$, $r(f^- + g^-)$ exist, it follows from (ii) and (iii) that

$$(3.39) \qquad \begin{aligned} s[(f + g)^+] &\leq s(f^+ + g^+) \leq s(f^+) + s(g^+), \\ r[(f + g)^-] &\leq r(f^- + g^-) \leq s(f^-) + r(g^-). \end{aligned}$$

Thus $s(f + g)$ exists under the hypotheses of (a). A dual argument shows that $r(f + g)$ exists in (b).

COROLLARY 3.1.   *Let $f \in \mathscr{C}$.*

(a) *If $s(f)$ exists, $|s(f)| \leq s(|f|)$.*

(b) *If $r(f)$ exists, $|r(f)| \leq s(|f|)$.*

Define sets K, $K^+ \subset R^n$ as follows:

$$(3.40) \qquad \begin{aligned} K &= \{x \in R^n : x_1 \geq 0, x_2 \geq 0, \cdots, x_n \geq 0\}, \\ K^+ &= \{x \in R^n : x_1 > 0, x_2 > 0, \cdots, x_n > 0\}. \end{aligned}$$

PROPOSITION 3.13.   *Let $h : K \to R$ be continuous and concave in K and such that $h(x) > 0$ and $h(\lambda x) = \lambda h(x)$ for every $x \in K^+$ and $\lambda \geq 0$. Let $f_1, f_2, \cdots, f_n \in \mathscr{C}^+$ be such that $s(f_i) < \infty$ for $1 \leq i \leq n$. Then*

$$(3.41) \qquad s[h(f_1(t), \cdots, f_n(t))] \leq h[s(f_1), \cdots, s(f_n)].$$

PROOF.   By Propositions 3.9 and 3.12, $s(\cdot)$ is an increasing gauge on $\mathscr{C}^+$. The theorem follows from a general result due to Bourbaki (see Berge [2], p. 212).

If $f \in \mathscr{C}$, define $\|f\|_p$ by

$$(3.42) \qquad \begin{aligned} \|f\|_p &= [s(|f|^p)]^{1/p}, &\qquad 1 \leq p < \infty, \\ \|f\|_\infty &= \sup\{z : v(|f(t)| > z) > 0\}. \end{aligned}$$

PROPOSITION 3.14. *Let $f, g \in \mathscr{C}$. Then*

(a) $0 \leqq \|f\|_p \leqq \|f\|_q$ *if* $1 \leqq p \leqq q \leqq \infty$,

(b) $v(|f(t)| \neq 0) = 0$ *if and only if* $\|f\|_p = $ *for* $1 \leqq p \leqq \infty$,

(c) $\|af\|_p = a\|f\|_p$ *if* $a \in R^+$ *and* $1 \leqq p \leqq \infty$,

(d) $\|fg\|_r \leqq \|f\|_p\|g\|_q$ *if* $1 \leqq p, q, r \leqq \infty$ *and* $r^{-1} = p^{-1} + q^{-1}$,

(e) $\|f + g\|_p \leqq \|f\|_p + \|g\|_p$ *if* $1 \leqq p \leqq \infty$.

PROOF. Assertions (b) and (c) are immediate. Apart from special cases, (d) and (e) follow from Proposition 3.13, or may be proved from the Hölder and Minkowski inequalities along the lines of Proposition 3.12. Assertion (a) is a consequence of (d).

4. Minimax decisions

Conditions for the existence of minimax decisions, as defined in Section 2, are provided by the theorem below. Examples of minimax procedures for a distribution free estimation problem follow.

PROPOSITION 4.1. *Let T, D be compact metric spaces and let $\ell : T \times D \to R$ be continuous. Then*

(a) *$s(\ell, d)$ and $r(\ell, d)$ are uniformly continuous on D,*

(b) *the suprema and infima over D of $r(\ell, d)$ are attained.*

PROOF. Let $m(\cdot, \cdot)$ denote the metric on D. Since $T \times D$ is compact metric, $\ell(t, d)$ is uniformly continuous on $T \times D$. Therefore, to every $\varepsilon > 0$ there corresponds an $\eta > 0$ such that

$$(4.1) \qquad m(d, d') < \eta \Rightarrow |\ell(t, d) - \ell(t, d')| < \varepsilon$$

for every $t \in T$. Applying Proposition 3.9, parts (b), (c), (d), (e), to the right side of (4.1) establishes

$$(4.2) \qquad |s(\ell, d) - s(\ell, d')| < \varepsilon, \qquad |r(\ell, d) - r(\ell, d')| < \varepsilon,$$

hence (a) and (b).

EXAMPLE. An example of the statistical model described in Section 2 is the nonparametric version of the two sample location shift model. If $(x_1, \cdots, x_m)$ are the observations of the first sample and $(y_1, \cdots, y_n)$ are the observations of the second sample, the model can be written in the form

$$(4.3) \qquad \begin{aligned} x_i &= F^{-1}(u_i), & 1 \leqq i \leqq m, \\ y_j &= \mu + F^{-1}(u_{m+j}), & 1 \leqq j \leqq n, \end{aligned}$$

where $(u_1, \cdots, u_{m+n})$ are realizations of independent, identically distributed random variables, each uniformly distributed on $[0, 1]$, $F \in \mathscr{F}$, the class of all continuous distribution functions on the real line, $\mu \in \Omega = (-\infty, \infty)$, and (μ, F) is the unknown parameter. Equations (4.3) are of the general form (2.1).

Let $\{d_{i,j} = y_j - x_i, 1 \leq i \leq m, 1 \leq j \leq n\}$ and let $a_1 < a_2 < \cdots < a_{M-1}$, where $M = mn + 1$, denote the ordered $\{d_{i,j}\}$. Under the original model, the strict ordering will be possible with probability one. Let $\Omega_1 = (-\infty, a_1)$, let $\Omega_i = (a_{i-1}, a_i)$ for $2 \leq i \leq M - 1$, and let $\Omega_M = (a_{M-1}, \infty)$. For arbitrary $A \subset \Omega$, define

$$(4.4) \qquad \delta_v(i, A) = \begin{cases} 1 & \text{if } A \cap \Omega_i \neq \varnothing \\ 0 & \text{otherwise,} \end{cases}$$

and

$$(4.5) \qquad \delta_u(i, A) = \begin{cases} 1 & \text{if } \Omega_i \subset A \\ 0 & \text{otherwise.} \end{cases}$$

Then, as shown in [1], for arbitrary $A \subset \Omega$,

$$(4.6) \qquad \begin{aligned} v(A \times \mathscr{F}) &= \frac{1}{M} \sum_{i=1}^{M} \delta_v(i, A), \\ u(A \times \mathscr{F}) &= \frac{1}{M} \sum_{i=1}^{M} \delta_u(i, A). \end{aligned}$$

This collection of upper and lower probabilities determines the upper and lower risks if the loss function does not depend upon F. For example, suppose that it is desired to estimate μ, that $h: R^+ \to R^+$ is strictly monotone increasing, and that the loss function of interest is

$$(4.7) \qquad \ell(\mu, d) = \begin{cases} h(|\mu - d|) & \text{if } |\mu - d| \leq c \\ h(c) & \text{if } |\mu - d| > c, \end{cases}$$

where $c > a_{M-1} - a_1$. Let $B_1 = [a_{M-1} - c, \frac{1}{2}(a_1 + a_2)]$, let $B_i = [\frac{1}{2}(a_{i-1} + a_i), \frac{1}{2}(a_i + a_{i+1})]$ for $2 \leq i \leq M - 2$, and let $B_{M-1} = [\frac{1}{2}(a_{M-2} + a_{M-1}), a_1 + c]$. Then for ℓ defined by (4.7),

$$(4.8) \qquad s(\ell, d) = \frac{1}{M} \Big[\sum_{j \neq i} h(|a_j - d|) + 2h(c) \Big]$$

if $d \in B_i$ for $1 \leq i \leq M - 1$, and

$$(4.9) \qquad \begin{aligned} s(\ell, d) &> s(\ell, a_{M-1} - c) & \text{if } d < a_{M-1} - c, \\ s(\ell, d) &> s(\ell, a_1 + c) & \text{if } d > a_1 + c. \end{aligned}$$

Similar expressions may be found for $r(\ell, d)$.

In particular, suppose that $h(x) = x$. Then, if M is even, $s(\ell, d)$ is minimized by any $d \in [\frac{1}{2}(a_{(M/2)-1} + a_{M/2}), \frac{1}{2}(a_{M/2} + a_{(M/2)+1})]$, while if M is odd, the minimizing value is $d = \frac{1}{2}(a_{(M-1)/2} + a_{[(M-1)/2]+1})$. This class of minimax estimates for μ includes the Hodges-Lehmann estimate median $\{a_1, \cdots, a_{M-1}\}$.

If $h(x) = x^2$, the minimax estimate for μ can be described as follows. Let $m_i = M^{-1} \sum_{j \neq i} a_j$. If there exists a k, $1 \leq k \leq M - 1$, such that $m_k \in B_k$, $s(\ell, d)$ is minimized by $d = m_k$. Otherwise, there will exist a k, $1 \leq k \leq M - 1$

such that $m_k > \frac{1}{2}(a_k + a_{k+1}) > m_{k+1}$; in this event $s(\ell, d)$ is minimized by $d = \frac{1}{2}(a_k + a_{k+1})$. Viewed as functions of $(x_1, \cdots, x_m, y_1, \cdots, y_n)$, these minimax decisions are minimax procedures in the sense of Definition 2.2.

REFERENCES

[1] R. J. BERAN. "On distribution-free statistical inference with upper and lower probabilities," *Ann. Math. Statist.*, Vol. 42 (1971), pp. 157–168.

[2] C. BERGE, *Topological Spaces*, Edinburgh and London, Oliver and Boyd, 1963.

[3] G. CHOQUET, "Theory of capacities," *Ann. Inst. Fourier, (Grenoble)*, Vol. 5 (1953/54), pp. 131–295.

[4] A. P. DEMPSTER, "On direct probabilities," *J. Roy. Statist. Soc. Ser. B.*, Vol. 25 (1963), pp. 102–107.

[5] ———, "New methods for reasoning toward posterior distributions based on sample data," *Ann. Math. Statist.*, Vol. 37 (1966), pp. 355–374.

[6] ———, "Upper and lower probabilities induced by a multivalued mapping," *Ann. Math. Statist.*, Vol. 38 (1967), pp. 325–339.

[7] ———, "Upper and lower probability inferences based on a sample from a finite univariate population," *Biometrika*, Vol. 54 (1967), pp. 515–528.

[8] ———, "Upper and lower probabilities generated by a random closed interval," *Ann. Math. Statist.*, Vol. 39 (1968), pp. 957–966.

[9] ———, "A generalization of Bayesian inference," *J. Roy. Statist. Soc. Ser. B.*, Vol. 30 (1968), pp. 205–247.

[10] ———, "Upper and lower probability inferences for families of hypotheses with monotone density ratios," *Ann. Math. Statist.*, Vol. 40 (1969), pp. 953–969.

[11] D. A. S. FRASER, "The fiducial method and invariance," *Biometrika*, Vol. 48 (1961), pp. 261–280.

[12] ———, "On fiducial inference," *Ann. Math. Statist.*, Vol. 32 (1961), pp. 661–676.

[13] ———, "On the consistency of the fiducial method," *J. Roy. Statist. Soc. Ser. B.*, Vol. 24 (1962), pp. 425–434.

[14] ———, "Structural probability and a generalization," *Biometrika*, Vol. 53 (1966), pp. 1–9.

[15] ———, "Data transformations and the linear model," *Ann. Math. Statist.*, Vol. 38 (1967), pp. 1456–65.

[16] ———, *The Structure of Inference*, New York, Wiley, 1968.

[17] I. HACKING, *Logic of Statistical Inference*, Cambridge, Cambridge University Press, 1965.

[18] P. J. HUBER, *Théorie de l'Inférence Statistique Robuste*, Montréal, Les Presses de l'Université de Montréal, 1969.

[19] V. STRASSEN. "Messfehler und Information." *Z. Wahrscheinlichkeitstheorie und Verw. Gebiete.* Vol. 2 (1964), pp. 273–305.

ON INEQUALITIES OF CRAMÉR-RAO TYPE AND ADMISSIBILITY PROOFS

COLIN R. BLYTH[1] and DONALD M. ROBERTS

UNIVERSITY OF ILLINOIS

1. Introduction and Summary

This paper is a discussion of the Hodges-Lehmann [7] method of proving admissibility for quadratic loss.

Section 2 compares the inequality $EU^2 \geq (EU)^2$ with the best Schwarz inequality $EU^2 \geq (EU)^2 + \{\text{Cov}(U, V)\}^2/\text{Var } V$ obtainable using linear functions of U and V, and considers invariance properties of these inequalities.

In Section 3, for a random variable X with possible distributions indexed by θ, we define a Cramér-Rao type inequality as one giving a lower bound on Var $T(X)$ in terms of $ET(X)$. Theorem 2 shows that for the best Schwarz inequality Var $T \geq \{\text{Cov}(T, V)\}^2/\text{Var } V$ using $V = V(X, \theta)$ to be of Cramér-Rao type, it is necessary that V depend on X only through a minimal sufficient statistic; this condition is also sufficient when there is a sufficient statistic with a complete family of possible distributions. In this case of completeness, it follows that the Cramér-Rao and Bhattacharyya inequalities require no regularity conditions beyond existence and nonconstancy of the derivatives involved.

Section 4 describes the Hodges-Lehmann method of proving an estimator T^* admissible for quadratic loss. In this method, the inequality showing that T makes T^* inadmissible is relaxed using the Cramér-Rao type inequality Var $(T - T^*) \geq \{\text{Cov}(T - T^*, T^*)\}^2/\text{Var } T^*$, and the relaxed inequality is shown to have no nontrivial solutions. In all examples known to us, this use of a Cramér-Rao type inequality can be replaced by a use of the weaker result Var $(T - T^*) \geq 0$; we suppose there are examples in which this cannot be done, but we have no such example.

Section 5 consists of several examples illustrating this method of proving admissibility.

2. Schwarz's inequality

For real valued random variables U and V, Schwarz's inequality

$$(2.1) \qquad \{EU^2\}\{EV^2\} \geq \{EUV\}^2$$

means that if EU^2 and EV^2 both exist, then EUV also exists and its square does not exceed their product.

[1] Work supported by NSF Grant GP 23835.

17

If V is a nonzero constant, Schwarz's inequality reduces to the Jensen inequality

$$(2.2) \qquad\qquad EU^2 \geqq \{EU\}^2.$$

This inequality is invariant under nonzero constant multiplications, and under translations: replacing U by cU just multiplies each side of (2.2) by c^2; and replacing U by $U - a$ just adds $a^2 - 2aEU$ to each side of (2.2).

If V is not a constant, Schwarz's inequality can be written

$$(2.3) \qquad\qquad EU^2 \geqq \frac{\{EUV\}^2}{EV^2}.$$

This inequality is invariant under nonzero constant multiplications, but is not invariant under translations: replacing U by cU, and V by dV, just multiplies each side by c^2; but replacing U by $U - a$, and V by $V - b$, changes the inequality to

$$(2.4) \qquad\qquad E(U - a)^2 \geqq \frac{\{E(U - a)(V - b)\}^2}{E(V - b)^2},$$

that is,

$$(2.5) \qquad EU^2 \geqq \{EU\}^2 - (EU - a)^2 + \frac{\{E(U - a)(V - b)\}^2}{E(V - b)^2}.$$

THEOREM 1.   *The strongest result obtainable by considering the Schwarz inequality under all possible translations is*

$$(2.6) \qquad\qquad EU^2 \geqq \{EU\}^2 + \frac{\{\mathrm{Cov}\,(U, V)\}^2}{\mathrm{Var}\,V}.$$

PROOF.   To show that the inequality

$$(2.6') \quad EU^2 \geqq \{EU\}^2 + \sup_{a,b}\left[-(EU - a)^2 + \frac{\{E(U - a)(V - b)\}^2}{E(V - b)^2}\right]$$

reduces to (2.6), notice that the second term on the right in (2.6') is translation invariant: translations on U and V inside the square brackets just change a and b, and supremum over all a and b is being taken. In particular, (2.6') is left unchanged when we replace, inside the square brackets only, U by $U_1 = U - EU$ and V by $V_1 = V - EV$. The result of this, after some easy simplifications, is

$$(2.7) \quad EU^2 \geqq \{EU\}^2 + \sup_{a,b}\left[\frac{-a^2 EV_1^2 + 2ab EU_1V_1 + (EU_1V_1)^2}{EV_1^2 + b^2}\right].$$

Here the quantity in square brackets has, for each b, its maximum value $(EU_1V_1)^2/EV_1^2$ when $a = b(EU_1V_1)/EV_1^2$. So the inequality (2.6') becomes

$$(2.8) \qquad\qquad EU^2 \geqq \{EU\}^2 + \frac{(EU_1V_1)^2}{EV_1^2},$$

which, written in terms of U and V, is (2.6).

This best Schwarz inequality (2.6) is invariant under nonzero constant multi-plications, under translations, and under replacement of U by $U - V$: replacing U by cU, and V by dV, just multiplies each side of (2.6) by c^2; replacing U by $U - a$, and V by $V - b$, just adds $a^2 - 2aEU$ to each side of (2.6); and replacing U by $U - V$ just adds $EV^2 - 2EUV$ to each side of (2.6).

Comparing the Jensen inequality (2.2) and the best Schwarz inequality (2.6), we see that (2.6) is a stronger result unless U and V are uncorrelated. Both in-equalities reduce to the same equality when U is a constant. The Jensen in-equality is an equality if and only if U is a constant. The best Schwarz inequality is an equality if and only if U and V are linearly related (which is true in particular if U is a constant).

3. Inequalities of Cramér-Rao type

Let X be a random variable with possible probability measures P_θ, $\theta \in \Omega$ on subsets of $\mathscr{X}$. There are no restrictions here on the space $\mathscr{X}$ or on the family of probability measures: any such restrictions will be stated where needed. For

$T = T(X)$ any real valued statistic, and
$V = V(X, \theta)$ any real valued random variable,

the best Schwarz inequality (2.6) gives, at every θ for which Var $V > 0$,

$$(3.1) \qquad ET^2 \geqq \{ET\}^2 + \frac{\{\mathrm{Cov}\,(T, V)\}^2}{\mathrm{Var}\,V},$$

that is,

$$(3.2) \qquad \mathrm{Var}\,T \geqq \frac{\{\mathrm{Cov}\,(T, V)\}^2}{\mathrm{Var}\,V},$$

or equivalently, from the invariance properties of (2.6),

$$(3.3) \qquad E\{T - g(\theta)\}^2 \geqq \{ET - g(\theta)\}^2 + \frac{\{\mathrm{Cov}\,(T, V)\}^2}{\mathrm{Var}\,V},$$

for any real valued function g defined on Ω. This inequality appears to give a lower bound on the risk of T as an estimator of $g(\theta)$ for squared error loss $\{T - g(\theta)\}^2$, and obviously also for quadratic loss $a(\theta)\{T - g(\theta)\}^2$ with $a(\theta) > 0$. But the apparent bound is useless because it depends on T: rather than compute Cov (T, V) to get a lower bound on the risk, we would compute Var T to get the risk itself.

However, when V is such that Cov (T, V) depends on T only through ET; that is, when V has the property

$$(3.4) \qquad ET_1 \underset{\theta}{\equiv} ET_2 \Rightarrow \mathrm{Cov}\,(T_1, V) \underset{\theta}{\equiv} \mathrm{Cov}\,(T_2, V),$$

then the best Schwarz inequality (3.1) takes the very useful form of a lower bound on the risk of T in terms of ET:

$$(3.5) \qquad ET = m(\theta) \Rightarrow ET^2 \geqq \{m(\theta)\}^2 + b_m(\theta),$$

that is,

$$(3.6) \qquad ET = m(\theta) \Rightarrow \operatorname{Var} T \geqq b_m(\theta),$$

or equivalently,

$$(3.7) \qquad ET = m(\theta) \Rightarrow E\{T - g(\theta)\}^2 \geqq \{m(\theta) - g(\theta)\}^2 + b_m(\theta),$$

where $b_m(\theta) = \{\operatorname{Cov}(T, V)\}^2/\operatorname{Var} V$. We will refer to (3.5) as an *inequality of Cramér-Rao type*. The question as to what functions V satisfy condition (3.4) and therefore give Cramér-Rao type inequalities has the following partial answer.

THEOREM 2. *A necessary condition for V to give a Cramér-Rao type inequality is that V depend on X only through a minimal sufficient statistic. This condition is also sufficient, when the minimal sufficient statistic has a complete family of possible distributions.*

PROOF. *Necessity.* For every statistic T and every sufficient statistic S, notice that $E(T|S)$ is a statistic and has the same expectation as T. The property (3.4) for V therefore implies

$$(3.8) \qquad \operatorname{Cov}\{E(T|S), V\} \equiv \operatorname{Cov}(T, V),$$

that is,

$$(3.9) \qquad E\{[E(T|S)]V\} - [E\{E(T|S)\}]EV \equiv E(TV) - (ET)(EV),$$

that is,

$$(3.10) \qquad E\{[E(T|S)]V\} \equiv E(TV),$$

that is,

$$(3.11) \qquad E[E\{[E(T|S)]V|S\}] \equiv E[E(TV|S)],$$

that is,

$$(3.12) \qquad E[E(T|S)E(V|S)] \equiv E[E(TV|S)].$$

This is true for every T. In particular, for $T = V(X, \theta_0)$ the above identity gives, at $\theta = \theta_0$,

$$(3.13) \qquad E_{\theta_0}[\{E(V|S)\}^2] = E_{\theta_0}[E(V^2|S)].$$

This shows, for every θ_0 in Ω, that the distribution of V given S must be concentrated on one point, that is, V must be a function of S.

Sufficiency. If S is a sufficient statistic and $V = V(S, \theta)$ we have

(3.14)
$$\text{Cov}\,(T,\,V) = E\{TV(S,\,\theta)\} - (ET)(EV)$$
$$= E\{[E(T|S)]\,V(S,\,\theta)\} - (ET)(EV).$$

When S has a complete family of possible distributions (making S minimal), we see that if $ET_1 \equiv ET_2$ then $E(T_1|S)$, $E(T_2|S)$ are two functions of S which have the same expectation and are therefore (with probability one) the same function, giving us $\text{Cov}\,(T_1,\,V) \equiv \text{Cov}\,(T_2,\,V)$.

REMARK. When the minimal sufficient statistic S does not have a complete family of possible distributions, $ET_1 \equiv ET_2$ does not imply $\text{Cov}\,\{T_1,\,V(S,\,\theta)\} \equiv \text{Cov}\,\{T_2,\,V(S,\,\theta)\}$. A counterexample is provided by the usual example X rectangular $(\theta - \frac{1}{2},\,\theta + \frac{1}{2})$, $-\infty < \theta < \infty$, in which X is minimal sufficient but not complete. Take $V = X$, $T_1 = 0$, $T_2 = \pm 1$ according as the integer nearest to X is above or below X. Then $ET_1 \equiv 0$ and $ET_2 \equiv 0$ but $\text{Cov}\,(T_1,\,V) \neq \text{Cov}\,(T_2,\,V)$.

COROLLARY 1. *When X has a complete family of possible measures, X is itself a minimal sufficient statistic, and the theorem reduces to this simple form: every best Schwarz inequality is of Cramér-Rao type; that is, every random variable V has property* (3.4).

Without appealing to the theorem, it is clear that ET determines T and therefore determines $\text{Cov}\,(T,\,V)$.

COROLLARY 2. *Every Cramér-Rao type inequality is invariant under a sufficiency reduction. If instead of working with X, we work only with the sufficient statistic S, then any Cramér-Rao type bound that was available is still available, and unchanged, because the V involved is a function of X through S only.*

For the standard particular Cramér-Rao type inequalities, this invariance under a sufficiency reduction is usually proved using the factorization theorem for densities.

3.1. *Families of variables V.* In all of this section, we could have taken a family $V_\alpha = V_\alpha(X,\,\theta)$, $\alpha \in A$ of such V and replaced b_m by $\sup_\alpha [\{\text{Cov}\,(T,\,V_\alpha)\}^2/\text{Var}\,V_\alpha]$. This is a commonly used way of getting a good bound. We have not done this here, because we are considering applications in which an obvious best V is available.

3.2. *Particular inequalities.* Particular choices of the V give the following particular Cramér-Rao type inequalities:

(a) the Cramér-Rao inequality [4] uses

(3.15)
$$V = \frac{1}{p_\theta(X)}\left\{\frac{\partial}{\partial \theta}\,p_\theta(X)\right\};$$

(b) the kth Bhattacharyya inequality [2] uses

(3.16)
$$V_\alpha = \frac{1}{p_\theta(X)}\left\{\sum_{i=1}^{k} c_{i\alpha}\,\frac{\partial^i}{\partial \theta^i}\,p_\theta(X)\right\},$$

where the $c_{i\alpha}$ range over all real numbers,

(c) the Barankin inequality [1] uses

$$(3.17) \qquad V_\alpha = \frac{1}{p_\theta(X)} \left\{ \sum_{i=1}^{n_\alpha} c_{i\alpha} p_{\theta_i}(X) \right\},$$

where n_α ranges over the positive integers, the $c_{i\alpha}$ range over the reals, and the θ_i range over Ω;

(d) the Chapman-Robbins inequality [3] uses

$$(3.18) \qquad V_\alpha = \frac{1}{p_\theta(X)} \{ p_{\theta_\alpha}(X) \},$$

where θ_α ranges over Ω;

(e) the Kiefer inequality [9] uses a continuous mixture, where Barankin uses a discrete mixture, of $p_{\theta_1}(X)$ over points θ_1 of Ω.

In all of these, the family of distributions is taken to be dominated, with P_θ having density p_θ relative to a fixed measure μ. The V used are all of the form $p_{\theta_1}(X)/p_\theta(X)$ where θ indexes the true distribution and θ_1 is any other point of Ω, or linear combinations over values of θ_1 in Ω, or limits of such linear combinations.

Existence of $V = p_{\theta_1}(X)/p_\theta(X)$ requires only that P_θ dominate P_{θ_1}; we can use $\frac{1}{2}(P_\theta + P_{\theta_1})$ for μ and will have $p_{\theta_1}(x) = 0$ whenever $p_\theta(x) = 0$. Such a V satisfies property (3.4) because for $ET = m(\theta)$ we have

$$(3.19) \qquad \mathrm{Cov}\,(T, V) = E(TV) - (ET)(EV)$$

$$= \int t(x) \frac{p_{\theta_1}(x)}{p_\theta(x)} \, p_\theta(x) \, d\mu(x) - m(\theta) \int \frac{p_{\theta_1}(x)}{p_\theta(x)} \, p_\theta(x) \, d\mu(x)$$

$$= m(\theta_1) - m(\theta),$$

which depends on T only through ET. And such a V will not be a constant, provided P_{θ_1} and P_θ are not the same measure. Thus, for the Barankin and Chapman-Robbins inequalities, all that is needed is that the P_{θ_1} measures used differ from and be dominated by P_θ.

For the Cramér-Rao and Bhattacharyya inequalities, θ must be real valued (easily extended to linear spaces [2]) and the derivatives involved must exist and not be constants. In addition property (3.4), obvious for the Barankin V, must survive the limit operation: this is assured by requiring differentiability under the expectation sign. However, this additional regularity condition is unnecessary when there is sufficient statistic S with a complete family of possible measures, because the Barankin variables V and therefore limits of them depend on X only through S, and all functions of S, θ have property (3.4) by Theorem 2.

Barankin [1] shows that when $p_\theta(x) = 0$ implies $p_{\theta_1}(x) = 0$ for all $\theta_1 \in \Omega$, it is enough to consider variables V of his form, because they give an achievable bound. It is unnecessary to use any other V because they cannot give a better bound; but it may be convenient to use other V that give the best bound more easily.

Cramér-Rao type inequalities have two uses in proving something good about T as an estimator of $g(\theta)$ with quadratic loss: (i) in proving minimum risk, either locally or uniformly, for given expectation, and (ii) in proving admissibility.

The second use (ii) is the subject of the rest of this paper; the first use (i) is as follows.

3.3. *Local use* (i).   Suppose T_0, with $ET_0 = m(\theta)$, achieves equality in (3.5) at $\theta = \theta_0$:

$$(3.20) \qquad E_{\theta_0}\{T_0 - g(\theta_0)\}^2 = \{m(\theta_0) - g(\theta_0)\}^2 + b_m(\theta_0).$$

Then every other estimator T with the same expectation $ET = m(\theta)$ has

$$(3.21) \qquad E_{\theta_0}\{T - g(\theta_0)\}^2 \geqq E_{\theta_0}\{T_0 - g(\theta_0)\}^2.$$

Therefore among all estimators of $g(\theta)$ having expectation $m(\theta)$, the unique (with P_{θ_0} probability one) one with minimum risk for $\theta = \theta_0$ is T_0; this for every $g(\theta)$ and every quadratic loss. Uniqueness because if T_1 were another estimator with expectation $m(\theta)$ and achieving equality in (3.5) for $\theta = \theta_0$, then $\frac{1}{2}(T_0 + T_1)$ would also have expectation $m(\theta)$ and would violate (3.5) at θ_0 unless $P_{\theta_0}(T = T_0) = 1$.

In particular, taking $g(\theta) = m(\theta)$, we see that T_0 is the unique (with P_{θ_0} probability one) unbiased estimator of $m(\theta)$ with minimum risk (minimum variance) at $\theta = \theta_0$.

3.4. *Uniform use* (i).   Suppose T^*, with $ET^* = m^*(\theta)$, achieves equality in (3.5) for all θ:

$$(3.22) \qquad E\{T^* - g(\theta)\}^2 \underset{\theta}{\equiv} \{m^*(\theta) - g(\theta)\}^2 + b_{m^*}(\theta).$$

Then every other estimator T with the same expectation $ET = m^*(\theta)$ has $E\{T - g(\theta)\}^2 \geqq E\{T^* - g(\theta)\}^2$, all θ. Therefore among all estimators of $g(\theta)$ having expectation $m^*(\theta)$, the unique (with probability one, all θ) one with uniformly minimum risk (or variance) is T^*; this for every $g(\theta)$ and every quadratic loss.

In particular, taking $g(\theta) = m^*(\theta)$, we see that T^* is the unique (with probability one) unbiased estimator of $m^*(\theta)$.

This use (i) is vacuous when there is a sufficient statistic S with a complete family of measures. The Rao-Blackwell theorem enables us to restrict attention to functions of S, and if T^* has expectation $m^*(\theta)$, then no other T has this same expectation and so the stated uniformly minimum risk properties are obvious.

4. Admissibility proofs for quadratic loss

An estimator T^* is inadmissible for estimating $g(\theta)$, with quadratic loss, if there is an estimator T that is a nontrivial solution of the inadmissibility inequality

$$(4.1) \qquad E\{T - g(\theta)\}^2 \leqq E\{T^* - g(\theta)\}^2,$$

or equivalently, writing $m(\theta)$ for ET,

$$(4.1') \qquad \text{Var } T + \{m(\theta) - g(\theta)\}^2 \leqq E\{T^* - g(\theta)\}^2,$$

or, again equivalently,

$$(4.1'') \qquad E\{T - T^*\}^2 + 2E\{T - T^*\}\{T^* - g(\theta)\} \leqq 0.$$

By a nontrivial solution (there is always the trivial solution $T = T^*$) we mean one for which the inequality is strict for at least one θ.

Hodges and Lehmann [7] introduced the following method, also used in [6] and [10], of proving admissibility. Using any Cramér-Rao type inequality (3.5) we see that if T is a nontrivial solution of the inadmissibility inequality (4.1) then T is also a nontrivial solution of the relaxed inequality

$$(4.2) \qquad b_m(\theta) + \{m(\theta) - g(\theta)\}^2 \leqq E\{T^* - g(\theta)\}^2.$$

This inequality is a relaxation of (4.1), the left side of (4.1) having been replaced by something at least as small. If it can be shown that this relaxed inequality (4.2) has no nontrivial solution m that is the expectation of some T, it therefore follows that (4.1) can have no nontrivial solutions, and so T^* is proved admissible for every quadratic loss.

The Hodges-Lehmann method works directly with the definition of admissibility, using the standard mathematical technique of replacing a complicated expression by a simple bound for it. This replaces an integral inequality in T by an easier inequality in m. When the Cramér-Rao inequality is used, as in the examples of [6], [7], [10], we have $b_m(\theta) = [m'(\theta)]^2/\text{Var } V$, so that (4.2) is always a differential inequality, which in those examples is easily shown to have no nontrivial solutions. (Any best Schwarz inequality could be used on (4.1) in the same way, but failing property (3.4) the relaxed inequality would still involve T so would be no improvement on (4.1).)

4.1. *Which Cramér-Rao type inequality to use.* The Hodges-Lehmann method cannot succeed unless T^* achieves equality in the Cramér-Rao type inequality (3.5) for all θ: otherwise $m^*(\theta) = ET^*$ would be a nontrivial solution of the relaxed inequality (4.2). So for the method to succeed the V used in (3.5) must be a linear function $\alpha(\theta)T^* + \beta(\theta)$ of T^*, or equivalently $V = T^*$ because of the invariance of (2.6) under linear transformations; there is no point in trying special cases of (3.5) such as the Cramér-Rao inequality.

For a Hodges-Lehmann proof of the admissibility of T^*, we check that T^* has property (3.4) [no check is needed when the family of distributions of X is complete], and write down the relaxed inequality (4.2) that results from using $V = T^*$ in (3.5):

$$(4.3) \qquad \frac{\{\text{Cov } (T, T^*)\}^2}{\text{Var } T^*} + \{m(\theta) - g(\theta)\}^2 - E\{T^* - g(\theta)\}^2 \leqq 0.$$

Because of the invariance of (2.6) under replacement of U by $U - V$, the application of (3.5) with $V = T^*$ to $E(T - T^*)^2$ in (4.1″) gives the same inequality as (4.3):

$$(4.3') \quad \frac{\{\text{Cov } T - T^*, T^*\}^2}{\text{Var } T^*} + \{m(\theta) - m^*(\theta)\}^2 + 2E\{T - T^*\}\{T^* - g(\theta)\} \leqq 0.$$

NOTE 1. The relaxed inequality (4.3′) can be further relaxed to

$$(4.4) \qquad \{m(\theta) - m^*(\theta)\}^2 + 2E\{T - T^*\}\{T^* - g(\theta)\} \leqq 0.$$

This amounts to using the Jensen inequality (2.2) instead of the Cramér-Rao type inequality (3.5) on $E(T - T^*)^2$ in (4.1″). If it can be shown that the further relaxed inequality (4.4) has no nontrivial solution m that can be the expectation function of some T, then the inadmissibility inequality (4.1) can have no nontrivial solutions, and T^* is proved admissible. All the examples of [6], [7], [10] can be worked using (4.4) instead of (4.3).

NOTE 2. Can the Hodges-Lehmann method fail to work? That is, can it happen that T^* is admissible so (4.1) has no nontrivial solutions, but (4.3) does have nontrivial solutions? And can it happen that neither (4.1) nor (4.3) has nontrivial solutions, but (4.4) does have nontrivial solutions? We do not yet have answers to these questions, except that it is easy to give examples (see Example 4) in which T^* is a constant and admissible, but (4.4) does have nontrivial solutions (the relaxed inequality (4.3) is not available when T^* is a constant). Negative answers for a particular T^* or class of T^* would permit use of the Hodges-Lehmann method in proving such a T^* inadmissible.

NOTE 3. When X, or a sufficiency reduction of X, has a complete family of distributions, why use a relaxed inequality? After all, the reason given for using (4.3) instead of (4.1) was to get an inequality in m instead of T, and under completeness m determines T so that (4.1) itself can be written in terms of m. The answer is that (4.1) in terms of m may involve an integral that can be eliminated by changing to (4.3). This is what happens in Example 1, where we begin by trying to work with (4.1) without relaxations. Inequality (4.4) has no such essential advantage over (4.3), but may be somewhat simpler and easier to work with than (4.3).

NOTE 4. It is often convenient to carry out these proofs in terms of $Z = T - T^*$ and $\zeta(\theta) = EZ = m(\theta) - m^*(\theta)$. In terms of Z and ζ, the inadmissibility inequality (4.1) is

$$(4.5) \quad \{\zeta(\theta)\}^2 + 2\zeta(\theta)\{m^*(\theta) - g(\theta)\} + 2 \operatorname{Cov}(Z, T^*) + \operatorname{Var} Z \leqq 0,$$

and the relaxed inequality (4.3) is

$$(4.6) \quad \{\zeta(\theta)\}^2 + 2\zeta(\theta)\{m^*(\theta) - g(\theta)\} + 2 \operatorname{Cov}(Z, T^*) + \frac{\{\operatorname{Cov}(Z, T^*)\}^2}{\operatorname{Var} T^*} \leqq 0,$$

and the further relaxed inequality (4.4) is

$$(4.7) \qquad \{\zeta(\theta)\}^2 + 2\zeta(\theta)\{m^*(\theta) - g(\theta)\} + 2 \operatorname{Cov}(Z, T^*) \leqq 0,$$

and the Hodges-Lehmann proof consists of showing that $\zeta(\theta) \equiv 0$ is the only solution of (4.6) or (4.7) that can be the expectation of some T, so that no nontrivial solution exists, and T^* must be admissible.

5. Examples

EXAMPLE 1. *Uniform* $(0, \theta)$. For $Y_1, \cdots, Y_n$ independent, each uniformly distributed on $(0, \theta)$, the sufficient statistic $X = \max Y_i$ has the complete family of possible densities

$$(5.1) \qquad\qquad \frac{n}{\theta^n} x^{n-1}, \qquad\qquad 0 \leqq x \leqq \theta, \theta > 0.$$

For estimating θ with quadratic loss, the uniformly best constant multiple of X is

$$(5.2) \qquad\qquad T^* = \frac{n+2}{n+1} X,$$

which has expectation

$$(5.3) \qquad\qquad m^*(\theta) = ET^* = \frac{n(n+2)}{(n+1)^2} \theta$$

and risk

$$(5.4) \qquad\qquad E(T^* - \theta)^2 = \frac{\theta^2}{(n+1)^2}.$$

This estimator T^* was proved admissible by Karlin [8]. For an estimator $T(X)$ with expectation $m(\theta)$ we have $T(X) = m(X) + (X/n)m'(X)$. Because of this simple inversion, T^* provides an easy illustration of the advantage of working with a relaxation instead of with the inadmissibility inequality itself (see Note 3, Section 4).

We begin by trying to work directly with the inadmissibility inequality (4.1) in the estimator T whose expectation is m:

$$(5.5) \qquad E\{T(X) - m(\theta)\}^2 + \{m(\theta) - \theta\}^2 \leqq \frac{\theta^2}{(n+1)^2}.$$

Because of the simple inversion, this can easily be written as an inequality in m:

$$(5.5') \qquad E\{m(X) + \frac{X}{n} m'(X) - m(\theta)\}^2 + \{m(\theta) - \theta\}^2 \leqq \frac{\theta^2}{(n+1)^2}.$$

Multiplying out the square in the first term and integrating one of the resulting terms, $EXm(X)m'(X)$, by parts, this inequality simplifies to

$$(5.5'') \qquad \frac{1}{n\theta^n} \int_0^\theta [m'(x)]^2 x^{n+1} \, dx + \{m(\theta) - \theta\}^2 \leqq \frac{\theta^2}{(n+1)^2}.$$

We want to prove that $m = m^*$ is the unique solution of this inequality, which is hard to work with because of the integral in the first term. An application of Jensen's inequality to that integral gives a relaxed inequality

$$(5.6) \qquad \frac{n+2}{n} \left\{ m(\theta) - \frac{n+1}{\theta^{n+1}} \int_0^\theta m(x) x^n \, dx \right\}^2 + \{m(\theta) - \theta\}^2 \leqq \frac{\theta^2}{(n+1)^2}.$$

A check shows that this is exactly the same relaxed inequality (4.3) as results from applying to Var T the Cramér-Rao type inequality using T^*. This inequality (5.6) still contains an integral, but written in terms of $r(\theta) = (1/\theta^n) \int_0^\theta m(x)x^n \, dx$, for which $m(\theta) = (n/\theta)r(\theta) + r'(\theta)$, it does not:

$$(5.6') \qquad \frac{n+2}{n}\left\{r'(\theta) - \frac{r(\theta)}{\theta}\right\}^2 + \left\{\frac{n}{\theta}r(\theta) + r'(\theta) - \theta\right\}^2 \leqq \frac{\theta^2}{(n+1)^2}.$$

We want to show that $r(\theta) = \theta^2 n/(n+1)^2$, corresponding to $m = m^*$, is the unique solution of (5.6'). For convenience, we now write $\theta^2 s(\theta) = r(\theta) - \theta^2 n/(n+1)^2$ and use a typical Hodges-Lehmann argument to show that $s(\theta) \equiv 0$ is the unique solution of (5.6') written in terms of $s(\theta)$ and slightly rearranged

$$(5.6'') \qquad \frac{n+2}{n}\{\theta s'(\theta) + (n+1)s(\theta)\}^2 + \left\{\theta s'(\theta) + \frac{1}{n+1}\right\}^2 \leqq \frac{1}{(n+1)^2}.$$

This inequality shows that $\theta s'(\theta) \leqq 0$; and that $\theta s'(\theta)$ and $\theta s'(\theta) + (n+1)s(\theta)$ are both bounded, making $s(\theta)$ also bounded. Now $\theta s'(\theta)$ cannot be bounded away from zero as $\theta \to 0$ or as $\theta \to \infty$, because either of these would make $s(\theta)$ unbounded. So there must be a sequence of θ values tending to zero and a sequence of θ values tending to infinity along which $\theta s'(\theta) \to 0$, and (5.6'') shows that $s(\theta) \to 0$ along these sequences. This together with $s'(\theta) \leqq 0$ implies $s(\theta) \equiv 0$ and proves T^* admissible.

Admissibility of T^* can also be proved using only the further relaxed inequality (4.4):

$$(5.7) \qquad \{\theta s'(\theta) + (n+2)s(\theta)\}^2 + \frac{2}{n+1}\theta s'(\theta) \leqq 0.$$

This inequality requires $\theta s'(\theta) \leqq 0$, and at this point we could use the inadmissibility inequality (5.5) with its first term omitted to see that $s(\theta)$ must be bounded, and admissibility would follow as above. But a proof can be given as below using only the inequality (5.7). (Note that $s(\theta) = a/\theta^{n+2}$ is a nontrivial solution of (5.7), but this solution is ruled out as not corresponding to any m, because $s(\theta) = (1/\theta^{n+2})\int_0^\theta m(x)x^n \, dx - n(n+1)^2$ which requires $\theta^{n+2}s(\theta) \to 0$ as $\theta \to 0$.)

First show that $\theta s'(\theta)$ cannot be bounded away from zero as $\theta \to \infty$; for if it were we would have $s(\theta) \to -\infty$, and so would have for all θ sufficiently large $s(\theta) < 0$ and therefore $\{\theta s'(\theta)\}^2 + \{2/(n+1)\}\theta s'(\theta) \leqq 0$, which requires $\theta s'(\theta)$ bounded as $\theta \to \infty$, which together with $s(\theta) \to -\infty$ as $\theta \to \infty$ would violate (5.7).

Thus there must be a sequence of θ values tending to infinity along which $\theta s'(\theta) \to 0$, and (5.7) shows that $s(\theta) \to 0$ along this sequence. From $s'(\theta) \leqq 0$ we can now conclude $s(\theta) \to 0$ as $\theta \to \infty$, and $s(\theta) \geqq 0$ for all θ.

Next show that $\theta s'(\theta)$ cannot be bounded away from zero as $\theta \to 0$; for if it were we would have $s(\theta) \to \infty$ as $\theta \to 0$. And in terms of $t(\theta) = \theta^{n+2}s(\theta)$, the inequality (5.7) is

$$(5.7') \qquad \left\{ \frac{t'(\theta)}{\theta^{n+1}} \right\}^2 + \frac{2}{n+1} \left\{ \frac{t'(\theta)}{\theta^{n+1}} \right\} - 2 \frac{n+2}{n+1} \frac{t(\theta)}{\theta^{n+2}} \leq 0.$$

We saw above that $t(\theta) \to 0$ as $\theta \to 0$, so $s(\theta) = t(\theta)/\theta^{n+2} \to \infty$ would imply (by L'Hospital's rule) $t'(\theta)/\theta^{n+1} \to \infty$, and therefore for all θ sufficiently small $t'(\theta)/\theta^{n+1} > 0$ and so

$$(5.8) \qquad \left\{ \frac{t'(\theta)}{\theta^{n+1}} \right\}^2 \leq 2 \frac{n+2}{n+1} \frac{t(\theta)}{\theta^{n+1}}.$$

This gives, for all x sufficiently small,

$$(5.9) \qquad \frac{t'(x)}{2(t(x))^{1/2}} \leq \text{const } x^{n/2}.$$

Integrating both sides from 0 to θ now gives, for all θ sufficiently small,

$$(5.10) \qquad (t(\theta))^{1/2} \leq \text{const } \theta^{n/2+1}$$

that is,

$$(5.11) \qquad t(\theta) \leq \text{const } \theta^{n+2},$$

which would contradict the fact that $t(\theta)/\theta^{n+2} \to \infty$ as $\theta \to 0$.

Thus there must be a sequence of θ values $\to 0$ along which $\theta s'(\theta) \to 0$, and (5.7) shows that $s(\theta) \to 0$ along this sequence. We can now conclude that $s(\theta) \equiv 0$, and the admissibility of T^* is proved.

EXAMPLE 2. *Exponential family*. Let X be a real valued random variable with the following exponential family (which is complete) of possible densities relative to a fixed σ-finite measure μ:

$$(5.12) \qquad \beta(\omega) \cdot e^{\omega x}, \qquad -\infty < x < \infty, \qquad a < \omega < b.$$

Here the interval (a, b) must be contained in the interval Ω of ω values for which $\int_{-\infty}^{\infty} e^{\omega x} d\mu(x) < \infty$. Consider $T^* = (X + k\lambda)/(1 + \lambda)$, where $\lambda > -1$ and k are constants, as an estimator of $EX = -\beta'(\omega)/\beta(\omega)$ with quadratic loss.

Hodges and Lehmann [7] proved admissibility for $(a, b) = \Omega$ of particular estimators T^* for the specific exponential families binomial, Poisson, normal $(\omega, 1)$, and gamma with scale parameter ω, given by specific choices of μ. Their proofs use the relaxed inequality (4.3); different proofs can be given using only (4.4).

Girshick and Savage [6] proved $T^* = X$ admissible for $(a, b) = \Omega = (-\infty, \infty)$. Their proof uses (4.3) which they further relax to (4.4), so can be carried out using only (4.4).

Karlin [8], using the limiting Bayes method, gave sufficient conditions for T^*, with $k = 0$, to be admissible for $(a, b) = \Omega$.

Ping [10] gave a Hodges-Lehmann type proof of the following extension of Karlin's theorem: T^* is admissible provided

(5.13) $\displaystyle\lim_{\omega \to a} \int_{\omega}^{\cdot \omega_0} [\beta(\xi)]^{-\lambda} e^{-k\lambda\xi} \, d\xi \; = \; \infty \; = \; \lim_{\omega \to b} \int_{\omega_0}^{\cdot \omega} [\beta(\lambda)]^{-\lambda} e^{-k\lambda\xi} \, d\xi.$

Ping writes down the relaxed inequality (his formula 1.4) but then replaces this by the further relaxed inequality (his formula 1.5), so that his proof actually uses only (4.4).

   EXAMPLE 3.   *Gamma with scale parameter.* For $X_1, \cdots, X_n$ independent, each with density $\lambda e^{-\lambda x}$, $x \geq 0$, $\lambda > 0$, the sufficient statistic $X = \Sigma X_i$ has exponential family (complete) of possible densities

(5.14) $$\frac{\lambda^n}{\Gamma(n)} x^{n-1} e^{-\lambda x}, \qquad\qquad\qquad x \geq 0, \qquad \lambda > 0.$$

For this family with $n > 2$ (not necessarily integer valued), for estimating $\lambda = 1/EX$ with quadratic loss, the uniformly best constant multiple of X is

(5.15) $$T^* = \frac{n-2}{X},$$

which has expectation

(5.16) $$m^*(\lambda) = ET^* = \frac{n-2}{n-1} \lambda$$

and risk

(5.17) $$E(T^* - \lambda)^2 = \frac{\lambda^2}{n-1}.$$

This extimator T^* was proved admissible by Ghosh and Singh [5] using a limiting Bayes argument. Here it provides an example, for an exponential family, of using the Hodges-Lehmann method to prove admissibility of an estimator of something other than EX. (Example 2 is restricted to estimation of EX.)

   For an estimator $T(X)$ with expectation $m(\lambda)$ and finite variance, Schwarz's inequality for T, $1/X$ proves the existence of

(5.18) $$r(\lambda) = E \frac{T}{X} = \frac{\lambda^n}{\Gamma(n)} \int_0^\infty \frac{t(x)}{x} x^{n-1} e^{-\lambda x} \, dx,$$

and we have

(5.19) $$m(\lambda) = \frac{n}{\lambda} r(\lambda) - r'(\lambda).$$

Written in terms of

(5.20) $$s(\lambda) = \frac{r(\lambda)}{\lambda^2} - \frac{1}{n-1},$$

the relaxed inequality (4.3) is

(5.21) $(n - 1)\{\lambda s'(\lambda)\}^2 + 2\{\lambda s'(\lambda)\} + (n - 1)(n - 2)\{s(\lambda)\}^2 \leqq 0,$

and the further relaxed inequality (4.4) is

(5.22) $\{(n - 2)s(\lambda) - \lambda s'(\lambda)\}^2 + 2\lambda s'(\lambda) \leqq 0.$

For each of these inequalities, it is easy to prove that $s(\lambda) \equiv 0$, corresponding to $m(\lambda) = m^*(\lambda)$, is the only solution that corresponds to some $m(\lambda) = ET$, so either inequality can be used to prove T^* admissible. The proofs are the same as those given for the corresponding inequalities in Example 1, except that the roles of $\lambda \to 0$ and $\lambda \to \infty$ are reversed.

EXAMPLE 4.   *Normal* $(\theta, 1)$. For X normal $(\theta, 1)$, $-\infty < \theta < \infty$, the estimator $T^* \equiv 0$ is obviously admissible for estimating θ with quadratic loss, but the Hodges-Lehmann method fails to prove this. Because T^* is a constant, the relaxed inequality (4.3) is not available; the further relaxed inequality (4.4) is

(5.23) $[m(\theta)]^2 - 2\theta m(\theta) \leqq 0.$

This inequality has the nontrivial solution $m(\theta) = \theta$, corresponding to $T(X) = X$. The same happens for any constant T^* in the exponential family problem of Example 2.

REFERENCES

[1] E. W. BARANKIN, "Locally best unbiased estimates," *Ann. Math. Statist.*, Vol. 20 (1949), pp. 477–501.
[2] A. BHATTACHARYYA, "On some analogues of the amount of information and their use in statistical estimation," *Sankhyā*, Vol. 8 (1946–1948), pp. 1–14, 201–218, 315–328.
[3] D. G. CHAPMAN and H. ROBBINS, "Minimum variance estimation without regularity assumptions," *Ann. Math. Statist.*, Vol. 22 (1951), pp. 581–586.
[4] H. CRAMÉR, *Mathematical Methods of Statistics*, Princeton, Princeton University Press, 1946.
[5] J. K. GHOSH and R. SINGH, "Estimation of the reciprocal of scale parameter of a Gamma density," *Ann. Inst. Statist. Math.*, Vol. 22 (1970), pp. 51–55.
[6] M. A. GIRSHICK and L. J. SAVAGE, "Bayes and minimax estimates for quadratic loss functions," *Proceedings of the Second Berkeley Symposium on Mathematical Statistics and Probability*, Berkeley and Los Angeles, University of California Press, 1951, pp. 53–73.
[7] J. L. HODGES and E. L. LEHMANN, "Some applications of the Cramér-Rao inequality," *Proceedings of the Second Berkeley Symposium on Mathematical Statistics and Probability*, Berkeley and Los Angeles, University of California Press, 1951, pp. 13–22.
[8] S. KARLIN, "Admissibility for estimation with quadratic loss,' *Ann. Math. Statist.*, Vol. 29 (1958), pp. 406–436.
[9] J. KIEFER, "On minimum variance estimates," *Ann. Math. Statist.*, Vol. 23 (1952), pp. 627–629.
[10] C. PING, "Minimax estimates of parameters of distributions belonging to the exponential family," *Chinese Math.*, Vol. 5 (1964), pp. 277–299.

THE LIKELIHOOD RATIO TEST
FOR THE
MULTINOMIAL DISTRIBUTION

J. OOSTERHOFF

UNIVERSITY OF NIJMEGEN

and

W. R. VAN ZWET

UNIVERSITY OF LEIDEN

1. Introduction and summary

Let $X^{(N)} = (X_1^{(N)}, \cdots, X_k^{(N)})$ be a random vector having a multinomial distribution with parameters N and $p = (p_1, \cdots, p_k)$,

(1.1)
$$P(X^{(N)} = x \,|\, p) = \frac{N!}{x_1! \cdots x_k!} p_1^{x_1} \cdots p_k^{x_k},$$

where $x = (x_1, \cdots, x_k)$ is a vector with nonnegative integer components with sum N, and p is any point in the simplex

(1.2)
$$\Omega = \left\{ (y_1, \cdots, y_k) \,\Big|\, \sum_{i=1}^{k} y_i = 1, y_i \geqq 0 \text{ for } i = 1, \cdots, k \right\}.$$

By $Z^{(N)} = (Z_1^{(N)}, \cdots, Z_k^{(N)})$ we denote the random vector with components

(1.3)
$$Z_i^{(N)} = \frac{X_i^{(N)}}{N}, \qquad\qquad i = 1, \cdots, k.$$

For $N = 1, 2, \cdots$, consider tests based on $Z^{(N)}$ for the hypothesis $H: p \in \Lambda_0$ against the alternative $K: p \in \Lambda_1$, where Λ_0 and Λ_1 are disjoint subsets of Ω and $\Lambda = \Lambda_0 \cup \Lambda_1$ may be a proper subset of Ω. It is assumed that the sizes α_N of the tests depend on N in such a way that $\alpha_N \to 0$ for $N \to \infty$. The likelihood ratio test based on $Z^{(N)}$ for H against K rejects H for large values of the statistic

(1.4)
$$\inf_{p \in \Lambda_0} \sup_{\pi \in \Lambda} \sum_{i=1}^{k} Z_i^{(N)} \log \frac{\pi_i}{p_i},$$

possibly with randomization on the set where the statistic assumes its critical value.

In [2] W. Hoeffding considered a special case of this situation where $\Lambda = \Omega$, in which case the likelihood ratio statistic (1.4) reduces to

Report SW 3/70 Mathematisch Centrum, Amsterdam.

31

$$(1.5) \qquad \inf_{p \in \Lambda_0} \sum_{i=1}^{k} Z_i^{(N)} \log \frac{Z_i^{(N)}}{p_i}.$$

The paper [2] is devoted to making precise the following proposition in this case: "*If a given test of size α_N is 'sufficiently different' from a likelihood ratio test, then there is a likelihood ratio test of size $\leq \alpha_N$ which is considerably more powerful than the given test at 'most' points p in the set of alternatives when N is large enough, provided that $\alpha_N \to 0$ at a suitable rate.*" By "considerably more powerful" is meant that the ratio of the error probabilities of the second kind at p of the two tests tends to zero more rapidly than any power of N. The condition that "$\alpha_N \to 0$ at a suitable rate" will typically imply that α_N tends to zero more rapidly than any power of N, that is, that $-\log \alpha_N / \log N \to \infty$.

If the likelihood ratio test is much better than a given test for most alternatives, it is natural to ask how much worse it can be for the remaining alternatives or sequences of alternatives. Let β_N denote the power function of the size α_N likelihood ratio test based on $Z^{(N)}$ for H against K and let β_N^+ be the size α_N envelope power for testing H, that is, $\beta_N^+(p)$ is the power at p of the size α_N most powerful test based on $Z^{(N)}$ for H against the simple alternative p. The shortcoming of the size α_N likelihood ratio test for a given N is defined by

$$(1.6) \qquad R_N(p) = \beta_N^+(p) - \beta_N(p), \qquad\qquad p \in \Lambda_1.$$

The main purpose of this paper is to show that for a simple hypothesis H and under a condition concerning the speed of convergence of α_N to zero, the shortcoming of the likelihood ratio test converges to zero uniformly on the set of alternatives. We note that for testing the simple hypothesis $H: p = p^0, p^0 \in \Lambda$ against $K: p \in \Lambda_1 = \Lambda - \{p^0\}$ the likelihood ratio statistic (1.4) reduces to

$$(1.7) \qquad \sup_{\pi \in \Lambda} \sum_{i=1}^{k} Z_i^{(N)} \log \frac{\pi_i}{p_i^0}.$$

THEOREM 1.1.   *Let Λ be an arbitrary subset of Ω, p^0 an arbitrary point of Λ and let R_N denote the shortcoming of the size α_N likelihood ratio test based on $Z^{(N)}$ for $H: p = p^0$ against $K: p \in \Lambda_1 = \Lambda - \{p^0\}$. If*

$$(1.8) \qquad \lim_{N \to \infty} \alpha_N = 0, \qquad -\log \alpha_N = o(N) \qquad \text{for} \quad N \to \infty,$$

then

$$(1.9) \qquad \lim_{N \to \infty} \sup_{p \in \Lambda_1} R_N(p) = 0.$$

Although Hoeffding's result and Theorem 1.1 are complementary in the sense mentioned above, we wish to point out that they are of an entirely different nature. Hoeffding's theorem concerns fixed alternatives and the performance of the likelihood ratio test is compared to that of a fixed sequence of tests by considering the ratio of error probabilities of the second kind. The alternatives at which the likelihood ratio test is considerably more powerful in Hoeffding's sense are necessarily alternatives where the power of the likelihood ratio test tends to one very rapidly. Since also the convergence of α_N to zero is assumed to be fast, the probabilities to be considered under the hypothesis as well as under the alternative are all probabilities of large deviations. The tools used to estimate these probabilities are Theorems 2.1 and A.1 in [2] which are reproduced here as Lemma 2.6.

In Theorem 1.1 on the other hand the performance of the likelihood ratio test is compared at each alternative to that of the most powerful test for that alternative. The comparison is in terms of power difference and the result is uniform on the set of alternatives. Alternatives or sequences of alternatives for which the power of the likelihood ratio test tends to one play a role only in so far as uniformity is concerned and the theorem is basically concerned with sequences of alternatives for which the power of the likelihood ratio test remains bounded away from one. Under alternatives we only have to compute probabilities of small deviations which is done by applying the central limit theorem. As α_N is allowed to tend to zero either slowly or fast, we are dealing with intermediate as well as large deviations under the hypothesis. In the former case where $-\log \alpha_N = o(N^{1/6})$, Theorem 1.1 was first proved by using classical limit theorems by J. Oosterhoff in [3] under the additional assumptions that $\Lambda = \Omega$ and that p^0 is an interior point of Ω. We shall use this result (Lemma 3.1) as a starting point for our investigation in the case where α_N tends to zero slowly. In the case where α_N tends to zero fast the resulting probabilities of large deviations are dealt with in the same manner as is done in [2].

The condition $-\log \alpha_N = o(N)$ in Theorem 1.1 is unduly restrictive and occurs there only for the sake of simplicity. In fact we shall show that it may be replaced by the assumption that there exists $\varepsilon > 0$ such that for all sufficiently large N

$$(1.10) \qquad\qquad \alpha_N \geqq \left(1 - p_m^0\right)^N e^{N\varepsilon},$$

where p_m^0 is the smallest positive coordinate of p^0. Moreover, further refinements of this condition are possible.

The reason that we need an assumption of this type at all, is to avoid complications arising from the fact that under sequences of alternatives converging sufficiently fast to certain boundary points of Ω, the distribution of the likelihood ratio statistic degenerates too rapidly. The nature of these complications is most easily made clear for alternatives located at the extreme points of Ω (that is, the points with a coordinate equal to one).

EXAMPLE 1.1.   Take for p^0 the point with coordinates $p_i^0 = k^{-1}, i = 1, \cdots, k$, and suppose that Λ contains all extreme points of Ω. Choose $\alpha_N = k^{-N}$. The statistic (1.7) assumes its maximum value if $Z_i^{(N)} = 1$ for some i. Since $P\left(Z_i^{(N)} = 1 \mid p^0\right) = k^{-N}$ for each i, the size α_N likelihood ratio test rejects $H : p = p^0$ with probability k^{-1} if $Z_i^{(N)} = 1$ for some i and hence its power at each of the extreme points of Ω is equal to k^{-1}. For each i, the size α_N most powerful test for $H : p = p^0$ against the simple alternative $p_i = 1$ rejects H if $Z_i^{(N)} = 1$ and has power one at $p_i = 1$. The shortcoming of the likelihood ratio test at each of the extreme points of Ω is therefore equal to $1 - k^{-1}$ for every N.

It is of course easy to modify this example in such a way that no randomization occurs.

Whereas Hoeffding's result is restricted to the case where $\Lambda = \Omega$ but allows a composite hypothesis H, Theorem 1.1 places no restriction on Λ but deals only with a simple hypothesis H. In Section 4 we shall show by means of a counter-example that even for the case where $\Lambda = \Omega$ Theorem 1.1 does not hold in general for a composite hypothesis H.

Section 2 of this paper contains some preliminary results on the multinomial distribution. In Section 3 we prove Theorem 1.1 and show that the condition $-\log \alpha_N = o(N)$ may be replaced by (1.10). Section 4 is devoted to the case where the hypothesis H is composite.

2. Preliminary results

For any set $A \subset \Omega$ we shall denote by A^N the set of all $y \in A$ for which Ny has integer coordinates.

LEMMA 2.1.   *For any $A \subset \Omega$ for which A^N is nonempty, the function $f(p = P(Z^{(N)} \in A|p)$ assumes its maximum value only at points p in the convex hull of A^N.*

PROOF.   Let π be a point in the complement of the convex hull of A^N. Since A^N contains only finitely many points its convex hull is closed and hence there exists a hyperplane separating π and A^N, that is, there exists a vector $a = (a_1, \cdots, a_k)$ such that $\Sigma a_i(z_i - \pi_i) > 0$ for all $z \in A^N$. Because $\Sigma z_i = \Sigma \pi_i = 1$, we may choose a in such a way that $\Sigma a_i \pi_i = 0$ and $\Sigma a_i z_i > 0$ for all $z \in A^N$. As $\Sigma a_i \pi_i = 0$ and $a_i = \pi_i = 0$ whenever $\pi_i = 0$, the points with coordinates $\pi_i + \varepsilon a_i \pi_i$ are points of Ω for all sufficiently small $\varepsilon > 0$. Hence

$$(2.1) \qquad \sum_{i=1}^{k} a_i \pi_i \frac{\partial}{\partial p_i} f(p)\big|_{p=\pi} = \sum_{i=1}^{k} a_i \sum_{z \in A^N} P(Z^{(N)} = z|\pi) N z_i$$

$$= N \sum_{z \in A^N} P(Z^{(N)} = z|\pi) \sum_{i=1}^{k} a_i z_i$$

is a directional derivative of f at π in a direction in Ω multiplied by a nonnegative constant. Note, however, that $a_i \pi_i$ may be equal to zero for all i if $\pi_i = 0$ for some i.

If $f(\pi) > 0$, then (2.1) is positive because $\Sigma a_i z_i > 0$ for all $z \in A^N$ and consequently f does not have a maximum at π. If $f(\pi) = 0$ the same conclusion holds since A^N is nonempty. Q.E.D.

For $z, p \in \Omega$ we define

$$(2.2) \qquad I(z, p) = \sum_{i=1}^{k} z_i \log \frac{z_i}{p_i},$$

where $z_i \log (z_i/p_i) = 0$ by definition if $z_i = 0$. It is well known that for fixed p this function is convex in z, positive unless $z = p$ and finite if $p_i \neq 0$ for all i. In Lemma 2.2 we show that under p the random variable $I(Z^{(N)}, p)$ is of order at most N^{-1} in probability uniformly in p.

LEMMA 2.2. *For every $\varepsilon > 0$ there exists $A > 0$ such that for all N*

$$(2.3) \qquad \sup_{p \in \Omega} P\left(I(Z^{(N)}, p) \geq \frac{A}{N} \Big| p\right) \leq \varepsilon.$$

PROOF. For $0 \leq z_i \leq 1, 0 < p_i \leq 1$,

$$(2.4) \qquad z_i \log \frac{z_i}{p_i} = z_i \log \left(1 + \frac{z_i - p_i}{p_i}\right) \leq z_i \frac{z_i - p_i}{p_i} = (z_i - p_i) + \frac{(z_i - p_i)^2}{p_i}.$$

Since uncer p, $Z_i^{(N)} = Z_i^{(N)} \log\left(Z_i^{(N)}/p_i\right) = 0$ a.s. if $p_i = 0$, we have under p

$$(2.5) \qquad 0 \leq I(Z^{(N)}, p) \leq \sum_{p_i \neq 0} \frac{(Z_i^{(N)} - p_i)^2}{p_i}$$

with probability one. It follows that

$$(2.6) \qquad E\left(I(Z^{(N)}, p)|p\right) \leq \sum_{p_i \neq 0} \frac{p_i(1 - p_i)}{Np_i} \leq \frac{k - 1}{N}.$$

Application of Markov's inequality completes the proof.

Let $\mathring{\Omega}$ denote the interior of Ω,

$$(2.7) \qquad \mathring{\Omega} = \{(y_1, \cdots, y_k)| \sum_{i=1}^{k} y_i = 1, y_i > 0 \text{ for } i = 1, \cdots, k\}$$

and define for $p^0 \in \mathring{\Omega}$, $p \in \Omega$,

$$(2.8) \qquad \sigma^2(p, p^0) = \sum_{i=1}^{k} p_i \left(\log \frac{p_i}{p_i^0}\right)^2 - \left(\sum_{i=1}^{k} p_i \log \frac{p_i}{p_i^0}\right)^2.$$

We shall have to consider the asymptotic distribution of

$$(2.9) \qquad T_p^{(N)} = \sum_{i=1}^{k} Z_i^{(N)} \log \frac{p_i}{p_i^0}$$

under p for fixed $p^0 \in \mathring{\Omega}$ and varying $p \in \Omega$. The distribution of $T_p^{(N)}$ under p is degenerate if and only if the positive coordinates of p are proportional to the corresponding coordinates of p^0 (as before we take $0 \log 0 = 0$ by definition). For $p \neq p^0$ and $p \geq \varepsilon > 0$ (that is, $p_i \geq \varepsilon$ for all $i = 1, \cdots, k$) the following lemma provides a uniform normal approximation. By Φ we denote the standard normal distribution function.

LEMMA 2.3. *For any fixed $p^0 \in \mathring{\Omega}$ and $\varepsilon > 0$,*

$$(2.10) \qquad \lim_{N \to \infty} P\left(\frac{T_p^{(N)} - I(p, p^0)}{(p, p^0)} N^{1/2} \leq a \Big| p\right) = \Phi(a)$$

uniformly for all a and all $p \in \Omega$ with $p \neq p^0$ and $p \geq \varepsilon$.

PROOF. Under p the distribution of $NT_p^{(N)}$ is the same as that of $\Sigma_{j=1}^{N} Y_j$, where $Y_1, \cdots, Y_N$ are independent and identically distributed random variables with

$$(2.11) \qquad\qquad P\left(Y_j = \log \frac{p_i}{p_i^0}\right) = p_i, \qquad\qquad i = 1, \cdots, k.$$

Hence

$$(2.12) \qquad\qquad E(T_p^{(N)}|p) = I(p, p^0),$$

$$(2.13) \qquad\qquad \sigma^2(T_p^{(N)}|p) = N^{-1}\sigma^2(p, p^0).$$

Let $F_{N, p}$ be the distribution function of

$$(2.14) \qquad\qquad \frac{T_p^{(N)} - I(p, p^0)}{\sigma(p, p^0)} N^{1/2}$$

and for $m = 2, 3$, let $v_{m, p}$ denote the mth absolute central moment of Y_j. Since the distribution of Y_j is degenerate only if the positive coordinates of p are proportional to the corresponding coordinates of p^0, $v_{m, p}$ is positive and finite if $p \neq p^0$ and $p \in \mathring{\Omega}$. Hence by the Berry-Esseen theorem (see [1]) we have for all a and N and for all $p \in \mathring{\Omega}$, $p \neq p^0$,

$$(2.15) \qquad\qquad |F_{N, p}(a) - \Phi(a)| \leq cv_{3, p}v_{2, p}^{-3/2} N^{-1/2},$$

where c is a constant independent of a, N and p. By (2.11)

$$(2.16) \qquad v_{m, p} = \sum_{j=1}^{k} p_j\eta_j^m, \qquad \eta_j = \left| \log \frac{p_j}{p_j^0} - \sum_{i=1}^{k} p_i \log \frac{p_i}{p_i^0} \right|;$$

if $p \neq p^0$ and $p \geq \varepsilon$ then $p_t\eta_t^3 = \max_j p_j\eta_j^3$ is positive and finite and as a result

$$(2.17) \qquad v_{3, p}v_{2, p}^{-3/2} \leq kp_t\eta_t^3 (p_t\eta_t^2)^{-3/2} = kp_t^{-1/2} \leq k\varepsilon^{-1/2}.$$

Together with (2.15) this proves the lemma.

LEMMA 2.4.   *For every fixed $p^0 \in \mathring{\Omega}$ and $\varepsilon > 0$ there exist $0 < M_1 < M_2 < \infty$ such that*

$$(2.18) \qquad\qquad M_1 I(p, p^0) \leq \sigma^2(p, p^0) \leq M_2 I(p, p^0)$$

for all $p \in \Omega$ with $p \geq \varepsilon$.

PROOF.   By expanding the logarithms involved we find that for $p \in \mathring{\Omega}$ with $\max|p_i - p_i^0| < \delta$,

$$I(p, p^0) = \frac{1}{2} \sum_{i=1}^{k} \frac{(p_i - p_i^0)^2}{p_i^0} + O(\delta^3),$$

$$(2.19)$$

$$\sigma^2(p, p^0) = \sum_{i=1}^{k} \frac{(p_i - p_i^0)^2}{p_i^0} + O(\delta^3).$$

The proof is completed by noting that for p outside a neighborhood of p^0 and $p \geq \varepsilon$, both $I(p, p^0)$ and $\sigma^2(p, p^0)$ are bounded away from zero and infinity.

For $p^0 \in \Lambda \subset \Omega$ we shall have to consider

$$(2.20) \qquad\qquad \sup_{\pi \in \Lambda} \sum_{i=1}^{k} Z_i^{(N)} \log \frac{\pi_i}{p_i^0},$$

where $z_i \log (\pi_i/p_i^0) = 0$ by definition if $z_i = 0$. Note that under $p \in \Lambda$ this random variable is defined (possibly $+\infty$) with probability one.

LEMMA 2.5. *Let Λ be an arbitrary subset of Ω, p^0 an arbitrary point of Λ and define $\Lambda_1 = \Lambda - \{p^0\}$. Furthermore, let c_N and a_N, $N = 1, 2, \cdots$, be sequences of nonnegative real numbers such that*

$$(2.21) \qquad \lim_{N \to \infty} c_N = 0, \qquad \lim_{N \to \infty} N c_N = \infty, \qquad \lim_{N \to \infty} \frac{N a_N^2}{c_N} = 0.$$

Then

$$(2.22) \qquad \sup_{p \in \Lambda_1} P \left(\sup_{\pi \in \Lambda} \sum_{i=1}^{k} Z_i^{(N)} \log \frac{\pi_i}{p_i^0} \leq c_N + a_N, I(Z^{(N)}, p^0) \geq c_N - a_N \big| p \right)$$

tends to zero for $N \to \infty$.

PROOF. Under $p \in \Lambda_1$,

$$(2.23) \qquad \sup_{\pi \in \Lambda} \sum_{i=1}^{k} Z_i^{(N)} \log \frac{\pi_i}{p_i^0} \geq \sum_{i=1}^{k} Z_i^{(N)} \log \frac{p_i}{p_i^0} = I(Z^{(N)}, p^0) - I(Z^{(N)}, p)$$

a.s. since under p, $0 \leq I(Z^{(N)}, p) < \infty$ a.s. Hence the lemma is proved if we show that

$$(2.24) \qquad \sup_{p \in \Lambda_1} P\big(c_N - a_N \leq I(Z^{(N)}, p^0) \leq c_N + a_N + I(Z^{(N)}, p)\big| p\big)$$

tends to zero for $N \to \infty$. By Lemma 2.2 it suffices to show that for every $A > 0$,

$$(2.25) \qquad \sup_{p \in \Omega} P \left(c_N - a_N \leq I(Z^{(N)}, p^0) \leq c_N + a_N + \frac{A}{N} \big| p \right) \to 0$$

for $N \to \infty$. We consider three cases.

(i) Suppose that $p^0 \in \overset{\circ}{\Omega}$. Since $c_N + a_N + A N^{-1} \to 0$ for $N \to \infty$, there exists $\varepsilon > 0$ such that for all sufficiently large N the set $\{z | z \in \Omega, I(z, p^0) \leq c_N + a_N + A N^{-1}\}$ is contained in the convex set $\{z | z \in \Omega, z_i \geq \varepsilon$ for $i = 1, \cdots, k\}$. By Lemma 2.1 the supremum over Ω in (2.25) may therefore be replaced by the supremum over the set of all $p \in \Omega$ with $p \geq \varepsilon$. Furthermore, we may again use the fact that under p

$$(2.26) \qquad I(Z^{(N)}, p^0) = \sum_{i=1}^{k} Z_i^{(N)} \log \frac{p_i}{p_i^0} + I(Z^{(N)}, p) \qquad \text{a.s.}$$

and $0 \leq I(Z^{(N)}, p) < \infty$ a.s. It follows from Lemma 2.2 that to prove (2.25) it is sufficient to show that for every $A > 0$ and $\varepsilon > 0$,

$$(2.27) \qquad \sup_{p \geq \varepsilon} P \left(c_N - a_N - \frac{A}{N} \leq \sum_{i=1}^{k} Z_i^{(N)} \log \frac{p_i}{p_i^0} \leq c_N + a_N + \frac{A}{N} \big| p \right)$$

tends to zero for $N \to \infty$.

The condition $Nc_N \to \infty$ implies that c_N is positive for all sufficiently large N; together with the condition $Na_N^2 c_N^{-1} \to 0$ it also yields

$$(2.28) \qquad a_N + \frac{A}{N} = o\left(\left(\frac{c_N}{N}\right)^{1/2}\right) = o(c_N)$$

for $N \to \infty$. Hence $c_N - a_N - AN^{-1} > 0$ for all sufficiently large N. As for $p = p^0$ the random variable in (2.27) is equal to 0 a.s., the supremum in (2.27) may be restricted to the set of all $p \neq p^0$ with $p \geq \varepsilon$. Applying Lemma 2.3 we find that it suffices to show that for every $A > 0$ and $\varepsilon > 0$

$$(2.29) \qquad \Phi\left(\frac{c_N + a_N + AN^{-1} - I(p, p^0)}{\sigma(p, p^0)} N^{1/2}\right)$$
$$- \Phi\left(\frac{c_N - a_N - AN^{-1} - I(p, p^0)}{\sigma(p, p^0)} N^{1/2}\right)$$

tends to zero for $N \to \infty$, uniformly for all $p \neq p^0$ with $p \geq \varepsilon$.

Define, for $N = 1, 2, \cdots$,

$$(2.30) \qquad \Omega_{N,1} = \left\{p \,\middle|\, p \in \Omega, p \neq p^0, p \geq \varepsilon, I(p, p^0) \leq \frac{c_N}{2}\right\},$$
$$\Omega_{N,2} = \left\{p \,\middle|\, p \in \Omega, p \neq p^0, p \geq \varepsilon, I(p, p^0) > \frac{c_N}{2}\right\}.$$

For $p \in \Omega_{N,1}$, (2.29) is bounded above by

$$(2.31) \qquad 1 - \Phi\left(\frac{\frac{1}{2}c_N - a_N - AN^{-1}}{\sigma(p, p^0)} N^{1/2}\right)$$

and by (2.28) and Lemma 2.4

$$(2.32) \qquad \frac{\frac{1}{2}c_N - a_N - AN^{-1}}{\sigma(p, p^0)} N^{1/2} \sim \frac{c_N N^{1/2}}{2\sigma(p, p^0)} \geq \frac{c_N N^{1/2}}{2[M_2 I(p, p^0)]^{1/2}}$$
$$\geq \left(\frac{Nc_N}{2M_2}\right)^{1/2} \to \infty \text{ for } N \to \infty.$$

For $p \in \Omega_{N,2}$, (2.29) is bounded above by

$$(2.33) \qquad \frac{a_N + AN^{-1}}{\sigma(p, p^0)} N^{1/2} \leq (a_N + AN^{-1})\left(\frac{N}{M_1 I(p, p^0)}\right)^{1/2}$$
$$\leq (a_N + AN^{-1})\left(\frac{2N}{M_1 c_N}\right)^{1/2} \to 0$$

by the mean value theorem, Lemma 2.4 and (2.28). Hence the suprema of (2.29) over both $\Omega_{N,1}$ and $\Omega_{N,2}$ tend to zero which proves the lemma for $p^0 \in \mathring{\Omega}$.

(ii) Suppose that p^0 is a boundary point but not an extreme point of Ω; without loss of generality we assume that for some $2 \leq m \leq k - 1$, $p_i^0 \neq 0$ for

$i = 1, \cdots, m$ and $p_i^0 = 0$ for $i = m + 1, \cdots, k$. Since $I(z, p^0) = \infty$ if $z_i \neq 0$ for some $m + 1 \leq i \leq k$, the set $\{z \,|\, z \in \Omega, I(z, p^0) \leq c_N + a_N + AN^{-1}\}$ is contained in the convex set $\{z \,|\, z \in \Omega, z_i = 0 \text{ for } i = m + 1, \cdots, k\}$. By Lemma 2.1 the supremum over Ω in (2.25) may therefore be replaced by the supremum over all $p \in \Omega$ with $p_i = 0$ for $i = m + 1, \cdots, k$. But under any p with $p_i = 0$ for $i = m + 1, \cdots, k$,

$$(2.34) \qquad I(Z^{(N)}, p^0) = \sum_{i=1}^{m} Z_i^{(N)} \log \frac{Z_i^{(N)}}{p_i^0} \qquad \text{a.s.}$$

and $(Z_1^{(N)}, \cdots, Z_m^{(N)})$ has a multinomial distribution with parameters N and $(p_1, \cdots, p_m)$. Thus we have reduced the problem of proving (2.25) to the same problem in a lower dimensional parameter space where $(p_1^0, \cdots, p_m^0)$ is now an interior point. This has been dealt with in (i).

(iii) Suppose that p^0 is an extreme point of Ω. This implies that $I(Z^{(N)}, p^0)$ can only assume the values 0 and ∞. Since $c_N - a_N > 0$ for all sufficiently large N, (2.25) is immediate. $Q.E.D.$

We remark that in the proof of Lemma 2.5 we have made use of the condition $c_N \to 0$ only to ensure that in case (i), for every $A > 0$

$$(2.35) \qquad \{z \,|\, z \in \Omega, I(z, p^0) \leq c_N + a_N + AN^{-1}\} \subset \{z \,|\, z \in \Omega, z \geq \varepsilon\}$$

for some $\varepsilon > 0$ for all sufficiently large N, whereas in case (ii) it is needed that the same condition holds for the reduced lower dimensional problem. As $a_N + AN^{-1} = o(c_N)$ by (2.14), Lemma 2.5 will continue to hold if we replace the condition $c_N \to 0$ by the following assumption. For all sufficiently large N the set $\{z \,|\, z \in \Omega, I(z, p^0) \leq c_N\}$ remains bounded away from the set of all points $z \in \Omega$ that have $z_i = 0$ for all i for which $p_i^0 = 0$ but also for at least one i with $p_i^0 \neq 0$. This extension of Lemma 2.5 is the main step in relaxing the condition $-\log \alpha_N = o(N)$ in Theorem 1.1 (see Section 3).

We complete this section by stating the result on large deviations of W. Hoeffding in [2] that we already referred to in Section 1. For a nonempty set $A \subset \Omega$ and $p \in \Omega$, define

$$(2.36) \qquad I(A, p) = \inf_{z \in A} I(z, p) = \inf_{z \in A} \sum_{i=1}^{k} z_i \log \frac{z_i}{p_i}.$$

If A is empty we take $I(A, p) = +\infty$. We recall that for any $A \subset \Omega$, A^N denotes the set of all $z \in A$ for which Nz has integer coordinates.

LEMMA 2.6 (Hoeffding).   *Uniformly for all $A \subset \Omega$ and all $p \in \Omega$,*

$$(2.37) \qquad P(Z^{(N)} \in A \,|\, p) = \exp \{-NI(A^N, p) + O(\log N)\}.$$

Moreover, for any $p \in \Omega$ and any sequence $A_N \subset \Omega$ with complex complements,

$$(2.38) \qquad I(A_N^N, p) = I(A_N, p) + O(N^{-1} \log N),$$

hence

$$(2.39) \qquad P(Z^{(N)} \in A_N \,|\, p) = \exp \{-NI(A_N, p) + O(\log N)\}.$$

3. Proof of Theorem 1.1

The size α_N likelihood ratio test based on $Z^{(N)}$ for $H: p = p^0$ against $K: p \neq p^0$ rejects H if

$$(3.1) \qquad I(Z^{(N)}, p^0) = \sum_{i=1}^{k} Z_i^{(N)} \log \frac{Z_i^{(N)}}{p_i^0} \geq c_N$$

with possible randomization if equality occurs. For this case, where $\Lambda = \Omega$, Oosterhoff [3] showed that Theorem 1.1 holds under the additional assumptions that $p^0 \in \overset{\circ}{\Omega}$ and that α_N tends to zero slowly. In his proof he found that under his conditions $-\log \alpha_N \sim N c_N$ for $N \to \infty$, which implies the conclusions concerning c_N in the following lemma.

LEMMA 3.1 (Oosterhoff).   *Let p^0 be an arbitrary point of $\overset{\circ}{\Omega}$ and let R_N denote the shortcoming of the size α_N likelihood ratio test (3.1) for $H: p = p^0$ against $K: p \in \Omega - \{p^0\}$. If*

$$(3.2) \qquad \lim_{N \to \infty} \alpha_N = 0, \qquad -\log \alpha_N = o(N^{1/6}) \qquad for \quad N \to \infty,$$

then

$$(3.3) \qquad \lim_{N \to \infty} \sup_{p \neq p^0} R_N(p) = 0,$$

and $N c_N \to \infty$, $c_N \to 0$ for $N \to \infty$.

We begin by removing, as far as possible, the restriction $p^0 \in \overset{\circ}{\Omega}$ in Lemma 3.1.

LEMMA 3.2.   *Let p^0 be an arbitrary point of Ω and let R_N denote the shortcoming of the size α_N likelihood ratio test (3.1) for $H: p = p^0$ against $K: p \in \Omega - \{p^0\}$. If*

$$(3.4) \qquad \lim_{N \to \infty} \alpha_N = 0, \qquad -\log \alpha_N = o(N^{1/6}) \qquad for \quad N \to \infty,$$

then

$$(3.5) \qquad \lim_{N \to \infty} \sup_{p \neq p^0} R_N(p) = 0.$$

Moreover, $N c_N \to \infty$, $c_N \to 0$ for $N \to \infty$ unless p^0 is an extreme point of Ω.

PROOF.   If $p^0 \in \overset{\circ}{\Omega}$ Lemma 3.2 is merely a repetition of Lemma 3.1. If p^0 is an extreme point of Ω, then the likelihood ratio test is uniformly most powerful and hence its shortcoming is identically equal to zero for all N. We may therefore suppose that p^0 is a boundary point but not an extreme point of Ω; without loss of generality we assume that for some $2 \leq m \leq k - 1$, $p_i^0 \neq 0$ for $i = 1, \cdots, m$ and $p_i^0 = 0$ for $i = m + 1, \cdots, k$.

In this case any admissible size α_N test for $H: p = p^0$ against $K: p \neq p^0$ rejects H with probability one if $Z_i^{(N)} \neq 0$ for at least one $i = m + 1, \cdots, k$, and with probability $\phi_N(z_1, \cdots, z_m)$ if $Z_i^{(N)} = z_i$ for $i = 1, \cdots, m$ and $Z_i^{(N)} = 0$ for $i = m + 1, \cdots, k$. The size α_N likelihood ratio test (3.1) is of this type with

$$(3.6) \qquad \phi_N(z_1, \cdots, z_m) = \begin{cases} 1 & \text{if} \quad \sum_{i=1}^{m} z_i \log \dfrac{z_i}{p_i^0} > c_N, \\[2ex] \delta & \text{if} \quad \sum_{i=1}^{m} z_i \log \dfrac{z_i}{p_i^0} = c_N, \\[2ex] 0 & \text{if} \quad \sum_{i=1}^{m} z_i \log \dfrac{z_i}{p_i^0} < c_N, \end{cases}$$

where $0 < \delta \leq 1$.

Let us introduce an auxiliary random vector $\tilde{Z}^{(N)} = (\tilde{Z}_1^{(N)}, \cdots, \tilde{Z}_m^{(N)})$ such that $N\tilde{Z}^{(N)}$ has a multinomial distribution with parameters N and $\tilde{p} = (\tilde{p}_1, \cdots, \tilde{p}_m)$, where $\tilde{p}$ is any point in

$$(3.7) \qquad \tilde{\Omega} = \{(y_1, \cdots, y_m) \mid \sum_{i=1}^{m} y_i = 1, y_i \geq 0 \text{ for } i = 1, \cdots, m\}.$$

Since $P\left(Z_{m+1}^{(N)} = \cdots = Z_k^{(N)} = 0 \mid p^0\right) = 1$, we have for the size α_N likelihood ratio test as well as for any admissible size α_N test

$$(3.8) \qquad \alpha_N = E\big(\phi_N(\tilde{Z}^{(N)}) \mid \tilde{p}^0\big),$$

where $\tilde{p} = (p_1^0, \cdots, p_m^0)$. For the power of such a test at $p \neq p^0$ we have

$$(3.9) \qquad \beta_N(p) = \begin{cases} 1 & \text{if} \quad p_1 = \cdots = p_m = 0, \\ 1 - \pi^N + \pi^N E\big(\phi_N(\tilde{Z}^{(N)}) \mid \tilde{p}\big) & \text{otherwise}, \end{cases}$$

where

$$(3.10) \qquad \pi = \sum_{i=1}^{m} p_i, \qquad \tilde{p}_i = \frac{p_i}{\pi} \qquad \text{for} \quad i = 1, \cdots, m.$$

For the random vector $\tilde{Z}^{(N)}$, consider the auxiliary problem of testing $\tilde{H} : \tilde{p} = \tilde{p}^0$ against $K : \tilde{p} \neq \tilde{p}^0$, where $\tilde{p}$ denotes the parameter vector of the distribution of $\tilde{Z}^{(N)}$. A test for this problem will reject $\tilde{H}$ with probability $\phi_N(z)$ if $\tilde{Z}^{(N)} = z$. Such a test has size α_N if and only if ϕ_N satisfies (3.8), and its power at $\tilde{p}$ is given by

$$(3.11) \qquad \tilde{\beta}_N(\tilde{p}) = E\big(\phi_N(\tilde{Z}^{(N)}) \mid \tilde{p}\big).$$

Thus there exists a one to one correspondence between the class of size α_N tests for H based on $Z^{(N)}$ that reject H with probability one if $Z_i^{(N)} \neq 0$ for at least one $i = m + 1, \cdots, k$ and the class of all size α_N tests for $\tilde{H}$ based on $\tilde{Z}^{(N)}$. Here corresponding tests have the same function ϕ_N and hence by (3.9) and (3.11) we find that for all p with $p_i \neq 0$ for at least one $i = 1, \cdots, m$, their power functions satisfy

$$(3.12) \qquad \beta_N(p) = 1 - \pi^N + \pi^N \tilde{\beta}_N(\tilde{p}),$$

where π and $\tilde{p}$ are defined by (3.10). Let β_N^+ and $\tilde{\beta}_N^+$ denote the size α_N envelope power functions for testing H on the basis of $Z^{(N)}$ and $\tilde{H}$ on the basis of $\tilde{Z}^{(N)}$, respectively. Since only admissible size α_N tests for H enter into the determination of β_N^+, it follows from (3.9) and (3.12) that

$$(3.13) \qquad \beta_N^+(p) = \begin{cases} 1 & \text{if} \quad p_1 = \cdots = p_m = 0, \\ 1 - \pi^N + \pi^N \tilde{\beta}_N^+(\tilde{p}) & \text{otherwise,} \end{cases}$$

where π and $\tilde{p}$ are defined by (3.10).

The likelihood ratio test for the auxiliary problem of testing $\tilde{H}$ against $\tilde{K}$ is based on the statistic $I(\tilde{Z}^{(N)}, \tilde{p}^0)$. As the function ϕ_N for the size α_N likelihood ratio test given by (3.6) satisfies (3.8), this function is also the test function of the size α_N likelihood ratio test for $\tilde{H}$ against $\tilde{K}$. In the first place this implies that the critical values of the two size α_N likelihood ratio tests are both equal to the same number c_N. In the second place it means that (3.13) will continue to hold if the envelope power functions β_N^+ and $\tilde{\beta}_N^+$ are replaced by the power functions β_N and $\tilde{\beta}_N$ of the size α_N likelihood ratio tests. Hence, if R_N and $\tilde{R}_N$ denote the shortcomings of the size α_N likelihood ratio tests for H against K and for $\tilde{H}$ against $\tilde{K}$, respectively, then

$$(3.14) \qquad R_N(p) = \begin{cases} 0 & \text{if} \quad p_1 = \cdots = p_m = 0, \\ \pi^N \tilde{R}_N(\tilde{p}) & \text{otherwise,} \end{cases}$$

where π and $\tilde{p}$ are defined by (3.10). Since $\pi \leq 1$ and $\tilde{R}_N(\tilde{p}^0) = 0$,

$$(3.15) \qquad \sup_{p \neq p^0} R_N(p) \leq \sup_{\tilde{p} \neq \tilde{p}^0} \tilde{R}_N(\tilde{p}).$$

As $\tilde{p}^0$ is an interior point of $\tilde{\Omega}$ we may apply Lemma 3.1 to the auxiliary testing problem to conclude that the right side of (3.15) tends to zero and that $Nc_N \to \infty$, $c_N \to 0$ for $N \to \infty$. Q.E.D.

Our next step will be to remove the restriction $\Lambda = \Omega$.

LEMMA 3.3. *Let Λ be an arbitrary subset of Ω, p^0 an arbitrary point of Λ and let R_N denote the shortcoming of the size α_N likelihood ratio test based on $Z^{(N)}$ for $H: p = p^0$ against $K: p \in \Lambda_1 = \Lambda - \{p^0\}$. If*

$$(3.16) \qquad \lim_{N \to \infty} \alpha_N = 0, \qquad -\log \alpha_N = o(N^{1/6}) \qquad \text{for} \quad N \to \infty,$$

then

$$(3.17) \qquad \lim_{N \to \infty} \sup_{p \in \Lambda_1} R_N(p) = 0.$$

PROOF. If p^0 is an extreme point of Ω, the likelihood ratio test for H against K is uniformly most powerful against K and hence its shortcoming is equal to zero for all $p \in \Lambda_1$ and all N. We may therefore suppose that p is not an extreme point of Ω.

The size α_N likelihood ratio test for H against K rejects H if

$$(3.18) \qquad \sup_{\pi \in \Lambda} \sum_{i=1}^{k} Z_i^{(N)} \log \frac{\pi_i}{p_i^0} \geq c_N^*,$$

possibly with randomization if equality occurs. Let us compare this test with the size α_N likelihood ratio test (3.1) for H against $p \neq p^0$. By Lemma 3.2 the short-coming of the latter test vanishes uniformly for all $p \neq p^0$ for $N \to \infty$ and hence Lemma 3.3 will be proved if we show that

$$(3.19) \qquad \sup_{p \in \Lambda_1} P\left(\sup_{\pi \in \Lambda} \sum_{i=1}^{k} Z_i^{(N)} \log \frac{\pi_i}{p_i^0} \leq c_N^*, I(Z^{(N)}, p^0) \geq c_N \middle| p \right)$$

tends to zero for $N \to \infty$, where c_N is the constant that occurs in (3.1). As p^0 is not an extreme point of Ω, Lemma 3.2 also ensures that $c_N \to 0$ and $Nc_N \to \infty$ for $N \to \infty$. Furthermore we note that under any $p \in \Lambda$

$$(3.20) \qquad \sup_{\pi \in \Lambda} \sum_{i=1}^{k} Z_i^{(N)} \log \frac{\pi_i}{p_i^0} \leq I(Z^{(N)}, p^0) \qquad\qquad \text{a.s.}$$

Since the tests (3.1) and (3.18) have the same size it follows that c_N^* and c_N may be chosen in such a way that $c_N^* \leq c_N$. To prove Lemma 3.3 it is therefore sufficient to show that

$$(3.21) \qquad \sup_{p \in \Lambda_1} P\left(\sup_{\pi \in \Lambda} \sum_{i=1}^{k} Z_i^{(N)} \log \frac{\pi_i}{p_i^0} \leq c_N, I(Z^{(N)}, p^0) \geq c_N \middle| p \right)$$

tends to zero for $N \to \infty$. As $c_N \to 0$ and $Nc_N \to \infty$ for $N \to \infty$, this is the content of Lemma 2.5 for $a_N = 0$. Q.E.D.

We now turn to the case where α_N tends to zero fast.

LEMMA 3.4.   *Lemma 3.3 holds if the conditions* (3.16) *concerning* α_N *are replaced by*

$$(3.22) \qquad \lim_{N \to \infty} \frac{-\log \alpha_N}{(\log N)^2} = \infty, \qquad -\log \alpha_N = o(N) \qquad \text{for} \quad N \to \infty.$$

PROOF.   For the same reason as in the proof of Lemma 3.3 we may restrict attention to the case where p^0 is not an extreme point of Ω. Consider the size α_N likelihood ratio test (3.1) for $H : p = p^0$ against $p \neq p^0$. The convexity of $I(z, p^0)$ in z ensures that the sets

$$(3.23) \qquad \begin{aligned} A_N &= \{z \,|\, z \in \Omega, I(z, p^0) \geq c_N\}, \\ B_N &= \{z \,|\, z \in \Omega, I(z, p^0) > c_N\} \end{aligned}$$

have convex complements. By the second part of Lemma 2.6

$$(3.24) \qquad \begin{aligned} \alpha_N &\leq P\big(I(Z^{(N)}, p^0) \geq c_N \,\big|\, p^0\big) \\ &= \exp\{-NI(A_N, p^0) + O(\log N)\} = \exp\{-Nc_N + O(\log N)\}, \end{aligned}$$

or $Nc_N \leq -\log \alpha_N + O(\log N)$. This implies that $c_N \to 0$ for $N \to \infty$ by the second part of (3.22). For $z \in \Omega$, the function $I(z, p^0)$ assumes all values in the interval $[0, -\log p_m^0]$ where p_m^0 is the smallest positive coordinate of p^0. As p^0 is not an extreme point of Ω, $-\log p_m^0 > 0$ and hence $0 \leq c_N < -\log p_m^0$ for all sufficiently large N. For these values of N, $I(B_N, p^0) = c_N$ and by the second part of Lemma 2.6

$$(3.25) \qquad \alpha_N \geqq P\big(I(Z^{(N)}, p^0) > c_N | p^0\big) = \exp\{-Nc_N + O(\log N)\}.$$

Hence

$$(3.26) \qquad \alpha_N = \exp\{-Nc_N + O(\log N)\}, \qquad Nc_N = -\log \alpha_N + O(\log N);$$

together with (3.22) this yields

$$(3.27) \qquad \lim_{N \to \infty} \frac{Nc_N}{(\log N)^2} = \infty, \qquad \lim_{N \to \infty} c_N = 0.$$

By the first part of Lemma 2.6 there exists a number $0 \leqq a < \infty$ independent of N, such that for every N and every $z^{(N)} \in \Omega^N$ with $I(z^{(N)}, p^0) < c_N - a(\log N)/N$,

$$(3.28) \qquad P\big(Z^{(N)} = z^{(N)} | p^0\big) \geqq \exp\{-Nc_N + a\log N + O(\log N)\} \geqq N\alpha_N.$$

Obviously, any size α_N test for $H: p = p^0$ cannot reject H with probability larger than N^{-1} if $Z^{(N)}$ assumes one of these values $z^{(N)}$. Hence the size α_N envelope power β_N^+ for testing H satisfies

$$(3.29) \qquad \beta_N^+(p) \leqq N^{-1} + P\big(I(Z^{(N)}, p^0) \geqq c_N - a_N | p\big)$$

for all $p \neq p^0$, where

$$(3.30) \qquad a_N = \frac{a\log N}{N}, \qquad\qquad 0 \leqq a < \infty.$$

We note that (3.29) is a slightly modified form of a conclusion due to W. Hoeffding in [2].

It follows from (3.29) that the shortcoming $R_N(p)$ at p of the size α_N likelihood ratio test (3.18) for H against K is bounded above by

$$(3.31) \qquad P\bigg(\sup_{\pi \in \Lambda} \sum_{i=1}^{k} Z_i^{(N)} \log \frac{\pi_i}{p_i^0} \leqq c_N^*, I(Z^{(N)}, p^0) \geqq c_N - a_N \Big| p\bigg) + \frac{1}{N}.$$

By the reasoning given in the proof of Lemma 3.3 we may assume that $c_N^* \leqq c_N$ and hence Lemma 3.4 is proved if we show that

$$(3.32) \qquad \sup_{p \in \Lambda_1} P\bigg(\sup_{\pi \in \Lambda} \sum_{i=1}^{k} Z_i^{(N)} \log \frac{\pi_i}{p_i^0} \leqq c_N, I(Z^{(N)}, p^0) \geqq c_N - a_N \Big| p\bigg)$$

tends to zero for $N \to \infty$. By (3.27) and (3.30), $c_N \to 0$, $Nc_N \to \infty$ and $Na_N^2/c_N \to 0$ for $N \to \infty$. Application of Lemma 2.5 completes the proof.

PROOF OF THEOREM 1.1.   The theorem is proved by splitting up the sequence α_N into two subsequences satisfying (3.16) and (3.22), respectively, and applying Lemmas 3.3 and 3.4.

In Section 1 we claimed that the condition $-\log \alpha_N = o(N)$ in Theorem 1.1 may be relaxed. To see how this can be achieved we obviously need not consider the proof of Theorem 1.1 for the case where $\alpha_N \to 0$ slowly; we only have to inspect the proof of Lemma 3.4.

In proving Lemma 3.4 we have made use of the condition $-\log \alpha_N = o(N)$ only to conclude that $c_N \to 0$ for $N \to \infty$, provided that p^0 is not an extreme point of Ω. This fact was needed on two occasions. In the first place it was used to ensure that, if p^0 is not an extreme point of Ω, we have $0 \leqq c_N < -\log p_m^0$ for all sufficiently large N, where p_m^0 denotes the smallest positive coordinate of p^0. As the function $I(z, p^0)$ assumes its largest finite value $-\log p_m^0$ at those extreme points $z \in \Omega$ for which $z_i = 1$ for some i with $p_i^0 = p_m^0$, the assertion $0 \leqq c_N < -\log p_m^0$ is equivalent to saying that the set

$$(3.33) \qquad C_N = \{z \,|\, z \in \Omega, \, I(z, p^0) \leqq c_N\}$$

does not contain these specific extreme points of Ω. We recall that C_N is the closure of the acceptance region of the size α_N likelihood ratio test (3.1) for $H: p = p^0$ against $p \neq p^0$.

In the second place, the fact that $c_N \to 0$ was used to ensure applicability of Lemma 2.5. However, in the remark following the proof of this lemma we pointed out that the lemma remains valid if the condition $c_N \to 0$ is replaced by the following assumption.

ASSUMPTION 1. *For all sufficiently large N the sets C_N defined in (3.33) remain bounded away from the set D_{p^0} of all points $z \in \Omega$ that have $z_i = 0$ for all i for which $p_i^0 = 0$ but also for at least one i with $p_i^0 \neq 0$.*

This assumption obviously implies that, for all sufficiently large N, the set C_N does not contain any extreme points of Ω, unless p^0 itself is an extreme point. It follows that Theorem 1.1 will continue to hold if the condition $-\log \alpha_N = o(N)$ is replaced by Assumption 1. Note that Assumption 1 imposes no restriction if p^0 is an extreme point of Ω.

One easily verifies that for $p^0 < 1$ (that is, $p_i^0 < 1$ for all i),

$$(3.34) \qquad \inf_{z \in D_{p^0}} I(z, p^0) = -\log (1 - p_m^0),$$

where p_m^0 is defined as above. Since $I(z, p^0)$ is convex and uniformly continuous on the set of all z that have $z_i = 0$ for all i with $p_i^0 = 0$, Assumption 1 is equivalent to the requirement that if $p^0 < 1$, there exists $\varepsilon > 0$ such that for all sufficiently large N, $c_N \leqq -\log (1 - p_m^0) - \varepsilon$. Going over the proof of Lemma 3.4 we find that this, in turn, is equivalent to

ASSUMPTION 2. *There exists $\varepsilon > 0$ such that for all sufficiently large N, $-\log \alpha_N \leqq N\big(-\log (1 - p_m^0) - \varepsilon\big)$, where p_m^0 denotes the smallest positive coordinate of p^0.*

Note that if p^0 is an extreme point of Ω, Assumption 2 imposes no restriction on the sequence α_N. As Assumptions 1 and 2 are equivalent, the condition $-\log \alpha_N = o(N)$ in Theorem 1.1 may be replaced by the obviously weaker Assumption 2.

By sharpening Lemmas 2.3 and 2.5 one can show that Theorem 1.1 will still continue to hold if C_N does approach D_{p^0} for $N \to \infty$, but does so sufficiently slowly. In Assumption 2 this corresponds to allowing ε to tend to zero for $N \to \infty$, provided that this convergence is sufficiently slow.

4. The case of a composite hypothesis

In this section we show by means of a counterexample that Theorem 1.1 breaks down in the case of a composite hypothesis H even when $\Lambda = \Omega$. We consider the binomial case $k = 2$ and write $Z^{(N)} = Z_1^{(N)}$, $1 - Z^{(N)} = Z_2^{(N)}$, $p = p_1$ and $1 - p = p_2$. Thus $NZ^{(N)}$ has a binomial distribution with parameters N and p where p is an arbitrary point in $[0, 1]$. For $z \in [0, 1]$, $p \in [0, 1]$ and $\Lambda_0 \subset [0, 1]$ we define

$$(4.1) \qquad I(z, p) = z \log \frac{z}{p} + (1 - z) \log \frac{1 - z}{1 - p},$$

and

$$(4.2) \qquad I(z, \Lambda_0) = \inf_{p \in \Lambda_0} I(z, p).$$

If Λ_0 is a proper subset of $[0, 1]$, one may consider the problem of testing $H: p \in \Lambda_0$ against $K: p \notin \Lambda_0$. A nonrandomized likelihood ratio test for H against K rejects H if

$$(4.3) \qquad I(Z^{(N)}, \Lambda_0) \geqq \tilde{c}_N;$$

the size of this test is
$$(4.4) \qquad \alpha_N = \sup_{p \in \Lambda_0} P\big(I(Z^{(N)}, \Lambda_0) \geqq \tilde{c}_N | p\big).$$

Consider any fixed sequence of positive numbers $\tilde{c}_N$ such that

$$(4.5) \qquad \lim_{N \to \infty} \tilde{c}_N = 0, \qquad \lim_{N \to \infty} N\tilde{c}_N = \infty.$$

We choose two positive integers a and b and a sequence d_N such that $0 < d_N < \tilde{c}_N$ for all N and $Nd_N \to 0$ for $N \to \infty$. Next we construct a set $\Lambda_0 \subset [0, 1]$ with the following property: there exists an infinite sequence of positive integers $N_1 < N_2 < \cdots$ such that for every i the following conditions are satisfied:

(i) Λ_0 contains points $p_{i, 1} < p_{i, 2}$ with

$$(4.6) \qquad I\left(\frac{a}{N_i}, p_{i, j}\right) = \tilde{c}_{N_i} - d_{N_i} \qquad \text{for} \quad j = 1, 2.$$

(ii) Λ_0 contains points $p_{i, 3} < p_{i, 4}$ with

$$(4.7) \qquad I\left(1 - \frac{b}{N_i}, p_{i, j}\right) = \tilde{c}_{N_i} \qquad \text{for} \quad j = 3, 4.$$

(iii) Λ_0 does not contain points in $(p_{i, 1}, p_{i, 2}) \cup (p_{i, 3}, p_{i, 4})$.

To see that this construction is possible we note that for sufficiently large N, $1 \leqq a, b \leqq N - 1$ and hence

$$(4.8) \qquad I\left(\frac{a}{N}, 0\right) = I\left(\frac{a}{N}, 1\right) = I\left(1 - \frac{b}{N}, 0\right) = I\left(1 - \frac{b}{N}, 1\right) = \infty.$$

Thus, for any sequence $N_1 < N_2 < \cdots$ with $N_1 - 1 \geqq \max(a, b)$, points $p_{i,j}$ with properties (i) and (ii) exist for every i. Notice that obviously $0 < p_{i,1} < aN_i^{-1} < p_{i,2}$ and $p_{i,3} < 1 - bN_i^{-1} < p_{i,4} < 1$. Since $\tilde{c}_N \to 0$ for $N \to \infty$, we can also ensure for every $0 < \varepsilon < \frac{1}{2}$ that $p_{1,2} < \varepsilon < 1 - \varepsilon < p_{1,3}$ by choosing N_1 large enough. Having chosen $N_1 \geqq \max(a, b) + 1$ in such a way that the above holds for some $0 < \varepsilon < \frac{1}{2}$, we proceed to choose N_i for $i = 2, 3, \cdots$ sequentially in such a way that

$$(4.9) \qquad \begin{aligned} \frac{a}{N_i} &< p_{i-1,1}, & I\left(\frac{a}{N_i}, p_{i-1,1}\right) &> \tilde{c}_{N_i} - d_{N_i}, \\ 1 - \frac{b}{N_i} &> p_{i-1,4}, & I\left(1 - \frac{b}{N_i}, p_{i-1,4}\right) &> \tilde{c}_{N_i}. \end{aligned}$$

This is clearly possible as $p_{i-1,1} > 0, I(0, p_{i-1,1}) > 0, p_{i-1,4} < 1, I(1, p_{i-1,4}) > 0$ for all $i \geqq 2$ and $\tilde{c}_N \to 0$ for $N \to \infty$. However, this implies that $p_{i,2} < p_{i-1,1}$ and $p_{i,3} > p_{i-1,4}$ for every $i \geqq 2$. Because we already made sure that $p_{1,2} < \varepsilon < 1 - \varepsilon < p_{1,3}$, condition (iii) will be satisfied if Λ_0 does not contain other points in an ε neighborhood of 0 and 1 besides the points $p_{i,j}$.

For an arbitrary sequence $\tilde{c}_N$ satisfying (4.5) and for a corresponding set Λ_0 that we have just constructed, we consider the sequence of likelihood ratio tests (4.3) for $H: p \in \Lambda_0$ against $K: p \notin \Lambda_0$. We shall show that α_N defined by (4.4) satisfies the conditions $\alpha_N \to 0$ and $-\log \alpha_N = o(N)$ of Theorem 1.1, but that the shortcoming of this sequence of likelihood ratio tests does not tend to zero uniformly for all $p \notin \Lambda_0$.

By (4.2) and (4.4)

$$(4.10) \qquad \alpha_N \leqq \sup_{p \in \Lambda_0} P\big(I(Z^{(N)}, p) \geqq \tilde{c}_N | p\big),$$

and since $N\tilde{c}_N \to \infty$, $\alpha_N \to 0$ for $N \to \infty$ by Lemma 2.2. Let p_0 be an isolated point of Λ_0 with $0 < p_0 < 1$, for example, $p_0 = p_{1,1}$. For z in a sufficiently small neighborhood of p_0, $I(z, \Lambda_0) = I(z, p_0)$ and the absolute value of the derivative of this function is smaller than δ. Since $\tilde{c}_N \to 0$, the set

$$(4.11) \qquad \tilde{A}_N = \{z | 0 \leqq z \leqq 1, I(z, \Lambda_0) \geqq \tilde{c}_N\}$$

will contain, for all sufficiently large N, a point $z^{(N)}$ for which $Nz^{(N)}$ is an integer and $I(z^{(N)}, p_0) \leqq \tilde{c}_N + \delta N^{-1}$. Hence by Lemma 2.6

$$(4.12) \qquad \alpha_N \geqq P\big(I(Z^{(N)}, \Lambda_0) \geqq \tilde{c}_N | p_0\big) \geqq \exp\{-N\tilde{c}_N - \delta + O(\log N)\},$$

and as $\tilde{c}_N \to 0$, $-\log \alpha_N = o(N)$ for $N \to \infty$.

For $N = N_i$ we need a sharper asymptotic lower bound for α_N. By properties (ii) and (iii) of the set Λ_0

$$(4.13) \qquad I\left(1 - \frac{b}{N_i}, \Lambda_0\right) = I\left(1 - \frac{b}{N_i}, p_{i,3}\right) = \tilde{c}_{N_i}$$

for all i. It follows that for every $\varepsilon > 0$ we have for all sufficiently large i

$$(4.14) \qquad \alpha_{N_i} \geq P\left(Z^{(N_i)} = 1 - \frac{b}{N_i}\Big|p_{i,3}\right)$$

$$= \exp\{-N_i \tilde{c}_{N_i}\}\, P\left(Z^{(N_i)} = 1 - \frac{b}{N_i}\Big|1 - \frac{b}{N_i}\right)$$

$$\geq (1 - \varepsilon)\, e^{-b}\frac{b^b}{b!}\exp\{-N_i \tilde{c}_{N_i}\}.$$

Also, by properties (i) and (iii) of the set Λ_0

$$(4.15) \qquad I\left(\frac{a}{N_i}, \Lambda_0\right) = I\left(\frac{a}{N_i}, p_{i,j}\right) = \tilde{c}_{N_i} - d_{N_i}$$

for $j = 1, 2$ and all i. Because $Nd_N \to 0$ for $N \to \infty$ this implies that for every $\varepsilon > 0$

$$(4.16) \qquad \sup_{p\in\Lambda_0} P\left(Z^{(N_i)} = \frac{a}{N_i}\Big|p\right) = \max_{j=1,2} P\left(Z^{(N_i)} = \frac{a}{N_i}\Big|p_{i,j}\right)$$

$$= \exp\{-N_i(\tilde{c}_{N_i} - d_{N_i})\}\, P\left(Z^{(N_i)} = \frac{a}{N_i}\Big|\frac{a}{N_i}\right)$$

$$\leq (1 + \varepsilon)\, e^{-a}\frac{a^a}{a!}\exp\{-N_i \tilde{c}_{N_i}\}$$

for all sufficiently large i. Together with (4.14) this implies that there exists a number $0 < \phi \leq 1$ such that the test T_N that rejects H with probability ϕ if $Z^{(N)} = aN^{-1}$ has size at most α_N whenever $N = N_i$ and i is sufficiently large. Hence, if β_N^+ denotes the size α_N envelope power for testing H, we have shown that for every $\varepsilon > 0$

$$(4.17) \qquad \beta_{N_i}^+\left(\frac{a}{N_i}\right) \geq \phi P\left(Z^{(N_i)} = \frac{a}{N_i}\Big|\frac{a}{N_i}\right) \geq \phi(1 - \varepsilon)\, e^{-a}\frac{a^a}{a!},$$

for all sufficiently large i. On the other hand, property (i) of the set Λ_0 ensures that for $N = N_i$ the critical region $\tilde{A}_N$ of the likelihood ratio test does not contain points in the interval $[p_{i,1}, p_{i,2}]$. If β_N denotes the power of the size α_N likelihood ratio test, this means that for all i

$$(4.18) \qquad \beta_{N_i}\left(\frac{a}{N_i}\right) \leq \beta_{N_i}(p_{i,1}) + \beta_{N_i}(p_{i,2}) \leq 2\alpha_{N_i},$$

where the right side tends to zero for $i \to \infty$. Together with (4.17) this proves that the shortcoming of the likelihood ratio test does not tend to zero uniformly for all $p \notin \Lambda_0$.

REFERENCES

[1] W. FELLER, *An Introduction to Probability Theory and Its Applications*, Vol. 2, New York, Wiley, 1966.
[2] W. HOEFFDING, "Asymptotically optimal tests for multinomial distributions," *Ann. Math. Statist.*, Vol. 36 (1965), pp. 369–401.
[3] J. OOSTERHOFF, *Combination of One-sided Statistical Tests*, Amsterdam, Mathematisch Centrum, 1969.

ON THE EXISTENCE OF PROPER BAYES MINIMAX ESTIMATORS OF THE MEAN OF A MULTIVARIATE NORMAL DISTRIBUTION

WILLIAM E. STRAWDERMAN
STANFORD UNIVERSITY

1. Introduction

Consider the problem of estimating the mean of a multivariate normal distribution with convariance matrix the identity and sum of squared errors loss. In an earlier paper [5] the author showed that if the dimension p is 5 or greater, then proper Bayes minimax estimators do exist. We review this result briefly in Section 2.

The main purpose of the present paper is to show that for p equal to 3 or 4, there do not exist spherically symmetric proper Bayes minimax estimators. The author has been unable, thus far, to disprove the existence of a nonspherical proper Bayes minimax estimator for p equal 3 or 4. Of course, for $p = 1, 2$, the usual estimator $\overline{X}$ is unique minimax but not proper Bayes.

In Section 3 we derive bounds for the possible bias of a minimax estimator. This result should be of some interest independent of its use in proving the main result of the paper. Section 4 is devoted to the proof of the main result.

2. The case of five and higher dimensions

In [5] the author produced a class of estimators for $p \geq 5$ which are proper Bayes minimax. This was done without loss of generality, in the case of a single observation. For completeness we briefly describe this result.

Let X be a p dimensional random vector distributed according to the multinormal distribution with mean θ and covariance matrix I.

The prior distribution on θ is given as follows: conditional on λ, where $0 < \lambda \leq 1$, let the distribution of θ be multinormal with mean zero and covariance matrix $[(1 - \lambda)/\lambda]I$. The unconditional density of λ with respect to Lebesgue measure is given by $\lambda^{-a}/(1 - a)$ for any a where $0 \leq a < 1$.

The Bayes estimator with respect to the above prior distribution on θ is given by

$$(1) \qquad \delta(X) = \left[1 - \left(\frac{p + 2 - 2a}{\|X\|^2} - \frac{2 \exp\{-\frac{1}{2}\|X\|^2\}}{\|X\|^2 \int_0^1 \lambda^{p/2-a} \exp\{-\lambda\|X\|^2\}} \right) \right] X.$$

It then follows from a simple extension of a result due to Baranchik [1], [2] that the estimator $\delta(X)$, which is proper Bayes by definition, is minimax for $p \geq 6$. In addition for $\frac{1}{2} \leq a < 1$, $\delta(X)$ is minimax for $p = 5$.

3. The bias of spherically symmetric minimax estimators

In this section we derive upper and lower bounds on the possible bias of spherically symmetric minimax estimators, that is, minimax estimators of the form $\delta(X) = h(\|X\|^2)X$. For such estimators, the ith component of the bias is given by $E_\theta h(\|X\|^2)X_i - \theta_i = -\varphi(\|\theta\|^2)\theta_i$. Using the multivariate information inequality as in Stein [3], p. 202, we have

$$(2) \qquad E_\theta(\|h(\|X\|^2)X - \theta\|^2)$$
$$\geq p + \|\theta\|^2 \varphi^2(\|\theta\|^2) - 2p\varphi(\|\theta\|^2) - 4\|\theta\|^2 \varphi'(\|\theta\|^2).$$

If $\delta(X) = h(\|X\|^2)X$ is minimax, (2) becomes

$$(3) \qquad 0 \geq \|\theta\|^2 \varphi^2(\|\theta\|^2) - 2p\varphi(\|\theta\|^2) - 4\|\theta\|^2 \varphi'(\|\theta\|^2).$$

Letting $t = \|\theta\|^2$ and $\psi(t) = t\,\varphi(t)$, we have

$$(4) \qquad 0 \geq \psi^2(t)/t - 2p\psi(t)/t - 4\psi'(t) - 4\psi(t)/t$$

or

$$(5) \qquad 4\psi'(t) \geq \psi(t)[\psi(t) - (2p - 4)]/t.$$

We prove the following lemma.

LEMMA 1.   *For* $p \geq 2$, $0 \leq \psi(t) \leq 2(p - 2)$, $0 < t < \infty$.

PROOF.   Suppose that for some $t_0 > 0$, $\psi(t_0) < 0$. This implies by (5) that $\psi(t) < 0$ for all $0 < t < t_0$, and hence

$$(6) \qquad \psi'(t)[\psi(t)(\psi(t) - 2p - 4)]^{-1} \geq 1/4t.$$

Integrating from $t(0 < t < t_0)$ to t_0

$$(7) \qquad \log\left[\frac{-\psi(t_0) + (2p - 4)}{-\psi(t_0)}\right]\left[\frac{-\psi(t)}{-\psi(t) + (2p - 4)}\right] \geq \frac{p - 2}{2}\log\left(\frac{t_0}{t}\right).$$

As $t \to 0$ the right side of (7) approaches $+\infty$ while the left side remains bounded, a contradiction. Hence it cannot happen that $\psi(t_0) < 0$.

Assume next for some $t_0 > 0$ that $\psi(t_0) > 2p - 4$. From (5) it follows that $\psi(t) > 2p - 4$ for all $t > t_0$. Proceeding as above except integrating the inequality (6) from t_0 to t, we obtain

$$(8) \qquad \log\left[\frac{\psi(t) - (2p - 4)}{\psi(t)}\right]\left[\frac{\psi(t_0)}{\psi(t_0) - (2p - r)}\right] \geq \frac{p - 2}{2}\log\left(\frac{t_0}{t}\right).$$

Again the right side approaches $+\infty$ as t approaches $+\infty$ while the left side remains bounded. This contradiction establishes the lemma.

From the lemma and the definition of $\psi(t)$ we get immediately the following result.

THEOREM 1. *If* $\delta(x) = h(\|X\|^2)X$ *is a minimax estimator, and*

$$(9) \qquad E_\theta(h(\|X\|^2)X_i - \theta_i) = -\varphi(\|\theta\|^2)\theta_i.$$

Then $0 \leq \varphi(\|\theta\|^2) \leq 2(p-2)/\|\theta\|^2$.

In other words, the bias of a spherically symmetric minimax estimator is always towards the origin but not by a factor larger than $2(p-2)/\|\theta\|2$.

4. The nonexistence of spherically symmetric proper Bayes minimax estimators in three and four dimensions

In this section we apply Theorem 1 to show that spherically symmetric proper Bayes estimators cannot be minimax for $p = 3$ and 4.

We note first that if $\delta(X)$ is a generalized Bayes estimator relative to the prior $dG(\theta)$,

$$(10) \qquad \delta_i(X) - X_i = \frac{\partial}{\partial X_i} \log \int_{R^p} \exp\left\{-\tfrac{1}{2}\|X - \theta\|^2\right\} dG(\theta).$$

The ith components of the $\delta(\cdot)$ is given by

$$(11) \quad A \int_{R^p} \left[\frac{\partial}{\partial x_i} \log \int_{R^p} \exp\left\{-\tfrac{1}{2}\|x - \theta\|^2\right\} dG(\theta)\right] \exp\left\{-\tfrac{1}{2}\|x - u\|^2\right\} \prod_{i=1}^p dx_i$$

$$= -A \int_{R^p} \left[\log \int_{R^p} \exp\left\{-\tfrac{1}{2}\|x - \theta\|^2\right\} dG(\theta)\right] \frac{\partial}{\partial x_i} \exp\left\{-\tfrac{1}{2}\|x - u\|^2\right\} \prod_{i=1}^p dx_i$$

$$= \frac{\partial}{\partial u_i} A \int_{R^p} \left[\log \int_{R^p} \exp\left\{-\tfrac{1}{2}\|x - \theta\|^2\right\} dG(\theta)\right] \exp\left\{-\tfrac{1}{2}\|x - u\|^2\right\} \prod_{i=1}^p dx_i$$

where $A = (2\pi)^{-p/2}$.

We have used integration by parts with respect to x_i to attain the first equality. To get the second equality we use the fact that the partial derivative of $\exp\{\tfrac{1}{2}\|x - u\|^2\}$ with respect to x_i is the negative of the partial derivative of the same quantity with respect to u_i. The partial with respect to u_i is then taken outside the integral.

If we now assume $\delta(\cdot)$ is minimax and spherically symmetric (which is equivalent to $dG(\theta)$ being orthogonally invariant [4], p. 42), we have from Theorem 1 that for $u_i > 0$

$$(12) \quad -\varphi(\|u\|^2)u_i$$

$$= \frac{\partial}{\partial u_i} A \int_{R^p} \left[\log \int_{R^p} \exp\left\{-\tfrac{1}{2}\|x - \theta\|^2\right\} dG(\theta)\right]$$

$$\exp\left\{-\tfrac{1}{2}\|x - u\|^2\right\} \prod_{i=1}^p dx_i$$

$$\geq \frac{-2(p-2)u_i}{\|u\|^2}.$$

For any vector $u = (u_1, u_2, \cdots, u_p)$, $\|u\| \geqq \varepsilon > 0$, $u_i \geqq 0$, we define vectors u^i, $i = 0, \cdots, p$ as follows. Let $u^0 = (u_1^0, u_2^0, \cdots, u_p^0)$ be any vector such that $0 \leqq u_i^0 \leqq u_i$ and $\|u^0\| = \eta < \varepsilon$, $\eta > 0$ fixed. Next let

$$(13) \qquad u_i^j = \begin{cases} u_i & \text{if } j \geqq i, \\ u_i^0 & \text{if } j < i, \end{cases}$$

and $u^j = (u_1^j, u_2^j, \cdots, u_p^j)$, $j = 1, \cdots, p$.

Note that we are merely changing the jth coordinate of u^{j-1} to that of u by the above construction. Hence $u^p = u$. Integrating the expression (12) from u^{i-1} to u^i with respect to u_i, $i = 1, \cdots, p$, and adding the results, we obtain by collapsing successive terms,

$$(14) \quad A \int_{R^p} \left[\log \int_{R^p} \exp\left\{ -\tfrac{1}{2}\|x - \theta\|^2 \right\} dG(\theta) \right] \exp\left\{ -\tfrac{1}{2}\|x - u\|^2 \right\} \prod_{i=1}^{p} dx_i$$

$$- A \int_{R^p} \left[\log \int_{R^p} \exp\left\{ -\tfrac{1}{2}\|x - \theta\|^2 \right\} dG(\theta) \right] \exp\left\{ -\tfrac{1}{2}\|x - u^0\|^2 \right\} \prod_{i=1}^{p} dx_i$$

$$\geqq -2(p - 2) \log \|u\| - 2(p - 2) \log \|u^0\|.$$

Because of the orthogonal invariance of $dG(\theta)$ and the fact that $\|u^0\| = \eta$, the two terms in the above inequality which depend on u^0 do so only through η. Hence, these two terms are constants. By Jensen's inequality applied to (14) it follows that

$$(15) \qquad \log A \left[\int_{R^p} \left(\int_{R^p} \exp\left\{ -\tfrac{1}{2}\|x - \theta\|^2 \right\} dG(\theta) \right) \exp\left\{ -\tfrac{1}{2}\|x - u\|^2 \right\} \prod_{i=1}^{p} dx_i \right]$$

or

$$\geqq -2(p - 2) \log \|u\| + c$$

$$(16) \qquad A \int_{R^p} \left(\int_{R^p} \exp\left\{ -\tfrac{1}{2}\|x - \theta\|^2 \right\} dG(\theta) \right) \exp\left\{ -\tfrac{1}{2}\|x - u\|^2 \right\} \prod_{i=1}^{p} dx_i$$

$$\geqq e^c \|u\|^{-2(p-2)}.$$

Note that by the orthogonal invariance of G, the above inequality holds for all u such that $\|u\| \geqq \varepsilon$.

The left side of (16) is essentially the (density of) the convolution of the standard normal with the convolution of $dG(\theta)$ and the standard normal. Hence the quantity on the left represents the Radon-Nikodym derivative (with respect to Lebesgue measure on R^p) of a measure which is finite if and only if $\int_{R^p} dG(\theta) < \infty$. However, integrating the right side of (16) over the sphere $\|u\| \geqq \varepsilon$ we see that the result can only be finite if $2(p - 2) - (p - 1) > 1$, or equivalently if $p > 4$. We therefore have the following result.

THEOREM 2. *Let $\delta(X)$ be a spherically symmetric minimax estimator which is generalized Bayes with respect to the (generalized) prior $dG(\theta)$. If $p \leqq 4$, $dG(\theta)$ cannot be a proper prior distribution.*

5. Remarks

We have shown that there do not exist spherically symmetric proper Bayes minimax estimators in four or lower dimensions. In a previous paper we demonstrated the existence of such estimators for $p \geq 5$. We are thus far unable to rule out the possibility that nonspherically symmetric proper Bayes minimax estimators exist in three and four dimensions although it seems highly unlikely that such estimators do exist.

Also we have been unable thus far to say anything concrete about the situation where some part of the covariance structure of the problem is unknown.

L. Brown has recently found a proof of the nonexistence of nonspherically symmetric proper Bayes estimators for $p = 3$ and 4.

The author wishes to thank L. Brown for bringing this problem to his attention and C. Stein for much helpful discussion.

REFERENCES

[1] A. J. BARANCHIK, "Multiple regression and estimation of the mean of a multivariate normal distribution," Stanford University, Technical Report No. 51 (1964).

[2] ———, "A family of minimax estimators of the mean of a multivariate normal distribution," *Ann. Math. Statist.*, Vol. 41 (1970), pp. 642–645.

[3] CHARLES STEIN, "Inadmissibility of the usual estimator for the mean of a multivariate normal distribution," *Proceedings of the Third Berkeley Symposium on Mathematical Statistics and Probability*, Berkeley and Los Angeles, University of California Press, 1955, pp. 197–206.

[4] W. E. STRAWDERMAN, "Generalized Bayes estimators and admissibility of estimators of the mean vector of a multivariate normal distribution with quadratic loss," Ph.D. thesis, Rutgers University, 1969.

[5] ———, "Proper Bayes minimax estimators of the multivariate normal mean," *Ann. Math. Statist.*, Vol. 42 (1971), pp. 385–388.

ON THE WIENER PROCESS APPROXIMATION TO BAYESIAN SEQUENTIAL TESTING PROBLEMS

P. J. BICKEL[1]

UNIVERSITY OF CALIFORNIA, BERKELEY

and

J. A. YAHAV[2,3]

UNIVERSITY OF TEL AVIV

1. Introduction and summary

In 1959 Chernoff [7] initiated the study of the asymptotic theory of sequential Bayes tests as the cost of observation tends to zero. He dealt with the case of a finite parameter space. The definitive generalization of the line of attack initiated in that paper was given by Kiefer and Sacks in [13]. Their work as well as that of Chernoff, the intervening papers of Albert [1], Bessler [3], and Schwarz [19], and the subsequent work of the authors [4] used implicitly or explicitly the theory of large deviations and applied only to situations where hypothesis and alternative were separated or at least an indifference region was present.

In the meantime in 1961 Chernoff [8] began to study the problem of testing $H : \theta \leq 0$ versus $K : \theta > 0$ on the basis of observation of a Wiener process with drift θ per unit time as an approximation to the discrete time normal observations problem. Having made the striking observation that study of the asymptotic behavior of the Bayes procedures for any normal prior was in this case equivalent to the study of the Bayes procedure with Lebesgue measure as prior and unit cost of observation, he reduced this problem for suitable loss functions to the solution of a free boundary problem for the heat equation. In subsequent work ([2], [9], [10] and [16]) the nature of this solution was investigated by Chernoff and others.

In this paper we are concerned with the problem of testing $H : \theta \leq 0$ versus $K : \theta > 0$ by sampling sequentially from a member of one parameter exponential (Koopman-Darmois) family of distributions (see equation (3.1)) at cost c per observation. We will assume the simple zero-one loss structure in which an error in decision costs one unit while being right costs nothing.

[1] Prepared with the partial support of the Office of Naval Research, Contract NONR N00014–69–A–0200–1038.

[2] Prepared with the partial support of U.S. Public Health Grant GM–10525(07).

[3] This research was done while the author was visiting the University of California, Berkeley.

Our main result, Theorem 4.2, states that if we assume a bounded continuous prior density ψ on the parameter space and that an observation has mean zero and variance one if $\theta = 0$, then our problem is asymptotically equivalent to the analogous Wiener process problem with drift θ per unit time, the same loss and cost structure and prior "density" $\equiv \psi(0)$. Chernoff's observation applies here also and this asymptotic problem is equivalent to the problem for fixed cost. A formal result in this direction was obtained for the special case of Bernoulli trials by Moriguti and Robbins [18]. Our technique may be viewed as an extension to the sequential case of an approach of Wald [21] and LeCam [14]. It is clearly applicable to other testing, estimation, and general decision problems.

We begin by examining the Wiener process problem and the embedded discrete time normal observation problem for a general continuous and bounded prior density ψ. Our first two results, Lemmas 1.1 and 2.2, establish the asymptotic relation between the Wiener process problem with prior density ψ and the same problem with prior density $\equiv \psi(0)$. Our basic tool is the similarity transform used by Chernoff in [8] and a weak compactness theorem which is a special case of an unpublished result of LeCam. A statement and proof of the latter for our special case is given in the Appendix (Theorem A.1). The validity of this result requires the use of randomized procedures. These are employed throughout the paper, despite the fact that the Bayes procedures for all our problems are non-randomized. Randomization also plays an important role in considering the relation between the discrete and continuous time problems where we make heavy use of sufficiency. Reference to Chapter 7 of Ferguson [12] may prove helpful.

In Section 3 we show essentially that the exponential family problem is asymptotically at least as hard as the Wiener process problem. To do this we successively, without substantial loss, reduce the problem to one in which observation is carried out in blocks, the parameter space is shrunk to a neighborhood of zero, and the time of observation is truncated. At this stage we use a Berry-Esseen type bound essentially due to Petrov [19] to show that the normal approximation is valid and then apply the results of Section 2. This approximation theorem is given as Lemma 3.3 and its proof is given in the Appendix.

Finally, in the fourth section we show that the Wiener process problem is at least as difficult asymptotically as the exponential family problem. In doing so, we exhibit implicitly a sequence of procedures, independent of ψ, for which the bound of Section 3 is achieved.

Some concluding remarks and statements of open problems are given in the last section.

2. The normal theory problem

In this section we shall describe randomized sequential procedures in continuous and discrete time and derive asymptotic results for the Wiener process problem and its discrete time approximations.

Let $\tilde{C}[0, \infty)$ be the set of all continuous functions defined on $[0, \infty)$ such that $\lim_{t \to \infty} x(t)/t^2 = 0$ endowed with the norm $\|x\| = \sup_t |x(t)|/(1 + t^2)$. The space $\tilde{C}$ is complete separable and metric. Let $\mathcal{B}$ denote the class of Borel sets on $\tilde{C}[0, \infty)$ (the product sigma field) and let $\mathcal{B}_t$ denote the Borel field generated by the maps $x \to x(s)$ for $0 \leq s \leq t$.

Let $\Omega = \tilde{C}[0, \infty) \times [0, 1]$, $\mathcal{A}$ be the product Borel field and Q_θ, $-\infty < \theta < \infty$, be the probability measure on $(\Omega, \mathcal{A})$ such that the stochastic process W and random variable U given by $W(x, z) = x$, $U(x, z) = z$ are independent and respectively a Wiener process with drift θ per unit time and a uniformly distributed variable on $[0, 1]$. The subscript θ will be used in this section when calculating expectations with respect to those measures or related measures of the discrete time problem. We are interested in testing $H: \theta \leq 0$ versus $K: \theta > 0$ with zero-one loss and cost c per unit time. A sequential procedure $\pi = (\delta, \tau)$ for this problem consists of a randomized stopping time τ and a randomized rule δ. Rigorously τ is a measurable map from Ω to $[0, \infty)$ such that for every $z \in [0, 1]$ and $t \leq \infty$ the event $[\tau(\cdot, z) < t] \in \mathcal{B}_t$. To describe δ we begin by defining the pre τ field $\tilde{\mathcal{B}}_\tau$. This is simply the class of all events $A \in \mathcal{A}$ such that for every $z \in [0, 1]$ and every $t \leq \infty$ the z section of $A \cap [\tau < t]$. that is, $\{x: (x, z) \in A \cap [\tau < t]\}$, is $\mathcal{B}_t$ measurable. Given τ, δ is any map from Ω to $[0, 1]$ which is $\tilde{\mathcal{B}}_\tau$ measurable. The use of these procedures should be clear. Having observed $U = z$, we employ $\tau(\cdot, z)$ and on stopping reject with probability $\delta(\cdot, z)$ and accept otherwise.

If $\ell(d, \theta)$ is our zero-one loss function we write the contitional risk of π given θ for observation cost c as,

$$(2.1) \qquad R_\theta(\pi, c) = E_\theta[\ell(\delta, \theta)] + cE_\theta(\tau)$$
$$= \varepsilon(\theta) E_\theta(\delta) + [1 - \varepsilon(\theta)]E_\theta(1 - \delta) + cE_\theta(\tau),$$

where $\varepsilon(\theta) = 1$ if $\theta \leq 0$ and 0 otherwise.

Let M_c be the bimeasurable transformation of Ω onto itself given by,

$$(2.2) \qquad M_c(x, z)(t) = \left[\frac{1}{\sqrt{c}} x(ct), z \right].$$

This is the similarity transformation suitable for this problem. Then,

$$(2.3) \qquad P_\theta M_c^{-1} = P_{\theta\sqrt{c}}.$$

M_c induces a mapping of the space of decision procedures onto itself as follows:

$$(2.4) \qquad M_c: \pi \to \pi_c = (\tau_c, \delta_c)$$

where

$$(2.5) \qquad \tau_c(x, z) = c\tau[M_c(x, z)],$$

$$(2.6) \qquad \delta_c(x, z) = \delta[M_c(x, z)].$$

Then,

$$(2.7) \qquad E_\theta(\delta_c) = E_{\theta\sqrt{c}}(\delta), \qquad E_\theta(\tau_c) = cE_{\theta\sqrt{c}}(\tau),$$

so that

$$(2.8) \qquad R_\theta(\pi_c, 1) = R_{\theta\sqrt{c}}(\pi, c).$$

Let ψ be any nonnegative measurable function on R. Define

$$(2.9) \qquad R(\psi, c) = \inf_\pi \int_{-\infty}^\infty R_\theta(\pi, c)\psi(\theta)\, d\theta.$$

LEMMA 2.1.

$$(2.10) \qquad \frac{1}{\sqrt{c}} R(\psi, c) = \underline{R}(\psi(\cdot\sqrt{c}), 1).$$

PROOF.   By (2.8),

$$(2.11) \qquad \int_{-\infty}^\infty R_\theta(\pi, c)\psi(\theta)\, d\theta = \sqrt{c} \int_{-\infty}^\infty R_{\theta\sqrt{c}}(\pi, c)\psi(\theta\sqrt{c})\, d\theta$$

$$= \sqrt{c} \int_{-\infty}^\infty R_\theta(\pi_c, 1)\psi(\theta\sqrt{c})\, d\theta.$$

Since the correspondence between π and π_c is one to one onto, the result follows by taking the infima over π on both sides.

All limits in the sequel are taken as $c \to 0$.

LEMMA 2:2.   *Let ψ be as above, bounded and continuous at zero. Then,*

$$(2.12) \qquad \lim \frac{1}{\sqrt{c}} R(\psi, c) = \psi(0)R^*(1)$$

where $R^*(1) = R(1, 1) = \inf_\pi \int_{-\infty}^\infty R_\theta(\pi, 1)\, d\theta$.

PROOF.   Note that $R^*(1)$ is finite (see, for example, the procedure of [5]). By Lemma 1.1, our hypothesis, and the dominated convergence theorem, we must have,

$$(2.13) \qquad \lim \frac{1}{\sqrt{c}} R(\psi, c) = \lim R[\psi(\cdot\sqrt{c}), 1]$$

$$\leqq \psi(0)R^*(1).$$

On the other hand by Theorem A.1 there exists a procedure $\pi(c)$ such that $R(\psi(\cdot\sqrt{c}), 1) = \int_{-\infty}^\infty R_\theta(\pi, 1)\psi(\theta\sqrt{c})\, d\theta$. Further given any sequence $c_n \downarrow 0$ there exists a procedure π and a subsequence $\{n_k\}$ such that,

$$(2.14) \qquad R_\theta(\pi, 1) \leqq \lim_k R_\theta(\pi(c_{n_k}), 1).$$

Then by Fatou's lemma, and the continuity of ψ,

$$(2.15) \qquad \liminf_k R[\psi(\cdot, \sqrt{c_{n_k}}), 1] \geqq \psi(0) \int_{-\infty}^{\infty} R_\theta(\pi, 1)\, d\theta$$

$$\geqq \psi(0) R^*(1).$$

The lemma follows.

Consider our problem with the modification that if you sample beyond time T then there is no terminal loss and no cost for additional sampling. Let $\underline{R}_\theta(\pi, c, T)$ denote the conditional risk given θ for the modified problem. Formally,

$$(2.16) \qquad \underline{R}_\theta(\pi, c, T) = E_\theta(\delta I[\tau \leqq T])\varepsilon(\theta) + [1 - \varepsilon(\theta)]$$
$$E_\theta[(1 - \delta)I[\tau \leqq T]] + cE_\theta[\min(\tau, T)],$$

where $I[A]$ is the indicator of A.

Given any procedure π, let $\pi_T = (\delta_T, \tau_T)$ and a truncation of π be defined by $\tau_T = \min(\tau, T)$, $\delta_T = \delta$ if $\tau < T$ and δ_T minimizes the posterior Bayes risk given $\mathscr{B}_T$.

Let

$$(2.17) \qquad \bar{R}^*(1, T) = \inf_\pi \int_{-\infty}^{\infty} R_\theta(\pi_T, 1)\, d\theta,$$

$$(2.18) \qquad \underline{R}^*(1, T) = \inf_\pi \int_{-\infty}^{\infty} \underline{R}_\theta(\pi, 1, T)\, d\theta.$$

LEMMA 2.3. *Let $K_c \to \infty$ as $c \to 0$, and let $\psi(\theta)$ be as in Lemma 2.2. Then*

$$(2.19) \qquad \liminf_\pi \frac{1}{\sqrt{c}} \int_{-K_c\sqrt{c}}^{K_c\sqrt{c}} \underline{R}_\theta\left(\pi, c, \frac{T}{c}\right)\psi(\theta)\, d\theta = \psi(0)\underline{R}^*(1, T),$$

$$(2.20) \qquad \liminf_\pi \frac{1}{\sqrt{c}} \int_{-\infty}^{\infty} R_\theta(\pi_{T/c}, c)\psi(\theta)\, d\theta = \psi(0)\bar{R}^*(1, T).$$

PROOF. By arguing as in the proof of Lemma 2.1 we have

$$(2.21) \quad \inf_\pi \frac{1}{\sqrt{c}} \int_{-K_c\sqrt{c}}^{K_c\sqrt{c}} \underline{R}_\theta\left(\pi, c, \frac{T}{c}\right)\psi(\theta)\, d\theta = \inf_\pi \int_{-K_c}^{K_c} \underline{R}_\theta(\pi, 1, T)\psi(\theta\sqrt{c})\, d\theta.$$

By arguing as in the proof of Lemma 2.2, we get that the right side of (2.21) converges as $c \to 0$ to the right side of (2.19) which proves (2.19). Exactly the same type of arguments prove (2.20) which completes the proof.

LEMMA 2.4.

$$(2.22) \qquad \lim_{T\to\infty} \underline{R}^*(1, T) = R^*(1),$$

$$(2.23) \qquad \lim_{T\to\infty} \bar{R}^*(1, T) = R^*(1).$$

PROOF. Clearly we have

$$(2.24) \qquad \underline{R}^*(1, T) \leqq R^*(1) \leqq \bar{R}^*(1, T).$$

By a weak compactness argument there exists for fixed T a $\tilde{\pi}(T)$ such that $\underline{R}^*(1, T) = \underline{R}(\tilde{\pi}, 1, T)$. Hence

$$(2.25) \quad \bar{R}^*(1, T) - \underline{R}^*(1, T) = \bar{R}^*(1, T) - \underline{R}(\tilde{\pi}, 1, T)$$

$$\leqq R(\tilde{\pi}_T, 1) - \underline{R}(\tilde{\pi}, 1, T)$$

$$\leqq \int_{-\infty}^{0} P_\theta\{W(T) > 0\}\, d\theta + \int_{0}^{\infty} P_\theta\{W(T) < 0\}\, d\theta.$$

The right side of (2.25) converges to zero as $T \to \infty$ which completes the proof of the lemma.

Before giving our final lemma we review two ways of defining sequential procedures for discrete time problems. Let $\mathcal{X} = R^\infty \times [0, 1]$ be the product of a countable number of copies of R and $[0, 1]$, and let $\mathcal{D}$ be the Borel field on this space. A randomized stopping time τ is now a measurable map from $\mathcal{X}$ to the natural numbers $\{0, 1, 2, \cdots, \infty\}$ such that the event $[\tau(\cdot, z) \leqq n]$ is, for every z and n, measurable with respect to the σ-field $\mathcal{B}_n^*$ generated by the map $(x_1, x_2, \cdots) \to (x_1, x_2, \cdots, x_n)$ on R^∞. We shall always suppose that the probability measure on $\mathcal{D}$ is such that the random sequence $(X_1, X_2, \cdots)$ and the random variable U given by $(X_1, X_2, \cdots)(x_1, x_2, \cdots, z) = (x_1, x_2, \cdots)$ and $U(x_1, x_2, \cdots, z) = z$ are independent and U is uniform on $[0, 1]$. Similarly a decision rule δ is any measurable map from $\mathcal{X}$ to $[0, 1]$ which is measurable with respect to $\tilde{\mathcal{B}}_\tau$ the ψ-field of all events $A \in \mathcal{D}$ such that the z section of $A \cap [\tau \leqq n]$ is in $\mathcal{B}_n^*$ for every n and z.

In this formulation (which we refer to as I) a procedure $\pi = (\delta, \tau)$ has the same interpretation as in the continuous time problem. On the other hand, following Ferguson [12] we can define a stopping rule τ by a sequence of functions $(\psi_0, \psi_1, \psi_2, \cdots)$ where ψ_j is a $\mathcal{B}_j^*$ measurable function from R^∞ to $[0, 1]$ and $\Sigma_{j=0}^\infty \psi_j \leqq 1$. If τ is a stopping time in the sense of (I), then the ψ_j are given by

$$(2.26) \qquad \psi_j(x_1, x_2, \cdots) = \lambda[z: \tau(x_1, x_2, \cdots, z) = j],$$

where λ is Lebesgue measure. Conversely, it is a well-known result of Wald and Wolfowitz [23] that given a stopping time in this second mode as $(\psi_0, \psi_1, \cdots)$ there is a stopping time in the sense of I satisfying (2.26) (see the proof of Theorem A.1). Similarly, a terminal decision rule is specified in the second mode as a sequence $(\delta_0, \delta_1, \cdots)$ of functions from R^∞ to $[0, 1]$ such that δ_j is measurable $\mathcal{B}_j^*$. Again given δ of type I,

$$(2.27) \qquad \delta_j(x_1, x_2, \cdots) = \int_0^1 \delta(x_1, x_2, \cdots, z)\, dz$$

and by [23] to any policy $((\delta_0, \delta_1, \cdots), (\psi_0, \psi_1, \cdots))$ there corresponds a policy $\pi = (\delta, \tau)$ satisfying (2.26) and (2.27). Now suppose that Q_θ (abusing notation) makes the X_i in $(X_1, X_2, \cdots)$ independent normal random variables with mean θ and variance one. If π is a policy as above we write the conditional risk given θ for the usual sequential testing problem as

$$(2.28) \quad R_\theta(\pi, c) = \varepsilon(\theta)E_\theta(\delta) + [1 - \varepsilon(\theta)]E_\theta(1 - \delta) + cE_\theta(\tau)$$

$$= \varepsilon(\theta) \sum_{j=0}^{\infty} E_\theta(\psi_j \delta_j) + [1 - \varepsilon(\theta)] \sum_{j=0}^{\infty} E_\theta[\psi_j(1 - \delta_j)]$$

$$+ c \sum_{j=0}^{\infty} jE_\theta(\psi_j),$$

if $\Sigma_{j=0}^{\infty} \psi_j = 1$ a.s. Q_θ, and $= \infty$ otherwise. We shall refer to this as the discrete time normal problem. Evidently any policy π as above for the given Q_θ may be considered as a policy in Wiener process problem with the same risk. We shall want to consider the normal block problem in which we are permitted to sample in blocks of size N only and are told only the block sums $S_N = \Sigma_{i=1}^{N} X_i$, $S_{2N} = \Sigma_{i=1}^{2N} X_i$, and so forth. Of course statistically, because of sufficiency, this last restriction has no effect on the difficulty of the problem. Let $\mathscr{B}(S_N, S_{2N}, \cdots, S_{jN})$ be the σ-field induced on R^∞ by the maps $(x_1, x_2, \cdots) \to (\Sigma_{i=1}^{N} x_i, \Sigma_{i=N+1}^{2N} x_i, \cdots, \Sigma_{i=(j-1)N+1}^{jN} x_i)$. Formally a block procedure π is any procedure in the discrete time problem such that τ only takes on the values $0, N, 2N, \cdots, jN, \cdots$ with probability one, for every z, j.

$$(2.29) \quad [\tau(\cdot, z) = jN] \in \mathscr{B}(S_N, S_{2N}, \cdots, S_{jN}),$$

and for every c, z, j,

$$(2.30) \quad [\delta(\cdot, z) \leqq c] \cap [\tau(\cdot, z) = jN] \in \mathscr{B}(S_N, \cdots, S_{jN}).$$

We can now state:

LEMMA 2.5. *For every procedure π in the Wiener process problem there exists a normal block procedure $\pi^{(N)}$ such that,*

$$(2.31) \quad |R_\theta(\pi, c) - R_\theta(\pi^{(N)}, c)| \leqq Nc.$$

PROOF. In view of our remarks we can give $\pi^{(N)}$ in the second formulation. Define

$$(2.32) \quad \psi_j^{(N)} = 0 \qquad \qquad \text{for} \quad j \neq iN,$$

$$(2.33) \quad \psi_{iN}^{(N)} = Q_\theta[(i - 1)N < \tau \leqq iN/W(N), W(2N), \cdots, W(iN)].$$

(The / indicates a suitable version of the conditional probability.) Note that since $W(iN)$ is sufficient for $\mathscr{B}_{iN}$, the $\psi_{iN}^{(N)}$ may be chosen independent of θ. Strictly speaking $\psi_{iN}^{(N)}$ is a function on $\tilde{C}[0, \infty]$ not R^∞. However by the usual arguments we may in fact take $\psi_{iN}^{(N)}$ to be a function of the variables $W(N), W(2N), \cdots, W(iN)$ only. It is clear that if the definition (2.33) defines a stopping time at all, it will be a block time. To check that it is a stopping time we need only to show that,

$$(2.34) \qquad \sum_{j=0}^{\infty} \psi_j^{(N)} \leqq 1.$$

It is enough to show that $\Sigma_{j=0}^{iN} \psi_j^{(N)} \leqq 1$ for every i. We have (for a suitable choice of the conditional probability)

$$(2.35) \quad 1 \geqq Q_\theta[\tau \leqq iN/W(N), W(2N), \cdots, W(iN)]$$

$$= \psi_0^{(N)} + \sum_{j=1}^{iN} Q_\theta[(j-1)N < \tau \leqq jN/W(N), W(2N), \cdots, W(iN)]$$

$$= \psi_0^{(N)} + Q_\theta[0 < \tau \leqq N/W(N), W(2N) - W(N),$$
$$W(3N) - W(N), \cdots, W(iN) - W(N)]$$

$$+ Q_\theta[N < \tau \leqq 2N/W(N), W(2N), W(3N) - W(2N), \cdots,$$
$$W(iN) - W(2N)]$$

$$+ \cdots + Q_\theta[(i-1)N < \tau \leqq iN/W(N), W(2N), W(3N), \cdots, W(iN)]$$

$$= \psi_0^{(N)} + Q_\theta[0 < \tau \leqq N/W(N)] + Q_\theta[N < \tau \leqq 2N/W(N), W(2N)]$$

$$+ \cdots + Q_\theta[(i-1)N < \tau \leqq iN/W(N). W(2N), \cdots, W(iN)]$$

$$= \sum_{j=0}^{iN} \psi_j^{(N)}.$$

Now define the $\delta_i^{(N)}$ by,

$$(2.36) \quad \psi_{iN}^{(N)} \delta_{iN}^{(N)} = E_\theta\{\delta I[(i-1)N < \tau \leqq iN]/W(N), \cdots, W(iN)\}$$
$$\text{for} \quad i = 0, 1, \cdots,$$

and $\delta_i^{(N)} = 0$ otherwise.

It is clear that $\pi^{(N)} = ((\psi_0^{(N)}, \cdots), (\delta_0^{(N)}, \delta_1^{(N)}, \cdots))$ is a block procedure and,

$$(2.37) \qquad \sum_{j=0}^{\infty} E_\theta(\psi_j^{(N)} \delta_j^{(N)}) = E_\theta(\delta)$$

while

$$(2.38) \qquad \sum_{j=0}^{\infty} jE_\theta(\psi_j^{(N)}) = \sum_{i=0}^{\infty} iN E_\theta(\psi_{iN}^{(N)})$$

$$= \sum_{i=0}^{\infty} iN Q_\theta[(i-1)N < \tau \leqq iN]$$

$$\leqq E_\theta(\tau) + N.$$

The lemma follows.

3. The exponential family problem: lower bound

In this section we introduce the exponential family model and derive a lower bound to the Bayes risk of the testing problem in terms of the Wiener process problem of the previous section. Without loss of generality we shall throughout this section suppose that $X_1, X_2, \cdots$ are the coordinate projections of R^∞ and are thus defined on the space of the previous section. We let P_θ be a probability measure on $\mathscr{X}$ which makes the X_i independent and identically distributed with density f_θ, with respect to some nondegenerate σ-finite measure μ on R. We take f_θ to be the function

$$(3.1) \qquad f_\theta(x) = e^{\theta x - b(\theta)},$$

where θ ranges over a set Θ such that zero is an interior point of Θ. (As in the previous section whatever be θ, U is independent of the X_i and uniformly distributed on $[0, 1]$.) Let ψ be as in Lemmas 2.3 and 2.4 a. bounded probability density (with respect to Lebesgue measure) on Θ and continuous at zero. As before we wish to test $H: \theta \leq 0$ versus $K: \theta > 0$ with zero-one loss, and at cost c per observation. Evidently the definitions of sequential procedure introduced in connection with the normal discrete time problem are appropriate for this exponential family problem also, the only difference being that risks must be calculated under P_θ rather than Q_θ. Since we shall occasionally have to talk about both problems we shall use the superscripts P, Q on expectations where this is necessary to avoid ambiguity.

Note that

$$(3.2) \qquad E_\theta^{(P)}(X_1) = b'(\theta), \qquad \mathrm{Var}_\theta^{(P)}(X_1) = b''(\theta).$$

We shall suppose that $b(0) = b'(0) = 0$ and $b''(0) = 1$. The general case reduces to this special one. To see this consider $Y_i = [X_i - b'(0)]/[b''(0)]^{1/2}$. The Y_i are a sequence of observations distributed according to an exponential family with density

$$(3.3) \qquad g_\theta(y) = \exp\{\theta[b''(0)]^{1/2}y - c(\theta)\},$$

with respect to a suitable underlying measure.

If we change parameters to $\eta = \theta[b''(0)]^{1/2}$ we are back in the previous case although this does, of course, give the prior density $[b''(0)]^{1/2}\psi\{\cdot[b''(0)]^{-1/2}\}$ for η. Also note that there is no loss of generality in assuming that the X_i are real valued. If X takes vector values (or even abstract values) and follows a one parameter exponential family with density of the form,

$$(3.4) \qquad f_\theta(x) = e^{\theta t(x) - b(\theta)}$$

then $t(X_1)$, $t(X_1) + t(X_2)$, $\cdots$ is a sequence of transitive sufficient statistics (see [12], Chapter 7) for the problem and of course $t(X_i)$ is a random variable following an exponential family probability law of the original form.

For any procedure π (in form I and II) define $B_\theta(\pi, c)$ to be the conditional risk of π given θ. Define the average risk of π, as usual, by

(3.5)
$$B(\pi, c, \psi) = \int_{-\infty}^{\infty} B_\theta(\pi, c)\psi(\theta)\, d\theta,$$

and let $\pi^*(c, \psi)$ denote the Bayes procedure for this problem which minimizes $B(\pi, c, \psi)$ over all π. For convenience we refer to these procedures as π^* in the sequel.

We shall prove

THEOREM 3.1. *Under the conditions of this section,*

(3.6)
$$\liminf_{c \to 0} \frac{1}{\sqrt{c}} B(\pi^*, c, \psi) \geqq \psi(0) R^*(1).$$

The proof proceeds by a series of lemmas. Let block procedures be defined as in the previous section.

LEMMA 3.1. *For every π, there exists a block procedure $\pi^{(N)}$ such that,*

(3.7)
$$\left| B_\theta(\pi^{(N)}, c) - B_\theta(\pi, c) \right| \leqq Nc$$

for every θ.

PROOF. The method of proof is the same as that of Lemma 2.5. Define $\pi = \big((\psi_0, \psi_1, \cdots), (\delta_0, \delta_1, \cdots)\big)$,

(3.8)
$$\psi_j^{(N)} = 0, \qquad j \neq iN, \qquad \psi_{iN}^{(N)} = \sum_{j=(i-1)N+1}^{iN} E_\theta^{(P)}(\psi_j | S_N, \cdots, S_{iN}),$$

and

(3.9)
$$\delta_{iN}^{(N)} \psi_{iN}^{(N)} = \sum_{j=(i-1)N+1}^{iN} E_\theta^{(P)}[\delta_j \psi_j | S_N, \cdots, S_{iN}].$$

Crucial use is made as before of the sufficiency of S_{iN} for P_θ on $\mathscr{B}_{iN}^*$ and the independence of the increments of the S_n process.

For any π let $\underline{B}_\theta(\pi, T)$ denote the conditional risk of π given θ for a modified version of the exponential family problem in which there is neither terminal loss nor additional cost of observation incurred after time T. Thus,

(3.10)
$$\underline{B}_\theta(\pi, c, T) = \varepsilon(\theta) E_\theta^{(P)}(\delta I[\tau \leqq T]) + c E_\theta^{(P)}[\min(\tau, T)]$$
$$+ [1 - \varepsilon(\theta)] E_\theta^{(P)}[(1 - \delta) I[\tau \leqq T]].$$

Let $\underline{R}_\theta(\pi, c, T)$ denote the same conditional expectation when the observations come from the normal distribution with mean θ and variance one, that is, when the expectation is taken with respect to Q_θ rather than P_θ. We shall also consider truncated procedures π_T defined in the natural way.

LEMMA 3.2. *For every π, there exists a block procedure $\pi^{(N)}$ such that,*

(3.11)
$$\left| \underline{B}_\theta(\pi^{(N)}, c, T) - \underline{B}_\theta(\pi, c, T) \right| \leqq Nc.$$

PROOF. As in Lemma 3.1.

Note that both lemmas apply to R_θ, $\underline{R}_\theta$ as a special case.

Let $P_{\theta, n}$ be the measure corresponding to the distribution of $S_n = \sum_{i=1}^n X_i$ where the X_i are independent and identically distributed according to f_θ. Let

$Q(\xi, \sigma^2)$ be the measure corresponding to the normal distribution with mean ξ and variance σ^2. Given a signed measure R defined on a σ-field $\mathscr{A}$ let $\|R\| = \sup_{A \in \mathscr{A}} |R(A)|$. Recall that if P, Q are probability measures dominated by a σ-finite measure μ then

$$(3.12) \qquad \|P - Q\| = \frac{1}{2} \int \left| \frac{dP}{d\mu} - \frac{dQ}{d\mu} \right| d\mu.$$

We need the following lemma which may be derived in the same fashion as a known result of Petrov [19].

LEMMA 3.3. *Let $\mathscr{F}$ be a family of densities (with respect to Lebesgue measure) on R. Suppose that $Z_1, \cdots, Z_n$ are independent and identically distributed according to $f \in \mathscr{F}$. Let $U_{f,n}$ be the probability induced by $n^{-1/2} \Sigma_{i=1}^n X_i$ and let Φ be the standard normal measure on R. Suppose that $\mathscr{F}$ satisfies the following conditions:*

(i) *$\mathscr{F}$ is precompact when considered as a subset of L_1 with the usual topology;*

(ii) *$c_1(\mathscr{F}) = \sup \{f(x): x \in R, f \in \mathscr{F}\} < \infty$;*

(iii) *$\int_{-\infty}^{\infty} xf(x)\, dx = 0. \qquad \int_{-\infty}^{\infty} x^2 f(x)\, dx = 1 \quad$ for every $f \in \mathscr{F}$;*

(iv) *$c_2(\mathscr{F}) = \sup \left\{ \int_{-\infty}^{\infty} |x|^3 f(x)\, dx : f \in \mathscr{F} \right\} < \infty$.*

Then,

$$(3.13) \qquad \sup \{ \|U_{f,n} - \Phi\| : f \in \mathscr{F} \} \leqq \frac{c(\mathscr{F})}{\sqrt{n}}.$$

PROOF. See Appendix.

LEMMA 3.4. *There exist $K_1, K_2(M)$ such that*

$$(3.14) \qquad \|Q_{(\xi, c^2)} - Q_{(0, 1)}\| \leqq K_1 |\xi| + K_2 |\sigma^2 - 1|$$

for all ζ and σ^2 such that $|\sigma^2 - 1| \leqq M$.

PROOF.

$$(3.15) \qquad \|Q_{(\xi, \sigma^2)} - Q_{(0, 1)}\| \leqq \|Q_{(\xi, \sigma^2)} - Q_{(\xi, 1)}\| + \|Q_{(\xi, 1)} - Q_{(0, 1)}\|$$
$$= \|Q_{(0, \sigma^2)} - Q_{(0, 1)}\| + \|Q_{(\xi, 1)} - Q_{(0, 1)}\|.$$

By (2.12) for $\xi > 0$,

$$(3.16) \qquad \|Q_{(\xi, 1)} - Q_{(0, 1)}\|$$
$$= \frac{1}{2} \left\{ \int_{-\infty}^{\xi/2} [\phi(t) - \phi(t - \xi)]\, dt + \int_{\xi/2}^{\infty} [\phi(t - \xi) - \phi(t)]\, dt \right\}$$
$$= \Phi(\xi/2) - \Phi(-\xi/2).$$

So, in general we get

$$(3.17) \qquad \|Q_{(\xi, 1)} - Q_{(0, 1)}\| \leqq K_1|\xi|.$$

Similarly for $|\sigma^2 - 1| \leqq M_2$,

$$(3.18) \qquad \|Q_{(0, \sigma^2)} - Q_{(0, 1)}\| \leqq K_2|\sigma^2 - 1|.$$

From Lemmas 3.3, 3.4 we obtain

LEMMA 3.5. *Suppose that $\{Z_i\}$ are independent and identically distributed with density g_θ (with respect to Lebesgue measure) where, the set $\{g_\theta: |\theta| \leqq \varepsilon\}$ satisfies conditions* (i), (ii) *and* (iv) *of Lemma 3.3 for some $\varepsilon > 0$, and further*

$$(3.19) \qquad \begin{aligned} e(\theta) &= E_\theta(Z_1) = \theta + O(\theta^2) \\ v(\theta) &= V_\theta(Z_1) = 1 + O(|\theta|). \end{aligned}$$

Let U_{θ_n} denote the distribution of $(1/\sqrt{n})\Sigma_{i=1}^n Z_i$.

Then there exists a $\delta > 0$ and constants d_1, d_2, d_3 such that

$$(3.20) \qquad \sup \{\|U_{\theta_n} - Q_{\theta_n}\| : |\theta| \leqq \delta\} \leqq \frac{d_1}{\sqrt{n}} + d_2|\theta| + d_3\theta^2\sqrt{n}.$$

PROOF.

$$(3.21) \qquad \|U_{\theta_n} - Q_{\theta_n}\| \leqq \|U_{\theta_n} - Q_{(ne(\theta), nv(\theta))}\| \\ + \|Q_{(ne(\theta), nv(\theta))} - Q_{(n\theta, n)}\|.$$

By Lemma 3.3 and our assumptions on $\{g_\theta: |\theta| \leqq \delta\}$,

$$(3.22) \qquad \sup \{\|U_{\theta_n} - Q_{(ne(\theta), nv(\theta))}\| : |\theta| \leqq \delta\} \leqq \frac{c(\delta)}{\sqrt{n}}$$

for $\delta \leqq \varepsilon$.

On the other hand, by Lemma 3.4,

$$(3.23) \qquad \|Q_{(ne(\theta), nv(\theta))} - Q_{(n\theta, n)}\| = \|Q_{(\sqrt{n}(e(\theta) - \theta), v(\theta))} - Q_{(0, 1)}\| \\ \leqq K(\delta)[\sqrt{n}|e(\theta) - \theta| + |v(\theta) - 1|]$$

for δ sufficiently small. The result follows by (3.19).

REMARK. If μ is dominated by Lebesgue measure and

$$(3.24) \qquad \sup \left\{ e^{\theta x}\frac{d\mu}{dx} : |\theta| \leqq M, x \in R \right\} < \infty$$

for some $M > 0$, then we may apply Lemma 3.5 to the exponential family and deduce that

$$(3.25) \qquad \|P_{\theta, n} - Q_{\theta, n}\| \leqq \frac{d_1}{\sqrt{n}} + d_2|\theta| + d_3\theta^2\sqrt{n}$$

if $|\theta| \leqq M$ for suitable d_1, d_2, d_3.

The following result is well known and is stated without proof.

LEMMA 3.6. *Let $P_1, P_2, \cdots, P_n; Q_1, Q_2, \cdots, Q_n$ be probability measures defined on the real line and let $P^{(n)}, Q^{(n)}$ be the corresponding n dimensional product measures. Then,*

$$(3.26) \qquad \|P^{(n)} - Q^{(n)}\| \leq \sum_{i=1}^{n} \|P_i - Q_i\|.$$

LEMMA 3.7. *Suppose that μ is dominated by Lebesgue measure and $\sup\{e^{\theta x}\, d\mu/dx : x \in R, |\theta| \leq M\} < \infty$. If $|\theta| \leq M$, then for any $N_c \geq 1$*

$$(3.27) \qquad \max\left\{\left|\underline{B}_\theta\left(\pi, c, \frac{T}{c}\right) - \underline{R}_\theta\left(\pi, c, \frac{T}{c}\right)\right|; |B_\theta(\pi_{T/c}, c) - R_\theta(\pi_{T/c}, c)|\right\}$$

$$\leq 2cN_c + \frac{T}{cN_c}(2 + T)\left\{\frac{d_1}{\sqrt{N_c}} + d_2|\theta| + d_3\theta^2\sqrt{N_c}\right\}.$$

PROOF. We give the argument for $\underline{B}_\theta$, that for B_θ is identical.

$$(3.28) \qquad \left|\underline{B}_\theta\left(\pi, c, \frac{T}{c}\right) - \underline{R}_\theta\left(\pi, c, \frac{T}{c}\right)\right| \leq \left|\underline{B}_\theta\left(\pi^{(N_c)}, c, \frac{T}{c}\right) - \underline{B}_\theta\left(\pi, c, \frac{T}{c}\right)\right|$$

$$+ \left|\underline{R}_\theta\left(\pi^{(N_c)}, c, \frac{T}{c}\right) - \underline{R}_\theta\left(\pi, c, \frac{T}{c}\right)\right| + \left|\underline{B}_\theta\left(\pi^{(N_c)}, c, \frac{T}{c}\right) - \underline{R}_\theta\left(\pi^{(N_c)}, c, \frac{T}{c}\right)\right|.$$

By Lemma 3.2 the first two terms on the right side of (3.28) are each bounded by cN_c. Now,

$$(3.29) \qquad \left|\underline{B}_\theta\left(\pi^{(N_c)}, c, \frac{T}{c}\right) - \underline{R}_\theta\left(\pi^{(N_c)}, c, \frac{T}{c}\right)\right|$$

$$\leq 2\left|E_\theta^{(P)}(\delta^{(N_c)}I[\tau^{(N_c)} \leq T/c]) - E_\theta^{(Q)}(\delta^{(N_c)}I[\tau^{(N_c)} \leq T/c])\right|$$

$$+ c\left|E_\theta^{(P)}\left(\min\left(\tau^{(N_c)}, \frac{T}{c}\right)\right) - E_\theta^{(Q)}\left(\min\left(\tau^{(N_c)}, \frac{T}{c}\right)\right)\right|$$

$$\leq 2\|P_\theta \underline{S}^{-1} - Q_\theta \underline{S}^{-1}\| + T\|P_\theta \underline{S}^{-1} - Q_\theta \underline{S}^{-1}\|,$$

where $\underline{S}$ maps $(x_1, x_2, \cdots, z)$ into

$$(3.30) \qquad \left(\sum_{i=1}^{N_c} x_i, \sum_{i=N_c+1}^{2N_c} x_i, \cdots, \sum_{i=(I_c-1)N_c+1}^{I_cN_c} x_i\right)$$

and $I[c] = [T/cN_c] + 1$.

Applying Lemma 3.5 (and the following remark) and Lemma 3.6 to (3.29), the result follows.

We are now able to prove Theorem 3.1. We begin by proving the theorem in the case μ is dominated by Lebesgue measure and $e^{\theta x}\, d\mu/dx$ is bounded in x for θ in some neighbourhood of zero.

Let K_c, N_c be positive numbers to be determined below. We have by the previous lemmas the following relations

$$(3.31) \qquad B(\pi^*, c, \psi) = \int_{-\infty}^{\infty} B_\theta(\pi^*, c)\psi(\theta) \, d\theta$$

$$\geqq \int_{-K_c\sqrt{c}}^{K_c\sqrt{c}} B_\theta(\pi^*, c)\psi(\theta) \, d\theta$$

$$\geqq \int_{-K_c\sqrt{c}}^{K_c\sqrt{c}} \underline{B}_\theta\left(\pi^*, c, \frac{T}{c}\right)\psi(\theta) \, d\theta$$

$$\geqq \int_{-K_c\sqrt{c}}^{K_c\sqrt{c}} \underline{R}_\theta\left(\pi^*, c, \frac{T}{c}\right)\psi(\theta) \, d\theta$$

$$- \int_{-K_c\sqrt{c}}^{K_c\sqrt{c}} \left\{2cN_c + \frac{T}{cN_c}\,(2 + T)\left(\frac{d_1}{\sqrt{N_c}} + d_2|\theta| \right.\right.$$

$$\left.\left. + d_3\theta^2\sqrt{N_c}\right)\right\}\psi(\theta) \, d\theta.$$

Now since $\psi(\theta) \leqq F$ ($\psi(\theta)$ is assumed to be bounded),

$$(3.32) \quad \frac{1}{\sqrt{c}} \int_{-K_c\sqrt{c}}^{K_c\sqrt{c}} \left\{2cN_c + \frac{T}{cN_c}\,(2 + T)\left(\frac{d_1}{\sqrt{N_c}} + d_2|\partial| + d_3\theta^2\sqrt{N^c}\right)\right\}\psi(\theta) \, d\theta$$

$$\leqq 2FK_c\left\{2cN_c + \frac{d_1 T(2 + T)}{cN_c^{3/2}} + \frac{T(2 + T)d_2 K_c\sqrt{c}}{cN_c} + \frac{T(2 + T)d_3 2K_c^2}{4N_c^{1/2}3}\right\}.$$

The right side converges to zero for $K_c = c^{-1/8 - 3\varepsilon}$, $N_c = c^{-7/8 + 4\varepsilon}$, and $0 < 176\varepsilon < 1$.

On the other hand considering π^* as a procedure for the Wiener process we have

$$(3.33) \quad \int_{-K_c\sqrt{c}}^{K_c\sqrt{c}} \underline{R}\left(\pi^*, c, \frac{T}{c}\right)\psi(\theta) \, d\theta \geqq \inf_\pi \int_{-K_c\sqrt{c}}^{K_c\sqrt{c}} \underline{R}\left(\pi, c, \frac{T}{c}\right)\psi(\theta) \, d\theta$$

and the result follows by Lemmas 2.3 and 2.4.

To prove the general case, that is, where μ is not dominated by Lebesgue measure consider the following problem.

We observe $Y_1, Y_2, \cdots$ where

$$(3.34) \qquad\qquad\qquad Y_i = (X_i, Q_i)$$

with X_i as before and $\{Q_i\}$ a sequence of independent identically distributed normal random variables independent of the $\{X_i\}$ with mean $\varepsilon\theta$ and variance ε. Let $W_i = X_i + Q_i$. The sequence $\{\Sigma_{i=1}^n W_i\}$ is sufficient and transitive for this new problem. The W_i are independent identically distributed according to a one parameter exponential family of the form (3.1) with $b'(0) = 0$, $b''(0) = 1 + \varepsilon$. Furthermore the underlying measure μ of this new family satisfies the condition of the remark following Lemma 3.5.

If we let $B^\varepsilon(c, \psi)$ be the Bayes risk of the best procedure for the new problem when the cost of observation (per vector) is c and ψ is the prior density on θ, then our initial discussion leads to

$$(3.35) \qquad \liminf_{c \to 0} \frac{B^\varepsilon(c, \psi)}{\sqrt{c}} \geqq \frac{\psi(0)}{\sqrt{(1 + \varepsilon)}} R^*(1).$$

Of course, $B(\pi^*, c, \psi) \geqq B^\varepsilon(c, \psi)$ for every $\varepsilon > 0$. The theorem follows.

4. The exponential family problem: upper bound

The basic result of this section is:

THEOREM 4.1. *Under the conditions of Section 2*

$$(4.1) \qquad \limsup \frac{1}{\sqrt{c}} B(\pi^*, c, \psi) \leqq \psi(0)R^*(1).$$

In fact, there exists a sequence of procedures $\{\pi^{**}\}$ which is independent of ψ such that

$$(4.2) \qquad \lim \frac{1}{\sqrt{c}} B(\pi^{**}, c, \psi) = \psi(0)R^*(1).$$

(Dependence on c in π^{**} is suppressed for brevity.)

From Theorems 4.1 and 3.1, we derive immediately our main result:

THEOREM 4.2. *Under the conditions of Section 2,*

$$(4.3) \qquad \lim \frac{1}{\sqrt{c}} B(\pi^*, c, \psi) = \psi(0)R^*(1).$$

We shall give the proof of (4.1) in detail for the case where μ satisfies the conditions of the remark following Lemma 3.5 and sketch the additional remarks needed for the general case and the construction of π^{**} at the end.

PROOF. Let π be any procedure for the Wiener process problem and $\pi_{T/c}$ be its truncation at T/c as in Section 2. By Lemma 2.5 there exists a block procedure $\pi_{T/c}^{(N_c)}$ which by construction is truncated at $[T/c] + N_c$ such that

$$(4.4) \qquad \left| R_\theta(\pi_{T/c}^{(N_c)}, c) - R_\theta(\pi_{T/c}, c) \right| \leqq cN_c.$$

Consider the following discrete time rule which we shall denote by $\pi^{(e)}$. Take $n_c^{(1)}$ observations. Stop and reject H if $\Sigma_i\{X_i; 1 \leqq i \leqq n_c^{(1)}\} > A_c^{(1)}$, stop and accept H if $\Sigma_i\{X_i; 1 \leqq i \leqq n_c^{(1)}\} < -A_c^{(1)}$. If $\left| \Sigma_i\{X_i; 1 \leqq i \leqq n_c^{(1)}\} \right| \leqq A_c^{(1)}$, take $n_c^{(2)}$ further observations and stop and reject H if

$$(4.5) \qquad \sum_i \{X_i; n_c^{(1)} + 1 \leqq i \leqq n_c^{(1)} + n_c^{(2)}\} > A_c^{(2)}$$

stop and reject H if

$$(4.6) \qquad \left| \sum_i \{X_i; n_c^{(1)} + 1 \leqq i \leqq n_c^{(1)} + n_c^{(2)}\} \right| < -A_c^{(2)}.$$

If

$$(4.7) \qquad \left| \sum_i \{X_i; \, n_c^{(1)} + 1 \leq i \leq n_c^{(1)} + n_c^{(2)}\} \right| \leq A_c^{(2)},$$

then disregard the first $n_c^{(1)} + n_c^{(1)}$ observations and follow the procedure $\pi_{T/c}^{(N_c)}$. Let $n_c^{(1)} = c^{-1/2+\varepsilon}$, $A_c^{(1)} = c^{-1/4}$, $n_c^{(2)} = c^{-3/4+3\varepsilon}$, $A_c^{(2)} = c^{-3/8+\varepsilon}$, and $N_c = c^{-7/8+3\varepsilon}$, where $176\varepsilon < 1$. In that case for absolutely continuous μ as above, we shall show that

$$(4.8) \qquad \limsup \frac{1}{\sqrt{c}} \left[B(\pi^{(e)}, c, \psi) - R(\pi_{T/c}, c, \psi) \right] \leq 0.$$

Given (4.8) it follows that

$$(4.9) \qquad \limsup \frac{1}{\sqrt{c}} B(\pi^*, c, \psi) \leq \limsup \frac{1}{\sqrt{c}} \inf_\pi \int_{-\infty}^{\infty} R_\theta(\pi_{T/c}, c)\psi(\theta) \, d\theta$$

$$= \psi(0)\bar{R}^*(1, T)$$

by Lemma 2.3. An application of Lemma 2.4 will then complete the proof of Theorem 4.1. To begin the proof of (4.8) note that, for arbitrary K_c,

$$(4.10) \qquad B(\pi^{(e)}, c, \psi)$$

$$= \int_{-\infty}^{\infty} B_\theta(\pi^{(e)}, c)\psi(\theta) \, d\theta$$

$$\leq cn^{(1)} + \int_{-\infty}^{0} P_\theta[S_{n_c^{(1)}} > A_c^{(1)}]\psi(\theta) \, d\theta$$

$$+ \int_{0}^{\infty} P_\theta[S_{n_c^{(1)}} < -A_c^{(1)}]\psi(\theta) \, d\theta$$

$$+ cn_c^{(2)} \int_{-\infty}^{\infty} P_\theta[|S_{n_c^{(1)}}| \leq A_c^{(1)}]\psi(\theta) \, d\theta$$

$$+ \int_{-\infty}^{0} P_\theta[S_{n_c^{(2)}} > A_c^{(2)}]\psi(\theta) \, d\theta + \int_{0}^{\infty} P_\theta[S_{n_c^{(2)}} < -A_c^{(2)}]\psi(\theta) \, d\theta$$

$$+ \int_{-K_c\sqrt{c}}^{K_c\sqrt{c}} B_\theta(\pi_{T/c}^{(N_c)}, c)\psi(\theta) \, d\theta$$

$$+ \int_{|\theta| > K_c\sqrt{c}} P_\theta[|S_{n_c^{(2)}}| \leq A_c^{(2)}](1 + T + cN_c)\psi(\theta) \, d\theta.$$

Since $P_\theta[S_n > A]$ is increasing in θ we may for arbitrary $H_c > 0$ bound the right side of (4.10) by

$$(4.11) \quad cn_c^{(1)} + P_0[|S_{n_c^{(1)}}| > A_c^{(1)}]$$

$$+ cn_c^{(2)} \{P_{-H_c\sqrt{c}}[S_{n_c^{(1)}} \geq -A_c^{(1)}] + P_{H_c\sqrt{c}}[S_{n_c^{(1)}} \leq A_c^{(1)}] + 2FH_c\sqrt{c}\}$$

$$+ P_0 \big[|S_{n_c^{(2)}}| > A_c^{(2)} \big] + \int_{-K_c\sqrt{c}}^{K_c\sqrt{c}} B_\theta(\pi_{T/c}^{(N_c)}, c)\psi(\theta)\,d\theta$$

$$+ (1 + T + cN_c)\{ P_{K_c\sqrt{c}}[S_{n_c^{(2)}} \leq A_c^{(2)}] + P_{-K_c\sqrt{c}}[S_{n_c^{(2)}} \geq -A_c^{(2)}]\},$$

where F is our bound on ψ. The idea now is to show that for suitable choices of K_c, H_c all of the above are negligible save $\int_{-K_c\sqrt{c}}^{K_c\sqrt{c}} B_\theta(\pi_{T/c}^{(N_c)}, c)\psi(\theta)\,d\theta$ and that this expression can be well approximated by $\int_{-K_c\sqrt{c}}^{K_c\sqrt{c}} R_\theta(\pi_{T/c}^{(N_c)}, c)\psi(\theta)\,d\theta$. We collect the estimates we need in three propositions. All of these employ the well-known inequality (see, for example, Chernoff [6]),

$$(4.12) \qquad P_\theta[S_n \geq A] \leq \min_{t \geq 0} E_\theta^{(P)}(e^{t(S_n - A)})$$

$$= \min_{t \geq 0} e^{n[b(t+\theta) - b(\theta)] - tA}.$$

PROPOSITION 4.1.

$$(4.13) \qquad \lim \frac{1}{\sqrt{c}} P_0\big[|S_{n_c^{(1)}}| \geq A_c^{(1)}\big] = 0,$$

$$(4.14) \qquad \lim \frac{1}{\sqrt{c}} P_0\big[|S_{n_c^{(2)}}| \geq A_c^{(2)}\big] = 0.$$

PROOF.   We prove (4.13), and (4.14) is argued similarly. By (4.12),

$$(4.15) \qquad \log P_0\big[S_{n_c^{(1)}} \geq A_c^{(1)}\big] = \min_{t > 0} \{n_c^{(1)}b(t) - tA_c^{(1)}\}.$$

Since $b(0) = b'(0) = 0$ and $b''(0) = 1$ for t sufficiently small $b(t) \leq \frac{2}{3}t^2$. Take $t = A_c^{(1)}/n_c^{(1)}$ to get

$$(4.16) \qquad \log P_0[S_{n_c^{(1)}} \geq A_c^{(1)}] \leq -\frac{1}{3}c^{-\varepsilon} \to -\infty.$$

Applying a similar argument to $\log P_0[S_{n_c^{(1)}} \leq A_c^{(1)}]$, the result follows.

PROPOSITION 4.2.   If $H_c = c^{-1/4 - 2\varepsilon}$,

$$(4.17) \qquad \lim \frac{1}{\sqrt{c}} cn_c^{(2)} P_{-H_c\sqrt{c}}\big[S_{n_c^{(1)}} \geq -A_c^{(1)}\big] = 0,$$

$$(4.18) \qquad \lim \frac{1}{\sqrt{c}} cn_c^{(2)} P_{H_c\sqrt{c}}\big[S_{n_c^{(1)}} \leq A_c^{(1)}\big] = 0.$$

PROOF.   By (4.12),

$$(4.19) \qquad \log P_{-H_c\sqrt{c}}\big[S_{n_c^{(1)}} \geq A_c^{(1)}\big]$$

$$= \min_{t > 0} n_c^{(1)}\{[b(t - H_c\sqrt{c}) - b(-H_c\sqrt{c})] + tA_c^{(1)}\}.$$

For c sufficiently small, expanding b about $-H_c\sqrt{c}$ and b' about zero, we get

$$(4.20) \qquad \log P_{-H_c\sqrt{c}}[S_{n_c^{(1)}} \geqq -A_c^{(1)}] \leqq -\tfrac{1}{3}n_c^{(1)}\left(\frac{Hc\sqrt{c} - A_c^{(1)}}{n_c^{(1)}}\right)^2$$

$$= -\tfrac{1}{3}c^{-\varepsilon}(c^{-\varepsilon} - 1)^2 \to -\infty.$$

The result follows and a similar argument establishes (4.18).

In an entirely analogous fashion, we have

PROPOSITION 4.3. *If* $K_c = c^{-1/8 - 3\varepsilon}$,

$$(4.21) \qquad \lim \frac{1}{\sqrt{c}} P_{-K_c\sqrt{c}}[S_{n_c^{(2)}} \geqq -A_c^{(2)}] = 0.$$

$$(4.22) \qquad \lim \frac{1}{\sqrt{c}} P_{K_c\sqrt{c}}[S_{n_c^{(2)}} \leqq A_c^{(2)}] = 0.$$

As a consequence of Propositions 4.1 through 4.3, to prove (4.8) we need only show that

$$(4.23) \qquad \limsup \frac{1}{\sqrt{c}} \int_{-K_c\sqrt{c}}^{K_c\sqrt{c}} [B_\theta(\pi_{T/c}^{(N_c)}, c) - R_\theta(\pi_{T/c}, c)]\psi(\theta)\, d\theta \leqq 0.$$

Now in view of (4.4)

$$(4.24) \qquad \frac{1}{\sqrt{c}} \int_{-K_c\sqrt{c}}^{K_c\sqrt{c}} |R_\theta(\pi_{T/c}, c) - R_\theta(\pi_{T/c}^{(N_c)}, c)|\psi(\theta)\, d\theta \leqq 2FK_c c N_c \to 0.$$

Finally,

$$(4.25) \qquad \frac{1}{\sqrt{c}}\left|\int_{-K_c\sqrt{c}}^{K_c\sqrt{c}} B_\theta(\pi_{T/c}^{(N_c)}, c) - R_\theta(\pi_{T/c}^{(N_c)}, c)\right|\psi(\theta)\, d\theta$$

$$\leqq 2FK_c \sup\left\{|B_\theta(\pi_{T/c}^{(N_c)}, c) - R_\theta(\pi_{T/c}^{(N_c)}, c)|: |\theta| \leqq K_c\sqrt{c}\right\} \to 0$$

by using Lemma 3.7 and the estimates (3.32). Combining (4.24), (4.25) and (4.23), (4.8) follows.

In the general case proceed as follows. Let $Q_1, Q_2, \cdots$ be a sequence of random variables (measurable functions) defined on the unit interval such that if we put the uniform distribution on $[0, 1]$ the Q_i are independent and normally distributed with mean zero and variance one. We may, of course, think of the Q_i as being defined on $\mathscr{X}$, depending on $(x_1, x_2, \cdots, z)$ through z only. Define $\pi_\varepsilon^{(e)}$ as follows; $\pi_\varepsilon^{(e)}$ agrees with $\pi^{(e)}$ for the first two stages of $\pi^{(e)}$. If

$$(4.26) \qquad \left|\sum_i \{X_i; 1 \leqq i \leqq n_c^{(1)}\}\right| \leqq A_c^{(1)},$$

$$\left|\sum_i \{X_i; n_c^{(1)} + 1 \leqq i \leqq n_c^{(1)} + n_c^{(2)}\}\right| \leqq A_c^{(2)},$$

then apply $\pi^{(e)}$ to the sequence

$$(4.27) \qquad X_{n_c^{(1)} + n_c^{(2)} + 1} + Q_1, \qquad X_{n_c^{(1)} + n_c^{(2)} + 2} + Q_2, \cdots.$$

Formally,

(4.28) $\quad \pi_\varepsilon^{(e)}(x_1, x_2, \cdots, z)$

$$= \pi^e(x_1, \cdots, x_{n_c^{(1)} + n_c^{(2)}}, \quad x_{n_c^{(1)} + n_c^{(2)} + 1} + Q_1(z), \cdots).$$

Arguing as before but now applying Lemma 3.7 to the variables

(4.29) $\qquad Z_i = (X_{n_c^{(1)} + n_c^{(2)} + i} + Q_i)$

which are readily seen to satisfy the condition of that lemma, we find that

(4.30) $\qquad \limsup \dfrac{1}{\sqrt{c}} B(\pi^*, c, \psi) \leqq \limsup \dfrac{1}{\sqrt{c}} B(\pi_\varepsilon^{(e)}, c, \psi)$

$$\leqq \sqrt{1 + \varepsilon} \, \psi(0) \bar{R}^*(1, T).$$

Letting $T \to \infty$ and $\varepsilon \to 0$ the result follows.

To construct a sequence of procedures which achieves the bound a slightly more involved argument is needed. First of all, arguing as before in Section 2, we show that in the Wiener process problem if $cN_cK_c \to 0$ and $T_c \to \infty$ then

(4.31) $\qquad \lim \dfrac{1}{\sqrt{c}} \displaystyle\int_{-K_c\sqrt{c}}^{K_c\sqrt{c}} R_\theta(\tilde{\pi}_{T/c}^{(N_c)}, c)\psi(\theta) \, d\theta = \psi(0)R^*(1)$

where $\tilde{\pi}$ is such that $\int_{-\infty}^{\infty} R_\theta(\tilde{\pi}, 1) \, d\theta = R^*(1)$. Choose $T_c \uparrow \infty$ so that $T_c^2 \, c^{1/16 - 11\varepsilon} \to 0$, and consider the procedures

(4.32) $\qquad\qquad\qquad \left(\tilde{\pi}_{T_c/c}^{(N_c)}\right)^{(e)}$

corresponding to $\tilde{\pi}_{T/c}^{(N_c)}$ defined in the proof of Theorem 4.1 for T_c varying as above. It is easy to check that if μ satisfies the conditions of the remark following Lemma 3.5, then

(4.33) $\qquad \limsup \dfrac{1}{\sqrt{c}} \displaystyle\int_{-\infty}^{\infty} \{B_\theta[(\tilde{\pi} \, {}_{T_c/c}^{(N_c)})^{(e)}, c] - R_\theta\{\tilde{\pi}_{T_c/c}, c\} \psi(\theta) \, d\theta \leqq 0.$

If μ does not satisfy the conditions following Lemma 3.5 the construction is even less explicit. We construct procedures $\pi_c^{(e)}$ corresponding to

(4.34) $\qquad\qquad\qquad \tilde{\pi}_{(T_c/c, \, \varepsilon_c)}^{(N_c)}$

to be defined below with variables $Q_i^{(c)}$ which are independent normal with mean zero and variance $\varepsilon_c \to 0$. It is necessary to examine the proof of Lemma 3.5 carefully since now d_1 will depend on c and n. It is easy to show that there exists a constant d_1^0 independent of n such that if $Z_i = X_i + Q_i^{(c)}$ then

(4.35) $\qquad d_1(c) \leqq d_1^0 \dfrac{n}{\sqrt{\varepsilon_c}} \exp \{-\gamma^2 \varepsilon_c n/2\},$

and $d_1(c)$ will remain bounded above for $n = N_c$ provided that $\varepsilon_c \geqq 3 \log N_c / \gamma^2 N_c$, say. For T_c, ε_c as above we have for any sequence of procedures $\{\pi\}$

$$(4.36) \qquad \limsup \frac{1}{\sqrt{c}} \int_{-\infty}^{\infty} [B_\theta(\pi_{\varepsilon_c}^{(e)}, c) - R_{\theta, \varepsilon_c}(\pi, c)] \psi(\theta) \, d\theta \leqq 0,$$

where $R_{\theta, \varepsilon}$ is the risk of π for the problem in which we observe the Wiener process with drift θ per unit time and variance $1 + \varepsilon$ per unit time. Finally, it follows from the results of Section 2 that

$$(4.37) \qquad \lim \frac{1}{\sqrt{c}} \inf_\pi \int_{-\infty}^{\infty} R_{\theta, \varepsilon_c}(\pi, c) \psi(\theta) \, d\theta = \psi(0) R^*(1).$$

Therefore if we take

$$(4.38) \qquad \tilde{\pi}_{(T_c/c, \, \varepsilon_c)}^{(N_c)}$$

to be the truncated block policy corresponding in the sense of Lemma 2.5 to the procedure $\tilde{\pi}_c$ which achieves $\min_\pi \int_{-\infty}^{\infty} R_{\theta, \varepsilon_c}(\pi, c) \, d\theta$ then

$$(4.39) \qquad \left(\tilde{\pi}_{(T/c, \, c)}^{(N_c)}\right)^{(e)}$$

achieve the bound. The theorem is proved.

5. Concluding remarks and open problems

The techniques of this paper are evidently not limited to the zero-one loss function considered. For different bounded loss functions we must use a different similarity transform, make different choices of K_c, H_c, N_c, and so on, obtain a different rate of convergence, but arrive at similar results. For example, if $\ell(\theta, d) = 0$ when d is the right decision and if $\ell(\theta, d) = \min\{|\theta|, 1\}$ when d is the wrong decision, then the Bayes risk of our problem is of the order of $c^{2/3}$ and the limiting coefficients of $c^{2/3}$ is $\psi(0)$ times the Bayes risk of Chernoff's problem [8] with unit cost and Lebesgue prior. We can also treat the problem of testing with shrinking indifference regions, say, of the form $[-A\sqrt{c}, B\sqrt{c}]$ for zero-one loss. The Bayes risk is of order $\sqrt{c}$ again and the coefficient is $\psi(0)$ times the risk of the Wiener process problem with unit cost, Lebesgue prior and indifference region $[-A, B]$. On the other hand if one permits ψ to vary with c, say, $\psi_c(t) = (1/\sqrt{c}) \psi(t/\sqrt{c})$ for a fixed prior density, one can under suitable regularity conditions for zero-one loss obtain an asymptotic risk of order $\sqrt{c}$ with coefficient the risk of the Wiener problem with unit cost and prior density ψ. Of course such densities presupposing more and more surety that the parameter is near zero with decreasing cost are not usually reasonable.

It seems that these techniques should also apply to other decision problems for the exponential family at least locally and should prove useful in non-Bayesian problems as well.

The result may also be generalized to nonexponential families by considering, under suitable regularity conditions the variables

$$(5.1) \qquad T_i = \left. \frac{\partial \log f_\theta(X_i)}{\partial \theta} \right|_{\theta = 0}.$$

To what extent an ambitious program such as that of LeCam [14] is possible in the sequential case is, however, unclear to us at present.

A great difficulty of the asymptotic theory of this paper is that in general it leads to problems for the Wiener process which, as the works of Chernoff indicate, can be solved at best approximately. In fact, from a (machine) computational point of view it might be easier, for example, to try to calculate the boundary for the Bernoulli process as an approximation to the Wiener boundary. The results of Moriguti and Robbins [17] as well as our paper indicate that such "boundary convergence" as in Schwarz [20] should hold. However, no proof is known to us.

APPENDIX

We retain the notation of Section 1. Our first aim is to prove the following weak compactness theorem.

THEOREM A.1. *Let* $\pi_n = (\delta_n, \tau_n)$ *be a sequence of procedures in the Wiener process problem. Then, there exists a subsequence* $\{n_k\}$ *and a procedure* $\pi = (\delta, \tau)$ *such that,*

(A.1)
$$\lim_k E_\theta(\delta_{n_k}) = E_\theta(\delta)$$

whenever $\limsup_k E_\theta(\tau_{n_k}) < \infty$ *and*

(A.2)
$$\liminf_k E_\theta(\tau_{n_k}) \geqq E_\theta(\tau)$$

for every θ. (E_θ *are taken with respect to* Q_θ *throughout.*)

The proof proceeds by a series of lemmas.

The following lemma is essentially a special case of Wald's theorem [22].

LEMMA A.1. *Suppose that all of the* τ_n *have common finite range* $\{t_1 < \cdots < t_s\}$. *Then the result of Theorem A.1 holds for suitable* $\{n_k\}$ *and for* $\pi = (\delta, \tau)$ *such that* τ *has the same range with* Q_θ *probability one. Furthermore, if* $\pi'_n = (\delta'_n, \tau'_n)$ *is another sequence of procedures with* τ'_n *having the same range and* $\tau'_n \leqq \tau_n$ *for all n, then we may choose* $\{n_k\}$ *to be the same for both sequences and choose the "limiting"* $\pi' = (\delta', \tau')$ *such that* $\tau' \leqq \tau$.

PROOF. We write the (δ_n, τ_n) in the second form of Section 3, $\tau_n = (\psi_{0n}, \psi_{1n}, \cdots, \psi_{sn})$, $\delta_n = (\delta_{0n}, \delta_{1n}, \cdots, \delta_{sn})$ with

(A.3)
$$\psi_{in}(x) = \lambda[z : \tau_n(x, z) = t_i]$$

(A.4)
$$\delta_{in}(x) = \int_0^1 \delta_{in}(x, z) \, dz.$$

Apply the weak compactness theorem (for tests) to $L_1(\Omega, \mathscr{B}_{t_j}, Q_0)$ (see Lehmann [15], p. 354) and the diagonal process to obtain a sequence $\{n'_k\}$ and $\mathscr{B}_{t_j}$ measurable functions ψ_j measurable functions $\psi_j, j = 1, \cdots, s$ such that

(A.5) $$\iint \psi_{jn_k}(x)g(x)Q_0(dx, dz) \rightarrow \iint \psi_j(x)g(x)Q_0(dx, dz)$$

for every g which is $\mathscr{B}_{t_j}$ measurable and such that $\iint |g(x)|Q_0(dx, dz) < \infty$. (The theorem is applicable since Ω is a complete separable metric space.) The ψ_j are evidently nonnegative. Further, if g is measurable and $Q_0 W^{-1}$ integrable,

(A.6) $$\iint \psi_{jn_k}(x)g(x)Q_0(dx, dz) = E_0[\psi_{jn_k}(W)g(W)]$$

$$= E_0\{\psi_{jn_k}(W)E_0[g(W)|\beta_{t_j}]\} \rightarrow E_0\{\psi_j(W)E_0[g(W)|\mathscr{B}_{t_j}]\}$$

$$= E_0[\psi_j(W)g(W)]$$

by (A.5).

Therefore,

(A.7) $$Q_0\left[\sum_{j=1}^{s} \psi_j(W) > 1\right] = E_0\left\{\left[\sum_{j=1}^{s} \psi_{jn_k}(W)\right]I\left[\sum_{j=1}^{s} \psi_j(W) > 1\right]\right\}$$

$$\rightarrow E_0\left\{\sum_{j=1}^{s} \psi_j(W)I\left[\sum_{j=1}^{s} \psi_j(W) > 1\right]\right\}.$$

By the same argument $E_0[\Sigma_{j=1}^{s}\psi_j(W)] = 1$.

Hence, since on $\mathscr{B}_{t_s}$ the $Q_\theta W^{-1}$ are equivalent

(A.8) $$Q_\theta\left[\sum_{j=1}^{s} \psi_j(W) = 1\right] = 1.$$

Evidently we may choose versions of the ψ_j such that $\psi_j \geqq 0$ and $\Sigma_{j=1}^{s}\psi_j = 1$ for all θ. Finally we conclude that $\psi = (\psi_1, \cdots, \psi_s)$ is a stopping time and

(A.9) $$E_\theta(\tau) = \sum_{j=1}^{s} t_j E_\theta(\psi_j)$$

$$= \sum_{j=1}^{s} t_j E_0\{\psi_j(W) \exp[\theta W(t_j) - \tfrac{1}{2}\theta^2 t_j]\}$$

$$= \lim_k \sum_{j=1}^{s} t_j E_0\{\psi_{jn_k}(W) \exp[\theta W(t_j) - \tfrac{1}{2}\theta^2 t_j]\}$$

$$= \lim_k E_\theta(\tau_{n_k}).$$

Now we can by diagonalization and a similar argument obtain a further subsequence $\{n_k\}$ and $\mathscr{B}_{t_j}$ measurable functions γ_j such that,

(A.10) $$E_0[\delta_{jn_k}(W)\psi_{jn_k}(W)g(W)] \rightarrow E_0[\lambda_j(W)g(W)]$$

for every integrable function g on $\tilde{C}$. Let,

(A.11) $$\delta_j = \lambda_j/\psi_j.$$

Since,

(A.12)
$$E_0[\lambda_j(W)g(W)] \leqq E_0[\psi_j(W)g(W)]$$

for every integrable g, we can select δ_j so that $0 \leqq \delta_j \leqq 1$ and, of course, δ_j is $\mathscr{B}_{t_j}$ measurable. Evidently, $((\psi_1, \cdots, \psi_s), (\delta_1, \cdots, \delta_s))$ a policy in the second form and $\{n_k\}$ satisfy (A.1) and (A.2). To obtain the procedure in form I simply define (following Wald and Wolfowitz [23]),

$$\tau(x, z) = t_r \quad \text{if} \quad \sum_{j=1}^{t_r-1} \psi_j(x) < z \leqq \sum_{j=1}^{t_r} \psi_j(x),$$

(A.13)
$$\delta(x, z) = \delta_j(x) \quad \text{on the set} \quad [\tau(x, z) = t_j].$$

Since $\tau_n \leqq \tau'_n$ the statement of the lemma leads to limiting times (in the second form) with $\Sigma_{j=1}^{\ell} \psi'_j(x) \geqq \Sigma_{j=1}^{\ell} \psi_j(x)$ for every ℓ and x and our second assertion follows from (A.13). The lemma is proved.

LEMMA A.2. *The theorem is valid if it is true that there exists a T such that $Q_0[\tau_n \leqq T] = 1$ for all n. Furthermore, order is preserved in the limit as in Lemma A.1.*

PROOF. Consider a grid 0, $T/2^m$, $2T/2^m$, $\cdots$, T. Define $\tau_n^{(m)} = kT/2^m$ if $(k-1)T/2^m < \tau_n \leqq kT/2^m$ for $k = 0, 1, \cdots, 2^m$.

Let $\pi_n^{(m)} = (\tau_n^{(m)}\delta_n)$. (Note that δ_n is $\tilde{\mathscr{B}}_{\tau^{(m)}}$ measurable.) Then,

(A.14)
$$E_\theta(\tau_n^{(m)}) - E_\theta(\tau_n) \leqq T/2^m$$

and

(A.15)
$$R_\theta(\pi_n^{(m)}) - R_\theta(\pi_n) \leqq T/2^m.$$

Extract a subsequence $\{n_k\}$ and limits in the sense of Lemma A.1 $\tau^{(m)}$, $\delta^{(m)}$ for each of the sequences $\pi_{n_k}^{(m)}$. Since $\tau_n^{(m)} \geqq \tau_n^{(m+1)}$ for every n, we may suppose that $\tau^{(m)} \geqq \tau^{(m+1)}$ for every m. Let $\tau = \lim_m \tau^{(m)}$. Note that $\tilde{\mathscr{B}}_{\tau^{(m)}} \subset \tilde{\mathscr{B}}_{\tau^{(m-1)}}$ for every m and,

(A.16)
$$\tilde{\mathscr{B}}_\tau = \bigcap_m \tilde{\mathscr{B}}_{\tau^{(m)}}.$$

Consider the functions $\{\delta^{(m)}\}$. These are $\tilde{\mathscr{B}}_{\tau^{(j)}}$ measurable for $m \geq j$. Extract a subsequence $\{m_k\}$ by the diagonal process and $\tilde{\mathscr{B}}_{\tau^{(j)}}$ measurable functions $\tilde{\delta}^{(j)}$ such that

(A.17)
$$E_\theta[\delta^{(m)}(W, U)g_j(W, U)] \to E_\theta[\tilde{\delta}^{(j)}(W, U)g_j(W, U)],$$

for every g_j which is $\tilde{\mathscr{B}}_{\tau^{(j)}}$ measurable and bounded for every θ. This follows by the weak compactness theorem for test functions applied to $\tilde{\mathscr{B}}_{\tau^{(j)}}$ successively since the Q_θ are all equivalent on $\tilde{\mathscr{B}}_{\tau^{(j)}}$ and the space Ω is complete separable metric. By construction for every θ the $\tilde{\delta}^{(j)}$ form a martingale and in view of (A.16) and by the martingale convergence theorem,

(A.18)
$$\tilde{\delta}^{(j)} \to E_\theta[\tilde{\delta}^{(1)}|\tilde{\mathscr{B}}_\tau]$$

a.s. Q_θ for every θ. Let,

(A.19) $\delta = E_0[\tilde{\delta}^{(1)}|\tilde{\mathscr{B}}_\tau]$.

Then δ is $\tilde{\mathscr{B}}_\tau$ measurable and

(A.20) $E_\theta(\delta) = E_\theta(\tilde{\delta}^{(1)}) = \lim_k E_\theta(\delta^{(m_k)}) = \lim E_\theta(\delta_{n_r})$

while

(A.21) $E_\theta(\tau) = \lim_k E(\tau^{(m_k)}) = \lim_k \lim_r E_\theta(\tau_{n_r}^{(m_k)})$

$$\leq \lim_r E_\theta(\tau_{n_r})$$

by (A.4). The lemma follows.

We complete the proof of the theorem. Given τ_n let $(\tau^{(T)}, \delta^{(T)})$ be the limits guaranteed by Lemma A.2 for a subsequence of the procedures $\pi_n^{(T)} = (\tau_n^{(T)}, \delta_n^{(T)})$ given by

(A.22) $\tau_n^{(T)} = \min(\tau_n, T)$,

$$\delta_n^{(T)} = \begin{cases} \delta & \text{if } \tau \leq T \\ 0 & \text{otherwise.} \end{cases}$$

By Lemma A.2 we can find a subsequence $\{n_k\}$ which works for every $T = 1$, $2, \cdots$ and such that $\tau^{(j)} \leq \tau^{(j+1)}$ for every j. Let $\tau = \lim_j \tau^{(j)}$. By the monotone convergence theorem,

(A.23) $E_\theta(\tau) = \lim_j E_\theta(\tau^{(j)}) \leq \lim_k \inf E_\theta(\tau_{n_k})$.

Consider the sequence $\delta^{(j)}$. Tracing back its construction via Lemmas A.1 and A.2 it is easy to see that the ordering $\delta_n^{(j)} \leq \delta_n^{(j+1)}$ is preserved with Q_θ probability one in the limit. Let $\delta = \sup_j \delta^{(j)}$. Clearly δ is $\tilde{\mathscr{B}}_\tau$ measurable and by the monotone convergence theorem.

(A.24) $E_\theta(\delta) = \lim_j \lim_k E_\theta(\delta_{n_k}^{(j)})$.

Therefore,

(A.25) $\lim_k \sup |E_\theta(\delta) - E_\theta(\delta_{n_k})|$

$$\leq \lim_j \sup \lim_k \sup Q_\theta[\tau_{n_k} > j]$$

$$\leq \lim_j \sup \frac{1}{j} \lim_k \sup E_\theta(\tau_{n_k}) = 0,$$

if $\lim \sup E_\theta(\tau_{n_k}) < \infty$. The theorem follows.

PROOF OF LEMMA 3.3.   We proceed as in [18]

(A.26) $\|U_{f,n} - \Phi\| = \frac{1}{2} \int_{-\infty}^{\infty} |f_n(t) - \phi(t)| \, dt$

where $f_n = dU_{f,n}(t)/dt$ and ϕ is the standard normal density. By the Schwarz and Minkowski inequalities,

$$(A.27) \qquad \|U_{f,n} - \phi\| \leq \frac{1}{2}\left[\int (1 + x)^{-2} dx\right]^{1/2}$$

$$\left\{\int (1 + x)^2 [f_n(x) - \phi(x)]^2 dx\right\}^{1/2}$$

$$\leq C_1\left(\left\{\int [f_n(x) - \phi(x)]^2 dx\right\}^{1/2}\right.$$

$$\left. + \left\{\int [xf_n(x) - x\phi(x)]^2 dx\right\}^{1/2}\right)$$

where C is a numerical constant. Since $C_1(\mathscr{F}) < \infty$ we may apply the Plancherel theorem to obtain

$$(A.28) \qquad \int [f_n(x) - \phi(x)]^2 dx = 2\pi \int\left[\lambda^n\left(\frac{t}{\sqrt{n}}\right) - e^{-t^2/2}\right]^2 dt,$$

where $\lambda(t) = \int_{-\infty}^{\infty} e^{itx} f(x) dx$. Similarly,

$$(A.29) \qquad \int [xf_n(x) - x\phi(x)]^2 dx$$

$$= 2\pi \int\left\{\left[\lambda^2\left(\frac{t}{\sqrt{n}}\right)\right]' - \left[e^{-t^2/2}\right]'\right\}^2 dt.$$

It is well known that

$$(A.30) \qquad \left|\lambda_n\left(\frac{t}{\sqrt{n}}\right) - e^{-t^2/2}\right| \leq \frac{C_3\{C_2(\mathscr{F})\}}{\sqrt{n}}\{|t|^3 + |t|^2\}e^{-t^2/4},$$

$$(A.31) \qquad \left|\left[\lambda_n\left(\frac{t}{\sqrt{n}}\right)\right]' - [e^{-t^2/2}]'\right| \leq \frac{C_3\{C_2(\mathscr{F})\}}{\sqrt{n}}\{|t|^3 + |t|^4\}e^{-t^2/4},$$

for

$$(A.32) \qquad |t| \leq \frac{C_4\sqrt{n}}{C_2^{1/2}(\mathscr{F})},$$

and

$$(A.33) \qquad \left|\left[\lambda^n\left(\frac{t}{\sqrt{n}}\right)\right]'\right| \leq C_5 n^{1/2} C_2^{1/3}(\mathscr{F})\left|\lambda^{n-1}\left(\frac{t}{\sqrt{n}}\right)\right|$$

where $C_1 - C_5$ are numerical constants.

Finally note that since the Riemann Lebesgue lemma holds uniformly on compact sets of L_1, we have

$$(A.34) \qquad \sup\{|\lambda(t)|:|t| \geq C_4/C_2^{1/2}(\mathscr{F}), f \in \mathscr{F}\} = C_3(\mathscr{F}) < 1.$$

(To prove this note that the map $(f, t) \to |\lambda(t)|$ is continuous on $L_1 \times [-\infty, \infty]$ with $\lambda(-\infty) = \lambda(+\infty) = 0$. Since $|\lambda(t)| < \int |f(t)| \, dt$ for every $t \neq 0$, (A.31) follows.) Now,

$$
(\text{A.35}) \qquad \int \left| \lambda^n \left(\frac{t}{\sqrt{n}} \right) - e^{-t^2/2} \right|^2 dt
$$

$$
\leqq (C_3^2/n) C_2^2(\mathscr{F}) \int_{|t| > C_4 n^{1/2} C^{-1/2}(\mathscr{F})} \{|t|^3 + |t|^2\}^2 \, e^{-t^2/2} \, dt
$$

$$
+ \int_{|t| > C_4 n^{1/2} C^{-1/2}(\mathscr{F})} e^{-t^2/2} \, dt
$$

$$
+ C_3^{n-2}(\mathscr{F}) \int_{|t| > C_4 n^{1/2} C^{-1/2}(\mathscr{F})} \left| \lambda^n \left(\frac{t}{\sqrt{n}} \right) \right| \, dt
$$

$$
\leqq C_7 \{ C_2^2(\mathscr{F})/n + C_3^{n-2}(\mathscr{F}) \} \leqq C^2(\mathscr{F})/n
$$

since $C_3 < 1$. A similar estimate can be given for the second term on the right of (A.27). The result follows.

REFERENCES

[1] A. E. ALBERT, "The sequential design of experiments for infinitely many states of nature," *Ann. Math. Statist.*, Vol. 32 (1961), pp. 774–799.

[2] J. A. BATHER, "Bayes procedures for deciding the sign of a normal mean," *Proc. Cambridge Philos. Soc.*, Vol. 58 (1962), pp. 599–620.

[3] S. A. BESSLER, "Theory and applications of the sequential design of experiments k-actions, and infinitely many experiments," Department of Statistics, Stanford University, Technical Report No. 55 (1960).

[4] P. J. BICKEL and J. A. YAHAV, "Asymptotically pointwise optimal procedures in sequential analysis," *Proceedings of the Fifth Berkeley Symposium on Mathematical Statistics and Probability*, Berkeley and Los Angeles, University of California Press, 1967, Vol. 1, pp. 401–413.

[5] ———, "On testing sequentially the mean of a normal distribution," Stanford Technical Report N.S.F. No. 26 (1967).

[6] H. CHERNOFF, "A measure of asymptotic efficiency for tests of a hypothesis based on the sum of observations," *Ann. Math. Statist.* Vol. 23 (1952), pp. 493–507.

[7] ———, "Sequential design of experiments," *Ann. Math. Statist.*, Vol. 30 (1959), pp. 755–770.

[8] ———, "Sequential tests for the mean of a normal distribution," *Proceedings of the Fourth Berkeley Symposium on Mathematical Statistics and Probability*, Berkeley and Los Angeles, University of California Press, 1961, Vol. 1, pp. 79–91.

[9] ———, "Sequential test for the mean of a normal distribution III (small t)," *Ann. Math. Statist.*, Vol. 36 (1965), pp. 28–54.

[10] ———, "Sequential test for the mean of a normal distribution IV (discrete case)," *Ann. Math. Statist.*, Vol. 36 (1965), pp. 55–68.

[11] A. DVORETZKY, A. WALD, and J. WOLFOWITZ, "Elimination of randomization in certain statistical decision procedures and zero-sum two-person games," *Ann. Math. Statist.*, Vol. 22 (1951), pp. 1–21.

[12] T. S. FERGUSON, *Mathematical Statistics*, New York, Academic Press, 1967.

[13] J. KIEFER and J. SACKS, "Asymptotically optimum sequential inference and design," *Ann. Math. Statist.*, Vol. 34 (1963), pp. 705–750.

[14] L. LeCam, "On the asymptotic theory of estimation and testing hypotheses," *Proceedings of the Third Berkeley Symposium on Mathematical Statistics and Probability*, Berkeley and Los Angeles, University of California Press, 1956, Vol. 1, pp. 129–156.

[15] E. L. Lehmann, *Testing Statistical Hypotheses*, New York, Wiley, 1959.

[16] D. V. Lindley and B. N. Barnett, "Sequential sampling: two decision problems with linear losses for binomial and normal random variables," *Biometrika*, Vol. 52 (1965), pp. 507–532.

[17] G. Lorden, "Integrated risk of asymptotically Bayes sequential tests," *Ann. Math. Statist.*, Vol. 38 (1967), pp. 1399–1422.

[18] S. Moriguti and H. Robbins, "A Bayes test of '$p \leq 1/2$' versus '$p > 1/2$'," *Rep. Statist. Appl. Res. Un. Japan Sci. Engrs.*, Vol. 9 (1962), pp. 39–60.

[19] V. V. Petrov, "Asymptotic analysis of some limit theorems in probability," *Vestnik Leningrad Univ.*, Vol. 16 (1961), pp. 51–61. (Also in *Selected Transl. Math. Statist. and Prob.*, Vol. 5 (1965), pp. 179–190.)

[20] G. Schwarz, "Asymptotic shapes of Bayes sequential testing regions," *Ann. Math. Statist.*, Vol. 33 (1962), pp. 224–236.

[21] A. Wald, "Tests of statistical hypotheses concerning several parameters when the number of observations is large," *Trans. Amer. Math. Soc.*, Vol. 54 (1943), pp. 426–482.

[22] ———, *Statistical Decision Functions*, New York, Wiley, 1950.

[23] A. Wald and J. Wolfowitz, "Two methods of randomization in statistics and the theory of games," *Ann. Math.*, Vol. 22 (1951), pp. 581–586.

SOME NEW RESULTS IN SEQUENTIAL ESTIMATION THEORY

YU. V. LINNIK and I. V. ROMANOVSKY
MATHEMATICAL INSTITUTE OF THE ACADEMY OF SCIENCES, LENINGRAD

1. The setup of the problem; some general remarks

The problem of sequential estimation attracted some fresh interest recently. We can note, basically, two directions in the corresponding recent literature: asymptotic investigations (see for instance [1], [2], [3], [4]) and exact formulas for "small samples" (see [5], [6], [7], [8]). We shall give here an account of some recent results in both these directions.

In articles [1] to [4] a Bayesian approach to the sequential estimation of parameters is considered. Here we use another, non-Bayesian approach.

We can consider sufficiently large families of stochastic processes with independent increments (and discrete or continuous time). We shall always study scalar processes unless stipulated otherwise.

In the case of discrete time, we shall suppose it integer valued so that our process will be reduced to the repeated sample of a certain population with a distribution in a family $\mathscr{P}_\theta$ characterized by a density $f(x, \theta)$ with respect to the Lebesgue or the counting measure. The parameter θ will be always scalar, unless stipulated otherwise.

We shall consider here mostly the processes with discrete time, addressing ourselves, say, to the standard Poisson process only in Section 3.

However, many of the results expounded here can be transferred to the continuous time case.

In both cases we shall consider Markov stopping times τ (see [9] for the definition) and scalar statistics T_τ that are unbiased estimates of a scalar function $g(\theta)$ of the parameter θ

$$(1.1) \qquad E_\theta(T_\tau) = g(\theta).$$

Moreover, we must choose among such unbiased estimates a statistic $\tilde{T}_\tau$ with the minimal variance $D_\theta(\tilde{T}_\tau)$ under the condition

$$(1.2) \qquad E_\theta(\tau) \leqq n,$$

where n is a given number. From (1.2) it follows that $P_\theta(\tau < \infty) = 1$.

Note that if such a statistic exists, this statistic and the corresponding stopping rule may depend upon θ. Thus a statistic which is optimal in the above sense uniformly with respect to θ does not exist in general.

However, if instead of exact optimality we consider asymptotic optimality, for $n \to \infty$ (which, of course, corresponds to the individual cost of an experiment converging to zero), the situation changes: the existence of the asymptotically optimal stopping rule and statistic T_τ is proved for the Bayes setup in the works of Bickel and Yahav [1] to [4].

In the present article we shall indicate asymptotically optimal estimates T_τ and stopping rules in the above mentioned setup. The presence or absence of the discontinuities in the corresponding information quantities will be very essential for the subject. An exact quantitative formulation will be given later.

We shall give some results for "small samples" about existence conditions for sampling plans which are efficient in the Rao-Cramér sense, about the types of such plans, and about the characterization of first hit plans by the simplest properties of Markov stopping times.

2. Asymptotic results in the case of the absence of discontinuities in the information quantities

In the present section we want to propound the point of view which indicates that, roughly speaking, if there are no discontinuities in the information quantities and the cost of the experiment is small, the sequential estimation in the setup described above can give only an infinitely small gain in comparison with the fixed sample method of estimation. We shall consider the family $\mathscr{P}_\theta$ of distributions with the density $f(x, \theta)$ with respect to the Lebesgue measure (some other requirements will be imposed later on).

Here we shall restrict our attention to the processes with independent increments and discrete time, that is, the simplest case of a repeated sample $x_1, x_2, \cdots$. For the continuous time case the corresponding theory is not yet worked out; to all appearances it can be made by analogy.

We shall look for an asymptotically unbiased estimate $\tilde{T}_\tau$ of the parameter θ which minimizes the variance $D_\theta(T_\tau)$ up to the infinitely small quantities of a given order under conditions $E_\theta \tau \leqq n$ and $n \to \infty$.

We shall show that, roughly speaking, in the absence of discontinuities in the information quantities such a statistic T_t can be constructed in the class of trivial stopping rules ($\tau = $ constant).

In the proof, we shall rely upon some theorems of Ibragimov and Hasminski on the asymptotic behavior of generalized Bayes estimates for constant sample sizes [12].

By the absence of discontinuities in the information quantities we mean the fulfillment of the following conditions of Ibragimov and Hasminski which we shall strengthen somewhat for the sake of convenience in the exposition.

CONDITION 1.   *The density $f(x, \theta)$ with respect to the Lebesgue measure is measurable in both arguments, and $\iint |\theta| f(x, \theta) f(x, \theta_0) \, dx d\theta < \infty$ for each point θ_0 in the interval of the values of the argument Θ.*

CONDITION 2.   *As $|\theta| \to \infty$, the integral $\int f(x, \theta) f(x, \theta_0) \, dx \to 0$.*

CONDITION 3. *As $\varepsilon \to 0$, the integral $\int f^{1-\varepsilon}(x, \theta_0)\, dx \to 1$.*

CONDITION 4. *For each set of real numbers $\theta_1, \cdots, \theta_s$ and for appropriately chosen intervals $[0, T_1], \cdots, [0, T_s]$, we shall have for all $t_j \in [0, T_j]$ and $\xi \downarrow 0$*

$$(2.1) \qquad \int \prod_{j=1}^{s} \left(\frac{f(x, \theta_0 + \xi\theta_j)}{f(x, \theta_0)} \right)^{t_j} f(x, \theta_0)\, dx$$

$$= 1 + \xi^{\alpha} a(t_1, \cdots, t_s; \theta_1, \cdots, \theta_s) + o(\xi^{\alpha}),$$

where the number α does not depend upon $\theta_1, \cdots, \theta_s$ and the function $a(\cdot) \not\equiv 0$.

CONDITION 5. *For all θ_1, θ_2 with $|\theta_j| \leq H < \infty$, $i = 1, 2$, we have for $\xi \to 0$*

$$(2.2) \qquad \left| \int \left\{ \log \frac{f(x, \theta_0 + \xi\theta_1)}{[f(x, \theta_0 + \xi\theta_2)]} \right\} f(x, \theta_0 + \xi\theta_2)\, dx \right| \leq |\xi|^{\alpha} c_1(|\theta_2 - \theta_1|),$$

$$(2.3) \qquad \left| \int \left\{ \log^2 \frac{f(x, \theta_0 + \xi\theta_1)}{[f(x, \theta_0 + \xi\theta_2)]} \right\} f(x, \theta_0 + \xi\theta_2)\, dx \right| \leq |\xi|^{\alpha} c_2(|\theta_2 - \theta_1|),$$

where $[f(x, \theta_0 + \xi\theta_2)]$ coincides with $f(x, \theta_0 + \xi\theta_2)$ if $f(\cdot, \cdot) \neq 0$ and equals 1 if $f(\cdot, \cdot) = 0$. The functions $c_i(h)$, $i = 1, 2$, depend upon H, and $c_i(h) \to 0$ for $h \to 0$.

Under Conditions 1 to 5 the theorems of Ibragimov and Hasminski [12] can be applied to the study of the asymptotic behavior of the variance of the Pitman estimate for a repeated sample.

We form the Pitman estimates $\tilde{\theta}_n$ for the parameter θ, as

$$(2.4) \qquad \tilde{\theta}_n = \int \theta p_n(\theta)\, d\theta,$$

where

$$(2.5) \qquad p_n(\theta) = \frac{\displaystyle\prod_{i=1}^{n} f(x_i, \theta)}{\displaystyle\int \prod_{i=1}^{n} f(x_i, \theta)\, d\theta}.$$

First consider the case of the location parameter, that is, $f(x, \theta) = f(x - \theta)$. Apart from Conditions 1 to 5, we also assume the finiteness of the following quantities (a) $\int x^2 f(x)\, dx$, (b) the information type quantities $E_0 |\ell^{(n)}(x_i)|^s$ with $n \leq 5$ and $s \leq 20$, and (c) $E_0 \max_{|\theta| \leq \varepsilon} |\ell^{(s)}(x_i - \theta)|^s$ with $s \leq 20$ and $\varepsilon > 0$, where ε is a small given number. Here $\ell(x) = \log f(x)$, and the above include the Fisher information quantity $I = E_0(\partial \log f / \partial x)^2$.

Then, according to [12] we can assert that the unbiased Pitman estimate (2.4) for the location parameter θ_0 has a finite variance; moreover

$$(2.6) \qquad E_\theta(\tilde{\theta}_n - \theta)^2 = \frac{1}{I_n} + \frac{c_1}{n^2} + o\left(\frac{1}{n^2}\right)$$

where c_1 is a constant depending upon $f(x)$ only.

In the general case, let $\ell(x, \theta) = \log f(x, \theta)$ and let $I_\theta = E(\partial \ell / \partial \theta)^2$ be the Fisher information number. If there is an $\varepsilon > 0$ such that the moments $E_\theta |\ell_\theta^{(n)}(x, \theta_0)|^s$, with $n \leq 5$ and $s \leq 20$, and $E_\theta \max_{|\theta| \leq \varepsilon} |\ell_\theta^s(x_1, \theta_0 - \theta)|^s$, with

$s \leqq 20$, are finite and if $\max_\theta E_\theta \tilde{\theta}_n^2 < \infty$, we have

$$(2.7) \qquad E_\theta(\tilde{\theta}_n - \theta) = \frac{c_0(\theta)}{n} + o\left(\frac{1}{n}\right),$$

$$(2.8) \qquad E_\theta(\tilde{\theta}_n - \theta)^2 = \frac{1}{I_\theta n} + \frac{c_1(\theta)}{n^2} + o\left(\frac{1}{n^2}\right),$$

where the $c_i(\theta)$ for $i = 0, 1$, are certain constructively given functions.

We return to the location parameter and formula (2.6). Here $\tilde{\theta}_n$ is an unbiased estimate of the parameter θ which possesses a variance. The relation (2.6), by comparison with the well known Rao-Cramér inequality shows that asymptotically, for large values of n, the estimate $\tilde{\theta}_{[n]}$ is very good.

Consider a Markov stopping rule τ with $E_\theta \tau < \infty$, and an unbiased estimate T_τ of the location parameter θ.

According to the well known Wolfowitz inequality (see [13]), we have

$$(2.9) \qquad D(T_\tau) \geqq \frac{1}{IE_\theta(\tau)}.$$

If the condition

$$(2.10) \cdot \qquad E_\theta \tau \leqq n$$

holds. Then (2.6) implies directly that

$$(2.11) \qquad D(T_\tau) \geqq D(\tilde{\theta}_{[n]})\left[1 + o\left(\frac{1}{n}\right)\right].$$

We can formulate (2.11) as the following theorem.

THEOREM 1.   *Assume that the Ibragimov-Hasminski conditions for the absence of discontinuities in the information quantities are satisfied. Then for any unbiased sequential estimate of the location parameter θ subject to the restriction $E_\theta(\tau) \leqq n$, the relative improvement of the variance over that of the constant sample size method is at most of the order $o(1/n)$.*

For the general case, taking into account the bias (see Equations (2.7) and (2.8)), we get a similar result.

In that case we must take an estimate $\tilde{\theta}_n$ having bias $o(1/n)$. We obtain Theorem 2.

THEOREM 2.   *Assume that the Ibragimov-Hasminski conditions for the absence of discontinuities in the information quantities are satisfied. Then for any unbiased sequential estimate of the location parameter θ subject to the restriction $E_\theta(\tau) \leqq n$, the relative improvement of the mean quadratic deviation is at most $o(1/n)$.*

For the proof of Theorem 2, we use (2.7) and (2.8). As an asymptotically optimal estimate for the sample size n, we take $\tilde{\theta}_{[n]}$. In forming the Wolfowitz inequality, we use (2.8). In other respects we argue as before.

Note that in the Theorems 1 and 2 we consider the mean quadratic deviation from the parameter value of the estimate $\tilde{\theta}_{[n]}$ itself, not that of the normed limit distribution. The computation of such a deviation often presents considerable difficulties (for analogous computations relating to the maximum likelihood estimates, see, for instance, [14]).

3. Sequential estimation in the case of discontinuities in the information quantities

In this section we shall give certain examples where the presence of discontinuities in the information quantities leads to the existence of sequential estimation procedures giving a considerable gain in the variance of the estimate over that of the fixed sample size method. Our examples relate to the location parameter θ and the density $f(x - \theta)$ which is continuous for $|x - \theta| < \frac{1}{2}$ and has the carrier $|x - \theta| \leq \frac{1}{2}$.

As $f(x - \theta) = 0$ for $|x - \theta| > \frac{1}{2}$, we have discontinuity in the expressions for the Fisher information quantity and cannot use the Rao-Cramér and Wolfowitz inequalities.

We assume that f satisfies conditions insuring, for sample sizes $n \geq 2$, the existence of the Pitman estimate

$$(3.1) \qquad \tilde{\theta}_n = \int \theta p_n(\theta) \, d\theta,$$

where

$$(3.2) \qquad p_n(\theta) = \prod_{i=1}^{n} f(x_i - \theta) \left[\int \prod_{i=1}^{n} f(x_i - \theta) \, d\theta \right]^{-1}.$$

As is well known (see [10] where the literature is indicated), the Pitman estimate $\tilde{\theta}_n$ will be unbiased and optimal with respect to the variance estimate of θ in the class of all "regular" estimates T_n (that is, such that $T_n(x_1 + c, \cdots, x_n + c) = T(x_1, \cdots, x_n) + c$). Therefore we can let this estimate exemplify the constant sample size n and compare its variance with that of estimates given by sequential methods for $E_\theta \tau \leq n$.

For the simplest case of the uniform distribution $f(x) = 1$ with $|x| \leq \frac{1}{2}$, we have (see [10]) $\tilde{\theta}_n = \frac{1}{2}(x_{max} - x_{min})$ and $D(\tilde{\theta}_n) = [2(n + 1)(n + 2)]^{-1}$. Choose the stopping rule $\tau = \min_n \{n; x_{max} - x_{min} > 1 - \varepsilon(n)\}$ and use the estimate $\tilde{\theta}_\tau$. For a suitable $\varepsilon(n)$ it will be unbiased such that $E_\theta(\tau) = n$ and $D(\tilde{\theta}_\tau) = \frac{1}{6}n^2$. It follows that

$$(3.3) \qquad D(\tilde{\theta}_\tau)/D(\theta_n) \to \frac{1}{3}$$

for $n \to \infty$ and that the sequential estimation improves the asymptotic variance by a factor of three.

Note that in this case for any strictly monotonic function W the stopping times $\tau = \min\{n; |x_{max} - x_{min}| > 1 - \alpha\}$ minimize the risk function $W(|x - \theta|) + c\tau$ for an appropriate choice of α.

It is interesting that a similar result holds for much more general distributions. I. I. Iaura and A. N. Shalyt were kind enough to make a calculation requested by the present authors. It was based upon the article of Ibragimov and Hasminski and led to the following theorem.

THEOREM 3.   *Let the density $f(x)$ having the carrier $|x| < \frac{1}{2}$, be symmetric and continuous together with its first derivative for $|x| < \frac{1}{2}$. Assume $f(x) = 1$ for $0 < \frac{1}{2} - |x| \leq \varepsilon$ where $\varepsilon > 0$ is a constant. Then the stopping rule $\tau = \min\{v: x_{\max}^{v} - x_{\min}^{v} > 1 - 2/n\}$ leads to the unbiased estimate $\theta_{\tau} = \frac{1}{2}(x_{\max}^{\tau} - x_{\min}^{\tau})$ and gives a variance asymptotically three times smaller than the fixed sample size method.*

Thus we see that the presence of discontinuities in the information quantities may lead to a considerable gain in variance by use of sequential instead of constant sample size estimation. The study of such a connection between discontinuities in the information quantities and the improvement of the estimation quality in the sequential case seems to be interesting.

We must remark that these discontinuities do not always lead to an asymptotic gain in the variance. Thus, Shalyt informed the authors that for the density $f(x - \theta) = \exp\{-(x - \theta)\}$ for $x \geq \theta$ and $f(x - \theta) = 0$ for $x < \theta$, there is no such gain.

4. Binomial and multinomial processes: the Poisson process

In this section we shall first consider estimation plans for processes with discrete time and a finite number of states. Such processes, inasmuch as the structure of the set of values of the corresponding random variable is not essential, are described by the multinomial scheme. The frequency vector describing the appearance of different states is a sufficient statistic in this case.

The estimation plans for the binomial case were studied in [5] and [6]. In particular, in [5] a statistic is given which is an unbiased estimate of the vector of probabilities of the states or of a given polynomial function of these probabilities. It is natural to indicate the plans for which such an estimate is unique.

A Markov stopping rule τ is called complete if the only unbiased estimate of zero is the trivial statistic $T_{\tau} \equiv 0$. In the case of bounded binomial schemes a necessary and sufficient condition for completeness of a Markov stopping rule is indicated in the work of De Groot [5]. In the multinomial case the following theorem of Zaidman holds.

THEOREM 4A.   *For the completeness of a Markov stopping rule it is necessary that it be nonrandomized.*

This means that all the points of the phase space are subdivided into three groups: the set of boundary points B, the set of transition points, and the set of unattainable points. A plan with such a Markov stopping rule and with the statistic obtained by the Rao-Blackwell process will be called a first hit plan in what follows.

A plan called "finite for θ" if $E_{\theta}\tau < \infty$, and "finite" if it is finite for all $\theta \in \Theta$. If no proper subset $B' \subset B$ is a set of boundary points of a finite plan, the plan is called a minimal plan.

THEOREM 4B.   *For the completeness of a first hit plan minimality is necessary.*

The proof of this theorem as well as Theorem 4A is based upon the construction of a nontrivial unbiased estimate of zero when the conditions of the theorem are violated.

Necessary and sufficient conditions of a geometrical nature for the completeness of multinomial first hit plans have not yet been found. However there are sufficient conditions valid for a large class of plans. In particular, the following theorem holds.

THEOREM 5. *Suppose that for an n-normal bounded plan, for each $x \in B$ there is a number i such that all the points $y = (y_1, \cdots, y_n)$ satisfying the conditions*

$$(4.1) \qquad y_i > x_i; \qquad y_j \leqq x_j, \quad j \neq i; \qquad \sum_j y_j = \sum x_j + 1,$$

are unattainable. Then the plan is complete.

The proof of this theorem is effected by induction. All the points of the boundary B are subdivided into the sets $B_k, B_{k+1}, \cdots, B_\ell$ on which the statistic τ has constant values (k being the minimal value). Further construct a new plan S' in which the set B_k is included in the set of transition points, whereas B_{k+1} is extended to $B'_{k+1} = B_k + B_\Delta$ so as to make the new plan S' closed and minimal. The induction hypothesis is that S' is a complete plan (the induction basis is the constant sample size plan $\tau = \ell$). Under these conditions we must prove that if $\phi(x)$ is a function on B and

$$(4.2) \qquad E_p \phi \equiv 0$$

p being the vector of the state probabilities, then $\phi \equiv 0$. Now the process of induction requires proving that $\phi = 0$ on B_k. The completeness of S' requires proving that for each $y \in S'$ we must have the equality

$$(4.3) \qquad E K(0, x) \phi(x) = 0,$$

where the summation extends over all the points $x \in S_k$ which can be reached starting from the point y, and where $K(0, x)$ is the number of trajectories reaching x and starting from the origin. Under the conditions of the theorem, the system (4.3) proves to have as many equations as variables and its matrix is triangular. This follows from the possibility of indexing the points of B_k and B_Δ in such a way that passage from the ith point of B_k to the jth point of B_Δ is impossible for $j > i$ (this indexing will be needed later also). Moreover as the matrix of the system (4.3) is nonsingular, we have $\phi \equiv 0$ on B_k which terminates the induction.

Note that for the binomial case condition, (4.1) is necessary because it is fulfilled only for the plans described by De Groot [5]. The plans described in Theorem 5 have another interesting property, which is in a sense the inverse of the completeness property.

THEOREM 6. *If S and S' are two different first hit plans under the conditions of Theorem 5, then*

$$(4.4) \qquad E_p \tau \neq E_p \tau'.$$

This theorem is proved by the same process of induction as the previous one.

We prove step by step the coincidence of the sets B and B'. Having proved for an appropriate set R, $R \subset B$ and $R \subset B'$, we study $E_p(\tau, B\backslash R)$ and $E_p(\tau, B'\backslash R)$ where

$$(4.5) \qquad E_p(\tau, M) = P_p(M)E_p(\tau|x \in M),$$

where $P_p(M)$ is the probability of reaching M from the origin, and where $E_p(\tau|x \in M)$ is the conditional expectation of τ given that we start from $x \in M$. The functions $E_p(\tau, B\backslash R)$ and $E_p(\tau, B'\backslash R)$ are polynomials in P_i.

Let plan S have the corresponding boundary set $B = \{B_k, \cdots, B_\ell\}$ and let plan S' have the set $B' = \{B'_{k'}, \cdots, B'_{\ell'}\}$ (see the proof of Theorem 5). Without loss of generality suppose that $k \leq k'$. We shall prove that $k = k'$ and $B_k = B_{k'}$. The vectors of B_k can be subdivided into several subsets, each indexed as in the proof of Theorem 5, by writing the components of vectors in a sequence fixed for each subset separately and indexing in decreasing lexicographical order, beginning with the vector $(k, 0, \cdots, 0)$ in the sequence selected for this set.

We take the first of these subsets of B_k and the first of its vectors x^0 and make the first component of the vector of the state probabilities tend to 1. We get $\lim E_p\tau' = K(0, x^0)k$ and so the vector x^0 must belong to B'. Then we continue the process in the same way with the stipulation that before making the first component of the probability vector tend to 1, we must divide $E_p(\tau, B\backslash R)$ and $E_p(t, B'\backslash R)$, which are equal, by $\Pi P x_i^0$.

The proof does not hold for an arbitrary complete plan (except for the binomial case) but one can conjecture that not only complete, but even simply minimal plans have this property. However, this question remains open. For nonminimal plans there are examples of families of plans with the same $E_p\tau$.

For sampling without repetition there is no separate completeness problem because the following theorem holds.

THEOREM 7.   *For a plan S to be complete for sampling without repetition it is necessary and sufficient that it be complete for the repeated sample method.*

We pass now to the Poisson process. This process can be treated as the limit case of the binomial process and the results relating to it are limit cases of the analogous facts for the Bernoulli scheme. For instance, the analogues of Theorems 5 and 6 are as follows.

THEOREM 8.   *For the completeness of a bounded first hit plan S given by a set of boundary points, consisting of a finite number of segments parallel to the time axis, in the space of the sufficient statistics $(t, \eta(t))$, it is necessary and sufficient that the common length of these segments be equal to the essential value of the maximum stopping time corresponding to this plan.*

THEOREM 9.   *A complete bounded first hit plan with a finite number of boundary segments is determined by the values of the mean stopping time of this process as a function of the intensity λ.*

The proofs of these theorems are analogous to the proofs for the binomial case and use induction on the length of the maximum stopping time. (See also [7], [8].)

5. Description of efficient sequential plans for the renewal process

In the present section we shall consider plans of sequential estimation for parameters of stochastic processes with independent increments which are renewal processes.

We shall consider sequential plans which are efficient in the sense of Wolfowitz's identity on a certain interval of parameter values. Investigations of this kind were started by De Groot [5] in 1959 and developed considerably by Trybula [6] in 1968. Trybula considered the Poisson process, Brownian motion, and so forth. In Section 7 of his interesting work he makes some general remarks on efficient plans of sequential estimation for homogeneous processes with independent increments.

One can consider homogeneous processes with independent increments and with discrete or continuous time.

Consider the case of discrete time. This will correspond to the repeated sample $x_1, x_2, \cdots$.

We shall suppose that the quantities x_i have a density $f(x, \theta) > 0$ with respect to the Lebesgue measure or the counting measure with carrier consisting of integer numbers. The parameter θ will lie in a certain interval Θ of the real axis. Let τ be a Markov stopping time and let $L(x, \tau, \theta)$ be the likelihood function. We shall suppose that requirements sufficient for fulfillment of the Rao-Cramér information inequalities hold (regarding these requirements, see for instance [15], [16]). Let T_τ be an estimate which is unbiased and efficient in the sense of the variance for a given function $g(\theta)$ of the parameter in a certain interval of parameter values. In this case

$$(5.1) \qquad T_\tau - g(\theta) = h(\theta) \frac{\partial \log L}{\partial \theta}.$$

Suppose now that the function $g(\theta)$ has a derivative $g'(\theta)$ having no zeros on a certain interval $I_1 \subset \Theta$. Then we can replace the parameter θ by the parameter $\theta_1 = g(\theta)$ and use the following theorem proved by Kagan in [17].

THEOREM 10. (Kagan). *Assume that the following conditions are satisfied:*

(1) *the density $f(x, \theta)$ satisfies the standard regularity conditions (see, for instance, [17]);*

(2) *the Markov stopping time τ is such that for a certain $n > 2$ the set $\{\tau = n\} = M_n$ contains $\overline{\Delta} = \Delta_1 \times \Delta_2 \times \cdots \times \Delta_n$, where the Δ_i are the intervals of R^1;*

(3) *the statistic $[T_\tau, \tau]$ is sufficient for θ in a certain interval;*

(4) *the mapping $T_n \to R^1$ is nontrivial on $\overline{\Delta}$ and for each pair of values θ, θ' in the interval mentioned above, there exists an i, $1 \leq i \leq n$, such that for $x \in \Delta_i$, $f(x, \theta)$ is not a multiple of $f(x, \theta')$.*
Then $f(x, \theta)$ is of exponential type.

Thus, under the conditions of the theorem the density $f(x, \theta)$ corresponds to the exponential family of distributions

$$(5.2) \qquad f(x, \theta) = \exp\{\theta T(x) - \rho(\theta)\}$$

for a certain interval of values of θ. Here $\rho(\theta)$ is the norming function and $T(x)$ is a statistic such that for each value $\tau = n$, $T_\tau = T_n = \Sigma_{i=1}^n T(x_i)$ gives the value of the sufficient statistic. An analogous result will hold also for densities with respect to the counting measure with carrier consisting of integers.

Without restricting the generality, we can suppose that relations (5.1) and (5.2) hold in an interval of values of θ where we can choose the values θ_1, and θ_2, and we can write equation (5.1) for them. Following articles [5] and [6] we can subtract one equality from the other. Then we shall get the linear relation

$$(5.3) \qquad aT_\tau + b\tau + c = 0$$

so that a sequential estimation plan which is efficient in the sense of the variance and corresponds to τ must be a first hit plan determined by the linear relation (5.3).

Let us suppose now that the sufficient statistic T_τ is integer valued and non-negative so that the process $\{T_t, t = 1, 2, \cdots\}$ is a renewal process. Let us mark the integer values time on the x axis and the values of the statistic T_t on the y axis. Then our process will be represented in the first quadrant.

In order that the plan with the boundary (5.3) be closed for a certain interval of values of the parameter θ the fulfillment of the following relation is necessary:

$$(5.4) \qquad \tau = AT_\tau + B,$$

A and B being positive integers.

Denote by $Cl(S)$ the set on which the plan S is closed (see [9], [10]), that is, the set of values θ for which $E_\theta \tau < \infty$.

Suppose that on an interval $I \subset \Theta$, $E_\theta \tau$ and $E_\theta \tau^2$ exist. Let $K(\tau, T_\tau)$ be the number of "trajectories" starting from the point $(0, 0)$ and reaching the point (τ, T_τ) on the plan boundary (it is finite because we consider only renewal processes). Then we have

$$(5.5) \qquad \sum_{\partial s} K(\tau, T_\tau) \exp\{\theta T_\tau - \tau\rho(\theta)\} = 1.$$

Differentiating this identity with respect to θ for $\theta \in I$, we get under certain natural conditions the Wald identity $E_\theta T_\tau = \rho'(\theta)E_\theta\tau$. From this and from (5.4), it follows that

$$(5.6) \qquad E_\theta \tau = B[1 - A\rho'(\theta)]^{-1}.$$

Then (5.4) implies

$$E_\theta \tau = A_\tau + B,$$
$$(5.7) \qquad E_\theta t^2 - AE_\theta T_\tau + BE_\theta \tau = 0,$$
$$E_\theta T_\tau \tau - AE(T_\tau^2) - BE_\theta T_\tau = 0$$

and a straightforward calculation gives

$$(5.8) \qquad D_\theta \tau = A^2 B\rho''(\theta)[1 - A\rho'(\theta)]^3.$$

(The calculation was made by Nz. M. Halfina to whom the authors wish to express their gratitude.)

This expression coincides with I_θ^{-1}. By dint of Wolfowitz's inequality this implies the efficiency of the unbiased estimate of the function (5.6) obtained from plan S, in the interval $I \subset Cl(S)$. All other efficiently estimable functions can be obtained from (5.6) with the help of linear transformations with constant coefficients (see [16]).

If $f(x, \theta)$ is the density with respect to the Lebesgue measure we shall again suppose that the statistic $T(x)$ is nonnegative. It is assumed to have a continuous density. In that case only constant sample size plans can be efficient, as a rule. Some particular cases of the processes with independent increments were considered by Trybula [6]. To study them systematically as we did for the discrete time case a generalization of Kagan's theorem from [17] for the continuous time case is needed. It has not yet been obtained.

6. Unsolved problems

We shall indicate here some unsolved problems of the theory of sequential estimation which seem to present a certain interest:

(1) generalize the results of Ibragimov and Hasminski to the case of multivariate distributions and formulate corresponding consequences for sequential analysis;

(2) compare the asymptotic properties of sequential estimation and fixed sample size method for nonquadratic losses and in the absence of discontinuities in the analogue of the information quantities;

(3) formulate and extend the above problems in the continuous time case;

(4) study the appearance of discontinuities in the information quantities and the possibilities of applying sequential analysis to improve the quality of the estimate in the case of discrete time and scalar or vector parameters for experiments which have a small cost;

(5) study the same question for the case of continuous time and scalar or vector parameter;

(6) find the geometric conditions for completeness of unbounded binomial plans which are closed only on a proper subset of the set $(0, 1)$;

(7) find the geometric conditions for completeness of multinomial plans;

(8) in the multinomial case, investigate the possibility that a bounded minimal plan be determined by the values of $E_p\tau$;

(9) investigate the most general conditions for the existence of linear optimal estimation plans for homogeneous processes with independent increments and discrete or continuous time.

REFERENCES

[1] P. J. BICKEL and J. A. YAHAV, "Asymptotically pointwise optimal procedures in sequential analysis," *Proceedings of the Fifth Berkeley Symposium on Mathematical Statistics and Probability*, Berkeley and Los Angeles, University of California Press, 1967, Vol. 1, pp. 401–413.

[2] ———, "Asymptotically optimal Bayes and minimax procedures in sequential estimation," *Ann. Math. Statist.*, Vol. 39 (1968), pp. 442–456.

[3] ——, "Some contributions to the asymptotic theory of Bayes solutions," *Z. Wahrschein-lichkeitstheorie und Verw. Gebiete.*, Vol. 2 (1969), pp. 257–276.

[4] ——, "On an A. P. O. rule in sequential estimation with quadratic loss," *Ann. Math. Statist.*, Vol. 40 (1969), pp. 417–427.

[5] M. H. DeGroot, "Unbiased sequential estimation for binomial population," *Ann. Math. Statist.*, Vol. 30 (1959), pp. 80–101.

[6] S. Trybula, "Sequential estimation in processes with independent increments," *Dissertationes Math. Rozprawy Mat.*, Vol. 60 (1968), pp. 1–50.

[7] R. A. Zaidman, Yu. V. Linnik, and S. V. Romanovsky, "Sequential estimation plans and Markov stopping times," *Dokl. Akad. Nauk SSSR*, Vol. 185 (1969), pp. 1222–1225. (In Russian.)

[8] R. A. Zaidman, Yu. V. Linnik, and V. N. Sudakov, "On sequential estimation and Markov stopping times for processes with independent increments," *USSR–Japan Symposium on Probability, Habarovsk, August* 1969, Novosibirsk, "Nauke," 1969, pp. 127–143. (In Russian.)

[9] A. N. Shiryayev, *Statistical Sequential Analysis*, "Nauke," 1969.

[10] A. N. Shalyt, "Some results on the sequential estimation of the location parameter," *Dokl. Akad. Nauk SSSR*, Vol. 189 (1969), pp. 57–58.

[11] ——, "On the optimal behavior of the location parameter in the class of invariant procedures," *Dokl. Akad. Nauk SSSR* (1970). (In Russian.)

[12] I. A. Ibragimov and R. Z. Hasminski, "On the asymptotic behavior of generalized Bayes estimates," *Dokl. Akad. Nauk SSSR*, Vol. 194 (1970), pp. 257–260. (In Russian.)

[13] J. Wolfowitz, "The efficiency of sequential estimation and Wald's equation for sequential processes," *Ann. Math. Statist.*, Vol. 18 (1947), pp. 215–230.

[14] Yu. V. Linnik and N. M. Mitrofanova, "Some asymptotic expansions for the distribution of the maximum likelihood estimate," *Contributions to Statistics Presented to Professor P. C. Mahalanobis on the Occasion of his Seventieth Birthday* (edited by C. R. Rao), Oxford and New York, Pergamon Press, 1965, pp. 229–238.

[15] L. N. Bolshev, "A refinement of the Cramér-Rao inequality," *Teor. Verojatnost. i Primenen.*, Vol. 6 (1961), pp. 319–326.

[16] Yu. V. Linnik, "Some remarks on the inequalities of Rao-Cramér, and Bhattacharya," *Matem. Zametki* (1970).

[17] A. M. Kagan, "Remarks on the sufficiency in a sequential process," *Matem. Zametki* (1970). (In Russian.)

[18] R. A. Zaidman, "Completeness of Markov stopping rules," *Zap. Naučn. Sem. Leningrad Otdel. Mat. Inst. Steklov* (1970). (In Russian.)

[19] Yu. V. Linnik, *Statistical Problems with Nuisance Parameters*, Moscow, Nauke, 1966. (English translation, Providence, American Mathematical Society, 1968.)

SEQUENTIAL RANK TESTS—
ONE SAMPLE CASE

RUPERT G. MILLER, JR.
STANFORD UNIVERSITY

1. Introduction

Let $X_1, X_2, \cdots$ be a sequence of independent randon variables, identically distributed according to the continuous c.d.f. F. The null hypothesis is H_0: $F(-x) = 1 - F(x)$, $0 \leq x < +\infty$; that is, the random variables are symmetrically distributed about zero. Sequential tests of this hypothesis which are based on the signs and ranks of the X_i are studied in this paper.

Sequential rank tests should be particularly useful in medical clinical trials. The one sample case arises naturally when patients are paired for similarity of influential physical traits, and are randomly assigned to one of two possible treatments so that each treatment is given to one member of each pair. The variable X_i is the difference between the treatment effects measured on the ith patient pair. Sequential binomial trials have been valuable in this context and will continue to be so. Rank tests, however, can take advantage of quantitative (non-dichotomous) information in each treatment comparison while at the same time making only minimal assumptions about the form of the distribution.

In 1969 Weed, Bradley, and Govindarajulu [4] proposed a sequential likelihood ratio test for this problem. Let $G(x) = P\{|X| < x | X < 0\}$ and $H(x) = P\{X < x | X > 0\}$. They considered the family of distributions whose left and right tails are related by $1 - H(x) = (1 - G(x))^A$, $A > 0$, and $F(0) = A/(1 + A)$. For this family the null hypothesis becomes $H_0: A = 1$, and an alternative hypothesis is $H_1: A = B$ where B is a specified constant. The likelihood for the signs and rank order of the absolute values of $X_1, \cdots, X_n$ is

$$(1) \qquad \binom{n}{n^-} \left(\frac{A}{1 + A}\right)^{n^-} \left(\frac{1}{1 + A}\right)^{n^+} n^-! \, n^+!$$

$$\int \cdots \int_{0 < x_1 < \cdots < x_n} \prod_{i=1}^{n} \{dG(x_i)\}^{\delta_i} \{dH(x_i)\}^{1 - \delta_i},$$

where $\delta_i = 1$ if the X_i with the ith smallest absolute value is negative, $\delta_i = 0$ if it is positive. For $H_0: A = 1$ and $H_1: A = B$, the likelihood ratio simplifies to

$$(2) \qquad LR_n = \frac{(B/1 + B)^n 2^n n!}{\prod_{i=1}^{n} [n_i^- + B n_i^+]},$$

where n_i^- is the number of X_j such that $X_j < 0$ and $|X_j| \geq |X_i|$ and n_i^+ is the

97

number of X_j such that $X_j > 0$ and $X_j \geqq |X_i|$. The Wald type sequential test is to continue sampling as long as $a < LR_n < b$, where $0 < a < 1 < b$. If LR_n exceeds the upper bound for some n, terminate sampling and decide in favor of H_1; if LR_n crosses the lower bound, terminate the test and decide H_0. Weed *et al* proved that this test terminates with probability one so the bounds can be approximated by $a = \beta/(1 - \alpha)$ and $b = (1 - \beta)/\alpha$, where α and β are the specified probabilities of errors of the first and second kind, respectively.

Weed *et al* also considered another model with $H(x) = G^A(x)$ and $F(0)$ arbitrary, but in this paper attention will be restricted to the model cited above.

Also in 1969, Miller [3] proposed an *ad hoc* sequential test based on the Wilcoxon signed rank statistic. Let $R_1 < \cdots < R_{n-}$ be the ranks of the negative X_i, and $S_1 < \cdots < S_{n+}$ the ranks of the positive X_i in the ordered sequence of the absolute values of the X_i. The Wilcoxon signed rank statistic is $SR_n = \Sigma_{i=1}^{n^+} S_i - \Sigma_{i=1}^{n^-} R_i$. Under H_0 the first two moments of SR_n are $E(SR_n) = 0$ and $\mathrm{Var}\,(SR_n) = n(n + 1)(2n + 1)/6$. The fixed sample size test suggests sampling as long as

$$(3a) \qquad |SR_n| \leqq |z|_N^\alpha [n(n + 1)(2n + 1)/6]^{1/2}$$

and

$$(3b) \qquad n < N.$$

If for some n prior or equal to N, $|SR_n|$ exceeds the bound in (3a), reject H_0. If n reaches N without (3a) being violated, accept H_0. The investigator selects the truncation point N and the probability of a type I error α, which determines the critical constant $|z|_N^\alpha$. A table of $|z|_N^\alpha$ for $\alpha = 0.10, 0.05, 0.01$, and $N = 10(5)30(10)60$ is given in [3]. The percentile points $|z|_N^\alpha$ were estimated by Monte Carlo simulation.

This test is computationally easy to perform since the SR_n can be computed sequentially.

$$(4) \qquad SR_n = \sum_{i=1}^{n} \sum_{j=1}^{i} \mathrm{sgn}\,(X_i + X_j)$$

$$= SR_{n-1} + \sum_{i=1}^{n} \mathrm{sgn}\,(X_i + X_n),$$

where

$$(5) \qquad \mathrm{sgn}\,(X_i + X_j) = \begin{cases} +1 & \text{if } X_i + X_j > 0, \\ -1 & \text{if } X_i + X_j < 0. \end{cases}$$

As the next observation X_n is obtained, it is easy to compare it with the preceding observations in order to compute the term $\Sigma_{i=1}^{n} \mathrm{sgn}\,(X_i + X_n)$. The addition of this sum and the previous SR_{n-1} yields SR_n.

Since $\Sigma_{i=n+1}^{N} i = (N - n)(N + n + 1)/2$, it is possible for the test to terminate sampling prior to N with the acceptance of H_0. If for any n

$$(6) \qquad |SR_n| \leqq |z|_N^\alpha [N(N + 1)(2N + 1)/6]^{1/2} - (N - n)(N + n + 1)/2,$$

then it will be impossible for $|SR_n|$ to reach the rejection boundary by time N, and the test can stop with acceptance of H_0. Expression (6) creates an inner acceptance boundary inside the outer rejection boundary (3a).

An analogous one sided test would be to continue sampling as long as $SR_n \leq z_N^\alpha [n(n+1)(2n+1)/6]^{1/2}$ and $n < N$. However, the critical constants z_N^α have not been computed. For the one sided test the acceptance boundary $SR_n \leq z_N^\alpha [N(N+1)(2N+1)/6]^{1/2} - (N-n)(N+n+1)/2$ is more apt to create substantial savings in the number of observations than for the two sided test.

In this paper a third test is presented. It is similar to the preceding test, but it employs a linear barrier instead of a square root barrier. Namely, continue sampling as long as

(7a) $$|SR_n| \leq |w|_N^\alpha n,$$

(7b) $$n < N$$

for the two sided test, or

(7a') $$SR_n \leq w_N^\alpha n$$

for the one sided test. The investigator specifies N and α, which determine $|w|_N^\alpha$ (Table Ia, b) or w_N^α (Table IIa, b). As in the previous test acceptance of H_0 can occur prior to N if

(8) $$|SR_n| \leq |w|_N^\alpha N - (N-n)(N+n+1)/2$$

in the two sided case, or if

(8') $$SR_n \leq w_N^\alpha N - (N-n)(N+n+1)/2$$

in the one sided case.

The linear barrier in (7a) or (7a') can be motivated in two ways. Suppose that the X_i are independently, identically distributed according to $F(x - \Delta)$, which has density $f(x - \Delta)$ symmetric about Δ. For $H_0 : \Delta = 0$ versus $H_1 : \Delta = \Delta_1$ the likelihood ratio for the signs and rank order of the absolute values of $X_1, \cdots, X_n$ can be written

(9) $$LR_n = 1 + \Delta_1 \left[\sum_{i=1}^{n^+} E\left(\frac{-f'(U_{(s_i)})}{f(U_{(s_i)})} \right) - \sum_{i=1}^{n^-} E\left(\frac{-f'(U_{(r_i)})}{f(U_{(r_i)})} \right) \right] + O(\Delta_1^2)$$

as $\Delta_1 \to 0$, where $U_{(1)} < \cdots < U_{(n)}$ are the order statistics from a sample generated by the density $2f(u)$ for $u > 0$. If F is chosen to be the logistic distribution, then

(10) $$E\left(\frac{-f'(U_{(s_i)})}{f'(U_{(s_i)})} \right) = \frac{s_i}{n+1},$$

and similarly for r_i. Thus,

(11) $$LR_n = 1 + \Delta_1 SR_n/(n+1) + O(\Delta_1^2),$$

and a linear barrier seems appropriate for local shift alternatives. This approach is analogous to the one employed for the two sample problem by W. J. Hall in unpublished work.

A second justification for a linear barrier arises from the approximate normality of SR_n. For moderate or large values of n, SR_n is approximately normally distributed with mean $\mu_n = n(n-1)\theta_2 + n\theta_1$ and variance $\sigma_n^2 = n(n-1)(n-2)\eta_3 + n(n-1)\eta_2 + n\eta_1$. The constants $\theta_1, \theta_2, \eta_1, \eta_2, \eta_3$ depend on F, but not n. In particular, $\theta_1 = P\{X_1 > 0\} - P\{X_1 < 0\}$, and $\theta_2 = [P\{X_1 + X_2 > 0\} - P\{X_1 + X_2 < 0\}]/2$. For translation alternatives $F(x - \Delta)$ the mean μ_n under $H_0: \Delta = -\Delta_1$ is the negative of the mean under $H_1: \Delta = +\Delta_1$; the variance σ_n^2 is the same under both H_0 and H_1. Thus, if it is assumed that SR_n is approximately normally distributed, the likelihood ratio for SR_n under $H_0: \Delta = -\Delta_1$ and $H_1: \Delta = +\Delta_1$ is

$$(12) \qquad L_n = \exp\left\{ \frac{1}{2\sigma_n^2}(SR_n + \mu_n)^2 - \frac{1}{2\sigma_n^2}(SR_n - \mu_n)^2 \right\},$$

and

$$(13) \qquad \log L_n = \frac{2\mu_n SR_n}{\sigma_n^2} \sim \left(\frac{2\theta_2}{\eta_3}\right)\frac{SR_n}{n},$$

which again suggests a linear barrier.

2. Percentile points for linear barriers

Define $Y_n = SR_n/n$, and

$$(14) \qquad W_N = \max\{Y_1, \cdots, Y_N\}, \qquad |W|_N = \max\{|Y_1|, \cdots, |Y_N|\}.$$

In order for the test defined by (7a), (7b) to have size α, the constant $|w|_N^\alpha$ must be the upper α-percentile point of the distribution of $|W|_N$; that is, $P\{|W|_N > |w|_N^\alpha\} = \alpha$. Similarly, w_N^α is defined by $P\{W_N > w_N^\alpha\} = \alpha$.

Both W_N and $|W|_N$ have discrete distributions which have not been treated analytically. However, their distributions can be estimated easily through Monte Carlo simulation. Two thousand W_N and $|W|_N$ were randomly generated for $N = 10(5)30(10)50$. The sample α-percentile points for $\alpha = 0.10, 0.50, 0.01$ are displayed in Table Ia and Table IIa. A pseudorandom number generator [2] was used to generate sequences of uniform variables, which were transformed into W_N or $|W|_N$. Different sequences were used for different values of N.

The stochastic behavior of the signed ranks SR_n, $n = 1, \cdots, N$, can be approximated by a Wiener process. It is well known that under H_0

$$(15) \qquad E(Y_n) = 0, \qquad \text{Var}(Y_n) = \frac{n(n+1)(2n+1)}{6n^2} \cong \frac{n}{3}.$$

For $m < n$ the covariance between Y_m and Y_n is

(16)
$$\text{Cov}\,(Y_m, Y_n) = \frac{m(n+1)(2m+1)}{6mn} \cong \frac{m}{3},$$

which is easily proved using the representation

(17)
$$\sum_{j=1}^{n} \sum_{i=1}^{j} \text{sgn}\,(X_i + X_j) = \sum_{j=1}^{m} \sum_{i=1}^{j} \text{sgn}\,(X_i + X_j) + \sum_{j=m+1}^{n} \sum_{i=1}^{j} \text{sgn}\,(X_i + X_j)$$

and the covariances

(18)
$$\text{Cov}\,[\text{sgn}\,(X_i + X_j),\, \text{sgn}\,(X_k + X_j)] = \tfrac{1}{3},$$
$$\text{Cov}\,[\text{sgn}\,(X_j + X_j),\, \text{sgn}\,(X_i + X_j)] = \tfrac{1}{2},$$

for $i, j, k,$ unequal. The approximate normality of Y_n and the variance-covariance in (15) and (16) suggest that the Y_n are behaving like $W(n)/\sqrt{3}$, where $W(t)$ is a standard Wiener process with zero mean and variance t.

This approximation gives

(19)
$$P\{W_N > c\} \cong P\{\max_{0 \le t \le N} W(t)/\sqrt{3} > c\}$$

$$= 2P\{W(N) > c\sqrt{3}\}$$

$$= 2[1 - \Phi(c\sqrt{3}/\sqrt{N})],$$

where the second line follows from the reflection principle and Φ is the unit normal c.d.f. The approximation (19) implies $w_N^\alpha \cong \sqrt{N}\, g^{\alpha/2}/\sqrt{3}$, where $g^{\alpha/2}$ is the upper $\alpha/2$-percentile point of the unit normal c.d.f. Values of $\sqrt{N}\, g^{\alpha/2}/\sqrt{3}$ appear in Table IIb.

If the probability of crossing both the positive and negative boundaries by time N is negligible, then $|w|_N^\alpha \cong \sqrt{N}\, g^{\alpha/4}/\sqrt{3}$. The latter values are presented in Table Ib. The probability of a double crossing appears to be negligible for N in the range of the tables, but it will increase with increasing N.

Comparison of Table Ia with Ib and Table IIa with IIb reveals close agreement between the Monte Carlo percentile points and the Wiener approximations. This seems remarkable since a discrete process in discrete time is being approximated by a continuous process in continuous time.

TABLE Ia

VALUES OF $|w|_N^\alpha$ BY MONTE CARLO APPROXIMATION

α \ N	10	15	20	25	30	40	50
.01	5.00	6.23	6.90	7.59	8.50	10.05	11.08
.05	4.10	5.07	5.70	6.50	6.93	7.97	8.82
.10	3.50	4.36	5.06	5.64	6.15	6.90	7.75

TABLE Ib

VALUES OF $|w|_N^\alpha$ BY WIENER APPROXIMATION

α \ N	10	15	20	25	30	40	50
.01	5.12	6.28	7.25	8.10	8.88	10.25	11.46
.05	4.09	5.01	5.79	6.47	7.09	8.18	9.15
.10	3.58	4.38	5.06	5.66	6.20	7.16	8.00

TABLE IIa

VALUES OF w_N^α BY MONTE CARLO APPROXIMATION

α \ N	10	15	20	25	30	40	50
.01	4.50	5.73	6.30	7.32	7.83	9.38	10.73
.05	3.70	4.30	5.00	5.50	5.96	7.00	7.82
.10	3.10	3.70	4.25	4.61	5.06	5.79	6.61

TABLE IIb

VALUES OF w_N^α BY WIENER APPROXIMATION

α \ N	10	15	20	25	30	40	50
.01	4.70	5.76	6.65	7.44	8.15	9.41	10.52
.05	3.58	4.38	5.06	5.66	6.20	7.16	8.00
.10	3.00	3.68	4.25	4.75	5.20	6.01	6.72

3. Power and expected sample size

The power and expected stopping time of the test using SR_n with linear barriers was investigated for shift alternatives in the double exponential distribution. Let

$$(20) \qquad f(x) = \tfrac{1}{2}e^{-|x-\Delta|}, \qquad -\infty < x < +\infty.$$

This distribution has mean Δ, variance 2 (standard deviation 1.41). The tails of (20), which are heavier than the normal, behave like the logistic distribution for which the Wilcoxon statistic is known to have good local properties. In [3] this distribution was used to examine the power and stopping time for SR_n with square root barriers.

For $N = 20$ the performance of the test was studied at $\Delta = 0, 0.5, 1, 1.5$; for $N = 50$, at $\Delta = 0, 0.25, 0.5, 0.75, 1$. The error probability α was taken to be either 0.05 or 0.01. For each combination of N, α, Δ, five hundred sequences of

TABLE IIIa

MONTE CARLO POWER AND EXPECTED SAMPLE
SIZE FOR TWO SIDED TEST, $N = 20$

Power	Δ			
	0	.5	1	1.5
$\alpha = .05$	.046	.39	.88	.99
$\alpha = .01$	.014	.19	.71	.96
Expected n				
$\alpha = .05$	19.8	18.5	14.8	13.0
$\alpha = .01$	20.0	19.5	17.4	15.5

TABLE IVa

MONTE CARLO POWER AND EXPECTED SAMPLE SIZE
FOR TWO SIDED TEST, $N = 50$

Power	Δ				
	0	.25	.5	.75	1
$\alpha = .05$	.062	.30	.82	.98	1.0
$\alpha = .01$	.010	.13	.59	.93	1.0
Expected n					
$\alpha = .05$	49.4	46.6	37.2	29.0	24.2
$\alpha = .01$	49.9	48.9	43.5	35.8	29.9

TABLE Va

MONTE CARLO POWER AND EXPECTED SAMPLE
SIZE FOR ONE SIDED TEST, $N = 20$

Power	Δ			
	0	.5	1	1.5
$\alpha = .05$	.052	.51	.94	.99
$\alpha = .01$	.020	.32	.78	.98
Expected n				
$\alpha = .05$	14.7	16.1	13.3	11.5
$\alpha = .01$	13.5	16.3	15.7	14.0

double exponential random variables were substituted into the test. For the
two sided test the results are displayed in Tables IIIa, IVa; the one sided test
results appear in Tables Va, VIa.

TABLE VIa

Power	Δ				
	0	.25	.5	.75	1
$\alpha = .05$	.052	.41	.87	.99	1.0
$\alpha = .01$	.006	.14	.59	.92	1.0
Expected n					
$\alpha = .05$	41.9	41.6	32.9	25.7	21.5
$\alpha = .01$	38.8	43.3	41.6	34.3	29.5

For the two sided test the power is the probability of exceeding the rejection boundary in the correct direction. Under these alternative hypotheses just one Monte Carlo sequence ever crossed the rejection boundary in the wrong direction. This occurred for $N = 50$, $\alpha = 0.05$, $\Delta = 0.25$.

For the two sided test the inner acceptance boundary was not used so that the expected sample sizes could be compared with those in [3] for SR_n with square root boundaries. The power of the test is unaffected by the inclusion, or exclusion, of the inner boundary, but the expected sample sizes would be slightly smaller with the inner boundary included. The early acceptance boundary was included for the one sided test.

Comparison of the entries in Tables IIIa, IVa with the corresponding values in [3] for SR_n with square root boundaries leads to the following conclusion. The test with linear barriers has slightly better power than the square root barrier test, but the reverse holds for expected sample sizes. The square root barrier gives expected sample sizes which are slightly smaller than those for the linear barrier. This means that if a sequence $X_1, X_2, \cdots$ is going to reject H_0, it will stop sooner with the square root barrier than the linear one.

As in the null case, the behavior of SR_n, $n = 1, 2, \cdots, N$, can be approximated by a Wiener process. Let $W(t)$ be a Wiener process with drift μt and variance $\sigma^2 t$. Dinges [1] proved that

$$(21) \qquad P\{\max_{0 \leq t \leq T} W(t) > c\} = \left(e^{c\mu/\sigma^2}\right)^2 \Phi\left(\frac{-\mu\sqrt{T}}{\sigma} - \frac{\sigma}{\sigma\sqrt{T}}\right)$$

$$+ \Phi\left(\frac{\mu\sqrt{T}}{\sigma} - \frac{c}{\sigma\sqrt{T}}\right).$$

For the distribution (20)

$$(22) \qquad E(Y_n) = nE[\text{sgn}(X_i + X_j)] + O(1),$$

$$= \tfrac{1}{2}[1 - (1 + \Delta)e^{-2\Delta}]n + O(1),$$

TABLE IIIb

WIENER POWER FOR TWO SIDED TEST, $N = 20$

Power	Δ			
	0	.5	1	1.5
$\alpha = .05$	.055	.37	.86	1.00
$\alpha = .01$	.015	.19	.64	.95

TABLE IVb

WIENER POWER FOR TWO SIDED TEST, $N = 50$

Power	Δ				
	0	.25	.5	.75	1
$\alpha = .05$	.061	.32	.80	.98	1.0
$\alpha = .01$	.013	.14	.58	.92	1.0

TABLE Vb

WIENER POWER FOR ONE SIDED TEST, $N = 20$

Power	Δ			
	0	.5	1	1.5
$\alpha = .05$	.053	.50	.93	1.00
$\alpha = .01$	.015	.27	.76	.98

TABLE VIb

WIENER POWER FOR ONE SIDED TEST, $N = 50$

Power	Δ				
	0	.25	.5	.75	1
$\alpha = .05$	.055	.42	.87	.99	1.0
$\alpha = .01$	.009	.16	.62	.92	1.0

$$(23) \qquad \mathrm{Var}\,(Y_n) = n\,\mathrm{Cov}\left[\mathrm{sgn}\,(X_i + X_j),\,\mathrm{sgn}\,(X_k + X_j)\right] + O(1),$$
$$= \tfrac{1}{3}\left[5e^{-2\Delta} - (4 + 6\Delta + 3\Delta^2)e^{-4\Delta}\right]n + O(1).$$

If μ and σ^2 are taken to be the coefficients of n in (22) and (23), expression (21) gives the entries in Tables IIIb, IVb, Vb, VIb.

Comparison of the a and b parts of Tables III, IV, V, VI reveals a remarkable agreement between the Monte Carlo approximation to the power and the Wiener approximation.

4. Comparison of tests

An interesting question is how to compare the three tests mentioned in Section 1. One approach is to adjust the structures of the tests so that they have the same power at a fixed alternative. The expected sample sizes can then be directly compared. This cannot be done analytically because workable expressions for the power are not available for the SR_n tests (except for Wiener approximation in the linear case) and for the expected sample sizes nothing is available for any of the tests. However, the comparison can be carried out by Monte Carlo simulation.

Let

$$(24) \qquad f(x) = \begin{cases} \dfrac{A}{1 + A} \, e^x, & x < 0, \\[2ex] \dfrac{A}{1 + A} \, e^{-Ax}, & x > 0. \end{cases}$$

This family of unsymmetric double exponential distributions constitutes Lehmann alternatives to which the likelihood ratio test (2) applies. For $N = 20$, $\alpha = 0.05$ five hundred sequences of random variables from the distribution (24) with $A = 1, 0.75, 0.5, 0.25$ were substituted into the one sided SR_n test with linear barrier. The resulting Monte Carlo power and expected sample sizes appear in Table VIIa in the M (Miller) linear line. For $B = 0.5$ the error probability β is 0.432. These constants were then used to define the sequential likelihood ratio test based on (2). An identical number of Monte Carlo sequences were substituted into this test, and the results appear in the W–B–G (Weed–Bradley–Govindarajulu) lines of the table.

TABLE VIIa

COMPARISON OF THE MILLER TESTS
($N = 20$, $\alpha = 0.05$) WITH THE WEED–BRADLEY–
GOVINDARAJULU TEST ($\alpha = 0.048$, $\beta = 0.432$,
$B = 0.5$) FOR LEHMANN ALTERNATIVES

Power	A			
	1	.75	.5	.25
M square root	.048	.18	.51	.95
M linear	.048	.18	.57	.97
W–B–G	.048	.21	.61	.94
Expected n				
M square root	13.4	14.3	13.9	10.1
M linear	14.7	15.9	15.9	12.9
W–B–G	8.4	12.9	14.5	11.3

To prevent unlimited sampling the W–B–G test was truncated at fifty observations. Just a few sequences needed to be truncated. For $A = 1$ there were three out of five hundred, $A = 0.75$ nine, and $A = 0.5$ six. For shift alternatives with $\Delta = 0$ there were two, and with $\Delta = 0.5$ eight.

The critical constant z_N^α for the one sided SR_n test with square root barrier when $N = 20$, $\alpha = 0.05$ was simulated in a special run so that this type of test could be compared with the previous two. (Complete tables of z_N^α have not been computed.) The estimated percentile point was 2.17. The Monte Carlo results for the square root barrier test appear in the M square root line. Both tests based on SR_n used the early acceptance boundary for H_0.

Since the likelihood ratio test is designed for Lehmann alternatives, it seemed fair to evaluate how it performs for shift alternatives where the Wilcoxon statistic is better adapted. Monte Carlo sequences from the distribution (20) were substituted into the three tests, and the outcome is displayed in Table VIIb.

Table VIIb

COMPARISON OF THE MILLER TESTS
($N = 20$, $\alpha = 0.05$) WITH THE WEED–BRADLEY–
GOVINDARAJULU TEST ($\alpha = 0.048$, $\beta = 0.432$,
$B = 0.5$) FOR SHIFT ALTERNATIVES

Power	Δ			
	0	.5	1	1.5
M square root	.048	.48	.90	.98
M linear	.052	.51	.94	.99
W–B–G	.034	.51	.87	.96
Expected n				
M square root	13.5	13.9	10.2	8.3
M linear	14.7	16.1	13.3	11.5
W–B–G	8.0	14.3	12.8	10.5

Examination of the values in the tables leads to the following conclusions. The M linear test has slightly better power than the M square root, but the M square root has smaller expected sample sizes. This agrees with the conclusion reached earlier in the two sided case. For both power and expected sample size the W–B–G test performs better than the M square root test for near alternatives whereas the M square root is better for far alternatives. The expected sample size under H_0 is very much smaller for the W–B–G test. The choice of alternative distribution, Lehmann or shift, appears to have little effect.

These conclusions are also reflected in the smallest value of n for which acceptance of H_0 or H_1 can occur. For M square root the earliest time at which acceptance of H_0 can happen is 10 and for H_1 it is 6. With M linear the values are 10 for both H_0 and H_1, and the W–B–G test needs only 2 for H_0 but 9 for H_1. Thus, the square root boundary allows earlier rejection of H_0 than the other two, and the likelihood ratio test permits earlier acceptance of H_0.

Which test should be used? The SR_n tests are much simpler computationally, and they are very simple to explain to the investigator. In medical applications selection of an alternative hypothesis and its associated power is often difficult or extremely arbitrary. A bound on the amount of sampling is usually easier to determine due to limitations of money, time, and so forth. The likelihood ratio test stops very early when H_0 is true, but in a medical setting the far alternatives seem more important. If there is little or no difference between treatments, then continuation of the trial is relatively unimportant from the ethical point of view, but if one treatment is much better than the other, you want to stop the trial as soon as possible.

The author would like to thank Elizabeth Hinkley for expertly programming the Monte Carlo computations. She proposed the use of the pseudorandom generator [2], which produces better sequences than other generators familiar to the author.

REFERENCES

[1] H. DINGES, "Ein verallgemeinestes Spiegelungsprinzip für den Prozess der Brownschen Bewegung," *Z. Wahrscheinlichkeitstheorie und Verw. Gebiete*, Vol. 1 (1962), pp. 177–196.
[2] P. A. W. LEWIS, A. S. GOODMAN, and J. M. MILLER, "A pseudo-random number generator for the System/360," *IBM Systems J.*, Vol. 8 (1969), pp. 136–146.
[3] R. G. MILLER, JR., "Sequential signed-rank test," *J. Amer. Statist. Assoc.*, Vol. 65 (1970), pp. 1554–1561.
[4] H. D. WEED, JR., R. A. BRADLEY, and Z. GOVINDARAJULU, "Stopping times of two rank order sequential probability ratio tests for symmetry based on Lehmann alternatives," *Florida State University Statistics Report*, No. 148 (1969).

EXAMPLES OF EXPONENTIALLY BOUNDED STOPPING TIME OF INVARIANT SEQUENTIAL PROBABILITY RATIO TESTS WHEN THE MODEL MAY BE FALSE

R. A. WIJSMAN

UNIVERSITY OF ILLINOIS

1. Summary and introduction

Two examples are presented (one of them the sequential χ^2 test) of parametric models in which the invariant Sequential Probability Ratio Test has exponentially bounded stopping time N, that is, satisfies $P(N > n) < \rho^n$ for some $\rho < 1$, where the true distribution P may be completely arbitrary except for the exclusion of a certain class of degenerate distributions. Another example demonstrates the existence of P under which N is not exponentially bounded, but even for those P we have $P(N < \infty) = 1$. In the last section a proof is given of the representation (2.1), (2.2) of the probability ratio R_n as a ratio of two integrals over the group G if G consists of linear transformations and translations.

Let $Z_1, Z_2, \cdots$ be independent, identically distributed (i.i.d.) random vectors which take their values in d dimensional Euclidean space E^d and possess distribution P. The symbol P will also be used for the probability of an event that depends on all the Z_i. Let Θ be an index set (parameter space) such that for each $\theta \in \Theta$, P_θ is a probability distribution on E^d. We shall say "the model is true" if the true distribution P is one of the P_θ, $\theta \in \Theta$, but it should be kept in mind throughout that we shall also consider the possibility that the model is false, that is, that P is not one of the P_θ. In the latter case we shall also speak of P being outside the model as opposed to P being in the model. Let Θ_1, Θ_2 be two disjoint subsets of Θ. It is not assumed that their union is Θ. The problem is to test sequentially H_1 versus H_2, where H_j is the hypothesis: $P = P_\theta$ for some $\theta \in \Theta_j$, $j = 1, 2$.

If the hypotheses H_j are simple, that is, $\Theta_j = \{\theta_j\}$, $j = 1, 2$, then Wald's Sequential Probability Ratio Test (SPRT) [23] computes the sequence of probability ratios

$$(1.1) \qquad R_n = \prod_{i=1}^{n} \frac{p_2(Z_i)}{p_1(Z_i)}, \qquad n = 1, 2, \cdots,$$

Research supported in part by National Science Foundation Grant GP-23835.

p_j being the density of P_{θ_j} with respect to some common dominating measure; stopping bounds $B < A$ are chosen and sampling continues until the first n when R_n is either $\geq A$ (accept H_2) or $\leq B$ (accept H_1). More convenient for our purpose is to consider $L_n = \log R_n$. Put $\log B = \ell_1$, $\log A = \ell_2$, N the random stopping time, then N is the first n when

$$(1.2) \qquad \qquad \ell_1 < L_n < \ell_2$$

is violated.

If we put $Y_i = \log\left[p_2(Z_i)/p_1(Z_i)\right]$ it follows from (1.1) that $L_n = \Sigma_1^n Y_i$, that is, L_n is a random walk on the real line, starting from zero. Wald [22] used this to prove $P(N < \infty) = 1$ for every P except if $P(Y_1 = 0) = 1$, which can also be written

$$(1.3) \qquad \qquad P\{p_1(Z_1) = p_2(Z_1)\} = 1.$$

Under the same assumption $P(Y_1 = 0) < 1$, Stein [21] obtained a much stronger result

$$(1.4) \qquad \qquad P(N > n) < \rho^n, \qquad \qquad n \geq r,$$

in which $0 < \rho < 1$ and r is some suitably chosen integer. Stein's proof relies on the demonstration that there exists an integer r and $p > 0$ such that

$$(1.5) \qquad P\{L_{n+r} \leq \ell_1 \quad \text{or} \quad \geq \ell_2 | \ell_1 < L_n < \ell_2\} > p, \qquad n = 1, 2, \cdots.$$

The property $P(N < \infty) = 1$ is usually referred to as "termination with probability one." For short, the property of N expressed by (1.4) will be called exponential boundedness of N (it would of course be more correct to call it exponential boundedness of the distribution of N; the shorter terminology follows Berk [3], [4]). The nondegeneracy condition that P not satisfy (1.3) is obviously not only sufficient but also necessary for N to be exponentially bounded. Thus, Stein's result is the most complete general result possible for the Wald SPRT.

If the hypotheses H_j are composite, a SPRT can be formulated only if somehow the composite hypotheses can be reduced to simple ones. In this paper we shall consider only the reduction that results from applying the principle of invariance (for the theory of invariance see, for example, [15]; for an extensive discussion of weight function and invariance reduction SPRT and their possible relation see Berk [4]). Suppose $X^{(n)}$ is the sample space of $(Z_1, \cdots, Z_n)$. Let G be a group of invariance transformations on $X^{(n)} \times \Theta$. The principle of invariance restricts decisions at the nth stage of sampling to those that depend on $x \in X^{(n)}$ only through its orbit Gx. At the same time, the distribution of orbits in $X^{(n)}$ depends on $\theta \in \Theta$ only through the orbit $G\theta$ of θ. (An alternative way of expressing these facts is by means of so called maximal invariants.) Suppose now that for $j = 1, 2$, Θ_j is a single orbit, that is, there exists $\theta_j \in \Theta_j$ such that $\Theta_j = G\theta_j$. Then as far as the distribution of orbits in $X^{(n)}$ is concerned, the hypotheses H_j are simple.

The probability ratio R_n at the nth stage may then be taken as the ratio of (random) orbit densities, one for θ_1 and one for θ_2. The resulting sequence $\{R_n : n = 1, 2, \cdots\}$ is used for testing in the same way as in the Wald SPRT. Such a test will be called an *invariant* SPRT. Many well known sequential tests of composite hypotheses are of this form: sequential t test, F test, and so forth.

The questions of termination with probability one and of exponential boundedness of N can also be asked of invariant SPRT but are much harder to answer. The results obtained so far are much less complete than for the Wald SPRT. In fact, the history of these endeavors has had a very modest beginning with only the question of termination with probability one being attacked and then only for P in the parametric model: Johnson [13], David and Kruskal [7], Ray [18], Jackson and Bradley [12], Wirjosudirjo [30], Ifram [9]. Berk [2] was the first explicitly to consider P outside the model. This was followed by Wijsman in [28] where the question of termination with probability one was also treated for a class of distributions larger than given by the model.

Study of the exponential boundedness of N started much later. Ifram [11] still restricted P within the parametric model. Later studies of parametric models by Wijsman [29], Abu-Salih [1] and Berk [4] allowed P outside the model. The first nonparametric model (sequential two sample rank order test) was treated by Savage and Sethuraman [19] and sharper results were obtained by Sethuraman [20]. In both [19] and [20] P was allowed outside the model.

The only result in the literature to date for invariant SPRT that approximates the completeness of Stein's result for the Wald SPRT is Sethuraman's [20] since he shows for the particular rank order test in question that N is exponentially bounded for every P, except for P in a certain class of degenerate distributions (but whether the P in this class actually misbehave, that is, cause N not to be exponentially bounded, is not known). In all parametric problems studied the results have been considerably less complete. In order to obtain termination with probability one a minimal assumption, made by all authors, has been that under P certain random variables (the X_i of (2.4)) have first moments, and for exponential boundedness of N the existence of their moment generating functions. In addition, nearly all authors have been plagued by exceptional P (where certain moments have certain exceptional values, not to be confused with degenerate P) under which only weaker results or even no results could be obtained (see, for example, the discussion in [29] Section 1). This is not typical for parametric as opposed to nonparametric problems, for Savage and Sethuraman [19] encountered the same difficulty. The only authors whose methods have made it possible to avoid having to exclude exceptional distributions are Berk [4] and Sethuraman [20]. The latter author revived the Stein type of proof in which (1.5) has to be established (this is of course much harder if L_n is not a random walk). Berk [4] followed suit and used a modification of Stein's method to prove exponential boundedness of N. Judging from these examples it is not unreasonable to guess that for sharpest results it is necessary to use the Stein type of proof, centering around (1.5). (However, the absence of exceptional distributions in

Berk's work must be due to his method [3] and assumptions in [3], [4], since his proof of termination with probability one has nothing to do with Stein's type of proof.)

There is a certain amount of overlap in results and in assumptions between [4] on one hand and [28], [29] on the other. There is a discussion of this in [4] Section 2. However, there are also differences. Berk obtains the strongest results whenever his method (relying on [3]) applies. Thus, he obtains exponential boundedness of N in the sequential t test also for those exceptional P for which only weaker results could be obtained in [29]. Moreover, [28] and [29] apply only to invariant SPRT and assume a multivariate normal model whereas Berk's method applies to a wider class of models and to weight function SPRT that are not necessarily invariant SPRT. On the other hand there are some intrinsic limitations in Berk's method that prevent the applicability of [3], [4] to certain cases that are amenable to [28], [29]. Specifically, Assumption 2.4(d) in [4] (which already assumes the existence of $E_P X_1$) implies in our notation (see (2.3), (2.4)) that $\lambda'(\theta) E_P X_1 + \log c(\theta)$ has a unique maximum as θ varies through $\Theta_j, j = 1, 2$. This condition is violated in Example 2, Section 5, if P is such that $E_P Z_1 Z_1'$ is a constant times the identity matrix because then $\lambda'(\theta) E_P X_1 + \log c(\theta)$ is constant on each Θ_j. The same happens in Example 3, Section 6, if $E_P Z_1 = 0$.

The main purpose of this paper is to provide two examples (2 and 3), one of them being the sequential χ^2 test, in which N is shown to be exponentially bounded for every P, excluding a certain class of degenerate P. In particular, it is not assumed that any of the random variables has finite expectation under P. We rely heavily on the Stein type of proof. To the best of our knowledge this is the first time such examples have been given in parametric models. These examples give hope that in the future other (perhaps all?) invariant SPRT may be shown to have exponentially bounded N for all but certain degenerate P.

In Section 3 a general discussion is given on distributions P for which N is not exponentially bounded. Such P will be called *obstructive*. That such P do indeed exist is illustrated in Example 1, Section 4. This example is nontrivial but so simple that the behavior of $\{L_n\}$ can be studied easily. It seems to be the first example of any nontrivial invariant SPRT—parametric or nonparametric— where the existence of obstructive P has been demonstrated. Berk [4], Section 5, gives examples of weight function SPRT where the tests do not even terminate with probability one for certain P. However, those tests do not seem to be invariant SPRT and fall therefore outside the scope of this paper.

2. Representation of R_n and approximation of L_n by Φ_n

We assume that for each $\theta \in \Theta$, P_θ has a density p_θ with respect to Lebesgue measure on E^d. Let G be a group of invariance transformations $Z_i \to g Z_i$, $i = 1, 2, \cdots$, where to each $g \in G$ is associated a d^2 nonsingular matrix C and a vector $b \in E^d$, and the action is defined by $g Z_i = C(Z_i + b)$. The precise assumptions on G are set forth in Section 7, Theorem 7.1. The hypotheses H_j for

$j = 1, 2$, are of the form: P is one of the P_θ with $\theta \in G\theta_j$, where θ_1 and θ_2 are not on the same orbit so that their orbits $G\theta_j$ are disjoint. It will be proved in Section 7 that the probability ratio R_n at the nth stage for the invariant SPRT can be represented in the form

$$(2.1) \qquad R_n = J_n(\theta_2)/J_n(\theta_1), \qquad n = d + 1, d + 2, \cdots,$$

in which

$$(2.2) \qquad J_n(\theta) = \int \prod_{i=1}^{n} p_{g\theta}(Z_i) v_G(dg).$$

In (2.2) v_G is right Haar measure on G and we assume $n \geq d + 1$. Note that the right invariance of v_G guarantees that (2.2) is constant on each orbit in Θ, so that R_n does not depend on the particular choice of θ_1 and θ_2. The representation (2.1), (2.2) bears a strong resemblance to (3) in [27], but in [27] the group G acted linearly whereas in the present paper the action of G may include translations. Furthermore, (2.2) is written in terms of G acting on Θ, whereas (3) in [27] was expressed in terms of G acting on the sample space.

From now on we shall assume that the model is exponential, that is, the density p_θ of the Z_i is expressible as

$$(2.3) \qquad p_\theta(z) = c(\theta)e^{\lambda'(\theta)s(z)}h(z),$$

in which $\lambda(\theta), s(z) \in E^k$ for some $k \geq 1$, $c(\theta), h(z) > 0$ and prime denotes transposition. It is convenient to denote

$$(2.4) \qquad X_i = s(Z_i), \qquad i = 1, 2, \cdots$$

so that $X_1, X_2, \cdots$ are i.i.d. vectors in E^k. Furthermore we introduce the function

$$(2.5) \qquad \psi_\theta(g, x) = \lambda'(g\theta)x + \log c(g\theta), \qquad x \in E^k,$$

and write $\bar{X}_n$ for $(1/n) \sum_1^n X_i$. After substitution of (2.3) to (2.5) into (2.2) we obtain

$$(2.6) \qquad J_n(\theta) = \int \exp \{n\psi_\theta(g, \bar{X}_n)\} v_G(dg).$$

If in (2.6) the integration were over a Euclidean space instead of over a group, the method of Laplace (see, for example, [6]) could be used to approximate $J_n(\theta)$ for large n. Assuming that Laplace's method is applicable even if the integration is over a group (provided certain conditions are fulfilled) we expect the main contribution to the integral to come from values of g that come close to maximizing ψ_θ. This suggests putting

$$(2.7) \qquad \phi_\theta(x) = \max_{g \in G} \psi_\theta(g, x),$$

assuming the maximum exists, and putting

(2.8) $$\Phi(x) = \phi_{\theta_2}(x) - \phi_{\theta_1}(x).$$

Then we may expect that asymptotically $R_n \sim \exp\{n\Phi(\bar{X}_n)\}$. Taking log and writing

(2.9) $$\Phi_n = n\Phi(\bar{X}_n)$$

we may expect then that Φ_n is in some sense an approximation to $L_n = \log R_n$. This approach was used in [28], [29].

Suppose now that the approximation of L_n by Φ_n has the following form: there is a constant B such that for all n

(2.10) $$|L_n - \Phi_n| < B.$$

Let $\ell_1' = \ell_1 - B, \ell_2' = \ell_2 + B$, where ℓ_1 and ℓ_2 are the stopping bounds on L_n (see (1.2)). In analogy to N defined in (1.2), define N' to be the first n for which $\ell_1' < \Phi_n < \ell_2'$ is violated. Then clearly $N \leq N'$ so that if the latter is exponentially bounded, so is the former. Since our aim is to prove exponential boundedness of N for all choices of ℓ_1, ℓ_2, the problem then becomes to prove exponential boundedness of N' for all choices of ℓ_1', ℓ_2'. This shifts therefore the original problem to a similar one, the only difference being that the original sequence $\{L_n\}$ has been replaced by a new sequence $\{\Phi_n\}$. The advantage is that $\{\Phi_n\}$ is usually much more tractable than $\{L_n\}$. In Examples 2 and 3 it is indeed possible first to show (2.10) and then to prove exponential boundedness of N'. In the proofs the primes on ℓ_1, ℓ_2 and N will be dropped for notational convenience.

3. Obstructive distributions

For the purpose of this paper we shall call a distribution P *obstructive* if for some stopping bounds (1.4) is not true, that is, N is not exponentially bounded when the true distribution is P. In Wald's SPRT Stein's result shows that P is obstructive if and only if it satisfies (1.3). In the case of invariant SPRT we know much less about the characterization of obstructive P. In fact, there is not even one example where this has been carried out. All that has been achieved so far is that in a few isolated examples classes of P that contain the obstructive P have been characterized. For instance, Sethuraman [20] shows in a sequential two sample rank test that the obstructive P satisfy (in his notation) $P(V(X_1, Y_1) = 0) = 1$, but it is not known whether all P in this class are obstructive. In Examples 2 and 3, Sections 5 and 6, certain classes of P will be characterized and shown to include the obstructive P. Again it is not known whether the inclusion is proper but this seems likely in the light of Example 1.

In the exponential model there is some reason to conjecture that any obstructive P satisfies

(3.1) $$P(v'X_1 = \text{constant}) = 1 \qquad \text{for some } v \in E^k, \qquad v \neq 0,$$

X_1 being defined in (2.4). That is, X_1 is confined to a $(k-1)$ dimensional hyperplane, a.e. P. However, very likely the class defined by (3.1) is usually too big.

At least in Examples 2 and 3 the class of P defined by (3.1) is larger than the class of obstructive P. A smaller class than defined by (3.1) is the one defined by Berk in his Assumption 2.4(c) of [4] (in different notation):

$$(3.2) \qquad P\{p_{\theta'}(Z_1) = p_{\theta''}(Z_1)\} = 1 \qquad \text{for some } \theta' \in \Theta_1, \theta'' \in \Theta_2.$$

This is the composite hypotheses analogue to (1.3). It is immediately seen that in the exponential model (3.2) implies (3.1). In Examples 2 and 3 it is true that the obstructive P, if any, are contained in the class defined by (3.2), but again the latter is too big. It turns out in those examples that obstructive P satisfy (3.2) only for certain pairs of (θ', θ''). One can make a conjecture that the pairs (θ', θ'') for which (3.2) may give obstructive P are those pairs that minimize the Kullback-Leibler "divergence between $p_{\theta'}$ and $p_{\theta''}$" (see (2.6) in [14]). It should be mentioned that in Examples 2 and 3 no P are actually demonstrated to be obstructive.

Example 1, Section 4 demonstrates for an invariant SPRT the existence of obstructive distributions P. They do satisfy (3.2), with the further restriction to pairs (θ', θ'') that minimize the Kullback-Leibler divergence; but even in this smaller class not all P are obstructive. In Example 1 we would have a complete characterization of obstructive P if we would know that every P outside the class (3.2) is not obstructive. Unfortunately, this is not known at present.

There is a qualitative difference between obstructive P for Wald's SPRT and obstructive P for invariant SPRT, at least for those of Example 1. In Wald's SPRT P is obstructive if and only if (1.3) is satisfied and in that case $L_n = 0$ for all n a.e. P so that not only is N not exponentially bounded but $N = \infty$ a.e. P. The obstructive P in Example 1 behave much more mildly: $\{L_n\}$ is truly random and $P(N < \infty) = 1$. The only thing that goes wrong is that the distribution of N does not have nice properties. Not only is N not exponentially bounded but the stopping bounds of the test can be chosen so that not even the first moment of N is finite.

4. Example 1

Let $Z_1, Z_2, \cdots$ be i.i.d. normal with mean ζ variance σ^2, both unknown. $H_j: \sigma = \sigma_j, j = 1, 2$, where the σ_j are given and distinct. Thus, ζ is a nuisance parameter. The problem is invariant under the transformation $Z_i \to Z_i + b$, $i = 1, 2, \cdots, \zeta \to \zeta + b, -\infty < b < \infty, \sigma \to \sigma$. The group G is therefore the group of reals b under addition, and right ($=$ left) Haar measure $\nu_G(dg)$ can be taken simply as db. The points θ_1 and θ_2 on the orbits Θ_j can be chosen as $\theta_j = (0, \sigma_j), j = 1, 2$. Then

$$(4.1) \qquad P_{g_j}^{\theta}(Z_i) = \sqrt{2\pi}\sigma_j)^{-1} \exp\{-(2\sigma_j^2)^{-1}(Z_i - b)^2\},$$

where b corresponds to g. Substituting this into (2.1) and (2.2), integrating with respect to b, and taking log yields

$$(4.2) \qquad L_n = \left[(2\sigma_1^2)^{-1} - (2\sigma_2^2)^{-1}\right] \sum_{i=1}^n (Z_i - \bar{Z}_n)^2 + (n-1) \log \frac{\sigma_1}{\sigma_2}$$

in which $\bar{Z}_n = (1/n)\Sigma_1^n Z_i$.

Now for convenience choose $\sigma_1 < \sigma_2$ in such a way that

$$(4.3) \qquad \sigma_1^{-2} + 2 \log \sigma_1 = \sigma_2^{-2} + 2 \log \sigma_2.$$

It is easily checked that with this choice the two normal densities with mean zero and variance σ_1^2, σ_2^2, respectively, are equal at ± 1, that is,

$$(4.4) \qquad p_{\theta_1}(\pm 1) = p_{\theta_2}(\pm 1).$$

Take the distribution P as follows:

$$(4.5) \qquad P(Z_1 = -1) = P(Z_1 = 1) = \tfrac{1}{2},$$

so that $\Sigma_1^n Z_i^2 = n$ a.e. P. Substituting this into (4.2) and using (4.3) we get

$$(4.6) \qquad L_n = \left[\log \frac{\sigma_2}{\sigma_1}\right](1 - n\bar{Z}_n^2) \qquad \text{a.e. } P.$$

Suppose ℓ_2 is chosen $> \log (\sigma_2/\sigma_1)$ then $L_n < \ell_2$ for all n a.e. P. The double inequality (1.2) reduces then to the single inequality $\ell_1 < L_n$, which can be put in the form

$$(4.7) \qquad \left| \sum_{i=1}^n Z_i \right| < a\sqrt{n}$$

for some $a > 0$ depending on σ_1, σ_2 and ℓ_1. The law of the iterated logarithm guarantees that for every $a > 0$, we have $P(N < \infty) = 1$, but if $a \geqq 1$ then $E_P N = \infty$ [5].

It follows from (4.4) and (4.5) that P satisfies (3.2) for the couple $(\theta', \theta'') = (\theta_1, \theta_2)$, where $\theta_j = (0, \sigma_j)$. It is clear from (4.2) that the behavior of L_n is unchanged if P puts probability $\tfrac{1}{2}$ on $c \pm 1$ for arbitrary c, so that such a P is also obstructive. It can be shown that this class of P, with $-\infty < c < \infty$, constitutes the class of obstructive P among all P satisfying (3.2) (or even among all P for which Z_1^2 has finite moment generating function). At the same time, a P with support on $c \pm 1$ satisfies (3.2) with

$$(4.8) \qquad \theta' = (c, \sigma_1), \qquad \theta'' = (c, \sigma_2),$$

and a simple computation shows that the pairs (θ', θ'') given by (4.8) are the pairs that minimize the Kullback-Leibler divergence. But note that not even all P satisfying (3.2) with θ', θ'' satisfying (4.8) are obstructive, since for this property it is not only required that P is supported on $c \pm 1$ but also that the probabilities in these two points are $\tfrac{1}{2}$.

5. Example 2

Let $Z_1, Z_2, \cdots$ be i.i.d. bivariate normal with mean 0 and covariance matrix Σ. Put $Z_i = (x_i, y_i)'$. We want to test the hypotheses $H_j, j = 1, 2$, that the characteristic roots of Σ^{-1} are σ_j, τ_j, with $\sigma_j > \tau_j$, where the σ_j and τ_j are given positive numbers such that $(\sigma_1, \tau_1) \neq (\sigma_2, \tau_2)$. Consider the invariance transformation $Z_i \to \Omega Z_i, i = 1, 2, \cdots, \Sigma \to \Omega \Sigma \Omega'$, with Ω 2^2 orthogonal. It is sufficient and more convenient to restrict G to all Ω having determinant $+1$. Then we may take Ω in the form

$$(5.1) \qquad \Omega = \begin{bmatrix} \cos \omega & -\sin \omega \\ \sin \omega & \cos \omega \end{bmatrix}, \qquad 0 \leqq \omega < 2\pi,$$

and right ($=$ left) Haar measure $\nu_G(d\Omega)$ can be taken as $d\omega$. The density p_θ of Z_i with respect to Lebesgue measure in E^2 is

$$(5.2) \qquad p_\theta(Z_i) = (2\pi)^{-1} |\Sigma|^{-1/2} \exp\left\{ -\tfrac{1}{2} \operatorname{tr} \Sigma^{-1} Z_i Z_i' \right\}.$$

We may identify θ with Σ^{-1} and take $\theta_j = \operatorname{diag}(\sigma_j, \tau_j), j = 1, 2$. In order to evaluate R_n, using (2.1), (2.2), we have to replace in (5.2) Σ^{-1} by $\Omega \operatorname{diag}(\sigma_j, \tau_j) \Omega'$, $j = 1, 2$, substitute into (2.1), (2.2), and integrate. The integrals involve expressions of the form

$$(5.3) \qquad \tfrac{1}{2}\pi \int_0^{2\pi} \exp\left\{ x \cos t \right\} dt = I_0(x),$$

where I_0 is one of the Bessel functions of imaginary argument ([25], p. 77). The random variables $Z_i = (x_i, y_i)'$ enter into the result only in the following combinations:

$$(5.4) \qquad U_n = \sum_{i=1}^{n} (x_i^2 + y_i^2),$$

$$(5.5) \qquad V_n = \left[\left(\sum_1^n x_i^2 - \sum_1^n y_i^2 \right)^2 + \left(2 \sum_1^n x_i y_i \right)^2 \right]^{1/2}.$$

In this notation we obtain for $L_n = \log R_n$ the result

$$(5.6) \qquad \begin{aligned} L_n = {} & \tfrac{1}{4}(\sigma_1 + \tau_1 - \sigma_2 - \tau_2) U_n + \frac{n}{2} \log \frac{\sigma_2 \tau_2}{\sigma_1 \tau_1} \\ & + \log I_0\left[\tfrac{1}{4}(\sigma_2 - \tau_2) V_n \right] - \log I_0\left[\tfrac{1}{4}(\sigma_1 - \tau_1) V_n \right]. \end{aligned}$$

On the other hand, performing the maximization in (2.7) and substituting into (2.8) we find

$$(5.7) \qquad \begin{aligned} \Phi_n = {} & \tfrac{1}{4}(\sigma_1 + \tau_1 - \sigma_2 - \tau_2) U_n + \frac{n}{2} \log \frac{\sigma_2 \tau_2}{\sigma_1 \tau_1} \\ & + \tfrac{1}{4}(\sigma_2 - \tau_2 - \sigma_1 + \tau_1) V_n. \end{aligned}$$

From the continuity of I_0 and its asymptotic behavior,

$$(5.8) \qquad \lim_{x \to \infty} (2\pi x)^{1/2} \exp\{-x\} I_0(x) = 1$$

(see [25], p. 203, but this also follows directly from the integral representation of $I_0(x)$ and Laplace's method), we deduce that for any given $\alpha, \beta > 0$ there exists B such that for all $x \geqq 0$

$$(5.9) \qquad \left| \log I_0(\alpha x) - \log I_0(\beta x) - (\alpha - \beta) x \right| < B.$$

Using this to compare (5.6) and (5.7) we see that (2.10) is satisfied. Hence for the question of exponential boundedness of N we may replace L_n by Φ_n, as discussed in Section 2.

It turns out that the distributions P, for which we cannot prove exponential boundedness of N, satisfy (3.2) with the pairs (θ', θ'') of the form $(g\theta_1, g\theta_2)$, $g \in G$ (and these are precisely the pairs that minimize the Kullback-Leibler divergence). That is, the two values of Σ^{-1} (under θ' and θ'') are Ω diag $(\sigma_1, \tau_1)\Omega'$ and Ω diag $(\sigma_2, \tau_2)\Omega'$, where Ω is any matrix of the form (5.1). Substituting (5.2) for Z_1 into (3.2), with the above values of Σ^{-1} and putting $\cos 2\omega = -u_1$, $\sin 2\omega = -u_2$, the negation of (3.2) then has the form

$$(5.10) \qquad P\{\tfrac{1}{2}(\sigma_2 - \tau_2 - \sigma_1 + \tau_1)[u_1(x_1^2 - y_1^2) + u_2(2x_1y_1)]$$
$$+ \tfrac{1}{2}(\sigma_1 + \tau_1 - \sigma_2 - \tau_2)(x_1^2 + y_1^2) + \log \frac{\sigma_2 \tau_2}{\sigma_1 \tau_1} = 0\} < 1.$$

Suppose first $\sigma_2 - \tau_2 - \sigma_1 + \tau_1 = 0$. Using (5.7) and (5.4) it is seen now that Φ_n is of the form $\Phi_n = \Sigma_1^n W_i$, with $W_1, W_2, \cdots$ i.i.d. and

$$(5.11) \qquad W_i = \tfrac{1}{4}(\sigma_1 + \tau_1 - \sigma_2 - \tau_2)(x_i^2 + y_i^2) + \tfrac{1}{2} \log (\sigma_2 \tau_2 / \sigma_1 \tau_1),$$

and thus by (5.10) $P(W_1 = 0) < 1$. Exponential boundedness follows from [21].

Assuming from now on that $\sigma_2 - \tau_2 - \sigma_1 + \tau_1 \neq 0$ we shall put

$$(5.12) \qquad a = \frac{\sigma_1 + \tau_1 - \sigma_2 - \tau_2}{\sigma_2 - \tau_2 - \sigma_1 + \tau_1}, \qquad b = \frac{2}{\sigma_2 - \tau_2 - \sigma_1 + \tau_1} \log \frac{\sigma_2 \tau_2}{\sigma_1 \tau_1},$$

so that (5.10) can be written

$$(5.13) \qquad P\{u_1(x_1^2 - y_1^2) + u_2(2x_1y_1) + a(x_1^2 + y_1^2) + b = 0\} < 1.$$

For the purpose of proving exponential boundedness of N we may divide Φ_n by any nonzero constant. Dividing in (5.7) by the coefficient of V_n we get

$$(5.14) \qquad \Phi_n = V_n + aU_n + bn.$$

A further simplification in notation will be made. Consider the vectors $Y_1, Y_2, \cdots$, where $Y_i = (x_i^2 - y_i^2, 2x_iy_i)'$, then $x_i^2 + y_i^2 = \|Y_i\|$ and (5.13), (5.14) can be put in the form

$$(5.15) \qquad P\{u'Y_1 + a\|Y_1\| + b = 0\} < 1$$

for every $u \in U$,

(5.16) $$\Phi_n = \| \sum_1^n Y_i \| + a \sum_1^n \| Y_i \| + bn.$$

in which $U = \{u \in E^2 : \| u \| = 1\}$. In (5.15), (5.16) there are some restrictions on a and b that follow from (5.12) and $\sigma_j \geqq \tau_j, j = 1, 2$. For instance, if $b = 0$ we must have $-1 < a < 0$. However, these restrictions are of no help in the proof of exponential boundedness of N and we shall ignore them. Also, the fact that the dimension of the sample space of the Y_i is two plays no role. We shall give the proof for $Y_i \in E^k$, for any positive integer k. This necessitates a redefinition of the set of unit vectors

(5.17) $$U = \{u \in E^k : \| u \| = 1\}.$$

LEMMA 5.1. *Let* $Y_1, Y_2, \cdots$ *be i.i.d. random vectors taking their values in* E^k *for some* $k \geqq 1$, *their common distribution* P *satisfying* (5.15) *with* U *defined in* (5.17). *Let* N *be the first integer* n *such that* $\ell_1 < \Phi_n < \ell_2$ *is violated, where* Φ_n *is given by* (5.16) *and* ℓ_1, ℓ_2 *are arbitrary real numbers. Then* N *is exponentially bounded, that is, satisfies* (1.4).

Before proving Lemma 5.1 it is convenient to prove

LEMMA 5.2. *Let* $\{X_t, t \in T\}$ *be a family of real valued random variables,* T *a compact index set, such that for every* $t \in T$, (i) $P(X_t < 0) > 0$, *and* (ii) $X_s \to X_t$ *in law as* $s \to t$. *Then there exists* $\delta, \varepsilon > 0$ *such that* $P(X_t < -\delta) > \varepsilon$ *for every* $t \in T$.

PROOF. Let F_t be the distribution function of X_t, and $x_t < 0$ a continuity point of F_t such that $F_t(x_t) = 2\varepsilon_t$ for some $\varepsilon_t > 0$ (this can be done by assumption (i)). For any $t \in T$ let $s \to t$ then $F_s(x) \to F_t(x)$ for every continuity point x of F_t, using assumption (ii). In particular, $F_s(x_t) \to F_t(x_t) = 2\varepsilon_t$ so that there exists a neighborhood U_t of t such that $s \in U_t$ implies $F_s(x_t) > \varepsilon_t$. Since T is compact, we can cover T with a finite number of such U_t, say $U_{t_1}, \cdots, U_{t_n}$. Put $\varepsilon = \min (\varepsilon_{t_i}, i = 1, \cdots, n)$, $\delta = \min (-x_{t_i}, i = 1, \cdots, n)$ so that $\varepsilon, \delta > 0$. Take an arbitrary $t \in T$ then t is in one of the U_{t_i}, say in U_{t_j}. We have then $F_t(-\delta) \geqq F_t(x_{t_j}) > \varepsilon_{t_j} \geqq \varepsilon$.

PROOF OF LEMMA 5.1. The case $P(Y_1 = 0) = 1$ is trivial for .1en by (5.15), (5.16) $\Phi_n = bn$, $b \neq 0$ so that N is constant. Henceforth we shall assume $P(Y_1 = 0) < 1$. We shall distinguish two cases.

Case 1. There exists $u \in U$ (defined in (5.17)) such that

(5.18) $$P\{u'Y_1 + a\| Y_1 \| + b \geqq 0\} = 1.$$

We define

(5.19) $$W_i = u'Y_i + a\| Y_i \| + b$$

so that $W_1, W_2, \cdots$ are i.i.d. real valued random variables. Equation (5.18) states that $P(W_1 \geqq 0) = 1$, and (5.15) in addition implies that $P(W_1 > 0) > 0$.

Thus, the random walk $T_n = \Sigma_1^n W_i$ takes only nonnegative steps, and the steps are positive with positive probability. Let N' be the first integer n such that $\ell_1 < T_n < \ell_2$ is violated, then N' is exponentially bounded according to [21]. Define

$$(5.20) \qquad\qquad S_n = \sum_{i=1}^n Y_i$$

then $\|S_n\| \geqq u'S_n$ and therefore, using (5.16) and (5.19),

$$(5.21) \qquad \Phi_n \geqq u'S_n + a \sum_1^n \|Y_i\| + bn = \sum_1^n W_i = T_n.$$

This, together with the fact that T_n is nondecreasing, implies $N \leqq N'$ so that N is also exponentially bounded.

Case 2. For all $u \in U$, $P(u'Y_1 + a\|Y_1\| + b \geqq 0) < 1$, which can also be written in the form

$$(5.22) \qquad P\{u'Y_1 + a\|Y_1\| + b < 0\} > 0$$

for every $u \in U$. We shall prove that there exists a positive integer r and $p > 0$ such that

$$(5.23) \qquad P\{\ell_1 < \Phi_{n+i} < \ell_2, i = 1, \cdots, r | \ell_1 < \Phi_n < \ell_2\} < 1 - p,$$
$$n = 1, 2, \cdots.$$

This statement is weaker than (1.5) but exponential boundedness of N follows from (5.23) in the same way as it does from (1.5).

We apply Lemma 5.2 to $X_t = t'Y_1 + a\|Y_1\| + b, t \in U$. Obviously, U is compact. Assumption (i) in Lemma 5.2 is (5.22) and assumption (ii) follows immediately from the fact that $X_s \to X_t$ everywhere as $s \to t$. Using Lemma 5.2, there exists $\delta_1, \varepsilon_1 > 0$ such that

$$(5.24) \qquad P\{u'Y_1 + a\|Y_1\| + b < -2\delta_1\} > 2\varepsilon_1$$

for every $u \in U$. Now take B_1 so large that $P(\|Y_1\| \geqq B_1) < \varepsilon_1$ and combine with (5.24), then

$$(5.25) \qquad P\{\|Y_1\| < B_1, u'Y_1 + a\|Y_1\| + b < -2\delta_1\} > \varepsilon_1$$

for every $u \in U$. For any vectors $s, y \in E^k$ we compute

$$(5.26) \qquad \|s + y\| - \|s\| \leqq \|s\|^{-1}(s'y + \tfrac{1}{2}\|y\|^2).$$

With S_n defined in (5.20) define $u_n = S_n/\|S_n\|$ if $S_n \neq 0$, and if $S_n = 0$ take u_n equal to any fixed vector in U. Let i be any positive integer. Since $S_{n+i} = S_{n+i-1} + Y_{n+i}$, (5.26) states that

$$(5.27) \qquad \|S_{n+i}\| - \|S_{n+i-1}\| \leqq u'_{n+i-1}Y_{n+i} + \frac{\|Y_{n+i}\|^2}{2\|S_{n+i-1}\|}.$$

From (5.16) we obtain

(5.28) $\Phi_{n+i} - \Phi_{n+i-1} = \|S_{n+i}\| - \|S_{n+i-1}\| + a\|Y_{n+i}\| + b.$

Substitution of (5.27) into (5.28) gives

(5.29) $\Phi_{n+i} - \Phi_{n+i-1}$

$$\leqq u'_{n+i-1} Y_{n+i} + a\|Y_{n+i}\| + b + (2\|S_{n+i-1}\|)^{-1}\|Y_{n+i}\|^2.$$

Let B_2 be so large that $B_1^2/(2B_2) < \delta_1$, then

(5.30) $[\|Y_{n+i}\| < B_i, \|S_{n+i-1}\| > B_2] \Rightarrow [(2\|S_{n+i-1}\|)^{-1}\|Y_{n+i}\|^2 < \delta_1].$

Together with (5.29) this gives

(5.31) $[\|Y_{n+i}\| < B_1, \|S_{n+i-1}\| > B_2, u'_{n+i-1} Y_{n+i} + a\|Y_{n+i}\| + b < -2\delta_1]$
$$\Rightarrow [\Phi_{n+1} - \Phi_{n+i-1} < -\delta_1].$$

Choose integer n_1 so that $n_1\delta_1 > d = \ell_2 - \ell_1$ and put $B = B_2 + n_1 B_1$. Then

(5.32) $[\|S_n\| > B, \|Y_{n+i}\| < B_1, i = 1, \cdots, n_1]$
$$\Rightarrow [\|S_{n+i-1}\| > B_2, i = 1, \cdots, n_1].$$

Define the events

(5.33) $A_i = [\|Y_i\| < B_1, u'_{i-1} Y_i + a\|Y_i\| + b < -2\delta_1], \qquad i = 1, 2, \cdots,$

then given $\|S_n\| > B, A_{n+i}$ implies the left side of (5.31) for $i = 1, \cdots, n_1$ (making use of (5.32)) and therefore

(5.34) $\left[\|S_n\| > B, \bigcap_{i=1}^{n_1} A_{n+i}\right] \Rightarrow [\Phi_{n+i} - \Phi_{n+i-1} < -\delta_1, i = 1, \cdots, n_1]$
$$\Rightarrow [\Phi_{n+n_1} - \Phi_n < -n_1\delta_1]$$
$$\Rightarrow [\Phi_{n+n_1} - \Phi_n < -d].$$

From (5.25) and the independence of the Y_i it follows that $P(A_i|S_{i-1}) > \varepsilon_1$ for every given value of S_{i-1}. (Note that the A_i are not independent, but A_i depends on $Y_1, \cdots, Y_{i-1}$ only through u_{i-1}, that is, through S_{i-1}.) Therefore, given $\|S_n\| > B$, the extreme left side of (5.34) has probability $\geqq \varepsilon_1^{n_1}$, and consequently

(5.35) $P\{\Phi_{n+n_1} - \Phi_n < -d \,|\, S_n\} > \varepsilon_1^{n_1}$ if $\|S_n\| > B.$

Since $P(Y_1 = 0) < 1$, we can choose $u \in U$ so that $P(u'Y_1 > 0) > 0$. Then there exists $\delta_2, \varepsilon_2 > 0$ such that $P(u'Y_1 > \delta_2) > \varepsilon_2$. Choose integer n_2 such that $n_2\delta_2 > 2B$. Then

(5.36) $[\|S_n\| \leqq B, u'Y_{n+i} > \delta_2, i = 1, \cdots, n_2] \Rightarrow [\|S_{n+n_2}\| > B].$

Therefore

(5.37) $P\{\|S_{n+n_2}\| > B \,|\, S_n\} > \varepsilon_2^{n_2}$ if $\|S_n\| \leqq B.$

Now put $r = n_1 + n_2$ and consider

(5.38) $\qquad P\{\Phi_{n+i} \leqq \ell_1 \quad \text{for some} \quad i = 1, \cdots, r \,|\, \ell_1 < \Phi_n < \ell_2, S_n\}.$

If $\|S_n\| > B$ it follows from (5.35) that (5.38) is $> \varepsilon_1^{n_1}$. If $\|S_n\| \leqq B$ it follows from (5.37) and from (5.35) with n replaced by $n + n_2$, that (5.38) is $> \varepsilon_2^{n_2}\varepsilon_1^{n_1}$. Taking $p = \varepsilon_2^{n_2}\varepsilon_1^{n_1}$, we have then, whether $\|S_n\|$ is $> B$ or $\leqq B$,

(5.39) $\qquad P\{\Phi_{n+i} \leqq \ell_1 \quad \text{for some} \quad i = 1, \cdots, r \,|\, \ell_1 < \Phi_n < \ell_2\} > p$

and (5.24) is an immediate consequence.

6. Example 3: sequential χ^2 test

Let $Z_1, Z_2, \cdots$ be i.i.d. d-variate normal with identity covariance matrix and unknown mean vector ζ. We want to test the hypotheses H_j that $\|\zeta\| = \gamma_j$, where $\gamma_1 \neq \gamma_2$ are given positive numbers. (Note that we exclude the null hypothesis $\zeta = 0$.) Let G be the group of all d^2 orthogonal matrices Ω and consider the invariance transformations $Z_i \to \Omega Z_i, i = 1, 2, \cdots, \zeta \to \Omega\zeta$. The parameter θ in Section 2 may be identified with ζ and we may take $\theta_j = \gamma_j u, j = 1, 2$, where u is any fixed element of U defined in (5.17) with $k = d$. The density of Z_i is given by

(6.1) $\qquad p_\theta(Z_i) = (2\pi)^{-d/2} \exp\{-\tfrac{1}{2}(Z_i - \zeta)'(Z_i - \zeta)\}.$

Replacing in (6.1) ζ by $\Omega\theta_j = \gamma_j \Omega u, j = 1, 2$, substituting into (2.1), (2.2) and taking log results in

(6.2) $\qquad L_n = \log f(\gamma_2 \|S_n\|) - \log f(\gamma_1 \|S_n\|) + (n/2)(\gamma_1^2 - \gamma_2^2),$

in which

(6.3) $\qquad\qquad\qquad S_n = \sum_{i=1}^{n} Z_i$

and

(6.4) $\qquad\qquad f(x) = \int e^{xu'\Omega u} \mu(d\Omega), \qquad\qquad [x \geqq 0, u \in U].$

In (6.4) μ is Haar measure on G. Clearly, f does not depend on the particular choice of $u \in U$. The function f satisfies the differential equation $f''(x) + (d - 1)x^{-1}f'(x) = f(x)$ and can be related to the hypergeometric function $_0F_1$ [24], [10]: $f(x) = {_0F_1}[\tfrac{1}{2}d(\tfrac{1}{2}x)^2]$ provided μ is normalized. This permits writing down the asymptotic behavior of f; we have $\lim_{x\to\infty} x^{(d-1)/2}e^{-x}f(x) = \text{constant}$. We can obtain this also directly from (6.4) using Laplace's method after writing $f(x)$ in the form $c\int_0^\pi \exp\{x\cos\omega\}(\sin\omega)^{d-2}\,d\omega$. A third way is provided by verifying that $f(x) = cx^{-\nu}I_\nu(x)$, with $\nu = \tfrac{1}{2}d - 1$ and consulting [25], p. 203 for the asymptotic expansion of I_ν. It follows that for any $\alpha, \beta > 0$ there exists B such that

(6.5) $\qquad\qquad |\log f(\alpha x) - \log f(\beta x) - (\alpha - \beta)x| < B.$

Maximizing the integrand in (6.4) in order to obtain Φ_n defined in (2.9) we get

$$(6.6) \qquad \Phi_n = (\gamma_2 - \gamma_1)\|S_n\| + (n/2)(\gamma_1^2 - \gamma_2^2)$$

and comparison between (6.2) and (6.6), using (6.5), shows that (2.10) is satisfied so that we may replace L_n by Φ_n in order to investigate exponential boundedness of N.

As in Example 2, in order to prove exponential boundedness of N it suffices to exclude distributions P satisfying (3.2) with (θ', θ'') of the form $(g\theta_1, g\theta_2)$, $g \in G$ (again, these are the pairs that minimize the Kullback-Leibler divergence). Equating $p_\theta(Z_1)$ for these pairs, using (6.1), yields $2(\gamma_2 - \gamma_1)u'Z_1 + \gamma_1^2 - \gamma_2^2 = 0$, with $u \in U$. If this is not to hold a.e. P then P must satisfy

$$(6.7) \qquad P(u'Z_1 + b = 0) < 1$$

for every $u \in U$ in which $b = -\frac{1}{2}(\gamma_1 + \gamma_2)$. On the other hand, (6.6), after dividing by the immaterial factor $\gamma_2 - \gamma_1$, may be written

$$(6.8) \qquad \Phi_n = \|S_n\| + bn.$$

It is seen that (6.7), (6.8) is a special case of (5.15), (5.16), with $a = 0$, and therefore exponential boundedness of N follows from Lemma 5.1.

7. Representation of orbit density ratio as ratio of integrals over the group G

Suppose X is a Euclidean space carrying a family of probability densities $p_\theta(x)$ with respect to Lebesgue measure in X, $\theta \in \Theta$. Suppose G is a group of invariance transformations, implying (among other things) that for every $g \in G$, $\theta \in \Theta$ and measurable set $A \subset X$

$$(7.1) \qquad \int_{gA} p_{g\theta}(x)\, dx = \int_A p_\theta(x)\, dx$$

in which dx is Lebesgue measure in X. Let Q_θ be the distribution on the orbit space X/G and assume it has a density q_θ with respect to some sigma-finite measure (since we shall need q_θ only for two values of θ, this assumption is justified), where $q_\theta(x)$ depends on x only through its orbit Gx. Let θ_1, θ_2 have distinct orbits $G\theta_1, G\theta_2$, and consider the orbit density ratio $R(x) = q_{\theta_2}(x)/q_{\theta_1}(x)$. If G acts on X linearly, it was shown in [27], Equation (3) that for all x, except possibly for x in a set of Lebesgue measure zero, $R(x)$ can be expressed as

$$(7.2) \qquad R(x) = J(\theta_1, x)/J(\theta_2, x)$$

with

$$(7.3) \qquad J(\theta, x) = \int p_\theta(gx)|g|\mu_G(dg),$$

in which $|g|$ is the Jacobian of the transformation $x \to gx$ and μ_G is left Haar measure on G. Using (7.1) it is easy to establish that

(7.4) $p_\theta(gx)|g| = p_{g^{-1}\theta}(x)$

and substitution of (7.4) into (7.3) gives

(7.5) $J(\theta, x) = \int p_{g^{-1}\theta}(x)\mu_G(dg).$

The measure ν_G defined by $\nu_G(dg) = \mu_G(dg^{-1})$ is right Haar measure on G and therefore (7.5) can be put in the form

(7.6) $J(\theta, x) = \int p_{g\theta}(x)\nu_G(dg).$

This leads immediately to (2.1), (2.2) provided (7.2), (7.3) can be justified also for G whose action on X consists not only of linear transformations but also of translations. The derivation of (3) in [27] depended on the nature of the action of G only insofar the existence of a flat local cross section at almost every x had been shown only for linear Cartan G spaces [26], Theorem 2. We shall extend this result now and show the existence of a flat local cross section at almost all x when G consists of linear transformations as well as translations. Basic to the proof are certain results of Palais [17]. Definitions in [17] and [26] will be used freely. No use will be made of Lemma 5 in [26].

THEOREM 7.1.   *Let $X_1, \cdots, X_n$ be copies of E^d, where $n > d \geqq 1$, and let $X = \prod_1^n X_i$. Let $G = LH$, in which L is a Lie group of linear transformations on E^d (with the usual topology as a matrix group) and H is the group of all translations of a subspace $B \subset E^d$ such that $LB = B$. Let the action of G on E^d be defined as follows: if $g = \ell h$, $\ell \in L$ corresponding to the d^2 nonsingular matrix C, $h \in H$ corresponding to the vector $b \in B$, then $gz = C(z + b)$, $z \in E^d$. Let the action of G on X be defined by $g(x_1, \cdots, x_n) = (gx_1, \cdots, gx_n)$, $x_i \in X_i$, $i = 1, \cdots, n$. Then there is an open subset X_0 of X, invariant under G and $X - X_0$ having zero Lebesgue measure, such that each point $x \in X_0$ admits a flat local cross section.*

Before proving the theorem we shall prove a lemma. Also, the definition of G needs discussion. As an analytic manifold G is defined as $L \times H$. Now there are two ways of defining the multiplication in order to make G into a group. The first one of these is consistent with the way we have defined the action of G on E^d (that is, $gz = C(z + b)$) and is given by

(7.7) $(C_1, b_1)(C_2, b_2) = (C_1 C_2, b_2 + C_2^{-1}b_1),$

$C_1, C_2 \in L$, $b_1, b_2 \in B$, where it should be observed that $b_2 + C_2^{-1}b_1 \in B$ by virtue of the assumption $LB = B$. From (7.7) follows that the elements of G of the form $(C, 0)$ form a subgroup isomorphic to L, and the elements of the form (I, b) (I = identity matrix) form a subgroup isomorphic to H. Furthermore, also from (7.7), $(C, 0)(I, b) = (C, b)$ so that the group multiplication (7.7) is consistent with writing the elements $g \in G$ in a unique way as $g = \ell h$, $\ell \in L$, $h \in H$. This suggests the notation $G = LH$. It is also easily verified that $g^{-1}hg \in H$, $g \in G$, $h \in H$, so that H is normal in G. The second way of defining the group multiplication, which also results in H being normal and which is consistent with the group action $gz = Cz + b$, is given by $(C_1, b_1)(C_2, b_2) = (C_1 C_2, b_1 + C_1 b_2)$. This is consistent with writing $g \in G$ in a unique way as $g = h\ell$, $h \in H$, $\ell \in L$, and we would have denoted $G = HL$. Either way of defining G from L and H is called a *semidirect product* (see (2.6), (6.20) in [8] and 30D in [16]). It is immaterial which of the two ways one chooses, as long as the action of G on E^d is defined in accordance. We have arbitrarily chosen the first way. After the proof of Theorem 7.1 we shall deal with Haar measure on G.

LEMMA 7.1. *Let $V_1, \cdots, V_d$ be copies of E^d and let L be a Lie group of d^2 matrices (with the usual topology) acting linearly on E^d. Let $V = \Pi_1^d V_i$, consider the points $v \in V$ as d^2 matrices $v = [v_1, \cdots, v_d]$, $v_i \in V_i$ being a $d \times 1$ column vector, and let $V_0 = \{v \in V : v \text{ is a nonsingular matrix}\}$. Then V_0 is a linear Cartan L space.*

PROOF. Since L is contained in the group $GL(d, R)$ of all real d^2 nonsingular matrices and L has the relative topology of $GL(d, R)$, the conclusion of the lemma is true for L if it is true for $GL(d, R)$. We shall therefore proceed to prove the lemma for $L = GL(d, R)$. Note that V_0 is a copy of L. Hence the task is to prove that L is a linear Cartan L space.

In order to show that every $\ell \in L$ has a thin neighborhood it is sufficient to do this for the d^2 identity matrix I. Let E be the d^2-dimensional Euclidean space in which L is embedded and observe that L has the relative topology of E. Define $U_0 = \{M \in E : \text{tr } MM' \leq c\}$ in which $c > 0$ is chosen so that $I + M$ is nonsingular for every $M \in U_0$ (it can be shown that any $c < 1$ will do). Then $U = I + U_0 \subset L$ is a compact neighborhood of I. It follows from the continuity of inversion and group multiplication that UU^{-1} is compact. Let $A = \{\ell \in L : \ell U \cap U \neq \emptyset\}$ then $A \subset UU^{-1}$ so that A has compact closure and therefore U is a thin neighborhood of I. Q.E.D.

PROOF OF THEOREM 7.1. Choose any n^2 orthogonal matrix Ω whose last column has all its entries equal to $n^{-1/2}$. Write $X = [X_1, \cdots, X_n]$ and let $W = [W_1, \cdots, W_n]$ be defined by $W = X\Omega$. It suffices to prove the conclusion of the theorem for W instead of X. The action of G on W is determined by the action of G on X. An easy calculation shows that L acts on the W_i in the same way as on the X_i. On the other hand, H acts trivially on $W_1, \cdots, W_{n-1}$ whereas the action on W_n is given by $hw_n = w_n + bn^{1/2}$ if h corresponds to $b \in B$. Write $W_n = W_n^{(1)} \times W_n^{(2)}$, where $W_n^{(1)} = B$ and $W_n^{(2)}$ its orthogonal complement in W_n. Thus, $LW_n^{(1)} = W_n^{(1)}$ and H is transitive on $W_n^{(1)}$ and acts trivially on $W_n^{(2)}$. It is easily established from the invariance of $W_n^{(1)}$ under L that the action of L on W_n induces an action of L on $W_n^{(2)}$ so that $W_n^{(2)}$ is a linear L space. Put $V = \Pi_1^d W_i$ and let V_0 be the subspace of V as defined in Lemma 7.1 so that $V - V_0$ is an invariant nullset. Write Y for $(\Pi_{d+1}^{n-1} W_i) \times W_n^{(2)}$, or $Y = W_n^{(2)}$ if $n = d + 1$, and Z for $W_n^{(1)}$, then Y is a linear L space and

$$(7.8) \qquad W_0 = V_0 \times Y \times Z$$

differs from W by an invariant nullset. Since by Lemma 7.1 V_0 is a Cartan L

space and Y is also an L space, it follows from [17] Proposition 1.3.3 that $V_0 \times Y$ is a Cartan L space (it is also linear but that is not being used). Using the fact that H is a group of translations of Z it is elementary to establish that W_0 is a Cartan G space. Furthermore, from the nature of L acting on V_0 it is obvious that the isotropy group is trivial at each point of V_0 and it follows that G acting on W_0 has trivial isotropy group G_x at each $x \in W_0$. Using [17] Lemma 2.2 there is at $x \in W_0$ a flat near slice (take as S^* in that lemma a translate through x of any linear complement to $(Gx)_x$, taking into account that G_x is trivial). By [17] Proposition 2.1.7 this near slice at x contains a slice S at x. Using [26] Lemma 3 and the fact that G_s is trivial for every $s \in S$, we conclude that S is a local cross section. This concludes the proof of Theorem 7.1.

We conclude this section by showing how Haar measure on $G = LH$ can be obtained from the Haar measures on L and H. Our G is a special case of the following general semidirect product. Given Lie groups L and H such that for each $\ell \in L$ there is an automorphism σ_ℓ of H satisfying $\sigma_{\ell_1 \ell_2} = \sigma_{\ell_2} \sigma_{\ell_1}, \ell_1, \ell_2 \in L$, (in our G, σ_ℓ is the automorphism $b \to C^{-1}b$ if C corresponds to ℓ), and moreover the function $(\ell, h) \to \sigma_\ell(h)$ is an analytic mapping of $L \times H$ onto H. Then G is defined as the analytic manifold $L \times H$ with the group multiplication $(\ell_1, h_1)(\ell_2, h_2) = (\ell_1 \ell_2, \sigma_{\ell_2}(h_1)h_2)$ (see 30D in [16]). (With this group multiplication and the obvious identification of L and H with subgroups of G it is readily verified that $\sigma_\ell(h) = \ell^{-1}h\ell$.) Let μ_L, μ_H be left Haar measure on L, H, respectively. Then the product measure $\mu_L \times \mu_H$ on $L \times H$ is left Haar measure on G (see 30D, E in [16]). This can also be expressed as follows: let v_L, v_H be right Haar measure on L, H, respectively, and let A be any measurable set in G, then the measure $\mu_G(v_G)$ defined by

$$(7.9) \qquad \mu_G A = \int\int_{\ell h \in A} \mu_L(d\ell)\mu_H(dh),$$

$$(7.10) \qquad v_G A = \int\int_{h\ell \in A} v_L(d\ell)v_H(dh)$$

is left (right) Haar measure on G. Here (7.10) follows from (7.9) using the familiar fact that for any group K, if μ_K is a left invariant measure on K then v_K defined by $v_K A = \mu_K A^{-1}$ is right invariant. We can write (7.9), (7.10) also in the following form: let f be integrable on G with respect to $\mu_G(v_G)$, then

$$(7.11) \qquad \int f(g)\mu_G(dg) = \int\int f(\ell h)\mu_L(d\ell)\mu_H(dh),$$

$$(7.12) \qquad \int f(g)v_G(dg) = \int\int f(h\ell)v_L(d\ell)v_H(dh).$$

The fact that the integral on the right side of (7.11) (of (7.12)) is unchanged if $f(g)$ is replaced by $f(g_1 g)$ (by $f(gg_1)$), for any $g_1 \in G$, can of course also be checked directly (as in (15.29)(a) of [8]).

APPLICATIONS. Using (7.11), the function $\psi(h)$ in Equation (10) of [27] could have been omitted by writing on the right side $p(hgx_0)$ instead of $p(ghx_0)$.

Using (7.12) and observing that right (= left) Haar measure on H can simply be taken as Lebesgue measure db on B, we can write (2.2) now in the form

$$(7.13) \qquad J_n(\theta) = \int \int \prod_{i=1}^{n} p_{h\ell\theta}(Z_i) v_L(d\ell)\, db$$

and (2.6) in the form

$$(7.14) \qquad J_n(\theta) = \int \int \exp\{n\psi_\theta(h\ell, \bar{X}_n)\} v_L(d\ell)\, db.$$

$$\diamond \qquad \diamond \qquad \diamond \qquad \diamond \qquad \diamond$$

I would like to thank R. H. Berk for several valuable comments.

REFERENCES

[1] M. S. ABU-SALIH, "On termination with probability one and bounds on sample size distribution of sequential probability ratio tests where the underlying model is an exponential family," unpublished Ph.D. thesis, University of Illinois, 1969.

[2] R. H. BERK, "Asymptotic properties of sequential probability ratio tests," unpublished Ph.D. thesis, Harvard University, 1964.

[3] ———, "Consistency a posteriori," *Ann. Math. Statist.*, Vol. 41 (1970), pp. 894–906.

[4] ———, "Stopping time of SPRTS based on exchangeable models," *Ann. Math. Statist.*, Vol. 41 (1970), pp. 979–990.

[5] D. BLACKWELL and D. FREEDMAN, "A remark on the coin tossing game," *Ann. Math. Statist.*, Vol. 35 (1964), pp. 1345–1347.

[6] N. G. DE BRUIJN, *"Asymptotic methods in analysis,"* Amsterdam, North-Holland Publishing Co., 1958.

[7] H. T. DAVID and W. H. KRUSKAL, "The WAGR sequential t-test reaches a decision with probability one," *Ann. Math. Statist.*, Vol. 27 (1956), pp. 797–805 and Vol. 29 (1958), p. 936.

[8] E. HEWITT and K. A. ROSS, *Abstract Harmonic Analysis*, Berlin-Heidelberg-New York, Springer, 1963.

[9] A. F. IFRAM, "On the asymptotic behavior of densities with applications to sequential analysis," *Ann. Math. Statist.*, Vol. 36 (1965), pp. 615–637.

[10] ———, "Hypergeometric functions in sequential analysis," *Ann. Math. Statist.*, Vol. 36 (1965), pp. 1870–1872.

[11] ———, "On the sample size and simplification of a class of sequential probability ratio tests," *Ann. Math. Statist.*, Vol. 37 (1966), pp. 425–434.

[12] J. E. JACKSON and R. A. BRADLEY, "Sequential χ^2- and T^2-tests," *Ann. Math. Statist.*, Vol. 32 (1961), pp. 1063–1077.

[13] N. L. JOHNSON, "Some notes on the application of sequential methods in the analysis of variance," *Ann. Math. Statist.*, Vol. 24 (1953), pp. 614–623.

[14] S. KULLBACK and R. A. LEIBLER, "On information and sufficiency," *Ann. Math. Statist.*, Vol. 22 (1951), pp. 79–86.

[15] E. L. LEHMANN, *Testing Statistical Hypotheses*, New York, Wiley, 1959.

[16] L. H. LOOMIS, *An Introduction to Abstract Harmonic Analysis*, New York, Van Nostrand, 1953.

[17] R. S. PALAIS, "On the existence of slices for actions of non-compact Lie groups," *Ann. Math.*, Vol. 73 (1961), pp. 295–323.

[18] W. D. RAY, "A proof that the sequential probability ratio test (S.R.P.T.) of the general linear hypothesis terminates with probability unity." *Ann. Math. Statist.*, Vol. 28 (1957), pp. 521–523.

[19] I. R. SAVAGE and J. SETHURAMAN, "Stopping time of a rank-order sequential probability ratio test based on Lehmann alternatives," *Ann. Math. Statist.*, Vol. 37 (1966), pp. 1154–1160 and Vol. 38 (1967), p. 1309.

[20] J. SETHURAMAN, "Stopping time of a rank-order sequential probability ratio test based on Lehmann alternatives—II," *Ann. Math. Statist.*, Vol. 41 (1970), pp. 1322–1333.

[21] C. STEIN, "A note on cumulative sums," *Ann. Math. Statist.*' Vol. 17 (1946), pp. 498–499.

[22] A. WALD, "On cumulative sums of random variables," *Ann. Math. Statist.*, Vol. 15 (1944), pp. 283–296.

[23] ——, *Sequential Analysis*, New York, Wiley, 1947.

[24] G. N. WATSON, "Asymptotic expansions of hypergeometric functions," *Cambridge Phil. Soc. Trans.*, Vol. 22 (1912), pp. 277–308.

[25] ——, *A Treatise on the Theory of Bessel Functions*, Cambridge, Cambridge University Press, 1922.

[26] R. A. WIJSMAN, "Existence of local cross-sections in linear Cartan *G*-spaces under the action of noncompact groups," *Proc. Amer. Math. Soc.*, Vol. 17 (1966), pp. 295–301.

[27] ——, "Cross-sections of orbits and their application to densities of maximal invariants," *Proceedings of the Fifth Berkeley Symposium on Mathematical Statistics and Probability*, Berkeley and Los Angeles, University of California Press, 1967, Vol. 1, pp. 389–400.

[28] ——, "General proof of termination with probability one of invariant sequential probability ratio tests based on multivariate normal observations," *Ann. Math. Statist.*, Vol. 38 (1967), pp. 8–24.

[29] ——, "Bounds on the sample size distribution for a class of invariant sequential probability ratio tests," *Ann. Math. Statist.*, Vol. 39 (1968), pp. 1048–1056.

[30] S. WIRJOSUDIRJO, "Limiting behavior of a sequence of density ratios," unpublished Ph.D. thesis, University of Illinois, 1961. (Abstract in *Ann. Math. Statist.*, Vol. 33 (1962), *pp.* 296–297.)

SOME ASYMPTOTIC PROPERTIES OF LIKELIHOOD RATIOS ON GENERAL SAMPLE SPACES

R. R. BAHADUR[1]

UNIVERSITY OF CHICAGO

and

M. RAGHAVACHARI[2]

CARNEGIE-MELLON UNIVERSITY

1. Introduction

It is shown in [2] (see also [3], [4], [14]) under certain conditions that in the case of independent and identically distributed observations likelihood ratios are asymptotically optimal test statistics in the sense of exact slopes. The present paper points out that many of the arguments and conclusions of the papers just cited extend to general sampling frameworks, and also develops certain refinements of these conclusions. The present generalizations and refinements seem worthwhile for the following reasons. They enable us to construct asymptotically optimal tests in problems such as testing independence in Markov chains and in exchangeable sequences. Secondly, they provide a useful method of finding the exact slope of a statistic which is equivalent, on some sample space, to the likelihood ratio on that sample space. It suffices in this case to evaluate the limit, in the nonnull case, of the normalized log likelihood ratio; it is not necessary to obtain estimates of the relevant large deviation probabilities in the null case; indeed, the latter estimates are implicit in the initial evaluation. Finally, the present elaborations throw some light on what sort of conditioning is advantageous in making conditional tests. It seems that a conditioning statistic is helpful if it produces an exact conditional null distribution for the contemplated test statistic *and* if in the testing problem on hand the conditioning statistic is useless by itself.

The following Sections 2 to 4 describe the general theory. Some examples illustrative of the theory are given in Sections 5 to 7.

[1] Research supported in part by National Science Foundation Research Grant NSF FP 16071 and the Office of Naval Research Grant N00014-67-A-028-009.

[2] Research supported by National Science Foundation Research Grant NSF GP 22595.

2. Notation and preliminary lemmas

Let S be a space of points s, and let $\mathscr{A}$ be a σ-field of sets of S. Let P and Q be probability measures on $\mathscr{A}$, and consider testing the null hypothesis that P obtains against the alternative that Q obtains.

In typical cases (see Sections 5 to 7 below) all $\mathscr{A}$ measurable procedures are not available to the statistician. Suppose then that we are given a σ-field $\mathscr{B} \subset \mathscr{A}$. and that we are restricted to $\mathscr{B}$ measurable procedures. Suppose for simplicity that P and Q are mutually absolutely continuous on $\mathscr{B}$, and let $r(s)$ be a $\mathscr{B}$ measurable function on S such that $0 < r(s) < \infty$ and

$$(2.1) \qquad dQ = r(s)\, dP \text{ on } \mathscr{B}.$$

This r is, of course, the likelihood ratio statistic for testing P against Q when the sample space is $(S, \mathscr{B})$.

Throughout the paper, if $\pi(s)$ is a statement which is either true or false for any given s in S and M is a measure on $\mathscr{A}$, "$\pi(s)$ a.e. $[M]$" will mean that there exists an $\mathscr{A}$ measurable set N of M measure zero such that $\{s: \pi(s) \text{ is false}\} \subset N$.

Let there be given a σ-field $\mathscr{C} \subset \mathscr{B}$, and let $\rho(s)$ be a $\mathscr{C}$ measurable function, $0 < \rho < \infty$, such that

$$(2.2) \qquad dQ = \rho(s)\, dP \text{ on } \mathscr{C},$$

so that ρ is the likelihood ratio statistic when the sample space is $(S, \mathscr{C})$. If $E_P(r(s)|\mathscr{C})$ is a version of the conditional expectation function of r given $\mathscr{C}$ when P obtains, then $\rho(s) = E_P(r(s)|\mathscr{C})$ a.e. P.

In applications, the field $\mathscr{C}$ just introduced is the σ-field induced by a (not necessarily real valued) statistic, say $y = U(s)$, and $\mathscr{C}$ plays two distinct roles. We may be studying the loss of information, if any, when the available sample space is that of the statistic, that is, $(S, \mathscr{C})$, rather than that of s, that is, $(S, \mathscr{B})$. Or we may wish to make conditional tests given y based on s, that is, $\mathscr{C}$ may be the conditioning field. In the remainder of this section $\mathscr{C}$ plays this second role of a conditioning field.

Let $T(s)$ be a real valued $\mathscr{B}$ measurable function, to be thought of as a test statistic, large values of T being significant. The conditional level attained by T given $\mathscr{C}$, say L, is defined as follows. Let $F(t, s)$ be a function defined for $-\infty < t < \infty$ and s in S such that $F(\cdot, s)$ is a left continuous probability distribution function for each s, and $F(t, \cdot)$ is $\mathscr{C}$ measurable for each t, and $F(t, s)$ is a version of the conditional probability function $P(T(s) < t|\mathscr{C})$ for each t. Then

$$(2.3) \qquad L(s) = 1 - F(T(s), s).$$

For a given s, L as just defined is the conditional probability given $\mathscr{C}$ of T being as large or larger than the observed value $T(s)$ if the hypothesis P is true. We shall usually refer to L as the level attained by $T|\mathscr{C}$. It is readily seen that if L and L^0 are two versions of the level attained by $T|\mathscr{C}$ then $L(s) = L^0(s)$ a.e. P.

LEMMA 1.   *The level L is a $\mathscr{B}$ measurable function of s, $0 \leq L \leq 1$.*
LEMMA 2.   *For $a > 0$, $P\big(L(s) < a|\mathscr{C}\big) \leq a$ a.e. P.*

The proofs of Lemmas 1 and 2 are virtually the same as the proofs of Propositions 3 and 4, respectively, of [4], and so are omitted.

We are interested mainly in the behavior of L in the nonnull case, that is, under Q. The following lemma gives a crude bound for the distribution of $L(r/\rho)$; this bound is of interest only because it is valid for all $\mathscr{B}$ measurable statistics T.

Let

$$(2.4) \qquad\qquad \lambda(s) = \min\{1, \rho(s)/r(s)\}$$

and

$$(2.5) \qquad\qquad D^2 = E_P\{[r(s) - \rho(s)]^2/\rho(s)\},$$

and assume for the moment that $D^2 < \infty$. Neither this condition nor Lemma 3 are used in subsequent sections.

LEMMA 3.   *The inequality $Q\big(L(s) < a\lambda(s)\big) \leq 3(a[1 + D^2])^{1/3}$ holds for $a > 0$.*

PROOF.   Let ε and δ be positive constants. Then by (2.4)

$$(2.6) \qquad Q(L < a\lambda) \leq Q(L < a\rho/r)$$

$$\leq Q(r < \varepsilon\rho) + Q(r > \delta\rho) + Q(L < a\rho/r,\ \varepsilon\rho \leq r \leq \delta\rho).$$

Now, according to (2.1) and (2.2),

$$(2.7) \qquad Q(r < \varepsilon\rho) = \int \{r < \varepsilon\rho\}\, r\, dP$$

$$\leq \varepsilon \int \{r < \varepsilon\rho\}\, \rho\, dP \leq \varepsilon.$$

Next,

$$(2.8) \qquad Q(r > \delta\rho) \leq \delta^{-1} E_Q(r/\rho) = \delta^{-1} E_P(r^2/\rho)$$

$$= \delta^{-1}[1 + D^2]$$

by Markov's inequality and (2.5). Finally, using the definitions (2.1), (2.2) and Lemma 2,

$$(2.9) \qquad Q(L < a\rho/r,\ \varepsilon\rho \leq r \leq \delta\rho) \leq Q(L < a/\varepsilon,\ r \leq \delta\rho)$$

$$= \int \{L < a/\varepsilon,\ r \leq \delta\rho\}\, r\, dP \leq \delta \int \{L < a/\varepsilon\}\, \rho\, dP$$

$$= \delta \int_S \rho P(L < a/\varepsilon\,|\,\mathscr{C})\, dP \leq (\delta a/\varepsilon) \int_S \rho\, dP = \delta a/\varepsilon.$$

It follows from (2.6) to (2.9) that $Q(L < a\lambda) \leqq \varepsilon + (\delta a)\varepsilon^{-1} + [1 + D^2]\delta^{-1}$ for all ε and $\delta > 0$. Choosing $\varepsilon = (\delta a)^{1/2}$ and then minimizing the bound with respect to δ completes the proof.

Now regard $r(s)$ as a test statistic and let $\hat{L}(s)$ be the level attained by $r | \mathscr{C}$.

LEMMA 4.   *The inequality $\hat{L}(s) \leqq \lambda(s)$ holds a.e. Q.*

PROOF.   Let $\hat{F}(t, s)$ be a version of the left continuous distribution function of r given $\mathscr{C}$ when P obtains. Consider a fixed $t > 0$, and let C be a $\mathscr{C}$ measurable set. Then from (2.1) and (2.2),

$$(2.10) \qquad \int_C [1 - \hat{F}(t, s)] \, dP = \int_C P(r \geqq t | \mathscr{C}) \, dP = \int \{C \cap \{r \geqq t\}\} \, dP$$

$$\leqq \int_C (r/t) \, dP = t^{-1} Q(C) = \int_C (\rho/t) \, dP.$$

Since $1 - \hat{F}$ and ρ/t are $\mathscr{C}$ measurable functions and since C is arbitrary it follows from (2.10) that $1 - \hat{F}(t, s) \leqq \rho(s)/t$ a.e. P. Since t is arbitrary, it now follows by a familiar argument that

$$(2.11) \qquad 1 - \hat{F}(t, s) \leqq \rho(s)/t \quad \text{for all} \quad t > 0 \text{ a.e. } P.$$

It follows from (2.11) that $\hat{L}(s) \equiv 1 - \hat{F}(r(s), s) \leqq \rho(s)/r(s)$ a.e. P. Hence the lemma, since $\hat{L} \leqq 1$ in any case, and $Q \ll P$ on $\mathscr{B}$.

Lemmas 3 and 4 suggest that if Q obtains, the conditional level attained by an optimal statistic is likely to be of the order $\rho(s)/r(s)$. If the sample space $(S, \mathscr{C})$ is itself highly informative for discriminating between P and Q, ρ is likely to be much larger than 1 when Q obtains; then $\rho(s)/r(s)$ is much larger than $1/r(s)$. By temporarily putting $\mathscr{C} = $ the trivial field in Lemmas 3 and 4, it is seen that the unconditional level attained by an optimal statistic on $(S, \mathscr{B})$ is likely to be of the order $1/r(s)$. These heuristic considerations suggest that conditioning is harmful, unless the sample space $(S, \mathscr{C})$ is useless for discriminating between P and Q. This suggestion is discussed more precisely in asymptotic terms in the following sections.

It may be worthwhile to look at some of the details of the special case when $\mathscr{C}$ is induced by a statistic and there exist regular conditional probabilities on $\mathscr{B}$ given $\mathscr{C}$. Suppose then that $y = U(s)$ is a measurable transformation of $(S, \mathscr{B})$ into a space $(Y, \mathscr{D})$. Assume that P admits a regular conditional probability measure on $\mathscr{B}$ given $U(s) = y$, that is, there exists a function $P_y(B)$ on $\mathscr{B} \times Y$ such that $P_y(\cdot)$ is a probability measure on $\mathscr{B}$ for each y and such that, for each B in $\mathscr{B}$, $P_y(B)$ is a version of $P(B | U(s) = y)$. Let $\mathscr{C} = U^{-1}(\mathscr{D})$. Then, for any statistic T, the level attained by $T | \mathscr{C}$ may be defined as $\{P_y(T(s) \geqq t)\}_{t = T(s), y = U(s)}$. In particular, the level attained by $r | \mathscr{C}$ is

$$(2.12) \qquad \hat{L}(s) = \{P_y(r(s) \geqq t)\}_{t = r(s), y = U(s)}.$$

Now let

$$(2.13) \qquad r_y(s) = r(s) \bigg/ \int_S r(s) \, dP_y. \qquad\qquad 0 \leqq r_y(s) \leqq \infty,$$

and for each y in Y, let the measure $Q_y(\cdot)$ be defined by $Q_y(B) = \int_B r_y(s)\, dP_y$ for B in $\mathscr{B}$. It is readily seen that Q_y is a regular conditional probability on $\mathscr{B}$ given $U(s) = y$ when Q obtains. Consequently, $r_y(s)$ is the conditional likelihood ratio given $U(s) = y$. It is also readily seen that, as an alternative to formula (2.12), we have

$$(2.14) \qquad \hat{L}(s) = \{\{P_y(r_y(s) \geq t)\}_{t = r_y(s)}\}_{y = U(s)}.$$

In subsequent sections, phrases such as "$\mathscr{C}$ is induced by U" or "$\mathscr{C}$ is induced by the mapping: $s \to U(s)$" mean that $\mathscr{C} = U^{-1}(\mathscr{D})$ where $\mathscr{D}$ is determined by the context as follows. If Y is a finite set, $\mathscr{D}$ is the field of all sets of Y; if Y is (or may be taken to be) k dimensional Euclidean space, $\mathscr{D}$ is the class of Borel sets of Y.

3. Limit theorems in the simplest case

In this section and the following ones we consider a space S of points s, a σ-field $\mathscr{A}$ of sets of S, and two sequences $\{\mathscr{B}_n : n = 1, 2, \cdots\}$ and $\{\mathscr{C}_n : n = 1, 2, \cdots\}$ of σ-fields such that

$$(3.1) \qquad\qquad \mathscr{C}_n \subset \mathscr{B}_n \subset \mathscr{A}, \qquad\qquad n = 1, 2, \cdots.$$

Let P and Q be probability measures on $\mathscr{A}$. We assume that P and Q are mutually absolutely continuous on $\mathscr{B}_n$, and let

$$(3.2) \qquad dQ = r_n(s)\, dP \text{ on } \mathscr{B}_n, \qquad dQ = \rho_n(s)\, dP \text{ on } \mathscr{C}_n,$$

where r_n is $\mathscr{B}_n$ measurable, ρ_n is $\mathscr{C}_n$ measurable, and $0 < r_n, \rho_n < \infty$; $n = 1, 2, \cdots$.

Let

$$(3.3) \qquad K_n^{(1)}(s) = n^{-1} \log r_n(s), \qquad K_n^{(2)}(s) = n^{-1} \log \rho_n(s),$$

and let

$$(3.4) \qquad\qquad \Delta_n(s) = K_n^{(1)}(s) - K_n^{(2)}(s).$$

Let

$$(3.5) \qquad \underline{\Delta}(s) = \liminf_{n \to \infty} \Delta_n(s), \qquad \overline{\Delta}(s) = \limsup_{n \to \infty} \Delta_n(s).$$

THEOREM 1. *The function $\underline{\Delta}$ satisfies $0 \leq \underline{\Delta}(s) \leq \infty$ a.e. Q.*

PROOF. Choose $\varepsilon > 0$. Since the event $\Delta_n < -\varepsilon$ is, by (3.3) and (3.4), identical with the event $r_n < \rho_n \exp\{-n\varepsilon\}$, we have by (3.2)

$$(3.6) \qquad Q(\Delta_n < -\varepsilon) = \int_{\Delta_n < -\varepsilon} r_n\, dP$$

$$\leq \exp\{-n\varepsilon\} \int_{\Delta_n < -\varepsilon} \rho_n\, dP$$

$$\leq \exp\{-n\varepsilon\} \int_S \rho_n\, dP = \exp\{-n\varepsilon\}.$$

Hence $\Sigma_n Q(\Delta_n < -\varepsilon) < \infty$. Hence $\underline{\Delta}(s) \geq -\varepsilon$, a.e. Q. Since ε is arbitrary, $\underline{\Delta}(s) \geq 0$ a.e. Q. Q.E.D.

Let

$$(3.7) \qquad \underline{K}^{(i)}(s) = \liminf_{n \to \infty} K_n^{(i)}(s), \qquad \bar{K}^{(i)}(s) = \limsup_{n \to \infty} K_n^{(i)}(s), \qquad i = 1, 2.$$

COROLLARY 1. *The inequalities* $0 \leq \underline{K}^{(2)}(s) \leq \underline{K}^{(1)}(s) \leq \infty$ *and* $0 \leq \bar{K}^{(2)}(s) \leq \bar{K}^{(1)}(s) \leq \infty$ *hold a.e.* Q.

PROOF. In view of Theorem 1, we need only show that $0 \leq \underline{K}^{(2)}(s)$ a.e. Q. To this end, for each n let $\mathscr{C}_n$ = the trivial field in Theorem 1. Then $\rho_n(s) \equiv 1$, $K_n^{(2)}(s) \equiv 0$; hence $\underline{K}^{(1)}(s) \geq 0$ a.e. Q, by (3.3) to (3.7) and Theorem 1. Since $\{\mathscr{B}_n\}$ is an arbitrary sequence, this last conclusion continues to hold when $\{\mathscr{B}_n\}$ is replaced by the initially given $\{\mathscr{C}_n\}$, so $\underline{K}^{(2)}(s) \geq 0$ a.e. Q. Q.E.D.

It is pointed out at the outset of Section 7 below that $\underline{K}^{(i)}$, $\bar{K}^{(i)}$, $i = 1, 2$, are, roughly speaking, generalizations of the Kullback-Leibler information numbers. and that Corollary 1 is a generalization of a well-known inequality concerning these numbers.

For each n let T_n be a real valued $\mathscr{B}_n$ measurable function, and let L_n be the level attained by $T_n | \mathscr{C}_n$. The following theorem is essentially an extension and refinement of Theorem 1 of [2] and of the main theorems of [4], [14].

THEOREM 2. *The inequalities* $\liminf_{n \to \infty} \{n^{-1} \log L_n(s)\} \geq -\bar{\Delta}(s)$ *and* $\limsup_{n \to \infty} \{n^{-1} \log L_n(s)\} \geq -\underline{\Delta}(s)$ *hold a.e.* Q.

PROOF. Choose and fix $\varepsilon > 0$. Then, for any constant $b \geq 0$ and any n,

$$(3.8) \qquad Q(L_n < \exp(-n[\Delta_n + 3\varepsilon]), b - \varepsilon < \Delta_n < b + \varepsilon)$$

$$\leq Q(L_n < \exp(-n[b + 2\varepsilon]), \Delta_n < b + \varepsilon)$$

$$(3.9) \qquad = \int \{L_n < \exp(-n[b + 2\varepsilon]), \Delta_n < b + \varepsilon\} r_n \, dP$$

$$(3.10) \qquad = \int \{L_n < \exp(-n[b + 2\varepsilon]), \Delta_n < b + \varepsilon\} \exp(n\Delta_n)\rho_n \, dP$$

$$\leq \exp(n[b + \varepsilon]) \int \{L_n < \exp(-n[b + 2\varepsilon]), \Delta_n < b + \varepsilon\}\rho_n \, dP$$

$$\leq \exp(n[b + \varepsilon]) \int \{L_n < \exp(-n[b + 2\varepsilon])\}$$

$$(3.11) \qquad \leq \exp(-n\varepsilon).$$

Here (3.9) follows from (3.2), (3.10) from (3.3) and (3.4), and (3.11) from Lemma 2, as in (2.9).

Now let $b_1, b_2, \cdots$ be an enumeration of the rational points of $[0, \infty)$, and let

$$(3.12) \qquad A_n(i) = \{s \colon L_n < \exp\{-n[\Delta_n + 3\varepsilon]\}, b_i - \varepsilon < \Delta_n < b_i + \varepsilon\},$$

(3.13) $$B(i) = \limsup_{n \to \infty} A_n(i)$$

and

(3.14) $$C = \bigcup_{i=1}^{\infty} B(i).$$

It follows from (3.8) to (3.11) with $b = b_i$ that $\Sigma_n \, Q\big(A_n(i)\big) < \infty$; hence $Q\big(B(i)\big) = 0$; hence $Q(C) = 0$. Let $D_\varepsilon = S - C - \{s : \underline{\Delta}(s) < 0\}$. Then $Q(D_\varepsilon) = 1$. We shall show that, for every s in D_ε,

(3.15) $$\liminf_{n \to \infty} \{n^{-1} \log L_n(s)\} \geqq -\bar{\Delta}(s) - 3\varepsilon$$

and

(3.16) $$\limsup_{n \to \infty} \{n^{-1} \log L_n(s)\} \geqq -\underline{\Delta}(s) - 3\varepsilon.$$

This will establish the theorem, since ε is arbitrary. Choose and fix an s in D_ε.

To establish (3.15) for the given s, we may suppose $\bar{\Delta} < \infty$ for otherwise (3.15) holds trivially. Let v denote the left side of (3.15) and let $j_1 < j_2 < \cdots$ be a sequence of positive integers such that $j_m^{-1} \log L_{j_m} \to v$ as $m \to \infty$. Since $0 \leqq \underline{\Delta} \leqq \bar{\Delta} < \infty$, there exists a subsequence $k_1 < k_2 < \cdots$ of $\{j_m\}$ such that $\Delta_{k_m} \to$ a limit Δ, say, where $0 \leqq \Delta < \infty$. There exists a rational $b_i \geqq 0$ such that $b_i - \varepsilon < \Delta < b_i + \varepsilon$; hence $b_i - \varepsilon < \Delta_{k_m} < b_i + \varepsilon$ for all sufficiently large m. It now follows from (3.12) to (3.14) and the choice of s that $L_{k_m} \geqq \exp \{-k_m(\Delta_{k_m} + 3\varepsilon)\}$ for all sufficiently large m. Since $\{k_m\}$ is a subsequence of $\{j_m\}$ it now follows from the choice of $\{j_m\}$ and $\{k_m\}$ that

(3.17) $$v \geqq \lim_{m \to \infty} (-\Delta_{k_m} - 3\varepsilon) = -\Delta - 3\varepsilon \geqq -\bar{\Delta} - 3\varepsilon.$$

so (3.15) holds.

To establish (3.16) we may suppose that $0 \leqq \underline{\Delta} < \infty$. There exists a sequence $j_1 < j_2 < \cdots$ such that $\Delta_{j_m} \to \underline{\Delta}$ as $m \to \infty$. Hence there exists a rational $b_i \geqq 0$ such that $b_i - \varepsilon < \Delta_{j_m} < b_i + \varepsilon$ for all sufficiently large m. Hence, by (3.12) to (3.14) and the choice of s, $L_{j_m} > \exp \{-j_m[\Delta_{j_m} + 3\varepsilon]\}$ for all sufficiently large m. Since the left side of (3.16) is not less than $\limsup_{m \to \infty} \{j_m^{-1} \log L_m\}$, it now follows from the present choice of $\{j_m\}$ that (3.16) holds. $Q.E.D.$

Now let $\hat{L}_n(s)$ be the level attained by $r_n | \mathscr{C}_n$. The following theorem is a partial generalization and refinement of Theorem 2 of [2].

THEOREM 3. *The equalities* $\liminf_{n \to \infty} \{n^{-1} \log \hat{L}_n(s)\} = -\bar{\Delta}(s)$ *and* $\limsup_{n \to \infty} \{n^{-1} \log \hat{L}_n(s)\} = -\underline{\Delta}(s)$ *hold a.e. Q.*

PROOF. It follows from (3.3) and (3.4) by Lemma 4 that

(3.18) $$n^{-1} \log \hat{L}_n(s) \leqq -\Delta_n(s) \quad \text{for all } n \text{ a.e. } Q.$$

It follows from (3.5) and (3.18) that $\lim \inf_{n \to \infty} \{n^{-1} \log \hat{L}_n\} \leq -\bar{\Delta}$ and $\lim \sup_{n \to \infty} \{n^{-1} \log \hat{L}_n\} \leq -\underline{\Delta}$ a.e. Q. Theorem 3 now follows from Theorem 2 applied to $\{r_n | \mathscr{C}_n\}$. $Q.E.D.$

We shall say that the sequence $\{T_n | \mathscr{C}_n\}$ has exact slope $c(s)$ when Q obtains (see [3]) if $n^{-1} \log L_n(s) \to -\frac{1}{2}c(s)$ as $n \to \infty$ a.e. Q. Let $\Delta(s)$ be an $\mathscr{A}$ measurable function, $0 \leq \Delta \leq \infty$. As an immediate consequence of Theorems 2 and 3 we have,

COROLLARY 2. *The sequence $\{r_n | \mathscr{C}_n\}$ has exact slope $2\Delta(s)$ when Q obtains if and only if $\lim_{n \to \infty} \Delta_n(s) = \Delta(s)$ a.e. Q. In the latter case, if $c(s)$ is the exact slope of any sequence $\{T_n | \mathscr{C}_n\}$ then $c(s) \leq 2\Delta(s)$ a.e. Q.*

Corollary 2 is perhaps the main conclusion of this paper. The elaborations given in Theorems 2 and 3 are, however, useful on occasion.

Suppose that $\lim_{n \to \infty} \Delta_n(s) = \Delta(s)$ *and* $\Delta(s) < \infty$ a.e. Q. Then, by Corollary 2,

$$(3.19) \qquad \hat{L}_n(s) = \exp\{-n\Delta_n(s) + o(n)\} \quad \text{as} \quad n \to \infty \text{ a.e. } Q.$$

The estimate $\exp\{-n\Delta_n\}$ is $\mathscr{B}_n$ measurable, that is, based on the same data as $\hat{L}_n$, but is often much easier to compute. The formulation (3.19) makes sense in the general case, but (3.19) is not valid unless the conditions stated are satisfied. To consider an example, suppose that $s = (x_1, x_2, \cdots)$ where the x_n are independent real valued random variables; under P each x_n is $N(0, 1)$; under Q, x_n is $N(\mu_n, 1)$ where $\mu_n = \exp\{n^2\}$. For each n let $\mathscr{B}_n$ be the field induced by the mapping: $s \to x_n$, and let $\mathscr{C}_n$ be the trivial field. Then $n^{-1} \log \hat{L}_n(s) + \Delta_n(s) \to -\infty$ a.e. Q. The verification is omitted.

Suppose for the moment that $K_n^{(i)}(s) \to K^{(i)}(s)$ as $n \to \infty$ a.e. Q where $0 \leq K^{(i)}(s) < \infty$, $i = 1, 2$. It then follows from Corollary 2 that the unconditional exact slope of $\{r_n\}$ is $2K^{(1)}(s)$, that the exact slope of $\{r_n | \mathscr{C}_n\}$ is $2[K^{(1)}(s) - K^{(2)}(s)]$, and that these are the maximum available unconditional and conditional slopes, respectively. Since $K^{(2)}(s) \geq 0$, it is plain that there is certainly no advantage in conditioning and that there is no disadvantage if and only if $K^{(2)}(s) = 0$ a.e. Q. Now, $2K^{(2)}(s)$ is the maximum available unconditional exact slope when $\{\mathscr{B}_n\}$ is replaced by $\{\mathscr{C}_n\}$, that is, $2K^{(2)}$ is the exact slope of $\{\rho_n\}$. It is thus seen that conditioning is asymptotically harmless if and only if the conditioning σ-field or statistic is asymptotically useless for testing P against Q. However, as is pointed out in the following section, if we are testing a *composite* null hypothesis there may exist an asymptotically harmless conditioning which has the following feature: the conditional distribution of a contemplated test statistic does not depend on which null P obtains. This last feature is very convenient, for practical as well as theoretical purposes.

Suppose that, for given $\{\mathscr{B}_n\}$ and $\{\mathscr{C}_n\}$, the assumptions of the preceding paragraph are satisfied and that $0 < K^{(1)}(s)$ a.e. Q. Since the exact slope of the optimal $\mathscr{B}_n$ measurable sequence $\{r_n\}$ is $2K^{(1)}$, and that of the optimal $\mathscr{C}_n$ measurable sequence $\{\rho_n\}$ is $2K^{(2)}$, and since the ratio of slopes is a measure of asymptotic efficiency (see [3]), it is seen that the asymptotic efficiency of $(S, \mathscr{C}_n)$ relative to

$(S, \mathscr{B}_n)$ in testing P against Q is $K^{(2)}(s)/K^{(1)}(s)$. As may be seen from examples in Sections 5 to 7 the $K^{(i)}(s)$ are usually, but not always, independent of s.

In concluding this section we state a partial analogue of Corollary 2 in terms of the size of the optimal test of P which has a given power against Q. (See [3], Section 5, for a description of the main relation between exact slopes and power.) Let β be given, $0 < \beta < 1$, and for each n let $\hat{\alpha}_n = \hat{\alpha}_n(\beta)$ be the size of the (possibly randomized) test of P based on r_n which has power $1 - \beta$ against Q.

THEOREM 4. *Suppose that* $K_n^{(1)}(s) \to K^{(1)}$ *as* $n \to \infty$ *a.e.* Q, *where* $K^{(1)}$ *is a constant,* $0 < K^{(1)} < \infty$. *Then, for each* β, $n^{-1} \log \hat{\alpha}_n(\beta) \to -K^{(1)}$ *as* $n \to \infty$.

This theorem is a generalization of a lemma of Stein. The proof is essentially the same as the proof of Stein's lemma on pp. 316–317 of [3] and so is omitted. It should be noted that, under the hypothesis of the theorem, the exact slope of $\{r_n\}$ against Q is $2K^{(1)}$, by Corollary 2. It should also be noted that (the dual of) Theorem 4 can be used to obtain the Hodges-Lehmann index [11] of likelihood ratio tests on an arbitrary sequence of sample spaces. The theorem is valid provided that $K_n^{(1)}(s) \to K^{(1)}$ in Q probability.

4. The general case

In this section we consider the framework $S = \{s\}$, $\mathscr{A}$, $\{\mathscr{B}_n\}$, and $\{\mathscr{C}_n\}$ of the preceding section, and suppose that we are given two disjoint sets $\mathscr{P}_0$ and $\mathscr{P}_1$ of probability measures on $\mathscr{A}$. The null hypothesis is that some P in $\mathscr{P}_0$ obtains; the alternative is that some Q in $\mathscr{P}_1$ obtains.

We assume that, for given $P \in \mathscr{P}_0$, and $Q \in P_1$, P and Q are mutually absolutely continuous on $\mathscr{B}_n$, and we write $r_n(s)$, $\rho_n(s)$, and $K_n^{(i)}(s)$ of the preceding section as $r_n(s: Q, P)$, $\rho_n(s: Q, P)$, and $K_n^{(i)}(s: Q, P)$ to indicate their dependence on P and Q, $i = 1, 2; n = 1, 2, \cdots$.

As a matter of economy, and without much loss of generality, we assume throughout this section that the following condition is satisfied: Given $P \in \mathscr{P}_0$ and $Q \in \mathscr{P}_1$, there exists a *constant* $K^{(1)}(Q, P)$, $0 \leq K^{(1)} \leq \infty$, such that

$$(4.1) \qquad \lim_{n \to \infty} K_n^{(1)}(s: Q, P) = K^{(1)}(Q, P) \text{ a.e. } Q.$$

Let

$$(4.2) \qquad J^{(1)}(Q) = \inf \{K^{(1)}(Q, P): P \in \mathscr{P}_0\}, \qquad 0 \leq J^{(1)} \leq \infty.$$

Now let T_n be a real valued $\mathscr{B}_n$ measurable statistic, and let $L_n(s; P)$ be the level attained by $T_n | \mathscr{C}_n$ in testing a given P. Let

$$(4.3) \qquad L_n^*(s) = \sup \{L_n(s; P): P \in \mathscr{P}_0\}.$$

Then L_n^* is, by definition, the level attained by $T_n | \mathscr{C}_n$ in testing that some P in $\mathscr{P}_0$ obtains. As noted on p. 29 of [4], it is not necessary to assume that L_n^* is $\mathscr{A}$ measurable or even that there exists an $\mathscr{A}$ measurable version thereof.

COROLLARY 3.   *For each Q in $\mathscr{P}_1$,*

(4.4) $$\liminf_{n \to \infty} \{n^{-1} \log L_n^*(s)\} \geqq -J^{(1)}(Q) \text{ a.e. } Q.$$

This corollary is a straightforward consequence of Theorem 2, Corollary 1, (4.1), (4.2), and (4.3). In view of Corollary 3, we shall say that, $\{T_n | \mathscr{C}_n\}$ is an asymptotically optimal sequence (for testing $\mathscr{P}_0$ against $\mathscr{P}_1$) if, for each Q in $\mathscr{P}_1$, it has exact slope $2J^{(1)}(Q)$ against Q; more precisely, if there exists a version of $\{L_n^*\}$ such that $n^{-1} \log L_n^*(s) \to -J^{(1)}(Q)$ as $n \to \infty$ a.e. Q, for each Q.

Let us say that $T_n | \mathscr{C}_n$ has an exact null distribution if there exists $F_n(t, s)$ such that F_n is a left continuous distribution function for each s and such that, for each P in $\mathscr{P}_0$ and for each t, $F_n(t, s)$ is a version of $P(T_n(s) < t | \mathscr{C}_n)$. In this case $1 - F_n(T_n(s), s)$ is a version of $L_n^*(s)$, and the maximization in (4.3) is avoided; this maximization is often inconvenient or impractical. The following Corollary 4 shows in part that a conditioning which produces an exact null distribution for T_n might also have the theoretical advantage of reducing the testing problem to the case considered in Section 3 and thereby producing an optimal testing sequence $\{T_n | \mathscr{C}_n\}$.

COROLLARY 4.   *Suppose that $T_n | \mathscr{C}_n$ has an exact null distribution, $n = 1, 2, \cdots$. Suppose also that for each $Q \in \mathscr{P}_1$ there exists a P_Q in $\mathscr{P}_0$ such that (a) $J^{(1)}(Q) = K^{(1)}(Q, P_Q)$, (b) $\lim_{n \to \infty} K_n^{(2)}(s: Q, P_Q) = 0$, a.e. Q, and (c) for each n, $r_n(s: Q, P_Q)$ is a strictly increasing function of $T_n(s)$. Then $\{T_n | \mathscr{C}_n\}$ is an asymptotically optimal sequence.*

PROOF.   For each n, let F_n be a function such that the conditions stated in the paragraph preceding Corollary 4 are satisfied, and let $L_n^*(s) = 1 - F_n(T_n(s), s)$. Choose and fix $Q \in \mathscr{P}_1$. By assumption there exists $P_Q \in \mathscr{P}_0$ such that the stated conditions (a) to (c) are satisfied. It follows from the present definition of L_n^* that, for each n, $L_n^*(s)$ is a version of $L_n(s; P_Q)$. It follows from condition (c) that, with $r_n = r_n(s: Q, P_Q)$, and with $\hat{L}_n(s; P_Q)$ and $L_n(s; P_Q)$ the levels attained by $r_n | \mathscr{C}_n$ and $T_n | \mathscr{C}_n$, respectively, in testing P_Q, we have $L_n(s; P_Q) = \hat{L}_n(s; P_Q)$, a.e. P_Q. Hence $L_n^*(s) = \hat{L}_n(s: P_Q)$ a.e. Q for all n. It now follows from Corollary 2 with $P = P_Q$, (4.1), and condition (b) that $n^{-1} \log L_n^*(s) \to -K^{(1)}(Q, P_Q)$ a.e. Q. It follows hence by condition (a) that $\{T_n | \mathscr{C}_n\}$ has exact slope $2J^{(1)}(Q)$ against Q. Q.E.D.

Although Corollary 4 is phrased in terms of asymptotic optimality, it can sometimes be used to find the exact slope of a given sequence $\{T_n | \mathscr{C}_n\}$ against a given Q by *defining* $\{\mathscr{B}_n\}$ suitably. To consider a special case, suppose we are given a measurable space $(S, \mathscr{A})$, a null set $\mathscr{P}_0$ of probability measures on $\mathscr{A}$, and a single nonnull probability measure Q on $\mathscr{A}$. For each n let T_n be an $\mathscr{A}$ measurable function on S into the real line such that $P(T_n < t) = F_n(t)$ for all P in $\mathscr{P}_0$ and all real t. For each n let $\mathscr{B}_n$ be the σ-field induced by T_n. Suppose that, with the present definition of $\mathscr{B}_n$, (4.1) holds for each P in $\mathscr{P}_0$, with some $K^{(1)}$. Let $J^{(1)}$ be defined by (4.2). Suppose there exists P_Q in $\mathscr{P}_0$ such that conditions (a) and (c) of Corollary 4 are satisfied. Since condition (b) is automatically satisfied when $\mathscr{C}_n$ is the trivial field for each n, we conclude that $n^{-1} \log [1 - F_n(T_n(s))] \to$

$-J^{(1)}(Q)$ as $n \to \infty$ a.e. Q, that is, with the present definition of $J^{(1)}$, $\{T_n\}$ has exact slope $2J^{(1)}$ against Q.

The mechanism described in Corollary 4 is rather special and is likely to be available only in rather special cases. Corollary 5 below describes a related but different method of constructing optimal sequences. For simplicity, we consider only unconditional test procedures.

CONDITION 1.   *For each Q in $\mathscr{P}_1$,*

(4.5) $$\liminf_{n \to \infty} T_n(s) \geqq J^{(1)}(Q) \text{ a.e. } Q.$$

CONDITION 2.   *Given ε and τ, $0 < \varepsilon < 1$ and $0 < \tau < 1$, for each n there exists a positive constant $k_n(\varepsilon, \tau)$ such that*

(4.6) $$P\big(T_n(s) \geqq t\big) \leqq \exp\{-n\tau t\}\,(1 + \varepsilon)^n\,k_n(\varepsilon, \tau)$$

for all $t > 0$ and all P in $\mathscr{P}_0$, and such that

(4.7) $$n^{-1} \log k_n(\varepsilon, \tau) \to 0 \quad as \quad n \to \infty.$$

COROLLARY 5.   *Suppose that $\{T_n\}$ satisfies Conditions 1 and 2. Then (i) $\{T_n\}$ is an asymptotically optimal sequence, (ii) for each Q in $\mathscr{P}_1$,*

(4.8) $$\lim_{n \to \infty} T_n(s) = J^{(1)}(Q) \text{ a.e. } Q.$$

and (iii) with $G_n(t) = \inf\{P\big(T_n(s) < t\big): P \in \mathscr{P}_0\}$,

(4.9) $$\lim_{n \to \infty} n^{-1} \log[1 - G_n(t)] = -t$$

for each $t \in \text{int}\{J^{(1)}(Q): Q \in \mathscr{P}_1\}$.

PROOF.   Choose ε and τ, $0 < \varepsilon < 1$ and $0 < \tau < 1$. By replacing k_n with $\max\{k_n, 1\}$ we may suppose that (4.6) holds for all real t, all P in $\mathscr{P}_0$, and all n, and that (4.7) is still satisfied. Hence, by the definition of G_n,

(4.10) $$L_n^*(s) = 1 - G_n\big(T_n(s)\big) \leqq (1 + \varepsilon)^n k_n(\varepsilon, \tau) \exp\{-n\tau T_n(s)\}$$

for all n and s. It follows from (4.5), (4.7), and (4.10) that, for any Q in $\mathscr{P}_1$,

(4.11) $$\limsup_{n \to \infty} n^{-1} \log L_n^*(s) \leqq \tau J^{(1)}(Q) + \log(1 + \varepsilon) \text{ a.e. } Q,$$

and that

(4.12) $$\liminf_{n \to \infty} n^{-1} \log L_n^*(s) \leqq \log(1 + \varepsilon) - \tau \limsup_{n \to \infty} T_n(s).$$

Since ε, τ and Q are arbitrary, it follows from (4.11), (4.12) and Corollary 3 that parts (i) and (ii) of Corollary 5 are valid.

Part (iii) follows from parts (i) and (ii) as a special case of the following proposition. Suppose that, for each Q in $\mathscr{P}_1$, $\{T_n\}$ has exact slope $c(Q)$ against Q, and that $T_n(s) \to b(Q)$ as $n \to \infty$ a.e. Q, where b and c are (possibly infinite) constants. Let G_n be defined as in part (iii) of Corollary 5. Then, for any finite t such that

$$f(t) = \sup \{\tfrac{1}{2}c(Q): Q \in \mathscr{P}_1, b(Q) < t\}$$

(4.13)

$$g(t) = \inf \{\tfrac{1}{2}c(Q): Q \in \mathscr{P}_1, b(Q) > t\}$$

are well defined, we have

(4.14)
$$-g(t) \leqq \liminf_{n \to \infty} \{n^{-1} \log [1 - G_n(t)]\}$$

$$\leqq \limsup_{n \to \infty} \{n^{-1} \log [1 - G_n(t)]\} \leqq -f(t).$$

This proposition is a straightforward consequence of the monotonicity of G_n for each n; we omit the details.

It seems that Conditions 1 and 2 are satisfied in a great variety of examples by the likelihood ratio statistic

(4.15)
$$\hat{T}_n(s) = \inf_{P \in \mathscr{P}_0} \sup_{Q \in \mathscr{P}_1} \{K_n^{(1)}(s: Q, P)\}$$

of Neyman and Pearson. The underlying reason is the following. *Conditions 1 and 2 are always satisfied by* $\{\hat{T}_n\}$ *if* $\mathscr{P}_0 \cup \mathscr{P}_1$ *is a finite set.* In many cases, although $\mathscr{P}_0 \cup \mathscr{P}_1$ is infinite, compactification devices which reduce the problem to the finite case are applicable, so that Conditions 1 and 2 do hold. The proof in [2] of Theorem 2 consists in verifying that, in the context of [2], Conditions 1 and 2 are satisfied by $\{\hat{T}_n\}$ provided a suitable compactification of $\mathscr{P}_0 \cup \mathscr{P}_1$ exists and certain other conditions are satisfied. In fact, Corollary 5 is a statement in general terms of the essential elements of the proof just cited. We think this generalized statement, however trivial or even tautological it may seem, is useful; see Sections 5 and 6.

We conclude this section with a sort of converse to Corollary 4.

COROLLARY 6.   *Suppose that there exists an asymptotically optimal sequence* $\{T_n | \mathscr{C}_n\}$. *Then* $Q \in \mathscr{P}_1$, $P \in \mathscr{P}_0$, $K^{(1)}(Q, P) = J^{(1)}(Q) < \infty$ *imply that* $K_n^{(2)}(s: Q, P) \to 0$ *as* $n \to \infty$, *a.e.* Q.

PROOF.   Let $\{T_n | \mathscr{C}_n\}$ be optimal, and for each n let $L_n^*(s)$ be a version of the level attained by $T_n | \mathscr{C}_n$ such that $n^{-1} \log L_n^*(s) \to -J^{(1)}(Q)$ a.e. Q, for each $Q \in \mathscr{P}_1$. Choose and fix Q and P such that the conditions stated are satisfied. For each n let $L_n(s)$ be the level attained by $T_n | \mathscr{C}_n$ in testing the simple hypothesis P. Then $L_n^*(s) \geqq L_n(s)$ a.e. P; hence $L_n^*(s) \geqq L_n(s)$ a.e. Q, for all n. Consequently,

(4.16) $$-K^{(1)}(Q, P) = -J^{(1)}(Q)$$

$$\geqq \limsup_{n \to \infty} \{n^{-1} \log L_n(s)\}$$

$$\geqq -\liminf_{n \to \infty} \{K_n^{(1)}(s: Q, P) - K_n^{(2)}(s: Q, P)\}$$

$$= -K^{(1)}(s: Q, P) + \bar{K}^{(2)}(s: Q, P) \text{ a.e. } Q$$

by Theorem 2 and (4.1). Hence $\bar{K}^{(2)}(s: Q, P) \leqq 0$ a.e. Q, so $\bar{K}^{(2)}(s: Q, P) = 0$ a.e. Q. Q.E.D.

5. Examples concerning exchangeable sequences

Let X be a finite set, say $X = \{a_1, \cdots, a_k\}$, $k \geq 2$, and let $\mathcal{M}$ be the field of all subsets of X. Let $S = X^{(\infty)}$ be the set of all sequences $s = (x_1, x_2, \cdots)$ with each $x_n \in X$, and let $\mathcal{A} = \mathcal{M}^{(\infty)}$. For each n, let $\mathcal{B}_n$ be the field induced by the mapping: $s \to (x_1, \cdots, x_n)$.

Let V be the set of all points $v = (v_1, \cdots, v_k)$ with $v_i \geq 0$ and $\Sigma_1^k v_i = 1$. For any v in V, let $M(\cdot|v)$ be the probability measure on $\mathcal{A}$ such that $M(s: x_1 = a_{i_1}, \cdots, x_n = a_{i_n}|v) = v_{i_1} \cdots v_{i_n}$ for all n and $a_{i_1}, \cdots, a_{i_n}$ in X. Let θ be a probability measure on V and let P_θ be defined by $P_\theta(A) = \int_V M(A|v)\, d\theta$ for $A \in \mathcal{A}$.

EXAMPLE 5.1. Let $\pi = (\pi_1, \cdots, \pi_k)$ be a given point in V with $\pi_i > 0$ for $i = 1, \cdots, k$, let θ_0 be the probability measure degenerate at π, and let Θ be the set of all measures θ such that $P_\theta\{s: x_1 = a_i\} = \pi_i$ for $i = 1, \cdots, k$. Let $\mathcal{P}_0$ consist of the one measure P_{θ_0}, and let $\mathcal{P}_1 = \{P_\theta : \theta \in \Theta\} - \{P_{\theta_0}\}$. In other words, we wish to test independence against exchangeability, the common marginal distribution of the x_n being known.

In the following, for $v = (v_1, \cdots, v_k)$ and $u = (u_1, \cdots, u_k)$ in V let

$$(5.1) \qquad \phi(v, u) = \sum_{i=1}^{k} v_i \log (v_i/u_i),$$

with $0/0 = 1$ (say) and $0 \log 0 = 0$. Then $0 \leq \phi \leq \infty$, and $\phi = 0$ if and only if $v = u$.

For each $s = (x_1, x_2, \cdots)$ let $f_i^{(n)}(s)$ denote the number of $x_j = a_i$ for $1 \leq j \leq n$, $i = 1, \cdots, k$; $n = 1, 2, \cdots$, and let $\xi^{(n)}(s) = (f_1^{(n)}(s)/n, \cdots, f_k^{(n)}(s)/n)$. Let $z(s) = \lim_{n \to \infty} \xi^{(n)}(s)$ if the limit exists and let $z(s) = (1/k, \cdots, 1/k)$ otherwise. Then z is an $\mathcal{A}$ measurable function on S into V. It is known that, for each θ,

$$(5.2) \qquad \lim_{n \to \infty} \xi^{(n)}(s) = z(s) \text{ a.e. } P_\theta,$$

and that $P_\theta(z(s) \in B) = \theta(B)$ for all Borel sets $B \subset V$. Now let

$$(5.3) \qquad T_n(s) = \phi(\xi^{(n)}(s), \pi).$$

We shall show that $\{T_n\}$ is an optimal sequence and that its exact slope is $2\phi(z(s), \pi)$. It is interesting to note that here the optimal slope is a random variable, and that this slope depends on which alternative P_θ obtains only to the extent that the probability distribution of z is then θ.

It follows from the easy part of Sanov's theorem (see [12]) that, for $t \geq 0$,

$$(5.4) \qquad P_{\theta_0}(T_n \geq t) = P_{\theta_0}(\phi(\xi^{(n)}(s), \pi), \geq t) \leq n^k \exp\{-n\alpha(t)\},$$

where $\alpha(t) = \inf\{\phi(v, \pi): v \in V, \ \phi(v, \pi) \geq t\} \geq t$; thus (see Condition 2 of Corollary 5), $1 - G_n(t) \leq n^k \exp\{-nt\}$. Now choose and fix a nonnull θ. A straightforward calculation shows that

$$(5.5) \qquad r_n(s: P_\theta, P_{\theta_0}) = \exp\{nT_n(s)\} \int_V \exp\{-n\phi(\xi^{(n)}(s), v)\}\, d\theta.$$

Since $\phi \geq 0$, it follows from (3.3) and (5.5) that

$$(5.6) \qquad\qquad K_n^{(1)}(s: P_\theta, P_{\theta_0}) \leq T_n(s)$$

for all s and n. Since $\phi(v, \pi)$ is continuous in v, it follows from (5.2) and (5.3) that

$$(5.7) \qquad\qquad \lim_{n \to \infty} T_n(s) = \phi\big(z(s), \pi\big) \text{ a.e. } P_\theta.$$

It follows from the estimate of the null distribution of T_n obtained at the outset of this paragraph that $n^{-1} \log L_n(s) \leq -T_n(s) + kn^{-1} \log n$ for all n and s. It follows hence from (5.6) and (5.7) that

$$(5.8) \qquad \liminf_{n \to \infty} \{n^{-1} \log L_n(s)\} \leq -\phi\big(z(s), \pi\big) \leq -\bar{K}^{(1)}(s: P_\theta, P_{\theta_0}) \text{ a.e. } P_\theta$$

and

$$(5.9) \qquad\qquad \limsup_{n \to \infty} \{n^{-1} \log L_n(s)\} \leq -\phi\big(z(s), \pi\big) \text{ a.e. } P_\theta.$$

It follows from Theorem 2 with $P = P_{\theta_0}$, $Q = P_\theta$, and $\mathscr{C}_n$ the trivial field that

$$(5.10) \qquad \liminf_{n \to \infty} \{n^{-1} \log L_n(s)\} \geq -\bar{K}^{(1)}(s: P_\theta, P_{\theta_0}) \text{ a.e. } P_\theta$$

and

$$(5.11) \qquad \limsup_{n \to \infty} \{n^{-1} \log L_n(s)\} \geq -\underline{K}^{(1)}(s: P_\theta, P_{\theta_0}) \text{ a.e. } P_\theta.$$

It follows from (5.8) to (5.11) first that

$$(5.12) \qquad\qquad \lim_{n \to \infty} K_n^{(1)}(s: P_\theta, P_{\theta_0}) = \phi\big(z(s), \pi\big) \text{ a.e. } P_\theta.$$

and next that

$$(5.13) \qquad\qquad \lim_{n \to \infty} \{n^{-1} \log L_n(s)\} = -\phi\big(z(s), \pi\big) \text{ a.e. } P_\theta.$$

Since θ is arbitrary, it is plain from (5.13) that $\{T_n\}$ has slope 2ϕ and from (5.12) and Corollary 2 with $\mathscr{C}_n$ trivial that $\{T_n\}$ is optimal.

The preceding argument, which is an elaboration of the argument of Corollary 5, could be greatly simplified if we could deduce (5.12) directly from (5.5) and (5.7), but this direct deduction seems difficult for arbitrary θ.

It may be noted that the T_n discussed above is not quite the statistic $\hat{T}_n$ defined by (4.15). It can be shown that in the present case $\{\hat{T}_n\}$ is also an optimal sequence, but we are unable to compute $\hat{T}_n$ explicitly.

EXAMPLE 5.2.   In the same framework as that of the preceding example, let Θ_0 be the set of all θ_0 which are degenerate at some point in the interior of V, and let Θ be the set of all measures on V such that $P_\theta\{s: x_1 = a_i\} > 0$ for $i = 1, \cdots, k$. Let $\mathscr{P}_0 = \{P_{\theta_0}: \theta_0 \in \Theta_0\}$ and $\mathscr{P}_1 = \{P_\theta: \theta \in \Theta\} - \mathscr{P}_0$. In other words, we wish to test independence versus exchangeability, the common marginal distribution of the x_n being unknown.

For each n let T_n be any $\mathscr{B}_n$ measurable statistic, let $\mathscr{C}_n$ be any field $\subset \mathscr{B}_n$, and let L_n^* be the level attained by $T_n | \mathscr{C}_n$, $n = 1, 2, \cdots$. We shall show that, for each nonnull P_θ,

$$(5.14) \qquad \lim_{n \to \infty} n^{-1} \log L_n^*(s) = 0 \text{ a.e. } P_\theta.$$

Thus there exists no test sequence for which the level attained goes to zero exponentially fast. It would be interesting to know whether there exists a sequence for which $\lim_{n \to \infty} L_n^*(s) = 0$ a.e. P_θ, for all nonnull P_θ, and if so, to determine the fastest possible rate of convergence.

To establish (5.14), let $\{v^{(1)}, v^{(2)}, \cdots\}$ be a countable everywhere dense subset of V with each $v^{(j)}$ in the interior of V. Let θ_j be the probability measure degenerate at $v^{(j)}$. Then $\{\theta_1, \theta_2, \cdots\} \subset \Theta_0$. For each n and j, let $L_n(s; P_{\theta_j})$ be the level attained by $T_n | \mathscr{C}_n$ in testing P_{θ_j}. Then, by (4.3), $L_n^*(s) \geqq L_n(s; P_{\theta_j})$ for all s, n, and j. Now choose and fix a nonnull P_θ. Since the proof of (5.12) involves no conditions on θ (except perhaps the mutual absolute continuity of P_θ and P_{θ_0} on $\mathscr{B}_n$ for $n = 1, 2, \cdots$), (5.12) holds with $\theta_0 = \theta_j$ and $\pi = v^{(j)}$. It now follows from Theorem 2 with $P = P_{\theta_j}$, $Q = P_\theta$, and trivial $\mathscr{C}_n$ that

$$(5.15) \qquad \liminf_{n \to \infty} \{n^{-1} \log L_n(s; P_{\theta_j})\} \geqq -\phi(z(s), v^{(j)}) \text{ a.e. } P_\theta,$$

for each j. Hence $\liminf_{n \to \infty} \{n^{-1} \log L_n^*(s)\} \geqq -\phi(z(s), v^{(j)})$ for all j a.e. P_θ. Since $0 = \phi(v, v) = \inf \{\phi(v, v^{(j)}) : j = 1, 2, \cdots\}$ for each $v \in V$, we conclude that $\liminf_{n \to \infty} n^{-1} \log L_n^*(s) \geqq 0$ a.e. P_θ. Since $L_n^* \leqq 1$, it follows that (5.14) holds.

6. Examples concerning Markov chains

Let $X = \{a_1, \cdots, a_k\}$ be a finite set, and let $S = X^{(\infty)}$ and $\mathscr{A}$ be defined as in the preceding section. It is assumed now that $s = (x_1, x_2, \cdots)$ is a Markov chain. Let $\mathscr{B}_n$ be the field induced by the mapping: $s \to (x_1, \cdots, x_{n+1})$, $n = 1, 2, \cdots$. In order to effect certain simplifications we shall assume that with probability one the sequence s starts off with $x_1 = $ a given point of X, say a_1. In effect, then, we shall be considering conditional tests given x_1, but this conditioning will not require explicit attention.

EXAMPLE 6.1. Let $\theta = \{\theta_{i,j}\}$ denote a $k \times k$ matrix with $\theta_{i,j} > 0$ and $\Sigma_{j=1}^k \theta_{i,j} = 1$ for $i = 1, \cdots, k$. Let P_θ be the measure on $\mathscr{A}$ determined by $P_\theta\{s : x_1 = a_1\} = 1$ and

$$(6.1) \qquad P_\theta\{s : x_2 = a_{i_2}, \cdots, x_{n+1} = a_{i_{n+1}}\} = \theta_{1, i_2} \theta_{i_2, i_3} \cdots \theta_{i_n, i_{n+1}}$$

for all n and all $a_{i_2}, \cdots, a_{i_{n+1}}$ in X. Let Θ be a given set of transition probability matrices θ; let Θ^0 be a given subset of Θ; let $\mathscr{P}_0 = \{P_\theta : \theta \in \Theta^0\}$, and let $\mathscr{P}_1 = \{P_\theta : \theta \in \Theta^1\}$ where $\Theta^1 = \Theta - \Theta^0$.

For each $s = (x_1, x_2, \cdots)$ and n, let $g_{i,j}^{(n)}(s)$ denote the number of indices m with $1 \leqq m \leqq n$ such that $x_m = a_i$ and $x_{m+1} = a_j$, for $i, j = 1, \cdots, k$. The matrix $g^{(n)}(s) = \{g_{i,j}^{(n)}(s)\}$ is called the transition count matrix. Let $f_i^{(n)}(s) = \Sigma_{j=1}^k g_{i,j}^{(n)}(s)$ be the total frequency of a_i in $\{x_1, \cdots, x_n\}$. Then $\Sigma_{i=1}^k f_i^{(n)}(s) = n$. Define $\gamma_{i,j}^{(n)}(s) = g_{i,j}^{(n)}(s)/f_i^{(n)}(s)$ if $f_i^{(n)}(s) > 0$ and $\gamma_{i,j}^{(n)}(s) = 1/k$ (say) otherwise, for $i, j = 1, \cdots, k$. Let $\gamma^{(n)}(s) = \{\gamma_{i,j}^{(n)}(s)\}$. The matrix $\gamma^{(n)}$ is an estimate of θ; in fact, it is a maximum likelihood estimate based on $(x_1, \cdots, x_{n+1})$ if θ is entirely unknown. Let $\xi^{(n)}(s) = (f_1^{(n)}(s), \cdots, f_k^{(n)}(s))/n$. It is known that, for each θ,

$$(6.2) \qquad \lim_{n \to \infty} \gamma^{(n)}(s) = \theta \text{ a.e. } P_\theta,$$

and that

$$(6.3) \qquad \lim_{n \to \infty} \xi^{(n)}(s) = \pi(\theta) \text{ a.e. } P_\theta,$$

where $\pi(\theta) = (\pi_1(\theta), \cdots, \pi_k(\theta))$ is the stationary distribution over X corresponding to θ, that is, $\pi(\theta)$ is the unique solution of $\Sigma_{i=1}^k \pi_i \theta_{i,j} = \pi_j, j = 1, \cdots, k$, with $\pi_i > 0$, $\Sigma_{i=1}^k \pi_i = 1$.

In the following let $\gamma_i^{(n)}(s) = (\gamma_{i,1}^{(n)}(s), \cdots, \gamma_{i,k}^{(n)}(s))$ and $\theta_i = (\theta_{i,1}, \cdots, \theta_{i,k})$ denote the ith rows of $\gamma^{(n)}(s)$ and θ, respectively, and let

$$(6.4) \qquad U_n(s; \theta) = \sum_{i=1}^k \xi_i^{(n)}(s) \phi(\gamma_i^{(n)}(s), \theta_i),$$

where ϕ is given by (5.1), and $\xi_i^{(n)} = f_i^{(n)}/n$ is the ith component of $\xi^{(n)}$. Then U_n is a sort of squared distance between $\gamma^{(n)}$ and θ. It is readily seen that, for any θ and θ^0,

$$(6.5) \qquad K_n^{(1)}(s: P_\theta, P_{\theta^0}) = U_n(s; \theta^0) - U_n(s; \theta).$$

It follows from (6.2), (6.3), (6.4), and (6.5) that (4.1) holds, with

$$(6.6) \qquad K^{(1)}(P_\theta, P_{\theta^0}) = \sum_{i=1}^k \pi_i(\theta) \phi(\theta_i, \theta_i^0).$$

Hence

$$(6.7) \qquad J^{(1)}(P_\theta) = \inf \left\{ \sum_{i=1}^k \pi_i(\theta) \phi(\theta_i, \theta_i^0): \theta^0 \in \Theta^0 \right\}.$$

Now let

$$(6.8) \qquad T_n(s) = \inf \{U_n(s: \theta^0): \theta^0 \in \Theta^0\},$$

with U_n defined by (6.4). We shall show, by means of Corollary 5, that $\{T_n\}$ is asymptotically optimal.

To show that Condition 1 of Corollary 5 is satisfied, choose and fix P_θ. Let s be a point such that $\gamma^{(n)}(s) \to \theta$ and $\xi^{(n)}(s) \to (\pi_1(\theta), \cdots, \pi_k(\theta))$. In view of (6.2) and (6.3) it will suffice to show that $\alpha(s) \equiv \lim\inf_{n \to \infty} T_n(s) \geqq J^{(1)}(P_\theta)$ for this s. We may suppose that $\alpha < \infty$. For each n, let $\delta^{(n)}$ be a point in Θ^0 such that $T_n \geqq U_n(s; \delta^{(n)}) - n^{-1}$; such a $\delta^{(n)}$ exists, by (6.8). There exists a sequence $m_1 < m_2 < \cdots$, of positive integers m_r and a probability matrix δ such that $T_{m_r} \to \alpha$ and $\delta^{(m_r)} \to \delta$ as $r \to \infty$. Let n be restricted to $\{m_r\}$. Then

$$(6.9) \qquad \alpha = \lim_{n \to \infty} T_n \geq \liminf_{n \to \infty} \{U_n(s; \delta^{(n)})\} \geq \sum_{i=1}^{k} \pi_i(\theta) \liminf_{n \to \infty} \phi(\gamma_i^{(n)}, \delta_i^{(n)})$$

by (6.4). Since $\pi_i(\theta) > 0$ and $\theta_{i,j} > 0$ for all i and j, since $\alpha < \infty$, and since $\gamma^{(n)} \to \theta$ and $\delta^{(n)} \to \delta$, it follows first that $\delta_{i,j} > 0$ for all i and j and next that $\alpha \geq \Sigma_{i=1}^{k} \pi_i(\theta)\phi(\theta_i, \delta_i)$. This last lower bound cannot be less than $J^{(1)}(P_\theta)$ defined by (6.6), (6.7) since δ is in the closure of Θ^0. Thus $\alpha \geq J^{(1)}(P_\theta)$.

In order to show that Condition 2 of Corollary 5 also holds, let M_n be the set of all $k \times k$ matrices $m = \{m_{i,j}\}$ with $m_{i,j} = 0, 1, 2, \cdots$ and $\Sigma_{i=1}^{k} \Sigma_{i=1}^{k} m_{i,j} = n$. It is readily seen that for any $m \in M_n$ and any θ, $P_\theta(g^{(n)}(s) = m) = v(m) \Pi_{i,j=1}^{k} [\theta_{i,j}]^{m_{i,j}}$, where $v(m)$ is the number of distinct sequences $(x_1, \cdots, x_{n+1})$, possibly zero, with $x_1 = a_1$ and transition count $g^{(n)} = m$. Now, $U_n(s; \theta)$ defined by (6.4) depends on s only through $g^{(n)}(s)$, say $U_n(s; \theta) \equiv U(g^{(n)}(s); \theta)$, and $\gamma^{(n)}(s)$ is also a function of $g^{(n)}(s)$, say $\gamma^{(n)}(s) \equiv G(g^{(n)}(s))$. An easy calculation shows that

$$(6.10) \qquad P_\theta(g^{(n)}(s) = m) = \exp\{-nU(m; \theta)\} P_{G(m)}(g^{(n)}(s) = m)$$
$$\leq \exp\{-nU(m; \theta)\}.$$

Now choose $t \geq 0$ and $\theta^0 \in \Theta^0$, and let $A = A(n; t; \theta^0)$ be the set $\{m : m \in M_n, U(m, \theta^0) \geq t\}$. Then

$$(6.11) \qquad P_{\theta^0}(U_n(s; \theta^0) \geq t) = P_{\theta^0}(U(g^{(n)}(s); \theta^0) \geq t)$$
$$= \sum_A P_{\theta_0}(g^{(n)} = m) \leq \sum_A \exp\{-nU(m; \theta^0)\}$$
$$\leq \exp\{-nt\} \sum_A 1 \leq \exp\{-nt\} \sum_{M_n} 1$$
$$\leq n^{k^2} \exp\{-nt\}.$$

It now follows from (6.8) that $P_{\theta^0}(T_n \geq t) \leq n^{k^2} \exp\{-nt\}$. Since θ^0 is arbitrary, $1 - G_n(t) \leq n^{k^2} \exp\{-nt\}$ for all n and t, and Condition 2 is satisfied.

It is shown in [8] that if Θ^0 consists of a single point θ^0 then T_n is asymptotically optimal even in the sense of [12].

The sequence $\{T_n\}$ considered above does not depend on what the given set Θ is. We now show that, with $T_n(s; \Theta^0)$ defined by the right side of (6.8), and with

$$(6.12) \qquad \hat{T}_n(s) = T_n(s; \Theta^0) - T_n(s; \Theta),$$

$\{\hat{T}_n\}$ is also asymptotically optimal. For any θ, $0 \leq T_n(s; \Theta) \leq U_n(s; \theta)$, and $U_n(s; \theta) \to 0$ a.e. P_θ. Hence, by (6.8) and (6.12), $T_n(s) - \hat{T}_n(s) \to 0$ a.e. P_θ. Secondly, $\hat{T}_n(s) \leq T_n(s)$ for all n and s, by (6.4) and (6.12). Since $\{T_n\}$ satisfies Conditions 1 and 2 of Corollary 5, we see that $\{\hat{T}_n\}$ also satisfies these conditions. In general, that is, for arbitrary Θ^0 and $\Theta^1 = \Theta - \Theta^0$, T_n and $\hat{T}_n$ are quite different. Presumably $\hat{T}_n$ is not only asymptotically optimal but actually better than T_n (see [2], pp. 16–17) for testing Θ^0 against Θ^1 in cases where $T_n \not\equiv \hat{T}_n$.

EXAMPLE 6.2.   Suppose now that Θ is the set of all θ with positive elements, and that Θ^0 is the set of all θ in Θ with identical rows. In other words, we wish to test independence against stationary Markovian dependence, the actual distribution in either case being unspecified.

In the present case, $T_n(s; \Theta) \equiv 0$, so the statistics $T_n(s)$ and $\hat{T}_n(s)$ defined by (6.8) and (6.12) are identical. In order to express $\{\hat{T}_n\}$ and its slope in explicit form we require the following easily verified proposition. Let $v = (v_1, \cdots, v_k)$ be a fixed point in V (that is, $v_i \geq 0$, $\Sigma_{i=1}^k v_i = 1$); let $\theta_1, \cdots, \theta_k$ be fixed points in V, say $\theta_i = (\theta_{i,1}, \cdots, \theta_{i,k})$; and let u be a variable point in V. Then, with the convention that $0 \cdot \infty = 0$, $\Sigma_{i=1}^k v_i \phi(\theta_i, u)$ is minimized by $u = (\Sigma_{i=1}^k v_i \theta_{i,1}, \cdots, \Sigma_{i=1}^k v_i \theta_{i,k})$. Now for each n let $h_i^{(n)}(s) =$ the number of $x_j = a_i$ for $2 \leq j \leq n+1$, and let $\eta^{(n)}(s) = n^{-1}(h_1^{(n)}(s), \cdots, h_k^{(n)}(s))$, that is, $\eta^{(n)}$ is the vector of relative frequency counts in $\{x_2, \cdots, x_{n+1}\}$. It follows from the stated proposition that

$$(6.13) \qquad \hat{T}_n(s) = \sum_{i=1}^k \xi_i^{(n)}(s) \phi(\gamma_i^{(n)}(s), \eta^{(n)}(s)).$$

It also follows that

$$(6.14) \qquad J^{(1)}(P_\theta) = \sum_{i=1}^k \pi_i(\theta) \phi(\theta_i, \pi(\theta)).$$

According to Example 6.1, $\{\hat{T}_n\}$ is asymptotically optimal and has exact slope $2J^{(1)}$.

Now for $n \geq 2$ let $W^{(n)}(s) = (W_1^{(n)}(s), \cdots, W_k^{(n)}(s))$ be the vector of frequency counts in $\{x_2, \cdots, x_n\}$ and let $\mathscr{C}_n$ be the field induced by the mapping: $s \to (x_1; W^{(n)}(s); x_{n+1})$. It is known (see [7]) that the conditional distribution of $(x_1, \cdots, x_{n+1})$ given $\mathscr{C}_n$ is the same for every θ in Θ^0. In particular, $\hat{T}_n | \mathscr{C}_n$ has an exact null distribution function, say $\hat{F}_n(t, s)$. We proceed to show, essentially by an extension of Corollary 5 to conditional tests, that $\{\hat{T}_n | \mathscr{C}_n\}$ is also an optimal sequence. It is plain from (6.13) and (6.14) that $\hat{T}_n(s) \to J^{(1)}(P_\theta)$ a.e. P_θ, for every θ. We shall show that $1 - \hat{F}_n(t, s) \leq \exp\{-nt\} f_n(s)$ for all n, t, and s, where $n^{-1} \log f_n(s) \to 0$ a.e. P_θ, for each θ. It will then follow from Corollary 3, as desired, that with $L_n^*(s) = 1 - \hat{F}_n(\hat{T}_n(s), s)$, $n^{-1} \log L_n^*(s) \to -J^{(1)}(P_\theta)$ a.e. P_θ for every nonnull θ.

Nonnull sets in $\mathscr{C}_n$ are unions of sets of the form

$$(6.15) \qquad \{s: x_1 = a_1, W^{(n)}(s) = b, x_{n+1} = a_j\} = C_n(b; a_j),$$

say, where a_j is a point of X and b is $(b_1, \cdots, b_k)$ with each b_i a nonnegative integer and $\Sigma_{i=1}^k b_i = n - 1$. For $t \geq 0$ let $B_n(t) = \{s: \hat{T}_n(s) \geq t\}$. Let θ^0 be a point in Θ^0. Then $1 - \hat{F}_n(t, s)$ equals $P_{\theta^0}(B_n(t) | C_n(b, a_j))$ evaluated at $b = W^{(n)}(s)$ and $a_j = x_{n+1}$. Now, $P_{\theta^0}(B_n | C_n) \leq P_{\theta^0}(B_n) / P_{\theta^0}(C_n)$. We have seen in Example 6.1 that $P_{\theta^0}(B_n) \leq n^{k^2} \exp\{-nt\}$. If θ^0 has $(\pi_1, \cdots, \pi_k)$ as each row, then

$$(6.16) \qquad P_{\theta^0}(C_n(b; a_j)) = \pi_j \left[\prod_{i=1}^k \left(\frac{\pi_i^{b_i}}{b_i!} \right) \right] (n-1)! = \pi_j \psi_n(b; \theta^0),$$

say. Let $\delta(\theta^0) = \min\{\pi_1, \cdots, \pi_k\}$. Then $\delta > 0$, and $P_{\theta^0}(C_n(b; a_j)) \geq \delta(\theta^0) \psi_n(b; \theta^0)$ for all j. Since θ^0 is arbitrary, we see that $1 - \hat{F}_n(t, s) \leq \exp\{-nt\} f_n(s)$, where

$$(6.17) \qquad f_n(s) = n^{k^2} \inf \{[\delta(\theta^0) \psi_n(W^{(n)}(s); \theta^0)]^{-1} : \theta^0 \in \Theta^0\}.$$

Suppose now that P_θ obtains. Since $W^{(n)}(s) = f^{(n)}(s) - (1, 0, \cdots, 0)$, it follows from (6.3) that $n^{-1} W^{(n)}(s) \to \pi(\theta)$ a.e. P_θ. Let θ^* be the point in Θ^0 which has $\pi(\theta)$ for each row. It follows from Stirling's formula that $n^{-1} \log \psi_n(W^{(n)}(s); \theta^*) \to 0$ a.e. P_θ, as $n \to \infty$. Since $f_n(s) \geq 1$, and since θ^* is a point in Θ^0, it follows from (6.17) that $n^{-1} \log f_n(s) \to 0$ a.e. P_θ.

EXAMPLE 6.3. Let Θ^0 be the same set as in the preceding example, let θ be a given transition probability matrix, $\theta \notin \Theta^0$, and suppose that we wish to test Θ^0 against θ. Then Examples 6.1 and 6.2 already provide three different asymptotically optimal test sequences. Another optimal sequence for the present case is $\{T_n^* | \mathscr{C}_n\}$, where $\mathscr{C}_n$ is the field considered above, and

$$(6.18) \qquad T_n^*(s) = \sum_{i=1}^{k} \xi_i^{(n)}(s) \left[\phi(\gamma_i^{(n)}(s), \pi(\theta)) - \phi(\gamma_i^{(n)}(s), \theta_i) \right].$$

That $\{T_n^* | \mathscr{C}_n\}$ has slope $2J^{(1)}(P_\theta)$ against θ may be seen from Corollary 4, as follows. Since T_n^* is $\mathscr{B}_n$ measurable, $T_n^* | \mathscr{C}_n$ has an exact null distribution. Let θ^* be the matrix with $\pi(\theta)$ as each row. Then $K^{(1)}(P_\theta, P_{\theta^*}) = J^{(1)}(P_\theta)$, and it follows from (6.4), (6.5), and (6.18) that $T_n^*(s) \equiv n^{-1} \log r_n(s: P_\theta, P_{\theta^*})$. It remains therefore to verify that condition (b) of Corollary 4 is satisfied with $Q = P_\theta$ and $P_Q = P_{\theta^*}$. This verification can be made by a direct calculation, but is immediately available from Corollary 6 since $\{\hat{T}_n | \mathscr{C}_n\}$ is an optimal sequence.

7. Examples concerning independent and identically distributed observations

In this section X is a Borel set of a Euclidean space of points x, $\mathscr{M}$ is the field of Borel sets of X, $S = X^{(\infty)}$ is the space of points $s = (x_1, x_2, \cdots)$, and $\mathscr{A} = \mathscr{M}^{(\infty)}$. For each n, $\mathscr{B}_n$ is the σ-field induced by the mapping: $s \to (x_1, \cdots, x_n)$. The set Θ is an index set of points θ, and Θ_0 is a subset of Θ. For each θ in Θ, p_θ is a probability measure on $\mathscr{M}$, and $P_\theta = P_\theta^{(\infty)}$. The null set of measures is $\mathscr{P}_0 = \{P_\theta : \theta \in \Theta_0\}$; the nonnull set is $\mathscr{P}_1 = \{P_\theta : \theta \in \Theta\} - \mathscr{P}_0$. It is assumed that, for any θ and θ_0 in Θ, p_θ and p_{θ_0} are mutually absolutely continuous on $\mathscr{M}$. Consequently, P_θ and P_{θ_0} are mutually absolutely continuous on $\mathscr{B}_n$ for each n, and (4.1) is satisfied with

$$(7.1) \qquad K^{(1)}(P_\theta, P_{\theta_0}) = \int_X \log (dp_\theta/dp_{\theta_0}) \, dp_\theta.$$

In accordance with the notation of [2], [3] we shall write the integral in (7.1) as $K(\theta, \theta_0)$, and write $J^{(1)}(P_\theta)$ defined by (4.2) as $J(\theta)$.

Let $y(x)$ be a Borel measurable transformation of X into a Euclidean space Y and for each n let $\mathscr{C}_n$ be the σ-field induced by: $s \to (y(x_1), \cdots, y(x_n))$. It is then readily seen that $K_n^{(2)}(s: P_\theta, P_{\theta_0}) \to K^*(\theta, \theta_0)$ a.e. P_θ, where

$$(7.2) \qquad K^* = \int_Y \log (dp_\theta y^{-1}/dp_{\theta_0} y^{-1}) \, dp_\theta y^{-1}.$$

With the present choice of $\mathscr{C}_n$ Corollary 1 reduces to the statement that $0 \leqq K^*(\theta, \theta_0) \leqq K(\theta, \theta_0) \leqq \infty$. This choice of $\mathscr{C}_n$ is not used elsewhere in this section.

In the following examples we give the exact slopes and related large deviation probability estimates for various likelihood ratio statistics. Most of these results have been obtained previously by other methods; our main object in reconsidering these examples here is to point out that the first part of Corollary 2 offers a simple method for all such examples. This method does not require explicit estimation of large deviation probabilities in the null case; indeed the estimates referred to can be obtained, if needed, after the exact slope is found.

EXAMPLE 7.1. Suppose X is the real line and p_0 is a nondegenerate probability measure on X. Suppose that the moment generating function $\phi(\theta) = \int_X \exp\{\theta x\} dp_0$ is finite for $\theta \in \Theta = [0, \delta)$ where $0 < \delta \leqq \infty$. For each θ in Θ let p_θ be defined by $dp_\theta = [\phi(\theta)]^{-1} \exp\{\theta x\} dp_0$ and let Θ_0 consist of the single point $\theta_0 = 0$. For $0 < \theta < \delta$ let $b(\theta) = E_\theta(x)$. Then $-\infty < b(\theta) < \infty$, and $K(\theta, \theta_0) = \theta b(\theta) - \log \phi(\theta) = J(\theta)$. For each n, let $T_n(s) = (x_1 + \cdots + x_n)/n$. Consider a particular $\theta > 0$. Then $T_n(s)$ is equivalent to $r_n(s: P_\theta, P_{\theta_0})$. Since (4.1) holds with $K^{(1)} = J$, it follows from Corollary 2 that $\{T_n\}$ has exact slope $2J(\theta)$ when θ obtains.

We observe next that $T_n(s) \to b(\theta)$ a.e. P_θ. Since b and J are continuous and strictly increasing over $(0, \delta)$ it follows that for given $t \in \{b(\theta): 0 < \theta < \delta\}$ there exists a unique θ_t such that $b(\theta_t) = t$, and f and g of (4.13) are both equal to $J(\theta_t) = \theta_t t - \log \phi(\theta_t)$. It follows hence from the conclusion of the preceding paragraph that, for each $t \in \{b(\theta): 0 < \theta < \delta\}$,

$$(7.3) \qquad n^{-1} \log P_0(x_1 + \cdots + x_n \geqq nt) \to -J(\theta_t)$$

as $n \to \infty$. It is thus seen that Chernoff's theorem [9] is deducible from Corollary 2 and the law of large numbers. A different proof, based on the central limit theorem, is given in [5].

EXAMPLE 7.2. Let $x = (y, z)$ where y and z are zero-one variables. Each possible value of x has positive probability but the distribution of x is otherwise entirely unknown. The null hypothesis is that y and z are independent. For each n let $\hat{T}_n$ be the likelihood ratio statistic (4.15). It follows from Theorem 2 of [2] that $\{\hat{T}_n\}$ is optimal. Write $x_n = (y_n, z_n)$, and let $\mathscr{C}_n$ be the field induced by $U_n(s) = (\sum_{i=1}^n y_i, \sum_{i=1}^n z_i)$. Then $\{\hat{T}_n | \mathscr{C}_n\}$ is also optimal. The level attained by $\hat{T}_n | \mathscr{C}_n$ equals the level attained by $T_n | \mathscr{C}_n$, where $T_n = |\sum_{i=1}^n y_i z_i - M_n(s)|$ and M_n is a complicated function of U_n and $\Sigma_1^n y_i z_i$;

$$(7.4) \qquad M_n \doteq n^{-1} \left(\sum_{i=1}^n y_i \right) \left(\sum_{i=1}^n z_i \right).$$

We omit the verification.

EXAMPLE 7.3. Suppose X is the k dimensional Euclidean space of points $x = (y_1, \cdots, y_k)$, and that Θ is the space of all points $\theta = (\mu_1, \cdots, \mu_k; \sigma)$ with $-\infty < \mu_i < \infty$ for each i and $0 < \sigma < \infty$. Suppose that when θ obtains

$y_1, \cdots, y_k$ are independent normally distributed variables with $E_\theta(y_i) = \mu_i$ and $\mathrm{Var}_\theta(y_i) = \sigma^2$ for $i = 1, \cdots, k$. The null hypothesis is that $\mu_1 = \cdots = \mu_k = 0$. It is readily seen that in this case

$$(7.5) \qquad J(\theta) = \tfrac{1}{2}k \log\left[1 + k^{-1}\delta^2(\theta)\right], \qquad \delta^2(\theta) = \sum_{i=1}^{k} \left(\frac{\mu_i}{\sigma}\right)^2.$$

Now for each $n \geq 2$, let $T_n(s)$ be n^{-1} times the appropriate F statistic based on $(x_1, \cdots, x_n)$. It has been shown by Abrahamson [1] that $\{T_n\}$ has exact slope $2J(\theta)$ against every θ and so is asymptotically optimal. The method used in [1] is to note that

$$(7.6) \qquad T_n(s) \to k^{-1}\delta^2(\theta) \text{ a.e. } P_\theta,$$

and then to show that, with $G_n(t)$ the null distribution function of T_n,

$$(7.7) \qquad n^{-1} \log\left[1 - G_n(t)\right] \to -\tfrac{1}{2}k \log\left[1 + t\right],$$

for each $t > 0$. Since the limit in (7.7) is continuous in t, it follows (see [3], pp. 309–310) from (7.6) that $\{T_n\}$ has slope $2J$ defined by (7.5). A second possible method of establishing Abrahamson's result is by means of Theorem 2 of [2] since T_n is equivalent to $\hat{T}_n$, but the verifications required seem formidable. We now show that Corollary 2 can be used to obtain first the slope of $\{T_n\}$ and then (7.7).

Choose and fix a nonnull $\theta = (\mu_1, \cdots, \mu_k; \sigma)$. Let $\theta^0 = (0, \cdots, 0; \sigma)$. Then $J(\theta) = K(\theta, \theta^0)$. Let $f_n(t)$ and $g_n(t)$ be the probability densities of $T_n(s)$ under θ and θ^0 respectively, and let $h_n(t) = f_n(t)/g_n(t)$ for $t > 0$. Let $\mathscr{C}_n$ be the σ-field induced by $T_n(s)$. Then

$$(7.8) \qquad \rho_n(s : P_\theta, P_{\theta^0}) = h_n\big(T_n(s)\big).$$

It follows from known results (see [13], p. 312) that

$$(7.9) \qquad h_n(t) = \sum_{j=0}^{\infty} \gamma_j(n)\pi_j(\tfrac{1}{2}n\delta^2)\left[\frac{nt}{n(1+t)-1}\right]^j$$

where

$$(7.10) \qquad \gamma_j(n) = \frac{\Gamma(\tfrac{1}{2}nk + j)\Gamma(\tfrac{1}{2}k)}{\Gamma(\tfrac{1}{2}nk)\Gamma(\tfrac{1}{2}k + j)},$$

and $\pi_j(\lambda)$ denotes the Poisson probability $\exp\{-\lambda\} \lambda^j/j!$. Since h_n is a strictly increasing function of t, we see from (7.8) that T_n and ρ_n are equivalent statistics. Consequently, $\{T_n\}$ has exact slope $2J(\theta)$ against θ if $\{\rho_n\}$ has exact slope $2J(\theta)$. We shall show that this last is the case by showing that, with $K_n^{(2)} = n^{-1} \log \rho_n$,

$$(7.11) \qquad K_n^{(2)}(s) \to J(\theta) \text{ a.e. } P_\theta,$$

for then Corollary 2 (with $\mathscr{B}_n$ and $\mathscr{C}_n$ of the corollary replaced by the present $\mathscr{C}_n$ and the trivial field, respectively) applies.

For each n let j_n be the positive integer such that $\frac{1}{2}n\delta^2 < j_n \leq \frac{1}{2}n\delta^2 + 1$. Since each term in the series in (7.9) is positive,

$$(7.12) \qquad h_n(t) \geq \gamma_{j_n}(n)\pi_{j_n}(\tfrac{1}{2}n\delta^2)[nt/(n(1+t)-1)]^{j_n}.$$

It now follows from (7.10) by an application of Stirling's formula that, for each t,

$$(7.13) \qquad \liminf_{n\to\infty} n^{-1} \log h_n(t) \geq \tfrac{1}{2}k \log (1 + k^{-1}\delta^2)$$

$$+ \tfrac{1}{2}\delta^2 \log \left[\frac{kt(1 + k^{-1}\delta^2)}{\delta^2(1 + t)} \right].$$

Since the left side of (7.13) is nondecreasing in t and the right side is continuous in t, it follows from (7.6), (7.8) and (7.13) that $\underline{K}^{(2)}(s) \geq \tfrac{1}{2}k \log (1 + k^{-1}\delta^2)$ a.e. P_θ. But $\tfrac{1}{2}k \log (1 + k^{-1}\delta^2) = J(\theta) = K(\theta, \theta^0) = K^{(1)}(P_\theta, P_{\theta^0})$. It therefore follows from Corollary 1, as desired, that (7.11) holds. Since $\{T_n\}$ is shown to have exact slope $2J(\theta)$ against each θ, it follows from (7.5) and (7.6) that (7.7) holds for each $t > 0$. Incidentally, it follows from (7.5), (7.6), (7.8), and (7.11) that $n^{-1} \log h_n(t) \to \tfrac{1}{2}k \log (1 + t)$ for each $t > 0$ and $\delta^2 > 0$.

EXAMPLE 7.4.   Let (a, b) be an interval on the real line, $-\infty \leq a < b \leq \infty$, and let $\mathscr{F}$ denote the set of all probability distribution functions F on the real line such that F assigns probability 1 to (a, b) and $F'(t)$ exists and is continuous and positive over (a, b). Let $\theta = (F, G)$ be a point of $\mathscr{F} \times \mathscr{F}$. Let X be the set of all points $x = (y, z)$ with real y and z and let p_θ be the measure on X corresponding to y and z being independent random variables with distribution functions F and G, respectively. Let $\Theta_0 = \{(H, H): H \in \mathscr{F}\}$ be the null set. The nonnull set is a single point $\theta = (F, G)$ with $F \not\equiv G$.

It is readily seen that, with $\theta_0 = (H, H)$ and $K(\theta, \theta_0)$ defined by the right side of (7.1),

$$(7.14) \qquad K(\theta, \theta_0) = \int_{-\infty}^{+\infty} \log (dF/dH)\, dF + \int_{-\infty}^{+\infty} \log (dG/dH)\, dG.$$

It follows easily from (7.14) that

$$(7.15) \qquad J(\theta) = K(\theta, \theta^0),$$

where $\theta^0 = (\bar{H}, \bar{H})$ and

$$(7.16) \qquad \bar{H}(t) = \tfrac{1}{2}[F(t) + G(t)].$$

For each n let $x_n = (y_n, z_n)$, let N denote $2n$, and let $u_{1,N}(s) \leq \cdots \leq u_{N,N}(s)$ be the ordered values in the set $\{y_1, z_1; \cdots; y_n, z_n\}$. Let $v_{i,N}(s) = 1$ if $u_{i,N}(s) = z_j$ for some $j = 1, \cdots, n$ and $v_{i,N}(s) = 0$ otherwise, and let $V_n(s) = (v_{1,N}(s), \cdots, v_{N,N}(s))$.

It has been shown by Hájek [10] that the rank statistic vector V_n is asymptotically fully informative in the following sense: there exist weights $\{a_{i,N}: i = 1, \cdots, N: N = 2, 4, \cdots\}$ such that, with $T_n(s) = \Sigma_{i=1}^N a_{i,N}v_{i,N}(s)$, $\{T_n\}$ has exact slope $2J(\theta)$ against the given θ; the weights depend of course on θ. As noted in

[10], this remarkable result implies that the likelihood ratio statistic for testing Θ_0 against θ based on V_n also has slope $2J(\theta)$.

Let $\mathscr{C}_n$ be the field induced by V_n. Then the likelihood ratio statistic based on V_n is $\rho_n(s) = \rho_n(s : P_\theta, P_{\theta^0})$. We have

$$(7.17) \qquad \rho_n(s) = \binom{N}{n} \{P_\theta(V_n(s) = b_n)\}_{b_n = V_n(s)},$$

where b_n denotes an N long vector of zeros and ones. It follows from Corollary 2 (with $\mathscr{B}_n$ and $\mathscr{C}_n$ of the corollary replaced with the present $\mathscr{C}_n$ and the trivial field, respectively) that the second part of Hájek's result is equivalent to

$$(7.18) \qquad \lim_{n \to \infty} K_n^{(2)}(s) = J(\theta) \text{ a.e. } P_\theta.$$

where $K_n^{(2)} = n^{-1} \log \rho_n$. That (7.18) does hold can be seen from [6]. It is shown in [6] that $K_n^{(2)}(s) \to I^*$ a.e. P_θ where I^* is a constant, and it follows from the formulae for I^* given in [6], [15] that in fact $I^*(\theta) = \inf \{K(\theta, \theta_0) : \theta_0 \in \Theta_0\} = J(\theta)$; we omit the details.

Now let $r_n(s) = r_n(s : P_\theta, P_{\theta^0})$ be the density of P_θ with respect to P_{θ^0} on $\mathscr{B}_n$. Let $\mathscr{D}_n$ be the σ-field induced by $U_n(s) = (u_{1,N}(s), \cdots, u_{N,N}(s))$, where the $u_{i,N}$ are the order statistics as above. We shall show that $\{r_n | \mathscr{D}_n\}$ is also an optimal sequence. For each null θ_0, the conditional distribution of $(y_1, z_1; \cdots; y_n, z_n)$ given $U_n = (a_1, \cdots, a_N)$, with $a_1 < \cdots < a_N$, is uniform over the permutations of $(a_1, \cdots, a_N)$; hence $r_n | \mathscr{D}_n$ has an exact null distribution. It follows from (7.15) that conditions (a) and (c) of Corollary 4 (with $\mathscr{C}_n$ of the corollary replaced by $\mathscr{D}_n$) are satisfied with $Q = P_\theta$ and $P_Q = P_{\theta^0}$. It therefore remains to show that condition (b) is satisfied by $\{\mathscr{D}_n\}$. Let $\xi_n(s)$ be a $\mathscr{D}_n$ measurable function such that $dP_\theta = \xi_n(s) \, dP_{\theta^0}$ on $\mathscr{D}_n$, $0 < \xi_n < \infty$, and let $K_n^{(3)}(s) = n^{-1} \log \xi_n(s)$. We have to show that

$$(7.19) \qquad K_n^{(3)}(s) \to 0 \text{ a.e. } P_\theta.$$

It seems difficult to establish (7.19) directly, but one can argue as follows. In the null case, U_n and V_n are independent random vectors. Since ρ_n is a function of V_n, it follows that the level attained by ρ_n is a version of the level attained by $\rho_n | \mathscr{D}_n$. Consequently $\{\rho_n | \mathscr{D}_n\}$ is an asymptotically optimal sequence. Since $J < \log 4$, it follows from Corollary 6 with $P = P_{\theta^0}$ and $Q = P_\theta$ (with $\mathscr{C}_n$ of the corollary replaced by $\mathscr{D}_n$) that (7.19) holds.

Let $f = F'$, $g = G'$, and $h = \frac{1}{2}(f + g)$. Then

$$(7.20) \qquad r_n(s) = \prod_{i=1}^{n} f(y_i) g(z_i) \bigg/ \prod_{i=1}^{n} h(y_i) h(z_i).$$

It follows from this formula, as is well known, that if f is a normal density and $g(t) = f(t - c)$ with $c > 0$, then the level attained by $r_n | \mathscr{D}_n$ equals the level attained by the Fisher-Pitman test $\sum_1^n z_i - \sum_1^n y_i | \mathscr{D}_n$. Since the latter test does not depend on $\theta = (F, G)$, we conclude that the Fisher-Pitman test is asymptotically optimal against all one sided normal translation alternatives.

REFERENCES

[1] I. G. ABRAHAMSON, "On the stochastic comparison of tests of hypotheses," Ph.D. dissertation, University of Chicago, 1965.

[2] R. R. BAHADUR, "An optimal property of the likelihood ratio statistics," *Proceedings of the Fifth Berkeley Symposium on Mathematical Statistics and Probability*, Berkeley and Los Angeles, University of California Press, 1967, Vol. 1, pp. 13–26.

[3] ———, "Rates of convergence of estimates and test statistics," *Ann. Math. Statist.*, Vol. 38 (1967), pp. 303–324.

[4] R. R. BAHADUR and P. J. BICKEL, "On conditional test levels in large samples," *Essays in Prob. Statist.*, University of North Carolina Monograph Series No. 3 (1970), pp. 25–34.

[5] R. R. BAHADUR and R. RANGA RAO, "On deviations of the sample mean," *Ann. Math. Statist.*, Vol. 31 (1960), pp. 1015–1027.

[6] R. H. BERK and I. R. SAVAGE, "The information in rank-order and the stopping time of some associated SPRT's," *Ann. Math. Statist.*, Vol. 39 (1968), pp. 1661–1674.

[7] P. BILLINGSLEY, "Statistical methods in Markov chains," *Ann. Math. Statist.*, Vol. 32 (1961), pp. 12–40.

[8] L. B. BOZA, "Asymptotically optimal tests for finite Markov chains," Ph.D. thesis, University of California, Berkeley, 1970.

[9] H. CHERNOFF, "A measure of asymptotic efficiency for tests of a hypothesis based on the sum of observations," *Ann. Math. Statist.*, Vol. 23 (1952), pp. 493–507.

[10] J. HÁJEK, "Asymptotic sufficiency of the vector of ranks in the Bahadur sense," *Ann. Math. Statist.*, Vol. 42 (1971).

[11] J. L. HODGES, JR. and E. L. LEHMANN, "The efficiency of some nonparametric competitors of the *t*-test," *Ann. Math. Statist.*, Vol. 27 (1956), pp. 324–335.

[12] W. HOEFFDING, "Asymptotically optimal tests for multinomial distributions," *Ann. Math. Statist.*, Vol. 36 (1965), pp. 369–408.

[13] E. L. LEHMANN, *Testing Statistical Hypotheses*, New York, Wiley, 1959.

[14] M. RAGHAVACHARI, "On a theorem of Bahadur on the rate of convergence of test statistics," *Ann. Math. Statist.*, Vol. 41 (1970), pp. 1695–1699.

[15] I. R. SAVAGE, "Nonparametric statistics: a personal view," *Sankhyā, Ser. A*, Vol. 31 (1969), pp. 107–144.

ON THE NORMAL APPROXIMATION FOR A CERTAIN CLASS OF STATISTICS

D. M. CHIBISOV

STEKLOV MATHEMATICAL INSTITUTE

1. Summary and introduction

We consider the class $C(\alpha)$ of asymptotic tests proposed by Neyman [6]. The term of order $n^{-1/2}$ in the normal approximation for the distributions of the test statistics is obtained. Moreover several theorems on conditional distributions are proved. They are used in deriving the main result but they also seem to be of independent interest.

Neyman [6] proposed a class of asymptotic tests called $C(\alpha)$ for the following statistical problem. Let a random variable (r.v.) X have a distribution depending on parameters $\theta = (\theta_1, \cdots, \theta_s)$ and ξ which take their values in open sets $\Theta \subset R^s$ and $\Xi \subset R^1$ respectively. (We denote by R^s, $s = 1, 2, \cdots$, the space of real row vectors $x = (x_1, \cdots, x_s)$ with the Euclidean norm $\|x\| = (xx')^{1/2}$, a prime denoting the transposition.) The hypothesis $H : \xi = \xi_0$, where $\xi_0 \in \Xi$ is a specified value, is to be tested on the basis of n independent observations $X_1, \cdots, X_n$ of the r.v. (In the sequel, we put $\xi_0 = 0$.) The distribution of X is assumed to have a density $f(x; \theta, \xi)$ (with respect to an appropriate measure) which satisfies certain regularity conditions. The $C(\alpha)$ tests are constructed as follows. Let a function $g(x, \theta)$ be such that

$$(1.1) \qquad E_{\theta, 0}\, g(X, \theta) \equiv 0, \qquad E_{\theta, 0}\, g^2(X, \theta) = \sigma^2 < \infty, \qquad \theta \in \Theta.$$

(The first assumption can always be satisfied by considering $g(x, \theta) - E_{\theta, 0}\, g(X, \theta)$ instead of $g(x, \theta)$.) Form the function

$$(1.2) \qquad Z_n(\theta) = \frac{1}{\sigma(\theta)\sqrt{n}} \sum_{i=1}^{n} g(X_i, \theta)$$

and let $\hat{\theta}_n$ be a locally root n consistent estimator of θ (which means that $\sqrt{n}(\hat{\theta}_n - \theta_0)$ is bounded in probability where θ_0 is the true value of θ; for a precise definition see [6]). It was shown in [6] that $Z_n(\hat{\theta}_n)$ is asymptotically $(0, 1)$ normally distributed whatever be the true value of θ if and only if $g(x, \theta)$ is orthogonal to the logarithmic derivatives

$$(1.3) \qquad h_j(x, \theta) = \frac{\partial}{\partial \theta_j} \log f(x; \theta, 0), \qquad j = 1, \cdots, s,$$

153

in the sense that

(1.4) $$E_{\theta, 0}\, g(X, \theta) h_j(X, \theta) \equiv 0, \qquad \theta \in \Theta, \quad j = 1, \cdots, s.$$

This implies that the test with the critical region $\{|Z_n(\hat\theta_n)| > z_{\alpha/2}\}$, where $z_{\alpha/2}$ is defined by $\mathcal{N}(z_{\alpha/2}) = 1 - \alpha/2$, $\mathcal{N}(z)$ being the $(0, 1)$ normal distribution function (d.f.), has the limiting significance level α whatever be the true value of θ. The class of tests of this form was called $C(\alpha)$.

A further result of [6] gives a rule for constructing an asymptotically optimal test of class $C(\alpha)$. Namely, let

(1.5) $$h_0(x, \theta) = \frac{\partial}{\partial \xi} \log f(x; \theta, \xi)\big|_{\xi = 0}$$

and let $g(x, \theta)$ be obtained from $h_0(x, \theta)$ by the orthogonalization process,

(1.6) $$g(x, \theta) = h_0(x, \theta) - \sum_1^s a_j(\theta) h_j(x, \theta)$$

to satisfy (1.4). Then the test of class $C(\alpha)$ with this $g(x, \theta)$ is an asymptotically optimal one.

Several examples of the use of $C(\alpha)$ tests in applied problems are given in [7].

When applying an asymptotic test one always encounters the question of the accuracy of the normal approximation. The standard methods related to the sums of independent random variables are inapplicable for the $C(\alpha)$ test statistics since the use of an estimate instead of θ in (1.2) makes the terms of the sum dependent. In the present paper the correction term of order $n^{-1/2}$ to the normal approximation under the hypothesis H is obtained when $g(X, \theta)$ and some related random variables have densities with respect to the Lebesgue measure. This result is contained in Theorem 2.1 stated in Section 2.

In Section 3, we give two theorems on conditional distributions. Namely, the normalized sum of independent random vectors is considered. Each of the vectors consists of two subvectors and hence so does their sum. The theorems concern the conditional distribution of the first subvector of the sum given the second one. The only paper which we know to deal with the conditional distributions of this kind is that of Steck [8]. In this paper some theorems on the convergence of conditional distributions to the normal have been proved.

In the theorems of Section 3 we restrict ourselves to the case of identically distributed summands with a one dimensional conditioning subvector which is sufficient for the proof of Theorem 2.1. Theorem 3.1 establishes a Lipschitz property for the dependence in variation of the conditional distribution on the value of the conditioning variable.

Theorem 3.2 gives an asymptotic estimate for tail probabilities of conditional distributions.

The proofs of theorems of Sections 2 and 3 are given in Sections 4 to 6. Section 7 contains some concluding remarks.

2. The main theorem

Let the hypothesis H be true and $\theta_0 \in \Theta$ be the true value of θ. Thus we have n independent random variables $X_1, \cdots, X_n$ with a common d.f., $F(x) = F(x; \theta_0, 0)$. (Since the values $\theta = \theta_0$ and $\xi = \xi_0$ will be fixed we shall omit them in our notations.) Denote

$$g(x) = g(x, \theta_0), \qquad g_j(x) = \frac{\partial g(x, \theta)}{\partial \theta_j}\bigg|_{\theta = \theta_0},$$

(2.1)
$$g_{i,j,\ell}(x, \theta) = \frac{\partial^2 g(x, \theta)}{\partial \theta_i \, \partial \theta_j}, \qquad g_{i,j}(x) = g_{i,j}(x, \theta_0),$$

$$g_{i,j,\ell}(x, \theta) = \frac{\partial^3 g(x, \theta)}{\partial \theta_i \, \partial \theta_j \, \partial \theta_\ell}, \qquad\qquad i, j, \ell = 1, 2, \cdots, s.$$

We shall state now the assumptions to be used in the theorem.

ASSUMPTION 1. $Eg(X) = 0, \qquad Eg^2(X) = 1, \qquad E|g(X)|^3 < \infty$.

ASSUMPTION 2. $Eg_j(X) = 0$ for $j = 1, 2, \cdots, s$.

REMARK. The variance of $g(X)$ may be taken equal to 1 by considering $g(x, \theta)/\sigma(\theta)$ instead of $g(x, \theta)$. Under certain regularity conditions Assumption 2 is equivalent to (1.4) in view of the equation

(2.2)
$$0 = \frac{\partial}{\partial \theta_j} \int g(x, \theta) f(x, \theta) \, dx = \int \frac{\partial g}{\partial \theta_j} \, dF + \int g \frac{\partial \log f}{\partial \theta_j} \, dF.$$

We prefer to use Assumption 2 directly because it does not refer to the dependence of the distribution on θ.

ASSUMPTION 3. $E|g_j(X)|^3 < \infty, \qquad j = 1, \cdots, s$.

ASSUMPTION 4. $E|g_{i,j}(X)|^{3/2+\delta} < \infty$ for some $\delta > 0$; $i, j = 1, \cdots, s$.

ASSUMPTION 5. There exist a neighborhood $U \subset \Theta$ of θ_0 and a function $K(x)$ such that $|g_{i,j,\ell}(x, \theta)| \leq K(x)$ for all $\theta \in U, i, j, \ell = 1, \cdots, s$; $E(K(x))^{1+\delta} < \infty$ for some $\delta > 0$.

For the convenience of notation, we introduce the following symbol.

DEFINITION 2.1. Let $\zeta_1, \zeta_2, \cdots$ be a sequence of random variables. We shall write $\zeta_n = \omega(a)$ if for any $c > 0$, $P\{|\zeta_n| > cn^a\} = o(n^{-1/2})$ as $n \to \infty$.

Now we assume that the estimator $\hat{\theta}_n$ is expressible in the form

(2.3)
$$\sqrt{n}\,(\hat{\theta}_n - \theta_0) = \frac{1}{\sqrt{n}} \sum_1^n h(X_i) + \eta_n$$

where $h(x) = (h_1(x), \cdots, h_s(x)), \eta_n = (\eta_{n,1}, \cdots, \eta_{n,s})$.

ASSUMPTION 6. $Eh_j(X) = 0, \qquad E|h_j(X)|^3 < \infty, \qquad j = 1, \cdots, s$.

ASSUMPTION 7. $\eta_{n,j} = \omega(-\delta)$ for some $\delta > 0$; $j = 1, \cdots, s$.

Note that δ in Assumptions 4, 5 and 7 need not be the same.

The remaining assumptions concern the joint distribution of the $(2s + 1)$ dimensional random vector $(Y_0, Y) = (Y_0, Y_1, \cdots, Y_{2s})$ where

(2.4) $Y_0 = g(X), \qquad Y_j = g_j(X), \qquad Y_{s+j} = h_j(X), \qquad j = 1, \cdots, s.$

Let $\varphi(\tau, t)$ be the characteristic function (c.f.) of (Y_0, Y),

(2.5) $\varphi(\tau, t) = E \exp \{\tau Y_0 + tY'\}, \qquad\qquad \tau \in R^1, t \in R^{2s}.$

ASSUMPTION 8. *For some $\gamma > 0$, $\varphi(\tau, t) = O(\|\tau, t\|^{-\gamma})$ as $\|\tau, t\| \to \infty$ (we write $\|\tau, t\| = (|\tau|^2 + \|t\|^2)^{1/2})$.*

ASSUMPTION 9. $E(\Pi_{j=1}^{2s} |Y_j|^{\varepsilon_j}) < \infty$ *for any combination of $\varepsilon_j = 0$ or 1, $j = 1, \cdots, 2s$.*

ASSUMPTION 10. *Put $\chi_j(\tau) = E[|Y_j|^3 \exp \{i\tau Y_0\}]$. There exists an n_1 such that the $\chi_j^{n_1}(\tau)$ are absolutely integrable on R^1, $j = 1, \cdots, 2s$.*

Denote by B the matrix with elements $b_{i,j} = Eg_{i,j}(X)$, $i, j = 1, \cdots, s$, and by Σ the covariance matrix of (Y_0, Y) with elements

(2.6) $\sigma_{i,j} = EY_iY_j, \qquad\qquad i, j = 0, 1, \cdots, 2s.$

Note that (Y_0, Y) has zero mean.

Let $(V_0, V_1, \cdots, V_{2s})$ be a normally distributed random vector with zero mean and covariance matrix Σ. Put

(2.7) $\begin{aligned} S &= (V_1, \cdots, V_s), \qquad\qquad T = (V_{s+1}, \cdots, V_{2s}), \\ W &= ST' + \tfrac{1}{2}TBT', \qquad \mu(x) = E[W | V_0 = x]. \end{aligned}$

Since $\sigma_{0,0} = 1$ (see Assumption 1) we have by the well-known formulae

(2.8) $E(V_j | V_0 = x) = \sigma_{0,j}x,$

(2.9) $\mathrm{Cov}\,(V_i, V_j | V_0 = x) = \sigma_{i,j} - \sigma_{0,i}\sigma_{0,j}, \qquad i, j = 1, \cdots, 2s.$

Therefore one can easily derive that

(2.10) $\mu(x) = \sum_{j=1}^{s} \sigma_{j,s+j} + \tfrac{1}{2} \sum_{j,\ell=1}^{s} b_{j\ell}\sigma_{s+j,s+\ell}$

$$- (1 - x^2)\left[\sum_{j=1}^{\ell} \sigma_{0,j}\sigma_{0,s+j} + \tfrac{1}{2} \sum_{j,\ell=1}^{s} b_{j,\ell}\sigma_{0,s+j}\sigma_{0,s+\ell}\right].$$

Denote by $\hat{\phi}_n(x)$ the d.f. of $Z_n(\hat{\theta}_n)$ where $Z_n(\theta)$ is defined by (1.2). We shall write $\mathcal{N}(x)$ and $n(x)$ for the $(0, 1)$ normal d.f. and its density respectively.

THEOREM 2.1. *Let the Assumptions 1 through 10 be satisfied. Then*

(2.11) $\hat{\phi}_n(x) = \phi_n(x) + \varepsilon_n(x)$

where

(2.12) $\phi_n(x) = \mathcal{N}(x) + n^{-1/2} n(x)[(\mu_3/6)(1 - x^2) - \mu(x)],$

with $\mu_3 = EY_0^3$ and $n^{1/2}\varepsilon_n(x) \to 0$ as $n \to \infty$ uniformly in $x \in A$ for any bounded $A \in R^1$.

3. Some theorems on conditional distributions

Let $Y_{0,i}, Y_i), Y_i = (Y_{i,1}, \cdots, Y_{i,k}), i = 1, \cdots, n$, be n independently identically distributed random vectors in R^{k+1} and

$$(3.1) \qquad \varphi(\tau, t) = E \exp \{\tau Y_{0,1} + t Y_1'\}, \qquad\qquad t \in R^k,$$

their common c.f. Let

$$(3.2) \qquad Z_n = \frac{1}{\sqrt{n}} \sum_{i=1}^{n} Y_{0,i}, \qquad S_n = (S_{n,1}, \cdots, S_{n,k}) = \frac{1}{\sqrt{n}} \sum_{i=1}^{n} Y_i.$$

In the following theorem the distribution of $(Y_{0,i}, Y_i)$ will be assumed to satisfy Assumptions 8 and 9 (which should be read in this case with $2s$ replaced by k). The c.f. of (Z_n, S_n) is

$$(3.3) \qquad \varphi^n(n^{-1/2}\tau, n^{-1/2}t) = \varphi_n(\tau, t),$$

say. Under Assumption 8, for n sufficiently large, this c.f. is absolutely integrable and (Z_n, S_n) has a joint $(k + 1)$ dimensional density. Denote this density by $p_n(z, x)$ and the marginal density of Z_n by $p_n(z)$. Then

$$(3.4) \qquad p_n(x|z) = \frac{p_n(z, x)}{p_n(z)}$$

is the conditional density of S_n given Z_n.

THEOREM 3.1. *Let the Assumptions 8 and 9 (with k instead of $2s$) be satisfied. Then for any bounded $A \subset R^1$ there exist a $K > 0$ and a finite N such that*

$$(3.5) \qquad \int_{R^k} \left| p_n(x|z_1) - p_n(x|z_2) \right| dx \le K|z_1 - z_2|$$

for all $z_1, z_2 \in A$ and $n \ge N$.

For the next theorem, consider n independent identically distributed two dimensional random vectors $(X_1, Y_1), \cdots, (X_n, Y_n)$. Denote their joint d.f. by $P(x, y)$ and their marginal distribution function by $P(x)$ and $Q(y)$, respectively. Let

$$(3.6) \qquad Z_n = \frac{1}{\sqrt{n}} \sum_{1}^{n} X_i, \qquad S_n = \frac{1}{\sqrt{n}} \sum_{1}^{n} Y_i.$$

Denote

$$(3.7) \qquad M_r = E|Y_1|^r, \qquad M_r(x) = E(|Y_1|^r | X_1 = x).$$

For two absolutely integrable functions $f_1(x)$ and $f_2(x)$, denote by $f_1 * f_2(x)$ their convolution,

$$(3.8) \qquad f_1 * f_2(x) = \int f_1(x - v) f_2(v) \, dv,$$

and by $f_1^{n*}(x)$ an n fold convolution of $f_1(x)$ with itself.

THEOREM 3.2.   *Assume that* (i) $EX_1 = EY_1 = 0$, (ii) $P(x)$ *has a density* $p(x)$, (iii) $M_r < \infty$ *for some* $r > 2$, *and* (iv) *the functions* $p^{v*}(x)$ *and* $[M_r(x)p(x)]^{v*}$ *are bounded for some finite* v. *Then*

$$(3.9) \qquad P\{|S_n| > x \,|\, Z_n = z\} = o(x^{-r}n^{1-r/2})$$

as $n \to \infty$, $x/\log n \to \infty$ *uniformly in* $z \in A$ *for any bounded* $A \subset R^1$.

For an unconditional counterpart of this theorem see Lemma 4.2 (iv) below.

4. Proof of Theorem 2.1

We establish first several lemmas. Denote

$$(4.1) \quad Z_{n,0} = \frac{1}{\sqrt{n}} \sum_1^n g(X_i), \qquad Z_{n,j} = \frac{1}{\sqrt{n}} \sum_1^n g_j(X_i), \qquad Z_{n,s+j} = \frac{1}{\sqrt{n}} \sum_1^n h_j(X_i)$$
$$j = 1, \cdots, s,$$

$$(4.2) \qquad S_n = (Z_{n,1}, \cdots, Z_{n,s}), \qquad T_n = (Z_{n,s+1}, \cdots, Z_{n,2s}).$$

Let $p_n(z)$ and $p_n(z, x)$, $x \in R^{2s}$, denote the marginal density of $Z_{n,0}$ and the joint density of $(Z_{n,0}, Z_{n,1}, \cdots, Z_{n,2s})$ respectively. Under Assumption 8 they exist for n sufficiently large. Denote by $\varphi_n(\tau, t) = \varphi^n(n^{-1/2}\tau, n^{-1/2}t)$ the c.f. of $(Z_{n,0}, Z_{n,1}, \cdots, Z_{n,2s})$ and by $\varphi_n(\tau) = \varphi_n(\tau, 0)$ the c.f. of $Z_{n,0}$.

LEMMA 4.1.   *Under Assumption* 8

(i) $p_n(z) \to \varkappa(z)$ *as* $n \to \infty$ *uniformly in* $z \in R^1$, *and*

(ii) $p_n(z)$ *has a derivative* $p'_n(z)$ *for* n *sufficiently large and* $\lim \sup_{n \to \infty} \sup_{z \in R^1} |p'_n(z)| < \infty$.

PROOF.   For part (i) see, for example, Feller [2], Theorem 2 in Chapter XV.5. Since

$$(4.3) \qquad \sup_z |p'_n(z)| \leqq \frac{1}{2\pi} \int |\tau \varphi_n(\tau)| \, d\tau,$$

one can get the proof of (ii) from (5.12), (5.17), (5.18) and (5.19) below. (Actually, $p'_n(z)$ converges to $\varkappa'(z)$ but we state in the lemma only what we need in the proof of the theorem.)

LEMMA 4.2.   *Let* $Y_1, \cdots, Y_n$ *be independent identically distributed random variables and* $S_n = n^{-1/2} \sum_1^n Y_i$. *Assume that* $E|Y_i|^r < \infty$ *for some* $r > 0$. *Then*

$$(4.4) \qquad P\{|S_n| > x\} = o(x^{-r}n^{1-r/2}) \qquad as \ n \to \infty,$$

provided one of the following conditions is satisfied: (i) $0 < r < 1$, (ii) $1 \leqq r < 2$, $EY_1 = 0$, (iii) $r = 2$, $EY_1 = 0$, $x \to \infty$, (iv) $r > 2$, $EY_1 = 0$, $x/\log n \to \infty$.

PROOF.   For the parts (i) and (ii) see Binmore and Stratton [1] (note that $E|Y_1|^r < \infty$ implies $P\{|Y_1| > x\} = o(x^{-r})$ as $x \to \infty$). Let $F_n(x)$ be the d.f. of S_n. Part (iii) follows from the inequality

$$(4.5) \qquad x^2 P\{|S_n| > x\} \leqq \int_{|y| \geqq x} y^2 \, dF_n(y).$$

and the uniform integrability of y^2 in $F_n(y)$ (see Loève [4], Theorem 11.4.A (iii)). Part (iv) follows from Theorem 1 of Nagaev [5]. This theorem actually provides an inequality which implies (4.4) only with O instead of o. However we shall indicate below (see 6.25) and the subsequent paragraph) a modification of Nagaev's proof which gives o in (4.4).

In terms of the symbol ω (Definition 2.1) we have,

COROLLARY 4.1. *Under the conditions of Lemma 4.2*

$$\text{(i)} \quad S_n = \omega\left(\frac{3-r}{2r}\right) \quad \text{if } r < 3,$$

$$\text{(ii)} \quad S_n = \omega(\varepsilon) \quad \text{for any } \varepsilon > 0 \text{ if } r \geq 3.$$

With the notation (4.1), (4.2), let

$$(4.6) \quad \begin{aligned} W_n &= S_n T_n' + \tfrac{1}{2} T_n B T_n', & \mu_n(x) &= E[W_n | Z_{n,0} = x], \\ P\{W_n < x\} &= G_n(x), & P\{W_n < x | Z_{n,0} = z\} &= G_n(x|z). \end{aligned}$$

Denote $G(x|z) = P\{W < x | V_0 = z\}$ (see (2.7)).

LEMMA 4.3. *Let Assumptions 8 and 9 be satisfied. Then, for any bounded $A \subset R^1$*

(i) *there exist a $K > 0$ and a finite N such that*

$$(4.7) \quad \sup_x |G_n(x|z_1) - G_n(x|z_2)| \leq K|z_1 - z_2|$$

for all $z_1, z_2 \in A$ and $n \geq N$;

(ii) $\sup\limits_{x \in R^1} |G_n(x|z) - G(x|z)| \to 0$ *uniformly in $z \in A$;*

(iii) $\mu_n(z) \to \mu(z)$ *uniformly in $z \in A$.*

PROOF. Since the inequality $\{W_n < x\}$ determines a Borel set in the sample space of (S_n, T_n), assertion (i) follows from Theorem 3.1. The convergence in assertion (ii) for any fixed z follows directly from Theorem 2.4 of Steck [8]; together with (i), this implies the asserted uniform convergence. Otherwise one could find an $\varepsilon > 0$, a subsequence $\{m\} \subset \{n\}$ and a sequence $\{z_m\}$ approaching a finite limit z_0, say, such that

$$(4.8) \quad \sup_x |G_m(x|z_m) - G(x|z_m)| > \varepsilon \quad \text{for all } m.$$

Then $\sup\limits_x |G_m(x|z_m) - G_m(x|z_0)|$ would not tend to zero which would contradict (i).

Denote by $a(z)$ and $\Sigma(z)$ the conditional mean and covariance matrix of $(V_1, \cdots, V_{2s})$ given $V_0 = z$ (see (2.7) above) and by $a_n(z)$ and $\Sigma_n(z)$ those of $(Z_{n,1}, \cdots, Z_{n,2s})$ given $Z_{n,0} = z$. Then (iii) follows from the convergence

$$(4.9) \quad a_n(z) \to a(z), \quad \Sigma_n(z) \to \Sigma(z) \quad \text{as } n \to \infty$$

uniformly in $z \in A$. This latter convergence can be proved by taking the first and second derivatives of

$$(4.10) \qquad \omega_n(t, z) = \frac{\int e^{-itz} \varphi_n(\tau, t) \, d\tau}{\int e^{-itz} \varphi_n(\tau, 0) \, d\tau},$$

the conditional c.f. of (S_n, T_n), given $Z_{n,0} = z$, at $t = 0$. The technique is quite similar to that used in [8]. It is easily verified that the derivatives of $\varphi_n(\tau, t) = \varphi^n(n^{-1/2}\tau, n^{-1/2}t)$ at $t = 0$ and any fixed τ converge to the corresponding derivatives of the limiting normal c.f. Then the passage to the limit under the integral sign which is justified by the bounded convergence theorem in the same way as in [8] leads to (4.9).

REMARK. The assertions (ii) and (iii) are valid actually under the conditions of Theorem 2.4 in [8] which are weaker than those used here. The proof of (iii) sketched above remains valid in this case; the proof of (ii) requires some standard but rather cumbersome technique.

The following lemma states some properties of the symbol ω (Definition 2.1). The proof is obvious and will be omitted.

LEMMA 4.4.
  (i) *If $\zeta_n = \omega(a)$ then $\zeta_n = \omega(a')$ for any $a' > a$.*
  (ii) *If $\zeta_n = \omega(a)$ and $\eta_n = \omega(b)$ then $\zeta_n + \eta_n = \omega(\max(a, b))$.*
  (iii) *If $\zeta_n = \omega(a)$ and $\eta_n = \omega(b)$ then $\zeta_n \eta_n = \omega(a + b)$.*

Now we proceed to the proof of the theorem. For notational convenience, let $\theta_0 = 0$. Expanding $g(X_i, \hat{\theta}_n)$ by the Taylor formula, we have

$$(4.11) \qquad Z_n(\hat{\theta}_n) = \frac{1}{\sqrt{n}} \sum_{i=1}^{n} g(X_i) + \frac{1}{\sqrt{n}} \sum_{j=1}^{s} \hat{\theta}_{n,j} \sum_{i=1}^{n} g_j(X_i)$$

$$+ \frac{1}{2\sqrt{n}} \sum_{j,\ell} \hat{\theta}_{n,j} \hat{\theta}_{n,\ell} \sum_{i=1}^{n} g_{j,\ell}(X_i) + \frac{1}{6\sqrt{n}} \sum_{j,k,\ell} \hat{\theta}_{n,j} \hat{\theta}_{n,k} \hat{\theta}_{n,\ell} \sum_{i=1}^{n} g_{j,k,\ell}(X_i, t_{n,i}\hat{\theta}_n),$$

where $0 \leqq t_{n,i} \leqq 1$. Denote by B_n the matrix with elements

$$(4.12) \qquad b_{n,j,\ell} = \frac{1}{n} \sum_{i=1}^{n} g_{j,\ell}(X_i), \qquad\qquad j, \ell = 1, \cdots, s.$$

Then using also (4.6) we can write

$$(4.13) \qquad Z_n(\hat{\theta}_n) = Z_{n,0} + n^{-1/2} W_n + n^{-1/2} R_n$$

where

$$(4.14) \qquad R_n = \eta_n S_n' + \tfrac{1}{2}(T_n + \eta_n)(B_n - B)(T_n + \eta_n)' + (T_n B \eta_n' + \tfrac{1}{2}\eta_n B \eta_n')$$

$$+ \tfrac{1}{6} \sum_{j,k,\ell} \hat{\theta}_{n,j} \hat{\theta}_{n,k} \hat{\theta}_{n,\ell} \sum_{i=1}^{n} g_{j,k,\ell}(X_i, t_{n,i}\hat{\theta}_n) = R_{n,1} + \tfrac{1}{2}R_{n,2} + R_{n,3} + \tfrac{1}{6}R_{n,4},$$

say. We shall show now that $R_n = \omega(0)$. By Lemma 5.2 (ii), it is sufficient to show that $R_{n,i} = \omega(0)$, $i = 1, \cdots, 4$.

By Assumptions 4, 5 and 7 we can find an α, $0 < \alpha < \frac{1}{3}$, such that

(4.15) $\qquad E|g_{i,j}(X)|^{3/2(1-\alpha)} < \infty, \qquad E(K(X))^{1/(1-\alpha)} < \infty, \qquad \eta_{n,j} = \omega\left(-\frac{\alpha}{2}\right).$

We have $R_{n,1} = \Sigma_{j=1}^{s} \eta_{n,j} Z_{n,j}$. By Assumption 3 and Corollary 4.1 (ii), we have $Z_{n,j} = \omega(\alpha/2)$. Therefore, $\eta_{n,j} Z_{n,j} = \omega(0)$ by Lemma 4.4, and $R_{n,1} = \omega(0)$.

Consider a term $R_{n,2}^{j,\ell} = (b_{n,j,\ell} - b_{j,\ell})(T_{n,j} + \eta_{n,j})(T_{n,\ell} + \eta_{n,\ell})$ of $R_{n,2}$. By Assumption 4 and Corollary 4.1 (i)

(4.16) $\qquad b_{n,j,\ell} - b_{j,\ell} = n^{-1/2} \cdot \frac{1}{\sqrt{n}} \sum_{i=1}^{n} (g_{j,\ell}(X_i) - b_{j,\ell}) = \omega\left(-\frac{1}{2} + \frac{3-r}{2r}\right)$

with $r = 3/2(1-\alpha)$ (see (4.15)). Thus $b_{n,j,\ell} - b_{j,\ell} = \omega(-\alpha)$. By Assumption 6 $T_{n,j} = \omega(\alpha/2)$ and by Lemma 4.4 $T_{n,j} + \eta_{n,j} = \omega(\alpha/2)$, $R_{n,2}^{j,\ell} = \omega(0)$, and $R_{n,2} = \omega(0)$.

In a similar way we obtain $R_{n,3} = \omega(0)$.

Now consider a term

(4.17) $\qquad R_{n,4}^{j,k,\ell} = \hat{\theta}_{n,j} \hat{\theta}_{n,k} \hat{\theta}_{n,\ell} \sum_{i=1}^{n} g_{j,k,\ell}(X_i, t_{n,i} \hat{\theta}_n)$

of $R_{n,4}$. We have $\hat{\theta}_{n,j} = n^{-1/2}(T_{n,j} + \eta_{n,j}) = \omega((\alpha-1)/2)$ and, by Lemma 4.4, $|\hat{\theta}_n| = \omega((\alpha-1)/2)$. Take a $\delta > 0$ such that $\{\theta : |\theta| \leq \delta\} \subset U$ (see Assumption 5). Then

(4.18) $\quad P\{|R_{n,4}^{j,k,\ell}| > c\} \leq P\{|R_{n,4}^{j,k,\ell}| > c, \hat{\theta}_n \in U\} + P\{|\hat{\theta}_n| > \delta\}$

$\qquad\qquad\qquad \leq P\{\sqrt{n}|\hat{\theta}_{n,j} \hat{\theta}_{n,k} \hat{\theta}_{n,\ell}| n^{-1/2} \sum_{i=1}^{n} K(X_i) > c\} + o(n^{-1/2}).$

We have

(4.19) $\qquad \sqrt{n}|\hat{\theta}_{n,j} \hat{\theta}_{n,k} \hat{\theta}_{n,\ell}| = \omega\left(\frac{1}{2} + 3\frac{\alpha-1}{2}\right) = \omega\left(\frac{3\alpha}{2} - 1\right).$

Hence it remains to show that

(4.20) $\qquad\qquad n^{-1/2} \sum_{i=1}^{n} K(X_i) = \omega\left(1 - \frac{3\alpha}{2}\right).$

Denote $\kappa = EK(X)$. By Assumption 5 (see (4.15)) and Corollary 4.1 (i),

(4.21) $\quad n^{-1/2} \sum_i K(X_i) - \kappa\sqrt{n} = n^{-1/2} \sum_i (K(X_i) - \kappa) = \omega\left(1 - \frac{3\alpha}{2}\right).$

Since α was chosen to be less than $1/3$, $\kappa\sqrt{n} = \omega(1 - 3\alpha/2)$. Therefore (4.21) implies (4.20).

We shall show now that the term $n^{-1/2} R_n$ in (4.13) may be neglected. Denote

(4.22) $\qquad\quad Z_n^* = Z_{n,0} + n^{-1/2} W_n, \qquad\qquad \phi_n^*(x) = P\{Z_n^* < x\}.$

Let (2.11) hold for $\phi_n^*(x)$. Then for an arbitrary $\delta > 0$,

$$(4.23) \qquad \hat{\phi}_n(x) = P\{Z_n^* + n^{-1/2} R_n < x\}$$
$$\leqq P\{Z_n^* < x + n^{-1/2}\delta\} + P\{|R_n| > \delta\}$$
$$= \phi_n(x + n^{-1/2}\delta) + \varepsilon_n(x + n^{-1/2}\delta) + o(n^{-1/2}).$$

A similar estimate from below gives

$$(4.24) \qquad \hat{\phi}_n(x) \geqq \phi_n(x - n^{-1/2}\delta) + \varepsilon_n(x - n^{-1/2}\delta) + o(n^{-1/2}).$$

The function $\phi_n(x)$ has a derivative bounded uniformly in n and x, $\phi_n'(x) < C$, say. Therefore

$$(4.25) \qquad |\hat{\phi}_n(x) - \phi_n(x)| \leqq n^{-1/2} C\delta + \bar{\varepsilon}_n(x) + o(n^{-1/2})$$

where $\bar{\varepsilon}_n(x) = \sup\left[|\varepsilon_n(x + u)|; |u| \leqq \delta\right]$. Since $\delta > 0$ is arbitrary, (4.25) implies the assertion of Theorem 2.1 for $\hat{\phi}_n(x)$.

Writing $P_n(x)$ for the d.f. of $Z_{n,0}$ we have from (4.6) and (4.22)

$$(4.26) \qquad \phi_n^*(x) = \int G_n\big((x - z)\sqrt{n}\,|\,z\big)\, dP_n(z).$$

The next step of the proof will be to show that $\phi_n^*(x)$ may be replaced by

$$(4.27) \qquad \phi_n^{**}(x) = \int G_n\big((x - z)\sqrt{n}\,|\,x\big)\, dP_n(z),$$

that is, for any $a > 0$

$$(4.28) \qquad \sup_{x \in [-a, a]} |\phi_n^*(x) - \phi_n^{**}(x)| = o(n^{-1/2}) \qquad \text{as } n \to \infty.$$

Put $\delta = n^{-3/8}$ and write the difference of (4.26) and (4.27) as

$$(4.29) \qquad \phi_n^*(x) - \phi_n^{**}(x)$$
$$= \left(\int_{x-\delta}^{x+\delta} + \int_{-\infty}^{x-\delta} + \int_{x+\delta}^{\infty}\right) \left[G_n\big((x - z)\sqrt{n}\,|\,z\big) - G_n\big((x - z)\sqrt{n}\,|\,x\big)\right] dP_n(z)$$
$$= I_1(x) + I_2(x) + I_3(x),$$

say. Applying Lemma 4.3 (i) with $A = [-a - 1, a + 1]$ and using the fact that $p_n(z)$ is bounded (Lemma 4.1 (i)), we obtain

$$(4.30) \qquad \max_{x \in [-a, a]} |I_1(x)| \leqq K \max p_n(x) \int_{x-\delta}^{x+\delta} |x - z|\, dz = O(\delta^2) = o(n^{-1/2}).$$

Further, $|I_2(x)| \leqq I_{2,1}(x) + I_{2,2}(x)$ where

$$(4.31) \qquad I_{2,1}(x) = \int_{-\infty}^{x-\delta} \left[1 - G_n\big((x - z)\sqrt{n}\,|\,x\big)\right] p_n(z)\, dz,$$

$$(4.32) \qquad I_{2,2}(x) = \int_{-\infty}^{x-\delta} \left[1 - G_n\big((x - z)\sqrt{n}\,|\,z\big)\right] p_n(z)\, dz.$$

Using the inequality

$$(4.33) \qquad 1 - G_n\big((x - z)\sqrt{n}\,|y\big) \leqq P\{W_n > n^{1/8} | Z_{n,0} = y\} \qquad \text{for } x - z > \delta$$

with $y = x$, we obtain

$$(4.34) \qquad\qquad I_{2,1}(x) \leqq P\{W_n > n^{1/8} | Z_{n,0} = x\}.$$

Assumptions $3, 6, 8,$ and 10 assure the fulfillment of the conditions of Theorem 3.2 for the vectors $(g_j(X_i), g(X_i))$ and $(h_j(X_i), g(X_i)), j = 1, \cdots, s,$ with $r = 3$. (One should only note that $\chi_j(\tau)$ in Assumption 10 is the Fourier transform of $M_3(x)p(x)$, and the integrability of $\chi_j^{n_1}(\tau)$ implies the boundedness of $[M_3(y)p(y)]^{n_1*}$.) Therefore

$$(4.35) \quad P\{|Z_{n,j}| > n^{1/16} | Z_{n,0} = x\} = o(n^{-1/2+3/16}) = o(n^{-1/2}), \quad j = 1, \cdots, 2s.$$

(In relations of this kind we mean that $o(n^{-1/2})$ is uniform in $x \in [-a, a]$ without stating it explicitly.) Now we obtain from the definition of W_n (see (4.6)) and Lemma 4.4 (applied to conditional probabilities) that

$$(4.36) \qquad\qquad P\{|W_n| > n^{1/8} | Z_{n,0} = x\} = o(n^{-1/2}).$$

In view of (4.34) this implies

$$(4.37) \qquad\qquad\qquad I_{2,1}(x) = o(n^{-1/2}).$$

Using (4.33) with $y = z$, we have

$$(4.38) \qquad I_{2,2}(x) \leqq \int_{-\infty}^{x-\delta} P\{W_n > n^{1/8} | Z_{n,0} = z\}\, dP_n(z)$$

$$\leqq \int_{-\infty}^{\infty} P\{W_n > n^{1/8} | Z_{n,0} = z\}\, dP_n(z) = P\{W_n > n^{1/8}\}.$$

This probability is estimated in the same way as (4.34) but Lemma 4.2 is used, rather than Theorem 3.2, which gives

$$(4.39) \qquad\qquad\qquad I_{2,2}(x) = o(n^{-1/2}).$$

The relations (4.29), (4.30), (4.37) and (4.39) together with a similar estimate for $I_3(x)$ prove (4.28).

Thus we are to prove the theorem for $\phi_n^{**}(x)$ defined by (4.27). Rewrite it in the form

$$(4.40) \qquad\qquad \phi_n^{**}(x) = \int P_n(x - z)\, dG_n(z\sqrt{n}\,|x).$$

Using (4.36) we have

$$(4.41) \int_{|z| > n^{-3/8}} P_n(x - z)\, dG_n(z\sqrt{n}\,|x) \leqq P\{W_n > n^{1/8} | Z_{n,0} = x\} = o(n^{-1/2}).$$

Therefore (writing again $\delta = n^{-3/8}$)

$$(4.42) \qquad\qquad \phi_n^{**}(x) = \int_{-\delta}^{\delta} P_n(x - z)\, dG_n(z\sqrt{n}\,|x) + o(n^{-1/2}).$$

By Taylor's formula we obtain

$$(4.43) \quad \phi_n^{**}(x) = \int_{-\delta}^{\delta} P_n(x - n^{-1/2}\mu_n(x)) \, dG_n(z\sqrt{n}\,|x)$$

$$+ \int_{-\delta}^{\delta} (n^{-1/2}\mu_n(x) - z) \, p_n(x - n^{-1/2}\mu_n(x)) \, dG_n(z\sqrt{n}\,|x)$$

$$+ \int_{-\delta}^{\delta} (n^{-1/2}\mu_n(x) - z)[p_n(x - z^*) - p_n(x - n^{-1/2}\mu_n(x))] \, dG_n(z\sqrt{n}\,|x)$$

$$+ o(n^{-1/2})$$

$$= J_1(x) + J_2(x) + J_3(x) + o(n^{-1/2}),$$

say, where z^* lies between z and $n^{-1/2}\mu_n(x)$. By virtue of (4.36)

$$(4.44) \quad J_1(x) = P_n(x - n^{-1/2}\mu_n(x))[1 - P\{|W_n| > n^{1/8}\,|Z_{n,0} = x\}]$$

$$= P_n(x - n^{-1/2}\mu_n(x)) + o(n^{-1/2}).$$

Now we shall show that $J_2(x)$ and $J_3(x)$ are $o(n^{-1/2})$. First, by Theorem 11.4.A (iii) of Loève [4], the assertions (ii) and (iii) of Lemma 4.3 imply

$$(4.45) \quad \int_{|z|>\delta} z \, dG_n(z\sqrt{n}\,|x) = n^{-1/2} \int_{|y|>n^{1/8}} y \, dG_n(y\,|x) = o(n^{-1/2}).$$

Up to the factor $p_n(x - n^{-1/2}\mu_n(x))$, which is bounded by Lemma 4.1 (i), $J_2(x)$ is equal to

$$(4.46) \quad n^{-1/2}\mu_n(x)\left[1 - \int_{|z|>\delta} dG_n(z\sqrt{n}\,|x)\right]$$

$$- \left[n^{-1/2}\mu_n(x) - \int_{|z|>\delta} z \, dG_n(z\sqrt{n}\,|x)\right],$$

and by (4.36) and (4.45) we get $J_2(x) = o(n^{-1/2})$.
Finally,

$$(4.47) \quad \begin{aligned} n^{-1/2}\mu_n(x) - z &= O(n^{-3/8}), \\ n^{-1/2}\mu_n(x) - z^* &= O(n^{-3/8}) \text{ for } z \in [-\delta, \delta], \end{aligned}$$

and making use of Lemma 4.1 (ii) we obtain $J_3(x) = O(n^{-3/4}) = o(n^{-1/2})$.

Thus

$$(4.48) \quad \phi_n^{**}(x) = P_n(x - n^{-1/2}\mu_n(x)) + o(n^{-1/2}),$$

or since $\mu_n(x) \to \mu(x)$ and $p_n(x)$ is bounded,

$$(4.49) \quad \phi_n^{**}(x) = P_n(x - n^{-1/2}\mu(x)) + o(n^{-1/2}).$$

By virtue of the well known expansion

$$(4.50) \quad P_n(x) = \mathcal{N}(x) + n^{-1/2}\tfrac{1}{6}\mu_3(1 - x^2)n(x) + o(n^{-1/2})$$

(see, for example, [2], Ch. XVI, § 4), the assertion of the theorem follows from (4.49).

5. Proof of Theorem 3.1

We have

$$(5.1) \qquad \int_{R^k} \left| p_n(x|z_1) - p_n(x|z_2) \right| dx = \int_{R^k} \left| \frac{p_n(z_1, x)}{p_n(z_1)} - \frac{p_n(z_2, x)}{p_n(z_2)} \right| dx \leqq I_1 + I_2,$$

where

$$I_1 = \int_{R^k} \frac{p_n(z_1, x)|p_n(z_2) - p_n(z_1)|}{p_n(z_1)\,p_n(z_2)}\, dx = \frac{|p_n(z_2) - p_n(z_1)|}{p_n(z_2)},$$

$$(5.2) \qquad I_2 = \int_{R^k} \frac{|p_n(z_1, x) - p_n(z_2, x)|p_n(z_1)}{p_n(z_1)\,p_n(z_2)}\, dx$$

$$= \frac{1}{p_n(z_2)} \int_{R^k} |p_n(z_1, x) - p_n(z_2, x)|\, dx = \frac{I}{p_n(z_2)},$$

say. By Lemma 4.1, $|p_n(z_2) - p_n(z_1)| \leqq C|z_2 - z_1|$ and $p_n(z_2)$ is bounded away from zero for $z_2 \in A$ and n sufficiently large. Thus we need only to obtain an inequality for I similar to (3.5). This will be based on the following lemma, which is an immediate multidimensional extension of Lemma 1.5.1 from Ibragimov and Linnik [3].

LEMMA 5.1. *Let a function $f(x)$, $x \in R^k$, be absolutely integrable in R^k, with Fourier transform*

$$(5.3) \qquad \psi(t) = \int_{R^k} e^{itx'} f(x)\, dx, \qquad\qquad t \in R^k,$$

which has derivatives

$$(5.4) \qquad \mathscr{D}_{\varepsilon_1, \cdots, \varepsilon_k} \psi(t) = \frac{\partial^{\varepsilon_1 + \cdots + \varepsilon_k}}{\partial t_1^{\varepsilon_1} \cdots \partial t_k^{\varepsilon_k}} \psi(t), \qquad \varepsilon_j = 0 \text{ or } 1, j = 1, \cdots, k,$$

with these derivatives (including $\mathscr{D}_{0, \cdots, 0} \psi(t) = \psi(t)$) being square integrable in R^k. Then

$$(5.5) \qquad \int_{R^k} |f(x)|\, dx \leqq \frac{1}{2^{k/2}} \left(\sum \int_{R^k} |\mathscr{D}_{\varepsilon_1, \cdots, \varepsilon_k} \psi(t)|^2\, dt \right)^{1/2},$$

where the summation is over all possible combinations of $\varepsilon_1, \cdots, \varepsilon_k = 0$ or 1.

Now let $\psi_n(z, \cdot)$ denote the Fourier transform of $p_n(z, \cdot)$. We have

$$(5.6) \qquad \int e^{itz} \psi_n(z, t)\, dz = \int e^{itz} \left(\int_{R^k} e^{itx'} p_n(z, x)\, dx \right) dz = \varphi_n(\tau, t).$$

Therefore, for sufficiently large n,

$$(5.7) \qquad \psi_n(z, t) = \frac{1}{2\pi} \int e^{-itz} \varphi_n(\tau, t)\, d\tau,$$

the application of the inversion formula being justified by Assumption 8. Let $\mathscr{D}$ stand for one of the operators $\mathscr{D}_{\varepsilon_1, \ldots, \varepsilon_k}$ from (5.4). Then

$$(5.8) \qquad \mathscr{D}\psi_n(z, t) = \frac{1}{2\pi} \int e^{-i\tau z} \mathscr{D}\varphi_n(\tau, t) \, d\tau.$$

(The differentiation under the integral sign is justified by Assumption 9 and relations (5.23), (5.24) below.) Furthermore, (5.8) implies

$$(5.9) \qquad \left| \mathscr{D}\big(\psi_n(z_1, t) - \psi_n(z_2, t)\big) \right| \leq \frac{|z_1 - z_2|}{2\pi} \int \left| \tau \mathscr{D}\varphi_n(\tau, t) \right| d\tau,$$

and in order to obtain the required inequality for I we need by Lemma 5.1 to show that

$$(5.10) \qquad \int_{R^k} a^2_{\varepsilon_1, \ldots, \varepsilon_k}(t) \, dt \leq K \qquad \text{for all } \varepsilon_1, \cdots, \varepsilon_k = 0 \text{ or } 1,$$

where

$$(5.11) \qquad a_{\varepsilon_1, \ldots, \varepsilon_k}(t) = \int \left| \tau \mathscr{D}_{\varepsilon_1, \ldots, \varepsilon_k} \varphi_n(\tau, t) \right| d\tau$$

and K is a constant which does not depend on n, τ and t.

Consider first $a(t) = a_{0, \ldots, 0}(t)$,

$$(5.12) \qquad a(t) = \int \left| \tau \varphi_n(\tau, t) \right| d\tau.$$

In view of (3.3) we need an estimate for $\varphi(\tau, t)$. By Assumption 8 we can find $C > 0$ and $B > 0$ such that

$$(5.13) \qquad |\varphi(\tau, t)| \leq C \|\tau, t\|^{-\gamma} \qquad \text{for } \|\tau, t\| \geq B.$$

We take B large enough to satisfy the inequality

$$(5.14) \qquad CB^{-\gamma} < 1.$$

Furthermore, it follows from Assumption 8 that Σ, the covariance matrix of $(Y_{0,1}, Y_1)$, is nondegenerate (otherwise there would exist $(\tau_0, t_0) \neq (0, 0)$ such that $E(\tau_0 Y_{0,1} + t_0 Y_1')^2 = 0$ and $\varphi(u\tau_0, ut_0) \equiv 1, -\infty < u < \infty$). Since $E(Y_{0,1}, Y_1) = 0$, one can find $\lambda > 0$ and $\delta > 0$ such that

$$(5.15) \qquad |\varphi(\tau, t)| \leq e^{-\lambda \|\tau, t\|^2} \qquad \text{for } \|\tau, t\| < \delta.$$

Moreover, $\sup_{\|\tau, t\| \geq \delta} |\varphi(\tau, t)| < 1$. Therefore, reducing λ if necessary, we get

$$(5.16) \qquad |\varphi(\tau, t)| \leq e^{-\lambda \|\tau, t\|^2} \qquad \text{for } \|\tau, t\| \leq B.$$

Setting $\tau(t) = \max \left[(B^2 n - \|t\|^2)^{1/2}, 0 \right]$, write $a(t)$ as

$$(5.17) \qquad a(t) = \left(\int_{|\tau| \leq \tau(t)} + \int_{|\tau| > \tau(t)} \right) \left| \tau \varphi_n(\tau, t) \right| d\tau = a^{(1)}(t) + a^{(2)}(t),$$

say. Then we have from (3.3), (5.13) and (5.16)

$$(5.18) \qquad a^{(1)}(t) \leqq 2 \int_0^{\tau(t)} \tau e^{-\lambda\|\tau, t\|^2} \, d\tau \leqq 2 \int_0^\infty \tau e^{-\lambda\|\tau, t\|^2} \, d\tau,$$

$$(5.19) \qquad a^{(2)}(t) \leqq 2 \int_{\tau(t)}^\infty \tau \frac{C^n n^{n\gamma/2}}{(\tau^2 + \|t\|^2)^{n\gamma/2}} \, d\tau = \frac{2C^n n^{n\gamma/2}}{(n\gamma - 2)[\tau^2(t) + \|t\|^2]^{(n\gamma/2)-1}}$$

$$= \frac{2C^n n}{(n\gamma - 2)B^{n\gamma - 2}} \qquad \text{for } \|t\|^2 \leqq B^2 n,$$

$$= \frac{2C^n n^{n\gamma/2}}{(n\gamma - 2)\|t\|^{n\gamma - 2}} \qquad \text{for } \|t\|^2 \geqq B^2 n.$$

We can estimate the integrals $\int [a^{(i)}(t)]^2 \, dt$, $i = 1, 2$. From (5.18),

$$(5.20) \qquad \int [a^{(1)}(t)]^2 \, dt$$

$$\leqq 4 \int_{R^k} \int_0^\infty \int_0^\infty \tau_1 \tau_2 \exp\{-\lambda(\|\tau_1, t\|^2 + \|\tau_2, t\|^2)\} \, d\tau_1 \, d\tau_2 \, dt < \infty.$$

Let $V_n(B)$ denote the volume of the k dimensional sphere $\|t\|^2 \leq B^2 n$; then for B fixed, $V_n(B) = O(n^{k/2})$. Hence we obtain from (5.19)

$$(5.21) \qquad \int [a^{(2)}(t)]^2 \, dt \leqq \frac{4C^{2n} n^2}{(n\gamma - 2)^2 B^{2n\gamma - 4}} V_n(B) + \frac{4C^{2n} n^{n\gamma}}{(n\gamma - 2)^2} \int_{\|t\| \geqq Bn^{1/2}} \frac{dt}{\|t\|^{2n\gamma - 4}}$$

$$= O(n^{k/2}(CB^{-\gamma})^{2n}) \to 0$$

in view of (5.14). Now (5.17), (5.20) and (5.21) imply (5.10) for $a(t) = a_{0, \cdots, 0}(t)$.

Consider now the general case of (5.10). Suppose without loss of generality that $\varepsilon_1 = \cdots = \varepsilon_\ell = 1$, $\varepsilon_{\ell+1} = \cdots = \varepsilon_k = 0$, $0 < \ell \leq k$, that is, that $\mathscr{D}_{\varepsilon_1, \cdots, \varepsilon_k} = \partial^\ell/\partial t_1 \cdots \partial t_\ell$ in (5.11). Let $T = \{j_1, \cdots, j_r\}$ be a subset of $\{1, \cdots, \ell\}$. Denote

$$(5.22) \qquad \varphi_T(\tau, t) = \frac{\partial^r}{\partial t_{j_1} \cdots \partial t_{j_r}} \varphi(\tau, t).$$

As is easily seen, $(\partial^\ell/\partial t_1 \cdots \partial t_\ell)\varphi_n(\tau, t)$ is the sum of the following terms

$$(5.23) \qquad n(n - 1) \cdots (n - m + 1) \, \varphi^{n-m}\left(\frac{\tau}{\sqrt{n}}, \frac{t}{\sqrt{n}}\right) \varphi_{T_1}\left(\frac{\tau}{\sqrt{n}}, \frac{t}{\sqrt{n}}\right)$$

$$\cdots \varphi_{T_m}\left(\frac{\tau}{\sqrt{n}}, \frac{t}{\sqrt{n}}\right) n^{-k/2}$$

where $T_1, \cdots, T_m$ is a partition of the set $\{1, \cdots, \ell\}$ into nonempty disjoint subsets, and the summation is over all possible partitions. Every term (5.23) is estimated as in the case of $a(t)$ above, and we shall only indicate the distinctions which arise.

Split the integral of τ times (5.23) into two parts as in (5.17) and call them again $a^{(1)}(t)$ and $a^{(2)}(t)$. We have for $T = \{j_1, \cdots, j_r\}$

$$(5.24) \qquad \left| \varphi_T(\tau, t) \right| \leq E \left| Y_{1, j_1} \cdots Y_{1, j_r} \right| = M_T,$$

say, which is finite by Assumption 9. Hence the modulus of (5.23) is bounded from above by

$$(5.25) \qquad n^{m-k/2} M_{T_1} \cdots M_{T_m} \left| \varphi \left(\frac{\tau}{\sqrt{n}}, \frac{t}{\sqrt{n}} \right) \right|^{n-m}.$$

This enables us to estimate $a^{(2)}(t)$ just as above. The only difference from (5.21) occurs in the factor of order of $n^{m-k/2}$. This does not matter in the presence of a geometric series term.

Concerning $a^{(1)}(t)$, (5.25) is sufficient when $m < k/2$. Consider the case $m > k/2$. We have

$$(5.26) \qquad \frac{\partial}{\partial t_j} \varphi(0, t) \bigg|_{t=0} = i E Y_{1, j} = 0, \qquad\qquad j = 1, \cdots, k,$$

and, writing for a moment t_0 instead of τ,

$$(5.27) \qquad \left| \frac{\partial^2}{\partial t_i \, \partial t_j} \varphi(t_0, t) \right| \leq E \left| Y_{1, i} Y_{1, j} \right|, \qquad i, j = 0, 1, \cdots, k.$$

Hence we can find a constant L such that, for all τ, t

$$(5.28) \qquad \left| \frac{\partial}{\partial t_j} \varphi(\tau, t) \right| \leq L \left\| \tau, t \right\|, \qquad\qquad j = 1, \cdots, k.$$

Note now that there are at least $2m - k$ sets among $T_1, \cdots, T_m$ containing just one element. In fact, if r is the number of such sets, then the remaining $m - r$ sets contain not less than $2(m - r)$ elements, that is, $k - r \geq 2(m - r)$ whence $r \geq 2m - k$. Suppose, to be definite, that $T_1, \cdots, T_{2m-k}$ contain one element each. Then by (5.28)

$$(5.29) \qquad \left| \varphi_{T_j} \left(\frac{\tau}{\sqrt{n}}, \frac{t}{\sqrt{n}} \right) \right| \leq L \left\| \tau, t \right\| n^{-1/2}, \qquad j = 1, \cdots, 2m - k.$$

Applying (5.24) to the remaining φ_T in (5.23), we obtain for (5.23), up to a constant factor, an upper bound $\left\| \tau, t \right\|^{2m-k} \varphi^{n-m} (\tau n^{-1/2}, t n^{-1/2})$. Therefore, proceeding as in (5.16), (5.18) and (5.20), we arrive at an inequality whose right side differs from that of (5.20) by some power of $\left\| \tau_1, t \right\| \left\| \tau_2, t \right\|$ under the integral sign, which does not affect the convergence of the integral. The proof is thus completed.

6. Proof of Theorem 3.2

Without loss of generality assume $\operatorname{Var} X_1 = \operatorname{Var} Y_1 = 1$. Put $u = xn^{1/2}$. All limits will be taken as $n \to \infty$, $x/\log n \to \infty$ (or, equivalently, as $n \to \infty$, $u/\sqrt{n} \log n \to \infty$) unless otherwise stated. We shall prove (3.8) with S_n rather than $|S_n|$. Since the same will be true for $-S_n$, this will imply (3.8). We take and fix an arbitrary bounded $A \subset R^1$ for reference when dealing with the uniformity of convergence.

Let $A_{n,u}$ denote the event

$$(6.1) \qquad \{Y_i < u \qquad \text{for all } i = 1, \cdots, n\}.$$

Writing $\bar{A}_{n,u}$ for its complement, we have

$$(6.2) \quad P\{S_n > x \,|\, Z_n = z\} \leqq P\{S_n > x, A_{n,u} \,|\, Z_n = z\} + P\{\bar{A}_{n,u} \,|\, Z_n = z\}.$$

Estimate first the last term in (6.2). We have obviously

$$(6.3) \qquad P\{\bar{A}_{n,u} \,|\, Z_n = z\} \leqq nP\{Y_n > u \,|\, Z_n = z\}.$$

We show now that

$$(6.4) \qquad \sup_z P\{Y_n > u \,|\, Z_n = z\}\, p_n(z) \leqq a_n P\{Y_n > u\}$$

where $a_n \to (2\pi)^{-1/2}$ and hence is bounded. Let $\Delta \subset R^1$ be an arbitrary bounded Borel set. Then

$$(6.5) \quad \int_\Delta P\{Y_n > u \,|\, Z_n = z\}\, p_n(z)\, dz = P\{Y_n > u, Z_n \in \Delta\}$$

$$= \int_{-\infty}^\infty \int_u^\infty P\{Z_n \in \Delta \,|\, X_n = x, Y_n = y\}\, dP(x, y)$$

$$= {}'\!\int_{-\infty}^\infty \int_u^\infty P\left\{Z_{n-1} \in \left[\left(\frac{n}{n-1}\right)^{1/2} \Delta - \frac{x}{(n-1)^{1/2}}\right] \middle|\, X_n\right.$$

$$\left. = x, Y_n = y\right\} dP(x, y)$$

$$\leqq \sup_t P\left\{Z_{n-1} \in \left(\frac{n}{n-1}\right)^{1/2} \Delta - t\right\} P\{Y_n > u\}$$

$$\leqq |\Delta| \left(\frac{n}{n-1}\right)^{1/2} \max_z p_n(z)\, P\{Y_n > u\},$$

where $|\Delta|$ is the Lebesgue measure of Δ. With $a_n = [n/(n-1)]^{1/2} \max_z p_n(z)$ this implies (6.4). The assumption that $p^{v*}(z)$ is bounded assures the convergence

$$(6.6) \qquad p_n(z) \to n(z) \qquad \text{uniformly in } z \in R^1.$$

This implies that $a_n \to (2\pi)^{-1/2}$ and moreover that $p_n(z)$ is bounded away from zero on A for n large enough. Since $P\{Y_n > u\} = o(u^{-r})$ we obtain from (6.3) and (6.4) that

(6.7) $P\{\bar{A}_{n,u}|Z_n = z\} = o(nu^{-r})$ uniformly in $z \in A$,

that is, this term has the required order.

Consider now the first term in the right side of (6.2). We shall write $\tilde{S}_n$, $\tilde{Z}_n$ for $S_n\sqrt{n}$, $Z_n\sqrt{n}$ (nonnormalized sums). For an event B, let I_B denote the indicator function; instead of $I_{A_{n,u}}$, we shall write $I_{n,u}$. Then for any $h > 0$,

(6.8) $P\{S_n > x, A_{n,u}|Z_n = z\} = P\{\tilde{S}_n > u, A_{n,u}|Z_n = z\}$
$$\leq e^{-hu} E[\exp\{h\tilde{S}_n\} I_{n,u}|Z_n = z].$$

Put

(6.9) $$d_r(v) = \int_{1/v}^{\infty} y^r \, dQ(y),$$

(6.10) $$c_n = \max\left(n^{-1/2}, d_r(\sqrt{n})\right),$$

(6.11) $$h_{n,u} = -u^{-1} \log(c_n n u^{-r}).$$

Writing $u = \lambda\sqrt{n} \log n$ where $\lambda \to \infty$, we see that

(6.12) $$h_{n,u} = \frac{-\log c_n n + r \log \lambda + r \log(n^{1/2} \log n)}{\lambda\sqrt{n} \log n} = o(n^{-1/2}).$$

Since $c_n \to 0$, we have from (6.11)

(6.13) $$\exp\{-h_{n,u}u\} = o(nu^{-r}).$$

We shall show that

(6.14) $$E[\exp\{h_{n,u}\tilde{S}_n\} I_{n,u}|Z_n = z] \, p_n(z)$$

is bounded uniformly in z and sufficiently large n. Then the theorem will follow from (6.1), (6.7), (6.8), (6.13) and (6.6).

Denote the density of $\tilde{Z}_n$ by $\tilde{p}_n(z)$; $\tilde{p}_n(z\sqrt{n})\sqrt{n} = p_n(z)$. We can rewrite (6.14) as

(6.15) $$E[\exp\{h_{n,u}\tilde{S}_n\} I_{n,u}|\tilde{Z}_n = z\sqrt{n}] \, \tilde{p}_n(z\sqrt{n}) \sqrt{n}.$$

This expression will be estimated with the help of the following lemma.

LEMMA 6.1.   *Let (U_1, V_1), (U_2, V_2) be independent random vectors, V_i having a density $p_i(v)$, $i = 1, 2$, and let $p(v)$ be the density of $V_1 + V_2$. Put*

(6.16) $$f_i(v) = E[U_i|V_i = v] \, p_i(v), \qquad\qquad i = 1, 2,$$
$$f(v) = E[U_1 U_2|V_1 + V_2 = v] \, p(v).$$

Then

(6.17) $$f(v) = f_{1*}f_2(v).$$

PROOF.   For any Borel set $B \subset R^1$, put

(6.18) $$\pi_i(B) = E[U_i I_B(V_i)], \qquad\qquad i = 1, 2,$$
$$\pi(B) = E[U_1 U_2 I_B(V_1 + V_2)].$$

Let $P_i(u, v)$ be the d.f. of (U_i, V_i), $i = 1, 2$. Then

$$(6.19) \qquad \pi(B) = \int_{R^2 \times R^2} u_1 u_2 I_B(v_1 + v_2) \, dP_1(u_1, v_1) \, dP_2(u_2, v_2)$$

$$= \int_{R^2} I_B(v_1 + v_2) \, d\pi_1 \, d\pi_2 = \pi_1 * \pi_2(B).$$

On the other hand,

$$(6.20) \qquad \int_B f_i(v) \, dv = \int_B E[U_i | V_i = v] \, p_i(v) \, dv$$

$$= E[U_i I_B(V_i)] = \pi_i(B), \qquad i = 1, 2,$$

That is, $f_i(v)$ is the density of π_i with respect to the Lebesgue measure. Similarly, $f(v)$ is the density of π. Thus (6.19) implies (6.17).

This lemma can be extended in an obvious way to any finite number of vectors (U_i, V_i). Denote

$$(6.21) \qquad f(z; h, u) = E[\exp \{hY_1\} I_{\{Y_1 < u\}} | X_1 = z].$$

Putting $U_i = \exp \{hY_i\} I_{\{Y_i < u\}}$, $V_i = X_i$ for $i = 1, \cdots, n$, we obtain

$$(6.22) \qquad E[\exp \{h\tilde{S}_n\} I_{n,u} | \tilde{Z}_n = z] \, \tilde{p}_n(z) = [f(z : h, u) \, p(z)]^{n*}.$$

Comparing this with (6.15), we see that all we need to show is

$$(6.23) \qquad \sup_z [f(z; h_{n,u}, u) \, p(z)]^{n*} = O(n^{-1/2}).$$

Note that

$$(6.24) \qquad \int f(z; h, u) \, p(z) \, dz = E[\exp \{hY_1\} I_{\{Y_1 < u\}}] = R(h, u),$$

say. The proof of Theorem 1 in [5] contains the following estimate

$$(6.25) \qquad R(h, u) = 1 + hm_1 + 2\theta_1 h^2 m_2 + \theta_2 K_r d_r(h) \, e^{hu} u^{-r},$$

where $|\theta_1| < 1$, $0 < \theta_2 < 1$, $m_k = EY_1^k$, $k = 1, 2$ and K_r is a function of r only. Actually in that proof $M_r = E|Y_1|^r$ rather than $d_r(h)$ is used, but it appears there in the inequality $1 - Q(x) \leq M_r x^{-r}$ (in our notation) which is used only for $x > 1/h$ and therefore holds true when M_r is replaced by $d_r(h)$. This is the modification of the proof we referred to in the proof of Lemma 4.2. It follows from (6.10) and (6.12) that $d_r(h_{n,u}) \leq c_n$ for n large enough. In view of (6.11), (6.12) and $m_1 = EY_1 = 0$, we obtain from (6.25)

$$(6.26) \qquad R(h_{n,u}, u) = 1 + O(n^{-1}).$$

Put

$$(6.27) \qquad r_{n,u}(z) = \frac{f(z; h_{n,u}, u) \, p(z)}{R(h_{n,u}, u)}.$$

Since $r_{n,u}$ is nonnegative and integrates to unity by (6.23), it is a probability density. By virtue of (6.26), (6.23) is equivalent to

$$(6.28) \qquad \sup_z \left[r_{n,u}(z) \right]^{n*} = O(n^{-1/2}).$$

First we shall show that

$$(6.29) \qquad \limsup_{n,u} \sup_z \left[r_{n,u}(z) \right]^{2\nu*} < \infty.$$

On applying (6.25) to the conditional distribution of Y_1 given $X_1 = z$ (see (6.21) and (6.24)), we obtain

$$(6.30) \qquad f(z; h, u) = 1 + hm_1(z) + 2\theta_1 h^2 m_2(z) + \theta_2 K_r M_r(z) e^{hu} u^{-r}$$

where $m_i(z) = E[Y_1^i | X_1 = z]$, $i = 1, 2$. (Actually, (6.30) corresponds to the version of (6.25) with M_r instead of $d_r(h)$.) It follows from (6.11) and (6.10) that

$$(6.31) \qquad \exp\{h_{n,u} u\} u^{-r} = 1/c_n n \leqq n^{-1/2}.$$

Moreover,

$$(6.32) \qquad m_i(z) \leqq \left(M_r(z) \right)^{i/r} \leqq 1 + M_r(z), \qquad\qquad i = 1, 2.$$

Therefore

$$(6.33) \quad f(z; h, u) \leqq 1 + h + 2h^2 + (h + 2h^2 + K_r n^{-1/2}) M_r(z) = \bar{f}(z; h, u),$$

say.

Define $\bar{r}_{n,u}(z)$ by (6.27) with f replaced by $\bar{f}$. Then we have from (6.12), (6.26) and (6.33)

$$(6.34) \qquad r_{n,u}(z) \leqq \bar{r}_{n,u}(z) = x_{n,u} p(z) + \beta_{n,u} M_r(z) p(z),$$

where $\alpha_{n,u} \to 1$, $\beta_{n,u} \to 0$. Denote by $\rho_{n,u}(t)$, $\bar{\rho}_{n,u}(t)$, $\varphi(t)$ and $\varphi_r(t)$ the c.f. of $r_{n,u}(z)$, $\bar{r}_{n,u}(z)$, $p(z)$ and $M_r(z) p(z)/M_r$, respectively. Put

$$(6.35) \qquad L = \sup_z p^{\nu*}(z), \qquad L_r = \sup_z \left[M_r(z) p(z)/M_r \right]^{\nu*}.$$

(This is finite by assumption (iv) of Theorem 3.2.) By the Plancherel identity (see, for example, Feller [2], Chapter XV, equation (3.8)),

$$(6.36) \qquad \frac{1}{2\pi} \int |\varphi(t)|^{2\nu} \, dt = \int \left[p^{\nu*}(z) \right]^2 dz \leqq L \int p^{\nu*}(z) \, dz = L.$$

Similarly

$$(6.37) \qquad \frac{1}{2\pi} \int |\varphi_r(t)|^{2\nu} \, dt \leqq L_r.$$

Using Minkowski's inequality, we have from (6.34), (6.36) and (6.37)

$$(6.38) \qquad \frac{1}{2\pi} \int \left| \bar{\rho}_{n,u}(t) \right|^{2\nu} dt \leq \left[\alpha_{n,u} L^{1/2\nu} + \beta_{n,u} M_r L_r^{1/2\nu} \right]^{2\nu} = L_{n,u},$$

say, and $\alpha_{n,u} \to 1$, $\beta_{n,u} \to 0$ imply $L_{n,u} \to L$. Furthermore

$$(6.39) \qquad \sup_z \left[\bar{r}_{n,u}(z) \right]^{2\nu*} \leq \frac{1}{2\pi} \int \left| \bar{\rho}_{n,u}(t) \right|^{2\nu} dt \leq L_{n,u}.$$

Obviously, $r_{n,u}^{k*}(z) \leq \bar{r}_{n,u}^{k*}(z)$ for any $k = 1, 2, \cdots$, whence $\sup_z \left[r_{n,u}(z) \right]^{2\nu*} \leq L_{n,u}$ which proves (6.29). Applying again the Plancherel identity we obtain

$$(6.40) \qquad \frac{1}{2\pi} \int \left| \rho_{n,u}(t) \right|^{4\nu} dt \leq L_{n,u}.$$

By virtue of the inequality

$$(6.41) \qquad \sup_z \left[r_{n,u}(z) \right]^{n*} \leq \frac{1}{2\pi} \int \left| \rho_{n,u}(t) \right|^n dt = \frac{1}{2\pi} J_{n,u},$$

say, in order to prove (6.28) we need to show that $J_{n,u} = O(n^{-1/2})$. The relations (6.26) and (6.30) give

$$(6.42) \qquad \int \left| r_{n,u}(z) - p(z) \right| dz \to 0.$$

For any density $r(z)$, we shall write $r^{(2)}(z) = r * r^-(z)$ where $r^-(z) = r(-z)$. Then (6.42) implies

$$(6.43) \qquad \int \left| r_{n,u}^{(2)}(z) - p^{(2)}(z) \right| dz \to 0.$$

Take an arbitrary $\delta > 0$ and put

$$(6.44) \qquad b_{n,u} = \int_{|z| \leq 1/\delta} z^2 r_{n,u}^{(2)}(z) \, dz.$$

Then (6.43) implies

$$(6.45) \qquad b_{n,u} \to \int_{|z| \leq 1/\delta} z^2 p^{(2)}(z) \, dz.$$

Taking into account that the c.f. of $r_{n,u}^{(2)}(z)$ is $\left| \rho_{n,u}(t) \right|^2$ and using the first of the truncation inequalities [4], 12.4.B', we obtain for $|t| \leq \delta$

$$(6.46) \qquad \left| \rho_{n,u}(t) \right|^2 \leq 1 - \tfrac{1}{3} t^2 b_{n,u} \leq \exp \left\{ - \tfrac{1}{3} t^2 b_{n,u} \right\}$$

or

$$(6.47) \qquad \left| \rho_{n,u}(t) \right|^n \leq \exp \left\{ - \tfrac{1}{6} n t^2 b_{n,u} \right\}.$$

Further, (6.42) implies that $\rho_{n,u}(t) \to \varphi(t)$ uniformly in $t \in R^1$, whence

$$(6.48) \qquad \sup_{|t| \geq \delta} \left| \rho_{n,u}(t) \right| \to \sup_{|t| \geq \delta} \left| \varphi(t) \right| < 1.$$

Split the integral $J_{n,u}$ in (6.41) into the integrals over $\{|t| \leqq \delta\}$ and $\{|t| > \delta\}$ and use (6.47) in the first and the inequality

$$(6.49) \qquad |\rho_{n,u}(t)|^n \leqq \left[\sup_{|t| \geqq \delta} |\rho_{n,u}(t)|\right]^{n-4\delta} |\rho_{n,z}(t)|^{4\nu}$$

in the second of them. Then by virtue of (6.40), (6.45) and (6.48) we obtain $J_{n,u} = O(n^{-1/2})$ which was to be proved.

7. Concluding remarks

After strengthening certain assumptions in Theorem 2.1, the same proof, somewhat refined, could give for $\varepsilon_n(x)$ an estimate $O(n^{-\beta})$ with $\frac{1}{2} < \beta < 1$. However it is impossible to obtain the naturally expected order n^{-1} by the present method. For this reason we restrict ourselves to the assertion that $\varepsilon_n(x) = o(n^{-1/2})$.

Though it is not explicitly stated in (2.12), the function $\phi_n(x)$ depends on θ_0 because μ_3 and $\mu(x)$ do, thus $\phi_n(x) = \phi_n(x, \theta_0)$, say. There is no such dependence (and the d.f. of $Z_n(\hat{\theta}_n)$ does not depend on θ_0 at all) when θ is the location scale parameter and has an appropriate invariance property. In the general case we cannot determine the critical value $z_{n,\alpha}$ from the equation $\phi_n(z, \theta_0) = 1 - \alpha$ since θ_0 is unknown. It may be shown, however, that under certain smoothness of the dependence of μ_3 and $\mu(x)$ on θ the critical value $\hat{z}_{n,\alpha}$, determined from the equation $\phi_n(z, \hat{\theta}_n) = 1 - \alpha$, has the property that

$$(7.1) \qquad P\{Z_n(\hat{\theta}_n) > \hat{z}_{n,\alpha}\} = \alpha + o(n^{-1/2}),$$

that is, it provides the same order of approximation as with known θ_0.

REFERENCES

[1] K. G. Binmore and H. H. Stratton, "A note on characteristic functions," *Ann. Math. Statist.*, Vol. 40 (1969), pp. 303–307.
[2] W. Feller, *An Introduction to Probability Theory and its Applications*, Vol. 2, New York, Wiley, 1966.
[3] I. A. Ibragimov and Yu. V. Linnik, *Independent and Stationarily Connected Random Variables*, Moscow, "Nauka," 1965.
[4] M. Loève, *Probability Theory*, Van Nostrand, 1960.
[5] S. V. Nagaev, "Some limit theorems on large deviations," *Teor. Verojatnost. i Primen.*, Vol. 10 (1965), pp. 231–254.
[6] J. Neyman, "Optimal asymptotic tests of composite statistical hypotheses," *Probability and Statistics*, Uppsala, Almquist and Wiksells, (The Harald Cramér Volume), 1959, pp. 213–234.
[7] J. Neyman and E. L. Scott, "On the use of $C(\alpha)$ optimal tests of composite hypotheses," *Bull. Inst. Internat. Statist.*, Vol. 41 (1965), pp. 476–495.
[8] G. P. Steck, "Limit theorems for conditional distributions," *Univ. California Publ. Statist.*, Vol. 2 (1957), pp. 237–284.

LOCAL ASYMPTOTIC MINIMAX
AND ADMISSIBILITY IN ESTIMATION

JAROSLAV HÁJEK

FLORIDA STATE UNIVERSITY

and

CHARLES UNIVERSITY, PRAGUE

1. Introduction

In their vigorous search for an adequate asymptotic theory of estimation, statisticians have tried almost all their methodological tools: prior densities, minimax, admissibility, large deviations, restricted classes of estimates (invariant, unbiased), contiguity, and so forth. The resulting body of knowledge is somewhat atomized and a certain synthetic work seems to be needed. In this section, let us try to single out several pieces of the jigsaw puzzle and combine them into a logically connected theory. Along with this we shall criticize some other approaches and make a few historical remarks.

Consider a fixed parametric space θ and a sequence of experiments described by families of densities $p_n(x_n, \theta)$, say $p_n(x_n, \theta) = \Pi_{i=1}^n f(y_i, \theta)$, where $x_n = (y_1, \cdots, y_n)$. First of all, it is necessary to single out "regular cases." This should not be done only formally, for example, only in terms of θ derivatives of $f(y, \theta)$. The statistical essence of regularity consists in the possibility of replacing the family of distributions by a normal family in a local asymptotic sense. Loosely speaking, given a point $t \in \theta$ and a small vicinity V_t of t, the quantity

$$(1.1) \qquad \Delta_{n,t} = n^{-1/2} \frac{\partial}{\partial \theta} \log p_n(x_n, \theta)\big|_{\theta=t}$$

should be approximately sufficient and normal (with constant covariance and expectation linear in θ) for $\theta \in V_t$ and n large. See Section 3 for a precise definition. The idea of approximating a general family by a normal family was first formulated by A. Wald [19], and then sophisticatedly developed by L. LeCam [12], [13], [15]. In spite of its importance, the idea has not yet found its way into current textbooks.

The next step is to get rid of ill behaved estimates and to characterize optimum ones. This may be achieved by scrutinizing an arbitrary sequence of estimates T_n from the point of view of minimax and admissibility, again in a *local* asymptotic sense. Theorem 4.1 below entails that there is a lower bound for asymptotic local maximum risk and that this bound may be achieved only if

Research sponsored in part by National Science Foundation Grant # Gu 2612 to the Florida State University and in part by the Army, Navy and Air Force under Office of Naval Research Contract Number NONR-988(08), Task Order NR 042-004.

$$(1.2) \qquad \left[\sqrt{n} \, (T_n - t) - \frac{\Delta_{n,t}}{\Gamma_t} \right] \to 0, \qquad [p_{n,t}]$$

where Γ_t does not depend on the observation nor on the choice of $\{T_n\}$.

Once the condition (1.2) has been established, we can develop particular methods providing estimates satisfying (1.2) for every $t \in \theta$. These methods would include Bayes estimates with respect to diffuse priors, maximum likelihood estimates, maximum probability estimates, RBAN estimates, "nonparametric" estimates, and so forth. The justification for a diffuse prior in Bayes estimation—and probably the only sound one— is that the resulting estimates satisfy (1.2), and, under additional conditions, have satisfactory local asymptotic minimax properties. Minimax scrutiny of Bayes estimates seems to be even more important in cases when we estimate a scalar parametric function $\tau = \tau(\theta)$ of a parameter whose dimensionality increases with n. Then it may be far from obvious what a "diffuse" prior means. Unfortunately local asymptotic minimax results are not systematically available for this case.

In maximum likelihood and Bayes estimation, we are troubled by the behavior of likelihood tails. This may be avoided by considering only some vicinity of a consistent estimate. This leads to consistent estimates considered as a starting point for arriving at better estimates.

If the families $p_n(\cdot, \theta)$ are well behaved, then $\mathscr{L}_\theta(\Delta_{n,\theta})$ will be smooth in θ. In turn, the distribution of any estimate T_n for which (1.2) is satisfied will be smooth in θ. Assuming that the laws $\mathscr{L}_\theta[\sqrt{n}(T_n - \theta)]$ converge continuously in θ to some laws L_θ, it was proved in [6] that L_t may be decomposed as a convolution

$$(1.3) \qquad L_t = \phi_t * G_t$$

where ϕ_t is a normal distribution and G_t is a distribution depending on the choice of T_n. It also may be shown that the best possible G_t is degenerate at the point zero. If G_t is normal, the L_t is also normal with larger covariance matrix than ϕ_t. In such a case, the ratio of the generalized variances of ϕ_t and L_t may serve as a measure of the efficiency of T_n. A simple short proof of (1.3) has been recently suggested by P. Bickel (personal communication). LeCam [16] provided still another proof and extended the result to families which may not be locally asymptotically normal. Then, of course, ϕ_t will not be a normal distribution, but, for example, the one sided exponential distribution.

Some points in the above exposition need comment. First, let us explain what we understand by local asymptotic minimax and admissibility. Roughly speaking, by saying that $\{T_n\}$ is locally asymptotically minimax (admissible) we shall mean that for any open interval $V \subset \theta$ the estimate is approximately minimax (admissible) on V if n is large. Actually, for n large there is little justification for relating the minimax criterion and admissibility to the whole parametric space, since after having obtained our observations, we are able to locate the unknown parameter with considerable precision.

It is important to note that an estimate may be exactly minimax for every n, but may fail to be *locally* asymptotically minimax. For example, this is true about the minimax estimator of the binomial p from X successes in n trials:

$$(1.4) \qquad\qquad t_n(X) = \frac{X + \sqrt{n}/2}{n + \sqrt{n}}.$$

This estimate is obviously not approximately minimax relative to any open interval V not containing $p = 1/2$, consequently, it is not locally asymptotically minimax. Actually, the maximum mean square error of t_n on V is $[4n(1 + n^{-1/2})^2]^{-1}$, whereas the minimax on V is smaller than $n^{-1} \max_{p\epsilon V}\{p(1 - p)\}$.

Also admissibility for all n does not entail local asymptotic admissibility. Estimate (1.4) may illustrate this point as well. Another example may be given from the field of sample surveys, where Joshi and Godambe (see [11]) proved that the sample average is an admissible estimate for the population average no matter what the sample design is. Again this estimate fails to be locally asymptotically admissible if the concept is properly defined for this situation.

According to Chernoff [5] the idea of local asymptotic minimax is due to C. Stein and H. Rubin, who showed that Fisher's programs could be rescued by it. The proof that local asymptotic minimax implies local asymptotic admissibility was first given by LeCam ([12], Theorem 14). There, he proved that superefficiency at one point entails bad risk values in the vicinity of this point, that is, that superefficiency excludes the local asymptotic minimax property. Apparently not many people have studied LeCam's paper so far as to read this very last theorem, and the present author is indebted to Professor LeCam for giving him the reference. Theorems 4.1 and 4.2 below may be regarded as extensions of the above mentioned result by LeCam and a related theorem by Huber. Huber, [10] in addition to LeCam's statement, proves that a locally asymptotically minimax estimate must be asymptotically normal with a given variance. This is made more precise by (1.2), since (1.2) entails asymptotic normality of $\sqrt{n}(T_n - t)$ on the basis of assumed asymptotic normality of $\Delta_{n,t}$. From the methodological point of view, LeCam's and the present paper both use the Blyth [4] approach to admissibility based on a normal prior, whereas Huber used the Hodges-Lehmann [9] approach based on the Cramér-Rao inequality.

In his subsequent papers [13], [15] LeCam perfected the idea of a locally asymptotically normal family. We shall base our proof on these papers, rather than on his 1953 paper [12] which contains some omissions. When approximating a general family by a normal one, the basic point is that the distance should be expressed in terms of the L_1 norm (variation) given by (3.5) below. The Lévy or Prohorov distance would not do in proving (4.8) below. Of course, we may not expect that, for example, the distribution of $\Delta_{n,t}$ approaches its normal limit in the L_1 norm. However, as was shown by LeCam [13] and as is proved here under different assumptions in Lemmas 3.2 and 3.3, it approaches in the L_1 norm a slightly deformed normal law with slightly deformed likelihood function, which is sufficient for our purposes. Another important point provided in LeCam [15] was a general definition of locally asymptotically normal families, which is not restricted to partial sequences in an infinite sequence of independent replications of some basic experiment. We take this attitude also, and our local asymptotic normality (LAN) conditions of Section 3 represent a selection from LeCam's [15] conditions DN1–7.

Another comment is invited by (1.2) in connection with the approach suggested by C. R. Rao ([17], p. 285) in his admirable book. He uses property (1.2) directly as a definition of asymptotic efficiency, which seems to be justified by the fact that the ratio of Fisher's information for T_n and for the whole observation, respectively, approaches unity under (1.2) and some additional assumptions. However, since Fisher's information is invariant under one-to-one transformations, I doubt whether giving it a central position in estimation problems can be rationally explained, even if used jointly with consistency. For example, in estimating the location parameter from n independent observations if the density is otherwise known, the maximun likelihood estimator contains all of the Fisher information (the apparently lost part of it may be "recovered"), but it is not the best estimate, not even the best location invariant estimate. The extent to which the maximum likelihood estimate lags behind the best one depends on how much the likelihoods are irregularly shaped, asymmetric, for example. Theorem 4.1 below provides a minimax justification for (1.2), which is somewhat less mystical.

To be more precise, Rao calls an estimate efficient, if (1.2) holds with Γ_t^{-1} replaced by any function β_t not involving the observations, but possibly depending on the estimate T_n. However, Theorem 4.1 implies that T_n cannot be locally asymptotically minimax, if $\beta_t \neq \Gamma_t^{-1}$. Rao is aware of bad consequences of $\beta_t \neq \Gamma_t^{-1}$ and proves in [18] that then the law of $\sqrt{n}(T_n - \theta)$ does not converge uniformly in θ, so that the limiting laws cannot be used to provide approximate confidence intervals.

Continuing our comments on approaches not embodied in the above exposition, let us mention another paper by LeCam [14]. There he proved that under certain conditions Bayes estimates provide asymptotic risks that can be improved on a set of measure zero only. Thus, if the prior is diffuse and the asymptotic risk of a Bayes estimate is continuous in θ, as it usually is, then it dominates the asymptotic risk of any other estimate having continuous asymptotic risk. A disadvantage of this approach is that it defines lower bounds in terms of certain estimates—Bayes estimates—and not in terms of intrinsic properties of the families of distributions involved. Also, a verification of assumptions entails difficult consistency investigations. On the other hand, it is the first paper of a very general scope, for example, the estimates are not assumed to be necessarily asymptotically normal. In regular cases, on which the present paper is focused, Bayes estimates for diffuse priors will satisfy (1.2), so that we get a good agreement with our previous considerations.

Investigations of estimates which are not necessarily asymptotically normal were started from a different angle by J. Wolfowitz [22]. He again defined the best possible asymptotic behavior in terms of certain special estimates—generalized maximum likelihood estimates. In order to be comparable in his sense, the estimates must be "regular," so that his approach cannot be applied to arbitrary estimates. However, for regular estimates we can under LAN conditions establish the decomposition (1.3), which provides the conclusions obtained by J. Wolfowitz more directly. As was shown by LeCam [16] similar decomposition as (1.3) may be obtained also for the "nonregular case" investigated by Weiss and Wolfowitz [20].

The property of being a generalized maximum likelihood estimate is implied by (1.2) and in the usual situation also the converse is true. The same holds about maximum probability estimators.

All of the above approaches are based on "ordinary" deviations of the estimate from the parameter, which are typically of order $n^{-1/2}$. A concept of asymptotic efficiency based on large deviations has been suggested by D. Basu [3] and R. R. Bahadur [1] and [2]. Fixing θ and some $\varepsilon > 0$, we define $\tau_n(\varepsilon, \theta)$ as the standard deviation of a normal distribution under which $|T_n - \theta| \geq \varepsilon$ would have the same probability as under $P_{n,\theta}$. Formally

$$(1.5) \qquad P_\theta(|T_n - \theta| \geq \varepsilon) = 2\Phi\left[-\frac{\varepsilon}{\tau_n(\varepsilon, \theta)}\right].$$

Now, under regularity conditions somewhat stronger than LAN below it may be proved that for any consistent estimate

$$(1.6) \qquad \lim_{\varepsilon \to 0} \lim_{n \to \infty} \{n\tau_n^2(\varepsilon, \theta)\} \geq \Gamma_\theta^{-1}$$

and that the equality holds for the maximum likelihood estimator. It will generally hold also for estimates satisfying (1.2). The only trouble with this approach is that (1.2) is no longer necessary and that too many estimates satisfy equality in (1.6), for example, the "superefficient" Hodges estimator, which must clearly be rejected by the minimax and ordinary deviations point of view. However, equality in (1.6) certainly may be used as an additional requirement for good estimates. Our main proposition is partially proved by a different method in [16].

2. Normal families of distributions

Consider the estimation of $\theta \in R$ by a random variable Z which has the normal distribution $N(\theta, 1)$ given θ. We shall assume that the loss function is $\ell(\hat{\theta} - \theta)$, where ℓ satisfies the following conditions:

$$(2.1) \qquad \ell(y) = \ell(|y|),$$

$$(2.2) \qquad \ell(y) \leq \ell(z) \qquad\qquad |y| \leq |z|,$$

(2.3) $$\int_{-\infty}^{\infty} \ell(y) \exp\left\{-\tfrac{1}{2}\lambda y^2\right\} dy < \infty \qquad\qquad \lambda > 0,$$

(2.4) $$\ell(0) = 0.$$

Condition (2.3) entails

(2.5) $$\int_{-\infty}^{\infty} \ell(y) y^2 \exp\left\{-\tfrac{1}{2}\lambda y^2\right\} dy < \infty, \qquad\qquad \lambda > 0.$$

We shall also introduce a truncated version of ℓ:

(2.6) $$\ell_a(y) = \min\left(\ell(y), a\right), \qquad\qquad 0 < a \leqq \infty,$$

(2.7) $$r = (2\pi)^{-1/2} \int_{-\infty}^{\infty} \ell(y) \exp\left\{-\tfrac{1}{2}y^2\right\} dy,$$

(2.8) $$r_a = (2\pi)^{-1/2} \int_{-\infty}^{\infty} \ell_a(y) \exp\left\{-\tfrac{1}{2}y^2\right\} dy$$

and

(2.9) $$r_a(b) = (2\pi)^{-1/2} \int_{-\sqrt{b}}^{\sqrt{b}} \ell_a(y) \exp\left\{-\tfrac{1}{2}y^2\right\} dy.$$

The estimator will be considered randomized, that is,

(2.10) $$\hat{\theta} = \xi(Z, U)$$

where U is a randomized variable. For convenience, we shall assume that U is uniformly distributed on $(0, 1)$. As always, U is independent of Z and θ. The introduction of randomized estimates is justified since our loss function $\ell(y)$ may not be convex.

The risk function corresponding to ℓ_a and ξ will be denoted as follows:

(2.11) $$R_a(\theta; \xi) = (2\pi)^{-1/2} \int_{-\infty}^{\infty} \int_0^1 \ell_a[\xi(z, u) - \theta] \exp\left\{-\tfrac{1}{2}(z - \theta)^2\right\} du\, dz.$$

The following is an extension of a result of Blyth [4] to situations involving truncation and randomization. The important feature in the lemma is the independence of the numbers a, α, b on ξ if (2.12) holds. Since the lemma follows the pattern of Blyth [4], the proof will be condensed.

LEMMA 2.1. *Under the above notations and assumptions, for any $\varepsilon > 0$ there exist positive numbers, a, b, α and a prior density $\pi(\theta)$, all depending on ε only, with the following property:*

For any randomized estimator $\xi(Z, U)$ such that

(2.12) $$P\left(\left|\xi(Z, U) - Z\right| > \varepsilon \middle| \theta = 0\right) > \varepsilon$$

then

(2.13) $$\int_{-b}^{b} \pi(\theta) R_a(\theta; \xi)\, d\theta > r + \alpha.$$

PROOF. It suffices to show that

(2.14)
$$\int_{-b}^{b} \pi(\theta) R_a(\theta; \xi) \, d\theta > r_a(b) + 2\alpha$$

for a, b sufficiently large and for an α which is independent of a, b. Note that

(2.15)
$$(2\pi)^{-1/2} \int_{-\sqrt{b}}^{\sqrt{b}} \ell_a(y - \beta) \exp\left\{-\tfrac{1}{2}y^2\right\} dy$$

$$\geqq r_a(b) + (2\pi)^{-1/2} \int_0^{\sqrt{b}} \left[\int_{-x}^{x} - \int_{-x+\beta}^{x+\beta}\right] \exp\left\{-\tfrac{1}{2}y^2\right\} dy \, d\ell_a(x)$$

$$\geqq r_a(b) + \delta, \quad \text{if} \quad |\beta| > \tfrac{1}{2}\varepsilon$$

where $\delta > 0$ depends only on ε, but not on a, b when they are sufficiently large.

Next, following the idea of Blyth [4], we shall assume θ to be distributed with density

(2.16)
$$\pi(\theta) = \frac{1}{\sigma(2\pi)^{1/2}} \exp\left\{-\frac{\theta^2}{2\sigma^2}\right\}$$

where σ will be appropriately chosen to depend on ε later. Then the conditional density of θ given $Z = z$ is

(2.17)
$$\psi(\theta|z) = \frac{(1 + \sigma^2)^{1/2}}{\sigma(2\pi)^{1/2}} \exp\left\{-\frac{1 + \sigma^2}{2\sigma^2}\left(\theta - \frac{z\sigma^2}{1 + \sigma^2}\right)^2\right\}$$

and the overall density of Z will be

(2.18)
$$f(z) = \frac{1}{[(1 + \sigma^2)(2\pi)]^{1/2}} \exp\left\{-\frac{z^2}{2(1 + \sigma^2)}\right\}.$$

In what follows we shall assume that

(2.19)
$$|z| \leqq b - \sqrt{b}.$$

Then

(2.20)
$$\int_{-b}^{b} \ell_a[\xi(z, u) - \theta] \psi(\theta|z) \, d\theta$$

$$\geqq (2\pi)^{-1/2} \int_{-\sqrt{b}}^{\sqrt{b}} \ell_a(y) \exp\left\{-\frac{y^2(1 + \sigma^2)}{2\sigma^2}\right\} dy$$

$$\geqq r_a(b) - \frac{K}{\sigma^2}$$

where K does not depend on a, b, σ^2. We have used the inequality

(2.21)
$$\exp\left\{-\frac{y^2(1 + \sigma^2)}{2\sigma^2}\right\} > \left[1 - \frac{y^2}{\sigma^2}\right] \exp\left\{-\tfrac{1}{2}y^2\right\}.$$

On the other hand, if

$$(2.22) \qquad\qquad |\xi(z, u) - z| > \varepsilon, \qquad\qquad\qquad |z| < M$$

we shall have

$$(2.23) \qquad\qquad \left| \xi(z, u) - \frac{z\sigma^2}{1 + \sigma^2} \right| > \tfrac{1}{2}\varepsilon \qquad \text{for} \quad (1 + \sigma^2) > \frac{2M}{\varepsilon}.$$

Consequently, in view of (2.15), we have

$$(2.24) \qquad (2\pi)^{-1/2} \int_{-b}^{b} \ell_a[\xi(z, u) - \theta]\psi(\theta|z)\, d\theta$$

$$\geqq (2\pi)^{-1/2} \int_{-\sqrt{b}}^{\sqrt{b}} \ell_a(y + \tfrac{1}{2}\varepsilon) \exp\left\{ -\frac{y^2(1 + \sigma^2)}{2\sigma^2} \right\} dy$$

$$\geqq (2\pi)^{-1/2} \int_{-\sqrt{b}}^{\sqrt{b}} \ell_a(y + \tfrac{1}{2}\varepsilon) \exp\left\{ -\tfrac{1}{2}y^2 \right\} dy - \frac{K}{\sigma^2}$$

$$\geqq r_a(b) + \delta - \frac{K}{\sigma^2},$$

if (2.19) and (2.23) hold.

Altogether, we have

$$(2.25) \qquad \int_{-b}^{b} \pi(\theta) R_a(\theta; \xi)\, d\theta = \int_0^1 \int_{-\infty}^{\infty} \int_{-b}^{b} \ell_a[\xi(z, u) - \theta]\psi(\theta|z) f(z)\, d\theta dz du$$

$$\geqq r_a(b) P(|Z| < b - \sqrt{b}) - \frac{K}{\sigma^2} + \delta P(|\xi(Z, U) - Z| > \varepsilon, |Z| < M).$$

From (2.12) we see that

$$(2.26) \qquad P(|\xi(Z, U) - Z| > \varepsilon, |Z| < M | \theta = 0) > \tfrac{1}{2}\varepsilon$$

for M sufficiently large. Now under $\theta = 0$ the density of Z is $(2\pi)^{-1/2} \exp\{-\tfrac{1}{2}z^2\}$, whereas the overall density is given by $f(z)$ of (2.18). The likelihood ratio for the two densities is for $|z| < M$ greater than

$$(2.27) \qquad\qquad \frac{1}{(1 + \sigma^2)^{1/2}} \exp\left\{ -\tfrac{1}{2}M^2 \right\}.$$

Thus

$$(2.28) \quad P(|\xi(Z, U) - Z| > \varepsilon, |Z| < M) > \frac{\varepsilon}{2(1 + \sigma^2)^{1/2}} \exp\left\{ -\tfrac{1}{2}M^2 \right\}.$$

Now it suffices to put

$$(2.29) \qquad\qquad 3\alpha = \frac{\delta\varepsilon}{2(1 + \sigma^2)^{1/2}} \exp\left\{ -\tfrac{1}{2}M^2 \right\} - \frac{K}{\sigma^2}$$

which is positive for σ sufficiently large. In the last step we choose a and b in such a way that

(2.30) $$r_a(b) > r - \tfrac{1}{2}\alpha$$

and given the above chosen σ^2,

(2.31) $$rP(|Z| > b - \sqrt{b}) < \tfrac{1}{2}\alpha.$$

Now (2.25) to (2.31) yield (2.14). $Q.E.D.$

Of course, usually there will be nuisance parameters, and then the following k-dimensional version of the preceding lemma is useful.

LEMMA 2.2. *Let $Z = (Z_1, \cdots, Z_k)$ be a normal random vector with expectation $\theta = (\theta_1, \cdots, \theta_k) \in R^k$ and fixed positive definite covariance matrix $\Sigma = \{\sigma_{i,j}\}_{i,j=1}^k$. Consider a function ℓ satisfying (2.1) to (2.4). Then for every $\varepsilon > 0$ there exist positive numbers a, b, α and a one-dimensional density $\pi(\cdot)$ with the following property:*

For any randomized estimator $\xi(Z_1, \cdots, Z_k, U)$ of θ_1 such that

(2.32) $$P(|\xi(Z_1, \cdots, Z_k, U) - Z_1| > \varepsilon | \theta_1 = \cdots = \theta_k = 0) > \varepsilon$$

then

(2.33) $$\int_{-b}^b \pi(\theta_1) E\{\ell_a[\xi(Z_1, \cdots, Z_k, U) - \theta_1]|\theta_i = \theta_1\sigma_{1,i}\sigma_{1,1}^{-1}, 1 \le i \le k\} \, d\theta_1$$

$$\ge (2\pi)^{-1/2} \int_{-\infty}^{\infty} \ell[y\sigma_1] \exp\{-\tfrac{1}{2}y^2\} \, dy + \alpha$$

where $\sigma_1 = (\sigma_{1;1})^{1/2}$

PROOF. For the submodel $\theta_i = \theta_1\sigma_{1,i}\sigma_{1,1}^{-1}, 1 \le i \le k, -\infty < \theta_1 < \infty, Z_1$ is a sufficient statistic, and the result follows from the preceeding lemma.

LEMMA 2.3. *Let Z be the same vector as in Lemma 2.2. Then for every $\delta > 0$ there exist positive numbers a, b and a density $\pi(\cdot)$ such that for any randomized estimator $\xi(Z_1, \cdots, Z_k, U)$*

(2.34) $$\int_{-b}^b \pi(\theta_1) E\{\ell_a[\xi(Z_1, \cdots, Z_k, U) - \theta_1]|\theta_i = \theta_1\sigma_{1,i}\sigma_{1,1}^{-1}, 1 \le i \le k\} \, d\theta_1$$

$$\ge (2\pi)^{-1/2} \int_{-\infty}^{\infty} \ell[y\sigma_1] \exp\{-\tfrac{1}{2}y^2\} \, dy - \delta$$

where σ_1 has the same meaning as in Lemma 2.2.

PROOF. The proof follows the same lines as the proofs of the previous lemmas.

3. Locally asymptotically normal families of distributions

The essence of the definition of a locally asymptotically normal family is that the log likelihood ratio is asymptotically normally distributed with a covariance matrix, which is locally constant, and with an expectation which is locally a linear function of θ. For our purpose—to obtain lower bounds for risks and necessary conditions for allowing equality in corresponding inequalities—we need the following version, employed already in [6]: consider a sequence of statistical experiments $(\mathscr{X}_n, \mathscr{A}_n, P_n(\cdot, \theta))$, $n \geq 1$, where θ runs through an open subset Θ of R^k. Take a point $t \in \Theta$ and assume it to be the true value of the parameter θ. We shall abbreviate $P_n = P_n(\cdot, t)$ and $P_{n,h} = P_n(\cdot, t + n^{-1/2}h)$. If appropriate, we could use in place of $\sqrt{n}$ some more general norming numbers $k(n) \to \infty$, or even matrices as in [6].

The norm of a point $h = (h_1, \cdots, h_k)$ from R^k will be denoted by $|h| = \max_{1 \leq i \leq k}|h_i|$, and $h'v$ will denote the scalar product of two vectors.

Given two probability measures P and Q, denote by dQ/dP the Radon-Nikodym derivative of the *absolutely continuous* part of Q with respect to P. Introduce the family of likelihood ratios

$$(3.1) \qquad r_n(h, x_n) = \frac{dP_{n,h}}{dP_n}(x_n), \qquad h \in R^k, n \geq n_h, x_n \in \mathscr{X}_n,$$

where n_h denotes the smallest integer such that $n \geq n_h$ entails $t + n^{-1/2}h \in \Theta$. In what follows the argument x_n will usually be omitted.

ASSUMPTION 3.1.   LAN *(local asymptotic normality) at* $\theta = t$. *Assume that*

$$(3.2) \qquad r_n(h) = \exp\{h'\Delta_{n,t} - \tfrac{1}{2}h'\Gamma_t h + Z_n(h, t)\}, \qquad h \in R^k, n \geq n_h,$$

where the random vector $\Delta_{n,t}$ *satisfies* $\mathscr{L}(\Delta_{n,t}|P_n) \to N(0, \Gamma_t)$, *and* $Z_n(h, t) \to 0$ *in* P_n *probability for every* $h \in R^k$. *Further assume that* $\det \Gamma_t > 0$.

EXAMPLE.   Consider the case $k = 1$, $x_n = (y_1, \cdots, y_n) \in R^n$ and $p_n(x_n, \theta) = \Pi_{i=1}^n f(y_i, \theta)$.

Then the existence of Fisher's information, its positivity and continuity at $\theta = t$, entails LAN. Of course, in order for Fisher's information to be well defined, $f(y, \theta)$ must be absolutely continuous in θ in a vicinity of t, and the derivative $\dot{f}(y, t) = (\partial/\partial\theta)f(y, \theta)|_{\theta=t}$ must exist for almost all y. Then the information equals

$$(3.3) \qquad I_\theta = \int_{-\infty}^\infty \left\{\frac{[\dot{f}(y, \theta)]^2}{f(y, \theta)}\right\} dy.$$

Satisfaction of LAN for this case may be proved by methods developed in [8] as is shown in the Appendix below. Specifically (3.2) will be satisfied for

$$(3.4) \qquad \Delta_{n,t} = n^{-1/2} \sum_{i=1}^n \frac{\dot{f}(y_i, t)}{f(y_i, t)}, \qquad \Gamma_t = I_t.$$

Let $\|P - Q\|$ be the L_1 norm of two probability measures. If p and q are densities of P and Q with respect to μ, then

$$(3.5) \qquad \|P - Q\| = \int |p - q| \, d\mu.$$

The following lemmas are essentially contained in LeCam [13], and form a bridge between exactly normal models and locally asymptotically normal models.

LEMMA 3.1. *For any sequence of statistics* $\{S_n\}$ *put*

$$(3.6) \qquad s_n(x, u) = \inf \{y : P_n[S_n \leq y | \Delta_{n,t} = x] \geq u\}, \qquad x \in R, 0 < u < 1,$$

and denote by $F_{n,h}$ *the distribution of* S_n *under* $P_{n,h}$ *and by* $F^*_{n,h}$ *the distribution of* $s_n(\Delta_{n,t}, U)$ *also under* $P_{n,h}$, *if* U *is uniformly distributed on* $(0, 1)$ *and independent of* $\Delta_{n,t}$.

Then, under Assumptions LAN

$$(3.7) \qquad \lim_{n \to \infty} \| F_{n,h} - F^*_{n,h} \| = 0, \qquad h \in R.$$

PROOF. We see from (3.6) that $F_{n,h} = F^*_{n,h}$ if $h = 0$. Now, as shown in the proof of the theorem in [6], $P_{n,h}$ may be approximated by $Q_{n,h}$ such that $\Delta_{n,t}$ is a sufficient statistic for the pair $(P_n, Q_{n,h})$ and $\| Q_{n,h} - P_{n,h} \| \to 0$. That yields (3.7).

In the next two lemmas we shall treat the cases $k = 1$ and $k > 1$ separately, because for $k = 1$ we are able to reach a greater degree of explicitness.

LEMMA 3.2. *Assume* $k = 1$. *Denote* $G_{n,h}(x) = P_{n,h}(\Delta_{n,t} \leq x)$ *and* $\Phi(x) = (2\pi) \int_{-\infty}^x \exp \{-\frac{1}{2}y^2\} \, dy$. *Let* $G^*_{n,h}$ *be the distribution of* $G^{-1}_{n,0} \Phi(Z\Gamma_t^{-1/2})$, *if* Z *is normal* $(h\Gamma_t, \Gamma_t)$.

Then, under Assumption LAN,

$$(3.8) \qquad \lim_{n \to \infty} \| G_{n,h} - G^*_{n,h} \| = 0, \qquad h \in R.$$

PROOF. We again have $G_{n,h} = G^*_{n,h}$ for $h = 0$. Since $G_{n,0}(x) \to \Phi(x\Gamma_t^{-1/2})$ uniformly under Assumptions LAN we have

$$(3.9) \qquad G^{-1}_{n,0} \Phi(x\Gamma_t^{-1/2}) \to x \text{ uniformly on compacts.}$$

Further, referring again to the proof of the theorem of [6], we may approximate $G_{n,h}$ by $\bar{G}_{n,h}(x) = Q_{n,h}(\Delta_{n,t} \leq x)$ satisfying $\| G_{n,h} - \bar{G}_{n,h} \| \to 0$. The rest follows by showing that

$$(3.10) \qquad \frac{d\bar{G}_{n,h}}{dG_{n,0}}(x) \to \exp \{-hx - \tfrac{1}{2}h^2\Gamma_t\},$$

$$(3.11) \qquad \frac{dG^*_{n,h}}{dG_{n,0}}(x) \to \exp \{-hx - \tfrac{1}{2}h^2\Gamma_t\}.$$

LEMMA 3.3. *Assume* $k \geq 1$. *Denote* $G_{n,h}(x) = P_{n,h}(\Delta_{n,t} \leq x)$, $x \in R^k$, *and denote by* $Z = (Z_1, \cdots, Z_k)$ *a random vector such that* $\mathcal{L}(Z) = N(\Gamma_t h, \Gamma_t)$, *if* $\theta = t + n^{-1/2}h$. *Then there exists a sequence of functions* $\phi_n(x)$ *such that*

$$(3.12) \qquad \lim_{n \to \infty} \sup_{x \in R^k} |\phi_n(x) - x| = 0$$

and the distribution of $\phi_n(Z)$ *under* $\theta = t + n^{-1/2}h$, *say* $G^*_{n,h}$, *satisfies*

$$(3.13) \qquad \lim_{n \to \infty} \| G^*_{n,h} - G_{n,h} \| = 0.$$

PROOF.   Consider a sequence of cubes

$$(3.14) \qquad C_j = \{(x_n, \cdots, x_k): |x_i| \leq j, 1 \leq i \leq k\}, \qquad j = 1, 2, \cdots.$$

Partition each cube C_j into j^{2k} subcubes $C_{j,i}$, $1 \leq i \leq j^{2k}$, all of equal volume $(2/j)^k$. Since $G_{n,0} \to N(0, \Gamma_t)$, we have

$$(3.15) \qquad \int_{C_{j,i}} dG_{n,0} \to \int_{C_{j,i}} dN(0, \Gamma_t).$$

Let $\phi_{n,j}$ be a function mapping $C_{j,i}$ into $C_{j,i}$,

$$(3.16) \qquad \phi_{n,j}: C_{j,i} \to C_{j,i}, \qquad 1 \leq j \leq \infty, 1 \leq i \leq j^{2k}$$

and such that

$$(3.17) \qquad \phi_{n,j}(x) = x \quad \text{for} \quad x \notin C_j.$$

Furthermore, we choose $\phi_{n,j}$ so that the difference between the $G^*_{n,j,0} = \mathscr{L}(\phi_{n,j}(Z)|N(0,\Gamma_t))$ and $G_{n,0}$ tends to zero in the L_1 norm, that is,

$$(3.18) \qquad \|G_{n,j,0} - G_{n,0}\| < \frac{1}{j} \quad \text{as} \quad n \geq n_j.$$

This is possible in view of (3.15). Now define $j(n)$ by $n_{j(n)} \leq n < n_{j(n)+1}$ and put

$$(3.19) \qquad \phi_n(x) = \phi_{n, j(n)}(x).$$

Then $\phi_n(x)$ satisfies (3.12) because of (3.17) and (3.16) and the fact that diameters of the cubes $C_{j,i}$ converge to 0 as $j \to \infty$. Furthermore, (3.18) entails $\|G^*_{n,0} - G_{n,0}\| \to 0$ since $G^*_{n,0} = G^*_{n,j(n),0}$.

The proof may then be concluded by showing that (3.10) holds, referring again to the theorem of [6], and that (3.11) is true. The last statement follows from (3.12).

4. The main proposition

We shall generalize and modify theorems by LeCam [12] and P. Huber [10], treating the cases $k = 1$ and $k \geq 1$ separately, again.

THEOREM 4.1.   *Assume $k = 1$. Under Assumptions* LAN *of Section 3, any sequence of estimates $\{T_n\}$ for θ satisfies for ℓ of (2.1) to (2.4)*

$$(4.1) \qquad \lim_{\delta \to 0} \lim_{n \to \infty} \inf \sup_{|\theta - t| < \delta} E_\theta \{\ell[\sqrt{n}(T_n - \theta)]\}$$

$$\geq (2\pi)^{-1/2} \int_{-\infty}^{\infty} \ell(y\Gamma_t^{-1/2}) \exp\{-\tfrac{1}{2}y^2\} \, dy.$$

Furthermore, we can have for a nonconstant ℓ

$$(4.2) \qquad \lim_{\delta - 0} \lim_{n \to \infty} \sup_{|\theta - t| < \delta} E_\theta \{\ell[\sqrt{n}(T_n - \theta)]\}$$

$$= (2\pi)^{-1/2} \int_{-\infty}^{\infty} \ell(y\Gamma_t^{-1/2}) \exp\{-\tfrac{1}{2}y^2\} \, dy$$

only if

(4.3)
$$\sqrt{n}(T_n - t) - \Gamma_t^{-1}\Delta_{n,t} \to 0$$

in P_n probability.

PROOF. We shall first prove (4.1). Introducing local coordinates by $\theta = t + hn^{-1/2}$ and using $E_\theta(\cdot)$ and $E(\cdot|\theta)$ interchangeably, we may write for n sufficiently large

(4.4)
$$\sup_{|\theta - t| < \delta} E_\theta\{\ell[\sqrt{n}(T_n - \theta)]\}$$

$$\geqq \int_{-b}^{b} \pi(h)E\{\ell_a[\sqrt{n}(T_n - t) - h]|t + n^{-1/2}h\}\,dh,$$

whatever the constants a, b and density $\pi(\cdot)$ may be. We fix some $\bar{\delta} > 0$ and choose a, b and π in such a way that

(4.5)
$$\int_{-b}^{b} \pi(h)E\{\ell_a[\xi(Z, U) - h]|t + n^{-1/2}h\}\,dh$$

$$\geqq (2\pi)^{-1/2}\int_{-\infty}^{\infty} \ell(y\Gamma_t^{-1/2})\exp\{-\tfrac{1}{2}y^2\}\,dy - \bar{\delta}$$

for any estimator $\xi(Z, U)$, provided that $\mathcal{L}(Z|t + n^{-1/2}h) = N(\Gamma_t h, \Gamma_t)$. This is possible according to Lemma 2.3, if applied to $\Gamma_t^{-1}Z$ which is normal (h, Γ_t^{-1}).

Next we identify $S_n = \lambda_n(T_n - t)$ in Lemma 3.1 and conclude (ℓ_a is bounded!) that for every $h \in R$

(4.6)
$$\big|E\{\ell_a[\sqrt{n}(T_n - t) - h]|t + n^{-1/2}h\}$$
$$- E\{\ell_a[s_n(\Delta_{n,t}, U) - h]|t + n^{-1/2}h\}\big| \to 0.$$

Furthermore, by Lemma 3.2, putting

(4.7)
$$\xi_n(Z, U) = s_n(G_{n,0}^{-1}\phi(Z\Gamma_t^{-1/2}), U)$$

we obtain for every $h \in R$

(4.8)
$$\big|E\{\ell_a[s_n(\Delta_{n,t}, U) - h]|t + n^{-1/2}h\}$$
$$- E\{\ell_a[\xi_n(Z, U) - h]|t + n^{-1/2}h\}\big| \to 0.$$

Consequently

(4.9)
$$\int_{-b}^{b} \pi(h)E\{\ell_a[\sqrt{n}(T_n - t) - h]|t + n^{-1/2}h\}\,dh$$

$$\geqq \int_{-b}^{b} \pi(h)E\{\ell_a[\xi_n(Z, U) - h]|t + n^{-1/2}h\}\,dh - \bar{\delta}, \quad n > n(a, b, \pi, \bar{\delta}).$$

Combining (4.4), (4.9), and (4.5), we obtain (4.1).

Now we shall prove the necessity of (4.3) for (4.2). Assuming that (4.3) does not hold we shall contradict (4.2). Again putting $S_n = \lambda_n(T_n - t)$, and recalling Lemma 3.1, we can see that $\lambda_n(T_n - t) - \Gamma_t^{-1}\Delta_{n,t}$ has the same distribution under P_n as

(4.10) $$s_n(\Delta_{n,t}, U) - \Gamma_t^{-1}\Delta_{n,t}.$$

Furthermore, in view of (3.9), if expression (4.10) fails to converge to zero in probability, then also

(4.11) $$\xi_n(Z, U) - \Gamma_t^{-1}Z$$

fails to do so. But then, there is an $\varepsilon > 0$ such that for every n there exists an $m > n$ such that

(4.12) $$P_m(|\xi_m(Z, U) - \Gamma_t^{-1}Z| > \varepsilon) > \varepsilon.$$

Therefore, according to Lemma 2.2, applied to $\Gamma_t^{-1}Z$ and $k = 1$, we choose a, b, α, and $\pi(\cdot)$ such that

(4.13) $$\int_{-b}^{b} \pi(h)E\{\ell_a[\xi_m(Z, U) - h]|t + n^{-1/2}h\}\,dh$$

$$> (2\pi)^{-1/2}\int_{-\infty}^{\infty} \ell(y\Gamma_t^{-1/2})\exp\left\{-\tfrac{1}{2}y^2\right\}dy + \alpha.$$

This in connection with (4.4) and (4.9) contradicts (4.2), since $\bar{\delta}$ can be made smaller than α. Q.E.D.

Without proof let us also formulate a k-dimensional version of the previous theorem. The parametric function of interest will be the first coordinate, θ_1, while the other coordinates will be regarded as nuisance parameters. The proof could be based on Lemma 2.2.

THEOREM 4.2.   *Under Assumptions* LAN *above, any sequence of estimates* T_n *for the first coordinate* θ_1 *of* θ *satisfies*

(4.14) $$\lim_{\delta \to 0}\lim_{n \to \infty}\inf\;\sup_{|\theta - t| < \delta} E_\theta\{\ell[\sqrt{n}(T_n - \theta_1)]\}$$

$$\geqq (2\pi)^{-1/2}\int_{-\infty}^{\infty} \ell(\sigma_1^t y)\exp\left\{-\tfrac{1}{2}y^2\right\}dy$$

where $\sigma_1^t = (\sigma_{t,1,1})^{1/2}$ *and* $\{\sigma_{t,i,j}\}_{i,j=1}^{k} = \Gamma_t^{-1}$.

We can have

(4.15) $$\lim_{\delta \to 0}\lim_{n \to \infty}\;\sup_{|\theta - t| < \delta} E_\theta\{\ell[\sqrt{n}(T_n - \theta_1)]\}$$

$$= (2\pi)^{-1/2}\int_{-\infty}^{\infty} \ell(\sigma_1^t y)\exp\left\{-\tfrac{1}{2}y^2\right\}dy$$

only if

(4.16) $$\{\sqrt{n}(T_n - t_1) - (\Gamma_t^{-1}\Delta_{n,t})_1\} \to 0$$

in P_n *probability, where* $(\cdot)_1$ *denotes the first coordinate in the corresponding vector.*

REMARK 1.   In (4.1) we could replace the left side by

(4.17) $\lim_{a \to \infty} \liminf_{n \to \infty} \sup_{\sqrt{n}|\theta - t| < a} E_\theta\{\ell[\sqrt{n}T_n - \theta)]\}.$

REMARK 2.   (4.16) in connection with asymptotic normality entails that $\sqrt{n}(T_n - t_1)$ is asymptotically normal $N(0, \sigma_{t,1,1})$, as is proved in Huber [9].

REMARK 3.   Condition (4.16) is only necessary. In order that there exists an estimate for which (4.15) holds for all $t \in \theta$, additional global as well as local conditions on the underlying family of distributions are necessary. If ℓ is not bounded, we also need to know something about how fast probabilities of moderate deviations of T_n from t approach zero.

REMARK 4.   If we are interested in a general scalar function $\sigma = \tau(\theta)$, we may reduce the problem to one considered in Theorem 4.2 by introducing new local coordinates $(\tau_1, \cdots, \tau_k)$ such that $\tau_1 \equiv \tau$.

APPENDIX

We shall here prove that the LAN conditions are satisfied under assumptions of Example in Section 3. To this aim it will be sufficient to adapt Theorem VI.2.1 of [8] for the present situation (see also Problem 7 of Chapter VI 1.c.).

Let us restate our conditions carefully:

A.1.   In some vicinity of $\theta = t$ the functions $f(y, \theta)$ are absolutely continuous in θ for all $y \in R$.

A.2.   For every θ in some vicinity of t the θ derivative $\dot{f}(y, \theta) = (\partial/\partial\theta)f(y, \theta)$ exists for almost all (Lebesgue measure) $y \in R$.

A.3.   The Fisher information

(A.1) $I_\theta = \int_{-\infty}^{\infty} \left\{ \frac{[\dot{f}(y, \theta)]^2}{f(y, \theta)} \right\} dy$

exists, is continuous at $\theta = t$ and $I_t > 0$. (The integrand in (A.1) is to be interpreted as zero if $f(y, \theta) = 0$).

Our preliminary goal is to show that

(A.2) $s(y, \theta) = [f(y, \theta)]^{1/2}$

is also absolutely continuous and has a mean square derivative at $\theta = t$.

LEMMA A.1.   *If $g(\theta) \geqq 0$ is absolutely continuous on (a, b) and its derivative $\dot{g}(\theta)$ satisfies*

(A.3) $\int_a^b \frac{|\dot{g}(\theta)|}{[g(\theta)]^{1/2}} d\theta < \infty,$

then $[g(\theta)]^{1/2}$ is also absolutely continuous on (a, b).

PROOF.   If $g(\theta) > 0$ and $\dot{g}(\theta)$ exists for some point θ, then it is well known from calculus that

(A.4) $$\frac{\partial}{\partial\theta}[g(\theta)]^{1/2} = \frac{\dot{g}(\theta)}{2[g(\theta)]^{1/2}}.$$

Furthermore, if $a \leq \alpha < \beta \leq b$ and g is positive on $[\alpha, \beta]$, it is easy to see that $[g(\alpha)]^{1/2}$ is absolutely continuous on $[\alpha, \beta]$ and

(A.5) $$[g(\beta)]^{1/2} - [g(\alpha)]^{1/2} = \tfrac{1}{2}\int_\alpha^\beta \frac{\dot{g}(u)}{[g(u)]^{1/2}}\,du.$$

In view of the continuity of $g(\theta)$ and in view of (A.3), (A.5) extends to $\alpha < \beta$ such that g is positive on (α, β), that is, even if possibly $g(\alpha) = 0$ or $g(\beta) = 0$. Now for any c, $a < c \leq b$ the interval (a, c) may be decomposed as follows:

(A.6) $$(a, c) = \left[\bigcup_{i=1}^\infty (\alpha_i, \beta_i)\right] \bigcup A$$

where (α_i, β_i) are disjoint intervals such that $g(\theta)$ is positive on them, $g(\alpha_i) = 0$ if $\alpha_i \neq a$ and $g(\beta_i) = 0$ if $\beta_i \neq c$, and $g(\theta) = 0$ for $\theta \in A$. Then, interpreting $\dot{g}(\theta)[g(\theta)]^{-1/2}$ as zero when $g(\theta) = 0$, we may write, in view of (A.3),

(A.7) $$\int_a^c \frac{\dot{g}(\theta)}{[g(\theta)]^{1/2}}\,d\theta = \sum_{i=1}^\infty \int_{\alpha_i}^\beta \frac{\dot{g}(\theta)}{[g(\theta)]^{1/2}}\,d\theta = [g(c)]^{1/2} - [g(a)]^{1/2},$$

because the summands with endpoints satisfying $0 = g(\alpha_i) = g(\beta_i)$ vanish according to (A.5). Relation (A.7) holding for all $c \in (a, b)$ proves absolute continuity.

LEMMA A.2.   *Under Assumptions A.1–A.3 the functions $s(y, \theta)$ are absolutely continuous in some vicinity of $\theta = t$ for almost all y.*

PROOF.   Continuity of I_θ and the Fubini theorem imply for some $\varepsilon > 0$

(A.8) $$\infty > \int_{t-\varepsilon}^{t+\varepsilon} I_\theta\,d\theta = \int_{-\infty}^\infty \int_{t-\varepsilon}^{t+\varepsilon} \left(\frac{[\dot{f}(y\ \theta)]^2}{f(y, \theta)}\right)\,d\theta\,dy.$$

Consequently, for almost all y

(A.9) $$\int_{t-\varepsilon}^{t+\varepsilon} \left(\frac{[\dot{f}(y, \theta)]^2}{f(y, \theta)}\right)\,d\theta < \infty$$

and, in turn, for almost all y,

(A.10) $$\int_{t-\varepsilon}^{t+\varepsilon} \frac{|\dot{f}(y, \theta)|}{[f(y, \theta)]^{1/2}}\,d\theta < \infty.$$

Thus it suffices to apply Lemma A.1 for $g(\theta) = f(y, \theta)$, for those y that satisfy (A.10).

LEMMA A.3. *Under A.1–A.3 the function $\dot{s}(y, t)$ defined by*

(A.11)
$$\dot{s}(y, t) = \begin{cases} \dfrac{\dot{f}(y, t)}{2[f(y, t)]^{1/2}} & \text{if} \quad f(y, t) > 0 \quad \text{and} \quad \dot{f}(y, t) \quad \text{exists} \\ 0 & \text{otherwise} \end{cases}$$

is the mean square derivative of $s(y, \theta)$ at $\theta = t$, that is,

(A.12)
$$\lim_{\Delta \to 0} \int_{-\infty}^{\infty} \left\{ \frac{1}{\Delta} [s(y, t + \Delta) - s(y, t)] - \dot{s}(y, t) \right\}^2 dy = 0.$$

PROOF. Lemma A.2 entails

(A.13)
$$\left\{ \frac{1}{\Delta} [s(y, t + \Delta) - s(y, t)] \right\}^2 = \left(\frac{1}{\Delta} \right)^2 \left(\int_0^{\Delta} \dot{s}(y, t + \lambda) \, d\lambda \right)^2$$
$$\leqq \frac{1}{\Delta} \int_0^{\Delta} [\dot{s}(y, t + \lambda)]^2 \, d\lambda.$$

Consequently, in view of continuity of I_θ,

(A.14)
$$\int_{-\infty}^{\infty} \left\{ \frac{1}{\Delta} [s(y, t + \Delta) - s(y, t)] \right\}^2 dy \leqq \frac{1}{\Delta} \int_{-\infty}^{\infty} \int_0^{\Delta} [\dot{s}(y, t + \lambda)]^2 \, d\lambda \, dy$$
$$= \frac{1}{4} \frac{1}{\Delta} \int_0^{\Delta} I_{t+\lambda} \, d\lambda \to \frac{1}{4} I_t$$
$$= \int_{-\infty}^{\infty} [\dot{s}(y, t)]^2 \, dy, \quad \text{as} \quad \Delta \to 0.$$

Put $M = \{y : s(y, t) > 0\}$. Then $(1/\Delta)[s(y, t + \Delta) - s(y, t)]$ converges to $\dot{s}(y, t)$ almost everywhere (Lebesgue measure) on M, and (A.14) entails

(A.15)
$$\limsup_{\Delta \to 0} \int_M \left\{ \frac{1}{\Delta} [s(y, t + \Delta) - s(y, t)] \right\}^2 dy \leqq \int_{-\infty}^{\infty} [\dot{s}(y, t)]^2 \, dy$$
$$= \int_M [\dot{s}(y, t)]^2 \, dy.$$

Utilizing Theorem V.1.3 of [8], we conclude that

(A.16)
$$\lim_{\Delta \to 0} \int_M \left\{ \frac{1}{\Delta} [s(y, t + \Delta) - s(y, t)] - \dot{s}(y, t) \right\}^2 dy = 0$$

and

(A.17)
$$\lim_{\Delta \to 0} \int_M \left\{ \frac{1}{\Delta} [s(y, t + \Delta) - s(y, t)] \right\}^2 dy = \int_{-\infty}^{\infty} [\dot{s}(y, t)]^2 \, dy.$$

However (A.17) and (A.14) are compatible only if

$$(A.18) \qquad \lim_{\Delta \to 0} \int_{M^c} \left\{ \frac{1}{\Delta} \left[s(y, t + \Delta) - s(y, t) \right] \right\}^2 dy = 0$$

where $M^c = R - M$. Now (A.17) and (A.18) together are equivalent to (A.12) if $\dot{s}(y, t)$ is defined by (A.11).

REMARK A.1.   Since $s(y, t) = 0$ on M^c, (A.18) is equivalent to

$$(A.19) \qquad \int_{\{y: f(y, t) = 0\}} f(y, \theta) \, dy = o[(\theta - t)^2]$$

which describes how large the singular part of $f(y, \theta)$ relative to $f(y, t)$ may be. Satisfaction of (A.18) is necessary for $q_n(x) = \Pi_{i=1}^n f(y_i, t + n^{-1/2}h)$ to be contiguous with respect to $p_n(x) = \Pi_{i=1}^n f(y_i, t)$.

THEOREM A.4.   *Under A.1–A.3 the* LAN *conditions defined in Section 3 are satisfied at* $\theta = t$ *with* $\Gamma_t = I_t$ *and* $\Delta_{n,t}$ *of* (3.4).

PROOF.   Introduce

$$(A.20) \qquad \begin{aligned} L_{n,h} &= \sum_{i=1}^n \log \frac{f(Y_i, t + n^{-1/2}h)}{f(Y_i, t)} \\ W_{n,h} &= 2 \sum_{i=1}^n \left\{ \frac{s(Y_i, t + n^{-1/2}h)}{s(Y_i, t)} - 1 \right\}. \end{aligned}$$

We need to prove that for every $h \in R$

$$(A.21) \qquad (L_{n,h} - h\Delta_{n,t} + \tfrac{1}{2}h^2 I_t) \overset{P_n}{\to} 0$$

in P_n probability, P_n referring to $\theta = t$. According to LeCam's second lemma in [8] (A.21) is equivalent to

$$(A.22) \qquad (W_{n,h} - h\Delta_{n,t} + \tfrac{1}{4}h^2 I_t) \overset{P_n}{\to} 0$$

if we show that $\mathscr{L}(W_{n,h}|P_n) \to N(-\tfrac{1}{4}h^2 I_t, h^2 I_t)$. All this will be accomplished, if we prove the following relations:

$$(A.23) \qquad E(\Delta_{n,t}|P_n) = 0,$$

$$(A.24) \qquad E(W_{n,h}|P_n) \to -\tfrac{1}{4}h^2 I_t,$$

$$(A.25) \qquad \text{Var}\,(W_{n,h} - h\Delta_{n,t}|P_n) \to 0,$$

$$(A.26) \qquad \mathscr{L}(\Delta_{n,t}|P_n) \to N(0, I_t).$$

We shall start with (A.23). We have

$$(A.27) \qquad 0 = \int_{-\infty}^{\infty} \frac{1}{\Delta} [f(y, t + \Delta) - f(y, t) \, dy$$

$$= \int_{-\infty}^{\infty} \frac{1}{\Delta} \left[s^2(y, t + \Delta) - s^2(y, t) \right] dy$$

$$= \int_{-\infty}^{\infty} \frac{1}{\Delta} \left[s(y, t + \Delta) - s(y, t) \right]^2 dy$$

$$+ 2 \int_{-\infty}^{\infty} \frac{1}{\Delta} \left[s(y, t + \Delta) - s(y, t) \right] s(y, t) \, dy$$

$$\to 0 + 2 \int_{-\infty}^{\infty} s(y, t) \dot{s}(y, t) \, dy = \int_{-\infty}^{\infty} \dot{f}(y, t) \, dy$$

$$\text{as} \quad \Delta \to 0.$$

The last statement follows from (A.11) and (A.12). Consequently

(A.28)
$$E(\Delta_{n, t} | P_n) = n^{-1/2} \prod_{i=1}^{n} E\left[\frac{\dot{f}(Y_i, t)}{f(Y_i, t)} \Big| P_n \right]$$

$$= n^{1/2} \int_{-\infty}^{\infty} \dot{f}(y, t) \, dy = 0.$$

In order to prove (A.24) and (A.25) we employ the same idea as in Lemmas VI.2.1a and VI.2.1b in [8]:

(A.29)
$$E(W_{n, h} | P_n) = 2 \sum_{i=1}^{n} E\left[\frac{s(Y_i, t + n^{-1/2}h)}{s(y_i, t)} - 1 \right]$$

$$= -h^2 \int_{-\infty}^{\infty} \left[\frac{s(y, t + n^{-1/2}h) - s(y, t)}{n^{-1/2}h} \right]^2 dy$$

$$\to -h^2 \int_{-\infty}^{\infty} [\dot{s}(y, t)]^2 \, dy = -\tfrac{1}{4} h^2 I_t.$$

Furthermore,

(A.30) $\text{Var} (W_{n, h} - h\Delta_{n, t})$

$$= 4 \sum_{i=1}^{n} \text{Var}\left[\frac{s(Y_i, t + n^{-1/2}h)}{s(Y_i, t)} - 1 - \tfrac{1}{2} n^{-1/2} h \frac{\dot{f}(Y_i, t)}{f(Y_i, t)} \right]$$

$$\leqq 4 \sum_{i=1}^{n} E\left[\frac{s(Y_i, t + n^{-1/2}h)}{s(Y_i, t)} - 1 - \tfrac{1}{2} n^{-1/2} h \frac{\dot{f}(Y_i, t)}{f(Y_i, t)} \right]^2$$

$$\leqq 4h^2 \int_{-\infty}^{\infty} \left[\frac{s(y, t + n^{-1/2}h) - s(y, t)}{n^{-1/2}h} - \dot{s}(y, t) \right]^2 dy \to 0.$$

Finally, (A.26) follows easily from (3.4) by the central limit theorem. *Q.E.D.*

REFERENCES

[1] R. R. BAHADUR, "On the asymptotic efficiency of tests and estimates," *Sankhyā*, Vol. 22 (1960), pp. 229–252.

[2] ———, "Rates of convergence of estimates and test statistics," *Ann. Math. Statist.*, Vol. 38 (1967), pp. 303–324.

[3] D. BASU, "The concept of asymptotic efficiency,' *Sankyā*, Vol. 17 (1956), pp. 193–196.

[4] C. R. BLYTH, "On minimax statistical decision procedures and their admissibility," *Ann. Math. Statist.*, Vol. 22 (1951), pp. 22–42.

[5] H. CHERNOFF, "Large sample theory: Parametric case," *Ann. Math. Statist.*, Vol. 27 (1956), pp. 1–22.

[6] J. HÁJEK, "A characterization of limiting distributions of regular estimates, *Z. Wahrschein-lichkeitstheorie und Verw. Gebiete,* Vol. 14 (1970), pp. 323–330.

[7] ———, "Limiting properties of likelihoods and inference," *Foundations of Statistical Inference* (edited by V. P. Godambe and D. A. Sprott), Toronto, Holt, Rinehart and Winston, 1971.

[8] J. HÁJEK and Z. ŠIDÁK, *Theory of Rank Tests*, New York, Academic Press, 1967.

[9] J. L. HODGES and E. L. LEHMANN, "Some applications of the Cramér-Rao inequality," *Proceedings of the Second Berkeley Symposium on Mathematical Statistics and Probability*, Berkeley and Los Angeles, University of California Press, 1950, pp. 13–22.

[10] P. HUBER, "Strict efficiency excludes superefficiency," (abstract), *Ann. Math. Statist.*, Vol. 37 (1966), p. 1425, proof unpublished but available.

[11] V. M. JOSHI, "Admissibility of estimates of the mean of a finite population," *New Developments in Survey Sampling*, New York, Wiley, 1969.

[12] L. LECAM, "On some asymptotic properties of maximum likelihood estimates and related Bayes' estimates," *Univ. California Publ. Statist.*, Vol. 1 (1953), pp. 277–330.

[13] ———, "On the asymptotic theory of estimation and testing hypotheses," *Proceedings of the Third Berkeley Symposium on Mathematical Statistics and Probability*, Berkeley and Los Angeles, University of California Press, 1956, Vol. 1, pp. 129–156.

[14] ———, "Les propriétés asymptotiques des solutions de Bayes," *Publ. Inst. Statist. Univ. Paris*, Vol. 7, fasc. (3–4)(1958), pp. 17–35.

[15] ———, "Locally asymptotically normal families of distributions," *Univ. California Publ. Statist.* Vol. 3 (1960), pp. 27–98.

[16] ———, "Limits of experiments," *Proceedings of the Sixth Berkeley Symposium on Mathematical Statistics and Probability*, Berkeley and Los Angeles, University of California Press, 1971, pp. 245–261.

[17] C. R RAO, *Linear Statistical Inference and Its Applications*, New York, Wiley, 1965.

[18] ——— -, "Criteria of estimation in large samples," *Sankhyā Ser. A*, Vol. 25 (1963), pp. 189–206.

[19] A. WALD, "Tests of statistical hypothesis concerning several parameters when the number of observations is large," *Trans. Amer. Math. Soc.*, Vol. 54 (1943), pp. 426–482.

[20] L. WEISS and J. WOLFOWITZ, "Generalized maximum likelihood estimators. *Teor. Verojatnost. i Primenen.*, Vol. 11 (1966), pp. 68–93.

[21] ———, "Maximum probability estimators," *Ann. Inst. Statist. Math.*, Vol. 19 (1967), pp. 193–206.

[22] J. WOLFOWITZ, "Asymptotic efficiency of the maximum likelihood estimator," *Teor. Verojatnost. i Primenen.*, Vol. 10 (1965), pp. 267–281.

APPLICATIONS OF CONTIGUITY TO MULTIPARAMETER HYPOTHESES TESTING

R. A. JOHNSON
and
G. G. ROUSSAS
UNIVERSITY OF WISCONSIN, MADISON

1. Summary and introduction

Consider a Markov process whose probability law depends on a k dimensional ($k \geqq 2$) parameter θ. The parameter space Θ is assumed to be an open subset of R^k. For each positive integer n, we consider the surface E_n defined by $(z - \theta_0)'\Gamma(z - \theta_0) = d_n$ for some sequence $\{d_n\}$ with $0 < d_n = O(n^{-1})$; Γ is a certain positive definite matrix.

For testing the hypothesis $H: \theta = \theta_0$ against the alternative $A: \theta \neq \theta_0$, a sequence of tests is constructed which, asymptotically, possesses the following optimal properties within a certain class of tests. It has best average power over E_n with respect to a certain weight function; it has constant power on E_n and is most powerful within the class of tests whose power is (asymptotically) constant on E_n. Finally, it enjoys the property of being asymptotically most stringent.

In this paper, we are dealing with the problem of testing the hypothesis $H: \theta = \theta_0$ when the underlying process is Markovian. The parameter θ varies over a k dimensional open subset of R^k denoted by Θ. Since the alternatives consist of all $\theta \in \Theta$ which are different from θ_0, one would not possibly expect to construct a test whose power would be "best" for each particular alternative. Therefore interest is centered on tests whose power is optimal over suitably chosen subsets of Θ. The class of subsets of Θ considered here consists of the surfaces of ellipsoids centered at θ_0. The question then arises as to which restricted class of tests one could search and still obtain an optimal test. The discussion detailed in Section 5 produces a class of tests, denoted by $\mathscr{F}$, which consists of those tests each of which is the indicator function of the complement of a certain closed, convex set. The precise definition of $\mathscr{F}$ is given in (4.4) and the arguments leading to it are due to Birnbaum [1] and Matthes and Truax [14]. The main steps of these arguments are summarized in an appendix for easy reference.

This research was supported by the National Science Foundation, Grant GP–20036.

The main results derived in this paper are of a local character. In order to give a brief description of them, we introduce the following notation. For each positive integer n, consider the surface E_n defined by

$$(1.1) \qquad E_n = \{z \in R^k; (z - \theta_0)' \Gamma (z - \theta_0) = d_n\},$$

where $0 < d_n = O(n^{-1})$ and Γ is a certain positive definite matrix. Also let ζ be a positive valued function defined on $R^k - \{\theta_0\}$ whose (surface) integral over each E_n is equal to 1. This function is given in (3.10) and (3.9). Then the test ϕ defined below by (4.5) has the following optimal properties. The power function of ϕ, weighted by the function ζ and integrated over E_n, is asymptotically largest when the competitor tests lie in $\mathscr{F}$. This is made precise in Theorems 4.1 and 6.1, the latter being a certain uniform version of the former. Next, the power of ϕ is asymptotically constant on E_n and the test ϕ is asymptotically most powerful on E_n among those tests in $\mathscr{F}$ which have asymptotically constant power on E_n. This is the content of Theorem 4.2. Finally, the test ϕ is asymptotically most stringent according to Theorems 8.1 and 8.2. Again, the latter of these theorems is a certain uniform version of the former.

In Section 9, the extra Assumption 5 is added under which the test ϕ is globally optimal in the sense that its power tends to 1, as $n \to \infty$, under nonlocal alternatives.

The hypothesis testing problem considered here has been considered by Wald [24] who also provided a solution to it (see Theorems I, II and III in Wald's paper). However, the discussion and solution to be presented here differ from those of Wald in the following respects. The assumptions made here are substantially weaker than those used by Wald. In particular, while Wald's results are formulated in terms of the maximum likelihood estimate, the present paper makes no reference to its existence. The present method of attacking the problem is that of utilizing available results obtained in Roussas [19], [20] and Johnson and Roussas [8], [9], which in turn were derived by exploiting the concept of contiguity introduced by LeCam [10]. (See also LeCam [11] and Hájek and Šidák [7], Chapter VI, and Roussas [21], [22].) As a consequence, the approach employed here is different and much less cumbersome than that of Wald. Finally, the present results also include the Markov case, whereas Wald's results were established for the independent, identically distributed case only. However, the classical methods have been used by Wald [24] and also Neyman [15] for testing composite hypotheses.

A test statistic similar to the one used here was also proposed by Rao [16] for the independent, identically distributed case and under the standard assumptions (of pointwise differentiability and so forth). However, no asymptotically optimal properties of the test were discussed except for its asymptotic distribution under the hypothesis being tested. Finally, some general results of a similar nature have been obtained by Chibisov [2]. The same author (Chibisov [3]) also obtained some average power type asymptotically optimal results in connection with the problem of testing a distribution function. An earlier version of the present paper appeared as a Technical Report, Roussas [23].

The relevant notation and assumptions are presented in Section 2. Auxiliary results necessary for the formulation of the main results are obtained in Section 3 and subsequent sections.

2. Notation and assumptions

Let Θ be a k dimensional open subset of R^k and for each $\theta \in \Theta$. consider the probability space $(\mathscr{X}, \mathscr{A}, P_\theta)$; $(\mathscr{X}, \mathscr{A}) = \Pi_{j=0}^\infty (R_j, \mathscr{B}_j)$, where $(R_j, \mathscr{B}_j) = (R, \mathscr{B})$ denotes the Borel real line and P_θ is the probability measure induced on $\mathscr{A}$ by a probability measure $p_\theta(\cdot)$ on $\mathscr{B}$ and a transition probability measure $p_\theta(\cdot\,;\,\cdot)$ defined on $R \times \mathscr{B}$. For each $\theta \in \Theta$, the coordinate process $\{X_n\}$, $n \geq 0$, n an integer, is a Markov process with initial measure $p_\theta(\cdot)$ and transition measure $p_\theta(\cdot\,;\,\cdot)$.

Let $\mathscr{A}_n$ denote the σ-field induced by the random variables $X_0, X_1, \cdots, X_n$ and let $P_{n,\theta}$ denote the restriction of P_θ to $\mathscr{A}_n$. By the assumptions to be made below, the following quantities exist and are well defined up to null sets. For $\theta, \theta^* \in \Theta$, let

(2.1)
$$\frac{dP_{0,\theta^*}}{dP_{0,\theta}} = q(X_0; \theta, \theta^*),$$

$$\frac{dP_{1,\theta^*}}{dP_{1,\theta}} = q(X_0, X_1; \theta, \theta^*).$$

Also set

(2.2) $$q(X_j | X_{j-1}; \theta, \theta^*) = q(X_{j-1'}, X_j; \theta, \theta^*)/q(X_{j-1}; \theta, \theta^*)$$

and

(2.3) $$\phi_j(\theta, \theta^*) = [q(X_j | X_{j-1}; \theta, \theta^*)]^{1/2}.$$

It follows that

(2.4) $$\frac{dP_{n,\theta^*}}{dP_{n,\theta}} = q(X_0; \theta, \theta^*) \prod_{j=1}^n \phi_j^2(\theta, \theta^*).$$

The Assumptions stated below are extracted from those used by one of the present authors in another paper (see Roussas [19]) and are stated here for the sake of completeness.

ASSUMPTION 1. *For each $\theta \in \Theta$, the Markov process $\{X_n\}$, $n \geq 0$ is (strictly) stationary and metrically transitive (ergodic).* (See, for example, Doob [4], pp. 191, 460).

ASSUMPTION 2. *The probability measures $\{P_{n,\theta}; \theta \in \Theta\}$ are mutually absolutely continuous for all $n \geq 0$.*

ASSUMPTION 3.   (i) *For each $\theta \in \Theta$, the random function $\phi_1(\theta, \theta^*)$ is differentiable in quadratic mean (q.m.) with respect to θ^* at the point (θ, θ) when P_θ is employed.* (See, for example, Loève [13] or LeCam [10].)

Let $\dot{\phi}_1(\theta)$ be the derivative in q.m. of $\phi_1(\theta, \theta^*)$ with respect to θ^* at (θ, θ). Then (ii) *$\dot{\phi}_1(\theta)$ is $\mathscr{A}_1 \times \mathscr{C}$ measurable, where $\mathscr{C}$ is the σ-field of Borel subsets of Θ.*

Let $\Gamma(\theta)$ be the covariance function defined by

$$(2.5) \qquad \Gamma(\theta) = 4\mathscr{E}_\theta[\dot{\phi}_1(\theta)\dot{\phi}_1'(\theta)].$$

Then, (iii) *$\Gamma(\theta)$ is positive definite for every $\theta \in \Theta$.*

ASSUMPTION 4.   *For each $\theta \in \Theta$, $q(X_0, X_1; \theta, \theta^*) \to 1$ in $P_{1,\theta}$ probability as $\theta^* \to \theta$.*

REMARK 2.1.   In the independent, identically distributed case, Assumption 1 is automatically satisfied (see, for example, Doob [4], p. 460). The random function $\phi_1(\theta, \theta^*)$ is equal to $[q(X_1; \theta, \theta^*)]^{1/2}$ and Assumption 4 is redundant (following from Assumption 3 (i)).

For later reference, we now introduce the k dimensional random vector $\Delta_n(\theta)$ which plays a fundamental role in this paper. Actually, $\Delta_n(\theta_0)$ replaces the maximum likelihood estimate, as will become apparent in the sequel.

$$(2.6) \qquad \Delta_n(\theta) = \frac{2}{\sqrt{n}} \sum_{j=1}^{n} \dot{\phi}_j(\theta),$$

where $\dot{\phi}_j(\theta)$ is given in Assumption 3.

In closing this section, we should like to mention that all results in this paper (except for Theorem 9.1) will be derived under the basic Assumptions 1 to 4 and this will not be mentioned again explicitly. Also the notation $P_{n,\theta}$ and P_θ will be used interchangeably and all limits will be taken as $\{n\}$, or subsequences thereof, converges to infinity unless otherwise specified.

3. Further notation and preliminary results

We recall that the problem of interest is that of testing $H: \theta = \theta_0$. In the sequel, dependence of various quantities on θ_0 will not be explicitly indicated. For instance, we shall write Γ, Δ_n rather than $\Gamma(\theta_0)$, $\Delta_n(\theta_0)$ and so forth.

Since the matrix Γ is positive definite there exists a nonsingular matrix M such that

$$(3.1) \qquad M'M = \Gamma.$$

From (3.1), it immediately follows that

$$(3.2) \qquad (M^{-1})'\Gamma M^{-1} = M\Gamma^{-1}M' = I,$$

where I is the $k \times k$ unit matrix.

Consider the following transformation of R^k onto itself

$$(3.3) \qquad M(z - \theta_0) = v - \theta_0,$$

where $v = u + \theta_0 - M\theta_0$ and $u = Mz$. For $c > 0$, let $E(c)$ be the surface (of an ellipsoid) defined by

(3.4) $$E(c) = \{z \in R^k; (z - \theta_0)' \Gamma(z - \theta_0) = c\}.$$

Then the transformation (3.3) sends the surface $E(c)$ onto the surface (of a sphere) $S(c)$, where

(3.5) $$S(c) = \{z \in R^k; (z - \theta_0)'(z - \theta_0) = c\}.$$

For $z \in R^k$, with $z \neq \theta_0$, set $c(z) = (z - \theta_0)'\Gamma(z - \theta_0)$. Then this quantity is positive, since Γ is positive definite by assumption. By means of $E(c(z))$ and $S(c(z))$, define the function ξ as in Wald [24], p. 445. Namely, for any $\rho > 0$, define $\omega(z, \rho)$ by

(3.6) $$\omega(z, \rho) = \{u \in E(c(z)); \|u - z\| \leq \rho\}$$

and let $\omega'(z, \rho)$ be the image of the set $\omega(z, \rho)$ under the transformation (3.3). Also denote by $A(\omega(z, \rho))$ and $A(\omega'(z, \rho))$ the areas of the sets $\omega(z, \rho)$ and $\omega'(z, \rho)$, respectively. Then the function ξ is defined as follows

(3.7) $$\xi(z) = \lim \frac{A(\omega'(z, \rho))}{A(\omega(z, \rho))} \qquad \text{as} \quad \rho \to 0.$$

Thus one has the positive valued function ξ defined on $R^k - \{\theta_0\}$ and it can be seen that its explicit form is

(3.8) $$\xi(z) = \frac{[|\Gamma|(z - \theta_0)'\Gamma(z - \theta_0)]^{1/2}}{\|\Gamma(z - \theta_0)\|}, \qquad z \neq \theta_0.$$

REMARK 3.1.   The significance of the function ξ defined by (4.6) may be seen from the relation

(3.9) $$\int_{E(c)} \xi(z) \, dA = \text{area of } S(c) \text{ to be denoted by } A(c),$$

where $E(c)$ and $S(c)$ are defined by (3.4) and (3.5), respectively, and the integral in (3.9) is a surface integral.

By setting

(3.10) $$\zeta(c; z) = \frac{\xi(z)}{A(c)}, \qquad z \in E(c),$$

one obtains a weight function $\zeta(c; \cdot)$ (integrating to 1) over each one of the surfaces $E(c)$.

In the sequel, we will be interested in parameter points θ_n of the form

(3.11) $$\theta_n = \theta_0 + \frac{h}{\sqrt{n}}, \qquad h \in R^k.$$

Also the h eventually, will be required to satisfy the condition $h'\Gamma h = c_n, c_n > 0$, so that $(\theta_n - \theta_0)'\Gamma(\theta_n - \theta_0) = c_n/n$ which we denote by d_n. Since $\theta_0 \in \Theta$ and

Θ is open, there exists a $d_0 > 0$ such that the surface $(\theta - \theta_0)' \Gamma (\theta - \theta_0) = d_0$ lies in Θ. We have

$$(3.12) \qquad E_n = E(d_n) = \{z \in R^k; (z - \theta_0)' \Gamma (z - \theta_0) = d_n\}$$

with $d_n = c_n/n$, and let

$$(3.13) \qquad \ddot{E}_n^* = E^*(c_n) = \{z \in R^k; z' \Gamma z = c_n\}.$$

Choose c_n satisfying the requirement

$$(3.14) \qquad\qquad 0 < c_n \leqq d_0 \text{ for all } n, \qquad 0 < d_n \leqq d_0 \text{ for all } n.$$

Then, with θ_n being the form (3.11), it follows that

$$(3.15) \qquad \theta_n \in E_n \quad \text{ if and only if } h \in E_n^*, E_n \subset \Theta \qquad \text{ for all } n.$$

REMARK 3.2. Let $z_n = \theta_0 + z/\sqrt{n}$. Then from (3.8) it follows that $\xi(z_n) = \xi(z)$. In particular, if θ_n is given by (3.11), then $\xi(\theta_n) = \xi(h)$, where $\theta_n \in E_n \subset \Theta$, so that $h \in E_n^*$. (See relations (3.12) to (3.15).)

4. Formulation of some of the main results

In the present paper, the problem we are interested in is that of testing the hypothesis $H: \theta = \theta_0$, for some fixed parameter point θ_0, against the alternative $A: \theta \neq \theta_0$. To this end, let

$$(4.1) \qquad\qquad \mathscr{L}_h = N(\Gamma h, \Gamma), \qquad\qquad\qquad h \in R^k$$

and define the set D as follows

$$(4.2) \qquad D = \{z \in R^k; z' \Gamma^{-1} z \leqq d\}, \qquad \mathscr{L}_0(D) = 1 - \alpha, \quad 0 < \alpha < 1.$$

Also define the class $\mathscr{C}$ by

$$(4.3) \qquad\qquad \mathscr{C} = \{C \in \mathscr{B}^k; C \text{ is closed and convex}\},$$

and set

$$(4.4) \qquad\qquad \mathscr{\bar{F}} = \{\psi; \psi = I[C^c], C \in \mathscr{C}\},$$

where I is the indicator of the set in the brackets.

In particular, set

$$(4.5) \qquad\qquad\qquad \phi = I[D^c].$$

Then, clearly, $D \in \mathscr{C}$, so that $\phi \in \mathscr{\bar{F}}$.

All tests herein will depend on the random vector Δ_n defined by (2.6) for $\theta = \theta_0$ and for reasons to be explained in Section 5, we may confine ourselves to tests in $\mathscr{\bar{F}}$.

For $\psi \in \mathscr{\bar{F}}$, that is $\psi = I[C^c]$ for some $C \in \mathscr{C}$, set

$$(4.6) \qquad\qquad \beta_n(\theta; C^c) = \beta_n(\theta; \psi) = \mathscr{E}_\theta \psi(\Delta_n).$$

REMARK 4.1.   As will be seen in the theorems to be stated below, the sequence of tests $\{\phi(\Delta_n)\}$ possesses certain optimal asymptotic properties, where ϕ is defined by (4.5). Also with θ_n and $\mathscr{L}_h$ defined by (3.11) and (4.1), respectively, it will be shown later (see Lemma 6.1 (i)) that

$$(4.7) \qquad \mathscr{E}_{\theta_0}\phi(\Delta_n) = P_{\theta_0}(\Delta_n \in D^c) \to \mathscr{L}_0(D^c).$$

Thus, if we decide to restrict attention to tests in the class $\bar{\mathscr{F}}$ of asymptotic level of significance α—which we shall do—we must have $\mathscr{L}_0(D^c) = \alpha$. This fact provides the justification for the equation $\mathscr{L}_0(D) = 1 - \alpha$ employed in (4.2).

Unless otherwise explicitly specified in all that follows and for each n, we shall consider only parameter points θ_n of the form $\theta_n = \theta_0 + h/\sqrt{n}$ with $h \in E_n^*$, so that $\theta_n \in E_n$ (see (3.12) and (3.13) for the definition of E_n and E_n^*). Although θ_n and h_n would be a more appropriate notation, we shall simply write θ and h, when no confusion is possible, with the understanding that $\theta = \theta_0 + h/\sqrt{n}$ and $h \in E_n^*$, so that $\theta \in E_n$. This will somewhat simplify an already cumbersome notation.

The first main result in this section is presented in the following theorem.

THEOREM 4.1.   *Let E_n be defined by (3.12) with c_n satisfying (3.14) and let $\zeta_n = \zeta(d_n; \cdot)$ be defined by (3.10). Also let ϕ be given by (4.5) and let $\{\psi_n\}$ be any sequence of tests in $\bar{\mathscr{F}}$ of asymptotic level of significance α. Then one has*

$$(4.8) \qquad \liminf \left[\int_{E_n} \beta_n(\theta; \phi)\zeta_n(\theta)\, dA - \int_{E_n} \beta_n(\theta; \psi_n)\zeta_n(\theta)\, dA \right] \geqq 0,$$

where $\beta_n(\theta; \phi)$ and $\beta_n(\theta; \psi_n)$ are defined by (4.6).

It is clear that one can take the sup over c_n belonging to a compact set before taking the lim inf in (4.8). This follows immediately since one can obtain (6.11) in the proof of Theorem 4.1 by passing to a subsequence of ellipsoids. (For a uniform version of the result just presented and also its interpretation, the reader is referred to Theorem 6.1.)

The second main result herein is the following one.

THEOREM 4.2.   *Let E_n, ϕ, $\{\psi_n\}$, $\beta_n(\theta; \phi)$ and $\beta_n(\theta; \psi_n)$ be as in Theorem 4.1.*

Then one has   (i) $\lim \{\sup [\beta_n(\theta; \phi); \theta \in E_n] - \inf [\beta_n(\theta; \phi); \theta \in E_n]\} = 0$

and   (ii) $\liminf \{\inf [\beta_n(\theta; \phi) - \beta_n(\theta; \psi_n); \theta \in E_n]\} \geqq 0$

for any tests ψ_n as described above and for which $\beta_n(\theta; \psi_n)$ satisfies (i).

As in the previous theorem, lim inf may be replaced by lim inf sup in (ii), since by passing to a subsequence of ellipsoids we can obtain (7.14).

The interpretation of the theorem is clear. Part (i) states that the power of the test ϕ on the surfaces E_n is asymptotically constant. The second part asserts that, within the class of tests whose power on E_n is asymptotically constant, the test ϕ is asymptotically most powerful.

The formulation (and proof) of the third main result in the present paper is deferred to Section 8, since it requires substantial additional notation.

5. Restriction to the class of tests $\bar{\mathscr{F}}$

Suppose that we are interested in testing the hypothesis $H: \theta = \theta_0$ against the alternative $A: \theta \neq \theta_0$ at asymptotic level of significance α. All tests are to be based on the random vector Δ_n and power is to be calculated under P_{θ_n}, where $\theta_n = \theta_0 + h/\sqrt{n}$. This can be done without loss of generality by Theorem 6.1 in Johnson and Roussas [9]. By Theorem 6.3 in [9], one has that for any tests ψ_n (not necessarily in $\bar{\mathscr{F}}$) and any bounded subset B of R^k,

$$(5.1) \qquad \sup \left[|\mathscr{E}_{\theta_n} \psi_n(\Delta_n) - \mathscr{E}_{\theta_n} \psi_n(\Delta_n^*)| ; h \in B \right] \to 0,$$

where Δ_n^* is an appropriate truncated version of Δ_n defined by (4.6) in the last reference above. Therefore from an asymptotic point of view, it suffices to base any tests on the random vector Δ_n^* rather than Δ_n. This is so regardless of whether we are interested in pointwise power or average power (see Theorem 4.1). Power is still to be calculated under P_{θ_n}. Next, by virtue of (5.2) in Johnson and Roussas [9], one has

$$(5.2) \qquad dR_{n,h}/dP_{\theta_0} = \exp \left\{ -B_n(h) + h' \Delta_n^* \right\},$$

where $\exp \{B_n(h)\} = \mathscr{E}_{\theta_0}(\exp \{h' \Delta_n^*\})$, $h \in R^k$. In the family of probability measures $R_{n,h'}$, the parameter is h and its range is all of R^k. Since for any $\theta \in \Theta$, $\theta = \theta_0 + h/\sqrt{n}$ for some $h \in R^k$, namely, $h = \sqrt{n}(\theta - \theta_0)$, it follows that $\theta = \theta_0$ if and only if $h = 0$. For $h = 0$, it follows from (5.2) that $R_{n,0} = P_{\theta_0}$.

By Theorem 5.1 in Johnson and Roussas [9],

$$(5.3) \qquad \sup (\| P_{\theta_n} - R_{n,h} \| ; h \in B) \to 0,$$

where B is any bounded subset of R^k. Thus for any tests $\psi_{n'}$, one has

$$(5.4) \qquad \sup \{ |\mathscr{E}_{\theta_n} \psi_n(\Delta_n^*) - \mathscr{E}[\psi_n(\Delta_n^*) | R_{n,h}]| ; h \in B \}$$
$$\leq \sup (\| P_{\theta_n} - R_{n,h} \| ; h \in B) \to 0.$$

Therefore it follows that, from an asymptotic power viewpoint, powers may be calculated under $R_{n,h}$ rather than P_{θ_n}. Furthermore this is true regardless of whether our interest lies in pointwise power or average power in the sense of Theorem 4.1.

In order to summarize: the original hypothesis testing problem $H: \theta = \theta_0$ against $A: \theta \neq \theta_0$ at asymptotic level of significance α, where tests are to be based on the random vector Δ_n and power is to be calculated under P_{θ_n}, may be replaced by the equivalent hypothesis testing problem $H^*: h = 0$ against $A^*: h \neq 0$ at asymptotic level of significance α, in connection with the family of probability measures defined by (5.2), where tests are to be based on the random vector Δ_n^* and power is to be calculated under $R_{n,h}$.

Now we introduce the random vector Z_n^*, where

$$(5.5) \qquad Z_n^* = \Gamma^{-1} \Delta_n^*.$$

Also the following notation is needed.

(5.6) $\mathscr{L}_{n,\theta}^* = \mathscr{L}(\Delta_n^* | P_\theta)$, $\theta \in \Theta$, $\mathscr{L}_{n,h}^* = \mathscr{L}(\Delta_n^* | R_{n,h})$, $h \in R^k$

and

(5.7) $L_{n,h}^* = \mathscr{L}(Z_n^* | R_{n,h})$, $h \in R^k$.

The following result holds true.

LEMMA 5.1. (i) *Let $\mathscr{L}_{n,\theta_0}^*$ and $\mathscr{L}_{n,h}^*$ be defined by (5.6). Then for every $h \in R^k$, one has $\mathscr{L}_{n,h}^* \ll \mathscr{L}_{n,\theta_0}^*$ and*

(5.8) $$\frac{d\mathscr{L}_{n,h}^*}{d\mathscr{L}_{n,\theta_0}^*} = \exp\{-B_n(h) + h'z\},   z \in R^k.$$

(ii) *Let $L_{n,h}^*$ be defined by (5.7). Then for every $h \in R^k$, one has $L_{n,h}^* \ll L_{n,0}^*$ and*

(5.9) $$\frac{dL_{n,h}^*}{dL_{n,0}^*} = \exp\{-B_n(h) + h'\Gamma z\},   z \in R^k.$$

PROOF.   (i) For $A \in \mathscr{B}^k$ and by virtue of (5.6) and (5.2), one has

(5.10) $\mathscr{L}_{n,h}^*(A) = R_{n,h}(\Delta_n^* \in A) = \displaystyle\int_{(\Delta_n^* \in A)} \exp\{-B_n(h) + h'\Delta_n^*\} dP_{\theta_0}$

$\qquad\qquad\qquad = \displaystyle\int_A \exp\{-B_n(h) + h'z\} d\mathscr{L}(\Delta_n^* | P_{\theta_0})$

$\qquad\qquad\qquad = \displaystyle\int_A \exp\{-B_n(h) + h'z\} d\mathscr{L}_{n,\theta_0}^*,$

as was asserted.

(ii) With A as above and by virtue of (5.2), (5.5) and (5.6), one has

(5.11) $L_{n,h}^*(A) = R_{n,h}(Z_n^* \in A) = \displaystyle\int_{(Z_n^* \in A)} \exp\{-B_n(h) + h'\Delta_n^*\} dP_{\theta_0}$

$\qquad\qquad\qquad = \displaystyle\int_{(Z_n^* \in A)} \exp\{-B_n(h) + h'\Gamma Z_n^*\} dP_{\theta_0}$

$\qquad\qquad\qquad = \displaystyle\int_A \exp\{-B_n(h) + h'\Gamma z\} d\mathscr{L}(Z_n^* | P_{\theta_0})$

$\qquad\qquad\qquad = \displaystyle\int_A \exp\{-B_n(h) + h'\Gamma z\} d\mathscr{L}(Z_n^* | R_{n,0})$

because $R_{n,0} = P_{\theta_0}$ by means of (5.2). Now since $\mathscr{L}(Z_n^* | R_{n,0}) = L_{n,0}^*$, we have

(5.12) $L_{n,h}^*(A) = \displaystyle\int_A \exp\{-B_n(h) + h'\Gamma z\} dL_{n,0}^*,$

as was asserted.

From (5.5), it follows that, in testing the hypothesis last described, our tests may be based on the random vector Z_n^* rather than Δ_n^*.

Now for each n, the family of probability densities

$$(5.13) \qquad \frac{dL_{n,h}^*}{dL_{n,0}^*} = \exp\{-B_n(h) + h'\Gamma z\}, \qquad z \in R^k$$

is of the form (A.9) in the Appendix. Therefore, Corollary A.1 in the Appendix applies and we conclude that an arbitrary test ψ_n' based on Z_n^* may be replaced by a test ψ_n based on Z_n^* of the form (A.4). Thus for each n, we consider tests ψ_n based on Z_n^* such that

$$(5.14) \qquad \psi_n(z) = \begin{cases} 1 & \text{if} \quad z \in C_n^c \\ 0 & \text{if} \quad z \in C_n^0 \text{ for some } C_n \in \mathscr{C}, \end{cases}$$

where $\mathscr{C}$ is given by (4.3); the test may be arbitrary (measurable) on C_n^b, and $\mathscr{E}[\psi_n(Z_n^*)|R_{n,0}] \to \alpha$.

Now it would be convenient to avoid arbitrariness of tests ψ_n on C_n^b; for instance, it would be convenient to set $\psi_n(z) = 0$ for $z \in C_n^b$, so that $\psi_n(z) = I[C_n^c]$. In order for this modification to be valid, we would have to show that by changing the test ψ_n on C_n^b in any arbitrary (measurable) way, both its asymptotic power and size remain intact. That this is, in fact, the case is the content of Lemma 5.3. In order to be able to prove the lemma, some additional notation and some preliminary results are needed. To this end, let

$$(5.15) \qquad \mathscr{C}^* = \{C \in \mathscr{B}^k; C \text{ is convex}\}$$

and also set

$$(5.16) \qquad \mathscr{L}_{n,\theta} = \mathscr{L}(\Delta_n|P_\theta), \qquad \theta \in \Theta.$$

Then by Theorem 6.2 in Johnson and Roussas [9], for $\theta_n^* = \theta_0 + h_n/\sqrt{n}$ with $h_n \to h \in R^k$, $\mathscr{L}_{n,\theta_n^*} \Rightarrow \mathscr{L}_{h'}$, where $\mathscr{L}_h$ is defined by (4.1) and $\Rightarrow$ denotes weak convergence of probability measures. Also $P_{\theta_n^*}(\Delta_n^* \neq \Delta_n) \to 0$ by Proposition 4.1 in Johnson and Roussas [9]. Therefore $\mathscr{L}_{n,\theta_n^*}^* \Rightarrow \mathscr{L}_{h'}$, where $\mathscr{L}_{n,\theta_n^*}^*$ is given by (5.6). On the other hand, $\|P_{\theta_n^*} - R_{n,h_n}\| \to 0$, as was mentioned before, so that $\mathscr{L}_{n,h_n}^* \Rightarrow \mathscr{L}_h$. That is, we have

$$(5.17) \qquad \mathscr{L}_{n,\theta_n^*}^* \Rightarrow \mathscr{L}_h, \qquad \mathscr{L}_{n,h_n}^* \Rightarrow \mathscr{L}_h.$$

The lemma below shows that these convergences are uniform over the class $\mathscr{C}^*$. More precisely, we have the following result.

LEMMA 5.2. *Let* θ_n, $\mathscr{L}_h$, $\mathscr{L}_{n,\theta_n}^*$, $\mathscr{L}_{n,h}^*$ *and* $\mathscr{C}^*$ *be defined by* (3.11), (4.1), (5.6) *and* (5.15), *respectively. Then for any bounded subset* B *of* R^k, *one has*

(i) $\qquad \sup\{\sup[|\mathscr{L}_{n,\theta_n}^*(C) - \mathscr{L}_h(C)|; C \in \mathscr{C}^*]; h \in B\} \to 0$

and

(ii) $\qquad \sup\{\sup[|\mathscr{L}_{n,h}^*(C) - \mathscr{L}_h(C)|; C \in \mathscr{C}^*]; h \in B\} \to 0.$

PROOF.   (i) The proof is by contradiction. Set

$$(5.18) \qquad \delta_n(h) = \sup \left[\left| \mathscr{L}^*_{n,\theta_n}(C) = \mathscr{L}_h(C) \right|; C \in \mathscr{C}^* \right]$$

and suppose that $\sup \left[\delta_n(h); h \in B \right] \nrightarrow 0$. Then there is a subsequence $\{m\} \subseteqq \{n\}$ and $h_m \in B$ such that $\delta_m(h_m) \to \delta$, for some $\delta > 0$. Equivalently,

$$(5.19) \qquad \sup \left[\left| \mathscr{L}^*_{m,\theta^*_m}(C) - \mathscr{L}_{h_m}(C) \right|; C \in \mathscr{C}^* \right] \to \delta,$$

where $\theta^*_m = \theta_0 + h_m/\sqrt{m}$. Let $\{h_r\} \subseteqq \{h_m\}$ be such that $h_r \to t \in R^k$. Then one has, by virtue of (5.17), $\mathscr{L}^*_{r,\theta^*_r} \Rightarrow \mathscr{L}_t$. Thus Theorem 4.2 in Rao [18] applies and gives

$$(5.20) \qquad \sup \left[\left| \mathscr{L}^*_{r,\theta^*_r}(C) - \mathscr{L}_t(C) \right|; C \in \mathscr{C}^* \right] \to 0.$$

On the other hand, we clearly have

$$(5.21) \qquad \sup \left[\left| \mathscr{L}_{h_r}(C) - \mathscr{L}_t(C) \right|; C \in \mathscr{C}^* \right] \to 0.$$

Relations (5.20) and (5.21) then imply that

$$(5.22) \qquad \sup \left[\left| \mathscr{L}^*_{r,\theta^*_r}(C) - \mathscr{L}_{h_r}(C) \right|; C \in \mathscr{C}^* \right] \to 0.$$

However, this contradicts (5.19) with m replaced by r.

(ii) We have

$$(5.23) \quad \left\| \mathscr{L}^*_{n,\theta_n} - \mathscr{L}^*_{n,h} \right\| = 2 \sup \left[\left| \mathscr{L}^*_{n,\theta_n}(A) - \mathscr{L}^*_{n,h}(A) \right|; A \in \mathscr{B}^k \right]$$
$$= 2 \sup \left[\left| P_{\theta_n}(\Delta^*_n \in A) - R_{n,h}(\Delta^*_n \in A) \right|; A \in \mathscr{B}^k \right]$$
$$\leqq 2 \sup \left[\left| P_{\theta_n}(E) - R_{n,h}(E) \right|; E \in \mathscr{A}_n \right] = \left\| P_{\theta_n} - R_{n,h} \right\|.$$

But $\sup \left(\left\| P_{\theta_n} - R_{n,h} \right\|; h \in B \right) \to 0$. Therefore,

$$(5.24) \qquad \sup \left(\left\| \mathscr{L}^*_{n,\theta_n} - \mathscr{L}^*_{n,h} \right\|; h \in B \right) \to 0.$$

Clearly, (5.24) implies that

$$(5.25) \qquad \sup \left\{ \sup \left[\left| \mathscr{L}^*_{n,\theta_n}(C) - \mathscr{L}^*_{n,h}(C) \right|; C \in \mathscr{C}^* \right]; h \in B \right\} \to 0.$$

This last convergence together with the first part of the lemma yields the desired conclusion.

The result just obtained is a strengthening of Lemma 2.1 in Chibisov [2] in that taking the sup over h, we allow h to vary over bounded rather than compact sets.

LEMMA 5.3.   *Let* Z^*_n, $L^*_{n,h}$ *and* $\mathscr{C}^*$ *be defined by* (5.5), (5.7) *and* (5.15), *respectively. Then for any bounded subset* B *of* R^k *and any sets* $C_n \in \mathscr{C}^*$, *one has*

$$(5.26) \qquad \sup \left[R_{n,h}(Z^*_n \in C^b_n); h \in B \right] = \sup \left[L^*_{n,h}(C^b_n); h \in B \right] \to 0.$$

PROOF.   For any $A \in \mathscr{B}^k$ one has $Z^*_n \in A$ if and only if $\Delta^*_n \in \hat{A}$, where

$$(5.27) \qquad \hat{A} = \{ u \in R^k; u = \Gamma z, z \in A \}.$$

Let A^0 and $\bar{A}$ denote the interior and the closure, respectively, of the set A.

We then have, by means of (2.3) and (1.1)

$$(5.28) \qquad L_{n,h}^*(C_n^b) = \mathscr{L}_{n,h}^*(\hat{C}_n^b) = \mathscr{L}_{n,h}^*(\hat{C}_n^-) - \mathscr{L}_{n,h}^*(\hat{C}_n^0)$$
$$= [\mathscr{L}_{n,h}^*(\hat{C}_n^-) - \mathscr{L}_h(\hat{C}_n^-)] - [\mathscr{L}_{n,h}^*(\hat{C}_n^0) - \mathscr{L}_h(\hat{C}_n^0)]$$

for any $C_n \in \mathscr{C}_*^*$. The equality $\mathscr{L}_h(\hat{C}_n^-) = \mathscr{L}_h(\hat{C}_n^0)$ holds because $C_n \in \mathscr{C}^*$ if and only if $\hat{C}_n \in \mathscr{C}^*$ and the boundary of any convex set in R^k has k dimensional Lebesgue measure zero and hence $\mathscr{L}_h$ measure zero. It is also well known that both the closure and the interior of a convex set are also convex.

Taking the sup of both sides of (5.28) as h varies in B, one obtains the desired result from Lemma 5.2 (ii).

Returning now to the discussion following the definition of the test ψ_n by (5.14), we conclude that we may restrict ourselves to tests ψ_n based on Z_n^* and having the following form

$$(5.29) \qquad\qquad \psi_n = I[C_n^c] \qquad \text{for some } C_n \in \mathscr{C}$$

and

$$(5.30) \qquad\qquad \mathscr{E}[\psi_n(Z_n^*)|R_{n,0}] \to \alpha.$$

6. Proof of the first main result

For the proof of the first theorem, we shall need some additional notation and also some preliminary results. Set

$$(6.1) \qquad\qquad Z_n = \Gamma^{-1}\Delta_n$$

and

$$(6.2) \qquad\qquad L_{n,\theta} = \mathscr{L}(Z_n|P_\theta), \qquad \theta \in \Theta.$$

Then $Z_n \in A$ if and only if $\Delta_n \in \hat{A}$, where $\hat{A}$ is given by (5.27). Also set

$$(6.3) \qquad\qquad L_h = N(h, \Gamma^{-1}), \qquad h \in R^k.$$

One then has the following result.

LEMMA 6.1.   (i) *Let* $\mathscr{L}_h$, $\mathscr{C}^*$ *and* $\mathscr{L}_{n,\theta}$ *be defined by* (4.1), (5.15) *and* (5.16), *respectively. Then*

$$(6.4) \qquad \sup \{\sup [|\mathscr{L}_{n,\theta_n}(C) - \mathscr{L}_h(C)|; C \in \mathscr{C}^*]; h \in B\} \to 0,$$

where θ_n *is given by* (3.11) *and* B *is any bounded subset of* R^k.

(ii) *Let* θ_n, $\mathscr{C}^*$ *and* B *be as above and let* $L_{n,\theta}$ *and* L_h *be defined by* (6.2) *and* (6.3), *respectively. Then*

$$(6.5) \qquad \sup \{\sup [|L_{n,\theta_n}(C) - L_h(C)|; C \in \mathscr{C}^*]; h \in B\} \to 0.$$

PROOF.   (i) The proof is similar to that of Lemma 5.2 (i) and the details are left to the reader.

(ii) For $A \in \mathscr{B}_*^k$ we have $L_{n,\theta_n}(A) = \mathscr{L}_{n,\theta_n}(\hat{A})$, where $\hat{A}$ is given by (5.27) and

$A \in \mathscr{C}^*$ $(A \in \overline{\mathscr{C}})$ if and only if $\hat{A} \in \mathscr{C}^*$ $(\hat{A} \in \overline{\mathscr{C}})$. It is also readily seen that

$$(6.6) \qquad L_h(A) = \mathscr{L}_h(\hat{A}).$$

Therefore

$$(6.7) \quad \sup \left[\left| L_{n,\theta_n}(C) - L_h(C) \right|; C \in \mathscr{C}^* \right] = \sup \left[\left| \mathscr{L}_{n,\theta_n}(\hat{C}) - \mathscr{L}_h(\hat{C}) \right|; \hat{C} \in \mathscr{C}^* \right]$$
$$= \sup \left[\left| \mathscr{L}_{n,\theta_n}(C) - \mathscr{L}_h(C) \right|; C \in \mathscr{C}^* \right].$$

Then taking the sup of both sides of this last relation as h varies in B and utilizing the first part of the lemma we obtain the desired result.

From (6.1) it follows that $\Delta_n \in A$ if and only if $Z_n \in \tilde{A}$, where

$$(6.8) \qquad \tilde{A} = \{ u \in R^k; u = \Gamma^{-1}z, z \in A \}.$$

Therefore by setting

$$(6.9) \qquad \beta_n(\theta; A) = P_\theta(\Delta_n \in A), \qquad \tilde{\beta}_n(\theta; \tilde{A}) = P_\theta(Z_n \in \tilde{A}), A \in \mathscr{B}^k,$$

we have

$$(6.10) \qquad \beta_n(\theta; A) = \tilde{\beta}_n(\theta; \tilde{A}), \qquad\qquad \theta \in \Theta, A \in \mathscr{B}^k.$$

We may now proceed with the proof of the first main result.

PROOF OF THEOREM 4.1.   The proof is by contradiction. Suppose that (4.8) is not true. Then there is a subsequence $\{m\} \subseteq \{n\}$ for which

$$(6.11) \quad \int_{E_m} \beta_m(\theta; \phi)\zeta_m(\theta)\, dA - \int_{E_m} \beta_m(\theta; \psi_m)\zeta_m(\theta)\, dA \to \delta, \quad \text{for some } \delta < 0.$$

By employing the notation in (4.6), this is rewritten as follows

$$(6.12) \qquad \int_{E_m} \beta_m(\theta; D^c)\zeta_m(\theta)\, dA - \int_{E_m} \beta_m(\theta; C_m^c)\zeta_m(\theta)\, dA \to \delta,$$

or

$$(6.13) \qquad \int_{E_m} \beta_m(\theta; C_m)\zeta_m(\theta)\, dA - \int_{E_m} \beta_m(\theta; D)\zeta_m(\theta)\, dA \to \delta.$$

By virtue of (6.10), this becomes

$$(6.14) \qquad \int_{E_m} \tilde{\beta}_m(\theta; \tilde{C}_m)\zeta_m(\theta)\, dA - \int_{E_m} \tilde{\beta}_m(\theta; \tilde{D})\zeta_m(\theta)\, dA \to \delta.$$

Now set

$$(6.15) \qquad \hat{\beta}(h; A) = L_h(A), \qquad\qquad h \in R^k, A \in \mathscr{B}^k,$$

where L_h is given by (6.3).

Then on account of (6.9) and (6.15), Lemma 6.1 (ii) implies that for arbitrary sets $D_m \in \mathscr{C}^*$

$$(6.16) \qquad \sup \left[\left| \tilde{\beta}_m(\theta_m; D_m) - \hat{\beta}(h; D_m) \right|; h \in B \right] \to 0$$

for any bounded subset B of R^k. In particular,

$$(6.17) \qquad \sup\left[\left|\tilde{\beta}_m(\theta_m; D_m) - \hat{\beta}(h; D_m)\right|; h \in E_m^*\right] \to 0,$$

where E_m^* is given by (3.13).

At this point we set

$$(6.18) \qquad \hat{\beta}(h; A) = \hat{\beta}\left(\sqrt{m}(\theta_m - \theta_0); A\right) = \beta^*(\theta_m; A), \qquad\qquad A \in \mathscr{B}^k$$

and we recall that, by (3.15), $h \in E_m^*$ if and only if $\theta_m \in E_m$. The convergence in (6.17) then becomes

$$(6.19) \qquad \sup\left[\left|\tilde{\beta}_m(\theta_m; D_m) - \beta^*(\theta_m; D_m)\right|; \theta_m \in E_m\right] \to 0,$$

or

$$(6.20) \qquad \sup\left[\left|\tilde{\beta}_m(\theta; D_m) - \beta^*(\theta; D_m)\right|; \theta \in E_m\right] \to 0.$$

Utilizing (6.20) with D_m replaced by C_m and D successively, we obtain

$$(6.21) \qquad \int_{E_m} \tilde{\beta}_m(\theta; \tilde{C}_m)\zeta_m(\theta)\, dA - \int_{E_m} \beta^*(\theta; \tilde{C}_m)\zeta_m(\theta)\, dA \to 0$$

and

$$(6.22) \qquad \int_{E_m} \tilde{\beta}_m(\theta; \tilde{D})\zeta_m(\theta)\, dA - \int_{E_m} \beta^*(\theta; \tilde{D})\zeta_m(\theta)\, dA \to 0.$$

From (6.14), (6.21) and (6.22), we obtain

$$(6.23) \qquad \int_{E_m} \beta^*(\theta; \tilde{C}_m)\zeta_m(\theta)\, dA - \int_{E_m} \beta^*(\theta; \tilde{D})\zeta_m(\theta)\, dA \to \delta,$$

or equivalently,

$$(6.24) \qquad \int_{E_m} \beta^*(\theta; \tilde{D}^c)\zeta_m(\theta)\, dA - \int_{E_m} \beta^*(\theta; \tilde{C}_m^c)\zeta_m(\theta)\, dA \to \delta.$$

Thus for all sufficiently large m, $m \geqq m_1$, say, we have

$$(6.25) \qquad \int_{E_m} \beta^*(\theta; \tilde{D}^c\zeta_m(\theta)\, dA < \int_{E_m} \beta^*(\theta; \tilde{C}_m^c)\zeta_m(\theta)\, dA + \frac{\delta}{2}$$

(recall that $\delta < 0$), or by means of (6.18),

$$(6.26) \qquad \int_{E_m} \hat{\beta}\left(\sqrt{m}(\theta - \theta_0); \tilde{D}^c\right)\zeta_m(\theta)\, dA$$
$$< \int_{E_m} \hat{\beta}\left(\sqrt{m}(\theta - \theta_0); \tilde{C}_m^c\right)\zeta_m(\theta)\, dA + \frac{\delta}{2} \qquad \text{for all } m \geqq m_1.$$

Let $A_m = \int_{E_m} \xi(z)\, dA$ (see also (3.9)). Then, on account of (3.10), (6.26) becomes

(6.27) $\quad \int_{E_m} \hat{\beta}(\sqrt{m}(\theta - \theta_0); \tilde{D}^c)\xi(\theta)\,dA$

$$< \int_{E_m} \hat{\beta}(\sqrt{m}(\theta - \theta_0); \tilde{C}_m^c)\xi(\theta)\,dA + \frac{\delta A_m}{2} \qquad \text{for all } m \geqq m_1.$$

Set

(6.28) $\quad \sqrt{m}(\theta - \theta_0) = h \qquad \text{so that} \quad \theta = \theta_0 + h/\sqrt{m}, \qquad h \in E_m^*,$

where E_m^* is given by (3.13). Then by virtue of (6.18) and Remark 3.2, the inequality in (6.27) becomes

(6.29) $\quad \int_{E_m^*} \hat{\beta}(h; \tilde{D}^c)\xi(h)\,dA < \int_{E_m^*} \hat{\beta}(h; \tilde{C}_m^c)\xi(h)\,dA + \frac{\delta A_m}{2|J_m|} \qquad \text{for all } m \geqq m_1,$

where the scaling factor J_m results from the transformation in (6.28). It is not hard to show that $|J_m| = m^{-(k-1)/2}$, whereas A_m, which is the surface area of the sphere with radius $(c_m/m)^{1/2}$ corresponding to E_m in (3.12), is equal to

(6.30) $$\frac{2\pi^{k/2}}{\Gamma\left(\dfrac{k}{2}\right)} \frac{c_m^{(k-1)/2}}{m^{(k-1)/2}},$$

as is well known. Therefore

(6.31) $$\frac{A_m}{2|J_m|} = \frac{\pi^{k/2}}{\Gamma\left(\dfrac{k}{2}\right)} c_m^{(k-1)/2}.$$

Now, on the basis of (3.14) and by passing to a subsequence if necessary, we may assume that $c_m \to c \geqq 0$. First consider the case that $c > 0$. Then for all sufficiently large m, $m \geqq m_2$, say, we have $c_m \geqq c/2$, so that $A_m/2|J_m| \geqq \delta_1$, where

(6.32) $$\delta_1 = \frac{\pi^{k/2}}{\Gamma\left(\dfrac{k}{2}\right)} \left(\frac{c}{2}\right)^{(k-1)/2}$$

Hence for $m \geqq m_3 = \max\{m_1, m_2\}$, (6.29) becomes

(6.33) $\quad \int_{E_m^*} \hat{\beta}(h; \tilde{D}^c)\xi(h)\,dA < \int_{E_m^*} \hat{\beta}(h; \tilde{C}_m^c)\xi(h)\,dA + \delta_2,$

where

(6.34) $$\delta_2 = \delta\delta_1 < 0.$$

At this point, we recall that $\hat{\beta}(h; \tilde{C}_m^c) = L_h(\tilde{C}_m^c)$ by (6.15). On the other hand, it is clear from (6.8) that $\tilde{A}^c = A^{c\sim}$, whereas $A^{\sim\hat{}} = A$, as it follows in an obvious manner from (5.27) and (6.8). Therefore one has $L_h(\tilde{C}_m^c) = L_h(C_m^{c\sim})$ and, by (6.6), this is equal to $\mathscr{L}_h(C_m^c)^{\sim\hat{}} = \mathscr{L}_h(C_m^c)$. Summarizing

(6.35) $$\hat{\beta}(h; \tilde{C}_m^c) = L_h(\tilde{C}_m^c) = \mathscr{L}_h(C_m^c).$$

By (5.16) $\mathscr{L}_{m,\theta_m}(C_m^c) = P_{\theta_m}(\Delta_m \in C_m^c)$, so that $\mathscr{L}_{m,0}(C_m^c) = P_{\theta_0}(\Delta_m \in C_m^c)$ and this converges to α. Then Lemma 6.1(i), in conjunction with (6.36), gives $L_0(C_m^{c\sim}) \to \alpha$. Now let $C \in \bar{\mathscr{F}}$ be such that $L_0(C^c) = \alpha$. Then $L_0(C_m^{c\sim}) - L_0(C^c) \to 0$ and from this it also follows that

$$(6.36) \qquad \sup \left[\left| L_h(C_m^{c\sim}) - L_h(C^c) \right|; h \in B \right] \to 0,$$

for any bounded subset B of R^k.

Since E_m^* remains bounded as $m \to \infty$, it follows that for all sufficiently large $m, m \geqq m_4$, say, one has $L_h(C_m^{c\sim}) \leqq L_h(C^c) + \varepsilon$. This, together with (6.15), gives

$$(6.37) \qquad \int_{E_m^*} \hat{\beta}(h; C_m^{c\sim}) \xi(h) \, dA \leqq \int_{E_m^*} \hat{\beta}(h; C^c) \xi(h) \, dA + \varepsilon A_m, \qquad m \geqq m_4.$$

Combining this inequality with (6.33), we obtain that for $m \geqq m_5 = \max \{m_3, m_4\}$,

$$(6.38) \qquad \int_{E_m^*} \hat{\beta}(h; \tilde{D}^c) \xi(h) \, dA < \int_{E_m^*} \hat{\beta}(h; C^c) \xi(h) \, dA + \varepsilon A_m + \delta_2.$$

From the expression of A_m given above, it follows that $A_m \to 0$. This result, together with (6.34), implies that for $m \geqq m_6$, some m_6, and some $\delta_3 < 0$,

$$(6.39) \qquad \int_{E_m^*} \hat{\beta}(h; \tilde{D}^c) \xi(h) \, dA < \int_{E_m^*} \hat{\beta}(h; \tilde{C}^c \xi) \xi(h) \, dA + \delta_3.$$

By (6.15), $\hat{\beta}(h; A)$ is the power of the test $\psi = I[A]$, based on the random vector Z whose distribution, under h, is $L_h = N(h, \Gamma^{-1})$ (see (6.3)). On account of (6.8), the set $\tilde{D}$ is given by

$$(6.40) \qquad \tilde{D} = \{u \in R^k; u = \Gamma^{-1} z, z \in D\}.$$

By taking into consideration (4.2), one has $\tilde{D} = \{u \in R^k; u' \Gamma u \leqq d\}$. Applying (6.6) with $A = \tilde{D}$ and $h = 0$ and also utilizing (4.2), we obtain $L_0(\tilde{D}) = 1 - \alpha$. That is $\tilde{D} = \{u \in R^k; u' \Gamma u \leqq d\}$, $L_0(\tilde{D}) = 1 - \alpha$, and $L_0(C) = 1 - \alpha$. Also since $E_m^* = \{h \in R^k; h' \Gamma h = c_m\}$, we have that both $\tilde{D}$ and E_m^* (for $m \geqq m_2$) are of the type required by Proposition II in Wald [24] for testing the hypothesis $h = 0$ in connection with the distribution $L_h = N(h, \Gamma^{-1})$. Hence relation (6.39) cannot hold true. The desired result is then established.

In order to complete the proof of the theorem, we have to show that its conclusion is true if $c = 0$; that is, if $c_m \to 0$. In this case, by substituting h_m for h in $h' \Gamma h = c_m$, we find that $h_m \to 0$, or equivalently $\sqrt{m}(\theta_m - \theta_0) \to 0$. Then repeating the arguments employed in the last paragraph of the proof of Theorem 4.1 in Johnson and Roussas [8], we obtain $\|P_{\theta_m} - P_{\theta_0}\| \to 0$. From this and by a simple contradiction argument, we also get $\sup (\|P_{\theta_m} - P_{\theta_0}\|; \theta_m \in E_m) \to 0$. Therefore uniformly in ψ_m in $\bar{\mathscr{F}}$ and $\theta_m \in E_m$, one has $\beta_m(\theta_m; \psi_m) - \beta_m(\theta_0; \psi_m) \to 0$. Hence $\beta_m(\theta_m; \psi_m) \to \alpha$ uniformly in $\psi_m \in \bar{\mathscr{F}}$ and $\theta_m \in E_m$, since $\beta_m(\theta_0; \psi_m) \to \alpha$. Applying this result for $\psi_m = \phi$, we obtain $\beta_m(\theta_m; \phi) - \beta_m(\theta_m; \psi_m) \to 0$ uniformly in $\psi_n \in \bar{\mathscr{F}}$ and $\theta_m \in E_m$, so that

$$(6.41) \qquad \int_{E_m} \beta_m(\theta; \phi)\zeta_m(\theta)\, dA - \int_{E_m} \beta_m(\theta; \psi_m)\zeta_m(\theta)\, dA \to 0.$$

Thus the left side of (4.8) (with lim inf replaced by lim) is equal to zero. The proof is completed.

From Lemma 6.1(i), it follows that from asymptotic power viewpoint (both in the pointwise and the average power sense), rather than considering asymptotic level α tests, we may restrict ourselves, for each n, to tests lying in the class $\mathscr{F}_0$ defined below by (8.2). In this case, the proof of Theorem 4.1 is considerably simpler in that one may deduce the desired contradiction from (6.29). This is so because $\hat{\beta}(0; \tilde{C}_m^c) = \alpha$. Also, in this case, one may formulate and prove a uniform version of Theorem 4.1. More precisely, one has the following result.

THEOREM 6.1. *With the same notation as that employed in Theorem 4.1 one has*

$$(6.42) \quad \lim \inf \left\{ \inf \left[\int_{E_n} \beta_n(\theta; \phi)\zeta_n(\theta)\, dA - \int_{E_n} \beta_n(\theta; \psi)\zeta_n(\theta)\, dA\,;\, \psi \in \mathscr{F}_0 \right] \right\} = 0.$$

PROOF OF THEOREM 6.1. Suppose that (6.42) is not true and let the left side of it be equal to some $\delta < 0$. (Clearly, δ may not be positive.) Then there is a subsequence $\{m\} \subseteq \{n\}$ such that

$$(6.43) \qquad \inf \left[\int_{E_m} \beta_m(\theta; \phi)\zeta_m(\theta)\, dA - \int_{E_m} \beta_m(\theta; \psi)\zeta_m(\theta)\, dA\,;\, \psi \in \mathscr{F}_0 \right] = \delta.$$

From this it follows that there exists a sequence $\{\psi_m\}$ of tests in $\mathscr{F}_0$ such that

$$(6.44) \qquad \int_{E_m} \beta_m(\theta; \phi)\zeta_m(\theta)\, dA - \int_{E_m} \beta_m(\theta; \psi_m)\zeta_m(\theta)\, dA \to \delta.$$

This is the same as relation (6.11) and a repetition of the arguments used in the proof of Theorem 4.1, leads us to (6.29). The desired contradiction then follows as indicated above.

7. Proof of the second main result

The following inequalities will be useful in the proof of the second theorem.

Let $\{\alpha_j, j \in I\}$ and $\{\beta_j, j \in I\}$ be any collections of bounded real numbers and let I be any index set. Then the following inequalities hold.

$$(7.1) \qquad \left| \sup\,(\alpha_j; j \in I) - \sup\,(\beta_j, j \in I) \right| \leq \sup\,(|\alpha_j - \beta_j|; j \in I)$$

and

$$(7.2) \qquad \left| \inf\,(\alpha_j; j \in I) - \inf\,(\beta_j; j \in I) \right| \leq \sup\,(|\alpha_j - \beta_j|; j \in I).$$

A few more facts will be needed before we proceed with the proof of the theorem.

Let Δ stand for the identity mapping in R^k. Also $L_h = N(h, \Gamma^{-1})$ by (6.3). Therefore $\mathscr{L}(\Delta | L_h) = N(h, \Gamma^{-1})$ and hence (see, for example, Rao [17], p. 152)

$$(7.3) \qquad \mathscr{L}(\Delta'\Gamma\Delta \mid L_h) = \chi^2_{k,\delta(h)'},$$

where $\delta(h) = h'\Gamma h$. From the definition (6.8) of $\tilde{A}$, it is immediate that $\tilde{A}^c = A^{c\sim}$. On the other hand, as was mentioned in the proof of Theorem 4.1, $A^{\sim\wedge} = A$, where $\hat{A}$ is given by (5.27). Utilizing these facts with $A = D^{c\sim} = \tilde{D}^c$), relation (6.6) becomes

$$(7.4) \qquad L_h(D^{c\sim}) = \mathscr{L}_h(D^c).$$

The quantities $\mathscr{L}_h$ and D are defined by (4.1) and (4.2), respectively. From (7.3) and (7.4), one obtains

$$(7.5) \qquad \mathscr{L}_h(D^c) = 1 - P[\chi^2_{k,\delta(h)} \leqq d] = \text{constant on each } E_n^*,$$

where E_n^* is given by (3.13).

Finally, let $h \in E_n^*$ and let $\theta_n = \theta_0 + h/\sqrt{n}$, so that $\theta_n \in E_n$, where E_n is defined by (3.12). Then by virtue of (6.15) and (6.18), we have $L_h(A) = \beta^*(\theta_n; A)$. Taking $A = D^{c\sim}$ and employing (7.4), one obtains $\beta^*(\theta_n; D^{c\sim}) = \mathscr{L}_h(D^c)$. This, together with (7.5) implies then

$$(7.6) \qquad \beta^*(\theta; D^{c\sim}) = \text{constant on each } E_n.$$

We may now start with the proof of the result.

PROOF OF THEOREM 4.2.   (i) By employing (7.6), we have

$$(7.7) \qquad \left| \sup \left[\beta_n(\theta; \phi); \theta \in E_n \right] - \inf \left[\beta_n(\theta; \phi); \theta \in E_n \right] \right|$$

$$\leqq \left| \sup \left[\beta_n(\theta; \phi); \theta \in E_n \right] - \sup \left[\beta^*(\theta; D^{c\sim}); \theta \in E_n \right] \right|$$
$$+ \left| \inf \left[\beta_n(\theta; \phi); \theta \in E_n \right] - \inf \left[\beta^*(\theta; D^{c\sim}); \theta \in E_n \right] \right|$$

and by means of (7.1) and (7.2), the right side above is bounded above by

$$(7.8) \qquad 2 \sup \left[|\beta_n(\theta; \phi) - \beta^*(\theta; D^{c\sim})|; \theta \in E_n \right]$$
$$= 2 \sup \left[|\beta_n(\theta; D^c) - \beta^*(\theta; D^{c\sim})|; \theta \in E_n \right]$$
$$= 2 \sup \left[|\beta_n(\theta; D) - \beta^*(\theta; \tilde{D})|; \theta \in E_n \right].$$

Set $\theta_n = \theta_0 + h/\sqrt{n}$ with $h \in E_n^*$. Then on account of (6.9) and (5.16), one has

$$(7.9) \qquad \beta_n(\theta_n; D) = P_{\theta_n}(\Delta_n \in D) = \mathscr{L}_{n,\theta_n}(D).$$

From (5.27) and (6.8), if follows that $D^{\sim\wedge} = D$. Therefore, by virtue of (6.18), (6.15) and (6.6), we have

$$(7.10) \qquad \beta^*(\theta_n; \tilde{D}) = \hat{\beta}(h; \tilde{D}) = L_h(\tilde{D}) = \mathscr{L}_h(D).$$

Hence

$$(7.11) \qquad 2 \sup \left[|\beta_n(\theta; D) - \beta^*(\theta; \tilde{D})|; \theta \in E_n \right]$$
$$= 2 \sup \left[|\mathscr{L}_{n,\theta_n}(D) - \mathscr{L}_h(D)|; h \in E_n^* \right].$$

Thus

$$(7.12) \qquad \left| \sup \left[\beta_n(\theta; \phi); \theta \in E_n \right] - \inf \left[\beta_n(\theta; \phi); \theta \in E_n \right] \right|$$
$$\leqq 2 \sup \left[\left| \mathscr{L}_{n,\theta_n}(D) - \mathscr{L}_h(D) \right|; h \in E_n^* \right].$$

Since the expression on the right side above converges to zero by Lemma 6.1(i), the proof of part (i) is completed.

(ii) We first show that

$$(7.13) \qquad \lim \inf \{ \sup \left[\beta_n(\theta; \phi) - \beta_n(\theta; \psi_n); \theta \in E_n \right] \} \geqq 0.$$

The proof is by contradiction. Suppose that (7.13) is not true and let the left side of it be equal to some $\delta < 0$. Then there is a subsequence $\{m\} \subseteqq \{n\}$ such that for all sufficiently large m, $m \geqq m_1$, say, one has

$$(7.14) \qquad \sup \left[\beta_m(\theta; \phi) - \beta_m(\theta; \psi_m); \theta \in E_m \right] < \frac{\delta}{2}.$$

This is equivalent to

$$(7.15) \quad \beta_m(\theta; \phi) - \beta_m(\theta; \psi_m) < \frac{\delta}{2} \qquad \text{for all } \theta \in E_m \text{ and all } m \geqq m_1,$$

or

$$(7.16) \qquad [\beta_m(\theta; \phi) - \beta_m(\theta; \psi_m)] \zeta_m(\theta) < \frac{\delta}{2} \zeta_m(\theta)$$

$$\text{for all } \theta \in E_m \text{ and all } m \geqq m_1.$$

Hence

$$(7.17) \qquad \lim \inf \left[\int_{E_m} \beta_m(\theta; \phi) \zeta_m(\theta) \, dA - \int_{E_m} \beta_m(\theta; \psi_m) \zeta_m(\theta) \, dA \right] \leqq \frac{\delta}{2},$$

since $\int_{E_m} \zeta_m(\theta) \, dA = 1$, and this contradicts (4.8).

We now continue as follows

$$(7.18) \qquad \inf \left[\beta_n(\theta; \phi) - \beta_n(\theta; \psi_n); \theta \in E_n \right]$$

$$= \inf \{ \beta_n(\theta; \phi) + \left[-\beta_n(\theta; \psi_n) \right]; \theta \in E_n \}$$
$$\geqq \inf \left[\beta_n(\theta; \phi); \theta \in E_n \right] + \inf \left[-\beta_n(\theta; \psi_n); \theta \in E_n \right]$$
$$= \inf \left[\beta_n(\theta; \phi); \theta \in E_n \right] - \sup \left[\beta_n(\theta; \psi_n); \theta \in E_n \right].$$

Adding and subtracting appropriate quantities, the last expression on the right side above becomes

$$(7.19) \qquad \inf \left[\beta_n(\theta; \phi); \theta \in E_n \right] - \sup \left[\beta_n(\theta; \psi_n); \theta \in E_n \right]$$

$$= - \{ \sup \left[\beta_n(\theta; \phi); \theta \in E_n \right] - \inf \left[\beta_n(\theta; \phi); \theta \in E_n \right] \}$$
$$- \{ \sup \left[\beta_n(\theta; \psi_n); \theta \in E_n \right] - \inf \left[\beta_n(\theta; \psi_n); \theta \in E_n \right] \}$$
$$+ \{ \sup \left[\beta_n(\theta; \phi); \theta \in E_n \right] - \inf \left[\beta_n(\theta; \psi_n); \theta \in E_n \right] \}.$$

The third of these terms is further written as $\sup [\beta_n(\theta; \phi); \theta \in E_m] + \sup [-\beta_n(\theta; \psi_m); \theta \in E_n]$ and this is bounded below by $\sup [\beta_n(\theta; \phi) - \beta_n(\theta; \psi_n); \theta \in E_n]$. Combining these results, we obtain then

$$(7.20) \qquad \inf [\beta_n(\theta; \phi) - \beta_n(\theta; \psi_n); \theta \in E_n]$$
$$\geqq - \{\sup [\beta_n(\theta; \phi); \theta \in E_n] - \inf [\beta_n(\theta; \phi); \theta \in E_n]\}$$
$$- \{\sup [\beta_n(\theta; \psi_n); \theta \in E_n] - \inf [\beta_n(\theta; \psi_n); \theta \in E_n]\}$$
$$+ \sup [\beta_n(\theta; \phi) - \beta_n(\theta; \psi_n); \theta \in E_n].$$

Now letting $n \to \infty$ (7.20) we have that the limit of the first term on the right side is equal to zero by the first part of the theorem, the limit of the second term on the same side is equal to zero, by assumption, and the lim inf of the third term on the same side is $\geqq 0$ by (7.13). This establishes (ii) and hence the theorem itself.

8. Formulation and proof of the third main result

Up to this point we have dealt with tests ψ_n in $\bar{\mathscr{F}}$ defined by (4.4), depending on the random vector Δ_n and having asymptotic level of significance α. In this section, we are going to further restrict the class $\bar{\mathscr{F}}$ of tests by introducing another class contained in $\bar{\mathscr{F}}$ and denoted by $\mathscr{F}_0$. The reasons for this restriction are implicit in the definition of the envelope power functions by (8.13) and (8.17) and also Lemma 8.2 below which is needed in the proof of Theorem 8.1. However, in order for the restriction under question to be legitimate, we must show that, asymptotically, nothing is lost in the process, either in terms of power or in terms of asymptotic level of significance.

Set

$$(8.1) \qquad \mathscr{C}_0 = \{C \in \mathscr{B}^k; C \text{ is closed, convex and } \mathscr{L}_0(C) = 1 - \alpha\},$$

where $\mathscr{L}_0$ is given by (4.1), and define the class $\mathscr{F}_0$ by

$$(8.2) \qquad \mathscr{F}_0 = \{\psi; \psi = I[C^c], C \in \mathscr{C}_0\}.$$

Since $\mathscr{L}(\Delta_n | P_{\theta_0}) = \mathscr{L}_{n, \theta_0} \Rightarrow N(0, \Gamma) = \mathscr{L}_0$ by Theorem 3.2.1 in Roussas [19], we have that $\mathscr{L}_{n, \theta_0}(C_n) - \mathscr{L}_0(C_n) \to 0$ for any sets $C_n \in \mathscr{C}^*$; this is so by Theorem 4.2 in Rao [18]. Thus $\omega_n = I[C_n^c]$ in $\mathscr{F}_0$, implies that $\mathscr{L}_0(C_n) = 1 - \alpha$ and the last convergence above gives $\mathscr{L}_{n, \theta_0}(C_n^c) \to \alpha$. That is, tests in $\mathscr{F}_0$ are of asymptotic level of significance α. Thus it suffices for us to show that every test ψ_n in $\bar{\mathscr{F}}$ which is of asymptotic level of significance α, can be replaced, from asymptotic power point of view, by tests ω_n in $\mathscr{F}_0$. More precisely, it suffices to establish the following result.

LEMMA 8.1.   *For any sequence of tests* $\{\psi_n\}$ *in* $\bar{\mathscr{F}}$ *for which* $\mathscr{E}_{\theta_0}\psi_n(\Delta_n) \to \alpha$, *there is a sequence of tests* ω_n *in* $\mathscr{F}_0$ *such that*

$$(8.3) \qquad \sup [|\mathscr{E}_{\theta_n}\psi_n(\Delta_n) - \mathscr{E}_{\theta_n}\omega_n(\Delta_n)|; h \in B] \to 0,$$

where $\theta_n = \theta_0 + h/\sqrt{n}$ *and* B *is any bounded subset of* R^k.

PROOF. Since $\psi_n \in \mathscr{F}$ and $\mathscr{E}_{\theta_0}\psi_n(\Delta_n) \to \alpha$, we have $\psi_n = I[C_n^c]$ for some $C_n \in \mathscr{F}$ and $\mathscr{L}_{n,\theta_0}(C_n) \to 1 - \alpha$. Setting $h = 0$ in Lemma 6.1 (i), we obtain $\mathscr{L}_{n,\theta_0}(C_n) - \mathscr{L}_0(C_n) \to 0$, so that

$$(8.4) \qquad \mathscr{L}_0(C_n) \to 1 - \alpha.$$

Thus for all sufficiently large n, $n \geqq n_1$, say, $\mathscr{L}_0(C_n) > 0$. This implies that for $n \geqq n_1$, the sets C_n are k dimensional since otherwise their k dimensional Lebesgue measure and hence $\mathscr{L}_0$ measure would be zero.

Then for each $n \geqq n_1$, consider the following modification of the set C_n: if $\mathscr{L}_0(C_n) < 1 - \alpha$, enlarge C_n, so that it remains closed and convex and $\mathscr{L}_0(C_n) = 1 - \alpha$. If $\mathscr{L}_0(C_n) > 1 - \alpha$, shrink the set C_n until $\mathscr{L}_0(C_n) = 1 - \alpha$. If $\mathscr{L}_0(C_n) = 1 - \alpha$, the set C_n is left intact. Denote the resulting set by $C_{n,0}$. Then $C_{n,0}$ is closed and convex and

$$(8.5) \qquad \mathscr{L}_0(C_{n,0}) = 1 - \alpha.$$

Thus setting $\omega_n = I[C_{n,0}^c]$, we have $\omega_n \in \mathscr{F}_0$. From (8.4) and (8.5), it follows that

$$(8.6) \qquad \mathscr{L}_0(C_n) - \mathscr{L}_0(C_{n,0}) \to 0.$$

From the process of arriving at $C_{n,0}$, it follows that $C_{n,0} \subseteqq C_n$ or $C_{n,0} \supseteqq C_n$. Therefore for each $h \in R^k$, we have

$$(8.7) \quad \mathscr{L}_h(C_n \Delta C_{n,0}) = \mathscr{L}_h(C_n - C_{n,0}) = \mathscr{L}_h(C_n) - \mathscr{L}_h(C_{n,0}) \qquad \text{if } C_{n,0} \subseteqq C_n$$

and

$$(8.8) \quad \mathscr{L}_h(C_n \Delta C_{n,0}) = \mathscr{L}_h(C_{n,0} - C_n) = \mathscr{L}_h(C_{n,0}) - \mathscr{L}_h(C_n) \qquad \text{if } C_{n,0} \supseteqq C_n.$$

Thus

$$(8.9) \qquad \mathscr{L}_h(C_n \Delta C_{n,0}) = \left| \mathscr{L}_h(C_n) - \mathscr{L}_h(C_{n,0}) \right|.$$

For $h = 0$ $\mathscr{L}_h(C_n \Delta C_{n,0}) \to 0$ by (8.6). On the other hand, $\mathscr{L}_h \ll \mathscr{L}_0$ for every $h \in R^k$. Thus $\mathscr{L}_h(C_n \Delta C_{n,0}) \to 0$. It can be further seen that for any bounded subset B of R^k, one has

$$(8.10) \qquad \sup \left[\mathscr{L}_h(C_n \Delta C_{n,0}) ; h \in B \right] \to 0.$$

Therefore, (8.9) and (8.10) give

$$(8.11) \qquad \sup \left[\left| \mathscr{L}_h(C_n) - \mathscr{L}_h(C_{n,0}) \right| ; h \in B \right] \to 0.$$

Now, for any B as described above, one has

$$(8.12) \qquad \sup \left[\left| \mathscr{E}_{\theta_n}\psi_n(\Delta_n) - \mathscr{E}_{\theta_n}\omega_n(\Delta_n) \right| ; h \in B \right]$$
$$= \sup \left[\left| \mathscr{L}_{n,\theta_n}(C_n) - \mathscr{L}_{n,\theta_n}(C_{n,0}) \right| ; h \in B \right]$$
$$\leqq \sup \left[\left| \mathscr{L}_{n,\theta_n}(C_n) - \mathscr{L}_h(C_n) \right| ; h \in B \right]$$
$$+ \sup \left[\left| \mathscr{L}_{n,\theta_n}(C_{n,0}) - \mathscr{L}_h(C_{n,0}) \right| ; h \in B \right]$$
$$+ \sup \left[\left| \mathscr{L}_h(C_n) - \mathscr{L}_h(C_{n,0}) \right| ; h \in B \right],$$

and each one of the terms on the right side above converges to zero on account of Lemma 6.1(i) and (8.11). The proof of Lemma 8.1 is completed and we have the justification for confining ourselves to the class of tests, $\mathscr{F}_0$.

Before we are able to formulate the third main result, we shall have to introduce a further piece of notation. To this end, let $\beta_n(\theta; \psi) = \mathscr{E}_\theta \psi(\Delta_n)$, as given in (4.6) and suppose that the test ψ lies in $\mathscr{F}_0 = \{\psi; \psi = I[C^c], C \in \mathscr{C}_0\}$, where $\mathscr{C}_0 = \{C \in \mathscr{B}^k; C$ is closed, convex and $\mathscr{L}_0(C) = 1 - \alpha\}$; we also recall that $\mathscr{L}_0 = N(0, \Gamma)$. Next define the modified *envelope power function* $\beta_n(\theta; \alpha)$ by

$$(8.13) \qquad \beta_n(\theta; \alpha) = \sup \{\beta_n(\theta; \psi); \psi \in \mathscr{F}_0\}.$$

It is to be noted that the tests involved in the definition of the modified envelope power function $\beta_n(\theta; \alpha)$ are not those of exact level α for finite n but, as with ϕ, they have level α under the limit distribution $\mathscr{L}_0$.

Then the third main result in this paper is as follows.

THEOREM 8.1. *Let E_n and ϕ be defined by (3.12) and (4.5), respectively, and let ψ_n be any tests in $\mathscr{F}_0$, where $\mathscr{F}_0$ is given in (8.2). Also let $\beta_n(\theta; \phi)$ and $\beta_n(\theta; \psi_n)$ be defined by (4.6) and let $\beta_n(\theta; \alpha)$ be given by (8.13). Then one has*

$$(8.14) \qquad \lim \sup \{\sup [\beta_n(\theta; \alpha) - \beta_n(\theta; \phi); \theta \in E_n]$$
$$- \sup [\beta_n(\theta; \alpha) - \beta_n(\theta; \psi_n); \theta \in E_n]\} \leqq 0.$$

Again we could replace lim sup by lim sup sup since relation (8.28) in the proof is obtained by passing to a subsequence of ellipsoids.

The interpretation of the theorem is that within the class $\mathscr{F}_0$, the test ϕ is *asymptotically most stringent* on E_n. We recall that in the present framework and for each n, the test ϕ_n would be said to be *most stringent* on E_n within the class $\mathscr{F}_0$, if the quantity $\sup [\beta_n(\theta; \alpha) - \beta_n(\theta; \psi_n); \theta \in E_n]$ were minimized for $\psi_n = \phi_n$.

For the proof of Theorem 8.1, a couple of auxiliary results will be needed. For their formulation, let us recall once again that $\mathscr{L}_h = N(\Gamma h, \Gamma)$ and set

$$(8.15) \qquad \bar{\beta}(h; A) = \mathscr{L}_h(A), h \in R^k, A \in \mathscr{B}^k.$$

For each n, let $h \in E_n^*$ and transform h to $\theta \in E_n$ through the transformation $\theta = \theta_0 + h/\sqrt{n}$. We recall that E_n and E_n^* are given by (3.12) and (3.13), respectively. Set

$$(8.16) \qquad \bar{\beta}(h; A) = \bar{\beta}(\sqrt{n}(\theta - \theta_0); A) = \beta'(\theta; A).$$

Next by means of $\beta'(\theta; A)$, define the *envelop power function* $\beta'(\theta; \alpha)$ as follows

$$(8.17) \qquad \beta'(\theta; \alpha) = \sup [\beta'(\theta; C^c); C \in \mathscr{C}_0].$$

The first auxiliary result is given in the following lemma.

LEMMA 8.2. *The function $\beta'(\theta; \alpha)$ defined by (8.17) remains constant on each E_n, where E_n is given by (3.12).*

PROOF.   Clearly, $\beta'(\theta;\alpha) = \bar{\beta}(h;\alpha)$, where

(8.18) $\bar{\beta}(h;\alpha) = \sup\,[\bar{\beta};\,C^c);\,C \in \mathscr{C}_0],\,h = \sqrt{n}(\theta - \theta_0).$

Thus it suffices to show that $\bar{\beta}(h;\alpha)$ stays constant on each E_n^*, where E_n^* is given by (3.13). With M defined by (3.1), consider the transformation $t = Mh$. Then, by (3.2), E_n^* is transformed into $S_n^* = \{t \in R^k;\,t't = \|t\|^2 = c_n\}$. Also the class of sets $\mathscr{C}_0$ is transformed into the class of sets $\mathscr{C}_*$, where

(8.19) $\mathscr{C}_* = \{C \in \mathscr{B}^k;\,C$ is closed, convex and $N_0(C) = 1 - \alpha\}$

and

(8.20) $N_t = N(MM't, M\Gamma M'),$ $t \in R^k.$

Therefore, by setting

(8.21) $\beta^0(t;A) = N_t(A),$ $t \in R^k,\quad A \in \mathscr{B}^k,$

$\beta^0(t;\alpha) = \sup\,[\beta^0(t;C^c);\,C \in \mathscr{C}_*],$

it suffices to show that $\beta^0(t;\alpha)$ is constant on each S_n^*. To obtain a contradiction, suppose that this is not so. Then there exist $t_1,\,t_2 \in S_n^*$ for which $\beta^0(t_1;\alpha) \neq \beta^0(t_2;\alpha)$ and let

(8.22) $\beta^0(t_1;\alpha) < \beta^0(t_2;\alpha).$

From the definition of $\beta^0(t;\alpha)$, there exists a set $C = \mathscr{C}_*$ such that

(8.23) $\beta^0(t_1;\alpha) < \beta^0(t_2;C^c).$

Now, clearly, $\beta^0(t_2;C^c) = N_{t_2}(C^c)$ is equal to the $N(t_2,I)$ measure of the set $(MM')^{-1}C^c$, and by symmetry, this is equal to the $N(t_1,I)$ measure of D, where D is the symmetric image of $(MM')^{-1}C^c$ with respect to the hyperplane through the origin that is perpendicular to the line segment connecting the points t_1 and t_2. But the $N(t_1,I)$ measure of D is equal to $N_{t_1}\big((MM')D\big)$, and, clearly, $(MM')D$ is the complement of a closed convex set, C_0^c, say. Then by symmetry, one clearly has $N_0(C_0^c = 1 - \alpha$, so that $C_0 \in \mathscr{C}_*$, and also $\beta^0(t_2;C^c) = \beta^0(t_1;C_0^c)$. Then (8.23) gives $\beta^0(t_1;\alpha) < \beta^0(t_1;C_0^c)$. However, this contradicts the definition of $\beta^0(t_1;\alpha)$ by (8.21). We reach the same conclusion if the inequality in (8.22) is reversed. Thus the proof of the lemma is completed.

The second auxiliary result referred to above is the following lemma. This lemma, as well as the one just established, are of some interest in their own right.

LEMMA 8.3.   *Let* $\beta_n(\theta;\alpha)$ *and* $\beta'(\theta;\alpha)$ *be defined by* (8.13) *and* (8.17), *respectively. Then for each n, one has*

   (i) $\sup\,[\,|\beta_n(\theta;\alpha) - \beta'(\theta;\alpha)|;\,\theta \in E_n] \to 0$

and

   (ii) $\sup\,[\beta_n(\theta;\alpha);\,\theta \in E_n] - \inf\,[\beta_n(\theta;\alpha);\,\theta \in E_n] \to 0,$
where E_n *is given by* (3.12).

PROOF.   (i) In the first place, relation (7.1) justifies the inequality below

(8.24) $\left|\beta_n(\theta;\alpha) - \beta'(\theta;\alpha)\right|$

$$= \left|\sup\left[\beta_n(\theta;\psi);\psi \in \mathscr{F}_0\right] - \sup\left[\beta'(\theta;C^c);C \in \mathscr{C}_0\right]\right|$$
$$= \left|\sup\left[\beta_n(\theta;C^c);C \in \mathscr{C}_0\right] - \sup\left[\beta'(\theta;C^c);C \in \mathscr{C}_0\right]\right|$$
$$\leqq \sup\left[\left|\beta_n(\theta;C^c) - \beta'(\theta;C^c)\right|;C \in \mathscr{C}_0\right]$$
$$= \sup\left[\left|\beta_n(\theta;C) - \beta'(\theta;C)\right|;C \in \mathscr{C}_0\right],$$

and this last expression is equal to $\sup\left[\left|\mathscr{L}_{n,\theta}(C) - \mathscr{L}_h(C)\right|;C \in \mathscr{C}_0\right]$, where $h = \sqrt{n}(\theta - \theta_0)$, since $\beta_n(\theta;C) = P_\theta(\Delta_n \in C) = \mathscr{L}_{n,\theta}(C)$ and $\beta'(\theta;C) = \bar{\beta}(h;C) = \mathscr{L}_h(C)$. That is, with $h = \sqrt{n}(\theta - \theta_0)$,

(8.25) $\left|\beta_n(\theta;\alpha) - \beta'(\theta;\alpha)\right| = \sup\left[\left|\mathscr{L}_{n,\theta}(C) - \mathscr{L}_h(C)\right|;C \in \mathscr{C}_0\right].$

Hence

(8.26) $\sup\left[\left|\beta_n(\theta;\alpha) - \beta'(\theta;\alpha)\right|;\theta \in E_n\right]$
$$= \sup\left\{\sup\left[\left|\mathscr{L}_{n,\theta}(C) - \mathscr{L}_h(C)\right|;C \in \mathscr{C}_0\right];h \in E_n^*\right\},$$

and the expression on the right side converges to zero by Lemma 6.1(i).

(ii) Letting $\theta \in E_n$ and utilizing Lemma 8.2 and inequalities (7.1) and (7.2), one has

(8.27) $\left|\sup\left[\beta_n(\theta;\alpha);\theta \in E_n\right] - \inf\left[\beta_n(\theta;\alpha);\theta \in E_n\right]\right|$
$$\leqq \left|\sup\left[\beta_n(\theta;\alpha);\theta \in E_n\right] - \beta'(\theta;\alpha)\right| + \left|\inf\left[\beta_n(\theta;\alpha);\theta \in E_n\right] - \beta'(\theta;\alpha)\right|$$
$$= \left|\sup\left[\beta_n(\theta;\alpha);\theta \in E_n\right] - \sup\left[\beta'(\theta;\alpha);\theta \in E_n\right]\right|$$
$$+ \left|\inf\left[\beta_n(\theta;\alpha);\theta \in E_n\right] - \inf\left[\beta'(\theta;\alpha);\theta \in E_n\right]\right|$$
$$\leqq 2\sup\left[\left|\beta_n(\theta;\alpha) - \beta'(\theta;\alpha)\right|;\theta \in E_n\right],$$

and this last expression tends to zero by part (i). This establishes the lemma.

We may now proceed with the proof of the third main result.

PROOF OF THEOREM 8.1.   The proof is by contradiction. Suppose that the theorem is not true and let the left side of (8.14) be equal to $4\delta > 0$. Then there exists a subsequence $\{m\} \subseteq \{n\}$ for which

(8.28) $\sup\left[\beta_m(\theta;\alpha) - \beta_m(\theta;\phi);\theta \in E_m\right]$
$$> \sup\left[\beta_m(\theta;\alpha) - \beta_m(\theta;\psi_m);\theta \in E_m\right] + 3\delta$$

for all sufficiently large m, $m \geqq m_1$, say.

The left side of the inequality above is bounded from above by

(8.29) $\sup\left[\beta_m(\theta;\alpha);\theta \in E_m\right] - \inf\left[\beta_m(\theta;\phi);\theta \in E_m\right]$

and its right side is bounded from below by

(8.30) $\inf\left[\beta_m(\theta;\alpha);\theta \in E_m\right] - \inf\left[\beta_m(\theta;\psi_m);\theta \in E_m\right] + 3\delta$

by virtue of (7.2).

By means of (8.29) and (8.30), relation (8.28) gives that, for all $m \geq m_1$

(8.31) $\quad \sup \left[\beta_m(\theta; \alpha); \theta \in E_m\right] - \inf \left[\beta_m(\theta; \phi); \theta \in E_m\right]$
$$> \inf \left[\beta_m(\theta; \alpha); \theta \in E_m\right] - \inf \left[\beta_m(\theta; \psi_m); \theta \in E_m\right] + 3\delta,$$

or

(8.32) $\quad \sup \left[\beta_m(\theta; \alpha); \theta \in E_m\right] - \inf \left[\beta_m(\theta; \alpha); \theta \in E_m\right]$
$$> \inf \left[\beta_m(\theta; \phi); \theta \in E_m\right] - \inf \left[\beta_m(\theta; \psi_m); \theta \in E_m\right] + 3\delta.$$

By Lemma 8.3 (ii), the left side of (8.32) tends to zero and hence it remains less than δ for all sufficiently large m, $m \geq m_2$, say. On the other hand, Theorem 4.2(i) yields

(8.33) $\quad \inf \left[\beta_n(\theta; \phi); \theta \in E_m\right] > \sup \left[\beta_m(\theta; \phi); \theta \in E_m\right] - \delta$

for all sufficiently large m, $m \geq m_3$, say. On account of these facts, inequality (8.32) then becomes

(8.34) $\quad \delta > \sup \left[\beta_m(\theta; \phi); \theta \in E_m\right] - \inf \left[\beta_m(\theta; \psi_m); \theta \in E_m\right] + 2\delta,$

or

(8.35) $\quad \inf \left[\beta_m(\theta; \psi_m); \theta \in E_m\right] - \delta = \inf \left[\beta_m(\theta; \psi_m) - \delta; \theta \in E_m\right]$
$$> \sup \left[\beta_m(\theta; \phi); \theta \in E_m\right]$$

for all $m \geq m_4 \geq \max \{m_1, m_2, m_3\}$. Hence

(8.36) $$\beta_m(\theta; \phi_m) - \delta > \beta_m(\theta; \phi)$$

for all $m \geq m_4$ and every $\theta \in E_m$. Therefore

(8.37) $\quad \displaystyle\int_{E_m} \beta_m(\theta; \phi)\zeta_m(\theta) \, dA - \int_{E_m} \beta_m(\theta; \psi_m)\zeta_m(\theta) \, dA < -\delta$

for all $m \geq m_4$, and this implies that

(8.38) $\quad \displaystyle\liminf \left[\int_{E_m} \beta_m(\theta; \phi)\zeta_m(\theta) \, dA - \int_{E_m} \beta_m(\theta; \psi_m)\zeta_m(\theta) \, dA\right] < -\delta.$

However, this contradicts Theorem 4.1. The desired result follows.

The following uniform version of Theorem 8.1 is also true.

THEOREM 8.2. *With the same notation as that employed in Theorem 8.1, one has*

(8.39) $\quad \displaystyle\limsup \left\{\sup \left[\beta_n(\theta; \alpha) - \beta_n(\theta; \phi); \theta \in E_n\right]\right.$
$$\left. - \inf \left\{\sup \left[\beta_n(\theta; \alpha) - \beta_n(\theta; \psi); \theta \in E_n\right]; \psi \in \mathscr{F}_0\right\}\right\} = 0.$$

PROOF. Suppose that the theorem is not true and let the left side of (8.39) be equal to some $\delta > 0$. (Clearly, δ may not be negative.) Then there exists a subsequence $\{m\} \subseteqq \{n\}$ such that

$$(8.40) \quad \sup \left[\beta_m(\theta; \alpha) - \beta_m(\theta; \psi); \theta \in E_m \right]$$
$$- \inf \{\sup \left[\beta_m(\theta; \alpha) - \beta_m(\theta; \psi); \theta \in E_m \right]; \psi \in \mathscr{F}_0 \} \to \delta.$$

Thus for $\varepsilon > 0$ and all $m \geqq m_5$, say, one has

$$(8.41) \quad \sup \left[\beta_m(\theta; \alpha) - \beta_m(\theta; \phi); \theta \in E_m \right] - \delta - \varepsilon$$
$$< \inf \{\sup \left[\beta_m(\theta; \alpha) - \beta_m(\theta; \psi); \theta \in E_m \right]; \psi \in \mathscr{F}_0 \}$$
$$< \sup \left[\beta_m(\theta; \alpha) - \beta_m(\theta; \phi); \theta \in E_m \right] - \delta + \varepsilon.$$

Therefore, for each $m \geqq m_5$, there exists a test $\psi_m \in \mathscr{F}_0$ such that

$$(8.42) \quad \sup \left[\beta_m(\theta; \alpha) - \beta_m(\theta; \phi); \theta \in E_m \right] - \delta - \varepsilon$$
$$< \sup \left[\beta_m(\theta; \alpha) - \beta_m(\theta; \psi_m); \theta \in E_m \right]$$
$$< \sup \left[\beta_m(\theta; \alpha) - \beta_m(\theta; \phi); \theta \in E_m \right] - \delta + \varepsilon,$$

or equivalently

$$(8.43) \quad \delta - \varepsilon < \sup \left[\beta_m(\theta; \alpha) - \beta_m(\theta; \phi); \theta \in E_m \right]$$
$$- \sup \left[\beta_m(\theta; \alpha) - \beta_m(\theta; \psi_m); \theta \in E_m \right] < \delta + \varepsilon,$$

provided $m \geqq m_5$.

It follows that

$$(8.44) \quad \sup \left[\beta_m(\theta; \alpha) - \beta_m(\theta; \phi); \theta \in E_m \right]$$
$$- \sup \left[\beta_m(\theta; \alpha) - \beta_m(\theta; \psi_m); \theta \in E_m \right] \to \delta \ (> 0).$$

However, this result contradicts (8.14). The proof of the theorem is completed.

9. Behavior of the power under nonlocal alternatives

Recall that $\phi = I[D^c]$, where D is given by (4.2). Also recall that the power of the test ϕ, based on $\Delta_n = \Delta_n(\theta_0)$, has been denoted by $\beta_n(\theta; \phi)$, $\theta \in \Theta$. Then the theorems formulated and proved in the previous sections, provide us with some optimal properties of the test ϕ. However, these properties are local in character, since the alternatives are required to lie close to the hypothesis being tested; actually, they are required to converge to θ_0 and at a specified rate.

The underlying basic Assumptions 1 to 4 employed throughout this paper, do not suffice for establishing optimal properties of the power function at alternatives removed from θ_0 or not converging to it at the specified rate. This can be done, however, under the following additional condition, Assumption 5.

ASSUMPTION 5. *Consider a sequence* $\{\theta_n\}$ *with* $\theta_n \in \Theta$ *for all* n. *Then* $\|\Delta_n(\theta_0)\| \to \infty$ *in* P_{n, θ_n} *probability whenever* $\|\sqrt{n}(\theta_n - \theta_0)\| \to \infty$.

The following result can now be established.

THEOREM 9.1. *Under Assumptions 1 to 5, for testing the hypothesis* $H : \theta = \theta_0$ *against the alternative* $A : \theta \neq \theta_0$ *at asymptotic level of significance* α, *the test defined by* (4.5) *possesses the optimal properties mentioned in Theorems* 4.1, 4.2, 6.1, 8.1, 8.2 *and also has the property that its power converges to* 1, *that is,* $\beta_n(\theta_n; \phi) \to 1$, *whenever* $\|\sqrt{n}(\theta_n - \theta_0)\| \to \infty$.

PROOF. The proof is immediate. Since Γ is positive definite, so is Γ^{-1}. Thus there exists a positive number p such that $z'\Gamma^{-1}z \geq p\|z\|^2$ for all $z \in R^k$. Therefore

$$(9.1) \qquad \beta_n(\theta_n; \phi) = \mathscr{E}_{\theta_n}\phi(\Delta_n) = P_{\theta_n}(\Delta_n \in D^c) = P_{\theta_n}(\Delta_n'\Gamma^{-1}\Delta_n > d).$$

However, this last quantity is greater than or equal to $P_{\theta_n}(p\|\Delta_n\|^2 > d)$ which converges to 1 by Assumption 5. Thus $\beta_n(\theta_n; \phi) \to 1$, as was to be seen.

APPENDIX

At the beginning of Section 5, it was pointed out that for testing the hypothesis $H : \theta = \theta_0$ against the alternative $A : \theta \neq \theta_0$, it suffices to consider the class of tests based only on Δ_n^*. Furthermore, each such test function is essentially the indicator of the complement of a closed, convex set in R^k. The reason for this is that the distribution $\mathscr{L}_{n,h}^*$ of Δ_n^*, under $R_{n,h}$ (defined in (5.6)), is of the standard exponential form, so that results obtained in Birnbaum [1] and Matthes and Truax [14] apply (see also Theorems 1.1 and 1.2 in Chibisov [2]). The purpose of this appendix is to elaborate further on this point.

In order to simplify the notation, in all that follows we shall omit the subscript n, since there is no danger of confusion.

From Lemma 5.1(i), one has

$$(A.1) \qquad \frac{d\mathscr{L}_h^*}{d\mathscr{L}_0^*} = \exp\{-B(h) + h'z\}, \qquad\qquad z, h \in R^k,$$

where $\mathscr{L}_h^* = \mathscr{L}(\Delta^* | R_h)$. Then the hypothesis testing problem above, described in terms of the family (A.1), becomes

$$(A.2) \qquad\qquad H^* : h = 0 \qquad \text{against} \quad A^* : h \neq 0.$$

For any test ϕ, the associated risk corresponding to the usual zero-one loss function is

$$(A.3) \qquad R_\phi(0) = \beta_\phi(0), \qquad R_\phi(h) = 1 - \beta_\phi(h) \qquad \text{for} \quad h \neq 0,$$

where $\beta_\phi(0)$ is the size of the test ϕ and $\beta_\phi(h)$ is its power at h.

The Bayes, or global risk, with respect to any prior probability distribution W on $\mathscr{B}^k$ that is associated with the test ϕ is given by

(A.4)
$$r(\phi, W) = \int R_\phi(h) \, dW.$$

The following theorem is obtained in Birnbaum [1].

THEOREM A.1.   *In connection with the family (A.1), for testing the hypothesis* $H^*: h = 0$ *against the alternative* $A^*: h \neq 0$ *on the basis of* Δ^*, *any Bayes test* ϕ_W *with respect to the prior distribution* W, *is given by*

(A.5)
$$\phi_W(z) = \begin{cases} 1 & if \quad z \in C_W^c \\ 0 & if \quad z \in C_W^0. \end{cases}$$

where C_W *is a closed, convex set in* R^k. *The test may be defined in an arbitrary (but measurable) manner on the boundary* C_W^b *of* C_W.

PROOF.   If w_0 is the mass assigned to $\{0\}$ by W, one has from (A.4)

(A.6)
$$r(\phi, W) = \int R_\phi(h) \, dW = w_0 [2\beta_\phi(0) - 1] + \int [1 - \beta_\phi(h)] \, dW$$

$$= (1 - w_0) + 2w_0 \int \phi(z) \, d\mathscr{L}_0^*$$

$$- \int \left[\int \phi(z) \exp \{- B(h) + h'z\} \, d\mathscr{L}_0^* \right] dW$$

$$= (1 - w_0) + \int [2w_0 - \int \exp \{- B(h) + h'z\} \, dW] \phi(z) \, d\mathscr{L}_0^*.$$

Clearly, the Bayes risk is minimized by the test

(A.7)
$$\phi_W(z) = \begin{cases} 1 & if \; z \in A_1 = [z \in R^k; 2w_0 < \int \exp \{- B(h) + h'z\} \, dW] \\ 0 & if \; z \in A_2 = [z \in R^k; 2w_0 > \int \exp \{- B(h) + h'z\} \, dW]; \end{cases}$$

ϕ_W may be defined arbitrarily (but in a measurable way) on the set

(A.8)
$$A_3 = \left[z \in R^k; 2w_0 = \int \exp \{- B(h) + h'z\} \, dW \right]$$

which is the boundary of $A_2 \cup A_3$. From the definition of $\exp \{B(h)\}$ in (5.1), it easily follows that $\exp \{- B(h)\}$ is bounded over bounded sets of h in R^k. Therefore for any probability measure W on $\mathscr{B}^k$, one defines a σ-finite measure μ_W on $\mathscr{B}^k$ as follows

(A.9)
$$\mu_W(B) = \int_B \exp \{- B(h)\} \, dW.$$

Then

(A.10)
$$\int \exp \{- B(h) + h'z\} \, dW = \int \exp \{h'z\} \, d\mu_W = \int \exp \{z'h\} \, d\mu_W$$

and by Theorem 9 on p. 52 in Lehmann [12], it follows that $\int \exp \{- B(h) + h'z\} \, dW$ is continuous (as a function of z). Next by using the inequality

(A.11)
$$\exp \{\lambda u + (1 - \lambda)v\} \leqq \lambda \exp \{u\} + (1 - \lambda) \exp \{v\} \quad 0 < \lambda < 1$$

with strict inequality unless $u = v$, one has that

$$(A.12) \quad \int \exp\left\{- B(h) + h'[\lambda z_1 + (1 - \lambda)z_2]\right\} dW$$

$$\leqq \lambda \int \exp\left\{- B(h) + h'z_1\right\} dW + (1 - \lambda) \int \exp\left\{-B(h) + h'z_2\right\} dW$$

$$\leqq 2w_0 \qquad \text{whenever} \quad z_1, z_2 \in A_2 \cup A_3 = C_W.$$

Thus C_W is a convex set. It is also closed, since $\int \exp\left\{- B(h) + h'z\right\} dW$ is continuous, as was shown above.

According to the weak compactness theorem for tests, for any given sequence of tests $\{\phi_n\}$, there is a subsequence $\{\phi_m\}$, which converges weakly to a test ϕ in the sense that

$$(A.13) \qquad \int \phi_m g d\mathscr{L}_0^* \to \int \phi g d\mathscr{L}_0^*$$

for every $\mathscr{L}_0^*$ integrable function g defined on R^k into R. A proof of this theorem can be found in Lehmann [12], pp. 354–356.

REMARK A.1. As a consequence of the weak compactness theorem stated above, one has that, if $\{\phi_m\}$ converges weakly to ϕ, then $\beta_{\phi_m}(h) \to \beta_\phi(h)$ for every $h \in R^k$. This follows from (A.13) above by replacing $g(x)$ by the $\mathscr{L}_0^*$ integrable function $\exp\left\{-B(h) + h'z\right\}$.

Now consider the class of tests ϕ of the following form

$$(A.14) \qquad \phi(z) = \begin{cases} 1 & \text{if } z \in C^c \\ 0 & \text{if } z \in C^0, \end{cases}$$

where C is a closed, convex set in R^k. The test may be defined in an arbitrary (but measurable) manner on the boundary C^b of C. The Bayes tests given by (A.5) are also of the form (A.14). However, in the following, we will be interested in tests of the form (A.14) which may not correspond to any prior W on $\mathscr{B}^k$.

We shall show below that the weak limit of a sequence of tests, each one of which is of the form (A.5), is also of the same form. To this end, denote by S_r the closed, solid sphere of radius r centered at the origin, and for any two closed subsets of S_r, A and B, consider their Hausdorff distance, $d(A, B)$, defined as follows

$$(A.15) \qquad d(A, B) = \inf\left\{\varepsilon > 0; A \subset N_\varepsilon(B), B \subset N_\varepsilon(A)\right\}.$$

Here $N_\varepsilon(A) = \{y \in R^k; (y - z)'(y - z) < \varepsilon \text{ for some } z \in A\}$ and similarly for $N_\varepsilon(B)$.

The following standard result on convex sets, which is established in Eggleston [5], p. 64, will also be needed.

THEOREM A.2. (Blaschke selection theorem). *Given any sequence of closed, convex subsets of S_r, $\{C_n\}$, there exists a subsequence $\{C_m\}$ and a nonvoid, convex subset C of S_r such that $d(C_m, C) \to 0$.*

By utilizing Theorem A.2, one can establish the following result, as in Matthes and Truax [14], p. 684.

THEOREM A.3. *If $\{\phi_m\}$ is a sequence of tests of the form (A.14) which converges weakly to the test ϕ in the sense of (A.13), then ϕ is also of the same form a.s. $[\mathscr{L}_0^*]$.*

The following definition refers to the essential completeness of a class of tests, namely.

DEFINITION A.1. *A class of tests is said to be essentially complete for testing the hypothesis $H^*: h = 0$ against the alternative $A^*: h \neq 0$ in the family (A.1), if for any test ψ not in the class there is a test ϕ in the class for which $\beta_\psi(0) \geq \beta_\phi(0)$ and $\beta_\psi(h) \leq \beta_\phi(h)$ for $h \neq 0$.*

We now show that tests of the form (A.14) form an essentially complete class. More precisely,

THEOREM A.4. *For testing the hypothesis $H^*: h = 0$ against the alternative $A^*: h \neq 0$ in the family (A.1), the class of tests of the form (A.14) is essentially complete when W varies over the class of all probability distributions on $\mathscr{B}^k$.*

PROOF. Assume that the test ψ is not of the form (A.14) and define the new risk R_ϕ^* associated with a test ϕ as follows

$$(A.16) \qquad R_\phi^*(h) = R_\phi(h) - R_\psi(h), \qquad\qquad h \in R^k,$$

where R_ϕ and R_ψ are given by (A.3) (with ϕ replaced by ψ for the latter). Let $\{h_n\}$ be a dense sequence in R^k with $h_1 = 0$ and distinct terms. We claim that for each $j = 1, 2, \cdots$, there exists a test ϕ_j of the form (A.5) with

$$(A.17) \qquad R_{\phi_j}^*(h_i) \leq 0, \qquad\qquad i = 1, 2, \cdots, j.$$

Suppose for a moment that (A.17) has been established. Then by considering the sequence $\{\phi_j\}$ there is a subsequence $\{\phi_m\}$ which converges weakly to a test ϕ, by the weak compactness theorem. Furthermore, the test ϕ is also of the form (A.14) a.s. $[\mathscr{L}_0^*]$. This is so by Theorem A.3. Now according to (A.17), $R_{\phi_m}^*(h_i) \leq 0$ for all i and m, or equivalently, $R_{\phi_m}(h_i) \leq R_\psi(h_i)$ for all i and m, as follows from (A.16). From Remark A.1, one then has that $\beta_\psi(0) \geq \beta_\phi(0)$ and $\beta_\psi(h_i) \leq \beta_\phi(h_i)$ for $i = 2, 3, \cdots$. By the fact that the power function of any test is continuous (by Theorem 9 on p. 52 in Lehmann [12]) and the choice of the sequence $\{h_i\}$, one has that $\beta_\psi(h) \leq \beta_\phi(h)$ for all $h \in R^k$, $h \neq 0$.

Thus it suffices to establish (A.17). For each $j = 1, 2, \cdots$, consider the risk set, $C_j(R^*)$ consisting of points of the form $(R_\phi^*(h_1), \cdots, R_\phi^*(h_j))'$ for some test ϕ. It is then an easy matter to show that $C_j(R^*)$ is closed and convex. As is well known, the minimax test ϕ_j, say, is obtained by finding the point in $C_j(R^*)$ on the equiangular line with the smallest coordinates. To this end, define d_j by $d_j = \inf \{\delta; \delta(1, \cdots, 1)' \in C_j(R^*)\}$ and set

$$(A.18) \qquad C(d_j) = \{z = (x_1, \cdots, x_j)' \in R^j; x_i < d_j, i = 1, \cdots, j\}.$$

Then $C(d_j)$ and $C_j(R^*)$ are disjoint, convex sets. By the separating hyperplane theorem (see, for example, Ferguson [6], pp. 73–74), it follows that there exists hyperplane $w'z = a$ with $w'z \geqq a$ for $z \in C_j(R^*)$, $w'z < a$ for $z \in C(d_j)$ and $w'z = a$ for $z = d_j(1, \cdots, 1)'$. The coordinates w_i, $i = 1, \cdots, j$ of w are $\geqq 0$ because otherwise $x_i \to -\infty$ would lead to a contradiction. Therefore we may as well assume that $(w_i, \cdots, w_j)'$ is a probability distribution over $\{h_1, \cdots, h_j\}$. Since $w'z \geqq a$ for $z \in C_j(R^*)$ and $w'z = a$ for $z = d_j(1, \cdots, 1)'$, it follows that the minimax test ϕ_j, corresponding to the point $d_j(1, \cdots, 1)'$, is also Bayes relative to the distribution $(w_1, \cdots, w_j)'$ and hence ϕ_j is of the form (A.14). Next from (A.16) it follows that $R^*_\psi(h_i) = 0$, $i = 1, \cdots, j$, and since ϕ_j is minimax, one has that $R^*_{\phi_j}(h_i) \leqq 0$ for all $i = 1, \cdots, j$. The claim in (A.17) is established and the proof of the theorem is completed.

To Theorem A.4, there is the following corollary.

COROLLARY A.1. *For testing the hypothesis $H^*: h = 0$ against the alternative $A^*: h \neq 0$ in the family of probability densities given in Lemma 5.1(ii), namely*

$$\text{(A.19)} \qquad \frac{dL^*_h}{dL^*_0} = \exp\{-B(h) + h'\Gamma z\}, \qquad\qquad z, h \in R^k,$$

where $L^*_h = \mathscr{L}(\Gamma^{-1}\Delta^* \mid R_h)$, the class of tests of the form (A.14) is essentially complete.

PROOF. The transformation $t = \Gamma h$ brings the family (A.18) into the form (A.1) and then Theorem A.4 applies.

This appendix is closed with the following remark.

REMARK A.2. As follows from Theorem A.4 and also Corollary A.1, when testing the hypothesis $H^*: h = 0$ against $A^*: h \neq 0$, for any test ψ not of the form (A.14), there exists a test of that form with no smaller power and no larger size whatever the size of ψ. Consequently, whatever criterion is proposed in terms of power, it is possible to restrict ourselves to members of the class of tests of the form (A.14).

REFERENCES

[1] A. BIRNBAUM, "Characterization of complete classes of tests of some multiparametric hypotheses, with applications to likelihood ratio tests," *Ann. Math. Statist.*, Vol. 26 (1955), pp. 21–36.

[2] D. M. CHIBISOV, "A theorem on admissible tests and its application to an asymptotic problem of testing hypotheses," *Theor. Probability Appl.*, Vol. 12 (1967), pp. 90–103.

[3] ———, "Transition to the limiting process for deriving asymptotically optimal tests," *Sankhyā Ser. A*, Vol. 31 (1969), pp. 241–258.

[4] J. L. DOOB, *Stochastic Processes*, New York, Wiley, 1953.

[5] H. G. EGGLESTON, *Convexity*, Cambridge, Cambridge University Press, 1966.

[6] T. S. FERGUSON, *Mathematical Statistics*, New York, Academic Press, 1967.

[7] J. HÁJEK and Z. ŠIDÁK, *Theory of Rank Tests*, New York, Academic Press, 1967.

[8] R. A. JOHNSON and G. G. ROUSSAS, "Asymptotically most powerful tests in Markov processes," *Ann. Math. Statist.*, Vol. 40 (1969), pp. 1207–1215.

[9] ———, "Asymptotically optimal tests in Markov processes," *Ann. Math. Statist.*, Vol. 41 (1970), pp. 918–938.

[10] L. LECAM, "Locally asymptotically normal families of distributions," *Univ. Calif. Publ. Statist.*, Vol. 3 (1960), pp. 37–98.

[11] ———, "Likelihood functions for large numbers of independent observations," *Research Papers in Statistics*, edited by F. N. David, New York, Wiley, 1966.

[12] E. L. LEHMANN, *Testing Statistical Hypotheses*, New York, Wiley, 1959.

[13] M. LOÈVE, *Probability Theory*, 3rd ed., Princeton, Van Nostrand, 1963.

[14] T. K. MATTHES and D. R. TRUAX, "Tests of composite hypotheses for the multivariate exponential family," *Ann. Math. Statist.*, Vol. 38 (1967), pp. 681–697.

[15] J. NEYMAN, "Optimal asymptotic tests of composite statistical hypotheses," *Probability and Statistics* (edited by Ulf Grenander), New York, Wiley, 1959.

[16] C. R. RAO, "Large sample tests of statistical hypotheses concerning several parameters with applications to problems of estimation," *Proc. Cambridge Philos. Soc.*, Vol. 44 (1948), pp. 50–57.

[17] ———, *Linear Statistical Inference and Its Applications*, New York, Wiley, 1965.

[18] R. R. RAO, "Relations between weak and uniform convergence of measures with applications," *Ann. Math. Statist.*, Vol. 33 (1962), pp. 659–680.

[19] G. G. ROUSSAS, "Asymptotic inference in Markov processes," *Ann. Math. Statist.*, Vol. 36 (1965), pp. 978–992.

[20] ———, "Some applications of the asymptotic distribution of likelihood functions to the asymptotic efficiency of estimates," *Z. Wahrscheinlichkeitstheorie und Verw. Gebiete*, Vol. 10 (1968), pp. 252–260.

[21] ———, "On the concept of contiguity and related theorems," Aarhus Universitet Preprint Series 1968/69 No. 33, 28 pp.

[22] ———, "The usage of the concept of contiguity in discussing some statistical problems," Aarhus Universitet Preprint Series 1968/69 No. 34, 42 pp.

[23] ———, "Multiparameter asymptotically optimal tests for Markov processes," Aarhus Universitet Preprint Series 1968/69 No. 42, 26 pp.

[24] A. WALD, "Tests of statistical hypotheses concerning several parameters when the number of observations is large," *Trans. Amer. Math. Soc.*, Vol. 54 (1943), pp. 426–482.

ITERATED LOGARITHM ANALOGUES FOR SAMPLE QUANTILES WHEN $p_n \downarrow 0$

J. KIEFER

CORNELL UNIVERSITY

1. Introduction

This paper is concerned with behavior of the Law of Iterated Logarithm (LIL) type for sample p_n-tiles, $p_n > 0$, when $p_n \downarrow 0$. The results are all stated for uniformly distributed random variables, from which they may easily be translated into results for general laws.

Let $X_1, X_2, \cdots$ be independent identically distributed random variables, uniformly distributed on $[0, 1]$. Let $T_n(x) = \{\text{number of } X_i \leq x, 1 \leq i \leq n\}$, so that $n^{-1}T_n$ is the right continuous *sample distribution function* based on $X_1, X_2, \cdots, X_n$. Define the *sample p_n-tile* $Z_n(p_n)$ as $\min \{z \colon T_n(z) \geq np_n\}$. This makes $Z_n(p_n) = np_n$-th order statistic when np_n is a fixed integer. (When $np_n \to \infty$ our results do not depend on the choice of definition of $Z_n(p_n)$ in cases of ambiguity.)

The earliest nontrivial result in this area, due to Baxter [2], is that, for any positive constant c,

$$(1.1) \qquad \limsup_n T_n(c/n) \log \log \log n (\log \log n)^{-1} = 1, \qquad \text{wp } 1.$$

On the other hand, it is trivial (and a consequence of Theorem 2 herein, with $k = 1$) that

$$(1.2) \qquad \liminf_n T_n(c/n) = 0, \qquad \text{wp } 1.$$

We thus no longer have the symmetry in asymptotic behavior of positive and negative deviations of $T_n(\pi_n) - ET_n(\pi_n)$ that prevails when π_n is constant; indeed, why should we, when $nT_n(c/n)$ is asymptotically Poisson rather than normal?

This difference in behavior means we will have to state results for the two directions of oscillations separately, and (since the analogue of (1.2) will not always be so simple to state) dictates a choice of nomenclature which we had best introduce at the outset: to eliminate possible confusion with reference to the two *directions* of oscillation, we drop the usual "upper or lower class" LIL terms completely, replacing these by "outer or inner class" for sequences $\{f_n\}$ beyond which $T_n(\pi_n)$ moves (in a direction away from $ET_n(\pi_n)$) finitely or infinitely often

Research carried out under ONR contract Nonr 401 (50) and NSF Grant GP 9297.

with probability one. Then, "top or bottom" bounds will refer to the most or least positive oscillations of $T_n(\pi_n)$. Thus, for example, $\{f_n\}$ is a bound of top inner class if $f_n > n\pi_n$ and $T_n(\pi_n) > f_n$ i.o. with probability one. If $\{(1 \pm \varepsilon)f_n\}$ gives top bounds of the two classes, we shall for brevity simply call $\{f_n\}$ a top bound. (If, as when π_n is constant, this yields too gross a result for T_n, we would instead specify a top bound on $T_n(\pi_n) - n\pi_n$.) Thus, log log n/log log log n is a top bound for $T_n(c/n)$ in Baxter's case mentioned above. Bounds for $Z_n(p_n)$ are described similarly. Of course, top bounds for Z_n are related to bottom bounds for T_n and vice versa, by the well known relation

$$(1.3) \qquad\qquad T_n(\pi_n) \geqq k_n \Leftrightarrow Z_n(k_n/n) \leqq \pi_n,$$

which follows from the definitions.

The proofs of the present paper employ standard techniques and estimates of binomial probabilities, and we have sometimes introduced inessential assumptions to maintain simplicity and brevity. The results are mainly about first order deviations of T_n or Z_n from their expectations; while some "strong form" results are known, with few exceptions (for example, Theorem 1) they entail much longer proofs and I do not presently know the conclusions for the full spectrum of sequences $\{\pi_n\}$ considered herein. Theorems 1 and 2 cover the behavior of $Z_n(k/n)$ with k fixed; Theorem 5 covers the domain of normal limiting behavior and resulting classical LIL form for $T_n(\pi_n)$, and Theorems 3 and 4 cover the behavior in between these two extremes, including Baxter's case; Theorem 6 translates some of the conclusions for T_n into conclusions for Z_n.

We now mention work related to that of the present paper, other than that of Baxter described above. Bahadur [2] used his relation between the Z_n and T_n processes to obtain the LIL for $Z_n(p)$ with p fixed from the classical LIL for binomial $T_n(p)$. The same method can be used to obtain the strong form of inner and outer classes [6]. While the classical techniques of Theorems 5 and 6 herein also yield the LIL for $Z_n(p)$ (the main departure from the usual LIL proof for $T_n(p)$ being that one is now led by (1.3) to the LIL for $T_n(p_n)$ for varying p_n), Bahadur's technique provides a great saving of effort when it comes to the strong form.

Eicker [4] obtained top outer bounds (analogous to (1.1)) for $T_n(\pi_n)$ when $\pi_n \downarrow 0$ and $n\pi_n/\log \log n \to \infty$, but not bottom outer bounds, and also obtained both inner bounds. This is the domain treated in Theorem 5 of the present paper, where the usual binomial-normal LIL form holds. Eicker's inner class proof follows essentially the classical lines which are therefore sketched only in brief outline herein; this proof applies also to certain cases where π_n is bounded away from zero and one. His top outer bound proof uses fine estimates of the probability that the T_n process exceeds certain polygonal bounds; our proof, while much more routine, is considerably shorter, and treats also the bottom outer bound.

Robbins and Siegmund [10] have just announced the use of an interpolating process and the derivation of probability estimates for ever exceeding certain bounds (in the spirit of their earlier work with sums of random variables), in obtaining strong top bounds for the first order statistics $Z_n(1/n)$. Our Theorem 2 with $k = 1$ states only the first order term of this strong form.

In [1], [6], [7], and [8] relations between the $T_n(p)$ and $Z_n(p)$ processes were studied (with respect to varying p as well as n). For example, in the spirit of [1], one thereby obtains in [8] (where some of the results of the present paper were also described) the strong form for $\sup_{0 < p < 1} |Z_n(p) - p|$ oscillations from the corresponding sample of distribution function results of Chung [3]. Large deviations of either these processes or of their difference, over the domain $0 < p < p_n \downarrow 0$, are related to the present results and will be treated elsewhere; some such considerations have appeared in the work of Chibisov, LeCam, and others. Also related are the numerous papers on weak laws for order statistics, about which we mention only the appearance therein, not surprisingly, of the "fundamental equation" (2.24) which arises below. (See, for example, [9].)

Section 2 contains definitions and relevant binomial tail probability estimates. Statements of the main results are contained in Section 3, along with proofs of the simple first two theorems. The remainder of the paper contains the other proofs.

2. Preliminaries

We shall use i.o., f.o., and a.a.n. in their customary meaning of infinitely often, finitely often, and for almost all (all but finitely many) n. We treat events indexed by the natural numbers $\{n\}$ or a subsequence $\{n_j\}$, and the usual expressions of limiting behavior ($\rightarrow$, $\sim$) or of order (such as $O(z(n))$ or $o(g(j))$) refer to behavior as $n \rightarrow +\infty$ or $j \rightarrow +\infty$. The symbols $\uparrow$, $\downarrow$ are used for monotone, not strict, approach.

We let $\log_1$ denote the natural logarithm and $\log_{j+1} = \log \log_j$. Also, $\log_j^i x = (\log_j x)^i$. In summations and other appearances of such an expression as $\log_j n$, the domain of n is understood to begin where the expression is meaningful.

We shall try to reserve π_n for the argument of T_n and p_n for that of Z_n, with $k_n \geqq 0$ being used for bounds on T_n. We define $T_{n_1, n_2} = T_{n_2} - T_{n_1}$, the observation counter based on $X_{n_1 + 1}, \cdots, X_{n_2}$. Limiting behavior will often be conveniently described in terms of

$$(2.1) \qquad h_n = n\pi_n/\log_2 n, \qquad H_n = np_n/\log_2 n.$$

Either of the Borel-Cantelli lemmas is denoted by BC.

We use int $\{x\}$ to denote the largest integer $\leqq x$, and int$^+$ $\{x\}$ for the smallest integer $\geqq x$.

When we consider a subsequence $\{n_j\}$ of the natural numbers, we write

$$(2.2) \qquad I_j = \{n: n_j < n \leqq n_{j+1}\} \quad \text{and} \quad I_{\bar{j}} = \{n: n_j < n < n_{j+1}\}.$$

We write $m_j = n_j - n_{j-1}$. The two subsequences $\{n_j\}$ we shall consider, with typical estimates they imply, are

$$n_j \sim \lambda^j \qquad \text{where} \quad \lambda > 1, \, m_j/n_j \sim (\lambda - 1)/\lambda,$$

(2.3)
$$\log_2 n_j \sim \log j,$$

and, for $\alpha > 0$,

$$n_j \sim e^{\alpha j \log j} \qquad\qquad m_j/n_j = 1 - O(j^{-\alpha}),$$

(2.4)
$$\log_2 n_j \sim \log j.$$

Much of our treatment can be carried out in terms of very simple events, essentially as employed in [2] for top bounds on T_n. Suppose $\pi_n \downarrow 0$ and $k_n \uparrow$, and let

(2.5)
$$A_n = \{T_n(\pi_n) \geqq k_n\}.$$

Let $\{n_j\}$ be *any* increasing sequence of natural numbers, and define

(2.6)
$$B_j^* = \{T_{n_j, n_{j+1}}(\pi_{n_{j+1}}) \geqq k_{n_{j+1}}\}$$

and

(2.7)
$$C_j^* = \{T_{n_{j+1}}(\pi_{n_j}) < k_{n_j}\}.$$

Then, since $T_n(\pi)$ is nondecreasing in n and π, and $\pi_n \downarrow$, $k_n \uparrow$,

(2.8)
$$C_j^* \subset \{T_n(\pi_n) < k_n, \, n \in I_j\},$$

and hence

(2.9)
$$\{C_j^*, \text{a.a.} \, j\} \Rightarrow \{A_n \, \text{f.o.}\}, \qquad \text{wp 1}.$$

In the other direction, obviously

(2.10)
$$\{B_j^* \, \text{i.o.}\} \Rightarrow \{A_n \, \text{i.o.}\}, \qquad \text{wp 1},$$

the events of (2.6) of course being useful because they are independent.

We shall see that, in the top outer class proofs of Theorems 1 and 3, we can even avoid the use of subsequences as employed in [2] and in "normal case" proofs such as that of Theorem 5 below, and, as an alternative, work with the events

(2.11)
$$A_n' = \{X_n \leqq \pi_n, \, T_{n-1}(\pi_n) \geqq k_n - 1\}.$$

As long as $1 \leqq k_n \uparrow$ and $\pi_n \downarrow 0$, it is evident that

(2.12)
$$\{A_n \, \text{i.o.}\} \Rightarrow \{A_n' \, \text{i.o.}\}, \qquad \text{wp 1}.$$

Similarly, if $\pi_n \downarrow$ and $k_n \uparrow$, for bottom outer bounds on T_n we use

(2.13)
$$\{T_{n_j}(\pi_{n_{j+1}}) < k_{n_{j+1}} \, \text{f.o.}\} \Rightarrow \{T_n(\pi_n) < k_n \, \text{f.o.}\}, \qquad \text{wp 1}.$$

The bottom inner bound treatment is only slightly less simple. If

(2.14)
$$Q_j^* = \{T_{n_j, n_{j+1}}(\pi_{n_{j+1}}) \leqq k_{n_{j+1}} - \gamma_j\},$$
$$R_j^* = \{T_{n_j}(\pi_{n_{j+1}}) > \gamma_j\},$$

where the γ_j are arbitrary nonnegative values, then

(2.15) $\{Q_j^* \text{ i.o.}\} \cap \{Q_j^* R_j^* \text{ f.o.}\} \Rightarrow \{T_{n_{j+1}}(\pi_{n_{j+1}}) \leqq k_{n_{j+1}} \text{ i.o.}\}$,   wp 1.

In fact, in the simple first order proofs of Theorems 2 and 4 it suffices to take $\gamma_j = 0$, and to show $P\{R_j^* \text{ f.o.}\} = 1$ for the second half of the left side of (2.15).

In using the simple devices of (2.5) through (2.15) we require (in addition to $\pi_n \downarrow 0$) $k_n \uparrow$. This is not always too convenient: as described below the statement of Theorem 3, for given $\{\pi_n\}$ the natural formula for a bound k_n in terms of π_n in Theorems 3 and 4 may not yield a monotone k_n. However, in such cases we will be able to replace the nonmonotone $\{k_n\}$ by a sequence $\{k_n^*\}$ such that

(2.16) $k_n^* \uparrow, \; k_n^* = [1 + O(1)]k_n$,

and then use the appropriate device of (2.5) through (2.15) on k_n^*; by virtue of our considerations being first order (so that we prove $(1 \pm \varepsilon)k_n^*$ lies in the appropriate class), $\{k_n\}$ is then by definition in the same class as $\{k_n^*\}$.

The study of sequences $\{\pi_n\}$ for which π_n is not monotone, or for which the technique of (2.16) fails, is more complex, requiring in place of (2.5) through (2.15) calculations which are somewhat similar to those stemming from (4.18) and (4.19) in the proof of Theorem 5. We omit such cases. It is obvious that bounds for some $\{\pi_n\}$ for which the departure from monotonicity is slight enough, may be obtained from those of majorizing and minorizing monotone sequences when corresponding bounds for the latter sequences coincide.

To give a little relief from the burden of memory or page turning, we shall reserve further definitions until they are encountered in the proof of Theorem 5.

We now list our binomial estimates. In Theorems 1 and 2 we consider $Z_n(k/n)$ with k fixed, and require only the following simple and familiar estimates ([5], p. 140) for nonnegative integral $\bar{k}$:

(2.17) $\limsup_n n\pi_n < \bar{k} \Rightarrow$

$$\log P\{T_n(\pi_n) \geqq \bar{k}\} - O(1) = \log P\{T_n(\pi_n) = \bar{k}\}$$

and
$$= \bar{k} \log(n\pi_n) - n\pi_n - \log(\bar{k}!) + o(1).$$

(2.18) $\{\pi_n = o(n^{-1/2}), \liminf_n n\pi_n > \bar{k}\} \Rightarrow$

$$\log P\{T_n(\pi_n) \leqq \bar{k}\} - O(1) = \log P\{T_n(\pi_n) = \bar{k}\}$$
$$= \bar{k} \log(n\pi_n) - n\pi_n - \log(\bar{k}!) + o(1).$$

In the domain of Theorems 3 and 4, we must take account of the fact that $k_n \to \infty$. Writing

$$b(z, N, P) = \binom{N}{z} P^z (1 - P)^{N-z}, \qquad\qquad z \text{ integral,}$$

(2.19)

$$B(z, N, P) = \sum_{y \leqq z} b(y, N, P),$$

$$B^+(z, N, P) = \sum_{y \geqq z} b(y, N, P),$$

we state the required binomial estimate as

LEMMA 1.   *Suppose* $N \to \infty$, $P_N \to 0$, $z_N \to +\infty$. *If*

(2.20)
$$P_N = o(N^{-1/2}), \qquad z_N = o(N^{1/2}),$$
$$\lim_N \sup NP_N/z_N < 1,$$

then

(2.21) $\log B^+(z_N, N, P_N) = z_N[\log(NP_N/z_N) - (NP_N/z_N) + 1 + o(1)].$

Moreover, (2.21) *holds with* B *replacing* B^+, *provided the last condition of* (2.20) *is replaced by* $\lim \inf_N NP_N/z_N > 1$; *and the right side of* (2.21) *equals log* $b(int^+ z_N, N, P_N)$ *or* $\log b(int\, z_N, N, P)$, *even without any third condition of* (2.20).

PROOF.   The last condition of (2.20) implies that B^+/b is bounded ([5], p. 140). Putting $z' = int^+ z_N$ and writing out $\log b(z', N, P_N)$ and using Stirling's approximation, we see that the first condition of (2.20) allows us to neglect $NP_N + \log(1 - P_N)^N$, the second allows us to neglect $\log[N^{z'}/N(N-1)\cdots(N - z' + 1)]$, and the two together imply $z_N P_N = o(1)$ and thus allow us to neglect $z' \log(1 - P_N)$. Since $z_N \sim z'$ and also we can absorb $\log z'$ into the $o(1)z_N$ term, we obtain (2.21). The result for B is obtained in the same way.

We now specialize the parameter values in Lemma 1, as used in the proofs of Theorems 3 and 4. Firstly, as we shall explain after Theorems 1 and 2 where k_n is bounded, we subsequently insure that k_n is unbounded by assuming

(2.22) $\lim_n \inf[\log(n\pi_n)/\log_2 n] \geqq 0$

in the top bound considerations for $T_n(\pi_n)$ of Theorem 3, and

(2.23) $\lim_n \inf[n\pi_n/\log_2 n] > 1$

in the bottom bound considerations of Theorem 4.

Secondly, in the domain where (2.22) or (2.23) is satisfied and where also $\pi_n = O(n^{-1}\log_2 n)$, the bounds on $T_n(\pi_n)$ are described in terms of the solutions of a certain transcendental equation.

The fundamental equation.   We consider the solutions, for $c > 0$, of the equation

(2.24) $\beta(\log \beta - 1) = (1 - c)/c.$

The left side of (2.24) is convex in $\beta > 0$ and attains its minimum value -1 at $\beta = 1$. Hence, (2.24) has a solution $\beta'_c > 1$ if $c > 0$ and a second positive solution $\beta''_c < 1$ if $c > 1$. For future reference we note that β'_c is decreasing in c while β''_c is increasing, and that

(2.25) $c \to 0 \Leftrightarrow \beta'_c \to \infty \Rightarrow \beta'_c \sim c^{-1}/\log c^{-1} \to +\infty$, $c\beta'_c \sim 1/\log c^{-1} \to 0$;

$c \to +\infty \Leftrightarrow \beta'_c \to 1 \Rightarrow c\beta'_c \to +\infty$;

$$c \downarrow 1 \Leftrightarrow \beta_c'' \to 0 \;\Rightarrow\; \beta_c'' \sim (c-1)/\log(c-1)^{-1}, \qquad c\beta_c'' \to 0;$$

$$c \to +\infty \Leftrightarrow \beta_c'' \to 1, \qquad c\beta_c'' \to +\infty.$$

Finally, on either part $\beta < 1$ or $\beta > 1$ of (2.24), $d(c\beta)/dc = (\beta - 1)/\log \beta > 0$.

We now state the specialization of Lemma 1 used in proving Theorems 3 and 4. Recall the definition (2.1) of h_n.

LEMMA 2. *Suppose* $n \to \infty$, $\pi_n \to 0$, $h_n = O(1)$, *and that* (2.22) *is satisfied. Let* $\{\rho_n\}$ *and* $\{d_n\}$ *be sequences of positive values which are bounded away from 0 and* ∞ *and for which* $n\rho_n$ *is integral. Assume*

$$(2.26) \qquad \limsup_n \rho_n/d_n\beta_{h_n}' < 1,$$

where β_c' *is defined below* (2.24). *Then*

$$(2.27) \qquad \log P\{T_{n\rho_n}(\pi_n) \geqq d_n h_n \beta_{h_n}' \log_2 n\}$$
$$= \{-d_n + h_n[d_n - \rho_n - d_n\beta_{h_n}' \log(d_n/\rho_n)] + o(1)\}\log_2 n;$$

moreover, $\log P\{T_{n\rho_n}(\pi_n) = \mathrm{int}^+ [d_n h_n \beta_{h_n}' \log_2 n]\}$ *satisfies the same relation.*

If (2.22) *and* (2.26) *are replaced in the above by* (2.23) *and*

$$(2.28) \qquad \liminf_n \rho_n/d_n\beta_{h_n}'' > 1,$$

then

$$(2.29) \qquad \log P\{T_{n\rho_n}(\pi_n) \leqq d_n h_n \beta_{h_n}'' \log_2 n\}$$
$$= \{-d_n + h_n[d_n - \rho_n - d_n\beta_{h_n}'' \log(d_n/\rho_n)] + o(1)\}\log_2 n.$$

PROOF. We shall demonstrate (2.27); the proof of (2.29) is almost identical. We put $N = n\rho_n$, $P_N = \pi_n$, $z_N = d_n h_n \beta_{h_n}' \log_2 n$ in Lemma 1. Then $z_N \to +\infty$ unless there is a sequence $\{n_j\}$ for which $h_{n_j} \to 0$ and $h_{n_j}\beta_{h_n}' = O(1/\log_2 n_j)$, which by the first line of (2.25) would entail $1/\log h_{n_j}^{-1} = O(1/\log_2 n_j)$; this last is contradicted by the fact that (2.22) (with $h_{n_j} \to 0$) implies $\log h_{n_j}^{-1} = o(\log_2 n_j)$; we conclude that $z_N \to +\infty$. Next, $h_n = O(1)$ implies the first condition of (2.20) as well as $z_N = O(\log_2 n) = O(\log_2 N)$, and this last yields the second condition of (2.20). Also, $NP_N/z_N = \rho_n/d_n\beta_{h_n}'$, so that (2.26) implies the last condition of (2.20). Thus, Lemma 1 applies, and substitution into (2.21) gives

$$(2.30) \qquad d_n h_n \beta_{h_n}' \{-\log \beta_{h_n}' + 1 - 1/\beta_{h_n}' + \log(\rho_n/d_n)$$
$$+ [1 - \rho_n/d_n]/\beta_{h_n}' + o(1)\}\log_2 n.$$

Since the first three terms in braces in (2.30) sum to $-1/h_n\beta_{h_n}'$ by (2.24), and since $d_n h_n \beta_{h_n}' o(1) = o(1)$, we obtain (2.27).

When $h_n \to \infty$, the approximations of Lemma 2 are insufficient, and we need the normal approximation instead. This tool is also well known; a careful reading of Feller ([5], pp. 168–173, 178–181) shows that the development there actually applies with only minor and obvious modifications when $\pi_N \to 0$ sufficiently slowly, and we state this result as

LEMMA 3. *If* $x_N = (k_N - N\pi_N)[N\pi_N(1 - \pi_N)]^{-1/2}$, *then*

(2.31) $\{N \to \infty, x_N \to +\infty, x_N[N\pi_N(1 - \pi_N)]^{-1/2} \to 0\} \Rightarrow$

$$B^+(k_N, N, \pi_N) = (2\pi)^{-1/2} x_N^{-1} \exp\{-x_N^2[1 + o(1)]/2\}.$$

(Feller's expression [5], p. 181, (6.11) for the $\frac{1}{2}o(1)x_N^2$ term in (2.31) is also correct under the present conditions; it is conveniently expressed as

(2.32) $\frac{1}{2}o(1)x_N^2 = \frac{1}{6}(2\pi_N - 1)x_N^3[N\pi_N(1 - \pi_N)]^{-1/2} + O(x_N^4[N\pi_N(1 - \pi_N)]^{-1})$

for use in other limit laws, but this will not be required herein.)

Finally, we shall also use the elementary fact that

(2.33) $$\inf_{N, \pi} B^+(N\pi - 1, N, \pi) > 0;$$

the "bad case" where the $N\pi - 1$ is required rather than $N\pi$ is of course $N \to \infty$, $N\pi \downarrow 0$. Similarly,

(2.34) $$\inf_{N, \pi} B(N\pi + 1, N, \pi) > 0.$$

In each of these expressions we include zero in the domain of N; both probabilities are then one, corresponding to the interpretation $b(0, 0, \pi) = 1$ which is appropriate in the application.

It seems essential that the proofs, as carried out in the present paper, be divided into the several cases as treated. For, the estimates of Lemma 2 are useless in the "normal" case of Theorem 5, just as those of Lemma 3 are useless for proving Theorems 3 and 4. Again, the geometric $\{n_j\}$ of (2.3) is inadequate in the inner class bottom bound proofs of Theorems 2 and 4, where (2.4) is used, but the latter cannot be used in the corresponding outer class proofs; in Theorem 5 we again use geometric $\{n_j\}$ where, also, it is impossible to avoid using subsequences by using (2.11) and (2.12) as in parts of Theorems 1 and 3. (For certain strong form results, other sequences $\{n_j\}$ must of course be considered.)

3. Main results

The short proofs of Theorems 1 and 2 will be given in this section, but proofs of the other theorems stated in this section will be deferred in favor of discussion here. For *bounded* top bounds on $T_n(\pi_n)$ or corresponding bottom bounds on the kth order statistic $Z_n(k/n)$ (k fixed), the situation is completely known [8], and is elementary to verify. We forego the artificial generality of bounded but varying k_n, or oscillatory $n\pi_n$, and state the result simply as

THEOREM 1. *If* k *is a positive integer and* $n\pi_n \downarrow 0$, *then*

(3.1) $P\{T_n(\pi_n) \geq k \text{ i.o.}\}$

$$= P\{Z_n(k/n) \leq \pi_n \text{ i.o.}\} = \begin{cases} 0 \Leftrightarrow \infty > \\ 1 \Leftrightarrow \infty = \end{cases} \sum_n n^{k-1}\pi_n^k$$

(*so that* $\log[nZ_n(k/n)]$ *has*

(3.2) $$-k^{-1}\left\{\log_2 n + \log_3 n + \cdots + (1 + \varepsilon)\log_j n\right\}$$

as bottom outer or inner bound, depending on whether $\varepsilon > 0$ or $\varepsilon \leq 0$).

PROOF. *Outer class.* By (1.3) we are concerned with the events A_n of (2.5), with $k_n = k$. The geometric sequence $\{n_j\}$ of (2.3) can be used with (2.7) and (2.9) to give a proof, in standard manner. However, we emphasize the simplicity of the present case by giving a proof without a subsequence, using the events $\{A_n'\}$ of (2.11) with $k_n = k$ and the relation (2.12). If the series of (3.1) converges, then BC and the estimate (2.17) with $\bar{k} = k - 1$ imply $P\{A_n'$ i.o.$\} = 1$.

Inner class. Let $n_j = 2^j$. Since $n\pi_n \downarrow 0$, divergence of the series of (3.1) implies divergence of $\Sigma_j (n_j \pi_{n_j})^k$. The estimate (2.17) for $P\{B_j^*\}$, BC, and (2.10) complete the proof of Theorem 1.

In view of Theorem 1, our further concern with top bounds k_n on $T_n(\pi_n)$ is with the case $k_n \to +\infty$. Thus, π_n should be such that the series of (3.1) diverges for each fixed k. This is obviously the case if (2.22) holds, but not if $\limsup_n [\log(n\pi_n)/\log_2 n] < 0$. Hence, ignoring oscillatory behavior where neither of these holds, we hereafter assume π_n satisfies (2.22) in discussing these top bounds. This condition is used in order to apply Lemma 2 in the proof of Theorem 3.

Even the first order lower bounds on $Z_n(k/n)$ in Theorem 1 depend on k, and thereby exhibit quite a different behavior from the upper bounds, to which we now turn.

THEOREM 2. *If k is a positive integer, then*

(3.3) $$P\{nZ_n(k/n) > (1 + \varepsilon)\log_2 n \text{ i.o.}\} = \begin{cases} 0 & \text{if } \varepsilon > 0, \\ 1 & \text{if } \varepsilon \leq 0. \end{cases}$$

PROOF. *Outer class.* By BC, (1.3), and (2.13) with $k_n = k$, it suffices to show that, for $d > 1$, there is a $\lambda > 1$ for which $P\{T_{n_j}(dn_{j+1}^{-1}\log_2 n_{j+1}) \leq k - 1\}$ is summable, where $n_j = \text{int } \lambda^j$. But, by (2.3) and (2.18) this probability is

(3.4) $$\exp\left\{(k-1)\log\left[d\lambda^{-1}\log j\right] - d\lambda^{-1}\log j + O(1)\right\},$$

which is summable if $1 < \lambda < d$.

Inner class. We now put $n_j = \text{int }\{e^{\alpha j \log j}\}$ with $\alpha > 0$. Then, by (2.4) and (2.18),

(3.5) $$P\{T_{n_j, n_{j+1}}(n_{j+1}^{-1}\log_2 n_{j+1}) \leq k - 1\}$$
$$= \exp\left\{(k-1)\log_3 n_{j+1} - \left[1 - O(j^{-\alpha})\right]\log_2 n_{j+1} + O(1)\right\},$$

whose sum diverges. By (2.14) and (2.15) with $\gamma_j = 0$ and $k_n = k - 1$, the proof is completed upon computing, again from (2.4) and (2.18) (now with $\bar{k} = 0$),

(3.6) $$P\{T_{n_j}(n_{j+1}^{-1}\log_2 n_{j+1}) \geq 1\}$$
$$= 1 - \exp\left\{-O(1)j^{-\alpha}\log_2 n_{j+1} + O(1) = O(j^{-\alpha}\log j).\right.$$

REMARKS. In (2.15), $\{Q_j^*$ i.o. wp 1$\}$ is essentially automatic in this case, and it is the event $\{Q_j^* R_j^*$ f.o. wp 1$\}$ for which geometric $n_j \sim \lambda^j$ is inadequate. With

n_j as chosen in the proof and $\alpha > 1$, not merely $\{Q_j^* R_j^* \text{ f.o. wp } 1\}$, but even $\{R_j^* \text{ f.o. wp } 1\}$ is satisfied; however, in strong form analogues one cannot always be so cavalier.

In view of Theorem 2, our further investigation of bottom bounds k_n on $T_n(\pi_n)$, with $k_n \to +\infty$, will be made under the assumption (analogous to (2.22)) that (2.23) holds. This will be discussed further, just after the statement of Theorem 4.

We now turn to the domain of behavior between that of the kth order statistic (fixed k) and that of "normal" LIL. This includes Baxter's case. We recall the definition (2.1) of h_n.

THEOREM 3.   *Suppose* $\pi_n \downarrow 0$, *(2.22) is satisfied,* $h_n = O(1)$, *and that there is a* k_n^* *satisfying (2.16) for* $k_n = h_n \beta'_{h_n} \log_2 n$. *Then*

$$(3.7) \qquad \limsup_n T_n(\pi_n)/h_n \beta'_{h_n} \log_2 n = 1 \qquad \text{wp } 1.$$

In particular, if $\pi_n \downarrow 0$, *a top bound* k_n^* *on* $T_n(\pi_n)$ *is given in various ranges of* π_n *by*:

$$(3.8) \qquad h_n \to c > 0 \Rightarrow k_n^* = c\beta'_c \log_2 n;$$

$$(3.9) \qquad h_n \to 0 \quad \text{and} \quad \log_2 n/\log h_n^{-1} \sim g_n \uparrow +\infty \Rightarrow k_n^* = g_n;$$

in particular,

$$(3.10) \quad \begin{aligned} k_n^* &= \log_2 n/\log_3 n \quad \text{if} \quad \log(n\pi_n) = o(\log_3 n), \\ k_n^* &= \log_2 n/(B + 1)\log_3 n \quad \text{if} \quad \pi_n \sim An^{-1}\log_2^{-B} n, \quad B > -1, \\ k_n^* &= \log_2 n/\log(n\pi_n)^{-1} \quad \text{if} \quad \begin{cases} \log_2 n/\log(n\pi_n)^{-1} \uparrow +\infty, \\ \log_3 n/\log(n\pi_n)^{-1} \to 0. \end{cases} \end{aligned}$$

The use of (2.16) here and in Theorem 4 is not as unnatural as it may first appear. For example, with the positive constant c near zero in Theorem 3 or near one in Theorem 4, and $h_n = c + (-1)^n/3n \log n \log_2 n$, we have $\pi_n \downarrow$, $h_n \to c$ (and, in the case of Theorem 4, even $n\pi_n \uparrow$); but the "natural" k_n given in the denominator of (3.7) or (3.11) is not monotone, which of course $k_n^* = c\beta_c \log_2 n$ is, and the conclusion of each theorem is still valid. The particular cases of (3.8), (3.9), (3.10) have been stated in terms of a simple increasing k_n^* rather than $h_n \beta'_{h_n} \log_2 n \sim \log_2 n/\log h_n^{-1}$.

We also note that, in both Theorem 3 and Theorem 4, h_n of exactly order one does not imply that $\lim h_n$ exists. One can obtain $1 < \liminf_n h_n < \limsup_n h_n < +\infty$ while $\pi_n \downarrow$, $n\pi_n \uparrow$, and even the "natural" $k_n \uparrow$ (discussed in the previous paragraph), by letting $h_{n+1} - h_n$ take successive blocks of positive and negative steps of size $\varepsilon/n \log n \log_2 n$ with ε sufficiently small.

THEOREM 4.   *Suppose* $\pi_n \downarrow 0$, *(2.23) is satisfied,* $h_n = O(1)$, *and that there is a* k_n^* *satisfying (2.16) for* $k_n = h_n \beta''_{h_n} \log_2 n$. *Then*

$$(3.11) \qquad \liminf_n T_n(\pi_n)/h_n \beta''_{h_n} \log_2 \text{n} = 1 \qquad \text{wp } 1.$$

In particular, if $\pi_n \downarrow 0$,

(3.12) $$h_n \to c > 1 \Rightarrow k_n^* = c\beta_c'' \log_2 n$$

is a bottom bound on $T_n(\pi_n)$.

By putting $h_n = 1 + \delta$ with $\delta > 0$ in (3.12), and then letting $\delta \to 0$ and using the third line of (2.25), we obtain that (in a case where (2.23) is not satisfied)

(3.13) $$\lim_n \sup h_n \leq 1 \Rightarrow \lim_n \inf T_n(\pi_n)/\log_2 n = 0, \qquad \text{wp } 1.$$

In fact, when $h_n \equiv 1$ we can use (3.4) or (1.3) and (3.3) with $k = 1$ to obtain

(3.14) $$\pi_n = n^{-1} \log_2 n \Rightarrow \lim_n \inf T_n(\pi_n) = 0, \qquad \text{wp } 1.$$

Since we shall see that the behavior of $c\beta_c'$ as $c \to 0$ in (2.25) yields the top bounds (3.8), (3.9), (3.10) as $h_n \to 0$, it is tempting to try to use the third line of (2.25) to obtain seemingly analogous *bottom* bounds in terms of $h_n - 1 \downarrow 0$. However, the latter are less accurately viewed as first order results in $h_n - 1$ than as second order results in h_n, which cannot be obtained from the behavior of $c\beta_c''$ as $c \downarrow 1$ without more effort, because of the failure of (2.23). We do not have complete results in this domain, and shall not discuss it further except to mention here, as an example of what is involved in the subdomain of smallest values of $h_n - 1$ of interest, that the determination of which sequences $h_n - 1$ of the particular form $L \log_3 n/\log_2 n$ (L constant) continue to imply the conclusion of (3.14) when $h_n \downarrow 1$, requires a more delicate argument than that used in proving Theorems 2 and 4. (Of course, the strong top bounds on $Z_n(1/n)$ [10] imply this second order consequence, but yield nothing about sequences $h_n - 1$ which vanish more slowly.)

We now turn to sequences $\{\pi_n\}$ which vanish slowly enough that "normal" LIL behavior prevails. Of course, we may write $\pi_n(1 - \pi_n) \sim \pi_n$.

THEOREM 5. *If* $\pi_n \downarrow 0$, $h_n \to +\infty$ *and* $n\pi_n \uparrow$, *then, for either choice of sign,*

(3.15) $$\lim_n \sup \pm [T_n(\pi_n) - n\pi_n][2n\pi_n \log_2 n]^{-1/2} = 1, \qquad \text{wp } 1.$$

REMARKS. The assumption that $n\pi_n$ is nondecreasing, although natural enough, is not essential, but is used to simplify the outer class proofs. For example, if $n^L\pi_n$ is increasing for any positive value L, it is only necessary to put $n_j \sim (1 + \varepsilon/3)^{2j/(L+1)}$ to use the same proof. However, if $n\pi_n$ oscillates too much, the right side of (4.14) need not be bounded away from zero, and a longer proof is needed. It will be evident that only minor changes in the proof are required to cover various other cases, for example, $\pi_n \downarrow \pi_0 > 0$, in which case the coefficient of $\log_2 n$ in (3.15) must of course be altered to $2n\pi_0(1 - \pi_0)$. It will be seen that the inner class proof does not use monotonicity of π_n or of $n\pi_n$. Eicker points out that his inner class results apply to more general sequences of sets than $[0, \pi_n]$; here, if J_n is a subset of $[0, 1]$ of Lebesgue measure π_n, we consider $T_n(J_n) = \{$number of X_i in J_n, $1 \leq i \leq n\}$. One must note, however, that such inner bounds on $T_n(J_n)$ may not be sharp if J_n moves too rapidly, and may even give the wrong order. One need only cite the familiar example of J_n chosen so that the random variables $T_n(J_n)$ are independent, in which case $\pm \left[T_n(J_n) - n\pi_n\right]$ has $\left[2n\pi_n \log n\right]^{1/2}$ as top bound under our assumptions on π_n.

In (3.1) equivalent T_n and Z_n bounds were treated, and (3.14) gives a consequence of Theorem 2 for T_n. It remains to translate the results of Theorems 3 to 5 into conclusions for the sample quantiles $Z_n(p_n)$ beyond the domain of Theorems 1 and 2. For each positive value v, we define c'_v to be the positive value satisfying the pair of equations

$$(3.16) \qquad \beta'_{c'_v}(\log \beta'_{c'_v} - 1) = (1 - c'_v)/c'_v, \qquad c'_v \beta'_{c'_v} = v,$$

with $\beta' > 1$. We define c''_v by the analogue of (3.16) obtained by replacing β' by $\beta'' < 1$. The existence and uniqueness of c'_v and c''_v follows from (2.25) and the sentence following it. We recall the definition (2.1) of H_n.

THEOREM 6.   *Suppose* $p_n \downarrow 0$. *If* $H_n \uparrow + \infty$, *then, for either choice of sign,*

$$(3.17) \qquad \limsup_n \pm \left[Z_n(p_n) - p_n\right]\left[2p_n n^{-1} \log_2 n\right]^{-1/2} = 1, \qquad \text{wp 1}.$$

If $0 < v < \infty$ *and* $H_n \to v$, *then*

$$(3.18) \qquad \begin{aligned} c''_v &= \limsup_n n Z_n(p_n)/\log_2 n, \qquad \text{wp 1} \\ c'_v &= \liminf_n n Z_n(p_n)/\log_2 n, \qquad \text{wp 1} \end{aligned}$$

If $H_n \to 0$ *(and* $np_n \geq 1$ *to avoid trivialities), then*

$$(3.19) \qquad \limsup_n n Z_n(p_n)/\log_2 n = 1, \qquad \text{wp 1};$$

while if $h_n \downarrow 0$ *and* $np_n \uparrow + \infty$, *then*

$$(3.20) \qquad \liminf_n H_n \log \left[n Z_n(p_n)/\log_2 n\right] = -1, \qquad \text{wp 1}.$$

REMARKS.   As in the case of Theorem 5, the assumption $H_n \uparrow$ is stronger than needed, but it is made to simplify the proof. Also, if $p_n \downarrow p_0 > 0$, essentially the same proof yields (3.17) with $2p_n$ replaced by $2p_0(1 - p_0)$. (Compare the remarks below Theorem 5.) In particular, when $p_n = p_0$ we obtain the LIL for sample p_0-tiles, but not the strong form [6], which would require considerably more effort using the present route. Part of (3.17) was stated in [8] under unnecessary restrictions. More satisfactory forms than (3.19) and (3.20) are obviously related to strong forms of Theorems 3 and 4. As they stand, (3.19) and (3.20) are also correct for the case $p_n = k/n$ of Theorems 2 and 1, and the grossness of (3.20) as a description for the latter is evident.

Proofs of Theorems 3 through 6

PROOF OF THEOREM 3.   In each particular case of (3.8), (3.9), (3.10), the assumptions preceding (3.7) as well as the correctness of the stated k_n^* follow easily from the first line of (2.25) and from the validity of (3.7), and we turn to the proof of the latter.

Inner class. Let $\varepsilon > 0$ be specified, $\varepsilon < 1$. Let $n_j \sim \varepsilon^{-j}$. Then $m_j \sim (1 - \varepsilon)n_j$ by (2.3). Hence, we can compute

$$(4.1) \qquad \log P\{T_{n_{j-1}, n_j}(\pi_{n_j}) \geqq d_{n_j} h_{n_j} \beta'_{h_{n-j}} \log_2 n_j\}$$

by using (2.27) with $\rho_n \sim d_n \sim 1 - \varepsilon$ and with n replaced by n_j in (2.27); here d_n is chosen so that (2.16) is satisfied, where $k_n^* = (1 - \varepsilon)^{-1} d_n h_n \beta'_{h_{n_j}} \log_2 n$. Note that (2.26) is satisfied because (2.24) and (2.25) and $h_n = O(1)$ imply

$$(4.2) \qquad \liminf_n \beta'_{h_n} > 1.$$

Also, $h_n \beta'_{h_n}$ is bounded, by (2.25). Hence, from (2.27), the expression (4.1) equals $-(1 - \varepsilon)[1 + o(1)] \log_2 n_j$, so (2.10) and BC yield the desired result. (The last clause of Lemma 1 indicates that we do not need to verify (2.26) for the above half of the proof, but (4.1) is needed below, anyway.)

Outer class. As in the case of Theorem 1, there is a proof using (2.7) to (2.9) with the n_j of (2.3), which we omit in order to demonstrate the simplicity of the situation by working directly with (2.12) and the A'_n of (2.11) with k_n replaced there by

$$(4.3) \qquad k'_n = d_n h_n \beta'_{h_n} \log_2 n,$$

and $d_n - 1 \to \varepsilon$, small and positive, with d_n chosen so that $k'_n \uparrow$ and k'_n is integral. By (4.2), for small enough ε and with $\rho_n = 1$, (2.26) is satisfied. Because of (2.26) (as used in (2.20)) we have for the event of (2.11) with k'_n for k_n,

$$(4.4) \qquad \begin{aligned} \log P\{A'_n\} &= \log \left[\pi_n B^+(k'_n - 1, n - 1, \pi_n)\right] \\ &\sim \log \left[\pi_n b(k'_n - 1, n - 1, \pi_n)\right] \\ &= \log \left[k'_n n^{-1} b(k'_n, n, \pi_n)\right]. \end{aligned}$$

From the previously obtained boundedness of $h_n \beta'_{h_n}$, we have $\log k'_n = o(\log_2 n)$. Consequently, from (4.4) and (2.27),

$$(4.5) \qquad \log P\{A'_n\} = -\log n + \{-(1 + \varepsilon) + h_n[\varepsilon - (1 + \varepsilon)\beta'_{h_n} \log(1 + \varepsilon)] + o(1)\} \log_2 n.$$

By (4.2), the quantity in square brackets in (4.5) is negative for ε sufficiently small (and positive) and for all large n. This, (2.12), and BC complete the proof.

PROOF OF THEOREM 4. Equation (3.12) follows from (3.11).

Outer class. Given ε, small and positive, let d_n be chosen so that $d_n h_n \beta'_{h_n} \log_2 n \uparrow$ and $d_n \to 1 - \varepsilon$. Put $n_j \sim \lambda^j$. Because of (2.13), we compute, using (2.29) with n replaced by n_{j+1} and $\rho_{n_{j+1}} = n_j/n_{j+1} \sim \lambda^{-1}$,

$$(4.6) \qquad \begin{aligned} \log P\{T_{n_j}(\pi_{n_{j+1}}) &< d_{n_{j+1}} h_{n_{j+1}} h_{n_{j+1}} \beta''_{h_{n_{j+1}}} \log_2 n_{j+1}\} \\ &= \{-(1 - \varepsilon) + h_{n_{j+1}}[1 - \varepsilon - \lambda^{-1} - (1 - \varepsilon)\beta''_{h_n} \\ &\quad \log(\lambda(1 - \varepsilon))] + o(1)\} \log_2 n_{j+1}, \end{aligned}$$

provided (2.28) is satisfied. Write $\liminf_n h_n = \bar{h}$. Since $\bar{h} > 1$ by (2.23), the

structure (2.24)—(2.25) of the functional equation implies that $1 < 1 - \bar{h}\beta_{\bar{h}}''$
$\log \beta_{\bar{h}}'' = \bar{h}(1 - \beta_{\bar{h}}'')$. Moreover, by the comment following (2.25) and the fact
that $(\beta - 1)/\log \beta < 1$ for $0 < \beta < 1$, we conclude that $h - h\beta_{\bar{h}}''$ increases in h.
Hence, $h_n(1 - \beta_{h_n}'') > 1 + 2\delta$ for some $\delta > 0$ and all large n. Now let $\lambda > 1$ be
chosen so close to one that $\lambda d\beta_h < 1$ (which yields (2.28)) and such that $1 - \lambda^{-1} <$
$\varepsilon\delta/(1 + 2\delta)$. Regarding the expression in braces on the right side of (4.6), other
than the $o(1)$ term, as a function of $\varepsilon > 0$, upon expanding it in powers of ε
and noting that $\log \lambda = 1 - \lambda^{-1} + O(\varepsilon^2)$ we obtain

$$(4.7) \quad - (1 - \varepsilon) + h_{h_{j+1}}\big[1 - \varepsilon - \lambda^{-1}$$
$$- (1 - \varepsilon)\beta_{h_{n_{j+1}}}''(-\varepsilon + 1 + \lambda^{-1} + O(\varepsilon^2))\big]$$
$$= - (1 - \varepsilon) + h_{n_{j+1}}(1 - \beta_{h_{n_{j+1}}}'')(1 - \lambda^{-1} - \varepsilon) + O(\varepsilon^2)$$
$$< - (1 - \varepsilon) + (1 + 2\delta)\big(-\varepsilon + \varepsilon\delta/(1 + 2\delta)\big) + O(\varepsilon^2)$$
$$< - 1 - \delta\varepsilon + O(\varepsilon^2).$$

Thus, for ε sufficiently small and positive, the probability of (4.6) is summable.

Inner class. As in the proof of Theorem 2, we now use the $n_j = \text{int}\{e^{\alpha j \log j}\}$
with $\alpha > 0$, of (2.4). Hence, choosing $d_n - 1 \to \varepsilon$ small and positive and such
that $d_n h_n \beta_{h_n}'' \log_2 n \uparrow$, we obtain from (2.29) with n_{j+1} for n and $\rho_{n_j} = m_j/n_j =$
$1 - O(j^{-\alpha})$,

$$(4.8) \quad \log P\{T_{n_j, n_{j+1}}(\pi_{n_{j+1}}) \leqq d_{n_{j+1}} h_{n_{j+1}} \beta_{h_{n_{j+1}}}'' \log_2 n_{j+1}\}$$
$$= \{- (1 + \varepsilon) + h_{n_{j+1}}\big[\varepsilon - (1 + \varepsilon)\beta_{h_{n_{j+1}}}'' \log (1 + \varepsilon)\big]$$
$$+ o(1)\} \log_2 n_{j+1}.$$

(While the last comment in the statement of Lemma 1 implies that this half of
the proof does not require (2.28), the latter is in fact satisfied provided
$(1 + \varepsilon)^{-1} > \lim \sup_n \beta_{h_n}''$, the last quantity being less than one by (2.24) and
(2.25) since $h_n = O(1)$.) As in the outer class proof, we again have $h_n(1 - \beta_{h_n}'') >$
$1 + 2\delta$ for some $\delta > 0$ and all large n, and consequently the expression in braces
on the right side of (4.8) is greater than $-1 + \delta\varepsilon$ for ε sufficiently small and
positive and for all large n. Hence, the probabilities of (4.8) have divergent sum.
In view of (2.14) and (2.15), it remains to compute

$$(4.9) \quad P\{T_{n_j}(\pi_{n_{j+1}}) \geqq 1\} = 1 - (1 - \pi_{n_{j+1}})^{n_j}$$
$$\sim (n_j/n_{j+1})h_{n_{j+1}} \log_2 n_{j+1}$$
$$= O(j^{-\alpha} \log j),$$

which is summable if $\alpha > 1$. This completes the proof of Theorem 4.

PROOF OF THEOREM 5. *Outer class.* Given $\varepsilon > 0$, put $n_j \sim (1 + \varepsilon/3)^j$ and
modify the previous notation by writing

$$(4.10) \quad \begin{aligned} k_n(\varepsilon) &= n\pi_n + (1 + \varepsilon)[2n\pi_n \log_2 n]^{1/2}, \\ D_n(\varepsilon) &= \{T_n(\pi_n) \geqq k_n(\varepsilon)\}, \qquad D_j^*(\varepsilon) = \bigcup_{n \in I_j} D_n(\varepsilon). \end{aligned}$$

The desired outer top bound result is

(4.11) $$P\{D_n(\varepsilon) \text{ i.o.}\} = 0.$$

The first step is common in such proofs. If we show that

(4.12) $$\liminf_j P\{D_{n_{j+1}}(\varepsilon/3) \mid D_j^*(\varepsilon)\} > 0,$$

then $P\{D_n(\varepsilon) \text{ i.o.}\} = P\{D_j^*(\varepsilon) \text{ i.o.}\} = 0$, by BC if $P\{D_{n_{j+1}}(\varepsilon/3)\}$ is summable, which it is by Lemma 3.

For brevity, we hereafter write k_n for $k_n(\varepsilon)$ (never for $k_n(\varepsilon/3)$). For $v \in I_j$, we define

(4.13) $$G_v = \{T_v(\pi_{n_{j+1}}) \geqq k_v \pi_{n_{j+1}}/\pi_v - 1\}.$$

If the event $D_j^*(\varepsilon)$ of (4.10) occurs, define the random variable N_j^* by $N_j^* = \min\{n : D_n(\varepsilon) \text{ occurs}, n \in I_j\}$. Clearly,

(4.14) $$P\{D_{n_{j+1}}(\varepsilon/3) \mid D_j^*(\varepsilon)\} = EP\{D_{n_{j+1}}(\varepsilon/3) \mid D_j^*(\varepsilon); N_j^*; T_{N_j^*}(\pi_{N_j^*})\}$$
$$\geqq \inf_{z \geqq k, \, v \in I_j^-} P\{D_{n_{j+1}}(\varepsilon/3) \mid T_v(\pi_v) = z\}.$$

Since

(4.15) $$P\{D_{n_{j+1}}(\varepsilon/3) \mid T_v(\pi_v) = z\} \geqq P\{G_v D_{n_{j+1}}(\varepsilon/3) \mid T_v(\pi_v) = z\}$$
$$= P\{D_{n_{j+1}}(\varepsilon/3) \mid G_v; T_v(\pi_v) = z\}$$
$$\cdot P\{G_v \mid T_v(\pi_v) = z\},$$

we obtain (4.12) from (4.14) if there is a $\delta > 0$ such that, for all large j and $v \in I_j^-$,

(4.16) $$\inf_{z \geqq k_v} P\{D_{n_{j+1}}(\varepsilon/3) \mid G_v; T_v(\pi_v) = z\} > \delta$$

and

(4.17) $$\inf_{z \geqq k_v} P\{G_v \mid T_v(\pi_v) = z\} > \delta.$$

The conditional probability of (4.17) is clearly

(4.18) $$B^+(k_v \pi_{n_{j+1}}/\pi_v - 1, z, \pi_{n_{j+1}}/\pi_v),$$

which is a minimum for $z = \text{int}^+\{k_v\}$. This and (2.33) yield (4.17).

Since $\pi_{n_{j+1}} \leqq \pi_v$, the probability of $D_{n_{j+1}}(\varepsilon/3)$ conditioned on values of $T_v(\pi_v)$ and $T_v(\pi_{n_{j+1}})$ is the same as that conditioned only on the last. Hence abbreviating $\text{int}^+\{k_v \pi_{n_{j+1}}/\pi_v - 1\}$ by μ, we see that the left side of (4.16) is at least

(4.19) $$\inf_{z \geqq k_v, \, z \geqq y \geqq \mu} P\{D_{n_{j+1}}(\varepsilon/3) \mid T_v(\pi_{n_{j+1}}) = y; T_v(\pi_v) = z\}$$
$$= \inf_{y \geqq \mu} P\{D_{n_{j+1}}(\varepsilon/3) \mid T_v(\pi_{n_{j+1}}) = y\}$$
$$= \inf_{y \geqq \mu} P\{T_{v, n_{j+1}}(\pi_{n_{j+1}}) \geqq k_{n_{j+1}}(\varepsilon/3) - y\}$$
$$= B^+(k_{n_{j+1}}(\varepsilon/3) - \text{int}^+\{k_v \pi_{n_{j+1}}/\pi_v - 1\}, n_{j+1} - v, \pi_{n_{j+1}}).$$

By (2.33) and the fact that $\mathrm{int}^+ \{x\} \geqq x$, we will thus establish (4.16) if we show that

$$(4.20) \qquad k_{n_{j+1}}(\varepsilon/3) - [k_v(\varepsilon)\pi_{n_{j+1}}/\pi_v - 1] \leqq (n_{j+1} - v)\pi_{n_{j+1}} - 1$$

for all large j and $v \in I_j^-$. Dividing both sides of (4.20) by $\pi_{n_{j+1}}$ and using (4.10), we obtain that (4.20) is equivalent to

$$(4.21) \qquad (1 + \varepsilon/3)n_{j+1}[2(\log_2 n_{j+1})/n_{j+1}\pi_{n_{j+1}}]^{1/2} + 2\pi_{n_{j+1}}^{-1}$$
$$\leqq (1 + \varepsilon)v[2(\log_2 v)/v\pi_v]^{1/2}.$$

The ratio of $2\pi_{n_{j+1}}^{-1}$ to the term preceding it approaches zero. Also, $n_{j+1}\pi_{n_{j+1}} > v\pi_v$ and $n_{j+1}(\log_2^{1/2} n_{j+1})/v \log_2^{1/2} v < n_{j+1}(\log_2^{1/2} n_{j+1})/n_j \log_2^{1/2} n_j \sim 1 + \varepsilon/3$. We conclude that (4.21) is satisfied for all large j and $v \in I_j^-$, completing the proof of (4.11).

The proof of the outer bottom result is very similar, so we shall merely list the changes. In (4.10) we replace $(1 + \varepsilon)$ by $-(1 + \varepsilon)$ in the definition of $k_n(\varepsilon)$, and $\geqq$ by $\leqq$ in the definition of $D_n(\varepsilon)$, as well as in the domain of z in (4.14), (4.16), and (4.17). The event G_v of (4.13) is replaced by

$$(4.22) \qquad \{T_v(\pi_{n_{j+1}}) \leqq k_v\pi_{n_{j+1}}/\pi_v + 1\};$$

also, int^+ is replaced everywhere by int. The probability (4.18) is replaced by

$$(4.23) \qquad B(k_v\pi_{n_{j+1}}/\pi_v + 1, z, \pi_{n_{j+1}}/\pi_v),$$

whose minimum subject to $z \leqq k_v$ is at $z = \mathrm{int} \{k_v\}$; this minimum is bounded away from zero, by (2.34). Finally, in (4.19) μ becomes $\mathrm{int} \{k_v\pi_{n_{j+1}}/\pi_v + 1\}$, and the domain of the first infimum is $\{y \leqq z \leqq k_v, y \leqq \mu\}$; when we replace the resulting domain $\{y \leqq \min(k_v, \mu)\}$ by $\{y \leqq \mu\}$, we cannot increase the infimum, and the analogues of last two expressions of (4.19) give, for a lower bound on the analogue of (4.16),

$$(4.24) \qquad \inf_{y \leqq \mu} P\{T_{v, n_{j+1}}(\pi_{n_{j+1}}) \leqq k_{n_{j+1}}(\varepsilon/3) - y\}$$
$$= B(k_{n_{j+1}}(\varepsilon/3) - \mathrm{int} \{k_v\pi_{n_{j+1}}/\pi_v + 1\}, n_{j+1} - v, \pi_{n_{j+1}}).$$

By (2.34) and the fact that $\mathrm{int} \{x\} \leqq x$, we obtain as the analogue of (4.20),

$$(4.25) \qquad k_{n_{j+1}}(\varepsilon/3) - [k_v\pi_{n_{j+1}}/\pi_v + 1] \geqq (n_{j+1} - v)\pi_{n_{j+1}} + 1.$$

Recalling that $(1 + \varepsilon)$ has been replaced by $-(1 + \varepsilon)$ in the definition (4.10) of $k_n(\varepsilon)$, we see that (4.25) is again equivalent to (4.21).

Inner class. The proof follows usual LIL lines and has been given by Eicker [4] in essentially this form, so we only sketch it. For the top inner bound, one shows, with $n_j \sim \lambda^j$, that

$$(4.26) \qquad P\{T_{n_j}(\pi_{n_j}) > n_j\pi_{n_j} + (1 - \varepsilon)[2h_{n_j}]^{1/2} \log_2 n_j \text{ i.o.}\} = 1,$$

by showing that

(4.27)
$$P\{T_{n_{j-1}, n_j}(\pi_{n_j}) > [n_j - (1 - [4\lambda/h_{n_j}]^{1/2})n_{j-1}]\pi_{n_j}$$

and
$$+ (1 - \varepsilon)[2h_{n_j}]^{1/2} \log_2 n_j \text{ i.o.}\} = 1,$$

(4.28)
$$P\{T_{n_{j-1}}(\pi_{n_j}) > (1 - [4\lambda/h_{n_j}]^{1/2})n_{j-1}\pi_{n_j}, \text{ a.a. } j\} = 1.$$

Of these, (4.27) is a consequence of (2.31) and BC provided λ is chosen so large that $[\lambda/(\lambda - 1)]^{1/2}[(4/\lambda)^{1/2} + (1 - \varepsilon)2^{1/2}] < 2^{1/2}$. The probability complementary to (4.28) is proved summable by using the standard Markov-Cramér inequality with abbreviations $T = T_{n_{j-1}}(\pi_{n_j})$, $\delta = [4\lambda/h_{n_j}]^{1/2}$, to obtain

(4.29)
$$P\{T \leqq (1 - \delta)ET\} = P\{e^{-\delta T} \geqq e^{-(1-\delta)\delta ET}\}$$
$$\leqq e^{(1-\delta)\delta ET} Ee^{-\delta T}$$
$$= \exp\{n_{j-1}[\pi_{n_j}(1 - \delta)\delta$$
$$+ \log(1 - \pi_{n_j} + \pi_{n_j}e^{-\delta})]\}$$
$$< \exp\{-\delta^2 n_{j-1}\pi_{n_j}/3\},$$

the last inequality for δ and π_{n_j} sufficiently small. For the bottom inner bound, replace $>$ by $<$, $(1 - \varepsilon)$ by $-(1 - \varepsilon)$, and $-[4\lambda/h_{n_j}]^{1/2}$ by $[4\lambda/h_{n_j}]^{1/2}$ in (4.26), (4.27), (4.28); and replace (4.29) by

(4.30)
$$P\{T \geqq (1 + \delta)ET\} \leqq e^{-(1+\delta)\delta ET}Ee^{\delta T},$$

with the same final estimate as in (4.29).

REMARK. The coefficient δ of T and ET in the second expression of (4.29) and (4.30) is not the usual minimizing value for exponential binomial bounds, but is less cumbersome and is close enough to yield the desired conclusions.

PROOF OF THEOREM 6. If $H_n \uparrow +\infty$ and $h_n = H_n \pm \lambda(2H_n)^{1/2}$ where $\lambda = 1 + \varepsilon$ or $1 - \varepsilon$, then $h_n \uparrow$ for large n. Applying Theorem 5 for these four possible choices of h_n, and noting that $h_n \mp \lambda(2h_n)^{1/2} = H_n + O(1)$, yields (3.17). Similarly, (3.18) follows from Theorems 3 and 4. If $H_n \to 0$, we obtain (3.19) from Theorem 2 and from (3.18) for v small and positive; by (2.24) and (2.25), $\lim_{v \downarrow 0} c_v'' = 1$. Finally, (3.20) follows from (1.3) with $k_n = np_n$ and $h_n = \exp\{-(1 \pm \varepsilon)H_n^{-1}\}$ upon invoking Theorem 3, the condition $np_n \to +\infty$ of the latter then being equivalent to (2.22).

REFERENCES

[1] R. R. BAHADUR, "A note on quantiles in large samples," *Ann. Math. Statist.*, Vol. 37 (1966), pp. 577–580.
[2] G. BAXTER, "An analogue of the law of the iterated logarithm," *Proc. Amer. Math. Soc.*, Vol. 6 (1955), pp. 177–181.
[3] K. L. CHUNG, "An estimate concerning the Kolmogoroff limit distribution," *Trans. Amer. Math. Soc.*, Vol. 67 (1949), pp. 36–50.
[4] F. EICKER, "A log log law for double sequences," to appear.
[5] W. FELLER, *An Introduction to Probability Theory and its Applications*, Vol. 1, New York, Wiley, 1957 (2nd ed.).

[6] J. KIEFER, "On Bahadur's representation of sample quantiles," *Ann. Math. Statist.*, Vol. 38 (1967), pp. 1323–1342.

[7] ———, "Deviations between the sample quantile process and the sample df," *Proceedings of the First International Conference on Nonparametric Inference*, Cambridge, Cambridge University Press, 1969, pp. 299–319.

[8] ———, "Old and new methods for studying order statistics and sample quantiles," *Proceedings of the First International Conference on Nonparametric Inference*, Cambridge, Cambridge University Press, 1969, pp. 349–357.

[9] V. F. KOLCHIN, "On the limiting behavior of extreme order statistics in a polynomial scheme," *Theor. Probability Appl.*, Vol. 14 (1969), pp. 458–469.

[10] H. ROBBINS and D. SIEGMUND, "An iterated logarithm law for maxima and minima" (abstract), *Ann. Math. Statist.*, Vol. 41 (1970), p. 757.

LIMITS OF EXPERIMENTS

L. LE CAM
UNIVERSITY OF CALIFORNIA, BERKELEY

1. Introduction

In a recent paper J. Hájek [4] proved a remarkably simple result on the limiting distributions of estimates of a vector parameter θ. It turns out that this result, as well as many of the usual statements about asymptotic behavior of tests or estimates, can be obtained by a general procedure which consists roughly in passing to the limit first and then arguing the case for the limiting problem. This passage to the limit relies on some general facts which are perhaps not entirely elementary. They depend heavily on the techniques of L. LeCam [8]. However, these general facts are of interest by themselves. If they are taken for granted the basic result of Hájek [4] and many results of A. Wald [13] become available immediately.

The present paper is organized as follows. Section 2 recalls a number of definitions and theorems which are variations on those given by the author in [8]. We have used here again a simplified definition of "experiments" barely different from the one given in [8]. There is no essential difficulty in returning to the more usual description, at least under appropriate restrictions. However the simplified (or "more abstract" as is claimed by some) description avoids measure theoretic technicalities and makes the arguments more transparent.

Section 3 gives further theorems concerning experiments with a fixed set of indices. It uses the metric introduced in [8] to define a weak topology on the space of experiments indexed by a given set Θ. Although the compactness statements proved in this section are not absolutely essential to the remainder of the paper, they do produce a number of simplifications.

The metric of [8] was intended, in part, to insure a certain continuity of risk functions, at least if loss functions stay bounded. The purpose of Section 4 is to show that a similar type of lower semicontinuity still exists for the weak topology of Section 3, even if the loss functions are only bounded from below.

In Section 5 we consider two types of limits: (1) limits of experiments in the weak topology of Section 3, and (2) experiments formed by taking limiting distributions of certain statistics. The main result is that the experiments of second type are always weaker than those of the first type. Statistics systems for which the two coincide are characterized. Another result of Section 5 is the existence of transitions which are convolutions in the case of shift invariant experiments. For this see also E. Torgersen [12] and H. Heyer [6].

Section 6 elaborates a few examples indicating some of the results implied by the previous propositions. It reproduces partially some results of Hájek, [4] and

245

[5]. Also it improves some results of LeCam [9] on the misbehavior of super-efficient estimates.

In the normal situation a simplified direct proof of Hájek's convolution result was communicated to us by P. J. Bickel.

Any resemblance between our results and those of Hájek is not entirely accidental, since the present paper was greatly modified after Hájek's presentation during the Symposium.

2. Definitions relative to experiments

Let Θ be a set. An experiment indexed by Θ is usually represented by a σ-field $\mathscr{A}$ carried by a set $\mathscr{X}$ and a family $\{P_\theta : \theta \in \Theta\}$ of probability measures on $\mathscr{A}$. We shall use instead a description which is obtainable from the usual one by ignoring the set $\mathscr{X}$.

Recall that an L-space is a Banach lattice whose norm satisfies the relation $\|\mu + v\| = \|\mu\| + \|v\|$ if $\mu \geqq 0$ and $v \geqq 0$. The dual of an L-space L is another Banach lattice M whose norm satisfies the relation $\|f \vee g\| = \|f\| \vee \|g\|$ for $f \geqq 0$ and $g \geqq 0$.

DEFINITION 1. *Let Θ be a set. An experiment $\mathscr{E}$ indexed by Θ is a function $\theta \rightsquigarrow P_\theta$ from Θ to an L-space L. This function is subject to the restriction that $P_\theta \geqq 0$ and $\|P_\theta\| = 1$.*

Let S be a subset of the L-space L. The band generated by S is the smallest linear subspace L_0 of L such that: (1) $S \subset L_0$, (2) if $\mu = \Sigma_{j=1}^n c_j |\mu_j|$ with $\mu_j \in S$ and $|v| \leqq \mu$ then $v \in L_0$, and (3) L_0 is complete for the norm.

If the experiment $\mathscr{E} : \theta \rightsquigarrow P_\theta$ is such that $S = \{P_\theta ; \theta \in \Theta\}$ has for band the whole range space L, we shall say that $\mathscr{E}$ generates L.

Since the above definition ignores the set of $\mathscr{X}$ of the more usual description it is necessary to translate to the present language the definition of "statistics." This is done by the "transitions" described below.

DEFINITION 2. *Let L_1 and L_2 be two L-spaces A transition from L_1 to L_2 is a positive linear map A from L_1 to L_2 such that $\|A\mu^+\| = \|\mu^+\|$ for all $\mu \in L_1$.*

Suppose that Z is a completely regular topological space. Let $C^b(Z)$ be the Banach space of bounded continuous numerical functions defined on Z. Let $C^*(Z)$ be the dual of $C^b(Z)$. This space $C^*(Z)$ is an L-space for the natural order. Thus if $\mathscr{E}$ is an experiment generating an L-space L, one can consider "transitions" from L to $C^*(Z)$. It will be convenient to call these transitions *statistics with values in Z*. It is clear that they correspond very exactly to the ordinary idea of *randomized Z-valued statistics* except for the circumstance that the randomization distributions need not be countably additive.

Consider a given L-space L with dual M. The formula $I\mu = \|\mu^+\| - \|\mu^-\|$ defines an element of M. The unit ball B of M is the set $B = \{u \in M ; |u| \leqq I\}$. Statistical tests correspond to the positive part of this unit ball. The following result is the fundamental tool in [8].

THEOREM 1. *Let L be an L-space with dual M. Let H be a sublattice of M whose unit ball* $\{u\,;\,u \in H,\,|u| \leq I\}$ *is* $\sigma(M, L)$ *dense in the unit ball of M. Assume also* $I \in H$. *Let Z be a completely regular space. Let* $\mathcal{M}$ *be the set of all transitions from L to* $C^*(Z)$. *Let* $\mathcal{M}_0$ *be the subset of* $\mathcal{M}$ *consisting of transitions T such that*

(1) $\gamma T \in H$ *for each* $\gamma \in C^b(Z)$,

(2) *there is a finite set* $F \subset Z$ *such that for every* $\mu \in L$ *the image* T_μ *of* μ *by T is carried by F.*

In $C^b(Z) \times L$ *let* $\mathcal{K}$ *be the class of finite unions of rectangles* $K_1 \times K_2$ *such that either* (1) K_1 *is* $\sigma[C^b(Z), C^*(Z)]$ *compact and* K_2 *is norm compact in L or,* (2) K_1 *is norm compact and* K_2 *is* $\sigma(L, M)$ *compact.*

Then $\mathcal{M}_0$ *is dense in* $\mathcal{M}$ *for the topology of uniform convergence on the sets* $S \in \mathcal{K}$.

With a slight change in notation this is Theorem 1 in [8].

Some other definitions of [8] which will be needed below are as follows.

Let Θ be a given set. Let $\mathcal{E} : \theta \rightsquigarrow P_\theta$ be an experiment generating an L-space $L(\mathcal{E})$. Let $\mathcal{F} : \theta \rightsquigarrow Q_\theta$ be another experiment indexed by the same set Θ. It generates a space $L(\mathcal{F})$.

DEFINITION 3. *The deficiency of* $\mathcal{E}$ *relative to* $\mathcal{F}$ *is the number*

$$(2.1) \qquad \delta(\mathcal{E}, \mathcal{F}) = \inf_A \sup_\theta \|A P_\theta - Q_\theta\|,$$

where the infimum is taken over all transitions from $L(\mathcal{E})$ *to* $L(\mathcal{F})$. *The "distance" between* $\mathcal{E}$ *and* $\mathcal{F}$ *is the number*

$$(2.2) \qquad \Delta(\mathcal{E}, \mathcal{F}) = \max\,[\delta(\mathcal{E}, \mathcal{F}), \delta(\mathcal{F}, \mathcal{E})].$$

This "distance" is only a pseudometric. It becomes an actual distance if two experiments whose distance is zero are considered *equivalent*.

DEFINITION 4. *For a given set* Θ *the equivalence class of an experiment* $\mathcal{E}$ *will be called the* type *of* $\mathcal{E}$ *and denoted* $\dot{\mathcal{E}}$.

Let $\mathcal{E}$ be an experiment $\mathcal{E} : \theta \rightsquigarrow P_\theta$ indexed by a set Θ. Let S be a subset of Θ. The experiment $\mathcal{E}$ restricted to S, that is the function $\theta \rightsquigarrow P_\theta$ defined on S only, will be denoted $\mathcal{E}_S$. If $\mathcal{E}$ and $\mathcal{F}$ are two experiments indexed by Θ the deficiency $\delta(\mathcal{E}_S, \mathcal{F}_S)$ will also be called the deficiency of $\mathcal{E}$ relative to $\mathcal{F}$ on the set S.

According to Wald a statistical decision problem is given by a triplet $\{\mathcal{E}, D, W\}$ where $\mathcal{E}$ is an experiment indexed by a set Θ and W is a function from $\Theta \times D$ to $[-\infty, +\infty]$.

In all the situations encountered below D will be a completely regular space. We shall identify the decision functions, also called decision procedures to the transitions from $L(\mathcal{E})$ to $C^*(D)$. If for each $\theta \in \Theta$ the loss function $z \rightsquigarrow W_\theta(z)$ defined on D by W is an element of $C^*(D)$, and if T is a decision procedure, the *risk* at θ of the procedure T is defined by the value $W_\theta T P_\theta$.

In order to state the result which was the main object of [8], we need a very particular class of decision spaces (D, W). A more general situation will be described in Section 4.

DEFINITION 5. *Let $\mathscr{S}$ be the class of decision spaces (D, W) formed by pairs where*

(1) *D is a compact, convex subset of the space $\mathscr{F}\{\Theta, [-1, +1]\}$ of functions from Θ to $[-1, +1]$,*

(2) *the set D has finite linear dimension,*

(3) *the value $W_\theta(z)$ of the loss at (θ, z), $\theta \in \Theta$, $z \in D$ is the value of the element z of $\mathscr{F}\{\Theta, [-1, +1]\}$ at the point θ.*

Note that for each θ, the function $z \rightsquigarrow W_\theta(z)$ is linear in z. Also $\|W\| = \sup\{|W_\theta(z)|, \theta \in \Theta, z \in D\}$ is always finite since $|W_\theta(z)| < 1$.

The main result of [8] can then be phrased as follows.

THEOREM 2. *Let $\mathscr{E}: \theta \rightsquigarrow P_\theta$ and $\mathscr{F}: \theta \rightsquigarrow Q_\theta$ be two experiments indexed by Θ. For any given $\varepsilon \in [0, 2]$ the following statements are all equivalent.*

(1) *There is a transition K from $L(\mathscr{E})$ to $L(\mathscr{F})$ such that $\sup_\theta \|KP_\theta - Q_\theta\| \leq \varepsilon$.*

(2) *If T is a decision procedure relative to the experiment $\mathscr{F}$ and a decision space $(D, W) \in \mathscr{S}$ there is a procedure S of $\mathscr{E}$ to the same space (D, W) such that*

$$(2.3) \qquad W_\theta S P_\theta \leqq W_\theta T Q_\theta + \varepsilon \|W\|$$

for every $\theta \in \Theta$.

(3) *If $\varepsilon' > \varepsilon$, if μ is a probability measure with finite support on Θ, and T, D, W are as in (2), there is an S from $\mathscr{E}$ to (D, W) such that*

$$(2.4) \qquad \int (W_\theta S P_\theta)\mu(d\theta) \leqq \int^{\cdot} (W_\theta T Q_\theta)\mu(d\theta) + \varepsilon' \|W\|.$$

That $(1) \Rightarrow (2) \Rightarrow (3)$ is obvious. The implication $(3) \Rightarrow (1)$ is the subject of [8] where it is shown also that in (3) one can assume that the procedure T is "special restricted" in the sense described by the sublattice range and finite support of Theorem 1.

An immediate corollary of Theorem 2 is as follows.

COROLLARY. *The deficiency $\delta(\mathscr{E}, \mathscr{F})$ is equal to $\sup_S \delta(\mathscr{E}_S, \mathscr{F}_S)$ where the supremum extends to all finite subsets $S \subset \Theta$.*

Furthermore, there is a transition K which achieves the infimum in $\inf_A \sup_\theta \|AP_\theta - Q_\theta\|$.

It is perhaps appropriate to include here a few words of warning on the difference between the setup just described and the more usual one. In the latter an experiment indexed by Θ is a system $\{\mathscr{X}, \mathscr{A}, P_\theta; \theta \in \Theta\}$ consisting of a σ-field $\mathscr{A}$ carried by a set $\mathscr{X}$ and a family $\{P_\theta; \theta \in \Theta\}$ of probability measures on $\mathscr{A}$. What we have called "transitions" are replaced by Markov kernels. If $\{\mathscr{X}, \mathscr{A}\}$ and $\{\mathscr{Y}, \mathscr{B}\}$ are two measurable spaces, a Markov kernel from $\mathscr{X}$ to $\mathscr{Y}$ is a function $K(B, x)$ defined on $\mathscr{B} \times \mathscr{X}$ and such that it is a probability measure as function of B and measurable as function of x. It is well known that if $\mathscr{B}$ is the Borel field of an analytic set $\mathscr{Y}$ in a Polish space and if the P_θ are dominated by a finite measure every one of our transitions from $\mathscr{L}(\mathscr{E})$ to measures on $\mathscr{Y}$ can be represented by a Markov kernel. The general situation is quite different.

In fact J. Denny [3] has pointed out to us the following result.

Let $\mathscr{E} = \{P_\theta; \theta \in \Theta\}$ be an experiment given by probability measures P_θ on the Borel sets of the real line. Let $\mathscr{E}^n$ be the direct product of n copies of $\mathscr{E}$. That is, $\mathscr{E}^n$ corresponds to n independent identically distributed observations, each distributed according to some P_θ. Assume that P_θ is nonatomic.

Then there is an experiment $\mathscr{F} = \{Q_\theta; \theta \in \Theta\}$ given by probability measures P_θ on a certain σ-field of subsets of the real line which has the following properties:

(1) for every integer n the experiments $\mathscr{E}^n$ and $\mathscr{F}^n$ are equivalent (in our sense),

(2) for every n, the sum of the observations is a sufficient statistic for $\mathscr{F}^n$.

To obtain $\mathscr{F}$ one just restricts the P_θ to a Hamel base which has elements in common with every perfect set.

The apparent teratology does not come from our definition of experiment but from the fact that we have elected to work with a category in which the morphisms are "transitions" which may or may not be Markov kernels. If we had restricted ourselves to Markov kernels we would not even be able to cover the situations described by the usual (Halmos-Savage) definition of sufficiency. As shown below, the category used here is just the appropriate one to reflect those properties which can be expressed in terms of joint distributions of *finite* (or countable) sets of likelihood ratios. One can also put it differently by stating that the system described here relies on the feeling that Boolean algebras of events are a more primitive notion that the points of the families of sets by which one represents the algebras.

3. A weak topology in experiments indexed by a given set

In this section Θ will be a fixed set. We are interested in the class of all experiments which can be indexed by Θ. Since the range L-space is unspecified this "class" does not ordinarily qualify as a "set." We shall see however that the corresponding equivalence classes called experiment types form a set $E(\Theta)$. It is this set which will be topologized.

When Θ is *finite*, D. Blackwell [2] used certain canonical measures to characterize experiment types as follows. Let U be the unit simplex of the product space R^Θ. Specifically U is the set

$$(3.1) \qquad U = \{x; x \in R^\Theta, x_\theta \geqq 0, \Sigma_\theta x_\theta = 1\}.$$

A *canonical measure* on U is a positive measure μ such that $\int x_\theta \, d\mu = 1$ for each $\theta \in \Theta$. Each canonical measure μ on U defines an experiment $\theta \rightsquigarrow p_\theta$ where p_θ is the probability measure defined on U by $dp_\theta = x_\theta \, d\mu$.

Conversely, let $\mathscr{E}; \theta \rightsquigarrow P_\theta$ be an experiment indexed by the finite set Θ. Let $m = \Sigma_\theta P_\theta$. Consider the Radon-Nikodym densities u_θ defined by $dP_\theta = u_\theta \, dm$. (Here each u_θ is an element of the dual M of $L(\mathscr{E})$ and $0 \leqq u_\theta \leqq I$ with $\Sigma_\theta u_\theta = I$.) Let u be the vector $u = \{u_\theta; \theta \in \Theta\}$. Let γ be an element of $C^b(U)$. The lattice algebraic properties of M allow the definition of $\gamma(u)$ as element of M. Thus, there is a well defined transition S from $L(\mathscr{E})$ to $C^*(U)$ whose value at $(\gamma, \lambda) \in$

$C^b \times L$ is $\gamma S\lambda = \langle \gamma(u), \lambda \rangle$. In another notation if λ is a probability measure $\lambda \in L(\mathscr{E})$ the value $S\lambda$ is the joint distribution of the likelihood ratios $\{u_\theta; \theta \in \Theta\}$ for an initial distribution λ. Consider then the image $\mu = Sm$ of $m = \Sigma P_\theta$ by S. This is obviously a canonical measure on the simplex U.

DEFINITION 6.   *Provide the space R^Θ with the maximum coordinate norm,* $|x| = \sup_\theta |x_\theta|$. *Let μ be a signed measure on U. The Dudley norm of μ is the number*

$$(3.2) \qquad \|\mu\|_D = \sup \left\{ \left| \int f d\mu \right|; f \in \Lambda \right\},$$

where Λ is the set of functions f defined on U and such that $|f| \leqq 1$ and $|f(x) - f(x')| \leqq |x - x'|$ for all x and x' in U.

One basic simple fact concerning such canonical measures is as follows.

PROPOSITION 1.   *Let $\mathscr{E}$ and $\mathscr{F}$ be two experiments indexed by the finite set Θ. Let μ and v be the corresponding canonical measures on the simplex $U \subset R^\Theta$. Then $\Delta(\mathscr{E}, \mathscr{F}) \leqq \|\mu - v\|_D$. Also $\Delta(\mathscr{E}, \mathscr{F}) = 0$ if and only if $\mu = v$.*

COROLLARY.   *Let $E(\Theta)$ be the set of experiment types indexed by the finite set Θ. Metrize $E = E(\Theta)$ by the experiment distance Δ. Metrize the set K of canonical measures on $U \subset R^\Theta$ by the Dudley norm. Then both E and K are compact metric spaces and the canonical one to one correspondence between them preserves the corresponding topologies and uniformities.*

PROOF.   The statement that $\Delta(\mathscr{E}, \mathscr{F}) = 0$ if and only if $\mu = v$ was already proved by Blackwell in [2]. The inequality $\Delta(\mathscr{E}, \mathscr{F}) \leqq \|\mu - v\|_D$ can easily be obtained by applying the implication $(3) \Rightarrow (1)$ of Theorem 2. This is done for instance in LeCam [10]. For further results see Torgersen [11]. The corollary is immediate since U is a compact set.

Although the above proposition is not actually indispensible for the results of Sections 4 and 5, it does afford a certain convenience and allows statements of theorems in which existence of certain limits can be assumed without actual loss of generality. For the same reason it will be appropriate to state a related proposition for the case where Θ is an infinite set. For this purpose let us recall that if α is a subset of a set Θ the experiment $\mathscr{E}: \theta \rightsquigarrow P_\theta$ restricted to α is denoted $\mathscr{E}_\alpha$. If α is finite we shall topologize the set of experiment types $E(\alpha)$ by the distance Δ, equivalent to the Dudley distance of canonical measures.

DEFINITION 7.   *Let Θ be an arbitrary set. Let E be a set of experiment types indexed by Θ. By the weak topology of E will be meant the weakest topology which makes the map $\mathscr{E} \rightsquigarrow \mathscr{E}_\alpha$ continuous for all the finite subsets α of Θ.*

To study the class $E(\Theta)$ of experiment types indexed by Θ it is convenient to investigate first the relations between $E(\alpha)$ and $E(\beta)$ for two *finite* subsets $\alpha \subset \beta \subset \Theta$.

It is obvious from the definitions that the restriction map from $E(\beta)$ to $E(\alpha)$ is a continuous map of $E(\beta)$ onto $E(\alpha)$.

For each finite set $\alpha \subset \Theta$ let E'_α be an experiment type $E'_\alpha \in E(\alpha)$. Such a family $\{E'_\alpha\}$ will be called *compatible* if whenever $\alpha \subset \beta$ the experiment type E'_α is the restriction to α of the experiment type E'_β. Let $E'(\Theta)$ be the set of all compatible families of experiment types. This is in an obvious way the projective limit of the sets $E(\alpha)$. Since each $E(\alpha)$ is a compact Hausdorff space, the usual projective limit theorems insure that $E'(\Theta)$ is a compact Hausdorff space whose natural map into $E(\alpha)$ is in fact *onto*. This leads to the following statement.

PROPOSITION 2. *Let Θ be an arbitrary set. Then the class $E(\Theta)$ of experiment types indexed by Θ is in one to one correspondence with the space of compatible families of experiment types $E'(\Theta)$. For its weak topology $E(\Theta)$ is a compact Hausdorff space. For the metric Δ the space $E(\Theta)$ is a complete metric space.*

PROOF. If $E(\Theta)$ and the compatible families $E'(\Theta)$ are the same set then $E(\Theta)$ is compact for the weak topology because $E'(\Theta)$ is compact. The completeness statement for the distance Δ follows because the uniform structure induced by Δ is stronger than the weak structure and because the set of pairs $(\mathscr{E}, \mathscr{F})$ such that $\Delta(\mathscr{E}, \mathscr{F}) \leq \varepsilon$ is closed for the weak topology. Furthermore if a compatible family $\{E'_\alpha\} \in E'(\Theta)$ derives from an experiment $\mathscr{E}$ indexed by Θ, the type of $\mathscr{E}$ is well determined as can be seen from the corollary of Theorem 2. Thus the proposition will be proved if we show that any compatible family $\{E'_\alpha\}$ can be obtained from an experiment in the sense of Definition 1.

For this purpose consider two finite sets $\alpha \subset \beta \subset \Theta$. Let R^α be the Cartesian product corresponding to α. Let U_α be the unit simplex of R^α. Let $C_\alpha = C(U_\alpha)$ be the Banach space of continuous functions on U_α. Let C^*_α be the space of Radon measures on U_α. Finally let K_α be the subset of C^*_α formed by canonical measures on U_α. For $\alpha \subset \beta$ and $y \in U_\beta$ let $s_{\alpha,\beta}(y) = \Sigma_{\theta \in \alpha} y_\theta$. Finally, for all $y \in U_\beta$ such that $s_{\alpha,\beta}(y) > 0$ let $\Pi_{\alpha,\beta}(y) \in U_\alpha$ be defined by $[\Pi_{\alpha,\beta} y]_\theta = [s_{\alpha,\beta}(y)]^{-1} y_\theta$.

Let $\mathscr{E}_\beta : \theta \rightsquigarrow P_\theta$ be an experiment indexed by β. Let μ_β be the corresponding canonical measure image of $m_\beta = \Sigma_{\theta \in \beta} P_\theta$. Let $m_\alpha = \Sigma_{\theta \in \alpha} P_\theta$ and let μ_α be the corresponding canonical measure for the experiment $\mathscr{E}_\alpha = \{P_\theta; \theta \in \alpha\}$. The canonical form of $\mathscr{E}_\beta$ is given by the measures $p_{\theta,\beta}$ defined by $dp_{\theta,\beta} = y_\theta d\mu_\beta$. Similarly, the canonical form of $\mathscr{E}_\alpha$ is given by measures $dp_{\theta,\alpha} = x_\theta d\mu_\alpha$. For $\theta \in \alpha$ the above relation can be read $x_\theta = [\Pi_{\alpha,\beta} y]_\theta = [s_{\alpha,\beta}(y)]^{-1} y_\theta$. Starting from the transformation $\Pi_{\alpha,\beta}$ one can define a transformation $A_{\alpha,\beta}$ from C^*_β to C^*_α and its transpose $A'_{\alpha,\beta}$ from C_α to C_β according to the formula $[\phi A'_{\alpha,\beta}](y) = [s_{\alpha,\beta}(y)] \phi[\Pi_{\alpha,\beta}(y)]$ and $< \phi, A_{\alpha,\beta} \mu > = < \phi A'_{\alpha,\beta}, \mu >$ for $\phi \in C_\alpha$ and $\mu \in C^*_\beta$. The transformation $A'_{\alpha,\beta}$ is well defined if $[\phi A'_{\alpha,\beta}](y)$ is put equal to zero whenever $s_{\alpha,\beta}(y) = 0$. One verifies easily that it is a transformation from C_α to C_β. Furthermore its transpose $A_{\alpha,\beta}$ is a transformation of C^*_β into C^*_α which maps the canonical measures K_β onto K_α. Finally, by virtue of its construction, the transformation $A_{\alpha,\beta}$ is such that, with the notation used above, $A_{\alpha,\beta} p_{\theta,\beta} = p_{\theta,\alpha}$ for each $\theta \in \alpha$. This can be stated in a different way as follows. Let $\{E'_\alpha\}$, $\alpha \subset \Theta$ be a compatible family of experiment types. For each E'_α, let $\{p_{\theta,\alpha}; \theta \in \alpha\}$ be its canonical representative. If $\alpha \subset \beta$, the restriction of E'_β to E'_α has for canonical representative the linear operation $A_{\alpha,\beta}$. To associate to the compatible family $\{E'_\alpha\}$ an experiment in the sense of Definition 1 one can then proceed as follows.

For each $\alpha \subset \Theta$ take a $\phi \in C^\alpha$ and all its images $\phi A'_{\alpha, \gamma}$ for $\gamma \supset \alpha$. If $\phi \in C_\alpha$ and $\psi \in C_\beta$ are such that $\phi A'_{\alpha, \gamma} = \psi A'_{\alpha, \gamma}$ for $\gamma = \alpha \cup \beta$ call them equivalent. Call such an equivalence class positive if the generating ϕ is positive. Also if $\phi \in C_\alpha$, $\psi \in C_\beta$ define the sum of their classes as the class of $\phi A'_{\alpha, \gamma} + \psi A'_{\beta, \gamma}$ with $\gamma = \alpha \cup \beta$. It is readily verified that these operations on classes are well defined. The set H of classes so obtained is a vector space and in fact a vector lattice. A compatible family $\{E'_\alpha\}$ defines, for each $\theta \in \Theta$, a linear functional P_θ on H as follows: if $\phi \in C_\alpha$ and $\theta \in \alpha$ define $< \phi, P_\theta > = \int \phi \, dp_{\theta, \alpha}$. To obtain an experiment in the sense of Definition 1, it is sufficient to take, in the dual of H, the band generated by these functionals. This completes the proof of the proposition.

Let us mention now a corollary of Propositions 1 and 2 which is often useful in specific computations.

Let Θ be an arbitrary set. Let $\tilde{\Theta}$ be the set of probability measures with *finite* support on Θ. Each element $\tau \in \tilde{\Theta}$ can be given by a finite set $\alpha \in \Theta$ and numbers τ_θ, $\theta \in \alpha$ such that $\tau_\theta \geqq 0$ and $\Sigma_{\theta \in \alpha} \tau_\theta = 1$. Let $\mathscr{E}: \theta \rightsquigarrow P_\theta$ be an experiment indexed by Θ. For each $\tau \in \tilde{\Theta}$ define another element H_τ of $L(\mathscr{E})$ by $dH_\tau = \Pi_\theta (dP_\theta)^{\tau_\theta}$. This is a well defined Hellinger product. The map $\tau \rightsquigarrow \|H_\tau\|$ from $\tilde{\Theta}$ [0, 1] will be called the Hellinger transform of the experiment $\mathscr{E}$. It is easily seen that if μ is the canonical measure of $\mathscr{E}_\alpha$, restriction of $\mathscr{E}$ to the support α of τ, then $H_\tau = \int (\Pi_{\theta \in \alpha} x_\theta^{\tau_\theta}) \, d\mu$.

PROPOSITION 3.   *Let $E(\Theta)$ be the set of experiment types indexed by Θ. There is a one to one correspondence between experiment types and their Hellinger transforms. The weak topology of $E(\Theta)$ is the same as the topology of pointwise convergence of the Hellinger transforms.*

PROOF.   That the Hellinger transforms determines the type of an experiment results from the uniqueness of the Laplace transform. The standard arguments show that the uniform structure of $E(\Theta)$ can be described as follows. Take a particular finite set $\alpha \subset \Theta$ and an $\varepsilon > 0$. Consider the set $S(\alpha, \varepsilon)$ of $\tau \in \tilde{\Theta}$ such that τ is carried by α and min $\{\tau_\theta; \theta \in \alpha\} \geqq \varepsilon$. Call two Hellinger transforms H and H' close of order $(\alpha, \varepsilon, \delta)$ if $|H_\tau - H'_\tau| < \delta$ for all $\tau \in S(\alpha, \varepsilon)$. This defines a uniformity on the space of Hellinger transforms. It is precisely the uniformity induced by the topology of $E(\Theta)$. This results from the fact that the functions $x \rightsquigarrow \Pi_\theta x_\theta^{\tau_\theta}$ form, for $\tau \in S(\alpha, \varepsilon)$, an equicontinuous set of functions of U_α.

4. Lower semicontinuity of risk functions

The present section indicates some continuity relations for risk functions when the set of experiments is topologized by the weak topology. To state a decision problem we need, in addition to the experiment $\mathscr{E}$, a set D of possible decisions and a loss function W. We shall always assume below that the following condition is satisfied.

The decision space D is completely regular. For each $\theta \in \Theta$, the loss function W_θ is the pointwise supremum of a nonempty subset of $C^b(D)$. The risk at θ of a decision function ρ is defined by the equality

(4.1) $$R(\theta, \rho) = \sup_{\gamma} \{\gamma \rho P_{\theta}; \gamma \in C^b(D), \gamma \leqq W_{\theta}\}.$$

This condition will be labeled the lower semicontinuity assumption, or for short, the l.s.c. assumption.

For a given experiment $\mathscr{E}$ and a given pair (D, W), let $\mathscr{D} = \mathscr{D}(\mathscr{E}, D, W)$ be the set of all available decision functions, that is, by definition, the set of all transitions from $L(\mathscr{E})$ to $C^*(D)$. This set $\mathscr{D}(\mathscr{E})$ will be given the topology of pointwise convergence on $C^b(D) \times L(\mathscr{E})$.

Let $K = K(\Theta)$ be the set of all positive finite nonnull measures which have finite support on Θ. For each $\mu \in K(\Theta)$, let

(4.2) $$\chi(\mu) = \inf \left\{ \int R(\theta, \rho) \mu(d\theta); \rho \in \mathscr{D}(\mathscr{E}) \right\}.$$

This will be called the envelope of the risk functions. Finally, let $\mathscr{R}(\mathscr{E})$ be the set of functions f from Θ to $(-\infty, +\infty]$ which are such that $R(\theta, \rho) \leqq f(\theta)$ for all $\theta \in \Theta$ and some $\rho \in \mathscr{D}(\mathscr{E})$.

In all these definitions it may become necessary to indicate which experiment and which pair (D, W) is involved. This will be done by writing $\chi(\mu; \mathscr{E})$ or $\chi(\mu; \mathscr{E}, W)$ or $\chi(\mu; W)$ and similar expressions for R and $\mathscr{R}$ according to which sets need to be specified.

One of the basic results from which many of the usual general statements of decision theory can be derived is as follows.

PROPOSITION 4. *Assume that the l.s.c. condition is satisfied. Then $\mathscr{D}(\mathscr{E})$ is a compact Hausdorff space and the function $\rho \rightsquigarrow \mathscr{R}(\theta, \rho)$ is lower semicontinuous on $\mathscr{D}(\mathscr{E})$. Furthermore this function is also convex in ρ. Finally a function f from Θ to $(-\infty, +\infty]$ belongs to $\mathscr{R}$ if and only if*

(4.3) $$\chi(\mu) \leqq \int f(\theta) \mu(d\theta)$$

for every $\mu \in K(\Theta)$.

PROOF. The compactness and lower semicontinuity assertions are almost immediate consequences of the definitions. For the last assertion note that it is enough to consider the subset Θ_f of Θ where f is finite. Consider then a finite set $A \subset \Theta_f$ and an $\varepsilon > 0$. Let S be the set of measures $\mu \in K(\Theta)$ which are carried by A and such that inf $\{\mu[\{\theta\}], \theta \in A\} \geqq \varepsilon$. On such a set S, integrals of risk functions are either everywhere finite or everywhere infinite. The usual theorems on separation of convex sets show then that a function f form S to $(-\infty, +\infty]$ agrees on S with an element of $\mathscr{R}$ if and only if $\int f d\mu \geqq \chi(\mu)$ for all $\mu \in S$. A passage to the limit as $\varepsilon \rightarrow 0$ and another passage to the limit letting A increase to Θ_f gives the result.

PROPOSITION 5. *Let Θ and the pair (D, W) be fixed. Let $E(\Theta)$ be the set of all experiment types indexed by Θ. Suppose that the l.s.c. assumption is satisfied. For each $\mathscr{E} \in E(\Theta)$, let $\mu \rightarrow \chi(\mu, \mathscr{E})$ be the corresponding envelope function. Let S be an arbitrary subset of $K(\Theta)$. Then $\mathscr{E} \rightsquigarrow \sup \{\chi(\mu, \mathscr{E}); \mu \in S\}$ is a lower semicontinuous function of $\mathscr{E}$ for the weak topology of $E(\Theta)$.*

PROOF. It is obviously sufficient to prove the result for a set S reduced to a single probability measure μ.

Since μ has finite support, it is also sufficient to prove the result for Θ finite. Assuming this, note that $\chi(\mu, \mathscr{E}, W)$ is the supremum $\sup_V \chi(\mu, \mathscr{E}, V)$ where V runs through loss functions V such that $V_\theta \in C^b(D)$ and $V_\theta \leqq W_\theta$ for all θ. Indeed, there is a decision function σ such that $\chi(\mu, \mathscr{E}, W) = \int R(\theta, \sigma)\mu(d\theta)$. Suppose that $\mathscr{E} = \{P_\theta; \theta \in \Theta\}$. For this fixed σ and for any $b(\theta) < W_\theta \sigma P_\theta$ one can find a $V_\theta \leqq W_\theta$ such that $V_\theta \sigma P_\theta \geqq b(\theta)$. Let $\|V\| = \sup\{|V_\theta(t)|; \theta \in \Theta, t \in D\}$. This is finite since $V_\theta \in C^b(D)$ and since Θ is finite. However we have also

$$(4.4) \qquad |\chi(\mu, \mathscr{E}, V) - \chi(\mu, \mathscr{F}, V)| \leqq \|\mu\| \|V\| \Delta(\mathscr{E}, \mathscr{F}).$$

Thus $\mathscr{E} \rightsquigarrow \chi(\mu, \mathscr{E}, V)$ is continuous and the supremum $\chi(\mu, \mathscr{E}, W)$ is lower semicontinuous.

COROLLARY 1. *Let f be a function from Θ to $(-\infty, +\infty]$. Suppose that f does not belong to $\mathscr{R}(\mathscr{E})$. Then there is an $\alpha > 0$ and a weak neighborhood G of $\mathscr{E}$ in $E(\Theta)$ such that $f + \alpha$ does not belong to any $\mathscr{R}(\mathscr{F})$, $\mathscr{F} \in G$.*

PROOF. The relation $f \notin \mathscr{R}(\mathscr{E})$ implies $\int f d\mu < \chi(\mu, \mathscr{E})$ for some $\mu \in K(\Theta)$. Thus, there is an $\alpha > 0$ such that if $g = f + \alpha$ then $\int g d\mu < \chi(\mu, \mathscr{E})$. The neighborhood $G = \{\mathscr{F}; \chi(\mu, \mathscr{F}) > \int g d\mu\}$ satisfies the required conditions.

Some particular cases of this corollary will be of interest in the following section. We shall restate two of them in a slightly different language. Recall that $\mathscr{E}_S$ means the experiment $\mathscr{E}: \theta \rightsquigarrow P_\theta$ with θ restricted to the set $S \subset \Theta$.

COROLLARY 2. *Suppose that r is an admissible element of $\mathscr{R}(\mathscr{E})$. Let θ_0 be a given element of Θ and let b be a number $b < r(\theta_0)$. There is an $\varepsilon > 0$, a finite set $S \subset \Theta$ and an $\alpha > 0$ such that if $f \in \mathscr{R}(\mathscr{F})$ for some $\mathscr{F}$ such that $\Delta(\mathscr{F}_S, \mathscr{E}_S) < \varepsilon$ satisfies the inequality $f(\theta_0) \leqq b$ then $f(\theta) > r(\theta) + \alpha$ for some $\theta \in S$.*

COROLLARY 3. *Let $a = \sup\{\chi(\mu, \mathscr{E}); \mu \in K(\Theta), \|\mu\| = 1\}$. Let b be a number $b < a$. There is a finite set S and an $\varepsilon > 0$ such that if $\Delta(\mathscr{F}_S, \mathscr{E}_S) < \varepsilon$ then $\sup\{f(\theta); \theta \in S\} > b$ for every $f \in \mathscr{R}(\mathscr{F})$.*

Note that even when W is bounded, in which case $\mathscr{E} \rightsquigarrow \chi(\mu, \mathscr{E}, W)$ is continuous in $\mathscr{E}$ for each fixed μ, we cannot conclude that the minimax risk $\sup\{\chi(\mu, \mathscr{E}); \mu \in K(\Theta), \|\mu\| = 1\}$ is a continuous function of $\mathscr{E}$. This would be true if instead of the weak topology we used the metric Δ on $E(\Theta)$.

5. Limits of experiments; distinguished statistics

Let Z be a completely regular space with its Banach space of bounded continuous numerical functions $C^b(Z)$. Let $C^*(Z)$ be the dual of $C^b(Z)$. We shall often call the elements of $C^*(Z)$ "measures" on Z or integrals even though this is an abuse of language. The space C^* can be topologized by the weak topology $\sigma[C^*(Z), C^b(Z)]$. To distinguish it from other weak topologies we shall call that one the *vague* topology of $C^*(Z)$. Recall that the positive elements of norm unity of $C^*(Z)$ (abusively called probability measures or distributions here) form a vaguely compact set.

PROPOSITION 6. *Let N be a directed set and let Θ be another set. For each $n \in N$ let $\mathscr{E}_n : \theta \rightsquigarrow F_{\theta,n}$ be an experiment such that each $F_{\theta,n}$ belongs to $C^*(Z)$. Assume that for each $\theta \in \Theta$ the $F_{\theta,n}$ converge vaguely to a limit F_θ. Let $\mathscr{F}$ be the experiment $\mathscr{F} = \theta \rightsquigarrow F_\theta$. On the space of experiment types $E(\Theta)$ let $\mathscr{E}$ be a cluster point of the directed set of types $\dot{\mathscr{E}}_n$ for the weak topology of $E(\Theta)$.*

Then $\mathscr{F}$ is weaker than $\mathscr{E}$ or more precisely $\delta(\mathscr{E}, \mathscr{F}) = 0$.

PROOF. According to the corollary of Theorem 2 it is sufficient to prove the result assuming that Θ is *finite*. This will be assumed henceforth. According to Proposition 1 one can also assume without loss of generality that $\Delta(\mathscr{E}_n, \mathscr{E}) \to 0$.

Let K be a compact convex subset of some Euclidean space and let W be a loss function defined on $\Theta \times K$. Assume that $|W| \leq 1$ and that for each $\theta \in \Theta$ the map $t \rightsquigarrow W_\theta(t)$ is continuous. For any decision procedure σ provided by $\mathscr{F}$ let $R(\theta, \sigma; \mathscr{F})$ be the risk of σ at θ. Take an $\varepsilon > 0$ and suppose $n(\varepsilon)$ so large that $n \geq n(\varepsilon)$ implies $\Delta(\mathscr{E}_n, \mathscr{E}) < \varepsilon$.

According to Theorem 1 for a fixed σ there is a decision procedure ρ such that $\sup_\theta |R(\theta, \sigma; \mathscr{F}) - \mathscr{R}(\theta, \rho; \mathscr{F})| < \varepsilon$ and such that ρ is *continuous* in the sense that its transpose maps $C(K)$ into $C^b(Z)$. This procedure ρ may also be applied to $\mathscr{E}_n$ giving a risk $R_n(\theta, \rho) = W_\theta \rho F_{\theta,n}$. Since $W_\theta \rho \in C^b(Z)$ there is an $n(\varepsilon, \sigma)$ such that $n \geq n(\varepsilon\sigma)$ implies $|R_n(\theta, \rho) - R(\theta, \rho)| < \varepsilon$. This gives $R_n(\theta, \rho) \leq R(\theta, \sigma; \mathscr{F}) + 2\varepsilon$. However since $\Delta(\mathscr{E}_n, \mathscr{E}) < \varepsilon$ for $n \geq \max[n(\varepsilon), n(\varepsilon, \sigma)]$ there is a procedure of $\mathscr{E}$ such that

(5.1)
$$R(\theta, \sigma'; \mathscr{E}) \leq R_n(\theta, \rho) + \varepsilon \leq R(\theta, \sigma; \mathscr{F}) + 3\varepsilon.$$

This is true for every K, every W and every σ. Thus $\sigma(\mathscr{E}, \mathscr{F}) \leq 3\varepsilon$ according to Theorem 2. This proves the desired result.

It is easy to construct examples where $\delta(\mathscr{F}, \mathscr{E}) = 2$. In other words, it may happen that $\mathscr{F}$ is trivial but $\mathscr{E}$ is perfect. However our next proposition shows that under special circumstances one can obtain the equivalence of $\mathscr{E}$ and $\mathscr{F}$.

Suppose again that N is a directed set and that Θ and Z are given. *Assume that Θ is finite and that Z is completely regular.* For each $n \in N$ let $\mathscr{E}_n : \theta \rightsquigarrow P_{\theta,n}$ be an experiment indexed by Θ. Let U be the unit simplex of R^Θ and let S_n be the canonical transition from $L(\mathscr{E}_n)$ to $C^*(U)$. This is the usual likelihood ratio vector described in Section 3. The experiment $\theta \rightsquigarrow S_n P_{\theta,n}$ is therefore the canonical representative of $\mathscr{E}_n$.

Let T_n be any statistic from $\mathscr{E}_n$ to the completely regular space Z. (See Definition 2 and the remarks following it.) Let us recall that the distribution of T_n given θ, usually written $\mathscr{L}[T_n | \theta]$ is written $T_n P_{\theta,n}$ in the present notation.

PROPOSITION 7. *With the notation just described, assume that for each $\theta \in \Theta$ the distributions $T_n P_{\theta,n}$ converge vaguely to a limit F_θ. Assume also that the experiments $\mathscr{E}_n$ converge to a limit $\mathscr{E}$. Let $\mathscr{F}$ be the experiment $\mathscr{F} ; \theta \rightsquigarrow F_\theta$. The following conditions are equivalent:*

(a) $\Delta(\mathscr{E}, \mathscr{F}) = 0$,

(b) *for each $\varepsilon > 0$ there is a transition Γ_ε such that*

(5.2)
$$\limsup_n \sup_\theta \| S_n P_{\theta,n} - \Gamma_\varepsilon T_n P_{\theta,n} \|_D \leq \varepsilon$$

for the Dudley norm of measures on U. *Furthermore* Γ_ε *is such that its transpose maps* $C^b(U)$ *into* $C^b(Z)$.

If these conditions are satisfied then $\Delta(\mathscr{E}_n, \mathscr{F}_n) \to 0$ *for* $\mathscr{F}_n = \{T_n P_{\theta,n}; \theta \in \Theta\}$.

PROOF. Let $G_{\theta,n} = S_n P_{\theta,n}$ and let $F_{\theta,n} = T_n P_{\theta,n}$. Since $\mathscr{E}'_n = \theta \leadsto G_{\theta,n}$ is the canonical form of $\mathscr{E}_n$, the convergence of $\mathscr{E}_n$ to $\mathscr{E}$ implies that $\| G_{\theta,n} - G_\theta \|_D \to 0$ for some limit measure G_θ. Suppose now that (b) is satisfied. Then $\Gamma_\varepsilon F_{\theta,n}$ converges to $\Gamma_\varepsilon F_\theta$. Thus the inequality in (b) may be replaced by the relation

$$(5.3) \qquad \limsup_n \sup_\theta \| G_\theta - \Gamma_\varepsilon F_\theta \| \leqq \varepsilon.$$

In particular $G_\theta = \lim_{\varepsilon \to 0} \Gamma_\varepsilon F_\theta$.

Let $\mathscr{F}_\varepsilon$ be the experiment $\theta \leadsto \Gamma_\varepsilon F_\theta$. This is obviously weaker than $\mathscr{F}$. As $\varepsilon \to 0$ the family $\mathscr{F}_\varepsilon$ has some cluster point say $\mathscr{F}^+$. Since each $\delta(\mathscr{F}, \mathscr{F}_\varepsilon) = 0$ one has also $\delta(\mathscr{F}, \mathscr{F}^+) = 0$. However by Proposition 6, the experiment $\theta \leadsto G_\theta$ is weaker than $\mathscr{F}^+$, hence also weaker than $\mathscr{F}$. Since $\mathscr{E} = \theta \leadsto G_\theta$, this gives $\delta(\mathscr{F}, \mathscr{E}) = 0$. Thus $\Delta(\mathscr{E}, \mathscr{F}) = 0$, according to Proposition 6.

Conversely, assume $\Delta(\mathscr{E}, \mathscr{F}) = 0$. Let Λ be the set of Lipschitz functions used to define the Dudley norm on $C^*(U)$. The equality $\Delta(\mathscr{E}, \mathscr{F}) = 0$ implies the existence of a transition Γ such that $G_\theta = \Gamma F_\theta$. Since Λ is a compact subset of $C(U)$, Theorem 1 implies the existence, for each $\varepsilon > 0$, of a transition Γ_ε satisfying the continuity requirement of (b) and such that $|f\Gamma F_\theta - f\Gamma_\varepsilon F_\theta| \leqq \varepsilon$ for all $f \in \Lambda$ and all θ. This can be written $\| G_\theta - \Gamma_\varepsilon F_\theta \|_D \leqq \varepsilon$ and implies (b). The last statement is a consequence of the triangle inequality.

Let us note that the relation $\Delta(\mathscr{E}_n, \mathscr{F}_n) \to 0$ can be interpreted to mean that the statistics T_n are asymptotically sufficient. However $\Delta(\mathscr{E}_n, \mathscr{F}_n) = 0$ would not necessarily imply $\Delta(\mathscr{E}, \mathscr{F}) = 0$. For this reason we shall introduce a definition.

DEFINITION 8. *Let* $\{T_n, n \in N\}$ *be a net of statistics as described before the statement of Proposition 5. If* Θ *is finite and condition* (b) *of Proposition* (5) *is satisfied the net* $\{T_n; n \in N\}$ *will be called distinguished. If* Θ *is finite, "distinguished" will mean that condition* (b) *is satisfied for every finite subset of* Θ.

Consider again a fixed arbitrary set Θ and a directed set N. For each $n \in N$ let $\mathscr{E}_n: \theta \leadsto P_{\theta,n}$ be an experiment indexed by Θ. Let Z and Z' be two completely regular spaces. For each n let T_n be a statistic from $\mathscr{E}_n$ to Z and let T'_n be a statistic from $\mathscr{E}_n$ to Z'.

PROPOSITION 8. *Assume that the distributions* $T_n P_{\theta,n}$ *converge vaguely to a limit* F_θ *and that the distributions* $T'_n P_{\theta,n}$ *converge vaguely to a limit* F'_θ. *If* $\{T_n; n \in N\}$ *is distinguished, there is a transition* M *such that* $F'_\theta = MF_\theta$ *for every* $\theta \in \Theta$.

PROOF. When Θ is finite, this is an immediate consequence of the definitions. If Θ is infinite it is still true, according to Theorem 2, that $\theta \leadsto F'_\theta$ is a weaker experiment than $\theta \leadsto F_\theta$. Hence the result.

The theorem given by Hájek in [4] appears similar to the above except for the fact that where we obtain the existence of a general transition, Hájek obtains a transition representable by convolution. This more precise statement can be derived from the above Proposition 8 under suitable assumptions, as we shall now show.

Let $\mathscr{F} = \{F_\theta ; \theta \in \Theta\}$ and let $\mathscr{F}' = \{F'_\theta ; \theta \in \Theta\}$.

Consider a transition A from $L(\mathscr{F})$ to itself and a transition A' from $L(\mathscr{F}')$ to itself.

DEFINITION 9. *The pair* (A, A') *leaves the system* $(\mathscr{F}, \mathscr{F}')$ *invariant if*:

(1) *A restricted to* $\{F_\theta ; \theta \in \Theta\}$ *is a permutation*,

(2) *A' restricted to* $\{F'_\theta ; \theta \in \Theta\}$ *is a permutation*,

(3) *If* $AF_\theta = F_\xi$ *then* $A'F'_\xi = F'_\theta$.

NOTE. If the maps $\theta \leadsto F_\theta$ and $\theta \leadsto F'_\theta$ are one to one, each one of A and A' induces a permutation of the set Θ. Condition 3 in Definition 9 says then that the permutation induced by A' is the *inverse* of that induced by A. It is more customary to write the definition in such a way that A and A' induce the *same* permutation. However the present formulation is more convenient here.

Suppose that $(\mathscr{F}, \mathscr{F}')$ is invariant by (A, A') and suppose that K is a transition from $\mathscr{L}(\mathscr{F})$ to $\mathscr{L}(\mathscr{F}')$ such that $KF_\theta = F'_\theta$. Then $A'KA$ is also such a transition.

Let $\mathscr{M}$ be the set of all transitions K from $\mathscr{L}(\mathscr{F})$ to $L(\mathscr{F}')$ such that $KF_\theta = F'_\theta$ for all $\theta \in \Theta$. For the topology of pointwise convergence on $M(\mathscr{F}') \times \mathscr{L}(\mathscr{F})$, this is a compact convex set which is transformed into itself by the continuous linear transformation $\mathscr{L} \leadsto A'KA$. An application of the Markov-Kakutani fixed point theorem gives immediately the following result.

PROPOSITION 9. *Assume that the experiments* $\mathscr{F} = \{F_\theta ; \theta \in \Theta\}$ *and* $\mathscr{F}' = \{F'_\theta ; \theta \in \Theta\}$ *are such that*:

(1) *$\mathscr{F}'$ is weaker than $\mathscr{F}$*,

(2) *there is a family* $(A_g, A'_g), g \in G$, *of transition pairs leaving the system* $(\mathscr{F}, \mathscr{F}')$ *invariant such that the induced family of transformations on $\mathscr{M}$ is either abelian or a solvable group or more generally a semigroup which admits almost invariant means.*

Then there is a transition K from $L(\mathscr{F})$ to $L(\mathscr{F}')$ such that $F'_\theta = KF_\theta$ *for all θ and* $A'_g KA_g = K$ *for all $g \in G$*.

In asymptotic theory one often encounters the following situation which is an important special case of the one just described.

Suppose that $\mathscr{F}$ and $\mathscr{F}'$ are as in Proposition 8 but that the two spaces Z and Z' are one and the same. Suppose also that $Z = Z'$ is a locally compact group and that the measures F_θ and F'_θ are Radon measures on Z. For any finite Radon measures μ on Z and any element $\alpha \in Z$ define the measure $\alpha\mu$, called μ shifted by α, by the equality $\int f(z)[\alpha\mu](dz) = \int f(\alpha z)\mu(dz)$.

One can say that the pair $(\mathscr{F}, \mathscr{F}')$ is invariant by the (left) group shifts if for each $\alpha \in Z$ the operations $F_\theta \leadsto \alpha F_\theta$ and $F'_\theta \leadsto \alpha F'_\theta$ are permutations and if $\alpha F_\theta = F_\xi$ implies $\alpha F'_\theta = F'_\xi$. A transition K which is "invariant" can then be described as a transition which commutes with the group shifts. That is for every $\alpha \in Z$ and $\mu \in L(\mathscr{F})$ one has $\alpha K\mu = K\alpha\mu$. It is to be expected that such transitions will in fact turn out to be convolutions by a fixed probability measure. However we have been able to prove this only under special assumptions.

PROPOSITION 10. *Let Z be a locally compact group. Let* $\mathscr{F} = \{F_\theta ; \theta \in \Theta\}$ *be an experiment defined by Radon measures on Z. Let K be a transition from $L(\mathscr{F})$*

*to finite Radon measures on Z. Assume that K commutes with the group shifts.
Then there is a probability measure Q such that $K\mu$ is the convolution $K\mu = \mu * Q$
if and only if the transpose of K transforms $C^b(Z)$ into $C^b(Z)$. This happens in
particular if the F_θ are either all discrete or all absolutely continuous with respect
to one of the Haar measures of Z.*

PROOF. Consider the case where all the F_θ are absolutely continuous with
respect to a Haar measure. Let f be a bounded measurable function defined on Z.
Then $\int |f| d\mu = 0$ for all $\mu \in L(\mathscr{F})$ if and only if f is locally equivalent to zero for
the Haar measure on Z. Indeed, suppose that $f \geqq 0$ has compact support. By
Fubini's theorem

$$(5.4) \qquad \int \left\{ \int f(\alpha z)\mu(dz) \quad \lambda(d\alpha) = \int \mu(dz) \int f(\alpha z)\lambda(d\alpha) \right.$$

$$= \|\mu\| \int f(\alpha)\lambda(d\alpha)$$

for the right Haar measure λ on Z.

Let H_0 be the space of equivalence classes bounded measurable functions
on Z, for the local equivalence relation defined by the Haar measure. According
to [7] the space H_0 admits a lifting which commutes with the shifts. Let H be
the range of the lifting. That is H consists of the functions which have been
selected as representatives of classes in H_0. Note that if $g \in C^b(Z)$ the representative of the class of g is g itself.

Since K is a transition from $L(\mathscr{F})$ to finite Radon measures on Z, its transpose
K^t maps the dual of the space of finite Radon measures into H_0. However,
composing K^t with the lifting, we can instead define K^t as a map into H. With
this agreement consider $K^t g$ for some $g \in C^b(Z)$. The equality $\alpha K\mu = K\alpha\mu$,
$\mu \in L(\mathscr{F})$, $\alpha \in Z$, implies $< \alpha^t K^t g, \mu > = < K^t \alpha^t g, \mu >$ for $\mu \in L(\mathscr{F})$ and $\alpha \in Z$.
However $\alpha^t K^t g$ is also in the lifting and therefore the *almost* everywhere equality
just indicated implies that $\alpha^t K^t g = K^t \alpha^t g$ *everywhere*. The same conclusion would
be available if we had assumed that the F_θ are discrete or that $K^t g$ can be taken
equal to an element of $C^b(Z)$.

Let then $\mathscr{K}$ be the space of continuous functions with compact support on
Z. Fix a $z \in Z$ and evaluate $K^t g$ at z for each $g \in \mathscr{K}$. This gives a positive linear
functional Q_z on $\mathscr{K}$. The equality $\alpha^t K^t g = K^t \alpha^t g$ yields $Q_{\alpha z} = \alpha Q_z$, that is
$Q_z = zQ$ with $Q = Q_0$. Thus

$$(5.5) \qquad (K^t g)(z) = \int g(t)Q_z(dt) = \int g(zt)Q(dt).$$

The result follows.

6. Applications to standard examples

In many of the customary applications of large sample theory one is interested
in the asymptotic behavior of certain test or estimate in the vicinity of a given
value of the parameter. This can be formalized as follows. One is given a directed

set N, for instance the set of integers. For each $n \in N$, let $\hat{\mathscr{E}}_n = \{Q_{\xi, n}; \xi \in \Xi_n\}$ be an experiment indexed by a parameter space Ξ_n. One is given in addition, for each $n \in N$, a subset Θ_n of the k-dimensional Euclidean space Θ and a function $\theta \leadsto \xi_n(\theta)$ from Θ_n to Ξ_n. The small vicinity of interest is the set $\xi_n(\Theta_n)$. For instance Ξ_n may be the k-dimensional Euclidean space itself and the functions ξ_n may be of the type $\xi_n(\theta) = \xi_n(0) + \delta_n \theta$ with δ_n tending to zero.

In brief the experiments of interest are not really the $\hat{\mathscr{E}}_n$ but the experiments $\mathscr{E}_n = \{P_{\theta, n}; \theta \in \Theta_n\}$ with $P_{\theta, n} = Q_{\xi_n(\theta), n}$.

To avoid extra complications we shall consider only cases where the following conditions are satisfied.

(A1) *If m and n are elements of N such that $m < n$ then $\Theta_m \subset \Theta_n$. Furthermore* $\Theta = \cup_n \Theta_n$.

(A2) *For every finite subset $S \subset \Theta$ the experiments $\mathscr{E}_{n,S} = \{P_{\theta, n}; \theta \in S\}$ have a limit in the sense of Section 2 and 3.*

One can extend arbitrarily the map $\theta \leadsto P_{\theta, n}$ to the whole of Θ. The experiment $\{P_{\theta, n}; \theta \in \Theta\}$ will still be denoted $\mathscr{E}_n$.

According to Section 3 these experiments $\mathscr{E}_n$ converge weakly to a certain limit $\mathscr{E}$, which is, of course, independent of the manner in which the $P_{\theta, n}$ are defined outside Θ_n.

The following examples are taken from the standard statistical example list.

EXAMPLE 1. Let $r_n(t)$ be the likelihood ratio $r_n(t) = dP_{t, n}/dP_{0, n}$. One assumes that there are k-dimensional random vectors Y_n and a positive definite matrix Γ such that $r_n(t) - \exp\{tY_n - \frac{1}{2}t\Gamma t'\}$ converges in $P_{0, n}$ probability to zero for each $t \in \Theta$. Furthermore one assumes that the distribution $\mathscr{L}[Y_n|P_{0, n}]$ converges to a Normal distribution with mean zero and covariance matrix Γ.

The approximation to $r_n(t)$ can also be rewritten $\exp\{\frac{1}{2}X_n\Gamma X_n'\}\exp\{-\frac{1}{2}(X_n - t)\Gamma(X_n - t)'\}$ with $\Gamma X_n = Y_n$. The limit experiment $\mathscr{E}$ is the experiment $\{P_\theta; \theta \in \Theta\}$ with P_θ normal with mean θ and covariance matrix Γ^{-1}.

EXAMPLE 2. Let N be the set of integers. Consider n independent, identically distributed variables $U_1, U_2, \cdots, U_n$, which are uniformly distributed on the interval $[0, 1 + \delta_n\theta]$ with θ such that $1 + \delta_n\theta > 0$ and $\delta_n = 1/n$. Let Y_n be the maximum of the observations $U_1, \cdots, U_n$. Then $\mathscr{L}\{n[(1 + \delta_n\theta) - Y_n]|1 + \delta_n\theta\}$ converges to the exponential distribution which has density e^{-x} on the positive part of the line. The measures P_θ of the limit experiment $\mathscr{E} = \{P_\theta; \theta \in \Theta\}$ have densities $e^{-(x-\theta)}$, $x \geqq \theta$.

EXAMPLE 3. Let $U_{n,j}$, $j = 1, 2, \cdots, n$, be independent identically distributed with individual distribution uniform on the interval $[\theta/n, (\theta/n) + 1]$. Let $Z_n' = n \min_j U_{n,j}$ and let $Z_n'' = n \max_j [U_{n,j} - 1]$. The pair (Z_n', Z_n'') has a limiting distribution equal to that of a pair (S, T) with $S = \theta + X$ and $T = \theta - Y$ where X and Y are independent variables such that $P[X \geqq x] = P[Y \geqq x] = e^{-x}$ for $x > 0$. The limit experiment $\mathscr{E}$ is the experiment where $P_\theta = \mathscr{L}\{(\theta + X, \theta - Y)\}$.

In all three examples we have stated a description of the limit experiment $\mathscr{E}$. Of course one needs to prove that $\mathscr{E}_n$ tend to the limit. For normal Example 1 note that the statistics $\{X_n\}$ are "distinguished" in the sense of Proposition 7. For the other two examples it is enough to remark that the statistics $X_n = n(Y_n - 1)$ in Example 2 and (Z'_n, Z''_n) in Example 3 are sufficient statistics whose *densities* converge. Thus they are automatically distinguished.

For any one of these examples let $\{T'_n\}$ be some other family of statistics. Suppose for instance that for each $\theta \in \Theta$ the distributions $\mathscr{L}(T'_n|\theta)$ converge to a limit F'_θ and that F'_θ is F'_0 shifted by the amount θ. That is $\mathscr{L}[T'_n - \theta|\theta]$ tends to a limit F'_0.

In Examples 1 and 2, Proposition 9 and 10 insure the existence of some probability measure Q such that $F'_0 = Q*F_0$ for the distribution $F_0 = \lim \mathscr{L}[X_n|\theta = 0]$.

In Example 3 things are more complicated since the shift group does not operate transitively on the plane. Introducing new variables $\xi = \frac{1}{2}(S + T)$ and $\eta = S - T$, let $H_\theta = \mathscr{L}[(\xi, \eta)|\theta]$. Clearly $\mathscr{L}[(\xi, \eta)|\theta] = \mathscr{L}[(\xi + \theta, \eta)|0] = H$ say. Simple computation shows then that the limiting distribution $G = F'_0$ must be obtainable by the formula

$$(6.1) \qquad \int f(z)G(dz) = \iiint f(z + x)\mu(dz|y)C(dx|y)M(dy),$$

where C and M are the conditional and marginal distributions such that $H(dx, dy) = C(dx|y)M(dy)$ and where for each y the symbol $\mu(dz|y)$ represents a probability measure in z.

In other words, for each y one convolutes the conditional distribution $C(dx|y)$ with some probability measure μ which depends on y. Then one averages the result over all values of y according to the marginal distribution M.

Some other statements can be obtained by application of the results of Section 4. We shall state a few for the case of the normal Example 1. Take as loss function the quadratic $(\theta - t)\Gamma(\theta - t')$. Then the normal vector X of the limit experiment is a minimax estimate with risk identically equal to the dimension k of Θ. Suppose then that $\{T'_n\}$ is any sequence of estimates. Let $R_n(\theta)$ be the risk of T'_n at θ. Let $\varepsilon > 0$ be given. By Proposition 5, Corollary 3, there is some *finite* set $S \subset \Theta$, some n_0 such that sup $\{R_n(\theta); \theta \in S\} > k - \varepsilon$ for all $n \geq n_0$. Furthermore S and n_0 do not depend on the choice of T'_n. This gives part of a result of Hájek in [5].

If $k \leq 2$, the estimate X is also admissible. Corollary 2 of Proposition 5 says then that if $a = \sup_{n>m} R_n(\theta_0) < k$ for a given θ_0 there is a finite set S an n_0 and an $\alpha > 0$ such that sup $\{R_n(\theta); \theta \in S\} \geq k + \alpha$ for $n \geq n_0$. Here again α, n_0 and S depend only on θ_0 and $a < k$. This strengthens considerably a result of [9] according to which super efficiency at one point must imply misbehavior nearby. The result of [9] was proved only for $k = 1$. Of course the result does not extend to $k \geq 3$ since X is no longer admissible.

It was proved in [13] that the χ^2 test which rejects $\theta = 0$ if $X'\Gamma X$ is larger than a given c_0 has best minimum power over the surface $\theta\Gamma\theta' = c_1$. Suppose then that $\alpha_0 = P_0\{X'\Gamma X \geq c_0\}$, $\beta_0 = P_\theta[X'\Gamma X \geq c_0]$ for $\theta\Gamma\theta' = c_1$. Let $\alpha_1 \leq \alpha_0$

and $\beta_1 \geqq \beta_0$ be two numbers. Suppose for instance $\beta_1 > \beta_0$ and let ϕ_n be tests such that $\int \phi_n dP_{0,n} \leqq \alpha_1$. Then there is some finite subset S, $S \subset \{\theta; \theta \Gamma \theta' = c_1\}$, and some n_0 such that if $n \geqq n_0$ we have $\int \phi_n dP_{\theta,n} < \beta_1$ for at least some $\theta \in S$. This is another application of Proposition 5, Corollary 3.

For such tests one can even say more. A. Birnbaum in [1] proved that every convex subset of the Euclidean space is admissible as a test of $\theta = 0$ against $\theta \neq 0$. Thus suppose that $\alpha_1 \leqq \alpha_0$, $\beta_1 \geqq \beta_0$ with at least one of the inequalities holding strictly. Let t be a particular point such that $t \Gamma t' = c_1$. There is then an n_0, and $\varepsilon > 0$ and a finite set $S \subset \Theta$ such that if $n \geqq n_0$ the inequalities $\int \phi_n dP_{t,n} \leqq \alpha_1$ and $\int \phi_n dP_{\theta,n} \geqq \beta_1$ imply $\varepsilon + \int \phi_n dP_{\theta,n} \leqq P_\theta[X'\Gamma X \geqq c_1]$ for some $\theta \in S$.

Analogous statements can be made for the cases of Examples 2 and 3. We shall leave them to the care of the reader.

REFERENCES

[1] A. Birnbaum, "Characterizations of complete classes of tests of some multiparametric hypotheses, with applications to likelihood ratio tests," *Ann. Math. Statist.*, Vol. 26 (1955), pp. 21–36.

[2] D. Blackwell, "Comparison of experiments," *Proceedings of the Second Berkeley Symposium on Mathematical Statistics and Probability*, Berkeley and Los Angeles, University of California Press, 1951, pp. 93–102.

[3] J. L. Denny, telephone conversation, 1968.

[4] J. Hájek, "A characterization of limiting distributions of regular estimates," *Z. Wahrscheinlichkeitstheorie und Verw. Gebiete*, Vol. 14 (1970), pp. 323–330.

[5] ———, "Local asymptotic minimax and admissibility in estimation," *Proceedings of the Sixth Berkeley Symposium on Mathematical Statistics and Probability*, Berkeley and Los Angeles, University of California Press, 197x, Vol. 1, pp. xxx–xxx.

[6] H. Heyer, "Erschöpfheit und Invarianz beim Vergleich von Experimenten," *Z. Wahrscheinlichkeitstheorie und Verw. Gebiete*, Vol. 12 (1969), pp. 21–25.

[7] A. Ionescu Tulcea and C. Ionescu Tulcea, "On the existence of a lifting commuting with the left translations of an arbitrary locally compact group," *Proceedings of the Fifth Berkeley Symposium on Mathematical Statistics and Probability*, Berkeley and Los Angeles, University of California Press, 1967, Vol. 2, Part 1, pp. 63–97.

[8] L. LeCam, "Sufficiency and asymptotic sufficiency," *Ann. Math. Statist.*, Vol. 35 (1964), pp. 1419–1455.

[9] ———, "On some asymptotic properties of maximum likelihood estimates and related Bayes estimates," *Univ. California Publ. Statist.*, Vol. 1, (1953), pp. 277–330.

[10] ———, *Théorie Asymptotique de la Décision Statistique*, Montreal, University of Montreal Press, (1969).

[11] E. N. Torgersen, "Comparison of experiments when the parameter space is finite," Ph.D. thesis, University of California Berkeley, 1968.

[12] ———, "Comparison of translation experiments," unpublished manuscript, May 1969.

[13] A. Wald, "Tests of statistical hypotheses concerning several parameters when the number of observations is large," *Trans. Amer. Math. Soc.*, Vol. 54, (1943), pp. 426–482.

ON THE STRONG CONSISTENCY OF APPROXIMATE MAXIMUM LIKELIHOOD ESTIMATORS

MICHAEL D. PERLMAN
UNIVERSITY OF MINNESOTA

1. Introduction and statement of problem

Wald's general conditions for strong consistency of Approximate Maximum Likelihood Estimators (AMLE) [11] have been extended by several authors, notably LeCam [9], Kiefer and Wolfowitz [8], Huber [7], Bahadur [1], and Crawford [4]. Except for mild identifiability and local regularity conditions these papers (except [9]) share two critical global assumptions, global in the sense that they concern the behavior of the Log Likelihood Ratio (LLR) over the entire parameter space Θ (which may be infinite dimensional). Crudely stated these are (a) there exists a "suitable compactification" $\bar{\Theta}$ of Θ (see [1], p. 320) to which the LLR may be extended in a continuous manner without altering the value of its supremum, and (b) the supremum of the LLR is integrable (dominance). Condition (b), however, is not satisfied in many common problems, especially multiparametric ones, where AMLE are known to be consistent. Kiefer and Wolfowitz and later Berk [3] suggested a method which seemed to overcome this difficulty in special cases, namely: consider the observations pairwise, or in groups of k. In the more general context of "maximum w" estimation described below, however, this method fails (see Example 2). Noticing this, Huber proposed that the LLR be divided by a function $b(\theta)$ such that this normalized LLR satisfies (a) and (b).

In Section 2 of this paper we show that under an extended global dominance assumption the method of Kiefer, Wolfowitz, and Berk is precisely the correct one. This idea is then extended to include Huber's modification. In Section 5 we show that (generalized versions of) LeCam's conditions are equivalent to those based on dominance.

The methods described above have several drawbacks, however: they require determination of a suitable group size k, normalizing functions $b(\theta)$, and a compactification $\bar{\Theta}$. As demonstrated by several examples below, these are not always naturally occurring quantities and may be difficult to determine. In Section 3 we introduce a new condition for strong consistency, based on a global uniformity assumption rather than dominance, which seems to present a more natural and straightforward method for determining strong consistency.

This research was supported by the National Science Foundation under Grant No. GP-9593.

263

It has the advantage that it is intrinsic, that is, it does not require searching for quantities k, $b(\theta)$, or $\bar{\Theta}$ which are not specified in the original problem.

Necessary and sufficient conditions for strong consistency are discussed in Section 4.

Throughout this paper we treat the problem of strong consistency of AMLE in the following generalized context.

Let $\mathscr{P}$ be a set of distinct probability distributions on a measurable space $(\mathscr{X}, \mathscr{A})$. Let $\theta = \theta(P)$ be a mapping of $\mathscr{P}$ onto a Hausdorff topological space Θ which satisfies the first axiom of countability. Let $X_1, X_2, \cdots$ be a sequence of independent, identically distributed (i.i.d.) random variables assuming values in $\mathscr{X}$, each distributed according to P_0, and let $\theta_0 = \theta(P_0)$. The symbol P_0 is also used to denote the product probability measure on the infinite product space of all sequences $(x_1, x_2, \cdots)$, and $_*P_0$ denotes the induced inner measure on this space. Let $w(x, \theta)$ be a real-valued function defined on $\mathscr{X} \times \Theta$ such that for each fixed θ, $w(\cdot, \theta)$ is measurable, and for $n = 1, 2, \cdots$, let

$$(1.1) \qquad w_n(\theta) = w_n(x_1, \cdots, x_n, \theta) \equiv \frac{1}{n} \sum_{i=1}^{n} w(x_i, \theta).$$

(For any other function $y(x, \theta)$, $y_n(\theta)$ is defined in a similar manner.) In this paper we discuss the strong consistency of estimators which are based on maximizing $w_n(\theta)$.

Let $\mathscr{S}_1$ denote the class of all estimating sequences $\{T_n\} = \{T_n(x_1, \cdots, x_n)\}$ (T_n is a Θ-valued function and is *not necessarily measurable*) such that for all P_0 in $\mathscr{P}$,

$$(1.2) \qquad _*P_0\big[\sup_\Theta w_n(\theta) = w_n(T_n) \text{ a.a. } n\big] = 1$$

(all suprema in this paper are taken with respect to θ over the indicated subset), where, if $\{A_n\}$ is any sequence of sets, $\{A_n \text{ a.a. } n\}$ is the set

$$(1.3) \qquad \liminf_{n \to \infty} A_n = \bigcup_{n=1}^{\infty} \bigcap_{k \geq n} A_k.$$

If $\{T_n\}$ is in $\mathscr{S}_1$, we call it a *Maximum w Estimator* (MWE). Since $\mathscr{S}_1$ may be empty (the supremum may not be attained), we shall mainly consider the larger class $\mathscr{S}_2$ consisting of all estimating sequences such that for all P_0 in $\mathscr{P}$

$$(1.4) \qquad _*P_0\big[H\big(\sup_\Theta w_n(\theta), w_n(T_n)\big) \to 0\big] = 1,$$

where

$$(1.5) \qquad H(a, b) \equiv \begin{cases} a - b & \text{if } a < \infty \\ b^{-1} & \text{if } a = \infty \text{ and } b > 0 \\ 1 & \text{if } a = \infty \text{ and } b \leq 0. \end{cases}$$

If $\{T_n\}$ is in $\mathscr{S}_2$, it is called an *Approximate Maximum w Estimator* (AMWE).

EXAMPLE 1. Suppose that $\theta(P)$ is one-to-one and that each P has a density $f(x, \theta)$ with respect to some measure μ. If $w(x, \theta) = \log f(x, \theta)$, then $\mathscr{S}_2$ contains all AMLE (in the sense of Wald [11], p. 600, Theorem 2).

EXAMPLE 2.   Let $\mathscr{P}$ denote the set of all distributions on $(-\infty, \infty)$ which possess a unique population median. Let $\Theta = (-\infty, \infty)$ and $\theta(P) = $ median of P. If $w(x, \theta) = -|x - \theta|$, then $\mathscr{S}_1$ contains all sample medians (recall that the sample median may not be uniquely determined).

For each θ_0 in Θ let $\{V_r\} = \{V_r(\theta_0)\}$ be a decreasing sequence of neighborhoods of θ_0 which form a base for the neighborhood system at θ_0 (so $\cap \, V_r = \{\theta_0\}$), and let $\Omega_r = \Omega_r(\theta_0) = \Theta - V_r(\theta_0)$. Then an estimating sequence $\{T_n\}$ is *strongly consistent* if and only if for all P_0 in $\mathscr{P}$ and $r \geq 1$,

$$(1.6) \qquad\qquad {}_*P_0[T_n \text{ in } V_r(\theta_0) \text{ a.a. } n] = 1.$$

Note that if this is satisfied for one such sequence of neighborhoods $\{V_r\}$, it must be satisfied for any other such sequence.

A convenient starting point for this problem is the following obvious fact:

LEMMA 1.1.   *A sufficient condition for the strong consistency of every* AMWE *is that*

$$(1.7) \qquad\qquad {}_*P_0\big[\limsup_{n \to \infty} \sup_{\Omega_r} u_n(\theta; \theta_0) < 0\big] = 1$$

for every P_0 in $\mathscr{P}$ and $r \geq 1$, where

$$(1.8) \qquad \begin{aligned} u(x, \theta) &\equiv u(x, \theta; \theta_0) \equiv w(x, \theta) - w(x, \theta_0), \\ u_n(\theta) &\equiv u_n(\theta; \theta_0) \equiv w_n(\theta) - w_n(\theta_0). \end{aligned}$$

(Under some additional assumptions (1.7) is also a necessary condition, see Section 4.)

The earlier papers [1] (p. 320), [4], [7] (p. 222), [8] (p. 890), [9] (pp. 302–304), and [11] all present conditions which imply Conditions 1 or 2 below, and therefore imply (1.7). (See discussion preceding Theorem 2.3 and see Section 5.) If Θ (hence Ω_r) is compact, Theorem 2.4 below is applicable. In this paper we are mainly concerned with the more interesting and difficult situation where Θ is not compact. In this case earlier papers (except LeCam [9]) assume that a "suitable compactification" of Θ exists (see [1], p. 320). Such a compactification is not always apparent if it exists; the one-point compactification is often unsuitable (see Example 4 and subsequent discussion). In this paper, as in LeCam ([9] pp. 302–304), we attempt to avoid the need to extend the parameter space. The lim sup in (1.7) is studied directly, first with no assumptions on Ω_r (Section 2) and then assuming Ω_r is σ-compact (Section 3).

2. Conditions for strong consistency of AMWE based on dominance and semi-dominance

Let Γ be a subset of Θ (so Γ is first countable and Hausdorff) and let $y(x, \theta)$ be a real-valued function, defined on $\mathscr{X} \times \Gamma$, which is measurable in x for each fixed θ. Let the sequence of $\mathscr{X}$-valued i.i.d. random variables $X_1, X_2, \cdots$

have probability distribution P (which need not be in $\mathscr{P}$). (Later we shall take $\Gamma = \Omega_r$, $y(x, \theta) = u(x, \theta; \theta_0)$, and $P = P_0$).

DEFINITION 1. *The function $y(x, \theta)$ is dominated (dominated by 0) on Γ with respect to P if there is a positive integer k and a real valued function $s(x_1, \cdots, x_k)$ on $\mathscr{X} \times \cdots \times \mathscr{X}$, measurable with respect to the product σ-field $\mathscr{A} \times \cdots \times \mathscr{A}$, such that*

(i) $\sup_\Gamma y_k(\theta) \leqq s(x_1, \cdots, x_k)$ *for all $x_1, \cdots, x_k$ in a set of probability one, and*

(ii) $Es(X_1, \cdots, X_k) < \infty$ (< 0).

(The subscript k will be used exclusively to refer to this definition.)

REMARK. Note that if $\sup_\Gamma y_k(\theta)$ is measurable, it can be used in place of $s(x_1, \cdots, x_k)$. In any case, note that s can be chosen to be a symmetric function of $x_1, \cdots, x_k$, for we may replace s by

$$(2.1) \qquad s^*(x_1, \cdots, x_k) = \frac{1}{k!} \sum s(x_{i(1)}, \cdots, x_{i(k)})$$

where the sum is taken over all permutations of $(1, \cdots, k)$. Also, s can be chosen such that $Es(X_1, \cdots, X_k) > -\infty$ for, if not, replace s by $\max(s, M)$ for any number M (or any M such that $E \max(s, M) < 0$).

DEFINITION 2. *$y(x, \theta)$ is semidominated (semidominated by 0) on Γ with respect to P if there exists a function $b(\theta)$ defined on Γ, $0 < b(\theta) < \infty$, such that $y(x, \theta)/b(\theta)$ is dominated (dominated by 0) on Γ with respect to P and $\inf_\Gamma b(\theta) > 0$.*

REMARK. Note that if $b_1(\theta)$, with $0 < b_1(\theta) < \infty$, is such that $y(x, \theta)/b_1(\theta)$ is dominated by 0 on Γ, it does not necessarily follow that $y(x, \theta)$ is semidominated by 0 on Γ, since replacing $b_1(\theta)$ by $b(\theta) = \max(b_1(\theta), a)$ with $a > 0$ will not necessarily preserve dominance by 0.

We first investigate some implications among Definitions 1, 2 and the condition (recall (1.7))

$$(2.2) \qquad {}_*P[\limsup_{n \to \infty} \sup_\Gamma y_n(\theta) < 0] = 1.$$

THEOREM 2.1. *If y is dominated (or semidominated) by 0 on Γ then (2.2) holds.*

PROOF. We show that dominance by 0 implies (2.2) using an idea of Berk [3]. For any $n \geqq k$, let $\alpha = \{\alpha_1, \cdots, \alpha_k\}$ denote a selection of k indices from $\{1, 2, \cdots, n\}$. Then

$$(2.3) \qquad y_n(\theta) = \binom{n}{k}^{-1} \sum_\alpha \left[k^{-1} \sum_{i \in \alpha} y(x_i, \theta) \right]$$

so that for all $x_1, \cdots, x_n$ in a set of probability one

$$(2.4) \qquad \sup_\Gamma y_n(\theta) \leqq \binom{n}{k}^{-1} \sum_\alpha s(x_{\alpha_1}, \cdots, x_{\alpha_k}) \equiv S_{n,k}.$$

(We choose s to be a symmetric function of $x_1, \cdots, x_k$.) Berk ([3], pp. 55–56) shows that $\{S_{n,k}\}_{n=k}^\infty$ forms a reverse martingale sequence and $S_{n,k} \to Es < 0$ almost surely as $n \to \infty$, which implies (2.2).

Finally we show that semidominance by 0 implies (2.2). If y is semidominated by 0 on Γ, there is a function $b(\theta)$ such that, applying the above argument,

$$(2.5) \qquad {}_*P[\limsup_{n \to \infty} \sup_{\Gamma} (y_n(\theta)/b(\theta)) < 0] = 1.$$

Since $\inf_{\Gamma} b(\theta) > 0$, this implies (2.2). $Q.E.D.$

EXAMPLE 3. $X_1, X_2, \cdots$ are i.i.d. random variables, each with the normal distribution $N(-1, 1)$. Take $\Gamma = [1, \infty)$ and $y(x, \theta) = \theta x$, so $y_n(\theta) = \theta \bar{x}_n$. Since $\bar{X}_n \to -1$ a.s., $\sup_{\Gamma} y_n(\theta) \to -1$ a.s. so (2.2) is satisfied. However, for all n

$$(2.6) \qquad P[\sup_{\Gamma} y_n(\theta) = \infty] = P[\bar{X}_n > 0] > 0,$$

so y is not dominated on Γ. Choosing $b(\theta) = \theta$ we see that y is semidominated by 0 on Γ. Thus neither (2.2) nor semidominance by 0 necessarily implies dominance by 0.

A partial converse to Theorem 2.1 is presented in Theorem 2.2 (ii). Several preliminary results are needed.

LEMMA 2.1. *If X and Y are independent real valued random variables, then $E(X + Y)^+ < \infty \Rightarrow EX^+ < \infty$ and $EY^+ < \infty$.*

PROOF. Since $E(X + Y)^+ = E\{E[(X + Y)^+ \mid Y]\}$ it follows that $E(X + y)^+ < \infty$ for almost all y. But $X^+ \leq (X + y)^+ + |y|$ so $EX^+ < \infty$, and similarly $EY^+ < \infty$. $Q.E.D.$

LEMMA 2.2. *If y is dominated or semidominated on Γ then for every θ' in Γ, $E[y_1(\theta')]^+ < \infty$. Thus for every θ' and n $Ey_n(\theta') = Ey_1(\theta')$ is well defined (possibly $= -\infty$) and $\sup_{\Gamma} Ey_1(\theta) < \infty$.*

PROOF. If y is dominated on Γ, Definition 1 implies that

$$(2.7) \qquad y_k(\theta') \leq \sup_{\Gamma} y_k(\theta) \leq s(x_1, \cdots, x_k)$$

so $E[y_k(\theta')]^+ \leq Es^+ < \infty$. The result then follows from Lemma 2.1 (the semi-dominated case is treated similarly). $Q.E.D.$

LEMMA 2.3. *Suppose that for every θ' in Γ, $Ey_1(\theta')$ is well defined (possibly $\pm \infty$). Then*

$$(2.8) \qquad {}_*P[\sup_{\Gamma} Ey_1(\theta) \leq \liminf_{n \to \infty} \sup_{\Gamma} y_n(\theta)] = 1$$

PROOF. For each θ' in Γ and all n, $y_n(\theta') \leq \sup_{\Gamma} y_n(\theta)$. Letting $n \to \infty$ the result follows from the Strong Law of Large Numbers (SLLN).

REMARK. Lemma 2.3 implies that

$$(2.9) \qquad {}_*P[\sup_{\Gamma} Ey_1(\theta) \leq \limsup_{n \to \infty} \sup_{\Gamma} y_n(\theta)] = 1,$$

and that

$$(2.10) \qquad {}_*P[\sup_{\Gamma} y_n(\theta) \to \sup_{\Gamma} Ey_1(\theta)] = 1$$

if and only if equality holds in (2.9), in fact, if and only if

$$(2.11) \qquad {}_*P[\sup_{\Gamma} Ey_1(\theta) \geq \limsup_{n \to \infty} \sup_{\Gamma} y_n(\theta)] = 1.$$

LEMMA 2.4.   (i) *If* $Ey_1(\theta') > -\infty$ *for some* θ' *in* Γ, *then* $E[\sup_\Gamma y_n(\theta)]^- < \infty$ *for all n such that* $\sup_\Gamma y_n(\theta)$ *is measurable, in which case* $E\sup_\Gamma y_n(\theta)$ *is well defined (possibly* $= +\infty$).

(ii) *If y is dominated on* Γ, *then* $E[\sup_\Gamma y_n(\theta)]^+ < \infty$ *for all* $n \geq k$ *such that* $\sup_\Gamma y_n(\theta)$ *is measurable, in which case* $E\sup_\Gamma y_n(\theta)$ *is well defined (possibly* $= -\infty$).

PROOF.   Part (i) is obvious; part (ii) follows from

$$(2.12) \qquad \sup_\Gamma y_n(\theta) \leq \binom{n}{k}^{-1} \sum_\alpha \sup_\Gamma \left[k^{-1} \sum_{i \in \alpha} y(x_i, \theta) \right] \equiv Y_{n,k}(\Gamma),$$

(see the proof of Theorem 2.1). *Q.E.D.*

The result need not hold if only semidominance is assumed.

LEMMA 2.5.   *If* $Ey_1(\theta')$ *is well defined for all* θ' *in* Γ *and if* $\sup_\Gamma y_n(\theta)$ *is measurable and* $E\sup_\Gamma y_n(\theta)$ *is well defined for almost all n then*

$$(2.13) \qquad \sup_\Gamma Ey_1(\theta) \leq \downarrow \lim_{n \to \infty} E\sup_\Gamma y_n(\theta).$$

PROOF.   From (2.12) with n, k replaced by $n+1$, n,

$$(2.14) \qquad \sup_\Gamma Ey_1(\theta) = \sup_\Gamma Ey_{n+1}(\theta) \leq E\sup_\Gamma y_{n+1}(\theta) \leq E\sup_\Gamma y_n(\theta),$$

which implies (2.13). *Q.E.D.*

Under the hypothesis of Lemma 2.5, (2.9) is valid with $_*P$ replaced by P, and should be compared with (2.13). The relationship between (2.9) and (2.13) is now clarified.

THEOREM 2.2.   (i) *If* $\sup_\Gamma y_n(\theta)$ *is measurable and* $E\sup_\Gamma y_n(\theta)$ *well defined for almost all n then*

$$(2.15) \qquad P[\limsup_{n \to \infty} \sup_\Gamma y_n(\theta) \leq \downarrow \lim_{n \to \infty} E\sup_\Gamma y_n(\theta)] = 1.$$

(ii) *If y is dominated on* Γ *and* $\sup_\Gamma y_n(\theta)$ *is measurable for almost all n then*

$$(2.16) \qquad P[\limsup_{n \to \infty} \sup_\Gamma y_n(\theta) = \downarrow \lim_{n \to \infty} E\sup_\Gamma y_n(\theta)] = 1.$$

Therefore under this measurability assumption, y is dominated by 0 on Γ *if and only if* (2.2) *is satisfied and y is dominated on* Γ.

PROOF.   For part (i) assume that the right side of the inequality is $< \infty$ (in which case y is dominated on Γ; otherwise (2.15) is trivial). Referring to (2.12), for any q such that $E\sup_\Gamma y_q(\theta) < \infty$, $\{Y_{n,q}\}$ is a reverse martingale, $n = 1, 2, \cdots$ and

$$(2.17) \qquad Y_{n,q} \to E\sup_\Gamma y_q(\theta) \text{ a.s.} \qquad \text{as } n \to \infty,$$

so $\limsup_{n \to \infty} \sup_\Gamma y_n(\theta) \leq E\sup_\Gamma y_q(\theta)$ a.s. Letting $q \to \infty$ we obtain (2.15).

For part (ii) we use (2.12) and (2.17) to apply a well-known extension of the Fatou-Lebesgue theorem ([10], p. 162), obtaining

$$(2.18) \qquad \downarrow \lim_{n \to \infty} E\sup_\Gamma y_n(\theta) \leq E[\limsup_{n \to \infty} \sup_\Gamma y_n(\theta)].$$

Combining this with (2.15) yields (2.16). *Q.E.D.*

REMARK. By the Hewitt-Savage zero-one law under the measurability assumption of Theorem 2.2 (ii), $\lim \sup_{n \to \infty} \sup_\Gamma y_n(\theta)$ is a constant a.s. (also lim inf), whether or not dominance holds.

Returning to the problem of strong consistency, we state the global

DOMINANCE AND MEASURABILITY ASSUMPTION $\mathscr{D}$. *For every P_0 in $\mathscr{P}$ and $r \geq 1$, $u(x, \theta; \theta_0)$ is dominated on Ω_r and $\sup_{\Omega_r} u_n(\theta; \theta_0)$ is measurable for almost all n.*

Condition 1. For every P_0 and $r \geq 1$, $u(x, \theta; \theta_0)$ is dominated by 0 on Ω_r. Then Theorem 2.2 (ii) implies: *if $\mathscr{D}$ holds then Condition 1 is necessary and sufficient for (1.7), and is thus sufficient for the strong consistency of all AMWE.*

The integer k needed to verify dominance of $u(x, \theta; \theta_0)$ may be ≥ 2, especially in multiparameter cases, as pointed out by Kiefer and Wolfowitz ([8], p. 904), Huber, and others. For example, in the context of Example 1, let $\theta = (\mu, \sigma)$ and consider AMLE of (μ, σ) in the location and scale family of densities $\sigma^{-1} f(\sigma^{-1}(x - \mu))$, f specified. If $f(0) > 0$ then $E \sup_{\Omega_r} u_1(\theta) = +\infty$ for r sufficiently large so if $u(x, \theta; \theta_0)$ is dominated, k must be ≥ 2. For example in the normal case, that is, $f(z) = (2\pi)^{-1/2} \exp \{-\frac{1}{2} z^2\}$, $E \sup_\Theta u_2(\theta) < \infty$.

Theorem 2.2 (ii) and Lemma 1.1 also imply that *a sufficient condition for the strong consistency of all AMWE is*

Condition 2. For every P_0 in $\mathscr{P}$ and $r \geq 1$, $u(x, \theta; \theta_0)$ is semidominated by 0 on Ω_r.

The need for considering semidominance is illustrated by

EXAMPLE 2 (*continued*). Consistent estimation of the population median (see Huber [7], p. 223). Assuming without loss of generality that $\theta_0 = 0$, we have $u(x, \theta) = |x| - |x - \theta|$. With

$$(2.19) \qquad \Omega_r = (-\infty, -r^{-1}] \cup [r^{-1}, \infty),$$

it can be shown that $u(x, \theta)$ is not dominated on Ω_r: for all r and n

$$(2.20) \quad \sup_{\Omega_r} u_n(\theta) \geq h_r(x_1, \cdots, x_n) \equiv \begin{cases} n^{-1} \min |x_i| & \text{if } x_i \geq r^{-1}, i = 1, \cdots, n, \\ -r^{-1} & \text{otherwise}. \end{cases}$$

Let $X_1, X_2, \cdots$ be i.i.d. real random variables, each distributed on $(-\infty, \infty)$ symmetrically about 0 and each $|X_i|$ having a cumulative distribution function (c.d.f.)

$$(2.21) \qquad F(x) = \begin{cases} \dfrac{\log (1 + x)}{1 + \log (1 + x)} & \text{if } x \geq 0, \\ 0 & \text{if } x < 0. \end{cases}$$

Since sgn X_i is independent of $|X_i|$,

$$(2.22) \quad E\{\min |X_i| \, | \, X_i \geq r^{-1}, i = 1, \cdots, n\}$$
$$= E\{\min |X_i| \, | \, |X_i| \geq r^{-1}, i = 1, \cdots, n\}$$
$$= E\{\min |X_i| \, | \, \min |X_i| \geq r^{-1}\}$$
$$= +\infty,$$

so $Eh_r(X_1, \cdots, X_n) = +\infty$. Thus $u(x, \theta)$ is not dominated on Ω_r. Setting $b(\theta) = |\theta|$ (or $|\theta - \theta_0|$ if $\theta_0 \neq 0$), however, Huber's arguments can be used to show that Condition 2 holds (see also our discussion preceding Theorem 2.3 below). (Huber takes $b(\theta) = |\theta| + 1$ but this is not necessary since $|\theta|$ is bounded away from 0 on Ω_r.) Thus all AMWE are strongly consistent even though dominance fails in this example.

It was stated above that the conditions of [1], [4], [8], [11] in fact imply Condition 1 and the conditions of Huber [7] imply Condition 2. These authors essentially assume that a compactification $\bar{\Theta}$ of Θ and an extension of $u(x, \theta)$ to $\bar{\Theta}$ exist where every $\theta' \neq \theta_0$ in $\bar{\Theta}$ possesses a sufficiently small neighborhood $V(\theta')$ such that $u(x, \theta; \theta_0)$ is dominated (or semidominated) by 0 on $V(\theta')$. Then using the compactness of $\bar{\Omega}_r$ and a by now well known argument (see Theorem 2.4 below) they deduce that (1.7) holds. The essential idea is, of course, to note that (1.7) holds with Ω_r replaced by $V(\theta')$, obtain a finite subcover $V(\theta_1), \cdots, V(\theta_h)$ of Ω_r, and conclude that (1.7) holds for Ω_r. Part (ii) of Theorem 2.3 isolates the key feature of the above method and shows that the conditions of the papers listed above do in fact imply Conditions 1 or 2.

THEOREM 2.3.   (i) *If y is dominated on Γ and $\Gamma' \subset \Gamma$, then y is dominated on Γ'. This remains true if "dominated" is replaced by "semidominated," "dominated by 0," or "semidominated by 0."*

(ii) *If y is dominated on Γ_i, $i = 1, \cdots, h$, then y is dominated on $\cup \Gamma_i$. This remains true if "dominated" is replaced as above.*

PROOF.   Part (i) is trivial. We prove part (ii) for $h = 2$ with "dominated" replaced by "dominated by 0," as this is the more difficult case. Using part (i) if necessary, we may assume that Γ_1 and Γ_2 are disjoint. Let $\Gamma = \Gamma_1 \cup \Gamma_2$ and let k_i and $s_i(x_1, \cdots, x_{k_i})$ be as in Definition 1, $i = 1, 2$. Let $m = k_1 k_2$,

$$(2.23) \qquad S(i) = \frac{1}{k(i)} \sum_{j=0}^{k(i)-1} s_i(x_{jk_i+1}, \cdots, x_{(j+1)k_i})$$

where $k(1) = k_2$, $k(2) = k_1$, and

$$(2.24) \qquad y^*(\theta) = \begin{cases} S(1) & \text{if } \theta \in \Gamma_1 \\ S(2) & \text{if } \theta \in \Gamma_2. \end{cases}$$

Thus $y_m(\theta) \leqq y^*(\theta)$ and $y_{mn}(\theta) \leqq y_n^*(\theta)$. Since $ES(i) = Es_i < 0$,

$$(2.25) \qquad \sup_{\Gamma} y_n^*(\theta) = \max\left(S_n(1), S_n(2)\right) \overset{\text{a.s.}}{\to} \max\left(Es_1, Es_2\right) < 0.$$

Applying Theorem 2.2 (ii) with y replaced by y^*, this implies that $E \sup_{\Gamma} y_p^*(\theta) < 0$ for some p. Setting $k = mp$ and $s(x_1, \cdots, x_k) = \sup_{\Gamma} y_p^*(\theta)$, we conclude that y is dominated by 0 on Γ. Note that we cannot set $k = m$ and $s = \sup_{\Gamma} y^*(\theta) = \max\left(S(1), S(2)\right)$, since $ES(i) < 0$ need not imply that $E \max\left(S(1), S(2)\right) < 0$. Q.E.D.

The preceding result is now used to show that, under local regularity assumptions only, (2.10) holds if Γ is compact. In Section 3 this is extended to σ-compact sets by imposing a uniformity assumption. We say $y(x, \theta)$ is *locally dominated* on Γ if every θ' in Γ possesses a neighborhood on which $y(x, \theta)$ is dominated. Recall that Γ is a first countable Hausdorff space.

THEOREM 2.4. *Let Γ be a compact subspace of Θ such that $y(x, \theta)$ is locally dominated on Γ. For each θ' in Γ suppose that $y(x, \cdot)$ is upper semicontinuous at θ' except for x in a P-null set possibly depending on θ' and that for some decreasing sequence $\{G_m\} = \{G_m(\theta')\}$ of subsets of Θ forming a base for the neighborhood system at θ', $\sup_{G_m} y_n(\theta)$ is measurable for almost all m and almost all n. Then (2.11) holds, implying (2.10). Also $Ey_1(\theta)$ is an upper semicontinuous function of θ.*

PROOF. Local dominance and compactness imply y is dominated on Γ by Theorem 2.3. Let k be an integer such that $E\sup_\Gamma y_k(\theta) < \infty$. Since $\cap\, G_m = \{\theta'\}$, upper semicontinuity of $y_k(\theta)$ implies $\sup_{G_m} y_k(\theta) \downarrow y_k(\theta')$ as $m \to \infty$ for almost all $x_1, \cdots, x_k$. Then by Lemmas 2.2 and 2.4 and the monotone convergence theorem, for each θ' in Γ.

$$(2.26) \qquad E \sup_{G_m} y_k(\theta) \downarrow Ey_1(\theta') \qquad \text{as } m \to \infty.$$

Thus, given $\delta > 0$, there is an integer $\mu = \mu(\theta')$ such that

$$(2.27) \qquad E \sup_{G_\mu} y_k(\theta) \leqq \sup_\Gamma Ey_1(\theta) + \delta.$$

There is a finite subset $\{\theta'_1, \cdots, \theta'_h\}$ of Γ such that $F_1, \cdots, F_h$ covers Γ, where $F_i = G_{\mu(\theta_i)}$. Let

$$(2.28) \qquad y^*(x, \theta) = y(x, \theta) - \sup_\Gamma Ey_1(\theta) - 2\delta,$$

so y^* is dominated by 0 on each F_i. Then Theorem 2.3 implies that y^* is dominated by 0 on Γ. Applying Theorem 2.1 to y^* and then letting $\delta \to 0$ yields (2.11) and hence (2.10). Finally, since

$$(2.29) \qquad Ey_1(\theta') \leqq \sup_{G_m} Ey_1(\theta) \leqq E \sup_{G_m} y_k(\theta),$$

(2.26) implies that $Ey_1(\theta')$ is upper semicontinuous at each θ'. Q.E.D.

The measurability assumption in Theorem 2.4 is satisfied, for example, if Γ is a separable space, each G_m is open, and $y(x, \cdot)$ is lower semicontinuous on Γ for almost all x. (For common spaces Γ, other criteria for measurability may be more useful, such as right continuity if θ is a real-valued parameter.) In particular, if Γ is separable and $\{y(x, \cdot)\}$ is an equicontinuous family of functions on Γ (except for x in a P-null set) then all the assumptions of Theorem 2.4 are satisfied.

We conclude this section with several remarks concerning the determination of a normalizing function $b(\theta)$ such that $y(x, \theta)$ is semidominated by 0. If $y(x, \theta)$ is not itself dominated by 0 on Γ no general conditions guaranteeing the existence of such a function $b(\theta)$ are known to the author and if such a function does exist no general formula for $b(\theta)$ is known. *Necessary* conditions for the existence of $b(\theta)$ are that $\sup_\Gamma E y_1(\theta) < 0$ and $b(\theta) \leq \delta |E y_1(\theta)|$ for some $\delta > 0$ (by (2.2) and Lemma 2.3). This suggests choosing $b(\theta) = |E y_1(\theta)|$ (as in Example 3) or more generally choosing $b(\theta)$ such that $b(\theta)/|E y_1(\theta)|$ is bounded away from 0 and ∞ (as in Example 2, see the continuation of this example in Section 3). This "rule of thumb" seems to be satisfactory in most statistical applications but, at the level of generality of this paper, it is not universally valid. To see this we present an example where y is not dominated by 0 but is semidominated by 0, and where $b(\theta)$ cannot be $|E y_1(\theta)|$. Let $\Gamma = [1, \infty)$ and let W be a random variable assuming the values 2 and -2 with probabilities 1/4 and 3/4 respectively. Let U be a stochastic process with parameter space Γ, U independent of W, such that for each θ, $U(\theta)$ is uniformly distributed on the interval $(-2\theta^2, -\theta)$ and $\{U(\theta): \theta \text{ in } \Gamma\}$ is a set of mutually independent random variables. Let $V(\theta) = \theta W$ and let $X = \{X(\theta)\} \equiv \{U(\theta) + V(\theta)\}$. Let $X_1, X_2, \cdots$ be a sequence of i.i.d. stochastic processes, each having the same distribution as X. Let $\mathscr{X}$ be the set of all real-valued functions on Γ, and for $(x, \theta) \in \mathscr{X} \times \Gamma$ set $y(x, \theta) = x(\theta)$. Then $P[\sup_\Gamma y_k(\theta) = \infty] > 0$ for every k, so y is not dominated. Setting $b(\theta) = \theta$ we have $y(X, \theta)/\theta = W + U(\theta)/\theta$ and $\sup_\Gamma [y_1(\theta)/\theta] \leq W - 1$, so y is semidominated by 0. However, $2|E y_1(\theta)| = 3\theta + \theta^2$ and for every k, we have $\sup_\Gamma [y_k(\theta)/(3\theta + \theta^2)] \geq 0$ with probability 1.

This example shows that the choice of $b(\theta)$, if it exists, is very delicate: if $b(\theta)$ is too small, $y(x, \theta)/b(\theta)$ may not even be dominated, while if $b(\theta)$ is too large $y(x, \theta)/b(\theta)$ may be dominated but not dominated by 0. Also, there are situations where no such $b(\theta)$ exists: let $\Gamma = [1, \infty)$ and let X be a stochastic process with parameter space Γ such that $\{X(\theta): \theta \in \Gamma\}$ are mutually independent and each $X(\theta)$ is uniformly distributed on $(-1, -1/\theta)$. With $y(x, \theta)$ as defined in the preceding paragraph, $y(x, \theta)$ is dominated on Γ but not semidominated by 0 on Γ. Note that if $b(\theta) = 1/\theta$, $y(x, \theta)/b(\theta)$ is dominated by 0 but inf $b(\theta) = 0$. If the example is changed slightly so that $X(\theta)$ is uniformly distributed on $(-1, 0)$ for each θ then there is no function $b(\theta) > 0$ such that $y(x, \theta)/b(\theta)$ is dominated by 0, even if we do not require that inf $b(\theta) > 0$.

It should be clear by now that due to the attempt to achieve wide generality Conditions 1 and 2, although extending (perhaps "consolidating" is a better term) earlier conditions based on compactification of Θ, do not eliminate the usefulness of these conditions in actually verifying strong consistency. As seen by Example 2, direct determination of a suitable integer k to verify that $u(x, \theta; \theta_0)$ is dominated by 0 or semidominated by 0 on Ω_r may not be feasible. Also, no general method is known for determining suitable normalizing functions $b(\theta)$. Conditions 1 and 2 and earlier conditions do share a common drawback, however: these conditions all require global dominance or semidominance over the possibly noncompact set Ω_r which, it is felt, is not a natural restriction. We now amplify these remarks and introduce a new method based on a global uniformity assumption which requires neither dominance on Ω_r nor compactification.

3. A new condition for strong consistency of AMWE based on uniformity

Throughout this section and the next it is assumed that $E_0 u_1(\theta; \theta_0)$ is well defined (possibly infinite) for every θ in Θ and every P_0 in $\mathscr{P}$, where E_0 denotes expectation under P_0, and therefore that $Ey_1(\theta)$ is well defined for every θ in Γ. By Lemma 2.3 a *necessary condition for* (1.7), *therefore weaker than Conditions 1 or 2, is*

Condition 3. For every P_0 in $\mathscr{P}$ and $r \geq 1$

$$(3.1) \qquad {}_*P_0[\sup_{\Omega_r} u_n(\theta; \theta_0) < \infty \text{ a.a. } n] = 1$$

and

$$(3.2) \qquad \sup_{\Omega_r} E_0 u_1(\theta; \theta_0) < 0.$$

(Notice that $\sup E_0 u_1$ is obviously easier to compute than $E_0 \sup u_n$.)

This condition has a natural interpretation. In Example 1, for instance, $u_1(\theta; \theta_0)$ is the log likelihood ratio and (3.2) is simply an identifiability condition stating that the topology of the parametrization is suited to the underlying probabilistic model. More precisely it states that if the density $f(x, \theta)$ converges to $f(x, \theta_0)$ in terms of Kullback-Leibler distance (information), so that the associated distributions converge, then θ must converge to θ_0. Clearly this is a minimal assumption which must be imposed.

An approach to the problem of strong consistency which seems natural, therefore, is to replace the assumption of dominance or semidominance of $u(x, \theta; \theta_0)$ on Ω_r by an assumption which implies that

$$(3.3) \qquad {}_*P_0[\limsup_{n \to \infty} \sup_{\Omega_r} u_n(\theta; \theta_0) = \sup_{\Omega_r} E_0 u_1(\theta; \theta_0)] = 1$$

for every P_0 and r. (The uniformity assumption is stated after Theorem 3.1.) That is, we are now concerned with verifying (2.10) rather than (2.16). Recall that Theorem 2.4 gave such conditions for a compact set Γ. A useful extension is based on the following result concerning equality of iterated limits, the proof of which is straightforward. When we say a limit exists we allow it to assume an infinite value.

LEMMA 3.1. *Let* $\{\beta(n, m)\}$ *be a double sequence of extended real numbers such that* $\beta(n, m)$ *is increasing in* m. *Suppose that the limit* $\beta(n, \infty) \equiv \uparrow \lim_{m \to \infty} \beta(n, m)$ *satisfies* $-\infty \leq \beta(n, \infty) < \infty$ *for a.a.* n, *that the limit* $\beta(\infty, m) \equiv \lim_{n \to \infty} \beta(n, m)$ *exists for all* m, *and that* $-\infty \leq \uparrow \lim_{m \to \infty} \beta(\infty, m) < \infty$. *Then*

$$(3.4) \qquad \lim_{m \to \infty} \beta(\infty, m) = \lim_{n \to \infty} \beta(n, \infty)$$

if and only if

$$(3.5) \qquad \beta(n, m) \to \beta(\infty, m)$$

uniformly in m,

and both (3.4) *and* (3.5) *are implied by*

$$(3.6) \qquad\qquad \beta(n, m) \uparrow \beta(n, \infty)$$

uniformly in n *(a.a. n).*

If $-\infty < \lim_{m \to \infty} \beta(\infty, m) < \infty$ *then* (3.4), (3.5), *and* (3.6) *are mutually equivalent.*

THEOREM 3.1. *Suppose that* $\Gamma = \cup_{m=1}^{\infty} \Gamma_m$ *where* $\Gamma_m \subset \Gamma_{m+1}$ *for all m. Assume that*

$$(3.7) \qquad\qquad -\infty \leqq \sup_{\Gamma} Ey_1(\theta) < \infty,$$

$$(3.8) \qquad\qquad {}_*P\left[\sup_{\Gamma} y_n(\theta) < \infty \text{ a.a. } n\right] = 1,$$

and

$$(3.9) \qquad\qquad {}_*P\left[\sup_{\Gamma_m} y_n(\theta) \to \sup_{\Gamma_m} Ey_1(\theta)\right] = 1 \qquad\qquad \text{for all } m.$$

Then ${}_*P[\sup_{\Gamma} y_n(\theta) \to \sup_{\Gamma} Ey_1\theta] = 1$, *if and only if the convergence in* (3.9) *is uniform in m, that is,*

$$(3.10) \qquad\quad {}_*P\left[\sup_{\Gamma_m} y_n(\theta) \to \sup_{\Gamma_m} Ey_1(\theta) \text{ uniformly in } m\right] = 1,$$

and both (2.10) *and* (3.10) *are implied by*

$$(3.11) \qquad\quad {}_*P\left[\sup_{\Gamma_m} y_n(\theta) \uparrow \sup_{\Gamma} y_n(\theta) \text{ uniformly in } n \text{ (a.a. } n)\right] = 1.$$

If $-\infty < \sup_{\Gamma} Ey_1(\theta) < \infty$, *then* (2.10), (3.10), *and* (3.11) *are mutually equivalent.*

PROOF. Let $\beta(n, m) = \sup_{\Gamma_m} y_n(\theta)$ and apply Lemma 3.1. *Q.E.D.*

Using this theorem we can extend Theorem 2.4 to σ-compact sets.

THEOREM 3.2. *Let Γ be a σ-compact space, that is, $\Gamma = \cup_{m=1}^{\infty} \Gamma_m$ where each Γ_m is compact. Suppose that for each θ' in Γ the local dominance, upper semicontinuity, and measurability assumptions of Theorem 2.4 are satisfied, and that the conditions of* (3.7) *and* (3.8) *hold. Then all the conclusions of Theorem 3.1 hold, that is,* (3.11) $\Rightarrow$ (3.10) $\Leftrightarrow$ (2.10), *and these are mutually equivalent if $\sup_{\Gamma} Ey_1(\theta)$ is finite. Furthermore $Ey_1(\theta)$ is upper semicontinuous on Γ.*

PROOF. We can assume that $\Gamma_m \subset \Gamma_{m+1}$. Theorem 2.4 implies (3.9) since Γ_m is compact, so all hypotheses of Theorem 3.1 are satisfied. Also $Ey_1(\theta)$ is upper semicontinuous on each Γ_m, hence on Γ. *Q.E.D.*

REMARK. Suppose the hypotheses of Theorem 3.2 hold and that $\Gamma = \cup_{m=1}^{\infty} \Gamma_m'$ is another representation of Γ as an increasing union of compact sets, so that (3.9) holds for $\{\Gamma_m'\}$ as well as $\{\Gamma_m\}$. Since (2.10) does not depend on the decomposition of Γ, (3.10) holds for $\{\Gamma_m'\}$ if and only if it holds for $\{\Gamma_m\}$.

We can now state the global

UNIFORMITY ASSUMPTION $\mathscr{U}$. *For each P_0 in $\mathscr{P}$ and $r \geqq 1$, $\Omega_r = \Omega_r(\theta_0)$ can be expressed as $\bigcup_{m=1}^{\infty} \Omega_{r,m}$ where $\Omega_{r,m} \subset \Omega_{r,m+1}$ and*

$$(3.12) \qquad {}_*P_0\Big[\sup_{\Omega_{r,m}} y_n(\theta) \to \sup_{\Omega_{r,m}} E_0 u_1(\theta; \theta_0) \text{ uniformly in } m\Big] = 1.$$

Then Theorem 3.1 implies: *if $\mathscr{U}$ holds then Condition 3 is necessary and sufficient for (1.7) and is therefore sufficient for the strong consistency of all AMWE.*

In some cases it may be easier to verify a condition based on (3.11), rather than (3.12) which is based on (3.10). Notice also that the crucial aspect of (3.12) is the *uniformity* of the convergence: if $\Omega_{r,m}$ is compact the convergence itself is guaranteed by local regularity assumptions only, as in Theorem 3.2.

It is felt that this represents a more directly applicable approach to the problem of strong consistency than that contained in the statement following Condition 1. This is illustrated by Example 2 where it was shown earlier that dominance $\mathscr{D}$ fails but where we now show that $\mathscr{U}$ and Condition 3 hold.

EXAMPLE 2 (*continued*). Fix P_0 and r and assume $\theta_0 = 0$. With $\Gamma = \Omega_r$ as defined in Section 2 let

$$(3.13) \qquad \Gamma_m = \Omega_{r,m} = [-m, -r^{-1}] \cup [r^{-1}, m],$$

a compact set. Since $u_n(\theta) \leqq n^{-1} \Sigma_i |X_i|$, (3.1) is obviously true. Since $u(x, \theta)$ is continuous in θ (in fact equicontinuous) and $u_1(\theta) \leqq m$ on Γ_m, the hypotheses of Theorem 2.4 are satisfied so (3.8) holds. Thus to verify $\mathscr{U}$ we must only verify the uniformity of the convergence in (3.12).

Note that $u_n(\theta)$ is a unimodal function with mode at $X[(n + 1)/2]$ (the $(n + 1)/2$th order statistic from a sample of size n) if n is odd, and mode "plateau" on the interval $(X[n/2], X[(n/2) + 1])$ if n is even. Thus, if we choose δ such that

$$(3.14) \qquad P_0[-\delta \leqq X_1 \leqq \delta] > \tfrac{1}{2}$$

it follows that

$$(3.15) \qquad P_0[-\delta \leqq \text{mode (plateau) of } u_n(\theta) \leqq \delta \text{ a.a. } n] = 1,$$

which implies the uniformity in (3.11) or (3.12). It remains only to verify (3.2). If $\theta \geqq 0$ ($\theta \leqq 0$ is similar),

$$(3.16) \qquad E_0 u_1(\theta) = 2\Big\{\theta P_0[X_1 \geqq \theta] + \int_{0+}^{\theta-} x\, dP_0(x)\Big\} - \theta$$

$$\leqq \theta\{2P_0[X_1 > 0] - 1\}$$

$$\leqq 0.$$

The first inequality is strict if $P_0[0 < X_1 < \theta] > 0$ and the second is strict if $P_0[X_1 > 0] < 1/2$. Thus $P_0[X_1 > 0] < 1/2$ implies $\sup_{\Omega_r} E_0 u_1(\theta) < 0$. If $P_0[X_1 > 0] = 1/2$ then for all $\theta > 0$, $P_0[0 < X_1 < \theta] > 0$ since otherwise the population median would not be unique. This again implies $\sup_{\Omega_r} E_0 u_1(\theta) < 0$ since this expectation decreases monotonically in $|\theta|$, so (3.2) holds. Incidentally, $|E u_1(\theta)| \leqq |\theta|$ since $|u_1(\theta)| \leqq |\theta|$, which shows that if $b(\theta) = |\theta|$, then $b(\theta)/|E u_1(\theta)|$ is bounded away from 0 (and ∞, as seen in Section 2), thus verifying an earlier remark.

Of course we saw earlier that Condition 2, based on semidominance, and Huber's conditions, based on compactification, each apply in this example. The present method has the advantage, however, that we did not have to search for normalizing functions $b(\theta)$, integers k, or suitable compactifications, which seems to be a great advantage in general. Even when dominance and Condition 1 apply the present method retains these advantages, as illustrated by the next example.

EXAMPLE 4.   In the context of Example 1, let $\Theta = [0, \infty)$, $\mu = $ Lebesgue measure, and

$$(3.17) \qquad f(x, \theta) = (2\pi)^{-1}\left[1 + \frac{a\theta}{1+\theta}\sin{(x-\theta)}\right] \qquad \text{if } 0 \leqq x \leqq 2\pi,$$

$f(x, \theta) = 0$ otherwise, where $0 < a < 1$ is a constant. We wish to show that all approximate maximum likelihood estimators are strongly consistent, although neither Wald's, Kiefer and Wolfowitz's nor Huber's conditions are satisfied here. Fix θ_0 and r, and let

$$(3.18) \qquad\qquad \Omega_r = [0, \theta_0 - r^{-1}] \cup [\theta_0 + r^{-1}, \infty).$$

Since $u(x, \theta) = \log f(x, \theta) - \log f(x, \theta_0) \leqq \log 2(1 + \theta_0)$, (3.1) is satisfied, and in fact, u is dominated on Ω_r. Setting $\Gamma = \Omega_r$ and $\Gamma_m = \Omega_{r,m} = \Gamma \cap [0, m]$, Γ_m is compact, u is dominated on Γ_m, and u is continuous so Theorem 2.4 implies (3.9). To verify (3.12) note that $\log t$ is uniformly continuous on $[1 - a, 1 + a]$, that for any x_i

$$(3.19) \qquad \left|\left[1 + \frac{a\theta}{1+\theta}\sin{(x_i - \theta)}\right] - [1 + a\sin{(x_i - \theta)}]\right| \leqq \frac{1}{1+\theta},$$

and that both terms in square brackets lie in $[1 - a, 1 + a]$. Therefore, given any $\delta > 0$ there exists $M > 0$ such that $\theta \geqq M$ implies

$$(3.20) \qquad \left|\frac{1}{n}\log\prod_{i=1}^{n}\left[1 + \frac{a\theta}{1+\theta}\sin{(x_i - \theta)}\right] - \frac{1}{n}\log\prod_{i=1}^{n}[1 + a\sin{(x_i - \theta)}]\right| \leqq \delta$$

independently of n and $x_1, x_2, \cdots$. Thus for θ sufficiently large, $u_n(\theta)$ can be approximated arbitrarily closely by a periodic function, uniformly in n and $x_1, x_2, \cdots$, which implies (3.12). Finally (3.2) follows from the information inequality and an easy limiting argument as $\theta \to \infty$, so Condition 3 is satisfied and all the AMLE are strongly consistent. Note that since dominance $\mathscr{D}$ holds this implies Condition 1 holds, but this would have been difficult to demonstrate directly.

In Example 4, conditions (a), (b), (c) of Bahadur ([1], p. 320) are satisfied, but this is not immediately evident. The difficulty arises when trying to find a suitable compactification, for the obvious one-point compactification $[0, \infty]$ is not adequate. We must adjoin to $\Theta = [0, \infty)$ an entire interval of length 2π, say, $I = [-2\pi, 0)$. Any θ in Θ can be uniquely represented as $\theta = 2\pi m + r$ where m is an integer and $-2\pi \leq r < 0$. Then the topology in $\bar{\Theta} = \Theta \cup I$ must be defined so that if $\{\theta_n\} \subset \Theta$ and $\bar{\theta} \in I$, $\theta_n \to \bar{\theta}$ if and only if $m_n \to \infty$ and $r_n \to \bar{\theta}$.

However, it has been pointed out to the author by Professor Bahadur that if in this example (or more generally in Example 1) we redefine $\theta(P) \equiv P$ and $\Theta \equiv \mathscr{P}$ and consider the topology of weak convergence in $\mathscr{P}$, then there is a natural compactification of $\mathscr{P}$, namely the closure of $\mathscr{P}$ in the set of all measures on $(\mathscr{X}, \mathscr{A})$ with total mass ≤ 1. This is in fact the natural parameterization and topology to consider in Example 1 since we are interested primarily in estimating the underlying probability distribution. Using this parameterization Bahadur's conditions (a), (b), (c) are easily verified in Example 4.

A similar situation occurs in the context of Example 1, where $\theta = (\mu, \sigma)$ and $f(x, \theta)$ is the density of the normal distribution $N(\mu, \sigma^2)$ discussed earlier. Here it is again difficult to find a "suitable compactification" of the space $\Theta = $ the open half plane $\{\sigma > 0\}$. If, however, the natural parameterization $\theta(P) = P$ and the natural compactification described above are considered, Bahadur's conditions (a) and (c) are readily verified. Condition (b) fails, but as pointed out earlier it can be replaced by the assumption of dominance with $k = 2$.

4. Necessary and sufficient conditions for strong consistency of AMWE

We now add some mild local regularity assumptions to the underlying assumptions introduced in Section 1 and show that in this case our sufficient conditions become necessary as well. Recall that Θ is a first countable Hausdorff space.

LOCAL REGULARITY ASSUMPTIONS $\mathscr{L}$.

(a) Θ *is locally compact, so we can choose* $\{V_r(\theta_0)\}$ *to be a* compact *base for the neighborhood system at each* θ_0;

(b) *for every* P_0 *in* $\mathscr{P}$, $u(x, \theta; \theta_0)$ *is locally dominated on* Θ *with respect to* P_0;

(c) *for every* P_0 *in* $\mathscr{P}$ *and* θ' *in* Θ, $u(x, \theta; \theta_0)$ *is upper semicontinuous at* θ' *except for* x *in a* P_0-*null set possibly depending on* θ';

(d) *for each* θ' *in* Θ *there exists a decreasing sequence of subsets* $\{G_m\} = \{G_m(\theta')\}$ *forming a base for the neighborhood system at* θ' *such that* $\sup_{G_m} u_n(\theta; \theta_0)$ *is measurable for a.a.* m *and a.a.* n;

(e) *for every* P_0 *in* $\mathscr{P}$, $\sup_{V_r} E_0 u_1(\theta; \theta_0) = 0$ *for a.a.* r.

Assumptions (b), (c), and (d) already appeared in Theorem 2.4. Assumption (e) is a very weak local identifiability condition. Note that $\sup_{V_r} E_0 u_1 \geq 0$ in any case since $u_1(\theta_0; \theta_0) \equiv 0$. In the case of maximum likelihood estimation (Example 1) (e) is always satisfied because of the information inequality.

THEOREM 4.1.   *If assumptions $\mathscr{L}$ hold, then* (1.7) *is necessary as well as sufficient for the strong consistency of all* AMWE.

PROOF.   From Theorem 2.4 it follows that if $\mathscr{L}$ holds then for every P_0 and r

$$(4.1) \qquad {}_*P_0[\sup_{V_r} u_n(\theta;\theta_0) \to 0] = 1$$

so that $\sup_{V_r} w_n(\theta)$ is finite for almost all n with inner probability one and

$$(4.2) \qquad {}_*P_0\left[\limsup_{n\to\infty} \sup_{\Omega_r} u_n(\theta;\theta_0) < 0\right]$$

$$= {}_*P_0\left[\limsup_{n\to\infty} \{\sup_{\Omega_r} u_n(\theta;\theta_0) - \sup_{V_r} u_n(\theta;\theta_0)\} < 0\right]$$

$$= {}_*P_0\left[\limsup_{n\to\infty} \{\sup_{\Omega_r} w_n(\theta) - \sup_{V_r} w_n(\theta)\} < 0\right].$$

Suppose that (1.7) is not satisfied. Then for some $\hat{P}_0$ and $\hat{r}$ the first term in (4.2), hence the last term, is < 1. Let $\hat{\theta}_0 = \theta(\hat{P}_0)$ and $\Omega_{\hat{r}} = \Omega_{\hat{r}}(\hat{\theta}_0)$. Now for every sample sequence $(x_1, x_2, \cdots)$ such that

$$(4.3) \qquad \lambda \equiv \limsup_{n\to\infty} \{\sup_{\Omega_{\hat{r}}} w_n(\theta) - \sup_{V_{\hat{r}}} w_n(\theta)\} \geqq 0$$

there exists a sequence of integers $n_i \to \infty$ such that

$$(4.4) \qquad \sup_{\Omega_{\hat{r}}} w_{n_i}(\theta) - \sup_{V_{\hat{r}}} w_{n_i}(\theta) \to \lambda \geqq 0 \qquad\qquad \text{as } i \to \infty$$

so there exists a sequence of points $\{\theta_{n_i}\} \subset \Omega_{\hat{r}}$ such that (see (1.4)) $H(\sup_\Theta w_{n_i}(\theta), w_{n_i}(\theta_{n_i})) \to 0$ as $i \to \infty$. Now choose an AMWE sequence $\{T_n(x_1, \cdots, x_n)\}$ in such a way that $T_{n_i}(x_1, \cdots, x_{n_i}) = \theta_{n_i}$ for all sequences $(x_1, x_2, \cdots)$ for which $\lambda \geqq 0$ (such a choice is always possible since we are not concerned with measurability of T_n). Then $\{T_n\}$ is not strongly consistent under $\hat{P}_0$ since

$$(4.5) \qquad {}_*\hat{P}_0[T_n \text{ in } V_{\hat{r}} \text{ a.a. } n] \leqq {}_*\hat{P}_0[\lambda < 0] < 1.$$

Q.E.D.

This result therefore implies: *if assumptions $\mathscr{D}$ and $\mathscr{L}$ hold then Condition 1 is necessary and sufficient for strong consistency of all* AMWE, *and if assumptions $\mathscr{U}$ and $\mathscr{L}$ hold then Condition 3 is necessary and sufficient for strong consistency of all* AMWE.

These statements can be strengthened if we add the natural global

IDENTIFIABILITY ASSUMPTION $\mathscr{I}$.   *For every P_0 in $\mathscr{P}$, $E_0 u_1(\theta;\theta_0) < 0$ if $\theta \neq \theta_0$.*

Clearly this is weaker than (3.2) and is therefore a necessary condition for strong consistency of all AMWE. In the context of Example 1 it simply states that if $\theta \neq \theta_0$, then the two underlying distributions must be distinct. We now show that the necessary and sufficient condition (1.7) can be weakened to: for every P_0 there exists an $r_0 = r(P_0)$ such that

$$(1.7') \qquad {}_*P_0[\limsup_{n\to\infty} \sup_{\Omega_{r_0}} u_n(\theta;\theta_0) < 0] = 1.$$

THEOREM 4.2. *If assumptions $\mathscr{L}$ are satisfied then* (1.7) *holds if and only if* (1.7′) *and $\mathscr{I}$ hold.*

PROOF.   Assume $\mathscr{I}$ and (1.7′). For any P_0 and r choose an open neighborhood U of θ_0 such that $U \subseteq V_r \cap V_{r_0}$ and let $W = V_{r_0} \sim U$. Then

$$(4.6) \qquad \sup_{\Omega_r} u_n(\theta; \theta_0) \leqq \max \left[\sup_{\Omega_{r_0}} u_n(\theta; \theta_0), \sup_W u_n(\theta; \theta_0) \right].$$

Now W is compact so by Theorem 2.4

$$(4.7) \qquad {}_*P_0 \left[\sup_W u_n(\theta; \theta_0) \to \sup_W E_0 u_1(\theta; \theta_0) \right] = 1.$$

However $E_0 u_1(\theta; \theta_0)$ is upper semicontinuous and thus achieves its maximum over W so $\sup_W E_0 u_1(\theta; \theta_0) < 0$ $\big($by $\mathscr{I}$ which implies (1.4)$\big)$. The converse is obvious. Q.E.D.

Therefore if assumptions $\mathscr{L}$ and $\mathscr{I}$ hold, the phrase "for every P_0 in $\mathscr{P}$ and every $r \geqq 1$" can be replaced by the phrase "for every P_0 in $\mathscr{P}$ there exists an $r_0 = r(P_0)$ such that" in assumptions $\mathscr{D}$, $\mathscr{U}$, Conditions 1, 2, 3, and (1.7), everywhere they appear in this paper.

Lastly, notice that if $\mathscr{L}$ holds and in addition Θ has a compact countable base for its topology (another global assumption) so Θ is second countable, then any open subset of Θ is σ-compact, in particular each Ω_r. Using the ideas of Theorem 3.2 this enables us to weaken slightly the uniformity assumption $\mathscr{U}$: the *convergence* in (3.12) is satisfied, so only the *uniformity* must be verified.

5. Relation to LeCam's condition for strong consistency

In Section 2 it was shown that the conditions of [1], [4], [7], [8], and [11], based on compactification of Θ, all imply either Condition 1 or 2. The conditions of LeCam ([9], pp. 302–304) are not based on compactification but rather on a form of dominance by Bochner-integrable random variables. We generalize LeCam's conditions slightly by introducing the following definitions. (The notation is that introduced in Section 2.)

DEFINITION 3.   *Let $B(\Gamma)$ denote the Banach space of all bounded real-valued functions on Γ with the usual sup norm. We say $y(x, \theta)$ is Bochner-dominated on Γ with respect to P if there is a positive integer j and a function $v(\theta) \equiv v(x_1, \cdots, x_j, \theta)$ mapping $\mathscr{X} \times \cdots \times \mathscr{X}$ into $B(\Gamma)$ such that*

(i) *v is a strongly measurable mapping (with respect to the product σ-field $\mathscr{A} \times \cdots \times \mathscr{A}$),*

(ii) *$\|v\| \equiv \sup_\Gamma |v(\theta)|$ is integrable,*

(iii) *for all $(x_1, \cdots, x_j)$ in a set of probability one, $y_j(\theta) \leqq v(\theta)$ uniformly for θ in Γ.*

We say $y(x, \theta)$ is Bochner-dominated by 0 on Γ if in addition

(iv) *$\sup_\Gamma E v(\theta) < 0$.*

Note that (i) and (ii) together are equivalent to Bochner-integrability of v. (For a definition of the terms used here see [6], also [5] and [9].) Also note that v

can be assumed to be a symmetric function of $x_1, \cdots, x_j$ (see the remark after Definition 1). Further, notice that (ii) implies $-\infty < \sup_\Gamma Ev(\theta)$.

DEFINITION 4. $y(x, \theta)$ is semi-Bochner-dominated (semi-Bochner-dominated by 0) on Γ with respect to P if there exists a function $b(\theta)$ defined on Γ, $0 < b(\theta) < \infty$, such that $y(x, \theta)/b(\theta)$ is Bochner-dominated (Bochner-dominated by 0) on Γ and $\inf_\Gamma b(\theta) > 0$.

LeCam's conditions for strong consistency of AMWE, in generalized form, are obtained from our Conditions 1 and 2 by replacing "dominated" by "Bochner-dominated" throughout. The following result shows that these conditions are in fact equivalent:

THEOREM 5.1. $y(x, \theta)$ is Bochner-dominated on Γ if and only if it is dominated on Γ. Similarly, Bochner dominance by 0 $\Leftrightarrow$ dominance by 0, semi-Bochner dominance $\Leftrightarrow$ semidominance, and semi-Bochner dominance by 0 $\Leftrightarrow$ semidominance by 0.

PROOF. We prove the second equivalence only; the other proofs are similar. If y is dominated by 0 on Γ, let $j = k$ and $v(x_1, \cdots, x_j, \theta) = s(x_1, \cdots, x_k)$. Clearly v is a strongly measurable mapping into $B(\Gamma)$. Since s may be chosen such that $-\infty < Es < 0$, (ii), (iii), and (iv) are satisfied, so y is .Bochner-dominated by 0. Next suppose that y is Bochner-dominated by 0 on Γ. Letting

$$(5.1) \qquad v_n(\theta) = \frac{1}{n} \sum_{i=0}^{n-1} v(x_{i, j+1}, \cdots, x_{(i+1), j}, \theta)$$

it follows from the Strong Law of Large Numbers (SLLN) for Bochner-integrable random variables taking values in a Banach space S that

$$(5.2) \qquad P[\sup_\Gamma |v_n(\theta) - E_V(\theta)| \to 0] = 1.$$

(See Beck [2] or Hanš [5]). In stating the SLLN they assume that the Banach space S is separable. However, even if this is not the case—for example, if $S = B(\Gamma)$—strong measurability implies that v is almost separably-valued ([6], p. 72) so the range of v lies in a separable closed linear subspace of S.) Therefore

$$(5.3) \qquad P[\sup_\Gamma v_n(\theta) \to \sup_\Gamma E_V(\theta) < 0] = 1.$$

(By Criterion 4 of Hanš [5], strong measurability implies that $v_n(\theta)$ is Borel measurable so $\sup_\Gamma v_n(\theta)$, being a continuous function of $v_n(\theta)$, is also Borel measurable, that is, a random variable.) However, we apply Theorem 2.2 (ii) with $y_n(\theta)$ replaced by $v_n(\theta)$ to see that there is an integer $m \geq 1$ such that $E \sup_\Gamma v_m(\theta) < 0$. Thus Definition 1 is satisfied if we take $k = mj$ and $s(x_1, \cdots, x_k) = \sup_\Gamma v_m(\theta)$.

I wish to thank Professors R. R. Bahadur, Robert Berk, and Lucien LeCam for helpful suggestions at various stages of this work.

REFERENCES

[1] R. BAHADUR, "Rates of convergence of estimates and test statistics," *Ann. Math. Statist.*, Vol. 38 (1967), pp. 303–324.

[2] A. BECK, "On the strong law of large numbers," *Ergodic Theory* (edited by F. B. Wright), New York, Academic Press, 1963, pp. 21–53.

[3] R. H. BERK, "Limiting behavior of posterior distributions when the model is incorrect," *Ann. Math. Statist.*, Vol. 37 (1966), pp. 51–58.

[4] G. B. CRAWFORD, "Consistency of non-parametric maximum likelihood estimates," 1958, unpublished.

[5] O. HANŠ, "Generalized random variables," *Conference on Information Theory, Statistical Decision Functions, Random Processes*, Prague, Publishing House of the Czechoslovak Academy of Science, 1956, pp. 61–104.

[6] E. HILLE and R. S. PHILLIPS, *Functional Analysis and Semi-Groups*, American Mathematical Society Colloquium Publication, Vol. 31, New York, American Mathematical Society, 1957 (revised ed.).

[7] P. J. HUBER, "The behavior of maximum likelihood estimates under nonstandard conditions," *Proceedings of the Fifth Berkeley Symposium on Mathematical Statistics and Probability*, Berkeley and Los Angeles, University of California Press, 1967, Vol. 1, pp. 221–233.

[8] J. KIEFER and J. WOLFOWITZ, "Consistency of the maximum likelihood estimator in the presence of infinitely many incidental parameters," *Ann. Math. Statist.*, Vol. 27 (1956), pp. 884–906.

[9] L. LeCAM, "On some asymptotic properties of maximum likelihood estimates and related Bayes estimates," *Univ. California Publ. Statistics*, Vol. 1 (1953), pp. 277–328.

[10] M. LOÈVE, *Probability Theory*, Princeton, Van Nostrand, 1963 (3rd ed.).

[11] A. WALD, "Note on the consistency of the maximum likelihood estimate," *Ann. Math. Statist.*, Vol. 20 (1949), pp. 595–601.

EFFICIENCY ROBUSTNESS OF ESTIMATORS

PAUL SWITZER
STANFORD UNIVERSITY

1. Introduction

This introduction sets out some general available procedures to get large sample efficient estimates of a location parameter when the governing distribution f is well specified. The next section reviews some attempts to relax the f specification. The third section discusses how one chooses from a given repertoire of competing estimators. It is there advocated that their respective estimated standard errors be used to govern the choice and two methods are presented for estimating standard errors nonparametrically for this purpose. Some Monte Carlo comparisons are presented in Section 4 using sample sizes of 30, 60, and 120 together with a short tail and long tail f. The possibility of using sample determined weightings of selected estimators is also briefly explored.

Our discussion of efficiency robustness is set in the context of estimating a location parameter θ, say the center of a symmetric distribution on the real line. We use $\hat{\theta}$ to generically denote a translation invariant estimator of θ. If f is the density function of the distribution and is sufficiently well specified, then there are various general methods available to obtain large sample efficient estimators. Some of these are:

(A) the maximum likelihood estimator, that is, the value of θ which maximizes $L(\theta) = \Pi_1^N f(X_i - \theta)$ where $X_1, X_2, \cdots, X_N$ are a sample of size N from f. For example, if f is Laplace then $\hat{\theta}$ is the sample median; if f is normal then $\hat{\theta}$ is the sample mean; if f is logistic then the MLE is not easily obtained.

(B) the Pitman estimator, namely, $\hat{\theta} = \int \theta L(\theta) d\theta / \int L(\theta) d\theta$, where $L(\theta)$ is the sample likelihood as above. For example, if f is normal then $\hat{\theta}$ is the sample mean; if f is uniform on an interval of fixed length then $\hat{\theta}$ is the midrange.

(C) the midpoint of a symmetric confidence interval for θ based on the locally most powerful rank test for the specified translation family f. The confidence probability is a fixed α, and the LMPRT depends on f through $J(u) = f'(F^{-1}(u))/f(F^{-1}(u))$ for $0 < u < 1$. For example, if f is Laplace then $\hat{\theta}$ is the average of a symmetric pair of the ordered X values; if f is logistic then $\hat{\theta}$ is the average of a symmetric pair of the ordered Walsh averages. Walsh averages are averages of pairs of X values. With $\alpha \to 0$ this is the method of Hodges and Lehmann [5].

(D) a specified weighted average of the ordered X values, namely, $\hat{\theta} =$

283

$\Sigma_1^N w(i/N + 1)X_{(i)}$ where $w(u)$ is proportional to $J'(u) \int_0^u J(v) \, dv$ and $J(u)$ is as given in (C) above (see [3]). For example, if f is Laplace then $w(u)$ is zero except at $u = \frac{1}{2}$ and $\hat{\theta}$ is therefore the sample median; if f is logistic then $w(u) = 6u(1 - u)$.

Generally speaking, these four estimation methods require different regularity conditions on f to safeguard their large sample efficiency. However, the conditions overlap considerably and will not be a concern here. In small samples, they can be regarded as competitors when they do not coincide.

Now consider this embarrassing situation. I have advised a client to use a particular $\hat{\theta}$ based on his specification of the shape of the density function f from which his sample was to be drawn. This $\hat{\theta}$ was justified by showing it was derived by one of the four methods above. He draws his sample and dutifully computes $\hat{\theta}$ and an estimate of its standard error $s(\hat{\theta})$. Not completely trusting my advice, he also computes his pet estimate $\hat{\theta}'$ and notes that its estimated standard error $s(\hat{\theta}')$ is much smaller than $s(\hat{\theta})$. He then sues me for malpractice.

The apparent poor showing of my recommended asymptotically efficient $\hat{\theta}$ could be ascribed to several sources of which three are: (i) the sampling variability in $s(\hat{\theta})$ and $s(\hat{\theta})$ could make the event $s(\hat{\theta}') \gg s(\hat{\theta})$ not particularly surprising even when the actual standard error of $\hat{\theta}$ is the smaller; (ii) for the sample size at hand, say $N = 100$, the asymptotically efficient $\hat{\theta}$ is indeed inferior to $\hat{\theta}'$ for the given f family; (iii) the f family was inappropriately specified and $\hat{\theta}$ is not even an asymptotically efficient estimator. It is to the third of these that our attention will now be turned.

2. Relaxed model specifications as an approach to robustness

For some time now, and especially in recent years, there has been a search for estimating procedures which retain high efficiency simultaneously for widely differing specifications f. Commonly used estimators like the mean or the median certainly would not qualify, the former having low efficiency for long tail f and the latter for short tail f, generally speaking. Instead, a variety of uncommon procedures have been advocated, some of which are reviewed below, not necessarily in historical order.

The approaches taken by Gastwirth [3] and Hogg [6] can be illustrated with the following example. Suppose f is allowed to be either a normal or a Laplace density. Using approach (D), say, we can find the functions J_G and J_L, respectively, which yield efficient estimators for each of the two families. Gastwirth's idea is to look for something intermediate between J_G and J_L which will have high relative efficiency regardless of whether f is normal or Laplace (or some convex combination of the two). This idea can be extended to three or more f families, to include Cauchy and logistic f, say. If this is done, then Gastwirth suggests that $\hat{\theta}$ be a weighted average of the $33\frac{1}{3}$rd, 50th and $66\frac{2}{3}$rd percentiles of the sample, using weights 0.3, 0.4, 0.3, respectively. The relative efficiency of this $\hat{\theta}$ never falls below 80 per cent for any f belonging to one of the four mentioned families.

Hogg, on the other hand, feels that with samples of moderate size it should be possible to distinguish whether f is normal or Laplace from the sample itself. He suggests using the fourth sample moment to discriminate between the two families. Then, use either the J_G or J_L based estimator, whichever is indicated by the fourth moment. This approach, which can also be extended to three or more f families, yields estimators which have full asymptotic efficiency for any f belonging to one of the admitted families.

Full efficiency within the context of a few well-specified f families can also be achieved, in a straightforward manner, by a somewhat extended use of the maximum likelihood method. Once again if f is either normal or Laplace, then the estimator

$$(1) \qquad \hat{\theta} = \begin{cases} \text{mean}, & \text{if } \left(\sum |X_i - \text{median}|\right)^2 > \frac{n\pi}{2e}\left(\sum |X_i - \text{mean}|^2\right), \\ \text{median}, & \text{otherwise}, \end{cases}$$

is the MLE for θ when both families are entertained together and the scale parameter is unknown. But note, should the real f fall outside this pair of families, neither formula (1) nor Hogg's method will necessarily indicate whether the mean or median has higher relative efficiency.

A quite different approach is taken by Huber [7]. There it is assumed that a tightly specified f governs the bulk of the sample data, while a loosely specified f' governs the remainder. This is Tukey's contamination model, in particular when f is taken to be normal and f' has longer tails than f. The robustness of any estimator $\hat{\theta}$ was gauged by Huber to be its maximum asymptotic variance as the contaminating f' ranged over the class $\mathscr{F}$ of its admitted possibilities.

For f normal (scale unknown) and f' any symmetric density centered at θ, the most robust $\hat{\theta}$ is found by minimizing $\sum_1^N \rho(X_i - \theta)$ where $\rho(t) = \max\left(\frac{1}{2}t^2, \lambda t - \frac{1}{2}t^2\right)$ and λ is related to the contamination proportion ε. Huber has shown that taking $\lambda = 2.0$ will do quite well for any ε less than 20 per cent. Tukey, some time ago [11], suggested using either trimmed or Winsorized means, both of which are close to Huber's estimator. The trimmed mean ignores a specified number of extreme observations in computing a sample average, whereas the Winsorized mean pulls in the extreme observations in a specified way.

In 1955, Stein [10] suggested that large sample uniformly efficient estimating procedures could be concocted without specifying very much at all about f. Hájek [4] and van Eeden [12] carried through on this suggestion. Specifically, one can estimate the whole J function of (C) so that $\int_0^1 [\hat{J}(u) - J(u)]^2\, du \to 0$ in probability, with only mild restrictions on f. This enables one to act as though J were known and thereby, using (C) or (D), say, to construct a uniformly fully efficient estimator $\hat{\theta}$. It is not hard to believe that the J function will not be well estimated generally unless the sample size is huge. In the meantime, such estimators remain to be tried on samples of moderate size, and the more modest objectives of the preceding paragraphs still command attention.

3. Comparison of procedures as an approach to robustness

It is possible to make a good case *against* the use of a particular procedure $\hat{\theta}$. Specifically, if a competing procedure $\hat{\theta}'$ is much better for some f and never much worse for any f, then there is little point in using $\hat{\theta}$. For example, Hodges and Lehmann [5] showed that the median of the Walsh averages (H-L estimator) can be infinitely more efficient than the sample mean, if f has infinite variance. On the other hand, the sample mean is never more than 125/108 times as efficient as the H-L estimator for any continuous symmetric f. This pretty much rules out the sample mean. Indeed, there exist estimators which are never less efficient than $\bar{X}$ in large samples, for example, use procedure (C) with the normal scores test.

In a similar vein, Bickel [1] compared the Hodges-Lehmann estimator with both trimmed means and Winsorized means. He found that the H-L estimator retains its *relative* robustness against these two challengers as well.

It would seem from the preceding considerations that one might confidently use the H-L estimator in preference to the competitors entertained in Bickel's paper. Two concerns remain. First, we may regret somewhat the efficiency lost by not using the simple mean, say, in those few situations when it is actually somewhat better than H-L. Second, the H-L estimator could itself have near zero efficiency relative to a third estimator, even when it is distinctly better than the mean.

The proposal of this paper is that the sample itself should be used to distinguish which one of several competing estimators is most efficient, for the unknown f from which the sample was drawn. To be able to use the sample in this way requires that the competing θ estimators be such that their standard errors can also be estimated without making use of the unknown shape f. In spirit, the proposal is unlike those of the previous section where the object was to pin down f. Rather, it addresses itself to the malpractice suit of Section 1 by standing opposed to arguments which fix upon a procedure based exclusively on *a priori* considerations. However, such considerations could be profitably used to set up the collection of competing estimators; for example, the collection should contain only estimators whose efficiency relative to one another ranges widely from very small to very large numbers as f ranges over a set of reasonable possibilities.

Specifically, let $\hat{\theta}_1$, $\hat{\theta}_2$, $\hat{\theta}_3$ be three sequences of competing estimators defined for every sample size N. The dependence on N is suppressed in the notation. For example, these might be the H-L estimator, the median, and the midrange, respectively, or they might be three trimmed means with different trimming proportions. Let S_1, S_2, S_3 be sequences of nonparametric estimators of the standard errors of $\hat{\theta}_1, \hat{\theta}_2, \hat{\theta}_3$ (more will be said about the S in a moment). Then the recommended estimator of the location parameter θ is

$$(2) \qquad\qquad \hat{\theta} = \sum_1^3 I_i \hat{\theta}_i$$

where

$$(3) \qquad I_i = \begin{cases} 1 & \text{if} \quad \min(S_1, S_2, S_3) = S_i, \\ 0 & \text{otherwise.} \end{cases}$$

That is, $\hat{\theta}$ is equivalent to one of $\hat{\theta}_1$, $\hat{\theta}_2$, $\hat{\theta}_3$ depending on which of the three has the smallest estimated standard error.

Now, for convenience, assume that the $\hat{\theta}_i$ were chosen so that $\sqrt{N}(\hat{\theta}_i - \theta)$ has a limiting normal distribution with zero mean, for $i = 1, 2, 3$. If $\sigma_i^2(f)$ denotes the variance of each of the three limiting distributions, then the large sample efficiency of $\hat{\theta}_i$ relative to $\hat{\theta}_j$ is $\sigma_j^2(f)/\sigma_i^2(f) = e_{i,j}(f)$, say. Hence, if $e_{i,j}(f)$ is greater than one for some f, then $\hat{\theta}_i$ is relatively more efficient than $\hat{\theta}_j$ for that f family. It is now further assumed that the standard error estimates were chosen so that NS_i^2 consistently estimates $\sigma_i^2(f)$, for each i and any f belonging to a *large* class $\mathcal{F}$; it is in this sense that we referred to the S as nonparametric. For example, if $\hat{\theta}_1$ is the sample mean, then NS_1^2 is the sample variance and has this property for any f having a finite variance. Some general methods for obtaining nonparametric S will be given shortly.

Let $\hat{\theta}(f)$ denote the most efficient of the three competiors for a given f. That is, if $e_{1,2}(f) > 1$ and $e_{1,3}(f) > 1$, then $\hat{\theta}(f) = \hat{\theta}_1$ for that f. With the assumptions of the preceding paragraph, it follows that $\sqrt{N}(\hat{\theta} - \theta)$ has the same limiting distribution as $\sqrt{N}(\hat{\theta}(f - \theta)$, for every $f \in \mathcal{F}$. This makes the proposed estimator $\hat{\theta}$ of (2) as efficient in large samples as $\hat{\theta}(f)$. So $\hat{\theta}$ is never less efficient than any of the initial competitors $\hat{\theta}_1$, $\hat{\theta}_2$, $\hat{\theta}_3$, (be they H-L, van Eeden, or whatever), and it is always more efficient than some two of them. Should $e_{1,2}(f) \cong 1$ and $e_{1,3}(f) > 1$, then $\hat{\theta}$ will bounce between $\hat{\theta}_1$ and $\hat{\theta}_2$, but this is of no concern since both are equally good and better than $\hat{\theta}_3$.

We have thus established a desirable large sample property for $\hat{\theta}$, namely, we do as well as if we had known which of $\hat{\theta}_1$, $\hat{\theta}_2$, $\hat{\theta}_3$ was best to begin with. However, its desirability hinges on how large we can make $\mathcal{F}$ and how well it does in samples of moderate size. For if $\mathcal{F}$ is too restricted then our $\hat{\theta}$ will show little advantage over the estimators of Gastwirth and Hogg. On the other hand, even when $\mathcal{F}$ is quite large, if enormous sample sizes are needed, then we may as well try van Eeden's uniformly efficient estimator. Recall that $\hat{\theta}$ of (2) is *not* uniformly fully efficient in $\mathcal{F}$; it is merely as efficient in $\mathcal{F}$ as the best of a preselected small collection of trial estimators, and in small samples it is necessarily *less* efficient than the best of the trial estimators.

As we remarked earlier, the choice of initial trial competitors $\hat{\theta}_i$ will be necessarily influenced by the ability to get nonparametric estimates of their standard errors. Here are two general procedures:

(E) Suppose a trial estimator $\hat{\theta}_i$ is constructed by the method (C), that is, by taking the midpoint of a level α nonparametric confidence interval for θ. Under quite general conditions on f, $\sqrt{N}(\hat{\theta}_i - \theta)$ will have an asymptotic normal distribution with asymptotic variance consistently estimated by a constant times the squared length of the confidence interval. This constant depends on the chosen α and N, but not on f (see Sen [9]). For example, when $\hat{\theta}_i$ is based on the Wilcoxon test, then the length of the confidence interval has the required consistency property for any f such that $\int f^2$ is finite.

(F) Assume the sample can be divided into K blocks of equal size $n = N/K$. In each of these blocks compute your favorite $\hat{\theta}_i$ estimates based on samples of size n; denote these by $\hat{\theta}_i^k$, $k = 1, 2, \cdots, K$ and $i = 1, 2, 3$ say. Then take the overall $\hat{\theta}_i$ to be the average of the block estimates, that is, $\hat{\theta}_i = \Sigma_1^k \hat{\theta}_i^k / K$. Its standard error is estimated nonparametrically by taking the sample variance of the $\hat{\theta}_i^k$, that is $S_i^2 = \Sigma_1^k (\hat{\theta}_i^k - \hat{\theta}_i)^2 / K(K - 1)$. Such S_i are consistent in the required sense provided only that the individual $\hat{\theta}_i^k$ has finite variance and that $K \to \infty$ with N. Estimators based on dividing up the data have been proposed by Box [2].

The second of the above two general procedures should be preferred for its simplicity. There will be some loss in efficiency due to the partitioning of the data, but this is usually slight. If both n and K are allowed to grow with N, and if $\sqrt{N}(\hat{\theta}_i^k - \theta)$ are themselves asymptotically normal, then there is no efficiency loss in large samples by dividing up the data. This second procedure also resembles the jack-knife method of estimation which might well be used in its place because it, too, produces an estimate of the standard error. See Miller [8] and the references therein contained for some caveats on the use of the jack-knife.

4. Some small sample Monte Carlo calculations

Let the total sample size N be divisible by six for purposes of the succeeding illustration. Divide the data into $K = N/6$ equal groups (at random). In each group k we compute the three midranges, namely,

$$
\begin{aligned}
\hat{\theta}_i^k &= \tfrac{1}{2}[X_{(3)} + X_{(4)}], \\
\hat{\theta}_2^k &= \tfrac{1}{2}[X_{(2)} + X_{(5)}], \\
\hat{\theta}_3^k &= \tfrac{1}{2}[X_{(1)} + X_{(6)}], \qquad\qquad k = 1, 2, \cdots, K.
\end{aligned}
$$

(4)

The three competing trial estimators $\hat{\theta}_1$, $\hat{\theta}_2$, $\hat{\theta}_3$ are the respective averages of these midranges, as outlined in (F). The nonparametric estimates of their respective standard errors S_i are also as given in (F) and are proportional to $\Sigma_1^k (\hat{\theta}_i^k - \hat{\theta}_i)^2$, $i = 1, 2, 3$.

Samples of size $N = 30, 60$, and 120 were drawn from a short tail distribution (uniform), a normal distribution and a long tail distribution (contaminated normal). The contaminated normal was a 90 per cent to 10 per cent mixture of a standard normal and three times a standard normal, respectively. Each sample was replicated 800 times. Table I shows the resulting Monte Carlo estimates of the variances of $\hat{\theta}_1$, $\hat{\theta}_2$ and $\hat{\theta}_3$ for each of the three sample sizes and each of three parent distribution models. The table entries have been scaled for ease of presentation, so horizontal comparisons should not be made. The standard errors of the table entries themselves run about 5 per cent.

TABLE I

SCALED VARIANCES OF VARIOUS ESTIMATORS
These variances were each computed from 800 independent
Monte Carlo replications of each sample.

		Uniform		Normal		Contaminated Normal	
$N = 30$	$\hat{\theta}_1$	1.61	(7%)	1.16	(35%)	1.39	(45%)
	$\hat{\theta}_2$	1.05	(18%)	1.15	(34%)	1.55	(41%)
	$\hat{\theta}_3$	0.54	(75%)	1.46	(31%)	4.48	(14%)
	$\hat{\theta}$	0.74		1.27		1.52	
	$\hat{\theta}^*$	1.47		2.56		2.78	
$N = 60$	$\hat{\theta}_1$	1.62	(1%)	1.11	(34%)	1.36	(48%)
	$\hat{\theta}_2$	1.09	(6%)	1.15	(39%)	1.55	(46%)
	$\hat{\theta}_3$	0.58	(93%)	1.52	(27%)	4.34	(6%)
	$\hat{\theta}$	0.68		1.23		1.45	
	$\hat{\theta}^*$	0.82		1.28		1.69	
$N = 120$	$\hat{\theta}_1$	1.56	(0%)	1.23	(33%)	1.52	(51%)
	$\hat{\theta}_2$	1.07	(1%)	1.14	(44%)	1.55	(48%)
	$\hat{\theta}_3$	0.58	(99%)	1.46	(23%)	4.20	(1%)
	$\hat{\theta}$	0.58		1.22		1.54	
	$\hat{\theta}^*$	0.72		1.11		1.47	

The results are as expected. For example, $\hat{\theta}_3$ is strongly favored by the short tail distribution and strongly disfavored by the long tail distribution. The suggested procedure (2) was also applied to these three competing estimators using the S_i as computed from (F) for each sample. The variance of the resulting estimator $\hat{\theta}$ is shown in the table. It is necessarily larger than the variance of the best of $\hat{\theta}_1, \hat{\theta}_2, \hat{\theta}_3$, but for the sample sizes used here it performs nearly as well for each of the three f families. The percentage figures in parentheses indicate the relative frequencies with which each of the three competitors were used when they were thrown into procedure (2).

5. Miscellanea

In using the recommended estimator $\hat{\theta}$ of (2), several reasonable questions arise. What if this estimator, itself, was used as one of its component competitors? To use it in this way one would need a nonparametric estimator of *its* standard error, say S. But whatever value we give to S should not be less than min (S_1, S_2, S_3), because $\hat{\theta}$ is either equal to the best of the three competitors or something worse. It follows that the inclusion of $\hat{\theta}$, itself, does not affect the procedure (2).

The general problem of getting an estimate S of the standard error of $\hat{\theta}$ is not taken up here in detail. In a sample of moderate size, $S = \min (S_1, S_2, S_3)$

will be slightly, but optimistically, biased. Corrections could be obtained by using the approximate joint normality of the competing $\hat{\theta}_i$; this is likely to lead to complicated calculations involving the covariances, which must then also be estimated nonparametrically.

If data groupings as in (F) are used, then we can indeed estimate nonparametrically the covariance between $\hat{\theta}_i$ and $\hat{\theta}_j$ by using the sample covariance,

$$(5) \qquad S_{i,j} = \sum_{k=1}^{K} (\hat{\theta}_i^k - \hat{\theta}_i)(\hat{\theta}_j^k - \hat{\theta}_j)/K(K-1).$$

Having the $S_{i,j}$ in hand might also suggest that we can improve on the recommendation of (2) by using an "almost optimally weighted" linear combination of the initial competitors $\hat{\theta}_1$, $\hat{\theta}_2$, $\hat{\theta}_3$, rather than using weights which are always zero or one as in (2). Specifically, take

$$(6) \qquad \hat{\theta}^* = \sum \hat{w}_i \hat{\theta}_i, \qquad\qquad \sum w_i = 1,$$

where the weights $\hat{w}_i$, possibly negative, are estimates of the optimal weights w_i we would use if we know the covariances precisely. The w_i satisfy the following linear equations (using T initial competitors)

$$(7) \qquad \sum_{j=1}^{T-1} \text{Cov}\,(\hat{\theta}_T - \hat{\theta}_i,\, \hat{\theta}_T - \hat{\theta}_j)w_j = \text{Cov}\,(\hat{\theta}_T - \hat{\theta}_i,\, \hat{\theta}_T),$$

$$i = 1, 2, \cdots, T-1, \qquad w_T = 1 - \sum_{j=1}^{T-1} w_j.$$

These equations are symmetric in $w_1, w_2, \cdots, w_T$. Changing the w to $\hat{w}$, they can be solved by using the covariances estimated from the sample. Provided the $\hat{\theta}_i$ are linearly independent, the resulting $\hat{w}_i$ will consistently estimate the w_i, and therefore $\sqrt{N}(\hat{\theta}^* - \theta)$ will have the same asymptotic normal distribution as if we had known the w_i. In general, the asymptotic variance of $\sqrt{N}(\hat{\theta}^* - \theta)$ will be strictly less than that for any initial competitor regardless of the density f governing the data, unless the optimal weights really are zero-one.

However, one might expect the small sample behavior of $\hat{w}_i$ to be erratic causing $\hat{\theta}^*$ to be actually inferior even to the $\hat{\theta}$ of (2) which necessarily uses zero-one weights. This phenomenon is clearly demonstrated by the Monte Carlo results of Table I, particularly for $N = 30$ where only five data groups were available to estimate the three variances and three covariances. In the case of uniform f, where 0-0-1 weighting is actually optimal, $\hat{\theta}$ continues to best $\hat{\theta}^*$ even in moderately large samples. For the normal and contaminated normal f, $\hat{\theta}^*$ is about as good as $\hat{\theta}$ when $N = 60$ and is possibly somewhat superior when $N = 120$. A few unreported Monte Carlo calculations at $N = 240$ continue to show that $\hat{\theta}^*$ is slightly but noticeably better than $\hat{\theta}$.

Herman Rubin suggested in a private communication that it would be reasonable to let the number of trial competitors used with procedure (2) depend on the sample size N. For purposes of illustration we have worked here with three competitors, but Rubin's suggestion appears quite reasonable. It is in the spirit of van Eeden's proposal [12] and merits further study.

In conclusion it should be recalled that this discussion of efficiency robustness was carried through in the context of a specific one sample problem of estimating the center of a symmetric distribution. Our main interest has been to characterize various approaches to the robustness question rather than to provide specific suggestions to specific problems. The two sample shift problem, for example, could have been treated in an almost identical manner where the parameter θ is the amount of the shift.

Immediately prior to the presentation of this paper in July, 1970, the author became aware of a highly relevant unpublished thesis by L. Jaeckel dated December 1969, Statistics Department, Berkeley. It contains essentially among other things, the proposal made in our formula (2). To this extent, and to the extent of any other overlap with this paper, priority belongs to Dr. Jaeckel.

REFERENCES

[1] P. J. BICKEL, "On some robust estimates of location," *Ann. Math. Statist.*, Vol. 36 (1965), pp. 847–858.

[2] G. E. P. BOX, "Non-normality and tests on variances," *Biometrika*, Vol. 40 (1953), pp. 318–335.

[3] J. L. GASTWIRTH, "On robust procedures," *J. Amer. Statist. Assoc.*, Vol. 61 (1966), pp. 929–948.

[4] J. HÁJEK, "Asymptotically most powerful rank-order tests," *Ann. Math. Statist.*, Vol. 33 (1962), pp. 1124–1147.

[5] J. L. HODGES JR. and E. L. LEHMANN, "Estimates of location based on rank tests," *Ann. Math. Statist.*, Vol. 24 (1963), pp. 598–611.

[6] R. V. HOGG, "Some observations on robust estimation," *J. Amer. Statist. Assoc.*, Vol. 62 (1967), pp. 1179–1186.

[7] P. J. HUBER, "Robust estimation of a location parameter," *Ann. Math. Statist.*, Vol. 35 (1964), pp. 73–102.

[8] R. G. MILLER, JR., "Jackknifing variances," *Ann. Math. Statist.*, Vol. 39 (1968), pp. 567–582.

[9] P. K. SEN, "On a distribution-free method of estimating asymptotically efficiency of a class of non-parametric tests," *Ann. Math. Statist.*, Vol. 37 (1966), pp. 1759–1770.

[10] C. STEIN, "Efficient nonparametric testing and estimation," *Proceedings of the Third Berkeley Symposium on Mathematical Statistics and Probability*, Berkeley and Los Angeles, University of California Press, 1956, Vol. I, pp. 187–196.

[11] J. W. TUKEY, *Reports 31–34*, Statistics Research Group, Princeton University, 1949.

[12] C. VAN EEDEN, "Efficiency-robust estimation of location," *Ann. Math. Statist.*, Vol. 41 (1970), pp. 172–181.

ISOTONIC TESTS FOR CONVEX ORDERINGS

RICHARD E. BARLOW

and

KJELL A. DOKSUM

UNIVERSITY OF CALIFORNIA, BERKELEY

1. Introduction

The problem of testing the hypothesis that F is a negative exponential distribution with unknown scale parameter against the alternative that F has monotone increasing nonconstant failure rate (F has Increasing Failure Rate, IFR) has been studied by a number of authors, some of whom are Proschan and Pyke [18], Nadler and Eilbott [17], Barlow [1], Bickel and Doksum [7], and Bickel [6]. Bickel and Doksum show that the test proposed by Proschan and Pyke is asymptotically inadmissible. They then take an essentially parametric approach to the problem. In particular they obtain the studentized asymptotically most powerful linear spacings tests for selected parametric families of distributions which are IFR when the parameter $\theta > 0$ and exponential when $\theta = 0$. Bickel [6] proves that these tests are actually asymptotically equivalent to the level α tests which are most powerful among all tests which are similar and level α (for the associated parametric problems).

Since the problem is essentially nonparametric, we take a nonparametric approach similar to the one taken by Chapman [10] and Doksum [12] in studying the problem of testing for goodness of fit to a specified distribution against stochastically ordered alternatives. In addition, we consider a more general class of problems which includes the problem of testing for monotone failure rate. The setup is similar to that in Barlow and van Zwet [2].

Let $\mathscr{F}$ be the class of absolutely continuous distribution functions F such that $F(0) = 0$ with positive and right (or left) continuous density f on the interval where $0 < F < 1$. It follows that the inverse function F^{-1} is uniquely defined on $(0, 1)$. We take $F^{-1}(1)$ to be equal to the right endpoint of the support of F (possibly $+\infty$) and define $F^{-1}(0) = 0$. For $F, G \in \mathscr{F}$ we say that F is c-*ordered* (convex ordered) with respect to $G(F \underset{c}{\leq} G)$ if and only if $G^{-1}F$ is convex on the

This research has been partially supported by the Office of Naval Research under Contract N00014–69–A–0200–1036 with the University of California at Berkeley. Reproduction in whole or in part is permitted for any purpose of the United States Government.

Professor Barlow's research was completed while the author was at Stanford University and partially supported there by NSF-GP17172 at Stanford.

interval where $0 < F < 1$ (van Zwet [21]). Denoting the densities of F and G by f and g, we find that $F \underset{c}{\leq} G$ implies that

$$(1.1) \qquad r(x) = \frac{d}{dx} G^{-1} F(x) = \frac{f(x)}{g[G^{-1} F(x)]}$$

is nondecreasing in x on the interval where $0 < F < 1$. The problem of estimating $r(x)$ when G is known was considered by Barlow and van Zwet [2], [3]. When $G(x) = 1 - \exp\{-x\}$, it is easy to verify that $r(x) = f(x)/[1 - F(x)]$, the failure rate function of F.

We assume G known, $F \underset{c}{\leq} G$ and consider the problem of testing

$$(1.2) \qquad H_0 : F \underset{c}{=} G$$

(that is, $G^{-1} F$ is linear on the support of F) against the alternative

$$(1.3) \qquad H_1 : F \underset{c}{\leq} G \quad \text{and} \quad F \underset{c}{\neq} G,$$

given a random sample $X = (X_1, X_2, \cdots, X_n)$ from F.

We call (1.2) and (1.3) the problem of testing for c-equivalence versus c-ordering. We study tests based on the "total time on test" statistics for this problem (Section 3). In the cases when G is the uniform or exponential distribution we show that the tests corresponding to the "cumulative total time on test statistics" are asymptotically minimax over a class of alternatives based on the Kolmogorov distance (Sections 6, 7, and 8) and in each of the classes of statistics considered by Bickel and Doksum [7] (in the exponential case).

2. Preliminaries

We can simplify our problem by introducing the following transformation

$$(2.1) \qquad H_F^{-1}(t) = \int_0^{F^{-1}(t)} g[G^{-1} F(u)] \, du, \qquad\qquad 0 \leq t \leq 1.$$

Recall that G is always fixed in this discussion. Note that H is a distribution since H^{-1} (the inverse of H) is strictly increasing on $[0, 1]$. In particular, $H_G^{-1}(t) = t$ so that H_G is the uniform distribution on $[0, 1]$. When it is clear from the context which distribution we are transforming, we will simply write H^{-1} for H_F^{-1}.

By (1.1) $F \underset{c}{\leq} G$ implies $f(x)/g[G^{-1} F(x)]$ is nondecreasing in x for $0 < F(x) < 1$ or $g[G^{-1}(t)]/f[F^{-1}(t)]$ is nonincreasing in t, $0 \leq t \leq 1$. Since

$$(2.2) \qquad \frac{d}{dt} H^{-1}(t) = \frac{g[G^{-1}(t)]}{f[F^{-1}(t)]},$$

it follows that H^{-1} is concave on $[0, 1]$ or H is convex on the interval where $0 < H < 1$ if and only if $F \underset{c}{\leq} G$. Hence, using transformation (2.1) we can reduce our problem (1.2) and (1.3) to that of testing

(2.3) $H_0 : H(x)$ linear for $0 < H(x) < 1$

versus

(2.4) $H_1 : H(x)$ convex and not linear for $0 < H(x) < 1$.

The following result from Barlow and van Zwet [2] will be needed.

LEMMA 2.1.   *If* F, $G \in \mathscr{F}$, *if* gG^{-1} *is uniformly continuous on* $[0, 1)$, *if* $\int_0^\infty x\, dF(x) < \infty$, *and if* $F^{-1}(1) < \infty$, *or* $gG^{-1}(y)/(1 - y)$ *is bounded on* $(0, 1)$, *or* $F \underset{c}{\leqslant} G$, *then* $H^{-1}(1) < \infty$.

If $G(x) = 1 - \exp\{-x\}$ for $x \geqq 0$, then

(2.5) $$H^{-1}(t) = \int_0^{F^{-1}(t)} [1 - F(u)]\, du.$$

In this case $H^{-1}(1) = \int_0^\infty x\, dF(x)$. If $F \underset{c}{\leqslant} G$, then $\int_0^\infty x\, dF(x) < \infty$ is automatically satisfied.

In testing $H_0 : F \underset{c}{=} G$ versus $H_1 : F \underset{c}{\leqslant} G$ we will be interested in tests ϕ that have isotonic power with respect to c-ordering: that is $F_1 \underset{c}{\leqslant} F_2$ implies $\beta_\phi(F_1) \geqq \beta_\phi(F_2)$ where $\beta_\phi(F)$ is the power of the test ϕ when F is the true distribution. One advantage of the transformation H_F^{-1} is that it transforms c-ordering into stochastic ordering

THEOREM 2.1.   *If* $F_1 \underset{c}{\leqslant} F_2$, $F_1, F_2, G \in \mathscr{F}$ *and if Lemma* 2.1 *holds then*

(2.6) $$\frac{H_{F_1}^{-1}(t)}{H_{F_1}^{-1}(1)} \geqq \frac{H_{F_2}^{-1}(t)}{H_{F_2}^{-1}(1)} \qquad\qquad 0 \leqq t \leqq 1.$$

If, in addition $F_2 \underset{c}{\leqslant} G$, *then*

(2.7) $$\frac{H_{F_2}^{-1}(t)}{H_{F_2}^{-1}(1)} \geqq t, \qquad\qquad 0 \leqq t \leqq 1.$$

PROOF.   Note that $F_1 \underset{c}{\leqslant} F_2$ implies

(2.8) $$\frac{f_1(x)}{f_2[F_2^{-1}F_1(x)]} \quad \text{increasing in } x,$$

or

(2.9) $$\frac{f_1[F_1^{-1}(u)]}{f_2[F_2^{-1}(u)]} \quad \text{increasing in } u.$$

Hence,

$$
(2.10) \quad \frac{H_{F_1}^{-1}(t)}{H_{F_1}^{-1}(1)} - \frac{H_{F_2}^{-1}(t)}{H_{F_2}^{-1}(1)} = \int_0^t \left[\frac{1}{H_{F_1}^{-1}(1)} \frac{gG^{-1}(u)}{f_1[F_1^{-1}(u)]} - \frac{1}{H_{F_2}^{-1}(1)} \frac{gG^{-1}(u)}{f_2[F_2^{-1}(u)]} \right] du
$$

$$
= \int_0^t \left[\frac{1}{H_{F_1}^{-1}(1)} \frac{f_2 F_2^{-1}(u)}{f_1 F_1^{-1}(u)} - \frac{1}{H_{F_2}^{-1}(1)} \right] \frac{gG^{-1}(u)}{f_2 F_2^{-1}(u)} du
$$

$$
\overset{\text{def}}{=} \int_0^t h(u) \frac{gG^{-1}(u)}{f_2 F_2^{-1}(u)} du.
$$

Since $\int_0^1 h(u) \left(gG^{-1}(u) \right) / \left(f_2 F_2^{-1}(u) \right) du = 0$ and $h(u)$ changes sign at most once and from positive to negative values if at all, it follows that

$$
(2.11) \qquad \int_0^t h(u) \frac{gG^{-1}(u)}{f_2 F_2^{-1}(u)} du \geqq 0.
$$

The second inequality follows from

$$
(2.12) \qquad\qquad H_G^{-1}(t) = t.
$$

Q.E.D.

Since G is assumed known we can estimate H_F^{-1} by substituting the empirical distribution F_n for F; that is,

$$
(2.13) \qquad H_n^{-1}(t) = H_{F_n}^{-1}(t) \overset{\text{def}}{=} \int_0^{F_n^{-1}(t)} gG^{-1} F_n(u) \, du
$$

and

$$
(2.14) \quad H_n^{-1}\left(\frac{i}{n}\right) = \int_0^{X_{i:n}} gG^{-1} F_n(u) \, du = \sum_{j=1}^{i} gG^{-1}\left(\frac{j-1}{n}\right)(X_{j:n} - X_{j-1:n}),
$$

where $X_{i:n}$ is the ith order statistic in a sample of size n from F and $X_{0:n} \equiv 0$. If $G(x) = 1 - \exp\{-x\}$ for $x \geqq 0$, then

$$
(2.15) \qquad H_n^{-1}\left(\frac{i}{n}\right) = n^{-1} \sum_{j=1}^{i} (n - j + 1)(X_{j:n} - X_{j-1:n}),
$$

that is, n^{-1} times the "total time on test" until the ith ordered observation from F.

The following result was proved in Barlow and van Zwet [2].

THEOREM 2.2. *If $F, G \in \mathscr{F}$ and*

(i) $\int_0^\infty x \, dF(x) < \infty$,

(ii) gG^{-1} *is uniformly continuous on* $[0, 1)$,

(iii) *either* $F^{-1}(1) < \infty$, $gG^{-1}(y)/(1 - y)$ *is bounded on* $(0, 1)$, *or* $F \underset{c}{\leq} G$ *and there exists* η, $0 < \eta < 1$, *such that for* $\eta \leqq y < 1$, $gG^{-1}(y)$ *is nonincreasing and* $gG^{-1}(y)/(1 - y)$ *is nondecreasing in* y, *then for* $n \to \infty$

$$(2.16) \quad \sup_{x \geqq 0} \left| \int_0^x g[G^{-1}F_n(u)] \, du - \int_0^x g[G^{-1}F(u)] \, du \right| \to 0 \quad \text{almost surely.}$$

2.1. *Order statistics from H.* The "total time on test" statistics, $H_n^{-1}(1/n) \leqq H_n^{-1}(2/n) \leqq \cdots \leqq H_n^{-1}((n-1)/n)$, "behave" asymptotically like order statistics from H. To see this let $U_{i:n}$ be the ith order statistic from the uniform distribution on $[0, 1]$. Then

$$(2.17) \quad Z_{i:n} \underset{\text{st}}{=} H^{-1}(U_{i:n}) \overset{\text{def}}{=} \int_0^{F^{-1}(U_{i:n})} gG^{-1}F(u) \, du \underset{\text{st}}{=} \int_0^{X_{i:n}} gG^{-1}F(u) \, du$$

will be distributed as the ith order statistic in a random sample of size n from the distribution H. Since we do not know F, $Z_{i:n}$, $i = 1, 2, \cdots, n$, are unobservable except in the case $G(x) = x$ for $0 \leqq x \leqq 1$. From Theorem 2.2 we see that

$$(2.18) \quad \left| H_n^{-1}\left(\frac{i}{n}\right) - Z_{i:n} \right| \to 0$$

almost surely and uniformly in i/n, $1 \leqq i \leqq n$. This observation suggests various tests for our transformed problem

$$(2.19) \qquad H_0 : H(x) \quad \text{linear for} \quad 0 < H(x) < 1$$

versus

$$(2.20) \qquad H_1 : H(x) \quad \text{convex and not linear for} \quad 0 < H(x) < 1$$

based on the "total time on test" statistics. Since our problem is clearly scale invariant, we consider tests based on the studentized statistics

$$(2.21) \qquad W_{F_n}\left(\frac{i}{n}\right) \overset{\text{def}}{=} W_{i:n} \overset{\text{def}}{=} H_n^{-1}\left(\frac{i}{n}\right) \Big/ H_n^{-1}(1).$$

Distributions which are c-ordered have studentized statistics which are stochastically ordered. This result is the basis for the isotonicity of tests to be considered in this paper.

THEOREM 2.3. *If F, K, $G \in \mathscr{F}$ and $F \underset{c}{\leq} K \underset{c}{\leq} G$, then*

$$(2.22) \qquad W_{F_n}\left(\frac{i}{n}\right) \underset{\text{st}}{\geqq} W_{K_n}\left(\frac{i}{n}\right) \underset{\text{st}}{\geqq} W_{G_n}\left(\frac{i}{n}\right)$$

where $\underset{\text{st}}{\geqq}$ denotes stochastic ordering and F_n, K_n, G_n are empirical distributions corresponding to independent random samples of size n from F, K, and G, respectively.

PROOF. Let $X_{1:n} < X_{2:n} < \cdots < X_{n:n}$ be an ordered sample from F. Let $V_{i:n} = K^{-1}F(X_{i:n})$ and note that

$$(2.23) \qquad \frac{V_{i:n} - V_{i-1:n}}{X_{i:n} - X_{i-1:n}}$$

is nondecreasing in i since $K^{-1}F$ is convex. Hence,

$$(2.24) \qquad gG^{-1}\left(\frac{i-1}{n}\right)(V_{i:n} - V_{i-1:n})/gG^{-1}\left(\frac{i-1}{n}\right)(X_{i:n} - X_{i-1:n}) \overset{\text{def}}{=} \frac{\beta_i}{\alpha_i},$$

is also nondecreasing in i, where

$$\beta_i = gG^{-1}\left(\frac{i-1}{n}\right)(V_{i:n} - V_{i-1:n}),$$

(2.25)

$$\alpha_i = gG^{-1}\left(\frac{i-1}{n}\right)(X_{i:n} - X_{i-1:n}).$$

Define $\psi(0) = 0$, $\psi(\alpha_1 + \cdots + \alpha_i) = \beta_1 + \cdots + \beta_i$, $1 \leq i \leq n$. Define $\psi(x)$ elsewhere on $[0, \alpha_1 + \cdots + \alpha_n]$ by linear interpolation between successive points defined above. Note that

$$(2.26) \qquad \frac{\psi(\alpha_1 + \cdots + \alpha_i) - \psi(\alpha_1 + \cdots + \alpha_{i-1})}{(\alpha_1 + \cdots + \alpha_i) - (\alpha_1 + \cdots + \alpha_{i-1})} = \frac{\beta_i}{\alpha_i}$$

is increasing in i, so that ψ is a convex function on $[0, \alpha_1 + \cdots + \alpha_n]$. Since $\psi(0) = 0$, ψ is also starshaped, that is $\psi(x)/x$ is nondecreasing in x. Hence

$$(2.27) \qquad \frac{\psi[\sum_1^r \alpha_i]}{\sum_1^r \alpha_i} = \frac{\sum_1^r \beta_i}{\sum_1^r \alpha_i}$$

is nondecreasing in r.

Inequalities (2.22) follow by noting that

$$(2.28) \qquad H_{F_n}^{-1}\left(\frac{i}{n}\right) = \sum_{j=1}^{i} gG^{-1}\left(\frac{j-1}{n}\right)(X_{j:n} - X_{j-1:n}) = \sum_{j=1}^{i} \alpha_j$$

and

$$(2.29) \qquad H_{K_n}^{-1}\left(\frac{i}{n}\right) = \sum_{j=1}^{i} gG^{-1}\left(\frac{j-1}{n}\right)(V_{j:n} - V_{j-1:n}) = \sum_{j=1}^{i} \beta_j.$$

Hence,

$$(2.30) \qquad \frac{H_{K_n}^{-1}\left(\frac{i}{n}\right)}{H_{F_n}^{-1}\left(\frac{i}{n}\right)} \leq \frac{H_{K_n}^{-1}(1)}{H_{F_n}^{-1}(1)}$$

implies

$$(2.31) \qquad \frac{H_{F_n}^{-1}\left(\frac{i}{n}\right)}{H_{F_n}^{-1}(1)} \geq \frac{H_{K_n}^{-1}\left(\frac{i}{n}\right)}{H_{K_n}^{-1}(1)}.$$

Stochastic ordering follows by noting that $(V_{1:n}, \cdots, V_{n:n})$ is stochastically equal to an *independent* ordered sample from K. This establishes the first stochastic inequality in (2.22). The second inequality follows similarly. *Q.E.D.*

The above proof is similar to that for Lemma 3.7 (i) Barlow and Proschan [5].

DEFINITION. *A test* ϕ *based on* $X_1, X_2, \cdots, X_n$ *is monotonic if*

$$(2.32) \qquad \phi(X_1, \cdots, X_n) = \begin{cases} 1 & if \quad T(X_1, \cdots, X_n) > c_{n,\alpha} \\ 0 & otherwise, \end{cases}$$

where T is nondecreasing coordinatewise.

DEFINITION. *A test ϕ is* isotonic *with respect to c-ordering if $F_1 \underset{c}{\leq} F_2$ and $X = (X_1, \cdots, X_n)(Y = (Y_1, \cdots, Y_n))$ is a random sample from F_1 (F_2) implies $\phi(X) \underset{st}{\geqq} \phi(Y)$.*

THEOREM 2.4. *Monotonic tests based on $W_{1:n}, W_{2:n}, \cdots, W_{n:n}$ are* isotonic *tests with respect to c-ordering. Isotonic tests of c-ordering have isotonic power with respect to c-ordering; that is, $F_1 \underset{c}{\leq} F_2 \underset{c}{\leq} G$ implies*

$$(2.33) \qquad \beta_\phi(F_1) \geqq \beta_\phi(F_2) \geqq \beta_\phi(G),$$

where $\beta_\phi(F)$ is the power of ϕ when the true distribution is F.

PROOF. This is an immediate consequence of Theorem 2.3. Q.E.D.

3. Tests for convex orderings

Note that $F \in \mathscr{F}$ implies $F^{-1}(0) = 0$ which in turn implies $H^{-1}(0) = 0$. Under the conditions of Theorem 2.2, $H_n^{-1}(1) \to H^{-1}(1)$ almost surely as $n \to \infty$. For the purpose of asymptotic comparison of competing tests we may suppose that $H^{-1}(1) = 1$. This simplifies the discussion somewhat. The problem of testing for c-ordering becomes

$$(3.1) \qquad H_0 : H(t) = t \quad on \quad [0, 1]$$

versus

$$(3.2) \qquad H_1 : H \quad convex \ on \quad [0, 1].$$

We are in effect testing that H is the uniform distribution on $[0, 1]$ versus the alternative that H has an increasing density (when $H^{-1}(1)$ is known).

3.1. *General scores statistics.* If we consider the problem in which the alternative to the uniform distribution is specified, then one can maximize the power by using the Neyman-Pearson lemma. Let $h(t) = d\,H(t)/dt$. If $Z_{1:n} < Z_{2:n} < \cdots < Z_{n:n}$ are the order statistics from H, then the Most Powerful (MP) level α test would reject when

$$(3.3) \qquad \sum_{i=1}^{n} \log h(Z_{i:n}) > k_{n,\alpha}.$$

Since $W_{i:n}$, $1 \leqq i \leqq n$, "behave" asymptotically like order statistics from H we are led to consider statistics of the form

$$(3.4) \qquad T_n(J) = n^{-1} \sum_{i=1}^{n} J[W_{i:n}],$$

where J is an increasing function on $[0, 1]$. (Note that since H is convex, h is increasing and so is $J(x) = \log h(x)$.) The corresponding test would reject H_0

for large values of the statistic. Tests based on such statistics are isotonic and hence have isotonic power by Theorem 2.4.

DEFINITION.   *The test ψ corresponding to $J(x) = x$, for which*

$$
(3.5) \qquad \psi[W_{1:n}, \cdots, W_{n:n}] = \begin{cases} 1 & \text{if } n^{-1} \sum_{i=1}^{n} W_{i:n} > k_{n,\alpha} \\ 0 & \text{otherwise}, \end{cases}
$$

is called the uniform scores *test and $n^{-1} \sum_{i=1}^{n} W_{i:n}$ (or $n^{-1} \sum_{i=1}^{n-1} W_{i:n}$ since $W_{n:n} \equiv 1$) is called the* cumulative total time on test statistic.

REMARK 3.1.   Suppose $G(x) = 1 - \exp\{-x\}$ for $x \geq 0$. Then

$$
(3.6) \qquad W_{i:n} = \frac{\sum_{j=1}^{i} (n - j + 1)(X_{j:n} - X_{j-1:n})}{\sum_{j=1}^{n} (n - j + 1)(X_{j:n} - X_{j-1:n})}
$$

and $n^{-1} \sum_{i=1}^{n-1} W_{i:n}$ is the cumulative total time on test statistic *studied by Nadler and Eilbott [17], Bickel and Doksum [7] and Barlow and Proschan [4].*

Other general scores tests are Fisher's test for the problem of combining tests with

$$
(3.7) \qquad\qquad J(x) = \log x,
$$

the Pearson or exponential scores test with

$$
(3.8) \qquad\qquad J(x) = -\log(1 - x),
$$

and the normal scores test with

$$
(3.9) \qquad\qquad J(x) = \Phi^{-1}(x),
$$

where Φ is the $N(0, 1)$ distribution.

We will show that the uniform scores test (when G is uniform or exponential) is asymptotically minimax over a certain natural class of alternatives determined by the Kolmogorov distance and with respect to a class of tests including all of the above examples.

Tests based on general scores statistics where J is increasing on $[0, 1]$ are clearly unbiased since they have isotonic power as noted previously.

CONDITION 3.1.   *The following regularity conditions are assumed to hold for J: J has the continuous derivative J' on $(0, 1)$ and $\int_0^1 J^2(x)\, dx < \infty$.*

To show that such tests are consistent we need the following result.

THEOREM 3.1.   *If $F, G \in \mathscr{F}$, if the conditions of Theorem 2.2 hold, if J is uniformly continuous on $[0, 1]$, and if $\int_0^1 J[(H^{-1}(u))/[(H^{-1}(1))]\, du < \infty$, then*

$$
(3.10) \qquad\qquad n^{-1} \sum_{i=1}^{n} J[W_{i:n}] \to \int_0^1 J\left[\frac{H^{-1}(u)}{H^{-1}(1)}\right] du
$$

almost surely as $n \to \infty$.

PROOF.   Without loss of generality we may assume $H^{-1}(1) = 1$. Let $Z_{1:n} \leqq$ $\cdots \leqq Z_{n:n}$ be order statistics from H. By the strong law of large numbers,

$$(3.11) \qquad n^{-1} \sum_{i=1}^{n} J[Z_{i:n}] \to \int_0^1 J\left[\frac{H^{-1}(u)}{H^{-1}(1)}\right] du$$

almost surely as $n \to \infty$. Since J is uniformly continuous and $|W_{i:n} - Z_{i:n}| \to 0$ uniformly in i/n and almost surely as $n \to \infty$, by Theorem 2.2 we have that

$$(3.12) \qquad n^{-1} \sum_{i=1}^{n} \{J[W_{i:n}] - J[Z_{i:n}]\} \to 0$$

almost surely as $n \to \infty$. Q.E.D.

Consistency of general scores tests follows from Theorem 3.1 and the observation that $F \underset{c}{<} G$ and $F \neq G$ implies

$$(3.13) \qquad \int_0^1 J\left[\frac{H^{-1}(u)}{H^{-1}(1)}\right] du > \int_0^1 J(u)\, du$$

by Theorem 2.1 (if J is strictly increasing).

Note that $\mu(H) = \int_0^1 (H^{-1}(u))/(H^{-1}(1))\, du = 1/2$ when $F \underset{c}{=} G$.

3.2. *The integral criterion.*   Another class of tests that are natural for our problem are those based on one sided distance functions; that is, functions which measure the "distance" between $H_n^{-1}(x)/H_n^{-1}(1)$ and x. The integral criterion is one such statistic; that is,

$$(3.14) \qquad \int_0^1 \left[\frac{H_n^{-1}(u)}{H_n^{-1}(1)} - u\right] dM_n(u) \overset{\text{def}}{=} n^{-1} \sum_{i=1}^{n} \left[W_{i:n} - \frac{i}{n}\right] L\left(\frac{i}{n}\right)$$

where $L(u) \geqq 0$. The corresponding test would reject H_0 for large values of the statistic. An equivalent statistic is

$$(3.15) \qquad n^{-1} \sum_{i=1}^{n} L\left(\frac{i}{n}\right) W_{i:n}.$$

Such statistics are called systematic statistics. When $L(i/n) \equiv 1$ we have the cumulative total time on test statistic. Bickel and Doksum [7] studied selected types of such statistics for $G(x) = 1 - \exp\{-x\}$ for $x \geqq 0$. In Section 7 we prove the asymptotic equivalence of these statistics to certain general scores statistics when $G(x) = 1 - \exp\{-x\}$. It is clear that such statistics lead to isotonic and hence unbiased tests.

If L satisfies Condition 3.1, $H^{-1}(1) = 1$ and $\int_0^1 xL[H(x)]\, dH(x) < \infty$, then

$$(3.16) \qquad n^{-1} \sum_{i=1}^{n} L\left(\frac{i}{n}\right) W_{i:n} \to \int_0^1 xL[H(x)]\, dH(x)$$

almost surely as $n \to \infty$. This can be proved using Theorem 2.2 and the method of proof used by Moore [16] in proving his Theorem 1.1 (that is, Theorem 4.1 in this paper). Consistency follows from the observation that $F \underset{c}{<} G$ and $F \underset{c}{\neq} G$ imply

$$(3.17) \qquad \int_0^1 xL[H(x)]\,dH(x) > \int_0^1 xL(x)\,dx$$

by Theorem 2.1 since $L(x) \geq 0$.

3.3. *The D_n^+ test.* The one sided Kolmogorov statistic suggests the one sided distance function

$$(3.18) \qquad D_n^+ = \sup_{1 \leq i \leq n} \left[W_{i:n} - \frac{i}{n} \right]$$

for use in the convex ordering problem. The corresponding test ϕ, would reject H_0 if $D_n^+ > c_{n,\alpha}$ where $c_{n,\alpha}$ is determined by H_0. By Theorem 2.3 this test will have isotonic power, since $F_1 \underset{c}{<} F_2$ implies

$$(3.19) \qquad \beta_\phi(F_1) = P_{F_1}[D_n^+ > c_{n,\alpha}] \geq P_{F_2}[D_n^+ > c_{n,\alpha}] = \beta_\phi(F_2).$$

Intuitively, any test based on a one sided distance function will have isotonic power by Theorem 2.3.

When G is the exponential distribution, the distribution of D_n^+ under H_0 is the same as that of the one sided Kolmogorov statistic since

$$(3.20) \qquad W_{i:n} \underset{\text{st}}{=} U_{i:n-1}$$

where $U_{i:n-1}$ is the ith order statistic in a sample of size $n-1$ from a uniform distribution on $[0, 1]$. Birnbaum and Tingey [9] computed the exact distribution for D_n^+ under H_0. For large n we can use the well-known result

$$(3.21) \qquad \lim_{n \to \infty} P_G \left\{ n^{1/2} \sup_{1 \leq i \leq n} \left[W_{i:n} - \frac{i}{n} \right] \leq t \right\} = 1 - e^{-2t^2}$$

for $t \geq 0$.

Seshadri, Csörgö, and Stephens [19] consider the D_n^+ test among other omnibus tests for exponentiality.

4. Asymptotic distribution of the cumulative total time on test statistic: general G

The *cumulative total time on test statistic* is

$$(4.1) \qquad n^{-1} \sum_{i=1}^{n-1} \frac{H_n\left(\frac{i}{n}\right)}{H_n^{-1}(1)} \overset{\text{def}}{=} n^{-1} \sum_{i=1}^{n-1} W_{i:n}.$$

We reject the null hypothesis (that is, $F \underset{c}{=} G$) for large values of the statistic.

We seek the asymptotic distribution of

$$(4.2) \qquad T_n \overset{\text{def}}{=} n^{1/2} \left[n^{-1} \sum_{i=1}^{n-1} W_{i:n} - \mu(H) \right]$$

under the general alternative distribution F, where

$$(4.3) \qquad \mu(H) \overset{\text{def}}{=} \int_0^1 \frac{H^{-1}(u)}{H^{-1}(1)} \, du.$$

To obtain the asymptotic distribution of T_n we use the following result (see D. S. Moore [16]).

Let $X_{1:n} < X_{2:n} < \cdots < X_{n:n}$ be the order statistics from F and

$$(4.4) \qquad S_n = n^{-1} \sum_{i=1}^{n} L\left(\frac{i}{n}\right) X_{i:n},$$

$$(4.5) \qquad \sigma^2 = \sigma^2(F) = 2 \iint_{s<t} L[F(s)] L[F(t)] F(s) [1 - F(t)] \, ds \, dt.$$

THEOREM 4.1. (D. S. Moore [16]). *If $\sigma^2 < \infty$ and*

(i) $E|X| = \int_0^1 |F^{-1}(u)| \, du < \infty$,

(ii) *L is continuous on $[0, 1]$ except for jump discontinuities at $a_1, \cdots, a_M$, and L' is continuous and of bounded variation on $[0, 1] - \{a_1, \cdots, a_M\}$, then*

$$(4.6) \qquad \mathscr{L}\left\{ n^{1/2} \left[S_n - \int_{-\infty}^{\infty} x L[F(x)] \, dF(x) \right] \right\} \to N(0, \sigma^2).$$

Stigler [20], Corollary 4.1 gives weaker conditions for asymptotic normality of sums of the form $S_n = \Sigma_{i=1}^{n} c_{i,n} X_{i:n}$.)

To use this result note that

$$(4.7) \quad H_n^{-1}\left(\frac{i}{n}\right) \overset{\text{def}}{=} \int_0^{F_n^{-1}(i/n)} g G^{-1} F_n(u) \, du = \int_0^{x_{i:n}} g G^{-1} F_n(u) \, du$$

$$= \sum_{j=1}^{i} g G^{-1}\left(\frac{j-1}{n}\right) (X_{j:n} - X_{j-1:n}),$$

where $X_0 \equiv 0$. Using (4.7) we see that

$$(4.8) \qquad \sum_{i=1}^{n-1} H_n^{-1}\left(\frac{i}{n}\right)$$

$$= \sum_{i=1}^{n-2} \left\{ g G^{-1}\left(\frac{i-1}{n}\right) - (n-i-1)\left[g G^{-1}\left(\frac{i}{n}\right) - g G^{-1}\left(\frac{i-1}{n}\right) \right] \right\} X_{i:n}$$

$$+ g G^{-1}\left(\frac{n-2}{n}\right) X_{n-1:n}.$$

Let

$$(4.9) \qquad S_n = n^{-1} \sum_{i=1}^{n-1} H_n^{-1}\left(\frac{i}{n}\right) - \mu(H) H_n^{-1}(1).$$

Then, assuming $gG^{-1}(1) = 0$ $\left(\text{so that } gG^{-1}\left((n-1)/n\right) \to 0\right)$,

$$(4.10) \qquad S_n = n^{-1} \sum_{i=1}^{n} \left\{ gG^{-1}\left(\frac{i-1}{n}\right) \right.$$
$$\left. - n\left(1 - \frac{i+1}{n} - \mu(H)\right)\left[gG^{-1}\left(\frac{i}{n}\right) - gG^{-1}\left(\frac{i-1}{n}\right)\right] \right\} X_{i:n}.$$

Assuming that $\psi(u) = gG^{-1}(u)$ has a continuous derivative on $[0, 1]$, we may approximate S_n by $n^{-1} \sum_{i=1}^{n} L(i/n) X_{i:n}$ where $L(u) = \psi(u) - \left(1 - u - \mu(H)\right)\psi'(u)$.

To apply Theorem 4.1 we wish to show that for this weight function

$$(4.11) \qquad \int_0^\infty x L[F(x)]\, dF(x) = 0.$$

LEMMA 4.1.   *If F, $G \in \mathscr{F}$, if $\int_0^\infty x\, dF(x) < \infty$, $F \underset{c}{\leqslant} G$, $g(0) < \infty$, and if ψ' is continuous on $(0, 1)$, then*

$$(4.12) \qquad H^{-1}(1) = -\int_0^\infty x\psi'[F(x)]\, dF(x),$$

where $\psi(u) = gG^{-1}(u)$.

PROOF.   Recall that $H^{-1}(1) = \int_0^\infty gG^{-1}F(u)\, du$. Integrating the right expression by parts we find

$$(4.13) \qquad \int_0^\infty gG^{-1}F(x)\, dx = xgG^{-1}F(x)\Big|_0^\infty - \int_0^\infty x\psi'[F(x)]\, dF(x).$$

Now

$$(4.14) \qquad \lim_{x \to \infty} xgG^{-1}F(x) = \lim_{x \to \infty} x\frac{f(x)}{r(x)} = 0,$$

since $F \underset{c}{\leqslant} G$ implies $r(x) = f(x)/gG^{-1}F(x)$ is nondecreasing and $\int_0^\infty xf(x)\, dx < \infty$ by assumption. Q.E.D.

LEMMA 4.2.   *Under the conditions of Lemma 4.1, $gG^{-1}(1) < \infty$, and $F^{-1}(0) = 0$,*

$$(4.15) \qquad \int_0^\infty x L[F(x)]\, dF(x)$$

$$\overset{\text{def}}{=} \int_0^\infty x\{\psi[F(x)] - (1 - F(x) - \mu(H))\psi'[F(x)]\}\, dF(x) = 0.$$

PROOF.   By Lemma 4.1 and Equation (4.3) the definition of $\mu(H)$,

(4.16) $\displaystyle\int_0^\infty x L(x)\, dF(x)$

$$= \int_0^\infty x\{\psi[F(x)] - [1 - F(x)]\psi'[F(x)]\}\, dF(x) - \int_0^1 H^{-1}(u)\, du.$$

Integrating by parts, we find that

(4.17) $\displaystyle\int_0^\infty x[1 - F(x)]\psi'[F(x)]\, dF(x) = \int_0^\infty F^{-1}(u)[1 - u]\psi'(u)\, du$

$$= -\int_0^\infty [1 - F(x)]\psi[F(x)]\, dx$$

$$+ \int_0^\infty x\psi[F(x)]\, dF(x).$$

Hence,

(4.18) $\displaystyle\int_0^\infty x L[F(x)]\, dF(x) = \int_0^\infty [1 - F(x)] g G^{-1} F(x)\, dx - \int_0^1 H^{-1}(u)\, du.$

Now

(4.19) $\displaystyle\int_0^1 H^{-1}(u)\, du \stackrel{\text{def}}{=} \int_0^1 \left[\int_0^{F^{-1}(u)} g G^{-1} F(s)\, ds\right] du$

$$= \int_0^\infty [1 - F(x)] g G^{-1} F(x)\, dx$$

by another integration by parts. It follows that

(4.20) $\displaystyle\int_0^\infty x L[F(x)]\, dF(x) = 0$

as claimed. $Q.E.D.$

It follows from Theorem 4.1 and Lemma 4.2 that

(4.21) $\mathscr{L}\{n^{1/2} S_n\} \to N(0, \sigma^2(F)),$

where

(4.22) $\displaystyle\sigma^2(F) = 2 \int_0^1 \left[\int_0^t \frac{[\psi(s) - (1 - s - \mu(H))\psi'(s)]}{fF^{-1}(s)} s\, ds\right]$

$$\cdot \frac{[\psi(t) - (1 - t - \mu(H))\psi'(t)]}{fF^{-1}(t)} (1 - t)\, dt.$$

Since $T_n = S_n/H_n^{-1}(1)$, an application of Slutsky's theorem (Cramér [11]) gives us the following result.

THEOREM 4.2. *Assume the conditions of Theorem 2.2. In addition, assume* $\psi(u) = gG^{-1}(u)$ *has a continuous derivative on* $[0, 1]$, $\psi(1) = 0$, $F^{-1}(0) = 0$, *and*

$\sigma^2(F) < \infty$, *then*

$$(4.23) \qquad \mathscr{L}\left\{n^{1/2}\left[n^{-1}\sum_{i=1}^{n-1} W_{i:n} - \mu(H)\right]\right\} \to N\left(0, \frac{\sigma^2(F)}{[H^{-1}(1)]^2}\right),$$

where $\mu(H) = \int_0^1 H^{-1}(u)\, du / H^{-1}(1)$ *and* $\sigma^2(F)$ *is given by* (4.22).

EXAMPLE.   Let $G(x) = 1 - e^{-x}$ for $x \geqq 0$. Then

$$(4.24) \qquad L(u) = 2(1 - u) - \zeta, \psi(u) = 1 - u,$$

$$(4.25) \qquad n^{-1}\sum_{i=1}^{n-1} W_{i:n} - \mu(H) = n^{-1}\sum_{i=1}^{n} \frac{\left[2\left(1 - \dfrac{i}{n}\right) - \mu(H)\right]X_{i:n}}{H_n^{-1}(1)},$$

We reject the exponential null hypothesis for large values of the statistic. Under the null hypothesis

$$(4.26) \qquad \sum_{i=1}^{n-1} W_{i:n} \underset{st}{=} \sum_{i=1}^{n-1} U_i,$$

where U_i, $i = 1, \cdots, n - 1$, are independent uniform random variables on $[0, 1]$. It follows that

$$(4.27) \qquad \mathscr{L}\left[(12n)^{1/2}\left\{n^{-1}\sum_{i=1}^{n-1} W_{i:n} - 1/2\right\}\right] \to N(0, 1)$$

under the null hypothesis.

In general, if $F \underset{c}{\leq} G$ and $G(x) = 1 - e^{-x}$ for $x \geqq 0$, then

$$(4.28) \qquad \mathscr{L}\left\{n^{1/2}\left[n^{-1}\sum_{i=1}^{n-1} W_{i:n} - \mu(H)\right]\right\} \to N(0, \sigma^2(F)),$$

where

$$(4.29) \qquad \sigma^2(F) = 2\int_0^1\left[\int_0^v \frac{\{2(1 - u) - \mu(H)\}}{fF^{-1}(u)}\, u\, du\right]$$

$$\cdot \frac{[2(1 - v) - \mu(H)]}{fF^{-1}(v)}(1 - v)\, dv.$$

It can be verified that $\sigma^2(G) = 1/12$.

In the case $G(x) = x$ for $0 \leqq x \leqq 1$,

$$(4.30) \qquad \sigma^2(F) = 2\int_0^1\left[\int_0^t sdF^{-1}(s)\right](1 - t)\, dF^{-1}(t)$$

and again $\sigma^2(G) = 1/12$.

In both cases $n^{-1}\Sigma_{i=1}^{n-1} W_{i:n}$ is asymptotically equivalent in distribution to $n^{-1}\Sigma_{i=1}^{n-1} U_i$ when $F \underset{c}{=} G$. This is *not* true for arbitrary G.

The result for the exponential case, (4.28), was first obtained by Nadler and Eilbott [17] by a different and more tedious argument.

5. Alternative classes of distributions based on the Kolmogorov distance

For this discussion we assume that $H^{-1}(1) = 1$. Consider the problem $H_0 \colon H(t) = t$ versus $H_1 \colon H(t)$ convex for $t \in [0, 1]$. (Note that $H(t) \leq t$ under H_1.) If C is a class of level α tests for this problem and Ω is a class of alternatives H with H convex, the $\psi \in C$ is said to be *minimax* over Ω and C if and only if it maximizes the minimum power, that is, if and only if

$$(5.1) \qquad \inf_{H \in \Omega} \beta_\psi(H) = \sup_{\phi \in C} \left[\inf_{H \in \Omega} \beta_\phi(H) \right].$$

It is clear that Ω cannot be taken to be all H with H convex since for this class, the infima in (5.1) would be α and all tests in C would be minimax. Thus, the alternatives in Ω must be "separated." Birnbaum [8], Chapman [10], Doksum [12], and others have considered alternatives separated by the Kolmogorov distance, that is, alternatives H, with H convex in this case, and $\sup_{t \in [0, 1]} [t - H(t)] \geq \Delta$. Here, $\Omega(\Delta)$ will denote the class of H with H convex and $\sup_{t \in [0, 1]} [t - H(t)] \geq \Delta$.

5.1. Extremal classes. The following distributions have Kolmogorov distance Δ (that is, $\sup_{t \in [0, 1]} [t - H(t)] = \Delta$) and are convex on $[0, 1]$ with $H^{-1}(1) = 1$:

$$(5.2) \qquad H_{u, \Delta}(t) = \begin{cases} a_1 t, & 0 \leq t \leq u, \quad \Delta \leq u \leq 1, \\ 1 - a_2(1 - t), & u \leq t \leq 1, \quad \Delta \leq u \leq 1, \end{cases}$$

where $a_1 = (u - \Delta)/u$, $a_2 = 1 + \Delta/(1 - u)$. See Figure 1.

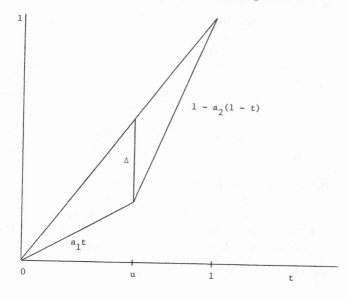

FIGURE 1
Graph of $H_{u, \Delta}$.

We will also need to use

$$(5.3) \qquad H_{u,\Delta}^{-1}(y) = \begin{cases} y/a_1, & 0 \leq y \leq u - \Delta, & \Delta \leq u \leq 1, \\ 1 - (1 - y)/a_2, & u - \Delta \leq y \leq 1, & \Delta \leq u \leq 1. \end{cases}$$

Let

$$(5.4) \qquad h_{u,\Delta}(t) = \begin{cases} a_1, & 0 \leq t \leq u \\ a_2, & u \leq t \leq 1 \end{cases}$$

denote the density of $H_{u,\Delta}(t)$. The distribution $F(F \underset{c}{<} G)$ corresponding to $H_{u,\Delta}$ has the form

$$(5.5) \qquad G_{u,\Delta}(x) = \begin{cases} G(a_1 x), & 0 \leq x \leq x_0 \\ G[a_1 x_0 + a_2(x - x_0)], & x_0 \leq x < \infty \end{cases}$$

where $x_0 = G^{-1}(u - \Delta)/a_1$. To verify this, compute

$$(5.6) \qquad H_{G_{u,\Delta}}^{-1}(y) = \int_0^{G_{u,\Delta}^{-1}(y)} g G^{-1} G_{u,\Delta}(x) \, dx$$

$$= \int_0^{G_{u,\Delta}^{-1}(y)} g(a_1 x) \, dx = \frac{G_{u,\Delta} G_{u,\Delta}^{-1}(y)}{a_1} = \frac{y}{a_1}$$

for $0 \leq y \leq u - \Delta$. A similar calculation verifies the assertion for $u - \Delta \leq y \leq 1$.

The following lemma is a consequence of the fact that

$$(5.7) \qquad \inf_{H \in \Omega(\Delta)} H(t) = \inf_{\Delta \leq u \leq 1} H_{u,\Delta}(t).$$

LEMMA 5.1.   *The distributions $\{H_{u,\Delta}\}$, $0 \leq \Delta \leq u \leq 1$, are least favorable in $\Omega(\Delta)$ for the class of monotone tests in the sense that if ϕ is a monotone test and $W_{i:n}$ is replaced by $Z_{i:n}$, then*

$$(5.8) \qquad \inf_{H \in \Omega(\Delta)} \beta_\phi(H) = \inf_{\Delta \leq u \leq 1} \beta_\phi(H_{u,\Delta}).$$

PROOF.   Suppose $H \in \Omega(\Delta)$ and $\Delta = u - H(u)$. Then $H_{u,\Delta}(x) \geq H(x)$ for $0 \leq x \leq 1$ which in turn implies $\beta_\phi(H_{u,\Delta}) \leq \beta_\phi(H)$ if ϕ is a monotone test. Q.E.D.

Let $r(x) = f(x)/g G^{-1} F(x)$ be the generalized failure rate function corresponding to F, so that $dH^{-1}(t)/dt = 1/r[F^{-1}(t)]$. We claim that the Kolmogorov distance applied to transforms of distributions provides a reasonable way of separating distributions having different failure rate variation. Suppose that F has transform $H_F \in \Omega(\Delta)$ and $\Delta = u - H(u)$. Then since $H^{-1}(t)$ is concave

$$(5.9) \qquad \sup_{0 \leq t \leq 1} r[F^{-1}(t)] - \inf_{0 \leq t \leq 1} r[F^{-1}(t)]$$

$$= \sup_{0 \leq t \leq 1} \left[\frac{d}{dt} H^{-1}(t) \right]^{-1} - \inf_{0 \leq t \leq 1} \left[\frac{d}{dt} H^{-1}(t) \right]^{-1}$$

$$\geq \sup_{0 \leq t \leq 1} \left[\frac{d}{dt} H_{u,\Delta}^{-1}(t) \right]^{-1} - \inf_{0 \leq t \leq 1} \left[\frac{d}{dt} H_{u,\Delta}^{-1}(t) \right]^{-1}$$

$$= a_2 - a_1 = \frac{\Delta}{u(1-u)} \geq 4\Delta.$$

Hence, large values of Δ correspond to large failure rate variation.

5.2. *Contiguity.* The concept of contiguous alternatives plays a crucial role in Sections 7 and 8. (See LeCam [14], Hájek [13].) Let $\{H_\nu, K_\nu\}\nu \geq 1$ be a sequence of similar testing problems. In this sequence the νth testing problem concerns n_ν observations $X_1, \cdots, X_{n_\nu}$ with $n_\nu \to \infty$. In our setup H_ν depends on ν through n_ν only, whereas K_ν depends on the parameters Δ_{n_ν}, in addition. Our problem is to determine

(5.10) $$\lim_{\nu \to \infty} \beta(\alpha, H_\nu, K_\nu) = \beta(\alpha), \qquad 0 \leq \alpha \leq 1,$$

where the sequence $\{\Delta_n\}$ will be chosen so that $\alpha < \beta(\alpha) < 1$. The concept of contiguous alternatives will be useful in computing (5.10) in Sections 7 and 8.

DEFINITION. *A sequence* $\{g_{u,\Delta_n}\}$ *is said to be* contiguous *to* $g_{u,0}$ *(in the sense of LeCam-Hájek) if for any sequence of random variables* $R_n(X_1, \cdots, X_n)$, $R_n \to 0$ *in* P_0 *probability implies* $R_n \to 0$ *in* P_{Δ_n} *probability where* P_θ *denotes the probability distribution of* $X_1, \cdots, X_n$ *if* $g_{u,\theta}$ *is true.*

The following conditions implying contiguity for sequences when $\lim_{n \to \infty} n^{1/2}\Delta_n = c$ for some $0 \leq c < \infty$ can be found in Bickel and Doksum [7]:

(5.11)

    (a) $\partial g_{u,\Delta}(x)/\partial\Delta \neq 0$ whenever $g_{u,\Delta}(x) > 0$,

    (b) $\displaystyle\int_0^\infty \sup \{[\partial g_{u,\Delta}(x)/\partial\Delta]^2 [g_{u,\Delta}(x)]^{-1} : 0 \leq \Delta \leq \delta\} \, dx < \infty$

for some $\delta > 0$.

It is easy to verify that (5.11) (b) holds for $g_{u,\Delta}(x) = \partial G_{u,\Delta}(x)/\partial x$ (where $G_{u,\Delta}$ is defined by (5.5) and $G(x) = 1 - e^{-x}$) for some $\delta > 0$ such that $0 < \delta < u < 1$. Hence, by condition (5.11) $\{g_{u,\Delta_n}\}$ are contiguous alternatives to $g_{u,0}$ if $0 < u < 1$. This fact will be used in Sections 7 and 8.

6. Asymptotic minimax property of the cumulative total time on test statistic: uniform case

In this section we assume that $G(x) = x$ for $0 \leq x \leq 1$ so that $H^{-1}(t) = F^{-1}(t)$. We also assume that $H^{-1}(1) = F^{-1}(1) = 1$. Our problem then is

(6.1) $$H_0 : H(t) = t, \qquad\qquad t \in [0, 1]$$

versus

(6.2) $$H_1 : H(t) \text{ convex}, \qquad\qquad t \in [0, 1].$$

Let $Z_1, Z_2, \cdots, Z_n$ be independent observations from H. We study statistics of the form

$$(6.3) \qquad\qquad n^{-1} \sum_{i=1}^{n} J[Z_i].$$

We show that the test ψ, corresponding to $J(x) = x$ is asymptotically minimax. The function $J(x) = x$ corresponds to the cumulative total time on test statistic.

6.1. *Asymptotic properties of the cumulative total time on test statistic.* Consider the test

$$(6.4) \qquad \psi(Z) = \begin{cases} 1 & \text{if } (12n)^{1/2} \left[n^{-1} \sum_{i=1}^{n} Z_i - 1/2 \right] > k_{n,\alpha}, \\ 0 & \text{otherwise,} \end{cases}$$

where $k_{n,\alpha}$ is defined by $\beta_\psi(U) \stackrel{\text{def}}{=} P_U[\psi(Z) > k_{n,\alpha}] = \alpha$ and $U(t) = t, t \in [0, 1]$.

Computing $\mu = E(Z)$ under $H_{u,\Delta}$ we find

$$(6.5) \qquad \mu = \mu(\Delta, u) = \int_0^u x \frac{u - \Delta}{u} dx + \int_u^1 x \left[\frac{1 - u + \Delta}{1 - u} \right] dx = \frac{1 + \Delta}{2}$$

and $\sigma^2 = \sigma^2(\Delta, u) = EZ^2 - \mu^2$ or

$$(6.6) \qquad\qquad \sigma^2 = \frac{1}{12} + \frac{\Delta}{3}(1 + u) - \frac{(2\Delta - \Delta^2)}{4} \geqq 0.01$$

for $0 \leqq \Delta \leqq 0.01$.

In order to compute asymptotic quantities such as

$$(6.7) \qquad\qquad \lim_{n \to \infty} \inf_{\Delta \leqq u \leqq 1} \beta_\psi(H_{u,\Delta})$$

the Berry-Esseen theorem (Loève, p. 288) will be needed. Doksum [12] made a similar application to a related problem. Applied to the random variables $Z_1, \cdots, Z_n$, it states that if $\mu = E(Z_i)$, $E(Z_i - \mu)^2 = \sigma^2$, $E|Z_i - \mu|^3 = \beta$, and H_n^* is the distribution of $\sum_{i=1}^{n}(Z_i - \mu)/n^{1/2}\sigma$, then there exists a constant $K < \infty$ such that for all x

$$(6.8) \qquad\qquad |H_n^*(x) - \Phi(x)| \leqq \frac{K\beta}{n^{1/2}\sigma^{3/2}}.$$

If H_n^* is the distribution of $\sum_{i=1}^{n}(Z_i - \mu)/n^{1/2}\sigma$ under $H_{u,\Delta}$, then $|Z_i - \mu| \leqq 1$ implies

$$(6.9) \qquad\qquad \beta = E|Z_i - \mu|^3 \leqq 1.$$

Then (6.6), (6.8) and (6.9) imply

$$(6.10) \qquad\qquad |H_n^*(x) - \Phi(x)| \leqq \frac{1000K}{n^{1/2}} \qquad \text{for all} \quad \Delta \in [0, 0.01],$$

for all $u \in [\Delta, 1]$, and for all x, where Φ is the $N(0, 1)$ distribution.

For the alternatives $H_{u,\Delta}$ and for the test corresponding to the statistic given by (6.3) we have

$$(6.11) \qquad \beta_\psi(H_{u,\Delta}) = P\left\{(12n)^{1/2}\left[n^{-1}\sum_{i=1}^{n}Z_i - 1/2\right] \geqq k_{n,\alpha}\middle| H_{u,\Delta}\right\}$$

$$(6.12) \qquad \beta_\psi(H_{u,\Delta}) = P\left\{n^{1/2}\left[n^{-1}\frac{\sum_{i=1}^{n}(Z_i-\mu)}{\sigma}\right] \geqq \frac{k_{n,\alpha}-(3n)^{1/2}\Delta}{(12)^{1/2}\sigma}\middle| H_{u,\Delta}\right\}.$$

This and (6.10) imply

$$(6.13) \qquad \left|\beta_\psi(H_{u,\Delta}) - \Phi\left(\frac{-k_{n,\alpha}+(3n)^{1/2}\Delta}{(12)^{1/2}\sigma}\right)\right| \leqq \frac{1000K}{n^{1/2}}$$

for all $\Delta \in [0, 0.01]$ and $u \in [\Delta, 1]$.

LEMMA 6.1. *The cumulative total time on test, ψ, satisfies*

(i) $\inf_{H\in\Omega(\Delta_n)} \beta_\psi(H)$ *tends to a limit between α and one as $n \to \infty$ if and only if* $\lim_{n\to\infty} n^{1/2}\Delta_n = c > 0$.

(ii) *For each sequence $\{\Delta_n\}$ such that $\lim_{n\to\infty} n^{1/2}\Delta_n = c > 0$ one has*

$$(6.14) \qquad \lim_{n\to\infty}\left[\inf_{H\in\Omega(\Delta_n)}\beta_\psi(H)\right] = \Phi(-k_\alpha + c\,3^{1/2}),$$

where k_α is defined by $\Phi(k_\alpha) = 1 - \alpha$.

PROOF. By Lemma 4.1

$$(6.15) \qquad \inf_{H\in\Omega(\Delta)}\beta_\psi(H) = \inf_{\Delta\leqq u\leqq 1}\beta_\psi(H_{u,\Delta}).$$

Let $u_0 = u_0(\Delta, n)$ be such that

$$(6.16) \qquad \beta_\psi(H_{u_0,\Delta}) = \inf_{\Delta\leqq u\leqq 1}\beta_\psi(H_{u,\Delta}) \overset{\text{def}}{=} \beta_\psi(\Delta).$$

Now (6.13) implies that

$$(6.17) \qquad \left|\beta_\psi(\Delta) - \Phi\left(\frac{-k_{n,\alpha}+(3n)^{1/2}\Delta}{12^{1/2}\sigma(\Delta,u_0)}\right)\right| \leqq \frac{1000K}{n^{1/2}}$$

where $\Delta \in [0, 0.01]$. From (6.6), one has that $\sigma^2(\Delta, u_0) \to 1/12$ as $\Delta \to 0$. Moreover, $-k_{n,\alpha} \to -k_\alpha$ with k_α satisfying $\Phi(-k_\alpha) = \alpha$. Thus, (6.17) implies that $\beta_\psi(\Delta_n)$ tends to a limit between α and one if and only if $(3n)^{1/2}\Delta_n \to c$ for some $c > 0$. This implies (i). Furthermore, when $n^{1/2}\Delta_n \to c$, then $\beta_\psi(\Delta_n) \to \Phi(-k_\alpha + 3^{1/2}c)$ which is (ii). Q.E.D.

6.2. *Asymptotic properties of general scores statistics.* The general scores statistics look like

$$(6.18) \qquad T_n(J) = n^{-1}\sum_{i=1}^{n}J(Z_i).$$

Consider the test

$$(6.19) \qquad \phi = \begin{cases} 1 & \text{if } n^{1/2}[T_n(J) - \mu_J]/\sigma_J > k_{n,\alpha}, \\ 0 & \text{otherwise}, \end{cases}$$

where $\mu_J = \int_0^1 J(x)\,dx$, $\sigma_J^2 = \int_0^1 J^2(x)\,dx - \mu_J^2$, and $k_{n,\alpha}$ is defined by $\beta_\phi(U) = \alpha$ where U is the uniform distribution on $[0, 1]$.

LEMMA 6.2. *If $\{\Delta_n\}$ satisfies $\lim_{n\to\infty} n^{1/2}\Delta_n = c > 0$, and J satisfies Condition 3.1, then*

$$(6.20) \qquad \lim_{n\to\infty} \beta_\phi(H_{u,\Delta_n}) = \Phi\left(-k_\alpha + \frac{c\left[-\dfrac{1}{u}\displaystyle\int_0^u J(x)\,dx + \dfrac{1}{1-u}\displaystyle\int_u^1 J(x)\,dx\right]}{\sigma_J}\right).$$

PROOF.   Let $E_J(\Delta, u)$ and $V_J(\Delta, u)$ denote $E[T_n(J)|H_{u,\Delta}]$ and $\text{Var}\,[T_n(J)|H_{u,\Delta}]$. Then by definition

$$(6.21) \qquad E_J(\Delta, u) = \left(\frac{u - \Delta}{u}\right)\int_0^u J(x)\,dx + \left(\frac{1 - u + \Delta}{1 - u}\right)\int_u^1 J(x)\,dx$$

by (5.2) and

$$(6.22) \qquad V_J(\Delta, u) = n^{-1}\left\{\left(\frac{u - \Delta}{u}\right)\int_0^u J^2(x)\,dx\right.$$

$$\left. + \left(\frac{1 - u + \Delta}{1 - u}\right)\int_u^1 J^2(x)\,dx - E_J^2(\Delta, u)\right\}.$$

Since in Condition 3.1 we assumed $\int_0^1 J^2(x)\,dx < \infty$, we see that $E_J(\Delta, u) < \infty$ and $V_J(\Delta, u) < \infty$ so long as $0 < u < 1$. Note that $V_J(\Delta, u) \to V_J(0, u) = n^{-1}\sigma_J^2 = n^{-1}[\int_0^1 J^2(x)\,dx - (\int_0^1 J(x)\,dx)^2]$ as $\Delta \to 0^+$. Thus, the central limit theorem implies that for each sequence $\{\Delta_n\}$ with $\Delta_n \to 0$ and $0 < u < 1$,

$$(6.23) \qquad \lim_{n\to\infty} P\left[n^{1/2}\left(\frac{T_n(J) - E_J(\Delta_n, u)}{\sigma_J}\right) \leq t\right] = \Phi(t).$$

From (6.19) we have

$$(6.24) \qquad \beta_\phi(H_{\Delta, u}) = P\left\{-n^{1/2}\left[\frac{T_n(J) - E_J(\Delta, u)}{\sigma_J}\right]\right.$$

$$\left. \leq -k_{n,\alpha} + \frac{n^{1/2}}{\sigma_J}[E_J(\Delta, u) - \mu_J]\right\}.$$

Using the definition of μ_J and $E_J(\Delta, u)$ we see that

$$(6.25) \qquad E_J(\Delta, u) - \mu_J = -\frac{\Delta}{u}\int_0^u J(x)\,dx + \frac{\Delta}{1-u}\int_u^1 J(x)\,dx.$$

Since $\lim_{n \to \infty} n^{1/2} \Delta_n = c > 0$ we see that

$$(6.26) \qquad \lim_{n \to \infty} n^{1/2} [E_J(\Delta_n, u) - \mu_J] = c \left[-\frac{1}{u} \int_0^u J(x) \, dx + \frac{1}{1-u} \int_u^1 J(x) \, dx \right]$$

which completes the proof of (6.20). $Q.E.D.$

We would like to show that

$$(6.27) \qquad \inf_{0 < u < 1} \left[\frac{-\dfrac{1}{u} \displaystyle\int_0^u J(x) \, dx + \dfrac{1}{1-u} \displaystyle\int_u^1 J(x) \, dx}{\sigma_J} \right]$$

is maximized for $J(x) = x$ since this would imply that the cumulative total time on test statistic maximizes the minimum power. (Note that (6.27) is unchanged if we replace J by $aJ + b$ when $a > 0$.)

The following lemma was communicated to the authors by W. R. van Zwet.

LEMMA 6.3.   (W. R. van Zwet)

$$(6.28) \qquad A_J = \inf_{0 < u < 1} \left[\frac{-u^{-1} \displaystyle\int_0^u J(x) \, dx + (1-u)^{-1} \displaystyle\int_u^1 J(x) \, dx}{\sigma_J} \right]$$

is maximized among all square integrable J on $(0, 1)$ by $J(x) = x$ where

$$(6.29) \qquad \sigma_J^2 = \int_0^1 J^2(x) \, dx - \left[\int_0^1 J(x) \, dx \right]^2.$$

PROOF.   Since the value of A_J remains unchanged if J is replaced by $aJ + b$, $a > 0$, we may assume

$$(6.30) \qquad \int_0^1 J(x) \, dx = \int_0^1 x \, dx = 1/2$$

and

$$(6.31) \qquad \int_0^1 J^2(x) \, dx = \int_0^1 x^2 \, dx = 1/3.$$

Let $J(x) = x + K(x)$. Then

$$(6.32) \qquad \int_0^1 K(x) \, dx = 0,$$

$$(6.33) \qquad \int_0^1 [K^2(x) + 2xK(x)] \, dx = 0,$$

and

$$(6.34) \qquad A_J = A_I + \inf_{0 < u < 1} \frac{-u^{-1} \displaystyle\int_0^u K(x) \, dx + (1-u)^{-1} \displaystyle\int_u^1 K(x) \, dx}{(1/12)^{1/2}},$$

where $I(x) = x$ and $A_I = 3^{1/2}$ (for $J = I$ the infimum is assumed at every u!). Suppose the proposition were false, then we would have for some K satisfying (6.32) and (6.33)

$$(6.35) \qquad \inf_{0 < u < 1} \left[-u^{-1} \int_0^u K(x) \, dx + (1 - u)^{-1} \int_u^1 K(x) \, dx \right]$$

$$= \inf_{0 < u < 1} \frac{1}{u(1 - u)} \int_u^1 K(x) \, dx > 0,$$

and hence

$$(6.36) \qquad \int_u^1 K(x) \, dx > 0 \qquad \text{for all} \quad 0 < u < 1.$$

However,

$$(6.37) \qquad \int_0^1 x K(x) \, dx = \int_0^1 dx \int_x^1 K(y) \, dy$$

and hence (6.36) would imply that $\int_0^1 x K(x) \, dx > 0$ which contradicts (6.33). Q.E.D.

We have proved a minimax result for testing that F is uniform versus F convex.

THEOREM 6.1. Let $G(x) = x$, $0 \leqq x \leqq 1$, in the problem (1.2) and (1.3). If J is square integrable, satisfies Condition 3.1 and ϕ is the level α general scores test associated with J, then

$$(6.38) \qquad \lim_{n \to \infty} \left[\inf_{H \in \Omega(\Delta_n)} \beta_\psi(H) \right] \geqq \left[\limsup_{n \to \infty} \inf_{H \in \Omega(\Delta_n)} \beta_\phi(H) \right];$$

that is, the level α cumulative total time on test statistic corresponding to $J(x) = x$ is minimax in the class of tests whose weight functions satisfy the conditions above.

Doksum [12] showed that $J(x) = x$ provides a minimax test over the class of those J satisfying Condition 3.1 and over the class of stochastically ordered alternatives determined by the Kolomogorov distance.

It follows from Theorem 6.1 that, in the minimax sense, the uniform scores test is better than tests based on Fisher's weights $\big(J(x) = \log x\big)$, better than the Pearson or exponential scores test and better than the normal scores test.

7. Asymptotic normality and efficiency of statistics based on total time on test statistics: exponential case

Bickel and Doksum [7] and Bickel [6] considered four classes of statistics for testing $H_0: F(x) = G_\lambda(x) = 1 - \exp\{-\lambda x\}$ against IFR alternatives. These four types of statistics were shown to be asymptotically equivalent and it was shown that each of the classes contains asymptotically most powerful statistics for parametric alternatives. We now show that the statistics

$$(7.1) \qquad T_n(J) = n^{-1} \sum_{i=1}^n J(W_{i:n})$$

based on the total time on test statistics are asymptotically equivalent to the four classes of statistics in [6] and [7]. Consequently, for a given parametric family $\{F_\theta\}$ of IFR distributions, it is possible to find a $J = J_{F_\theta}$ such that the test that rejects H_0 for large values of $T_n(J)$ is asymptotically most powerful.

Note that under H_0, $W_{1:n}, \cdots, W_{n-1:n}$ are distributed as the order statistics of a sample of size $n - 1$ from the uniform distribution on $[0, 1]$. Using well-known results (for example, [16]) on linear combinations of order statistics in reverse, we have that $T_n(J)$ is asymptotically equivalent to

$$(7.2) \qquad S_n(J) = n^{-1} \sum_{i=1}^{n} J'\left(\frac{i}{n+1}\right) W_{i:n}.$$

More precisely, if we define

$$(7.3) \qquad \mu_J = \int_0^1 J(x)\, dx, \quad \text{and} \quad \sigma_J^2 = \int_0^1 J^2(x)\, dx - \mu_J^2,$$

then we have the following Lemma.

LEMMA 7.1. *Suppose that $0 < \sigma^2 < \infty$, and that J' satisfies condition* (ii) *of Theorem 4.1, then*

$$(7.4) \qquad n^{1/2}\{[S_n(J) - (J(1) - \mu_J)] - [T_n(J) - \mu_J]\}$$

converges to zero in probability under H_0.

Next note that since $n\bar{X}\, W_{i:n} = \Sigma_{j=1}^{i}\, (n - j + 1)(X_{j:n} - X_{j-1:n})$ if we set $D_j = (n - j + 1)(X_{j:n} - X_{j-1:n})$ and

$$(7.5) \qquad V_n(J) = -(n\bar{X})^{-1} \sum_{i=1}^{n} J\left(\frac{i}{n+1}\right) D_j$$

then

$$(7.6) \qquad n^{1/2}[S_n(J) - [V_n(J) + J(1)]]$$

tends to zero in probability under H_0. For a given parametric family $\{F_\theta\}$ of distributions, it is shown in [6] that there exists a function $a(u) = a_{F_\theta}(u)$ (see [6], Equation (2.9)) such that the test that rejects H_0 for large values of $V_n(a)$ is asymptotically most powerful for $\{F_\theta\}$. Using this, Lemma 7.1, and the definition of contiguity, we have

THEOREM 7.1. *If $J = a$ satisfies the conditions of Lemma 7.1 and $\{F_\theta\}$ satisfies the conditions of Corollary 2.1 of Bickel [6], then the test that rejects H_0 for large values of $T_n(a)$ is asymptotically most powerful among all similar tests.*

Bickel and Doksum ([7], Section 7) show that $n^{1/2} \bar{X} V_n(J)$ can be approximated by a sum $\Sigma_{i=1}^{n} h(X_i)$, of independent, identically distributed random variables. We now proceed to give a similar approximation using a different derivation. Note that we can write

$$(7.7) \quad - \bar{X} V_n(J)$$

$$= n^{-1} \sum_{i=1}^{n} \left\{\left(1 - \frac{i-1}{n}\right)\left[n\left(J\left(\frac{i}{n+1}\right) - J\left(\frac{i+1}{n+1}\right)\right)\right] + J\left(\frac{i+1}{n+1}\right)\right\} X_{i:n}.$$

It follows that if we define

(7.8) $$L(u) = L_J(u) = (1 - u)J'(u) - J(u)$$

and

(7.9) $$W_n(J) = n^{-1} \sum_{i=1}^{n} L\left(\frac{i}{n+1}\right) X_{i:n},$$

then

(7.10) $$n^{1/2} \left(W_n(J) - \bar{X} V_n(J)\right)$$

and

(7.11) $$n^{1/2} \left(\frac{W_n(J)}{\bar{X}} - V_n(J)\right)$$

tend to zero in probability under H_0.

We will need

(7.12) $$\xi_W = \int_0^\infty x L[F(x)] \, dF(x)$$

and

(7.13) $$\tau_W^2 = 2 \iint_{0 < s < t < \infty} L[F(s)] L[F(t)] F(s) [1 - F(t)] \, ds \, dt.$$

If we apply Moore's approximation [16] to $W_n(J)$ we get

LEMMA 7.2. *If $\tau_W^2 < \infty$ if $\mu_{F^{-1}} < \infty$, and if L satisfies condition* (ii) *of Theorem 4.1, then $n^{1/2}\{[W_n(J) - \xi_W] - Q_n(J)\}$ tends to zero in probability, where*

(7.14) $$Q_n(J) = n^{-1} \sum_{i=1}^{n} B_F(X_i)$$

and

(7.15) $$B_F(x) = \int_0^x F(t) L[F(t)] \, dt - \int_x^\infty [1 - F(t)] L[F(t)] \, dt.$$

This result establishes the asymptotic equivalence under contiguous alternatives of all the statistics in this section with sums of independent, identically distributed random variables. For the computations of asymptotic power we need some lemmas.

LEMMA 7.3. *If the conditions of Lemma 7.2 hold, then $E[B_F(X)|F] = 0$.*
PROOF. If we define $I_x(t) = 1(0)$ if $x \leq t(x > t)$, then

(7.16) $$B_F(x) = \int_0^\infty [F(t) - I_x(t)] L[F(t)] \, dt.$$

The result follows since $E[I_X(t)|F] = F(t)$. Q.E.D.

LEMMA 7.4. *If $F(x) = 1 - \exp\{-x\}$, then*

$$(7.17) \qquad B_F(x) = J[F(x)] - \int_0^x J[F(t)]\,dt.$$

PROOF. Note that $d\{[1 - F(x)]J[F(x)]\}/dx = L[F(x)]f(x)$ and that $f(x) = 1 - F(x)$. Thus integrating by parts

$$(7.18) \qquad \int_0^x F(t)L[F(t)]\,dt$$

$$= \int_0^x F(t)[1 - F(t)]^{-1}\,d[(1 - F(t))J[F(t)]]$$

$$= F(x)J[F(x)] - \int_0^x [1 - F(t)]J[F(t)]\,d[F(t)[1 - F(t)]^{-1}]$$

$$= F(x)J[F(x)] - \int_0^x J[F(t)]\,dt,$$

where the last equality follows from $F(t)[1 - F(t)]^{-1} = e^t - 1$.

Similarly,

$$(7.19) \qquad \int_x^\infty [1 - F(t)]L[f(t)]\,dt$$

$$= \int_x^\infty d[(1 - F(t))J[F(t)]] = -(1 - F(x))J[F(x)].$$

Q.E.D.

LEMMA 7.5. (i) *If $F(x) = G_\lambda(x) = 1 - \exp\{-\lambda x\}$, then $B_{G_\lambda}(x) = B_{G_1}(\lambda x)/\lambda$.*
(ii) *If $J(u) = u$, then $B_{G_1}(x) = 2G_1(x) - x$.*

PROOF. Part (i) follows by setting $x = \lambda t$ in the definition of $B_{G_\lambda}(x)$. Part (ii) is immediate. *Q.E.D.*

LEMMA 7.6. *If $t[1 - F(t)]J[F(t)] \to 0$ as $t \to \infty$, then*
(i) $\xi_W = -\int_0^\infty [1 - F(t)]J[F(t)]\,dt$;
(ii) *if $F(x) = G_1(x)$, then $\xi_W = -\mu_J$.*

PROOF.

$$(7.20) \qquad \xi_W = \int_0^\infty t[1 - F(t)]J'[F(t)]\,dF(t) - \int_0^\infty tJ[F(t)]\,dF(t).$$

But

$$(7.21) \qquad \int_0^\infty t[1 - F(t)]J'[F(t)]\,dF(t) = \int_0^\infty t[1 - F(t)]\,dJ[F(t)]$$

$$= -\int_0^\infty [1 - F(t)]J[F(t)]\,dt$$

$$+ \int_0^\infty tJ[F(t)]\,dF(t)$$

by integration by parts. Part (i) follows.

To show (ii), note that if we set $u = F(t) = G_1(t)$, then

$$(7.22) \qquad \frac{du}{dt} = \exp\{-t\} = 1 - u;$$

thus

$$(7.23) \qquad \int_0^\infty [1 - F(t)]J[F(t)]\, dt = \int_0^1 J(u)\, du.$$

Q.E.D.

If we now put together the results of this section, we have that

$$(7.24) \qquad n^{1/2}[T_n(J) - \mu_J] - n^{1/2}[(\bar{X})^{-1}[Q_n(J) - \mu_J] + \mu_J]$$

converges to zero under G_1 and under contiguous alternatives, where

$$(7.25) \qquad Q_n(J) = n^{-1} \sum_{i=1}^n B_{G_1}(X_i).$$

Let

$$(7.26) \qquad \mu(F, J) = \lim_{n \to \infty} \{\bar{X}^{-1}[Q_n(J) - \mu_J] + \mu_J\}$$

$$= [\mu(F^{-1})]^{-1}\left[\int_0^\infty J[G_1(t)]\, dF(t)\right.$$

$$\left. - \int_0^\infty [1 - F(t)]J[G_1(t)]\, dt - \mu_J\right] + \mu_J.$$

We have shown the following theorem.

THEOREM 7.2.   *If the conditions of Lemma 7.1 and 7.2 are satisfied, and if $\{F_n\}$ is a sequence of alternative distributions contiguous to G_1, then*

$$(7.27) \qquad \frac{n^{1/2}}{\sigma_J}[T_n(J) - \mu_J - \mu(F_n, J)]$$

converges in law to a standard normal variable.

REMARK 7.1.   Theorem 7.2 can be used to obtain the results of Theorem 7.1. Let

$$(7.28) \qquad \phi = \begin{cases} 1 & \text{if } \dfrac{n^{1/2}}{\sigma_J}[T_n(J) - \mu_J] > k_{n,\alpha}, \\[2mm] 0 & \text{otherwise.} \end{cases}$$

and

$$(7.29) \qquad B_\phi(G_{u,\Delta}) = P_{G_{u,\Delta}}\left\{\frac{n^{1/2}}{\sigma_J}[T_n(J) - \mu_J - \mu(G_{u,\Delta}, J)]_\Delta\right.$$

$$\left. > k_{n,\alpha} - \frac{1}{\sigma_J}\mu(G_{u,\Delta}, J)\right\}$$

where $G_{u,\Delta}$ is defined by (5.5) and $G(x) = 1 - e^{-x}$. If $\lim_{n\to\infty} n^{1/2}\Delta_n = c > 0$, then using Theorem 7.2 it is easy to verify that

$$(7.30) \qquad \lim_{n \to \infty} \beta_\phi(G_{u, \Delta_n}) = \Phi \left\{ -k_\alpha + \frac{c}{\sigma_J} \left[-\frac{1}{u} \int_0^u J(x)\,dx + \frac{1}{1-u} \int_u^1 J(x)\,dx \right] \right\}.$$

8. Asymptotic minimax property of the cumulative total time on test statistic: exponential case

In this section we again consider the null hypothesis $H_0: F(x) = G_\lambda(x) = 1 - \exp\{-\lambda x\}$. Let

$$(8.1) \qquad T_n(J) = n^{-1} \sum_{i=1}^{n-1} J(W_{i:n})$$

where J is increasing, $\mu_J = \int_0^1 J(x)\,dx < \infty$, $\sigma_J^2 = \int_0^1 J^2(x)\,dx - \mu_J^2 < \infty$, and

$$(8.2) \qquad W_{i:n} = \frac{\displaystyle\sum_{j=1}^{i} (n - j + 1)(X_{j:n} - X_{j-1:n})}{\displaystyle\sum_{i=1}^{n} (n - j + 1)(X_{j:n} - X_{j-1:n})}$$

in this case. Consider the test

$$(8.3) \qquad \phi = \begin{cases} 1 & \text{if} \quad \dfrac{n^{1/2}}{\sigma_J} [T_n(J) - \mu_J] > k_{n, \alpha}, \\ 0 & \text{otherwise}, \end{cases}$$

where $k_{n, \alpha}$ is defined by $\beta_\phi(U) = \alpha$ and U is the uniform distribution on $[0, 1]$.

Let F have transform H (see (2.1)) and

$$(8.4) \qquad H_1^{-1}(t) = \frac{H^{-1}(t)}{H^{-1}(1)}.$$

Let $\Omega_1(\Delta)$ be the class of distributions $F \in \mathscr{F}$ for which

$$(8.5) \qquad \sup_{\Delta < u < 1} [u - H_1(u)] \geqq \Delta$$

and H is convex on $[0, 1]$. Note that if $F \in \Omega_1(\Delta)$ there does not necessarily exist $u \in [\Delta, 1]$ such that $F \underset{c}{\leqslant} G_{u, \Delta}$. If we restrict ourselves to totally ordered classes of distributions in $\Omega_1(\Delta)$ containing some $G_{u, \Delta}$, then we have that $F \in \Omega_1(\Delta)$ implies $F \underset{c}{\leqslant} G_{u, \Delta}$. The union of these classes is the class $\Gamma(\Delta)$ of F in $\Omega_1(\Delta)$ for which there exists as $u \in [\Delta, 1]$ such that $F \underset{c}{\leqslant} G_{u, \Delta}$. Suppose that $F \in \Gamma(\Delta)$. Since ϕ is an isotonic test we have by Theorem 2.4, that

$$(8.6) \qquad \beta_\phi(F) = P\left[\frac{n^{1/2}}{\sigma_J} [T_n(J) - \mu_J] > k_{n, \alpha} \middle| F \right]$$

$$\geqq P\left[\frac{n^{1/2}}{\sigma_J} [T_n(J) - \mu_J] > k_{n, \alpha} \middle| G_{u, \Delta} \right] = \beta_\phi(G_{u, \Delta})$$

for some $u \in [\Delta, 1]$. Hence,

$$(8.7) \qquad \inf [\beta_\phi(F): F \in \Gamma(\Delta)] = \inf [\beta_\phi(G_{u, \Delta}): u \in [\Delta, 1]].$$

Let $T_n^{(1)} = n^{-1} \sum_{i=1}^{n-1} W_{i:n}$ and

$$(8.8) \qquad \psi = \begin{cases} 1 & \text{if } (12n)^{1/2}[T_n^{(1)} - 1/2] \geqq k_{n, \alpha}, \\ 0 & \text{otherwise}, \end{cases}$$

that is, the test based on the cumulative total time on test statistic. Let $\mu(H_1) = [H^{-1}(1)]^{-1} \int_0^1 H^{-1}(u) \, du$ and $\sigma_1^2(H) = \sigma^2(F)[H^{-1}(1)]^{-2}$ where $\sigma^2(F)$ is given by (4.29).

LEMMA 8.1. *If* $\lim_{n \to \infty} n^{1/2}\Delta_n = c$, *then*

$$(8.9) \qquad \lim_{n \to \infty} [\inf \beta_\psi(F): F \in \Gamma(\Delta_n)] = \Phi(-k_\alpha + 3^{1/2}c).$$

PROOF.   Assume $H^{-1}(1) = 1$. By definition

$$(8.10) \quad \begin{aligned} \beta_\psi(G_{u, \Delta}) &= P_{u, \Delta}[(12n)^{1/2}(T_n^{(1)} - 1/2) \geqq k_{n, \alpha}] \\ &= P_{u, \Delta}[(12n)^{1/2}[(T_n^{(1)} - 1/2) - (\mu(H_{u, \Delta}) - 1/2)] \\ &\geqq k_{n, \alpha} - (12n)^{1/2}[\mu(H_{u, \Delta}) - 1/2]]. \end{aligned}$$

For the distribution G_{u, Δ_n}, $\Delta_n = c/n^{1/2}$, and the statistic

$$(8.11) \qquad S_n^{(1)} = n^{-1} \sum_{i=1}^n J_0\left(\frac{i}{n}\right) X_{i:n}, J_0(t) = 2(1 - t)$$

we find that the error term I_{2n} of Moore [16] is zero, while the second one, I_{3n}, tends to zero in probability uniformly in $u \in [\Delta_n, 1]$, that is,

$$(8.12) \qquad \sup_u P_{u, \Delta_n}(|I_{3n}| > \varepsilon) \to 0 \qquad \text{as} \quad n \to \infty$$

for each $\varepsilon > 0$. Thus, $n^{-1}[S_n^{(1)} - \mu(H_{u, \Delta})]$ can be expressed as a sum of independent, identically distributed random variables with third moments plus a term that tends to zero uniformly in u. Using the representation

$$(8.13) \qquad n^{1/2}[T_n^{(1)} - \mu(H_{u, \Delta})] = \frac{n^{1/2}}{\bar{X}} [S_n^{(1)} - \bar{X}\mu(H_{u, \Delta})].$$

we find that the same thing is true for

$$(8.14) \qquad n^{1/2}[T_n^{(1)} - \mu(H_{u, \Delta})].$$

If we now apply a uniform version of Slutsky's theorem and the Berry-Esseen theorem, we have that

$$(8.15) \qquad \sup_{u \in [\Delta_n, 1]} |\beta_\psi(G_{u, \Delta_n}) - \{1 - \Phi[k_{n, \alpha} - (12n)^{1/2}(\mu(H_{u, \Delta_n}) - 1/2)]\}| \to 0$$
$$\text{as} \quad n \to \infty.$$

The result now follows by the computations leading to Lemma 6.1. *Q.E.D.*

From Lemma 8.1, Equation (7.29), and Lemma 6.3, we obtain:

THEOREM 8.1. *If J and the conditions of Lemmas* 7.1 *and* 7.2 *are satisfied, then*

$$(8.16) \qquad \lim_{n \to \infty} \left[\inf \beta_\psi(F) \colon F \in \Gamma(\Delta_n)\right] \geqq \limsup_{n \to \infty} \left[\inf \beta_\phi(F) \colon F \in \Gamma(\Delta_n)\right]$$

for each sequence $\{\Delta_n\}$ *satisfying* $\lim_{n \to \infty} n^{1/2}\Delta_n = c$, *for some* $c \in [0, \infty)$.

Thus, we have shown that the cumulative total time on test statistic is asymptotically minimax over $\Gamma(\Delta)$ in each of the classes of statistics of Section 7.

REMARK 8.1. Bickel and Doksum [7] considered the problem of Section 7. exponentiality against four totally ordered families of distributions. They determined the studentized asymptotically most powerful linear spacings test for each family. The cumulative total time on test statistic is the asymptotically most powerful linear spacings test for the family of densities

$$(8.17) \qquad f_\theta(x) = \left[1 + \theta(1 - e^{-x})\right] \exp\left\{-\left[x + \theta(x + e^{-x} - 1)\right]\right\}$$

where $f_0(x) = e^{-x}$ and $f_\theta(x)$ has increasing failure rate for $\theta > 0$. Bickel [6] showed that this test is in fact asymptotically equivalent to the level α test which is most powerful against the family $\{f_\theta(x)\}$ among all tests which are similar and level α.

Bickel and Doksum [7] computed the asymptotic efficiency of each linear spacings test considered relative to each family of distributions for comparison purposes ([7], Table 6.1). Let $e(W_i, j)$ denote the asymptotic efficiency of Bickel and Doksum's test W_i relative to the jth family, $j = 1, \cdots, 4$, of distributions (W_1 corresponds to the cumulative total time on test statistic). It is easy to verify from ([7], Table 6.1) that

$$(8.18) \qquad \min_j e(W_1, j) = \max_i \min_j e(W_i, j);$$

that is, the cumulative total time on test statistic maximizes the minimum asymptotic efficiency. This observation suggested the minimax property proved in Theorem 8.1.

REMARK 8.2. It is possible to prove a minimax result for the cumulative total time on test statistic and the class of alternatives $\Omega_1(\Delta)$ rather than $\Gamma(\Delta)$ if we worked with $\inf_F \lim_{n \to \infty} \beta_\psi(F)$ rather than $\lim_{n \to \infty} \inf_F \beta_\psi(F)$. This is clear since the asymptotic power of

$$(8.19) \qquad T_n^{(1)} = n^{-1} \sum_{i=1}^{n-1} W_{i:n}$$

depends only on $\mu(H_1)$ and

$$(8.20) \qquad \inf_{F \in \Omega(\Delta)} \mu(H_1) = \inf_{u \in [\Delta, 1]} \mu(H_{u, \Delta}).$$

However, we would then have to assume that the distributions F in $\Omega(\Delta)$ satisfied conditions such that $T_n^{(1)}$ is asymptotically normal. Such conditions are given in [6] and [7] for parametric families of distributions.

9. Asymptotic minimax property of the cumulative total time on test statistic: general G

If we consider the problem of testing that $F = G$ where G is completely specified then a natural class of statistics are those of the form

$$(9.1) \qquad \phi^* = \begin{cases} 1 & \text{if} \quad \dfrac{n^{1/2}}{\sigma_J} \left[n^{-1} \sum_{i=1}^{n} J[G(X_i)] - \mu_J \right] > k_{n,\,\alpha}, \\[2mm] 0 & \text{otherwise.} \end{cases}$$

For $J(x) = x$, $\phi^* = \psi^*$ is the uniform scores test.

Suppose $G^{-1}(1) = F^{-1}(1) = 1$ and $F \underset{c}{\leq} G$. Then $F(x) \leq G(x)$ for $0 \leq x \leq 1$ and if J is increasing, ϕ^* will provide an isotonic test for our problem.

If we confine attention to the extremal class of alternative distributions defined in (5.5) then we can prove the following asymptotic minimax theorem.

THEOREM 9.1.   *If J is square integrable and*

$$(9.2) \qquad \frac{1}{gG^{-1}(y)} g\left[\left(1 - \frac{\Delta}{u}\right) G^{-1}(y) \right] \to 1$$

uniformly in $y \in [0, 1]$ as $\Delta \to 0$, then

$$(9.3) \qquad \lim_{n \to \infty} \left[\inf_{0 < u < 1} \beta_{\psi^*}(G_{u,\,\Delta_n}) \right] \geq \lim \sup_{n \to \infty} \left[\inf_{0 < u < 1} \beta_{\phi^*}(G_{u,\,\Delta_n}) \right],$$

where $\lim_{n \to \infty} n^{1/2} \Delta_n = c > 0$.

The proof is similar to that of Theorem 6.1. In this case, $E_J(\Delta, u) = E_{G_{u,\Delta}} [J[G(X)]]$ becomes

$$(9.4) \qquad E_J(\Delta, u) = \frac{(u - \Delta)}{u} \int_0^u J(y) \left\{ \frac{1}{gG^{-1}(y)} g\left[\left(\frac{u - \Delta}{u} \right) G^{-1}(y) \right] \right\} dy$$

$$+ \left[1 + \frac{\Delta}{1 - u} \right] \int_u^1 J(y) \left\{ \frac{1}{gG^{-1}(y)} g[a_1 x_0 + a_2(G^{-1}(y) - x_0)] \right\} dy$$

where $a_1 = (u - \Delta)/u$, $a_2 = 1 + \Delta/(1 - u)$ and $x_0 = G^{-1}(u - \Delta)/a_1$. As before

$$(9.5) \qquad \lim_{n \to \infty} n^{1/2} \left[E_J(\Delta_n, u) - \mu_J \right] = c \left[-\frac{1}{u} \int_0^u J(y)\, dy + \frac{1}{1 - u} \int_u^1 J(y)\, dy \right].$$

We would like to acknowledge the help of W. R. van Zwet who found a general proof of Lemma 6.3 which replaced many tedious arguments involving special cases.

REFERENCES

[1] R. E. BARLOW, "Likelihood ratio tests for restricted families of probability distributions," *Ann. Math. Statist.*, Vol. 39 (1968), pp. 547–560.

[2] R. E. BARLOW and W. R. VAN ZWET, "Asymptotic properties of isotonic estimators for the generalized failure rate function," *Proceedings of the First International Symposium on Nonparametric Techniques in Statistical Inference*, Cambridge, Cambridge University Press, 1970, pp. 159–176.

[3] ———, "Asymptotic properties of isotonic estimators for the generalized failure rate function. Part II: Asymptotic distributions," University of California, Berkeley, Operations Research Center Report ORC 69–10, 1970.

[4] R. E. BARLOW and F. PROSCHAN, "A note on tests for monotone failure rate based on incomplete data," *Ann. Math. Statist.*, Vol. 40 (1969), pp. 595–600.

[5] ———, "Inequalities for linear combinations of order statistics from restricted families," *Ann. Math. Statist.*, Vol. 37 (1966), pp. 1574–1592.

[6] P. BICKEL, "Tests for monotone failure rate II," *Ann. Math. Statist.*, Vol. 40 (1969), pp. 1250–1260.

[7] P. BICKEL and K. DOKSUM, "Tests for monotone failure rate based on normalized spacings," *Ann. Math. Statist.*, Vol. 40 (1969), pp. 1216–1235.

[8] Z. W. BIRNBAUM, "On the power of a one-sided test of fit for continuous distribution functions," *Ann. Math. Statist.*, Vol. 24 (1953), pp. 284–289.

[9] Z. W. BIRNBAUM and F. TINGEY, "One-sided confidence contours for probability distribution functions," *Ann. Math. Statist.*, Vol. 22 (1951), pp. 592–596.

[10] D. G. CHAPMAN, "A comparative study of several one-sided goodness-of-fit tests," *Ann. Math. Statist.*, Vol. 29 (1958), pp. 655–674.

[11] H. CRAMÉR, *Mathematical Methods in Statistics*, Princeton, Princeton University Press, 1946.

[12] K. DOKSUM, "Asymptotically minimax distribution-free procedures," *Ann. Math. Statist.*, Vol. 37 (1966), pp. 619–628.

[13] J. HÁJEK and Z. ŠIDÁK, *Theory of Rank Tests*, New York, Academic Press, 1967.

[14] L. LECAM, "Likelihood functions for large numbers of independent observations," *Festschrift for J. Neyman*, New York, Wiley, 1966, pp. 167–187.

[15] M. LOÈVE, *Probability Theory*, Princeton, Van Nostrand, 1963 (3rd ed.).

[16] D. S. MOORE, "An elementary proof of asymptotic normality of linear functions of order statistics," *Ann. Math. Statist.*, Vol. 39 (1968), pp. 263–265.

[17] J. NADLER and EILBOTT, "Testing for monotone failure rates," unpublished.

[18] F. PROSCHAN and R. PYKE, "Test for monotone failure rate," *Proceedings of the Fifth Berkeley Symposium on Mathematical Statistics and Probability*, Berkeley and Los Angeles, University of California Press, 1967, Vol. 3, pp. 293–312.

[19] V. SESHADRI, M. CSÖRGÖ, and M. A. STEPHENS, "Tests for the exponential distribution using Kolmogorov-type statistics," *J. Roy. Statist. Soc. Ser. B*, Vol. 31 (1969), pp. 499–509.

[20] S. M. STIGLER, "Linear functions of order statistics," *Ann. Math. Statist.*, Vol. 40 (1969), pp. 770–788.

[21] W. R. VAN ZWET, *Convex Transformations of Random Variables*, Amsterdam, Mathematical Centre, 1964.

ASYMPTOTICALLY DISTRIBUTION FREE STATISTICS SIMILAR TO STUDENT'S t

Z. W. BIRNBAUM

UNIVERSITY OF WASHINGTON

1. A class of statistics

1.1. Let X be a random variable with a continuous distribution function $F(x) = P\{X \leqq x\}$, $X_1, X_2, \cdots, X_n$ a random sample of X, and $X_{(1)} \leqq X_{(2)} \leqq \cdots X_{(n)}$ the corresponding ordered sample. For given $0 < \gamma < 1$, we consider the γ quantile of X

$$(1.1.1) \qquad \mu_\gamma = F^{(-1)}(\gamma)$$

and the corresponding sample quantile

$$(1.1.2) \qquad V_\gamma = X_{(k)},$$

where

$$(1.1.3) \qquad k = [\gamma n] + 1.$$

We now consider the statistic

$$(1.1.4) \qquad S_\gamma = \frac{V_\gamma - \mu_\gamma}{X_{(k+r_2)} - X_{(k-r_1)}},$$

where r_1, r_2 are integers such that $0 < r_1 < k$, $0 < r_2 \leqq n - k$.

1.2. The statistics of the form (1.1.4) have a structure somewhat similar to Student's t: the numerator is the difference between an estimate of a location parameter (sample quantile V_γ) and that location parameter (population quantile μ_γ), while the denominator is an estimate (sample interquantile range) of a scale parameter (population interquantile range). A more pertinent analogy with the t statistic is this: the statistic S_γ is invariant under linear transformations and hence, if the distribution function $F(\cdot)$ is given, S_γ has for fixed γ, n, r_1, r_2 a probability distribution independent of location and scale parameters; that is the same for all random variables with distribution functions $F((x - a)/b)$ with arbitrary real a and positive b. One could, therefore, choose, for example, $F(x) = (2\pi)^{-1/2} \int_{-\infty}^{x} e^{-y^2/2} \, dy$ and tabulate the probability distributions of S_γ for practically meaningful values of γ, n, r_1, r_2; these probability distributions

Research supported in part by the U.S. Office of Naval Research.

would be independent of the mean and the variance of X, and could be used in a manner analogous to that in which one uses the t statistic.

The S statistics defined by (1.1.4) can be computed and used when only the three order statistics $X_{(k)}$, $X_{(k+r_2)}$, $X_{(k-r_1)}$ are available, for example in situations when a number of more extreme order statistics have been "censored." This constitutes a possible practical advantage as compared with the t statistic, which can be computed only when the complete sample is available.

Another minor advantage of the S statistics is that, as soon as the order statistics $X_{(k)}$, $X_{(k+r_2)}$, $X_{(k-r_1)}$ are available, the calculation of S is quite simple.

1.3.   Among the statistics (1.1.4), the one corresponding to $\gamma = \frac{1}{2}$ and hence $\mu_{1/2}$ = median of X is closest in structure and possible applications to the t statistic. Using the notations

$$n = 2m + 1 \quad \text{(odd sample size)},$$
(1.3.1)
$$\mu_{1/2} = \mu = F^{(-1)}(1/2) = \text{median},$$
$$V_{1/2} = V = X_{(m+1)},$$

and, specializing even more, $r_1 = r_2 = r$,

we have the statistic

(1.3.2)
$$S = \frac{X_{(m+1)} - \mu}{X_{(m+1+r)} - X_{(m+1-r)}} \qquad 1 \leqq r \leqq m.$$

This statistic was considered in some detail in [1], while similar statistics were previously studied in [2]. From now on we shall limit our considerations to the statistic (1.3.2).

2. Exact distribution

2.1.   We assume in the following the existence everywhere of the probability density $f(x) = F'(x)$. Using well-known expressions for the joint probability distribution of the order statistics $X_{(m+1-r)}$, $X_{(m+1)}$, $X_{(m+1+r)}$, one obtains

(2.1.1) $P\{S > \lambda\} = P(\lambda) = \dfrac{(2m + 1)!}{[(m - r)! \, (r - 1)!]^2}$

$$\cdot \int_{v=0}^{+\infty} \int_{u=(\lambda-1)v/\lambda}^{v} \int_{w=v}^{u+v/\lambda} f(u)f(v)f(w)$$

$$\cdot F^{m-r}(u) [F(v) - F(u)]^{r-1} [F(w) - F(v)]^{r-1} [1 - F(w)]^{m-r} \, dw \, du \, dv$$

for $\lambda > 0$.

2.2.   In [1], expression (2.1.1) was used to prove the following statement: if the probability density $f(x)$ is bell shaped about its median μ, that is, $f(\mu - x) = f(\mu + x)$ for $x \geqq 0$ and $f(\mu + x)$ is nonincreasing for $x \geqq 0$, then

(2.2.1) $$P\{|S| > \lambda\} \leqq \binom{2m + 1}{m - r}\binom{2r}{r}[\lambda(\lambda - 1)]^{-r} 2^{-(m+r)}$$

for $\lambda > 1$.

2.3. For every family of probability distributions with a location and scale parameter determined, say, by the probability density $f(x)$ for the standardized random variable X, $E(X) = 0$, $\text{Var}(X) = 1$, expression (2.1.1) and a similar expression for $\lambda < 0$ could be used to compute numerically all probabilities needed for practical use. The evaluation of the triple integrals in (2.1.1), however, appears to be quite time consuming. Fortunately, the following simple intuitive argument makes it clear that for m large and r/m small the probability distribution of S will be practically independent of $f(x)$; for such values of m and r, all three order statistics $X_{(m+1-r)}$, $X_{(m+1)}$, $X_{(m+1+r)}$ fall with probability close to one very close to μ, where $F(x)$ is approximately equal to $1/2 + f(0)(x - \mu)$. Therefore S is approximately distributed as if the sample was obtained from a random variable with uniform distribution on $[\mu - 1/2f(0), \mu + 1/2f(0)]$, and since a change of location and of scale does not affect the probability distribution of S, it is approximately distributed as if the sample came from a random variable distributed uniformly on $[-1/2, 1/2]$.

3. Limiting distributions

3.1. THEOREM 3.1. *If $f(\mu) > 0$, and $f'(x)$ exists and is continuous in an interval $(\mu - \delta, \mu + \delta)$ for some $\delta > 0$ then, for r fixed, the probability distribution of the statistic (1.3.2) has the limit*

(3.1.1) $$\lim_{m \to \infty} P\left\{\sqrt{\frac{2}{m}} S \leqq s\right\} = \frac{1}{(2r - 1)!} \int_0^\infty \Phi(zs) z^{2r-1} e^{-z} \, dz$$

where $\Phi(\cdot)$ is the standardized normal distribution function.

An elementary proof of (3.1.1) appears in [3]. In view of the intuitive argument in Section 2.3, there is reason to believe that the right side of (3.1.1) is a good approximation, already for moderate values of m, when $f(x)$ is reasonably close to a linear function in a neighborhood of μ; by contrast, a good approximation need not be expected, for example, for $f(x) = \frac{1}{2}e^{-|x|}$.

3.2. When not only m, but also r is large, the following argument yields an even simpler result.

The right side of (3.1.1) is

(3.2.1) $$R(s) = \int_0^\infty \Phi(zs) \, \gamma_{2r}(z) \, dz,$$

where $\gamma_{2r}(z)$ is the probability density of a gamma distribution with parameter $2r$, hence with mode $2r - 1$ and variance $2r$. The change of variable

(3.2.2) $$y = \frac{z \cdot}{2r - 1}$$

yields

(3.2.3) $$R(s) = \int_0^\infty \Phi[(2r - 1)sy]\, \gamma_{2r}^*(y)\, dy,$$

where $\gamma_{2r}^*(y) = \gamma_{2r}[(2r - 1)y]\,(2r - 1)$ is a probability density with mode at $y = 1$ and variance $2r/(2r - 1)^2$, hence with its mass concentrated about $y = 1$. One concludes from this by a routine argument that $R(s)$ is approximated by $\Phi[2r - 1)s]$, hence

(3.2.4) $$P\left\{\sqrt{\frac{2}{m}}\, S \leq s\right\} \sim \Phi[(2r - 1)s]$$

for r large and r/m small.

4. Some numerical computations and comparisons

4.1.   One sided critical values for the limiting distribution (3.1.1) were calculated with the aid of a high speed computer by Richard Erickson [4] for the significance levels $\alpha = 0.10, 0.05, 0.025, 0.01, 0.005$, and $r = 1\,(1)10$, and for $\alpha = 0.001, 0.0005$ and selected values of r. It would be desirable to compute the corresponding critical values for a specific family of probability distributions, preferably the normal family, from the exact expression (2.1.1), and to determine how large m must be to make the difference between these exact critical values and those obtained from (3.1.1) negligible. As mentioned before, these calculations are quite time consuming, and have not yet been carried out.

4.2.   Monte Carlo estimates of the exact probabilities (2.1.1) were obtained by Professor Jean Tague at Memorial University, St. John's, Newfoundland, for X normally distributed and the following values of the arguments: $m = 1(1)10$, $r = 1(1)m$, $\lambda = 0.0(0.1)5.0$ and some selected large values of λ. These Monte Carlo values [5] are the best information available at the present on the exact values of (2.1.1).

4.3.   Let $\lambda_{\alpha, m, r}$ be the exact "critical value" determined by (2.1.1), under the assumption that X has normal distribution, so that for sample size $2m + 1$ one has $P\{S > \lambda_{\alpha, m, r}\} = \alpha$. Three different approximations to these exact values are available:

$\lambda_{\alpha, m, r}^* =$ the Monte Carlo estimates obtained by interpolation from [5],

$\lambda_{\alpha, m, r}^{**} =$ the approximations obtained in [4] from (3.1.1)

$\lambda_{\alpha, m, r}^{***} =$ the approximations obtained from (3.2.3).

From the manner in which these values have been obtained, one would expect that $\lambda_{\alpha, m, r}^*$ is generally a good approximation for $\lambda_{\alpha, m, r}$ while $\lambda_{\alpha, m, r}^{**}$ is good for m large, and $\lambda_{\alpha, m, r}^{***}$ only for r large and r/m small. Table I contains a comparison of the three approximations for $m = 10$ (sample size $2m + 1 = 21$), and $\alpha = 0.10$.

TABLE I

THREE APPROXIMATIONS TO VALUES $\lambda_{\alpha, m, r}$
For a sample of size $2m + 1$ from a normal population one has

$$P\left\{\frac{X_{(m+1)} - \mu}{X_{(m+1+r)} - X_{(m+1-r)}} > \lambda_{\alpha, m, r}\right\} = \alpha \quad \text{for} \quad m = 10, \alpha = 0.10.$$

r	λ^*	λ^{**}	λ^{***}
1	2.2045	1.5876	2.8657
2	.8537	.8745	.9552
3	.5223	.5335	.5731
4	.3765	.3937	.4094
5	.2904	.3113	.3184
6	.2381	.2563	.2605
7	.1919	.2173	.2204
8	.1658	.1887	.1910
9	.1386	.1668	.1686

REFERENCES

[1] Z. W. BIRNBAUM, "On a statistic similar to Student's t," *Nonparametric Techniques in Statistical Inferences*, London, Cambridge University Press, 1970, pp. 427–433.
[2] F. N. DAVID and N. L. JOHNSON, "Some tests of significance with ordered variables," *J. Roy. Statist. Soc. Ser. B*, Vol. 18 (1956), pp. 1–20.
[3] Z. W. BIRNBAUM and I. VINCZE, "Limiting distributions of statistics similar to Student's t," to be submitted for publication.
[4] R. A. ERICKSON, "On a Rényi type statistic for the median," unpublished M. S. thesis, University of Washington, 1970.
[5] JEAN TAGUE, "Monte Carlo tables for the S-statistic," unpublished, Memorial University of Newfoundland, 1969.

DECISION THEORY FOR SOME NONPARAMETRIC MODELS

KJELL A. DOKSUM

UNIVERSITY OF CALIFORNIA, BERKELEY

1. Introduction and summary

The problem considered in this paper is that of obtaining optimal decision rules when a parametric form of the distribution of the observations is not known exactly. Thus we assume that the underlying distribution function F of the X_i in the random sample $X = (X_1, \cdots, X_n)$ is in a class Ω of distribution functions, and Ω is not indexed in a natural way by a parameter θ in m dimensional Euclidean space R^m. Let $R(F, d)$ denote the risk of the decision rule $d = d(X)$ when F is the true distribution. Minimax procedures that minimize the maximum risk sup $\{R(F, d); F \in \Omega\}$ have been obtained in special cases by Hoeffding [8], Ruist [14], Huber [9], [10], [11], and Doksum [3]. In particular, Huber was able to show that if Ω is the class of all distributions in a neighborhood of a normal distribution, then the minimax procedures are based on statistics that are, approximately, trimmed means. Most stringent procedures that minimize the maximum shortcoming sup$_F$ $\{R(F, d) - \inf_d R(F, d)\}$ have been considered by Schaafsma [15].

Another approach would be to define a probability (weight function) P on Ω and then minimize the average (Bayes) risk $\int_\Omega R(F, d) P(dF)$, thereby obtaining what is called the Bayes solution. This approach has been taken by Kraft and van Eeden [13], Ferguson [6], and Antoniak [1], who were able to obtain explicit Bayes solutions for some probabilities P. Their work is closely related to the work of Fabius [5], who considered properties of posterior distributions for a class of probability measures P that essentially contains those of Kraft and van Eeden and of Ferguson. Fabius' work in turn is related to that of Freedman [7], who considered properties of Bayes procedures in the case where the X_i are discrete random variables. The relationship between these papers will be discussed further in Section 5.

In this paper, we introduce a criterion which involves minimizing a quantity between the maximum risk and the average risk. This criterion is appropriate when the probability P on Ω is not fully specified, but only the distribution of $F(t_1), \cdots, F(t_k)$ is known for some $t_1 < \cdots < t_k$. Thus past records may

Prepared with the partial support of the National Science Foundation under Grant GP–15283, and with the partial support of the Office of Naval Research under Contract N00014––69–A–0200–1036. Completed while the author was at the University of Oslo.

be available for $F(1)$, $F(2)$, and so on, but not for $F(e)$, $F(\pi)$ and so on. The criterion is to minimize the average maximum risk, where the average is computed with respect to the distribution λ of $F(t_1), \cdots, F(t_k)$. More precisely, let $t_1 < \cdots < t_k$ be fixed and let $\Omega(q, k)$ be the class of distribution functions in Ω that pass through $(t_1, q_1), (t_2, q_2), \cdots, (t_k, q_k)$, $0 \leqq q_1 \leqq \cdots \leqq q_k \leqq 1$. We define the average maximum risk (or mixed risk) as $\int_{R^k} [\sup_{F \in \Omega(q, k)} R(F, d)] \lambda(dq)$. The decision rule that minimizes this risk is called the mixed Bayes-minimax rule, or mixed rule. It will be shown in Sections 2 and 3 that, under certain conditions on the risk function, the mixed rule can be obtained by computing the posterior distribution of a multinomial parameter $p = (p_1, \cdots, p_{k-1})$ having as prior distribution the distribution μ of $F(t_2) - F(t_1), F(t_3) - F(t_2), \cdots, F(t_k) - F(t_{k-1})$.

The mixed Bayes-minimax rule d_k can be thought of as an approximation to the Bayes rule. In Section 4, Prohorov's theorem is used to show that if Ω is contained in $C[0, 1]$ or $D[0, 1]$, and if a Bayes solution d exists, then d_k converges to d in the sense of convergence of Bayes risks. Thus in situations where P is known, but the Bayes rule is hard to compute, one can use the mixed rule as an approximation. If the limit $\lim_{k \to \infty} d_k$ can be computed, then it gives a method of obtaining the Bayes solution. Note that $k = 0$ corresponds to the minimax problem.

2. The mixed Bayes-minimax problem

Let $X_1, \cdots, X_n$ be independent, identically distributed random variables with distribution function F, where F belongs to some specified class Ω of distribution functions. It will be convenient to assume that Ω is a measurable subset of some larger class Γ of functions with a σ-field $\mathscr{S}$. We further assume that there is a probability P on $(\Gamma, \mathscr{S})$, with $P(\Omega) = 1$.

Let $L(F, d)$ be a real valued function that denotes the loss of the real valued decision rule $d = d(X_1, \cdots, X_n)$ when $F \in \Omega$ is the true distribution. Let $X = (X_1, \cdots, X_n)$. Then the risk of d is

$$(2.1) \qquad R(F, d) = E_F L(F, d(X)).$$

If there is a rule (procedure) that minimizes the maximum risk

$$(2.2) \qquad R(d) = \sup_{F \in \Omega} R(F, d),$$

then it is called a minimax procedure. Similarly, if there is a d that minimizes the average risk

$$(2.3) \qquad r(P, d) = \int_{\Omega} R(F, d) P(dF),$$

then it is a Bayes procedure. Here F is thought of as a random distribution function with distribution P; that is, F is a stochastic process with sample paths $F(t)$, $t \in R$. Computing Bayes procedures will involve computing the posterior distribution of F given X.

The mixed Bayes-minimax procedure minimizes a function that is between $R(d)$ and $r(P, d)$. This function is the average maximum risk when a finite dimensional distribution corresponding to P is known, and the average is taken with respect to this distribution. We now proceed with the definition. The *carrier* of a given distribution is in general the smallest compact set whose probability under the given distribution is one. Let $C(F)$ denote the carrier of $F \in \Omega$; we define the *support* of Ω to be $S(\Omega) = \cup_{F \in \Omega} C(F)$. Let $t_1 < \cdots < t_k$ be k points in $S(\Omega)$. The distribution of $F(t_1), \cdots, F(t_k)$ under P will be denoted by λ, or $\lambda(\cdot; P, k)$. Thus, if we write $q = (q_1, \cdots, q_k)$, $0 \leqq q_1 \leqq \cdots \leqq q_k \leqq 1$, then

$$(2.4) \qquad \lambda(q; P, k) = P(F: F(t_1) \leqq q_1, \cdots, F(t_k) \leqq q_k).$$

The class of distributions in Ω whose value at t_i is q_i, $i = 1, \cdots, k$, will be denoted by $\Omega(q, k)$, that is,

$$(2.5) \qquad \Omega(q, k) = \{F \in \Omega: F(t_i) = q_i, i = 1, \cdots, k\}.$$

The *average maximum risk* of a decision rule d is now defined as

$$(2.6) \qquad r_k(P, d) = \int_{R^k} \big[\sup_{F \in \Omega(q, k)} R(F, d) \big] \lambda(dq; P, k).$$

If there is a d_0 that minimizes $r_k(P, d)$, d_0 is called a mixed Bayes-minimax (or mixed) procedure.

As we are dealing with functions of $X \in R^n$ and $F \in \Omega$, we need a joint distribution for X and a random element F of Ω with distribution P. If $\mathscr{B}^n$ denotes the σ-field of Borel sets in R^n, we define the probability $\tilde{P}$ on $(R^n \times \Gamma, \mathscr{B}^n \times \mathscr{S})$ by

$$(2.7) \qquad \tilde{P}(B \times S) = \int_S \mu_F(B)\, P(dF), \qquad B \in \mathscr{B}^n, \qquad S \in \mathscr{S},$$

where μ_F is the probability in R^n corresponding to the distribution of X, and it is assumed that μ_F is $\mathscr{S}$ measurable. We also assume that a conditional distribution of X given F exists and satisfies

$$(2.8) \qquad \tilde{P}(X_1 \leqq x_1, \cdots, X_n \leqq x_n | F) = \prod_1^n F(x_i).$$

Furthermore, we assume that F has a conditional distribution given $F(t_i) = q_i$, $i = 1, \cdots, k$; we denote this conditional distribution by P^q. For a further discussion of these definitions, see Fabius [5]. The assumptions involved in the definitions are satisfied for the complete, separable metric spaces considered in Section 4.

The following inequalities follow at once from the definitions. Note that when more than one set of $t_1, \cdots, t_k$ is considered, we will use double subscripts and write $t_{m,1}, \cdots, t_{m,k_m}$.

LEMMA 2.1. *The risks defined above satisfy the following relations:*

(i) $R(d) = r_0(P, d) \geqq r_k(P, d) \geqq r(P, d)$, $k \geqq 1$;

(ii) *if* $\{\prod_m: t_{m,1} < \cdots < t_{m,k_m}\}$, $m = 1, 2, \cdots$, *is a sequence of partitions such that each partition is a refinement of the previous one, then*

$$(2.9) \qquad r_{k_m}(P, d) \geqq r_{k_\ell}(P, d) \qquad for\ m < \ell.$$

We next give a parametric example in which the mixed risk r_k equals the Bayes risk r.

EXAMPLE 2.1.   Let $\Gamma = \Omega$ be the class of normal distribution functions F_θ with mean θ and variance unity. Suppose that P is the measure for which θ has a normal distribution with mean ξ and variance unity. Let $\mathscr{S}$ be the class of sets of the form $\{F_\theta: \theta \in B\}$, where B is a Borel subset of the reals. All the quantities of this section are now defined. Moreover, since $F(t_1)$ determines θ and θ determines $F(t_1)$, then $r_k(P, d) = r(P, d)$, $k \geq 1$.

Next we consider a "discrete" example in which the mixed risk eventually equals the Bayes risk.

EXAMPLE 2.2.   Let $\Omega = \Gamma$ be a countable class $\{F_1, F_2, \cdots\}$ and let $\mathscr{S}$ be the collection of all subsets of Ω. If Ω is discrete (that is, Ω has no limit points for the sup norm), and if $\{t_{m, j}\}$ of Lemma 2.1 (ii) becomes dense in $S(\Omega)$ as $m \to \infty$, then there exists $m_1 \geq 1$ such that $r_{k_m}(P, d) = r(P, d)$ for all $m \geq m_1$. To see this, note that, by our assumptions, there exists m_1 such that

$$(2.10) \qquad \big(F_i(t_{m, 1}), \cdots, F_i(t_{m, k_m})\big) \neq \big(F_j(t_{m, 1}), \cdots, F_j(t_{m, k_m})\big)$$

for $m \geq m_1$ and all $i \neq j$.

We now define a decision rule d_k that, in some cases, will be the mixed Bayes-minimax procedure. Let $t_1 = \inf \{t: t \in S(\Omega)\}$ and $t_k = \sup \{t: t \in S(\Omega)\}$. It follows that $F(t_k) = 1$. We will assume that $F(t_1) = 0$ a.s. (P), $-\infty < t_1 < t_k < \infty$, and that $k \geq 3$. Let $q = \{q_1, \cdots, q_k\}$ with $0 = q_1 \leq \cdots \leq q_k = 1$. Let $F_{q, k}$ be the polygonal distribution function that equals q_i at t_i, $i = 1, \cdots, k$, and is linear over each interval $[t_i, t_{i+1}]$, $i = 1, \cdots, k - 1$. Let F_k denote the random distribution function obtained by letting q in $F_{q, k}$ have distribution $\lambda = \lambda(\cdot; P, k)$. We assume that F_k is a measurable function on some measure space to $(\Gamma, \mathscr{S})$. Let P_k denote the distribution of F_k. Finally, d_k will denote the Bayes solution for $F_{q, k}$ when q has prior λ, that is, d_k minimizes

$$(2.11) \qquad r(P_k, d) = \int_{R_k} R(F_{q, k}, d)\lambda(dq),$$

THEOREM 2.1.   *If $F_{q, k} \in \Omega$ for almost all q in the carrier C_λ of λ, if d_k minimizing* (2.11) *exists, and if*

$$(2.12) \qquad L(F, d_k) = L(G, d_k)$$

for all F, $G \in \Omega(q, k)$ and almost all q in C_λ then d_k is the mixed Bayes-minimax procedure.

PROOF.   Let N_i denote the number of X in $(t_i, t_{i+1}]$, $i = 1, \cdots, k - 1$. Then $\mathbf{N} = (N_1, \cdots, N_{k-1})$ is sufficient for $F_{q, k}$, and d_k depends on the X only through $\mathbf{N}$. Thus the distribution of d_k as a function of the X is the same for all F in $\Omega(q, k)$. This and (2.12) imply that $R(F, d_k) = R(G, d_k)$ for all F, $G \in \Omega(q, k)$. Thus $r_k(P, d_k) = r(P_k, d_k)$, and since d_k is optimal for P_k, then $r(P_k, d_k) \leq r(P_k, d)$ for all other rules d. Finally, since $F_{q, k} \in \Omega$ for almost all q, $r(P_k, d) \leq r_k(P, d)$ and the results follow.

REMARK 2.1. From the above proof, it is clear that (2.12) can be replaced by the condition

$$(2.13) \qquad r_k(P, d_k) = r(P_k, d_k)$$

and that (2.13) is weaker than (2.12).

EXAMPLE 2.3. If $0 \in \{t_1, \cdots, t_k\}$ and $L(F, d) = [d - F(0)]^2$, then (2.12) is satisfied. Generalizing this, we have that (2.12) is satisfied for any loss function depending on F only through $F(t_1), \cdots, F(t_k)$; that is, the loss is defined through those points where we have information about F. Such a loss function corresponding to squared error when estimating the mean of F would be $[d - \mu(F, k)]^2$, where

$$(2.14) \qquad \mu(F, k) = \tfrac{1}{2} \sum_{i=1}^{k-1} (t_{i+1} + t_i)[F(t_{i+1}) - F(t_i)]$$

In the next section, we consider a testing problem for which (2.12) is satisfied.

3. A testing problem

Suppose $\Omega = \Omega_0 \cup \Omega_1$ with $\Omega_0 \cap \Omega_1$ empty, $\Omega_0, \Omega_1 \in \mathscr{S}$, and we want to test $H_0 \colon F \in \Omega_0$ against $H_1 \colon F \in \Omega_1$. $\varphi = \varphi(X)$ will denote a test function, and $L(F, \varphi)$ will be the usual loss for the testing problem; that is, $L(F, \varphi) = L_i$, a positive constant, if H_i is falsely rejected, $i = 0, 1$, and $L(F, \varphi) = 0$ otherwise. The Bayes risk is

$$(3.1) \qquad r(P, \varphi) = L_0 \int_{\Omega_0} E_F(\varphi) \, P(dF) + L_1 \int_{\Omega_1} [1 - E_F(\varphi)] \, P(dF)$$

and the average maximum risk is

$$(3.2) \qquad r_k(P, \varphi) = L_0 \int_{Q_0} \Big[\sup_{F \in \Omega(q, k)} E_F(\varphi) \Big] \lambda(dq; P, k)$$
$$+ L_1 \int_{Q_1} \Big[\sup_{F \in \Omega(q, k)} [1 - E_F(\varphi)] \Big] \lambda(dq; P, k),$$

where $Q_i = \{(F(t_1), \cdots, F(t_k)) \colon F \in \Omega_i\}$, $i = 1, 2$; that is, Q_0 and Q_1 are the sets in R^k corresponding to Ω_0 and Ω_1. We assume that $Q_0 \cap Q_1$ has λ probability zero. This assumption is needed to obtain (3.2) above. For this loss function it is clear that (2.12) of Theorem 2.1 is satisfied; if in addition $F_{q,k} \in \Omega$, then the result can be applied. If $F_{q,k} \notin \Omega$, then $L(F_{q,k}, \varphi)$ is not defined. However, there is a natural way of defining $L(F_{q,k}, \varphi)$ and making $F_{q,k}$ a member of $\Omega(q, k)$: for a given q not in $Q_0 \cap Q_1$, $L(F, \varphi)$ has the same value for each $F \in \Omega(q, k)$; define $L(F_{q,k}, \varphi)$ to be this value.

In what follows, it will be assumed that either $F_{q,k} \in \Omega$ or $L(F_{q,k}, \varphi)$ has been defined as above. Let $f(x|q)$ denote the density of $F_{q,k}$. Then from (3.2) and Theorem 2.1 we can conclude that the mixed Bayes-minimax procedure φ_k rejects H_0 when

$$(3.3) \quad L_1 \int_{Q_1} \prod_{i=1}^{k-1} (q_{i+1} - q_i)^{N_i} \lambda(dq) \geqq L_0 \int_{Q_0} \prod_{i=1}^{k-1} (q_{i+1} - q_i)^{N_i} \lambda(dq), \ i = 1, 2.$$

Let $p_i = q_{i+1} - q_i, i = 1, \cdots, k - 1$ and

(3.4) $$A_i = \{(F(t_2) - F(t_1)), \cdots, (F(t_k) - F(t_{k-1})) : F \in \Omega_i\}.$$

Then (3.3) becomes

(3.5) $$L_1 \int_{A_1} \prod_{i=1}^{k-1} p_i^{N_i} \pi(dp) \geqq L_0 \int_{A_0} \prod_{i=1}^{k-1} p_i^{N_i} \pi(dp),$$

where π is the distribution of $(F(t_2) - F(t_1), \cdots, F(t_k) - F(t_{k-1}))$ when F has distribution P. Note that (3.5) is the solution to a Bayesian multinomial testing problem, in which $(N_1, \cdots, N_{k-1} | P)$ is a multinomial variable with parameters n and p, and we are testing $H_0: p \in A_0$ versus $H_1: p \in A_1$. The solution is the test φ_k that rejects H_0 when the ratio of the posterior probability of H_1 to the posterior probability of H_0 exceeds the ratio of the losses L_0/L_1; that is, if p has the posterior density $g(p|\mathbf{N})$, $\mathbf{N} = (N_1, \cdots, N_{k-1})$, then the test φ_k rejects H_0 when

(3.6) $$\frac{\int_{A_1} g(p|\mathbf{N}) \, dp}{\int_{A_0} g(p|\mathbf{N}) \, dp} \geqq \frac{L_0}{L_1}.$$

EXAMPLE 3.1. Consider the goodness of fit problem where Ω_0 contains only the uniform distribution F_0 on $(0, 1)$ and

(3.7) $$\Omega_1 = \{F: C_F \subset [0, 1], F \leqq F_0\} - \Omega_0.$$

If P assigns probability $\frac{1}{2}$ to Ω_0, and if $t_1 = 0$, $t_2 = \frac{1}{2}$, $t_3 = 1$, then from (3.5)

(3.8) $$\varphi_k = \begin{cases} 1 & \text{if} \quad 2^{-n} \int_0^{1/2-0} p_1^{N_1}(1 - p_1)^{n-N_1} \pi_1(dp_1) \geqq \frac{L_1}{L_0} \\ 0 & \text{otherwise} \end{cases}$$

where $\pi_1(p_1) = P(F(\frac{1}{2}) \leqq p_1)$, so that $\pi_1(\frac{1}{2} - 0) = \frac{1}{2}$ and $\pi_1(\frac{1}{2}) = 1$. Thus φ_k is based on a decreasing function of $N_1 = $ number of X less than or equal to $\frac{1}{2}$. When $2\pi_1$ is the beta distribution, then φ_k can be obtained from the tables of the incomplete beta function.

REMARK 3.1. The loss function of this section is not the only one for which optimal mixed tests can be obtained. For instance, if Ω_0 and Ω_1 are as in Example 3.1, then the loss for deciding H_0 when $F \in \Omega_1$ could be defined to be $L_1(\frac{1}{2} - F(\frac{1}{2}))$. The conditions of Theorem 2.1 would then still be satisfied and the optimal test can be obtained from the corresponding multinomial problem.

4. Convergence of the mixed solutions to the Bayes solution

In this section, we will consider classes of distribution functions on $[0, 1]$, that is, Ω is such that its support $S(\Omega)$ is contained in $[0, 1]$. Then Ω is a subset of the class $D[0, 1]$ of right continuous functions on $[0, 1]$ with limits from the left. On $D[0, 1]$ we use the Skorohod topology with the modified Skorohod metric (see for example [2], p. 113) so that $D[0, 1]$ is a complete separable metric space. In the notation of Section 2, $\Gamma = D[0, 1]$ and $\mathscr{S}$ is the σ-field generated by the open sets, F is a random function (a measurable function on some measure space to $(\Gamma, \mathscr{S})$) with distribution P. We assume that the probability P on $(\Gamma, \mathscr{S})$ satisfies $P(F \in \Omega) = 1$. If $\{F_k\}$ is a sequence of random functions with distributions $\{P_k\}$, then F_k is said to *converge in distribution* to F if

$$(4.1) \qquad \int_\Gamma h(F) P_k(dF) \to \int_\Gamma h(F) P(dF)$$

for each continuous, bounded, real valued function h on Γ. If (4.1) holds, then P_k is also said to *converge weakly* to P.

If F_k is the polygonal random distribution function of Section 2, then F_k does not necessarily converge in distribution to F. However, we will show that the Bayes risk of the Bayes solution for the prior P_k converges to the Bayes risk of the Bayes solution for the prior P. To do this, we will make use of the random distribution function G_k that is constant over each interval $[t_i, t_{i+1})$, and for which the joint distribution of $G_k(t_1), \cdots, G_k(t_k)$ equals that of $F(t_1), \cdots, F(t_k)$. The symbol Q_k will denote the distribution of G_k. We will need double subscripts on the t and again write $t_{m,1}, \cdots, t_{m,k_m}$ instead of $t_1, \cdots, t_k$. We assume that $\{\Pi_m: t_{m,1} < \cdots < t_{m,k_m}\}$, $m = 1, \cdots, k_m$ is a sequence of partitions of $[0, 1]$ such that Π_{m+1} is a refinement of Π_m and $\max_i |t_{m,(i+1)} - t_{m,i}| \to 0$ as $m \to \infty$.

It is now easy to show using Prohorov's theorem that:

LEMMA 4.1. *If $\Omega \subset D[0, 1]$, then G_k converges in distribution to F.*

PROOF. For each $F_0 \in \Omega$ and $\delta \in (0, 1)$ define

$$(4.2) \qquad v(F_0, \delta) = \sup \min [F_0(t) - F_0(s), F_0(u) - F_0(t)],$$

where the sup is over $s \leqq t \leqq u$, $u - s = \delta$. Similarly, define

$$(4.3) \qquad w_0(F_0, \delta) = \sup [F_0(u) - F_0(s)],$$

where the sup is over $1 - \delta \leqq s < u < 1$. By Prohorov's theorem applied to $D[0, 1]$ it is enough to show that (i) the finite dimensional distributions of Q_k converge weakly to the finite dimensional distributions of P at points $s_1, \cdots, s_r$ in the set $\{t: P[F(t) \neq F(t^-)] = 0\} \cup \{0, 1\}$, and that (ii) for each $\varepsilon, \eta > 0$, there exists $\delta \in (0, 1)$ and an integer k_0 such that

$$(4.4) \qquad Q_k(v(G_k, \delta) < \varepsilon) > 1 - \eta \qquad \text{for} \quad k \geqq k_0,$$

and

$$(4.5) \qquad Q_k(w_0(G_k, \delta) < \varepsilon) > 1 - \eta \qquad \text{for} \quad k \geqq k_0.$$

(See for example [2], pp. 125–126.)

The convergence of the indicated finite dimensional distributions follows from the definition of G_k. The inequalities (4.4) and (4.5) are easy to establish using the tightness of **P**. We omit the first subscript on the t to simplify the notation. Let k_0 be such that $t_{(i+1)} - t_i < \delta$, $i = 1, \cdots, k$, for all $k \geqq k_0$. Then since G_k is constant between points t_i,

$$(4.6) \qquad v(G_k, \delta) \leqq \max \min \left[F(t_i) - F(t_j), F(t_j) - F(t_\ell) \right],$$

where the max is over $t_\ell \leqq t_j \leqq t_i$, $t_i - t_\ell \leqq 2\delta$. The right side is bounded above by $v(F, 2\delta)$. Thus

$$(4.7) \qquad v(G_k, \delta) \leqq v(F, 2\delta) \qquad \text{for} \quad k \geqq k_0.$$

Similarly, $w_0(G_k, \delta) \leqq w_0(F, 2\delta)$. Since P is tight, we can choose δ so that $P(v(F, 2\delta) < \varepsilon) > 1 - \eta$ and $P(w_0(F, 2\delta) < \varepsilon) > 1 - \eta$. This implies the result.

Let $G_{q,k}$ denote the distribution function that is constant on $[t_i, t_{i+1})$ and whose value at t_i is q_i, where $0 = q_1 \leqq \cdots \leqq q_k = 1$. Thus Q_k is the distribution of $G_{q,k}$ when q has distribution λ, and G_k is the random distribution function obtained by letting q in $G_{q,k}$ have distribution λ.

Recall that d_k is the decision rule that minimizes the Bayes risk $r(P_k, d)$.

THEOREM 4.1. *If $G_{q,k}$ is an element of Ω for almost all q in C_λ, if the conditions of Theorem 2.1 are satisfied, if a Bayes solution d exists, and if d has a continuous (in F) bounded risk $R(F, d)$, then for $\Omega \subset D[0, 1]$, d_k converges to d in the sense that the Bayes risk $r(P, d_k)$ of d_k converges to the Bayes risk $r(P, d)$ of d.*

PROOF. Consider the Bayes solution $\hat{d}_k$ for the prior Q_k. Since $G_{q,k}$ is constant between points t_i, $\hat{d}_k$ will depend on the X only through $\mathbf{S} = (S_1, \cdots, S_{k-1})$, where $S_i = $ number of X equal to t_{i+1}, $i = 1, \cdots, k - 1$. (Recall that $F(t_1) = F(0) = 0$ a.s. (P) by assumption.) Note that $\mathbf{S}$ and $\mathbf{N}$ have the same distribution under $G_{q,k}$ and that this is the distribution $\mathbf{N}$ has under $F_{q,k}$. This implies that d_k is also a Bayes solution for the prior Q_k. We now have

$$(4.8) \qquad r_k(P, d_k) = r(P_k, d_k) = r(Q_k, d_k) = r(Q_k, \hat{d}_k) \leqq r(Q_k, d).$$

By Lemma 4.1,

$$(4.9) \qquad \lim_{k \to \infty} r(Q_k, d) = r(P, d).$$

Equations (4.8) and (4.9) yield

$$(4.10) \qquad \limsup_{k \to \infty} r_k(P, d_k) \leqq r(P, d).$$

On the other hand, by Lemma 2.1,

$$(4.11) \qquad r_k(P, d_k) \geqq r(P, d_k).$$

Since d is the Bayes solution for P,

$$(4.12) \qquad r(P, d_k) \geqq r(P, d).$$

Putting these inequalities together, we get

(4.13) $\qquad \lim_{k \to \infty} r_k(P, d_k) = r(P, d), \qquad \lim_{k \to \infty} r(P, d_k) = r(P, d).$

If Ω is a class of continuous distribution functions on $[0, 1]$, then it is a subset of the class $C[0, 1]$ of continuous functions on $[0, 1]$. On $C[0, 1]$ we use the sup norm and the σ-algebra generated by the open sets.

LEMMA 4.2. *If $\Omega \subset C[0, 1]$, then F_k converges in distribution to F.*

PROOF. The convergence of finite dimensional distributions follows from the definition of F_k. For each of $F_0 \in \Omega$ and $\delta \in (0, 1)$, let

(4.14) $\qquad w(F_0, \delta) = \sup_{t - s = \delta} [F_0(t) - F_0(s)].$

Let k_0 be such that $t_{i+1} - t_i < \delta$, $i = 1, \cdots, k$, for all $k \geq k_0$. Now the tightness of $\{P_k\}$ follows from the inequality $w(F_k, \delta) \leq w(F, 3\delta)$ and the tightness of P. Thus the result follows from Prohorov's theorem applied to $C[0, 1]$.

We can now prove that d_k converges to d under fewer conditions than when $\Omega \subset D[0, 1]$.

THEOREM 4.2. *If $F_{q,k} \in \Omega$ for almost all q in C_λ, if d_k is a mixed Bayes-minimax solution, if a Bayes solution d exists, and if d has a continuous bounded risk $R(F, d)$, then for $\Omega \subset C[0, 1]$, $\lim_{k \to \infty} r(P, d_k) = r(P, d)$.*

PROOF. Since d_k is Bayes for P_k, then

(4.15) $\qquad r_k(P, d_k) = r(P_k, d_k) \leq r(P_k, d).$

By Lemma 4.2,

(4.16) $\qquad \lim_{k \to \infty} r(P_k, d) = r(P, d).$

The rest of the proof now follows on the lines of the proof of Theorem 4.1.

5. Examples of random distribution functions

In order to obtain the mixed Bayes-minimax solutions, we have to specify a distribution for the random distribution function F_k which is linear between the points $(t_1, q_1), \cdots, (t_k, q_k)$, $0 = q_1 \leq \cdots \leq q_k = 1$. Equivalently, we have to define a probability λ on

(5.1) $\qquad A_k = \{q \in R^k : 0 = q_1 \leq \cdots \leq q_k = 1\},$

or a probability π on

(5.2) $\qquad B_k = \left\{ p \in R^{k-1} : 0 \leq p_i \leq 1, \sum_{i=1}^{k-1} p_i = 1 \right\}.$

Here q_i is thought of as $F(t_i)$ and p_i as $F(t_{i+1}) - F(t_i)$. We say that $p = (p_1, \cdots, p_{k-1})$ has distribution π.

One way to obtain a class of distributions π of p (Freedman [7], Fabius [5], Connor and Mosimann [16]), is to let p have the same distribution as the vector whose ith coordinate is

$$(5.3) \qquad Z_i \prod_{j=1}^{i-1} (1 - Z_j), \qquad\qquad i = 1, \cdots, k - 1,$$

where $Z_1, \cdots, Z_{k-1}$ are independent random variables satisfying

$$(5.4) \qquad 0 < Z_i \leqq 1, \qquad Z_{k-1} = 1.$$

For each choice of distributions $H_1, \cdots, H_{k-2}$ of the Z, we obtain a probability π on B_k. For this class of probabilities, it is easy ([7], p. 1401 and [5], p. 848) to compute posterior probabilities $\pi(p|\mathbf{N})$ of p given $\mathbf{N} = (N_1, \cdots, N_{k-1})$. Such probabilities π are called *tailfree* by Freedman [7] and Fabius [5] and *neutral* by Connor and Mosimann [16]. If we let each Z_i have a beta distribution $B(r_i, s_i)$ with parameters r_i and s_i, then π is called the *generalized Dirichlet distribution* [16]. If in addition,

$$(5.5) \qquad s_i = \sum_{j=i+1}^{k-1} r_j, \qquad\qquad i = 1, \cdots, k - 2, \quad s_{k-1} = 0,$$

then π is called the *Dirichlet* distribution with parameters $r_1, \cdots, r_{k-1}$.

Extensions of the definition of π on B_k to

$$(5.6) \qquad B_\infty = \left\{ p \in R^\infty : 0 \leqq p_i \leqq 1, \sum_{i=1}^{\infty} p_i = 1 \right\}$$

are obtained by replacing (5.4) by

$$(5.7) \qquad 0 \leqq Z_i \leqq 1, \qquad \lim_{r \to \infty} \prod_{i=1}^{r} (1 - Z_i) = 0 \qquad \text{a.s.}$$

If the Z have beta distributions, then the resulting π on B_∞ is called the *infinite dimensional generalized Dirichlet* distribution (Freedman [7]).

Note that, if p has a Dirichlet distribution, then Cov $(p_i, p_j) < 0$. However, for the generalized Dirichlet distribution, it is possible to have Cov $(p_i, p_j), > 0$ (see [16], p. 198).

EXAMPLE 5.1. Consider the problem of estimating the mean $\mu(F) = E_F(X_1)$ when $p = [F(t_2) - F(t_1), \cdots, F(t_k) - F(t_{k-1})]$ has a generalized Dirichlet distribution π with parameters $(r_1, s_1), \cdots, (r_{k-1}, s_{k-1})$. If we consider the squared error loss function, then the Bayes estimate of

$$(5.8) \qquad \mu(F_k) = \tfrac{1}{2} \sum_{i=1}^{k-1} [F(t_{i+1}) - F(t_i)](t_{i+1} + t_i)$$

is

$$(5.9) \qquad \hat{\mu}_k = E_\pi(\mu(F_k)|X) = \tfrac{1}{2} \left(\sum_{i=1}^{k-1} [t_{i+1} + t_i] E_\pi(p_i|\mathbf{N}) \right),$$

where (see [5])

$$(5.10) \qquad E_\pi(p_i|\mathbf{N}) = \frac{r_i + N_i}{r_1 + s_1 + n} \prod_{\ell=1}^{i-1} \frac{s_\ell + n - \sum\limits_{j=1}^{\ell} N_j}{r_{\ell+1} + s_{\ell+1} + n - \sum\limits_{j=1}^{\ell} N_j}.$$

Following Ferguson [6], let α be a finite, finitely additive measure on R and let

$$(5.11) \qquad r_i = \alpha(t_i, t_{i+1}], \qquad s_i = \sum_{j=i+1}^{k-1} r_j, \qquad s_{k-1} = 0.$$

Assume that α assigns measure zero to the region outside $(t_1, t_k]$. Then

$$(5.12) \qquad E_\pi\big(\mu(F_k)\big|X\big) = \big(\alpha(R) + n\big)^{-1} \sum_{i=1}^{k-1} \tfrac{1}{2}(t_{i+1} + t_i)\big(\alpha(t_i, t_{i+1}] + N_i\big)$$

$$= \alpha_n E\big(\mu(F_k)\big) + (1 - \alpha_n)\bar{X}',$$

where $\alpha_n = \alpha(R)[\alpha(R) + n]^{-1}$ and $\bar{X}'$ is the average of the random variables X' obtained by replacing each X in the interval $(t_i, t_{i+1}]$ by the midpoint $\tfrac{1}{2}(t_{i+1} + t_i)$. Note that if the t become dense in $(t_1, t_k]$ as in Section 4, then from (5.12)

$$(5.13) \qquad \lim_{k \to \infty} E_\pi\big(\mu(F_k)\big|X\big) = \alpha_n \mu_0 + (1 - \alpha_n)\bar{X},$$

where $\mu_0 = \alpha(R)^{-1} \int x\alpha(dx)$. This is the estimate obtained by Ferguson [6]. Note that in addition to being the Bayes estimate of $\mu(F_k)$, the estimate (5.9) is the mixed Bayes-minimax estimate of $\mu(F)$ for the loss function $[d - \mu(F, k)]^2$ of Example 2.3.

EXAMPLE 5.2. Suppose again that p has a generalized Dirichlet distribution π. If $s \in \{t_1, \cdots, t_k\}$, the mixed Bayes-minimax estimate of $F(s)$ using squared error loss $[d - F(s)]^2$ is

$$(5.14) \qquad \hat{F}(s) = \sum_{j=1}^{i-1} E_\pi(p_j|\mathbf{N}),$$

where $s = t_i$ and $E_\pi(p_j|\mathbf{N})$ is given by (5.10). If in addition (5.11) is satisfied, then (5.14) becomes

$$(5.15) \qquad \alpha_n \alpha_0(s) + (1 - \alpha_n)F_n(s),$$

where $\alpha_0(s) = \alpha(-\infty, s]/\alpha(R)$ and $F_n(s)$ is the empirical distribution function of the sample. This is exactly the estimate obtained by Ferguson [6].

Next we consider the problem of defining a probability P on a set Ω of distribution functions F in such a way that it is possible to compute the posterior of F given X under the prior P. Ferguson [6] shows that for each finite, finitely additive measure α on R, it is possible to define a *Dirichlet process* F in such a way that $P_F(A_1), \cdots, P_F(A_m)$ has a Dirichlet distribution with parameters $\alpha_1, \cdots, \alpha_m$, where $d_i = \alpha(A_i)$, and $A_1, \cdots, A_m$ is a measurable partition of R.

He shows that the posterior of F given X is again a Dirichlet process with α replaced by $\alpha + \Sigma_{i=1}^{n} \delta(x_i)$, where $\delta(x)$ is the measure giving mass one to the point x. This makes it possible to compute the Bayes procedure for this prior. The estimate (5.15) in Example 5.2 above is both the Bayes and the mixed Bayes-minimax estimate for this prior.

Fabius ([5], p. 853) gives a general construction of probabilities on the set Ω of all distribution functions on $[0, 1]$ that include the Dirichlet process on $[0, 1]$, the processes of Kraft [12], Kraft and van Eeden [13], and those special cases of the processes of Dubins and Freedman that are contained in [13]. Kraft and van Eeden [13] compute the Bayes estimate for one of these processes for a problem in bioassay. Ferguson shows that if F is the Dirichlet process and Ω_1 is the class of discontinuous distribution functions, then $P(F \in \Omega_1) = 1$. Kraft [12] shows that it is possible to use the construction of Fabius to obtain a process F such that $P(F \in \Omega^*) = 1$, where Ω^* is the class of absolutely continuous distribution functions.

Using definitions (2.2) and (2.3) of Fabius [5], it is possible to check that the Dirichlet process is tailfree for all trees of partitions. Thus one can use expression (2.4) of [5] for the posterior distribution of a tailfree process to conclude that the posterior of a Dirichlet process is again Dirichlet.

I am indebted to Lucien LeCam, Thomas Ferguson, Jaap Fabius, and others for helpful discussions, and to Michael Stuart for a careful reading of the manuscript that led to many improvements.

REFERENCES

[1] C. Antoniak, "Mixtures of Dirichlet processes with application to some nonparametric problems," Ph.D. thesis, Mathematics Department, University of California, Los Angeles, 1969.

[2] P. Billingsley, *Convergence of Probability Measures*, New York, Wiley, 1968.

[3] K. A. Doksum, "Minimax results for IFRA scale alternatives, *Ann. Math. Statist.*, Vol. 40 (1969), pp. 1778–1783.

[4] L. E. Dubins and D. A. Freedman, "Random distribution functions," *Proceedings of the Fifth Berkeley Symposium on Mathematical Statistics and Probability*, Berkeley and Los Angeles, University of California Press, 1966, Vol. 2, pp. 183–214.

[5] J. Fabius, "Asymptotic behavior of Bayes' estimates," *Ann. Math. Statist.*, Vol. 35 (1964), pp. 846–856.

[6] T. S. Ferguson, "A Bayesian analysis of some nonparametric problems," 1970, submitted for publication.

[7] D. A. Freedman, "On the asymptotic behavior of Bayes' estimates in the discrete case," *Ann. Math. Statist.*, Vol. 34 (1963), pp. 1386–1403.

[8] W. Hoeffding, "'Optimum' nonparametric tests," *Proceedings of the Second Berkeley Symposium on Mathematical Statistics and Probability*, Berkeley and Los Angeles, University of California Press, 1951, pp. 83–92.

[9] P. J. HUBER, "A robust version of the probability ratio test," *Ann. Math. Statist.*, Vol. 36 (1965), pp. 1753–1758.

[10] ———, "Robust confidence limits," *Z. Wahrscheinlichkeitstheorie und Verw. Gebiete*, Vol. 10 (1968), pp. 269–278.

[11] ———, *Théorie de l'Inférence Statistique Robuste*, Montreal, Les Presses de l'Université de Montréal, 1969.

[12] C. H. KRAFT, "A class of distribution function processes which have derivatives," *J. Appl. Probability*, Vol. 1 (1964), pp. 385–388.

[13] C. H. KRAFT and C. VAN EEDEN, "Bayesian bio-assay," *Ann. Math. Statist.*, Vol. 35 (1964), pp. 886–890.

[14] E. RUIST, "Comparison of tests for non-parametric hypotheses," *Arkiv Math.*, Vol. 3 (1954), pp. 133–163.

[15] W. SCHAAFSMA, *Hypothesis Testing Problems with the Alternative Restricted by a Number of Inequalities*, Groningen, Noordhoff, 1966.

[16] R. J. CONNOR and J. E. MOSIMANN, "Concepts of independence for proportions with a generalization of the Dirichlet distribution," *J. Amer. Statist. Assoc.*, Vol. 64 (1969), pp. 194–206.

A FEW SEEDLINGS OF RESEARCH

J. M. HAMMERSLEY
UNIVERSITY OF OXFORD

1. Sowing

Graduate students sometimes ask, or fail to ask: "How does one do research in mathematical statistics?" It is a reasonable question because the fruits of research, lectures and published papers bear little witness to the ways and means of their germination and ripening. How did that author ever come to state this theorem so aptly, to snatch this neat proof from the thin air: surely it never sprang fully armed like Pallas Athene from the brow of Zeus? No, indeed not, rest reassured. But still the question is a hard one. The answer depends much upon the field of statistics, and even more upon the tastes and skills and prejudices of the researcher. The only means of appraisal is case study, and no author has the data for any case study but his own. Without more ado or apology, I shall speak of some of my own work and how it came about; and it will have to be a personal story, with all natural drawbacks, and of course to get a proper view of things at large, you will, nay must, look elsewhere for other accounts by other men of their own work.

The Committee on Support of Research in the Mathematical Sciences, in the introduction to a volume of essays [21], wrote:

"Our task was to assess the present status and the projected future needs, especially fiscal needs, of the mathematical sciences. . . . We realize that even scientific readers of our report, let alone nonscientists, may feel that they are not adequately informed about what mathematical research, especially modern mathematical research, consists of. Similarly, even professional mathematicians . . . may be unaware of the applications. . . . To provide additional background of factual information concerning the mathematical sciences, we are supplementing our report with the present collection of essays. . . . We believe that the mathematical community has no obligations more important than those concerned with education. . . ."

The essays in this collection provide an excellent account of the substance and applications of mathematical research: it deals, as the Committee intended, with the what. It does not try to deal with the how. Yet anyone agreeing with the last sentence in the above quotation, as I heartily do, must feel that the how also deserves educational coverage. So I have conceived the present paper as if it were *one* chapter for a companion volume, yet to be compiled, on the how of mathematical research.

The reader whom I have in mind, mainly though not exclusively, is a graduate student just about to embark on research in mathematical statistics. He will have had a thorough undergraduate education in mathematics, so he will have a good technical grasp of the subject; but probably he will also be limited by some of the unavoidable drawbacks inherent in any undergraduate course. In particular, he will suffer from a surfeit of knowledge and from overexposure to the *fait accompli* and to rigor. In the secondary school he will have done some nonrigorous mathematics (at least I hope so) if he has not squelched too deeply into the "New Math" mire; but he will not yet know how to argue nonrigorously at a more advanced level. So I have deliberately written a good deal of this paper in a nonrigorous style; for nonrigorous thought is an essential first stage in almost any piece of research. The rigor comes at the end of the process: undergraduate training serves the end, not the start.

Most of this paper deals with one particular problem: the problem itself, namely the what, is less important than its how. It would be impossibly tedious to give the how in full detail; and there is no need for me to say much about the false avenues explored or the mistakes made. Everybody commits these, and no one finds them hard to imagine. What I have done instead, and what I hope will suffice, is to leave the investigation of the problem in an unfinished state. This may show in snapshot form what research looks like while still in progress, not as it usually appears in the literature as a packaged end-product. The material in the later parts of the paper is at a more rudimentary stage of development than in the earlier parts, where I have ventured to formulate some of the results as theorems. This makes a virtue of necessity; for I have had to write for a publication deadline. The whole investigation has occupied eleven weeks—about three weeks to do the mathematics, interspersed throughout eight weeks of writing it up. It will be no bad thing for realism if this haste shows in the text.

There are two other things. First, I have emphasized, especially at the end of the paper, conjectures and unsolved problems. These, into which the graduate student may feel inclined to stick his teeth, serve the same purpose as exercises in a textbook. Second, I have posed a problem (on the interfusion of aluminium and copper) more or less in the raw state in which it comes from the scientist to the mathematician. This can serve as an exercise in constructing a suitable mathematical model (the one sketched in the text may *not* be suitable).

Some people would say that recent years have been a golden age for mathematical research, with funds and graduate students on an unprecedented scale. I thought so too once; but, having seen the age unfold, my views have changed. The plenitude has enlarged the quantity of mediocre research without enhancing quality. I regret the way a doctorate has become virtually a union ticket for university employment. Admittedly, my own experience colors these views: I was lucky enough never to have to write a dissertation. Having started before the time when doctorates were fashionable, I was able to jump in at the deep end straight away; and, instead of being restrained by the single theme of a dissertation, I could pick and choose from a wide range of scientific enquiries, of a richness which today's graduate students rarely meet or even visualize, and in the three years it would have taken me to write one doctoral thesis I published a dozen papers on all sorts of different topics.

Who sows the seed? Whence the subject matter? By inclination I am not a theorist, but a problem solver. When I first acquired my bachelor's degree in mathematics some twenty years ago, I had the good fortune to be appointed as a junior assistant in a small department at Oxford whose purpose and name was the Design and Analysis of Scientific Experiment. This meant it had to do the sums and headscratching and minister to the general statistical needs of biologists, foresters, and farmers in the university. There could have been no better place for source material. Scientists of all breeds would bring in a cornucopia of subjects, susceptible of mathematical treatment. This is not to claim that I regularly answered their questions: far from it, a mathematician in these circumstances will more likely convert the questions to his own purposes. For example, quite early on I managed to conjure a paper on nonharmonic Fourier series [10] out of the apparently unpromising raw material of carrot roots soaked in acetic acid, much to the consternation of the botanist concerned. (This may be the only occasion when the august pages of *Acta Mathematica* have talked of carrots and vinegar.) Later I moved towards theoretical physics, which is a chastening experience because theoretical physicists tend to be much better at mathematics than professional mathematicians themselves. (Sykes and Essam's work, to be mentioned in Section 20, is a good example of the mathematical virtuosity of physicists.) After some years, I got known for having solved a few of the problems put to me. This, let me explain with all due lack of modesty, is an unusual reputation for any mathematician to earn. Accredited mathematical statisticians never—what never? well, hardly ever— solve any problems at all: for, no sooner do they really acquire that knack, than they become known as theoretical physicists, or molecular biologists, or engineers, or the like; but I cannot claim to have been so successful as that. Nevertheless, over the years I have accumulated a stock of unsolved or partly solved mathematical problems.

So I could rummage in this stock for something to talk about when invited to speak at the Sixth Berkeley Symposium; and presently I came across the problem of what will (or might) happen if you clamp a flat sheet of copper closely against an aluminum one. As a simplified mathematical model, we can think of the copper sheet as occupying one half of space, and the aluminum the other half, with a plane boundary in between. Each metal consists of atoms positioned on a lattice: say, for simplicity, the lattice is the set of points with integer coordinates (x, y, z). Initially the copper atoms occupy the lattice points with $x \geqq 0$ while the aluminum occupies the lattice points with $x < 0$. However, a small percentage of the lattice points have no atom: these are called vacancies. The situation is a three-dimensional analogue of the child's toy with fifteen small numbered squares which can slide in a square box large enough for sixteen. Any of the six atoms next to a vacancy may by chance fall into it; and if some of these six atoms are copper and the rest aluminum, one kind may enjoy a preferential probability of jumping into the vacancy. Naturally, when an atom jumps into a vacancy, it leaves a new vacancy whence it came: or in other words, the vacancy jumps to a neighboring lattice point in exchange for an atom. Thus the vacancies describe random walks over the lattice. Yet this motion of the vacancies is only of intermediate concern: the ultimate interest is the motion they impart to the atoms. How will a marked atom move with the passage of time? Or better, what are the equations of the general atomic commotion caused by the randomly walking vacancies, and how fast will the copper plate diffuse into the aluminum plate and vice versa? How will this depend upon the percentage concentration of vacancies and any preferential probabilities postulated?

And now, dear graduate student, I pass this problem on to you as a possible topic for research, because with only two weeks to go before the Berkeley Symposium my own analysis of it was suddenly interrupted by that indefatigable beaver, Professor Gian-Carlo Rota of M.I.T. Might he jog my memory (he wrote firmly) of the collected works of Ulam under his editorship, and of its imminent dispatch to the printers, and of my promise to write some comments on certain papers he had sent me? And so I turned to these papers by Ulam, and at once a problem in one of them [25] intrigued and fascinated me.

That is how the seeds of research may be sown. Chance sows, and curiosity nurtures.

2. Ulam's problem

Ulam asks what is the distribution of the length (that is, number of terms) of a longest monotone subsequence of (not necessarily consecutive) terms in a random permutation of the first $n^2 + 1$ natural numbers. For example, if $n = 2$ and $n^2 + 1 = 5$, the permutation 51423 contains no monotone subsequence of length 4, but at least one (actually three) of length 3 (.1.23, 5.42., and 5.4.3), so any longest monotone subsequence of this particular permutation has length $3 = n + 1$. There are 120 possible permutations of 12345; and the longest length is 3 for 86 of them, 4 for 32 of them, and 5 for the remaining 2.

Ulam remarks that the longest length must always be at least $n + 1$ for any n, by virtue of what he calls "a well-known theorem." He then refers to a Monte Carlo study of the cases $n = 4$ to $n = 10$ by E. Neighbor: he cites no reference, and indeed Neighbor's work may be unpublished. For three of these values of n, he quotes the average longest lengths in the Monte Carlo samples, namely,

$$
\begin{aligned}
8.46 &= 1.69n &&\text{for} \quad n = 5 \\
14.02 &= 1.75n &&\text{for} \quad n = 8 \\
17.85 &= 1.78n &&\text{for} \quad n = 10.
\end{aligned}
$$

(2.1)

All the averages, he says, were about $1.7n$, while the distribution "turned out to have a Gaussian form starting at the guaranteed minimum [that is, $n + 1$], having its maximum at the average, and becoming vanishingly small at about 2.2 times the minimum." A careless, or perhaps a fancifully numerological reading of this quotation might suggest that the distribution was spread between $n\sqrt{1}$ and $n\sqrt{5}$ with mean $n\sqrt{3}$, thus giving a standard deviation also proportional to n. We shall discover that all these suggestions are wrong.

Now, a highly finished paper, with all its theorems carefully proved, all avenues explored, and all loose ends carefully snipped, may arouse one's admiration; but its very perfection drains it of vitality, and there is little one can do with it except file it. Papers are more entertaining if they are still rich in conjectures, with results unproved or even wrong: Ulam's paper is like this; and I shall do what I can to write here in a similar vein, incidentally saving space by avoiding rigor. As a high exemplar, we may recall that the most enigmatically stimulating communication in the whole history of mathematics was written under extreme restrictions on space—margin too small.

3. How well known is a well-known theorem!

Ulam's "well-known theorem" runs as follows:

THEOREM 1. (Ulam? or A.N. Other??) *Any real sequence of $n^2 + 1$ terms contains a monotone subsequence of $n + 1$ terms.*

Enterprising readers will doubtless wish to prove this theorem for themselves: so I defer the proof while they get to work. I regret to say this theorem was not well-known to me, indeed not known at all; and it took me a couple of days to invent my own proof, based on negative induction (if the theorem is false for n, it is also false for $n - 1$). Having found a proof, I cast around for colleagues who might know the theorem and be able to cite a reference (Ulam gives neither proof nor reference, thereby arousing his reader's curiosity). It was a long time before I met anyone able to answer: eventually, however, I turned up trumps in Professor Lincoln Moses. He believed there was a proof, based on the pigeonhole principle, in a collection of essays [21] prepared for the Committee on Support of Research in the Mathematical Sciences (COSRIMS); and he thought that Ulam himself had written the essay in question. Together Moses and I went to the Berkeley Library to verify this reference: but fortunately we found it was out on loan and so temporarily unavailable. I say "fortunately" because I now had the chance of reconstructing on my own a second proof, using the pigeonhole principle; and the reader has too.

A few days later I delivered the lecture on this paper at a session of the Sixth Berkeley Symposium; and I took the opportunity of doing a little operational research on the familiarity of well-known theorems. I asked the audience how many of them knew Ulam's "well-known theorem." Out of an audience of 59 accomplished mathematicians, just 3 knew it. Adding 1 to 59 to include my own ignorance, we conclude that 95 percent of mathematicians will be ignorant of a well-known theorem. But one theorem is a small sample of theorems.

Professor Milton Sobel was one of the three wise men in the audience at the Berkeley Symposium; and he told me that the theorem plus a proof—indeed a a third way of proving it—had appeared in Martin Gardner's column [7] in

Scientific American. It is astonishing that the theorem should be so little known after appearance in such justly famous and widely read columns.

Lastly, the COSRIMS essays [21] came back to the library, and I was able to consult them. And it came to pass as Moses had prophesied, that Ulam had indeed contributed an essay to this collection; but his essay made no mention of the theorem. Yet the collection contained another essay, which did mention it; and—curious coincidence?—the author of this essay was Gian-Carlo Rota, who, as we have already seen, is guilty of this whole diversion away from copper and aluminum.

The purpose of this chit-chat on the background to Theorem 1 has been to interpose a long enough piece of prose to discourage the reader's eye from straying from the statement of the theorem to its proof. But it also serves to introduce a more serious maxim, which every properly ambitious graduate student should know. For Rota's use of the pigeonhole principle, when I eventually read it, did not seem to me quite as satisfactory as the one I had reconstructed for myself: his argument is somewhat longer and proceeds by *reductio ad absurdum*, whereas my more direct line is a constructive proof. Moral: never read the literature before you absolutely have to (and not always, even then), for thus it will not cloud your imagination and sometimes you may be able to do better on your own.

Added in in proof. Theorems 1 and 2 are both due to Erdös and Szekeres *Compositio Math.*, Vol. 2 (1935), pp. 463–470 (in particular, page 468).

Theorem 1 is the particular case $a = d = n$ of the slightly more general Theorem 2 (which could be new, though I doubt it).

THEOREM 2. *Any real sequence of at least $ad + 1$ terms contains either an ascending subsequence of $a + 1$ terms or a descending subsequence of $d + 1$ terms.*

Here and later I interpret "ascending" to mean "nondecreasing," and "descending" to mean "strictly decreasing." (Theorem 2 is also true for "ascending" = "strictly increasing," and "descending" = "nonincreasing," the changes in the proof being trivial.)

Suppose we have a set of pigeonholes $P_1, P_2, \cdots$ and a given real sequence $X_N = \{x_1, x_2, \cdots, x_N\}$ with $N \geq ad + 1$. Place the terms of X_N successively in the pigeonholes: first put x_1 in P_1; then generally, with $x_1, x_2, \cdots, x_{i-1}$ already placed, place x_i in the pigeonhole P_j with the least value of j such that P_j already contains no term larger than x_i. Thus, at any stage of the procedure the contents of each pigeonhole comprises an ascending subsequence of X_N. Moreover (since j is least), when x_i goes into P_j, then P_{j-1} must already contain an earlier and greater term (that is, some x_h such that $h < i$ and $x_h > x_i$); and similarly P_{j-2} must contain an earlier and greater term than this x_h, and so on as far as P_1. Hence, there is a descending sequence with one term in each occupied pigeonhole (and it is explicitly recoverable by working backwards from any term in the last occupied pigeonhole). The required subsequence will have been constructed as soon as either some pigeonhole contains $a + 1$ terms or $d + 1$ pigeonholes become occupied. This event will occur because of the pigeonhole principle and because $N \geq ad + 1$.

4. Notation

The following notation will be useful. If r is any real number, define $\chi(r)$ to be the least integer such that $\chi(r) \geqq r$. We write

(4.1)
$$X = \{x_1, x_2, \cdots\}$$

for an infinite real sequence; and

(4.2)
$$X_N = \{x_1, x_2, \cdots, x_N\}$$

for the first N terms of X. We write

(4.3)
$$\ell_N = \ell(X_N)$$

for the length of a longest ascending subsequence of X_N, and

(4.4)
$$\ell'_N = \ell'(X_N)$$

for the length of a longest descending subsequence of X_N. Thus

(4.5)
$$\ell_N^* = \ell^*(X_N) = \max(\ell_N, \ell'_N)$$

denotes the length of a longest monotone subsequence of X_N.

With this notation we can now state Theorem 3.

THEOREM 3. *For any real sequence X,*

(4.6)
$$\ell^*(X_N) \geqq \chi(\sqrt{N}),$$

and this inequality is best possible.

Since $\ell^*(X_N)$ is a nondecreasing function of N, the inequality (4.6) follows at once from the particular case $N = n^2 + 1$ covered by Theorem 1. To see that (4.6) is best possible, we arrange the positive integers into successive parts of one, two, three, $\cdots$ terms; and we reverse each part to obtain (1) $(3\ 2)$ $(6\ 5\ 4)$ $(10\ 9\ 8\ 7) \cdots$. Then we interlace this sequence with its negative to yield the particular sequence $X = \{x_1, x_2, \cdots\}$ exhibited in (4.7).

(4.7)

N:	1		2	3	4		5		6	7	8	9		10		11		12	13	14	15	16		17 $\cdots$
x_N:	1		-1	3	2		-3		-2	6	5	4		-6		-5		-4	10	9	8	7		$-10 \cdots$
ℓ_N:	1		1	2	2		2		2	3	3	3		3		3		3	4	4	4	4		4 $\cdots$
ℓ'_N:	1		2	2	2		3		3	3	3	3		4		4		4	4	4	4	4		5 $\cdots$.

We can now verify for this particular sequence X that

(4.8)
$$\ell_N = \chi[(N + \tfrac{1}{4})^{1/2} - \tfrac{1}{2}] \leqq \ell'_N = \ell_N^* = \chi(\sqrt{N}).$$

5. Random sequences

From now on we shall suppose, unless the context specifically denies it, that X is a sequence of identically distributed independent random variables, whose common distribution is continuous. This assumption of a continuous distribution conveniently ensures that, apart from a set of measure zero, no two terms of X are equal. Sets of measure zero will be ignored without further comment; so we can presume that $x_i \neq x_j$ for all $i \neq j$. Instead of supposing that the x_i were independent, we might have supposed them symmetrically dependent in Sparre Andersen's sense, or exchangeable (see [4], p. 225 for definitions of this); but the extra generality would be more apparent than real in the present context, and we shall not entertain it.

Clearly $\ell(X_N)$, $\ell'(X_N)$, and $\ell^*(X_N)$ are distribution-free random variables (that is to say, their distributions do not depend upon the common distribution of the terms of X); and, by symmetry, $\ell(X_N)$ and $\ell'(X_N)$ have the same distribution. What can we say about the distributions of $\ell(X_N)$ and $\ell^*(X_N)$, as Ulam asks, and what practical applications are there for the results? Ulam and Neighbor's results suggest that $N^{-1/2}\ell^*(X_N)$ converges to some random variable: what is it? We shall discover that a stronger asymptotic result holds.

THEOREM 4. *If X is a random sequence of the type described above, then $N^{-1/2}\ell(X_N)$ and $N^{-1/2}\ell^*(X_N)$ both converge in probability to an absolute constant c as $N \to \infty$. They also converge in pth absolute mean for any p satisfying $0 < p < \infty$.*

It seems very likely that these two random variables also converge with probability 1, but I cannot yet prove this conjecture.

To prove Theorem 4, we first assemble some remarks on subadditive stochastic processes and stochastic summation.

6. Subadditive and superadditive stochastic processes

Subadditive stochastic processes were first invented [15] to deal with time-dependent percolation processes; but they have other applications, one of which is to Ulam's problem. They also afford a generalization of renewal processes. Much remains to be done on the theory of these processes; and the original paper [15] lists a number of conjectures (the authors of a paper which does not furnish ample conjectures may be suspected, rightly or wrongly, of not working to the limits of their capabilities). With one exception, all these conjectures remain open: the exception is due to Kingman [18], who proved the ergodic theorem for subadditive processes.

A subadditive stochastic process is a family of real random variables $\{w_{r,s}(\omega)\}$ defined on a probability space (Ω, B, P) and indexed by a pair of nonnegative integers r, s such that $r \leq s$. The process satisfies three postulates.

POSTULATE (i). *The process is stationary in the sense that its finite-dimensional distributions are the same as those of the shifted process*

$$(6.1) \qquad w'_{r,s} = w'_{r,s}(\omega) = w_{r+1,s+1}(\omega).$$

POSTULATE (ii). *Each random variable of the process has finite expectation. The stationarity then ensures that this expectation depends only upon the difference of the indices r and s:*

$$(6.2) \qquad E(w_{r,s}) = g_{s-r}, \; say.$$

POSTULATE (iii).   *For any $\omega \in \Omega$ and any three indices r, s, t such that $r \leqq s \leqq t$, we have the subadditive property*

(6.3) $$w_{r,t}(\omega) \leqq w_{r,s}(\omega) + w_{s,t}(\omega).$$

From (6.2) and (6.3), we deduce

(6.4) $$g_{u+v} \leqq g_u + g_v, \qquad\qquad u, v = 0, 1, 2 \cdots ;$$

and hence, by a standard result on subadditive functions,

(6.5) $$\lim_{t \to \infty} g_t/t = \inf_{t \geqq 1} g_t/t = c,$$

where c satisfies $-\infty \leqq c < \infty$. If g_t/t is bounded below then c is finite and is called the time constant of the process. In what follows, we always assume that the time constant exists.

A process $w_{r,s}$ is called superadditive if $-w_{r,s}$ is subadditive; that is, if the inequalities in (6.3) and (6.4) are reversed. Theorems for subadditive processes remain true for superadditive processes with obvious trivial modifications. If equality holds in (6.3) and (6.4), the theory reduces to the ordinary theory of additive processes. The postulates given above are those due to Kingman [18]; and are slightly more stringent than those in the original paper [15], where one-dimensional distributions were used in place of finite-dimensional ones in Postulate (i). The stricter requirement (i) is needed for the ergodic theorem, which runs as follows:

THEOREM 5.   (Kingman) *Let $w = \{w_{r,s}\}$ be a subadditive stochastic process with a time constant c; and let I be the σ-field of events defined in terms of w and invariant under the shift $w \to w'$. Then as $t \to \infty$, $t^{-1}w_{0,t}(\omega)$ converges almost surely to a random variable $W(\omega)$, which can be expressed as a conditional expectation*

(6.6) $$W(\omega) = \lim_{t \to \infty} t^{-1}E(w_{0,t}|I).$$

Moreover $W = c$ almost surely when I consists only of events of probability 0 or 1.

7. Stochastic summation

Many of the common procedures in the classical theory of summation [16] can be thought of in terms of discrete frequency distributions, though admittedly this is not the most usual way of looking at them. The purpose of summation is to assign a meaning to the statement

(7.1) $$s_n = \sum_{i=0}^{n} u_i \to s = \sum_{i=0}^{\infty} u_i \quad \text{as} \quad n \to \infty.$$

Consider a family of discrete probability distributions defined on the nonnegative integers, members of the family being indexed by some numerical parameter μ:

(7.2) $$p_\mu = \{p_{0,\mu}, p_{1,\mu}, \cdots\}, \qquad p_{n,\mu} \geqq 0, \qquad \sum_{n=0}^{\infty} p_{n,\mu} = 1.$$

Quite commonly, μ is the mean of the distribution P_μ. Suppose the series

$$(7.3) \qquad \sigma_\mu = \sum_{n=0}^{\infty} s_n p_{n, \mu}$$

converges in the ordinary sense; so σ_μ, the expectation of the partial sums s_n in (7.1), exists with respect to the distribution p_μ. If further σ_μ converges to a number s as $\mu \to \infty$, then we say (as the meaning to be assigned to (7.1)) that $\Sigma \, u_i$ is summable to s in the sense of p_μ; and we write (7.1) as

$$(7.4) \qquad s = \sum_{i=0}^{\infty} u_i \qquad (p_\mu),$$

or as

$$(7.5) \qquad s_n \to s \qquad (p_\mu).$$

For example, consider three famous particular cases of this procedure. First, if p_μ is the discrete uniform distribution

$$(7.6) \qquad p_{n, \mu} = \begin{cases} N^{-1}, & n = 0, 1, \cdots, N - 1, \\ 0, & n \geqq N, \end{cases} \qquad \mu = \tfrac{1}{2}(N - 1),$$

we say that s is summable in the sense of Césaro, written $s_n \to s$ (C, 1). Second, if p_μ is the geometric distribution

$$(7.7) \qquad p_{n, \mu} = p^n(1 - p), \qquad \mu = p/(1 - p),$$

we say that s is summable in the sense of Abel, written $s_n \to s$ (A). Third, if p_μ is the Poisson distribution

$$(7.8) \qquad p_{n, \mu} = e^{-\mu} \mu^n / n!,$$

we say that s is summable in the sense of Borel, written $s_n \to s$ (B).

Besides being summable (p_μ), the series (7.4) or the sequence (7.5) may or may not be convergent in the ordinary sense. Two important classes of theorems deal with this situation. First, the so-called Abelian theorems assert (roughly speaking) that ordinarily convergent series (or sequences) are summable (p_μ). Second, the so-called Tauberian theorems assert the converse, provided that the terms u_i satisfy an additional condition (called a Tauberian condition) whose effect is to exclude the possibility of anomalous individual terms. For example, the Tauberian condition $u_n = O(n^{-1})$ suffices in order that Abel-summability should imply convergence.

Now the classical theory of summation deals with real variables u_i or s_n; but there is no reason why it should not be extended to random variables. I do not know whether the literature contains a systematic extension along these lines, though certainly there are isolated cases dealing mainly with Césaro-summation of random variables: if no such systematic extension exists, it could perhaps provide a straightforward topic for a doctoral dissertation. Here I have raised the subject because in the next section I shall use a Tauberian argument on Borel-summation of random variables.

8. Asymptotic behavior in Ulam's problem

Consider all possible subsequences of length n contained in X_N, and let $m_{n,N}$ denote the expected number of these which are monotone.

THEOREM 6.   *If* $n = \chi(e\sqrt{N}) + t$ *and* $t \geqq 0$, *then*

$$(8.1) \qquad P(\ell_N^* \geqq n) \leqq m_{n,N} = \frac{2}{n!}\binom{N}{n} \leqq \frac{e^{-2t}}{\pi\sqrt{N}}.$$

If ρ_k is the probability that exactly k of these subsequences are monotone, then

$$(8.2) \qquad P(\ell_N^* \geqq n) = \sum_{k\geqq 1} \rho_k \leqq \sum_{k\geqq 0} k\rho_k = m_{n,N} = \frac{2}{n!}\binom{N}{n};$$

for there are $\binom{N}{n}$ such subsequences and $2/n!$ is the probability that any specified one of them will be monotone. Also, since $n \geqq \chi(e\sqrt{N})$,

$$(8.3) \qquad \frac{m_{n+1,N}}{m_{n,N}} = \frac{N-n}{(n+1)^2} \leqq \frac{N}{(e\sqrt{N})^2} = e^{-2};$$

and hence

$$(8.4) \qquad m_{n,N} \leqq e^{-2t} m_{\chi(e\sqrt{N}),N}.$$

Hence, to complete the proof of Theorem 6, we need only establish that

$$(8.5) \qquad m_{n,N} \leqq 1/\pi\sqrt{N}$$

holds in the special case

$$(8.6) \qquad n = \chi(e\sqrt{N}).$$

Now if $N < 8$, we find $n > N$ and therefore $m_{n,N} = 0$ and (8.5) is trivially true. For the cases $8 \leqq N < 16$, calculation yields

$$(8.7)$$

N:	8	9	10	11	12	13	14	15
n:	8	9	9	10	10	10	11	11
$11!\,m_{n,N}$:	990	110	1100	121	726	3146	364	1365.

Thus in any of these eight cases, we have

$$(8.8) \qquad m_{n,N} \leqq 3146/11! \leqq 1/4\pi < 1/\pi\sqrt{N}, \qquad 8 \leqq N < 16.$$

So it remains to prove (8.5) for $N \geqq 16$. Using Stirling's formula in the form

$$(8.9) \qquad \log a! = (a + \tfrac{1}{2})\log a - a + \tfrac{1}{2}\log(2\pi) + \frac{\theta}{12a}, \qquad 0 < \theta < 1,$$

we deduce

$$(8.10) \quad \log m_{n,N} \leqq \log 2 + (N + \tfrac{1}{2}) \log N - N + \tfrac{1}{2} \log (2\pi) + (12N)^{-1}$$

$$- (2n + 1) \log n + 2n - \log (2\pi)$$

$$- (N - n + \tfrac{1}{2}) \log (N - n) + (N - n) - \tfrac{1}{2} \log (2\pi)$$

$$\leqq n + n \log \frac{N}{n^2} - (N - n + \tfrac{1}{2}) \log \left(1 - \frac{n}{N}\right) - \log (\pi n) + \frac{1}{12N}.$$

But $N/n^2 \leqq e^{-2}$; and also

$$(8.11) \qquad - (N - n) \log \left(1 - \frac{n}{N}\right) \leqq (N - n) \sum_{r=1}^{\infty} \left(\frac{n}{N}\right)^r = n.$$

Hence

$$(8.12) \qquad \log m_{n,N} \leqq - \tfrac{1}{2} \log \left(1 - \frac{n}{N}\right) - \log (\pi n) + (12N)^{-1}.$$

With $N \geq 16$, we have

$$(8.13) \qquad \frac{n}{N} \leqq \frac{e\sqrt{N} + 1}{N} = \frac{1}{\sqrt{N}} \left(e + \frac{1}{\sqrt{N}}\right) \leqq \tfrac{1}{4}(e + \tfrac{1}{4});$$

and therefore

$$(8.14) \qquad \log m_{n,N} \leqq - \log (\pi n) - \tfrac{1}{2} \log \frac{15 - 4e}{16} + \frac{1}{192}$$

$$\leqq - \log (\pi e \sqrt{N}) - \tfrac{1}{2} \log \tfrac{1}{4} + \frac{1}{192} \leqq - \log (\pi \sqrt{N}).$$

This proves (8.5), as required.

Theorem 6 shows that the upper tail of the distribution of $\ell_N^*/\sqrt{N}$, that is to say from e upwards, has probability $O(N^{-1/2})$ as $N \to \infty$; and likewise, because $\Sigma_{t=0}^{\infty} t^p e^{-2t}$ converges, the contribution from this tail to any pth absolute moment of $\ell_N^*/\sqrt{N}$ is $O(N^{-(p+1)/2})$. Thus in proving Theorem 4, there is no essential loss of generality in treating $\ell_N^*/\sqrt{N}$ as though it were a bounded random variable, restricted to the closed interval $[0, e]$. Accordingly, convergence in pth absolute mean will follow if we can prove convergence in probability. The same remarks apply to $\ell_N/\sqrt{N}$ because $\ell_N \leqq \ell_N^*$. Lastly, ℓ_N and ℓ_N' have the same distribution; so $\ell_N^*/\sqrt{N} = \max (\ell_N/\sqrt{N}, \ell_N'/\sqrt{N})$ will converge in probability to c if $\ell_N/\sqrt{N}$ does.

Thus it remains to prove that

$$(8.15) \qquad \ell_N \sim c\sqrt{N}$$

in probability as $N \to \infty$. There are, I suppose, three stages in solving a problem.

The first stage is to get a clear idea of the essence of the problem, and to clear away the minor irrelevancies. I have just done this for Theorem 4: equation (8.15) contains the essence of Theorem 4, and the statements about convergence in pth absolute mean are minor irrelevancies. This concentration upon the main issue may easily, and here does, simplify things: the distinction between ℓ_N^* and ℓ_N is comparatively unimportant, but ℓ_N is an easier quantity to handle. The second stage is to conceive the basic ideas, on which the proof will rest; and the third stage is to work out the details of the proof. This third stage is normally quite easy, just a matter of craftsmanship. The second stage is by far the most difficult, since it requires a certain power of imagination coupled with a mathematical alertness. My own approach, for what it is worth, is to run quickly through a catalogue of available mathematical tools: such a catalogue is liable to be quite short. Here (8.15) is a fairly hard result to get at; and most of the familiar tools (for example, law of large numbers, Markov processes, renewal theory, and so forth) can be eliminated as insufficiently powerful tools for the job. The only two tools, which seemed to me to be strong enough for the task, were subadditive processes and submartingales. Being more familiar with the former, I started with them; and since they worked, I did not pursue the other alternative. However, devotees of submartingales will doubtless wish to explore the latter possibility. In considering each tool in the catalogue, one has to envisage various different ways of using it. One needs to think of the problem upside down and inside out, as it were, and to entertain unusual ways of handling the tool. This is where the alertness comes in; for otherwise one may miss the elusive idea that does the trick. The situation is rather like playing a game of chess: one follows certain strategies, but one always has to be alert for the position that conceals a winning combination. In the present case, the main difficulty to surmount is that subadditive processes are essentially linear, namely

(8.16)
$$w_{0,t} \sim ct,$$

whereas (8.15) is nonlinear. How does one introduce the square root? There is a clue in the fact that it is a *square* root: why not make use of the geometrical properties of a square? Readers, who would like to evaluate their sense of mathematical alertness, may care to pause at this point and ask themselves what is going to happen next, *given* that subadditive processes, Borel-summation, and the geometry of a square are the combination of ideas that will prove (8.15).

To return to the chess analogy, you may imagine that you are faced with a chess columnist's game position, for which you are *told* that a winning combination exists and, moreover, can be achieved by a queen sacrifice followed by a discovered check and a pawn promotion. Of course, all that sort of information makes things much easier than it would have been if you had actually been playing the game itself and had had to make yourself aware of the existence and nature of the combination. If you wish to think out for yourself how (8.15) may be proved, do not turn over the page yet—for to do so would give the game away too soon.

Consider a Poisson process of unit density in the Euclidean plane with a co-ordinate system (ξ, η). For nonnegative integers r, s $(r \leqq s)$ let $S_{r,s}$ be the square

$$(8.17) \qquad r \leqq \xi < s, \qquad r \leqq \eta < s,$$

with the convention that this square is the empty set if $r = s$. We say that a set of points (ξ_i, η_i), $i = 1, 2, \cdots, k$, in the plane form a *chain* if they can be connected together by a path that proceeds in a northeasterly direction, where "northeast" means any direction between north and east inclusive: that is to say, if and only if

$$(8.18) \qquad \xi_1 \leqq \xi_2 \leqq \cdots \leqq \xi_k \quad \text{and} \quad \eta_1 \leqq \eta_2 \leqq \cdots \leqq \eta_k.$$

The length of a chain is defined to be the number of points in the chain. Now consider the points of the Poisson process, which fall in the square $S_{r,s}$, and let $w_{r,s}$ be the length of a longest chain which can be formed from some subset of these points. Here $w_{r,s} = 0$ if $S_{r,s}$ is empty or contains no points of the Poisson process.

Since the squares $S_{r,s}$ and $S_{s,t}$ are contained in $S_{r,t}$, and any point in $S_{s,t}$ is to the northeast of any point in $S_{r,s}$, we have

$$(8.19) \qquad w_{r,t} \geqq w_{r,s} + w_{s,t}, \qquad\qquad r \leqq s \leqq t.$$

The finite-dimensional distributions of the process $\{w_{r,s}\}$ are invariant under the shift $w_{r,s} \to w'_{r,s} = w_{r+1,s+1}$ by the spatial homogeneity of the Poisson process.

Let τ be the number of points of the Poisson process in $S_{0,t}$; and suppose that these points have coordinates (ξ_i, η_i) where $i = 1, 2, \cdots, \tau$. Let us also order these points so that

$$(8.20) \qquad \xi_1 < \xi_2 < \cdots < \xi_\tau.$$

Here we may ignore the possibility that any two ξ or any two η are equal, for this event has zero probability. The corresponding sequence $\{\eta_1, \eta_2, \cdots, \eta_\tau\}$ will be a sequence of identically distributed independent random variables; and $w_{0,t}$ will be the length of the longest ascending subsequence in $\{\eta_1, \eta_2, \cdots, \eta_\tau\}$. So $w_{0,t}$ has the same distribution as ℓ_τ. The random variable τ has a Poisson distribution with parameter t^2, the area of $S_{0,t}$; and hence, using Theorem 6, we have

$$(8.21) \qquad Ew_{0,t} = \sum_{N=0}^{\infty} e^{-t^2} \frac{t^{2N}}{N!} E(\ell_N)$$

$$= O\left\{ \sum_{N=0}^{\infty} e^{-t^2} \frac{t^{2N}}{N!} \sqrt{N} \right\} = O(t) \qquad \text{as} \quad t \to \infty.$$

So $t^{-1} Ew_{0,t}$ is bounded. We have now verified all three postulates in Section 6, and we can conclude that $\{w_{r,s}\}$ is a superadditive stochastic process. By Theorem 5, there exists a random variable W such that

(8.22) $t^{-1}w_{0,t} \to W$ almost surely   as $t \to \infty$.

The points of the Poisson process which lie in the unit square $r \leqq \zeta < r + 1$, $s \leqq \eta < s + 1$ can be generated from a random variable v (having a Poisson distribution with parameter 1) and an infinite collection of independent observations from a uniform distribution on $[0, 1)$ from which we utilize the first $2v$ observations to yield the coordinates of v points in the square. We write $\zeta_{r,s}$ for the collection of all random variables associated in this way with this square. The family $\{\zeta_{r,s}; r, s = 0, 1, \cdots\}$ is a family of mutually independent collections; and $w_{r,s}$ can be written as a fixed function

(8.23) $w_{r,s} = F\{\zeta_{r,s}, \zeta_{r+1,s}, \cdots, \zeta_{s+1,r}, \cdots\}$.

In particular, $w_{r,s}$ is independent of those $\zeta_{\rho,\sigma}$ with $\rho < r$ and $\sigma < s$. So the invariant σ-field of the $w_{r,s}$-events can be embedded in the remote σ-field of the $\zeta_{r,s}$; and hence consists only of events of probability 0 or 1. Thus W is almost surely a constant c; and we conclude that

(8.24) $t^{-1}w_{0,t} \to c$ almost surely   as $t \to \infty$.

It only remains to find a Tauberian argument for unscrambling the Borel-summation induced by the Poisson process. To this end we extend the definition of $w_{0,t}$: for any real $t \geqq 0$, no longer an integer necessarily, we say that $w_{0,t}$ is the length of the longest chain amongst the points of the Poisson process in the square $0 \leqq \zeta < t, 0 \leqq \eta < t$. For each given realization of the Poisson process, $w_{0,t}$ is a nondecreasing function of t: this fact serves as the required Tauberian condition. In the first place it ensures that (8.24) remains true for any sequence of real t tending to infinity, and therefore for any sequence of real random variables t tending to infinity with probability 1. Next we define $t(N)$ by the requirement that it is the smallest value of t such that the square $0 \leqq \zeta < t, 0 \leqq \eta < t$ shall contain exactly N points of the Poisson process. Here $N = 1, 2, \cdots$; and for each given N, the distribution of $w_{0,t(N)}$ is the same as the distribution of ℓ_N in Ulam's problem. Moreover, from the properties of the Poisson process

(8.25) $t(N)/\sqrt{N} \to 1$ almost surely   as $N \to \infty$.

Putting $t = t(N)$ in (8.24) we deduce that

(8.26) $w_{0,t(N)}/\sqrt{N} \to c$ almost surely   as $N \to \infty$.

Although we have almost sure convergence in (8.26), we have not proved almost sure convergence in Theorem 4: the reason is that $w_{0,t(N)}$ in (8.26) is not associated with a sequence X in the way in which ℓ_N is in the statement of Theorem 4. It merely has the same distribution as ℓ_N for each *given* N. Therefore we have only proved convergence in probability in Theorem 4.

THEOREM 7.   *The constant c in Theorem 4 satisfies*

(8.27) $\frac{1}{2}\pi \leqq c \leqq e$.

Consider the points of the Poisson process (with unit parameter) and a square of area N. For any point P in the square, let $Q(P)$ denote the point of the Poisson process which is northeast of P and as close to P as possible. Let Q_0 be the southwest corner of the square, and from the sequence

$$(8.28) \qquad\qquad Q_{i+1} = Q(Q_i), \qquad\qquad i = 0, 1, 2, \cdots.$$

The expected value of the horizontal (or vertical) projection of $Q_i Q_{i+1}$ is

$$(8.29) \qquad\qquad \int_0^\infty dr \int_0^{\pi/2} d\theta (re^{-\pi r^2/4})(r \cos \theta) = \frac{2}{\pi}.$$

These projected lengths are independent; and so the strong law of large numbers shows that (with probability 1 as $N \to \infty$) $\frac{1}{2}\pi\sqrt{N} + o(\sqrt{N})$ terms of (8.28) can be formed before reaching the opposite boundary of the square. The sequence (8.28) provides a chain, not longer than a longest chain in the square. This proves the lower bound in (8.27); and the upper bound follows at once from Theorem 6. Professor Kingman has remarked to me that, if in the third line of the proof of Theorem 7 we interpret the word "close" in terms of distance measured parallel to the diagonal of the square, we get a different integral in (8.29) with a value $(\pi/8)^{1/2}$. Thus the lower bound in (8.27) can be raised from $\frac{1}{2}\pi = 1.57 \cdots$ to $(8/\pi)^{1/2} = 1.59 \cdots$. Professor Blackwell has also obtained this result independently.

9. Monte Carlo methods

In this section I shall describe how to study the behavior of ℓ_N by a Monte Carlo method, called dummy truncation. This device, originally introduced many years ago to deal with percolation processes [12], is really a very simple idea. In studying a quantity, such as $L_N = N^{-1/2}\ell_N$, which converges in probability as $N \to \infty$, it may be better to spend a given amount of computing time on a small number of samples with large N rather than a large number of samples with small N. Too small a value of N entails the risk of not reaching the region of asymptotic behavior; but a small sample size with large N need not, on the other hand, prejudice the accuracy of the estimation because the sampling variance of L_N becomes small as N increases on account of the convergence. Ulam's account of Neighbor's Monte Carlo work does not mention the sample size (though one may guess it was large since a computer was used), but Ulam does state that the values of N were small ($N \leq 101$); and we shall see presently that his N are all much too small to represent the true asymptotic behavior. How does one know when the asymptotic region is reached? There is no panacea, of course; but a reasonable procedure is to study successive values of N until L_N appears to settle down to some stable value. Thus we regard the whole vector

$$(9.1) \qquad\qquad \{\ell(X_1), \ell(X_2), \cdots, \ell(X_N), \cdots\},$$

as a single observation of the Monte Carlo sample, and the different observations of the sample come from taking different sequences X. The sample size will be small if we only consider a small number of different X. In practice we have to terminate (9.1) at some value of N, this value being chosen when $L(X_N) = N^{-1/2}\ell(X_N)$ reaches stability. Here we have to be cautious: since X is common to all terms in (9.1), the terms in (9.1) are highly correlated. Therefore the stability of $L(X_N)$ for each individual X is not enough: we must also check that the stable values for different X agree with each other.

The plan will be impractical unless we have an efficient computing algorithm for generating the successive coordinates of (9.1). To this end, suppose that

$$(9.2) \qquad\qquad X = \{x_1, x_2, \cdots\}$$

consists of independent identically distributed observations x_i from a uniform rectangular distribution on the interval $[0, 1]$; and introduce a real variable x, called the dummy truncator, which can take any value in $[0, 1]$. We define X_N^x to be the subsequence of

$$(9.3) \qquad\qquad X_N = \{x_1, x_2, \cdots, x_N\}$$

obtained by deleting from X_N all $x_i > x$. We write, as usual, $\ell(X_N^x)$ for the length of the longest ascending subsequence of X_N^x; and we now regard $\ell(X_N^x)$ as a function of x for each N. Of course

$$(9.4) \qquad\qquad \ell(X_N) = \ell(X_N^1);$$

so we can recover (9.1) if we have

$$(9.5) \qquad\qquad \{\ell(X_1^x), \ell(X_2^x), \cdots, \ell(X_N^x), \cdots\}$$

available in the computer. However, this is more than we need store in the computer. To see this, we note that $\ell(X_N^x)$ is an integer step function of x, satisfying the recurrence relation

$$(9.6) \qquad \ell(X_{N+1}^x) = \begin{cases} \ell(X_N^x), & 0 \le x < x_{N+1} \\ \max\{\ell(X_N^x), 1 + \ell(X_N^{x_{N+1}})\}, & x_{N+1} \le x \le 1. \end{cases}$$

This recurrence relation starts from

$$(9.7) \qquad\qquad \ell(X_1^x) = \begin{cases} 0, & 0 \le x < x_1 \\ 1, & x_1 \le x \le 1. \end{cases}$$

Moreover, $\ell(X_N^x)$ is completely specified by a statement of the positions of its steps: suppose these occur at $x = y_{i,N}$, $i = 1, 2, \cdots, I(N)$, where

$$(9.8) \qquad\qquad y_{1,N} < y_{2,N} < \cdots < y_{I(N),N}.$$

From (9.6), we see that

$$(9.9) \qquad\qquad y_{1,N+1} < y_{2,N+1} < \cdots < y_{I(N+1),N+1}$$

is obtained from (9.8) by adding x_{N+1} to the end of (9.8) if $x_{N+1} > y_{I(N),N}$, and otherwise by substituting x_{N+1} in place of the least $y_{i,N} \geqq x_{N+1}$. The recurrence from (9.8) to (9.9) starts from

(9.10)
$$y_{1,1} = x_1.$$

Thus all we need do is to store (9.8) in the computer and successively update it to (9.9).

This procedure looks formidable at first sight; but a numerical example will show that it is really very simple. Suppose that

(9.11)
$$X = \{0.23, 0.47, 0.14, 0.22, 0.96, 0.83, \cdots\}$$

Then

(9.12)
$$\{y_{1,1}\} = \{0.23\}$$
$$\{y_{1,2}, y_{2,2}\} = \{0.23, 0.47\}$$
$$\{y_{1,3}, y_{2,3}\} = \{0.14, 0.47\}$$
$$\{y_{1,4}, y_{2,4}\} = \{0.14, 0.22\}$$
$$\{y_{1,5}, y_{2,5}, y_{3,5}\} = \{0.14, 0.22, 0.96\}$$
$$\{y_{1,6}, y_{2,6}, y_{3,6}\} = \{0.14, 0.22, 0.83\}$$

For hand computing, this can be achieved by writing the terms of (9.11) in rows (pigeonholes, rather like those in the proof of Theorem 2 but with ascending and descending roles reversed); and ℓ_N will be the number of rows used, while the y are the last entries in each row. Successive appearances of this tableau will look like:

$0.23 \rightarrow 0.23 \rightarrow 0.23, 0.14 \rightarrow 0.23, 0.14 \rightarrow 0.23, 0.14 \rightarrow 0.23, 0.14$
$\qquad 0.47 \qquad 0.47 \qquad\qquad 0.47, 0.22 \quad 0.47, 0.22 \quad 0.47, 0.22$
$\qquad\qquad\qquad\qquad\qquad\qquad\qquad\qquad\qquad 0.96 \qquad\quad 0.96, 0.83$

Since we have

(9.13)
$$\ell(X_N) = I(N),$$

we shall generate (9.1) if we generate X from a sequence of pseudorandom numbers, which we then sort in an overwritten list (9.8), the length of the list at any instant n yielding $\ell(X_n)$. From Theorem 4, the length of the final list will be about $c\sqrt{N}$. Hence to generate (9.1) as far as N, the storage requirement will be approximately $c\sqrt{N}$ and the computing time will be proportional to $\Sigma_{n=1}^{N} c\sqrt{n} = \frac{2}{3} cN^{3/2}$. This is not excessive even for values of N as large as a million. (Programming experts will see that the foregoing computing time can be considerably shortened by appropriate block addressing and address modifiers; but I shall not go into that here.)

Dr. D. C. Handscomb and Mrs. L. Hayes were kind enough to program the foregoing algorithm on the computer at Oxford. The program was written to accept any value of $N \leqq 10^6$; but due to pressure of time (I was due to fly to Berkeley at the end of that week), we only ran the program up to $N = 10^4$, we only printed out values of $\ell(X_N)$ for $N = 100, 400, 1600, 4900, 10000$; and we only did this for 10 different sequences X. This, of course, is a mere sketch of a Monte Carlo calculation; and a full-sized calculation ought to be undertaken in due course of time. However, even these sketchy calculations with a sample size of 10 provide some interesting results. In Table I, the 10 lines correspond to the 10 different sequences X, the quantities tabulated being $\ell(X_N)$.

TABLE I

MONTE CARLO OBSERVATIONS OF $\ell(X_N)$

$N = 100$	$N = 400$	$N = 1600$	$N = 4900$	$N = 10000$
20	38	76	133	198
18	40	76	132	197
17	40	75	132	197
16	35	74	134	198
17	35	74	135	198
18	39	76	132	197
18	37	74	136	195
16	35	74	134	198
19	39	76	132	197
17	39	75	132	197

The consistency between sequences is impressive: indeed I have an uneasy feeling that it is fortuitously too good, and that a larger sample would reveal more scatter. However, for what these data are worth, we get the following

TABLE II

MONTE CARLO ESTIMATES

N	$E(\ell_N/\sqrt{N})$	Var ℓ_N
100	1.76 ± 0.04	1.6
400	1.88 ± 0.03	3.8
1600	1.88 ± 0.007	0.8
4900	1.90 ± 0.005	1.5
10000	1.97 ± 0.003	0.8

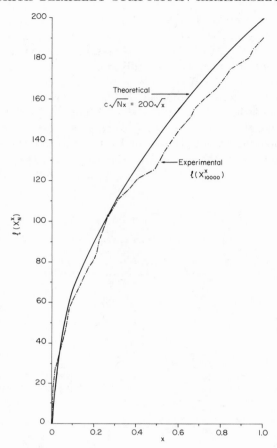

FIGURE 1
Monte Carlo experiment for $\ell(X_N^x)$ with $N = 10000$.

estimates. The $\pm$ entries in the second column are standard errors: these and the estimates of variance in the third column are certainly ragged and seem to me to be suspiciously small. Better estimates must await more extensive Monte Carlo calculations. The mean length 17.6 ± 0.4 for $N = 100$ agrees with the Ulam-Neighbor figure of 17.85 for $N = 101$ (his figures were for the slightly larger quantity ℓ^* in place of ℓ); but the second column suggests that $\ell_N/\sqrt{N}$ does not come close to c until N is substantially larger, say $N = 10000$ or more. The value of c seems to be near 2. Figures 1 and 2 show the results of an eleventh observation. Figure 1 gives a graph of $\ell(X_N^x)$ for $N = 10000$ and $0 \leqq x \leqq 1$. Figure 2 shows the convergence of $N^{-1/2}\ell(X_N)$ for $N = 1, 2, \cdots, 10000$. This eleventh observation is markedly smaller than the other 10 observations.

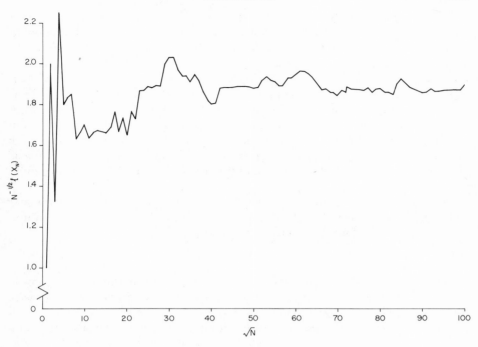

FIGURE 2
Monte Carlo experiment for $N^{-1/2}\ell(X_N)$ with $N = 1, 2, \cdots , 10000$.

10. Attack on c: first method

The arguments in the first eight sections of this paper are rigorous; or rather (since "rigorous" is a subjective term), people, finding the arguments non-rigorous, should at least have little difficulty in modifying them to accord with their own personal standards. However, as usual the price paid for rigor is conclusions which are rather insipid and general, which merely assert the existence of limits or the qualitative behavior of functions. This and the next two sections, on the other hand, deal with the harder problem of assigning a numerical value to the constant c, whose mere existence was established by Theorem 4. Here I jettison rigor without compunction, because a premature attempt to retain it would quite simply halt the work. Of course, I do not mean that rigor should be scorned for its own sake, but merely that the initial investigation is best done without the shackles of rigor and I have not yet got beyond this initial stage. A research worker needs to be able to think nonrigorously in order to get off the ground in the first place, and this ability is in fact rather more difficult to achieve than rigorous thought; for it requires a rather sophisticated sense of judgment over the adequacy of approximations, the plausibility of the reasoning, and the prospects for ultimately making it watertight. One should try to make many different attacks upon a difficult problem—wave after wave of attack if possible, from all sorts of angles, to test the weak points of the defenses and reveal the footholds from which a final assault may be launched. This preliminary reconnoitering and skirmishing should be highly informal, though not without a perceptive eye for routes able to support the weight of later formality. I shall try, rather sketchily, to illustrate this by presenting three separate nonrigorous determinantions of c. The first determination has a superficial attractiveness; but it is, I believe, thoroughly disreputable. The second and third arguments look progressively wilder but are, I think, more promising. But these assessments are matters of taste and experience, and the reader must make up his own mind about them.

Consider, as in the proof of Theorem 7, a Poisson process with unit parameter and a square of area N. Let $P_1, P_2, \cdots, P_\ell$ be the points of a longest chain in the square drawn from the points of the Poisson process. Here

$$(10.1) \qquad \ell = \ell_N = c\sqrt{N},$$

with (of course) a suitable nonrigorous interpretation upon the equality signs in (10.1). The average horizontal (or vertical) displacement between two successive points P_i, P_{i+1} is $1/c$ since the square has side $\sqrt{N}$; and hence the area of a rectangle R, having sides parallel to the square and opposite vertices at P_i and P_{i+2} (*not* P_{i+1}) is about $4/c^2$. However, the rectangle R may be expected to contain just one point (P_{i+1}) of the Poisson process (if it contained two or more points there would be at least a 50 per cent chance of two or more points of the chain in R). The expected number of Poisson points in an area $4/c^2$ is $4/c^2$. Hence

$$(10.2) \qquad 4/c^2 = 1,$$

which gives $c = 2$.

This looks like the right value of c, and at first glance the method has a pleasant appearance of simplicity. On closer examination, however, it becomes far less attractive. There is in fact a vicious circularity about the argument; and this is best exhibited by considering instead the rectangle R' with opposite vertices at P_i and P_{i+1}. The area of R' is about $1/c^2$, and R' certainly contains no points of the Poisson process. But $1/c^2$ is nothing like zero. The trouble arises because R' depends upon the properties of the chain, and hence upon the Poisson process. There seems little prospect of mending this first method.

This is a pity, because an extension of this method would have yielded not merely c, but also the values of

$$(10.3) \qquad \lambda(N) = E(\ell_N)$$

for finite N. To pass quickly over this, consider the rectangle R'' with opposite vertices at P_i and P_{i+k}. This has area k^2/c^2; and so the probability that there are n Poisson points in R'' is

(10.4)
$$e^{-k^2/c^2}(k^2/c^2)^n/n!$$

The expected length of longest chain from these n points in R'' is $\lambda(n)$ from (10.3); and R'' is known to contain a longest chain of length $k - 1$. Hence

(10.5)
$$\sum_{n=0}^{\infty} e^{-k^2/c^2} \frac{(k^2/c^2)^n}{n!} \lambda(n) = k - 1.$$

There are, of course, obvious queries over summing from $n = 0$ rather than from $n = k$, say; but let us ignore them.

The left side of (10.5) is what might be called a Borel transform, by analogy with Section 7. How does one invert a Borel transform? Putting

(10.6)
$$k^2/c^2 = \theta,$$

we get

(10.7)
$$\sum_{n=0}^{\infty} e^{-\theta} \frac{\theta^n}{n!} \lambda(n) = c\theta^{1/2} - 1;$$

and hence taking Laplace transforms with respect to θ, we have

(10.8)
$$\sum_{n=0}^{\infty} \frac{\lambda(n)}{(s+1)^n} = \frac{c\sqrt{\pi}}{2s^{3/2}} - \frac{1}{s}.$$

This is a trifle nonsensical: the right side has a branch point, and it becomes negative when $s \to 0$ while the left side remains obstinately positive. A source of this discrepancy is that, in taking Laplace transforms, we have treated θ as a continuous variable from 0 to ∞ although (10.6) restricts it to discrete values. There are infinitely many continuous functions of θ which agree with the right side of (10.7) for these discrete values: which of these functions can we choose to make the right side of (10.8) positive?

However, to continue, we put $t = 1/(s+1)$ in (10.8) and expand the right side in powers of t, obtaining

(10.9)
$$\sum_{n=0}^{\infty} \lambda(n)t^n = c \sum_{n=0}^{\infty} \frac{(n + \frac{1}{2})!}{n!} t^{n+3/2} - \sum_{n=1}^{\infty} t^n.$$

In this we want to equate coefficients. In ordinary circumstances we could replace a series like $\Sigma\, a_n t^{n+1}$ by $\Sigma\, a_{n-1} t^n$; and we extend this principle of shifting n by an integer to nonintegral shifts. This yields

(10.10)
$$\lambda(n) = c \frac{(n-1)!}{(n - \frac{3}{2})!} - 1.$$

For large n, this gives $\lambda(n) \sim c\sqrt{n}$ as it should. For small n, we insert $c = 2$ from (10.2) and calculate the right side of (10.10). Table III compares these results with the exact values of $\lambda(n)$ obtained later in Section 17.

TABLE III

VALUES OF $\lambda(N)$

N	Formula (10.10)	Exact $\lambda(N)$
2	1.27	1.50
3	2.01	2.00
4	2.61	2.42
5	3.13	2.79
6	3.59	3.14
7	4.00	3.47
8	4.39	3.77
9	4.75	4.06

The agreement is only moderate, perhaps all that could have been expected from an outlandish calculation like this one. However, some concealed ideas may possibly lurk in the foregoing, and perhaps they can be discovered by a little gentle cooking of the mathematics and a proper disregard for the occasional whiff of burning.

11. Attack on c: second method

It is more or less true (in some sense) that

$$(11.1) \qquad \ell(X_N) = c\sqrt{N} + o(\sqrt{N}).$$

If the error term $o(\sqrt{N})$ behaves smoothly enough, then

$$(11.2) \qquad E\{\ell(X_{N+1}) - \ell(X_N)|\ell(X_N)\} \sim c\sqrt{(N+1)} - c\sqrt{N} \sim \frac{c}{2\sqrt{N}}.$$

Now consider a square S with N points uniformly distributed in it; and consider adding one more point to S, this extra point being also uniformly distributed over S. We look at the conditional situation, given the positions of the original N points. The extra point can only increase a longest chain of length $\ell(X_N)$ by 1. Hence the expected conditional increase, namely (11.2), is equal to the probability that the new point will cause an increase of 1. Thus the area of the region in S, in which the new point will cause an increase, is $c/2\sqrt{N}$.

Consider a longest chain $P_1, P_2, \cdots, P_\ell$ from the original N points. Take P_0 and $P_{\ell+1}$ to be the southwest and northeast corners of S. Let R_i, $i = 0, 1, \cdots, \ell$, be the rectangle with opposite vertices P_i and P_{i+1}. The new point will cause an increase of 1 to *this* chain if and only if it falls into one of the rectangles R_0, $R_1, \cdots, R_\ell$. If these rectangles were squares, all of equal size, their total area would be $(\ell+1)/(\ell+1)^2 \sim 1/c\sqrt{N}$. Actually their total area will be larger than this (actually about twice as large), because they are not of equal size. There will be a similar chain of rectangles from any other longest chain $P_1', P_2', \cdots, P_\ell'$; and the rectangles from one chain will overlap those of another chain. Let us hope that the overlapping compensates more or less for the underestimation of the total area associated with an individual chain. Hence the average number of points which form an ith point of some longest chain is

(11.3) $$(c/2\sqrt{N})/(1/c\sqrt{N}) = \tfrac{1}{2}c^2.$$

This holds for each given value of i; and therefore (ignoring the distinction between arithmetic and geometric means) the total number of longest chains is about

(11.4) $$(\tfrac{1}{2}c^2)^\ell \sim (\tfrac{1}{2}c^2)^{c\sqrt{N}}.$$

On the other hand from Theorem 6, the expected number of increasing subsequences of length $c\sqrt{N}$ is

(11.5) $$\frac{1}{(c\sqrt{N})!}\binom{N}{c\sqrt{N}} \sim \left(\frac{e}{c}\right)^{2c\sqrt{N}}.$$

Equating (11.4) and (11.5) we get

(11.6) $$\tfrac{1}{2}c^2 = (e/c)^2$$

which leads to

(11.7) $$c = 2^{1/4}e^{1/2} = 1.961. \ldots$$

This method is admittedly very rough and ready: the saving grace is the fourth root at the end of the calculation, which reduces any relative error by a factor of 4. It is the only method I have been able to invent which involves a final fourth root: the other two methods presented here, as well as further methods which I shall not mention, end by taking a square root. The suggestion accordingly is that one should try to look for a method which ends by taking an arbitrarily high root, thus swamping any approximations in the early part of the calculation (Littlewood's "high indices" principle).

At first sight it seems odd that the number of longest chains is as large as (11.5). The reason becomes apparent when one considers three successive terms $y < y' < y''$ in a longest ascending subsequence of X. There is a reasonable chance that X contains another term z, say, such that $y < z < y''$ while $z > y'$. Thus the term y' may be replaced by z. This sort of substitution can take place along the whole length of the chain; and hence the number of longest chains will increase exponentially with the length of a longest chain. Actually I can prove rigorously that the expected number of longest chains is at least

(11.8) $$(e/c)^{2c\sqrt{N} + o(\sqrt{N})}.$$

The reason why I have to be content with this as a lower bound is that there is a small probability of the length of a longest chain being less than $c\sqrt{N}$; if a longest chain is shorter than $c\sqrt{N}$, the number of such chains increases exponentially over and above (11.8). This very large increase in numbers may override the small probability.

12. Attack on c: third method

This method is based on the principle that the only good Monte Carlo method is a dead one: to wit, having done the appropriate transformation to prepare a problem for Monte Carlo sampling (as by dummy truncation in Section 9), one then ought to abandon the proposed Monte Carlo sampling and instead investigate the transformed problem analytically. (If the analysis appears intractable, the transformation is not an adequate preparation for Monte Carlo sampling!)

The analytical idea adopted here is to calculate the area under the curve $\ell(X_N^x)$, defined in (9.6) and regarded as a function of x. Since the number of terms in the sequence X_N^x will be asymptotically Nx for large N, we have

$$(12.1) \qquad \ell(X_N^x) = c\sqrt{Nx} + o(\sqrt{N}).$$

The area under this curve is

$$(12.2) \qquad A = \int_0^1 \ell(X_N^x)\,dx.$$

We shall calculate A in two ways, one of which expresses it as a multiple of $1/c$. By equating these two results, we then determine c. The second calculation, leading to a multiple of $1/c$, might seem to conflict with the dogma that integration is a linear operation: the moral is that in searching for methods of calculating scale factors one should not be blinded by the dogmas of functional analysis.

The first calculation of A is straightforward: from (12.1) and (12.2) we have

$$(12.3) \qquad A = \tfrac{2}{3}c\sqrt{N} + o(\sqrt{N}).$$

For the second calculation of A, we note that the curve $\ell(X_N^x)$ can be built up stage by stage by the recurrence relation (9.6). Suppose that the recurrence has gone as far as $N = n$, and we look for the expected increase in area, say q_n, in going from $N = n$ to $N = n + 1$. We shall have

$$(12.4) \qquad A = \sum_{n=0}^{N-1} q_n.$$

The discontinuities of $\ell(X_n^x)$ occur at

$$(12.5) \qquad y_{1,n} < y_{2,n} < \cdots < y_{I(n),n}$$

in accordance with (9.8); and the recurrence arises from adding a new point x_{n+1} uniformly distributed over $[0, 1]$. The added area will be $(y_{j+1} - x_{n+1})$, where y_{j+1} is the first term in (12.5) which exceeds x_{n+1}. Given that x_{n+1} falls in (y_j, y_{j+1}) the conditional expected additional area is $\tfrac{1}{2}(y_{j+1} - y_j)$; and the probability that x_{n+1} does fall in (y_j, y_{j+1}) is $(y_{j+1} - y_j)$. Hence

$$(12.6) \qquad q_n = \sum_{j=0}^{I(n)} \tfrac{1}{2}(y_{j+1} - y_j)^2,$$

where we have written y_i for $y_{i,n}$ and taken $y_0 = 0$, $y_{I(n)+1} = 1$.

Now consider the contribution to q_n from an interval $(x, x + dx)$. Here dx is small; but we suppose that n is large enough to make v large, where v is the number of y_i in $(x, x + dx)$. It is reasonable to assume that the distribution of the discontinuities y_i is locally homogeneous and random: I shall refer to this as "assumption α," and will return later to discuss it. Thus, in $(x, x + dx)$ the y_i behave as though they were independently and uniformly distributed over this interval of length dx. However, it is known [19] that when v points are uniformly and independently distributed over a unit interval, thus dividing it into $v + 1$ subintervals, the sum of squares of the lengths of these subintervals has an expected value $2/(v + 2)$. Multiplying by $(dx)^2$ to allow for the scale factor between $(x, x + dx)$ and a unit interval, writing $2/v$ instead of $2/(v + 2)$ because v is large, and incorporating the factor $\frac{1}{2}$ in (12.6) we see that the contribution to q_n from $(x, x + dx)$ is

$$(12.7) \qquad\qquad\qquad (dx)^2/v.$$

However v is the number of (unit height) steps of $\ell(X_n^x)$ in $(x, x + dx)$; so

$$(12.8) \qquad\qquad\qquad v = \frac{\partial \ell(X_n^x)}{\partial x} dx.$$

Inserting this into (12.7) and integrating to collect together all possible intervals $(x, x + dx)$ we get

$$(12.9) \qquad\qquad\qquad q_n = \int_0^1 dx \left/ \frac{\partial \ell(X_n^x)}{\partial x} \right. .$$

Thus (12.1) and (12.9) yield

$$(12.10) \qquad\qquad q_n = \int_0^1 \frac{dx}{(c\sqrt{n})(\frac{1}{2}x^{-1/2})} = \frac{4}{3c\sqrt{n}}.$$

By (12.4) we get

$$(12.11) \qquad\qquad A \sim \sum_{n=n_0}^{N-1} \frac{4}{3c\sqrt{n}} = \frac{8\sqrt{N}}{3c} + o(\sqrt{N}).$$

Here the summation has to begin at some large value $n = n_0$, because we have assumed v large. Equating (12.3) and (12.11) we find that

$$(12.12) \qquad\qquad\qquad c = 2.$$

The mathematical novice will doubtless be appalled at this argument: he will complain that in (12.8) I have differentiated a step function, and in (12.9) I have integrated the reciprocal of this derivative; and that, if the reciprocal of the derivative of a step function has any meaning, it must be infinite everywhere, except for zero values at the positions of the steps. And he will also point to the looseness which sometimes envisages random values and sometimes their expected values and which slips carelessly from one to the other. Experienced mathematicians will feel no such qualms. With one exception, all the steps in the argument can be made rigorous. To sketch very briefly the necessary amendments we first choose an arbitrary $\varepsilon > 0$ and then find a partition of $[0, 1]$ into equal intervals of length $\delta = \delta(\varepsilon)$ such that the Darboux sum

$$(12.13) \qquad \sum_{j=1}^{1/\delta} (j\delta)^{1/2}\delta$$

approximates

$$(12.14) \qquad \int_0^1 x^{1/2}\,dx$$

to within ε. The intervals $(j\delta, j\delta + \delta)$ can then be used in place of $(x, x + dx)$. The derivative in (12.8) is replaced by a finite difference, which will behave itself if $n \geqq n_0(\varepsilon)$. And N in (12.11) can be chosen large in comparison with n. Eventually we shall find that c differs from 2 by a fixed multiple of ε; and (12.12) will result from the arbitrariness of ε.

The only gap in the argument which I cannot yet fill in rigorously is a justification of assumption α. Unfortunately this assumption is crucial to the argument; for, if we replace the 2 in the numerator of $2/(v + 2)$ by some other constant, the ultimate value of c will be altered correspondingly. Of course, Cauchy's inequality shows that the sum of squares in question cannot be less than $1/(v + 1)$: this leads to a rigorous proof that $c \geqq \sqrt{2}$, a result which is not as good as $c \geqq \frac{1}{2}\pi$ obtained in Theorem 7. Assumption α is actually a stronger assumption than one needs to arrive at $2/(v + 2)$; and, at one stage in the development, I thought that it might be possible to justify an adequate weaker assumption by entering Laplace transformed space at an appropriate moment and utilizing the results of Section 13. (This would have involved an appropriate transformation of the x-axis, and a more complicated integral instead of (12.4); but that would only call for a few technical adjustments of a fairly simple kind.) However, the manipulation in Laplace transformed space (see the end of Section 16) has proved more slippery and difficult than I first thought; and so far I cannot provide a rigorous proof of (12.12). However, I should be very surprised if (12.12) is false.

There is a rather treacherous variant of this method—treacherous because, unlike the foregoing, it hides an elusive mixture of conditional expectations that are not at all easily amenable to rigor. The conditional expectation of $\ell(X_{N+1})$ given $\ell(X_N^x)$ is $1 - y_{I(N), N}$, since $\ell(X_{N+1}) - \ell(X_N)$ can only take the values 0 and 1 and the latter value occurs if and only if $y_{I(N)N} \leqq x_{N+1} \leqq 1$. Hence

$$(12.15) \qquad E\ell(X_{N+1}) - E\ell(X_N) = E(1 - y_{I(N), N}).$$

The left side of (12.15) is the result of differencing $c\sqrt{N} + o(\sqrt{N})$ with respect to N; and ought to be about $\frac{1}{2}c/\sqrt{N}$ if the error term is smooth (this sort of difficulty ought to be surmountable by reversing the argument and summing the differences over N as in (12.4)). The right side of (12.15) is the displacement in x near $x = 1$ just sufficient to ensure unit increase in $\ell(X_N^x)$, and should be nearly equal to the reciprocal of $\partial\ell(X_N^x)/\partial x$ at $x = 1$, namely $2/c\sqrt{N}$ from (12.1). Thus

$$(12.16) \qquad \tfrac{1}{2}c/\sqrt{N} = 2/c\sqrt{N},$$

which once again yields (12.12).

This argument can be made a little bit more plausible by considering what happens at the discontinuity $x = y_{iN}$. By the same argument as leads to (12.15), we have

$$(12.17) \qquad E\ell(X_{N+1}^x) - \ell(X_N^x) = y_{i,N} - y_{i-1,N} \qquad (x = y_{i,N}),$$

where the expectation in (12.17) is conditional on $\ell(X_N^x)$ being given. Now $\ell(X_N^x)$ regarded as a function of x has a jump of height 1 at $x = y_{i,N}$. Hence summing (12.17) over $i = 1, 2, \cdots, I(N)$, we have the Stieltjes integral

$$(12.18) \qquad \int_{x=0}^{1} \{E\ell(X_{N+1}^x) - \ell(X_N^x)\}\, d\ell(X_N^x) = 1.$$

Here we have written 1 for the upper limit of integration and for the right side: strictly it should have been $y_{I(N),N}$; but this is very nearly equal to 1. If we now regard $\ell(X_N^x)$ as a continuous function of N, for example by making it piecewise linear between the original integer values of N, we can write (12.18) as

$$(12.19) \qquad E\int_{x=0}^{1} \frac{\partial\ell(X_N^x)}{\partial N}\, d\ell(X_N^x) = 1.$$

Suppose that somehow we can take expected values over X_N^x (and it is not clear how to do so) in such a way that $\ell(X_N^x)$ may be replaced by its asymptotic value $c\sqrt{N}x$. Then this would give

$$(12.20) \qquad \int_{x=0}^{1} \frac{\partial}{\partial N}(c\sqrt{Nx}) \frac{\partial}{\partial x}(c\sqrt{Nx})\, dx = 1;$$

whereupon an easy computation leads from (12.20) to (12.12).

In Section 1 I said that published proofs differ from the arguments of their gestation. Sections 10, 11, and 12 may illustrate the kind of preliminary thinking from which a finished proof might be derived. Each appears in a different stage of development and none has reached fruition. As a matter of fact I have presented them in reverse chronology. Section 12, apart from the much younger final variant, is about six weeks old and comes nearest to being the framework for a rigorous argument. Section 11 is about two weeks old, and Section 10 is only two days old. These ages are reflected in the relative coherence or incoherence of the text. If any one of them had reached the stage of a watertight argument, the others would have been discarded. Moreover the necessary epsilontics of a proof would have shrouded the underlying ideas and their origins.

13. The joint characteristic function

It is always worth trying to solve a problem by brute force. If the problem is a difficult one, brute force will have slender hopes of success; but, should it succeed, it is likely to yield far more detailed results than general theory ever does. Charles Darwin was fond of saying that scientists should perform damn-fool experiments from time to time: these usually fail, but are a triumph when they come off. He illustrated this precept once by playing the trombone at his tulips, with negative results.

We should be able to deduce almost anything we wanted to know about Ulam's problem if we could obtain a tractable expression for the joint characteristic function of the discontinuities in the Monte Carlo truncated function (9.6). The distribution of ℓ_N is nonparametric in the sense that it does not depend upon the common distribution of the x_i in X; and the distribution of ℓ_N is the same as that of ℓ'_N. There are technical simplifications in supposing that the x_i come from the common distribution

$$(13.1) \qquad\qquad P(x_i \leq x) = 1 - e^{-x}, \qquad\qquad 0 \leq x \leq \infty,$$

and in considering ℓ'_N instead of ℓ_N. The necessary changes to (9.6) and (9.8) are as follows. We define $\bar{X}^x_N$ to be the subsequence of X_N which contains only those $x_i \geq x$. Thus $\bar{X}^0_N = X_N$ in particular. We write $\ell'(\bar{X}^x_N)$ for the length of a longest descending subsequence in $\bar{X}^x_N$. Thus (9.6) becomes

$$(13.2) \qquad \ell'(\bar{X}^x_{N+1}) = \begin{cases} \ell'(\bar{X}^x_N), & x_{N+1} < x \leq \infty \\ \max\{\ell'(\bar{X}^x_N), 1 + \ell'(\bar{X}^{x_{N+1}}_N)\}, & 0 \leq x \leq x_{N+1}. \end{cases}$$

Suppose that the steps of $\ell'(\bar{X}^x_N)$ occur at

$$(13.3) \qquad\qquad y_{1,N} > y_{2,N} > \cdots > y_{I(N),N}.$$

For convenience, we drop the second suffix and extend the range of the first suffix: so (13.3) becomes

$$(13.4) \qquad\qquad y_1 \geq y_2 \geq \cdots \geq y_i \geq \cdots \geq 0.$$

In (13.4) all $y_i = 0$ when $i > I(N)$; and, with probability 1, strict inequality holds for $y_i > y_{i+1}$ when $i \leq I(N)$. The advantage of (13.4) lies in not having to bother about the value of $I(N)$ in the calculation, since the notation automatically takes care of it. Our aim is to find an expression for

$$(13.5) \qquad \phi_N(\mathbf{s}) = \phi_N(s_1, s_2, \cdots) = E \exp\left\{ - \sum_{i=1}^{\infty} s_i y_i \right\}$$

where the suffix N in (13.5) recalls the suppressed second suffix in the y_i. We note that the extended y_i, namely those with $i > I(N)$, being zero do not affect the value of $\phi_N(\mathbf{s})$. In (13.5), we would have a joint characteristic function if the s_i were all pure imaginary quantities: it is, however, more convenient to take the s_i to be real and nonnegative; and this will not affect the usefulness of $\phi_N(\mathbf{s})$. We now seek a recurrence relation between ϕ_N and ϕ_{N+1}.

To pass from (13.4), which relates to N, to the corresponding sequence relating to $N + 1$, we have to draw x_{N+1} from the distribution (13.1), and then insert x_{N+1} in place of y_j where j is the smallest integer such that $y_j > x_{N+1}$. If no such j exists, we simply augment the y by x_{N+1} as it stands. This will provide a new sequence

(13.6)
$$y_1' \geqq y_2' \geqq \cdots \geqq 0$$

in place of (13.4); from which

(13.7)
$$\phi_{N+1}(\mathbf{s}) = E \exp\left\{ -\sum_{i=1}^{\infty} s_i y_i' \right\}.$$

We shall, however, calculate (13.7) in two stages. In the first stage we calculate

(13.8)
$$E^* \exp\left\{ -\sum_{i=1}^{\infty} s_i y_i' \right\},$$

where E^* denotes the conditional expectation given the sequence (13.4). The second stage will complete the process by taking expectations over the sequence (13.4). For typographical convenience we write x in place of x_{N+1}. Thus drawing x from the distribution (13.1), we have

(13.9)
$$E^* \exp\left\{ -\sum_{i=1}^{\infty} s_i y_i' \right\}$$

$$= \int_{y_1}^{\infty} e^{-x} \exp\left\{ -s_1 x - \sum_{i=2}^{\infty} s_i y_i \right\} dx$$

$$+ \sum_{j=2}^{\infty} \int_{y_j}^{y_{j-1}} e^{-x} \exp\left\{ -\sum_{i=1}^{j-1} s_i y_i - s_j x - \sum_{i=j+1}^{\infty} s_i y_i \right\} dx$$

$$= \frac{1}{s_1 + 1} \exp\left\{ -(s_1 + 1)y_1 - \sum_{i=2}^{\infty} s_i y_i \right\}$$

$$+ \sum_{j=2}^{\infty} \frac{1}{s_j + 1} \left[\exp\left\{ -\sum_{i=1}^{j-1} s_i y_i - (s_j + 1)y_j - \sum_{i=j+1}^{\infty} s_i y_i \right\} \right.$$

$$\left. - \exp\left\{ -\sum_{i=1}^{j-1} s_i y_i - (s_j + 1)y_{j-1} - \sum_{i=j+1}^{\infty} s_i y_i \right\} \right].$$

Now we effect the second stage of the calculation. Using the definition (13.5) with appropriately adjusted values of $\mathbf{s}$, we find

(13.10)
$$\phi_{N+1}(\mathbf{s}) = \frac{1}{s_1 + 1} \phi_N(s_1 + 1, s_2, s_3, \cdots)$$

$$+ \sum_{j=2}^{\infty} \frac{1}{s_j + 1} \left[\phi_N(s_1, s_2, \cdots, s_{j-1}, s_j + 1, s_{j+1}, \cdots) \right.$$

$$\left. - \phi_N(s_1, s_2, \cdots, s_{j-2}, s_{j-1} + s_j + 1, 0, s_{j+1}, \cdots) \right].$$

This recurrence relation can be started by considering the case $N = 0$, in which $y_1 = y_2 = \cdots = 0$. Thus (13.10) holds for $N = 0, 1, 2, \cdots$ and starts from

$$(13.11) \qquad \phi_0(\mathbf{s}) = 1.$$

The next problem is to solve the functional recurrence relation (13.10), and a pretty unpleasant relation it looks. Functional equations in one variable can be troublesome, those in infinitely many variables are worse. The only hope, I felt, was to work out the first few values $\phi_0, \phi_1, \phi_2, \cdots$ from (13.10), to guess the general result, and then prove it by induction. This is a good example of brute force mathematics, and it needs plenty of courage and a lot of tedious elementary algebra. It is clear that ϕ_N will be a rational function of $s_1, s_2, \cdots, s_N$. The difficulty is that a rational function of several variables can be written in an enormous number of different ways, and one has to hit on just the right way of writing it before one has much chance of guessing the general result. If the reader does not believe this, let him try solving (13.10) for himself.

I spent two or three days over this job. The expressions ϕ_1 and ϕ_2 are quite easily found; but ϕ_3 was a lot messier, and I had to write it down in many different algebraic forms before I got it into a shape which looked like a reasonable extension of ϕ_1 and ϕ_2. I then guessed what ϕ_N should be like on the basis of ϕ_1, ϕ_2, and ϕ_3, and was able to confirm this guess by induction. This confirmation is quite easy and runs as follows.

Define

$$(13.12) \qquad S_1 = s_1, \quad S_2 = s_1 + s_2, \quad S_3 = s_1 + s_2 + s_3, \cdots$$

and

$$(13.13) \qquad S(\mathbf{a}) = \prod_{i=1}^{k} \frac{(S_i + a_{i-1})!}{(S_i + a_i)!},$$

where $\mathbf{a} = (a_0, a_1, \cdots, a_k)$ is a sequence of integers satisfying

$$(13.14) \qquad 0 = a_0 < a_1 < a_2 < \cdots < a_k.$$

Introduce the functional operators $J_1, J_2, \cdots$ by the definitions

$$(13.15) \qquad J_1 f(s_1, s_2, \cdots) = (s_1 + 1)^{-1} f(s_1 + 1, s_2, s_3, \cdots),$$

$$(13.16) \quad J_j f(s_1, s_2, \cdots) = (s_j + 1)^{-1} [f(s_1, \cdots, s_{j-1}, s_j + 1, s_{j+1}, \cdots)$$
$$- f(s_1, \cdots, s_{j-2}, s_{j-1} + s_j + 1, 0, s_{j+1}, \cdots)], j \geqq 2.$$

From (13.10), we have

$$(13.17) \qquad \phi_{N+1}(\mathbf{s}) = \sum_{j=1}^{\infty} J_j \phi_N(\mathbf{s}).$$

Now, from (13.13),

$$(13.18) \qquad J_1 S(a_0, a_1, \cdots, a_k) = S(a_0, a_1 + 1, a_2 + 1, \cdots, a_k + 1),$$

and

(13.19)
$$J_j S(a_0, a_1, \cdots, a_k) = 0, \qquad\qquad j \geqq k + 2.$$

To deal with the remaining cases $2 \leqq j \leqq k + 1$ we adopt the convention that an empty product, like $\Pi_{j=1}^0$, is interpreted as 1. We have, for $2 \leqq j \leqq k + 1$,

(13.20) $\quad J_j S(a_0, a_1, \cdots, a_k)$

$$= (s_j + 1)^{-1} \left[\left\{ \prod_{i=1}^{j-1} \frac{(S_i + a_{i-1})!}{(S_i + a_i)!} \right\} \left\{ \prod_{i=j}^{k} \frac{(S_i + a_{i-1} + 1)!}{(S_i + a_i + 1)!} \right\} \right.$$

$$\left. - \left\{ \prod_{i=1}^{j-2} \frac{(S_i + a_{i-1})!}{(S_i + a_i)!} \right\} \left\{ \frac{(S_j + a_{j-2} + 1)!}{(S_j + a_{j-1} + 1)!} \right\} \left\{ \prod_{i=j}^{k} \frac{(S_i + a_{i-1} + 1)!}{(S_i + a_i + 1)!} \right\} \right]$$

$$= \left\{ \prod_{i=1}^{j-2} \frac{(S_i + a_{i-1})!}{(S_i + a_i)!} \right\} \left\{ \prod_{i=j}^{k} \frac{(S_i + a_{i-1} + 1)!}{(S_i + a_i + 1)!} \right\}.$$

$$\frac{1}{S_j - S_{j-1} + 1} \left\{ \frac{(S_{j-1} + a_{j-2})!}{(S_j + a_{j-1})!} - \frac{(S_j + a_{j-2} + 1)!}{(S_j + a_{j-1} + 1)!} \right\}$$

$$= \left\{ \prod_{i=1}^{j-2} \frac{(S_i + a_{i-1})!}{(S_i + a_i)!} \right\} \left\{ \prod_{i=j}^{k} \frac{(S_i + a_{i-1} + 1)!}{(S_i + a_i + 1)!} \right\}.$$

$$\sum_{r=a_{j-2}+1}^{a_{j-1}} \frac{(S_{j-1} + a_{j-2})!}{(S_{j-1} + r)!} \frac{(S_j + r)!}{(S_j + a_{j-1} + 1)!}$$

$$= \sum_{r=a_{j-2}+1}^{a_{j-1}} S(a_0, a_1, \cdots, a_{j-2}, r, a_j + 1, a_{j+1} + 1, \cdots, a_k + 1).$$

Since

(13.21)
$$\phi_1(\mathbf{s}) = (s_1 + 1)^{-1} = S(0, 1)$$

we deduce from (13.17), (13.18), (13.9) and (13.20) that

(13.22)
$$\phi_N(\mathbf{s}) = \sum_{\mathbf{a}}^{N} c(\mathbf{a}) S(\mathbf{a}),$$

where $c(\mathbf{a})$ is a positive integer depending on the sequence $\mathbf{a}$, and $\Sigma_{\mathbf{a}}^{N}$ denotes summation over all integer sequences $\mathbf{a}$ satisfying

(13.23)
$$0 = a_0 < a_1 < a_2 < \cdots < a_k = N.$$

Here k may have any integer value provided $1 \leqq k \leqq N$. Thus there are 2^{N-1} sequences satisfying (13.23); and accordingly $\phi_N(\mathbf{s})$ is a linear combination of 2^{N-1} functions like (13.13), the coefficients of the linear combination being positive integers. This exhibits the functional form of ϕ_N; and to make further progress we need to study the coefficients $c(\mathbf{a})$.

14. Recurrence relations for the coefficients c (a)

The operators $J_1, J_2, \cdots$ are linear operators on the positive integral orthant of the vector space spanned by the functions (13.13). The vectors **a** can be taken as a natural representation of the basis of the orthant. In deriving recurrence relations for the coefficients $c(\mathbf{a})$, it suffices to study the mapping induced on the representation by the linear operator

$$(14.1) \qquad\qquad J = \sum_{j=1}^{\infty} J_j.$$

Let the set of vectors **a** which are sequences of integers satisfying

$$(14.2) \qquad\qquad 0 = a_0 < a_1 < \cdots < a_k = N$$

be denoted by A_N. Given a vector $\mathbf{a} \in A_N$ and an integer r satisfying $0 \leq r \leq N$, we define a mapping T_r from A_N to A_{N+1}:

$$(14.3) \qquad T_r \mathbf{a} = T_r(a_0, a_1, \cdots, a_k) = \mathbf{b} = (b_0, b_1, \cdots, b_{k'}),$$

where the b_i are obtained as follows. Determine the smallest integer j such that $r \leq a_j$. Then (ignoring any empty instruction like $0 \leq i < 0$) put

$$(14.4) \qquad\qquad \begin{aligned} b_i &= a_i, & 0 \leq i < j \\ b_j &= r, \\ b_i &= a_i + 1, & j < i \leq k. \end{aligned}$$

If this process results in $b_k = N + 1$, the process is complete; if it results in $b_k < N + 1$, put $k' = k + 1$ and $b_{k'} = N + 1$. A vector $\mathbf{a} \in A_N$ is an inverse image of $\mathbf{b} \in A_{N+1}$ if there exists an r, $0 \leq r \leq N$ such that $T_r \mathbf{a} = \mathbf{b}$. We write $T^{-1}(\mathbf{b})$ for the set of all inverse images of **b**. Comparison of (14.1), (14.3), and (14.4) with (13.18) and (13.20) now shows that

$$(14.5) \qquad\qquad c(\mathbf{b}) = \sum_{\mathbf{a} \in T^{-1}(\mathbf{b})} c(\mathbf{a}).$$

This is the desired recurrence relation for the coefficients. The recurrence starts from

$$(14.6) \qquad\qquad c(0, 1) = 1,$$

in accordance with (13.21).

The rules by which the recurrence (14.5) operates are somewhat complicated and a numerical example will help to clarify them. We use the rules to construct ϕ_2 from ϕ_1, ϕ_3 from ϕ_2, and so on. Suppose that we have got as far as ϕ_4, and we now wish to calculate ϕ_5 from ϕ_4. To simplify the notation we shall write typically

$$(14.7) \qquad\qquad S_{034} = S(0, 3, 4).$$

Here the inital zero suffix serves to distinguish $S_{01} = S(0, 1)$, for example, from the variable S_1. In this notation ϕ_4 is given by

$$(14.8) \qquad \phi_4(\mathbf{s}) = S_{04} + 5S_{014} + 5S_{024} + 3S_{034} + 3S_{0124}$$
$$+ 3S_{0134} + 3S_{0234} + S_{01234}.$$

The calculation proceeds by means of Table IV. The first column contains the

TABLE IV

CALCULATION OF ϕ_5 FROM ϕ_4

$\mathbf{a}$	$c(\mathbf{a})$	$T_0\mathbf{a}$	$T_1\mathbf{a}$	$T_2\mathbf{a}$	$T_3\mathbf{a}$	$T_4\mathbf{a}$
04	1	05	015	025	035	045
014	5	025	015	0125	0135	0145
024	5	035	015	025	0235	0245
034	3	045	015	025	035	0345
0124	3	0235	0135	0125	01235	01245
0134	3	0245	0145	0125	0135	01345
0234	3	0345	0145	0245	0235	02345
01234	1	02345	01345	01245	01235	012345

suffices $\mathbf{a} \in A_4$ appearing in ϕ_4, and the second column gives the corresponding coefficients $c(\mathbf{a})$ taken from (14.8). The last five columns tabulate $T_r\,\mathbf{a}$, for $r = 0, 1, \cdots, 4$. In general when calculating ϕ_{N+1} from ϕ_N the table will contain 2^{N-1} rows and $N + 3$ columns for $\mathbf{a}$, $c(\mathbf{a})$, and $T_r\,\mathbf{a}(r = 0, 1, \cdots, N)$. To calculate $T_2(034)$, for example, we note that 3 is the least integer in $\mathbf{a} = (034)$ which exceeds $r = 2$. The integer 3 is accordingly reduced to 2; all preceding integers are unaltered; and all succeeding integers are increased by 1. Thus $T_2(034) = (025)$. The exception to this rule arises when it would lead to a vector which did not end in $N + 1 = 5$. In that event $N + 1$ is added to the end of the vector as a final coordinate. Thus $T_2(014) = 0125$, and not 012 under the unamended rule. Examination of the various entries in Table IV should make everything clear. The entries in the body of the table are the vectors $\mathbf{b}$ in A_{N+1}; and to calculate $c(\mathbf{b})$ we add together the $c(\mathbf{a})$ entries, in the second column, for each row in which $\mathbf{b}$ occurs. For example, from rows 2, 5, and 6,

$$(14.9) \qquad c(0135) = 5 + 3 + 3 = 11.$$

This leads to

$$(14.10) \qquad \phi_5(\mathbf{s}) = S_{05} + 14S_{015} + 14S_{025} + 9S_{035} + 4S_{045} + 11S_{0125}$$
$$+ 11S_{0135} + 11S_{0145} + 11S_{0235} + 11S_{0245}$$
$$+ 6S_{0345} + 4S_{01235} + 4S_{01245} + 4S_{01345}$$
$$+ 4S_{02345} + S_{012345}.$$

In this way I calculated $\phi_1, \phi_2, \cdots, \phi_7$; and Mr. A. Izenman checked my calculations and extended them to ϕ_8 and ϕ_9. The size of the calculation doubles for each new ϕ; and ϕ_9 is about as far as one can go with paper and pencil.

Things could go further with a computer; but even a large computer would feel its resources strained round about ϕ_{20}; and the method is clearly impractical for $N \geq 30$.

What is needed is some algebraic apparatus, say generating functions, to carry the work to larger values of N. I have not yet succeeded in constructing such apparatus. However, examination of the first few values of ϕ_N led to the formulation of the functional form of ϕ, namely the functions (13.13); and, in the same spirit, we can look at the numerical properties of the first few coefficients $c(\mathbf{a})$ in the hope of spotting some general pattern. In combinatorial work especially, but also in other areas of mathematics, I find it very helpful to study the numerical properties of particular cases. If the numerical data are extensive, one must summarize in some way that will fruitfully reveal the intrinsic pattern. Research experience, rather than undergraduate learning, seems to be the only road to cultivating an instinct for the fruitful choice of good summarizing quantities.

15. Properties of the coefficients c (a)

Even for small values of N, there is an unwieldy amount of data associated with the coefficients $c(\mathbf{a})$; and some method of summarizing it is advisable. As an ad hoc device, guided by a mixture of instinct and experience, I decided to look at the quantities $\alpha_{p,q}^{(N)}$ defined by

$$(15.1) \qquad \alpha_{p,q}^{(N)} = \sum_{a_p = q, a_k = N} c(a_0, a_1, \cdots, a_k), \qquad 1 \leq p \leq q \leq N.$$

For example, from (14.10),

$$(15.2) \qquad \alpha_{24}^{(5)} = 11 + 11 + 6 = 28.$$

The results are tabulated in Tables V to IX, with row and column totals, for $N = 1, 2, \cdots, 7$.

TABLE V

VALUE OF $\alpha_{p,q}^{(1)}$

	$q = 1$	
$p = 1$	1	1
	1	

TABLE VI

VALUES OF $\alpha_{p,q}^{(2)}$

	$q = 1$	2	
$p = 1$	1	1	2
2		1	1
	1	2	

TABLE VII

VALUES OF $\alpha_{p,q}^{(3)}$

	$q = 1$	2	3	
$p = 1$	3	2	1	6
2		1	4	5
3			1	1
	3	3	6	

TABLE VIII

VALUES OF $\alpha_{p,q}^{(4)}$

	$q = 1$	2	3	4	
$p = 1$	12	8	3	1	24
2		4	6	13	23
3			1	9	10
4				1	1
	12	12	10	24	

TABLE IX

VALUES OF $\alpha_{p,q}^{(5)}$

	$q = 1$	2	3	4	5	
$p = 1$	60	40	15	4	1	120
2		20	30	28	41	119
3			5	12	61	78
4				1	16	17
5					1	1
	60	60	50	45	120	

Have you spotted any numerical patterns yet? If you wish to test your skill at pattern spotting, *do not turn over the page* until you have first had a very good look at tables V to IX inclusive and formed your own conjectures about the corresponding numerical patterns for the next two cases ($N = 6$ and $N = 7$). These next two cases are covered by Tables X and XI on the next page.

Scrutiny of these tables suggests certain interesting patterns. In the first place it appears to be true that

$$(15.3) \qquad \alpha_{p,q}^{(N)} = N\alpha_{p,q}^{(N-1)}, \qquad\qquad 1 \leqq p \leqq q \leqq N - 2.$$

On the other hand, (15.3) is certainly *not* true for $q = N - 1$ or $q = N$. Secondly, if we denote the row totals by

$$(15.4) \qquad \alpha_p^{(N)} = \sum_{q=p}^{N} \alpha_{p,q}^{(N)},$$

then it appears to be true that

$$(15.5) \qquad \alpha_1^{(N)} = N!$$

TABLE X

VALUES OF $\alpha_{p,q}^{(6)}$

	$q = 1$	2	3	4	5	6	
$p = 1$	360	240	90	24	5	1	720
2		120	180	168	120	131	719
3			30	72	105	381	588
4				6	20	181	207
5					1	25	26
6						1	1
	360	360	300	270	251	720	

TABLE XI

VALUES OF $\alpha_{p,q}^{(7)}$

	$q = 1$	2	3	4	5	6	7	
$p = 1$	2520	1680	630	168	35	6	1	5040
2		840	1260	1176	840	495	428	5039
3			210	504	735	830	2332	4611
4				42	140	276	1821	2279
5					7	30	421	458
6						1	36	37
7							1	1
	2520	2520	2100	1890	1757	1638	5040	

and

$$(15.6) \qquad \alpha_p^{(N)} = \alpha_{p-1}^{(N)} - \alpha_{p-1,N}^{(N)}, \qquad\qquad 2 \leq p \leq N.$$

We shall prove later that (15.3), (15.5), and (15.6) are indeed true in general. For the moment we only note that they were originally obtained on the empirical evidence of Tables V, VI, $\cdots$, XI; and that they have the following important implication: all the numbers $\alpha_{p,q}^{(N)}$ can be reconstructed from a knowledge of the last columns $\alpha_{p,N}^{(N)}$ only. For suppose that we are given $\alpha_{p,n}^{(n)}$ for all p, n satisfying $1 \leq p \leq n \leq N$, and suppose that we have so far managed from these to reconstruct $\alpha_{pq}^{(n)}$ for all p, q, n satisfying $1 \leq p \leq q \leq n \leq N - 1$. Then we can reconstruct the first $N - 2$ columns of the table $\alpha_{p,q}^{(N)}$ by use of (15.3). The Nth column of the table has been given us; and we can calculate the row totals of the table by successive use of (15.5) and (15.6). We can then fill in the $(N - 1)$th column, since it is the only missing column and we know the row totals. The assertion about reconstruction now follows by induction upon N.

Thus it is enough to study the quantities

$$(15.7) \qquad \beta_p^{(N)} = \alpha_{p,N}^{(N)},$$

which we now tabulate for $1 \leq p \leq N \leq 9$.

I have said that Table XII is sufficient for the reconstruction of the earlier tables; but it is much more important than this, and actually it contains the complete solution of Ulam's problem (or rather, we should possess the complete

TABLE XII

Values of $\beta_p^{(N)}$

$p = 1$	$N = 1$	2	3	4	5	6	7	8	9
1	1	1	1	1	1	1	1	1	1
2		1	4	13	41	131	428	1429	4861
3			1	9	61	381	2332	14337	89866
4				1	16	181	1821	17557	167080
5					1	25	421	6105	83029
6						1	36	841	16465
7							1	49	1513
8								1	64
9									1
	1	2	6	24	120	720	5040	40320	362880

solution if we knew the complete form of Table XII instead of its first 9 columns only). At first I did not realize the significance of Table XII: it merely evolved as a study of certain numerical patterns associated with the coefficient $c(\mathbf{a})$. It was not until I had calculated Tables V, VI, $\cdots$, IX and hence the first 5 columns of Table XII that I understood what Table XII meant. When its meaning dawned on me, I decided to calculate Tables X and XI, and Mr. Izenman extended this to $N = 8$ and $N = 9$, and at that stage Table XII emerged in its present form. I shall explain the meaning of Table XII in the next section: for the moment, let us look at the numerical patterns in Table XII.

Evidently we have for the first row

$$(15.8) \qquad \beta_1^{(N)} = 1, \qquad\qquad N \geqq 1.$$

The second row is not so simple; but it turns out that

$$(15.9) \qquad \beta_2^{(N)} = \frac{(2N)!}{N!(N+1)!} - 1, \qquad\qquad N \geqq 2,$$

and this can be proved in general. There is also a fairly clear pattern in diagonals near the bottom of the table:

$$(15.10) \qquad \beta_N^{(N)} = 1, \qquad\qquad N \geqq 1$$

$$(15.11) \qquad \beta_{N-1}^{(N)} = (N-1)^2, \qquad\qquad N \geqq 2.$$

Mr. Izenman discovered the formulae for the next two diagonals,

$$(15.12) \qquad \beta_{N-2}^{(N)} = \tfrac{1}{2} N(N-1)(N-2)(N-3) + 1, \qquad\qquad N \geqq 3$$

and

$$(15.13) \qquad \beta_{N-3}^{(N)} = 1 + \tfrac{1}{3} N(N-1)\big[-115 + 57N - 10N(N-1)$$
$$+ \frac{7}{4!} N(N-1)(N-2)(N-3)$$
$$+ \frac{1}{5!} N(N-1)(N-2)(N-3)(N-4)\big], \qquad N \geqq 4.$$

Equations (15.10), (15.11), and (15.12) can be proved in general; but (15.13) has not been proved in general and may not have been written in the most transparent form. This question of transparency is elusive. For example, when one knows the reason why (15.12) is true, it is appropriate to write (15.12) in the form

$$(15.14) \qquad \beta_{N-2}^{(N)} = \binom{N-1}{2} + \left[\binom{N-1}{2} - 1\right]\left[2\binom{N-1}{2} - 1\right], \quad N \geq 3;$$

and it seems to be more or less an algebraic accident that the right side of (15.14) happens to simplify to the right side of (15.12). If (15.13) really is as complicated as it looks, then an attempt to find exact general formulae for the quantities $\beta_p^{(N)}$ would seem out of the question, and we might have to be content with approximations or asymptotic formulae. This issue remains unsettled.

16. Interpretation of the coefficients c (a)

The following interpretation of the coefficients $c(\mathbf{a})$ emerged gradually from an attempt to prove (15.3), which at that early stage was merely a conjecture based on the numerical evidence of Tables V, VI, $\cdots$, IX. In deriving the formulae for ϕ_N we used descending subsequences of $\{x_1, x_2, \cdots\}$, where the x_i came from the exponential distribution (13.1). This, however, was merely a device for easing the analysis and obtaining manageable functions (13.13). However, the coefficients $c(\mathbf{a})$ are much more deeply implicated in the combinatorial structure of the problem; and, to interpret them, we return to the original formulation of the problem, namely ascending subsequences of random permutations of the integers $\{1, 2, \cdots, N\}$.

We set up a method of coding these permutations. Suppose that $\pi = \{\pi_1, \pi_2, \cdots, \pi_N\}$ is a given permutation of $\{1, 2, \cdots, N\}$. Define

$$(16.1) \qquad\qquad a_i = a_i(\pi), \qquad\qquad i = 1, 2, \cdots, \ell$$

to be the greatest integer j such that $\{\pi_1, \pi_2, \cdots, \pi_j\}$ has a longest ascending subsequence of length i. In the definition (16.1), $i = 1, 2, \cdots, \ell$ where ℓ is the length of a longest ascending subsequence in $\pi = \{\pi_1, \pi_2, \cdots, \pi_N\}$. Also define $a_0 = 0$; and write

$$(16.2) \qquad\qquad \mathbf{a}(\pi) = \left(a_0, a_1(\pi), \cdots, a_\ell(\pi)\right).$$

For example, if $N = 9$ and

$$(16.3) \qquad\qquad \pi = \{8\ 9\ 1\ 4\ 3\ 6\ 5\ 7\ 2\},$$

then

$$(16.4) \qquad\qquad \mathbf{a}(\pi) = (0\ 1\ 5\ 7\ 9).$$

It is evident that the final coordinate in $\mathbf{a}(\pi)$ must always equal N, the number of elements in π. For a given vector $\mathbf{a}$, let $\gamma(\mathbf{a})$ denote the number of permutations π such that $\mathbf{a}(\pi) = \mathbf{a}$. We are going to prove that

(16.5) $$\gamma(\mathbf{a}) = c(\mathbf{a}).$$

If $N = 1$, the only possible permutation is $\pi = \{1\}$ and $a(\pi) = (0, 1)$. Hence $\gamma(0, 1) = 1 = c(0, 1)$ and (16.5) is true for $N = 1$. Now assume that (16.5) is true for N. Consider a given permutation $\pi = \{\pi_1, \pi_2, \cdots, \pi_N\}$. This can be converted to a permutation of $\{1, 2, \cdots, N + 1\}$ by inserting $N + 1$ in $N + 1$ available places. We write these extended permutations as

(16.6) $$\begin{aligned} T_0\pi &= \{N + 1, \pi_1, \pi_2, \cdots, \pi_N\} \\ T_1\pi &= \{\pi_1, N + 1, \pi_2, \cdots, \pi_N\} \\ T_2\pi &= \{\pi_1, \pi_2, N + 1, \cdots, \pi_N\} \\ \overline{\phantom{T_2\pi = \{\pi_1, \pi_2, N}} \\ T_N\pi &= \{\pi_1, \pi_2, \cdots, \pi_N, N + 1\}. \end{aligned}$$

However, by (14.3) and (14.4), we have

(16.7) $$\mathbf{a}(T_r\,\pi) = T_r\,\mathbf{a}(\pi), \qquad\qquad r = 0, 1, \cdots, N.$$

This holds for all π; and hence (16.5) is true for $N + 1$ in place of N. Thus (16.5) is generally true by induction on N.

But now (15.1) and (15.7), taken together with (16.2), prove that $\beta_p^{(N)}$ is the number of permutations of $\{1, 2, \cdots, N\}$ which contain a longest ascending subsequence of length p. Hence

(16.8) $$P\{\ell(X_N) = p\} = \beta_p^{(N)}/N!,$$

which provides the distribution of the random variable $\ell(X_N)$. This explains the importance of Table XII.

With these preliminaries settled, we can now prove (15.3). Consider a given permutation $\pi = \{\pi_1, \pi_2, \cdots, \pi_N\}$. For $r = 1, 2, \cdots, N + 1$ define

(16.9) $$U_r\pi = \{\pi_1^*, \pi_2^*, \cdots, \pi_N^*, r\}$$

where $\pi_i^* = \pi_i$ or $\pi_i^* = \pi_i + 1$ accordingly as $\pi_i < r$ or $\pi_i \geqq r$. If

(16.10) $$\mathbf{a}(\pi) = (a_0, a_1, \cdots, a_{\ell-1}, N)$$

then

(16.11) $$\mathbf{a}(U_r\pi) = \begin{cases} (a_0, a_1, \cdots, a_{\ell-1}, N + 1), & r = 1, 2, \cdots, N \\ (a_0, a_1, \cdots, a_{\ell-1}, N, N + 1), & r = N + 1. \end{cases}$$

This gives the important identity

(16.12) $$\begin{aligned} (N &+ 1)c(a_0, a_1, \cdots, a_{\ell-1}, N) \\ &= c(a_0, a_1, \cdots, a_{\ell-1}, N + 1) + c(a_0, a_1, \cdots, a_{\ell-1}, N, N + 1), \end{aligned}$$

because π is an arbitrary permutation of $\{1, 2, \cdots, N\}$. Since $a_{\ell-1} \leqq N - 1$, we now see that

$$(16.13) \qquad \alpha_{p,q}^{(N+1)} = (N+1)\alpha_{p,q}^{(N)}, \qquad 1 \leq p \leq q \leq N-1,$$

which is (15.3) with $N+1$ in place of N.

A few of the coefficients $c(\mathbf{a})$ can be determined explicitly. For example, consider the permutations which are coded by the vector

$$(16.14) \qquad \mathbf{a} = (0, r, r+1, r+2, \cdots, N),$$

that is to say with $a_i = i + r - 1$, for $i = 1, 2, \cdots, N - r + 1$; all have

$$(16.15) \qquad \pi_1 > \pi_2 > \cdots > \pi_r = 1 < \pi_{r+1} < \pi_{r+2} < \cdots < \pi_N,$$

and *vice versa*. But we can choose any set of $r - 1$ elements from $\{2, 3, \cdots, N\}$ and arrange them in descending order to give $\pi_1, \pi_2, \cdots, \pi_{r-1}$. Therefore

$$(16.16) \qquad c(0, r, r+1, r+2, \cdots, N) = \binom{N-1}{r-1}.$$

From (16.12) and (16.16) we deduce

$$(16.17) \quad c(0, r, r+1, r+2, \cdots, N-2, N) = \left\{\frac{N^2 - N(r+1) + 1}{r-1}\right\}\binom{N-2}{r-2}.$$

For example,

$$(16.18) \qquad c(0345) = \binom{4}{2} = 6,$$

and

$$(16.19) \qquad c(035) = \left(\frac{25 - 20 + 1}{2}\right)\binom{3}{1} = 9;$$

and these confirm the coefficients of S_{0345} and S_{035} in (14.10).

The identity (16.12) shows that our problem would be solved, at least in principle, if for each N we knew the values of the 2^{N-2} coefficients $c(\mathbf{a})$ in which $a_{\ell-1} = N - 1, a_\ell = N$. This suggests that we ought to look for further identities like (16.12) which would successively reduce the problem to a determination of coefficients with

$$a_{\ell-1} = N - 1, a_\ell = N$$
$$(16.20) \qquad a_{\ell-2} = N - 2, a_{\ell-1} = N - 1, a_\ell = N$$

----

until we reach known coefficients of the form (16.16). However this attractive possibility has so far eluded me. Nor have I made any substantial progress towards an explanation of the spectrum of values assumed by the coefficients

$c(\mathbf{a})$. Table XIII shows the number of distinct values taken by $c(\mathbf{a})$; and this number is noticeably smaller than the number of coefficients.

TABLE XIII

DATA ON THE COEFFICIENT SPECTRUM

Value of N:	1	2	3	4	5	6	7	8	9
Number of coefficients $c(\mathbf{a})$	1	2	4	8	16	32	64	128	256
Number of distinct coefficients $c(\mathbf{a})$	1	1	2	3	6	9	16	29	55

For purposes of reference and to spare other investigators the labor of recalculating the coefficients $c(\mathbf{a})$ for $N \leqq 9$, I give in Table XIV a condensed list of the coefficients for $N = 9$. Coefficients for $N \leqq 8$ can be easily recovered from Table XIV by means of the identity (16.12). To save space the coefficients are simply listed in natural order beginning with $c(01)$ and ending with $c(0123456789)$; and an entry such as u^n means that the coefficient u occurs n times consecutively in this position of the list, and semicolons separate coefficients for different values of ℓ: thus, in this notation, (14.10) would take the compact form 1; 14^2, 9, 4; 11^5, 6; 4^4; 1.

TABLE XIV

LIST OF COEFFICIENTS FOR $N = 9$

1; 1430^2, 1001, 572, 275, 110, 35, 8; 6529^3, 6031, 5035, 4168, 2431, 6529^2, 6031, 5035, 3751, 2431, 3820, 3772, 3322, 2536, 1672, 1609^2, 1321, 913, 520^2, 400, 133^2, 28; 4364^4, 4280, 4028, 4364^3, 4280, 4028, 4364^2, 4280, 4028, 3812^2, 3644, 2876^2, 1475, 4364^3, 4280, 4028, 4364^2, 4280, 4028, 3812^2, 3644, 2876^2, 1892, 2339^3, 2255, 2339^2, 2303, 1871^2, 1271, 866^5, 650, 245^3, 56; 1405^{14}, 1363, 1405^9, 1363, 1405^5, 1363, 1201^3, 877, 1405^9, 1363, 1405^5, 1363, 1201^3, 877, 730^9, 568, 259^4, 70; 314^{34}, 266, 314^{14}, 266, 161^5, 56; 55^{27}, 28; 8^8; 1.

We now return to a further consideration of assumption α in Section 12, where we had to calculate the quantity q_n in (12.6). Actually we shall change the ground a little by supposing that, instead of sampling the x_i in X from the uniform distribution on $[0, 1]$, we are sampling from (13.1) and looking at the distribution of ℓ'_N. The necessary technical adaptation to pass from one form of the problem to the other is simple, merely a suitable transformation of the x-axis. Also (13.4) will hold in place of (12.5) and we write

(16.21) $$z_i = y_i - y_{i+1}.$$

We want to calculate

(16.22) $$q_N = \frac{1}{2} \sum_{i=1}^{\infty} z_i^2.$$

From (13.5) and (13.12) we have

$$(16.23) \qquad \phi_N(\mathbf{s}) = E \exp \left\{ - \sum_{i=1}^{\infty} s_i y_i \right\} = E \exp \left\{ - \sum_{i=1}^{\infty} S_i(y_i - y_{i+1}) \right\}$$

$$= E \exp \left\{ - \sum_{i=1}^{\infty} S_i z_i \right\}.$$

Hence, from (13.13) and (13.22),

$$(16.24) \qquad E \exp \left\{ - \sum_{i=1}^{\infty} S_i z_i \right\} = \sum_{\mathbf{a}} c(\mathbf{a}) \prod_{i=1}^{k} \frac{(S_i + a_{i-1})!}{(S_i + a_i)!}.$$

If the right side of (16.24) had just one term in the sum, this would establish the independence of the z_i required under assumption α. As things stand, it is no more than suggestive.

We can go a little further by differentiating (16.24) twice, and then putting $S_1 = S_2 = \cdots = 0$. Thus

$$(16.25) \qquad q_N = \frac{1}{2} \left[\sum_{i=1}^{\infty} \frac{\partial^2}{\partial S_i^2} E \exp \left\{ - \sum_{j=1}^{\infty} S_j z_j \right\} \right]_{\mathbf{s}=0}$$

$$= \frac{1}{N!} \sum_{\mathbf{a}} c(\mathbf{a}) v(\mathbf{a}),$$

where

$$(16.26) \qquad v(a_0, a_1, \cdots, a_k) = \sum_{1 \leq i < j \leq k} \frac{1}{a_i a_j}.$$

Maybe this can be manipulated further.

17. Distributional properties of $\ell(X_N)$ for small N

Table XV exhibits the principal statistics of the distribution of the random variable $\ell_N = \ell(X_N)$ for $N \leq 9$, calculated from Table XII.

TABLE XV

STATISTICS OF ℓ_N FOR $N \leq 9$

N	$E(\ell_N)$	$E(\ell_N)/\sqrt{N}$	Var ℓ_N	$\sqrt{(\text{Var } \ell_N)}$
1	1.00000	1.00000	0.00000	0.00000
2	1.50000	1.06066	0.25000	0.50000
3	2.00000	1.15470	0.33333	0.57735
4	2.41667	1.20830	0.41005	0.64035
5	2.79167	1.24844	0.49863	0.70614
6	3.14028	1.28201	0.57065	0.75541
7	3.46528	1.30975	0.63218	0.79510
8	3.77034	1.33302	0.69106	0.83130
9	4.05833	1.35278	0.74859	0.86521

I have tried extrapolating $E(\ell_N)/\sqrt{N}$ as $N \to \infty$ by calculating divided differences with $N^{-1/2}$ as argument; but this does not work at all well with the values of N available in Table XV, and yields to an estimate of c substantially less than 2.

18. Application to nonparametric testing of stationary sequences

As explained earlier, I took up Ulam's problem because it was a challenging mathematical problem, which aroused my curiosity. But, before long, of course, I asked myself if the mathematics might, by some happy accident, have applications.

One of the standard methods of testing whether a sequence of independent random variables is stationary, against the alternative that it has a trend, is to count the number of local maxima in the sequence. This test however has the drawback of being rather easily affected by local aberrations in the sequence: a test that took a more synoptic view of the whole sequence would be preferable. This situation is rather like that met in looking for periodicities in a stochastic process: the old-fashioned periodogram analysis suffers because genuine periodicities can be obscured by a few accidental phase-shifts; and the autocorrelation coefficient and its Fourier transform provide a better approach since they filter out these local irregularities.

The length of a longest ascending (or descending) subsequence should give quite a good synoptic nonparametric test statistic of stationarity. A sequence of length N will have a longest ascending subsequence of length about $2\sqrt{N}$ if it is stationary, but one of length proportional to N if it has an increasing trend. If the original sequence is reasonably long, so that N is much larger than $2\sqrt{N}$, the test will be very sensitive, especially because the Monte Carlo experiments suggest that ℓ_N has a small sampling variance (which might even be bounded as $N \to \infty$). The test statistic ℓ_N is also very easily computed by the algorithm in Section 9. But before the test can be put forward for practical use, we need to know more about $\mathrm{Var}\, \ell_N$.

19. Distribution of the number of ladder points

When I delivered the lecture on this paper at the Sixth Berkeley Symposium, Section 18 represented my thoughts on applications. But two days later Professor A. Dvoretzky, who had been in the audience, suggested to me that I should look into the corresponding nonparametric test based upon ladder points; he thought it likely that this would be both an easier mathematical problem and a more powerful test.

A point x_j in the sequence $X_N = \{x_1, x_2, \cdots, x_N\}$ is called a ladder point if $x_i \leqq x_j$ for all $i \leqq j$. What is the distribution of k_N, the number of ladder points in X_N, given that the elements of X are independently and identically distributed with a probability density function? Let

$$(19.1) \qquad f_N(t) = \sum_{n=1}^{N} P(k_N = n)t^n$$

be the generating function of the distribution. Let us pass from X_N to X_{N+1}; then there is probability $1/(N + 1)$ that x_{N+1} will add an extra ladder point to k_N, and probability $N/(N + 1)$ that it will not. Hence

$$(19.2) \qquad f_{N+1}(t) = f_N(t)(N + t)/(N + 1);$$

whence

$$(19.3) \qquad f_N(t) = (1/N!) \prod_{r=0}^{N-1} (r + t)$$

$$= (1/N!) \sum_{n=1}^{N} (-1)^{N+n} S_N^n t^n,$$

where S_N^n are the Stirling numbers of the first kind ([17], p. 22). Hence

$$(19.4) \qquad P(k_N = n) = (-1)^{N+n} S_N^n / N! = |S_N^n|/N!$$

gives the distribution of the number of ladder points. From (19.3) we can easily derive the mean and variance of k_N:

$$(19.5) \qquad E(k_N) = \sum_{r=1}^{N} r^{-1} \sim \log N + \gamma,$$

$$(19.6) \qquad \operatorname{Var}(k_N) = \sum_{r=1}^{N} (r - 1)/r^2 \sim \log N + \gamma - \frac{\pi^2}{6},$$

where γ is Euler's constant. For large N, we find that k_N has asymptotically a Poisson distribution with parameter $\log N$. Since $N^{-1} \log N$ is much smaller than $(2\sqrt{N})/N$, the test based on ladder points will be more sensitive than the one proposed in Section 18. It is one of those hard but sad facts of mathematics that the easier and less diverting mathematical problems are likely to be the more useful in practice. Now anything *both* useful *and* mathematically trivial will have been published several times over already; and one really ought to check the literature for references. I found papers by Chandler [2], Foster and Stuart [5], and Stuart [22], and—ironically enough—a couple of my own early papers [8], [9], which I had forgotten. (The conjectures in [9] were subsequently solved by Erdös [3] and by Moses and Wyman [20].)

20. Cross connections and conjectures

In Section 3 I said that it is better not to be influenced by reading the literature; and this should include forgetting about one's own work as well as ignoring other people's. Had I remembered my earlier work on Stirling's numbers, I would have been deprived of an important motive for thinking about Ulam's problem. Equally, it is difficult to escape from ideas and techniques that one has used before. The methods used in proving (8.15), in particular the introduction of a Poisson process and the associated Tauberian argument, originated from a paper [1] on the travelling salesman problem. There they sufficed for convergence with probability 1; but here, for various reasons, they only lead to convergence in probability. I want, of course, to show that (8.15) is also true with probability 1; and for this I believe that fresh ideas are needed, and I have tried hard to escape from the shackles of the earlier methods in [1], but without success. Preconceptions die hard.

But although the literature can be stultifying if one pays too much attention to proof, it can be stimulating if one concentrates upon conjectures. So I shall end by tracing some cross-connections between this paper and sundry problems and conjectures.

If N towns are distributed at random in a region of area A, and L_N is the length of the shortest journey that the travelling salesman must make to visit them all, then

$$(20.1) \qquad\qquad L_N \sim C\sqrt{NA} \qquad \text{as} \quad N \to \infty$$

with probability 1, where C is an absolute constant. The relations (8.15) and (20.1) are closely alike. In [1] it is shown that C satisfies various inequalities; but nobody has yet solved the problem of determining C exactly. What is wanted is theory for C along the lines of Sections 10, 11, and 12. There is also a similar problem for Steiner's network problem [11]. (Incidentally, I take this opportunity of correcting an error of calculation: in [1], p. 302, relation (7) should have read $2^{1/3}/3^{1/2} \leqq \alpha_3$, and consequently the relevant part of (8) should be $0.72742 \leqq \alpha_3$.)

I have written elsewhere [13] of the distinctions between "soft" and "hard" mathematics. One, though naturally not the only, distinction is that hard mathematics is often concerned with calculating the numerical value of a constant. To this extent Section 8 is soft mathematics dealing with generalities, while Sections 9 to 12 are hard mathematics aimed at determining the value of c. Of course, mathematical physicists are concerned with numerical values; and this is one of the reasons why their mathematical expertise tends to be sharper and stronger than that of pure mathematicians.

Rota's paper [21] has already featured in this story; and it has other cross connections which I shall mention briefly. He writes about the place of combinatorial analysis in mathematical research and he lists seven challenging problems: (i) the Ising problem; (ii) percolation theory; (iii) the number of necklaces, and Pólya's problem; (iv) self-avoiding random walks; (v) the travelling salesman problem; (vi) the coloring problem; and (vii) the pigeonhole principle and Ramsay's theorem. It is interesting to note how many of these topics are centered upon the determination of constants; and I am also pleased to find a high proportion of my own favourite problems in his list.

The Ising problem is one of the most celebrated problems in theoretical physics, nearly fifty years old now and still guarding its secrets about the numerical values of certain constants as well as more qualitative questions about the existence of singularities. Percolation theory is my own invention; and it has, as Rota explains, a close connection with the Ising problem. A general exposition together with a bibliography of percolation theory and its relation to the Ising problem and to various other questions in physics and chemistry appears in [6]: this also gives references to the self-avoiding walk problem, which is intimately connected with percolation theory. Another celebrated problem, closely associated with the Ising problem, is the monomer-dimer problem (see [14] for details and bibliography). This too asks for the value of a constant; but it also contains some "soft" mathematical problems, which would contribute greatly to our understanding of stochastic processes in more than one dimension if only we could solve them. The following problem is typical; and I am indebted to Professor David Blackwell for kindly supplying me with a translation of it into the language of modern mathematics.

"Denote by F the set of all functions f from the lattice points of the plane to $\{e, n, w, s\}$ such that

$$f(x, y) = w \Leftrightarrow f(x + 1, y) = e$$

and

$$f(x, y) = s \Leftrightarrow f(x, y + 1) = n.$$

For any finite set A of lattice points, denote by F_A the set of all restrictions of functions $f \in F$ to A. For any $B \supset A$, the uniform distribution on F_B induces a probability distribution $p(A, B)$ on F_A. Does $p(A, B)$ converge as B increases to the set of all lattice points, for every A?"

Percolation theory gave birth to subadditive stochastic processes, already discussed in Section 6. Despite considerable work on percolation problems, a great deal remains to be done. Rota [21] says that the percolation problem "was brilliantly solved by Michael Fisher, a British physicist now at Cornell University." This, however, is not quite correct, although Professor Fisher has done a great deal to advance our knowledge of these matters. I too was at one time under the misapprehension that the percolation problem was solved; in [6], p. 897, I wrote:

"Sykes and Essam have very recently (verbal communication to one of the authors) utilized somewhat similar conversions in a proof that the bond process critical probabilities of the triangular, square, and hexagonal lattices are respectively $2 \sin \pi/18$, $1/2$, and $1 - 2 \sin \pi/18$. Their brilliant solution of these three exceptionally difficult problems, all hitherto unsolved, is a most remarkable achievement."

Alas, these three problems are still unsolved; when Sykes and Essam published their work [23, 24], they wrote [24], p. 1125:

"We shall suppose, without offering a proof, that for real p ($0 \leq p \leq 1$) the function K is singular at $p = p_c$, but nowhere else. This is to be expected in the light of exact results for closely related problems, and in particular, for percolation problems on lattices of the Bethe type for which K has been given exactly."

I am in no doubt that Sykes and Essam have got the right numerical answers (which agree, for example, with results obtained from series expansions and by Monte Carlo methods); and their exploitation of "matching" graphs is a valuable new tool in the subject; but they do not claim to have found a rigorous proof, and their argument rests upon plausible assumptions such as the one quoted above.

The travelling salesman problem we have already noted in (20.1). The most famous case of the coloring problem is to evaluate a constant, known to be either 4 or 5. Rota's illustration of the pigeonhole principle in [21] was discussed in Section 3. On self-avoiding walks, where one of the issues is to determine the numerical value of the so-called connective constant, Rota writes: "it is likely that this problem will be at least partly solved in the next few years, if interest in it stays alive." And this is the rub: he qualifies his decent optimism; for today an increasing number of graduate students, reared on the deficient diet of modern mathematics, take fright at difficult specific problems which they have neither the courage nor the intellectual training to tackle, and they turn aside to a tedious retilling of easier soils.

REFERENCES

[1] J. E. BEARDWOOD, J. H. HALTON, and J. M. HAMMERSLEY, "The shortest path through many points," *Proc. Cambridge Philos. Soc.*, Vol. 55 (1959), pp. 299–327.

[2] K. N. CHANDLER, "The distribution and frequency of record values," *J. Roy. Statist. Soc. Ser. B.*, Vol. 14 (1952), pp. 220–228.

[3] P. ERDÖS, "On a conjecture of Hammersley," *J. London Math. Soc.*, Vol. 28 (1953), pp. 232–236.

[4] W. FELLER, *An Introduction to Probability Theory and its Applications*, Vol. 2, New York, John Wiley, 1966, p. 225.

[5] F. G. FOSTER and A. STUART, "Distribution-free tests in time series based on the breaking of records," *J. Roy. Statist. Soc. Ser. B*, Vol. 16 (1954), pp. 1–22.

[6] H. L. FRISCH and J. M. HAMMERSLEY, "Percolation processes and related topics," *J. Soc. Indust. Appl. Math.*, Vol. 11 (1963), pp. 894–914.

[7] M. GARDNER, "Mathematical games," *Scientific American*, Vol. 216, (March 1967), pp. 123–129 and Vol. 216, (April 1967), pp. 116–123.

[8] J. M. HAMMERSLEY, "On estimating restricted parameters," *J. Roy. Statist. Soc. Ser. B*, Vol. 12 (1950), pp. 192–229.

[9] ———, "The sums of products of the natural numbers," *Proc. London Math. Soc. Ser. 3*, Vol. 1 (1951), pp. 426–452.

[10] ———, "A non-harmonic Fourier series," *Acta Math.*, Vol. 89 (1953), pp. 243–260.

[11] ———, "On Steiner's network problem," *Mathematika*, Vol. 8 (1961), pp. 131–132.

[12] ———, "A Monte Carlo solution of percolation in the cubic crystal," *Methods in Computational Physics*, New York, Academic Press, 1963, Vol. 1 pp. 281–298.

[13] ———, "The enfeeblement of mathematical skills by modern mathematics and by similar soft intellectual trash in schools and universities," *Bull. Inst. Math. Appl.*, Vol. 4 (1968), pp. 66–85.

[14] J. M. HAMMERSLEY and V. V. MENON, "A lower bound for the monomer-dimer problem." *J. Inst. Math. Appl.*, Vol. 6 (1970), pp. 341–364.

[15] J. M. HAMMERSLEY and D. J. A. WELSH, "First-passage percolation, subadditive processes, stochastic networks, and generalized renewal theory," *Bernoulli-Bayes-Laplace Anniversary Volume*, Berlin, Springer, 1965, pp. 61–110.

[16] G. H. HARDY, *Divergent Series*, Oxford, Oxford University Press, 1949.

[17] C. JORDAN, *Calculus of Finite Differences*, third edition. New York, Chelsea Publishing Company, 1965.

[18] J. F. C. KINGMAN, "The ergodic theory of subadditive stochastic processes." *J. Roy. Statist. Soc. Ser B*, Vol. 30 (1968), pp. 499–510.

[19] P. A. P. MORAN, "The random division of an interval," *J. Roy. Statist. Soc. Ser. B*, Vol. 9 (1947), pp. 92–98 and Vol 13 (1951), pp. 147–150.

[20] L. MOSES and M. WYMAN, "Asymptotic development of the Stirling numbers of the first kind," *J. London Math. Soc.*, Vol. 33 (1958), pp. 133–146.

[21] G.-C. ROTA, "Combinatorial analysis," *The Mathematical Sciences: Essays for the Committee on Support of Research in the Mathematical Sciences*, Cambridge, M.I.T. Press, 1969, pp. 197–208.

[22] A. STUART, "The efficiency of the records test for trend in normal regression," *J. Roy. Statist. Soc. Ser. B*, Vol. 19 (1957), pp. 149–153.

[23] M. F. SYKES and J. W. ESSAM, "Some exact critical percolation probabilities for bond and site problems in two dimensions," *Phys. Rev. Lett.*, Vol. 10 (1963), pp. 3–4.

[24] ———, "Exact critical percolation probabilities for site and bond problems in two dimensions," *J. Mathematical Phys.*, Vol. 5 (1964), pp. 1117–1127.

[25] S. M. ULAM, "Monte Carlo calculations in problems of mathematical physics," *Modern Mathematics for the Engineer: Second Series* (edited by E. F. Beckenbach) New York, McGraw-Hill, 1961.

CLASSES OF DISTRIBUTIONS APPLICABLE IN REPLACEMENT WITH RENEWAL THEORY IMPLICATIONS

ALBERT W. MARSHALL
UNIVERSITY OF ROCHESTER
and
FRANK PROSCHAN
FLORIDA STATE UNIVERSITY

1. Introduction and summary

Age and block replacement policies are commonly used to diminish in-service failures. Unfortunately, for some items (say, those with decreasing failure rate), use of these policies may actually *increase* the number of in-service failures.

In this paper we determine the largest classes of life distributions for which age and block replacement diminishes, either stochastically or in expected value, the number of failures in service. We obtain bounds on survival probability, moment inequalities, and renewal quantity inequalities for distributions in these classes. We show that under certain reliability operations on components in a given class of life distributions (such as formation of systems, addition of life lengths, and mixtures of distributions), life distributions are obtained which remain within the class.

We consider items which perform a function that is to be continued over an indefinite period of time. To make this possible, an item which fails while in service is immediately replaced by a new item of the same kind.

Sometimes the interruption caused by an in-service failure is costly compared with the item replacement cost. If it is possible to make a "planned" replacement of an unfailed item, thus avoiding the high cost associated with a failure replacement, then planned replacements provide a practical means for avoiding reliance upon aged or worn items.

It has long been realized that for units with certain kinds of life distributions, planned replacements actually increase the frequency of failures. A goal of this paper is to identify the life distributions for which planned replacements are, or are not, beneficial.

We assume that the life lengths of all items to be placed in service are independent and have a common distribution F. Without further mention, we assume that $F(z) = 0$ for $z < 0$, and we denote the *survival function* by $\bar{F} \equiv 1 - F$.

The two planned replacement policies most commonly employed are age and block replacement. Under an *age replacement policy*, a unit is replaced upon failure or upon reaching a specified age T, whichever comes first. Under a *block replacement policy*, a replacement is made whenever a failure occurs and additionally at specified times $T, 2T, 3T, \cdots$. Age replacement results in fewer planned replacements, since replacements are planned according to a unit's age. On the other hand, block replacement can be scheduled in advance, and perhaps coordinated with the replacement of associated units.

Barlow and Proschan [3] have compared these replacement policies with respect to the number of failures, the number of planned replacements, and the total number of replacements by any time t. They reference several other authors who have studied these replacement policies.

Barlow and Proschan assume that the distribution F has an increasing failure rate. A distribution F is said to have an *Increasing Failure Rate* (IFR) if $\log \bar{F}$ is concave, that is, if for all $x > 0$, $\bar{F}(x + t)/\bar{F}(t)$ is decreasing in t such that $t \geqq 0$ and $\bar{F}(t) > 0$. If F has a density, this is equivalent to the condition that for some version f of the density, the failure rate $r(t) = f(t)/\bar{F}(t)$ is increasing in t for which $\bar{F}(t) > 0$.

We find two other classes of distributions important in the comparison of replacement policies.

A distribution F (or survival function $\bar{F}$) is said to be

(i) *New Better than Used* (NBU) if

$$(1.1) \qquad\qquad \bar{F}(x + y) \leqq \bar{F}(x)\bar{F}(y)$$

for all $x, y \geqq 0$,

(ii) *New Better than Used in Expectation* (NBUE) if the mean μ of F is finite and

$$(1.2) \qquad\qquad \mu \geqq \int_0^\infty \bar{F}(t + x)\, dx/\bar{F}(t)$$

for all $t \geqq 0$ such that $\bar{F}(t) > 0$. Notice that equation (1.1) can be interpreted as saying that the chance $\bar{F}(x)$ that a new unit will survive to age x is greater than the chance $\bar{F}(x + y)/\bar{F}(y)$ that an unfailed unit of age y will survive an additional time x. On the other hand, (1.2) says only that the expected life length of a new unit is greater than the expected remaining life of a used but unfailed unit.

Another class of distributions known to be important in reliability consists of the distributions with an *Increasing Failure Rate Average* (IFRA). A distribution F is said to be IFRA if $-[\log \bar{F}(t)]/t$ is increasing in $t > 0$, or equivalently, $[\bar{F}(s)]^{1/s} \geqq [\bar{F}(t)]^{1/t}$ for $0 < s < t$. Notice that if t is a multiple of s, this inequality is also satisfied when F is NBU.

The chain of implications

$$(1.3) \qquad\qquad \text{IFR} \Rightarrow \text{IFRA} \Rightarrow \text{NBU} \Rightarrow \text{NBUE}$$

is readily established. To prove the last implication, one needs the fact that NBU distributions have finite means (proved in Section 4).

Each of the above classes of distributions has a companion class defined by reversing the inequality of the definition. Thus, we define *Decreasing Failure Rate* (DFR), *Decreasing Failure Rate Average* (DFRA), *New Worse than Used* (NWU) and *New Worse than Used in Expectation* (NWUE). The corresponding chain of implications

(1.4) $$\text{DFR} \Rightarrow \text{DFRA} \Rightarrow \text{NWU}$$

holds. If F is NWU and if additionally it has a finite mean, then F is NWUE.

The classes NBU, NWU, NBUE and NWUE have received little attention in the literature in spite of their intuitive appeal. Their importance in replacement policy evaluation is demonstrated in Section 2, and their role in renewal theory is discussed in Section 3. Some basic inequalities for the classes are presented in Section 4, and the preservation of these classes under reliability operations is the subject of Section 5.

2. Replacement policy comparisons

In this section we obtain some comparisons for age and block replacement policies assuming that the underlying life distribution F belongs to an appropriate class. For these comparisons, we need the following notation:

$N(t)$ = number of failures (renewals) in $[0, t]$ for an ordinary renewal process, with no planned replacements. This quantity records the number of failures in $[0, t]$ if replacements are made only upon failure.

$N_A(t, T)$ = number of failures in $[0, t]$ under an age replacement policy with replacement age T.

$N_B(t, T)$ = number of failures in $[0, t]$ under a block replacement policy with replacement interval T.

These quantities do not record planned replacements, but only replacements due to an in-service failure. However, in-service failures are not recorded if they should happen to coincide with a planned replacement.

We caution the reader on one point: contrary to what is sometimes assumed in renewal theory (for example, by Feller, [8], p. 346), these quantities do *not* automatically count the origin as a renewal point (point of failure).

Y_1 = time of first failure when no planned replacements are made = $\inf \{t : N(t) \geq 1\}$.

Y_i = length of time between $(i - 1)$st and ith failure when no planned replacements are made = $\inf \{t : N(t) \geq i\} - \inf \{t : N(t) \geq i - 1\}, i = 2, 3, \cdots$. Similarly define

(2.1) $$Y_{i, A}(T), \qquad Y_{i, B}(T), \qquad i = 1, 2, \cdots$$

for the processes $N_A(t, T)$ and $N_B(t, T)$, respectively.

In the following theorems and lemma, we give the proof of the first of two parallel results. The second result in brackets in each case can be proved in the same way but with all inequalities reversed. (The symbol $\underset{st}{\leqq}$ will be used for "stochastically less than.")

THEOREM 2.1. $Y_i \underset{st}{\leqq} Y_{i,A}(T)$ $[Y_i \underset{st}{\geqq} Y_{i,A}(T)]$ for all $T > 0, i = 1, 2, \cdots \Leftrightarrow F$ is NBU [NWU].

PROOF.   Note that Y_i and $Y_{i,A}(T)$ have distributions independent of i. Clearly,

$$(2.2) \qquad P\{Y_1 > t\} = \bar{F}(t), \qquad P\{Y_{1,A}(T) > t\} = [\bar{F}(T)]^j \bar{F}(t - jT)$$
$$\text{for} \quad jT \leqq t < (j+1)T, \qquad j = 0, 1, \cdots.$$

If F is NBU, then $\bar{F}(t) \leqq [\bar{F}(T)]^j \bar{F}(t - jT)$ by a repeated application of the definition.

If $P\{Y_i > t\} \leqq P\{Y_{i,A}(T) > t\}$ for all t, T, take $T = \max(x, y)$, $t - T = \min(x, y)$ to obtain $\bar{F}(x + y) \leqq \bar{F}(x)\bar{F}(y)$. Q.E.D.

COROLLARY 2.1. $N(t) \underset{st}{\geqq} N_A(t, T)$ $[N(t) \underset{st}{\leqq} N_A(t, T)]$ for all $t, T > 0 \Leftrightarrow F$ is NBU [NWU].

PROOF.   If F is NBU, then since $Y_1, Y_2, Y_3, \cdots$ are independent and $Y_{1,A}, Y_{2,A}, \cdots$ are independent, it follows from Theorem 2.1 that

$$(2.3) \qquad P\{N(t) \geqq n\} = P\{Y_1 + \cdots + Y_n \leqq t\}$$
$$\geqq P\{Y_{1,A}(T) + \cdots + Y_{n,A}(T) \leqq t\} = P\{N_A(t, T) \geqq n\}.$$

If $N(t) \underset{st}{\geqq} N_A(t, T)$ then

$$(2.4) \qquad P\{Y_1 > t\} = P\{N(t) = 0\} \leqq P\{N_A(t, T) = 0\} = P\{Y_{1,A}(T) > t\},$$

and hence F is NBU by Theorem 2.1. Q.E.D.

THEOREM 2.2. $\mu = EY_i \leqq EY_{i,A}(T)$ $[\mu \geqq EY_{i,A}(T)]$ for all $T > 0, i = 1, 2, \cdots \Leftrightarrow F$ is NBUE [NWUE].

PROOF.   It is sufficient to prove the theorem for $i = 1$. We compute

$$(2.5) \quad EY_{1,A}(T) = \int_0^\infty P\{Y_{1,A}(T) > t\} \, dt$$

$$= \int_0^T \bar{F}(t) \, dt + \bar{F}(T) \int_T^{2T} \bar{F}(t - T) \, dt + \bar{F}^2(T) \int_{2T}^{3T} \bar{F}(t - 2T) \, dt + \cdots$$

$$= \int_0^T \bar{F}(t) \, dt / F(T).$$

(See [11].) But $\int_0^T \bar{F}(t) \, dt / F(T) \geqq \mu$ for all $T \Leftrightarrow$

$$(2.6) \qquad \bar{F}(T) \int_0^\infty \bar{F}(t) \, dt = \int_0^\infty \bar{F}(t) \, dt - F(T) \int_0^\infty \bar{F}(t) \, dt \geqq \int_T^\infty \bar{F}(t) \, dt$$

for all T, that is, $\Leftrightarrow F$ is NBUE. Q.E.D.

Because the process $N_B(t, T)$ has more dependencies than the process $N_A(t, T)$, results for block replacement are not quite as easily obtained as for age replacement. We require the following lemma which is of some independent interest.

LEMMA 2.1. *Let planned replacements occur at fixed time points* $0 < t_1 < t_2 < \cdots$ *under Policy 1, and at these and the additional point* $t_0 > 0$ *under Policy 2. Let* $N_i(t)$ *be the number of failures in* $[0, t]$ *under Policy* i, $i = 1, 2$. *Then* $N_1(t) \underset{st}{\geqq} N_2(t)$ $[N_1(t) \underset{st}{\leqq} N_2(t)]$ *for all* $t > 0$ *and all* $t_0, t_1, t_2, \cdots \Leftrightarrow F$ *is* NBU [NWU].

PROOF. Suppose first that F is NBU. For $t < t_0$, $N_1(t)$ and $N_2(t)$ have the same distribution. Next assume that $t_0 \leqq t \leqq t_k$ where t_k is the smallest $t_j > t_0$; take $t_k = \infty$ if $t_0 > t_j$ for all $j > 0$. Let Z be the age of the unit in operation at time t_0^-; the distribution of Z does not depend upon the policy. Let τ_i denote the interval between t_0 and the time of first failure subsequent to t_0^- under policy i, $i = 1, 2$. Since F is NBU,

$$(2.7) \qquad P\{\tau_1 > t \,|\, Z\} \leqq P\{\tau_2 > t \,|\, Z\} \qquad \text{for all} \quad t \geqq 0.$$

Let U_i be the number of failures in $[t_0, t]$ under policy i, $i = 1, 2$. If $X_1, X_2, \cdots$ are independent and have distribution F,

$$(2.8) \quad P\{U_1 \geqq n \,|\, Z\} = P\{t_0 + \tau_1 + X_1 + \cdots + X_{n-1} \leqq t \,|\, Z\}$$

$$\geqq P\{t_0 + \tau_2 + X_1 + \cdots + X_{n-1} \leqq t \,|\, Z\} = P\{U_2 \geqq n \,|\, Z\}$$

for $n = 0, 1, 2, \cdots$. By unconditioning on Z, we conclude that $P\{U_1 \geqq n\} \geqq P\{U_2 \geqq n\}$, $n = 0, 1, 2, \cdots$. Thus $N_1(t) \underset{st}{\geqq} N_2(t)$.

Finally, assume that $t > t_k$. Let $N_i(t_k, t)$ denote the number of failures in $(t_k, t]$ under Policy i, $i = 1, 2$. Then $N_i(t_k) + N_i(t_k, t) = N_i(t)$, with $N_i(t_k)$ and $N_i(t_k, t)$ independent, $i = 1, 2$. Since $N_1(t_k) \underset{st}{\geqq} N_2(t_k)$ and $N_1(t_k, t) \equiv N_2(t_k, t)$, we conclude that $N_1(t) \underset{st}{\geqq} N_2(t)$.

Now suppose $N_1(t) \underset{st}{\geqq} N_2(t)$ for all t and all $t_0, t_1, \cdots$. Choose $0 < t_0 < t_1$. Then

$$(2.9) \quad \bar{F}(t_1) = P\{N_1(t_1) = 0\} \leqq P\{N_2(t_1) = 0\} = \bar{F}(t_0)\bar{F}(t_1 - t_0).$$

Since $0 < t_0 < t_1$ are arbitrary, F is NBU. $Q.E.D.$

THEOREM 2.3. $Y_i \underset{st}{\leqq} Y_{i,B}(T)$ $[Y_i \underset{st}{\geqq} Y_{i,B}(T)]$ *for all* $T > 0, i = 1, 2, \cdots \Leftrightarrow$ F *is* NBU [NWU].

PROOF. For given $S_{i-1,B} = Y_{1,B}(T) + \cdots + Y_{i-1,B}(T)$, let k be the smallest integer for which $kT \geqq S_{i-1,B}$. Then $P\{Y_{i,B}(T) > t \,|\, S_{i-1,B}\} = P\{Y_1^* > t\}$, where Y_1^* is the time to first failure when planned replacements are made at $kT - S_{i-1,B} = \delta, \delta + T, \delta + 2T, \cdots$.

If F is NBU, apply Lemma 2.1 to compare successive pairs of a sequence of replacement policies, in which the ith policy calls for planned replacement at time points $\{0, \delta, T + \delta, 2T + \delta, \cdots, iT + \delta\}$. This comparison yields

$$(2.10) \qquad \bar{F}(t) = P\{Y_i > t\} = P\{Y_1 > t\} \leqq P\{Y_1^* > t\} \qquad \text{for all} \quad t \geqq 0,$$

and $Y_i \underset{\text{st}}{\leqq} Y_{i,B}(T)$ follows upon unconditioning.

The converse follows as in Theorem 2.1 since $Y_{1,B}(T)$ and $Y_{1,A}(T)$ have the same distribution. $Q.E.D.$

THEOREM 2.4. $N(t) \underset{\text{st}}{\geqq} N_B(t, T)$ $[N(t) \underset{\text{st}}{\leqq} N_B(t, T)]$ for all $t \geqq 0, T > 0 \Leftrightarrow$ F is NBU $[$NWU$]$.

PROOF. If F is NBU, apply Lemma 2.1 to compare successive pairs of a sequence of replacement policies in which the jth policy calls for planned replacement at time points $0, T, \cdots, (j - 1)T$. The converse follows as in the converse of Corollary 2.1 with $i = 1$. $Q.E.D.$

We remark that $N(t) \underset{\text{st}}{\geqq} N_B(t, T)$ is equivalent to $Y_1 + \cdots + Y_n \underset{\text{st}}{\leqq} Y_{1,B}(T) + \cdots + Y_{n,B}(T)$, $n = 1, 2, \cdots$. This result is not immediate from Theorem 2.3 because the $Y_{i,B}(T)$ are not independent.

THEOREM 2.5. $\mu = EY_i \leqq EY_{i,B}(T)$ $[EY_i \geqq EY_{i,B}(T)]$ for all $T > 0, i = 1, 2, \cdots \Leftrightarrow F$ is NBUE. $[$NWUE$]$.

PROOF. Suppose F is NBUE. Let $S_{i-1,B}$ and δ be as in the proof of Theorem 2.3. Then

$$(2.11)$$
$$E[Y_{i,B}(T)|S_{i-1,B}] = \int_0^\delta \bar{F}(t) \, dt + \bar{F}(\delta) \int_\delta^{\delta+T} \bar{F}(t - \delta) \, dt$$

$$+ \bar{F}(\delta)\bar{F}(T) \int_{\delta+T}^{\delta+2T} \bar{F}(t - \delta - T) \, dt + \cdots$$

$$+ \bar{F}(\delta)[\bar{F}(T)]^{j-1} \int_{\delta+jT}^{\delta+(j+1)T} \bar{F}(t - \delta - jT) \, dt + \cdots$$

$$= \int_0^\delta \bar{F}(t) \, dt + \bar{F}(\delta) \int_0^T \bar{F}(t) \, dt/F(T).$$

But for all $T, \delta > 0$,

$$(2.12) \qquad \int_0^\delta \bar{F}(t) \, dt/F(\delta) \geqq \mu \qquad \text{and} \qquad \int_0^T \bar{F}(t) \, dt/F(T) \geqq \mu.$$

Thus

$$(2.13) \qquad \int_0^\delta \bar{F}(t) \, dt + \bar{F}(\delta) \int_0^T \bar{F}(t) \, dt/F(T) \geqq \mu F(\delta) + \mu \bar{F}(\delta) = \mu.$$

It follows that $E[Y_{i,B}(T)|S_{i-1,B}] \geqq \mu$.

The converse follows as in Theorem 2.2 since $Y_{1,B}(T)$ and $Y_{1,A}(T)$ are identically distributed. $Q.E.D.$

The preceding results compare policies with and without planned replacements. We consider now the comparison of age replacement policies with differing replacement age T.

THEOREM 2.6. $N_A(t, T)$ *is stochastically increasing* [*decreasing*] *in* $T > 0$ *for each fixed* $t \Leftrightarrow F$ *is* IFR [DFR].

PROOF. Suppose F is IFR. For fixed $T > 0$, $\{N_A(t, T), t \geq 0\}$ is a renewal process with underlying distribution

$$(2.14) \quad S_T(x) = 1 - [\bar{F}(T)]^n \bar{F}(x - nT), \quad nT \leq x < (n + 1)T, n = 0, 1, \cdots.$$

Barlow and Proschan ([4], p. 61) show that $S_T(x)$ is increasing in $T > 0$ for fixed $x \geq 0$. Hence the nth convolution $S_T^{(n)}(x) = P\{N_A(t, T) \geq n\}$ is increasing in $T > 0$ for fixed $t \geq 0$.

Barlow and Proschan ([4], p. 61) also show that $P\{N_A(t, T) = 0\}$ decreasing in $T > 0$ for all $t \geq 0$ implies F is IFR, and this completes the proof. Q.E.D.

It is possible, under weaker conditions, to compare two age replacement policies for which the planned replacement age of one policy is a multiple of that of the other policy. This comparison is in fact a generalization of Corollary 2.1, obtained by setting $k > T/t$ below.

THEOREM 2.7. $N_A(t, kT) \underset{st}{\geq} N_A(t, T)$ [$N_A(t, kT) \underset{st}{\leq} N_A(t, T)$] *for all* $t \geq 0$, $T > 0, k = 1, 2, \cdots \Leftrightarrow F$ *is* NBU [NWU].

PROOF. Suppose first that F is NBU. With the notation introduced in the proof of Theorem 2.6, we have, for $nT \leq x < (n + 1)T$,

$$(2.15) \quad \bar{S}_T(x) - \bar{S}_{kT}(x) = [\bar{F}(T)]^n \bar{F}(x - nT) - [\bar{F}(kT)]^{[n/k]} \bar{F}(x - [n/k]kT).$$

Since F is NBU,

$$(2.16) \qquad [\bar{F}(kT)]^{[n/k]} \leq [\bar{F}(T)]^{[n/k]k},$$

and

$$(2.17) \qquad \bar{F}(x - [n/k]kT) \leq [\bar{F}(T)]^{n - k[n/k]} \bar{F}(x - nT).$$

Thus $\bar{S}_T(x) \geq \bar{S}_{kT}(x)$. Hence

$$(2.18) \qquad P\{N_A(t, kT) \geq n\} = S_{kT}^{(n)}(t) \geq S_T^{(n)}(t) = P\{N_A(t, T) \geq n\}$$

for $t \geq 0, T > 0$, that is, $N_A(t, kT) \underset{st}{\geq} N_A(t, T)$.

Next suppose $N_A(t, 2T) \underset{st}{\geq} N_A(t, T)$ for all $t, T > 0$. Take $T = \max(x, y)$ and $t - T = \min(x, y)$. Then $t \leq 2T$, so that

$$(2.19) \qquad \bar{F}(x)\bar{F}(y) = \bar{F}(T)\bar{F}(t - T) = P\{N_A(t, T) = 0\}$$
$$\geq P\{N_A(t, 2T) = 0\} = \bar{F}(t) = \bar{F}(x + y).$$

THEOREM 2.8. $EY_{i, A}(T)$ *is decreasing* [*increasing*] *in* $T > 0, i = 1, 2, \cdots \Leftrightarrow$

$$(2.20) \quad \int_0^T \bar{F}(t) \, dt/F(T) \text{ is decreasing [increasing] in } T \text{ such that } F(T) > 0.$$

PROOF. We have seen in the proof of Theorem 2.2 that

$$(2.21) \qquad EY_{i, A}(T) = \int_0^T \bar{F}(t) \, dt/F(T). \qquad Q.E.D.$$

The condition that $\int_0^T \bar{F}(t)\,dt/F(T)$ is decreasing in T has to our knowledge not been encountered previously in reliability theory. Its meaning or significance, apart from that given by Theorem 2.8, is not presently clear. One can easily show that F IFR $\Rightarrow$ (2.20) $\Rightarrow$ F NBUE, and that the conditions are distinct.

In view of Theorem 2.6, and because the condition there that F is IFR means $\bar{F}(x + t)/\bar{F}(t)$ is decreasing in t, one might have expected in place of (2.20) to have encountered in Theorem 2.8 the condition that

$$(2.22) \qquad \int_0^\infty \bar{F}(x + t)\,dx/\bar{F}(t) = \int_t^\infty \bar{F}(x)\,dx/\bar{F}(t)$$

is decreasing in $t \geqq 0$ such that $\bar{F}(t) > 0$. A distribution which satisfies (2.22) is said to have a *Decreasing Mean Residual Life* (DMRL). It is of some interest that (2.20) and (2.22) are not related. If $\bar{F}(x) = e^{-x}$ for $0 \leqq x < 1$ and $\bar{F}(x) = e^{-2x}$ for $x \geqq 1$, then F satisfies (2.20) (and also F is IFRA) but it is not DMRL because of the discontinuity at 1 within its interval of support. On the other hand, if $\bar{F}(x) = e^{-x}$ for $0 \leqq x < 1$, $\bar{F}(x) = e^{-1}$ for $1 \leqq x < 2$, and $\bar{F}(x) = 0$ for $x \geqq 2$, then F is DMRL, but F does not satisfy (2.20) because its support is not an interval (neither is F IFRA, although F is NBUE as a consequence of its being DMRL).

Results parallel to Theorems 2.6 and 2.8 for block replacement are unknown. A theorem identical with Theorem 2.7 except with block in place of age replacement is easily obtained using Lemma 2.1.

3. Renewal theory inequalities

We obtain here several results which hold for renewal processes when times between failures have a distribution that is NBU or NBUE. The first of these, like Theorems 2.1 and 2.2, provides a characterization of the NBU class. A similar characterization of the NBUE class is also obtained. The renewal theory implications obtained in Propositions 3.1 through 3.9 and Example 3.1 below are summarized in Figure 1. Following this, moment inequalities are obtained for the renewal quantity; in each case the Poisson process yields a bound.

In cases where there is a parallel result with inequalities reversed, we give only one proof, as in Section 2.

Again, we caution the reader. $N(t)$ does *not* count the origin as a renewal point (unless the initial item placed in service has life length zero).

For any two random variables U and V, dependent or not, we write $U * V$ to represent a random variable with a distribution that is the convolution of the distributions of U and V. For any distribution function F, $F^{(n)}$ denotes the nth convolution of F, and $F^{(0)}$ is degenerate at 0.

PROPOSITION 3.1. $N(s) * N(t) \underset{\text{st}}{\leqq} N(s + t)$ $\left[N(s) * N(t) \underset{\text{st}}{\geqq} N(s + t) \right]$ *for all s,* $t \underset{\text{st}}{\geqq} 0 \Leftrightarrow F$ *is* NBU [NWU].

PROOF. Suppose first that F is NBU. Then the result follows from Lemma 2.1 with $t_0 = s$, $t_1 > s + t$. Next, suppose that $N(s) * N(t) \leqq N(s + t)$. Then

$P\{N(s + t) = 0\} \leq P\{N(s) * N(t) = 0\} = P\{N(s) = 0\}P\{N(t) = 0\}$, which is the condition that F is NBU. $Q.E.D.$

Denote the distribution of time between u and the next following renewal by F_u; it is often called the distribution of residual life at time u. In order to write F_u in terms of F, it is convenient to use the standard notation $M(t) = EN(t)$.

PROPOSITION 3.2. F is NBU [NWU] $\Rightarrow \bar{F}_u(t) \leq \bar{F}(t)$ $[\bar{F}_u(t) \geq \bar{F}(t)]$ for all $t, u \geq 0$.

PROOF. If F is NBU, then

$$(3.1) \qquad \bar{F}_u(t) = \bar{F}(t + u) + \int_0^u \bar{F}(t + u - z) \, dM(z)$$

$$\leq \bar{F}(t)\bar{F}(u) + \int_0^u \bar{F}(t)\bar{F}(u - z) \, dM(z)$$

$$= \bar{F}(t)\bar{F}_u(0) \leq \bar{F}(t). \quad Q.E.D.$$

The process $\{N_u(t), t \geq 0\}$, where $N_u(t) = N(t + u) - N(u)$, is a modified renewal process in which the distribution of time to first renewal is F_u, and the distribution of time between successive renewals is F.

PROPOSITION 3.3. For each $u \geq 0$, $\bar{F}_u(t) \leq \bar{F}(t)$ $[\bar{F}_u(t) \geq \bar{F}(t)]$ for all $t \geq 0$ $\Leftrightarrow N(t) \underset{st}{\leq} N_u(t)$ $[N(t) \underset{st}{\geq} N_u(t)]$ for all $t \geq 0$.

PROOF. Suppose first that $\bar{F}_u(t) \leq \bar{F}(t)$ for all $t, u \geq 0$. Then for $n > 0$,

$$(3.2) \qquad P\{N(t) \geq n\} = \int_0^t F^{(n-1)}(t - x) \, dF(x)$$

$$\leq \int_0^t F^{(n-1)}(t - x) \, dF_u(x) = P\{N_u(t) \geq n\}.$$

Next, suppose that $N(t) \underset{st}{\leq} N_u(t)$ for all $t \geq 0$. Then $P\{N(t) = 0\} \geq P\{N_u(t) = 0\}$, that is, $\bar{F}_u(t) \leq \bar{F}(t)$. $Q.E.D.$

We have shown in Proposition 3.2 that F is NBU implies $\bar{F}_u(t) \leq \bar{F}(t)$ for all $t, u \geq 0$, but the truth of the converse has not been determined. The following proposition provides a weaker conclusion.

PROPOSITION 3.4. $\bar{F}_u(t) \leq \bar{F}(t)$ for all $t, u \geq 0 \Rightarrow F$ is NBUE.

PROOF. It is well known (see, for example, Feller (1968), p. 355) that $\lim_{u \to \infty} \bar{F}_u(t) = \mu^{-1} \int_t^\infty \bar{F}(z) \, dt$ if F has a finite mean μ, and otherwise $\lim_{u \to \infty} \bar{F}_u(t) \equiv 1$. Since $\bar{F}_u(t) \leq \bar{F}(t)$, we can choose t for which $\bar{F}(t) < 1$ to conclude that $\mu < \infty$. Then $\lim_{u \to \infty} \bar{F}_u(t) = \mu^{-1} \int_t^\infty \bar{F}(z) \, dz \leq \bar{F}(t)$ is the desired result. $Q.E.D.$

The parallel result, that $\bar{F}_u(t) \geq \bar{F}(t)$ for all $t, u \geq 0 \Rightarrow F$ is NWUE, is true only with the additional assumption that $\mu < \infty$.

The converse of Proposition 3.4 is false, as we shall later show in Example 3.1.

The class of NBUE distributions can also be characterized in terms of a renewal quantity stochastic ordering. Let $\hat{N}(t) =$ number of failures in $[0, t]$

for a stationary renewal process. The definition of this modified renewal process requires $\mu < \infty$. The distribution of time to first renewal has density $\bar{F}(t)/\mu$, $t > 0$, and the distribution of subsequent interrenewal times is F.

PROPOSITION 3.5.   *F is* NBUE [NWUE] iff $N(t) \underset{st}{\leqq} \hat{N}(t)$ $[N(t) \underset{st}{\geqq} \hat{N}(t)]$ *for all* $t > 0$.

This result is in fact a special case of Proposition 3.3, obtained by letting $u \to \infty$. Alternatively, it can be directly proved using the argument of Proposition 3.3 by taking $\hat{N}$ in place of N_u and $\mu^{-1} \int_0^\infty \bar{F}(t + x) \, dx$ in place of $\bar{F}_u(t)$. That F is NBUE implies $N(t) \underset{st}{\leqq} \hat{N}(t)$ was obtained by Barlow and Proschan ([3], Theorem 4.1).

Let us turn now to some results concerning the renewal function $M(t) = EN(t)$.

PROPOSITION 3.6. $\bar{F}_u(t) \leqq \bar{F}(t)$ $[\bar{F}_u(t) \geqq \bar{F}(t)]$ *for all* u, $t > 0 \Rightarrow M(s + t) \geqq M(s) + M(t)$ $[M(s + t) \leqq M(s) + M(t)]$ *for all* s, $t > 0$.

PROOF. $M(s + t) - M(t) = \sum_{k=0}^\infty (F_t * F^{(k)})(s) \geqq \sum_{k=0}^\infty (F * F^{(k)})(s) = M(s)$.

PROPOSITION 3.7. $M(s + t) \geqq M(s) + M(t) \Rightarrow M(t) \leqq t/\mu$.

PROOF.   Since $\lim_{s \to \infty} [M(s)/s] = 1/\mu$, this result is trivial. Q.E.D.

Again the parallel result, that $M(s + t) \leqq M(s) + M(t) \Rightarrow M(t) \geqq t/\mu$, requires the additional assumption that $\mu < \infty$.

PROPOSITION 3.8.   *F is* NBUE [NWUE] $\Rightarrow M(t) \leqq t/\mu$ $[M(t) \geqq t/\mu]$.

This result was obtained by Barlow and Proschan [3], and is immediate from Proposition 3.5 upon taking expectations. It is extended below in Proposition 3.10 and 3.11.

PROPOSITION 3.9.   *F is* NBU $\Rightarrow$ Var $N(t) \leqq M(t)$.

PROOF.   The argument given by Barlow and Proschan (1964, Theorem 4.2) applies, as their hypothesis that F is IFR is unnecessarily strong.

A weaker upper bound for Var $N(t)$ has been obtained by Esary, Marshall and Proschan ([7], Section 6) which holds for general interrenewal time distributions F. They show that

$$\text{(3.3)} \qquad\qquad \text{Var } N(t) \leqq M(t) + [M(t)]^2.$$

The following example is of interest, as it provides a counterexample to converses of several preceding propositions.

EXAMPLE 3.1.   Suppose that F places mass $\frac{1}{2}$ at 1 and mass $\frac{1}{2}$ at 3. Then F has mean $\mu = 2$ and $\int_0^\infty \bar{F}(t + x) \, dx/\bar{F}(t) \leqq 2$ for $t \geqq 0$, so that F is NBUE.

Observe that for this example, $M(1+) = \frac{1}{2}$ and $M(2+) = 1(\frac{1}{2})^2 + 2(\frac{1}{2})^2 = \frac{3}{4}$ so that $M(2+) < M(1+) + M(1+)$. Hence F is NBUE $\Rightarrow M(s + t) \geqq M(s) + M(t)$ for all s, $t \geqq 0$. Consequently it cannot be that F is NBUE $\Rightarrow \bar{F}_u(t) \leqq \bar{F}(t)$, because by Proposition 3.6, this would be an even stronger conclusion than $M(s + t) \geqq M(s) + M(t)$.

Of course for this F, we still have from Proposition 3.8 that $M(t) \leqq t/\mu$, so that the converse of Proposition 3.7 is false.

Figure 1 summarizes the results of Propositions 3.1 to 3.8 and Example 3.1. Also indicated are some dotted implications which have not yet been proved or disproved.

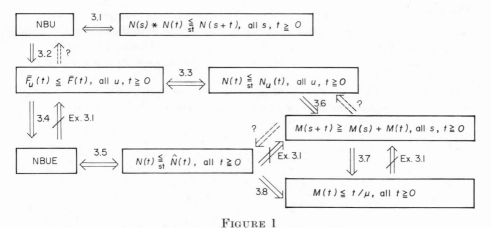

FIGURE 1

Summary of the results of Propositions 3.1 to 3.8 and Example 3.1.

Let us now consider a generalized renewal process in which the interrenewal times have not necessarily identical distributions. We denote the distribution of time to first renewal by F_1, and the distribution of time from the jth to $(j + 1)$st renewal by $F_{j+1}, j = 1, 2, \cdots$. For convenience we often write $F^{[j]}$ in place of $F_1 * \cdots * F_j$. Denote a generalized renewal process by $\{N_0(t), t \geq 0\}$, and let $M_0(t) = EN_0(t)$. Then M_0 is given by

$$(3.4) \qquad M_0(t) = \sum_{j=1}^{\infty} F^{[j]}(t), \ t \geq 0.$$

PROPOSITION 3.10. *If M_0 is a generalized renewal function in which each F_i has common mean μ and is* NBUE [NWUE], *then*

$$(3.5) \qquad M_0(t) \leq \frac{t}{\mu} \ \left[M_0(t) \geq \frac{t}{\mu} \right] \quad \text{for all} \quad t \geq 0.$$

PROOF. First, note that by Proposition 3.8, the result holds when the F_i are identical. We shall proceed to prove the result by induction.

Suppose that the result holds whenever $F_k \equiv F_{k+1} \equiv \cdots$ (that is, only the first k distributions can differ), and consider a renewal process where $F_{k+1} \equiv F_{k+2} \equiv \cdots$. Then by the induction hypothesis the result holds when interrenewal time distributions are $F_2, F_3, \cdots$, since only the first k distributions can differ. Thus,

$$(3.6) \qquad M_0(t) = \int_0^t \left[1 + \sum_{j=1}^{\infty} F_2 * \cdots * F_{j+1}(t - x) \right] dF_1(x)$$

$$\leq \int_0^t \left[1 + \frac{t - x}{\mu} \right] dF_1(x) = \frac{t}{\mu} + F_1(t) - \frac{1}{\mu} \int_0^t \bar{F}_1(x) \, dx \leq \frac{t}{\mu}$$

since F_1 is NBUE. The result follows by taking the limit as $k \to \infty$. Q.E.D.

The result of Proposition 3.10 can be viewed as a comparison with the Poisson process in which each of the interrenewal times is exponential with mean μ. For this process $M_0(t) \equiv t/\mu$. Other moment comparisons with the Poisson process were made by Barlow and Proschan [3] that hold also for generalized renewal processes.

PROPOSITION 3.11.   *If N_0 is a generalized renewal process in which each interrenewal time distribution F_i has a common mean μ and is NBUE, then*

$$(3.7) \qquad E\binom{N_0(t) + n}{m} \leq \sum_{j=0}^{\infty} \binom{j + n}{m} e^{-t/\mu} \left(\frac{t}{\mu}\right)^j \frac{1}{j!},$$

$$(3.8) \qquad EN_0^n(t) \leq \sum_{j=0}^{\infty} j^n e^{-t/\mu} \left(\frac{t}{\mu}\right)^j \frac{1}{j!},$$

for $m, n = 0, 1, 2, \cdots$, and $0 \leq t < \infty$. The reverse inequalities hold if F_i is NWUE.

PROOF OF (3.7).   First, assume that $n = 0$, $m = 0$; then both sides equal one. Assume that (3.7) holds for $n = 0$, and $m = 0, 1, \cdots, k - 1$. Then, with $G(t) = 1 - e^{-t/\mu}$,

$$(3.9) \qquad E\binom{N(t)}{k}$$

$$= \sum_{i=k}^{\infty} \binom{i}{k} [F^{[i]}(t) - F^{[i+1]}(t)] = \sum_{i=k}^{\infty} \binom{i-1}{k-1} F^{[i]}(t)$$

$$= \sum_{\ell=0}^{\infty} \binom{k+\ell-2}{k-2} \sum_{i=1}^{\infty} F^{[k+\ell-1+i]}(t)$$

$$= \sum_{\ell=0}^{\infty} \binom{k+\ell-2}{k-2} \sum_{i=1}^{\infty} \int F_{k+\ell} * \cdots * F_{k+\ell-1+i}(t-x) \, dF^{[k+\ell-1]}(x)$$

$$\leq \sum_{\ell=0}^{\infty} \binom{k+\ell-2}{k-2} \sum_{i=1}^{\infty} \int G^{(i)}(t-x) \, dF^{[k+\ell-1]}(x)$$

$$= \sum_{\ell=0}^{\infty} \binom{k+\ell-2}{k-2} \sum_{i=1}^{\infty} \int F^{[k+\ell-1]}(t-x) \, dG^{(i)}(x)$$

$$\leq \sum_{i=1}^{\infty} \int \sum_{\ell=0}^{\infty} \binom{k+\ell-2}{k-2} G^{(k+\ell-1)}(t-x) \, dG^{(0)}(x)$$

$$= \sum_{i=k}^{\infty} \binom{i-1}{k-1} G^{(i)}(t) = \sum_{j=0}^{\infty} \binom{j}{k} e^{-t/\mu} \left(\frac{t}{\mu}\right)^j \frac{1}{j!}.$$

The first inequality above follows from Proposition 3.10; the second inequality follows from the inductive hypothesis.

To establish (3.7) for $n = 1, 2, \cdots$, use induction and the identity

(3.10) $\qquad \dbinom{i + n}{m} = \dbinom{i + n - 1}{m - 1} + \dbinom{i + n - 2}{m - 1} + \cdots + \dbinom{m - 1}{m - 1}$

for $i + n > m$.

4. Moment inequalities

In this section we obtain some inequalities for distributions that are NBU, NWU, NBUE, or NWUE. We use the notation

(4.1) $\qquad \mu_r = \int x^r \, dF(x), \qquad \lambda_r = \mu_r/\Gamma(r + 1), \qquad\qquad r \geqq 0.$

Barlow, Marshall, and Proschan [2] have shown that if F is IFR, then λ_{r+s}/λ_r is decreasing in $r \geqq 0$ for fixed $s \geqq 0$. A somewhat weaker condition, that $\lambda_r^{1/r}$ is decreasing in $r > 0$, holds when F is IFRA [7]. These results can be expressed somewhat differently as:

(i) if $\log \bar{F}(t)$ is concave in $t \geqq 0$, then $\log \lambda_r$ is concave in $r \geqq 0$;

(ii) if $-\log \bar{F}(t)$ is starshaped in $t \geqq 0$, then $-\log \lambda_r$ is starshaped in $r \geqq 0$. (ϕ is *starshaped* on $[0, \infty)$ if $\phi(ax) \leqq a\phi(x)$ for all $x \geqq 0, 0 \leqq a \leqq 1$.)

The condition that F is NBU is just the condition that $-\log \bar{F}(t)$ is superadditive in $t \geqq 0$; consequently it is natural to conjecture that

(iii) if $-\log \bar{F}(t)$ is superadditive in $t \geqq 0$, then $-\log \lambda_r$ is superadditive in $r \geqq 0$. The truth of this conjecture is established in the following theorem.

THEOREM 4.1.   *If F is* NBU   [NWU], *then*

(4.2) $\qquad\qquad\qquad \lambda_{r+s} \leqq \lambda_r \lambda_s \qquad [\lambda_{r+s} \geqq \lambda_r \lambda_s]$

for all $r, s \geqq 0$.

PROOF.   If F is NBU, then $\bar{F}(x + y) \leqq \bar{F}(x)\bar{F}(y)$, and so

(4.3) $\qquad \dfrac{x^{r-1}}{\Gamma(r)} \dfrac{y^{s-1}}{\Gamma(s)} \bar{F}(x + y) \leqq \dfrac{x^{r-1}\bar{F}(x)}{\Gamma(r)} \dfrac{y^{s-1}\bar{F}(y)}{\Gamma(s)}, \qquad\qquad x, y \geqq 0.$

It follows that

(4.4) $\qquad \displaystyle\int_0^\infty \int_0^\infty \dfrac{x^{r-1}}{\Gamma(r)} \dfrac{x^{s-1}}{\Gamma(s)} \bar{F}(x + y) \, dx\, dy$

$$\qquad\qquad \leqq \int_0^\infty \frac{x^{r-1}\bar{F}(x)}{\Gamma(r)} \, dx \int_0^\infty \frac{y^{s-1}\bar{F}(y)}{\Gamma(s)} \, dy = \lambda_r \lambda_s.$$

The left member of this inequality is

$$(4.5) \quad \int_0^\infty \int_0^\infty \frac{x^{r-1}}{\Gamma(r)} \frac{y^{s-1}}{\Gamma(s)} \int_{x+y}^\infty dF(z)\, dx dy = \int_0^\infty dF(z) \int_0^z \frac{x^{r-x}}{\Gamma(r)} dx \int_0^{z-1} \frac{y^{s-1}}{\Gamma(s)} dy$$

$$= \int_0^\infty dF(z) \int_0^z \frac{x^{r-1}(z-x)^s}{\Gamma(r)\Gamma(s+1)} dx$$

$$= \int_0^\infty dF(z) \frac{z^{r+s}}{\Gamma(r+s+1)} = \lambda_{r+s}.$$

In case F is NWU, a proof is obtained by reversing the above inequalities.

THEOREM 4.2. *If F is NBUE [NWUE], then*

$$(4.6) \qquad\qquad \lambda_{r+1} \leqq \lambda_r \lambda_1 \qquad [\lambda_{r+1} \geqq \lambda_r \lambda_1]$$

for all $r > 0$.

PROOF.   If F is NBUE, then $\int_t^\infty \bar{F}(x)\, dx \leqq \mu_1 \bar{F}(t)$ so that

$$(4.7) \qquad\qquad \int_0^\infty \frac{t^{r-1}}{\Gamma(r)} \int_t^\infty \bar{F}(x)\, dx dt \leqq \mu_1 \int_0^\infty \frac{t^{r-1}}{\Gamma(r)} \bar{F}(t)\, dt = \lambda_1 \lambda_r.$$

By interchanging the order of integration, we easily compute that the left side of this inequality is λ_{r+1}. If F is NWUE, the proof is modified by reversing inequalities. *Q.E.D.*

Notice that with $r = 1$ in Theorem 4.2 we obtain that if F is NBUE[NWUE] then the *coefficient* of *variation* $\sigma/\mu \leqq 1$ $[\sigma/\mu > 1]$, where σ is the standard deviation of F.

We have in the previous theorems ignored an interesting question: are the moments finite?

PROPOSITION 4.1.   *If F is NBU, then $\mu_r < \infty$ for all $r > 0$.*

PROOF.   Choose $t < \infty$ such that $\bar{F}(t) < 1$. Since $\bar{F}$ is monotone and NBU, it follows that if x and k satisfy $kt \leqq x < (k+1)t$, then

$$(4.8) \qquad\qquad \bar{F}(x) \leqq \bar{F}(kt) \leqq [\bar{F}(t)]^k \leqq [\bar{F}(t)]^{(x/t)-1}$$

Thus

$$(4.9) \quad \mu_r = r \int_0^\infty x^{r-1} \bar{F}(x)\, dx$$

$$\leqq r \int_0^t x^{r-1} dx + r \int_t^\infty x^{r-1} [\bar{F}(t)]^{(x/t)-1} dx < \infty. \; Q.E.D.$$

If F is NWU, μ_r need not be finite for any $r > 0$. In fact, Barlow, Marshall and Proschan [2] have pointed out following their Theorem 6.2 that a DFR distribution (which, of course, is NWU) may have infinite moments of all positive orders.

We already know, as an integral part of the definition, that NBUE and NWUE distributions have a finite mean.

PROPOSITION 4.2. *If F is* NBUE, *then* $\mu_r < \infty$ *for all* $r > 0$.

PROOF. This is a consequence of $\mu_1 < \infty$ and Theorem 4.2.

Barlow and Marshall [1] have shown that if F is IFR and F has mean μ, then $F(t) \leq 1 - e^{-t/\mu}$ for all $t < \mu$. A somewhat weaker bound can be obtained if F is known only to be NBUE.

THEOREM 4.3. *If F is* NBUE *and* μ *is the mean of F, then* $F(t) \leq t/\mu$ *for all* $t \leq \mu$.

PROOF. If F is NBUE, then $\mu F(t) \leq \int_0^t \bar{F}(x)\,dx$. Trivially, $\int_0^t \bar{F}(x)\,dx \leq t$. Q.E.D.

The inequality of Theorem 4.3 is sharp in the sense that equality can be attained. Moreover, the bound cannot be improved even with the stronger condition that F is NBU. To see this, we exhibit an NBU distribution G which attains equality: let

$$(4.10) \qquad \log \bar{G}(x) = -k\Delta, \qquad kt \leq x < (k+1)t, \qquad k = 0, 1, \cdots.$$

With Δ chosen to satisfy

$$(4.11) \qquad \mu = \int_0^\infty \bar{G}(x)\,dx = \sum_{k=0}^\infty te^{-k\Delta} = t/(1 - e^{-\Delta}),$$

we have

$$(4.12) \qquad \bar{G}(t) = e^{-\Delta} = 1 - \frac{t}{\mu}.$$

Theorem 4.3 provides a lower bound on $\bar{F}(t)$ for $t \leq \mu$; the upper bound for $t \geq \mu$ provided by Markov's inequality can be improved under the assumption that F is NBUE. One way to do this for large t is to combine the result of Theorem 4.2 with the Markov inequality $\bar{F}(t) \leq \mu_r/t^r$. However, this does not provide a sharp bound. Sharp upper bounds for $\bar{F}(t)$, $t > \mu$, under the conditions that F is NBU or NBUE are not known.

Bounds for NBU survival functions $\bar{F}$ can be obtained in terms of a percentile.

THEOREM 4.4. *If F is* NBU *and* $\bar{F}(t) = \alpha$, *then*

$$(4.13) \qquad
\begin{aligned}
&\bar{F}(x) \geq \alpha^{1/k}, && \frac{t}{k+1} < x \leq \frac{t}{k}, && k = 1, 2, \cdots, \\
&\bar{F}(x) \leq \alpha^k, && kt \leq x < (k+1)t, && k = 0, 1, \cdots.
\end{aligned}$$

PROOF. For $t/(k+1) < x \leq t/k$, $\bar{F}(x) \geq \bar{F}(t/k) \geq [\bar{F}(t)]^{1/k}$; this establishes the lower bound. For $kt \leq x < (k+1)t$, $\bar{F}(x) \leq \bar{F}(kt) \leq [\bar{F}(t)]^k$, which is the upper bound. Q.E.D.

The upper bound of Theorem 4.4 is itself an NBU survival function, which, of course, attains equality and shows that the inequality is sharp. On the other hand, the lower bound is *not* an NBU survival function, and sharpness is not so trivially established.

Let $\bar{G}_k(x) = \alpha^{(j-1)/k}$ for $[(j-1)/(k+1)]t < x \leq [j/(k+1)]t, j = 1, 2, \cdots,$ $k + 1$, and let $\bar{G}_k(x) = 0$ for $x > t$. This survival function is NBU and attains equality for x in the interval $t/(k+1) < x \leq t/k$.

THEOREM 4.5.   *If F is NWU and $\bar{F}(t) = \alpha$, then*

$$(4.14) \quad \begin{aligned} \bar{F}(x) &\leq \alpha^{1/(k+1)}, & \frac{t}{k+1} &\leq x < \frac{t}{k}, & k &= 1, 2, \cdots, \\ \bar{F}(x) &\geq \alpha^{k+1}, & kt &\leq x < (k+1)t, & k &= 0, 1, \cdots. \end{aligned}$$

PROOF.   For $t/(k+1) \leq x \leq t/k$, $\bar{F}(x) \leq \bar{F}[t/(k+1)] \leq [\bar{F}(t)]^{1/(k+1)}$; this establishes the upper bound. For $kt \leq x < (k+1)t$, $\bar{F}(x) \geq \bar{F}((k+1)t) \geq [\bar{F}(t)]^{k+1}$, which is the lower bound. *Q.E.D.*

The lower bound of Theorem 4.5 is itself an NWU survival function, so that the inequality is sharp in the sense that equality can be attained. The upper bound is *not* an NWU survival function, but it is still true that equality can be attained, although not by the same distribution for each x.

Let $\bar{H}_k(x) = \alpha^{(j+1)/(k+1)}$ for $jt/k \leq x < [(j+1)/k]t$, $j = 0, 1, \cdots$. This survival function is NWU and attains equality for x in the interval $t/(k+1) \leq x < t/k$.

THEOREM 4.6.   *If F is NBUE [NWUE] and μ is the mean of F, then*

$$(4.15) \quad \mu e^{-t/\mu} \geq \int_t^\infty \bar{F}(x)\,dx \quad \left[\mu e^{-t/\mu} \leq \int_t^\infty \bar{F}(x)\,dx \right] \quad \text{for all} \quad t \geq 0.$$

PROOF.   Let F_1 be the distribution with density $f_1(x) = \bar{F}(x)/\mu$, $x \geq 0$. If F is NBUE, then

$$(4.16) \quad \bar{F}_1(z) = \int_z^\infty \bar{F}(x)/\mu\,dx \leq \mu \bar{F}(z)/\mu = \mu f_1(z),$$

or equivalently, $r_1(z) \equiv f_1(z)/\bar{F}_1(z) \geq 1/u$. Thus

$$(4.17) \quad \frac{1}{\mu} \int_t^\infty \bar{F}(x)\,dx = \bar{F}_1(t) \equiv \exp\left\{ -\int_0^t r_1(z)\,dz \right\}$$

$$\leq \exp\left\{ -\int_0^t \frac{1}{\mu}\,dz \right\} = e^{-t/\mu}. \quad Q.E.D.$$

The bound $\mu e^{-t/\mu}$ of Theorem 4.6 may be viewed as $\int_t^\infty \bar{G}(x)\,dx$, where $\bar{G}(x) = e^{-x/\mu}$, $x \geq 0$. Since F and G have the same mean, we obtain immediately the corollary:

COROLLARY 4.1.   *If F is NBUE and μ is the mean of F, then*

$$(4.18) \quad \mu(1 - e^{-t/\mu}) \leq \int_0^t \bar{F}(x)\,dx \quad \text{for all} \quad t \geq 0.$$

The inequality is reversed if F is NWUE.

Marshall and Proschan [10] have shown that the inequalities of Theorem 4.6 are preserved under the formation of parallel systems, and the inequalities of Corollary 4.1 are preserved under the formation of series systems. Moreover, all these inequalities are preserved under convolutions.

5. Class preservation under reliability operations

In this section we determine which of the classes of life distributions, NBU, NWU, NBUE, and NWUE, are preserved under the formation of (1) coherent systems, (2) convolutions, and (3) mixtures. These operations occur quite naturally in reliability models.

5.1. *Coherent systems.* Let $x_i = 1$ if the ith component in a system functions, 0 otherwise, $i = 1, \cdots, n$. Let $\phi(x) = 1$ if the system functions, 0 otherwise. The function ϕ is called the *structure function* of the system. A system is coherent if (a) its structure function ϕ is increasing and (b) ϕ is not identically 0 and not identically 1. See Birnbaum, Esary, and Saunders [5] and Barlow and Proschan [4].

THEOREM 5.1. *If each component of a coherent system of independent components has an* NBU *life distribution, then the system has an* NBU *life distribution.*

The proof is presented in Esary, Marshall, and Proschan [6].

NBUE *not preserved.* To show that the NBUE class need not be preserved under the formation of coherent systems, consider a series system of two independent components each having the life distribution F of Example 3.1. We have verified that F is NBUE, but not NBU. However, the mean life of a new system is

$$(5.1) \qquad v = \int_0^3 \bar{F}^2(x)\, dx = \tfrac{3}{2},$$

whereas the mean remaining life of a system of age one is two. Thus system life is *not* NBUE.

REMARK. We saw in the remark following Theorem 4.2 that an NBUE distribution has coefficient of variation $\leqq 1$. A simple counter example shows that a coherent system of independent components each having coefficient of variation $\leqq 1$ may itself have coefficient of variation > 1; form a series system of two independent components each having the distribution which places mass $\tfrac{1}{2}$ at 0 and 1.

NWU, NWUE *not preserved.* To show that neither the NWU nor the NWUE classes are preserved under the formation of coherent systems, consider a parallel system of two independent components each having life distribution $1 - e^{-t}$, which is both NWU and NWUE. Then system life distribution is given by

$$(5.2) \qquad F(t) = (1 - e^{-t})^2,$$

so that system failure rate is

$$(5.3) \qquad r(t) = 1 - \frac{1}{2e^t - 1},$$

a strictly increasing function. Thus system life is neither NWU nor NWUE.

5.2. *Convolutions.* Both the NBU and the NBUE classes are preserved under convolution, as shown in the next two theorems.

THEOREM 5.2.   *Let F_1, F_2 be* NBU *distributions. Then the convolution*

$$(5.4) \qquad\qquad F(t) = \int_0^t F_1(t - x) \, dF_2(x)$$

is NBU.

PROOF.

$$(5.5) \qquad \bar{F}(x + y) = \int_0^x \bar{F}_2(x + y - z) \, dF_1(z) + \int_0^\infty \bar{F}_2(y - z) \, d_z F_1(x + z).$$

But

$$(5.6) \qquad \int_0^x \bar{F}_2(x + y - z) \, dF_1(z) \leqq \bar{F}_2(y) \int_0^x \bar{F}_2(x - z) \, dF_1(z)$$
$$= \bar{F}_2(y) [\bar{F}(x) - \bar{F}_1(x)],$$

and, integrating by parts,

$$(5.7) \qquad \int_0^\infty \bar{F}_2(y - z) \, d_z F_1(x + z)$$
$$= \bar{F}_2(y)\bar{F}_1(x) + \int_0^\infty \bar{F}_1(x + z)[- d_z F_2(y - z)]$$
$$\leqq \bar{F}_2(y)\bar{F}_1(x) + \bar{F}_1(x) \int_0^\infty \bar{F}_1(z)[- d_z F_2(y - z)]$$
$$= \bar{F}_2(y)\bar{F}_1(x) + \bar{F}_1(x) [\bar{F}(y) - \bar{F}_2(y)].$$

Thus

$$(5.8) \quad \bar{F}(x + y) \leqq \bar{F}_2(y)\bar{F}(x) + \bar{F}_1(x)\bar{F}(y) - \bar{F}_1(x)\bar{F}_2(y)$$
$$= \bar{F}(x)\bar{F}(y) - [\bar{F}(x) - \bar{F}_1(x)][\bar{F}(y) - \bar{F}_2(y)] \leqq \bar{F}(x)\bar{F}(y).$$

THEOREM 5.3.   *Let F_1, F_2 be* NBUE. *Then the convolution* $F(t) = \int_0^t F_1(t - x) \, dF_2(x)$ *is* NBUE.

PROOF.

$$(5.9) \qquad \int_0^\infty \bar{F}(t + x) \, dx$$
$$= \int_0^\infty \int_0^t \bar{F}_1(t + x - u) \, dF_2(u) \, dx + \int_0^\infty \int_t^\infty \bar{F}_1(t + x - u) \, dF_2(u) \, dx.$$

But for $u \leqq t$, $\int_0^\infty \bar{F}_1(t - u + x) \, dx \leqq \mu_1 \bar{F}_1(t - u)$ by (1.2), where μ_i is the mean of F_i, $i = 1, 2$. Thus

$$(5.10) \qquad \int_0^\infty \int_0^t \bar{F}_1(t + x - u) \, dF_2(u) \, dx \leqq \mu_1 \int_0^t \bar{F}_1(t - u) \, dF_2(u)$$
$$= \mu_1 [\bar{F}(t) - \bar{F}_2(t)].$$

For $u > t$, $\int_0^\infty \bar{F}_1(t + x - u) \, dx = u - t + \mu_1$, so that

$$(5.11) \qquad \int_0^\infty \int_t^\infty \bar{F}_1(t + x - u) \, dF_2(u) \, dx = \int_t^\infty (u - t + \mu_1) \, dF_2(u).$$

Let $w = u - t$. Then

(5.12) $\displaystyle\int_t^\infty (u - t + \mu_1)\, dF_2(u) = \mu_1 \bar{F}_2(t) + \int_0^\infty w\, dF_2(w + t)$

$$= \mu_1 \bar{F}_2(t) + \int_0^\infty \bar{F}_2(w + t)\, dw$$

$$\leqq \mu_1 \bar{F}_2(t) + \mu_2 \bar{F}_2(t).$$

Thus

(5.13) $\displaystyle\int_0^\infty \bar{F}(t + x)\, dx \leqq \mu_1 \bar{F}(t) + \mu_2 \bar{F}_2(t) \leqq (\mu_1 + \mu_2)\bar{F}(t).$

REMARK. If F_1 and F_2 each have coefficient of variation $\leqq a$, then their convolution $F_0(t) = \int_0^\infty F_1(t - x)\, dF_2(x)$ has coefficient of variation $\leqq a$. To prove this, let F_i have mean μ_i, variance σ_i^2, $i = 0, 1, 2$. Then

(5.14) $\quad a^2 \mu_0^2 - \sigma_0^2 = a^2 (\mu_1 + \mu_2)^2 - (\sigma_1^2 + \sigma_2^2)$

$$= (a\mu_1^2 - \sigma^2) + (a_2 \mu_2^2 - \sigma^2) + 2a^2 \mu_1 \mu_2 \geqq 0.$$

NWU, NWUE *not preserved.* To show that neither the NWU nor the NWUE classes are preserved under convolution, let $F(t) = 1 - e^{-t}$, which is both NWU and NWUE. Then $F^{(2)}(t) = 1 - (1 + t)e^{-t}$, a gamma distribution of order two, which has strictly increasing failure rate, and thus is neither NWU nor NWUE.

5.3. *Mixtures.*

DEFINITION. *Let $\mathscr{F} = \{F_\alpha : \alpha \in \mathscr{A}\}$ be a family of probability distributions and G a probability distribution. Then*

(5.15) $$F(t) = \int F_\alpha(t)\, dG(\alpha)$$

is a mixture of probability distributions from $\mathscr{F}$.

As we shall see below, none of the classes, NBU, NBUE, NWU, NWUE, is preserved under mixtures. However, we can demonstrate preservation of a subclass of the NWU and of the NWUE classes under mixtures.

THEOREM 5.4. *Suppose F is the mixture of F_α, $\alpha \in \mathscr{A}$, with each F_α NWU [NWUE] and no two distinct F_α, $F_{\alpha'}$ crossing on $(0, \infty)$. Then F is NWU [NWUE].*

PROOF. NWU *case.* By the Chebyshev inequality for similarly ordered functions (Hardy, Littlewood, and Pólya [9], Theorem 43),

(5.16) $\quad \bar{F}(s)\bar{F}(t) \equiv \displaystyle\int \bar{F}_\alpha(s)\, dG(\alpha) \int \bar{F}_\alpha(t)\, dG(\alpha) \leqq \int \bar{F}_\alpha(s)\bar{F}_\alpha(t)\, dG(\alpha).$

By the NWU property

(5.17) $\quad \displaystyle\int \bar{F}_\alpha(s)\bar{F}_\alpha(t)\, dG(\alpha) \leqq \int \bar{F}_\alpha(s + t)\, dG(\alpha) \equiv \bar{F}(s + t).$

Thus F is NWU.

NWUE *case.* As above,

$$(5.18) \qquad \mu \bar{F}(t) \equiv \int \mu_\alpha \, dG(\alpha) \int \bar{F}_\alpha(t) \, dG(\alpha) \leqq \int \mu_\alpha \bar{F}_\alpha(t) \, dG(\alpha).$$

Using the NWUE defining property (1.2) first and the Fubini Theorem next, we have

$$(5.19) \qquad \int \mu_\alpha \bar{F}_\alpha(t) \, dG(\alpha) \leqq \int \int \bar{F}_\alpha(t + x) \, dx dG(\alpha)$$

$$= \int \int \bar{F}_\alpha(t + x) \, dG(\alpha) \, dx = \int \bar{F}(t + x) \, dx.$$

Thus by (1.2), F is NWUE. $Q.E.D.$

REMARK. We may readily show that if F is a mixture of F_α, $\alpha \in \mathscr{A}$, with each F_α having $CV \geqq a$, then F has $CV \geqq a$, $a \geqq 0$.

NBU, NBUE *classes not preserved*. To see that neither the NBU nor the NBUE class is preserved under mixtures, note that a mixture of nonidentical exponential distributions has a strictly decreasing failure rate, and thus cannot be NBU nor NBUE.

NWU *class not preserved*. To see that the NWU class is not preserved under mixtures, consider the following example. Let $\bar{F}_\delta(x) = e^{-k\delta}$ for $(k - 1)\delta < x \leqq k\delta$, $k = 1, 2, \cdots$. Then it is easy to verify that F is NWU.

Next we show that the mixture $F = \frac{1}{2}F_\delta + \frac{1}{2}F_\gamma$ does not satisfy the NWU property

$$(5.20) \qquad D \equiv \bar{F}(x + y) - \bar{F}(x)\bar{F}(y) \geqq 0$$

$$\text{for} \quad 0 < y < \delta < x < \gamma < 2\delta < x + y < 2\gamma$$

(for example, take $y = 3$, $\delta = 4$, $x = 6$, $\gamma = 7$). Then

$$(5.21) \qquad \begin{aligned} &\bar{F}_\delta(x) = e^{-2\delta}, \qquad \bar{F}_\delta(y) = e^{-\delta}, \qquad \bar{F}_\delta(x + y) = e^{-3\delta}; \\ &\bar{F}_\gamma(x) = e^{-\gamma}, \qquad \bar{F}_\gamma(y) = e^{-\gamma}, \qquad \bar{F}_\gamma(x + y) = e^{-2\gamma}. \end{aligned}$$

We compute

$$(5.22) \qquad \begin{aligned} D &= \tfrac{1}{2}(e^{-3\delta} + e^{-2\gamma}) - \tfrac{1}{4}(e^{-2\delta} + e^{-\gamma})(e^{-\delta} + e^{-\gamma}) \\ &= \tfrac{1}{4}(e^{-\gamma} - e^{-\delta})(e^{-\gamma} - e^{-2\delta}) < 0. \end{aligned}$$

We summarize the preservation results of this section in Table I.

TABLE I

PRESERVATION OF LIFE DISTRIBUTION CLASSES UNDER RELIABILITY OPERATIONS

	Formation of Coherent Systems	Convolutions	Arbitrary Mixtures	Mixtures of Distributions That Do Not Cross
NBU	preserved	preserved	not preserved	not preserved
NBUE	not preserved	preserved	not preserved	not preserved
NWU	not preserved	not preserved	not preserved	preserved
NWUE	not preserved	not preserved	?	preserved

REFERENCES

[1] R. E. BARLOW and A. W. MARSHALL, "Bounds for distributions with monotone hazard rate," *Ann. Math. Statist.*, Vol. 35 (1964), pp. 1234–1257.

[2] R. E. BARLOW, A. W. MARSHALL and F. PROSCHAN, "Properties of probability distributions with monotone hazard rate," *Ann. Math. Statist.*, Vol. 34 (1963), pp. 375–389.

[3] R. E. BARLOW and F. PROSCHAN, " Comparison of replacement policies and renewal theory implications," *Ann. Math. Statist.*, Vol. 35 (1964), pp. 577–589.

[4] ———, *Mathematical Theory of Reliability*, New York, Wiley, 1965.

[5] Z. W. BIRNBAUM, J. D. ESARY and S. C. SAUNDERS, "Multi-component systems and structures and their reliability," *Technometrics*, Vol. 3 (1961), pp. 55–77.

[6] J. D. ESARY, A. W. MARSHALL and F. PROSCHAN, "Some reliability applications of the hazard transform," *SIAM J. Appl. Math.*, Vol. 18 (1970), pp. 849–860.

[7] ———, "Shock models and wear processes," to appear in *Ann. Math. Statist.*

[8] W. FELLER, *An Introduction to Probability Theory and Its Applications*, Vol. 2, New York, Wiley, 1966.

[9] G. H. HARDY, J. E. LITTLEWOOD and G. PÓLYA, *Inequalities*, Cambridge, Cambridge University Press, 1952.

[10] A. W. MARSHALL and F. PROSCHAN, "Mean life of series and parallel systems," *J. Appl. Probability*, Vol. 7 (1970), pp. 165–174.

[11] G. WEISS, "On the theory of replacement of machinery with a random failure time," *Naval Res. Logist. Quart.*, Vol. 3 (1956), pp. 279–293.

SPACINGS REVISITED

RONALD PYKE

UNIVERSITY OF WASHINGTON

and

IMPERIAL COLLEGE

1. Introduction

This paper surveys some of the developments which have appeared in the literature on spacings during the five years since the presentation of two papers [16] and [17]. The first of these, by Proschan and this author, deals with the asymptotic theory of a class of tests for Increasing Failure Rate (IFR) which are based on spacings, whereas the second paper surveys the substantial literature on spacings that had appeared prior to 1965. In the present article we also set out some open problems which still remain in the asymptotic theory of tests based on spacings.

The general area of limit theorems for dependent random variables is broad and complex, with no unifying methodology. For example, problems related to rank statistics, linear combinations of order statistics and stationary sequences all require different approaches. Limit theorems for spacings represent some of the more challenging problems involving dependent variables, and the various approaches used provide interesting comparisons.

2. Basic formulations

By spacings we refer to the gaps or distances between successive points on a line. Let $\{T_n : n \geq 0\}$ be a sequence of random variables (r.v.) for which $T_0 \leq T_1 \leq T_2 \cdots$. The spacings are then the differences $\{T_i - T_{i-1}\}$. There is a basic ambiguity in the theory of spacings caused by the radically different assumptions which can be placed on the T process. These differences can clearly be seen for example between the three basic models outlined below.

Model I: order statistics. For fixed n, one is given independent random variables $X_1, X_2, \cdots, X_n$, with common distribution function (d.f.) F_X. One defines $T_1 \leq T_2 \leq \cdots \leq T_n$ to be the order statistics of the sample and considers the spacings $D_i = T_i - T_{i-1}$. The range for i is $2 \leq i \leq n$ unless the support of F_X indicates that spacings D_1 and/or D_{n+1} may be defined. The usual situation under Model I is a hypothesis testing one in which the two hypotheses are

$$(2.1) \qquad H_0 : F_X \in \mathscr{F}_0, \qquad H_1 : F_X \in \mathscr{F}_1.$$

This paper was prepared with the partial support of the National Science Foundation Grant GP–23653.

Usually $\mathscr{F}_0$ consists of a single continuous d.f. which can therefore be assumed to be uniform on $(0, 1)$.

Model II: point processes. In this model, the T process could be a general one sided point process with $T_0 = 0$, say. The most common situation is that of testing the null hypothesis that the T process is Poisson against some alternative involving non-Poissonian point processes. In this special case the spacings are taken to be $D_i = (T_i - T_{i-1})/T_{n+1}$ with $i = 1, 2, \cdots, n + 1$, where $T_0 \leqq T_1 \leqq T_2 \leqq \cdots$, are the successive points.

Model III: renewal processes. This is technically a special case of Model II. Assume that the T process is a renewal process with common d.f. F. Use the proportional spacings or interoccurrence times $D_i = (T_i - T_{i-1})/T_{n+1}$. The most common hypothesis testing problem within this context is to test the null hypothesis that F is exponential against some appropriate alternative for the common d.f. F.

Although the null hypothesis limit theory of statistics based on these spacings is the same (namely, that of uniform spacings) for all three of the specific models given above, the asymptotic theory under the alternatives is drastically different for each model. Consequently, particularly in the study of asymptotic power, asymptotic relative efficiency, and limiting distributions under contiguous alternatives must be made separately for each case and different techniques must be used.

3. Uniform spacings

The theory here is essentially complete (see [17]). Let $U_1 \leqq \cdots \leqq U_n$ be uniform $(0, 1)$ order statistics and let $\{D_{n,i} = n(U_i - U_{i-1}): 1 \leqq i \leqq n + 1\}$ be the set of $n + 1$ weighted spacings, with $U_0 = 0$ and $U_{n+1} = 1$. Let $\{Y_i: i \geqq 1\}$ be independent exponential r.v. of mean 1. Let

$$(3.1) \qquad\qquad S_n = n^{-1/2} \sum_{i=1}^{n+1} (Y_i - 1).$$

Write

$$(3.2) \qquad \mathbf{D}_n = (D_{n,1}, \cdots, D_{n,n+1}), \qquad \mathbf{Y}_n = (Y_1, \cdots, Y_{n+1}),$$

and for any Borel measurable function g on R_{n+1} let $G_n = g(\mathbf{D}_n)$ and $J_n = g(\mathbf{Y}_n)$. It is well known that $\mathscr{L}(G_n) = \mathscr{L}(J_n | S_n = 0)$. Functions of uniform spacings have been studied by several authors; the first general methodology was given by Darling [10]. The most general method available today for proving limit theorems for functions of spacings was introduced by LeCam [14], who applied classical limit theory to (J_n, S_n) and utilized the representation of spacings as exponential random variables conditioned by $S_n = 0$.

Using the same general approach, this result was generalized for a wider class of functions g by Pyke [17] in 1965 and by Wichura [25] in 1968. In the latter case, conditional distributions given $S_n = x \neq 0$ are also considered. Bickel

[2] in 1969 also generalized the theorem for the case of sums to include conditional distributions given $S_n = x \neq 0$ and obtained for this case uniformity in x.

Although the conditional construction $\mathscr{L}(D_n) = \mathscr{L}(Y_n | S_n = 0)$ provides a relatively simple collection of limit theorems for uniform spacings, there remains at least one striking open question.

Problem 1 (Uniform spacings under random sample size). Let $\{N_t : t \geq 0\}$ be a positive integer valued process for which $N_t/t \overset{P}{\to} 1$ as $t \to \infty$. For G_n as defined above, show that $G_n \overset{L}{\to} G$ implies that $G_{N_t} \overset{L}{\to} G$ for a large class of functions g.

Random sample size versions of limit theorems have previously been obtained, for example, for sums, maxima and empirical processes of independent, identically distributed random variables; but nothing has been obtained for spacings. For example, it is unknown whether or not G_{N_t} converges in law for such a simple statistic as $G_n = n^{-1/2} \sum_{i=1}^{n+1} (D_{n,i} - 1)^2$.

4. Limit theory under alternatives: Model III

As indicated above, several different models lead to test statistics which under the null hypothesis are functions of uniform spacings. However, when one turns to questions of distribution theory under alternative hypotheses, one finds that each model poses distinct problems. The easiest model to work under alternatives is Model III, for in the case of a renewal process the interoccurrence times $\{T_i - T_{i-1}\}$ remain independent and identically distributed under the alternatives. Consequently most results can be derived directly from standard theory. For example, consider the problem of establishing the weak convergence of the empirical process of the spacings. Under Model III, $D_{n,i} = X_i/\bar{X}_n$ for $1 \leq i \leq n+1$ where $\bar{X}_n = (X_1 + \cdots + X_{n+1})/n$ and where $\{X_i : i \geq 1\}$ are independent with common d.f. F. Assume that F is continuous, $E(X_1) = 1$ and $\mathrm{Var}(X_1) = \sigma^2 < \infty$. Let F_n denote the empirical d.f. of $\{X_1, \cdots, X_{n+1}\}$ and let H_n denote the same of $\{D_{n,1}, \cdots, D_{n,n+1}\}$. (Excuse the unconventional but convenient use of n for $n+1$ in the subscripts of $\bar{X}_n$, F_n, and H_n.) Then $H_n(x) = F_n(x\bar{X}_n)$ for all x. Write

$$(4.1) \qquad U_n = n^{1/2}(F_n \circ F^{-1} - e), \qquad V_n = n^{1/2}(H_n \circ F^{-1} - e)$$

for the empirical processes of $\{X_i\}$ and $\{D_{n,i}\}$, respectively, defined on $(0, 1)$, where e denotes the identity function $e(x) = x$. Elementary algebra shows that for all x

$$(4.2) \qquad V_n(F(x)) = U_n(F(x\bar{X}_n)) + \left\{ \frac{F(x\bar{X}_n) - F(x)}{x(\bar{X}_n - 1)} \right\} x Z_n$$

where $Z_n = n^{1/2}(\bar{X}_n - 1)$. (The second term is taken to be zero when $x = 0$.) Notice that

$$(4.3) \qquad Z_n = n^{1/2}(\bar{X}_n - 1) = -n^{1/2} \int_0^\infty [F_n(x) - F(x)]\, dx = -\int_0^1 U_n\, dF^{-1}.$$

Since it is known that $U_n \xrightarrow{L} U_0$ where U_0 is a Brownian bridge, one may, without loss of generality for our purposes, assume that U_n converges uniformly to U_0. (See Pyke [18] and Pyke and Shorack [19].) When this is the case it follows from (4.3) that $Z_n \xrightarrow{P} Z_0$ where

$$(4.4) \qquad Z_0 = - \int_0^1 U_0 \, dF^{-1}$$

is a $N(0, \sigma_1^2)$ r.v. with

$$(4.5) \qquad \sigma_1^2 = 2 \int_0^1 \int_0^v u(1 - v) \, dF^{-1}(u) \, dF^{-1}(v) < \infty.$$

To show this, use uniform convergence to get

$$(4.6) \qquad \int_\varepsilon^{1-\varepsilon} U_n \, dF^{-1} \to \int_\varepsilon^{1-\varepsilon} U_0 \, dF^{-1}$$

for any $\varepsilon > 0$. Then use Chebyshev's inequality to show that the remaining integrals can be made small in probability in view of the finiteness of σ_1^2, a direct consequence itself of the finiteness of $E(X_1^2)$. This establishes the fact that $(U_n, Z_n) \xrightarrow{L} (U_0, Z_0)$ on the natural product space. It is now convenient again to use equivalent constructions as in [18] for which $Z_n \to Z_0$ and U_n converges uniformly to U_0. The expression in (4.2) then suggests that under suitable smoothing and tail conditions on $f = F'$, one should obtain that V_n converges uniformly to

$$(4.7) \qquad V_0 = U_0 + (f \circ F^{-1})F^{-1} Z_0.$$

Hence V_0 is a Gaussian process of zero mean whose covariance function, by virtue of the special construction of Z_0, can be computed to be

$$(4.8) \qquad E[V_0(u)V_0(v)] = u(1 - v) + h(u)h(v)\sigma^2 + h(u)g(v) + h(v)g(u)$$

for $0 \leq u \leq v \leq 1$ where $h = (f \circ F^{-1})F^{-1}$ and

$$(4.9) \qquad g(u) = E[Z_0 U_0(u)] = - \int_0^1 (u \wedge w - uw) \, dF^{-1}(w).$$

Sufficient conditions for this to hold are given in the following result.

THEOREM 4.1. *If* (i) f *satisfies a Lipschitz condition on bounded intervals and* (ii) $\sup_{|x|>T} f(x)x \to 0$ *as* $T \to \infty$, *then* V_n *converges in law to* V_0, *a Gaussian process of mean zero and covariance given by* (4.8).

PROOF. Using the special constructions described above it suffices to establish that

$$(4.10) \qquad \sup_x \left| \frac{F(\bar{X}_n x) - F(x)}{(\bar{X}_n - 1)x} Z_n - f(x)Z_0 \right| |x| \to 0,$$

or equivalently that $\sup_x |f(\theta_{n,x})Z_n - f(x)Z_0||x| \to 0$, where $\theta_{n,x}$ is some value between x and $\bar{X}_n x$. Since $Z_n \to Z_0$, condition (ii) enables one to delete Z_n and

Z_0 from the expression. Then by (i), there exists for each T a constant M such that

$$(4.11) \qquad \sup_{|x| < T} |f(\theta_{n,x}) - f(x)\|x| \leq M|1 - \bar{X}_n|T^2 \to 0.$$

Condition (ii) guarantees that the supremum can be made small on the complement of $(-T, T)$. Q.E.D.

The above result does not use the assumption that $X_i \geq 0$ which would ordinarily be assumed within the context of Model III. Weak convergence could also be obtained relative to the stronger metrics ρ_q of [19], which are defined by $\rho_q(f, h) = \sup |f - h|/q$, for a wide class of weighting functions q. Along this line we quote the following result of Shorack [21] which involves, however, the inverse empirical (or quantile) process rather than the empirical process. The former has some implicit simplicities over the latter. Although results are given in [21] for the general case, for simplicity of notation we state the result only for the uniform case.

THEOREM 4.2 (Shorack).   *If F is an exponential d.f. of mean 1,*

$$(4.12) \qquad W_n(t) = n^{1/2}[D_{n,i} - F^{-1}(t)] \qquad \text{for} \quad i - 1 < (n+1)t \leq i,$$

and q is a nonnegative function which is nondecreasing (nonincreasing) on $(0, 1/2]$ $([1/2, 1))$ and whose reciprocal is square integrable over $(0, 1)$, then $W_n \overset{L}{\to} W$ relative to the metric $\rho_{q/(1-e)}$.

It should be remarked that if one deletes the factor $n^{1/2}$ from (4.2), it is obvious that the Glivenko-Cantelli theorem for H_n holds whenever $\sup |x|f(x) < \infty$. This observation is due to Bickel and provides a simpler proof of the Glivenko-Cantelli theorem given in [17].

5. Limit theory under alternatives: Model I

When describing alternatives for tests based on spacings, one should keep in mind that spacings tests are only appropriate for problems whose alternative hypotheses involve the shape of the density functions. Two suitable problems of interest might be

$$(5.1) \qquad H_0: F_X \text{ is uniform } (0, 1) \text{ versus } H_1: f_X \searrow \text{ on } (0, \infty)$$

and

$$(5.2) \qquad H_0: F_X \text{ is exponential, versus } H_1: F_X \text{ is IFR } \left(\text{that is } f_X(1 - F_X)^{-1} \searrow \right)$$

Both of these alternatives restrict the shape of the density function f_X; other examples might involve the unimodality or bimodality of f_X or the monotoneity of f_X'/f_X.

For the problem in (5.2), Proschan and Pyke [16] proposed a family of tests based on the normalized spacings $\bar{D}_{n,i} = (n - i + 1)(T_i - T_{i-1})$. Under H_0 these normalized spacings are independent and identically distributed whereas under H_1 they are stochastically decreasing. The test statistics were of the form

$$(5.3) \qquad G_n = \sum_{i<j} g(\bar{D}_{n,i}, \bar{D}_{n,j})$$

for bounded nonnegative g satisfying $g(\cdot, y) \searrow$ and $g(x, \cdot) \nearrow$ for all x, y. (It is well to emphasize that only nonhomogeneous functions of spacings should be considered in Model I.) Central limit theorems for these statistics are given in [16] and [17]. Of particular interest is the Wilcoxon-like statistic, $V_n =$ number of pairs $(\bar{D}_{n,i}, \bar{D}_{n,j})$ with $\bar{D}_{n,i} > \bar{D}_{n,j}$ for $i < j$. Similar statistics suggest themselves for testing (5.1), but in terms of the regular spacings $D_{n,i} = T_i - T_{i-1}$.

All approaches to limit theorems for alternatives under Model I can be said to depend upon Taylor's expansion methods. The dependency of the spacings causes considerable difficulties, but basically the approach is to observe that when $F_X = F$,

$$(5.4) \qquad nD_{n,i} \overset{L}{=} n[F^{-1}(Y_i/S_{n+1}) - F^{-1}(Y_{i-1}/S_{n+1})] = Y_i/f(F^{-1}(\theta_{n,i}))\bar{X}_n$$

or

$$(5.5) \qquad \bar{D}_{n,i} \overset{L}{=} (n - i + 1)[F^{-1} \circ H(Y_i^*) - F^{-1} \circ H(Y_{i-1}^*)] \overset{L}{=} Y_i r(A_{n,i}),$$

where $\{Y_i\}$ are independent exponential random variables with mean one, H is their common d.f., $\{Y_i^*\}$ denote exponential order statistics, $r = (1 - e)/f \circ F^{-1}$ is the reciprocal of the failure rate defined through F^{-1} on $(0, 1)$, and where $\{\theta_{n,i}\}$, $\{A_{n,i}\}$ are defined by appropriate Taylor expansions. Thus although the spacings are dependent, the above representations indicate their approximate independence and exponentiality. Except when working under contiguous alternatives, the errors in these approximations make a significant cumulative contribution. To illustrate this, consider somewhat simpler statistics of the form $G_n^* = \Sigma g(\bar{D}_{n,i}, i/n)$. In [17] it is shown that under regularity assumptions on g and f_X, $n^{1/2}(G_n^* - R_n) \overset{P}{\to} 0$ where

$$(5.6) \qquad R_n = \sum [g(Y_i r(i/n), i/n) + (Y_i - 1)C(i/n)]$$

and

$$(5.7) \qquad C(w) = (1 - w)^{-1} \int_w^1 (1 - u)[r'(u)/r(u)] \int_0^\infty e^{-y}(y - 1)g(yr(u), u)\, dy\, du.$$

This result shows how the errors cumulate to give the second summand in (5.6). It is interesting to compare the form of this result with that of Bickel and Doksum [3], derived under contiguous alternatives, which states that $n^{-1/2}(T_n - S_n) \overset{P}{\to} 0$ where $T_n = \Sigma h(X_i)$, $S_n = \Sigma a(i/n)(\bar{D}_{n,i} - 1)$, and h and a are functions satisfying certain regularity assumptions which are related by

$$(5.8) \qquad a(u) = (1 - u)^{-1} \int_{-\log(1-u)}^\infty h'(x)e^{-x}\, dx.$$

In [3], Bickel and Doksum show that under contiguous alternatives $\{f_{\theta_n}\}$ with $\theta_n = bn^{-1/2}$ and $b \geq 0$, the test based on V_n described above is asymptotically inadmissible. They do this by showing that V_n is asymptotically equivalent to

$\Sigma\, iR_i$ (where the R_i are the ranks of the normalized spacings) which in turn is always asymptotically inferior to $-\Sigma\, i\log\left[1\,-\,R_i/(n\,+\,1)\right]$. They show, furthermore, that among "studentized" linear functions of spacings or of their ranks there exist asymptotically most powerful ones. By analogy, the result suggests the following.

Problem 2. For the problem (5.1) of testing uniformity against monotoneity (or unimodality), do asymptotically most powerful tests exist among those based upon linear functions of (nonnormalized) spacings?

Other results involving limit theory under contiguous alternatives include the following one by Weiss [22]. Assume for X a sequence of distribution functions G_n with density functions $g_n\,=\,1\,+\,n^{-\delta}r$ on $(0,\,1)$, where $\delta\,>\,0$ and r is a function whose second derivative exists and is uniformly bounded in absolute value. Let $\mathbf{W}\,=\,(W_1,\,\cdots,\,W_{n+1})$ be independent exponential random variables with W_i having mean $g_n\circ G_n^{-1}(i/(n\,+\,1))$ and set $Z_i\,=\,W_i/(W_1\,+\,\cdots\,+\,W_{n+1})$. If $\mathbf{D}_n\,=\,(D_{n,\,1},\,\cdots,\,D_{n,\,n+1})$ are the spacings when $F_X\,=\,G_n$, then Weiss proves the following theorem.

THEOREM 5.1. (Weiss). *The ratio $f_{\mathbf{Z}_n}(\mathbf{D}_n)/f_{\mathbf{D}_n}(\mathbf{D}_n)$ tends to 1 in probability.*

It is further shown that in fact Theorem 5.1 holds when convergence in probability is replaced by convergence in 1-mean, thereby implying the contiguity of the corresponding measures.

In [23] this result is used to establish the asymptotic distribution of homogeneous functions of spacings under such contiguous alternatives. Although expressed in terms of spacings, Theorem 5.1 can equivalently be viewed as a result about the order statistics $T_1\,\leq\,T_2\,\leq\,\cdots\,\leq\,T_n$ in view of the one to one correspondence between the two. In fact, the proof in [22] notationally works with the equivalent densities of the order statistics. Thus if $V_i\,=\,Z_1\,+\,\cdots\,+\,Z_i$, then Theorem 5.1 is equivalent to stating that $f_{\mathbf{T}_n}(\mathbf{T}_n)/f_{\mathbf{V}_n}(\mathbf{T}_n)\overset{P}{\to}1$.

In a recent paper, Rao and Sethuraman [20], motivated by problems involving circular data, also develop limit theorems for spacings under contiguous Model I alternatives. They use the representation (5.4) and postulate, as in Weiss [22], a sequence of distribution functions G_n with densities $g_n\,=\,1\,+\,n^{-\delta}r$, $\delta\,\geq\,1/4$, where r possesses a uniformly continuous derivative r'. The approach of Rao and Sethuraman is to study the empirical processes of the spacings, and they begin by studying these in the approximate situation in which $\theta_{n,\,i}$ in representation (5.4) is a constant. Specifically, they study the empirical process of $\{Y_i/c_{n,\,i}\colon 1\,\leq\,i\,\leq\,n\}$ where $\{Y_i\}$ are independent exponential random variables of mean one, and where the constants are determined by $c_{n,\,i}\,=\,c_n(i/n)$ where

$$(5.9) \qquad c_n(u)\,=\,1\,+\,A(u)n^{-\delta}\,+\,R_n(u), \qquad\qquad \delta\,\geq\,1/4.$$

In this representation, the function A satisfies certain smoothness conditions and R_n is uniformly $o(n^{-1/2})$.

If H denotes the exponential d.f. of mean one, set

$$(5.10) \qquad U_i(y)\,=\,I_{[Y_i\,\leq\,y]}\,-\,H(y); \qquad W_n(y)\,=\,n^{-1/2}\sum_{i=1}^{n}U_i(yc_{n,\,i})$$

for $y \geq 0$. Then W_n is the empirical process of $\{Y_i/c_{n,i}\}$ and may be written as

$$(5.11) \qquad W_n(y) = n^{-1/2} \sum_{i=1}^{n} U_i(y) + n^{-1/2} \sum_{i=1}^{n} [U_i(yc_{n,i}) - U_i(y)]$$
$$= V_n(y) + Q_n(y), \qquad\qquad\qquad y \geq 0.$$

Here, V_n is the usual empirical process for an exponential sample whose weak convergence is known. The remainder term Q_n is bounded in absolute value by the modulus of continuity of the exponential empirical process over intervals of width

$$(5.12) \qquad \sup \{|1 - c_{n,i}| : 1 \leq i \leq n\} \leq n^{-\delta} \sup \{|A(u)| : 0 \leq u \leq 1\}.$$

However, only for $\delta > 1/2$ is the bound sufficient to enable one to deduce $Q_n \xrightarrow{P} 0$ directly from the weak convergence of V_n. To establish the limiting behavior of Q_n for general δ one must make better use of the fact that Q_n is a sum of n independent processes and apply the standard methods of weak convergence (see [4]). In passing, observe that for any $\delta > 0$, $Q_n(y) \xrightarrow{P} 0$ for each *fixed* y. To verify this, compute

$$(5.13) \qquad \text{Var } Q_n(y) \leq n^{-1} \sum_{i=1}^{n} |e^{-y} - e^{-c_{n,i}y}| = n^{-1} \sum_{i=1}^{n} u|1 - u^{c_{n,i}-1}|$$

for $u = e^{-y}$. Since $u|1 - u^b| \leq |1 - (1 + b)^{-1}|(1 + b)^{-1/b} \leq |1 - (1 + b)^{-1}|$ for $0 \leq u \leq 1$ and real b, it follows that

$$(5.14) \qquad \text{Var } Q_n(y) \leq n^{-1} \sum_{i=1}^{n} |1 - c_{n,i}^{-1}|$$

which converges to zero under even weaker assumptions than (5.9).

The study of empirical processes of "perturbed" random variables also has applications in the theory of rank statistics when the underlying independent distribution functions are not identical.

Problem 3. Let $\{F_{n,i} : 1 \leq i \leq n, n \geq 1\}$ be a triangular array of distribution functions and let $\{X_i : i \geq 1\}$ be a sequence of independent uniform $(0, 1)$ random variables. Find necessary and sufficient conditions on $\{F_{n,i}\}$ to ensure the weak convergence of the empirical processes

$$(5.15) \qquad b_n \sum_{i=1}^{n} \{I_{[X_i \leq F_{n,i}(y)]} - F_{n,i}(y)\}, \qquad -\infty < y < \infty,$$

for suitable constants $\{b_n\}$.

6. Limit theory under alternatives: Model II

This is a situation of considerable importance and the reader is referred to Cox and Lewis [9] for a review of the pertinent literature. Let it suffice here to mention the following general question, suggested by the results of Weiss [22], Bickel and Doksum [3], and Rao and Sethuraman [20] described above.

Problem 4.   Let $\{\lambda_\theta : \theta \geqq 0\}$ be a family of intensity functions for nonhomogeneous Poisson processes in which $\lambda_0 \equiv 1$. Find sufficient conditions of these functions to ensure that a sequence $\{P_{\theta_n}\}$ of corresponding measures is contiguous with P_0, the measure for the Poisson process of rate 1.

7. Other results

In this section, capsule descriptions of other recent references on spacings are given.

In [5] and [6], Blumenthal studies the limit theory of statistics of the form $\sum_{i=1}^{n} \{D_i^X / D_i^Y\}^r$ where $\{D_i^X\}$ and $\{D_i^Y\}$ are the spacings of two independent samples, with common d.f. F_X and F_Y, respectively, and where r is a constant. The proofs are difficult, indicating again the particular complexities associated with spacings and the need for more general methods. A study of the limiting behavior of statistics of the form $\sum_i g(D_i^X, D_i^Y)$ suggests itself. In [7], Blumenthal establishes a strong limit theorem in connection with the two sample spacings problem. Note the open problems stated on p. 112 of [7].

In [13], Kale derives some of the standard spacings' statistics as functions of the data that minimize certain distances between the empirical and null hypothesis d.f. In particular the statistics $\sum D_{n,i}^{-1}$, $\sum (D_{n,i} - 1/n)^2$, $\sum \log D_{n,i}$, $\min D_{n,i}$ and $\max D_{n,i}$ can be obtained in this way. See also [11] and [12].

The asymptotic distribution of the k smallest spacings, for fixed k, is obtained by Weiss [24] for the case of a continuous density f_X over $(0, 1)$ which is bounded away from zero.

In unpublished papers, Blumenthal [8] derives the asymptotic normality of $\sum \log D_{n,i}$ for Model I alternatives satisfying restrictive smoothness conditions, while Shorack [21] derives the weak convergence of the empirical processes of spacings relative to a wide class of metrics and applies his results to tests based on linear combinations of functions of ordered spacings.

8. Miscellaneous problems

Problem 5.   A study of rates of convergence for limiting distributions of spacings would be of interest. No results are presently available. In the case of uniform spacings presumably classical theory could be applied by making use of LeCam's approach [14].

Problem 6.   Show that

$$(8.1) \qquad \sum_{i,j} \int \int |f_{D_{n,i}, D_{n,j}}(x, y) - f_{D_{n,i}}(x) f_{D_{n,j}}(y)| dx dy = O(n)$$

for a large class of underlying densities f_X. This would give an applicable measure of the rate of asymptotic independence of spacings; (see [17], p. 434).

Problem 7.   If, as in [17], p. 420, one defines the empirical distribution functions of the spacings $D_{n,i}$ and $\bar{D}_{n,i}$, respectively, by

$$(8.2) \qquad L_n(x) = n^{-1} \sum_i I[D_{n,i} \leqq x], \qquad M_n(x) = n^{-1} \sum_i I[\bar{D}_{n,i} \leqq x]$$

and the associated processes by

$$(8.3) \qquad Y_n(x) = n^{1/2}[L_n(x) - L(x)], \qquad Z_n(x) = n^{1/2}[M_n(x) - M(x)]$$

where

$$(8.4) \qquad L(x) = 1 - \int_{-\infty}^{\infty} f(y)e^{-xf(y)}\,dy, \qquad M(x) = 1 - \int_0^1 e^{-x/r(u)}\,du,$$

can one verify the weak convergence of these processes? In [17] the convergence of the finite dimensional distribution functions is established rather straight-forwardly, but the weak convergence has not been shown. Such a result could possibly have application to the rank statistics of Bickel and Doksum [3].

ADDENDUM

The discussion in Section 5 of [20] was based on a preprint of the same title (Tech. Rpt. No. Math-Stat/11/69-Mar. 1969, Indian Statist. Instit.). Reference [20] is given since the results there are more general and accessible than those of the preprint.

REFERENCES

[1] L. ABRAMSON, "The distribution of the smallest sample spacing," (abstract), *Ann. Math. Statist.*, Vol. 37 (1966), p. 1421.

[2] P. J. BICKEL, "Tests for monotone failure rate, II," *Ann. Math. Statist.*, Vol. 40 (1969), pp. 1250–1260.

[3] P. J. BICKEL and K. A. DOKSUM, "Tests for monotone failure rate based on normalized spacings," *Ann. Math. Statist.*, Vol. 40 (1969), pp. 1216–1235.

[4] P. BILLINGSLEY, *Convergence of Probability Measures*, New York, Wiley, 1968.

[5] S. BLUMENTHAL, "Contributions to sample spacings theory, I: Limit distributions of sums of ratios of spacings," *Ann. Math. Statist.*, Vol. 37 (1966), pp. 904–924.

[6] ———, "Contributions to sample spacings theory, II: Tests of the parametric goodness of fit and two-sample problems," *Ann. Math. Statist.*, Vol. 37 (1966), pp. 925–939.

[7] ———, "Limit theorems for functions of shortest two-sample spacings and a related test," *Ann. Math. Statist.*, Vol. 38 (1967), pp. 108–116.

[8] ———, "Logarithms of sample spacings," *SIAM J. Appl. Math.*, Vol. 16 (1968), pp. 1184–1191.

[9] D. R. COX and P. A. W. LEWIS, *Statistical Analysis of Series of Events*, London, Methuen, 1966.

[10] D. A. DARLING, "On a class of problems related to the random division of an interval," *Ann. Math. Statist.*, Vol. 28 (1953), pp. 823–838.

[11] J. R. GEBERT and B. K. KALE, "Goodness of fit tests based on discriminatory information," *Statistiche Hefte*, Vol. 10 (1969), pp. 197–200.

[12] V. P. GODAMBE and B. K. KALE, "A test of goodness of fit," *Statistiche Hefte*, Vol. 8 (1967), pp. 165–172.

[13] B. K. KALE, "Unified derivation of tests of goodness of fit based on spacings," *Sankhyā*, Ser. A, Vol. 31 (1969), pp. 43–48.

[14] L. LeCam, "Un théorème sur la division d'un intervalle par des points pris au hasard," *Publ. Inst. Statist. Univ. Paris*, Vol. 7 (1958), pp. 7–16.

[15] J. I. Naus, "Some probabilities, expectations and variances for the size of largest clusters and smallest intervals," *J. Amer. Statist. Assoc.*, Vol. 61 (1966), pp. 1191–1199.

[16] F. Proschan and R. Pyke, "Tests for monotone failure rate," *Proceedings of the Fifth Berkeley Symposium on Mathematical Statistics and Probability*, Berkeley and Los Angeles, University of California Press, 1967, Vol. 3, pp. 293–312.

[17] R. Pyke, "Spacings," *J. Roy. Statist. Soc., Ser. B*, Vol. 7 (1965) pp. 395–449.

[18] ———, "Applications of almost surely convergent constructions of weakly convergent processes," *International Symposium on Probability and Information Theory, Lecture Notes on Mathematics*, 89 (edited by M. Behara, K. Krickeberg, and J. Wolfowitz), Berlin, Heidelberg, New York, Springer-Verlag, 1969.

[19] R. Pyke and G. R. Shorack, "Weak convergence of a two-sample empirical process and a new approach to Chernoff-Savage theorems," *Ann. Math. Statist.*, Vol. 39 (1968), pp. 755–771.

[20] J. S. Rao and J. Sethuraman, "Pitman efficiencies of tests based on spacings," *Proceedings of the First International Symposium on Non-Parametric Techniques in Statistical Inference*, New York, Cambridge University Press, 1970, pp. 405–415.

[21] G. R. Shorack, "Convergence of spacings' processes with applications," to be published.

[22] L. Weiss, "On asymptotic sampling theory for distributions approaching the uniform distribution," *Z. Wahrscheinlichkeitstheorie und Verw. Gebiete*, Vol. 4 (1965), pp. 217–221.

[23] ———, "Limiting distributions of homogeneous functions of sample spacings for distributions approaching the uniform distribution," *Z. Wahrscheinlichkeitstheorie und Verw. Gebiete*, Vol. 10 (1968), pp. 193–197.

[24] ———, "The joint asymptotic distribution of the k-smallest sample spacings," *J. Appl. Probability*, Vol. 6 (1969), pp. 442–448.

[25] M. Wichura, "On the weak convergence of non-Borel probabilities on a metric space," Ph.D. thesis, Columbia University, 1968.

[26] S. Blumenthal, "Tests of fit based on partial sums of the ordered spacings," *Ann. Inst. Statist. Math.*, Vol. 22 (1970), pp. 261–276.

ON LARGE SAMPLE PROPERTIES OF CERTAIN NONPARAMETRIC PROCEDURES

HERMAN RUBIN
PURDUE UNIVERSITY

1. Summary and introduction

Efficiencies of one sided and two sided procedures are considered from the standpoint of risk. It is shown that the two sided Kolmogorov-Smirnov (K-S) and Kuiper procedures, which were shown in [4] to be asymptotically equiefficient with the median for translation alternatives for symmetric unimodal distributions, have efficiencies for sample sizes in a wide range in the general vicinity of that of the median; but even if certain standard asymptotic approximations can be made, the efficiencies are not too close to that of the median, and in many cases the dominant asymptotic correction term does not even yield the sign of the deviation for samples of size 10^{20}.

A procedure briefly discussed in [1], for which the Pitman efficiency is zero, has good Bayes risk efficiency for translation alternatives for any distribution and merits further work for two sided testing.

In the one sided case, the one sided K-S procedure appears to be somewhat worse to much worse than a procedure introduced by the author in [3]. Also, the K-S procedure involves a choice of significance level which is highly distribution dependent.

We shall consider the "moderately large sample" efficiencies of certain well known and not sufficiently well known nonparametric procedures from a decision theoretic standpoint. By "moderately large sample" we shall mean that central limit type theorems yield adequate approximations to the distributions involved, but that the further asymptotic approximations of the type in [4] are not necessarily very good. We shall also assume that the samples under consideration are sufficiently large that the large sample form of the risk can be used.

That is, we shall carry out our computations as if the observations can be considered as a stochastic process on $[0, 1]$ such that

$$(1.1) \qquad X(t) = \theta h(t) + Y(t),$$

Research was supported in part by the Office of Naval Research Contract N00014-67-A-226-0014, project number NR042-216 at Purdue University. Reproduction in whole or in part is permitted for any purpose of the United States Government.

429

where Y is a separable Gaussian process with mean 0 and covariance function

$$(1.2) \qquad\qquad \sum (t, u) = 4[\min (t, u) - tu],$$

and $h(t)$ is a multiple of $fF^{-1}(t)$, chosen so that $h(0.5) = 1$. The choice of these normalizing factors is for computational convenience; the median corresponds to $-X(0.5)$ in standard units. For simulation purposes, we have chosen five distributions:

normal, with

$$(1.3) \qquad\qquad h_N(t) = \exp \left\{ -0.5 \left\{ [N(0, 1)(t)]^{-1} \right\}^2 \right\};$$

logistic, with

$$(1.4) \qquad\qquad h_L(t) = 4t(1 - t);$$

double exponential, with

$$(1.5) \qquad\qquad h_D(t) = \begin{cases} 2t, & t \leq 0.5, \\ 2(1 - t), & t > 0.5; \end{cases}$$

Cauchy, with

$$(1.6) \qquad\qquad h_C(t) = \sin^2 \pi t;$$

and a distribution with density $C(1 + \tau|x - \theta|)^{-10/9}$, in which case

$$(1.7) \qquad\qquad h_T(t) = \begin{cases} (2t)^{10}, & t \leq 0.5, \\ [2(1 - t)]^{10}, & t > 0.5. \end{cases}$$

The loss structure was taken to be $2|\theta|\,d\theta$ for a wrong decision in the one sided testing problem [2]—it can be strongly argued that for "reasonably large" samples no other loss function is reasonable for this problem. For the two sided problem the weight function was taken, as in [4], to be 1 if a type I error is made, and $|\theta|^k \, d\theta/\sqrt{2\pi} \, \mu_k$ for a type II error, where μ_k is the kth absolute moment of the normal distribution. The choice of multiplicative constants was chosen so that if Z is $N(\theta, \sigma^2)$, then it will never pay to accept the null hypothesis if $\sigma > 1$, but for Z sufficiently small it will pay if $\sigma < 1$.

In the two sided case these normalizations correspond to establishing a base for the sample size. In the one sided case, if the value of the translation parameter at which there is indifference is θ^*, the risk is $E(\theta^{*2})$.

The procedures we have evaluated by Monte Carlo are, for the two sided problem, Kolmogorov-Smirnov and Kuiper; and we have compared them to the median, for which it is known [4] that they are asymptotically equiefficient. For the one sided case, the Kolmogorov-Smirnov statistic has been compared with a symmetrized version introduced by the author in [3].

2. Two sided tests: asymptotic treatment

The procedures that we shall consider are the median, Kolmogorov-Smirnov, and Kuiper. We shall also consider a test, suggested in [2], for which the Pitman efficiency is 0, but whose Bayes risk efficiency is that of the best order statistic. For the median (in our approximation: $X(0.5) + \theta h(0.5)$) the probability of exceeding C under the null hypothesis is

$$(2.1) \qquad P_M = \frac{2}{\sqrt{2\pi}} \int_c^\infty e^{-t^2/2} \, dt \sim \frac{2}{\sqrt{2\pi}\, c} e^{-c^2/2}.$$

For the K-S statistic $(\sup |X(t) + \theta h(t)|)$ the corresponding probability is

$$(2.2) \qquad P_{K\text{-}S} = 2\Sigma(-1)^{n-1} e^{-c^2 n^2/2} \sim 2 e^{-c^2/2},$$

and for the Kuiper statistic, the probability is

$$(2.3) \qquad P_K = 2\Sigma(n^2 c^2 - 1) e^{-c^2 n^2/2} \sim 2 c^2 e^{-c^2/2}.$$

(Note that there is a scale factor of 2 in the expressions for P_K and $P_{K\text{-}S}$).

Now let us examine what happens under the alternative. Let $X^+(\theta) = \sup\big(X(t) + \theta h(t)\big)$. For θ reasonably large, if $h(t) \sim 1 - \lambda|t - \frac{1}{2}|^\gamma, \gamma > \frac{1}{2}, X^+(\theta)$ is approximately $\theta + \lambda^{-1/\gamma}\theta^{-1/\gamma}Y_\gamma - Z$, where Z is normal $(0, 1)$ and Y_γ is a positive random variable whose distribution is not known except for $\gamma = 1$. Hence that θ for which $X^+(\theta) = c$ is approximately

$$(2.4) \qquad \theta_c = c + Z - \lambda^{-1/\gamma}c^{-1/\gamma}Y_\gamma.$$

Therefore

$$(2.5) \qquad E_{K\text{-}S}(\theta_c^k) \sim c^k + \binom{k}{2} c^{k-2} - kK_\gamma c^{k-1-1/\gamma}.$$

For the Kuiper statistic we also need $X^-(\theta) = \inf\big(X(t) + \theta h(t)\big)$. Here if θ is reasonably large and $h(t) \sim pt^\beta, \beta > \frac{1}{2}, X^-(\theta) \sim -p^{-1/\beta}\theta^{-1/\beta}W_\beta$, and

$$(2.6) \qquad E_K(\theta_c^k) \sim c^k + \binom{k}{2} c^{k-2} - kK_\gamma c^{k-1-1/\gamma} - kH_\beta c^{k-1-1/\beta}.$$

For the distributions we are considering, the values of γ and β are shown in Table I. (For the normal, the tail behavior is slightly more complicated, but

TABLE I

VALUES OF γ AND β

	γ	β
normal	2	1
logistic	2	1
double exponential	1	1
Cauchy	2	2
long tailed	1	10

since for the Kuiper statistic the larger of β and γ is what counts, this is not a problem.)

Incidentally, in the case of the median,

$$(2.7) \qquad E_M(\theta_c^k) \sim c^k + \binom{k}{2} c^{k-2}.$$

Note that for the K-S and the Kuiper statistic, $E(\theta_c^k)$ is smaller than for the median. However, the c required to obtain a given type I error is somewhat larger.

Now let us investigate what happens for mth power loss for samples of size n if the cut off point is c. We obtain for the type II risk

$$(2.8) \qquad R_2 = \frac{2}{\sqrt{2\pi}\, \mu_m n^{(m+1)/2}} E\left(\frac{\theta_c^{m+1}}{m+1}\right).$$

Hence our combined risk is

$$(2.9) \qquad R = P_I(c) + \left(\frac{c^{m+1}}{m+1} + \frac{m}{2} c^{m-1} - Rc^{m-1/\gamma} + \cdots\right) Bn^{-(m+1)/2},$$

where $P_I(c) \sim Ac + e^{-c^2/2}$. Now a lengthy calculation shows that the dominant correction term to the asymptotic expression

$$(2.10) \qquad R \sim B(m+1)^{(m-1)/2}\left(\frac{\log n}{n}\right)^{(m+1)/2}$$

has the relative value

$$(2.11) \qquad C_1 = \frac{1}{2}\frac{(q+1-m)\log\log n}{[(m+1)\log n]^{1/2}},$$

which of course increases with q. Thus, for extremely large n, the median is better than the K-S test, which is better than the Kuiper test.

However, extremely large depends on $\log\log n$. Since $\log\log 10^{20} < 4$, for practical purposes the next term (which depends on A) comes into effect, and the $-Rc^{m-1/\gamma}$ term may actually be dominant.

3. Two sided tests: moderately large sample and empirical results

A computation based on the likelihood ratio shows that for small n the K-S and Kuiper statistics are approximately equivalent to the best procedure. (This requires the probability of type I error to be nearly 1.) Apparently this efficiency drops off rapidly. Let us look at the results of Monte Carlo computations (Table II).

The values are independent for the different distributions, but dependent within any one distribution. The standard deviations (estimated from 1000 sample processes) of these efficiencies are 1 to 2 per cent for samples of size 10

TABLE II

Efficiency (per cent)

		Kolmogorov-Smirnov			Kuiper		
		constant loss	absolute error	squared error	constant loss	absolute error	squared error
Normal	1	127	123	123	—	—	—
	2	129	126	119	70	72	72
	5	122	118	119	65	68	71
	10	119	119	118	68	70	73
	10^2	115	116	114	74	76	80
	10^3	117	114	113	75	81	85
	10^5	114	112	110	80	87	91
	10^{10}	112	109	107	88	93	96
	10^{20}	108	106	105	93	97	98
Logistic	1	111	114	116	—	—	—
	2	117	117	111	73	75	76
	5	114	112	113	69	72	74
	10	111	113	113	72	73	75
	10^2	110	112	111	75	78	81
	10^3	111	111	110	77	82	86
	10^5	110	109	108	82	88	91
	10^{10}	109	107	106	89	93	96
	10^{20}	107	105	104	94	97	98
Double exponential	1	—	—	—	—	—	—
	2	79	79	80	64	63	67
	5	78	80	82	60	61	62
	10	79	82	83	61	63	66
	10^2	82	86	87	64	68	72
	10^3	85	88	90	67	72	76
	10^5	88	91	93	72	78	82
	10^{10}	92	94	96	79	85	88
	10^{20}	94	96	97	86	90	92
Cauchy	1	—	—	—	—	—	—
	2	87	90	92	85	87	89
	5	88	91	91	80	84	87
	10	89	93	94	81	85	87
	10^2	93	97	98	84	89	92
	10^3	97	99	100	89	93	95
	10^5	98	101	101	92	96	98
	10^{10}	103	102	102	97	99	101
	10^{20}	102	102	102	99	101	102
Long tailed	1	—	—	—	100 +	105	112
	2	70	73	77	90	96	102
	5	66	65	65	78	86	91
	10	57	61	64	81	86	94
	10^2	61	68	73	81	93	100
	10^3	66	74	79	88	98	104
	10^5	74	82	86	94	103	107
	10^{10}	83	89	92	103	107	109
	10^{20}	93	94	95	106	108	108

and, with very few exceptions, 0.1 to 0.2 per cent for samples of size 10^{20}. Thus, while individual figures for small sample sizes are not too reliable, the general picture is clear: for the Kolmogorov-Smirnov test, the flatness at the median determines the efficiency, and for samples of size 10^{20} relative to the base, the dominant asymptotic error has yet to make its presence felt.

The results are also similar for the Kuiper statistic. Several cases also clearly show the dip for small samples in the efficiencies. These results also agree with the exact calculations for K-S with 0th power loss for the double exponential in [3]. The optimal significance levels also are not much affected by the test.

A test occasionally considered (see, for example [1]) is to use $T_n = \sqrt{n} \sup |(F_n - F)/[F(1 - F)]^{1/2}|$. The statistic T_n is more sensitive to deviations in the tails than the K-S statistic. Now examination of

$$(3.1) \qquad T_n(x) = \sqrt{n} \, \frac{F_n(x) - F(x)}{[F(x)(1 - F(x))]^{1/2}}$$

by the usual methods shows that

$$(3.2) \qquad T_n \sim (2 \log \log n)^{1/2}.$$

A further examination shows that the statistic cannot be very sensitive to Pitman alternatives since that x for which $T_n = |T_n(x)|$ is likely to be near 0 to 1. Of course, $(2 \log \log 10^{20})^{1/2} < 2.15 (2 \log \log 10)^{1/2}$, so that even this argument may not be too serious for reasonable sample sizes. But we note that for kth power loss, for the K-S test the critical deviation is approximately $\frac{1}{2}[(k + 1) \log n]^{1/2}$, which grows much more rapidly. However, if we break the ordered observations below the median into groups of size $1, 2, 4, 8, \cdots$, and examine the distribution of $T_n(x)$ in the corresponding intervals, we find that

$$(3.3) \qquad P(T_n > c) < K \log (n + 1) e^{-c^2/2}.$$

This shows that from the Bayes risk standpoint, this statistic bears much the same relationship to the best order statistic as the K-S or Kuiper statistic does to the median! This test consequently merits investigation.

4. One sided tests

If the weight function is $2|\theta| \, d\theta$ for a wrong decision, and if the structural model is such that the observation $Y = \phi(\theta, X)$, and for $\theta < \hat{\theta}(x)$ the decision is made that $\theta < 0$, the risk is $E[(\hat{\theta}(x))^2]$. This calculation can be applied in our model to the one sided K-S tests and also to the symmetric test given by the author in [3]. The symmetric test has similar properties to the median for all symmetric unimodal distributions; its reciprocal efficiency relative to the median is between $2 - \frac{1}{6}\pi^2 = 0.355$ and a number bounded by $(\frac{1}{3}\sqrt{\pi} + 1)^2 \sim 7.9$.

The one sided K-S test does not fare so well. From equation (1.1), note that if θ is large the maximum of $X(t)$ will be large with large probability since $X(\frac{1}{2}) = \theta + Y(\frac{1}{2})$, but the minimum may still be quite negative if θ is not very large, especially if h is small some distance away from 0. This is indeed borne out by the empirical results. Unfortunately, it was not anticipated just how bad things would get, and hence it is necessary to crudely estimate some numbers. The results are given in Table III with standard errors in parentheses.

TABLE III

VARIANCE OF INDIFFERENCE POINT

	Normal	Logistic	Double exponential	Cauchy	Long tail
Symmetric	.735(.033)	.796(.034)	1.405(.061)	1.386(.062)	6.13(.19)
K-S, one sided	.851(.027)	.942(.029	1.737(.057)	1.902(.073)	~11(~3)
K-S, one sided 50 per cent	.852(.017)	.950(.029)	1.931(.07)	~4.5	~160
Optimal level	.49	.47	.39	.28	.008

Again, the values for the one sided K-S at optimal level for the double exponential and the optimal level agree very well with the theoretical values of $\sqrt{14} - 2 = 1.74166$ and $e^{\sqrt{0.875}} = 0.39244$, respectively. Note that for not too bad distributions, the one sided K-S test is fairly good if used at the optimal level, which varies considerably with the distribution. If the 50 per cent level were used, as is optimal for any symmetric test statistic, the tail of the Cauchy is already bad enough to cause problems. It was not anticipated in the empirical procedure that when θ was chosen to make the maximum of $X(\theta)$ greater than 12.8 (or the minimum less than -12.8), which is beyond the 10^{-33} level for the Kuiper statistic that there would be any significant problems with the minimum (maximum). A few values for the Cauchy distribution were far enough out to give questionable accuracy at the 50th percentile; for the long tailed distribution the figures given are probably slightly conservative.

The author wishes to express his thanks to Glennis Cohen for her assistance in producing the programs for the simulation and processing of the numerical results, and to Arthur Rubin for his invaluable help in the debugging of these programs.

REFERENCES

[1] T. W. ANDERSON and D. A. DARLING, "Asymptotic theory of certain 'goodness of fit' criteria based on stochastic processes," *Ann. Math. Statist.*, Vol. 23 (1952) pp. 193–212.
[2] H. CHERNOFF, "Sequential tests for the mean of a normal distribution," *Proceedings of the Fourth Berkeley Symposium on Mathematical Statistics and Probability*, Berkeley and Los Angeles, University of California Press, 1961, Vol. 1, pp. 79–91.
[3] H. RUBIN, "Decision-theoretic evaluation of some non-parametric methods," *Nonparametric Techniques in Statistical Inference* (edited by M. L. Puri, London), Cambridge University Press, 1970, pp. 579–583.
[4] H. RUBIN and J. SETHURAMAN, "Bayes risk efficiency," *Sankhyā Ser. A.* Vol. 27 (1965) pp. 347–356.

ASYMPTOTIC DISTRIBUTION OF THE LOG LIKELIHOOD RATIO BASED ON RANKS IN THE TWO SAMPLE PROBLEM[1]

I. R. SAVAGE[2]
and
J. SETHURAMAN
FLORIDA STATE UNIVERSITY

1. Introduction

Let $X_1, X_2, \cdots, Y_1, Y_2, \cdots$, be independent random variables where the X (Y) have common *strictly increasing* and *continuous* distribution function F_1^* (F_2^*). Let $N = 2n$ and $W_{N,1} \leq W_{N,2} \leq \cdots \leq W_{N,N}$ be a rearrangement of $X_1, X_2, \cdots, X_n, Y_1, Y_2, \cdots, Y_n$ in increasing order of magnitude, $n = 1, 2, \cdots$. Define

$$(1.1) \qquad Z_{N,i} = \begin{cases} 0 & \text{if } W_{N,i} \text{ is an } X, \\ 1 & \text{if } W_{N,i} \text{ is a } Y, \end{cases}$$

$i = 1, \cdots, N$, and let $\mathbf{Z}_N = (Z_{N,1}, \cdots, Z_{N,N})$.

Let F_1 and F_2 be two arbitrary *strictly increasing continuous* distribution functions. Let

$$(1.2) \qquad L_N = L_N(\mathbf{z}_N) = \frac{P(\mathbf{Z}_N = \mathbf{z}_N | F_1^* = F_1, F_2^* = F_2)}{P(\mathbf{Z}_N = \mathbf{z}_N | F_1^* = F_2^* = F_1)}.$$

Note that the denominator in the above does not depend on F_1 and is equal to $1/\binom{N}{n}$ and that the numerator is unchanged if F_1 and F_2 are replaced by $F_1 K^{-1}$ and $F_2 K^{-1}$, where K is a strictly increasing continuous distribution function. Let

$$(1.3) \qquad \ell_N = \ell_N(\mathbf{z}_N) = \log L_N(\mathbf{z}_N).$$

From now on $P(E)$ will stand for the probability of the event E when the common distribution of X and Y are F_1^* and F_2^*, respectively. Our main aim is to prove the asymptotic normality of $\ell_N(\mathbf{Z}_N)$ under suitable conditions (see Theorems 5.1 and 5.2). The conditions imposed are $A1$, $A2$, $A3$, and B or $\bar{A}1$, $\bar{A}2$, and $\bar{B}$ (see

[1] Research supported by the Army, Navy and Air Force under the Office of Naval Research Contract No. NONR 988(08), Task Order NR 042-004. Reproduction in whole or in part is permitted for any purpose of the U.S. Government.

[2] Now visiting at the Center for Advanced Studies in the Behavioral Sciences, Stanford.

Sections 2 and 4). Conditions B and $\bar{B}$ are similar and in our opinion they form the crucial condition. In Section 6 we exhibit an example in which Condition $\bar{B}$ is not satisfied and the asymptotic distribution of ℓ_N is not normal.

In many statistical contexts it is of great importance to study the distribution of a likelihood ratio like L_N. It seems however that in nonparametric settings like ours other statistics like linear rank statistics have been studied more than the likelihood ratio. Laws of large numbers and bounds to the large deviation probabilities relating to linear rank statistics have been obtained earlier than those relating to ℓ_N. Problems of asymptotic normality of linear rank statistics have been well studied by now. We now describe some of the known results for ℓ_N.

Savage and Sethuraman [8] have shown, when $F_2 = F_1^A$ where A is a constant, that there is a number $I(F_1^*, F_2^*; F_1, F_1^A)$ such that for each $\varepsilon > 0$ there is a $\rho < 1$ with

$$(1.4) \qquad P\left\{\left|\frac{1}{N}\ell_N(\mathbf{Z}_N) - I(F_1^*, F_2^*; F_1, F_1^A)\right| > \varepsilon\right\} < \rho^N,$$

for all large N. More generally, under some mild conditions, Berk and Savage [1] showed that there is a constant $I(F_1^*, F_2^*, F_1, F_2)$ such that for each $\varepsilon > 0$ there is a $\rho < 1$ with

$$(1.5) \qquad P\left\{\left|\frac{1}{N}\ell_N(\mathbf{Z}_N) - I(F_1^*, F_2^*; F_1, F_2)\right| > \varepsilon\right\} < \rho^N,$$

for all large N. These results were used by these authors in their study of sequential tests based on ℓ_N. Some exact large deviation results for ℓ_N have recently been obtained by Hájek [5] when $F_1^* = F_2^*$ and F_1 and F_2 satisfy some mild conditions. The only results known about the asymptotic normality of ℓ_N were obtained simultaneously by Sethuraman [9] and Govindarajulu [4], for the special case when $F_2 = F_1^A$ where A is a constant. In this case $\sqrt{N}\,[\ell_N/N - I(F_1^*, F_2^*; F_1, F_1^A)]$ has a limiting normal distribution. This result was arrived at rather directly since in this special case $\ell_N(\mathbf{Z}_N)$ turned out to be a slightly modified Chernoff-Savage statistic.

In this paper we consider the asymptotic distribution of ℓ_N in the general case. We basically have two different proofs of the asymptotic distribution which are applicable to different situations. These two situations do not exhaust all possible cases and by the time we have imposed all of our conditions they do not cover the case treated in Sethuraman [9]. In the first situation it is assumed among other things that the density functions of F_1 and F_2 are bounded away from 0. In the second situation the density functions of F_1 and F_2 are histograms with a finite number of feet. There are some other conditions imposed on F_1^* and F_2^*. The proof of the asymptotic normality of ℓ_N uses some hints from game theory in the first situation and relies heavily on the results of Berk and Savage [1] in the second situation. A heuristic proof which applies to the first situation is sketched in Section 3 and we hope that these heuristics can be justified in situations more general than what we have been able to do.

Section 2 deals with the notation and various preliminaries. Section 4 contains some intermediate lemmas and theorems and the main results are in Section 5 in the form of Theorems 5.1 and 5.2.

It may be noted that we are trying to establish the asymptotic normality of ℓ_N in the two sample problem when the ratio of the first sample size to the combined sample size, λ_N, is identically equal to $\frac{1}{2}$. It is believed that our conclusions are valid at least when $\lambda_N = \lambda + o(1/\sqrt{N})$ where $0 < \lambda < 1$, but we do not treat these extensions here.

2. Definitions and notations

Let Δ be the class of all probability measures on $[0, 1]$. An element in Δ may also be viewed as a distribution function (left continuous or right continuous). Thus when we say that P belongs to Δ we may mean either that P is a probability measure on $[0, 1]$ or that $P = P(x)$ is a distribution function on $[0, 1]$. This ambiguity does not cause confusion and helps in reducing the amount of notation. We say that a sequence $\{P_n, n = 0, 1, \cdots\}$ in Δ converges to P_0 when P_n converges to P_0 weakly. Endowed with this topology, Δ becomes a compact metric space (for example, see Gnedenko and Kolmogorov [3], Chapter 2). If $\{\mu_n, n = 0, 1, \cdots\}$ is a sequence of probability measures on Δ (on the Borel σ-field generated by the open sets in Δ) we say that $\mu_n \to \mu_0$ if $\int_\Delta g \, d\mu_n \to \int_\Delta g \, d\mu_0$ for every bounded continuous function g on Δ. With this notation of convergence (which is the usual weak convergence for measures) the space of probability measures on Δ becomes another compact metric space (for instance, see Billingsley [2], Section 6).

When the probability density function (p.d.f.) of a distribution function $P(x)$ exists we denote it by $p(x)$, the corresponding lower case letter. Let Δ^* be the subset of Δ consisting of distribution functions $P(x)$, with continuous probability density functions $p(x)$ and with $p(x) \geqq \delta$ on $[0, 1]$ for some $\delta > 0$. Let Δ_N be the subset of Δ which consists of distributions that give masses in multiples of $1/N$ to at most N distinct points of $[0, 1]$, $N = 1, 2, \cdots$.

Now let $0 = a_0 < a_1 < \cdots < a_R = 1$ be fixed and generate a partition of $[0, 1]$. Relative to this partition we define $\bar{\Delta}$ to be the subset of Δ consisting of distribution functions $P(x)$ with p.d.f. $p(x)$ satisfying

$$(2.1) \qquad p(x) = p_r \geqq 0, \qquad x \in [a_{r-1}, a_r), \qquad r = 1, \cdots, R.$$

Let $\bar{\Delta}^*$ be the subset of $\bar{\Delta}$ consisting of distribution functions $P(x)$ with p.d.f. $p(x)$ satisfying

$$(2.2) \qquad p(x) = p_r > 0, \qquad x \in [a_{r-1}, a_r), \qquad r = 1, \cdots, R.$$

Finally let $\bar{\Delta}_N$ be the subset of $\bar{\Delta}$ consisting of distribution functions $P(x)$ with p.d.f. $p(x)$ satisfying

$$(2.3) \qquad p(x) = p_r, \qquad p_r(a_r - a_{r-1}) = \text{a multiple of } \frac{1}{N},$$
$$x \in [a_{r-1}, a_r), \qquad r = 1, \cdots, R.$$

Notice that when we talk of $\bar{\Delta}$ or $\bar{\Delta}^*$ or $\bar{\Delta}_N$, we always have the partition genera-ted by $\{a_0, a_1, \cdots, a_{R_j}\}$ in mind and that this partition is arbitrary but fixed throughout this paper.

We have remarked earlier that ℓ_N and L_N are unchanged if we replace F_1 and F_2 by $F_1 K^{-1}$ and $F_2 K^{-1}$, where K is strictly increasing continuous distribution function on $[0, 1]$. For the same reason the distributions of ℓ_N and L_N remain unchanged if we replace F_1^* and F_2^* by $F_1^* K^{-1}$ and $F_2^* K^{-1}$. We will therefore normalize $F_1, F_2, F_1^*,$ and F_2^* by appropriately choosing K as follows: put

$$H(x) = \tfrac{1}{2}[F_1(x) + F_2(x)],$$
(2.4)
$$U_j(t) = F_j(H^{-1}(t)),$$
$$H^*(x) = \tfrac{1}{2}[F_1^*(x) + F_2^*(x)],$$

and

(2.5)
$$U_j^*(t) = F_j^*(H^{*-1}(t)), \qquad\qquad 0 \leq t \leq 1, \quad j = 1, 2.$$

Note that

(2.6)
$$U_1(t) + U_2(t) = U_1^*(t) + U_2^*(t) = 2t, \qquad 0 \leq t \leq 1,$$

and that $U_1, U_2, U_1^*,$ and U_2^* have probability density functions satisfying

(2.7)
$$u_1(t) + u_2(t) = u_1^*(t) + u_2^*(t) = 2, \qquad 0 \leq t \leq 1.$$

When $u_1(t)$ and $u_2(t)$ do not vanish on $[0, 1]$, define

(2.8)
$$M(t) = \log \frac{u_1(t)}{u_2(t)}.$$

We now state conditions concerning $F_1, F_2, F_1^*,$ and F_2^*, in terms of $U_1, U_2, U_1^*,$ and U_2^*.

CONDITION A.

A1. $U_1^*(t)$ is strictly convex on $[0, 1]$ and $u_1^*(t)$ is continuous.

A2. U_1, U_2 belong to Δ^*.

A3. $M(t)$ is strictly increasing on $[0, 1]$.

Condition A3 is equivalent to $u_1(t)$ being strictly increasing on $[0, 1]$ which is the same as $U_1(t)$ being strictly convex. Conditions A2 and A3 together imply that M is of bounded variation on $[0, 1]$. A general example where A2 and A3 are satisfied is as follows: $F_2 = \phi(F_1)$ where ϕ is a distribution function on $[0, 1]$ with ϕ' strictly decreasing and continuous with $\phi' \geq \delta$ for some $\delta > 0$. As a further special case we can put $\phi(t) = (1 - \lambda)t + \lambda t^A$ for some $A < 1$ and $0 < \lambda < 1$.

Condition A1 is restrictive and is not satisfied if $U_1^*(t) = t, 0 \leq t \leq 1$, which corresponds to the null hypothesis in the two sample problem. Condition A2 is similarly restrictive. It is not satisfied in the case $F_2 = F_1^A$ though we know from Sethuraman [9] that ℓ_N is asymptotic normal in this case.

Note that the roles X and Y can be reversed without changing the asymptotic distribution of ℓ_N. Thus if U_1, U_2, U_1^*, and U_2^* do not satisfy Condition A one could reverse the roles of X and Y and verify whether U_2, U_1, U_2^*, and U_1^* satisfy Condition A.

CONDITION $\bar{\text{A}}$.

$\bar{\text{A}}1$. The p.d.f. u_1^* is continuous on $[0, 1]$.

$\bar{\text{A}}2$. There is a partition $0 = a_0 < a_1 < \cdots < a_R = 1$ of $[0, 1]$ and U_1, U_2 belong to $\bar{\Delta}^*$, that is

$$(2.9) \qquad u_j(t) = u_{j,r} > 0, \qquad t \in [a_{r-1}, a_r), \quad r = 1, \cdots, R, \quad j = 1, 2.$$

Condition $\bar{\text{A}}2$ is used by Berk and Savage [1] when obtaining bounds on large deviations probabilities of ℓ_N. They were successful in removing these restrictions later on in their paper. We have not been able to do this. However Condition $\bar{\text{A}}2$ is a useful point to start an investigation of the properties of ℓ_N.

Let $F_{1,N}(x)$, $F_{2,N}(x)$, and $H_N(x)$ be the right continuous empirical distribution functions of $(X_1, \cdots, X_n)$, $(Y_1, \cdots, Y_n)$ and $(X_1, \cdots, X_n, Y_1, \cdots, Y_n)$, respectively. Clearly $H_N(x) = \frac{1}{2}[F_{1,N}(x) + F_{2,N}(x)]$. For $t \in [0, 1]$ define

$$(2.10) \qquad H_N^{-1}(t) = \inf \{x : H_N(x) \geqq t\}.$$

Then H_N^{-1} is left continuous. Define

$$(2.11) \qquad U_{j,N}(t) = F_{j,N}\big(H_N^{-1}(t)\big), \qquad 0 \leqq t \leqq 1, \quad j = 1, 2.$$

Note that $U_{1,N}$ and $U_{2,N}$ are left continuous and

$$(2.12) \qquad U_{1,N}(t) = \frac{1}{n} \sum_{i < Nt+1} (1 - Z_{N,i}),$$

$$(2.13) \qquad U_{2,N}(t) = \frac{1}{n} \sum_{i < Nt+1} Z_{N,i},$$

and

$$(2.14) \qquad U_{1,N}(t) + U_{2,N}(t) = 2t, \qquad t = 0, \frac{1}{N}, \cdots, \frac{N}{N}.$$

With probability one, $U_{1,N}$ and $U_{2,N}$ assume $\binom{N}{n}$ different values, each, in Δ.

Let $P \in \Delta$ and $Q \in \Delta^*$. Equation (3.1) motivates the following definitions. Let

$$(2.15) \qquad i(P, Q; U_{1,N}) = \frac{1}{2} \sum_{j=1}^{2} \int_0^1 \log \frac{u_j(t)}{q(t)} \, dU_{j,N}\big(P(t)\big)$$

and

$$(2.16) \qquad i(P, Q; U_1^*) = \frac{1}{2} \sum_{j=1}^{2} \int_0^1 \log \frac{u_j(t)}{q(t)} \, dU_j^*\big(P(t)\big).$$

Notice that these integrals always exist under Condition A or $\bar{\text{A}}$ since $\log u_j/q$ is a bounded function, $j = 1, 2$. For any $P \in \Delta$, the element P_N of Δ_N is defined as

$$(2.17) \quad P_N(t) = \begin{cases} \dfrac{i}{N}, & P^{-1}\left(\dfrac{i}{N}\right) \leq t < P^{-1}\left(\dfrac{i+1}{N}\right), i = 0, 1, \cdots, N-1 \\ 1, & P^{-1}(1) \leq t \leq 1. \end{cases}$$

Then

$$(2.18) \quad P_N^{-1}\left(\dfrac{i}{N}\right) = P^{-1}\left(\dfrac{i}{N}\right), \qquad i = 0, 1, \cdots, N,$$

and

$$(2.19) \quad |P_N(t) - P(t)| \leq \dfrac{1}{N}, \qquad 0 \leq t \leq 1,$$

Thus

$$(2.20) \quad i(P, Q; U_{1,N}) =$$

$$\dfrac{1}{2n} \left\{ \sum_{i=1}^{N} \log \dfrac{u_1\left(P^{-1}\left(\dfrac{1}{N}\right)\right)}{q\left(P^{-1}\left(\dfrac{i}{N}\right)\right)} (1 - Z_{N,i}) + \sum_{i=1}^{N} \log \dfrac{u_2\left(P^{-1}\left(\dfrac{i}{N}\right)\right)}{q\left(P^{-1}\left(\dfrac{i}{N}\right)\right)} Z_{N,i} \right\}$$

$$= i(P_N, Q; U_{1,N}).$$

Let

$$(2.21) \quad W_N(t) = \sqrt{N}\,[U_{1,N}(t) - U_1^*(t)], \qquad 0 \leq t \leq 1$$

and

$$(2.22) \quad W_N^1(t) = \begin{cases} W_N(t) & \text{for } \dfrac{1}{N} \leq t \leq 1, \\ 0 & \text{for } 0 \leq t < \dfrac{1}{N}. \end{cases}$$

Let D^- be the space of left continuous function on $[0, 1]$ which are right continuous at 0. This space becomes a complete metric space under the Skorohod topology. If $u_1^*(t)$ is continuous (this is Condition $\bar{A}1$ and it is a part of Condition A1) it follows from Pyke and Shorack [7] (Theorem 4.1(a)) that the distributions of $\{W_N^1(t), 0 \leq t \leq 1\}$ in D^- converge weakly to the distribution of a Gaussian process $\{W(t), 0 \leq t \leq 1\}$ with mean function 0 and variance-covariance function

$$(2.23) \quad K(t, s) = \tfrac{1}{2}U_1^*(t)[1 - U_1^*(s)]u_2^*(t)u_2^*(s) + \tfrac{1}{2}U_2^*(t)[1 - U_2^*(s)]u_1^*(t)u_1^*(s).$$

Since u_1^* is continuous, it is easily seen from the form of $K(t, s)$ in (2.23) that $\{W(t), 0 \leq t \leq 1\}$ has continuous path functions with probability 1. We can therefore state Lemma 2.1 below without proof in which for any function h on

$[0, 1]$ and any $\delta > 0$,

$$(2.24) \qquad \omega(h, \delta) = \sup_{|t-s| \leq \delta,\, 0 \leq t,\, s \leq 1} |h(t) - h(s)|.$$

LEMMA 2.1. *Let u_1^* be continuous. Then*

(i) *the distributions of $\{W_N^1(t), 0 \leq t \leq 1\}$ converge weakly to the distribution of $\{W(t), 0 \leq t \leq 1\}$ and*

(ii) *for each $\varepsilon > 0$,*

$$(2.25) \qquad \lim_N \sup P\{\omega(W_N^1, \delta) \geq \varepsilon\} \to 0$$

as $\delta \to 0$.

We would next like to investigate the asymptotic distribution of $i(P, Q; U_{1,N})$. Let $\log q$ be of bounded variation and let either $\bar{A}2$ or A2 and A3 hold. Then, from (2.20),

(2.26)

$$\sqrt{N}\left[i(P, Q; U_{1,N}) - i(P, Q; U_1^*)\right]$$

$$= \sqrt{N}\left[i(P_N, Q; U_{1,N}) - i(P_N, Q; U_1^*)\right] + \sqrt{N}\left[i(P_N, Q; U_1^*) - i(P, Q; U_1^*)\right]$$

$$= \frac{1}{2}\sqrt{N} \sum_{j=1}^{2} \int \log \frac{u_j}{q}\, d(U_{j,N}P_N - U_j^*P_N) + \frac{1}{2}\sqrt{N} \sum_{j=1}^{2} \int \log \frac{u_j}{q}\, d(U_j^*P_N - U_j^*P)$$

$$= \frac{1}{2}\sqrt{N} \int \log \frac{u_1}{u_2}\, d(U_{1,N}P_N - U_1^*P_N) - \frac{1}{2}\sqrt{N} \sum_{j=1}^{2} \int (U_j^*P_N - U_j^*P)\, d\log\frac{u_j}{q}.$$

Using (2.19), this becomes

$$(2.27) \qquad -\frac{1}{2}\int W_N(P_N)\, dM + O\left(\frac{1}{\sqrt{N}}\right) = -\frac{1}{2}\int W_N^1(P_N)\, dM + O_p\left(\frac{1}{\sqrt{N}}\right).$$

LEMMA 2.2. *Let u_1^* be continuous and A2 and A3 hold. Let $P^{(N)}$ be a random element in Δ, $N = 1, 2, \cdots$, such that the distribution of $P^{(N)}$ converges to the degenerate distribution at P^*, where $P^* \in \Delta^*$. Let $Q \in \Delta^*$ and $\log q$ be of bounded variation. Then*

$$(2.28) \qquad \sqrt{N}\left[i(P^{(N)}, Q; U_{1,N}) - i(P^{(N)}, Q; U_1^*)\right]$$

has a limiting normal distribution with mean 0 and variance σ_{P*}^2 given by

$$(2.29) \qquad \sigma_{P*}^2 = \frac{1}{4}\iint K(t, s)\, dMP^{*-1}(t)\, dMP^{*-1}(s).$$

(Note that the P^* above and in Lemma 2.3 can be any element of Δ^* and not necessarily the P^* defined in Condition B which follows later.)

PROOF. Using the Skorohod representation for sequences of random elements in $\Delta \times D^-$ (as Pyke and Shorack [7] do in D^-), we can assume that $(P^{(N)}, W_N) \to (P^*, W)$ in $\Delta \times D^-$ with probability 1. We can then imitate the proof of Theorem 4.1(b) of Pyke and Shorack [7] for our case as follows. From (2.26)

$$(2.30) \qquad \sqrt{N} \left[i(P^{(N)}, Q; U_{1,N}) - i(P^{(N)}, Q; U_1^*) \right] + \frac{1}{2} \int W(t) \, dMP^{*-1}(t)$$

$$= -\frac{1}{2} \int \left[W_N^1(P_N^{(N)}) - W(P^*) \right] dM + O_p \left(\frac{1}{\sqrt{N}} \right).$$

Using Lemma 2.1, this becomes

$$(2.31) \qquad -\frac{1}{2} \int \left(W_N^1(P_N^{(N)}) - W(P_N^{(N)}) + W(P_N^{(N)}) - W(P^*) \right) dM + O_p \left(\frac{1}{\sqrt{N}} \right)$$

$$= o_p(1).$$

This completes the proof of Lemma 2.2.

COROLLARY. *Under the conditions of Lemma 2.2 the limiting distribution of*

$$(2.32) \qquad \sqrt{N} \left[i(P^*, Q; U_{1,N}) - i(P^*, Q; U_1^*) \right]$$

is normal with mean 0 and variance $\sigma_{P^*}^2$.

When $\bar{A}$ holds we can simplify the expressions for $i(P, Q; U_{1,N})$, and so on whenever $Q \in \bar{\Delta}^*$. Let $Q \in \bar{\Delta}^*$. Then

$$(2.33) \qquad i(P, Q; U_{1,N}) = \frac{1}{2} \sum_{j=1}^{2} \int \log \frac{u_j(t)}{q(t)} \, dU_{j,N}(P(t))$$

$$= \frac{1}{2} \sum_{j=1}^{2} \sum_{r=1}^{R} \log \frac{u_{j,r}}{q_r} \left[U_{j,N}(P(a_r)) - U_{j,N}(P(a_{r-1})) \right],$$

which depends on P only through its values $P(a_r)$, $r = 0, 1, \cdots, R$. Thus a general P in Δ may be replaced by the unique $\bar{P}$ in $\bar{\Delta}$ with $P(a_r) = \bar{P}(a_r)$, $r = 0, 1, \cdots, R$, in $i(P, Q; U_{1,N})$ without changing its value. Therefore it is natural to restrict P to belong to $\bar{\Delta}$ and Q to $\bar{\Delta}^*$ in dealing with $i(P, Q; U_{1,N})$, and so forth, when $(\bar{A})$ holds. Also note that with the convention $0 \log 0 = 0$, $i(P, P; U_{1,N})$ can be defined for all $P \in \bar{\Delta}$. The following lemma and its corollary are analogous to Lemma 2.2 and its corollary.

LEMMA 2.3. *Let $\bar{A}1$ and $\bar{A}2$ hold. Let $P^{(N)}$ be a random element of $\bar{\Delta}$, $N = 1, 2, \cdots$, such that the distributions of $P^{(N)}$ in Δ converge weakly to the degenerate distribution at P^* where $P^* \in \bar{\Delta}^*$. Then*

$$(2.34) \qquad \sqrt{N} \left[i(P^{(N)}, P^{(N)}; U_{1,N}) - i(P^{(N)}, P^{(N)}; U_1^*) \right]$$

has a limiting normal distribution with mean 0 and variance $\sigma_{P^*}^2$ given in (2.29).

PROOF. Using the Skorohod representation as in Lemma 2.2 we may assume that $(P^{(N)}, W_N^1) \to (P^*, W)$ with probability 1. Since $P^{(N)} \in \bar{\Delta}$ and $P^* \in \bar{\Delta}^*$, it follows that the total variation of $\log p^{(N)}$ is bounded in probability for all large N and converges to the total variation of $\log p^*$. Thus using a slight extension of (2.27) and Lemma 2.1,

(2.35)

$$\sqrt{N}\left[i(P^{(N)}, P^{(N)}; U_{1,N}) - i(P^{(N)}, P^{(N)}; U_1^*)\right] + \frac{1}{2}\int W(t)\, dMP^{*-1}(t)$$

$$= -\frac{1}{2}\int \left[W_N^1(P_N^{(N)}) - W(P^*)\right] dM + O_p\left(\frac{1}{\sqrt{N}}\right)$$

$$= -\frac{1}{2}\int \left[W_N^1(P_N^{(N)}) - W(P_N^{(N)}) + W(P_N^{(N)}) - W(P^*)\right] dM + O_p\left(\frac{1}{\sqrt{N}}\right)$$

$$= o_p(1).$$

This completes the proof of Lemma 2.3.

COROLLARY. *Let the conditions of Lemma 2.3 hold. The limiting distribution of*

(2.36)
$$\sqrt{N}\left[i(P^*, P^*; U_{1,N}) - i(P^*, P^*; U_1^*)\right]$$

is normal with mean 0 and variance σ_{P*}^2.

3. A heuristic proof of the asymptotic normality of ℓ_N

An expression for L_N may be written as below using a formula from Hoeffding [6]. Fix $Q \in \Delta^*$ and write

(3.1) $$L_N(\mathbf{Z}_N) = N! \int \cdots \int\limits_{0 \leq w_1 \leq \cdots \leq w_N \leq 1} \prod_{i=1}^N \{[u_1(w_i)]^{1-Z_{N,i}}[u_2(w_i)]^{Z_{N,i}}\, dw_1 \cdots dw_N$$

$$= N! \int \cdots \int\limits_{0 \leq w_1 \leq \cdots \leq w_N \leq 1} \exp\left\{\sum_{i=1}^N \left[(1 - Z_{N,i}) \log \frac{u_1(w_i)}{q(w_i)}\right.\right.$$

$$\left.\left. + Z_{N,i} \log \frac{u_2(w_i)}{q(w_i)}\right]\right\} dQ(w_1) \cdots dQ(w_N)$$

$$= N! \int \cdots \int\limits_{0 \leq w_1 \leq \cdots \leq w_N \leq 1} \exp\left\{Ni(P_{\mathbf{w}}, Q; U_{1,N})\right\} dQ(w_1) \cdots dQ(w_N),$$

where $P_{\mathbf{w}}$ is the empirical distribution function of $w_1, \cdots, w_N$ and $i(P_{\mathbf{w}}, Q; U_{1,N})$ is as defined in (2.15). Note that $P_{\mathbf{w}}$ is in Δ_N. Expression (3.1) is the first place we have written down L_N explicitly and is fundamental to this investigation. A simple consequence of (3.1) is

(3.2)
$$L_N(\mathbf{Z}_N) \leq \exp\left\{N \sup_{P \in \Delta_N} i(P, Q; U_{1,N})\right\}$$

since

(3.3)
$$N! \int \cdots \int\limits_{0 \leq w_1 \leq \cdots \leq w_N \leq 1} dQ(w_1) \cdots dQ(w_N) = 1.$$

Since Q is arbitrary, (3.2) implies that

$$(3.4) \qquad L_N(\mathbf{Z}_N) \leqq \exp\left\{N \inf_{Q \in \Delta^*} \sup_{P \in \Delta_N} i(P, Q; U_{1,N})\right\}$$

$$= \exp\left\{N \inf_{Q \in \Delta^*} \sup_{P \in \Delta^*} i(P, Q; U_{1,N})\right\}$$

in view of (2.20). Again, since the integrand in (3.1) is an exponential raised to the Nth power, we may for large N act as if

$$(3.5) \qquad \ell_N = \log L_N \sim N \inf_{P \in \Delta^*} \sup_{Q \in \Delta^*} i(P, Q; U_{1,N}),$$

where $\sim$ means that the expressions on both sides of it are equal in some asymptotic sense. Let us look at the expression on the right side with $U_{1,N}$ replaced by its limit which is U_1^*. We get

$$(3.6) \qquad N \inf_{Q \in \Delta^*} \sup_{P \in \Delta^*} i(P, Q; U_1^*)$$

which may be replaced by

$$(3.7) \qquad N \sup_{P \in \Delta^*} \inf_{Q \in \Delta^*} i(P, Q; U_1^*),$$

if the appropriate result concerning the equality of a min max to a max min holds. This last expression can be shown (see Lemma 4.1) to be equal to

$$(3.8) \qquad N \sup_{P \in \Delta^*} i(P, P; U_1^*).$$

Let us assume that there is a unique P^* in that Δ^* such that the above is equal to $Ni(P^*, P^*; U_1^*)$. This assumption will be made precise in Section 4 as Condition B. The simplification which occurs when we use the limit U_1^* leads us to write

$$(3.9) \qquad \ell_N \sim Ni(P^*, P^*; U_{1,N}).$$

If the above were true, then from the corollary to Lemma 2.2 it follows that

$$(3.10) \qquad \frac{1}{\sqrt{N}}[\ell_N - Ni(P^*, P^*; U_1^*)]$$

has a limiting normal distribution with mean 0 and variance σ_{P*}^2.

4. Some intermediate results for i (P, Q; ·)

This section deals with some properties of $i(P, Q; U_1^*)$ and $i(P, Q: U_{1,N})$. These results are used in the proof of the main theorems of Section 5.

LEMMA 4.1.   *Let Condition A2 or $\overline{A}2$ hold. Let $P \in \Delta^*$. Then*

$$(4.1) \qquad i(P, Q; U_1^*) > i(P, P; U_1^*)$$

for all $Q \in \Delta^*$ with $Q \neq P$.

PROOF. Condition A2 or $\bar{A}2$ implies that $i(P, Q; U_1^*)$ is finite for $P \in \Delta^*$ and $Q \in \Delta^*$. For $Q \in \Delta^*$ with $Q \neq P$,

$$(4.2) \qquad i(P, Q; U_1^*) - i(P, P; U_1^*) = \frac{1}{2} \sum_{j=1}^{2} \int \left(\log \frac{u_j}{q} - \log \frac{\dot{u}_j}{p} \right) dU_j^* P$$

$$= \int \log \frac{p}{q} \, dP > 0.$$

This proves (4.1).

LEMMA 4.2. *Let $Q \in \Delta^*$. Let A2 hold. Then $i(P, Q; U_1^*)$ is a continuous function of P with P varying in Δ.*

PROOF. Let $P^m \in \Delta$, $m = 1, 2, \cdots$, and $P^m \to P$. Then

$$(4.3) \qquad i(P^m, Q; U_1^*) - i(P, Q; U_1^*) = \frac{1}{2} \sum_{j=1}^{2} \int \log \frac{u_j}{q} \, d(U_j^* P^m - U_j^* P) \to 0,$$

since $\log u_j/q$ is bounded and continuous and $U_j^* P^m \to U_j^* P$ in Δ, $j = 1, 2$.

COROLLARY. *Under the conditions of Lemma 4.1*

$$(4.4) \qquad \sup_{P \in \Delta^*} i(P, Q; U_1^*) = \sup_{P \in \Delta} i(P, Q; U_1^*).$$

REMARK. Let (V_1, V_2) be a pair of distribution functions on $[0, 1]$. Define

$$(4.5) \qquad i(P, Q; V_1, V_2) = \frac{1}{2} \sum_{j=1}^{2} \int \log \frac{u_j}{q} \, dV_j(P),$$

whenever $P \in \Delta$ and $Q \in \Delta^*$. One can see in a manner similar to the proof of Lemma 4.2 that for fixed $Q \in \Delta^*$, $i(P, Q; V_1, V_2)$ is a continuous function of $\big(V_1(P), V_2(P) \big)$.

LEMMA 4.3. *Let $Q \in \Delta^*$. Let A1, A2, and A3 hold. Then $i(P, Q; U_1^*)$ is a strictly concave function of P with P varying in Δ.*

PROOF. Let $P_1, P_2, \in \Delta$, $P_1 \neq P_2$, $0 < \alpha < 1$. Then

$$(4.6) \qquad i\big(\alpha P_1 + (1 - \alpha)P_2, Q; U_1^*\big) - \alpha i(P_1, Q; U_1^*) - (1 - \alpha)i(P_2, Q; U_1^*)$$

$$= \frac{1}{2} \sum_{j=1}^{2} \int \log \frac{u_j}{q} \, d\big[U_j^*(\alpha P_1 + (1 - \alpha)P_2) - \alpha U_j^*(P_1) - (1 - \alpha)U_j^*(P_2) \big]$$

$$= \frac{1}{2} \int \log \frac{u_1}{u_2} \, d\big[U_1^*(\alpha P_1 + (1 - \alpha)P_2) - U_1^*(P_1) - (1 - \alpha)U_1^*(P_2) \big]$$

$$= -\frac{1}{2} \int \big[U_1^*(\alpha P_1 + (1 - \alpha)P_2) - \alpha U_1^*(P_1) - (1 - \alpha)U_1^*(P_2) \big] \, dM$$

$$> 0,$$

since U_1^* is strictly convex and M is strictly increasing.

COROLLARY. *If $Q \in \Delta^*$, U_1^* is convex, $\log u_1/u_2$ is nondecreasing and bounded then $i(P, Q; U_1^*)$ is concave in P.*

We now state one form of the critical and hard to verify condition of this paper.

CONDITION B.   *There exists a unique P^* in Δ^* such that*

$$(4.7) \qquad \max_{P \in \Delta^*} i(P, P; U_1^*) = i(P^*, P^*; U_1^*)$$

and $\log p^*$ *is of bounded variation on* $[0, 1]$.

Notice that $i(P, P; U_1^*)$ may be written as $\int_0^1 \xi(t, P, p)\, dt$ with

$$(4.8) \qquad \xi(t, P, p) = \tfrac{1}{2} \sum_{j=1}^{2} [\log u_j - \log p] u_j^*(P) p.$$

Thus one might use the Euler formula of the calculus of variations to verify B. The following lemma also can be useful in verifying Condition B.

LEMMA 4.4.   *Let U_1^* be convex and $\log u_1/u_2$ be nondecreasing. Let A2 hold. Then $i(P, P; U_1^*)$ is a strictly concave function of P with P varying in Δ^*. Thus if there is a P_1 with*

$$(4.9) \qquad i(P_1, P_1; U_1^*) = \max_{P \in \Delta^*} i(P, P; U_1^*),$$

then P_1 is unique.

PROOF.   Let $P_1, P_2, \in \Delta^*, P_1 \neq P_2$ and $0 < \alpha < 1$. Then $\alpha P_1 + (1 - \alpha)P_2 \in \Delta^*$ and

$$(4.10) \quad i(\alpha P_1 + (1 - \alpha)P_2, \alpha P_1 + (1 - \alpha)P_2; U_1^*)$$
$$- \alpha i(P_1, P_1; U_1^*) - (1 - \alpha)i(P_2, P_2; U_1^*)$$
$$= -\frac{1}{2} \int [U_1^*(\alpha P_1 + (1 - \alpha)P_2) - \alpha U_1^*(P_1) - (1 - \alpha)U^*(P_2)]\, dM$$
$$- \int [(\alpha p_1 + (1 - \alpha)p_2) \log (\alpha p_1 + (1 - \alpha)p_2) - \alpha p_1 \log p_1$$
$$- (1 - \alpha)p_2 \log p_2]\, dt > 0,$$

The first term is ≥ 0 since U_1^* is convex and M is nondecreasing and the second term is > 0 since $x \log x$ is strictly convex.

LEMMA 4.5.   *Let Condition B hold. Let U_1^* be convex, M be nondecreasing and Condition A2 hold. Then*

$$(4.11) \qquad \max_{P \in \Delta \text{ or } \Delta^*} \min_{Q \in \Delta^*} i(P, Q; U_1^*) = \min_{Q \in \Delta^*} \max_{P \in \Delta \text{ or } \Delta^*} i(P, Q; U_1^*)$$

and

$$(4.12) \qquad \max_{P \in \Delta \text{ or } \Delta^*} i(P, P^*; U_1^*) = i(P^*, P^*; U_1^*),$$

where the P^* is as specified in Condition B.

PROOF.   We first establish (4.12). Only this conclusion of the lemma will be used later. The equality in (4.11) is aesthetically pleasing and is implied by (4.12) in the presence of (4.1) and (4.4) as we shall see later.

Let $P \in \Delta^*$. From Lemma 4.1

(4.13) $$i(P, P; U_1^*) = \min_{Q \in \Delta^*} i(P, Q; U_1^*).$$

From Condition B

(4.14) $$i(P^*, P^*; U_1^*) = \max_{P \in \Delta^*} i(P, P; U_1^*) = \max_{P \in \Delta^*} \min_{Q \in \Delta^*} i(P, Q; U_1^*).$$

Now, let $0 < \alpha < 1$, $P \in \Delta^*$, $P \neq P^*$. Then for $Q \in \Delta^*$, $i(P', Q; U_1^*)$ is a concave function of P' from the corollary to Lemma 4.3. Thus

(4.15) $$i((1 - \alpha)P^* + \alpha P, Q; U_1^*) \geqq (1 - \alpha)i(P^*, Q; U_1^*) + \alpha i(P; Q; U_1^*)$$
$$\geqq (1 - \alpha)i(P^*, P^*; U_1^*) + \alpha i(P, Q; U_1^*)$$

from (4.1). Put $Q = Q_1$ where $Q_1 = (1 - \alpha)P^* + \alpha P$. The above reduces to

(4.16) $$i(Q_1, Q_1; U_1^*) \geqq (1 - \alpha)i(P^*, P^*; U_1^*) + \alpha i(P, Q_1; U_1^*).$$

In view of Condition B we must have

(4.17) $$i(P, Q_1; U_1^*) \leqq i(P^*, P^*; U_1^*)$$

for all $P \in \Delta^*$. Now let $\alpha \to 0$. We show later in this proof that

(4.18) $$i(P, Q_1; U_1^*) \to i(P, P^*; U_1^*).$$

Thus

(4.19) $$i(P, P^*; U_1^*) \leqq i(P^*, P^*; U_1^*)$$

for all $P \in \Delta^*$. The conclusion (4.12) of Lemma 4.5 follows from (4.19) and (4.4).

Again, from (4.12) and (4.14)

(4.20) $$\min_{Q \in \Delta^*} \max_{P \in \Delta \text{ or } \Delta^*} i(P, Q; U_1^*) \leqq \max_{P \in \Delta \text{ or } \Delta^*} i(P, P^*; U_1^*)$$
$$= i(P^*, P^*; U_1^*)$$
$$\leqq \max_{P \in \Delta \text{ or } \Delta^*} \min_{Q \in \Delta^*} i(P, Q; U_1^*).$$

But a min max is always greater than or equal to a max min. Thus there is equality throughout (4.20) which now establishes (4.11).

To make this proof complete it therefore remains to establish (4.18) which can be done as follows:

(4.21) $$i(P, Q_1; U_1^*) - i(P, P^*; U_1^*) = \frac{1}{2} \sum_{j=1}^{2} \int \left(\log \frac{u_j}{q_1} - \log \frac{u_j}{p^*} \right) dU_j^* P$$
$$= \int \log \frac{p^*}{(1 - \alpha)p^* + \alpha p} \, p \, dt,$$

which tends to 0 as α tends to 0 by the dominated convergence theorem since the integrand in the last expression tends to 0 and is bounded in modulus for P and P^* belonging to Δ^*.

COROLLARY.   *Let Condition* B *hold. Let* A1, A2, *and* A3 *hold. Then*

$$(4.22) \qquad i(P, P^*; U_1^*) < i(P^*, P^*; U_1^*)$$

for all $P \neq P^*$, $P \in \Delta$. *Further if* P^{**} *is any random element in* Δ *with*

$$(4.23) \qquad i(P^{**}, P^*; U_1^*) \geq i(P^*, P^*; U_1^*)$$

with probability 1 *then* $P^{**} = P^*$ *with probability* 1.

This corollary follows readily from (4.12) in Lemma 4.5 and the strict concavity of $i(P, P^*; U_1^*)$ in P established in Lemma 4.3.

THEOREM 4.1.   *Let Conditions* B, A1, A2, *and* A3 *hold. Then there exist random elements* $P^{(N)}$ *in* Δ_N, $N = 1, 2, \cdots$, *such that* (i)

$$(4.24) \qquad i(P^{(N)}, P^*; U_{1,N}) \geq \sup_{P \in \Delta_N} i(P, P^*; U_{1,N}) - \frac{1}{N}$$

and such that (ii) *the probability measures induced by* $P^{(N)}$ *converge to the degenerate probability measure at* P^*. *Here* P^* *is as given in Condition* B.

PROOF.   With probability 1, $U_{1,N}$ is one of the $\binom{N}{n}$ distribution functions that give masses $1/n$ to n of the N points $1/N, 2/N, \cdots, N/N$. For each value of $U_{1,N}$ we can fix a $P^{(N)}$ in Δ_N such that

$$(4.25) \qquad i(P^{(N)}, P^*; U_{1,N}) \geq \sup_{P \in \Delta_N} i(P, P^*; U_{1,N}) - \frac{1}{N}.$$

Notice that this supremum is always finite since

$$(4.26) \qquad |i(P, P^*; U_{1,N})| \leq \frac{1}{2} \sum_{j=1}^{2} \sup_t \left| \log \frac{u_j(t)}{p^*(t)} \right|.$$

Thus we obtain random elements $P^{(N)}$ in Δ_N, $N = 1, 2, \cdots$, satisfying (4.24). Since Δ is compact and metric, the probability measures induced by $\{P^{(N)}\}$ have limit points. Let the probability measure induced by P^{**} be one such limit point. Since $\sup_t |U_{1,N}(t) - U_1^*(t)| \to 0$ with probability 1, the probability measure of $U_1^*(P^{**})$ is a limit point of the probability measures of $\{U_{1,N}(P^{(N)})\}$. (All these random elements are in Δ.) From the remark following Lemma 4.2, the distribution of $i(P^{**}, P^*; U_1^*)$ is a limit point of the distributions of $\{i(P^{(N)}, P^*; U_{1,N})\}$. But, from (2.20),

$$(4.27) \qquad i(P^{(N)}, P^*; U_{1,N}) \geq \sup_{P \in \Delta N} i(P, P^*; U_{1,N}) - \frac{1}{N}$$
$$= \sup_{P \in \Delta} i(P, P^*; U_{1,N}) - \frac{1}{N}$$
$$\geq i(P^*, P^*; U_{1,N}) - \frac{1}{N}.$$

Again $i(P^*, P^*; U_{1,N})$ converges in probability to the constant $i(P^*, P^*; U_1^*)$. Hence

$$(4.28) \qquad i(P^{**}, P^*; U_1^*) \geq i(P^*, P^*; U_1^*)$$

with probability 1. From the corollary to Lemma 4.5, $P^{**} = P^*$ with probability 1. This establishes the second part of Theorem 4.1.

REMARK.   Under the conditions of Theorem 4.1.

(4.29) $$\sup_{P \in \Delta_N} i(P, P^*; U_{1, N}) \to i(P^*, P^*; U_1^*)$$

in probability.

COROLLARY.   *Under the conditions of Theorem 4.1 and with the sequence* $\{P^{(N)}\}$ *as defined there*

(4.30) $$\sqrt{N}\left[i(P^{(N)}, P^*; U_{1, N}) - i(P^{(N)}, P^*; U_1^*)\right]$$

has a limiting normal distribution with mean 0 *and variance* σ_{P*}^2 *as given in* (2.29).

The above corollary follows from an application of Lemma 2.2.

We now state a condition analogous to the Condition B which will be used when Conditions $\bar{A}1$ and $\bar{A}2$ hold.

CONDITION $\bar{B}$.   *There exists a unique* P^* *in* $\bar{\Delta}^*$ *such that*

(4.31) $$i(P^*, P^*; U_1^*) = \max_{P \in \bar{\Delta}} i(P, P; U_1^*).$$

THEOREM 4.2.   *Let Conditions* $\bar{B}$, $\bar{A}1$, *and* $\bar{A}2$ *hold. Then there exist random elements* $P^{(N)}$ *in* $\bar{\Delta}_N$, $N = 1, 2, \cdots$, *such that* (i)

(4.32) $$i(P^{(N)}, P^{(N)}; U_{1, N}) = \max_{P \in \bar{\Delta}_N} i(P, P; U_{1, N}),$$

and such that (ii) *the probability measures induced by* $\{P^{(N)}\}$ *converge to the degenerate probability measure at* P^* *where* P^* *is as given in Condition* $\bar{B}$.

PROOF.   Notice that $\bar{\Delta}_N$ contains only a finite number of elements. For each one of the $\binom{N}{n}$ possible values of $U_{1, N}$ we can fix a $P^{(N)}$ in $\bar{\Delta}_N$ to satisfy (4.32). These $P^{(N)}$ are random elements in $\bar{\Delta}_N$, $N = 1, 2, \cdots$. Since Δ is compact and metric, the probability measures of $\{P^{(N)}\}$ have limit points. Notice that these limit points can only be probability measures in $\bar{\Delta}$. Let the probability measure of P^{**} be one such limit point. In a fashion similar to the proof of Theorem 4.1, the distribution of $i(P^{**}, P^{**}; U_1^*)$ is a limit point of the distributions of $i(P^{(N)}, P^{(N)}; U_{1, N})$. From Berk and Savage [1] (page 1668, line 9)

(4.33) $$\max_{P \in \bar{\Delta}_N} i(P, P; U_{1, N}) \geqq \sup_{P \in \bar{\Delta}} i(P, P; U_{1, N}) + O\left(\frac{\log N}{N}\right)$$

$$\geqq i(P^*, P^*; U_{1, N}) + O\left(\frac{\log N}{N}\right).$$

Combining these facts,

(4.34) $$i(P^{**}, P^{**}; U_1^*) \geqq i(P^*, P^*; U_1^*)$$

with probability 1. From Condition $\bar{B}$ it now follows that $P^{**} = P^*$ with probability 1. This establishes the second part of Theorem 4.2.

COROLLARY.   *Under the conditions of Theorem 4.2 and with the sequence as defined there*

(4.35) $$\sqrt{N}\left[i(P^{(N)}, P^{(N)}; U_{1, N}) - i(P^{(N)}, P^{(N)}; U_1^*)\right]$$

has a limiting normal distribution with mean 0 *and variance* σ_{P*}^2 *as given in* (2.29).

This corollary follows immediately by an application of Lemma 2.3.

5. The proof of the asymptotic normality of ℓ_N

We present two theorems on the asymptotic normality of ℓ_N, one when B, A1, A2, and A3 hold and the other when $\bar{\text{B}}$, $\bar{\text{A}}1$, and $\bar{\text{A}}2$ hold. It may be noted once again that Sethuraman [9] has established the asymptotic normality of ℓ_N for a case not covered by the two theorems alluded to above. In Section 6 we have an example where $\bar{\text{A}}1$ and $\bar{\text{A}}2$ hold but $\bar{\text{B}}$ does not hold and the asymptotic distribution of ℓ_N is not normal.

The proof of the asymptotic normality when $\bar{\text{B}}$, $\bar{\text{A}}1$, and $\bar{\text{A}}2$ hold is short and is facilitated considerably by the results of Berk and Savage [1]. The proof when B, A1, A2, and A3 hold is more lengthy and uses the results of Section 4 heavily. We feel that it might be possible to generalize this method of proof to include the case when $\bar{\text{B}}$, $\bar{\text{A}}1$ and $\bar{\text{A}}2$ hold and also to include other cases of interest which are now excluded by the restrictive conditions B, A1, A2, and A3. We have not been able to do this so far.

THEOREM 5.1.   *Let B, A1, A2, and A3 hold. Then*

$$(5.1) \qquad \frac{1}{\sqrt{N}} \left[\ell_N - Ni(P^*, P^*; U_1^*) \right]$$

has a limiting normal distribution with mean 0 and variance σ_{P*}^2 as given in (2.29). Here P^* is as specified by Condition B.

PROOF.   Substituting $Q = P^*$ in (3.2) we obtain

$$(5.2) \qquad \ell_N \leqq N \sup_{P \in \Delta_N} i(P, P^*; U_{1,N}).$$

Let $P^{(N)}$ be the random element in Δ_N given by Theorem 4.1. Using equalities (4.12) and then (4.24), we have

$$(5.3) \qquad \frac{1}{\sqrt{N}} \left[\ell_N - Ni(P^*, P^*; U_1^*) \right]$$

$$\leqq \sqrt{N} \left[\sup_{P \in \Delta_N} i(P, P^*; U_{1,N}) - \sup_{P \in \Delta} i(P, P^*; U_1^*) \right]$$

$$\leqq \sqrt{N} \left[i(P^{(N)}, P^*; U_{1,N}) - i(P^{(N)}, P^*; U_1^*) \right] + \frac{1}{\sqrt{N}}.$$

The corollary to Theorem 4.1 states that the above expression has a limiting normal distribution with mean 0 and variance σ_{P*}^2. Thus

$$(5.4) \qquad \liminf_N P\{ [\ell_N - Ni(P^*, P^*; U_1^*)]/\sqrt{N} \leqq x \} \geqq \Phi\left(\frac{x}{\sigma_{P*}} \right),$$

where $\Phi(x)$ is the distribution function of the standard normal random variable.

For any function h on $[0, 1]$, define

$$(5.5) \qquad \|h\| = \sup_{0 \leqq t \leqq 1} |h(t)|.$$

Substituting $Q = P^*$ in (3.1) and letting $S = \{0 \leqq w_1 \leqq \cdots \leqq w_N \leqq 1 : \|P_{\mathbf{w}} -$

$P^*|| \leqq \varepsilon/\sqrt{N}\}$ we have, for any $\varepsilon > 0$,

$$(5.6) \qquad L_N \geqq N! \int \cdots \int_S \exp \{Ni(P_\mathbf{w}, P^*, U_{1,N})\} \, dP^*(w_1) \cdots dP^*(\omega_N)$$

$$\geqq K_N(\varepsilon) \exp \{N \inf_P [i(P, P^*; U_{1,N}): P \in \Delta_N, ||P - P^*|| \leqq \varepsilon/\sqrt{N}]\},$$

where

$$(5.7) \qquad K_N(\varepsilon) = \int \cdots \int dP^*(w_1) \cdots dP^*(w_N)$$

with the integral taken over the set $\{w_1, \cdots, w_N : \|P_\mathbf{w} - P^*\| \leqq \varepsilon/\sqrt{N}\}$. From the Kolmogorov-Smirnov theorem (Billingsley [2], Section 16)

$$(5.8) \qquad K_N(\varepsilon) \to K(\varepsilon) > 0.$$

Consider

(5.9)

$$\sqrt{N}\Big|\inf_{P \in \Delta_N, \, \|P - P^*\| \leqq \varepsilon/\sqrt{N}} i(P, P^*; U_{1,N}) - i(P^*, P^*; U_{1,N})\Big|$$

$$\leqq \sqrt{N} \sup_{P \in \Delta, \, \|P - P^*\| \leqq \varepsilon/\sqrt{N}} \big|i(P, P^*; U_{1,N}) - i(P^*, P^*; U_{1,N})\big|$$

$$= \sqrt{N} \sup_{\|P - P^*\| \leqq \varepsilon/\sqrt{N}} \big|i(P, P^*; U_{1,N}) - i(P, P^*; U_1^*)$$

$$- i(P^*, P^*; U_{1,N}) + i(P^*, P^*; U_1^*) + i(P, P^*; U_1^*) - i(P^*, P^*; U_1^*)\big|$$

$$\leqq \sup_{\|P - P^*\| \leqq \varepsilon/\sqrt{N}} \tfrac{1}{2}\Big| \int [W_N^1(P_N) - W_N^1(P_N^*)] \, dM\Big|$$

$$+ \sup_{\|P - P^*\| \leqq \varepsilon/\sqrt{N}} \frac{\sqrt{N}}{2} \Big| \int [U_1^*(P) - U_1^*(P^*)] \, dM \Big| + O_p\Big(\frac{1}{\sqrt{N}}\Big)$$

Thus using (2.29) in which P_N and P_N^* are also defined, we have

$$(5.10) \qquad \leqq \tfrac{1}{2} C \, \omega\Big(W_N^1, \frac{\varepsilon}{\sqrt{N}} + \frac{2}{N}\Big) + C\varepsilon + O_p\Big(\frac{1}{\sqrt{N}}\Big),$$

where $C = \mathrm{Var}\,(M)$ and in which we have used the fact that $\|P_N - P_N^*\| \leqq \varepsilon/\sqrt{N} + 2/N$ and $\|u_1^*\| \leqq 2$. From Lemma 2.1 (ii), the first term in the above is $o_p(1)$. Thus from (5.6), (5.8), and (5.10), we have

(5.11)

$$\frac{1}{\sqrt{N}} [\ell_N - Ni(P^*, P^*; U_1^*)]$$

$$\geqq \sqrt{N} [i(P^*, P^*; U_{1,N}^*) - i(P^*, P^*; U_1^*)] + o_p(1) - C\varepsilon.$$

Using the corollary to Lemma 2.2 in the above and noting that $\varepsilon > 0$ is arbitrary, we have

$$(5.12) \qquad \limsup_N P\{[\ell_N - Ni(P^*, P^*; U_1^*)]/\sqrt{N} \leq x\} \leq \Phi\left(\frac{x}{\sigma_{P*}}\right).$$

Theorem 5.1 now follows from (5.4) and (5.12).

The next theorem applies to the case when u_1 and u_2 are histograms.

THEOREM 5.2.   *Let $\bar{B}$, $\bar{A}1$, and $\bar{A}2$ hold. Then $[\ell_N - Ni(P^*, P^*; U_1^*)]/\sqrt{N}$ has a limiting normal distribution with mean 0 and variance σ_{P*}^2 as given in (2.29). Here P^* is as specified in Condition $\bar{B}$.*

PROOF.   The expression for L_N in (3.1) can be simplified in this case. Putting $Q = $ the uniform distribution on $[0, 1]$ in (3.1),

$$(5.13) \quad L_N = N! \int\cdots\int_{0 \leq w_1 \leq \cdots \leq w_N \leq 1} \prod_{i=1}^{N} \{[u_1(w_i)]^{1-Z_{N,i}}[u_2(w_i)]^{Z_{N,i}}\}\, dw_1 \cdots dw_N$$

$$= N! \sum_{P \in \Delta_N} \prod_{r=1}^{R} \left\{\prod_{j=1}^{2} \bar{u}_{j,r}^{N[U_{j,N}(P(a_r)) - U_{j,N}(P(a_{r-1}))]}\right\} (N[P(a_r) - P(a_{r-1})])!$$

$$= N! \sum_{P \in \Delta_N} \exp\{Ni(P, P; U_{1,N})\} \prod_{r=1}^{R} \frac{[P(a_r) - P(a_{r-1})]^{N[P(a_r) - P(a_{r-1})]}}{(N[P(a_r) - P(a_{r-1})])!},$$

where $\bar{u}_{j,r} = u_{j,r}(a_r - a_{r-1})$, $j = 1, 2$, $r = 1, \cdots, R$. Using Stirling's formula and simplifying (5.13), Berk and Savage [1] (page 1667, line 17 and page 1668, line 9) have shown that

$$(5.14) \qquad \left|\ell_N - N \sup_{P \in \Delta_N} i(P, P; U_{1,N})\right| = O(\log N)$$

and

$$(5.15) \qquad \left|\sup_{P \in \bar{\Delta}_N} i(P, P; U_{1,N}) - \sup_{P \in \bar{\Delta}} i(P, P; U_{1,N})\right| = O\left(\frac{\log N}{N}\right).$$

Now, using (5.14) and (5.15),

$$(5.16) \qquad \sqrt{N}\left[i(P^*, P^*; U_{1,N}) - i(P^*, P^*; U_1^*)\right]$$

$$\leq \sqrt{N}\left[\sup_{P \in \bar{\Delta}} i(P, P; U_{1,N}) - i(P^*, P^*; U_1^*)\right]$$

$$\leq \frac{1}{\sqrt{N}}\left[\ell_N - Ni(P^*, P^*; U_1^*)\right] + O\left(\frac{\log N}{\sqrt{N}}\right)$$

$$= \sqrt{N}\left[\sup_{P \in \bar{\Delta}_N} i(P, P; U_{1,N}) - \sup_{P \in \bar{\Delta}} i(P, P; U_1^*)\right] + O\left(\frac{\log N}{\sqrt{N}}\right)$$

$$\leq \sqrt{N}\left[i(P^{(N)}, P^{(N)}; U_{1,N}) - i(P^{(N)}, P^{(N)}; U_1^*)\right] + O\left(\frac{\log N}{\sqrt{N}}\right),$$

where $P^{(N)}$ is a random element in $\bar{\Delta}_N$ chosen as in Theorem 4.2. The top expression in (5.16) has a limiting normal distribution with mean 0 and variance σ_{P*}^2 from the corollary to Lemma 2.3. From the corollary to Theorem 4.2 the last expression in (5.16) has the same limiting distribution. This establishes that $[\ell_N - Ni(P^*, P^*; U_1^*)]/\sqrt{N}$ has a limiting normal distribution with mean 0 and variance σ_{P*}^2.

6. An example where ℓ_N has a nonnormal asymptotic distribution

An example is given below where with the usual normalization ℓ_N has an asymptotic distribution which is not normal. In this example U_1 and U_2 have probability density functions which are histograms and U_1^* and U_2^* have continuous probability density functions. Thus Conditions $\bar{A}1$ and $\bar{A}2$ are automatically satisfied. We establish that $\bar{B}$ is not satisfied. Finally we evaluate the asymptotic distribution of ℓ_N.

Let $R = 2$, $a_0 = 0$, $a_1 = \frac{1}{2}$, $a_2 = 1$. Let $\alpha > 0$, $\beta > 0$ and $\alpha + \beta = 2$. Let $\lambda = \log (\alpha/\beta) > 0$, $\mu = \log (\frac{1}{4}\alpha\beta)^{1/2}$. Let

$$(6.1) \qquad u_1(t) = \begin{cases} \alpha & \text{if} \quad 0 \leq t \leq \frac{1}{2} \\ \beta & \text{if} \quad \frac{1}{2} < t \leq 1, \end{cases}$$

$$(6.2) \qquad u_2(t) = 2 - u_1(t), \qquad\qquad 0 \leq t \leq 1.$$

For $P \in \bar{\Delta}$ set $P(\frac{1}{2}) = \pi$. Then

$$(6.3) \qquad p(t) = \begin{cases} 2\pi & \text{if} \quad 0 \leq t \leq \frac{1}{2} \\ 2(1 - \pi) & \text{if} \quad \frac{1}{2} < t < 1. \end{cases}$$

Also, $0 \leq \pi \leq 1$ and π parametrizes the class $\bar{\Delta}$. Now, for any U_1^*, U_2^*,

$$(6.4) \qquad i(\pi) = i(P, P; U_1^*)$$

$$= \frac{1}{2} \left\{ U_1^*(\pi) \log \frac{\alpha}{2\pi} + [1 - U_1^*(\pi)] \log \frac{\beta}{2(1 - \pi)} \right.$$

$$\left. + U_2^*(\pi) \log \frac{\beta}{2\pi} + [1 - U_2^*(\pi)] \log \frac{\alpha}{2(1 - \pi)} \right\}$$

$$= [U_1^*(\pi) - \pi] \log \frac{\alpha}{\beta}$$

$$+ [-\pi \log \pi - (1 - \pi) \log (1 - \pi)] + \log \left(\frac{\alpha\beta}{4}\right)^{1/2}$$

$$= \lambda[U_1^*(\pi) - \pi] + L(\pi) + \mu,$$

where $L(\pi) = -\pi \log \pi - (1 - \pi) \log (1 - \pi)$. Notice that $\ell(\pi) = L'(\pi) = \log (1 - \pi)/\pi$. Let $\lambda_0 = 1/(1 + e^\lambda)$, $\lambda^0 = (1 - \lambda_0) = e^\lambda/(1 + e^\lambda)$. Then $0 < \lambda_0 < \frac{1}{2} < \lambda^0 < 1$ and $\ell(\lambda_0) = -\ell(\lambda^0) = \lambda$. We choose U_1^* and U_2^* as follows.

$$(6.5) \qquad u_1^*(t) = \begin{cases} 2 - 2t/\lambda_0 & \text{if} \quad 0 \leq t \leq \lambda_0, \\ 1 - \ell(t)/\lambda & \text{if} \quad \lambda_0 \leq t \leq \lambda^0, \\ 2 - 2(t - \lambda^0)/(1 - \lambda^0) & \text{if} \quad \lambda^0 \leq t \leq 1, \end{cases}$$

and

$$(6.6) \qquad\qquad u_2^*(t) = 2 - u_1^*(t), \qquad\qquad 0 \leq t \leq 1.$$

Then u_1^* is continuous on $[0, 1]$, $0 \leq u_1^*(t) \leq 2$, and $\int_0^1 u_1^*(t)\, dt = 1$. Thus u_1^* and u_2^* are continuous probability density functions. An explicit form for U_1^* is

$$(6.7) \qquad U_1^*(t) = \begin{cases} 2t - t^2/\lambda_0 & \text{if} \quad 0 \leq t \leq \lambda_0, \\ t + c - L(t)/\lambda & \text{if} \quad \lambda_0 \leq t \leq \lambda^0, \\ 2t - \lambda^0 - (t - \lambda^0)^2/(1 - \lambda^0) & \text{if} \quad \lambda^0 \leq t \leq 1, \end{cases}$$

where $c = L(\lambda_0)/\lambda = L(\lambda^0)/\lambda$. Substituting this U_1^* in (6.4) we obtain

$$(6.8) \quad i(\pi) - \mu = \begin{cases} \lambda(\pi - \pi^2/\lambda_0) + L(\pi) & \text{if} \quad 0 \leq \pi \leq \lambda_0, \\ \lambda c & \text{if} \quad \lambda_0 \leq \lambda \leq \lambda^0, \\ \lambda\big(\pi - \lambda^0 - (\pi - \lambda^0)^2/(1 - \lambda^0)\big) + L(\pi) & \text{if} \quad \lambda^0 \leq \pi \leq 1. \end{cases}$$

It can be shown (for instance by the sign of the derivatives, and so on) that $i(\pi) - \mu < c\lambda$ for $\pi < \lambda_0$ and $\pi > \lambda^0$, and that for any $\theta > 0$ there exists a $\delta > 0$ such that

$$(6.9) \qquad i(\pi) - \mu \leq c\lambda - \delta \qquad \text{for all} \quad \pi < \lambda_0 - \theta,\ \pi > \lambda^0 + \theta.$$

Thus

$$(6.10) \qquad\qquad \max_{0 \leq \pi \leq 1} i(\pi) = \lambda c + \mu = i(t) \qquad \text{for any} \quad t \in [\lambda_0, \lambda^0].$$

This means that Condition $\bar{\text{B}}$ is not satisfied.

For $\theta > 0$ and x real let

$$(6.11) \qquad\qquad G_\theta(x) = P\big\{ \sup_{\pi \in [\lambda_0 - \theta,\, \lambda^0 + \theta]} W(\pi) \leq x \big\},$$

where $\{W(t), 0 \leq t \leq 1\}$ is the Gaussian process in D^- defined with mean function 0 and variance-covariance $K(t, s)$ as defined in (2.23).

Recalling relation (5.14) we have

$$(6.12) \qquad\qquad \big| \ell_N - N \sup_{P \in \Delta N} i(P, P; U_{1,N}) \big| = O(\log N).$$

Further, by a simplification similar to (6.4)

$$(6.13) \qquad\qquad i(P, P; U_{1,N}) = \lambda(U_{1,N}(\pi) = \pi) + L(\pi) + \mu,$$

whenever $\pi = P(\tfrac{1}{2})$ is a multiple of $1/N$. Thus, from (5.14) and (6.13),

$$(6.14) \qquad\qquad \sqrt{N}\,\big[i(P, P; U_{1,N}) - i(P, P; U_1^*) \big] = \lambda W_N(\pi),$$

whenever $P \in \bar{\Delta}_N$, that is when $\pi = P(\tfrac{1}{2})$ is a multiple of $1/N$.

Now

(6.15)

$$\sqrt{N}\left[\sup_{P \in \Delta_N} i(P, P; U_{1, N}) - (\lambda c + \mu)\right]$$

$$\geq \sqrt{N} \sup_P \{[i(P, P; U_{1, N}) - i(P, P; U_1^*)]; \lambda_0 \leq \pi \leq \lambda^0, P \in \bar{\Delta}_N\}$$

$$= \lambda \sup_P \{W_N(\pi); \lambda_0 \leq \pi \leq \lambda^0, \pi \text{ a multiple of } 1/N\},$$

using (6.14). From Lemma 2.1 (i), comparing (5.14) and (6.15), we have that

(6.16) $$\limsup_N P\{(\ell_N - N[\lambda c + \mu])/\sqrt{N} \leq x\} \leq G_0(x).$$

Again, for any $\theta > 0$, $\sqrt{N}[\sup_{P \in \bar{\Delta}_N} i(P, P; U_{1, N}) - (\lambda c + \mu)]$ is equal to the maximum of

(6.17) $$\sqrt{N}\left[\sup_P \{i(P, P; U_{1, N}) - (\lambda c + \mu): P \in \bar{\Delta}_N, \pi \notin [\lambda_0 - \theta, \lambda^0 + \theta]\}\right]$$

and

(6.18) $$\sqrt{N} \sup \{i(P, P; U_{1, N}) - (\lambda c + \mu): P \in \bar{\Delta}_N, \pi \in [\lambda_\theta - \theta, \lambda^\theta + \theta]\}$$

which is smaller than the maximum of

(6.19) $$\sqrt{N} \sup \{[i(P, P; U_{1, N}) - i(P, P; U_1^*)]: P \in \bar{\Delta}_N, \pi \notin [\lambda_0 - \theta, \lambda^0 + \theta]\}$$
$$- \delta\sqrt{N}$$

and

(6.20) $$\sqrt{N} \sup_P \{[i(P, P; U_{1, N}) - i(P, P; U_1^*)]: P \in \bar{\Delta}_N, \pi \in [\lambda_0 - \theta, \lambda^0 + \theta]\},$$

in view of (6.9) and (6.10). The first term in the above tends to $-\infty$ in probability. The limiting distribution function of the second term is $G_\theta(x)$ by an application of Lemma 2.1 (i). Thus

(6.21) $$\liminf_N P\{[\ell_N - N(\lambda c + \mu)]/\sqrt{N} \leq x\} \geq G_\theta(x).$$

For the Gaussian process W with continuous path functions one can show that

(6.22) $$\lim_{\theta \to 0} G_\theta(x) = G_0(x).$$

Since $\theta > 0$ is arbitrary in (6.21), using (6.22) and then comparing it with (6.16) we find that

(6.23) $$\lim_N P\{\ell_N - N(\lambda c + \mu)]/\sqrt{N} \leq x\} = G_0(x)$$

which is the probability that the supremum of the Gaussian process W on $[\lambda_0, \lambda^0]$ is less than or equal to x. The distribution G_0 is clearly not normal. This completes the example.

$$\diamond \quad \diamond \quad \diamond \quad \diamond \quad \diamond$$

We thank Professors R. R. Bahadur and J. Hájek for several helpful remarks and comments.

REFERENCES

[1] R. BERK and I. R. SAVAGE, "The information in a rank-order and the stopping time of some associated SPRT's," *Ann. Math. Statist.*, Vol. 39 (1968), pp. 1661–1674.

[2] P. BILLINGSLEY, *Convergence of Probability Measures*, New York, Wiley, 1968.

[3] B. V. GNEDENKO and A. N. KOLMOGOROV, *Limit Distributions for Sums of Independent Random Variables* (translated by K. L. Chung), Cambridge, Addison-Wesley, 1954.

[4] Z. GOVINDARAJULU, "Asymptotic normality and efficiency of two sample rank order sequential probability ratio tests based on a Lehmann alternative," unpublished, 1968.

[5] J. HÁJEK, oral communication.

[6] W. HOEFFDING, "Optimum nonparametric tests," *Proceedings of the Second Berkeley Symposium on Mathematical Statistics and Probability*, Berkeley and Los Angeles, University of California Press, 1951, pp. 83–92.

[7] R. PYKE and G. SHORACK, "Weak convergence of a two-sample empirical process and a new proof to the Chernoff-Savage theorem," *Ann. Math. Statist.*, Vol. 39 (1968), pp. 755–771.

[8] I. R. SAVAGE and J. SETHURAMAN, "Stopping time of a rank-order sequential probability ratio test based on Lehmann alternatives," *Ann. Math. Statist.*, Vol. 37 (1966), pp. 1154–1160. (Correction note, *Ann. Math. Statist.*, Vol. 38 (1967), p. 1309.)

[9] J. SETHURAMAN, "Stopping time of a rank-order sequential probability ratio test based on Lehmann alternatives—II," *Ann. Math. Statist.*, Vol. 41 (1970), pp. 1322–1333.

ON SOME RESULTS AND PROBLEMS IN CONNECTION WITH STATISTICS OF THE KOLMOGOROV-SMIRNOV TYPE

I. VINCZE
MATHEMATICAL INSTITUTE
HUNGARIAN ACADEMY OF SCIENCES

1. Introduction

Since Kolmogorov and Smirnov established their limiting distribution theorems concerning maximal deviations between empirical and theoretical distributions, an increasing amount of scientific work has been done by statisticians in this field. Practical importance and theoretical interest give the motivation. Researchers have worked on distribution laws, power considerations and the limiting process in the last five years with considerable results. The present paper will consider only a few results, those nearest to the author's work and interest of the past few years. The first part, Section 2, concerns the case when the parent distribution is noncontinuous, the third and fourth sections consider the one and two sample problem, in the fifth section an analogous question for a density function due to Révész is discussed, while in the last section the two dimensional problem is considered.

2. The Gnedenko-Korolyuk distribution for discontinuous random variables

In his paper Schmid [12] has given the limiting distribution law of the Kolmogorov and of the Smirnov statistics, that is, of

(2.1)
$$D_n = \sup_{(x)} |F_n(x) - F(x)|,$$
$$D_n^+ = \sup_{(x)} [F_n(x) - F(x)]$$

for discontinuous $F(x)$, where $F_n(x)$ denotes the empirical distribution function of a sample of size n from a population distributed according to $F(x)$. Using the ballot lemma Csáki [2] determined the exact distribution of D_n^+ for finite n which corresponds to the well known Smirnov-Birnbaum-Tingey distribution for continuous $F(x)$. His formula has a fairly complicated form.

The two sample case for discontinuous $G(x) \equiv F(x)$ was considered by the author [20], who determined for finite $m = n$ the exact distribution of

$$(2.2) \quad \begin{aligned} D_{n,n}^+ &= \max_{(x)} [F_n(x) - G_n(x)], \\ D_{n,n} &= \max_{(x)} |F_n(x) - G_n(x)|, \end{aligned}$$

as well as their limiting forms as $n \to \infty$.

Let $F(x)$ have jumps at $x_1, \cdots, x_r$, with $x_i < x_{i+1}$, and be continuous otherwise. Let $F(x)$ be left continuous satisfying the following relations: with $x_0 = -\infty, x_{r+1} = +\infty$,

$$(2.3) \quad \begin{aligned} F(x_i) - F(x_{i-1} + 0) &= p_i, & i = 1, \cdots, r+1 \\ F(x_i + 0) - F(x_i) &= q_i, & i = 1, \cdots, r, \end{aligned}$$

where $\Sigma_{i=1}^{r+1} p_i + \Sigma_{i=1}^{r} q_i = 1$.

For the limiting distributions we have for $y \geqq 0$,

$$(2.4) \quad \lim_{n \to \infty} P\left[\left(\frac{n}{2}\right)^{1/2} D_{n,n}^+ < y \right]$$

$$= \frac{1}{(2\pi)^r p_{r+1}} \int \cdots \int_{G+} \prod_{i=1}^{r+1} \left[1 - \exp\left\{ -\frac{2}{p_i} (y - S_{i-1} - T_{i-1})(y - S_i - T_{i-1}) \right\} \right]$$

$$\exp\left\{ -\frac{1}{2} \sum_{i=1}^{r} (u_i^2 + w_i^2) - \frac{1}{2p_{r+1}} (S_r + T_r)^2 \right\} \prod_{i=1}^{r} du_i dw_i,$$

where

$$(2.5) \quad \begin{aligned} S_0 &= 0, & S_i &= \sum_{j=1}^{i} u_j p_j^{1/2}, & i = 1, \cdots, r+1, \\ T_0 &= 0, & T_i &= \sum_{j=1}^{i} v_j q_j^{1/2}, & i = 1, \cdots, r \end{aligned}$$

and for the domain of integration we have

$$(2.6) \quad G^+ = \{ S_{i-1} + T_{i-1} < y, S_i + T_{i-1} < y, i = 1, \cdots, r \}.$$

In the two sided case the following form holds, for $y > 0$,

$$(2.7) \quad \lim_{n \to \infty} P\left[\left(\frac{n}{2}\right)^{1/2} D_{n,n} < y \right]$$

$$= \frac{1}{(2\pi)^r p_{r+1}} \int \cdots \int_{G} \prod_{i=1}^{r+1} \left(\sum_{\gamma=-\infty}^{\infty} \left[\exp\left\{ -\frac{2}{p_i} (2\gamma y - u_i p_i^{1/2}) 2\gamma y \right\} \right. \right.$$

$$\left. \left. - \exp\left\{ -\frac{2}{p_i} [(2\gamma + 1) y + S_i + T_{i-1}][(2\gamma + 1) y + S_{i-1} + T_{i-1}] \right\} \right] \right)$$

$$\exp\left\{ -\frac{1}{2} \sum_{i=1}^{r} (u_i^2 + w_i^2) - \frac{1}{2p_{r+1}} (S_r + T_r)^2 \right\} \prod_{i=1}^{r} du_i dw_i$$

with the same S_i and T_i as above, while for the domain of integration we have

$$(2.8) \quad G = \{- y < S_{i-1} + T_{i-1} < y_1 - y < S_i + T_{i-1} < y, i = 1, \cdots, r\}.$$

As in the one sample case the distributions are no longer independent of $F(x)$; they depend on the values $F(x_i)$ and $F(x_i + 0)$ at the points of discontinuity, but only on them.

The case of $P_i = 0$ for each i, that is, the case of a discrete distribution was considered by Š. Šujan [16]. For finite n the corresponding distributions can be obtained from the formulas given in the paper cited, but the above relations do not work for $p_i = 0$. According to the calculation of Šujan for purely discrete random variables, with our above notations the following hold

$$(2.9) \quad \lim_{n \to \infty} P\left[\left(\frac{n}{2}\right)^{1/2} D_{n,n}^+ < y\right]$$

$$= \frac{1}{(2\pi)^{r-1} q_r^{1/2}} \int_{G_{r-1}} \cdots \int \exp\left\{-\frac{1}{2} \sum_{i=1}^{r} w_i^2 - \frac{1}{2g_r} T_{r-1}^2\right\} \prod_{i=1}^{r-1} dw_i,$$

where $G_{r-1}^+ = \{T_i < y, i = 1, \cdots, r - 1\}$.

The corresponding limit relation for $D_{n,n}$ has the same form with the single exception that the domain of integration has the two sided form

$$(2.10) \qquad G_{r-1} = \{- y < T_i < y, i = 1, \cdots, r - 1\}.$$

It was pointed out by Kolmogorov and proved by Noether [9] that for a critical value c the relation

$$(2.11) \qquad P(D_n > c \,|\, F \text{ is discrete}) \leqq P(D_n > c \,|\, F \text{ is continuous})$$

holds. This means that for given size α of a test based on the Kolmogorov statistic

$$(2.12) \qquad D_n = \sup_{(x)} |F_n(x) - F(x)|$$

the critical region must be at least as large as in the continuous case. Table I gives numerical calculations carried out by Šujan showing how pessimistic the Kolmogorov-Smirnov two sample test may be in the discrete case.

TABLE I

NUMERICAL EXAMPLE OF PESSIMISM OF
KOLMOGOROV-SMIRNOV TWO SAMPLE TEST FOR THE CASE
$n = 3$, $r = 2$, $c = 0$, $q_2 = 1 - q_1$

q_1	$P(D_{3,3} > 0)$
1/4	0.308
1/3	0.332
1/2	0.363
F continuous	0.75

The last row corresponds to the continuous case. The example chosen is a very extreme case. It can be shown that the less pessimistic case is $p_1 = p_2 = \frac{1}{2}$.

3. On distributions of statistics connected with the two sample problem

In the last two decades a lot of effort has been expended to determine the exact probabilities of the Kolmogorov-Smirnov two sample statistic, which was known for special sample sizes only ($m = kn$, k positive integer). In this respect, the paper of Steck [15] can be considered as containing far reaching results. Using ideas of Maag and Stephens, [7] and also of Lehmann, the Smirnov statistics were expressed in terms of the ranks of one sample. Steck then expressed explicitly the distribution in the form of a determinant, when one underlying distribution is the power of the other, $G(x) = [F(x)]^k$. Further, by giving determinant formulas for the frequency content under the null hypothesis of any parallelepiped in the sample space of the ranks of one sample, he obtained the null joint distribution of the one sided statistic and thus the null distribution of the two sided statistic, for arbitrary sample sizes.

Since the application of a pair of statistics as test statistic was introduced by Vincze [19], some authors have determined joint distribution laws considering the question from this point of view. V. Sujan [17] obtained the joint distribution of the maximum deviation and the number of runs; she, among others, proved that these two statistics are asymptotically independent. Mohanty and Pestros [8] considered joint distributions belonging to different rank statistics. In my paper the test was examined when the alternative is specified, in which case the likelihood ratios are completely ordered. The test for one sided alternatives was constructed by the authors mentioned using a partial ordering in the range space of the joint statistic based on certain relations of likelihood ratios.

The considerations mentioned led to a development in the theory of simple random walks, and an interesting new method was constructed by Dwass [4] in the technique of generating functions.

For the former point we refer to the paper of Sen [13] in which, among other results, a systematic treatment is given of certain useful path transformations.

In [4] the random walk $\{\vartheta_1, \vartheta_2, \cdots\}$ is considered with $P(\vartheta_i = +1) = p$, $P(\vartheta_i = -1) = 1 - p = q$, the ϑ_i being independent. Assuming that $p > q$ the random walk returns to the origin at most finitely often with probability one. Denote by T the last index for which the partial sum $s_i = \vartheta_1 + \vartheta_2 + \cdots + \vartheta_i$ vanishes (that is, $s_T = 0$, T even). Let U be a function defined on the random walk which is completely determined by the first T of the ϑ_j. Let us define

$$(3.1) \qquad U_n(\vartheta_1, \vartheta_2, \cdots, \vartheta_{2n}) = U(\vartheta_1, \vartheta_2, \cdots, \vartheta_T), \qquad \text{when} \quad T = 2n.$$

But the U_n can be considered as those rank statistics which play a role in the two sample problems. The relation

$$(3.2) \qquad E(U) = \sum_{n=0}^{\infty} E(U_n) P(T = 2n) = (1 - 2p) \sum_{n=0}^{\infty} \binom{2n}{n} (pq)^n E(U_n)$$

given by Dwass [4] makes it possible for him to derive previously known and also new distributions in a very simple way.

An example in [19] shows that when we use a pair of statistics instead of one statistic in the case of a given simple alternative, the probability of the error of second kind is reduced to one half or one quarter of its previous value. In any case, depending on the alternative, the second kind error can be diminished from extremely high values to acceptable ones from the practical point of view. The examples mentioned concern the pair $(D_{n,n}^+, R_{n,n}^+)$, where $D_{n,n}^+$ is the one sided maximal deviation between the two empirical distribution functions, while $R_{n,n}^+$ is the index of the sample element in the ordered union of two samples for which the maximum $D_{n,n}^+$ first occurs. The test based on $(D_{n,n}^+, R_{n,n}^+)$ is compared with the test based on $D_{n,n}^+$ only for the equal sample sizes, $n = 10, 30, 50$. We thought that the power of the Kolmogorov-Smirnov test would be improved in this way for arbitrary sample sizes, that is, for the tests based on $(D_{n,m}^+, R_{n,m}^+)$ and $D_{n,m}^+$, respectively. Surprisingly, Steck in his paper [15] showed that the situation is not so simple; it depends on the relationship between m and n. If they are relatively prime then there is precisely one value of $R_{n,m}^+$ which is associated with a given value of $D_{n,m}^+$, hence any test based on $(D_{n,m}^+, R_{n,m}^+)$ is equivalent to the one based on $D_{n,m}^+$. The interesting and surprising consequence is the following: in a given case with $\sup_{(x)} [F(x) - G(x)] = 0.2$ using $(D_{50,50}^+, R_{50,50}^+)$ the second kind error turns out to be 0.047, while by using only $D_{50,50}^+$ the corresponding value is 0.173, nearly four times as much. When we turn to samples $n = 50$ and $m = 51$ my method does not lead to any improvement of the two sample Smirnov test. What is the probability of the second kind error, how does it relate to the above two values, what is the asymptotic relation of the corresponding power functions? These are questions of interest.

I should like to mention that in our paper with Reimann [10], statistics of the following type were considered

$$(3.3) \qquad nF_n(x) - mG_m(x),$$

the distribution of which can be determined easily, and does not show the same irregularity for m and n relatively prime as shown by Steck.

4. Some questions connected with the one sample problem

While an exact formula for the distribution of the Smirnov statistic

$$(4.1) \qquad D_n^+ = \sup_{(x)} [F_n(x) - F(x)]$$

was known very early (Smirnov 1944, independently Birnbaum and Tingey 1951), the distribution of the two sided statistic

$$(4.2) \qquad D_n = \sup_{(x)} |F_n(x) - F(x)|$$

was obtained only within the last five years. Durbin [3] derived the generating function for the probabilities that the empirical distribution function lies between two parallel straight lines. He obtained recursion formulas and for a particular case, that is, for a certain integral value of a parameter, he has exact formulas. Epanechnikov [5] by means of a recurrence relation determined the exact probabilities. Independently Steck in his paper [14] gives, as he says, "a neat determinant for the probability that the order statistics for a sample of uniform random variables all lie in a multidimensional rectangle. An immediate application of this result gives the probability that the empirical distribution function lies between two other distribution functions." These authors have obtained a result for which a great deal of effort was expended in the last two decades.

Turning to the one sided case, the use of the ballot lemma (see Takács [18]) enables us to get very simple derivations of the distribution of D_n^+ and of distributions of a number of related statistics.

The following modified and extended form of the ballot lemma suggested by me leads almost immediately to the Smirnov-Birnbaum-Tingey theorem.

Let $A_0, A_1, A_2, \cdots, A_n$ be a complete system of events such that $P(A_0) = p$, $P(A_i) = q, i = 1, \cdots, n, p + nq = 1$. Denoting by v_i the frequency of the event A_i in n trials, the following relation holds

$$(4.3) \qquad P\left(\sum_{j=1}^{i} v_j < i, \qquad i = 1, \cdots, n \right) = p.$$

It is easy to see that an elementary proof can be obtained for this theorem from the following nice lemma due to and proved by Tusnády [2].

Let the points $P_1, P_2, \cdots, P_n$ be given on a directed circle of unit circumference and let us choose the positive number q such that $0 < nq < 1$. To an arbitrary point Q of the circle, construct the points $Q_1, Q_2, \cdots, Q_n$ consecutively in the positive direction with $\widehat{QQ_i} = iq$. Let the point Q be called a point of first category if the arc QQ_k contains less than k of points $P_1, P_2, \cdots, P_n$ for $k = 1, \cdots, n$. Then the measure of the set of points of first category is $1 - nq$.

PROOF (Tusnády). To each point P_i a chain

$$(4.4) \qquad C_i = \{ R_{i,1}, R_{1,2}, \cdots, R_{i,v(i)} \}$$

will be ordered in the following way: these points are consecutive but in the negative direction on the circle; further the arc $P_i R_{i,j}$ contains at least j of the points $P_1, P_2, \cdots, P_n$, including the point P_i for $j = 1, \cdots, v(i)$, while less than j for $j = v(i) + 1$.

It can easily be seen that if a chain covers the point P_j then it covers C_j as well.

A chain is called maximal if no other chain covers it. Two maximal chains are disjoint, hence the total length of all maximal chains is nq.

A point Q on the circle is now of the first category if and only if no chain covers it. Consequently the measure of all points of first category is $1 - nq$, as stated above.

Unfortunately a "two sided ballot lemma," which would be a tool for the derivation of the distribution of D_n, for example, does not yet exist.

Added in proof. See S. G. Mohanty, "Combinatorial methods in probability and statistics," lecture presented at the 58th Session of the Indian Science Congress Association, 1971.

A number of interesting articles have appeared in recent years which concern the power or asymptotic power of Kolmogorov-Smirnov tests. It is beyond our scope to mention these, but as a nice summarization of certain results in this field, I would mention the book of Hájek and Šidák [6].

5. Distribution of a Rényi type statistic due to Révész

Let the null hypothesis be that the density function is fully specified by $f(x)$, a function satisfying certain conditions (for example, $f'(x)$ exists and $|f'(x)|$ is bounded).

Let

$$(5.1) \qquad x_0 < x_1 < x_2 < \cdots < x_{a_n}$$

be a division of the real axis or of the interval where $f(x) > 0$. The a_n, $n = 1, 2, \cdots$, are restricted by the relations

$$(5.2) \qquad n^{1/3} \log n < a_n < n^{1-\varepsilon}, \qquad\qquad \varepsilon > 0.$$

Denoting by k_{i+1} the number of elements out of the sample $(X_1, X_2, \cdots, X_n)$ in the interval (x_i, x_{i+1}), $i = 1, \cdots, a_n - 1$, the empirical density function is defined by

$$(5.3) \qquad f_n(x) = \frac{k_{i+1}}{n(x_{i+1} - x_i)}, \quad x_i \leqq x < x_{i+1}, i = 0, 1, \cdots, a_n - 1.$$

Révész [11] proved the following limiting relations which are *distribution free*

$$(5.4) \quad \lim_{n \to \infty} P\left[\left(\frac{n}{a_n}\right)^{1/2} \sup_{x_\alpha < x < x_\beta} \frac{f_n(x) - f(x)}{f(x)} < (2 \log a_n - \log \log a_n + y)^{1/2}\right]$$

$$= \exp\{-\exp\{-y/2\}/2\pi^{1/2}\},$$

$$(5.5) \quad \lim_{n \to \infty} P\left[\left(\frac{n}{a_n}\right)^{1/2} \sup_{x_\alpha < x < x_\beta} \left|\frac{f_n(x) - f(x)}{f(x)}\right| < (2 \log a_n - \log \log a_n + y)^{1/2}\right]$$

$$= \exp\{-\exp\{-y/2\}/\pi^{1/2}\}.$$

The values $x_\alpha = x(\alpha, n)$ and $x_\beta = x(\beta, n)$ form an interval for which $f(x) \geqq (\log n)^{-1/3}$.

Further

$$(5.6) \qquad \int_{x_\alpha}^{x_\beta} f(x)\, dx \to 1, \qquad\qquad \text{as} \quad n \to \infty.$$

6. On a two dimensional analogue of the Gnedenko-Korolyuk distribution

The difficulty encountered in constructing distribution free methods for samples taken on two or more variate random variables is well known. Recently Bickel [1] has given a distribution free version of the Smirnov two sample statistic in the p variate case. We shall consider the original Smirnov statistics in two dimensions

(6.1)
$$D_{n,n}^+ = \sup_{(x,y)} \left[F_n(x, y) - G_n(x, y) \right],$$

$$D_{n,n} = \sup_{(x,y)} \left| F_n(x, y) - G_n(x, y) \right|$$

in case of equal sample sizes. Our consideration leads to an immediate extension of the random walk model. As is known and will be illustrated below, the distribution of $D_{n,n}^+$ or $D_{n,n}$ does depend on the common theoretical continuous distribution function $G(x, y) \equiv F(x, y)$ of the two samples. Our aim is to propose a problem which will concern the independent case $F(x, y) = H_1(x) H_2(y)$ and which shows the difficulties even in this simple—distribution free—case.

Let (X_i, Y_i) and (X_i', Y_i'), $i = 1, \cdots, n$, be two samples with

(6.2) $$P(X_i < x, Y_i < y) = P(X_i' < x, Y_i' < y) = F(x, y).$$

Since $F(x, y)$ is continuous there is a "two dimensional ordering" of the two samples with probability one in the following way: let $\eta_1^* < \eta_2^* < \cdots < \eta_{2n}^*$ be the ordered union of the samples $(Y_1, Y_2, \cdots, Y_n)$ and $(Y_1', Y_2', \cdots, Y_n')$ and let us denote by ξ_i the corresponding X or X' of η_i^*, that is, we have

(6.3) $$(\xi_1, \eta_1^*), (\xi_2, \eta_2^*), \cdots, (\xi_{2n}, \eta_{2n}^*).$$

Taking now the ordered version of the ξ_i

(6.4) $$\xi_1^* < \xi_2^* < \cdots < \xi_{2n}^*,$$

the following random variables will be introduced

(6.5) $$\vartheta_{i,j} = \begin{cases} +1 & \text{if} \quad \eta_i^* = Y_h \quad \text{and} \quad \xi_i = \xi_j^* = X_h \\ -1 & \text{if} \quad \eta_i^* = Y_\ell' \quad \text{and} \quad \xi_i = \xi_j^* = X_h' \\ 0 & \text{otherwise.} \end{cases}$$

Now we have an arrangement of $+1$ and of -1, each n in number, in a $2n \times 2n$ table containing in each row and in each column exactly one element. This corresponds to and is analogous with the random walk used first by Gnedenko and Korolyuk and independently by Drion. Let us introduce the "partial" sums in the following way

(6.6) $$s_{0,0} = 0, \quad s_{k,\ell} = \sum_{i \leq k} \sum_{j \leq \ell} \vartheta_{i,j}, \ 1 \leq k \leq 2n, 1 \leq \ell \leq 2n, \quad s_{2n,2n} = 0.$$

As can be seen very easily

$$(6.7) \qquad D_{n,n}^{+} = \frac{1}{n} \max_{(k,\ell)} s_{k,\ell}, \qquad D_{n,n} = \frac{1}{n} \max_{(k,\ell)} |s_{k,\ell}|.$$

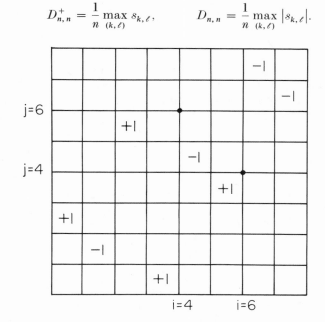

FIGURE 1

Array of $\vartheta_{i,j}$ for $n = 4$.

For example, in Figure 1 we have $n = 4$ and as can be justified very simply the following relations hold

$$(6.8) \qquad D_{4,4}^{+} = \tfrac{1}{4} s_{4,6} = \tfrac{1}{4} s_{6,4} = \tfrac{1}{2}.$$

There are altogether $(2n)!$ possible arrangements within the square and each of them allows $\binom{2n}{n}$ possible allocations of the $+1$ and the -1, that is, the number of all possible configurations is $(2n)! \binom{2n}{n} = [(2n)!/n!]^2$.

Unfortunately the different arrays may have different probabilities depending on $F(x, y)$.

Consider the two extreme cases, $Y = X$ and $Y = -X$.

If $Y = X$, then $(\vartheta_{1,1}, \vartheta_{2,2}, \cdots, \vartheta_{2n,2n})$ are the nonzero terms in the $2n \times 2n$ square, where each of the $\binom{2n}{n}$ possible arrays of the $+1$ and the -1 is of equal probability. In this case the relations hold

$$(6.9) \qquad P\left(D_{n,n}^{+} < \frac{k}{n}\right) = 1 - \frac{\binom{2n}{n-k}}{\binom{2n}{n}}, \qquad k = 0, 1, 2, \cdots, n,$$

and

(6.10) $$P\left(D_{n,n} < \frac{k}{n}\right) = \frac{1}{\binom{2n}{n}} \sum_{j=-\infty}^{\infty} - 1^j \binom{2n}{n - jk}, \quad k = 1, \cdots, n,$$

that is, the Gnedenko-Korolyuk distributions are valid.

In the second case, when $Y = -X$, the nonzero terms are $(\vartheta_{1,\,2n}, \vartheta_{2,\,2n-1}, \cdots, \vartheta_{2n,\,1})$. We mention without proof that in this case the distribution of the Kuiper statistics determined by Maag and Stephens [7] is valid. In the two sided case we have

(6.11)

$$P\left(D_{n,n} < \frac{k}{n}\right) = 1 - \frac{2}{\binom{2n}{n}}\left[k \sum_{j=1}^{\infty} \binom{2n}{n - jk} - (k + 1) \sum_{j=1}^{\infty} \binom{2n}{n - j(k + 1)}\right],$$

$$k = 2, 3, \cdots, n.$$

This was derived for finding the distribution of the maximum deviation when two samples of the same size n are distributed uniformly on the circumference of a circle.

These two cases already show the dependence on the theoretical distribution function. Let us turn now to the independent case. The following problem is raised.

The distribution of the maximum deviation and of the absolute maximum deviation is to be determined when the two random variables are independent

(6.12)
$$P\left[D_{n,n}^+ < \frac{k}{n} \middle| F(x, y) = H_1(x)H_2(y)\right] = ?$$

$$P\left[D_{n,n} < \frac{k}{n} \middle| F(x, y) = H_1(x)H_2(y)\right] = ?$$

In this case each array has the same probability, as given above, and the determination of the probabilities can be reduced to the enumeration of those paths, that is, arrays, for which the maximum is k/n or $\max_{(i,\,j)} s_{i,\,j} = k$. As the above example shows, the Markovian property does not hold. There does not exist a "first" maximum, instead simultaneously two places with $s_{i,\,j} = k$. In this way, however, a reflection can be made and a path with $\{s_{0,\,0} = 0, s_{2n,\,2n} = 0\}$ can be transformed into a path with $\{s_{0,\,0} = 0, \quad s_{2n,\,2n} = 2k\}$; however this is not a one to one mapping of the paths.

Concerning the reflection, let us consider the pair of indices (i^*, j^*) for which $s_{i^*,\,j^*} = k = \max s_{i,\,j}$ and change $\vartheta_{i,\,j}$ into $-\vartheta_{i,\,j}$, for which either $i > i^*$

or $j > j^*$ or both. Denote by α and β the number of $+1$ and -1, respectively, in the set $\{\vartheta_{i,j}, i \leqq i^*, j \leqq j^*\}$. Then $\alpha - \beta = k$, and the $+1$, of which there are $n - \alpha$, will be replaced by -1, and the -1, of which there are $n - \beta$, will be replaced by $+1$. In this way the number of $+1$ will amount to $\alpha + n - \beta$, while the number of -1 will be $\beta + n - \alpha$. Consequently

$$(6.13) \qquad s_{2n, 2n} = \alpha + n - \beta - (\beta + n - \alpha) = 2(\alpha - \beta) = 2k.$$

Now the number of paths with $s_{0,0} = 0$, $s_{2n, 2n} = 2k$ is $(2n)! \binom{2n}{n-k}$ which must be smaller than the number of arrays with $D_{n,n}^+ \geqq k/n$. Consequently the relation holds

$$(6.14) \qquad P\left(D_{n,n}^+ < \frac{k}{n} \right) \leqq 1 - \frac{\binom{2n}{n-k}}{\binom{2n}{n}}.$$

This relation follows also from a consideration of Nedoma (personal communication).

For $n = 2$ the number of possible arrays is 144, the distribution of $D_{4,4}^+$ is contained in Table II.

TABLE II

DISTRIBUTION OF $D_{4,4}^+$

k	$P\left(D_{n,n} < \dfrac{k}{n} \right)$	$1 - \dfrac{\binom{2n}{n-k}}{\binom{2n}{n}}$
1	$\dfrac{15}{144}$	$\dfrac{1}{3} = \dfrac{48}{144}$
2	$\dfrac{92}{144}$	$\dfrac{5}{6} = \dfrac{120}{144}$
3	1	1

In the case of $n = 3$ the number of possible arrangements is 14,400 for which a computer is needed.

REFERENCES

[1] P. J. BICKEL, "A distribution free version of the Smirnov two sample test in the p-variate case," *Ann. Math. Statist.*, Vol. 40 (1969), pp. 1–23.
[2] E. CSÁKI and G. TUSNÁDY, "On the number of intersections and the ballot theorem," *Studia Sci. Math. Hungar.*, to be published.
[3] J. DURBIN, "The probability that the sample distribution function lies between two parallel straight lines," *Ann. Math. Statist.*, Vol. 39 (1968), pp. 398–411.

[4] M. Dwass, "Simple random walk and rank order statistics," *Ann. Math. Statist.*, Vol. 38 (1967), pp. 1042–1053.

[5] V. A. Epanechnikov, "The significance level and power of the two-sided Kolmogorov test in the case of small samples," *Teor. Verojatnost. i Primenen.*, Vol. 13 (1968), pp. 725–730.

[6] J. Hájek and Z. Šidák, *Theory of Rank Tests*, New York, Academic Press, 1967.

[7] U. R. Maag and M. A. Stephens, "The V_{NM} two sample test," *Ann. Math. Statist.*, Vol. 39 (1968), pp. 923–935.

[8] S. G. Mohanty and C. I. Petros, "Joint distributions of several nonparametric statistics and tests based on them," unpublished.

[9] G. E. Noether, "Note on the Kolmogorov statistics in the discrete case," *Metrika*, Vol. 7 (1969), pp. 115–116.

[10] J. Reimann and I. Vincze, "On the comparison of two samples with slightly different sizes," *Publ. Math. Inst. Hungar. Acad. Sci.*, Vol. 5 (1960), pp. 293–309.

[11] P. Révész, "Testing of density functions," *Mathematica (Cluj)*, 1971, in print.

[12] P. Schmid, "On the Kolmogorov and Smirnov limit theorems for discontinuous distribution functions," *Ann. Math. Statist.*, Vol. 29 (1958), pp. 1011–1027.

[13] K. Sen, "Paths of an odd number of steps with final position unspecified," *J. Indian Statist. Assoc.*, Vol. 7 (1969), pp. 107–135.

[14] G. P. Steck, "Rectangle probabilities for uniform order statistics and the probability that the empirical distribution lies between two distribution functions," *Ann. Math. Statist.*, Vol. 42 (1971), pp. 1–11.

[15] ———, "The Smirnov two sample tests as rank tests," *Ann. Math. Statist.*, Vol. 40 (1969), pp. 1449–1466.

[16] Š. Šujan, "On some problems concerning Kolmogorov-Smirnov test in the case of discrete theoretical distribution," thesis, Comenius University, Bratislava, 1970.

[17] V. Sujan, "The Kolmogorov-Smirnov statistics and the runs," unpublished.

[18] L. Takács, *Combinatorial Methods in the Theory of Stochastic Processes*, New York, Wiley, 1967.

[19] I. Vincze, "Some questions connected with two sample tests of Smirnov type," *Proceedings of the Fifth Berkeley Symposium on Mathematical Statistics and Probability*, Berkeley and Los Angeles, University of California Press, Vol. 1 (1967), pp. 654–666.

[20] ———, "On Kolmogorov-Smirnov type distribution theorems," *Proceedings of the First International Symposium on Nonparametric Techniques*, New York, Cambridge University Press (1970), pp. 385–401.

EFFICIENT ESTIMATION OF REGRESSION COEFFICIENTS IN TIME SERIES

T. W. ANDERSON
STANFORD UNIVERSITY

1. Introduction

This paper deals with estimating regression coefficients in the usual linear model. Let $\mathbf{y}$ be a T-component random vector with expected value

$$(1.1) \qquad \mathscr{E}\mathbf{y} = \mathbf{Z}\boldsymbol{\beta},$$

where $\mathbf{Z}$ is a $T \times p$ matrix of numbers and $\boldsymbol{\beta}$ is a p-component vector of parameters. (All vectors are column vectors.) For convenience we assume that the rank of $\mathbf{Z}$ is the number of columns, p. The covariance matrix of $\mathbf{y}$ is

$$(1.2) \qquad \mathscr{C}(\mathbf{y}) = \mathscr{E}(\mathbf{y} - \mathbf{Z}\boldsymbol{\beta})(\mathbf{y} - \mathbf{Z}\boldsymbol{\beta})' = \boldsymbol{\Sigma}.$$

(Transposition of a vector or matrix is denoted by a prime.) Again for convenience we shall assume that $\boldsymbol{\Sigma}$ is positive definite. The problem is to estimate $\boldsymbol{\beta}$ on the basis of one observation on $\mathbf{y}$ when $\mathbf{Z}$ is known.

When $\boldsymbol{\Sigma}$ is known or is known to within a constant multiple, the Markov or Best Linear Unbiased Estimate (BLUE) is given by

$$(1.3) \qquad \mathbf{b} = (\mathbf{Z}'\boldsymbol{\Sigma}^{-1}\mathbf{Z})^{-1}\mathbf{Z}'\boldsymbol{\Sigma}^{-1}\mathbf{y}.$$

The least squares estimate is given by

$$(1.4) \qquad \mathbf{b}^* = (\mathbf{Z}'\mathbf{Z})^{-1}\mathbf{Z}'\mathbf{y}.$$

The covariance matrix of the Markov estimate is

$$(1.5) \qquad \mathscr{C}(\mathbf{b}) = (\mathbf{Z}'\boldsymbol{\Sigma}^{-1}\mathbf{Z})^{-1}.$$

The covariance matrix of the least squares estimate is

$$(1.6) \qquad \mathscr{C}(\mathbf{b}^*) = (\mathbf{Z}'\mathbf{Z})^{-1}\mathbf{Z}'\boldsymbol{\Sigma}\mathbf{Z}(\mathbf{Z}'\mathbf{Z})^{-1}.$$

Both of the estimates are linear and unbiased.

Research supported by the Office of Naval Research under Contract Number N00014–67–A–00112–0030. Reproduction in whole or in part permitted for any purpose of the United States Government.

471

The optimality property of the Markov estimate implies that $\mathscr{C}(\mathbf{b}^*) - \mathscr{C}(\mathbf{b})$ is positive semidefinite; that is, any linear function of the Markov estimate has a variance no larger than the variance of that linear function of the least squares estimate. Since the least squares estimate can always be calculated, but the Markov estimate is unavailable if the covariance matrix $\boldsymbol{\Sigma}$ is not known to within a constant of proportionality, an interesting problem is to find the conditions under which the least squares estimate is identical to the Markov estimate. It will be noted that they are identical when $\boldsymbol{\Sigma}$ is a multiple of the identity, $\mathbf{I}$. The general answer is given by the following theorem.

THEOREM 1. *The least squares estimate* (1.4) *is identical to the best linear unbiased estimate* (1.3) *if and only if* $\mathbf{Z} = \mathbf{V}^* \mathbf{C}$, *where the p columns of* $\mathbf{V}^*$ *are p linearly independent characteristic vectors of* $\boldsymbol{\Sigma}$ *and* $\mathbf{C}$ *is a nonsingular matrix.*

The sufficiency of the condition was essentially given by myself in 1948 in *Skandinavisk Aktuarietidskrift* [1]. In that paper I showed that if $\mathbf{y}$ is normally distributed, then the least squares estimate is identical to the maximum likelihood estimate; under normality, of course, the maximum likelihood estimate is best linear unbiased. Watson [9] studied a measure of efficiency of estimates from which the necessity of the condition can be deduced; Magness and McGuire [7] explicitly deduced the necessity (while Zyskind [11] announced it).

A problem that is more explicitly and specially a time series problem occurs in the case where the residuals constitute a stationary stochastic process. The property $\sigma_{st} = \sigma(s - t)$, where $\boldsymbol{\Sigma} = (\sigma_{st})$, denotes stationarity in the wide sense. In general, the least squares estimate and the best linear unbiased estimate will be different. The characteristic vectors of $\boldsymbol{\Sigma}$ depend on the values of the serial or lag covariances and hence the best linear unbiased estimate depends on these parameters, which are generally unknown.

In this case we consider the covariance matrices of the estimates, normalize them suitably and identically, and consider the limits of them as $T \to \infty$. Grenander in [4], Rosenblatt in the Third Berkeley Symposium [6], and these two authors in [5] found conditions for which the two limiting covariance matrices are the same. They did not indicate that their results were asymptotic analogues of the result for a finite sample, and the statement of their results and their methods of proof do not make it easy to see the relationship.

In this paper I shall prove the results for the finite dimensional case and the limiting case in a similar fashion in order that the relationship between the results be clearer and that the asymptotic results be more easily understood. The emphasis here is on the linear algebra; the rigorous derivation of the limits, which is rather involved, is omitted (but is given in Section 10.2 of [3]).

The method of proof is not the most direct for Theorem 1, because the proof uses covariance matrices instead of the structure of the estimates themselves. On the other hand, the asymptotic results must be derived in terms of the covariance matrices because the order of the observation vector increases, and thus the structure of the estimate changes. To obtain comparable proofs, covariance matrices must be used throughout. A byproduct of my proof of the theorems is a different statement of the conditions of Grenander and Rosenblatt, which, I hope, is more enlightening than the original. Watson [9] related the two sets of results by considering the finite sample case in the framework of the approach of Grenander and Rosenblatt.

2. The finite sample case

We shall now proceed to prove Theorem 1 by considering the conditions for which the two covariance matrices, (1.5) and (1.6), are identical. To study this problem it will be convenient to transform the coordinate system in the T-dimensional space to the coordinate system defined by the characteristic vectors of the covariance matrix Σ. Let

$$(2.1) \qquad \Lambda = \begin{pmatrix} \lambda_1 & 0 & \cdots & 0 \\ 0 & \lambda_2 & \cdots & 0 \\ \vdots & \vdots & & \vdots \\ 0 & 0 & \cdots & \lambda_T \end{pmatrix},$$

where $\lambda_1 \geqq \lambda_2 \geqq \cdots \geqq \lambda_T (>0)$ are the characteristic roots of Σ. Let $\mathbf{V}$ be a $T \times T$ matrix with columns as corresponding normalized characteristic vectors. These properties can be summarized in the two matrix equations

$$(2.2) \qquad \Sigma \mathbf{V} = \mathbf{V}\Lambda,$$

$$(2.3) \qquad \mathbf{V}'\mathbf{V} = \mathbf{I},$$

which imply $\Sigma = \mathbf{V}\Lambda\mathbf{V}'$ and $\mathbf{I} = \mathbf{V}\mathbf{V}'$. We can refer the matrix of independent variables to this coordinate system. Then

$$(2.4) \qquad \mathbf{Z} = \mathbf{V}\mathbf{G},$$

where

$$(2.5) \qquad \mathbf{G}' = (\mathbf{g}_1, \cdots, \mathbf{g}_T),$$

and $\mathbf{g}_t$ is a p-component vector, $t = 1, \cdots, T$. The two covariance matrices depend on three matrices involving $\mathbf{Z}$ and Σ. These can be written in terms of Λ and $\mathbf{G}$ as

$$(2.6) \qquad \mathbf{Z}'\mathbf{Z} = \mathbf{G}'\mathbf{V}'\mathbf{V}\mathbf{G} = \mathbf{G}'\mathbf{G} = \sum_{t=1}^{T} \mathbf{g}_t\mathbf{g}_t',$$

$$(2.7) \qquad \mathbf{Z}'\Sigma\mathbf{Z} = \mathbf{G}'\mathbf{V}'\Sigma\mathbf{V}\mathbf{G} = \mathbf{G}'\Lambda\mathbf{G} = \sum_{t=1}^{T} \lambda_t\mathbf{g}_t\mathbf{g}_t',$$

$$(2.8) \qquad \mathbf{Z}'\Sigma^{-1}\mathbf{Z} = \mathbf{G}'\mathbf{V}'\Sigma^{-1}\mathbf{V}\mathbf{G} = \mathbf{G}'\Lambda^{-1}\mathbf{G} = \sum_{t=1}^{T} \frac{1}{\lambda_t} \mathbf{g}_t\mathbf{g}_t'.$$

The columns of $\mathbf{V}$ are characteristic vectors of Σ^{-1} corresponding to roots which are the reciprocals of the characteristic roots of Σ. We shall follow these matrices along.

The characteristic roots may not all be different. Let us indicate the multiplicity of the roots by writing the diagonal matrix $\mathbf{\Lambda}$ in the partitioned form

$$(2.9) \qquad \mathbf{\Lambda} = \begin{pmatrix} v_1 \mathbf{I} & \mathbf{0} & \cdots & \mathbf{0} \\ \mathbf{0} & v_2 \mathbf{I} & \cdots & \mathbf{0} \\ \vdots & \vdots & & \vdots \\ \mathbf{0} & \mathbf{0} & \cdots & v_H \mathbf{I} \end{pmatrix},$$

where $v_1 > v_2 > \cdots > v_H \; (>0)$ are the different characteristic roots. The orders of the diagonal blocks are the multiplicities of the corresponding roots, say $m_1, m_2, \cdots, m_H \; (\Sigma_{h=1}^{H} m_h = T)$. We partition $\mathbf{V}$ and $\mathbf{G}$ similarly,

$$(2.10) \qquad \mathbf{V} = (\mathbf{V}^{(1)}, \mathbf{V}^{(2)}, \cdots, \mathbf{V}^{(H)}),$$

$$(2.11) \qquad \mathbf{G} = \begin{pmatrix} \mathbf{G}^{(1)} \\ \mathbf{G}^{(2)} \\ \vdots \\ \mathbf{G}^{(H)} \end{pmatrix}.$$

Now let us go back to the matrices we considered previously, and express them in these new terms. $\mathbf{Z}$ is written as

$$(2.12) \qquad \mathbf{Z} = \sum_{h=1}^{H} \mathbf{V}^{(h)} \mathbf{G}^{(h)}.$$

The three matrices appearing in the covariance matrices are

$$(2.13) \qquad \mathbf{Z}' \mathbf{Z} = \sum_{h=1}^{H} \mathbf{G}^{(h)\prime} \mathbf{G}^{(h)},$$

$$(2.14) \qquad \mathbf{Z}' \mathbf{\Sigma} \mathbf{Z} = \sum_{h=1}^{H} v_h \mathbf{G}^{(h)\prime} \mathbf{G}^{(h)},$$

$$(2.15) \qquad \mathbf{Z}' \mathbf{\Sigma}^{-1} \mathbf{Z} = \sum_{h=1}^{H} \frac{1}{v_h} \mathbf{G}^{(h)\prime} \mathbf{G}^{(h)}.$$

The definition of a submatrix of $\mathbf{V}$ may have some indeterminacy in it. We can replace $\mathbf{V}^{(h)}$ by $\mathbf{V}^{(h)} \mathbf{Q}^{(h)}$ and replace $\mathbf{G}^{(h)}$ by $\mathbf{Q}^{(h)\prime} \mathbf{G}^{(h)}$, where $\mathbf{Q}^{(h)}$ is an orthogonal matrix of order m_h. Such a transformation leaves each of the last four equations invariant.

Theorem 1 shall be shown to be equivalent to the following theorem.

THEOREM 2. $\mathscr{C}(\mathbf{b}) = \mathscr{C}(\mathbf{b}^*)$ *if and only if*

$$(2.16) \qquad \sum_{h=1}^{H} \rho(\mathbf{G}^{(h)}) = p,$$

where $\rho(\mathbf{G}^{(h)})$ *denotes the rank of* $\mathbf{G}^{(h)}$.

In order to simplify the study of the conditions for the equality of the covariance matrices, it is convenient to transform the matrices again. Let $\mathbf{P}$

be a nonsingular matrix such that

$$(2.17) \qquad \mathbf{P}'(\mathbf{Z}'\mathbf{Z})\mathbf{P} = \mathbf{I},$$

$$(2.18) \qquad \mathbf{P}'(\mathbf{Z}'\mathbf{\Sigma}\mathbf{Z})\mathbf{P} = \mathbf{D},$$

where $\mathbf{D}$ is a diagonal matrix with $d_{11} \geqq d_{22} \geqq \cdots \geqq d_{pp} > 0$. (These are the characteristic roots of $\mathbf{Z}'\mathbf{\Sigma}\mathbf{Z}(\mathbf{Z}'\mathbf{Z})^{-1}$.) Let us also make the transformation of the other matrix, $\mathbf{P}'(\mathbf{Z}'\mathbf{\Sigma}^{-1}\mathbf{Z})\mathbf{P}$. The covariance matrix of $\mathbf{P}^{-1}\mathbf{b}$ is the inverse of this last matrix. The covariance matrix of $\mathbf{P}^{-1}\mathbf{b}^*$ is $\mathbf{D}$. (This can be seen from the original expression for the covariance matrix of $\mathbf{b}^*$, (1.6), by multiplication on the left by $\mathbf{P}^{-1}$ and on the right by $\mathbf{P}'^{-1}$ and with use of the properties of the matrices we have just discussed.) The question of equality of the original covariance matrices has now been reduced to the problem of when the covariance matrix of $\mathbf{P}^{-1}\mathbf{b}$ is $\mathbf{D}$.

The three matrices in $\mathscr{C}(\mathbf{b})$ and $\mathscr{C}(\mathbf{b}^*)$ can be written

$$(2.19) \qquad \mathbf{I} = \mathbf{P}'\mathbf{Z}'\mathbf{Z}\mathbf{P} = \sum_{h=1}^{H} \mathbf{C}^{(h)},$$

$$(2.20) \qquad \mathbf{D} = \mathbf{P}'\mathbf{Z}'\mathbf{\Sigma}\mathbf{Z}\mathbf{P} = \sum_{h=1}^{H} v_h \mathbf{C}^{(h)},$$

$$(2.21) \qquad \mathbf{P}'\mathbf{Z}'\mathbf{\Sigma}^{-1}\mathbf{Z}\mathbf{P} = \sum_{h=1}^{H} \frac{1}{v_h} \mathbf{C}^{(h)},$$

where $\mathbf{C}^{(h)} = \mathbf{P}'\mathbf{G}^{(h)'}\mathbf{G}^{(h)}\mathbf{P}$. Note that $\rho(\mathbf{C}^{(h)}) = \rho(\mathbf{G}^{(h)})$. Let us consider the diagonal elements of each of the last three equations. They are

$$(2.22) \qquad 1 = \sum_{h=1}^{H} c_{ii}^{(h)},$$

$$(2.23) \qquad d_{ii} = \sum_{h=1}^{H} v_h c_{ii}^{(h)},$$

$$(2.24) \qquad \sum_{h=1}^{H} \frac{1}{v_h} c_{ii}^{(h)}.$$

Since the matrix $\mathbf{C}^{(h)}$ is positive semidefinite, each diagonal element is non-negative. For each i the sum of these nonnegative components is 1; hence, the elements in the ith diagonal position can be considered as probabilities. Let X_i be a random variable that takes on the value v_h with probability $c_{ii}^{(h)}$, $h = 1, \cdots, H$. Then d_{ii} is the expected value of this random variable. The last expression is the expected value of the reciprocal of this positive random variable. If the two covariance matrices are to be the same, the ith diagonal element of the last matrix must be the reciprocal of that diagonal element of the second matrix. Thus, the random variable just defined can take on only one value with probability 1. (This is basically the condition for equality in the Cauchy-Schwarz inequality.) This implies that for each i, $c_{ii}^{(h)} = 1$ for one index

h and is 0 for other values of h because the v_h are distinct. These facts imply that the diagonal elements of the matrices $\mathbf{C}^{(h)}$ are 1's and 0's. The matrices $\mathbf{C}^{(h)}$ have diagonal elements as follows:

(2.25)

$$
\mathbf{C}^{(1)} = \begin{bmatrix} 1 & & & & & & \\ & \ddots & & & & & \\ & & 1 & & & & \\ & & & 0 & & & \\ & & & & \ddots & & \\ & & & & & 0 & \\ & & & & & & 0 \\ & & & & & & & \ddots \\ & & & & & & & & 0 \end{bmatrix}, \qquad \mathbf{C}^{(2)} = \begin{bmatrix} 0 & & & & & & \\ & \ddots & & & & & \\ & & 0 & & & & \\ & & & 1 & & & \\ & & & & \ddots & & \\ & & & & & 1 & \\ & & & & & & 0 \\ & & & & & & & \ddots \\ & & & & & & & & 0 \end{bmatrix}, \cdots.
$$

If a matrix $\mathbf{C}^{(h)}$ has 1 in the ith diagonal position, the other matrices have 0 in that position. (Then $d_{ii} = v_h$. Since the v_h and d_{ii} are numbered in descending order, the 1's in $\mathbf{C}^{(1)}$ are in the upper left corner, and so on.) Some matrices may only have 0's on the main diagonal. Since $\mathbf{C}^{(h)}$ is positive semidefinite, a diagonal element of 0 implies that the entire corresponding row and column are 0. Thus

(2.26)

$$
\mathbf{C}^{(1)} = \begin{bmatrix} 1 & & & \\ & \ddots & & \mathbf{0} \\ & & 1 & \\ \mathbf{0} & & & 0 \end{bmatrix}, \qquad \mathbf{C}^{(2)} = \begin{bmatrix} \mathbf{0} & & \mathbf{0} & & \mathbf{0} \\ & 1 & & & \\ \mathbf{0} & & \ddots & & \mathbf{0} \\ & & & 1 & \\ \mathbf{0} & & \mathbf{0} & & \mathbf{0} \end{bmatrix}, \cdots.
$$

Since the $\mathbf{C}^{(h)}$ sum to $\mathbf{I}$, and the nonzero blocks are not overlapping,

(2.27)

$$
\mathbf{C}^{(1)} = \begin{bmatrix} \mathbf{I} & \mathbf{0} \\ \mathbf{0} & \mathbf{0} \end{bmatrix}, \qquad \mathbf{C}^{(2)} = \begin{bmatrix} \mathbf{0} & \mathbf{0} & \mathbf{0} \\ \mathbf{0} & \mathbf{I} & \mathbf{0} \\ \mathbf{0} & \mathbf{0} & \mathbf{0} \end{bmatrix}, \cdots.
$$

We have then $\mathbf{C}^{(1)}$ with an identity in the upper left corner and so on. The rank of each $\mathbf{C}^{(h)}$ is equal to the number of diagonal elements that are 1. Thus, the sum of the ranks is equal to p. Therefore, the equality of the covariance matrices implies that the sum of the ranks is p.

The converse can be obtained by use of Cochran's theorem. (See Lemma 7.4.1 of [2], for example.) However, we shall use a simplified proof of a generalization of one part of Cochran's theorem due to Styan [8]. We assume the sum of the ranks of the $\mathbf{C}^{(h)}$'s is p. Let the nonnull $\mathbf{C}^{(h)}$'s be $\mathbf{L}_1, \cdots, L_K, K \leq p$, and let the ranks of these matrices be $r_1, \cdots, r_K$, respectively. Then $\mathbf{L}_j$ can be written $\mathbf{A}'_j \mathbf{A}_j$, where $\mathbf{A}_j$ is $r_j \times p, j = 1, \cdots, K$. Let $\mathbf{U}$ be the diagonal matrix with jth diagonal block of order r_j consisting of $u_j \mathbf{I}$, respectively, where u_j is the jth value of $v_1, \cdots, v_H$ corresponding to a nonnull $\mathbf{C}^{(h)}, j = 1, \cdots, K$. Let

$$(2.28) \qquad \mathbf{A} = \begin{pmatrix} \mathbf{A}_1 \\ \vdots \\ \mathbf{A}_K \end{pmatrix}.$$

Then (2.19) and (2.20) are

$$(2.29) \qquad \mathbf{I} = \mathbf{A}'\mathbf{A},$$

$$(2.30) \qquad \mathbf{D} = \mathbf{A}'\mathbf{U}\mathbf{A}.$$

Equation (2.29) shows that $\mathbf{A}$ is orthogonal as $\sum_{j=1}^{K} r_j = p$, and so it follows from (2.30) that

$$(2.31) \qquad \mathbf{D}^{-1} = \mathbf{A}'\mathbf{U}^{-1}\mathbf{A},$$

which is (2.21). Since

$$(2.32) \qquad \sum_{h=1}^{H} \rho(\mathbf{C}^{(h)}) = \sum_{j=1}^{K} r_j = p,$$

Theorem 2 is proved. (That equality of covariance matrices implies the rank condition can be proved by the method used in the converse, but it does not generalize directly to the case of stationary residuals.)

As was indicated earlier, $\mathbf{G}^{(h)}$ in $\mathbf{Z} = \sum_{h=1}^{H} \mathbf{V}^{(h)}\mathbf{G}^{(h)}$ can be replaced by $\mathbf{Q}^{(h)\prime}\mathbf{G}^{(h)}$ where $\mathbf{Q}^{(h)}$ is orthogonal. In particular, $\mathbf{Q}^{(h)}$ can be chosen so that $\mathbf{G}^{(h)}$ has as many nonzero rows as its rank. (For the nonnull $\mathbf{C}^{(h)}$ or $\mathbf{G}^{(h)}$, the resulting matrices are $\mathbf{A}_1, \cdots, \mathbf{A}_K$.) This proves Theorem 1 for the finite-dimensional case.

3. Large sample theory for stationary residuals

We now turn to the problem involving stationary time series. The elements of the covariance matrix of $\mathbf{y}$ are

$$(3.1) \qquad \sigma_{st} = \sigma(s - t) = \int_{-\pi}^{\pi} e^{i(s-t)\lambda} f(\lambda) \, d\lambda,$$

where $f(\lambda)$ is the spectral density, which is assumed to exist. Also we assume that the spectral density satisfies the inequalities

$$(3.2) \qquad 0 < \frac{m}{2\pi} \leqq f(\lambda) \leqq \frac{M}{2\pi},$$

when m and M are some positive constants. In developing the asymptotic theory, I shall not attempt to state all of the conditions. (They are given in Section 10.2.3 of [3].) We write

$$(3.3) \qquad f(\lambda) = \frac{1}{2\pi} \sum_{h=-\infty}^{\infty} e^{i\lambda h} \sigma(h).$$

Let the diagonal matrix $\mathbf{D}_T$ be defined by

$$(3.4) \qquad \qquad \text{diag} \left(\mathbf{Z}'\mathbf{Z}\right) = \text{diag} \left(\mathbf{D}_T^2\right),$$

where we take the positive square roots. Since we are interested in $T \to \infty$, we shall use the index T when convenient to emphasize that we have a sequence of estimates. The suitable normalization of the estimates is multiplication by this matrix $\mathbf{D}_T$. We consider the limits of the covariance matrices of $\mathbf{D}_T\mathbf{b}$ and of $\mathbf{D}_T\mathbf{b}^*$. The question is what are necessary and sufficient conditions on the independent variables and the spectral density such that

$$(3.5) \qquad \qquad \lim_{T \to \infty} \mathscr{C}(\mathbf{D}_T\mathbf{b}) = \lim_{T \to \infty} \mathscr{C}(\mathbf{D}_T\mathbf{b}^*).$$

Let

$$(3.6) \qquad \qquad \mathbf{Z}' = (\mathbf{z}_1, \cdots, \mathbf{z}_T).$$

Consider the sum on t of $\mathbf{z}_{t+h}\mathbf{z}_t'$ and multiply on each side by $\mathbf{D}_T^{-1}$ to obtain the matrix of lagged correlations of order h. Let the limit of this matrix as $T \to \infty$ be

$$(3.7) \qquad \qquad \mathbf{R}(h) = \lim_{T \to \infty} \mathbf{D}_T^{-1} \sum_t \mathbf{z}_{t+h}\mathbf{z}_t'\mathbf{D}_T^{-1}.$$

We assume that these limits exist for $t = 0, \pm 1, \pm 2, \cdots$. Then this sequence of matrices has the spectral representation

$$(3.8) \qquad \qquad \mathbf{R}(h) = \int_{-\pi}^{\pi} e^{i\lambda h} \, d\mathbf{M}(\lambda),$$

where $\mathbf{M}(\lambda)$ has complex-valued elements, is Hermitian, and has increments that are positive semidefinite.

We shall now consider the limits of the covariance matrices of the normalized estimates. Those covariance matrices involve the limits of the matrices $\mathbf{D}_T^{-1}\mathbf{Z}'\mathbf{Z}\mathbf{D}_T^{-1}$, $\mathbf{D}_T^{-1}\mathbf{Z}'\boldsymbol{\Sigma}\mathbf{Z}\mathbf{D}_T^{-1}$, and $\mathbf{D}_T^{-1}\mathbf{Z}\boldsymbol{\Sigma}^{-1}\mathbf{Z}\mathbf{D}_T^{-1}$. In fact,

$$(3.9) \qquad \qquad \lim_{T \to \infty} \mathscr{C}(\mathbf{D}_T\mathbf{b}) = \lim_{T \to \infty} (\mathbf{D}_T^{-1}\mathbf{Z}'\boldsymbol{\Sigma}^{-1}\mathbf{Z}\mathbf{D}_T^{-1})^{-1},$$

$$(3.10) \qquad \lim_{T \to \infty} \mathscr{C}(\mathbf{D}_T\mathbf{b}^*) = \mathbf{R}^{-1}(0) \lim_{T \to \infty} \mathbf{D}_T^{-1}\mathbf{Z}'\boldsymbol{\Sigma}\mathbf{Z}\mathbf{D}_T^{-1}\mathbf{R}^{-1}(0).$$

The second matrix is

$$(3.11) \qquad \lim_{T \to \infty} \mathbf{D}_T^{-1} \sum_{h=-(T-1)}^{T-1} \sum_t \mathbf{z}_{t+h}\mathbf{z}_t'\sigma(h)\mathbf{D}_T^{-1} = \sum_{h=-\infty}^{\infty} \mathbf{R}(h)\sigma(h)$$

$$= \int_{-\pi}^{\pi} \sum_{h=-\infty}^{\infty} \sigma(h)e^{i\lambda h} \, d\mathbf{M}(\lambda)$$

$$= \int_{-\pi}^{\pi} 2\pi f(\lambda) \, d\mathbf{M}(\lambda).$$

Of course, these operations need to be justified to give a rigorous proof, but that requires considerable detail. The full proof is given in Section 10.2.3 of my book [3] and is along the lines indicated by Grenander and Rosenblatt [5]. The three matrices we are interested in can be written

$$(3.12) \qquad \lim_{T \to \infty} \mathbf{D}_T^{-1} \mathbf{Z}' \mathbf{Z} \mathbf{D}_T^{-1} = \int_{-\pi}^{\pi} d\mathbf{M}(\lambda),$$

$$(3.13) \qquad \lim_{T \to \infty} \mathbf{D}_T^{-1} \mathbf{Z}' \boldsymbol{\Sigma} \mathbf{Z} \mathbf{D}_T^{-1} = \int_{-\pi}^{\pi} 2\pi f(\lambda) \, d\mathbf{M}(\lambda),$$

$$(3.14) \qquad \lim_{T \to \infty} \mathbf{D}_T^{-1} \mathbf{Z}' \boldsymbol{\Sigma}^{-1} \mathbf{Z} \mathbf{D}_T^{-1} = \int_{-\pi}^{\pi} \frac{1}{2\pi f(\lambda)} \, d\mathbf{M}(\lambda).$$

The derivation for the third matrix is an involved demonstration also given in [3]. These three expressions are the analogues of (2.6), (2.7), and (2.8) in the finite-dimensional case. Carrying the analogy to the finite-dimensional case further, we shall write these integrals in another manner to resemble (2.13), (2.14), and (2.15). Let

$$(3.15) \qquad S(u) = \{\lambda \,|\, 2\pi f(\lambda) \leqq u\}, \qquad\qquad m \leqq u \leqq M,$$

$$(3.16) \qquad \mathbf{T}(u) = \int_{S(u)} d\mathbf{M}(\lambda).$$

The component functions of $\mathbf{T}(u)$ are real. Then our three matrices can be written as

$$(3.17) \qquad \lim_{T \to \infty} \mathbf{D}_T^{-1} \mathbf{Z}' \mathbf{Z} \mathbf{D}_T^{-1} = \int_m^M d\mathbf{T}(u),$$

$$(3.18) \qquad \lim_{T \to \infty} \mathbf{D}_T^{-1} \mathbf{Z}' \boldsymbol{\Sigma} \mathbf{Z} \mathbf{D}_T^{-1} = \int_m^M u \, d\mathbf{T}(u),$$

$$(3.19) \qquad \lim_{T \to \infty} \mathbf{D}_T^{-1} \mathbf{Z} \boldsymbol{\Sigma}^{-1} \mathbf{Z} \mathbf{D}_T^{-1} = \int_m^M \frac{1}{u} \, d\mathbf{T}(u).$$

Similar to the finite-dimensional case we let $\mathbf{P}$ be a nonsingular matrix such that

$$(3.20) \qquad \mathbf{P}' \int_m^M d\mathbf{T}(u) \, \mathbf{P} = \mathbf{I},$$

$$(3.21) \qquad \mathbf{P}' \int_m^M u \, d\mathbf{T}(u) \, \mathbf{P} = \mathbf{D}.$$

where $\mathbf{D}$ is diagonal and $d_{11} \geqq d_{22} \geqq \cdots \geqq d_{pp} > 0$. The same transformation is applied to the third matrix, $\int_m^M u^{-1} \, d\mathbf{M}(u)$, which is the inverse of $\lim_{T \to \infty} \mathscr{C}(\mathbf{D}_T \mathbf{b})$. The other limiting covariance matrix is $\lim_{T \to \infty} \mathscr{C}(\mathbf{D}_T \mathbf{b}^*) = \mathbf{D}$.

If we let

$$(3.22) \qquad \mathbf{L}(u) = \mathbf{P}' \mathbf{T}(u) \mathbf{P},$$

then the three matrices of interest are

$$(3.23) \qquad \mathbf{I} = \lim_{T \to \infty} \mathbf{P}' \mathbf{D}_T^{-1} \mathbf{Z}' \mathbf{Z} \mathbf{D}_T^{-1} \mathbf{P} = \int_m^M d\mathbf{L}(u),$$

$$(3.24) \qquad \mathbf{D} = \lim_{T \to \infty} \mathbf{P}' \mathbf{D}_T^{-1} \mathbf{Z}' \boldsymbol{\Sigma} \mathbf{Z} \mathbf{D}_T^{-1} \mathbf{P} = \int_m^M u \, d\mathbf{L}(u),$$

$$(3.25) \qquad \lim_{T \to \infty} \mathbf{P}' \mathbf{D}_T^{-1} \mathbf{Z}' \boldsymbol{\Sigma}^{-1} \mathbf{Z} \mathbf{D}_T^{-1} \mathbf{P} = \int_m^M \frac{1}{u} \, d\mathbf{L}(u).$$

A diagonal element of $\mathbf{L}(u)$, say $l_{ii}(u)$, has the properties of a cumulative distribution function. The corresponding diagonal elements of (3.24) and (3.25) are $\int_m^M u \, dl_{ii}(u)$ and $\int_m^M u^{-1} \, dl_{ii}(u)$, which are the expected values of the random variable with this distribution and its reciprocal. Thus, if the two limiting covariance matrices are equal, the matrix (3.25) is the inverse of (3.24) and

$$(3.26) \qquad \int_m^M u \, dl_{ii}(u) = \left[\int_m^M \frac{1}{u} dl_{ii}(u) \right]^{-1};$$

this implies that $l_{ii}(u)$ has one point of increase and the increase is 1 at this point. Let the points of increase be $u_1 \geqq u_2 \cdots \geqq u_K \geqq 0$, and let $\mathbf{L}_j$ be the increase of $\mathbf{L}(u)$ at u_j, $j = 1, \cdots, K$. Then the three matrices can be written

$$(3.27) \qquad \mathbf{I} = \lim_{T \to \infty} \mathbf{P}' \mathbf{D}_T^{-1} \mathbf{Z}' \mathbf{Z} \mathbf{D}_T^{-1} \mathbf{P} = \sum_{j=1}^K \mathbf{L}_j,$$

$$(3.28) \qquad \mathbf{D} = \lim_{T \to \infty} \mathbf{P}' \mathbf{D}_T^{-1} \mathbf{Z}' \boldsymbol{\Sigma} \mathbf{Z} \mathbf{D}_T^{-1} \mathbf{P} = \sum_{j=1}^K u_j \mathbf{L}_j,$$

$$(3.29) \qquad \lim_{T \to \infty} \mathbf{P}' \mathbf{D}_T^{-1} \mathbf{Z}' \boldsymbol{\Sigma}^{-1} \mathbf{Z} \mathbf{D}_T^{-1} \mathbf{P} = \sum_{j=1}^K \frac{1}{u_j} \mathbf{L}_j.$$

We are now back to the same forms that we had for the finite-dimensional case, (2.19), (2.20), (2.21). The only difference is that in the earlier case we had not culled out the vacuous matrices $\mathbf{C}^{(h)}$. From this point on the reasoning is the same. The matrices $\mathbf{L}_1, \mathbf{L}_2, \cdots, \mathbf{L}_K$ have the form of (2.27); that is, the diagonal blocks are $\mathbf{I}$'s and $\mathbf{0}$'s and off diagonal blocks are $\mathbf{0}$'s.

The converse is similar to the finite-dimensional case. If $\mathbf{L}(u)$ has K points of increase and the sum of the ranks of the increases is p (and the increases are positive semidefinite with sum of $\mathbf{I}$ and weighted sum of $\mathbf{D}$), then by the previous reasoning, they are of the form (2.27) and (3.29) is $\mathbf{D}^{-1}$. We put these properties in terms of $\mathbf{M}(\lambda)$ and summarize them in a theorem.

THEOREM 3. *The limiting covariances of $\mathbf{D}_T \mathbf{b}$ and $\mathbf{D}_T \mathbf{b}^*$ are identical if and only if $f(\lambda)$ takes on no more than p values on the set of λ for which $\mathbf{M}(\lambda)$ increases and the sum of the ranks of $\int d\mathbf{M}(\lambda)$ over the sets of λ for which $f(\lambda)$ takes on these values is p.*

The set of λ for which $\mathbf{M}(\lambda)$ increases is called the *spectrum* of $\mathbf{M}(\lambda)$. The sets of λ for which $f(\lambda)$ assumes its values are called *the elements of the spectrum*. The properties of $\mathbf{L}_1, \cdots, \mathbf{L}_K$ (idempotent and orthogonal) determine these sets; Grenander and Rosenblatt used them, though indirectly.

When the residuals are uncorrelated, $f(\lambda) = \sigma(0)/(2\pi)$ and the conditions of Theorem 3 are satisfied. However, we may be interested in conditions on the independent variables also which insure that least squares be asymptotically efficient regardless of $f(\lambda)$.

THEOREM 4. *The limiting covariances of $\mathbf{D}_T\mathbf{b}$ and $\mathbf{D}_T\mathbf{b}^*$ are identical for all stationary processes with spectral densities which are bounded and bounded away from 0 if and only if $\mathbf{M}(\lambda)$ increases at not more than p values of λ, $0 \leq \lambda \leq \pi$, and the sum of the ranks of the increase in $\mathbf{M}(\lambda)$ is p.*

If the number of points at which $\mathbf{M}(\lambda)$ increases is at most p, $0 \leq \lambda \leq \pi$, then the spectrum of $\mathbf{M}(\lambda)$ consists of these p points and their corresponding negative values. The spectral density (which is symmetric) can then take on at most p values, namely, its values at these p points, $0 \leq \lambda \leq \pi$. On the other hand if $\mathbf{M}(\lambda)$ increases at more than p points, $0 \leq \lambda \leq \pi$ then an $f(\lambda)$ can be constructed so that it takes on no more than p values.

An example of independent variables $\{z_{jt}\}$ such that $\mathbf{M}(\lambda)$ has one point of increase is $z_{jt} = t^{j-1}, j = 1, \cdots, p, t = 1, 2, \cdots$; the jump is at 0 and the increase in $\mathbf{M}(\lambda)$ at $\lambda = 0$ is a positive definite matrix

$$(3.30) \qquad \mathbf{M}_0 = \left(\frac{(2j-1)^{1/2} \, (2k-1)^{1/2}}{j+k-1} \right).$$

In this case $\mathbf{R}(h) = \mathbf{M}_0$, $h = 0, \pm 1, \cdots$. If

$$(3.31) \qquad \mathbf{z}_t = \boldsymbol{\alpha}_0 + \sum_{j=1}^{H} (\boldsymbol{\alpha}_j \cos v_j t + \boldsymbol{\beta}_j \sin v_j t),$$

then $\mathbf{M}(\lambda)$ has an increase of rank 1 at $\lambda = 0$ and an increase of rank 2 at $\lambda = v_j$ (with $0 < v_j < \pi$), $j = 1, \cdots, H$. In these examples the spectral distribution function of each independent variable is a pure jump function, which can be considered as the opposite of a density. Trigonometric functions act like characteristic vectors of a covariance matrix in the sense that they are involved in spectral representation. Comparison of $\boldsymbol{\Sigma} = \mathbf{V}\boldsymbol{\Lambda}\mathbf{V}'$ and (3.1) suggests that columns of $\mathbf{V}$ correspond to functions $e^{i\lambda s}$, the diagonal components of $\boldsymbol{\Lambda}$ correspond to the values of $2\pi f(\lambda)$, and summation with respect to the index of diagonal components of $\boldsymbol{\Lambda}$ corresponds to integration with respect to $\lambda/(2\pi)$. The analogue of $\mathbf{V}'\boldsymbol{\Sigma}\mathbf{V} = \boldsymbol{\Lambda}$ is (3.3), which involves a limiting procedure.

The author is indebted to George Styan for helpful discussions.

REFERENCES

[1] T. W. ANDERSON, "On the theory of testing serial correlation," *Skand. Aktuarietidskr.*, Vol. 31 (1948), pp. 88–116.

[2] ———, *An Introduction to Multivariate Statistical Analysis*, New York, Wiley, 1958.

[3] ———, *The Statistical Analysis of Time Series*, New York, Wiley, 1971.

[4] U. GRENANDER, "On the estimation of regression coefficients in the case of an autocorrelated disturbance," *Ann. Math. Statist.*, Vol. 25 (1954), pp. 252–272.

[5] U. GRENANDER and M. ROSENBLATT, *Statistical Analysis of Stationary Time Series*, New York, Wiley, 1957.

[6] M. ROSENBLATT, "Some regression problems in time series analysis," *Proceedings of the Third Berkeley Symposium on Mathematical Statistics and Probability*, Berkeley and Los Angeles, University of California Press, 1956, Vol. 1, pp. 165–186.

[7] T. A. MAGNESS and J. B. McGUIRE, "Comparison of least squares and minimum variance estimates of regression parameters," *Ann. Math. Statist.*, Vol. 33 (1962), pp. 462–470.

[8] G. P. H. STYAN, "Notes on the distribution of quadratic forms in singular normal variables," *Biometrika*, Vol. 57 (1970), pp. 567–572.

[9] G. S. WATSON, "Serial correlation in regression analysis, I," Mimeo Series No. 49, Institute of Statistics, University of North Carolina, Chapel Hill.

[10] ———, "Linear least squares regression," *Ann. Math. Statist.*, Vol. 38 (1967), pp. 1679–1699.

[11] G. ZYSKIND, "On conditions of equality of best and simple linear least squares estimation" (abstract), *Ann. Math. Statist.*, Vol. 33 (1962), pp. 1502–1503.

THE SPECTRAL ANALYSIS OF STATIONARY INTERVAL FUNCTIONS

DAVID R. BRILLINGER

UNIVERSITY OF CALIFORNIA, BERKELEY

1. Introduction and summary

We consider stationary, additive, interval functions $\mathbf{X}(\Delta)$. These are vector valued stochastic processes having real intervals $\Delta = (\alpha, \beta]$ as domain, having finite dimensional distributions invariant under time translation and satisfying

$$(1.1) \qquad \mathbf{X}(\Delta_1 \cup \Delta_2) = \mathbf{X}(\Delta_1) + \mathbf{X}(\Delta_2),$$

for disjoint intervals Δ_1, Δ_2. Such processes are considered in some detail in Bochner [5]. Setting

$$(1.2) \qquad \mathbf{X}(t) = \mathbf{X}(0, t],$$

$-\infty < t < \infty$, and in the reverse direction setting

$$(1.3) \qquad \mathbf{X}(\alpha, \beta] = \mathbf{X}(\beta) - \mathbf{X}(\alpha),$$

we see that a consideration of stationary interval functions is equivalent with a consideration of processes $\mathbf{X}(t)$, $-\infty < t < \infty$, having stationary increments. These last are discussed in Yaglom [24] for example. Important examples of processes of the type under consideration are provided by the point processes. Here the components of $\mathbf{X}(\Delta)$ give the number of events of various sorts that occur in the interval Δ. A variety of properties and applications of point processes may be found in Cox and Lewis [11], Bartlett [4], and Srinivasan [21].

The paper is divided into various sections. In Section 2 we introduce a key assumption for the processes; specifically we require that all the moments of $\mathbf{X}(\Delta)$ exist and have particular integral representations. We are then able to define

$$(1.4) \qquad f_{a_1, \cdots, a_k}(\lambda_1, \cdots, \lambda_k),$$

$-\infty < \lambda_j < \infty, a_1, \cdots, a_k = 1, \cdots, r$, the cumulant spectra of order k of the r vector valued $\mathbf{X}(\Delta)$. These turn out to be generalizations of the cumulant spectra of order k of a continuous time series discussed in Brillinger and Rosenblatt [9]. We then present a spectral representation for $\mathbf{X}(\Delta)$. This representation was introduced in Kolmogorov [17] for real valued processes with stationary increments. It takes the form

$$(1.5) \qquad \mathbf{X}(0, t] = \int_{-\infty}^{\infty} \left[\frac{\exp\{i\lambda t\} - 1}{i\lambda} \right] d\mathbf{Z}_X(\lambda),$$

with $\mathbf{Z}_X(\lambda)$, $-\infty < \lambda < \infty$, a vector valued process. The process $\mathbf{Z}_X(\lambda)$ relates to the cumulant spectrum (1.4) through the expression

$$(1.6) \qquad \text{cum} \{dZ_{a_1}(\lambda_1), \cdots, dZ_{a_k}(\lambda_k)\}$$

$$= \delta\left(\sum_1^k \lambda_j\right) f_{a_1, \cdots, a_k}(\lambda_1, \cdots, \lambda_k) \, d\lambda_1, \cdots, d\lambda_k$$

with $\delta(\lambda)$ denoting the Dirac delta function.

In Section 3 of the paper we indicate how the theory developed applies to integrated continuous time series, to point processes, and to processes that are hybrids of these last two. In the case of point processes we relate the cumulant spectra to important parameters that have been introduced by Bahba [1] and by Kuznetsov and Stratonovich [18].

Section 4 of the paper discusses various asymptotic properties of the statistic

$$(1.7) \qquad \mathbf{d}_X^{(T)}(\lambda) = \int_0^T \exp\{-i\lambda t\} \, d\mathbf{X}(t)$$

based on an observed stretch of an $\mathbf{X}(\Delta)$ process. It will be seen to behave in a similar manner to the finite Fourier transform of a stretch of a continuous stationary series. It follows that the estimates of the various cumulant spectra of $\mathbf{X}(\Delta)$ may be formed in the manner of Brillinger and Rosenblatt [9] and that the properties developed in that paper, such as asymptotic normality, will continue to hold. A selection of results that therefore become available is provided. In particular results relating to the linear time invariant regression of one stationary interval function on another are given. Because point processes are particular cases of the processes under consideration it follows that an asymptotic theory for the spectral estimates of order two of point processes has now been provided.

In Section 5 we apply the previously mentioned asymptotic results to develop estimates of the parameters suggested by Bahba and by Kuznetsov and Stratonovich for point processes. In Section 6 we consider the problem of the estimation of the second order spectra of a continuous time series when its values are available only for random times that are the occasions of events of an independent point process.

Section 7 discusses briefly some practical implications and extensions of the results of the paper. The proofs of the various lemmas and theorems of the paper are given in Section 8.

I would like to thank P. A. W. Lewis for a variety of helpful comments on the point process sections of the paper.

2. Random interval functions

Let Δ denote the collection of finite intervals of the form $\Delta = (\alpha, \beta]$. We consider r vector valued stochastic processes $\mathbf{X}(\Delta)$, $\Delta \in \Delta$ with the additivity property

(2.1) $$\mathbf{X}(\Delta) = \mathbf{X}(\Delta_1) + \mathbf{X}(\Delta_2),$$

for $\Delta, \Delta_1, \Delta_2 \in \Delta$ with $\Delta = \Delta_1 \cup \Delta_2, \Delta_1 \cap \Delta_2 = \varnothing$. Such a process will be called an *r vector valued additive stochastic interval function*. As one example we mention

(2.2) $$\mathbf{X}(\Delta) = \int_{\Delta} \mathbf{Y}(t)\, dt,$$

where $\mathbf{Y}(t)$, $-\infty < t < \infty$, is a continuous r vector valued time series. As a second example we consider $\mathbf{X}(\Delta) = \mathbf{N}(\Delta)$ where $\mathbf{N}(\Delta)$ is an r vector valued point process with $N_a(\Delta)$ giving the number of events of the ath component of the process that occur in the interval Δ. As a final example we mention

(2.3) $$\mathbf{X}(\Delta) = \int_{\Delta} \mathbf{Y}(t)\, dN(t),$$

where $\mathbf{Y}(t)$, $-\infty < t < \infty$, is an r vector valued continuous time series and $N(\Delta)$, $\Delta \in \Delta$, is a point process. If $\tau_1, \cdots, \tau_n$ denote the times of events of the process $N(\Delta)$ in the interval Δ, then $\mathbf{X}(\Delta)$ equals

(2.4) $$\mathbf{Y}(\tau_1) + \cdots + \mathbf{Y}(\tau_n),$$

in this case.

In connection with the process $\mathbf{X}(\Delta)$, $\Delta \in \Delta$, we set down

ASSUMPTION 2.1.   *The process* $\mathbf{X}(\Delta)$, $\Delta \in \Delta$, *is an r vector valued stochastic interval function possessing moments of all orders such that for* $\Delta_1, \cdots, \Delta_k \in \Delta$; $a_1, \cdots, a_k = 1, \cdots, r$; $k = 1, 2, \cdots$,

(2.5) $$E\{X_{a_1}(\Delta_1) \cdots X_{a_k}(\Delta_k)\} = \int_{\Delta_1} \cdots \int_{\Delta_k} dM_{a_1, \cdots, a_k}(t_1, \cdots, t_k)$$

for some function $M_{a_1, \cdots, a_k}(t_1, \cdots, t_k)$, $-\infty < t_j < \infty$, *of bounded variation in finite intervals.*

In the case that $\mathbf{X}(\Delta)$ satisfies this assumption and $\phi_a(t)$ is bounded and continuous for t in some interval of Δ and 0 outside the interval, we may define stochastic integrals of the form

(2.6) $$\int \phi_a(t)\, dX_a(t)$$

as the limit in mean (of any order $\nu > 0$) of approximating Riemann sums

(2.7) $$\sum_{j=1}^{n} \phi_a(t_j)\, X_a(\Delta_j),$$

where $t_j \in \Delta_j$ and $\Delta_1 \cup \cdots \cup \Delta_n$ is a partition of the support of $\phi_a(t)$, $a = 1, \cdots, r$. (See Cramér and Leadbetter [12], p. 86 for the case $\nu = 2$.) These integrals have the property

(2.8) $E\left\{\int \phi_{a_1}(t_1)\, dX_{a_1}(t_1) \cdots \int \phi_{a_k}(t_k)\, dX_{a_k}(t_k)\right\}$

$$= \int \cdots \int \phi_{\sigma}(t_1) \cdots \phi_{a_k}(t_k)\, dM_{a1, \cdots, a_k}(t_1, \cdots, t_k),$$

for $a_1, \cdots, a_k = 1, \cdots, r$.

For $\Delta = (\alpha, \beta]$ in Δ we denote the translated interval $(\alpha + t, \beta + t]$ by $\Delta + t$ for $-\infty < t < \infty$. We will now say that an r vector valued additive stochastic interval function is *stationary* if the joint distributions of all finite collections of variates

(2.9) $\mathbf{X}(\Delta_1 + t), \cdots, \mathbf{X}(\Delta_k + t),$

$\Delta_1, \cdots, \Delta_k \in \Delta$, $-\infty < t < \infty$, and $k = 1, 2, \cdots$, do not depend on t. In this connection we have

LEMMA 2.1.   *If* $\mathbf{X}(\Delta)$, $\Delta \in \Delta$, *is a stationary r vector valued additive interval function satisfying Assumption* 2.1, *then for* $a_1, \cdots, a_k = 1, \cdots, r; k = 1, 2, \cdots,$

(2.10) $E\{X_{a_1}(\Delta_1) \cdots X_{a_k}(\Delta_k)\}$

$$= \int_{\Delta_1} \cdots \int_{\Delta_k} dM'_{a_1, \cdots, a_k}(t_1 - t_k, \cdots, t_{k-1} - t_k)\, dt_k,$$

for some function $M'_{a_1, \cdots, a_k}(u_1, \cdots, u_{k-1})$, $-\infty < u_j < \infty$, *of bounded variation in finite intervals.*

In the case $k = 1$, the lemma indicates that

(2.11) $EX_a(\Delta) = C'_a|\Delta|,$

for some constant C'_a, $a = 1, \cdots, r$ with $|\Delta|$ denoting the length of the interval Δ.

It follows from this lemma that one can write

(2.12)   cum $\{X_{a_1}(\Delta_1), \cdots, X_{a_k}(\Delta_k)\}$

$$= \int_{\Delta_1} \cdots \int_{\Delta_k} dC'_{a_1, \cdots, a_k}(t_1 - t_k, \cdots, t_{k-1} - t_k)\, dt_k,$$

for a function $C'_{a_1, \cdots, a_k}(u_1, \cdots, u_{k-1})$, of bounded variation in finite intervals. In differential notation we may write this last as

(2.13)   cum $\{dX_{a_1}(u_1 + t), \cdots, dX_{a_{k-1}}(u_{k-1} + t), dX_{a_k}(t)\}$

$$= dC'_{a_1, \cdots, a_k}(u_1, \cdots, u_{k-1})\, dt.$$

Taking note of Lemma 2.1 and (2.12) we set down the key assumption of our work. It is

ASSUMPTION 2.2.   *The process* $\mathbf{X}(\Delta)$, $\Delta \in \Delta$, *is a stationary r vector valued additive interval function satisfying Assumption* 2.1 *and such that* $C'_{a_1, \cdots, a_k}(u_1, \cdots, u_{k-1})$ *of* (2.12) *satisfies*

(2.14) $\int_{-\infty}^{\infty} \cdots \int |u_j|\, d|C'_{a_1, \cdots, a_k}(u_1, \cdots, u_{k-1})| < \infty,$

for $j = 1, \cdots, k;$ $a_1, \cdots, a_k = 1, \cdots, r;$ $k = 2, 3, \cdots.$

This assumption has the nature of a mixing condition on the increments of $\mathbf{X}(t)$, that is, increments that are well separated in time are only weakly dependent.

In view of condition (2.13) we can define the Fourier transforms

$$(2.15) \qquad f_{a_1, \cdots, a_k}(\lambda_1, \cdots, \lambda_k)$$

$$= (2\pi)^{-k+1} \int_{-\infty}^{\infty} \cdots \int \exp\left\{ -i \sum_{1}^{k-1} \lambda_j u_j \right\} dC'_{a_1, \cdots, a_k}(u_1, \cdots, u_{k-1}),$$

for $-\infty < \lambda_1, \cdots, \lambda_k < \infty$, where we understand $\lambda_1 + \cdots + \lambda_k = 0$ in the definition. For completeness we set

$$(2.16) \qquad f_a(\lambda) = C'_a,$$

where C'_a was defined in (2.11), $a = 1, \cdots, r$. The transform $f_{a_1, \cdots, a_k}(\lambda_1, \cdots, \lambda_k)$ is called a *cumulant spectrum of order k* of the process $\mathbf{X}(\Delta)$, $\Delta \in \Delta$. We will sometimes find it convenient to adopt the unsymmetric notation

$$(2.17) \qquad f'_{a_1, \cdots, a_k}(\lambda_1, \cdots, \lambda_{k-1}) = f_{a_1, \cdots, a_k}(\lambda_1, \cdots, \lambda_k).$$

The second order cumulant spectra, $f'_{a,b}(\lambda)$, $-\infty < \lambda < \infty$, are of particular importance. It is convenient to collect them together into the $r \times r$ *spectral density matrix*

$$(2.18) \qquad \mathbf{f}'_{X,X}(\lambda) = [f'_{a,b}(\lambda)].$$

We also collect the first order spectra together into the r vector

$$(2.19) \qquad \mathbf{f}'_X = [f'_a].$$

There is an intimate connection between stationary interval functions and stationary series. Suppose that, $\mathbf{X}(\Delta)$, $\Delta \in \Delta$, satisfies Assumption 2.2. and has cumulant spectra

$$(2.20) \qquad f_{a_1, \cdots, a_k}(\lambda_1, \cdots, \lambda_k).$$

Suppose the real valued, $\phi_a(t)$, $-\infty < t < \infty$, satisfies

$$(2.21) \qquad \int |t| \|\phi_a(t)\| \, dt < \infty,$$

for $a = 1, \cdots, r$. Then the r vector valued times series, $\mathbf{Y}(t)$, $-\infty < t < \infty$, with components

$$(2.22) \qquad Y_a(t) = \int \phi_a(t-u) \, dX_a(u),$$

$a = 1, \cdots, r$ may be seen to be stationary and such that

$$(2.23) \qquad \mathrm{cum}\, \{Y_{a_1}(t+t_1), \cdots, Y_{a_{k-1}}(t+t_{k-1}), Y_{a_k}(t)\}$$

$$= \int \cdots \int \phi_{a_1}(t_1 - u_1) \cdots \phi_{a_{k-1}}(t_{k-1} - u_{k-1}) \, \phi_{a_k}(u_k)$$

$$dC'_{a_1, \cdots, a_k}(u_1, \cdots, u_{k-1}) \, du_k.$$

Taking a Fourier transform, we see that the cumulant spectra of $\mathbf{Y}(t)$, in the sense of Brillinger and Rosenblatt [9], are given by

$$(2.24) \qquad \Phi_{a_1}(\lambda_1) \cdots \Phi_{a_k}(\lambda_k) f_{a_1, \cdots, a_k}(\lambda_1, \cdots, \lambda_k),$$

where

$$(2.25) \qquad \Phi_a(\lambda) = \int \exp\{-i\lambda t\}\, \phi_a(t)\, dt,$$

for $a = 1, \cdots, r$.

A variety of authors (including Kolmogorov [17], Doob [14] p. 551, Ito [16], Yaglom [24], [25] p. 86, Bochner [5] p. 159) have given spectral representations for stationary interval functions (or processes with stationary increments). In this connection we mention

THEOREM 2.1. *Let the process* $X(\Delta)$, $\Delta \in \Delta$, *satisfy Assumption 2.2. Let*

$$(2.26) \qquad \mathbf{Z}_X^{(T)}(\lambda) = (2\pi)^{-1} \int_{-T}^{T} \left[\frac{1 - \exp\{-i\lambda t\}}{-it} \right] d\mathbf{X}(t),$$

for $-\infty < \lambda < \infty$. *Then there exists,* $\mathbf{Z}_X(\lambda)$, $-\infty < \lambda < \infty$, *such that* $\mathbf{Z}_X^{(T)}(\lambda)$ *tends to* $\mathbf{Z}_X(\lambda)$ *in mean order* v, *for any* $v > 0$. $\mathbf{Z}_X(\lambda)$ *satisfies*

$$(2.27) \qquad \text{cum}\{Z_{a_1}(\lambda_1), \cdots, Z_{a_k}(\lambda_k)\}$$

$$= \int_0^{\lambda_1} \cdots \int_0^{\lambda_k} \delta\left(\sum_1^k \alpha_j\right) f_{a_1, \cdots, a_k}(\alpha_1, \cdots, \alpha_k)\, d\alpha_1 \cdots d\alpha_k,$$

for $a_1, \cdots, a_k = 1, \cdots, r$; $k = 1, 2, \cdots$. *Also*

$$(2.28) \qquad \mathbf{X}(\Delta) = \int_{-\infty}^{\infty} \left[\int_\Delta \exp\{i\lambda t\}\, dt \right] d\mathbf{Z}_X(\lambda),$$

with probability one.

In differential notation particular cases of (2.27) include:

$$(2.29) \qquad E\, d\mathbf{Z}_X(\lambda) = \delta(\lambda)\mathbf{f}_X'\, d\lambda;$$

$$(2.30) \qquad \text{Cov}\{d\,\mathbf{Z}_X(\lambda), d\,\mathbf{Z}_X(\mu)\} = \delta(\lambda - \mu)\mathbf{f}_{X,X}'(\lambda)\, d\lambda\, d\mu;$$

$$(2.31) \qquad \text{cum}\{dZ_{a_1}(\lambda_1), \cdots, dZ_{a_k}(\lambda_k)\}$$
$$= \delta(\lambda_1 + \cdots + \lambda_k) f_{a_1, \cdots, a_k}'(\lambda_1, \cdots, \lambda_{k-1})\, d\lambda_1 \cdots d\lambda_k.$$

Also if we set $\mathbf{X}(t) = \mathbf{X}(0, t]$, then (2.28) takes the form

$$(2.32) \qquad \mathbf{X}(t) = \int \left[\frac{\exp\{i\lambda t\} - 1}{i\lambda} \right] d\mathbf{Z}_X(\lambda),$$

for $-\infty < t < \infty$.

The representation (2.28) is useful for displaying the effect of linear time invariant operations on the process, $\mathbf{X}(\Delta)$, $\Delta \in \Delta$. Suppose $\mathbf{a}(\Delta)$, $\Delta \in \Delta$, is an $s \times r$ matrix valued interval function of bounded variation satisfying

$$(2.33) \qquad \int |t|\, d|\mathbf{a}(t)| \,<\, \infty.$$

Set

$$(2.34) \qquad \mathbf{A}(\lambda) \,=\, \int \exp\left\{-i\lambda t\right\} d\mathbf{a}(t),$$

for $-\infty < \lambda < \infty$. The s vector valued interval function

$$(2.35) \qquad \mathbf{Y}(\Delta) \,=\, \mathbf{a} * \mathbf{x}(\Delta)$$
$$\,=\, \int \mathbf{a}(\Delta - u)\, d\mathbf{X}(u),$$

$\Delta \in \Delta$, may be seen to satisfy Assumption 2.2. Also the process $\mathbf{Z}_Y(\lambda)$, $-\infty < \lambda < \infty$, of its spectral representation may be seen to satisfy,

$$(2.36) \qquad d\mathbf{Z}_Y(\lambda) \,=\, \mathbf{A}(\lambda)\, d\mathbf{Z}_X(\lambda).$$

We may infer from this last that the spectral density matrices of $\mathbf{X}(\Delta)$ and $\mathbf{Y}(\Delta)$ are related by

$$(2.37) \qquad \mathbf{f}'_{Y,Y}(\lambda) \,=\, \mathbf{A}(\lambda)\mathbf{f}'_{X,X}(\lambda)\overline{\mathbf{A}(\lambda)}^{\tau}.$$

This last relation has the identical form with that giving the effect of linear time invariant operations on the spectral density matrices of time series.

We conclude this section by remarking that the function $M'_{a_1,\cdots,a_k}(u_1,\cdots,u_{k-1})$ of (2.10) may be determined as

$$(2.38) \qquad M'_{a_1,\cdots,a_k}(u_1,\cdots,u_{k-1})$$
$$= E\left\{T^{-1}\int_0^T X_{a_1}(t,\,t\,+\,u_1]\cdots X_{a_{k-1}}(t,\,t\,+\,u_{k-1}]\,dX_{a_k}(t)\right\}.$$

3. Some examples

EXAMPLE 3.1. Suppose that $\mathbf{Y}(t)$, $-\infty < t < \infty$, is an r vector valued stationary time series possessing moments of all orders. If

$$(3.1) \quad c'_{a_1,\cdots,a_k}(u_1,\cdots,u_{k-1}) \,=\, \operatorname{cum}\left\{Y_{a_1}(t+u_1),\cdots,Y_{a_{k-1}}(t+u_{k-1}),Y_{a_k}(t)\right\},$$

satisfies

$$(3.2) \qquad \int\cdots\int |u_j|\big|c'_{a_1,\cdots,a_k}(u_1,\cdots,u_{k-1})\big|\, du_1\cdots du_{k-1} \,<\, \infty.$$

the cumulant spectra of the series, $\mathbf{Y}(t)$, $-\infty < t < \infty$, are given by

$$(3.3) \qquad f_{a_1, \cdots, a_k}(\lambda_1, \cdots, \lambda_k)$$

$$= (2\pi)^{-k+1} \int \cdots \int \exp\left\{-i \sum_1^{k-1} \lambda_j u_j\right\} c'_{a_1, \cdots, a_k}(u_1, \cdots, u_{k-1})$$

$$du_1 \cdots du_{k-1},$$

understanding $\lambda_1 + \cdots + \lambda_k = 0$ (see Brillinger and Rosenblatt [9]). Also the Cramér representation of $\mathbf{Y}(t)$ is given by

$$(3.4) \qquad \mathbf{Y}(t) = \int \exp\{i\lambda t\}\, d\mathbf{Z}_Y(\lambda),$$

where

$$(3.5) \qquad \mathbf{Z}_Y(\lambda) = \underset{T \to \infty}{\text{l.i.m.}}\ (2\pi)^{-1} \int_{-T}^{T} \left[\frac{\exp\{-i\lambda t\} - 1}{-it}\right] \mathbf{Y}(t)\, dt.$$

Suppose we construct the interval process

$$(3.6) \qquad \mathbf{X}(\Delta) = \int_\Delta \mathbf{Y}(t)\, dt,$$

then we quickly see that this process satisfies Assumption 2.2 with

$$(3.7) \qquad C'_{a_1, \cdots, a_k}(u_1, \cdots, u_{k-1}) = \int_0^{u_1} \cdots \int_0^{u_{k-1}} c'_{a_1, \cdots, a_k}(u_1, \cdots, u_{k-1})$$

$$du_1 \cdots du_{k-1}.$$

The cumulant spectra of the interval process, $\mathbf{X}(\Delta)$, $\Delta \in \Delta$, are therefore the same as the cumulant spectra of the time series $\mathbf{Y}(t)$, $-\infty < t < \infty$.

A comparison of expressions (3.5) and (2.26) indicates that $\mathbf{Z}_X(\lambda)$ of the spectral representation of $\mathbf{X}(\Delta)$ is equivalent with $\mathbf{Z}_Y(\lambda)$ of the Cramér representation of $\mathbf{Y}(t)$.

In a later section we will see that our proposed empirical analysis of the process, $\mathbf{X}(\Delta)$, $\Delta \in \Delta$, reduces to the usual empirical analysis of the continuous series $\mathbf{Y}(t)$, $-\infty < t < \infty$.

EXAMPLE 3.2.   Consider an r vector valued point process, $\mathbf{N}(\Delta)$, $\Delta \in \Delta$. Here $N_a(\Delta)$ represents the number of events of the ath sort that occur in the interval Δ. If we let $\mathbf{1}_a$ denote a vector with 1 as its ath component and 0 elsewhere, then we may set down

ASSUMPTION 3.1.   *The point process, $\mathbf{N}(\Delta)$, $\Delta \in \Delta$, possesses moments of all orders and is such that if $\Delta_1, \cdots, \Delta_k$ are disjoint intervals with $|\Delta_1|, \cdots |\Delta_k| \leq \delta < \infty$,*

$$(3.8) \qquad P\{\mathbf{N}(\Delta_1) = \mathbf{n}_1, \cdots, \mathbf{N}(\Delta_k) = \mathbf{n}_k\} < K_\delta |\Delta_1|^{|\mathbf{n}_1|} \cdots |\Delta_k|^{|\mathbf{n}_k|}$$

for some finite K_δ and for $\mathbf{n}_1, \cdots, \mathbf{n}_k$ having nonnegative integral coordinates.

Also if $t_j \in \Delta_j$, *for such* $\Delta_1, \cdots, \Delta_k$, *there is a function* $p_{a_1, \cdots, a_k}(t_1, \cdots, t_k)$, *bounded in finite intervals, such that*

$$(3.9) \qquad \lim_{|\Delta_j| \to 0} |\Delta_1|^{-1} \cdots |\Delta_k|^{-1} P\{\mathbf{N}(\Delta_1) = \mathbf{1}_{a_1}, \cdots, \mathbf{N}(\Delta_k) = \mathbf{1}_{a_k}\}$$
$$= p_{a_1, \cdots, a_k}(t_1, \cdots, t_k),$$

uniformly in $t_1, \cdots, t_k$.

The functions $p_{a_1, \cdots, a_k}(t_1, \cdots, t_k)$ have been called *product density functions*, see Srinivasan [21]. The function $p_a(t)$, $-\infty < t < \infty$, is called the *density of events* of the ath sort at time t, $a = 1, \cdots, r$.

We note that the process satisfies

$$(3.10) \qquad P\left\{\sum_{a=1}^{r} N_a(\Delta) > 1\right\} = O(|\Delta|^2),$$

and is therefore *orderly* (events tend not to happen simultaneously). Also we have

$$(3.11) \qquad P\{N_a(\Delta) = 1\} = p_a(t)|\Delta| + o(|\Delta|),$$

and

$$(3.12) \qquad P\{N_a(\Delta) = 0\} = 1 - p_a(t)|\Delta| + o(|\Delta|).$$

In the theorem below we let $\delta\{x\}$ denote the Kronecker delta $\delta\{x\} = 1$ if $x = 0$ and $\delta\{x\} = 0$ otherwise. We let $\chi_\Delta(\tau)$ denote the indicator function $\chi_\Delta(\tau) = 1$ if $\tau \in \Delta$, $\chi_\Delta(\tau) = 0$ otherwise. We have

THEOREM 3.1. *Let the* r *vector valued point process* $\mathbf{N}(\Delta)$, $\Delta \in \Delta$, *satisfy Assumption* 3.1. *Then*

$$(3.13)$$

$$E\{N_{a_1}(\Delta_1) \cdots N_{a_k}(\Delta_k)\} = \sum_{\ell=1}^{k} \sum_{a_1, \cdots, a_\ell = 1}^{r} [\prod_{j \in v_1} \delta\{\alpha_1 - a_j\}] \cdots [\prod_{j \in v_\ell} \delta\{\alpha_\ell - a_j\}]$$

$$\int \cdots \int [\prod_{j \in v_1} \chi_{\Delta_j}(\tau_1)] \cdots [\prod_{j \in v_\ell} \chi_{\Delta_j}(\tau_\ell)] p_{a_1, \cdots, a_\ell}(\tau_1, \cdots, \tau_\ell) \, d\tau_1 \cdots d\tau_\ell$$

with the sum extending over all partitions $(v_1, \cdots, v_\ell)$ *of the set* $(1, \cdots, k)$.

We see that the moments of $\mathbf{N}(\Delta)$, $\Delta \in \Delta$, have the integral representation required in Assumption 2.1. Particular cases of this theorem include

$$(3.14) \qquad EN_a(\Delta) = \int_\Delta p_a(\tau) \, d\tau,$$

$$(3.15) \qquad E\{N_a(\Delta_1)N_a(\Delta_2)\} = \int_{\Delta_1} \int_{\Delta_2} p_{a,a}(\tau_1, \tau_2) \, d\tau_1 \, d\tau_2 + \int_{\Delta_1 \cap \Delta_2} p_a(\tau) \, d\tau,$$

$$(3.16) \qquad E\{N_a(\Delta_1)N_b(\Delta_2)\} = \int_{\Delta_1} \int_{\Delta_2} p_{a,b}(\tau_1, \tau_2) \, d\tau_1 \, d\tau_2 \qquad \text{if} \quad a \neq b,$$

and

$$(3.17) \qquad E\{N_a(\Delta)^k\} = \sum_{\ell=1}^{k} \mathscr{S}_\ell^{(k)} \int_\Delta \cdots \int_\Delta p_{a,\cdots,a}(\tau_1, \cdots, \tau_\ell) \, d\tau_1 \cdots d\tau_\ell,$$

where $\mathscr{S}_\ell^{(k)}$ denotes a Stirling number of the second kind. Expression (3.17) was set down by Ramakrishnan [20] and Kuznetsov and Stratonovich [18].

Kuznetsov and Stratonovich [18] remarked that it might prove more useful to consider the cumulant functions

$$(3.18) \qquad q_{a_1, \cdots, a_k}(t_1, \cdots, t_k)$$
$$= \sum_{\ell=1}^{k} (-1)^{\ell-1} (\ell-1)! \, p_{a_j; \, j \in v_1}(t_j; j \in v_1) \cdots p_{a_j; \, j \in v_\ell}(t_j; j \in v_\ell),$$

where the summation extends over all partitions $(v_1, \cdots, v_\ell)$ of $(1, \cdots, k)$. These functions have the property of tending to 0 as $|t_i - t_j| \to \infty$ in the case that the increments of $\mathbf{N}(t)$ are tending to become independent as they separate in time. Particular cases of the functions include

$$(3.19) \qquad\qquad q_a(t) = p_a(t),$$

$$(3.20) \qquad\qquad q_{a,b}(t_1, t_2) = p_{a,b}(t_1, t_2) - p_a(t_1)p_b(t_2).$$

The inverse relation to (3.18) is

$$(3.21) \qquad p_{a_1, \cdots, a_k}(t_1, \cdots, t_k) = \sum_{\ell=1}^{k} q_{a_j; \, j \in v_1}(t_j; j \in v_1) \cdots q_{a_j; \, j \in v_\ell}(t_j; j \in v_\ell).$$

We have

THEOREM 3.2. *Let the r vector valued point process* $\mathbf{N}(\Delta)$, $\Delta \in \Delta$, *satisfy Assumption 3.1. Then*

$$(3.22)$$
$$\operatorname{cum}\{N_{a_1}(\Delta_1), \cdots, N_{a_k}(\Delta_k)\} = \sum_{\ell=1}^{k} \sum_{\alpha_1, \cdots, \alpha_\ell = 1}^{r} \left[\prod_{j \in v_1} \delta\{\alpha_1 - a_j\}\right] \cdots \left[\prod_{j \in v_\ell} \delta\{\alpha_\ell - a_j\}\right]$$
$$\int \cdots \int \left[\prod_{j \in v_1} \chi_{\Delta_j}(\tau_1)\right] \cdots \left[\prod_{j \in v_\ell} \chi_{\Delta_j}(\tau_\ell)\right] q_{\alpha_1, \cdots, \alpha_\ell}(\tau_1, \cdots, \tau_\ell) \, d\tau_1 \cdots d\tau_\ell$$

where the summation extends over all partitions $(v_1, \cdots, v_\ell)$ *of the set* $(1, \cdots, k)$.

The relation (3.22) has the same form as the relation (3.13). As particular cases we mention

$$(3.23) \qquad\qquad EN_a(\Delta) = \int_\Delta q_a(\tau) \, d\tau = \int_\Delta p_a(\tau) \, d\tau,$$

$$(3.24) \quad \operatorname{Cov}\{N_a(\Delta_1), N_a(\Delta_2)\} = \int_{\Delta_1} \int_{\Delta_2} q_{a,a}(\tau_1, \tau_2) \, d\tau_1 \, d\tau_2 + \int_{\Delta_1 \cap \Delta_2} q_a(\tau) \, d\tau,$$

$$(3.25) \qquad \operatorname{Cov}\{N_a(\Delta_1), N_b(\Delta_2)\} = \int_{\Delta_1} \int_{\Delta_2} q_{a,b}(\tau_1, \tau_2) \, d\tau_1 \, d\tau_2,$$

and

$$(3.26) \qquad \mathrm{cum}_k \left\{ N_a(\Delta) \right\} = \sum_{\ell=1}^{k} \mathscr{S}_\ell^{(k)} \int_\Delta \cdots \int_\Delta q_{a, \cdots, a}(\tau_1, \cdots, \tau_\ell) \, d\tau_1 \cdots d\tau_\ell.$$

This last expression was given by Kuznetsov and Stratonovich [18]. We remark that in differential notation, (3.22) has the form

$$(3.27) \quad \mathrm{cum} \left\{ dN_{a_1}(t_1), \cdots, dN_{a_k}(t_k) \right\} = q_{a_1, \cdots, a_k}(t_1, \cdots, t_k) \, dt_1 \cdots dt_k,$$

if the t_j are distinct. As a further implication of the theorem we have

COROLLARY 3.1. *Under the conditions of the theorem and if $\phi_a(t)$ is continuous with finite support, $a = 1, \cdots, k$, then*

$$(3.28) \quad \mathrm{cum} \left\{ \int \phi_1(t) \, dN_{a_1}(t), \cdots, \int \phi_k(t) \, dN_{a_k}(t) \right\}$$

$$= \sum_{\ell=1}^{k} \sum_{\alpha_1, \cdots, \alpha_\ell = 1}^{r} \left[\prod_{j \in v_1} \delta\{\alpha_1 - a_j\} \right] \cdots \left[\prod_{j \in v_\ell} \delta\{\alpha_\ell - a_j\} \right] \int \cdots \int \left[\prod_{j \in v_1} \phi_j(\tau_1) \right]$$

$$\cdots \left[\prod_{j \in v_\ell} \phi_j(\tau_\ell) \right] q_{\alpha_1, \cdots, \alpha_\ell}(\tau_1, \cdots, \tau_\ell) \, d\tau_1 \cdots d\tau_\ell$$

where the summation is over all partitions $(v_1, \cdots, v_\ell)$ of $(1, \cdots, k)$.

If the point process $\mathbf{N}(\Delta)$, $\Delta \in \Delta$, is stationary, then

$$(3.29) \qquad p_{a_1, \cdots, a_k}(t + t_1, \cdots, t + t_k) = p_{a_1, \cdots, a_k}(t_1, \cdots, t_k),$$

and

$$(3.30) \qquad q_{a_1, \cdots, a_k}(t + t_1, \cdots, t + t_k) = q_{a_1, \cdots, a_k}(t_1, \cdots, t_k),$$

for all real $t, t_1, \cdots, t_k$. In this case we set

$$(3.31) \qquad r_{a_1, \cdots, a_k}(u_1, \cdots, u_{k-1}) = q_{a_1, \cdots, a_k}(u_1, \cdots, u_{k-1}, 0).$$

The parameter r_a is called the *mean intensity* of the process $N_a(\Delta)$, $\Delta \in \Delta$; $r_{a,a}(u)$ is called the *covariance density* of the process $N_a(\Delta)$, $\Delta \in \Delta$; and $r_{a,b}(u)$, for $a \neq b$, is called the *cross covariance density* of the component $N_a(\Delta)$ with the component $N_b(\Delta)$.

We now set down

ASSUMPTION 3.2. $\mathbf{N}(\Delta)$, $\Delta \in \Delta$, *is an r vector valued stationary point process satisfying Assumption 3.1 and such that*

$$(3.32) \qquad \int \cdots \int |u_j| |r_{a_1, \cdots, a_k}(u_1, \cdots, u_{k-1})| \, du_1 \cdots du_{k-1} < \infty$$

for $a_1, \cdots, a_k = 1, \cdots, r$; $k = 2, 3, \cdots$.

If the process $\mathbf{N}(\Delta)$, $\Delta \in \Delta$, satisfies this assumption, then we may define the

Fourier transforms

$$(3.33) \quad g_{a_1, \cdots, a_k}(\lambda_1, \cdots, \lambda_k)$$

$$= \int \cdots \int \exp \left\{ -i \sum_1^{k-1} \lambda_j u_j \right\} r_{a_1, \cdots, a_k}(u_1, \cdots, u_{k-1}) \, du_1 \cdots du_{k-1},$$

understanding $\lambda_1 + \cdots + \lambda_k = 0$. For completeness we set

$$(3.34) \quad g_a(\lambda) = r_a = q_a(t) = p_a(t),$$

in the case $k = 1$. We now have

THEOREM 3.3.   *Let the point process* $\mathbf{N}(\Delta)$, $\Delta \in \Delta$, *satisfy Assumption 3.2. Then the process satisfies Assumption 2.2. Its cumulant spectra are given by*

$$(3.35) \quad f_{a_1, \cdots, a_k}(\lambda_1, \cdots, \lambda_k)$$

$$= (2\pi)^{-k+1} \sum_{\ell=1}^{k} \sum_{\alpha_1, \cdots, \alpha_\ell = 1}^{r} \left[\prod_{j \in v_1} \delta\{\alpha_1 - a_j\} \right]$$

$$\cdots \left[\prod_{j \in v_\ell} \delta\{\alpha_\ell - a_j\} \right] g_{\alpha_1, \cdots, \alpha_\ell} \left[\sum_{j \in v_1} \lambda_j, \cdots, \sum_{j \in v_\ell} \lambda_j \right],$$

with the summation extending over all partitions $(v_1, \cdots, v_\ell)$ *of* $(1, \cdots, k)$.

As particular cases of the cumulant spectra we mention

$$(3.36) \quad f'_a = r_a,$$

$$(3.37) \quad f'_{a, a}(\lambda) = (2\pi)^{-1} [g'_{a, a}(\lambda) + g_a]$$

$$= (2\pi)^{-1} \left[\int \exp \{-i\lambda t\} \, r_{a, a}(t) \, dt + r_a \right],$$

in agreement with Bartlett [4], p. 183. Also

$$(3.38) \quad f'_{a, b}(\lambda) = (2\pi)^{-1} \int \exp \{-i\lambda t\} \, r_{a, b}(t) \, dt \qquad \text{if} \quad a \neq b,$$

and

$$(3.39) \quad f'_{a, a, a}(\lambda_1, \lambda_2) = (2\pi)^{-2} [g'_{a, a, a}(\lambda_1, \lambda_2) + g'_{a, a}(\lambda_1) + g'_{a, a}(\lambda_2)$$

$$+ g'_{a, a}(-\lambda_1 - \lambda_2) + g'_a],$$

We have the following relation, inverse to (3.35),

$$(3.40) \quad g_{a_1, \cdots, a_k}(\lambda_1, \cdots, \lambda_k)$$

$$= \sum_{\ell=1}^{k} \sum_{\alpha_1, \cdots, \alpha_\ell = 1}^{r} (-1)^{\ell-1} (\ell - 1)! \, (2\pi)^{k-\ell} \left[\prod_{j \in v_1} \delta\{\alpha_1 - a_j\} \right]$$

$$\cdots \left[\prod_{j \in v_\ell} \delta\{\alpha_\ell - a_j\} \right] \cdot f_{\alpha_1, \cdots, \alpha_\ell} \left(\sum_{j \in v_1} \lambda_j, \cdots, \sum_{j \in v_\ell} \lambda_j \right),$$

where the summation is again over all partitions $(v_1, \cdots, v_\ell)$ of the integers $(1, \cdots, k)$.

In Section 2 we discussed a class linear time invariant operations on stationary interval processes. It may be of interest to indicate a subclass of these operations which carry point processes over into point processes. Let $\sigma_j, j = 0, \pm 1, \cdots,$ be a sequence of real numbers. Let

(3.41) $\qquad\qquad a(\Delta) = \text{the number of } \sigma_j \in \Delta,$

then.

(3.42) $\qquad\qquad Y(\Delta) = a * N(\Delta)$

$$= \int a(\Delta - u) \, dN(u)$$

will be a real valued point process in the case that $N(\Delta)$, $\Delta \in \Delta$, is one. If τ_j, $j = 0, \pm 1, \cdots$, denote the times of events of a realization of $N(\Delta)$, then events of this $Y(\Delta)$ occur at the times $\tau_j + \sigma_k, j, k = 0, \pm 1, \cdots$.

Daley [13] discusses the second order spectral theory of point processes, considers operations on point processes, and presents a variety of examples.

EXAMPLE 3.3. Suppose that $\mathbf{Y}(t)$, $-\infty < t < \infty$, is an r vector valued stationary time series satisfying the conditions of Example 3.1 and having Cramér representation

(3.43) $\qquad\qquad \mathbf{Y}(t) = \int \exp\{i\lambda t\} \, d\mathbf{Z}_Y(\lambda).$

Suppose $N(\Delta)$, $\Delta \in \Delta$, is an independent stationary point process satisfying Assumption 3.2 and having spectral representation

(3.44) $\qquad\qquad N(\Delta) = \int \left[\int_\Delta \exp\{i\lambda t\} \, dt \right] dZ_N(\lambda).$

In Section 6 of the paper we will consider the process

(3.45) $\qquad\qquad \mathbf{X}(\Delta) = \int_\Delta \mathbf{Y}(t) \, dN(t)$

$$= \mathbf{Y}(\tau_1) + \cdots + \mathbf{Y}(\tau_n),$$

if $\tau_1, \cdots, \tau_n$ are the events of $N(\Delta)$ in the interval Δ. One can check that this process satisfies Assumption 2.2. If its spectral representation is

(3.46) $\qquad\qquad \mathbf{X}(\Delta) = \int \left[\int_\Delta \exp\{i\lambda t\} \, dt \right] d\mathbf{Z}_X(\lambda),$

then we see directly that

(3.47) $\qquad\qquad d\mathbf{Z}_X(\lambda) = \int \left[d\mathbf{Z}_Y(\lambda - \alpha) \right] dZ_N(\alpha),$

for $-\infty < \lambda < \infty$. Expression (3.47) may be used to determine the cumulant spectra of $\mathbf{X}(\Delta)$ in terms of those of $\mathbf{Y}(t)$ and $N(\Delta)$.

We mention that Walker suggested the consideration of real valued processes of the form (3.45) in the discussion of Bartlett [3].

4. Stochastic properties of finite Fourier transforms

We now turn to an investigation of certain statistics useful in the estimation of the cumulant spectra of a stationary interval function $\mathbf{X}(\Delta)$, $\Delta \in \Delta$. We will suppose that the values of $\mathbf{X}(\Delta)$ are available for Δ contained in the support of a function $h(t/T)$, $T = 1, 2, \cdots$. We set down,

ASSUMPTION 4.1.   *The function $h(t)$, $-\infty < t < \infty$, is measurable in t, bounded, zero for $|t| > 1$ and there exists a finite K such that*

$$(4.1) \qquad \int \left| h(t + u) - h(t) \right| dt < K|u|$$

for all real u.

The inequality (4.1) will be satisfied if $h(t)$ is of bounded variation, for example. For given T, the function $h(t/T)$ has been called a *taper* by Tukey [22]. It has also been called a *data window*.

The principal statistics of our analysis of interval processes are the finite Fourier transforms,

$$(4.2) \qquad d_a^{(T)}(\lambda) = \int h_a(t/T) \exp\left\{ -i\lambda t \right\} dX_a(t),$$

$a = 1, \cdots, r$, $-\infty < \lambda < \infty$. In the case of Example 3.1, the statistic (4.2) takes the form

$$(4.3) \qquad d_a^{(T)}(\lambda) = \int h_a(t/T) \exp\left\{ -i\lambda t \right\} Y_a(t)\, dt,$$

that is, it is the Fourier transform of the tapered values that was considered in Brillinger and Rosenblatt [9]. In the case of Example 3.2, if we let $\tau_a(1), \cdots, \tau_a(n_a)$ denote the times of events of the ath sort that occur in the support of $h_a(t/T)$, then the statistic (4.2) has the form

$$(4.4) \qquad \sum_{j=1}^{n_a} h_a(\tau_a(j)/T) \exp\left\{ -i\lambda\tau_a(j) \right\}.$$

This statistic, excluding the taper, was considered in Bartlett [3] for the case $r = 1$ and suggested for the case of general r by Jenkins in the discussion of that paper. In the case of Example 3.3, the statistic has the form

$$(4.5) \qquad \sum_{j=1}^{n} h_a(\tau_j/T) \exp\left\{ -i\lambda\tau_j \right\} Y_a(\tau_j),$$

if $\tau_1, \cdots, \tau_n$ denote the times of events of the process $N(\Delta)$ in the support of $h_a(t/T)$.

We next present a basic theorem indicating the asymptotic joint cumulants of the Fourier transform (4.2). In the theorem we let

$$(4.6) \qquad H_{a_1, \cdots, a_k}(\lambda) = \int h_{a_1}(t) \cdots h_{a_k}(t) \exp\left\{ -i\lambda t \right\} dt.$$

THEOREM 4.1. *Let the process $X(\Delta)$, $\Delta \in \Delta$, satisfy Assumption 2.2. Let $h_a(t)$, $a = 1, \cdots, r$, $-\infty < t < \infty$, satisfy Assumption 4.1. Then as $T \to \infty$*

$$(4.7) \qquad \text{cum } \{d_{a_1}^{(T)}(\lambda_1), \cdots, d_{a_k}^{(T)}(\lambda_k)\}$$

$$= TH_{a_1, \cdots, a_k}\left(T \sum_1^k \lambda_j\right)(2\pi)^{k-1} f_{a_1, \cdots, a_k}(\lambda_1, \cdots, \lambda_{k-1}) + O(1)$$

for $a_1, \cdots, a_k = 1, \cdots, r$; $k = 1, 2, \cdots$. The $O(1)$ term is uniformly bounded in $\lambda_1, \cdots, \lambda_k$.

We see that the joint cumulants are of reduced order unless $\sum_1^k \lambda_j$ is near zero. We see from (4.6) and (4.7) that the joint cumulants based on disjoint stretches of data are of reduced order as well.

If $h_a(t) = 1$ for $0 \leqq t \leqq 1$ and $h_a(t) = 0$ otherwise, then this theorem has identical nature with the key theorem used in Brillinger and Rosenblatt [9], Brillinger [6], Brillinger [7] to develop properties of spectral estimates. The results of these papers therefore become directly available. We indicate a selection of results that now hold.

We begin by considering the asymptotic distribution of the finite Fourier transform. Let $N_r^C(\mu, \Sigma)$ denote the complex r variate normal distribution with mean μ and covariance matrix Σ. We have

THEOREM 4.2. *Let $\mathbf{X}(\Delta)$, $\Delta \in \Delta$, be an r vector valued interval process satisfying Assumption 2.2. Let $s_j(T)$ be an integer with $\lambda_j(T) = 2\pi s_j(T)/T \to \lambda_j$ as $T \to \infty$ for $j = 1, \cdots, J$. Suppose $\lambda_j(T) \pm \lambda_k(T) \neq 0$ for $j, k = 1, \cdots, J$. Let*

$$(4.8) \qquad \mathbf{d}_X^{(T)}(\lambda) = \int_0^T \exp\{-i\lambda t\}\, d\mathbf{X}(t)$$

for $-\infty < \lambda < \infty$. Then $\mathbf{d}_X^{(T)}(\lambda_j(T))$, $j = 1, \cdots, J$ are asymptotically independent $N_r^C(\mathbf{0}, 2\pi T\mathbf{f}'_{X,X}(\lambda))$ variates, respectively. Also $\mathbf{d}_X^{(T)}(0) = \mathbf{X}(0, T]$ is asymptotically $N_r(T\mathbf{f}'_X, 2\pi T\mathbf{f}'_{X,X}(0))$ independently of the previous variates.

This theorem has the nature of a central limit theorem. Let $W_r^C(n, \Sigma)$ denote the complex Wishart distribution of dimensions $r \times r$, degrees of freedom n and covariance matrix Σ. Define the matrix of periodograms

$$(4.9) \qquad \mathbf{I}_{X,X}^{(T)}(\lambda) = (2\pi T)^{-1}\,\mathbf{d}_X^{(T)}(\lambda)\,\overline{\mathbf{d}_X^{(T)}(\lambda)}^{\mathsf{t}}.$$

We have the following corollary.

COROLLARY 4.1. *Under the conditions of Theorem 4.2, if $\lambda_1 = \cdots = \lambda_J = \lambda$ and if*

$$(4.10) \qquad \mathbf{f}_{X,X}^{(T)}(\lambda) = J^{-1}\sum_{j=1}^J \mathbf{I}_{X,X}^{(T)}(\lambda_j(T)),$$

$\mathbf{f}_{X,X}^{(T)}(\lambda)$ *is asymptotically $J^{-1}W_r^C(J, \mathbf{f}'_{X,X}(\lambda))$ as $T \to \infty$.*

This corollary makes precise the chi square approximation for the distribution of second order spectral densities of point processes suggested by Bartlett [3].

We next construct consistent asymptotically normal estimates of the cumulant spectra of different orders of an interval process $\mathbf{X}(\Delta)$, $\Delta \in \Delta$. We begin by letting $W(u_1, \cdots, u_k)$ be a weight function satisfying

ASSUMPTION 4.2.   *The function* $W(u_1, \cdots, u_k)$, $-\infty < u_j < \infty$, *is symmetric in* $u_1, \cdots, u_k$, *is concentrated on the plane* $\Sigma_1^k u_j = 0$, *and is such that*

$$(4.11) \qquad \int_{-\infty}^{\infty} \cdots \int W(u_1, \cdots, u_k) \, \delta\left(\sum_1^k u_j\right) du_1 \cdots du_k = 1$$

and

$$(4.12) \qquad \left| W\left(u_1, \cdots, u_{k-1}, -\sum_1^{k-1} u_j\right)\right|, \left|\frac{\partial}{\partial u_\ell} W\left(u_1, \cdots, u_{k-1}, -\sum_1^{k-1} u_j\right)\right|$$
$$\leq A\left(1 + \left[\sum_1^{k-1} u_j^2\right]^{1/2}\right)^{-k-\varepsilon+1},$$

for some $A, \varepsilon > 0$, $\ell = 1, \cdots, k$.

Given the sequence of nonnegative numbers $B_T^{(k)}$, $T = 2, 3, \cdots$, we set

$$(4.13) \qquad W_T(u_1, \cdots, u_k) = \left(B_T^{(k)}\right)^{-k+1} W\left(B_T^{(k)-1} u_1, \cdots, B_T^{(k)-1} u_k\right).$$

We suppose $B_T^{(2)} \leq B_T^{(3)} \leq \cdots$. Next we set $\Psi(u_1, \cdots, u_k) = 1$ if $\Sigma_1^k u_j = 0$ but no proper subset of the u_j has sum 0, and set it $= 0$ otherwise. Let

$$(4.14) \qquad \mathbf{d}_X^{(T)}(\lambda) = \int_0^T \exp\{-i\lambda t\} \, d\mathbf{X}(t).$$

Finally set

$$(4.15) \qquad I_{a_1, \cdots, a_k}^{(T)}(\lambda_1, \cdots, \lambda_k) = (2\pi)^{-k+1} T^{-1} \prod_{j=1}^k d_{a_j}^{(T)}(\lambda_j).$$

As an estimate of $f_{a_1, \cdots, a_k}(\lambda_1, \cdots, \lambda_k)$ we now take

$$(4.16) \qquad f_{a_1, \cdots, a_k}^{(T)}(\lambda_1, \cdots, \lambda_k)$$

$$= \left(\frac{2\pi}{T}\right)^{k-1} \sum_{s_1} \cdots \sum_{s_k} W_T\left(\lambda_1 - \frac{2\pi s_1}{T}, \cdots, \lambda_k - \frac{2\pi s_k}{T}\right) \cdot$$
$$\cdot \Psi(s_1, \cdots, s_k) I_{a_1, \cdots, a_k}^{(T)}\left(\frac{2\pi s_1}{T}, \cdots, \frac{2\pi s_k}{T}\right).$$

In connection with this estimate we have the theorem,

THEOREM 4.3.   *Let* $\mathbf{X}(\Delta)$, $\Delta \in \Delta$, *satisfy Assumption 2.2. Let* $W(u_1, \cdots, u_k)$ *satisfy Assumption 4.2. Let* $f_{a_1, \cdots, a_k}^{(T)}(\lambda_1, \cdots, \lambda_k)$ *be given by* (4.16). *Let* $B_T^{(k)} \to 0$, $\left(B_T^{(k)}\right)^{k-1} T \to \infty$ *as* $T \to \infty$, *then*

$$(4.17) \qquad E f_{a_1, \cdots, a_k}^{(T)}(\lambda_1, \cdots, \lambda_k)$$

$$= \int \cdots \int W_T(\lambda_1 - \alpha_1, \cdots, \lambda_k - \alpha_k) f_{a_1, \cdots, a_k}(\alpha_1, \cdots, \alpha_k)$$

$$\cdot \delta(\alpha_1 + \cdots + \alpha_k) \, d\alpha_1 \cdots d\alpha_k + O\big(B_T^{(k)^{-1}} T\big)$$

$$= f_{a_1, \cdots, a_k}(\lambda_1, \cdots, \lambda_k) + O\big(B_T^{(k)}\big) + O\big(B_T^{(k)^{-1}} T\big),$$

(4.18) $\quad \lim_{T \to \infty} \big(B_T^{(k)}\big)^{k-1} T \, \mathrm{Cov} \, \{f_{a_1, \cdots, a_k}^{(T)}(\lambda_1, \cdots, \lambda_k), f_{a_1', \cdots, a_k'}^{(T)}(\mu_1, \cdots, \mu_k)\}$

$$= 2\pi \sum_P \delta\{\lambda_1 - \mu_{P,1}\} \cdots \delta\{\lambda_k - \mu_{P,k}\} f'_{a_1 a_{P,1}'}(\lambda_1) \cdots f'_{a_k a_{P,k}'}(\lambda_k)$$

$$\int \cdots \int W(u_1, \cdots, u_k)^2 \delta\left(\sum_1^k u_j\right) du_1 \cdots du_k,$$

where the summation is over all permutations P of the integers $1, \cdots, k$. Collections of spectral estimates are asymptotically jointly normally distributed as $T \to \infty$ with estimates of different orders asymptotically independent and estimates of the same order having covariance structure given by (4.18).

We next turn to the development of an empirical analysis of the linear time invariant model,

(4.19) $$\mathbf{Y}(\Delta) = \mathbf{a} * \mathbf{X}(\Delta) + \boldsymbol{\varepsilon}(\Delta)$$

$$= \int \mathbf{a}(\Delta - u) \, d\mathbf{X}(u) + \boldsymbol{\varepsilon}(\Delta),$$

with $\mathbf{X}(\Delta)$, $\boldsymbol{\varepsilon}(\Delta)$, $\Delta \in \Delta$, independent stationary interval processes and

(4.20) $$\int |u| \, d|\mathbf{a}(u)| < \infty.$$

In differential notation we may write (4.19) as

(4.21) $$d\mathbf{Y}(t) = \int \big[d\mathbf{a}(t - u)\big] \, d\mathbf{X}(u) + d\boldsymbol{\varepsilon}(t).$$

Denote the cross spectral density matrix of the process

(4.22) $$\begin{bmatrix} \mathbf{X}(\Delta) \\ \mathbf{Y}(\Delta) \end{bmatrix},$$

$\Delta \in \Delta$, by

(4.23) $$\begin{bmatrix} \mathbf{f}'_{X,X}(\lambda) & \mathbf{f}'_{X,Y}(\lambda) \\ \mathbf{f}'_{Y,X}(\lambda) & \mathbf{f}'_{Y,Y}(\lambda) \end{bmatrix},$$

and that of $\boldsymbol{\varepsilon}(\Delta)$, $\Delta \in \Delta$, by $\mathbf{f}'_{\varepsilon, \varepsilon}(\lambda)$. Set

(4.24) $$\mathbf{A}(\lambda) = \int \exp \{-i\lambda t\} \, d\mathbf{a}(t).$$

Then (4.19) gives

(4.25)
$$\mathbf{f}'_{Y,X}(\lambda) = \mathbf{A}(\lambda)\mathbf{f}'_{X,X}(\lambda),$$

(4.26)
$$\mathbf{f}'_{Y,Y}(\lambda) = \mathbf{A}(\lambda)\mathbf{f}'_{X,X}(\lambda)\overline{\mathbf{A}(\lambda)}^{\tau} + \mathbf{f}'_{\varepsilon,\varepsilon}(\lambda).$$

These last suggest that we may base estimates of $\mathbf{A}(\lambda)$ and $\mathbf{f}'_{\varepsilon,\varepsilon}(\lambda)$ on an estimate of the spectral density matrix (4.23). We could construct an estimate of this last in the manner of (4.16); however, in order to display an alternate form of spectral estimate of order two we proceed slightly differently.

In constructing this alternate estimate we let $h(t)$, $-\infty < t < \infty$, be a tapering function satisfying Assumption 4.1. We then set

(4.27)
$$H_k(\lambda) = \int h(t)^k \exp\{-i\lambda t\}\, dt,$$

for $-\infty < \lambda < \infty$. We next set

$$\mathbf{d}_X^{(T)}(\lambda) = \int h(t/T) \exp\{-i\lambda t\}\, d\mathbf{X}(t),$$

(4.28)

$$\mathbf{d}_Y^{(T)}(\lambda) = \int h(t/T) \exp\{-i\lambda t\}\, d\mathbf{Y}(t);$$

and we let $W(\alpha)$ be a weight function satisfying

ASSUMPTION 4.3. $W(\alpha)$, $-\infty < \alpha < \infty$, *is real valued, even, absolutely integrable, has an absolutely integrable first derivative, and*

(4.29)
$$\int_{-\infty}^{\infty} W(\alpha)\, d\alpha = 1.$$

The variate (4.22) has mean

(4.30)
$$\begin{bmatrix} \mathbf{f}'_X \\ \mathbf{f}'_Y \end{bmatrix} |\Delta|.$$

Estimates of $\mathbf{f}'_X$, $\mathbf{f}'_Y$ based on tapered values are provided by

$$\mathbf{f}_X^{(T)} = \int h(t/T)\, d\mathbf{X}(t) / \int h(t/T)\, dt = \mathbf{d}_X^{(T)}(0)/[TH_1(0)],$$

(4.31)

$$\mathbf{f}_Y^{(T)} = \int h(t/T)\, d\mathbf{Y}(t) / \int h(t/T)\, dt = \mathbf{d}_Y^{(T)}(0)/[TH_1(0)],$$

respectively. The Fourier transform of the process (4.22) corrected for its sample mean is then given by

(4.32)
$$\mathbf{e}_X^{(T)}(\lambda) = \int \exp\{-i\lambda t\} h(t/T)\big[d\mathbf{X}(t) - \mathbf{f}_X^{(T)}\, dt\big]$$
$$= \mathbf{d}_X^{(T)}(\lambda) - \mathbf{d}_X^{(T)}(0) H_1(T\lambda)/H_1(0),$$

$$\mathbf{e}_Y^{(T)}(\lambda) = \int \exp\{-i\lambda t\}h(t/T)\left[d\mathbf{Y}(t) - \mathbf{f}_Y^{(T)}\,dt\right]$$

$$= \mathbf{d}_Y^{(T)}(\lambda) - \mathbf{d}_Y^{(T)}(0)H_1(T\lambda)/H_1(0).$$

Let B_T be a sequence of nonnegative numbers tending to 0 as $T \to \infty$. Set

$$(4.33) \qquad\qquad W^{(T)}(\alpha) = B_T^{-1}W(B_T^{-1}\alpha).$$

As an estimate of the cross spectral density matrix (4.23) we now propose

$$(4.34) \quad \begin{bmatrix} \mathbf{f}_{X,X}^{(T)}(\lambda) & \mathbf{f}_{X,Y}^{(T)}(\lambda) \\ \mathbf{f}_{Y,X}^{(T)}(\lambda) & \mathbf{f}_{Y,Y}^{(T)}(\lambda) \end{bmatrix} = \int W^{(T)}(\lambda - \alpha)(2\pi T)^{-1}\begin{bmatrix} \mathbf{e}_X^{(T)}(\alpha) \\ \mathbf{e}_Y^{(T)}(\alpha) \end{bmatrix}\overline{\begin{bmatrix} \mathbf{e}_X^{(T)}(\alpha) \\ \mathbf{e}_Y^{(T)}(\alpha) \end{bmatrix}}^\tau d\alpha.$$

As estimates of $\mathbf{A}(\lambda)$, $\mathbf{f}_{\varepsilon,\varepsilon}'(\lambda)$, we then take

$$(4.35) \qquad\qquad \mathbf{A}^{(T)}(\lambda) = \mathbf{f}_{Y,X}^{(T)}(\lambda)\mathbf{f}_{X,X}^{(T)}(\lambda)^{-1},$$

$$(4.36) \qquad\qquad \mathbf{g}_{\varepsilon,\varepsilon}^{(T)}(\lambda) = \mathbf{f}_{Y,Y}^{(T)}(\lambda) - \mathbf{f}_{Y,X}^{(T)}(\lambda)\mathbf{f}_{X,X}^{(T)}(\lambda)^{-1}\mathbf{f}_{X,Y}^{(T)}(\lambda).$$

We can now state the following theorem.

THEOREM 4.4. *Let the process* $\mathbf{X}(\Delta)$, $\Delta \in \Delta$, *satisfy Assumption 2.2. Suppose* $\mathbf{f}_{X,X}'(\lambda)$ *is nonsingular. Let the process* $\varepsilon(\Delta)$, $\Delta \in \Delta$, *satisfy Assumption 2.2., have mean* $\mathbf{0}$ *and be statistically independent of the process* $\mathbf{X}(\Delta)$, $\Delta \in \Delta$. *Let* $\mathbf{a}(\Delta)$ *satisfy* (4.20). *Let* $\mathbf{Y}(\Delta)$ *be given by* (4.19). *Let* $W(\alpha)$ *satisfy Assumption 4.3 and* $h(t)$ *satisfy Assumption 4.1. Then if* $B_T \to 0$, $B_T T \to \infty$ *as* $T \to \infty$,

$$(4.37) \qquad\qquad \lim_{T\to\infty} \overrightarrow{\mathrm{ave}}\,\mathbf{A}^{(T)}(\lambda) = \mathbf{A}(\lambda),$$

$$(4.38) \qquad \lim_{T\to\infty} B_T T\,\overrightarrow{\mathrm{Cov}}\{\mathrm{vec}\,\mathbf{A}^{(T)}(\lambda),\,\mathrm{vec}\,\mathbf{A}^{(T)}(\mu)\}$$

$$= 2\pi H_4(0)H_2(0)^{-2}\delta\{\lambda - \mu\}\mathbf{f}_{\varepsilon,\varepsilon}'(\lambda) \otimes \mathbf{f}_{X,X}'(\lambda)^{-1}\int W(\alpha)^2\,d\alpha,$$

$$(4.39) \qquad\qquad \lim_{T\to\infty} \overrightarrow{\mathrm{ave}}\,\mathbf{g}_{\varepsilon,\varepsilon}^{(T)}(\lambda) = \mathbf{f}_{\varepsilon,\varepsilon}'(\lambda),$$

$$(4.40) \quad \lim_{T\to\infty} B_T T\,\overrightarrow{\mathrm{Cov}}\{g_{j,k}^{(T)}(\lambda),\,g_{m,n}^{(T)}(\mu)\}$$

$$= 2\pi H_4(0)H_2(0)^{-2}\big(\delta\{\lambda - \mu\}\big[\mathbf{f}_{\varepsilon,\varepsilon}'(\lambda)\big]_{j,m}\big[\mathbf{f}_{\varepsilon,\varepsilon}'(-\lambda)\big]_{k,n}$$

$$+ \delta\{\lambda + \mu\}\big[\mathbf{f}_{\varepsilon,\varepsilon}'(\lambda)\big]_{j,n}\big[f_{\varepsilon,\varepsilon}'(-\lambda)\big]_{k,m}\big)\int W(\alpha)^2\,d\alpha,$$

$$(4.41) \qquad\qquad \lim_{T\to\infty} B_T T\,\overrightarrow{\mathrm{Cov}}\{\mathrm{vec}\,\mathbf{A}^{(T)}(\lambda),\,g_{j,k}^{(T)}(\mu)\} = 0,$$

for $j,\,k,\,m,\,n = 1,\,\cdots,\,s$. *Also the variates* $\mathbf{A}^{(T)}(\lambda)$, $g_{\varepsilon,\varepsilon}^{(T)}(\mu)$ *are asymptotically jointly normal with the above covariance structure.*

(In this theorem $\overrightarrow{\text{ave}}$, $\overrightarrow{\text{Cov}}$ have technical definitions allowing the use of Taylor series expansions in determining asymptotic moments. See Brillinger and Tukey [10].)

In the case $r = s = 1$, we may define the *coherency* of $X(\Delta)$ with $Y(\Delta)$ at frequency λ by

$$(4.42) \qquad |R_{Y,X}(\lambda)|^2 = \frac{|f'_{Y,X}(\lambda)|^2}{[f'_{X,X}(\lambda)f'_{Y,Y}(\lambda)]}.$$

As an estimate of the coherency we consider the statistic

$$(4.43) \qquad |R^{(T)}_{Y,X}(\lambda)|^2 = \frac{|f^{(T)}_{Y,X}(\lambda)|^2}{[f^{(T)}_{X,X}(\lambda)f^{(T)}_{Y,Y}(\lambda)]}.$$

We then have from the theorem

COROLLARY 4.2. *Under the conditions of the theorem and if $f'_{X,X}(\lambda), f'_{Y,Y}(\lambda) \neq 0$, $|R^{(T)}_{Y,X}(\lambda)|^2$ is asymptotically normal with*

$$(4.44) \qquad \lim_{T \to \infty} \overrightarrow{\text{ave}} |R^{(T)}_{Y,X}(\lambda)|^2 = |R_{Y,X}(\lambda)|^2$$

and

$$(4.45) \qquad \lim_{T \to \infty} B_T T \overrightarrow{\text{Cov}} \{|R^{(T)}_{Y,X}(\lambda)|^2, |R^{(T)}_{Y,X}(\mu)|^2\}$$
$$= 4\pi H_4(0) H_2(0)^{-2} [\delta\{\lambda - \mu\}$$
$$+ \delta\{\lambda + \mu\}]|R_{Y,X}(\lambda)|^2 [1 - |R_{Y,X}(\lambda)|^2]^2 \int W(\alpha)^2 \, d\alpha.$$

A comparison of the results of this theorem and its corollary, with the corresponding results for the regression of one vector valued stationary time series on another, shows that they are identical. This will also be the case for the interval process extension of many of the asymptotic results of the analysis of stationary time series.

5. Estimation of product densities

Let $\mathbf{N}(\Delta)$, $\Delta \in \Delta$, be a stationary point process satisfying Assumption 3.2. We have defined various characteristics of such a process. These may be summarized as follows:

$$(5.1) \qquad p_{a_1, \cdots, a_k}(t_1, \cdots, t_k)$$
$$= \lim_{dt_j \to 0} p\{dN_{a_1}(t_1) = 1, \cdots, dN_{a_k}(t_k) = 1\}/(dt_1 \cdots dt_k)$$

for $t_1, \cdots, t_k$ distinct;

$$(5.2) \qquad q_{a_1, \cdots, a_k}(t_1, \cdots, t_k)$$
$$= \sum_{\ell=1}^{k} (-1)^{\ell-1} (\ell - 1)! p_{a_j; j \in v_1}(t_j; j \in v_1) \cdots p_{a_j; j \in v_\ell}(t_j; j \in v_\ell);$$

(5.3) $p_{a_1,\cdots,a_k}(t_1,\cdots,t_k)$

$$= \sum_{\ell=1}^{k} q_{a_j;\,j\in v_1}(t_j;j\in v_1)\cdots q_{a_j;\,j\in v_\ell}(t_j;j\in v_\ell);$$

(5.4) $r_{a_1,\cdots,a_k}(u_1,\cdots,u_{k-1}) = q_{a_1,\cdots,a_k}(u_1,\cdots,u_{k-1},0);$

(5.5) $g'_{a_1,\cdots,a_k}(\lambda_1,\cdots,\lambda_{k-1})$

$$= \int\cdots\int \exp\left\{-i\sum_{1}^{k-1}\lambda_j u_j\right\} r_{a_1,\cdots,a_k}(u_1,\cdots,u_{k-1})\,du_1\cdots du_{k-1};$$

and if $g'_{a_1,\cdots,a_k}(\lambda_1,\cdots,\lambda_{k-1})$ is integrable

(5.6)

$$r_{a_1,\cdots,a_k}(u_1,\cdots,u_{k-1})$$

$$= (2\pi)^{-k+1}\int\cdots\int \exp\left\{i\sum_{1}^{k-1}\lambda_j u_j\right\} g'_{a_1,\cdots,a_k}(\lambda_1,\cdots,\lambda_{k-1})\,d\lambda_1\cdots d\lambda_{k-1}.$$

Also

(5.7) $$g_{a_1,\cdots,a_k}(\lambda_1,\cdots,\lambda_k) = g'_{a_1,\cdots,a_k}(\lambda_1\cdots,\lambda_{k-1})$$

understanding $\Sigma_1^k \lambda_j = 0$. Continuing

(5.8) $f_{a_1,\cdots,a_k}(\lambda_1,\cdots,\lambda_k)$

$$= (2\pi)^{-k+1}\sum_{\ell=1}^{k}\sum_{\alpha_1,\cdots,\alpha_\ell=1}^{r} \Big[\prod_{j\in v_1}\delta\{\alpha_1-a_j\}\Big]\cdots\Big[\prod_{j\in v_\ell}\delta\{\alpha_\ell-a_j\}\Big]$$

$$\cdot g_{\alpha_1,\cdots,\alpha_\ell}\Big[\sum_{j\in v_1}\lambda_j,\cdots,\sum_{j\in v_1}\lambda_j\Big];$$

(5.9) $g_{a_1,\cdots,a_k}(\lambda_1,\cdots,\lambda_k)$

$$= \sum_{\ell=1}^{k}\sum_{\alpha_1,\cdots,\alpha_\ell=1}^{r}(-1)^{\ell-1}(\ell-1)!\Big[\prod_{j\in v_1}\delta\{\alpha_1-a_j\}\Big]\cdots\Big[\prod_{j\in v_\ell}\delta\{\alpha_\ell-a_j\}\Big]$$

$$\cdot (2\pi)^{k-\ell}f_{\alpha_1,\cdots,\alpha_\ell}\Big[\sum_{j\in v_1}\lambda_j,\cdots,\sum_{j\in v_\ell}\lambda_j\Big].$$

The summations in (5.2), (5.3), (5.8), (5.9) are over all partitions $(v_1,\cdots,v_\ell)$ of the integers $(1,\cdots,k)$.

In the previous section we developed an estimate of $f_{a_1,\cdots,a_k}(\lambda_1,\cdots,\lambda_k)$. Let us now put this work to use in developing estimates of the various functions listed above. As an estimate of $g_{a_1,\cdots,a_k}(\lambda_1,\cdots,\lambda_k)$, in the light of (5.9), we may consider

(5.10) $g_{a_1,\cdots,a_k}^{(T)}(\lambda_1,\cdots,\lambda_k)$

$$= \sum_{\ell=1}^{k}\sum_{\alpha_1,\cdots,\alpha_\ell=1}^{r}(-1)^{\ell-1}(\ell-1)!\Big[\prod_{j\in v_1}\delta\{\alpha_1-a_j\}\Big]\cdots\Big[\prod_{j\in v_\ell}\delta\{\alpha_\ell-a_j\}\Big]$$

$$\cdot (2\pi)^{k-\ell}f_{\alpha_1,\cdots,\alpha_\ell}^{(T)}\Big[\sum_{j\in v_1}\lambda_j,\cdots,\sum_{j\in v_\ell}\lambda_j\Big],$$

where the $f_{a_1, \cdots, a_\ell}^{(T)}(\mu_1, \cdots, \mu_\ell)$. $\ell = 1, \cdots, k$, are formed in the manner of Theorem 4.3. From that theorem we see that

$$(5.11) \qquad Eg_{a_1, \cdots, a_k}^{(T)}(\lambda_1, \cdots, \lambda_k)$$
$$= g_{a_1, \cdots, a_k}(\lambda_1, \cdots, \lambda_k) + O(B_T^{(k)}) + O(B_T^{(k)^{-1}} T^{-1});$$

and because estimates of order less than k have asymptotic variance of smaller order than that of estimates of order k, the covariance of $g_{a_1, \cdots, a_k}^{(T)}(\lambda_1, \cdots, \lambda_k)$ with $g_{b_1, \cdots, b_\ell}^{(T)}(\mu_1, \cdots, \mu_\ell)$ will be asymptotically equivalent to that of $f_{a_1, \cdots, a_k}^{(T)}$ $(\lambda_1, \cdots, \lambda_k)$ with $f_{b_1, \cdots, b_\ell}^{(T)}(\mu_1, \cdots, \mu_\ell)$ as given in Theorem 4.3. Also the estimates will be asymptotically normal and estimates of different orders will be asymptotically independent.

Suppose next that $g_{a_1, \cdots, a_k}'(\lambda_1, \cdots, \lambda_{k-1})$ vanishes for $|\lambda_j| > \Lambda$. As an estimate of $r_{a_1, \cdots, a_k}(u_1, \cdots, u_{k-1})$ we can then consider

$$(5.12) \qquad r_{a_1, \cdots, a_k}^{(T)}(u_1, \cdots, u_{k-1})$$
$$= (2\pi)^{-k+1} \int_{|\lambda_j| \le \Lambda} \cdots \int \exp\left\{i \sum_1^{k-1} \lambda_j u_j\right\} g_{a_1, \cdots, a_k}^{(T)}\left(\lambda_1, \cdots, \lambda_{k-1}, -\sum_1^{k-1} \lambda_j\right)$$
$$d\lambda_1 \cdots d\lambda_{k-1}.$$

From (5.11) this estimate will be asymptotically unbiased.

By analogy with Theorem 5.2 of Brillinger [6], we would expect, for example, that

$$(5.13) \qquad \lim_{T \to \infty} T \operatorname{Cov}\{r_{a_1, b_1}^{(T)}(u_1), r_{a_2, b_2}^{(T)}(u_2)\}$$
$$= \int_{-\Lambda}^{\Lambda} \exp\{i\alpha(u_1 - u_2)\} f_{a_1, a_2}'(\alpha) f_{b_1, b_2}'(-\alpha)\, d\alpha$$
$$+ \int_{-\Lambda}^{\Lambda} \exp\{i\alpha(u_1 + u_2)\} f_{a_1, b_2}'(\alpha) f_{b_1, a_2}'(-\alpha)\, d\alpha$$
$$+ 2\pi \iint_{-\Lambda}^{\Lambda} \exp\{i(\alpha_1 u_1 + \alpha_2 u_2)\} f_{a_1, b_1, a_2, b_2}'(\alpha_1, -\alpha_1, \alpha_2)\, d\alpha_1 d\alpha_2,$$

in the case $k = 2$ and $a_j \ne b_j$.

Next one can take

$$(5.14) \qquad q_{a_1, \cdots, a_k}^{(T)}(t_1, \cdots, t_k) = r_{a_1, \cdots, a_k}^{(T)}(t_1 - t_k, \cdots, t_{k-1} - t_k)$$

as an estimate of $q_{a_1, \cdots, a_k}(t_1, \cdots, t_k)$ and

$$(5.15) \qquad p_{a_1, \cdots, a_k}^{(T)}(t_1, \cdots, t_k) = \sum_{\ell=1}^{k} q_{a_j; \, j \in \nu_1}^{(T)}(t_j; j \in \nu_1) \cdots q_{a_j; \, j \in \nu_\ell}^{(T)}(t_j; j \in \nu_\ell)$$

as an estimate of $p_{a_1, \cdots, a_k}(t_1, \cdots, t_k)$.

In the case $k = 1$, we would estimate r_a by $N_a(0, T]/T$. In Theorem 4.2 we saw that this statistic was asymptotically normal with mean r_a and variance $2\pi T^{-1} f_{a, a}'(0)$.

6. Estimation of second order spectra from sampled values

Let $Y(t)$, $-\infty < t < \infty$, be a real valued time series satisfying the conditions of Example 3.1, having mean c'_Y and autocovariance function $c'_{Y,Y}(u)$, $-\infty < u < \infty$. Let $N(\Delta)$, $\Delta \in \Delta$, be an independent real valued point process satisfying Assumption 3.2, having mean intensity r_N and autocovariance density $r_{N,N}(u)$, $-\infty < u < \infty$. Suppose that events of a realization of the process $N(\Delta)$ occur at the times $\tau_1, \cdots, \tau_n$ in the interval $(0, T]$. Consider the problem of estimating the autocovariance $c'_{Y,Y}(u)$, $-\infty < u < \infty$, and power spectrum $f'_{Y,Y}(\lambda)$, $-\infty < \lambda < \infty$, of the series $Y(t)$ from the values

(6.1) $$\tau_1, \cdots, \tau_n$$

and

(6.2) $$Y(\tau_1), \cdots, Y(\tau_n).$$

We can construct a stationary interval process $X(\Delta)$, $\Delta \in \Delta$, in the manner of Example 3.3 by setting

(6.3) $$X(\Delta) = \int_\Delta Y(t)\,dN(t),$$

or, in differential notation, by setting

(6.4) $$dX(t) = Y(t)\,dN(t).$$

The first and second order measures of this process satisfy

(6.5) $$C'_X\,dt = c'_Y r_N\,dt,$$

and

(6.6) $$\begin{aligned} dC'_{X,X}(u)\,dt = \big(c'_{Y,Y}(u)r_{N,N}(u) &+ c'_{X,Y}(u)r_N\delta(u) + c'_{Y,Y}(u)r_N^2 \\ &+ (c'_Y)^2 r_{N,N}(u) + (c'_Y)^2 r_N\delta(u)\big)\,du\,dt. \end{aligned}$$

The measure $C'_{X,X}(u)$ is seen to have absolutely continuous part and an atom of mass $c'_{Y,Y}(0)r_N + (c'_Y)^2 r_N$ at $u = 0$. If we let $r_{X,X}(u)$ denote the derivative of the absolutely continuous part of $C'_{X,X}(u)$ then, from (6.6),

(6.7) $$r_{X,X}(u) = c'_{Y,Y}(u)r_{N,N}(u) + c'_{Y,Y}(u)r_N^2 + (c'_Y)^2 r_{N,N}(u)$$

for $-\infty < u < \infty$. For convenience set

(6.8) $$h(u) = r_{X,X}(u) - (c'_Y)^2 r_{N,N}(u).$$

If

(6.9) $$r_N^2 + r_{N,N}(u) \neq 0,$$

then, from (6.7),

(6.10) $$c'_{Y,Y}(u) = \frac{h(u)}{\big[r_{N,N}(u) + r_N^2\big]}.$$

We see, from (6.8) and (6.10), that an estimate of $c'_{Y,Y}(u)$ may be constructed from estimates of $r_{X,X}(u)$, c'_Y, $r_{N,N}(u)$, r_N. One can then proceed to form an estimate of $f'_{Y,Y}(\lambda)$.

Alternatively we could proceed directly to the frequency domain and note that the power spectrum of the process $X(\Delta)$ is given by

$$(6.11) \quad f'_{X,X}(\lambda) = (2\pi)^{-1} \left[\int r_{X,X}(u) \exp\{-i\lambda u\}\, du + c'_{Y,Y}(0)r_N + (c'_Y)^2 r_N \right]$$

$$= \int f'_{Y,Y}(\lambda - \alpha) g'_{N,N}(\alpha)\, d\alpha + f'_{Y,Y}(\lambda)r_N^2 + (c'_Y)^2 f'_{N,N}(\lambda)$$
$$+ (2\pi)^{-1} c'_{Y,Y}(0)r_N,$$

for $-\infty < \lambda < \infty$. If we rewrite this in the form

$$(6.12) \quad f'_{Y,Y}(\lambda)r_N^2 + \int f'_{Y,Y}(\alpha)g'_{N,N}(\lambda - \alpha)\, d\alpha$$

$$= f'_{X,X}(\lambda) - (c'_Y)^2 f'_{N,N}(\lambda) - (2\pi)^{-1}c'_{Y,Y}(0)r_N$$
$$= H(\lambda),$$

then we have an integral equation for $f'_{Y,Y}(\lambda)$. This equation may be solved for $f'_{Y,Y}(\lambda)$, under the condition (6.9), as follows: set

$$(6.13) \quad P(\lambda) = (2\pi)^{-1} \int \exp\{-i\lambda u\} r_{N,N}(u)/[r_N^2 + r_{N,N}(u)]\, du,$$

then

$$(6.14) \quad f'_{Y,Y}(\lambda) = r_N^{-2} H(\lambda) - 2\pi r_N^{-2} \int P(\lambda - \alpha)H(\alpha)\, d\alpha.$$

Once estimates of r_N, $r_{N,N}(u)$, c'_Y, $c'_{Y,Y}(0)$, $f'_{N,N}(\lambda)$, $f'_{X,X}(\lambda)$, are available an estimate of $f'_{Y,Y}(\lambda)$ may be constructed from (6.14). The estimates may be determined as follows:

$$(6.15) \qquad r_N^{(T)} = n/T;$$

$$(6.16) \qquad c_Y^{(T)} = [Y(\tau_1) + \cdots + Y(\tau_n)]/n;$$

$$(6.17) \qquad m_{Y,Y}^{(T)}(0) = [Y(\tau_1)^2 + \cdots + Y(\tau_n)^2]/n;$$

$$(6.18) \qquad c_{Y,Y}^{(T)}(0) = m_{Y,Y}^{(T)}(0) - c_Y^{(T)2};$$

and finally estimates $f_{N,N}^{(T)}(\lambda)$, $f_{X,X}^{(T)}(\lambda)$ may be constructed in the manner of (4.16) or (4.34).

A problem related to the one just considered is that of obtaining as estimate of the cross spectrum $f'_{Y_1,Y_2}(\lambda)$ of a series $Y_1(t)$ with a series $Y_2(t)$ from the values

$$(6.19) \qquad \tau_1, \cdots, \tau_n,$$

$$(6.20) \qquad Y_1(\tau_1), \cdots, Y_1(\tau_n),$$

and

(6.21) $$Y_2(\tau_1), \cdots, Y_2(\tau_n).$$

In this case the expression (6.11) is replaced by

(6.22) $$f'_{X_1, X_2}(\lambda) = \int f'_{Y_1, Y_2}(\lambda - \alpha) g'_{N, N}(\alpha) \, d\alpha + f'_{Y_1, Y_2}(\lambda) r_N^2$$
$$+ (c'_{Y_1})(c'_{Y_2}) f'_{N, N}(\lambda) + (2\pi)^{-1} c'_{Y_1, Y_2}(0) r_N.$$

A second related problem would be to construct an estimate of $f'_{Y_1, Y_2}(\lambda)$ from the values

(6.23) $$\sigma_1, \cdots, \sigma_m,$$

(6.24) $$\tau_1, \cdots, \tau_n,$$

(6.25) $$Y_1(\sigma_1), \cdots, Y_1(\sigma_m),$$

(6.26) $$Y_2(\tau_1), \cdots, Y_2(\tau_n),$$

where $\sigma_1, \cdots, \sigma_m$ are the times of events in $(0, T]$ of a point process $N_1(\Delta)$ and $\tau_1, \cdots, \tau_n$ are the times of events in $(0, T]$ of a related point process $N_2(\Delta)$ with the bivariate point process satisfying Assumption 3.2. In this case expression, (6.11) is replaced by the simpler expression

(6.27) $$f'_{X_1, X_2}(\lambda) = \int f'_{Y_1, Y_2}(\lambda - \alpha) g_{N_1, N_2}(\alpha) \, d\alpha + f'_{Y_1, Y_2}(\lambda) r_{N_1} r_{N_2}$$
$$+ c_{Y_1} c_{Y_2} f'_{N_1, N_2}(\lambda).$$

7. Further considerations

We next discuss briefly some practical implications and extensions of the previous results. We saw, in Section 2, that if $\mathbf{X}(\Delta)$, $\Delta \in \Delta$, was a stationary interval process with cumulant spectra

(7.1) $$f_{X_{a_1}, \cdots, X_{a_k}}(\lambda_1, \cdots, \lambda_k),$$

then

(7.2) $$Y_a(t) = \int \phi_a(t - u) \, dX_a(u),$$

$a = 1, \cdots, r, -\infty < t < \infty$, was a stationary time series with cumulant spectra

(7.3) $$f_{Y_{a_1}, \cdots, Y_{a_k}}(\lambda_1, \cdots, \lambda_k) = \Phi_{a_1}(\lambda_1) \cdots \Phi_{a_k}(\lambda_k) f_{X_{a_1}, \cdots, X_{a_k}}(\lambda_1, \cdots, \lambda_k).$$

This suggests that one might estimate the spectrum (7.1) by a statistic of the form

(7.4) $$f_{Y_{a_1}, \cdots, Y_{a_k}}^{(T)}(\lambda_1, \cdots, \lambda_k) / [\Phi_{a_1}(\lambda_1) \cdots \Phi_{a_k}(\lambda_k)],$$

having formed

$$(7.5) \qquad f_{Y_a, \cdots, Y_a}^{(T)} (\lambda_1, \cdots, \lambda_k)$$

in the manner of Brillinger and Rosenblatt [9]. (In the case $k = 2$ this suggestion was made by Priestly in the discussion of Bartlett [3].) This procedure is seen to be analogous with the technique of prewhitening a time series prior to estimating its spectrum. This analogy suggests that we should choose the $\phi_a(\lambda)$ so that the spectrum (7.3) is near constant for λ_j in some finite region. The estimate (7.4) is seen to have the important advantage of allowing the use of existing spectral programs and also of allowing a simultaneous prewhitening of the data.

The proposed analysis may be related to the analysis of a continuous time series in another way. The basic statistic of our analysis is

$$(7.6) \qquad \int_0^T \exp\{-i\lambda t\} h(t/T) \, d\mathbf{X}(t).$$

If we approximate (7.6) by a Stieltjes sum, then we obtain

$$(7.7) \qquad \sum_{t=0}^{T-1} \exp\{-i\lambda t\} h(t/T) [\mathbf{X}(t+1) - \mathbf{X}(t)].$$

An examination of expression (7.7) shows that it corresponds to carrying out an empirical spectral analysis on the time series of first differences. This procedure is common in the analysis of economic time series.

Computations involved in forming (7.6) may be prohibitive. Therefore there is much to be said for a procedure involving splitting the data into N segments of length S, forming an estimate

$$(7.8) \qquad f_{a_1, \cdots, a_k}^{(S)} (\lambda_1, \cdots, \lambda_k)_n$$

for the nth segment, $n = 1, \cdots, N$, and taking

$$(7.9) \qquad N^{-1} \sum_{n=1}^N f_{a_1, \cdots, a_k}^{(S)} (\lambda_1, \cdots, \lambda_k)_n$$

as a final estimate. Authors recommending such a procedure include: Bartlett [2], Welch [23], Lewis [19], and Huber *et al* [15]. The asymptotics of such estimates are directly determinable from the results of Theorem 4.3 because, following the remark after Theorem 4.1, Fourier transforms based on disjoint stretches of data are asymptotically independent. A variety of further remarks concerning practical aspects of the calculations in the case of a point process are made in Lewis [19].

We remark that the calculations proposed in this paper reduce, in the case that the interval process $\mathbf{X}(\Delta)$, $\Delta \in \Delta$, is an integral of a continuous time series, to the usual calculations of the frequency analysis of time series.

Extensions of the definitions and theorems of this paper to a case in which t is vector valued, $t \in R^p$, appear fairly immediate if one takes the approach of

Brillinger [7]. A different sort of extension would result from a consideration of processes whose differences of higher order than the first are stationary (see Yaglom [24]).

8. Proofs

PROOF OF LEMMA 2.1. If $M_{a_1,\cdots,a_k}(t_1,\cdots,t_k)$ corresponds to the measure determined by the coordinates $t_1,\cdots,t_k$, let $N_{a_1,\cdots,a_k}(u_1,\cdots,u_{k-1},t_k)$ correspond to the measure determined by the coordinates $u_1 = t_1 - t_k, \cdots, u_{k-1} = t_{k-1} - t_k, t_k$. The initial measure is invariant under the transformation $t_1,\cdots,t_k \to t_1 + t,\cdots,t_k + t$. The second measure is therefore invariant under the transformation $t_k \to t_k + t$. We see therefore that

$$(8.1) \qquad N_{a_1,\cdots,a_k}(u_1,\cdots,u_{k-1},t_k) - N_{a_1,\cdots,a_k}(u_1,\cdots,u_{k-1},0)$$
$$= N_{a_1,\cdots,a_k}(u_1,\cdots,u_{k-1},t_k + t) - N_{a_1,\cdots,a_k}(u_1,\cdots,u_{k-1},t).$$

Suppressing $a_1,\cdots,a_k, u_1,\cdots,u_{k-1}$ this last may be written

$$(8.2) \qquad N(t_k + t) = N(t_k) + N(t) - N(0).$$

Under the given conditions, all solutions of this functional equation have the form

$$(8.3) \qquad N(t_k) = M't_k + N(0),$$

giving the indicated result.

PROOF OF THEOREM 2.1. Assume the results of Theorem 4.1 hold. It will be proved later. Set

$$(8.4) \qquad \mathbf{d}_X^{(T)}(\lambda) = \int_{-T}^{T} \exp\{-i\lambda t\}\, d\mathbf{X}(t),$$

using the notation of Section 4 with $h(t) = 1$ for $|t| \leq 1$ and $h(t) = 0$ otherwise. One has therefore

$$(8.5) \qquad \mathbf{Z}_X^{(T)}(\lambda) = (2\pi)^{-1} \int_0^{\lambda} d_X^{(T)}(\alpha)\, d\alpha.$$

One now uses expression (4.7) to see that

$$(8.6) \qquad E\big|Z_a^{(T)}(\lambda) - Z_a^{(S)}(\lambda)\big|^{2k} \to 0$$

as $S, T \to \infty$ for $k = 1, 2, \cdots; a = 1, \cdots, r$. It follows that there exists $\mathbf{Z}_X(\lambda)$ such that $\mathbf{Z}_X^{(T)}(\lambda) \to \mathbf{Z}_X(\lambda)$ in mean of order v for any $v > 0$.

One next checks that

$$(8.7) \qquad \text{cum}\,\{Z_{a_1}^{(T)}(\lambda_1),\cdots,Z_{a_k}^{(T)}(\lambda_k)\}$$
$$\to \int_0^{\lambda_1}\cdots\int_0^{\lambda_k} \delta\left(\sum_1^k \alpha_j\right) f_{a_1,\cdots,a_k}(\alpha_1,\cdots,\alpha_k)\, d\alpha_1\cdots d\alpha_k$$

as $T \to \infty$, again using expression (4.7). This gives (2.27).

Finally one checks that

$$(8.8) \qquad E|X_a(\Delta) - \int_{-\infty}^{\infty} \left[\int_{\Delta} \exp\{i\lambda t\}\, dt \right] dZ_a^{(T)}(\lambda)|^2 \to 0$$

as $T \to \infty$. This gives (2.28).

PROOF OF THEOREM 3.1.   We first state and prove a lemma.

LEMMA 8.1.   *With the conditions and notation of Assumption 3.1,*

$$(8.9) \qquad \lim_{|\Delta_j| \to 0} |\Delta_1|^{-1} \cdots |\Delta_k|^{-1} E\{N_{a_1}(\Delta_1) \cdots N_{a_{m_1}}(\Delta_1)] \cdots$$

$$\cdots [N_{a_{m_{k-1}+1}}(\Delta_k) \cdots N_{a_{m_k}}(\Delta_k)]\}$$

$$= \sum_{\alpha_1, \cdots, \alpha_k = 1}^{r} \left[\prod_{j=1}^{m_1} \delta\{a_j - \alpha_1\} \right]$$

$$\cdots \left[\prod_{j=m_{k-1}+1}^{m_k} \delta\{a_j - \alpha_k\} \right] p_{\alpha_1, \cdots, \alpha_k}(t_1, \cdots, t_k)$$

uniformly in $t_1, \cdots, t_k$ *for integers* $1 \leqq m_1 < m_2 < \cdots < m_{k-1} < m_k$.

PROOF.   Suppose first that

$$(8.10) \qquad \begin{matrix} a_1, \cdots, a_{m_1} = \alpha_1, \\ \vdots \\ a_{m_{k-1}+1}, \cdots, a_{m_k} = \alpha_k. \end{matrix}$$

Now

$$(8.11) \qquad E\{N_{\alpha_1}(\Delta_1)^{m_1} \cdots N_{\alpha_k}(\Delta_k)^{m_k - m_{k-1}}\}$$

$$= \sum_{n_j \geqq 1} n_1^{m_1} \cdots n_k^{m_k - m_{k-1}} P[N_{\alpha_1}(\Delta_1) = n_1, \cdots, N_{\alpha_k}(\Delta_k) = n_k]$$

$$= P[N_{\alpha_1}(\Delta_1) = 1, \cdots, N_{\alpha_k}(\Delta_k) = 1] + \sum n_1^{m_1} \cdots n_k^{m_k - m_{k-1}} L(n_1, \cdots, n_k),$$

with the second summation extending over some $n_j \geqq 2$ and with $|L(n_1, \cdots, n_k)| \leqq K_\delta |\Delta_1|^{n_1} \cdots |\Delta_k|^{n_k}$ from (3.8), and so

$$(8.12) \qquad \lim_{|\Delta_j| \to 0} |\Delta_1|^{-1} \cdots |\Delta_k|^{-1} E\{N_{\alpha_1}(\Delta_1)^{m_1} \cdots N_{\alpha_k}(\Delta_k)^{m_k - m_{k-1}}\}$$

$$= p_{\alpha_1, \cdots, \alpha_k}(t_1, \cdots, t_k),$$

uniformly in $t_1, \cdots, t_k$ from (3.9). Continuing if (8.10) is not satisfied for some $\alpha_1, \cdots, \alpha_k$, then one can see from (3.8) that the limit in (8.9) is 0 uniformly in $t_1, \cdots, t_k$. This completes the proof of the lemma.

Turning to the proof of the theorem; let $\phi_j(t)$ be continuous is some interval of Δ and 0 elsewhere for $j = 1, \cdots, k$. We have

$$(8.13) \qquad \int \phi_j(t)\, dN_{a_j}(t) = \lim_{\varepsilon \to 0} \sum_i \phi_j(i\varepsilon) N_{a_j}(i\varepsilon, i\varepsilon + \varepsilon].$$

By bounded convergence,

$$(8.14) \quad E\left\{ \int \phi_1(t)\, dN_{a_1}(t) \cdots \int \phi_k(t)\, dN_{a_k}(t) \right\}$$

$$= \lim_{\varepsilon \to 0} \sum_{i_1} \cdots \sum_{i_k} \phi_1(i_1\varepsilon) \cdots \phi_k(i_k\varepsilon) E\{N_{a_1}(i_1\varepsilon, i_1\varepsilon + \varepsilon] \cdots N_{a_k}(i_k\varepsilon, i_k\varepsilon + \varepsilon]\}$$

$$= \lim_{\varepsilon \to 0} \sum_{\ell=1}^{k} \sum_{\alpha_1, \cdots, \alpha = 1}^{r} \left[\prod_{j \in v_1} \delta\{a_j - \alpha_1\} \right] \cdots \left[\prod_{j \in v_\ell} \delta\{a_j - \alpha_\ell\} \right]$$

$$\cdot \sum_{i_1} \cdots \sum_{i_\ell} \left[\prod_{j \in v_1} \phi_j(i_1\varepsilon) \right] \cdots \left[\prod_{j \in v_\ell} \phi_j(i_\ell\varepsilon) \right] p_{\alpha_1, \cdots, \alpha}(i_1\varepsilon, \cdots, i_\ell\varepsilon)\varepsilon^\ell,$$

where the summation extends over partitions $(v_1, \cdots, v_\ell)$ of $(1, \cdots, k)$ if we separate out terms in (8.14) with the same argument and use Lemma 8.1. We now see that expression (8.14) equals

$$(8.15) \quad \sum_{\ell=1}^{k} \sum_{\alpha_1, \cdots, \alpha_\ell = 1}^{r} \left[\prod_{j \in v_1} \delta\{a_j - \alpha_1\} \right] \cdots \left[\prod_{j \in v_\ell} \delta\{a_j - \alpha\} \right]$$

$$\cdot \int \cdots \int \left[\prod_{j \in v_1} \phi_j(\tau_1) \right] \cdots \left[\prod_{j \in v_\ell} \phi_j(\tau_\ell) \right] p_{\alpha_1, \cdots, \alpha_\ell}(\tau_1, \cdots, \tau_\ell)\, d\tau_1 \cdots d\tau_\ell.$$

Expression (3.13) now follows from (8.15) taking the $\phi_j(t)$ to be indicator functions.

PROOF OF THEOREM 3.2. One proves (3.28) from (8.15) and then obtains (3.22) by taking the $\phi_j(t)$ to be indicator functions.

PROOF OF THEOREM 3.3. This follows directly from (3.28).

PROOF OF THEOREM 4.1. Let $h_a^{(T)}(t) = h_a(t/T)$. The cumulant at issue is given by

$$(8.16) \quad \int \cdots \int h_{a_1}^{(T)}(t_1) \cdots h_{a_k}^{(T)}(t_k)$$

$$\cdot \exp\left\{ -i \sum_1^k \lambda_j t_j \right\} dC'_{a_1, \cdots, a_k}(t_1 - t_k, \cdots, t_{k-1} - t_k)\, dt_k$$

$$= \int \cdots \int \left[\int_t h_{a_1}^{(T)}(u_1 + t) \cdots h_{a_{k-1}}^{(T)}(u_{k-1} + t) h_{a_k}^{(T)}(t) \right.$$

$$\left. \cdot \exp\left\{ -i \sum_1^k \lambda_j t \right\} dt \right] \exp\left\{ -i \sum_1^{k-1} \lambda_j u_j \right\} dC'_{a_1, \cdots, a_k}(u_1, \cdots, u_{k-1}).$$

The indicated result now follows as

$$(8.17) \quad \left| \int_t \{ h_{a_1}^{(T)}(u_1 + t) \cdots h_{a_{k-1}}^{(T)}(u_{k-1} + t) h_{a_k}^{(T)}(t) \right.$$

$$\left. - h_{a_1}^{(T)}(t) \cdots h_{a_k}^{(T)}(t) \exp\{-i\lambda t\}\, dt \right| \leq C \sum_1^{k-1} |u_j|,$$

for some finite C following Assumption 4.1.

PROOF OF THEOREM 4.2.   This follows directly from Theorem 4.1 in the manner of corresponding results in Brillinger [7] and Brillinger [8].

PROOF OF THEOREM 4.3.   This follows directly from Theorem 4.1 in the manner of the principal theorems in Brillinger and Rosenblatt [9].

PROOF OF THEOREM 4.4.   This follows directly from Theorem 4.1 in the manner of corresponding results in Brillinger [7] and Brillinger [8].

REFERENCES

[1] H. J. BAHBA, "On the stochastic theory of continuous parametric systems and its application to electron cascades," *Proc. Roy. Soc. Ser. A*, Vol. 202 (1950), pp. 301–322.

[2] M. S. BARTLETT, "Smoothing periodograms from time-series with continuous spectra," *Nature*, Vol. 161 (1948), pp. 686–687.

[3] ———, "The spectral analysis or point processes," *J. Roy. Statist. Soc. Ser. B*, Vol. 25 (1963), pp. 264–296.

[4] ———, *Stochastic Processes*, Cambridge, Cambridge University Press, 1966.

[5] S. BOCHNER, *Harmonic Analysis and the Theory of Probability*, Berkeley and Los Angeles, University of California Press, 1960.

[6] D. R. BRILLINGER, "Asymptotic properties of spectral estimates of second order," *Biometrika*, Vol. 56 (1969), pp. 375–390.

[7] ———, "The frequency analysis of relations between stationary spatial series," *Proceedings Twelfth Biennial Seminar*, Montreal, Canadian Mathematical Congress, 1970, pp. 39–81.

[8] ———, *The Frequency Analysis of Vector-Valued Time Series*, New York, Holt, Rinehart, and Winston, to appear 1971.

[9] D. R. BRILLINGER and M. ROSENBLATT, "Asymptotic theory of estimates of kth order spectra," *Spectral Analysis of Time Series* (edited by B. Harris), New York, Wiley, 1967, pp. 153–188.

[10] D. R. BRILLINGER and J. W. TUKEY, *Asymptotic Variances, Moments, Cumulants, and Other Averaged Values*, 1964, unpublished.

[11] D. R. COX and P. A. W. LEWIS, *The Statistical Analysis of Series of Events*, London, Methuen, 1966.

[12] H. CRAMÉR and M. R. LEADBETTER, *Stationary and Related Stochastic Processes*, New York, Wiley, 1967.

[13] D. J. DALEY, "Spectral properties of weakly stationary point processes," *J. Roy. Statist. Soc. Ser. B*, Vol. 33 (1971).

[14] J. L. DOOB, *Stochastic Processes*, New York, Wiley, 1953.

[15] P. J. HUBER, B. KLEINER, TH. GASSER, G. DUMERMUTH, "Statistical methods for investigating phase relations in stationary stochastic processes," International Seminar on Digital Processing of Analog Signals, Zürich, 1970.

[16] K. ITO, "Stationary random distributions," *Mem. Coll. Sci. Univ. Kyoto Ser. A Math.*, Vol. 28 (1954), pp. 209–224.

[17] A. N. KOLMOGOROV, "Curves in Hilbert space invariant with regard to a one parameter group of motions," *Dokl. Akad. Nauk SSSR*, Vol. 26 (1940), pp. 6–9.

[18] P. I. KUZNETSOV and R. L. STRATONOVICH, "A note on the mathematical theory of correlated random points," *Non-Linear Transformations of Stochastic Processes* (edited by P. I. Kuznetsov, R. L. Stratonovich, and V. I. Tikhonov), Oxford, Pergamon, 1965, pp. 101–115.

[19] P. A. W. LEWIS, "Remarks on the theory, computation and application of the spectral analysis of series of events," *J. Sound Vib.*, Vol. 12 (1970), pp. 353–375.

[20] A. RAMAKRISHNAN, "Stochastic processes relating to particles distributed in a continuous infinity of states," *Proc. Cambridge Philos. Soc.*, Vol. 46 (1950), pp. 595–602.

[21] S. K. SRINIVASAN, *Stochastic Theory and Cascade Processes*, New York, Elsevier, 1969.

[22] J. W. TUKEY, "An introduction to the calculations of numerical spectrum analysis," *Spectral Analysis of Time Series* (edited by B. Harris), New York, Wiley, 1967, pp. 25–46.

[23] P. D. WELCH, "A direct digital method of power spectrum estimation," *IBM J. Res. Develop.*, Vol. 5 (1961), pp. 141–156.

[24] A. M. YAGLOM, "Correlation theory of processes with stationary *n*-th order increments," *Amer. Math. Soc. Transl.*, Vol. 8 (1958), pp. 87–142.

[25] ———, *An Introduction to the Theory of Stationary Random Functions*, Englewood Cliffs, Prentice-Hall, 1962.

CREDIBILITY PROCEDURES

HANS BÜHLMANN

SWISS FEDERAL INSTITUTE OF TECHNOLOGY

1. Historical review

The concept of credibility is used by actuaries to estimate expected values (net premiums) from statistical data. The first papers on the subject were written by Whitney [8] and Perryman [7]. In the 1950's Arthur Bailey [1] in two special parametric cases gave a mathematical model from which the credibility procedures could be justified. Only in the last few years a nonparametric credibility theory was developed [3] which is now being further refined [4], [5]. The technique derived from the theory is becoming a major actuarial tool in non-life insurance.

As will be apparent from the formulation below, the method of estimation which actuaries mean when referring to credibility procedures is of quite general interest and can easily be transcribed to other fields of application where it may be used for forecasting. For reasons of intuitive appeal I shall, however, restrict the terminology to the actuarial application. The presentation here given follows in many respects that in [4] (in German).

2. The problem

For $i = 1, \cdots, n; j = 1, \cdots, N$, we consider random variables $X_{i,j}$, non-negative real numbers $P_{i,j}$, and maps $\rho_{i,j}$ from R^∞ to R^1, where we think of j as indicating the risk (or risk group) within the collective of risks and i as indicating the period (say year) over which these risks (risk groups) can be observed. The above introduced abstract concepts have the following intuitive meaning:

$X_{i,j}$ is the observable risk performance (year i, risk j),

$P_{i,j}$ is the measure of exposure (year i, risk j), and

$\rho_{i,j}$ is the map assigning the risk performance $X_{i,j}$ to the doubly stochastic sequence of individual claims (year i, risk j). (We call $\rho_{i,j}$ the insurance conditions.)

Finally we introduce a parameter $\vartheta_{i,j}$ taking values in an abstract θ, $i = 1, \cdots, n$; $j = 1, \cdots, N$, and we think of it as characterizing the "quality" of the risk j in the year i. Using all symbols just introduced we write

515

(2.1) $F_\vartheta(x|P, \rho) = P[X_{i,j} \leq x|\vartheta_{i,j} = \vartheta, P_{i,j} = P, \rho_{i,j} = \rho]$

for the probability distribution function of $X_{i,j}$ given ϑ, P, ρ.

We now introduce the year $i = 0$. It is the one (just beginning now) for which a forecast (rating) has to be made. Actuaries would call it the rating year (rating period). On the other hand the positive values of the index i are thought of as referring to the past years (observation period) in the "reversed natural time order." We also reserve the index k for the one risk which we want to rate.

Mathematically we are faced with the following estimation problem: estimate $\mu(\vartheta_{0,k}, P_{0,k}, \rho_{0,k}) = E[X_{0,k}|\vartheta_{0,k}, P_{0,k}, \rho_{0,k}]$, where $P_{0,k}$ and $\rho_{0,k}$ (exposure and insurance conditions) are supposed to be *known* and $\vartheta_{0,k}$ (quality) is *unknown*. The data available for this estimation are:

the observations on $\mathcal{X} = \{(X_{i,j}, i = 1, \cdots, n; j = 1, \cdots, N)\}$,

the past exposures $\mathcal{P} = \{(P_{i,j}, i = 1, \cdots, n; j = 1, \cdots, N)\}$,

and the past insurance conditions $\mathcal{R} = \{(\rho_{i,j}, i = 1, \cdots, n; j = 1, \cdots, N)\}$.

Observe that we are treating here only the case of estimating the mean. In [5] this method is extended to estimating $\mu(\vartheta_{0,k}, P_{0,k}, \rho_{0,k}) + \beta\sigma^2(\vartheta_{0,k}, P_{0,k}, \rho_{0,k})$ and in principle one might try to estimate any functional of $F_{\vartheta_{0,k}}(x|P_{0,k}, \rho_{0,k})$ by a method similar to the one described here.

3. Assumptions

The following properties are assumed to hold throughout this paper.

(i) *Independence of risk performances.* Given the values of the parameter $\vartheta_{i,j}$ the risk performances $X_{i,j}$ shall be independent for all i and j.

(ii) *Homogeneity in time.* The quality of each risk shall not vary in time, that is, $\vartheta_{i,j} = \vartheta_j$ independent of i for all $j = 1, \cdots, N$.

(iii) *Independence of parameter values.* The parameters $\vartheta_j, j = 1, \cdots, N$, are independent random variables all obeying the same distribution v with $v(M) = P[\vartheta_j \in M]$ for all measurable subsets of $M \subset \theta$.

(iv) *Existence of as-if-statistics per risk.* This means that it shall be possible to reconstruct "artificial statistics" as if the insurance conditions of each risk had always been the same, that is, $\rho_{i,j} = \rho_j$ independent of i for all $j = 1, \cdots, N$ and in particular $\rho_{0,k} = \rho_k$ for the risk to be rated.

The reader is referred to [4] for a discussion of these assumptions from an actuarial viewpoint. In particular the distribution v is thought of as an idealized version of frequencies of risk qualities in the *collective* of risks from which the *portfolio* containing the risks j, where $j = 1, \cdots, N$, has been drawn. We therefore call $v(M)$ the *structural distribution* of the collective. Classical actuarial estimating methods implicitly or explicitly always assume this structural distribution to degenerate, which then leads to the fiction of a "homogeneous collective."

4. Equilibrium

Denote by X an element of $\mathscr{X}$, P an element of $\mathscr{P}$, and R an element of $\mathscr{R}$ and let $\bar{P} = (P, P_{0,k})$ and $\bar{R} = (\rho_1, \cdots, \rho_N)$. We then write $\hat{\mu}_k(X; \bar{P}; \bar{R})$ for the estimate of $\mu(\vartheta_k, P_{0,k}, \rho_k) = \mu_k$ and call μ_k the *correct premium* and $\hat{\mu}_k$ the *rating*. For our further investigations it is then important to distinguish the two cases and whether the structural distribution is *known* or *unknown*.

4.1. *Structural distribution known.* Given the structural distribution, we call a rating $\hat{\mu}_k$ *in equilibrium over a* (measurable) *set* $S \subset \mathscr{X} \times \theta^N$ if

$$(4.1) \qquad \int_S \hat{\mu}_k(X; \bar{P}; \bar{R})\, dP = \int_S \mu(\vartheta_k, P_{0,k}, \rho_k)\, dP$$

where, with the usual abuse of notation,

$$(4.2) \qquad dP = \prod_{j=1}^{N} \left(\prod_{i=1}^{n} dF_{\vartheta_j}\left[x_{i,j} \middle| P_{i,j}, \rho_j \right] \right) dv(\vartheta_j).$$

From this form of the probability P it follows in particular, that (ϑ_1, X_1), $(\vartheta_2, X_2), \cdots, (\vartheta_N, X_N)$ are independent random variables.

Observe that we write here X_k for $(X_{1,k}, X_{2,k}, \cdots, X_{n,k})$. Later on we shall also write $\bar{P}_k$ for $(P_{0,k}, P_{1,k}, \cdots, P_{n,k})$.

We now ask that the rating $\hat{\mu}_k$ be in equilibrium for all cylinders $S \subset \mathscr{X} \times \theta^N$ with basis in $\mathscr{X}$. This means intuitively that the difference between correct premium and rating should average out to zero over any part of the collective which is characterized by experience alone. If we require this, we know from the Radon-Nikodym theorem that we must have

$$(4.3) \qquad \hat{\mu}_k(X; \bar{P}; \bar{R}) = E\left[\mu(\vartheta_k, P_{0,k}, \rho_k) \middle| X_1, \cdots, X_N \right]$$

and by the independence of $\{(\vartheta_k, X_k), k = 1, \cdots, N\}$ we have

$$(4.4) \qquad \hat{\mu}(X; \bar{P}; \bar{R}) = E\left[\mu(\vartheta_k, P_{0,k}, \rho_k) \middle| X_k \right].$$

To express that the right side only depends on the observations made on the risk k alone, we write for the rating $\hat{\mu}_k$

$$(4.5) \qquad \hat{\mu}_k(X_k; \bar{P}_k; \rho_k) = E\left[\mu(\vartheta, P_{0,k}, \rho_k) \middle| X_k \right]$$

where we further drop the index k from ϑ_k to express that the structural distribution v is the same for all $\vartheta_j, j = 1, \cdots, N$. Observe that the rating formula (4.5) could also be derived from the postulate

$$(4.6) \qquad \int_{\mathscr{X} \times \theta^N} \left[\hat{\mu}_k(X; \bar{P}; \bar{R}) - \mu(\vartheta_k, P_{0,k}, \rho_k) \right]^2 dP = \text{minimum},$$

but we prefer the equilibrium argument which corresponds to the requirements for a rating in practical applications.

4.2. *Structural distribution unknown.* Under 4.1 we have derived the rating $\hat{\mu}_k$ and we have found that it is a function depending on the arguments $(X_k; \bar{P}_k; \rho_k)$.

The *form of this function* is however *undetermined* if the structural distribution is *not* given. In this case we follow the basic idea of Robbins [6] and use the full information $(X, \bar{P}, \bar{R})$ to *estimate* the form of the rating function. Instead of discussing this in full generality we will show how to proceed in the case of the "credibility rating" which we now want to present.

5. Credibility rating, generalities

In many actuarial applications it can at best be hoped that first and second moments relating to the structural distribution are known. We therefore want to develop suitable *approximations* to the rating $\hat{\mu}_k$, which itself can only be computed if the whole structural distribution is given. The easiest way to get such approximations is by restricting admissible predictors of the correct premium to the linear form. Following standard actuarial terminology we speak then of "credibility."

A *credibility rating* μ_k^* is defined as

(5.1) $$\mu_k^*(X; \bar{P}; \bar{R}) = \sum_{i,j} \alpha_{i,j} X_{i,j} + \beta_k$$

where $\alpha_{i,j}$ and β_k are constants, and we say that μ_k^* approximates $\hat{\mu}_k$ best if

(5.2) $$E[\mu_k^*(X; \bar{P}; \bar{R}) - \hat{\mu}_k(X_k; P_k; \rho_k)]^2$$

is smallest among all credibility ratings μ_k^*. The expected value operation over $\mathscr{X} \times \theta^N$ is denoted by $E[\cdot]$ as in Section 4. Since

(5.3) $$E[\mu_k^*(X; \bar{P}; \bar{R}) - \mu(\vartheta_k, P_{0,k}, \rho_k)]^2 = E[\mu_k^*(X; \bar{P}; \bar{R}) - \hat{\mu}_k(X_k; \bar{P}_k; \rho_k)]^2 + E[\hat{\mu}_k(X_k; \bar{P}_k; \rho_k) - \mu(\vartheta_k, P_{0,k}, \rho_k)]^2,$$

the best credibility rating can also be derived from having

(5.4) $$E[\mu_k^*(X; \bar{P}; \bar{R}) - \mu(\vartheta_k, P_{0,k}, \rho_k)]^2$$

smallest among all μ_k^*.

In order to render our computations easier we are in the following also assuming some special properties of the distributions relating to $X_{i,j}, i = 1, \cdots, n; j = 1, \cdots, N$. (a) $E[X_{i,j}|\vartheta_{i,j} = \vartheta, P_{i,j} = P, \rho_{i,j} = \rho] = \mu(\vartheta, \rho)$ independent of P and (b) $\text{Var}[X_{i,j}|\vartheta_{i,j} = \vartheta, P_{i,j} = P, \rho_{i,j} = \rho] = \sigma^2(\vartheta, \rho)/P$. These assumptions (expressing intuitively that $X_{i,j}$ is some "kind of an average") are discussed in [2]. Finally we introduce the observations $E[\mu(\vartheta, \rho_j)] = m_j$, $\text{Var}[\mu(\vartheta, \rho_j)] = w_j$, and $E[\sigma^2(\vartheta, \rho_j)] = v_j$.

6. Credibility rating if m_j, v_j, w_j are given

Since

(6.1) $$E[\mu_k^*(X; \bar{P}; \bar{R}) - \mu(\vartheta_k, \rho_k)]^2$$

$$= E\left[\sum_i \alpha_{i,k} X_{i,k} + \beta_k + \sum_{i,j \neq k} \alpha_{i,j} E(X_{i,j}) \right.$$

$$\left. + \sum_{i,j \neq k} \alpha_{i,j} [X_{i,j} - E(X_{i,j})] - \mu(\vartheta_k, \rho_k) \right]^2$$

$$= E\left[\sum_{i,j \neq k} \alpha_{i,j} [X_{i,j} - E(X_{i,j})] \right]^2 + E\left[\sum_i \alpha_{i,k} X_{i,k} + \beta_k' - \mu(\vartheta_k, \rho_k) \right]^2,$$

where for brevity $\beta_k' = \beta_k + \sum_{i,j \neq k} \alpha_{i,j} E(X_{i,j})$, the optimal μ_k^* can be searched for among those linear estimators for which $\alpha_{i,j} = 0$ for $j \neq k$. Hence we have

$$(6.2) \qquad E\left[\sum_i \alpha_{i,k} X_{i,k} + \beta_k - \mu(\vartheta_k, \rho_k) \right]^2 = E\left[\sum_i \alpha_{i,k} [X_{i,k} - \mu(\vartheta_k, \rho_k)] \right]^2$$

$$+ E\left[\left(\sum_i \alpha_{i,k} - 1 \right) \mu(\vartheta_k, \rho_k) + \beta_k \right]^2$$

$$= \sum_i \alpha_{i,k}^2 \frac{v_k}{P_{i,k}} + \left(\sum_i \alpha_{i,k} - 1 \right)^2 w_k.$$

Put $\beta_k = (1 - \Sigma_i \alpha_{i,k}) m_k$, which minimizes the second term on the right of (6.2). The minimum is achieved for

$$(6.3) \qquad \alpha_{i,k} = \frac{P_{i,k} w_k}{v_k + \Sigma_i P_{i,k} w_k}.$$

Hence we have for the credibility rating

$$(6.4) \qquad \mu_k^* = \frac{\sum_i P_{i,k} X_{i,k}}{\sum_i P_{i,k}} \frac{\sum_i P_{i,k} w_k}{v_k + \Sigma_i P_{i,k} w_k} + m_k \frac{v_k}{v_k + \Sigma_i P_{i,k} w_k}.$$

Observe that we have obtained a *weighted average* between the individual average experience and the theoretical mean over the whole collective. It is customary to call the weight attached to the individual average experience the *credibility factor* which thus turns out to be $\Sigma_i P_{i,k} w_k / (v_k + \Sigma_i P_{i,k} w_k)$.

7. Credibility rating if m_j, v_j, w_j are not given

As indicated in Section 4 we will now use all the observations (including those on the "other risks" $j \neq k$) to estimate these quantities. Of course if such a procedure should be meaningful the risks must be "comparable" in some sense. This means that we must have some knowledge about the functional relationship between the m_j, $j = 1, \cdots, N$ (and similarly for the v_j and the w_j). The presentation of the basic ideas becomes clearest if we assume that $m_j = m$, $v_j = v$ and $w_j = w$ for all j. (This is equivalent to the postulate of equal insurance conditions $\rho_j = \rho$ for all risks in all years under consideration.) We then try to find an estimate by again starting with a linear credibility rating

$$(7.1) \qquad \mu_k^* = \sum_{i,j} \alpha_{i,j} X_{i,j}$$

and we postulate

$$(7.2) \qquad E\left[\sum_{i,j} \alpha_{i,j} X_{i,j} - \mu(\vartheta_k, \rho)\right]^2$$

minimum under all credibility ratings of the form (7.1).

As it is by no means clear that the minimum solution of (7.2) is in equilibrium over $\mathscr{X} \times \theta^N$, the whole collective, (contrary to the case treated in Section 6 where this is automatically the case) we require this equilibrium property in addition to (7.2)

$$(7.3) \qquad E\left[\sum_{i,j} \alpha_{i,j} X_{i,j}\right] = m$$

which is equivalent to $\Sigma_{i,j}\, \alpha_{i,j} = 1$. Then formulate the Lagrangian

$$(7.4) \qquad \phi(\alpha_{1,1}, \cdots, \alpha_{n,1}, \alpha_{1,2}, \cdots, \alpha_{n,2}, \cdots, \alpha_{1,N}, \cdots, \alpha_{n,N}, \alpha)$$

$$= E\left[\sum_{i,j} \alpha_{i,j} X_{i,j} - \mu(\vartheta_k, \rho)\right]^2 - 2\alpha \sum_{i,j} \alpha_{i,j} E[X_{i,j}]$$

from which the system of equations is obtained

$$(7.5) \qquad \frac{\partial \phi}{\partial \alpha_{\ell,h}} = 2E\left[\left\{\sum_{i,j} \alpha_{i,j} X_{i,j} - \mu(\vartheta_k, \rho)\right\} X_{\ell,h}\right] - 2\alpha E[X_{\ell,h}] = 0,$$

$$\ell = 1, \cdots, n; h = 1, \cdots, N.$$

Hence using (7.3) and the conditional independence of all $X_{i,j}$

$$(7.6) \qquad E\left[\sum_{i,j} \alpha_{i,j}\{\mu(\vartheta_h, \rho)[\mu(\vartheta_j, \rho) - \mu(\vartheta_k, \rho)]\} + \alpha_{\ell,h}\frac{\sigma^2(\vartheta_h, \rho)}{P_{\ell,h}}\right] = \alpha m$$

and as

$$(7.7) \qquad E[\mu(\vartheta_h, \ell)\mu(\vartheta_j, \rho)] = m^2 + \delta_{h,j}w, \qquad \delta_{h,j} = \begin{matrix} 1, h = j \\ 0, h \neq j, \end{matrix}$$

we get

$$(7.8) \qquad \sum_{i=1}^n \alpha_{i,h}w + \alpha_{\ell,h}\frac{v}{P_{\ell,h}} = \delta_{h,k}w + \alpha m.$$

Hence

$$(7.9) \qquad \alpha_{\ell,h} = \frac{P_{\ell,h}}{v}\left[\left(\delta_{h,k} - \sum_{i=1}^n \alpha_{i,h}\right)w + \alpha m\right] = P_{\ell,h} \cdot C,$$

(where C is independent of ℓ). Summing over the first index we get

$$(7.10) \qquad \sum_{i=1}^n \alpha_{i,h} = \frac{\sum_{i=1}^n P_{i,h}}{v}\left[\left(\delta_{h,k} - \sum_{i=1}^n \alpha_{i,h}\right)w + \alpha m\right]$$

or

$$(7.11) \qquad \sum_{i=1}^{n} \alpha_{i,h} \left(1 + \frac{\sum_i P_{i,h} w}{v} \right) = \frac{\sum_{i=1}^{n} P_{i,h} w}{v} \delta_{h,k} + \frac{\sum_{i=1}^{n} P_{i,h}}{v} \alpha m$$

or equivalently `

$$(7.12) \qquad \sum_{i=1}^{n} \alpha_{i,h} = \frac{\sum_i P_{i,h} w}{v + \sum_i P_{i,h} w} \delta_{h,k} + \frac{\sum_i P_{i,h}}{v + \sum_i P_{i,h} w} \alpha m$$

from which

$$(7.13) \qquad \alpha_{\ell,h} = \frac{P_{\ell,h} w}{v + \sum_{i=1}^{n} P_{i,h} w} \delta_{h,k} + \frac{P_{\ell,h}}{v + \sum_{i=1}^{n} P_{i,h} w} \alpha m,$$
$$\ell = 1, \cdots, n; h = 1, \cdots, N.$$

Let us write $P_{\cdot h}$ for $\Sigma_{i=1}^{n} P_{i,h}$. Then $\Sigma_{i,h} \alpha_{i,h} = 1$ yields

$$(7.14) \qquad 1 = \frac{P_{\cdot k} w}{v + P_{\cdot k} w} + \sum_{h=1}^{N} \frac{P_{\cdot h}}{v + P_{\cdot h} w} \alpha m.$$

Hence

$$(7.15) \qquad \alpha m = \frac{\dfrac{v}{v + P_{\cdot k} w}}{\sum_{h=1}^{n} \dfrac{P_{\cdot h}}{v + P_{\cdot h} w}}.$$

We finally get

$$(7.16) \qquad \alpha_{\ell,h} = \frac{P_{\ell,k} w}{v + P_{\cdot k} w} \delta_{h,k} + \frac{\dfrac{P_{\ell,h}}{v + P_{\cdot h} w}}{\sum_{h=1}^{N} \dfrac{P_{\cdot h}}{v + P_{\cdot h} w}} \frac{v}{v + P_{\cdot k} w}$$

and

$$(7.17) \qquad \mu_k^* = \sum_{i,h} \alpha_{i,h} X_{i,h} = \frac{P_{\cdot k} w}{v + P_{\cdot k} w} \frac{\sum_{i=1}^{n} P_{i,k} X_{i,k}}{\sum_{i=1}^{n} P_{i,k}}$$

$$+ \frac{v}{v + P_{\cdot k} w} \frac{\sum_{h=1}^{N} \left(\dfrac{\sum_{i=1}^{n} P_{i,h} X_{i,h}}{\sum_i P_{i,h}} \right) \Pi_h}{\sum_{h=1}^{N} \Pi_h}$$

or in shorter form

$$(7.18) \qquad \mu_k^* = \frac{P_{\cdot k} w}{v + P_{\cdot k} w} \bar{X}_{\cdot k} + \frac{v}{v + P_{\cdot k} w} \bar{X}_{\cdot\cdot}.$$

with

$$\bar{X}_{\cdot k} = \frac{\sum_i P_{i,k} X_{i,k}}{\sum_i P_{i,k}}$$

(7.19)

$$\bar{X}_{\cdot\cdot} = \frac{\sum_h \Pi_h \bar{X}_{\cdot h}}{\sum_h \Pi_h}$$

$$\Pi_h = \frac{P_{\cdot h}}{v + P_{\cdot h} w}.$$

Observe the very close relationship between (6.4) and (7.18). In the latter the theoretical mean over the collective (which we denoted by m) is replaced by the estimate $\bar{X}_{\cdot\cdot}$. Note, however, that our formula still involves the unknowns v and w. (Actually only the proportion of the two plays a role.) We want to show in the sequel how these quantities as well may be estimated from the observations on $\mathcal{X} = \{X_{i,j}, i = 1, \cdots, n; j = 1, \cdots, N\}$.

8. Estimating v and w

We recall that $v = E[\sigma^2(\vartheta, \rho)]$ and $w = \mathrm{Var}[\mu(\vartheta, \rho)]$. Intuitively v measures the variability within each risk and w the variation between different risks. It is hence natural to consider the following statistics V and W to estimate the two above quantities

(a)
$$V = \frac{1}{N} \sum_{h=1}^{N} \frac{1}{n-1} \sum_{\ell=1}^{n} \frac{P_{\ell,h}}{P} (X_{\ell,h} - \bar{X}_{\cdot h})^2$$

$$= \frac{1}{N(n-1)} \sum_{\ell,h} \frac{P_{\ell,h}}{P} X_{\ell,h}^2 - \frac{1}{N(n-1)} \sum_{h=1}^{N} \frac{P_{\cdot h}}{P} \bar{X}_{\cdot h}^2,$$

with $P_{\cdot h} = \Sigma_{\ell=1}^{n} P_{\ell,h}$, $\Sigma_{h=1}^{N} P_{\cdot h} = P$, and $\bar{X}_{\cdot h} = \Sigma_{\ell=1}^{n} P_{\ell,h} X_{\ell,h}/P_{\cdot h}$,

and

(b) $$W = \frac{1}{N(n-1)} \sum_{\ell,h} \frac{P_{\ell,h}}{P} (X_{\ell,h} - \tilde{X})^2 = \frac{1}{N(n-1)} \sum_{\ell,h} \frac{P_{\ell,h}}{P} X_{\ell,h}^2 - \frac{1}{N(n-1)} \tilde{X}^2$$

with $\tilde{X} = \Sigma_{\ell,h} P_{\ell,h} X_{\ell,h}/P = \Sigma_{n=1}^{N} P_{\cdot h} \bar{X}_{\cdot h}/P$.
Compute then

(a) $$N(n-1)E[V|\vartheta_1, \vartheta_2, \cdots, \vartheta_N] = \sum_{\ell,h} \frac{P_{\ell,h}}{P} \left\{ \mu^2(\vartheta_h, \rho) + \frac{\sigma^2(\vartheta_h, \rho)}{P_{\ell,h}} \right\}$$

$$- \sum_{h=1}^{N} \frac{P_{\cdot h}}{P} \left\{ \mu^2(\vartheta_h, \rho) + \frac{\sigma^2(\vartheta_h, \rho)}{P_{\cdot h}} \right\}$$

$$= \frac{n-1}{P} \sum_{h=1}^{N} \sigma^2(\vartheta_h, \rho).$$

(Observe that we have used $\operatorname{Var}\left[\bar{X}._h|\vartheta_h\right] = \sigma^2(\vartheta_h, \rho)/P._h$ which follows from our assumptions.)

Hence

$$(8.1) \qquad E[V] = \frac{E\left[\sigma^2(\vartheta, \rho)\right]}{P} = \frac{v}{P}.$$

On the other hand

$$(b) \quad (nN - 1)E[W|\vartheta_1, \vartheta_2, \cdots, \vartheta_N] = \sum_{\ell,h} \frac{P_{\ell,h}}{P} \left\{ \mu^2(\vartheta_h, \rho) + \frac{\sigma^2(\vartheta_h, \rho)}{P_{\ell,h}} \right\}$$

$$- \left(\sum_{h=1}^{N} \frac{P._h}{P} \mu(\vartheta_h, \rho) \right)^2 - \sum_{h=1}^{N} \frac{P._h}{P^2} \sigma^2(\vartheta_h, \rho)$$

$$= \sum_{h=1}^{N} \frac{P._h}{P} \mu^2(\vartheta_h, \rho) - \left(\sum_{h=1}^{N} \frac{P._h}{P} \mu(\vartheta_h, \rho) \right)^2$$

$$+ \sum_{h=1}^{N} \left(\frac{n}{P} - \frac{P._h}{P^2} \right) \sigma^2(\vartheta_h, \rho).$$

Hence

$$(8.2) \qquad (nN - 1)E[w] = \sum_{n=1}^{N} \frac{P._h}{P} (m^2 + w) - m^2 - \sum_{n=1}^{N} \frac{P^2._h}{P^2} w + (nN - 1) \frac{v}{P}$$

or

$$(8.3) \qquad E[W] = \frac{v}{P} + \frac{1}{nN - 1} \sum_{h=1}^{N} \frac{P._h}{P} \left(1 - \frac{P._h}{P} \right) w$$

which we abbreviate

$$(8.4) \qquad E[W] = \frac{v}{P} + \tilde{\Pi}w \qquad \text{with} \quad \tilde{\Pi} = \frac{1}{nN - 1} \sum_{h=1}^{N} \frac{P._h}{P} \left(1 - \frac{P._h}{P} \right).$$

This leads to the linear equations

$$\frac{\hat{v}}{P} = V$$

$$(8.5)$$

$$\frac{\hat{v}}{P} + \tilde{\Pi}\hat{w} = W$$

for the unbiased and consistent estimators $\hat{v}$ and $\hat{w}$.

There is however one difficulty in using these estimates. The estimator $\hat{w}$ turns out negative whenever $W < V$. This is indeed possible and we therefore modify our estimates as follows: (1) If $W > V$ choose the estimator as given by (8.5).

(2) If $W < V$ put $\hat{w} = 0$, that is, use the credibility rating as for a homogeneous collective

$$(8.6) \qquad \mu_k^* = \tilde{X} = \sum_{\ell, h} \frac{P_{\ell, h}}{P} X_{\ell, h}.$$

We believe this procedure to be reasonable in spite of the fact that it destroys the unbiasedness of $\hat{w}$.

For a numerical example where this method of estimation is applied the reader is referred to [4]. We give the results of the computations as presented there. The data are shown in Table I. In the last row we have tabulated $P_{.h} = \sum_{\ell} P_{\ell, h}$ and $\bar{X}_{.h} = \sum_{\ell} P_{\ell, h} X_{\ell, h} / P_{.h}$. Using the estimates, formula (8.5), in Section 8 we get

$$(8.7) \qquad \begin{aligned} \hat{v} &= 209.0 \\ \hat{w} &= 12.1. \end{aligned}$$

TABLE I

DATA FOR EXAMPLE OF THE METHOD OF ESTIMATION

Year ℓ	Risk h													
	1		2		3		4		5		6		7	
	P	X	P	X	P	X	P	X	P	X	P	X	P	X
5	5	0·0	14	11.3	18	8.0	20	5.4	21	9.7	43	9.7	70	9.0
4	6	0.0	14	25.0	20	1.9	22	5.9	24	8.9	47	14.5	77	9.6
3	8	4.2	13	18.5	23	7.0	25	7.1	28	6.7	53	10.8	85	8.7
2	10	0.0	11	14.3	25	3.1	29	7.2	34	10.3	61	12.0	92	11.7
1	12	7.7	10	30.0	27	5.2	35	8.3	42	11.1	70	13.1	100	7.0
	41	3.1	62	19.5	113	5.0	131	7.0	149	9.5	274	12.1	424	9.2

On the basis of these values for $\hat{v}$ and $\hat{w}$ we tabulate

the credibility factors $\quad \gamma_h = \dfrac{P_{.h} w}{v + P_{.h} w} \quad$ using formula (6.4),

the average experience $\quad \bar{X}_{..} = \sum_h \dfrac{\prod_h}{\prod} \bar{X}_{.h} \quad$ using formula (7.18), and

the credibility rating $\quad \mu_h^* = \gamma_h \bar{X}_{.h} + (1 - \gamma_h) \bar{X}_{..} \quad$ using formula (7.18).

The resulting estimates are shown in Table II. It is instructive to compare the last two rows of the table, the credibility rating and the individual average experience.

TABLE II

RESULTING ESTIMATES

	Risk h						
	1	2	3	4	5	6	7
γ_h (in %)	70.4	78.2	86.7	88.4	89.6	94.1	96.1
$\bar{X}_{..}$	9.4	9.4	9.4	9.4	9.4	9.4	9.4
μ_h^*	5.0	17.3	5.6	7.3	9.5	11.9	9.2
$\bar{X}_{.h}$	3.1	19.5	5.0	7.0	9.5	12.1	9.2

REFERENCES

[1] A. L. BAILEY, "Credibility procedures: Laplace generalization of Bayes' rule and the combination of collateral knowledge with observed data," *Proceedings Casualty Actuarial Society*, Vol. 37 (1950), pp. 7–23.

[2] H. BÜHLMANN, "A distribution free method for general risk problems," *ASTIN-Bulletin*, Vol. 3 (1964), pp. 144–152.

[3] ———, "Experience rating and credibility," *ASTIN-Bulletin*, Vol. 4 (1967), pp. 199–227.

[4] H. BÜHLMANN and E. STRAUB, "Glaubwürdigkeit für Schadensätze," *Mitt. Verein. Schweiz. Versich.-math.*, Vol. 70 (1970), pp. 111–133.

[5] H. BÜHLMANN, *Mathematical Methods in Risk Theory*, Heidelberg, Springer Verlag, 1970.

[6] H. ROBBINS, "The empirical Bayes approach to statistical decision problems," *Ann. Math. Statist.*, Vol. 35 (1964), pp. 1–20.

[7] F. S. PERRYMAN, "Some notes on credibility," *Proceedings Casualty Actuarial Society*, Vol. 19 (1932), pp. 65–84.

[8] A. W. WHITNEY, "The theory of experience rating," *Proceedings Casualty Actuarial Society*, Vol. 4 (1918), pp. 274–292.

SOME EFFECTS OF ERRORS OF MEASUREMENT ON LINEAR REGRESSION

W. G. COCHRAN
HARVARD UNIVERSITY

1. Introduction

I assume a bivariate distribution of pairs (y, X) in which y has a linear regression on X

$$(1.1) \qquad y = \beta_0 + \beta_1 X + e,$$

where e, X are independently distributed and $E(e|X) = 0$. However, the measurement of X is subject to error. Thus we actually observe pairs (y, x), with $x = X + h$, where h is a random variable representing the error of measurement.

Given a random sample of pairs (y, x), previous writers have discussed various approaches to the problem of making inferences about the line $\beta_0 + \beta_1 X$, sometimes called the *structural* relation between y and X. In the present context this line might be called "the regression of y on the correct X" to distinguish it from "the regression of y on the fallible x." An obviously relevant question is: under assumption (1.1), what is the nature of the regression of y on x?

Lindley [5] gave the necessary and sufficient conditions that the regression of y on the fallible x be linear in the narrow sense. This means that $E(y|x)$ is linear in x, or equivalently that

$$(1.2) \qquad y = \beta_0' + \beta_1' x + e',$$

where $E(e'|x) = 0$. This definition does not require that e' and x be independently distributed. Lindley's proof assumes that the error of measurement h is distributed independently of X. His necessary and sufficient conditions are that Fisher's cumulant function (logarithm of the characteristic function) of h be a multiple of that of X. Roughly speaking, this implies that h and X belong to the same class of distributions. Thus if X is distributed as $\chi^2\sigma^2$, so is h, though the degrees of freedom can differ: if X is normal, h must be normal.

Several writers have discussed the corresponding necessary and sufficient conditions if we demand in addition that the residual e' in (1.2) be distributed independently of x. In particular, Fix [3] showed that if the second moment of

This work was supported by the Office of Naval Research through Contract N00014-56A-0298-0017, NR-042-097 with the Department of Statistics, Harvard University.

either X or h exists and X, h are independent as before, the conditions for linearity of regression in this fuller sense are that both X and h be normally distributed.

Thanks partly to computers, numerous regression studies are being done nowadays, particularly in the social sciences and medicine, in which all the variables are difficult to measure, and therefore are presumably measured with sizeable errors. In thinking about mathematical models appropriate to such uses, it seems clear that the forces which determine the nature of the distribution of h (the imperfections of the measuring instrument or process) are quite different from those that determine the nature of the distribution of the correct X. Consequently, my opinion is that in such applications even the Lindley conditions will not be satisfied, except perhaps by a fluke or as an approximation (for example, the cumulants of X and h might be similar in the sense that both distributions are close to normal).

This paper considers the regression of y on x when Lindley's conditions are not satisfied. There are at least two reasons for interest in this regression. The objective may be to obtain a consistent estimate of β_1 for purposes of interpretation or adjustment by covariance (Lord [6]). Secondly, the purpose may be to predict y from the fallible x by the regression technique in which case the shape of this regression is relevant.

The strategy used here is first to construct a straight line relation between y and the fallible x which may be called the *linear component* of the regression of y on x. This is the line that we are estimating, in some sense, when we compute a sample linear regression of y on x.

The paper then takes a look at the question: what is the nature of the departure from linearity when Lindley's conditions are not satisfied? In particular, does the linear component dominate? If it does, then as Kendall [4] remarks, "A slight departure from linearity will sometimes allow the ordinary theory to be used as an approximation." I have been unable to obtain any general results that are exact, but something can be learned by a combination of an approach *via* moments and the working out of some easy particular cases. These suggest, fortunately, that the linear component often dominates, even with measurements of rather poor reliability, but the issue needs more thorough investigation by someone with greater mathematical power.

2. The linear component

As stated, a linear regression of y on the correct X (in the fullest sense) is assumed, namely,

$$(2.1) \qquad\qquad y = \beta_0 + \beta_1 X + e,$$

where e, X are independently distributed and $E(e|X) = 0$. I also assume X scaled so that $E(X) = 0$.

As regards the error of measurement h, Lindley's result requires h, X to be independently distributed, but this assumption limits the range of applications of the result. Some measuring instruments or methods underestimate high values of X and overestimate low values. I have been unable to obtain conditions analogous to Lindley's when X, h follow a general bivariate distribution $\phi(X, h)$ but there is no difficulty in obtaining the linear component in this case. Denote $E(h)$ by μ_h, since measurements may be biased. It is assumed, as seems reasonable for most applications, that h and hence x are distributed independently of e. Hence the regression of y on x is, from (2.1),

$$(2.2) \qquad E(y|x) = \beta_0 + \beta_1 E(X|x) = \beta_0 + \beta_1 R(x),$$

say.

Thus we need to find $R(x)$. Let $\phi(X, h)$ be the joint frequency function of X, h. The marginal distribution of the fallible x is

$$(2.3) \qquad \psi(x) = \int \phi(X, x - X) \, dX,$$

while $R(x) = E(X|x)$ satisfies the equation

$$(2.4) \qquad R(x)\psi(x) = \int X\phi(X, x - X) \, dX.$$

The linear component of $R(x)$ can be defined by fitting the straight line $L(x) = C_0 + C_1 x$ to $R(x)$ by the population analogue of the method of least squares. That is, we choose C_0, C_1 to minimize

$$(2.5) \qquad \int \{R(x) - C_0 - C_1 x\}^2 \psi(x) \, dx.$$

Clearly,

$$(2.6) \qquad \int R(x)\psi(x) \, dx = \iint X\phi(X, h) \, dXdh = \mu_X = 0,$$

$$(2.7) \qquad \int xR(x)\psi(x) \, dx = \iint (X^2 + Xh)\phi(X, h) \, dXdh = \sigma_X^2 + \sigma_{Xh},$$

where σ_{Xh} is the population covariance of X and h. Hence the normal equations for C_0 and C_1 give

$$(2.8) \qquad C_0 = -C_1\mu_h, \quad C_1 = (\sigma_X^2 + \sigma_{Xh})/(\sigma_X^2 + \sigma_h^2 + 2\sigma_{Xh}).$$

From (2.2), the linear component of the regression of y on x is $L(x) = \beta_0 + \beta_1(C_0 + C_1 x)$. If we write this $\beta_0' + \beta_1' x$, we have

$$(2.9) \qquad \beta_0' = \beta_0 - \beta_1 C_1\mu_h, \quad \beta_1' = \beta_1 C_1 = \beta_1(\sigma_X^2 + \sigma_{Xh})/(\sigma_X^2 + \sigma_h^2 + 2\sigma_{Xh}).$$

Incidentally, expressions (2.9) for β'_0, β'_1 can be obtained directly by noting that our procedure is equivalent to defining $(\beta'_0 + \beta'_1 x)$ as the linear component of the regression of y on x if we write $y = \beta'_0 + \beta'_1 x + e'$ and determine β'_0, β'_1 so that the residuals e' satisfy the conditions

$$(2.10) \qquad E(e') = 0, \qquad \text{Cov } (e', x) = 0.$$

Formulas (2.9) for β'_0 and β'_1 agree with the well-known elementary results in the literature, usually obtained on the assumption that X, h follow a bivariate normal. Bias in the measurements affects the intercept β'_0 but not the slope β'_1. If h and X are uncorrelated, $\beta'_1 = \beta_1 \sigma_X^2/\sigma_x^2$, the factor σ_X^2/σ_x^2 being often called the *reliability* of the measurement x. For given σ_X^2, σ_h^2, positive correlation of the errors with X makes the underestimation of the slope worse, while negative correlation alleviates it if $\sigma_h^2 < \sigma_X^2$. In the Berkson case [1], the investigator plans to apply preselected amounts x of some agent or treatment in a laboratory experiment, but owing to errors in measuring out this amount, the amount X actually applied is different. Here, Cov $(x, h) = 0$, so that $\sigma_{Xh} = -\sigma_h^2$ and (2.9) shows that $\beta'_1 = \beta_1$. This situation also applies when large samples are grouped by their values of X into classes to facilitate the calculation of regression on a desk machine, provided that x is taken as the *mean* of X within each class. The common practice is of course to take x as the midpoint of the class. This makes Cov (x, h) slightly positive for most unimodal distributions of X, so that some residual inconsistency in β'_1 as an estimate of β_1 remains, though the inconsistency is in general trivial if at least ten classes are used.

Suppose now that y is also subject to an error of measurement d. If Y represents the correct value of y, we may rewrite the original model (1.1) as

$$(2.11) \qquad Y = \beta_0 + \beta_1 X + e, \qquad y = Y + d.$$

Hence

$$(2.12) \qquad E(y|x) = \beta_0 + \beta_1 R(x) + E(d|x).$$

If errors in y are independent of Y, X and h, then $E(d|x) = \mu_d$, the amount of bias in d, and we get the old result that such errors in y do not affect the slope of the regression line. If d is correlated with Y, the choice of an appropriate model requires care. Specification of the joint frequency functions $\phi(X, h)$, $\theta(Y, d)$ is not enough to determine $E(d|x)$; we need to know the relation between d and h. The following might serve for applications in which the process by which y is measured is independent of that by which x is measured. Noting that $E(X) = 0$, $E(Y) = \beta_0$, write

$$(2.13) \qquad d = \mu_d + \frac{\sigma_{Yd}}{\sigma_Y^2}(Y - \beta_0) + d', \qquad h = \mu_h + \frac{\sigma_{Xh}}{\sigma_X^2} X + h',$$

where h', d', with zero means, are assumed independent of each other and of X, Y, and e. This model does not imply that d and h are independent, since

$$(2.14) \qquad \text{Cov } (dh) = \frac{\sigma_{Yd}\sigma_{Xh}\sigma_{XY}}{\sigma_Y^2 \sigma_X^2},$$

but this correlation arises only as a consequence of the X, Y correlation.

In some applications there may be further correlation between d and h because the measuring processes are not independent. For instance, an individual pair (x, y) might be estimates of a town population five years apart, where the municipal statisticians use the same techniques in a town, the technique varying from town to town. In general, it will obviously be difficult to know which model to pursue, and to get data for verification of a model.

If (2.13) holds,

$$(2.15) \qquad E(d|x) = \mu_d + \frac{\sigma_{Yd}}{\sigma_Y^2} \beta_1 R(x).$$

From (2.8) and (2.12), we obtain for the linear component of the regression of y on x,

$$(2.16) \qquad \left[\beta_0 + \mu_d - \beta_1 C_1 \mu_h \frac{(\sigma_Y^2 + \sigma_{Yd})}{\sigma_Y^2} \right] + \beta_1 x \frac{(\sigma_Y^2 + \sigma_{Yd})(\sigma_X^2 + \sigma_{Xh})}{\sigma_Y^2(\sigma_X^2 + 2\sigma_{Xh} + \sigma_h^2)}.$$

As is obvious from graphical considerations, errors in y that are positively correlated with Y tend to increase the absolute value of β_1', whereas errors in x have the opposite effect. With errors in both y and x, β_1' may be either greater or less than β_1.

The method of obtaining the linear component extends naturally to a multiple linear regression of y on $x_1, x_2, \cdots, x_k$. Even when the errors of measurement h_i are independent of X_i and of each other—the simplest case—β_i' is a linear function of all β_j whose corresponding x_j are subject to errors of measurement (Cochran [2]). When the h_i and X_i are correlated, we again meet the problem of specifying the nature of the correlation between h_i and h_j.

One objective in working out the relations between β_i' and the β_j is as a possible means of estimating the coefficients β_j of the structural regression by using data from supplementary studies of the errors of measurement. With errors in more than one variate, however, the algebraic results suggest that the information needed about errors of measurement is more than we are likely to be able to obtain.

3. Polynomial approach by moments

Now consider the nature of $R(x)$ with errors in x only when Lindley's conditions are not satisfied. Like Lindley, I assume h and X independent, with $\mu_h = 0$. I first chose some simple forms for the frequency functions $f(X)$ of X and $g(h)$ of h for which $R(x)$ can be worked out exactly in closed form. Examples are the χ distribution with a small number of degrees of freedom, the normal, the uniform, and the exponential types like e^{-X}, with $X > 0$, or $\frac{1}{2}e^{-|x|}$, with $-\infty < X < \infty$. Inspection of a few cases indicated that if either X or h follows a skew distribution, the departure of $R(x)$ from linearity, in a region around the mean of x, is of the simple type that can be approximated by a quadratic curve (an example will be given in Section 4).

If, however, both h and X are symmetrically distributed about their means 0, then $\psi(x)$ is also symmetrical and $R(-x) = -R(x)$, which suggests a cubic approximation with a zero quadratic term. The equations of the approximating quadratic or cubic, by the least squares method, are obtained easily from the low moments of the distributions of h and X. (In the symmetric case it is possible that for some frequency functions a quadratic approximation in $|x|$, with reversal of sign when x is negative, might do better than the cubic, but the fitting requires calculation of some incomplete moments and this has not been pursued.)

In fitting a polynomial approximation of degree p to $R(x)$, we choose the coefficients C_i to minimize

$$(3.1) \qquad \int \{R(x) - \sum_{i=0}^{p} C_i x^i\}^2 \psi(x)\, dx.$$

The rth normal equation is

$$(3.2) \qquad \sum_{i=0}^{p} C_i \mu_{r+i, x} = \iint X(X + h)^r f(X) g(h)\, dX dh,$$

where the μ denote moments about the mean.

Here we are fitting only the simplest nonlinear approximations, say $Q(x)$.

Case 1.   *X or h skew.*

$$(3.3) \qquad Q(x) = C_1 x + C_2(x^2 - \mu_{2x}),$$

where

$$(3.4) \qquad \Delta = \mu_{4x}\mu_{2x} - \mu_{3x}^2 - \mu_{2x}^3,$$

$$(3.5) \qquad C_1 = (\mu_{4x}\mu_{2x} - \mu_{3x}\mu_{3X} - \mu_{2x}^2\mu_{2X})\Delta^{-1},$$

$$(3.6) \qquad C_2 = (\mu_{2x}\mu_{3X} - \mu_{3x}\mu_{2X})\Delta^{-1}.$$

Case 2.   *X and h both symmetrical.*

$$(3.7) \qquad C(x) = c_1 x + c_3 x^3,$$

where

$$(3.8) \qquad \Delta = \mu_{6x}\mu_{2x} - \mu_{4x}^2,$$

$$(3.9) \qquad c_1 = (\mu_{6x}\mu_{2X} - \mu_{4x}\mu_{4X} - 3\mu_{4x}\mu_{2X}\mu_{2h})\Delta^{-1},$$

$$(3.10) \qquad c_3 = (\mu_{4x}\mu_{2X} - \mu_{2x}\mu_{4X} - 3\mu_{2x}\mu_{2X}\mu_{2h})\Delta^{-1}.$$

There is obvious interest in seeing how well the quadratic and cubic approximations fit $R(x)$. The reduction in the variance of $R(x)$ due to these approximations can be obtained from the normal equations by the usual analysis of variance rule of multiplying the solutions C_i or c_i by the right sides of the normal equations. But I have been unable to obtain an exact expression for the variance of $R(x)$ which does not involve computing an integral, so that I do not have a general result for the closeness of fit in this case, though it can be obtained by numerical integration in specific examples, as discussed in Section 5.

This approach extends also to correlated errors, the expressions for the C_i involving joint moments of h and X which are easily obtained.

4. Two examples

As an example of the quadratic approximation I take

$$(4.1) \qquad f(X) = Xe^{-X^2/2}, \qquad X > 0, \qquad g(h) = N(0, \sigma^2).$$

Thus X is skew, essentially a χ variate with two degrees of freedom, mean $\sqrt{\pi/2}$ and variance $(2 - \pi/2)$, about 0.429, while h is normal. The reliability of the measurement is $(4 - \pi)/(4 - \pi + 2\sigma^2)$, so that measurements with different degrees of reliability from 92 per cent to 54 per cent are represented by taking $\sigma = 0.2(0.1)0.6$. Measurements of lower reliability have been reported, but this range should cover the great majority of applications. Reliability of 50 per cent is far from impressive: the measurement error has as big a variance as the correct measurement. It is not claimed that this example corresponds to any actual situation in practice: although X is essentially positive, x is not.

For this example, $R(x)$ works out as

$$(4.2) \qquad R(x) = \frac{\sigma}{(1 + \sigma^2)^{1/2}} \left\{ \frac{(u^2 + 1)P(u) + uz(u)}{uP(u) + z(u)} \right\}$$

where $u = x/\sigma(1 + \sigma^2)^{1/2}$, $z(u)$, and $P(u)$ are the ordinate and cumulative of $N(0, 1)$. (In this example X was not scaled so that $E(X) = 0$). Figure 1 presents points on $R(x)$ for $\sigma = 0.3$ and $\sigma = 0.6$, plus the line $R(x) = x$ that would apply if there were no error of measurement.

As suggested, the major departure from linearity of the points in Figure 1 near μ_x is a simple curvature that is well represented by a quadratic curve, though this quadratic would do very badly at points far away from μ_x. For instance, with x and u becoming large and positive, $P(u)$ tends to one and $z(u)$ to zero, so that $R(x)$ behaves like $x/(1 + \sigma^2)$, while for x negative $R(x)$ tends to become asymptotic to the origin. However, the approximation errors in the quadratic $Q(x)$ receive very little weight at these points, since they are far enough away from the mean of x so that $\psi(x)$ is tiny. The mean square error (MSE) of the quadratic approximation, the integral of $[Q(x) - R(x)]^2 \psi(x)\, dx$, is only around 0.0003 to 0.0007. The MSE of the linear approximation has the following values.

σ	.2	.3	.4	.5	.6
MSE$[L(x)]$	.0010	.0025	.0041	.0052	.0069

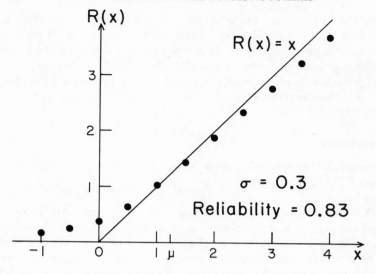

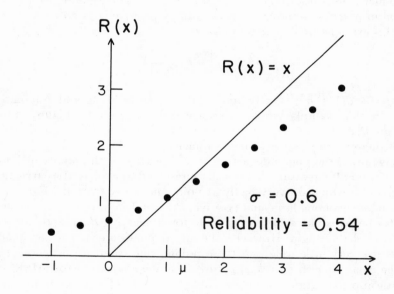

FIGURE 1

Regression $R(x)$ of X (correct) on x (fallible) measurement.
X is χ (2 d.f.), error of measurement is $N(0, \sigma^2)$.

The example of the symmetrical case has $X = N(0, 1)$, while the measurement error h is uniform between $-L$ and L. The reliability is $3/(3 + L^2)$ and with $L = 0.5(0.25)1.5$, it varies between 93 per cent and 57 per cent. Figure 2 shows $R(x)$ for $L = 1$, $L = 1.5$, with reliabilites 0.75 and 0.57. In this example

$$(4.3) \qquad R(x) = [z(x - L) - z(x + L)][C(x + L) - C(x - L)]^{-1}.$$

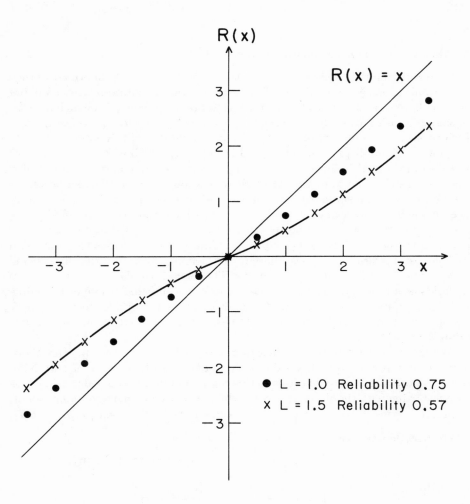

FIGURE 2

Regression of $R(x)$ of X (correct) on x (fallible) measurement.
X is $N(0, 1)$, error of measurement is uniform $(-L, L)$.

As x moves far away from its mean at 0, $R(x)$ tends to become $(x - L)$, again linear in x rather than cubic. The MSE of the cubic approximation is very small, the greatest value in the cases worked being 0.00054 at $L = 1.5$. For the linear approximation, the MSE are as follows.

L	0.5	0.75	1	1.25	1.5
$\text{MSE}[L(x)]$	$.0^4 5$	$.0^3 16$	$.0011$	$.0046$	$.0116$

5. Adequacy of the linear approximation

In considering the adequacy of the linear approximation, I am not concerned with applications in which the aim is to estimate the structural relationship, but only with those in which (i) the objective is to predict y from x, in which case the usual advice is to utilize the observed regression of y on x, or (ii) to test the null hypothesis $\beta_1' = 0$ by a t test, as a means of testing the null hypothesis $\beta_1 = 0$.

First, as Kendall [4] reminded us, Lindley's type of linearity falls short of the assumptions of independence of the residual e' and x and normality of e' that are needed for use of the standard regression formulas. The situation resembles that in the standard texts on sample surveys, in which attempts have been made to construct a theory of linear regression estimators without allotting any specific structure to a relation between y and x except that all values of y and x in the population are bounded. It is known in such cases that the usual sample estimate $\hat{\beta}_1'$ of β_1' is biased, the leading term in the bias being $-E(e'x^3)/n\sigma_x^4$, where n is the sample size. The usual formula for the variance of $\hat{\beta}_1'$ holds as a first approximation, but it too has a bias that becomes negligible in large samples, while the numerator and denominator of the t test of $\hat{\beta}_1'$ only become independent asymptotically. At best we can say that the usual methods apply asymptotically.

As regards the effect of errors in x on the precision of regression estimates of y, the most important quantity is the variance of the residuals from the regression of y on x, or from approximations to this regression that we consider using. It is worth starting with the simplest case in which X, h are normal and independent, so that the regression of y on x is linear in both senses. Here $\beta_1' = \beta_1 \sigma_X^2 / \sigma_x^2 = G\beta_1$,

where G is the coefficient of reliability. Further, since

$$(5.1) \qquad \sigma_y^2 = \sigma_e^2 + \beta_1^2 \sigma_X^2 = \sigma_{e'}^2 + \beta_1'^2 \sigma_x^2 = \sigma_{e'}^2 + G\beta_1^2 \sigma_X^2,$$

we have

$$(5.2) \qquad \sigma_{e'}^2 = \sigma_e^2 + (1 - G)\beta_1^2 \sigma_X^2 = (1 - \rho^2)\sigma_y^2 + (1 - G)\rho^2 \sigma_y^2.$$

This is the familiar result that, even when there is no problem about nonlinearity, errors in x increase the variance of the deviations from the linear prediction model by $(1 - G)\rho^2 \sigma_y^2$, an increase that hurts most, for given G, when the prediction formula is very good (ρ high).

Returning to the assumptions of this paper, suppose y is predicted from its regression on x. Since

(5.3) $$y = \beta_0 + \beta_1 X + e = \beta_0 + \beta_1 R(x) + e',$$

(5.4) $$\sigma_{e'}^2 = \sigma_e^2 + \beta_1^2(\sigma_X^2 - \sigma_R^2) = \sigma_e^2 + \beta_1^2\sigma_X^2(1 - \sigma_R^2/\sigma_X^2).$$

In the examples that I have worked, σ_R^2/σ_X^2 comes numerically to within one or two percentage points of $G = \sigma_X^2/\sigma_x^2$. Consequently, if the population regression of y on x is used for prediction, the loss of precision due to errors in x is about what we expect from the value of the reliability coefficient G.

To pass to the additional loss of precision if $L(x)$ instead of $R(x)$ is used in the prediction formula, write

(5.5) $$\beta_0 + \beta_1 R(x) + e' = \beta_0 + \beta_1 L(x) + e'',$$

giving

(5.6) $$\sigma_{e''}^2 = \sigma_{e'}^2 + \beta_1^2(\sigma_R^2 - \sigma_L^2) = \sigma_{e'}^2 + \beta_1^2 \text{ MSE } (L)$$

since the method of fitting L makes Cov $(L, R - L) = 0$. From the little tables of MSE (L) in Section 4, the highest ratios of MSE (L) to σ_X^2, which occur when reliability is lowest, are around 0.01 to 0.02. If these examples typify what happens in applications, it appears that errors in x create an increase of around $(1 - G)\rho^2\sigma_y^2$ in the variance of residuals, but that even when reliability is low, the additional increase in this variance due to use of $L(x)$ instead of $R(x)$ is unimportant.

Since e' and x are not independent, a look at the conditional distribution of e' for fixed x may be worthwhile. From (5.2),

(5.7) $$e' = e + \beta_1[X - R(x)].$$

By our hypothesis, the term e is independent of X, h, and hence x, so that this term has the same shape of distribution and variance in all arrays with x fixed. The second term is determined by the distribution of X for fixed x. In Example 1 this distribution works out as

(5.8) $$f(X|x) = \frac{(1 + \sigma^2)}{\sigma^2\sqrt{2\pi}} \frac{1}{K(x)} X \exp\left\{-\frac{1}{2}\frac{(1 + \sigma^2)}{\sigma^2}\left[X - \frac{x}{1 + \sigma^2}\right]^2\right\}$$

where, with $\mu = x/\sigma(1 + \sigma^2)^{1/2}$,

(5.9) $$K(x) = z(u) + uP(u).$$

This distribution changes in shape as x varies; its variance increases with x, the increase being small when the reliability G is high but more marked when G is lower.

In Example 2, $f(X|x)$ is simply the incomplete normal

(5.10) $$f(X|x) = \frac{1}{\sqrt{2\pi}} \frac{1}{K(x)} \exp\left\{-\tfrac{1}{2}X^2\right\}; \qquad x - L \leqq X \leqq x + L$$

with $K(x) = P(x + L) - P(x - L)$. When x is at its mean 0, this is symmetrical with its maximum variance, but changes from negative to positive skewness as x changes from negative to positive.

Thus the distribution of e' is a compound of two independent distributions, one unchanging, the other changing with x. As we have seen, the average variances of the two components are $\sigma_y^2(1 - \rho^2)$ and approximately $\rho^2\sigma_y^2(1 - G)$. With G high and ρ modest, the unchanging component should dominate, and the assumption that the distribution of e' is the same for different x may be a reasonable approximation, but with say $\rho = 0.8$, $G = 0.5$, the two components have variances $0.36\sigma_y^2$ and $0.32\sigma_y^2$ and are about equally important.

6. Summary and discussion

This paper deals with applications in which the standard linear regression model $y = \beta_0 + \beta_1 X + e$, with e, X independent and $E(e) = 0$, is assumed to apply to a bivariate sample of pairs (y, X). However, owing to difficulties in measuring the X values, we actually have a bivariate sample (y, x) where $x = X + h$, h being an error of measurement. My opinion is that in applications even Lindley's conditions for linearity of the regression of y on x in the narrow sense will not in general be satisfied.

Facing this situation we can define a linear relation $y = \beta_0' + \beta_1'x$ that may be called the linear component of the regression of y on x. The elementary results in the literature for the relations between β_0', β_1' and β_0, β_1 (usually derived on the assumption that h, X are normally distributed) hold for this linear component. The linear component can be obtained when h and X are correlated, whereas Lindley assumes h, X independent, and extends to errors in y also and to multiple linear regression, subject to problems about specification of the nature of correlated errors.

The next step was to work out the exact regression of y on x for a number of specific examples in which this regression has a closed form. These suggest that the departure from linearity can be approximated by a quadratic in x if either h or X is skew and by a cubic in x if h, X are symmetrical. The equations of the approximating quadratic or cubic are easily obtainable from the lower moments of the distributions of h and X. Further, in these examples the linear component dominates, in the sense that the mean square deviation of y from the linear component is only slightly larger than that from the exact regression of y on x, even for measurements of reliability not much more than 50 per cent.

A further result that holds well in these examples is of interest when the fallible x is used to predict y. When h and X are independent and normal, so that the regression of y on x is linear, it is known that the residual variance from the regression of y on x exceeds that for the regression of y on X by $(1 - G)\rho^2\sigma_y^2$, where G is the coefficient of reliability of x. This result remains a good approximation when X, h are independent but have different distributions so that Lindley's conditions do not hold.

Unfortunately the results here are only suggestive and leave unanswered the important questions. For example, (i) what is the analogous result to Lindley's when h and X are not independent? Some measuring instruments have the property of underestimating high values of X and overestimating low values, and *vice versa*. (ii) How far can the moments approach be trusted? Are there distributions for which the departure from linearity is more complex than a quadratic or cubic? (iii) Lindley's conditions guarantee linearity only in the narrow sense; the deviations e' from the regression of y on x are not independent of x. There is reason to believe that in this case linear regression theory can be used asymptotically in large samples, but more needs to be known about the practical importance of the disturbances present in small samples.

REFERENCES

[1] J. BERKSON, "Are there two regressions?," *J. Amer. Statist. Assoc.*, Vol. 45 (1950), pp. 164–180.

[2] W. G. COCHRAN, "Errors of measurement in statistics," *Technometrics*, Vol. 10 (1968), pp. 637–666.

[3] E. FIX, "Distributions which lead to linear regressions," *Proceedings of the Berkeley Symposium on Mathematical Statistics and Probability*, Berkeley and Los Angeles, University of California Press, 1949, pp. 79–91.

[4] M. G. KENDALL, "Regression, structure and functional relationship." *Biometrika*, Vol. 38 (1951), pp. 11–25.

[5] D. V. LINDLEY, "Regression lines and the linear functional relationship," *J. Roy. Statist. Soc. Ser. B*, Vol. 9 (1947), pp. 218–244.

[6] F. LORD, "Large-sample covariance analysis when the control variable is fallible," *J. Amer. Statist. Assoc.*, Vol. 55 (1960), pp. 307–321.

ESTIMATION FOR A REGRESSION MODEL WITH AN UNKNOWN COVARIANCE MATRIX

LEON JAY GLESER
THE JOHNS HOPKINS UNIVERSITY
and
INGRAM OLKIN
STANFORD UNIVERSITY

1. Summary and introduction

A linear regression model is considered under which the residual error vector is assumed to have a multivariate normal distribution with unknown covariance matrix Σ. To estimate Σ, it is assumed that the regression design can be given independent replications. This problem has been considered by Rao, who obtains a point estimator and suggests two classes of confidence regions for the vector β of regression parameters. In the present paper, we find the maximum likelihood estimators of β and of Σ, and derive their distributions. One of Rao's two classes of confidence regions for β had previously been inapplicable due to the lack of tables for upper tail values of the distribution of the pivotal quantity. These tables are now provided, and the performances of the two classes of confidence regions are compared in terms of their expected volumes.

In the classical linear regression model, the vector of observations $y = (y_1, y_2, \cdots, y_p)$ has the form

$$(1.1) \qquad y = \beta X + \varepsilon,$$

where $\beta: 1 \times q$ is an unknown vector of regression parameters, X is a known $q \times p$ matrix of rank $q \leq p$, and ε has a p variate normal distribution with mean vector zero and covariance matrix $\Sigma = \sigma^2 I$. Since the simple structure of the covariance matrix may not be valid for some problems, extensions of the results of the classical model to models where Σ has a more general structure have been considered. Such attempts can be classified in the following hierarchy of complexity:

  (i) Σ an arbitrary known matrix,
  (ii) Σ known up to a scale factor σ^2,

This work was supported in part by the National Science Foundation Grant GP-6681 at Stanford University.

(iii) Σ unknown but with some special structure,

(iv) Σ completely unknown and arbitrary.

The maximum likelihood estimators (MLE) for cases (i) and (ii) are well known (see Anderson [1]). In both of these cases, the MLE $\hat{\beta}(\Sigma)$ of β has the form $\hat{\beta}(\Sigma) = y\Sigma^{-1}X'(X\Sigma^{-1}X')^{-1}$ with covariance matrix $(X\Sigma^{-1}X')^{-1}$ yielding the minimum concentration ellipse among all linear unbiased estimators of β. (Note that in case (ii), $\hat{\beta}(\Sigma)$ is independent of the unknown scale factor σ^2.) Watson [22], [23] and Watson and Hannan [24] have investigated the errors involved when the assumptions made concerning Σ in cases (i) and (ii) are violated.

As an example of a model of the type considered in case (iii), assume that Σ has the intraclass correlation structure. This class of linear regression models has been considered by Halperin [10], by Geisser and Greenhouse [5], [6], and by other authors. Alternative possible special models for Σ include the models of autocorrelation, circular symmetry, and compound symmetry. In each of these special cases, as well as in cases (i) and (ii), inference concerning the parameters of the regression model is possible even when only one replication of the random vector y is available.

If, however, we are in complete ignorance of Σ, it is clear that more than one observation must be taken on y in order to estimate both β and Σ. In some problems, one may actually have independent replications of the y's: for example, (a) where each y vector represents a score vector on an examination and the replications are individuals from a particular homogeneous group, or (b) in the analysis of growth curves (see Rao [19], Pothoff and Roy [15], Gleser and Olkin [8], [9]). The replications on y enable us to simultaneously estimate β and Σ.

Versions of case (iv) have been considered by many authors. Cochran and Bliss [2] discuss a variant of this model in connection with the comparison of discriminant functions from two populations. Rao [16], [17], [18] considers the problem of testing the hypothesis that the vector β of regression parameters obeys certain linear constraints, derives the likelihood ratio test statistic for this problem, and obtains its null and nonnull distributions. Further distributional results for the likelihood ratio statistic are given by Narain [13], Olkin and Shrikhande [14], and Kabe [11].

Rao [16], [18], [20], [21] also considers the problem of estimating β. He obtains a certain "least squares" estimator for β which is, in fact, the MLE of β (Gleser and Olkin [7]). They find the MLE of β and Σ, give representations for their densities, and compare the covariance matrices of the MLE of β and the BLUE of β when Σ is known. The comparison shows that for even moderate sample sizes, there is little difference in the accuracies of the two estimators. (Similar results are also given by Rao [21] and Williams [25].) The above results, together with a new and very useful representation for the density of the MLE of β, appear in Section 2 and Appendix A.

Rao ([16]–[21]) has proposed two classes of confidence regions for (linear combinations of) the elements of the vector β—one class based on a statistic closely related to Mahalanobis's distance, the other on the likelihood ratio test statistic for testing that β obeys certain linear restraints. These two procedures are described in Section 3. Distributional difficulties with the former class of confidence regions have up to now severely limited its applicability, and have prevented comparisons with the class of regions based on the likelihood ratio statistic. In Appendix B of this paper, we provide the necessary tables for the application of this confidence procedure in certain cases, and indicate how these tables may be used (and extended) in more general contexts. The availability of these tables permits comparison of the two classes of confidence regions; these comparisons appear in Section 4. An illustrative example is given in Section 5.

2. The regression model: estimators of β and Σ

Let $y^{(1)}, \cdots, y^{(N)}$ be N independent random p dimensional row vectors, each having a multivariate normal distribution with mean vector $\mathscr{E}(y^{(j)}) = \beta X$ and covariance matrix Σ, where X is a known $q \times p$ matrix of rank $q \leqq p$, where β is a $1 \times q$ vector of unknown regression parameters, and where Σ is an unknown positive definite matrix.

We may immediately reduce the data to the sufficient statistic $(\bar{y}, S)$, where $\bar{y} = N^{-1} \Sigma_{j=1}^{N} y^{(j)}$ is the sample mean vector and $S = \Sigma_{i=1}^{N} (y^{(i)} - \bar{y})' (y^{(i)} - \bar{y})$ is the sample cross product matrix. Thus, $\bar{y}$ and S are independently distributed, $\bar{y}$ has a multivariate normal distribution with mean vector βX and covariance matrix $N^{-1} \Sigma$, (denoted $\bar{y} \sim N(\beta X, N^{-1} \Sigma)$), and S has the Wishart distribution with $n \equiv N - 1$ degrees of freedom and expectation $\mathscr{E}(S) = n\Sigma$, (denoted $S \sim W(\Sigma; p, n)$), S being $p \times p$. The joint density of $\bar{y}$ and S is given by

(2.1)
$$p(\bar{y}, S) = c|\Sigma|^{-N/2}|S|^{(n-p-1)/2} \exp \left\{ -\tfrac{1}{2} \text{tr } \Sigma^{-1}[S + N(\bar{y} - \beta X)'(\bar{y} - \beta X)] \right\},$$
where

(2.2)
$$c^{-1} = [2^N \pi^{(p+1)/2} N^{-1}]^{p/2} \prod_{i=1}^{p} \Gamma[\tfrac{1}{2}(n - i + 1)].$$

To obtain the MLE of β and Σ, first maximize $p(\bar{y}, S)$ with respect to Σ; this yields

(2.3)
$$N\hat{\Sigma}(\beta) = S + N(\bar{y} - \beta X)'(\bar{y} - \beta X)$$
(see, for example, Anderson [1], p. 46). Inserting $\hat{\Sigma}(\beta)$ for Σ in the joint density yields a constant multiple of

(2.4)
$$|S + N(\bar{y} - \beta X)'(\bar{y} - \beta X)|^{-N/2} = |S|^{-N/2}[1 + N(\bar{y} - \beta X)S^{-1}(\bar{y} - \beta X)']^{-N/2},$$
from which, maximizing with respect to β, we obtain the MLE of β to be

(2.5)
$$\hat{\beta} = \bar{y}S^{-1}X'(XS^{-1}X')^{-1}.$$

The MLE of Σ is then $\hat{\Sigma} \equiv \hat{\Sigma}(\hat{\beta})$.

The distribution of $\hat{\beta}$ is obtained in Appendix A. There, it is shown that the following result holds.

THEOREM 2.1.   *The probability density of $\hat{\beta}$ is*

$$(2.6)\qquad p(\hat{\beta}) = \sum_{j=0}^{\infty} c_j \frac{\left|NX\Sigma^{-1}X'\right|^{1/2}[Q(\hat{\beta})]^j \exp\left\{-\tfrac{1}{2}Q(\hat{\beta})\right\}}{(2\pi)^{q/2}2^j[\Gamma(\tfrac{1}{2}q + j)/\Gamma(\tfrac{1}{2}q)]} \equiv \sum_{j=0}^{\infty} c_j h_j(\hat{\beta}),$$

where $Q(\hat{\beta}) = N(\hat{\beta} - \beta)X\Sigma^{-1}X'(\hat{\beta} - \beta)'$,

$$(2.7)\qquad c_j = \frac{c_0}{j!} \frac{\Gamma(\tfrac{1}{2}(p - q) + j)}{\Gamma(\tfrac{1}{2}(p - q))} \frac{\Gamma(\tfrac{1}{2}q + j)}{\Gamma(\tfrac{1}{2}q)} \frac{\Gamma(\tfrac{1}{2}(n + q + 1))}{\Gamma(\tfrac{1}{2}(n - p + q) + j)},$$

the components of $\hat{\beta}$ range from $-\infty$ to ∞, and

$$(2.8)\qquad c_0 = \frac{\Gamma(\tfrac{1}{2}(n + 2q - p + 1))\Gamma(\tfrac{1}{2}(n + 1))}{\Gamma(\tfrac{1}{2}(n + q + 1))\Gamma(\tfrac{1}{2}(n - p + q + 1))}.$$

Note that $c_j \geq 0$ for all j. It can be shown that $\sum_{j=0}^{\infty} c_j = 1$ and that each $h_j(\hat{\beta})$, $j = 0, 1, \cdots$, is a q variate density. (Indeed, $h_0(\hat{\beta})$ is the density of a q variate normal distribution having mean vector β and covariance matrix $(NX\Sigma^{-1}X')^{-1}$.) Thus, (2.6) is a mixture of the densities $h_j(\hat{\beta})$. Using standard results concerning mixtures of densities, we can conclude that for any measurable set R in q dimensional space,

$$(2.9)\qquad c_0 P\{u \in R\} \leq P\{\hat{\beta} \in R\} \leq c_0 P\{u \in R\} + (1 - c_0),$$

where $u \sim N(\beta, (NX\Sigma^{-1}X')^{-1})$. From the fact that for fixed h.

$$(2.10)\qquad \frac{\Gamma(t + h)}{\Gamma(t)} = t^h[1 + o(1)], \qquad\qquad t \to \infty,$$

it follows that

$$(2.11)\qquad c_0 = 1 - \frac{q(p - q)}{2N} + O(N^{-2})$$

as $N \to \infty$. From (2.9) and (2.11), we see that $\sqrt{N}(\hat{\beta} - \beta)$ has an asymptotic q variate normal distribution with mean vector zero and covariance matrix $(X\Sigma^{-1}X')^{-1}$, and we also have a measure of the accuracy of the approximation involved in replacing the finite sample distribution of $\hat{\beta}$ with the asymptotic distribution.

Two alternative forms for the density (2.6) of $\hat{\beta}$ in terms of an integral representation and a hypergeometric series may prove helpful (Gleser and Olkin [7]). These are the following:

$$(2.12)\qquad p(\hat{\beta}) = h_0(\hat{\beta}) \int_0^1 \frac{g^{(p-q)/2 - 1}(1 - g)^{(n+2q-p+1)/2 - 1} \exp\left\{\tfrac{1}{2}gQ(\hat{\beta})\right\} dg}{B(\tfrac{1}{2}(p - q), \tfrac{1}{2}(n - p + q + 1))},$$

and

$$(2.13) \qquad p(\hat{\beta}) = c_0 h_0(\hat{\beta})_1 F_1\left(\tfrac{1}{2}(p - q), \tfrac{1}{2}(n + q + 1); \tfrac{1}{2}Q(\hat{\beta})\right),$$

where $_1F_1(a, b; z)$ is the confluent hypergeometric function

$$(2.14) \qquad {}_1F_1(a, b; z) = \sum_{j=0}^{\infty} \frac{\Gamma(a + j)}{\Gamma(a)} \frac{\Gamma(b)}{\Gamma(b + j)} \frac{z^j}{j!}.$$

From (2.12), a direct computation (involving an interchange of the order of integration between $\hat{\beta}$ and g) yields $\mathscr{E}(\hat{\beta}) = \beta$ (that is, $\hat{\beta}$ is unbiased) and

$$(2.15) \qquad N \operatorname{Cov}(\hat{\beta}) = \frac{n - 1}{n - p + q - 1} (X\Sigma^{-1}X')^{-1}.$$

We have derived the estimators $\hat{\beta}$ and $\hat{\Sigma}$ assuming that Σ is unknown. If Σ is known, then the estimator $\hat{\beta}(\Sigma) = \bar{y}\Sigma^{-1}X'(X\Sigma^{-1}X')^{-1}$ is the Gauss-Markov (BLUE) estimator of β— that is, among all unbiased linear estimators of β, $\hat{\beta}(\Sigma)$ has the smallest ellipsoid of concentration. The covariance matrix of $\hat{\beta}(\Sigma)$ is $(NX\Sigma^{-1}X')^{-1}$; from this fact and (2.15), it follows that for all Σ,

$$(2.16) \qquad \operatorname{Cov}(\hat{\beta}) = \left(1 + \frac{p - q}{n - p + q - 1}\right) \operatorname{Cov}[\hat{\beta}(\Sigma)].$$

For n moderately large with respect to $p - q$, $\operatorname{Cov}(\hat{\beta})$ and $\operatorname{Cov}[\hat{\beta}(\Sigma)]$ are nearly equal (more accurately, they are of the same order of magnitude in N). We thus have an estimator $\hat{\beta}$ for β which, regardless of the value Σ of the unknown covariance matrix, has for large enough N approximately the minimal ellipse of concentration achievable by the BLUE of β given that value of Σ. Comparisons similar to the above have been made in Gleser and Olkin [7], Rao [21], and Williams [25].

It is worth noting that as $N \to \infty$ both $\sqrt{N}(\hat{\beta} - \beta)$ and $\sqrt{N}[\hat{\beta}(\Sigma) - \beta]$ have the limiting distribution $N(0, (X\Sigma^{-1}X')^{-1})$. A measure of the error involved in assuming that $\hat{\beta}$ and $\hat{\beta}(\Sigma)$ have the same distribution in small samples can be obtained from (2.9) and (2.11).

The distribution of $\hat{\Sigma}$ is given in Appendix A.

3. Confidence regions for β

From (2.6), (2.12), or (2.13) it can be seen that the density of $\hat{\beta}$ is constant on ellipsoids that have the form

$$(3.1) \qquad Q(\hat{\beta}) = \text{constant},$$

where

$$(3.2) \qquad Q(\hat{\beta}) = N(\hat{\beta} - \beta)(X\Sigma^{-1}X')(\hat{\beta} - \beta)'.$$

The regions of form (3.1) are thus ellipsoids of concentration for the distribution of $\hat{\beta}$. Since $\hat{\beta}$ has approximately a q variate normal distribution with mean vector β and covariance matrix $(NX\Sigma^{-1}X')^{-1}$, this suggests using the ellipsoid $\{\beta: Q(\hat{\beta}) \leq \chi_q^2(\gamma)\}$, where $\chi_q^2(\gamma)$ is the upper tail of a χ_q^2 distribution, as a 100γ per cent confidence interval for β. Unfortunately, this region cannot be used since Σ is unknown. We can, however, replace Σ by its MLE $\hat{\Sigma}$, and form a confidence region for β based on the pivotal quantity

$$(3.3) \qquad \Delta = N(\hat{\beta} - \beta)(X\hat{\Sigma}^{-1}X')(\hat{\beta} - \beta)'.$$

Since $(N\hat{\Sigma})^{-1} = S^{-1} - N(1 + r)^{-1}S^{-1}(\bar{y} - \hat{\beta}X)'(\bar{y} - \hat{\beta}X)S^{-1}$, where

$$(3.4) \qquad r = N(\bar{y} - \hat{\beta}X)S^{-1}(\bar{y} - \hat{\beta}X)',$$

and since $XS^{-1}(\bar{y} - \hat{\beta}X) = 0$, it follows that $X\hat{\Sigma}^{-1}X' = NXS^{-1}X'$ and

$$(3.5) \qquad \Delta = N^2(\hat{\beta} - \beta)(XS^{-1}X')(\hat{\beta} - \beta)'.$$

Although the region $\{\beta: N^2(\hat{\beta} - \beta)(XS^{-1}X')(\hat{\beta} - \beta) \leq \chi_q^2(\gamma)\}$ has asymptotic confidence γ as $N \to \infty$, it is not an exact 100γ per cent confidence region for β. Thus for moderate sample sizes it may be of value to determine exact confidence regions for β based on the pivotal quantity Δ defined in (3.5).

The problem of finding the constant $b^{(\gamma)}$ for which the region

$$(3.6) \qquad E_1 = \{\beta: N(\hat{\beta} - \beta)(XS^{-1}X')(\hat{\beta} - \beta)' \leq b^{(\gamma)}\}$$

has *exact* confidence γ is quite difficult since $b^{(\gamma)}$ or equivalently $c^{(\gamma)} = b^{(\gamma)}/(1 + b^{(\gamma)})$ is obtained as the solution of the integral equation

$$(3.7) \qquad \int_0^1 dg \int_0^{c^{(\gamma)}} dh \, \frac{g^{a_1-1}(1 - g)^{a_2-1}h^{d_1-1}(1 - h)^{d_2-1}(1 - gh)^{-(d_1+d_2)}}{B(a_1, a_2 - d_1)B(d_1, d_2)} = \gamma,$$

where $a_1 = \frac{1}{2}(p - q)$, $a_2 = \frac{1}{2}(n + 2q - p + 1)$, $d_1 = \frac{1}{2}q$, and $d_2 = \frac{1}{2}(n - p + 1)$.

THEOREM 3.1. *If $c^{(\gamma)}$ is chosen to satisfy (3.7), then E_1 (with $b^{(\gamma)} = c^{(\gamma)}/(1 - c^{(\gamma)})$) is a 100γ per cent confidence region for β.*

PROOF. From (3.4) and Lemma 2 of Appendix A, $(n - p + 1)\Delta/q(1 + r)$ has, conditional upon r, Snedecor's F distribution with q and $n - p + 1$ degrees of freedom. Also $(1 + r)^{-1}$ has a Beta distribution with parameters $\frac{1}{2}(n - p + q + 1)$ and $\frac{1}{2}(p - q)$. It follows, therefore, that for $P\{\beta \in E_1\}$ to be equal to γ, we must have

$$(3.8) \quad \gamma = P\{\beta \in E_1\} = P\left\{\frac{(n - p + 1)\Delta}{q(1 + r)} \leq \left(\frac{n - p + 1}{q}\right)\frac{b^{(\gamma)}}{1 + r}\right\}$$

$$= \int_0^\infty \frac{r^{(p-q)/2-1} \, dr}{B(\frac{1}{2}(p - q), \frac{1}{2}(n + q - p + 1))(1 + r)^{(n+1)/2}}$$

$$\cdot \int_0^{b^{(\gamma)}/(1+r)} \frac{x^{q/2-1} \, dx}{B(\frac{1}{2}q, \frac{1}{2}(n - p + 1))(1 + x)^{(n-p+q+1)/2}}.$$

By a change of variables to $g = r/(1+r)$, $h = (x + xr)/(1 + x + xr)$, we obtain (3.7). *Q.E.D.*

Another expression for $P\{\beta \in E_1\}$ has been given by Rao [17] in terms of the hypergeometric function. However, in either form it is difficult to solve for the cutoff point $b^{(\gamma)}$. A computer program has been written utilizing a certain mixture representation for the integral (3.7). This program is described in Appendix B.

Notice that the statistic $r = N(\bar{y} - \hat{\beta}X)S^{-1}(\bar{y} - \hat{\beta}X)'$ is a function of the sufficient statistic $(\bar{y}, S)$ and has a distribution which is functionally independent of the parameters β and Σ under the model (1.1). Thus, r is an ancillary statistic. Indeed, the statistic r can be used to test the goodness of fit of the model (1.1) (see Rao [20]). Following a somewhat standard practice, we might agree to find a confidence region for β which has probability of coverage γ, conditional upon r for each possible value of r. Returning to the distributional fact used in the proof of Theorem 3.1, we see that one such region is

$$(3.9) \qquad E_2 = \left\{ \beta : \frac{(n - p + 1)N(\hat{\beta} - \beta)(XS^{-1}X')(\hat{\beta} - \beta)'}{q(1 + r)} \leqq F^{(\gamma)}_{q, n-p+1} \right\},$$

where $F^{(\gamma)}_{q, n-p+1}$ is the upper tail of Snedecor's F distribution with q and $n - p + 1$ degrees of freedom. Since E_2 has, conditional upon r, coverage γ for β, it is also a 100γ per cent unconditional confidence region for β. Because tables of the F distribution are easily available, the region E_2 has been preferred by statisticians. However, in certain circumstances the performance of region E_1 may be superior to that of region E_2. Without values of $b^{(\gamma)}$, comparisons of these two confidence regions are difficult, if not impossible, to do. Using the tables of $b^{(\gamma)}$, such comparisons can now be made.

Before leaving the present section, however, it is worth noting that the region E_2 is the set of all vectors β_0 in q dimensional space for which the null hypothesis $H: \beta = \beta_0$ is not rejected by the appropriate likelihood ratio test at level $\alpha = 1 - \gamma$. The likelihood ratio test of $H: \beta = \beta_0$ versus general alternatives has rejection region

$$(3.10) \qquad \frac{(n - p + 1)N(\hat{\beta} - \beta_0)(XS^{-1}X')(\hat{\beta} - \beta_0)'}{q(1 + r)} \geqq F^{(\gamma)}_{q, n-p+1}$$

(Rao [21]), so that for given values of $\bar{y}$ and S (and thus of $\hat{\beta}$, S, and r), we accept $H: \beta = \beta_0$ if and only if β_0 is in E_2.

4. Comparison of the two procedures

Historically, there have been two main sets of criteria for the comparison of confidence regions—those based on concepts of power and those based on volume considerations. Since every confidence region can generate a test for such hypotheses as $H: \beta = \beta_0$, it seems reasonable to apply power considerations in the comparison of confidence regions. However, the difficulty involved in obtaining and analyzing the nonnull distributions of Δ and $\Delta(1 + r)^{-1}$ discourage comparisons based on power concepts (see Rao [18], [21]).

Comparisons of confidence regions through consideration of their volumes also have intuitive appeal, since the volume of a region can be viewed as a measure of the "quantity" of models (parameters) which are accepted by (included in) the confidence procedure. For example, in the case of two confidence intervals A and B of confidence γ we would prefer interval A to interval B if the length of A were always less than the length of B, because intuitively we we would feel that A would give us a more precise picture of which models are reasonable, given the data.

In the present situation our regions are ellipsoids in q dimensional Euclidean space. Since the volume of an ellipsoid

$$(4.1) \qquad (u_1, u_2, \cdots, u_m) A^{-1}(u_1, u_2, \cdots, u_m)' \leq 1$$

is $c(m)|A|^{1/2}$, where $c(m) = (2\pi)^{m/2} \Gamma(m/2)$, we conclude that

$$
(4.2) \qquad
\begin{aligned}
&\text{volume } E_1 = c(q)|NXS^{-1}X'|^{-1/2}b_0^{q/2}, \\
&\text{volume } E_2 = c(q)|NXS^{-1}X'|^{-1/2}(1 + r)^{q/2}[qF_0/(n - p + 1)]^{q/2},
\end{aligned}
$$

where $F_0 = F_{q, n-p+1}^{(\gamma)}$ and $b_0 = b^{(\gamma)}$.

Since these volumes are random variables, we may compare their expected values. Thus, we say that region E_1 is preferable to region E_2 if and only if $\mathscr{E}[\text{volume } E_1] \leq \mathscr{E}[\text{volume } E_2]$, or equivalently if and only if the ratio

$$(4.3) \qquad I_{1,2} = \left[\frac{(n - p + 1)b_0}{qF_0}\right]^{q/2} \frac{\mathscr{E}[|NXS^{-1}X'|^{1/2}]}{\mathscr{E}[|NXS^{-1}X'|^{1/2}(1 + r)^{q/2}]}$$

is less than or equal to 1. By Lemma 2 of Appendix A, $XS^{-1}X'$ and r are independently distributed and $(1 + r)^{-1}$ has a Beta distribution with $\frac{1}{2}(n - p + q + 1)$ and $\frac{1}{2}(p - q)$ degrees of freedom. Thus (4.3) becomes

$$(4.4) \qquad I_{1,2} = \frac{\Gamma[\frac{1}{2}(n - q + 1)] \Gamma[\frac{1}{2}(n - p + q + 1)]}{\Gamma[\frac{1}{2}(n + 1)] \Gamma[\frac{1}{2}(n - p + 1)]} \left[\frac{(n - p + 1)b_0}{qF_0}\right]^{q/2}.$$

From equation 4.4 (and Table III of Appendix B), values of $I_{1,2}$ are computed for $n = 10(2)30(5)35$, $p = 2(1)\frac{1}{2}n$, $q = 1(1) p - 2$, and $\gamma = 0.90, 0.95, 0.975, 0.99$. In the resulting table we have observed certain patterns. (A selection from this table appears in Table I below.) First, if we fix n, p, and q, and allow γ to increase, then the ratio $I_{1,2}$ increases, becoming greater than 1 for large enough γ. The larger q is, the smaller the value of γ at which $I_{1,2}$ changes from less than 1 to greater than 1. Saying this another way, for fixed r, p, γ, the ratio $I_{1,2}$ is nearly monotonically decreasing in q (the decrease of $I_{1,2}$ in q is reversed in the third decimal place for $q \geq p - 4$).

TABLE I

Ratio of the Expected Volume of E_1 to E_2

		n = 14					n = 24 (continued)		
p	q	0.90	0.95	0.99	p	q	0.90	0.95	0.99
2	1	1.00+	1.00+	1.01	8	1	1.00+	1.01	1.02
						2	1.00+	1.01	1.03
3	1	1.00+	1.00+	1.01		3	1.00−	1.01	1.03
						4	0.99	1.00+	1.03
4	1	1.00+	1.01	1.02		5	0.98	1.00−	1.02
	2	1.00+	1.01	1.02		6	0.98	0.99	1.01
5	1	1.00+	1.01	1.03	9	1	1.01+	1.01	1.02
	2	1.00+	1.01	1.04		2	1.00	1.01	1.04
	3	0.99	1.00+	1.03		3	1.00−	1.01	1.04
						4	0.99	1.00+	1.04
6	1	1.01	1.02	1.04		5	0.98	0.99	1.03
	2	1.00+	1.02	1.05		6	0.97	0.98	1.00+
	3	0.99	1.01	1.04		7	0.97	0.98	1.00+
	4	0.98	0.99	1.02					
					10	1	1.00+	1.01	1.03
7	1	1.01	1.02	1.05		2	1.00+	1.02	1.05
	2	1.00+	1.02	1.07		3	1.00−	1.01	1.05
	3	0.98	1.01	1.06		4	0.98	1.00+	1.05
	4	0.96	0.98	1.03		5	0.97	0.99	1.03
	5	0.95	0.97	1.00+		6	0.96	0.97	1.01
						7	0.95	0.97	1.00−
						8	0.95	0.96	0.99

		n = 24							
					11	1	1.00	1.01	1.03
						2	1.00+	1.02	1.05
p	q	0.90	0.95	0.99		3	0.99	1.01	1.06
						4	0.98	1.00+	1.05
2	1	1.00	1.00+	1.00+		5	0.96	0.98	1.04
						6	0.94	0.97	1.01
3	1	1.00+	1.00+	1.00+		7	0.93	0.95	1.00−
						8	0.93	0.94	0.98
4	1	1.00+	1.00+	1.01		9	0.93	0.94	0.97
	2	1.00+	1.00+	1.01					
					12	1	1.01	1.01	1.04
5	1	1.00+	1.00+	1.01		2	1.00+	1.02	1.06
	2	1.00+	1.00+	1.01		3	0.99	1.02	1.07
	3	1.00−	1.00+	1.01		4	0.97	1.00−	1.06
						5	0.95	0.98	1.04
6	1	1.00+	1.00+	1.01		6	0.92	0.95	1.01
	2	1.00+	1.01	1.02		7	0.91	0.93	0.99
	3	1.00−	1.01	1.02		8	0.90	0.92	0.96
	4	1.00−	1.00+	1.01		9	0.90	0.91	0.95
						10	0.91	0.92	0.94
7	1	1.00+	1.01	1.02					
	2	1.00+	1.01	1.03					
	3	1.00−	1.01	1.03					
	4	0.99	1.00+	1.02					
	5	0.99	1.00−	1.01					

TABLE I (Continued)

RATIO OF THE EXPECTED VOLUME OF E_1 TO E_2

			$n = 35$							
p	q	0.90	0.95	0.99		p	q	0.90	0.95	0.99
2	1	1.00	1.00+	1.00+		10	1	1.00+	1.00+	1.01
							2	1.00+	1.01	1.02
3	1	1.00−	1.00+	1.00+			3	1.00−	1.01	1.02
							4	0.99	1.00+	1.02
4	1	1.00−	1.00+	1.00+			5	0.99	1.00−	1.02
	2	1.00−	1.00+	1.00+			6	0.98	0.99	1.02
							7	0.98	0.99	1.01
5	1	1.00+	1.00+	1.00+			8	0.98	0.99	1.00+
	2	1.00−	1.00+	1.00+						
	3	1.00−	1.00+	1.01		11	1	1.00+	1.00+	1.01
							2	1.00+	1.01	1.02
6	1	1.00+	1.00+	1.00+			3	1.00−	1.01	1.03
	2	1.00+	1.00+	1.01			4	0.99	1.00+	1.03
	3	1.00−	1.00+	1.01			5	0.99	0.99	1.02
	4	1.00−	1.00+	1.01			6	0.98	0.99	1.02
							7	0.98	0.99	1.01
7	1	1.00+	1.00+	1.01			8	0.97	0.98	1.00+
	2	1.00−	1.00+	1.01			9	0.97	0.98	1.00−
	3	1.00−	1.00+	1.01						
	4	1.00−	1.00+	1.01		12	1	1.00+	1.01	1.01
	5	1.00−	1.00−	1.01			2	1.00+	1.01	1.03
							3	1.00−	1.01	1.03
8	1	1.00+	1.00+	1.01			4	0.99	1.00+	1.03
	2	1.00+	1.00+	1.01			5	0.98	1.00−	1.03
	3	1.00−	1.00+	1.01			6	0.97	0.98	1.02
	4	1.00−	1.00+	1.01			7	0.97	0.98	1.02
	5	0.99	1.00−	1.01			8	0.96	0.98	1.01
	6	0.99	1.00−	1.01			9	0.96	0.97	1.00−
							10	0.97	0.97	0.99
9	1	1.00+	1.00+	1.01						
	2	1.00+	1.01	1.02						
	3	1.00−	1.01	1.02						
	4	0.99	1.01+	1.02						
	5	0.99	1.00+	1.02						
	6	0.99	1.00−	1.01						
	7	0.99	0.99	1.01+						

Second, if we fix p, q, and γ, and allow n to increase, then the ratio $I_{1,2}$ converges to 1. This result is not at all surprising since the pivotal quantities Δ and $\Delta(1 + r)^{-1}$ converge to one another in probability at an exponential rate as $n \to \infty$, regardless of the values of p, q, and γ.

Finally, if we fix n, q, and γ, and allow p to increase, then the ratio $I_{1,2}$ may increase or decrease depending on whether the initial value of $I_{1,2}$ in the series is greater or less than one. The actual pattern of movement of $I_{1,2}$ in p is probably a slowly undulating one, offering little practical guidance in the choice of procedure.

Recalling that values of $I_{1,2}$ greater than one favor procedure E_2, that values of $I_{1,2}$ less than one favor procedure E_1, and that a value of $I_{1,2}$ equal to one favors neither procedure, the patterns which we have noted in our table of $I_{1,2}$ suggest that the confidence region E_1 should be used if the requirements for probability of coverage are modest ($\gamma = 0.90$, or even 0.95), the number q of regression parameters is not much less than the dimension p of a single replication y of the model (1.1), and/or if N is of moderate size. However, it should be kept in mind that $I_{1,2}$ is a dimensionless quantity (a ratio of volumes), so that if a large saving in expected volume is of interest, the $I_{1,2}$ tells us little unless we also know the expected volume of one of the two confidence regions.

It should also be remarked that in our table of $I_{1,2}$, values very rarely are less than 0.88 or greater than 1.07. Thus, unless one is greatly concerned about keeping the expected volume of the region as low as possible, the choice between the regions E_1 and E_2 can be governed by computational convenience, by other aspects of the context of the given research problem, or by personal conviction.

REMARK.   One advantage in using the conditional region E_2 is that its conditional probability of coverage given r is independent of r. Since r is a monotone function of the likelihood ratio test statistic for the goodness of fit of model (1.1), one can perform a preliminary test for the fit of the model without affecting the coverage probability of the confidence region for the parameters of the model (assuming the model is accepted by the likelihood ratio test). The expected volume of E_2 would, of course, be affected by such a two stage procedure. A similar two stage procedure based on r and E_1 could be constructed, but this would require new tables of $b^{(\gamma)}$. To our knowledge, no satisfactory criterion for comparing such two stage procedures has yet been proposed, so that balancing this advantage of region E_2 against a possibly smaller expected volume for E_1 must be left entirely to the individual.

5. An illustrative example

To illustrate the computation of the point estimators of β and Σ and the construction of the two confidence regions E_1 and E_2 for β, we make use of the growth curve data reported earlier by Potthoff and Roy [15]. In a study performed at the University of North Carolina Dental School, measurements were made of the distance (in mm.) from the center of the pituitary to the pteryomaxillary fissure for eleven girls and sixteen boys at ages 8, 10, 12, and 14 years. The resulting data for the boys is given in Table II below.

TABLE II

DISTANCE IN MM FROM CENTER OF PITUITARY TO PTERYOMAXILLARY FISSURE

Subject Age	Age in Years			
	8	10	12	14
1	26	25	29	31
2	21.5	22.5	23	26.5
3	23	22.5	24	27.5
4	25.5	27.5	26.5	27
5	20	23.5	22.5	26
6	24.5	25.5	27	28.5
7	22	22	24.5	26.5
8	24	21.5	24.5	25.5
9	23	20.5	31	26
10	27.5	28	31	31.5
11	23	23	23.5	25
12	21.5	23.5	24	28
13	17	24.5	26	29.5
14	22.5	25.5	25.5	26
15	23	24.5	26	30
16	22	21.5	23.5	25

In the present analysis we adopt a linear model for the growth curve; namely,

$$(5.1) \qquad y_i = \beta_1 + \tfrac{1}{3}\beta_2(t_i - 11),$$

where y_i is the distance (in mm.) measured at time t_i with $i = 1, 2, 3, 4$. We have chosen to represent the model in terms of the orthogonal polynomials for the sake of computational convenience. In terms of the model (1.1),

$$(5.2) \qquad X = \begin{pmatrix} 1 & 1 & 1 & 1 \\ -3 & -1 & 1 & 3 \end{pmatrix},$$

with $p = 4$ and $q = 3$. The sample size $N = 16$, so that $n = 15$. Computation yields the following:

$$\bar{y} = (22.88, 23.81, 25.72, 27.47),$$

$$(5.3) \qquad S = \begin{pmatrix} 90.25 & 34.37 & 42.16 & 24.19 \\ 34.37 & 68.44 & 32.91 & 42.16 \\ 54.44 & 32.91 & 105.48 & 48.61 \\ 24.19 & 42.16 & 48.61 & 65.23 \end{pmatrix},$$

from which
$$\hat{\beta} = (25.00, 0.83),$$

(5.4)
$$\hat{\Sigma} = \begin{pmatrix} 5.78 & 2.02 & 2.59 & 1.50 \\ 2.02 & 4.40 & 2.10 & 2.65 \\ 2.59 & 2.10 & 6.61 & 3.04 \\ 1.50 & 2.65 & 3.04 & 4.08 \end{pmatrix}.$$

The 95 per cent confidence region of form E_1 for (β_1, β_2) is given by

(5.5) $\quad E_1 = \{(\beta_1, \beta_2) : 0.34(\beta_1 - 25.00)^2$
$$+ 0.10(\beta_1 - 25.00)(\beta_2 - 0.83) + 7.27(\beta_2 - 0.83)^2 \leqq 0.767\},$$

where $b^{(.95)} = 0.767$ is obtained by linearly interpolating the values of $c^{(.95)}$ for $n = 14$ and $n = 16$, and then from the resulting c forming $b = c(1 - c)^{-1}$. The 95 per cent confidence region of form E_2 for (β_1, β_2) is given by

(5.6) $\quad E_2 = \{(\beta_1, \beta_2) : 0.34(\beta_1 - 25.00)^2$
$$+ 0.10(\beta_1 - 25.00)(\beta_2 - 0.83) + 7.27(\beta_2 - 0.83)^2 \leqq 0.611\},$$

since $r = 0.144$, $F_{2,14}^{(.95)} = 3.74$. Notice that the volume of E_2 is less than the volume of E_1 for this example. Although this result will not always occur if this particular example is replicated (since r is a random variable), the tables of $I_{1,2}$ described in Section 4 would lead us to expect the result we have obtained (since for both $n = 14$ and $n = 16$, with $p = 4$, $q = 2$, and $\gamma = 0.95$, the value of $I_{1,2}$ is 1.01).

$$\diamond \quad \diamond \quad \diamond \quad \diamond \quad \diamond$$

APPENDIX A

DISTRIBUTIONAL RESULTS

A.1. Introduction

In this appendix we derive the distributions of $\hat{\beta}$ and $\hat{\Sigma}$ by means of a certain canonical distributional representation of these statistics. As a first step in obtaining this representation, note that $\hat{\beta}$ is invariant under the transformation $\tilde{y} = \bar{y}A$, $\tilde{S} = A'SA$, $\tilde{X} = XA$ for A nonsingular. Consequently, if we choose A so that $A'\Sigma A = I$ (that is, $A = \Sigma^{-1/2}$), then $\tilde{y} \sim N(\beta\tilde{X}, N^{-1}I)$, $\tilde{S} \sim W(I; p, n)$, $\tilde{y}$ and $\tilde{S}$ are independently distributed. In terms of $\tilde{y}$, $\tilde{S}$, and $\tilde{X}$,

(A.1)
$$\hat{\beta} = \tilde{y}\tilde{S}^{-1}\tilde{X}'(\tilde{X}\tilde{S}^{-1}\tilde{X}')^{-1},$$
$$N\hat{\Sigma} = \Sigma^{1/2}[\tilde{S} + N(\tilde{y} - \hat{\beta}\tilde{X})'(\tilde{y} - \hat{\beta}\tilde{X})]\Sigma^{1/2}.$$

Further simplification is possible. There exists a nonsingular $q \times q$ matrix $\mathbf{T}$ and a $p \times p$ orthogonal matrix Γ such that

(A.2)
$$\tilde{X} = T(I_q, 0)\Gamma'.$$

(MacDuffee p. 77 [12]), where I_q is the $q \times q$ identity matrix. This has the effect of reducing the dimensionality of the space as follows. Transform from $\tilde{y}, \tilde{S}$ to

$$\text{(A.3)} \qquad z = \sqrt{N}\tilde{y}\Gamma, \qquad V = \Gamma'\tilde{S}\Gamma.$$

Let $z = (\dot{z}, \ddot{z})$, where $\dot{z}$ consists of the first q components of z, and then partition V as

$$\text{(A.4)} \qquad V = \begin{pmatrix} V_{11} & V_{12} \\ V_{21} & V_{22} \end{pmatrix}, \qquad V_{11}: q \times q, \qquad V_{22}: (p - q) \times (p - q).$$

It is easily verified that $\dot{z}, \ddot{z}$, and V are stochastically independent, that $\dot{z} \sim N(\sqrt{N}\beta T, I_q)$, that $\ddot{z} \sim N(0, I_{p-q})$, and that $V \sim W(I; p, n)$. Furthermore,

$$\text{(A.5)} \qquad \begin{aligned} b &\equiv \sqrt{N}\,\hat{\beta}T = \dot{z} - \ddot{z}V_{22}^{-1}V_{21}, \\ \tilde{\Sigma} &\equiv \Gamma'\Sigma^{-1/2}\hat{\Sigma}\Sigma^{-1/2}\Gamma = V + \begin{pmatrix} V_{12}V_{22}^{-1}\ddot{z}' \\ \ddot{z}' \end{pmatrix}(\ddot{z}V_{22}^{-1}V_{21}, \ddot{z}), \end{aligned}$$

where

$$\text{(A.6)} \qquad \tilde{\Sigma} = \begin{pmatrix} \tilde{\Sigma}_{11} & \tilde{\Sigma}_{12} \\ \tilde{\Sigma}_{21} & \tilde{\Sigma}_{22} \end{pmatrix}$$

is partitioned in the manner of V. Let $\mu = \sqrt{N}\,\beta T$. The following lemma is known (and easily verified).

LEMMA A.1. *If V has a $W(I; p, n)$ distribution, then $M = V_{11} - V_{12}V_{22}^{-1}V_{21} \sim W(I_q; q, n - p + q)$, $V_{22} \sim W(I_{p-q}; p - q, n)$, and the $q(p - q)$ elements of $L = V_{22}^{-1/2}V_{21}$ are independently distributed as $N(0, 1)$. Furthermore, M, V_{22}, and L are mutually stochastically independent.*

A.2. The distribution of $\hat{\beta}$

Since $\sqrt{N}\,\hat{\beta} = bT^{-1}$, to obtain the distribution of $\hat{\beta}$ it is sufficient to find the distribution of b. From (A.5) and Lemma A.1, we see that

$$\text{(A.7)} \qquad b = \dot{z} - wL \equiv \dot{z} - \ddot{z}V_{22}^{-1/2}L,$$

where $w, \dot{z}$, and L are independent. Again from Lemma A.1, it follows that the conditional distribution of b given w is $N(\mu, (1 + ww')I_q)$. Let $r = ww'$ and note that $r = \ddot{z}V_{22}^{-1}\ddot{z}' = N(\bar{y} - \beta X)S^{-1}(\bar{y} - \beta X)'$. Since $\ddot{z} \sim N(0, I_{p-q})$ and $V_{22} \sim W(I; p - q, n)$ and $\ddot{z}$ and V_{22} are independent, it can be shown in a straightforward manner using Hsu's theorem (Anderson [1], p. 319) that r has the density

$$(A.8) \qquad p(r) = \frac{r^{(p-q)/2-1}}{B\left(\frac{1}{2}(p-q), \frac{1}{2}(n-p+q+1)\right)(1+r)^{(n+1)/2}}.$$

The distribution of b given w is the same as that of b given r (since the former conditional distribution depends upon w only through $r = ww'$), namely $N(\mu, (1+r)I_q)$, so that

$$(A.9) \qquad p(b, r) = p(b|r)p(r)$$
$$= \frac{r^{(p-q)/2-1}\exp\left\{-\frac{1}{2}\frac{(b-\mu)(b-\mu)'}{1+r}\right\}}{(2\pi)^{q/2} B\left(\frac{1}{2}(p-q), \frac{1}{2}(n-p+q+1)\right)(1+r)^{(n+q+1)/2}}.$$

Transforming from b to $\hat{\beta} = N^{-1/2}bT^{-1}$ and from r to $g = r/(1+r)$, noting that $TT' = X\Sigma^{-1}X'$, that $\mu = \sqrt{N}\beta T$, and integrating over g, where $0 \le g \le 1$, yields (2.12). The expansion of the integral form (2.12) of $p(\hat{\beta})$ in terms of the confluent hypergeometric function $_1F_1\left(\frac{1}{2}(p-q), \frac{1}{2}(n+q+1); Q(\hat{\beta})\right)$ (equation (2.13)) is well known (for example, see Erdélyi [4], p. 255). Finally by grouping terms appropriately in the infinite sum representation of $_1F_1$ in the representation (2.13) for $p(\hat{\beta})$, we obtain (2.11) and the result of Theorem 2.5.

REMARK. The representations (2.12) and (2.13) for $p(\hat{\beta})$ were obtained by a slightly more complicated proof in Gleser and Olkin [7]. The representation (2.11) is new. As demonstrated in Section 2, the new representation is useful in finding approximations to $p(\hat{\beta})$ for moderate values of the sample size N.

As a byproduct of the above derivations and from Lemma A.1, we have the following result which is useful in Sections 3 and 4.

LEMMA A.2. *The distribution of* $(n-p+1)N(\hat{\beta}-\beta)XS^{-1}X'(\hat{\beta}-\beta)'/q(1+r)$ *given* $r = N(\bar{y}-\hat{\beta}X)S^{-1}(\bar{y}-\hat{\beta}X)'$ *is* $F_{q,n-p+1}$. *Further,* r *and* $XS^{-1}X'$ *are stochastically independent,* $(1+r)^{-1}$ *has a beta distribution with parameters* $\frac{1}{2}(n-p+q+1)$ *and* $\frac{1}{2}(p-q)$, *and* $(XS^{-1}X')^{-1} \sim W\left((X\Sigma^{-1}X')^{-1}; q, n-p+q\right)$.

PROOF. From (A.5), $\sqrt{N}(\hat{\beta}-\beta) = (b-\mu)T^{-1}$. Since, as shown above, the conditional distribution of b given r is $N(\mu, (1+r)I_q)$, since from Lemma A.1, M is independent of $\dot{z}, \ddot{z}, L$, and V_{22} (thus of $\hat{\beta}$ and r), and since

$$(A.10) \qquad \Delta = N(\hat{\beta}-\beta)(XS^{-1}X')(\hat{\beta}-\beta)' = (b-\mu)M^{-1}(b-\mu)',$$

it follows that $(n-p+1)\Delta/q(1+r)$ given r has Snedecor's F distribution with q and $n-p+1$ degrees of freedom (see Anderson [1], Theorem 5.2.2). That $(1+r)^{-1}$ has the beta distribution with parameters $\frac{1}{2}(n-p+q+1)$ and $\frac{1}{2}(p-q)$ follows from (A.8). Finally

$$(A.11) \qquad (XS^{-1}X')^{-1} = (T')^{-1}MT^{-1},$$

and thus $(XS^{-1}X')^{-1} \sim W\left((TT')^{-1}; q, n-p+q\right)$. Since $(TT')^{-1} = (X\Sigma^{-1}X')^{-1}$, the proof of the lemma is completed.

A.3. The distribution of $\hat{\Sigma}$

From Equation (A.5),

$$(A.12) \quad \tilde{\Sigma} = \begin{pmatrix} \tilde{\Sigma}_{11} & \tilde{\Sigma}_{12} \\ \tilde{\Sigma}_{21} & \tilde{\Sigma}_{22} \end{pmatrix} = V + \begin{pmatrix} V_{12} V_{22}^{-1} \\ I \end{pmatrix} \tilde{z}' \tilde{z} (V_{22}^{-1} V_{21}, I),$$

where $V \sim W(I; p, n)$ is independently distributed of $\tilde{z} \sim N(0, I_{p-q})$. Let

$$(A.13) \quad \begin{aligned} M &= V_{11} - V_{12} V_{22}^{-1} V_{21}, \\ \tilde{\Sigma}_{12} &= V_{12} V_{22}^{-1} (V_{22} + \tilde{z}' \tilde{z}), \quad \tilde{\Sigma}_{22} = V_{22} + \tilde{z}' \tilde{z}, \end{aligned}$$

be a transformation from (V_{11}, V_{12}, V_{22}) to $(M, \tilde{\Sigma}_{12}, \tilde{\Sigma}_{22})$. Noting that

$$(A.14) \quad V_{12} V_{22}^{-1} V_{21} = \tilde{\Sigma}_{12} \tilde{\Sigma}_{22}^{-1} (\tilde{\Sigma}_{22} - \tilde{z}' \tilde{z}) \tilde{\Sigma}_{22}^{-1} \tilde{\Sigma}_{12}'$$

and that $\tilde{\Sigma}_{12} = \tilde{\Sigma}_{21}'$, it follows by a direct computation that

$$(A.15) \quad p(\tilde{z}, M, \tilde{\Sigma}_{12}, \tilde{\Sigma}_{22})$$

$$= \frac{C(p, n)}{(2\pi)^{(p-q)/2}} |\tilde{\Sigma}_{22}|^{-q} |\tilde{\Sigma}_{22} - \tilde{z}' \tilde{z}|^{(n+2q-p-1)/2} |M|^{(n-p-1)/2}$$

$$\cdot \exp \left\{ -\tfrac{1}{2} [\operatorname{tr} \tilde{\Sigma}_{22} + \operatorname{tr} M + \operatorname{tr} \tilde{\Sigma}_{12} \tilde{\Sigma}_{22}^{-1} (\tilde{\Sigma}_{22} - \tilde{z}' \tilde{z}) \tilde{\Sigma}_{22}^{-1} \tilde{\Sigma}_{21}] \right\},$$

where $M > 0$, $\tilde{\Sigma}_{22} - \tilde{z}' \tilde{z} > 0$,

$$(A.16) \quad C^{-1}(p, n) = 2^{np/2} \pi^{p(p-1)/4} \prod_{i=1}^{p} \Gamma\left(\tfrac{1}{2}(n - i + 1)\right),$$

and the elements of $\tilde{\Sigma}_{12}$ and $\tilde{z}$ are unrestricted.

Now let

$$(A.17) \quad v = \tilde{z} \tilde{\Sigma}_{22}^{-1/2}, \qquad \tilde{\Sigma}_{11} = M + \tilde{\Sigma}_{12} \tilde{\Sigma}_{22}^{-1} \tilde{\Sigma}_{21}$$

be a transformation from $(\tilde{z}, M)$ to $(v, \tilde{\Sigma}_{11})$. Then

$$(A.18) \quad p(\tilde{\Sigma}, v) = \frac{C(p, n)}{(2\pi)^{(p-q)/2}} |\tilde{\Sigma}_{22}|^{(n-p-1)/2} |\tilde{\Sigma}_{11} - \tilde{\Sigma}_{12} \tilde{\Sigma}_{22}^{-1} \tilde{\Sigma}_{21}|^{(n-p-1)/2}$$

$$\cdot |\tilde{\Sigma}_{22}|^{1/2} (1 - vv')^{(n+2q-p-1)/2}$$

$$\cdot \exp \left\{ -\tfrac{1}{2} [\operatorname{tr} \tilde{\Sigma}_{11} + \operatorname{tr} \tilde{\Sigma}_{22} - \operatorname{tr} v \tilde{\Sigma}_{22}^{-1/2} \tilde{\Sigma}_{21} \tilde{\Sigma}_{12} \tilde{\Sigma}_{22}^{-1/2} v'] \right\}$$

$$= \left[C(p, n) |\tilde{\Sigma}|^{(n-p-1)/2} \exp \left\{ -\tfrac{1}{2} \operatorname{tr} \tilde{\Sigma} \right\} \right]$$

$$\cdot \left[\frac{|\tilde{\Sigma}_{22}|^{1/2} (1 - vv')^{(n+2q-p-1)/2}}{(2\pi)^{(p-q)/2}} \exp \left\{ +\tfrac{1}{2} v \Xi(\tilde{\Sigma}) v' \right\} \right],$$

where $\Xi(H)$ is, for any positive definite

$$(A.19) \quad H = \begin{pmatrix} H_{11} & H_{12} \\ H_{21} & H_{22} \end{pmatrix},$$

defined by $\Xi(H) = H_{22}^{-1/2} H_{21} H_{12} H_{22}^{-1/2}$, and where the range of definition is $\tilde{\Sigma} > 0$, $vv' \leq 1$. We make use of the invariance of vv' under the transformation $v \to v\Gamma$, Γ orthogonal, to reduce the expression still further. Let U be the orthogonal matrix such that

(A.20) $$U\Xi(\tilde{\Sigma})U' = \text{diag}\,(v_1, \cdots, v_{p-q}) \equiv D_v,$$

where the values of v_i are the characteristic roots of $\Xi(\tilde{\Sigma})$. Hence, letting $s = vU'$ where $s: 1 \times (p - q)$, we obtain

(A.21) $$p(\tilde{\Sigma}) = \left[C(p, n) |\tilde{\Sigma}|^{(n-p-1)/2} \exp \left\{ -\tfrac{1}{2} \operatorname{tr} \tilde{\Sigma} \right\} \right]$$

$$\cdot \left[\frac{|\tilde{\Sigma}_{22}|^{1/2}}{(2\pi)^{(p-q)/2}} \int_{ss' \leq 1} (1 - ss')^{(n+2q-p-1)/2} \exp \left\{ \tfrac{1}{2} s D_v s' \right\} ds \right].$$

An alternative expression for $p(\tilde{\Sigma})$ may be obtained by noting that

(A.22)

$$\int_{ss' \leq 1} (1 - ss')^{(n+2q-p-1)/2} \exp \left\{ \tfrac{1}{2} s D_v s' \right\}$$

$$= 2^{q-p} \Gamma \left[\tfrac{1}{2}(n + 2q - 2 + 1) \right]$$

$$\cdot \sum_{j_1, \cdots, j_{p-q}=0}^{\infty} \left[\Gamma \left(\sum_{i=1}^{p-q} j_i + \tfrac{1}{2}(n + q + 1) \right) \right]^{-1} \prod_{i=1}^{p-q} \left(\frac{v_i}{2} \right)^{j_i} \frac{\Gamma(j_i + \tfrac{1}{2})}{j_i!}.$$

When $p - q \geq q$, some of the v_i are 0 with probability one, so that the expressions for $p(\tilde{\Sigma})$ can be somewhat simplified.

The distribution of $\hat{\Sigma}$ may be determined by making the transformation from $\tilde{\Sigma}$ to $\hat{\Sigma} = N^{-1} \Sigma^{1/2} \Gamma \tilde{\Sigma} \Gamma' \Sigma^{1/2}$.

APPENDIX B

TABLES FOR APPLYING CONFIDENCE REGION E_1

Table III gives values of $c^{(\gamma)}$ (see equation (3.7)) needed in order to construct 100γ per cent confidence regions of the form E_1 (see (3.6)). The present tables are calculated for $n = 10(2)30(5)35$, $p = 2(1)\tfrac{1}{2}n$, $q = 1(1)\,p - 2$, and $\gamma = 0.90$, 0.95, 0.975, 0.99. These values of n, p, q, and γ have been chosen as illustrative, but not exhaustive, examples of situations met in practice. For example, it is usually desirable for n to be somewhat larger than p so that sufficient degrees of freedom are available to accurately estimate Σ. For the distribution of $\hat{\Sigma}$ to be nonsingular, we must have $n \geq p + 1$; the assumption $n \geq 2p$ provides a comfortable number of degrees of freedom for $\hat{\Sigma}$. When n is large (say, over 40), this assumption is unnecessarily strict and can be replaced by the condition that $n - p - 1$ be of a reasonable magnitude.

The values of n given are not uncommon in practice. Values of n less than 10 are rarely practical (unless $p = 2$) for reasons already indicated. If n is larger than 35 or 40, large sample approximations may be appropriate (unless p is too large). Simple linear or quadratic interpolation in the tables should give enough accuracy in most situations for the application of the confidence region E_1 when n is odd, $11 \leqq n \leqq 34$.

The coverage probabilities γ chosen for Table III are those customarily given in standard tables for upper tail probabilities. Finally, the values of q which have been chosen reflect the fact that (at least in the context of growth curves) the most desirable models are those which require estimation of the fewest parameters.

Starting with equation (3.7), the table was constructed as follows. First the expansion

$$(B.1) \qquad (1 - gh)^{-(d_1 + d_2)} = \sum_{j=0}^{\infty} \frac{\Gamma(d_1 + d_2 + j)}{\Gamma(d_1 + d_2)} \frac{(gh)^j}{j!}$$

enables us to expand the double integral in (3.7) in the following infinite series:

$$(B.2) \qquad \gamma = \sum_{j=0}^{\infty} \frac{\Gamma(d_1 + d_2 + j)}{\Gamma(d_1 + d_2)j!}$$

$$\cdot \frac{\int_0^1 g^{a_1 + j - 1}(1 - g)^{a_2 - 1} \, dg \int_0^{c(\gamma)} h^{d_1 + j - 1}}{B(a_1, a_2 - d_1)B(d_1, d_2)} (1 - h)^{d_2^{-1}} \, dh$$

$$= \sum_{j=0}^{\infty} c_j I_{c(\gamma)}(d_1 + j, d_2),$$

where the values of c_j with $j = 0, 1, \cdots$ have already been defined in Theorem 2.1; where for constants $f_1, f_2 > 0, 0 \leqq z \leqq 1$,

$$(B.3) \qquad I_z(f_1, f_2) = \int_0^z \frac{w^{f_1 - 1}(1 - w)^{f_2 - 1} \, dw}{B(f_1, f_2)},$$

and where $a_1 = \frac{1}{2}(p - q), a_2 = \frac{1}{2}(n + 2q - p + 1), d_1 = \frac{1}{2}q, d_2 = \frac{1}{2}(n - p + 1)$. The interchange of summation and integration used to obtain equation (B.2) is readily justified from Fubini's theorem by noting that $c_j \geqq 0$, all j, $\Sigma_{j=0}^{\infty} c_j = 1$, and $0 \leqq I_z(f_1, f_2) \leqq 1$. These facts also support the following inequality:

$$(B.4) \qquad \sum_{j=0}^{M} c_j I_z(d_1 + j, d_2) \leqq \sum_{j=0}^{\infty} c_j I_z(d_1 + j, d_2)$$

$$\leqq \sum_{j=0}^{M} c_j I_z(d_1 + j, d_2) + \left(1 - \sum_{j=M+1}^{\infty} c_j\right),$$

which holds for all nonnegative integers M. This inequality permits evaluation of the error involved in truncating the infinite sum $\gamma(z) \equiv \Sigma_{j=0}^{\infty} c_j I_z(d_1 + j, d_2)$

after M terms have been computed. Using this inequality, a grid of values for the infinite sum $\gamma(z)$ was computed (to five place accuracy) for each n, p, q chosen, and for values of z ranging by jumps of 0.02 from 0.50 to 0.98. After such a grid was formed, a value of γ was chosen ($\gamma = 0.90, 0.95, 0.975, 0.99$) and the grid was searched for that value z^* of z which yielded a calculated value of $\gamma(z)$ closest to γ. Since $\gamma(z)$ is monotonic increasing in z, z was allowed to move in increments of 0.001 down or up from z^* depending on whether $\gamma(z^*)$ was greater or less than γ. This movement was terminated once the size of $\gamma(z) - \gamma$ reversed from that of $\gamma(z^*) - \gamma$. A similar incremental movement in steps of 0.0001 from this new value of z was terminated when once again $\gamma(z) - \gamma$ reversed sign. The value of z computed in this entire series for which $|\gamma(z) - \gamma|$ was a minimum was then chosen to be $c^{(\gamma)}$. The resulting values of $c^{(\gamma)}$ are accurate to within $\pm 5 \times 10^{-5}$—assuming that we want $c^{(\gamma)}$ to give us coverage γ up to an error of $\pm 5 \times 10^{-6}$ and ignoring errors in the calculation of the individual terms $c_j I_z(d_1 + j, d_2)$. The value of $c^{(\gamma)}$ was checked by evaluating (B.2) within a six place accuracy.

The computations are simplified by noting the recursion

$$\text{(B.5)} \qquad c_{j+1} = c_j \frac{(q + 2j)(p - q + 2j)}{(n + q + 1 + 2j)(2 + 2j)}.$$

Users of Table III should note that $n = N - 1$, where N is the sample size, and that for the value of γ selected, E_1 is to be used with $b^{(\gamma)} = c^{(\gamma)}/(1 - c^{(\gamma)})$.

TABLE III

TABLES OF CRITICAL VALUES $c^{(\gamma)}$ FOR CONFIDENCE REGION E_1

			$n = 10$						$n = 12$		
p	q	0.90	0.95	0.975	0.99	p	q	0.90	0.95	0.975	0.99
2	1	29496	39059	47606	57308	2	1	24342	32672	40351	49397
3	1	35507	46173	55314	65209	3	1	28535	37844	46197	55735
4	1	43063	54628	63985	73502	4	1	33711	44003	52913	62682
	2	54450	63812	71169	78562		2	44962	53881	61302	69251
5	1	52419	64336	73269	81642	5	1	40093	51253	60466	70060
	2	63314	72388	79077	85361		2	51696	60903	68255	75786
	3	67957	75669	81371	86783		3	57359	65398	71765	78283
						6	1	47901	59595	68679	77545
							2	59364	68491	75411	82124
							3	64550	72346	78246	84002
							4	67525	74504	79803	85010

TABLE III (Continued)

		n = 14						n = 16			
p	q	0.90	0.95	0.975	0.99	p	q	0.90	0.95	0.975	0.99
2	1	20693	28031	34938	43276	2	1	17984	24524	30771	38451
3	1	23765	31921	39457	48357	3	1	20322	27541	34346	42574
4	1	27477	36511	44656	54013	4	1	23097	31059	38435	47178
	2	38079	46318	53429	61362		2	32930	40479	47159	54827
5	1	31983	41909	50586	60213	5	1	26406	35162	43098	52279
	2	43220	51937	59250	67157		2	36931	44993	51982	59824
	3	49251	57076	63548	70497		3	42981	50394	56708	63708
6	1	37456	48211	57247	66846	6	1	30367	39938	48383	57856
	2	49136	58174	65494	73116		2	41527	50044	57246	65109
	3	55061	63005	69373	75981		3	47673	55363	61765	68688
	4	58707	65907	71661	77642		4	51664	58735	64581	70888
7	1	44077	55460	64555	73717	7	1	35118	45478	54313	63865
	2	55857	64953	72011	79036		2	46779	55638	63904	70583
	3	61453	69278	75315	81334		3	52891	60733	67084	73756
	4	64764	71780	77196	82622		4	56735	63864	69611	75644
	5	66952	73405	78398	83428		5	59396	66001	71320	76919
						8	1	40806	51832	60851	70173
							2	52726	61736	68862	76107
							3	58629	66442	72567	78787
							4	62215	69238	74736	80331
							5	64635	71100	76165	81343
							6	66372	72419	77164	82032

		$n = 18$						$n = 20$			
p	q	0.90	0.95	0.975	0.99	p	q	0.90	0.95	0.975	0.99
2	1	15893	21779	27464	34542	2	1	14238	19590	24806	31370
3	1	17729	24180	30347	37928	3	1	15718	21546	27182	34208
4	1	19873	26949	33626	41713	4	1	17420	23771	29850	37339
	2	28957	35872	42099	42092		2	25820	32179	37986	44899
5	1	22389	30141	37341	45904	5	1	19392	26317	32866	40826
	2	32137	39541	46107	53665		2	28397	35203	41343	48555
	3	38050	45004	51055	57928		3	34083	40582	46326	52965
6	1	25357	33834	41557	50542	6	1	21683	29231	36268	44676
	2	35768	43650	50511	58250		2	31317	38574	45030	52489
	3	41871	49158	55388	62329		3	37242	44086	50055	56855
	4	46002	52797	58552	64931		4	41388	47850	53423	59728
7	1	28870	38103	46317	55622	7	1	24360	32576	40103	48919
	2	39906	48222	55304	63096		2	34629	42334	49074	56719
	3	46134	53693	60025	66925		3	40755	47915	54061	60944
	4	50242	57215	63011	69303		4	44952	51647	57337	63668
	5	53177	59690	65081	70926		5	48042	54354	59688	65609
8	1	33035	43014	51636	61092	8	1	27497	36413	44407	53554
	2	44608	53279	60469	68160		2	38379	46504	53469	61201
	3	50860	58595	64919	71642		3	44651	52078	58335	65202
	4	54864	61918	67653	73730		4	48852	55729	61475	67753
	5	57672	64218	69526	75158		5	51888	58325	63678	69516
	6	59751	65899	70879	76170		6	54195	60276	65321	70819
9	1	37974	48629	57509	66882	9	1	31181	40800	49208	58558
	2	49917	58805	65947	73341		2	42619	51103	58209	65902
	3	56047	63818	69996	76366		3	48956	56577	62861	69603
	4	59856	66866	72420	78150		4	53093	60073	65791	71909
	5	62468	68928	74044	79335		5	56027	62514	67810	73472
	6	64371	70413	75200	80166		6	58226	64327	69301	74625
	7	65819	71533	76069	80787		7	59333	65718	70435	75488
						10	1	35513	45799	54517	63894
							2	47388	56131	63255	70746
							3	53677	61383	67582	74062
							4	57669	64648	70237	76071
							5	60447	66893	72046	77434
							6	62493	68526	73347	78398
							7	64066	69771	74337	79131
							8	65306	70743	75098	79680

TABLE III (Continued)

			$n = 22$						$n = 24$		
p	q	0.90	0.95	0.975	0.99	p	q	0.90	0.95	0.975	0.99
2	1	12890	17791	22597	28693	2	1	11775	16293	20748	26436
3	1	14103	19405	24570	31066	3	1	12791	17656	22428	28483
4	1	15488	21234	26790	33713	4	1	13937	19181	24295	30734
	2	23275	29142	34552	41068		2	21182	26623	31685	37842
5	1	17072	23306	29280	36648	5	1	15231	20889	26363	33193
	2	25400	31666	37390	44208		2	22965	28764	34119	40579
	3	30840	36915	42350	48720		3	28146	33836	38978	45074
6	1	18888	25652	32060	39859	6	1	16706	22818	28682	35927
	2	27795	34480	40522	47633		2	24953	31125	36772	43512
	3	33480	39886	45555	52119		3	30385	36387	41764	48083
	4	37573	43691	49042	55188		4	34377	40162	45278	51227
7	1	20983	28325	35187	43415	7	1	18387	24996	31272	38938
	2	30489	37598	43941	51299		2	27177	33738	39680	46686
	3	36409	43139	49022	55742		3	32852	39165	44764	51266
	4	40595	46967	52475	58720		4	36965	43008	48301	54397
	5	43749	49806	55007	60878		5	40122	45909	50937	56691
8	1	23410	31369	38693	47319	8	1	20308	27450	34150	42215
	2	33522	41052	47666	55207		2	29673	36637	42870	50130
	3	39653	46686	52747	59565		3	35575	42193	47996	54645
	4	43901	50500	56127	62416		4	39787	46070	51513	57705
	5	47055	53287	58568	64449		5	42983	48967	54115	59942
	6	49497	55418	60416	65965		6	45502	51222	56117	61636
9	1	26227	34834	42605	51561	9	1	22518	30234	37367	45815
	2	36941	44875	51720	59373		2	32466	39830	46330	53783
	3	43235	50533	56717	63551		3	38580	45489	51466	58224
	4	47504	54283	59975	66229		4	42864	49364	54928	61167
	5	50626	56987	62302	68131		5	46071	52223	57453	63297
	6	53011	59025	64031	69512		6	48572	54426	59380	64900
	7	54899	60624	65382	70587		7	50583	56181	60906	66163
10	1	29510	38781	46964	56164	10	1	25066	33387	40952	49744
	2	40787	49081	56091	63754		2	35594	43345	50073	57649
	3	47177	54680	60915	67665		3	41889	49060	55172	61974
	4	51417	58317	64009	70148		4	46209	52894	58534	64766
	5	54459	60888	66170	71859		5	49396	55680	60951	66757
	6	56747	62808	67767	73108		6	51853	57803	62776	68240
	7	58553	64293	68993	74055		7	53813	59484	64215	69413
	8	59999	65481	69969	74807		8	55408	60840	65364	70333
11	1	33342	43262	51787	61098						
	2	45095	53674	60748	68287						
	3	51497	59122	65320	71876						
	4	55635	62572	68180	74099						
	5	58550	64967	70141	75602						
	6	60722	66736	71579	76697						
	7	62402	68090	72672	77521						
	8	63738	69158	73525	78154						
	9	64830	70027	74219	78669						

TABLE III (Continued)

p	q	0.90	0.95	0.975	0.99	p	q	0.90	0.95	0.975	0.99
\multicolumn{6}{c}{$n = 24$ (continued)}	\multicolumn{6}{c}{$n = 26$}										

p	q	0.90	0.95	0.975	0.99	p	q	0.90	0.95	0.975	0.99
11	1	28009	36953	44922	53975	2	1	10836	15026	19178	24506
	2	29100	47207	54116	61738						
	3	45519	52906	59092	65850	3	1	11698	16188	20617	26272
	4	49831	56647	62305	68452						
	5	52965	59331	64592	70297	4	1	12661	17478	22205	28205
	6	55348	61347	66288	71641		2	19425	24489	29232	35041
	7	57227	62923	67609	72682						
	8	58750	64194	68670	73521	5	1	13743	18918	23967	30331
	9	60004	65230	69526	74181		2	20939	26321	31330	37422
							3	25870	31208	36066	41874
12	1	31420	40990	49313	58528						
	2	43017	51427	58439	66006	6	1	14959	20525	25917	32661
	3	49486	57020	63204	69816		2	22617	28336	33621	39998
	4	53737	60619	66229	72206		3	27786	33410	38494	44522
	5	56769	63151	68336	73856		4	31664	37135	42015	47745
	6	59048	65036	69891	75062						
	7	60824	66491	71083	75978	7	1	16331	22320	28072	35192
	8	62250	67651	72029	76700		2	24483	30557	36123	42778
	9	63420	68599	72799	77286		3	29892	35811	41117	47356
	10	64393	69381	73427	77758		4	33896	39613	44672	50560
							5	37028	42546	47387	52988
						8	1	17888	24340	30479	37997
							2	26557	32996	38839	45745
							3	32206	38422	43944	50372
							4	36328	42289	47516	53542
							5	39517	45236	50213	55914
							6	42071	47571	52330	57759
						9	1	19661	26613	33156	41070
							2	28868	35684	41799	48936
							3	34747	41253	46969	53544
							4	38975	45168	50546	56679
							5	42202	48110	53200	58970
							6	44764	50419	55265	60741
							7	46851	52282	56922	62151
						10	1	21679	29165	36114	44389
							2	31452	38651	45029	52378
							3	37538	44323	50211	56893
							4	41847	48251	53749	59937
							5	45098	51172	56335	62155
							6	47648	53436	58344	63829
							7	49709	55247	59931	65155
							8	51412	56736	61229	66235

TABLE III (Continued)

		n = 26 (*continued*)						n = 28			
p	q	0.90	0.95	0.975	0.99	p	q	0.90	0.95	0.975	0.99
11	1	23994	32043	39405	48021	2	1	10036	13943	17829	22841
	2	34331	41902	48512	56006						
	3	40604	47646	53674	60416	3	1	10775	14942	19071	24371
	4	44963	51550	57131	63326						
	5	48207	54415	59641	65423	4	1	11596	16047	20439	26045
	6	50726	56613	61550	66998		2	17936	22671	27130	32630
	7	52742	58356	63051	68226						
	8	54396	59777	64271	69221	5	1	12510	17272	21947	27882
	9	55779	60960	65283	70044		2	19234	24251	28950	34708
							3	23932	28955	33558	39102
12	1	26652	35289	43049	51951						
	2	37538	45460	52257	59818	6	1	13530	18630	23606	29884
	3	43961	51225	57347	64082		2	20667	25985	30937	36962
	4	48333	55062	60681	66825		3	25588	30873	35688	41446
	5	51538	57837	63069	68773		4	29339	34521	39178	44692
	6	53998	59944	64864	70220						
	7	55951	61602	66271	71352	7	1	14672	20140	25442	32078
	8	57538	62941	67399	72250		2	22250	27886	33098	39394
	9	58856	64046	68328	72986		3	27399	32955	37981	43947
	10	59967	64972	69101	73599		4	31282	36697	41531	47210
							5	34362	39620	44271	49702
13	1	29708	38941	47059	56159						
	2	41112	49345	56274	63824	8	1	15956	21824	27469	34474
	3	47623	55053	61206	67845		2	24002	29972	35451	42010
	4	51958	58774	64371	70384		3	29383	35218	40455	46618
	5	55090	61429	66614	72175		4	33393	39042	44047	49880
	6	57462	63416	68274	73481		5	36542	42001	46795	52346
	7	59323	64962	69556	74481		6	39100	44377	48985	54293
	8	60830	66209	70591	75295						
	9	62069	67226	71427	75939	9	1	17400	23697	29700	37067
	10	63109	68077	72126	76481		2	25944	32263	38010	44825
	11	63990	68793	72711	76928		3	31556	37673	43117	49463
							4	35682	41562	46726	52688
							5	38892	44543	49466	55115
							6	41476	46917	51629	57014
							7	43609	48860	53391	58556
						10	1	19037	25802	32184	39926
							2	28095	34769	40775	47089
							3	33937	40333	45968	52474
							4	38168	44270	49581	55649
							5	41422	47254	52288	58008
							6	44019	49611	54411	59844
							7	46146	51522	56122	61317
							8	47922	53108	57534	62526

TABLE III (Continued)

		n = 28 (continued)						n = 30			
p	q	0.90	0.95	0.975	0.99	p	q	0.90	0.95	0.975	0.99
11	1	20894	28160	34934	43045	2	1	09346	13004	16655	21382
	2	30484	37520	43775	51005						
	3	36540	43204	49010	55627	3	1	09987	13875	17741	22729
	4	40862	47173	52607	58750						
	5	44141	50135	55256	61012	4	1	10696	14834	18935	24204
	6	46735	52456	57319	62767		2	16656	21100	25305	30521
	7	48843	54325	58971	64165						
	8	50591	55865	60325	65306	5	1	11476	15885	20234	25791
	9	52066	57155	61454	66249		2	17784	22481	26906	32367
							3	22257	26995	31361	36649
12	1	23004	30795	37960	46398						
	2	33145	40542	47033	54433	6	1	12344	17045	21663	27527
	3	39388	46303	52251	58935		2	19020	23986	28641	34350
	4	43773	50265	55788	61950		3	23703	28683	33247	38746
	5	47061	53191	58372	64126		4	27324	32238	36681	41977
	6	49631	55455	60353	65777						
	7	51702	57263	61928	67087	7	1	13308	18329	23234	29426
	8	53407	58741	63207	68142		2	20376	25627	30518	36477
	9	54839	59977	64274	69021		3	25277	30505	35270	40971
	10	56057	61022	65172	69757		4	29030	34163	38779	44246
							5	32046	37060	41528	46784
13	1	25414	33758	41309	50040						
	2	36096	43837	50527	58026	8	1	14382	19750	24960	31490
	3	42501	49641	55694	62396		2	21871	27424	32562	38776
	4	46915	53556	59131	65262		3	26992	32478	37444	43345
	5	50180	56411	61610	67307		4	30876	36232	41017	46645
	6	52706	58598	63495	68851		5	33969	39178	43790	49178
	7	54722	60328	64978	70058		6	36508	41568	46020	51193
	8	56370	61732	66174	71025						
	9	57746	62900	67167	71828	9	1	15584	21328	26864	33745
	10	58909	63881	67997	72495		2	23519	29387	34780	41247
	11	59905	64718	68701	73058		3	28864	34613	39779	45866
							4	32873	38451	43399	49170
14	1	28170	37084	44997	53963		5	36038	41437	46185	51689
	2	39370	47423	54260	61780		6	38617	43844	48412	53683
	3	45892	53214	59325	65975		7	40765	45829	50234	55299
	4	50296	57041	62620	68658						
	5	53504	59791	64963	70547	10	1	16931	23082	28960	36198
	6	55958	61874	66727	71960		2	25338	31536	37184	43895
	7	57902	63513	68110	73068		3	30908	36925	42286	48549
	8	59475	64828	69209	73936		4	35035	40834	45937	51838
	9	60779	65914	70115	74651		5	38262	43846	48719	54320
	10	61878	66825	70873	75249						
	11	62814	67598	71515	75752	10	6	40869	46253	50922	56265
	12	63621	68259	72059	76171		7	43031	48232	52724	57852
							8	44852	49885	54220	59157

TABLE III (Continued)

		$n = 30$ (continued)						$n = 30$ (continued)			
p	q	0.90	0.95	0.975	0.99	p	q	0.90	0.95	0.975	0.99
11	1	18444	25029	31259	38840	15	1	26782	35392	43100	51915
	2	27349	33888	39790	46733		2	47771	45649	52390	59875
	3	33141	39426	44976	51394		3	44278	51484	57538	64172
	4	37373	43385	48628	54631		4	48732	55400	60948	66992
	5	40650	46409	51391	57065		5	52010	58243	63401	69004
	6	43275	48804	53559	58952		6	54533	60409	65254	70510
	7	45437	50760	55322	60485		7	56542	62120	66710	71685
	8	47247	52386	56777	61740		8	58181	63508	67886	72635
	9	48786	53759	57999	62782		9	59545	64657	68856	73411
							10	60696	65622	69669	74060
12	1	20155	27209	33810	41744		11	61682	66445	70359	74611
	2	29570	36452	42594	49728		12	62535	67156	70954	75085
	3	35579	42125	47846	54392		13	63277	67768	71459	75476
	4	39903	46118	51484	57565						
	5	43213	49131	54202	59919						
	6	45842	51498	56319	61735			$n = 35$			
	7	47988	53414	58024	63189						
	8	49778	55003	59432	64389	p	q	0.90	0.95	0.975	0.99
	9	51290	56336	60606	65381	2	1	79720	11125	14293	18425
	10	52587	57472	61598	66211						
						3	1	08440	11766	15099	19435
13	1	22092	29642	36623	44898						
	2	32029	39255	45623	42918	4	1	08950	12461	15972	20527
	3	38235	45026	50892	57515		2	14127	17972	21644	26246
	4	42635	49033	54498	60621						
	5	45957	52012	57145	62868	5	1	09504	13214	16913	21693
	6	48572	54332	59193	64594		2	14949	18990	22838	27644
	7	50690	56197	60830	65970		3	18935	23078	26937	31669
	8	52444	57733	62174	67099						
	9	53918	59014	63287	68019	6	1	10109	14034	17933	22950
	10	55178	60104	64232	68807		2	15839	20088	24119	29135
	11	56264	61039	65037	69468		3	19999	24335	28359	33275
							4	23305	27643	31613	36407
14	1	24285	32356	39705	48270						
	2	34757	42321	48898	56324	7	1	10773	14930	19046	24321
	3	41128	48144	54123	60777		2	16803	21271	25492	30717
	4	45573	52124	57651	63760		3	21147	25685	29881	34985
	5	48890	55054	60221	65913		4	24574	29097	33219	38178
	6	51469	57306	62177	67529		5	27396	31866	35901	40716
	7	53542	59102	63731	68810						
	8	55246	60570	64995	69848	8	1	11500	15907	20249	25790
	9	56671	61791	66042	70704		2	17853	22554	26975	32422
	10	57882	62823	66924	71423		3	22386	27134	31503	36792
	11	58923	63706	67676	72032		4	25936	30648	34923	40042
	12	59825	64468	68320	72547		5	28842	33485	37658	42616
							6	31283	35841	39909	44709
						9	1	12301	16978	21567	27390
							2	18997	23943	28574	34245
							3	23725	28691	33238	38709
							4	27400	32305	36735	42010
							5	30388	35204	39513	44607
							6	32884	37597	41785	46704
							7	35009	39618	43689	48453

TABLE III (Continued)

$n = 35$ (*continued*)

p	q	0.90	0.95	0.975	0.99	p	q	0.90	0.95	0.975	0.99
10	1	13182	18148	22994	29105	15	1	19271	26052	32422	40113
	2	20245	25450	30297	36197		2	28481	35154	41130	48098
	3	25174	30366	35092	40746		3	34455	40834	46425	52840
	4	28974	34077	38662	44089		4	38811	44890	50152	56134
	5	32040	37029	41469	46687		5	42179	47982	52967	58602
	6	34588	39458	43761	48792		6	44877	50434	55181	60528
	7	36746	41494	45669	50530		7	47096	52435	56980	62089
	8	38600	43231	47285	51988		8	48952	54095	58459	63353
							9	53534	55503	59710	64423
11	1	14157	19437	24558	30970		10	51895	56706	60773	65324
	2	21607	27083	32152	38279		11	53083	57755	61700	66113
	3	26745	32170	37080	42916		12	54126	58671	62504	66790
	4	30665	35968	40701	46267		13	55050	59480	63213	67387
	5	33808	38972	43542	48879						
	6	36401	41423	45836	50964	16	1	20940	28158	34861	42856
	7	38586	43469	47740	52682		2	30639	37626	43817	50955
	8	40456	45207	49347	54122		3	36806	43415	49159	55660
	9	42078	46707	50727	55353		4	41240	47488	52843	58867
							5	44633	50565	55612	61262
12	1	15238	20855	26268	32995		6	47328	52983	57710	63112
	2	23095	28855	34147	40491		7	49529	54945	59515	64605
	3	28444	34106	39195	45197		8	51359	56562	60942	65812
	4	32486	37989	42872	48568		9	52911	57928	62143	66826
	5	35698	41033	45724	51164		10	54240	59089	63158	67674
	6	38331	43500	48015	53226		11	55395	60095	64036	68410
	7	40537	45549	49907	54919		12	56407	60976	64804	69055
	8	42415	47281	51496	56329		13	57300	61748	65474	69611
	9	44038	48769	52856	57531		14	58093	62430	66063	70096
	10	45452	50058	54026	58559						
						17	1	22818	30498	37544	45824
13	1	16439	22418	28137	35182		2	33005	40298	46681	53940
	2	24727	30783	36306	42872		3	39354	46176	52030	58599
	3	30288	36190	41454	47613		4	43847	50246	55673	61711
	4	34444	40146	45166	50980		5	47245	53284	58371	64003
	5	37718	43219	48022	53549		6	49921	55653	60459	65766
	6	40383	45694	50302	55579		7	52087	57556	62128	67168
	7	42603	47737	52172	57238		8	53882	59123	63495	68311
	8	44481	49452	53729	58600		9	55394	60438	64641	69269
	9	46099	50923	55064	59773		10	56681	61547	65595	70052
	10	47502	52188	56203	60760		11	57797	62509	66428	70744
	11	48735	53298	57201	61628		12	58769	63341	67143	71327
							13	59626	64075	67772	71846
14	1	17777	24144	30184	37554		14	60386	64721	68324	72298
	2	26517	32878	38630	45405		15	61064	65297	68817	72697
	3	32286	38428	43860	50156						
	4	36548	42442	47590	53498						
	5	39876	45533	50435	56027						
	6	42565	48007	52694	58022						
	7	44787	50030	54527	59621						
	8	46660	51725	56054	60949						
	9	48263	53167	57349	62070						
	10	49649	54406	58454	63019						
	11	50864	55489	59422	63853						
	12	51933	56438	60264	64572						

REFERENCES

[1] T. W. ANDERSON, *Introduction to Multivariate Statistical Analysis*, New York, Wiley, 1958.
[2] W. G. COCHRAN and C. I. BLISS, "Discriminant functions with covariance," *Ann. Math. Statist.*, Vol. 19 (1948), pp. 151–176.
[3] R. C. ELSTON and J. E. GRIZZLE, "Estimation of time-response curves and their confidence bands," *Biometrics*, Vol. 18 (1962), pp. 148–159.
[4] A. ERDÉLYI, *et al.*, *Higher Transcendental Functions*, Vol. 1, Bateman Manuscript Project, New York, McGraw-Hill, 1953.
[5] S. GEISSER and S. W. GREENHOUSE, "An extension of Box's results on the use of the F distribution in multivariate analysis," *Ann. Math. Statist.*, Vol. 29 (1958), pp. 885–891.
[6] ———, "On methods in the analysis of profile data," *Psychometrika*, Vol. 24 (1959), pp. 95–112.
[7] L. J. GLESER and I. OLKIN, "Estimation for a regression model with covariance," Stanford University, Technical Report No. 15, August 14, 1964.
[8] ———, "A K-sample regression model with covariance," *Multivariate Analysis I* (edited by P. R. Krishnaiah), New York, Academic Press, 1966, pp. 59–72.
[9] ———, "Linear models in multivariate analysis," *Essays in Probability and Statistics* (edited by R. C. Bose, *et al.*), Chapel Hill, University of North Carolina Press, 1970, pp. 267–292.
[10] M. HALPERIN, "Normal regression theory in the presence of intra-class correlation," *Ann. Math. Statist.*, Vol. 22 (1951), pp. 573–580.
[11] D. G. KABE, "On the noncentral distribution of Rao's U statistic," *Ann. Inst. Statist. Math.*, Vol. 17 (1965), pp. 75–80.
[12] C. C. MACDUFFEE, *The Theory of Matrices*, New York, Chelsea, 1946 (2nd ed.).
[13] R. D. NARAIN, "Some results on discriminant functions," *J. Indian Soc. Agric. Statist.*, Vol. 2 (1950), pp. 49–59.
[14] I. OLKIN and S. S. SHRIKHANDE, "On a modified T^2 problem," *Ann. Math. Statist.*, Vol. 25 (1954), p. 808, no. 8.
[15] R. F. POTHOFF and S. N. ROY, "A generalized multivariate analysis of variance model useful especially for growth curve problems," *Biometrika*, Vol. 51 (1964), pp. 313–326.
[16] C. R. RAO, "Tests with discriminant functions in multivariate analysis," *Sankhyā*, Vol. 7 (1946), pp. 407–414.
[17] ———, "On some problems arising out of discrimination with multiple characters," *Sankhyā*, Vol. 9 (1949), pp. 343–366.
[18] ———, "A note on the distribution of $D_{p+q}^2 - D_q^2$ and some computational aspects of D^2 statistic and discriminant function," *Sankhyā*, Vol. 10 (1950), pp. 257–268.
[19] ———, "Some statistical methods for the comparison of growth curves," *Biometrics*, Vol. 14 (1958), pp. 1–17.
[20] ———, "Some problems involving linear hypotheses in multivariate analysis," *Biometrika*, Vol. 46 (1959), pp. 49–58.
[21] ———, "Least squares theory using an estimated dispersion matrix and its application to measurement of signals," *Proceedings of the Fifth Berkeley Symposium on Mathematical Statistics and Probability*, Berkeley and Los Angeles, University of California Press, 1967, Vol. 1, pp. 355–372.
[22] G. S. WATSON, "Serial correlation in regression analysis. I," *Biometrika*, Vol. 42 (1955), pp. 327–341.
[23] ———, "Linear least squares regression," *Ann. Math. Statist.*, Vol. 38 (1967), pp. 1679–1699.
[24] G. S. WATSON and E. J. HANNAN, "Serial correlation in regression analysis II," *Biometrika*, Vol. 43 (1956), pp. 436–445.
[25] J. S. WILLIAMS, "The variance of weighted regression estimators," *J. Amer. Statist. Assoc.*, Vol. 62 (1967), pp. 1290–1301.

ON THE STATISTICAL THEORY OF ANALYTIC GRADUATION

JAN M. HOEM

THE CENTRAL BUREAU OF STATISTICS OF NORWAY

1. Introduction; motivating examples

1.1. *Example 1. Rates of mortality.*   Standard methods for the investigation of human mortality will produce statistics such as those given in extract in Table I. The mortality rate at age x is interpreted as a measure of the mortality risk for women born in the year 1968 minus x, and the corresponding number "exposed to risk" in Column 2 is used as a measure of the accuracy of this rate. (This will be made clearer later on.) Unless the population is substantially larger than the one producing these data. the diagram of the sequence of rates, plotted against age, will have a rather rugged appearance. Figure 1, based on the same data as Table I, shows the typical form of such diagrams. There seems to be a universal conviction, however, that "real mortality" would be portrayed by a smooth curve, and that any irregularities of curves of observed mortality rates are due to accidental circumstances. The observed rates are then regarded as "raw" or primary estimates of the underlying "real" rates, and graduation is employed to get a smoother curve.

A number of techniques have been developed to graduate age specific mortality rates, as can be seen from any text on the subject. (See, for instance, [55]; [59], pp. 145–197; [83], pp. 216–237, 243–244, and 251–252.) Most of these methods have been developed by intuitive arguments, at least initially, but investigations of statistical properties of some of them have also appeared [1]; [2]; [43]; [44]; [46]; [61]; [69]; [71]; [76]; [83], p. 252. One class of such methods consists in fitting a parametric function to the observed rates. We shall call this the class of *analytic graduation methods*.

Quite a number of functions have been suggested for analytic graduation of mortality rates [45], pp. 236–238; [67], pp. 453–454; [79], pp. 79–85; [83], pp. 56–60 and 243–244. By far the most commonly used for the adult ages is the Gompertz-Makeham formula

$$(1.1) \qquad g_x(\alpha, \beta, c) = \alpha + \beta c^x \qquad \text{for} \quad \beta > 0, c > 1, \alpha > -\beta c^{xmin},$$

where x represents age attained. We have fitted this function to our data in

The present paper was written while the author visited the University of California, Berkeley, on a Ford Foundation fellowship.

This research was partially supported by National Science Foundation Grant GP 15283.

Figure 1 by minimum χ^2. Other common methods are least squares and some moments methods. We shall describe each of these in turn.

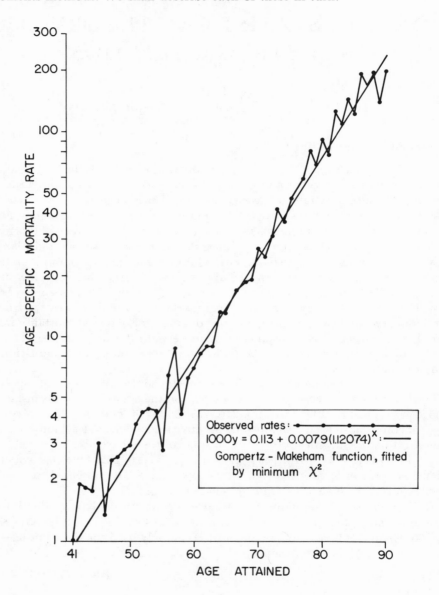

FIGURE 1
Age specific mortality rates per 1000.
Females, Oslo, 1968.

TABLE I

AGE SPECIFIC MORTALITY

FEMALES, MUNICIPALITY OF OSLO, NORWAY, 1968

Column 2 shows arithmetic mean of the number of persons at a given age as of January 1, 1968,
and the corresponding number as of December 31, 1968.
Column 3 shows age at death taken as 1968 minus year of birth.
Column 4 shows ratio between entries in columns 3 and 2, multiplied by 1000.
Source: Central Bureau of Statistics of Norway.

Age 1	Exposed to risk 2	Deaths 3	Mortality rate per thousand 4
40	2798	1	0.357
41	2924.5	3	1.025
42	3156	6	1.901
43	3272.5	6	1.833
44	3465.5	6	1.731
45	3639	11	3.022
46	3770	5	1.326
47	4057	10	2.464
48	3886.5	10	2.573
49	3650.5	10	2.739
..	......	..	.....
..	......	..	.....
80	1204.5	111	92.154
81	1064	81	76.127
82	930	118	126.881
83	835	90	107.784
84	709.5	101	142.353
85	615.5	74	120.227
86	502	97	193.227
87	408	68	166.666
88	341	66	193.548
89	378	53	140.211
90	218.5	43	196.796

1.2 *Example* 2. *Rates of fertility.* A standard investigation of age specific
human fertility would produce a table quite similar to Table I, except of course
that column 3 would contain numbers of births (or usually numbers of liveborn
children) by age of mother. A corresponding diagram would look something
like the one in Figure 2, and graduation would again give a smoother curve.

A fertility curve of this sort closely resembles certain density functions, and
one category of functions proposed for the analytic graduation of fertility curves
consists of densities from the Pearson family [13]; [27]; [45], pp. 140–169;
[53]; [56]; [73]; [77]; [80]; [81]; in particular Pearson type I, III, IV, VI, and
the normal density, multiplied by a constant.

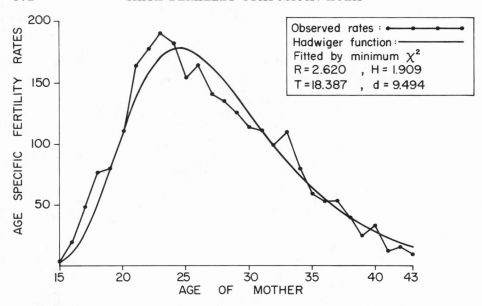

FIGURE 2
Age specific fertility rates per 1000.
Females, Stavanger (Norway), 1968.

Another category of graduating functions consists of polynomials in x [11], [27], such as

$$(1.2)\qquad b_x(a, b, c, d) = (x - \alpha + 1)(\beta - x)^2(a + bx + cx^2 + dx^3),$$

where x stands for age of mother at childbearing, and where $[\alpha, \beta\rangle$ is the fertile period for females. It is customary to take $\alpha = 15$ and $\beta = 45$ or $\beta = 50$, but in certain cases α and β occur as parameters which are estimated [13]; [54]; [77].

The Hadwiger function,

$$(1.3)\quad h_x(R, T, H, d) = \frac{RH}{T\sqrt{\pi}}\left(\frac{T}{x - d}\right)^{3/2} \exp\left\{- H^2\left(\frac{T}{x - d} + \frac{x - d}{T} - 2\right)\right\},$$

with $R > 0, T > 0, H > 0, d < \alpha$, is a third type of graduating formula [27]; [28]; [31]; [45], pp. 149–169; [80]; [81]. (We follow Yntema's notation.) Other functions have also been suggested [12], [48], [49], [50], [54], [72].

Naturally, the same type of functions will be used for the graduation of other vital rates whose diagrams have the same general form as fertility rates, such as marriage rates [22], pp. 99–101, and [49].

1.3. The above age specific rates of mortality and fertility are examples of the kind of vital rates which occur in fields such as actuarial science, biostatistics, and demography. In the present paper we shall make a contribution to the statistical theory of curve fitting as applied to such rates in general. We shall suggest a probabilistic model within which the rates appear as estimators for certain parameters called forces of transition, and shall show how the analytic graduation can be interpreted as a procedure used to further estimate a set of "more basic" parameters, namely those of the graduating function.

The model will be introduced in Section 2. In Sections 3 and 4, we describe how the rates appear within the model, and Sections 5 to 9 are devoted to the study of analytic graduation methods. We shall be concerned mainly with the asymptotic statistical properties (as the population size N increases) of the estimators. Most of our results are straightforward consequences of general asymptotic theory, and we shall often use standard theorems from that field, like those of Chapters 4 and 5 in [10] and Theorem 4.2.5 in [3], without explicit reference. We *shall* quote references whenever we use a deeper result.

Since we use standard theorems, it is not surprising that we can prove theorems which correspond to previous standard results. Thus (speaking informally here) we shall see that none of the general estimation procedures we study will be better than one of the maximum likelihood type, and that a (modified) minimum χ^2 procedure is equally good, while moment methods will usually give less favorable results.

We feel that there may be a need for some explanation why procedures of the type which we shall describe are preferred to certain others. Rather than breaking up our presentation of the techniques involved by giving parts of this explanation as we go along, we have preferred to include it all in Section 10.

Apart from what is contained in Sections 1.1 and 1.2 above, no numerical examples will be given in this paper. Numerical investigations are planned and will be reported at a later date.

2. A Markov process model

2.1. *The general model.* To describe the phenomena at hand we shall use a Markov process model. Let y_t be the sample function value at time t of a time inhomogeneous Markov process with a denumerable state space I and a continuous time parameter restricted to some finite time interval $[0, \zeta\rangle$. Let the transition probabilities be

$$(2.1) \qquad P_{i,J}(s, t) = P\{y_t \in J \,|\, y_s = i\},$$

for $0 \leqq s < t < \zeta$, and $i \in I$ and $J \subseteq I$, and assume that $P_{i,I}(s, t) \equiv 1$ and $\lim P_{i,j}(s, t) \equiv \delta_{i,j}$ (a Kronecker delta) as $t \downarrow s$. We introduce the *forces of transition*,

$$(2.2) \qquad \mu_{i,j}(s) = \lim_{t \downarrow s} P_{i,j}(s, t)/(t - s) \qquad \text{for} \quad i \neq j,$$

and the *forces of decrement*,

$$(2.3) \qquad \mu_i(s) = \lim_{t \downarrow s} \{1 - P_{i,i}(s, t)\}/(t - s),$$

for $0 \leqq s < \zeta$, and assume that all μ_i and $\mu_{i,j}$ are finite and integrable over $[0, \zeta)$. We also assume that

$$(2.4) \qquad \mu_i = \sum_{j \in I - i} \mu_{i,j} \qquad \text{for each} \quad i \in I.$$

We shall call a state i absorbing if $\mu_i = 0$.

The problem which leads us to study analytic graduation consists in finding a method for estimating one or more of the $\mu_{i,j}(\cdot)$ from data of the type which one encounters within the fields of application mentioned at the beginning of Section 1.3.

2.2. *Examples.* We shall give some examples to show how models in the applications appear as particular cases of the general model in Section 2.1 above.

(i) Our simplest example will be a model with only two states, called "alive" (state 1) and "dead" (state 2). State 2 is absorbing, and there is only one nonzero force of transition and of decrement, namely,

$$(2.5) \qquad \mu(\cdot) = \mu_1(\cdot) = \mu_{1,2}(\cdot),$$

called *the force of mortality*. The rates of Section 1.1 will be seen to appear within this model. The time parameter is represented by a person's age.

(ii) The age specific fertility rates of Section 1.2 can be interpreted within a model with a double infinity of states. A woman will be said to be in state $(k, 1)$ at age x if she is alive then and her parity is k, that is, she has had k births, $k = 0, 1, 2, \cdots$. She will be said to be in state $(k, 2)$ at age x if she has died within age x and her parity at death was k. All states of the form $(k, 2)$ are absorbing. We select two suitable functions, $\mu(\cdot)$ and $\varphi(\cdot)$, and set

$$(2.6) \qquad \mu_{(k, 1), (k, 2)}(\cdot) = \mu(\cdot),$$

and

$$(2.7) \qquad \mu_{(k, 1), (k+1, 1)}(\cdot) = \varphi(\cdot),$$

for $k = 0, 1, \cdots$, while all other $\mu_{i,j} = 0$. The function μ will be called the *force of mortality*, and φ will be called the *force of fertility*. Again the time parameter is represented by the woman's age. This model, which we have studied in some detail previously [36], is not particularly "realistic", but it is probably the simplest one in which the rates of Section 1.2 can be meaningfully discussed. More realistic fertility models of this type have appeared elsewhere [37], [38].

(iii) To describe *marriage formation and dissolution*, we suggest a model with five states, called "never married" (state 1), "married" (state 2), "widowed" (state 3), "divorced" (state 4), and "dead" (state 5). State 5 is absorbing. The following forces correspond to impossible direct transitions, and are therefore identically equal to zero: $\mu_{i,1}$ for $i > 1$, $\mu_{3,4}$, $\mu_{4,3}$, $\mu_{5,j}$ for $j < 5$. The model applies to one sex only, while the other sex appears only implicitly, as a kind of shadow factor. We have also looked at marriage models elsewhere [41].

Other models of this type have been studied by, for instance, Du Pasquier [21]; Sverdrup [70]; Simonsen [66]; Chiang [15], Chapters 4, 5, and 7; and Hoem [35]. Compare also [25] and [64].

In each of these models, a state $i \in I$ corresponds to some vital status, that is, a marital status, a social status, a birth parity, and so on. A transition similarly corresponds to a vital event, such as a death, a birth, a marriage, a divorce, and so on. An individual sample path will be visualized intuitively as a person (or sometimes a group of persons, such as a household or a family) moving through some of the statuses of the system specified. The sample paths will be taken as stochastically independent.

2.3. *Seniority.* In demographic models one often wishes to distinguish between an age parameter (which may be actual age obtained, duration of marriage, interval since last previous birth, or the like), calendar time, and observational time. It is the age parameter which corresponds to the time parameter in the Markov process of Section 2.1. In a general model it may be useful to have a separate name for this (unspecified) age parameter, covering all interpretations which it may have in the applications. Following Henry [33] who calls it *ancienneté*, we shall use the name *seniority* for it.

2.4. *Some basic assumptions about the forces of transition.* In what follows, we shall disregard the forces of transition which are identically equal to zero because they correspond to impossible direct transitions by the definition of the model. Even if the state space I can be (countably) infinite, there are many cases where only a finite number of the nonzero forces of transition are *distinct*. [Compare Example 2.2, (ii).] *We shall assume everywhere that there exist A nonnegative real functions,* $\lambda_1, \cdots, \lambda_A$, *such that each* $\mu_{i,j}$, *not identically zero, equals some* λ_a, *and such that for each* λ_a *there exists a* $\mu_{i,j} = \lambda_a$. By (2.4), we may then write

$$(2.8) \qquad \mu_i = \sum_{a=1}^{A} c_a(i)\lambda_a \qquad \text{for each} \quad i \in I,$$

where

$$(2.9) \qquad c_a(i) = \sum_{\{j \in I - i : \mu_{i,j} = \lambda_a\}} 1.$$

Each $c_a(i) < \infty$ because all $\mu_i < \infty$. Equation (2.8) shows that for any given $i \in I$, exactly $\Sigma_a c_a(i)$ of the $\mu_{i,j}$ can be positive, that is, a finite number of the $\mu_{i,j}$ only, while the rest are identically equal to zero.

Let us also assume everywhere that

$$(2.10) \qquad \sup\{\mu_i(s): 0 \leqq s < \zeta, i \in I\} < \infty.$$

This assumption is not necessary for what follows, and it can be relaxed [39]; [40], Section 5, but one will expect it to hold in practice and it will simplify our exposition. It follows from (2.10) [40], Section 3.1, that there only exists a finite number of *distinct* vectors $\mathbf{c}(i) = (c_1(i), \cdots, c_A(i))$. Let us call these $(c_{b,1}, \cdots, c_{b,A})$ for $b = 1, 2, \cdots, B$, and let

$$(2.11) \qquad \gamma_b = \sum_{a=1}^{A} c_{b,a} \lambda_a \qquad \text{for} \quad b = 1, 2, \cdots, B.$$

Then each nonzero μ_i equals some γ_b and for each γ_b there is a μ_i which equals it. There is, thus, only a finite number of distinct forces of decrement as well. (This need not be the case if (2.10) does not hold.) For each $i \in I$, let $b(i)$ be defined by $\mu_i = \gamma_{b(i)}$. Then, by (2.9),

$$(2.12) \qquad c_{b(i), a} = \sum_{\{j \in I - i : \mu_{i,j} = \lambda_a\}} 1.$$

Let

$$(2.13) \qquad c = \begin{pmatrix} c_{1,1}, & \cdots, & c_{1,A} \\ \cdots & \cdots & \cdots \\ c_{B,1}, & \cdots, & c_{B,A} \end{pmatrix}$$

We shall finally assume everywhere that the rank of c is B. The extension to rank $c < B$ is easy [40], Section 3.1, but it leads to slightly more complicated formulas.

3. The primary or "raw" estimates

3.1. *Approximation of the λ_a by step functions.* As a first step in our description of the kind of estimation methods which have produced the rates of Sections 1.1 and 1.2, we shall approximate the λ_a by step functions. The seniority interval $[0, \zeta\rangle$ is paritioned into D subintervals, $[\zeta_0, \zeta_1\rangle$, $[\zeta_1, \zeta_2\rangle$, $\cdots$, $[\zeta_{D-1}, \zeta_D\rangle$, with $\zeta_0 = 0$ and $\zeta_D = \zeta$. Let $I_d(\cdot)$ be the indicator function of the interval $[\zeta_{d-1}, \zeta_d\rangle$, and let

$$(3.1) \qquad \lambda_a^{\#}(\cdot) = \sum_{d=1}^{D} \lambda_{a,d} I_d(\cdot).$$

Here each $\lambda_{a,d}$ is a constant chosen in such a way that it can represent the values of λ_a in $[\zeta_{d-1}, \zeta_d\rangle$. If λ_a is assumed to be a nice and smooth function, with certain known monotonicity properties, say, then $\lambda_a^{\#}$ will inherit these properties, modified of course, by the fact that the latter is a step function.

In what follows, we shall assume that the $\lambda_a^{\#}$ give an adequate representation of the λ_a, and our calculations will be made as if we actually had $\lambda_a = \lambda_a^{\#}$ for $a = 1, 2, \cdots, A$.

3.2. *More about the ζ_d.* In this presentation, we use the same partitioning $\{\zeta_d : d = 0, 1, \cdots, D\}$ for all λ_a. In certain situations one would rather use different partitionings for different λ_a. The results of this paper will continue to hold for such cases with only quite obvious modifications [37].

The approach sketched in Section 3.1 is closely related to histogram methods for the estimation of a probability density or a generalized failure rate [6], [74], [75]. Although the lengths of the histogram intervals are often made to converge to zero as the number of observations increase, this is not the case for the seniority subintervals above. The ζ_d will typically be selected according to conventional rules established with different considerations in mind than statistical convergence properties. When ζ is of the size order of several decades, as is often the case, the seniority interval $[0, \zeta\rangle$ will usually be partitioned into one year or five year intervals, possibly with a longer "tail" interval at the upper end. There is a tendency to use shorter subintervals in a large population than in a small one, at least if the data are reliable, but an interval length shorter than one year is commonly used only in certain standardized contexts, such as in investigations of infant mortality, where ζ equals one year of age [8], p. 211; [30], Tables 38 to 42; [67], p. 84. There seems to be no tendency toward letting such interval lengths decrease to zero.

3.3. *On the observational plan.* There are a number of observational plans (or ascertainment methods) in use in the fields of application we have in mind, and one could construct others from ideas used in life testing. (See, for example, [20], [32], [65], [82].

In the present paper we shall only consider observational plans where a group of people are followed continuously over some time interval $[0, T\rangle$. The data collection will consist of noting what happens to each person while under observation, that is, which states of I he visits and just when vital events occur to him.

It is characteristic for the types of populations which occur in practice that some people enter them and others leave during the study period. They may also be heterogeneous with respect to seniority, in the sense that those who come under observation (whether they are in the population from the outset or enter later on) may have different seniorities at time T. We want to cover such possibilities. Let, therefore, N be the number of individuals ever observed. Let us say that person number k enters the population at some time $t_k \in [0, T\rangle$ with seniority x_k and a status corresponding to state r_k, and that he stays there at least until time $t_k + z_k \in [0, T]$, when observation is discontinued. We shall take the entrance time t_k, the initial seniority x_k, the initial state r_k, and the exposure time z_k to be preassigned, that is, not random. (Other possibilities are discussed in [40].) Any period spent in an absorbing state, for instance after the death of the individual, is included in the period of exposure $[t_k, t_k + z_k\rangle$, although of course no actual observation is made after a path has entered such a state.

We shall also take N to be nonrandom.

3.4. *Estimation of the* $\lambda_{a,d}$. We get an estimator for $\lambda_a^\#$ by plugging estimators $\hat{\lambda}_{a,d}$ for the $\lambda_{a,d}$ into the right side of (3.1). (This is how the rugged curves in Figures 1 and 2 have arisen.) Standard estimators used for this purpose are occurrence/exposure rates, like those in Sections 1.1 and 1.2 [62], [63]. We shall see how these arise.

Some of the $\lambda_{a,d}$ may be known to be zero because they correspond to vital events which are impossible during $[\zeta_{d-1}, \zeta_d\rangle$, such as births after menopause. We will take the other $\lambda_{a,d}$ to be strictly positive. Let

(3.2) $\mathscr{G} = \{(a, d): \lambda_{a,d} > 0\}.$

We can regard $\{\lambda_{a,d}: (a, d) \in \mathscr{G}\}$ as a point in the space

(3.3) $\Lambda_0 = \underset{(a,d)\in\mathscr{G}}{\times} \{x_{a,d} > 0\}.$

The situation in hand will usually restrict the possible points we actually can have to a proper subset Λ of Λ_0. We shall take Λ to be open.

Now let $M_k(a, d)$ be the number of transitions observed for path number k during the seniority interval $[\zeta_{d-1}, \zeta_d\rangle$, direct from any state i to any state j where $\mu_{i,j} = \lambda_a$. Let $U_k(i, d)$ be the total time spent in state i during $[\zeta_{d-1}, \zeta_d\rangle$ by this path, and let

(3.4) $V_k(b, d) = \sum_{\{i\in I: \mu_i = \gamma_b\}} U_k(i, d).$

Then $V_k(b, d)$ is the total time spent in any state i where $\mu_i = \gamma_b$ by path k during the interval mentioned. Finally, let

$$L_k(a, d) = \sum_{b=1}^{B} c_{b,a} V_k(b, d)$$
(3.5)
$$= \sum_{i\in I} c_{b(i),a} U_k(i, d),$$

and let us use the notation $X = \sum_k X_k$, where X_k is any quantity depending on k. Since the forces of transition are represented by step functions, we can then use the same method as in [37], Section 4.8 to write the likelihood in the form

(3.6) $\prod_{(a,d)\in\mathscr{G}} \lambda_{a,d}^{M(a,d)} \exp\left\{-\sum_{d=1}^{D}\sum_{b=1}^{B} \gamma_{b,d} V(b, d)\right\}$

$$= \prod_{(a,d)\in\mathscr{G}} \lambda_{a,d}^{M(a,d)} \exp\left\{-\sum_{(a,d)\in\mathscr{G}} \lambda_{a,d} L(a, d)\right\},$$

where $\gamma_{b,d} = \sum_{a=1}^{A} c_{b,a}\lambda_{a,d}$. (Compare (2.11).) Thus we are dealing with a Darmois-Koopman class of probability distributions, and one may show [40], Section 3.1, that $\{M(a, d), V(b, d): a = 1, 2, \cdots, A; b = 1, 2, \cdots, B; d = 1, 2, \cdots, D\}$ is minimal sufficient for the $\lambda_{a,d}$. An unrestricted maximization of the likelihood function would give the estimators

(3.7) $\hat{\lambda}_{a,d} = M(a, d)/L(a, d),$

which are the occurrence/exposure rates we mentioned. (We arbitrarily set $\hat{\lambda}_{a,d} = 0$ if $L(a, d) = 0$.) The point $\{\hat{\lambda}_{a,d}: (a, d) \in \mathscr{G}\}$ need not lie in Λ, nor even in Λ_0 if some $\hat{\lambda}_{a,d} = 0$. However, under certain conditions, spelt out in Theorem 1 below, the probability that the point lies in Λ increases to 1 as $N \to \infty$.

3.5. *Asymptotic properties of the* $\hat{\lambda}_{a,d}$. For each $(a, d) \in \mathscr{G}$, the variables $L_1(a, d), \cdots, L_N(a, d)$ will not generally be identically distributed unless all (x_k, z_k, r_k) are equal. Similarly for $M_1(a, d), \cdots, M_N(a, d)$. Nevertheless, one can prove the following consistency theorem [40], Section 4.2:

THEOREM 1. *Assume that* $P\{L(a, d) > 0\} \to 1$ *as* $N \to \infty$, *and that a finite positive limit*

$$(3.8) \qquad L_{a,d} = \lim_{N \to \infty} EL(a, d)/N$$

exists. Then $\hat{\lambda}_{a,d}$ *converges to* $\lambda_{a,d}$ *in probability as* $N \to \infty$.

To arrive at a theorem concerning the asymptotic distribution of $\hat{\lambda}_{a,d}$ as $N \to \infty$, we make an additional set of assumptions, which establish a grouping of the (x_k, z_k, r_k) at a finite set of strategic values. More precisely we make

ASSUMPTION 1. *There exists a finite set of possible initial seniorities,* $y_1, \cdots, y_H$, *a finite set of possible exposure times,* $w_1, \cdots, w_J$, *and a finite set of possible initial states,* $s_1, \cdots, s_Q$, *such that each* (x_k, z_k, r_k) *must equal some* (y_h, w_j, s_q). *We let*

$$(3.9) \qquad \mathscr{K}_{h,j,q} = \{k : (x_k, z_k, r_k) = (y_k, w_j, s_q)\},$$

and let $S_{h,j,q}(N)$ *be the number of elements in* $\mathscr{K}_{h,j,q}$. *We assume that*

$$(3.10) \qquad \alpha_{h,j,q} = \lim_{N \to \infty} S_{h,j,q}(N)/N$$

exists for each (h, j, q).

If $\varepsilon_{h,j,q}(a, d) = EL_k(a, d)$ for $k \in \mathscr{K}_{h,j,q}$, we get under Assumption 1 that the $L_{a,d}$ of (3.8) satisfy

$$(3.11) \qquad L_{a,d} = \sum_{h,j,q} \alpha_{h,j,q} \varepsilon_{h,j,q}(a, d).$$

We may then prove ([40], Section 4.2; compare [47]) Theorem 2.

THEOREM 2. *Under Assumption 1, the variables* $N^{\frac{1}{2}}(\hat{\lambda}_{a,d} - \lambda_{a,d})$ *for which* $P\{L(a, d) > 0\} \to 1$ *as* $N \to \infty$, *are asymptotically independent and normally distributed with means 0 and asymptotic variances*

$$(3.12) \qquad \sigma_{a,d}^2 = \text{as. var } N^{1/2}(\hat{\lambda}_{a,d} - \lambda_{a,d}) = \lambda_{a,d}/L_{a,d}.$$

We note that under the assumptions of Theorem 2,

$$(3.13) \qquad \hat{\sigma}_{a,d}^2 = N\hat{\lambda}_{a,d}/L(a, d)$$

is a consistent estimator for $\sigma_{a,d}^2$. Thus we see a justification of the use of the number exposed to risk [which is $L(a, d)$ here] as an intuitive measure of the accuracy of the corresponding rate of transition $\hat{\lambda}_{a,d}$, as mentioned in Section 1.1 for a special case.

We also note that we do not need to know the value of N in order to estimate the $\lambda_{a,d}$ and the asymptotic variance $\sigma_{a,d}^2/N$ of the $\hat{\lambda}_{a,d}$.

4. Non-observation of part of the state space

4.1. *A problem.* In Section 3 we assumed that one could observe what state a sample path visited at any time. This need not be the case in practice. Let us give two examples.

(i) Demographic studies will often be concerned with people living in a restricted area, such as a country or part of a country, and there will be some in and out migration. Say that a study of marriages is carried out, perhaps based on a model like the one in Section 2.2 (iii). If a person initially lives in the study area, then leaves and stays away for a while, and subsequently returns while the study is still being conducted, it rarely happens that his changes of marital status (if any) while outside the study area are traced. In many cases one will know his marital status on departure from the study area, as well as his status as he returns, but nothing more.

(ii) Similar problems occur in studies of the mortality of insured lives. A person may cancel his insurance policy and be uninsured for a while, then take out a new policy, which may be cancelled again after a while, and so on. The insurer will keep track of deaths among the persons covered by his policies, but will not usually know what happens to the uninsured.

The question is how one should take account of phenomena like these in the estimation procedures.

4.2. *Formalization of the two examples.*

(i) To describe the example in (i) above in terms of a probabilistic model, let $I_1 = \{1, 2, 3, 4, 5\}$ be the state space of the example in 2.2 (iii), let $I_2 = \{1, 2\}$, and let $I = I_1 \times I_2$. An individual with marital status j will be said to be in state $(j, 1)$ if he lives in the study area, and in state $(j, 2)$ if he lives outside it. A migration *out of* the study area will correspond to a transition from a state $(j_1, 1)$ to a state $(j_2, 2)$. A migration *into* the study area will correspond to a transition from a state $(j_2, 2)$ to a state $(j_1, 1)$. In most cases, $j_1 = j_2$. In any case, we shall take j_1 to be observable. Whatever moves the sample path otherwise makes while in the subspace $\{(j, 2): j \in I_1\}$ will not be observed.

(ii) To formalize the second example above, let us use four states, called "alive and insured" (state 1), "alive and uninsured" (state 2), "dead while insured" (state 3), and "dead while uninsured" (state 4). Which transitions are possible and which are not follows directly from the state names. Except for transitions from state 2 to state 4, all transitions (and the dates on which they occur) are recorded. (This is essentially the model studied by Du Pasquier [21], Fix and Neyman [25], and Sverdrup [70], except that they took all transitions as recorded. Recording problems different from the present one have been studied by Høyland [42], Kruopis [47], and others.)

Let us take all forces of transition to be constants. This will suffice for our purposes, which are those of illustration. Generalization to other cases is simple. We shall take the forces of mortality of the insured and the uninsured to be equal, and let $\mu = \mu_{1,3} = \mu_{2,4}$. Let $v = \mu_{1,2}$, $\rho = \mu_{2,1}$, $\alpha = \mu + v$,

$\beta = \mu + \rho$. Sample path number k is followed over the period $[0, z_k]$, and we say that $k \in \mathscr{K}$ if this sample path is in state 2 or 4 at time z_k, that is, if person number k is uninsured then. All N paths start in state 1, and they make a total number of $M_{i,j}$ jumps from state i to state j, for $(i, j) \in \{(1, 2), (1, 3), (2, 1)\}$. Let W denote the total time spent in state 2 by the paths $k \notin \mathscr{K}$, and let V be the total time spent in state 1 by all paths taken together. For $k \in \mathscr{K}$, let $z_k - U_k$ be the time of the last jump recorded from state 1 to state 2 for path k, that is, the time to last observed cancellation. Then the corresponding likelihood can be written as

$$(4.1) \qquad e^{-\alpha V - \beta W} v^{M_{1,2}} \rho^{M_{2,1}} \mu^{M_{1,3}} \beta^{-K} \prod_{k \in \mathscr{K}} (\mu + \rho e^{-\beta U_k}),$$

where K is the number of elements in $\mathscr{K}$. The maximum likelihood estimator of v turns out to be

$$(4.2) \qquad \hat{v} = M_{1,2}/V,$$

which is what (3.7) would have given. ($M_{1,2}$ is the number of cancellations observed.) Closed, explicit expressions for the maximum likelihood estimators of μ and ρ do not exist in this case. We can still get an estimator of μ, however, by letting

$$(4.3) \qquad \hat{\mu} = M_{1,3}/V.$$

($M_{1,3}$ is the number of insured deaths.) The properties of $\hat{v}$ and $\hat{\mu}$ will appear by specialization of the results in Section 4.3 below.

4.3. *The general case.* Consider now the general model with the assumptions made in Sections 2 and 3.1. Let the state space I be partitioned into two disjoint subsets, H and J, and assume that all transitions between states in H can be recorded, while no transitions between states in J are recorded. Any transition from a state in H to one in J is recorded, as are all jumps from J to H. For both kinds of jumps, one also records the state *to* which the jump is made.

We redefine the quantities $M(a, d)$ and $L(a, d)$, initially introduced in Section 3.4, as follows.

Let $\mathscr{A}$ be the set of the a for which there exists a $\mu_{i,j}$, with $i \in H$, such that $\mu_{i,j} = \lambda_a$. For each $a \in \mathscr{A}$ and each $d \in \{1, 2, \cdots, D\}$ let $M(a, d)$ now be the total number of transitions observed during the seniority interval $[\zeta_{d-1}, \zeta_d\rangle$, for all paths taken together, direct from any state $i \in H$ to any state $j \in I - i$ such that $\mu_{i,j} = \lambda_a$. Furthermore, let

$$(4.4) \qquad L(a, d) = \sum_{i \in H} c_{b(i), a} U(i, d),$$

with $c_{b(i), a}$ given by (2.12) and $U(i, d)$ defined as in Section 3.4. For $a \in \mathscr{A}$, let $\hat{\lambda}_{a,d}$ be given by (3.7) with the *new* definitions of $M(a, d)$ and $L(a, d)$. *Then Theorems 1 and 2 hold verbatim for the $a \in \mathscr{A}$*, even though the $\hat{\lambda}_{a,d}$ need not be maximum likelihood estimators, as demonstrated in the example in (ii) above. If there does not exist any $\mu_{i,j}$, with $i \in J$, such that $\mu_{i,j} = \lambda_a$ for any $a \in \mathscr{A}$, the $\hat{\lambda}_{a,d}$ *will* be maximum likelihood estimators, in the sense that they maximize the likelihood under free variation of the $\lambda_{a,d}$ in Λ_0.

If the state j cannot be recorded when there is a jump from a state $i \in H$ to a state $j \in J$, the results above continue to hold, provided we again redefine the quantities involved in a natural way. In the definition of $M(a, d)$, we must only include jumps from i to j where both i and $j \neq i$ belong to H, and where $\mu_{i,j} \doteq \lambda_a$. The set $\mathscr{A}$ is similarly reduced. This time we also redefine $c_{b,a}$ by letting

$$(4.5) \qquad c_{b(i),a} = \sum_{\{j \in H - i: \mu_{i,j} = \lambda_a\}} 1 \qquad \text{for} \quad i \in H, a \in \mathscr{A}.$$

Using (4.5), we define $L(a, d)$ for $a \in \mathscr{A}$ by (4.4).

5. Conventions and notation relating to analytic graduation

5.1. *Analytic graduation.*   Although an original λ_a is assumed to be a nice and smooth function, the estimators $\hat{\lambda}_{a,d}$ now in use, such as those in (3.7), will typically produce a $\hat{\lambda}_a^{\#}$ which is considered too irregular, except in large populations. (Compare the account on page 561 in [18].) Analytic graduation then consists in selecting some nice, parametric function $g_a(\cdot, \boldsymbol{\theta}_a)$ and some representative seniority ξ_d from each interval $[\zeta_{d-1}, \zeta_d\rangle$, and in getting an estimator $\hat{\boldsymbol{\theta}}_a$ for $\boldsymbol{\theta}_a$ by fitting the values $\{g_a(\xi_d, \boldsymbol{\theta}_a): d = 1, 2, \cdots, D\}$ to $\{\hat{\lambda}_{a,d}: d = 1, 2, \cdots, D\}$ by a suitable method. The function $g_a(\cdot, \hat{\boldsymbol{\theta}}_a)$, usually regarded as a function of a *continuous* seniority variable x, represents the final estimator for the function $\lambda_a(\cdot)$.

Most methods for constructing an estimator $\hat{\boldsymbol{\theta}}_a$ are based on analogies with estimation methods used in other contexts [1], [59], [83]. We shall study least squares and minimum χ^2 methods in Section 6 [7], [12], [16], [27], [28], [48], [49], [50], [61]. (See also [60].) In Section 7, we shall discuss moment methods [11]; [13]; [28]; [45], pp. 140–169; [53]; [56]; [73]; [77]; [80]; [81]; and in Section 8 we shall introduce a technique of the maximum likelihood type. Some authors have also used methods involving the minimization of sums of absolute deviations [17], [28].

5.2. *Further assumptions and conventions.* We shall be working with a single, fixed value of a, and shall therefore suppress this subscript except where it may cause confusion.

In what follows, we shall disregard the fact that some of the λ_d may be known to equal zero. The case where some λ_d actually do equal zero needs only trivial notational modifications.

Let

$$(5.1) \qquad g_d(\boldsymbol{\theta}) = g(\xi_d, \boldsymbol{\theta}),$$

$$(5.2) \qquad \mathbf{g}(\boldsymbol{\theta}) = (g_1(\boldsymbol{\theta}), \cdots, g_D(\boldsymbol{\theta}))'.$$

(The prime denotes a transpose.)

ASSUMPTION 2. *We assume that $\boldsymbol{\theta}$ varies in an open subset Θ of the G-dimensional Euclidean space $\mathbf{R}_G$, where $G < D$. Let $\mathbf{g}$ be a one to one, bicontinuous, continuously differentiable mapping of Θ into*

$$(5.3) \qquad \Lambda_0 = \underset{d=1}{\overset{D}{\times}} \{x_d > 0\}.$$

Define

$$(5.4) \qquad \mathbf{J}(\boldsymbol{\theta}) = \begin{pmatrix} \dfrac{\partial}{\partial \theta_1} g_1(\boldsymbol{\theta}), \cdots, \dfrac{\partial}{\partial \theta_G} g_1(\boldsymbol{\theta}) \\ \cdots\cdots\cdots\cdots\cdots \\ \dfrac{\partial}{\partial \theta_1} g_D(\boldsymbol{\theta}), \cdots, \dfrac{\partial}{\partial \theta_G} g_D(\boldsymbol{\theta}) \end{pmatrix},$$

and assume that $\mathbf{J}(\boldsymbol{\theta})$ has rank G for each $\boldsymbol{\theta} \in \Theta$.

We denote the true value of $\boldsymbol{\theta}$ by $\boldsymbol{\theta}^0$, and let

$$(5.5) \qquad \mathbf{J}_0 = \mathbf{J}(\boldsymbol{\theta}^0), \ \boldsymbol{\lambda}^0 = \mathbf{g}(\boldsymbol{\theta}^0), \quad \text{and} \quad L_d^0 = \lim_{N \to \infty} E_{\boldsymbol{\theta}^0} L(d)/N.$$

[Compare (3.8).] We also let

$$(5.6) \qquad \sigma_{d,0}^2 = \lambda_d^0/L_d^0 \quad \text{and} \quad \boldsymbol{\Sigma}^0 = \text{diag}\,(\sigma_{1,0}^2, \cdots, \sigma_{D,0}^2)$$

[compare (3.12)], with the convention that we write $\mathbf{M} = \text{diag}\,(m_1, \cdots, m_S)$ if $\mathbf{M}$ is a diagonal $S \times S$ matrix with the m_s as diagonal elements.

Let us denote it by a right superscript N if we want to stress that a quantity depends on N.

In Sections 3 and 4 we brought out some estimators $\hat{\boldsymbol{\lambda}}(N) = (\hat{\lambda}_1^{(N)}, \cdots, \hat{\lambda}_D^{(N)})'$ of the common occurrence/exposure type for the parameter $\boldsymbol{\lambda} = (\lambda_1, \cdots, \lambda_D)'$, and we stated some theorems concerning their asymptotic properties. In much of what follows, it is precisely these properties which are of interest, and not the form of the estimators themselves. In Sections 6 and 7, therefore, we shall take $\hat{\boldsymbol{\lambda}}^{(N)}$ to be any estimator for $\boldsymbol{\lambda}$, not necessarily the one given by (3.7), and we shall continuously make

ASSUMPTION 3.

$$(5.7) \qquad N^{1/2}(\hat{\boldsymbol{\lambda}}^{(N)} \mathscr{L}_{\boldsymbol{\theta}^0} \boldsymbol{\lambda}^0) \xrightarrow{\mathscr{L}_{\boldsymbol{\theta}^0}} \mathscr{N}(\mathbf{0}, \boldsymbol{\Sigma}_0),$$

where $\mathscr{N}(\mathbf{0}, \boldsymbol{\Sigma}_0)$ is the multinormal distribution with mean $\mathbf{0}$ and a positive definite covariance matrix $\boldsymbol{\Sigma}_0$, which need not be the same as $\boldsymbol{\Sigma}^0$.

6. Analytic graduation through minimization of a quadratic form

6.1. *The graduation method.* Let $\mathbf{M}$ be a positive definite, symmetric $D \times D$ matrix whose elements $m_{i,j}$ may (but need not) be random variables. Let

$$(6.1) \qquad Q(\boldsymbol{\theta}) = N(\hat{\boldsymbol{\lambda}} - \mathbf{g}(\boldsymbol{\theta}))' \mathbf{M}(\hat{\boldsymbol{\lambda}} - \mathbf{g}(\boldsymbol{\theta})).$$

Assume that there exists a $\boldsymbol{\theta}$, say $\hat{\boldsymbol{\theta}}$, which minimizes $Q(\boldsymbol{\theta})$. We shall then take $\hat{\boldsymbol{\theta}}$ to be our estimator for $\boldsymbol{\theta}$.

A whole class of graduation methods is generated by the various choices of the matrix $\mathbf{M}$. Thus if we take $\mathbf{M} = \mathbf{I}$, the identity matrix, we get

$$(6.2) \qquad Q(\boldsymbol{\theta}) = N \sum_{d=1}^{D} \left(\hat{\lambda}_d - g_d(\boldsymbol{\theta}) \right)^2,$$

and $\hat{\boldsymbol{\theta}}$ becomes a least squares estimator. An analogy with the modified minimum χ^2 method results from setting

$$(6.3) \qquad \mathbf{M} = \operatorname{diag}(1/\hat{\sigma}_1^2, \cdots, 1/\hat{\sigma}_D^2),$$

where the $\hat{\sigma}_d^2$ are given by (3.13). We then get

$$(6.4) \qquad Q(\boldsymbol{\theta}) = \sum_{d=1}^{D} \{M(d) - L(d)g_d(\boldsymbol{\theta})\}^2/M(d).$$

If, in particular, $\mathbf{g}(\boldsymbol{\theta})$ is a linear function of $\boldsymbol{\theta}$, say

$$(6.5) \qquad \mathbf{g}(\boldsymbol{\theta}) = \mathbf{J}_0 \boldsymbol{\theta} + \mathbf{g}_0,$$

where $\mathbf{J}_0$ is a known $D \times G$ matrix of rank G, and $\mathbf{g}_0$ is a known $D \times 1$ vector, we get

$$(6.6) \qquad \hat{\boldsymbol{\theta}} = (\mathbf{J}_0' \mathbf{M} \mathbf{J}_0)^{-1} \mathbf{J}_0' \mathbf{M}(\hat{\boldsymbol{\lambda}} - \mathbf{g}_0).$$

A particular case of (6.5) is given in (1.2).

6.2. *Asymptotic theory.* Let $\{\mathbf{M}^{(N)}\}$ be a sequence of positive definite, symmetric, possibly random, $D \times D$ matrices. For simplicity we assume that the $\mathbf{M}^{(N)}$ are not functions of $\boldsymbol{\theta}$. (This can be modified. Compare, for example, [14], Theorem 5.) Let $\hat{\boldsymbol{\theta}}^{(N)}$ be a value of $\boldsymbol{\theta}$. if any, which minimizes $Q(\boldsymbol{\theta})$ with $\mathbf{M} = \mathbf{M}^{(N)}$ and $\hat{\boldsymbol{\lambda}} = \hat{\boldsymbol{\lambda}}^{(N)}$. We can then prove the following theorem by the methods of general asymptotic statistical theory. (See [51]. All the hard parts of the proof can be handled by the argument in [9].)

THEOREM 3. *Make Assumptions 2 and 3, and assume also that*

$$(6.7) \qquad \operatorname{plim} \mathbf{M}^{(N)} = \mathbf{M}_0,$$

where $\mathbf{M}_0$ is positive definite. With a probability increasing to 1 as $N \to \infty$, there then exists a value $\hat{\boldsymbol{\theta}}^{(N)} \in \Theta$ which minimizes $Q(\boldsymbol{\theta})$, and

$$(6.8) \qquad N^{1/2}(\hat{\boldsymbol{\theta}}^{(N)} - \boldsymbol{\theta}^0) \xrightarrow{\mathscr{L}_{\theta^0}} \mathscr{N}(\mathbf{0}, \boldsymbol{\Sigma}),$$

where

$$(6.9) \qquad \boldsymbol{\Sigma} = (\mathbf{J}_0' \mathbf{M}_0 \mathbf{J}_0)^{-1} \mathbf{J}_0' \mathbf{M}_0 \boldsymbol{\Sigma}_0 \mathbf{M}_0 \mathbf{J}_0 (\mathbf{J}_0' \mathbf{M}_0 \mathbf{J}_0)^{-1}$$

is positive definite.

COROLLARY. $N^{1/2}\{\mathbf{g}(\hat{\boldsymbol{\theta}}^{(N)}) - \boldsymbol{\lambda}^0\} \xrightarrow{\mathscr{L}_{\theta^0}} \mathscr{N}(\mathbf{0}, \mathbf{J}_0' \boldsymbol{\Sigma} \mathbf{J}_0)$.

REMARK 1. If $\mathbf{M}_0 = \boldsymbol{\Sigma}_0^{-1}$, as is the case when we use (3.7) and (6.3), we get $\boldsymbol{\Sigma}$ equal to

$$(6.10) \qquad \boldsymbol{\Sigma}_{0,0} = (\mathbf{J}_0'\boldsymbol{\Sigma}_0^{-1}\mathbf{J}_0)^{-1}.$$

REMARK 2. Since $G < D$, $\mathbf{J}_0'\boldsymbol{\Sigma}\mathbf{J}_0$ is singular.

REMARK 3. If we regard $\hat{\boldsymbol{\theta}}^{(N)}$ as a mapping from $\mathbf{R}_D$ to $\mathbf{R}_G$ (that is, a function of $\hat{\boldsymbol{\lambda}}$), we obviously have

$$(6.11) \qquad \hat{\boldsymbol{\theta}}^{(N)}\big(\mathbf{g}(\boldsymbol{\theta})\big) = \boldsymbol{\theta} \qquad \text{for} \quad \boldsymbol{\theta} \in \Theta,$$

for any positive definite $\mathbf{M}^{(N)}$.

6.3. *The choice of* $\{\mathbf{M}^{(N)}\}$. Since different sequences $\{\mathbf{M}^{(N)}\}$ give rise to estimators $\{\hat{\boldsymbol{\theta}}^{(N)}\}$ which may have different asymptotic covariance matrices, one will want to know how to select a $\{\mathbf{M}^{(N)}\}$ so as to get a $\boldsymbol{\Sigma}$ which is as favorable as possible. Given two such matrices, $\boldsymbol{\Sigma}_1$ and $\boldsymbol{\Sigma}_2$, where $\boldsymbol{\Sigma}_2 - \boldsymbol{\Sigma}_1$ is positive semidefinite, we shall regard $\boldsymbol{\Sigma}_1$ as the more favorable, since each of the variances on its diagonal will be no greater than the corresponding variance on the diagonal of $\boldsymbol{\Sigma}_2$. At the same time, $\mathbf{J}_0'\boldsymbol{\Sigma}_1\mathbf{J}_0$ will be preferred to $\mathbf{J}_0'\boldsymbol{\Sigma}_2\mathbf{J}_0$ (compare the corollary to Theorem 3), since also $\mathbf{J}_0'(\boldsymbol{\Sigma}_2 - \boldsymbol{\Sigma}_1)\mathbf{J}_0$ will be positive semidefinite. The following theorem tells us that an $\{\mathbf{M}^{(N)}\}$ with $\mathbf{M}_0 = \boldsymbol{\Sigma}_0^{-1}$ will be optimal in this sense.

THEOREM 4. *Let $\boldsymbol{\Sigma}$ and $\boldsymbol{\Sigma}_{0,0}$ be given by (6.9) and (6.10), respectively, Then $\boldsymbol{\Sigma} - \boldsymbol{\Sigma}_{0,0}$ is positive semidefinite under the assumptions of Theorem 3.*

PROOF. (i) Let $\mathbf{A}$ be any $D \times G$ matrix of rank $G < D$. Then $\mathbf{A}(\mathbf{A}'\mathbf{A})^{-1}\mathbf{A}'$ is idempotent, so all its characteristic roots equal 0 or 1. Thus $\mathbf{I} - \mathbf{A}(\mathbf{A}'\mathbf{A})^{-1}\mathbf{A}'$ has only 0 and 1 as characteristic roots, and this matrix, therefore, is positive semidefinite.

(ii) Let us then prove that $\boldsymbol{\Sigma}_0 - \mathbf{J}_0(\mathbf{J}_0'\boldsymbol{\Sigma}_0^{-1}\mathbf{J}_0)^{-1}\mathbf{J}_0'$ is positive semidefinite. Let $\mathbf{B}$ be a nonsingular matrix such that $\mathbf{B}'\boldsymbol{\Sigma}_0\mathbf{B} = \mathbf{I}$. Let $\mathbf{v}$ be an arbitrary $D \times 1$ vector, and let $\mathbf{w} = \mathbf{B}^{-1}\mathbf{v}$. Then

$$(6.12) \qquad \mathbf{v}'\{\boldsymbol{\Sigma}_0 - \mathbf{J}_0(\mathbf{J}_0'\boldsymbol{\Sigma}_0^{-1}\mathbf{J}_0)^{-1}\mathbf{J}_0'\}\mathbf{v} = \mathbf{w}'\{\mathbf{I} - \mathbf{B}'\mathbf{J}_0(\mathbf{J}_0'\mathbf{B}\mathbf{B}'\mathbf{J}_0)^{-1}\mathbf{J}_0'\mathbf{B}\}\mathbf{w}$$
$$= \mathbf{w}'\{\mathbf{I} - \mathbf{A}(\mathbf{A}'\mathbf{A})^{-1}\mathbf{A}'\}\mathbf{w},$$

with $\mathbf{A} = \mathbf{B}'\mathbf{J}_0$. Our assertion then follows from step (i) above.

(iii) Finally, let $\mathbf{v}$ be as above, and let

$$(6.13) \qquad \mathbf{w} = \mathbf{M}_0\mathbf{J}_0(\mathbf{J}_0'\mathbf{M}_0\mathbf{J}_0)^{-1}\mathbf{v}.$$

Then $\mathbf{J}_0'\mathbf{w} = \mathbf{v}$ and so

$$(6.14) \qquad \mathbf{v}'\{\boldsymbol{\Sigma} - \boldsymbol{\Sigma}_{0,0}\}\mathbf{v} = \mathbf{w}'\{\boldsymbol{\Sigma}_0 - \mathbf{J}_0\boldsymbol{\Sigma}_{0,0}\mathbf{J}_0'\}\mathbf{w} \geqq 0$$

by step (ii) above. Thus $\boldsymbol{\Sigma} - \boldsymbol{\Sigma}_{0,0}$ is positive semidefinite. *Q.E.D.*

6.4. *The choice of* $\{\hat{\boldsymbol{\lambda}}^{(N)}\}$. In Sections 6.1 to 6.3 above, we have focused on a single estimator $\{\hat{\boldsymbol{\lambda}}^{(N)}\}$. Assume now that two such sequences are proposed, say $\{\hat{\boldsymbol{\lambda}}_1^{(N)}\}$ and $\{\hat{\boldsymbol{\lambda}}_2^{(N)}\}$, both satisfying the assumptions of Theorem 3, with asymptotic covariance matrices $\boldsymbol{\Sigma}_1/N$ and $\boldsymbol{\Sigma}_2/N$, respectively. Say that $\boldsymbol{\Sigma}_2 - \boldsymbol{\Sigma}_1$ is positive semidefinite. Intuitively one would expect $\{\hat{\boldsymbol{\theta}}_1^{(N)}\}$ to have a more favorable asymptotic covariance matrix than $\{\hat{\boldsymbol{\theta}}_2^{(N)}\}$, when $\{\hat{\boldsymbol{\theta}}_i^{(N)}\}$, for $i = 1, 2$, is produced from $\{\hat{\boldsymbol{\lambda}}_i^{(N)}\}$ by the method of Section 6.1 with a choice of $\mathbf{M}_i^{(N)}$ which is optimal according to Theorem 4. This turns out to be correct.

THEOREM 5. *If $\boldsymbol{\Sigma}_1$ and $\boldsymbol{\Sigma}_2$ are positive definite and $\boldsymbol{\Sigma}_2 - \boldsymbol{\Sigma}_1$ is positive semidefinite, then $\boldsymbol{\Sigma}_{0,2} - \boldsymbol{\Sigma}_{0,1}$ is positive semidefinite, where*

$$(6.15) \qquad \boldsymbol{\Sigma}_{0,i} = (\mathbf{J}_0' \boldsymbol{\Sigma}_i^{-1} \mathbf{J}_0)^{-1}$$

for $i = 1, 2$. Here $\mathbf{J}_0$ is any $D \times G$ matrix of rank $G < D$.

PROOF. If $\mathbf{A}$ and $\mathbf{B}$ are positive definite $D \times D$ matrices with positive semidefinite $\mathbf{A} - \mathbf{B}$, then $\mathbf{B}^{-1} - \mathbf{A}^{-1}$ will also be positive semidefinite [26], page 55, Theorem 2.5. From this the theorem easily follows. *Q.E.D.*

7. Moment methods

7.1. *The graduation method.* A moment method estimator $\tilde{\boldsymbol{\theta}}^{(N)}$ of $\boldsymbol{\theta}$ is defined as a solution of the system of equations

$$(7.1) \qquad \sum_{d=1}^{D} \xi_d^r \{\hat{\lambda}_d^{(N)} - g_d(\hat{\boldsymbol{\theta}}^{(N)})\} = 0 \qquad \text{for} \quad r = 0, 1, \cdots, G - 1,$$

if it exists. Let

$$(7.2) \qquad \mathbf{M} = \begin{pmatrix} 1, & 1, & \cdots, & 1 \\ \xi_1, & \xi_2, & \cdots, & \xi_D \\ \xi_1^2, & \xi_2^2, & \cdots, & \xi_D^2 \\ \cdot & \cdot & \cdots & \cdot \\ \xi_1^{G-1}, & \xi_2^{G-1}, & \cdots, & \xi_D^{G-1} \end{pmatrix}.$$

Then (7.1) can be rewritten as

$$(7.3) \qquad \mathbf{M}\{\hat{\boldsymbol{\lambda}}^{(N)} - \mathbf{g}(\tilde{\boldsymbol{\theta}}^{(N)})\} = 0.$$

We shall extend this definition, and shall call $\tilde{\boldsymbol{\theta}}^{(N)}$ a *generalized moment method estimator* for $\boldsymbol{\theta}$ if it is a solution of (7.3), where $\mathbf{M}$ here can be any $G \times D$ matrix, that is, $\mathbf{M}$ need not be given by (7.2).

To give an example of an estimator generated by (7.3) but not satisfying (7.1), we shall consider the King-Hardy method of estimating the three parameters, α, β, and c, of the Gompertz-Makeham function in (1.1). Say that we can take $[0, \zeta\rangle$ to be the age interval $[x_0, x_0 + 3h\rangle$ for some integer h, and that $\zeta_x = x_0 + x$ for $x = 0, 1, \cdots, 3h$, so that we have one year age intervals. Then the King-Hardy estimators are the solution $(\tilde{\alpha}, \tilde{\beta}, \tilde{c})$ of the equations

$$(7.4) \qquad \sum_{x=x_0+(k-1)h}^{x_0+kh-1} (\tilde{\alpha} + \tilde{\beta}\tilde{c}^x) = \tilde{H}_k \qquad \text{for} \quad k = 1, 2, 3,$$

where

$$\tilde{H}_k = \sum_{x = x_0 + (k-1)h}^{x_0 + kh - 1} \hat{\lambda}_x.$$

We get [59], p. 167,

(7.5) $$\hat{c}^h = (\tilde{H}_3 - \tilde{H}_2)/(\tilde{H}_2 - \tilde{H}_1),$$

and similar formulas for $\tilde{\alpha}$ and $\tilde{\beta}$. If we let $\mathbf{m}_x$ be a $1 \times h$ vector where all elements equal x, and let

(7.6) $$\mathbf{M} = \begin{pmatrix} \mathbf{m}_1 & \mathbf{m}_0 & \mathbf{m}_0 \\ \mathbf{m}_0 & \mathbf{m}_1 & \mathbf{m}_0 \\ \mathbf{m}_0 & \mathbf{m}_0 & \mathbf{m}_1 \end{pmatrix},$$

then (7.3) reduces to (7.4) in the case where

(7.7) $$g_x(\alpha, \beta, c) = \alpha + \beta c^{(x_0 + x)}.$$

In applications to analytic graduation, the matrix $\mathbf{M}$ is usually nonrandom and not a function of N or $\boldsymbol{\theta}$. For simplicity we shall only study this case, but generalization to possibly random $\mathbf{M}$, possibly depending on N and $\boldsymbol{\theta}$, can be made by standard methods [24], [78].

If, in particular, $\mathbf{g}(\boldsymbol{\theta})$ is given by (6.5), we get

(7.8) $$\tilde{\boldsymbol{\theta}}^{(N)} = (\mathbf{M}\mathbf{J}_0)^{-1}\mathbf{M}(\hat{\boldsymbol{\lambda}}^{(N)} - \mathbf{g}_0),$$

provided $\mathbf{M}\mathbf{J}_0$ is nonsingular.

7.2. *Asymptotic theory*.

THEOREM 6. *Make Assumptions 2 and 3, and assume also that $\mathbf{M}\mathbf{J}(\theta)$ is nonsingular for any $\boldsymbol{\theta} \in \Theta$. There then exists a neighborhood Ω of $\mathbf{g}(\Theta)$ and a one to one mapping $\tilde{\boldsymbol{\theta}}^{(N)}$ from $\mathbf{R}_D$ to $\mathbf{R}_G$, continuous in Ω, such that*

(7.9) $$\tilde{\boldsymbol{\theta}}^{(N)}\big(\mathbf{g}(\boldsymbol{\theta})\big) = \boldsymbol{\theta}$$

for $\boldsymbol{\theta} \in \Theta$ and (7.3) holds for all $\hat{\boldsymbol{\lambda}}^{(N)} \in \Omega$. Furthermore,

(7.10) $$N^{1/2}(\tilde{\boldsymbol{\theta}}^{(N)} - \boldsymbol{\theta}^0) \xrightarrow{\mathscr{L}_{\theta^0}} \mathscr{N}(\mathbf{0}, \boldsymbol{\Sigma}),$$

where

(7.11) $$\boldsymbol{\Sigma} = (\mathbf{M}\mathbf{J}_0)^{-1}\mathbf{M}\boldsymbol{\Sigma}_0\{(\mathbf{M}\mathbf{J}_0)^{-1}\mathbf{M}\}'.$$

$\boldsymbol{\Sigma} - \boldsymbol{\Sigma}_{0,0}$ is positive semidefinite. ($\boldsymbol{\Sigma}_{0,0}$ is given in (6.10).)

PROOF. By Theorem 1 in [24], p. 1054, we need only prove the final assertion above. Let $\mathbf{v}$ be an arbitrary $G \times 1$ vector, and let $\mathbf{w} = (\mathbf{J}_0'\mathbf{M}')^{-1}\mathbf{v}$. Then

(7.12) $$\mathbf{v}'\{\boldsymbol{\Sigma} - \boldsymbol{\Sigma}_{0,0}\}\mathbf{v} = (\mathbf{M}\mathbf{w})'\{\boldsymbol{\Sigma}_0 - \mathbf{J}_0\boldsymbol{\Sigma}_{0,0}\mathbf{J}_0'\}(\mathbf{M}\mathbf{w}) \geqq 0$$

by step (ii) of the proof of Theorem 4. *Q.E.D.*

REMARK 4. By the final assertion of the theorem, the generalized moment method can never give a more favorable asymptotic covariance matrix for the estimatior of θ than the corresponding "optimal" estimator found in Section 6.

REMARK 5. Since $\hat{\lambda}^{(N)}$ is $N^{1/2}$-consistent for λ^0,

$$(7.13) \qquad P\{\hat{\lambda}^{(N)} \in \Omega\} \to 1 \qquad \text{as} \quad N \to \infty.$$

REMARK 6. The analogues of Theorem 5 and the corollary to Theorem 3 hold in the present situation.

When the generalized moment method is applied to a particular case, it is frequently modified to suit the characteristics of the situation in hand. We shall give examples of this in Sections 7.3 and 7.4.

7.3. *Modifications, Example* 1: *Gompertz-Makeham graduation*. In mortality studies using the Gompertz-Makeham formula (1.1), one will frequently find that c is estimated by (7.5), but that estimators for α and β are subsequently found by some other method, for instance by minimizing

$$(7.14) \qquad \sum_{x=x_0}^{x_0+3h-1} (\hat{\lambda}_x^{(N)} - \alpha - \beta \tilde{c}^x)^2$$

[83], p. 225. Let us consider a slightly more general case, and let us estimate α and β by minimizing

$$(7.15) \qquad Q(\alpha, \beta) = N(\hat{\lambda}^{(N)} - \alpha e - \beta \tilde{\psi}^{(N)})' \mathbf{M}^{(N)} (\hat{\lambda}^{(N)} - \alpha e - \beta \tilde{\psi}^{(N)}).$$

Here $\{\mathbf{M}^{(N)}\}$ is a sequence of matrices of the kind studied in Section 6, e is a $3h \times 1$ vector where all elements equal 1, and

$$(7.16) \qquad \tilde{\psi}^{(N)} = (\tilde{c}^{x_0}, \tilde{c}^{x_0+1}, \cdots, \tilde{c}^{x_0+3h-1})'.$$

Assuming that $\mathbf{M}^{(N)}$ is positive definite, and letting

$$(7.17) \qquad \mathbf{K}^{(N)} = (e, \tilde{\psi}^{(N)}),$$

we get the estimators

$$(7.18) \qquad \begin{pmatrix} \hat{\alpha} \\ \hat{\beta} \end{pmatrix} = (\mathbf{K}^{(N)'} \mathbf{M}^{(N)} \mathbf{K}^{(N)})^{-1} \mathbf{K}^{(N)'} \mathbf{M}^{(N)} \hat{\lambda}^{(N)}.$$

Now let m_x be defined as below (7.5), let α_0, β_0, and c_0 be the true values of the parameters, let

$$(7.19) \qquad \begin{aligned} \psi_0 &= (c_0^{x_0}, \cdots, c_0^{x_0+3h-1})', \qquad \mathbf{K}_0 = (e, \psi_0), \\ \gamma &= (c_0 - 1)/\{h\beta_0 c_0^{x_0+h-1}(c_0^h - 1)^2\}, \end{aligned}$$

and

$$(7.20) \qquad \Phi = \begin{pmatrix} (\mathbf{K}_0' \mathbf{M}_0 \mathbf{K}_0)^{-1} \mathbf{K}_0 \mathbf{M}_0 \\ (m_0, m_{-\gamma}, m_\gamma) \end{pmatrix}.$$

We then get

THEOREM 7. *Let $p \lim \mathbf{M}^{(N)} = \mathbf{M}_0$, where $\mathbf{M}_0$ is positive definite, and make Assumption 3. Then*

$$(7.21) \qquad N^{1/2}\{(\hat{\hat{\alpha}}, \hat{\hat{\beta}}, \hat{c})' - (\alpha_0, \beta_0, c_0)'\} \xrightarrow{\mathscr{L}_{(\alpha_0, \beta_0, c_0)}} \mathscr{N}(\mathbf{0}, \mathbf{\Phi}\mathbf{\Sigma}_0\mathbf{\Phi}').$$

Stevens [68], Patterson [57], Lipton and McGilchrist [52], and others in their references have studied the estimation of the Gompertz-Makeham parameters in a regression model. Stevens [68] found that King-Hardy's method may be very inefficient there. The estimators developed for the regression model can also be used for purposes of analytic graduation, and it would be interesting to see an investigation of their merits in that context.

7.4. *Modifications, Example* 2: *Hadwiger graduation.* Consider now the problem of graduating a set $\{\hat{\lambda}_x : x = \alpha, \alpha + 1, \cdots, \beta - 1\}$ of female fertility rates calculated for single year age groups by fitting the Hadwiger function (1.3) to the rates. If we regard h_x as a function of a continuous x, and define

$$(7.22) \qquad R_k'(R, T, H, d) = \int_d^\infty x^k h_x(R, T, H, d)\, dx,$$

then

$$(7.23) \qquad R_0'(R, T, H, d) = R, \ R_1'(R, T, H, d) = R(T + d),$$

and the formulas for R_k' for $k \geqq 2$ can be found from the fact that the corresponding cumulants for $d = 0$ are

$$(7.24) \qquad \kappa_k = (1)(3)\cdots(2k - 3)2^{k-1}H^{2k-2}/T^{3k-4} \qquad \text{for} \quad k \geqq 2.$$

(Compare [45], pp. 150, 151, 160.) No such nice formulas are known for the discrete case, that is, for

$$(7.25) \qquad R_k(R, T, H, d) = \sum_{x=\alpha}^{\beta-1} x^k h_x(R, T, H, d),$$

where only integer values of x are used in the summation. Rather than attempting cumbersome calculations with the R_k, and acting on the analogy between the R_k and the R_k', Yntema [28], [81] has suggested an estimation procedure which amounts to the following: regard h_x as a function of a *continuous* x. Let $U = T + d$.

Then

$$(7.26) \qquad h_U(R, T, H, d) = \frac{RH}{T\sqrt{\pi}},$$

and the mode of the function is

$$(7.27) \qquad M = d + 3T\{(1 + 16H^4/9)^{1/2} - 1\}/(4H^2).$$

One easily sees that $M < U$. Solve (7.27) with respect to H, introduce the result into (7.26), let

$$(7.28) \qquad a = \tfrac{4}{3}\pi(Th_U/R)^2, \qquad b = (M-d)/T,$$

and get

$$(7.29) \qquad b = \{(1+a^2)^{1/2} - 1\}/a.$$

For the range of values in which a will usually lie, the right side here is approximately equal to $1 - a^{-1}$. Solving $b \approx 1 - a^{-1}$ with respect to T after substituting $U - T$ for d, we get

$$(7.30) \qquad T \approx R^2/\{\tfrac{4}{3}\pi(U - M)h_U^2\}.$$

Now introduce the estimators as follows: Let

$$(7.31) \qquad \hat{R} = \sum_{x=\alpha}^{\beta-1} \hat{\lambda}_x, \qquad \hat{U} = \sum_{x=\alpha}^{\beta-1} x\hat{\lambda}_x/\hat{R}.$$

(Compare (7.23).) If $[y]$ denotes the integer value of y, let

$$(7.32) \qquad V = [\hat{U} + \tfrac{1}{2}], \qquad \hat{h} = \hat{\lambda}_V, \qquad \hat{M} = \min\{x: \hat{\lambda}_x \geqq \hat{\lambda}_y \text{ for all } y\}$$

Finally, let

$$(7.33) \qquad \hat{T} = \hat{R}^2/\{\tfrac{4}{3}\pi(\hat{U} - \hat{M})\hat{h}^2\}, \qquad \hat{d} = \hat{U} - \hat{T},$$

[compare (7.30)], and let

$$(7.34) \qquad \hat{H} = \hat{h}\hat{T}\sqrt{\pi}/\hat{R} = \tfrac{3}{4}\hat{R}/\{\sqrt{\pi}\hat{h}(\hat{U} - \hat{M})\}.$$

(Compare (7.26).) Then $\hat{\boldsymbol{\theta}} = (\hat{R}, \hat{T}, \hat{H}, \hat{d})$ is an estimator for $\boldsymbol{\theta} = (R, T, H, d)$. To study its asymptotic properties, we let $\boldsymbol{\theta}^0 = (R^0, T^0, H^0, d^0)$ be the true value of $\boldsymbol{\theta}$, and introduce

$$R_0 = R_0(\boldsymbol{\theta}^0), \qquad U_0 = R_1(\boldsymbol{\theta}^0)/R_0, \qquad V_0 = [U_0 + \tfrac{1}{2}],$$
$$h_0 = h_{V_0}(\boldsymbol{\theta}^0),$$

$$(7.35) \qquad M_0 = \min\{x \in \{\alpha, \alpha+1, \cdots, \beta-1\}: h_x(\boldsymbol{\theta}^0) \geqq h_y(\boldsymbol{\theta}^0)$$
$$\text{for all } y \in \{\alpha, \alpha+1, \cdots, \beta-1\}\},$$

$$T_0 = R_0^2/\{\tfrac{4}{3}\pi(U_0 - M_0)h_0^2\}, \qquad d_0 = U_0 - T_0,$$
$$H_0 = h_0 T_0 \sqrt{\pi}/R_0.$$

Let $\mathbf{e}$ be a $(\beta - \alpha) \times 1$ vector where all elements equal 1, let

$$(7.36) \qquad \boldsymbol{\psi} = R_0^{-1}(\alpha - T_0, \alpha + 1 - T_0, \cdots, \beta - 1 - T_0)',$$

and let

$$(7.37) \qquad \boldsymbol{\Phi} = (\mathbf{e}, \mathbf{e}T_0/R_0, \mathbf{e}T_0 h_0\sqrt{\pi}/R_0^2, \boldsymbol{\psi}).$$

We then have

THEOREM 8. *Under Assumption* 3,

$$(7.38) \qquad N^{1/2}\{(\hat{R}, \hat{T}, \hat{H}, \hat{d})' - (R_0, H_0, T_0, d_0)'\} \xrightarrow{\mathscr{L}_{\theta^0}} \mathscr{N}(\mathbf{0}, \boldsymbol{\Phi}'\boldsymbol{\Sigma}_0\boldsymbol{\Phi}).$$

REMARK 7.   Note that $\boldsymbol{\Phi}'\boldsymbol{\Sigma}_0\boldsymbol{\Phi}$ is singular.

REMARK 8.   Note also that (R_0, T_0, H_0, d_0) is not the true value of the parameters here. No one seems to have looked into the difference (R^0, T^0, H^0, d^0) minus (R_0, T_0, H_0, d_0) in any detail.

8. A maximum likelihood method

In Sections 6 and 7, the estimators for θ appear as functions of the "raw" estimator $\hat{\lambda}$ for λ. If one may really assume that $\lambda_a = \mathbf{g}_a(\boldsymbol{\theta}_a)$ for $a = 1, 2, \cdots, A$, different approaches may be at least as efficient, however. One obvious possibility is to enter the $\mathbf{g}_a(\boldsymbol{\theta}_a)$ into the likelihood function and maximize with respect to the $\boldsymbol{\theta}_a$. In the situation of Section 3.4, this will amount to maximizing $\Sigma_a \hat{\eta}_a(\boldsymbol{\theta}_a)$, where

$$(8.1) \qquad \hat{\eta}_a(\boldsymbol{\theta}_a) = \sum_d M(a, d) \log g_{a,d}(\boldsymbol{\theta}_a) - \sum_d L(a, d) g_{a,d}(\boldsymbol{\theta}_a).$$

For simplicity, we shall assume that $\boldsymbol{\theta}_1, \boldsymbol{\theta}_2, \cdots, \boldsymbol{\theta}_A$ are functionally independent, so that we can maximize the likelihood function (if at all) by maximizing each $\hat{\eta}_a$ separately.

In the situation of Section 4.3, the log likelihood function is of a different form, but we shall still construct an estimator $\boldsymbol{\theta}_a^*$ for $\boldsymbol{\theta}_a$ by maximizing $\hat{\eta}_a$.

The following theorem holds.

THEOREM 9.   *Fix $a \in \{1, 2, \cdots, A\}$ and let the $M(a, d)$ and $L(a, d)$ be given as in Section 3.4 or 4.3. Assume that $P\{L(a, d) > 0\} \to 1$ as $N \to \infty$ for all d where $(a, d) \in \mathscr{G}$. Make Assumptions 1 and 2.*

With a probability increasing to 1 as $N \to \infty$, there will then exist a value $\boldsymbol{\theta}_a^{*(N)} \in \Theta_a$ which maximizes $\hat{\eta}_a(\boldsymbol{\theta}_a)$, and

$$(8.2) \qquad N^{1/2}(\boldsymbol{\theta}_a^{*(N)} - \boldsymbol{\theta}_a^0) \xrightarrow{\mathscr{L}_{\theta_a^0}} \mathscr{N}(\mathbf{0}, \boldsymbol{\Sigma}_{0,0}),$$

where $\boldsymbol{\Sigma}_{0,0}$ is given by (6.10), provided there exist constants k_a and k_a' such that

$$(8.3) \qquad k_a' \geqq g_{a,d}(\boldsymbol{\theta}_a) \geqq k_a > 0 \qquad \text{for all} \quad \boldsymbol{\theta}_a \in \Theta_a.$$

PROOF.   (i) *Preliminaries.*   Suppress the subscript a and fix the true value θ^0. Let

$$(8.4) \qquad \ell_d = \ell_d^{(N)} = E_{\theta^0} L(d), \qquad L_d = \lim_{N \to \infty} \ell_d^{(N)}/N,$$

and note that [40], (10),

$$(8.5) \qquad E_{\theta^0} M(d) = \lambda_d^0 \ell_d.$$

Let

$$(8.6) \qquad \hat{Q}(\lambda) = \sum_d M(d) \log \lambda_d - \sum_d L(d) \lambda_d,$$

so that

$$(8.7) \qquad \hat{\eta}(\boldsymbol{\theta}) = \hat{Q}(\mathbf{g}(\boldsymbol{\theta})),$$

and let

$$(8.8) \qquad Q(\boldsymbol{\lambda}) = E_{\boldsymbol{\theta}^0} \hat{Q}(\boldsymbol{\lambda}) = \sum_d \ell_d \{\lambda_d^0 \log \lambda_d - \lambda_d\},$$

and

$$(8.9) \qquad \eta(\boldsymbol{\theta}) = Q(\mathbf{g}(\boldsymbol{\theta})) = E_{\boldsymbol{\theta}^0} \hat{\eta}(\boldsymbol{\theta}).$$

Finally, let

$$(8.10) \qquad \Delta(\boldsymbol{\lambda}) = Q(\boldsymbol{\lambda}^0) - Q(\boldsymbol{\lambda}) = \sum_d \ell_d \{\lambda_d^0 (\log \lambda_d^0 - \log \lambda_d) - (\lambda_d^0 - \lambda_d)\}.$$

For large enough N, each ℓ_d will be positive. For such N, we will have $\Delta(\boldsymbol{\lambda}) > 0$ for all $\boldsymbol{\lambda} \neq \boldsymbol{\lambda}^0$, and $\Delta(\boldsymbol{\lambda})$ will strictly increase as each $|\lambda_d - \lambda_d^0|$ increases. For every $\varepsilon > 0$ there then exists a $\delta_1(\varepsilon)$ such that if $\Delta(\mathbf{g}(\boldsymbol{\theta})) \leqq \varepsilon$ then $|\mathbf{g}(\boldsymbol{\theta}) - \mathbf{g}(\boldsymbol{\theta}^0)| \leqq \delta_1(\varepsilon)$, and by the bicontinuity of $\mathbf{g}$ there further exists a $\delta(\varepsilon)$ such that $|\boldsymbol{\theta} - \boldsymbol{\theta}^0| \leqq \delta(\varepsilon)$. Conversely there exists a $\delta_0(\varepsilon)$ such that $\Delta(\mathbf{g}(\boldsymbol{\theta})) \leqq \delta_0(\varepsilon)$ if $|\boldsymbol{\theta} - \boldsymbol{\theta}^0| \leqq \varepsilon$. Let

$$(8.11) \qquad S_\varepsilon = \{\boldsymbol{\theta} \in \Theta : |\boldsymbol{\theta} - \boldsymbol{\theta}^0| \leqq \varepsilon\},$$

and choose ε. Choose $\varepsilon' > 0$, and let ε'' be so small that $\varepsilon'' \leqq \delta_0(\varepsilon)$, and $\delta(2\varepsilon'') \leqq \varepsilon'$, and

$$(8.12) \qquad 0 < 2\varepsilon'' < \eta(\boldsymbol{\theta}^0) - \inf\{\eta(\boldsymbol{\theta}) : \boldsymbol{\theta} \in \Theta\},$$

and let

$$(8.13) \qquad \Theta_{\varepsilon''} = \{\boldsymbol{\theta} \in \Theta : \Delta(\mathbf{g}(\boldsymbol{\theta})) \leqq 2\varepsilon''\}.$$

(ii) *Existence of* $\boldsymbol{\theta}^*$.   Let

$$(8.14) \qquad \gamma = \sup \sum_d \{|\log g_d(\boldsymbol{\theta})| + g_d(\boldsymbol{\theta})\}.$$

By (8.3), $\gamma < \infty$. Let $A_{\varepsilon''}^{(N)}$ be the event that

$$(8.15) \qquad |N^{-1} M(d) - \lambda_d^0 \ell_d| < \varepsilon''/\gamma, \qquad |N^{-1} L(d) - \ell_d| < \varepsilon''/\gamma.$$

Then $P_{\boldsymbol{\theta}^0}(A_{\varepsilon''}^{(N)}) \to 1$. Assume that (8.15) holds. Then

$$(8.16) \qquad |\hat{\eta}(\boldsymbol{\theta}) - \eta(\boldsymbol{\theta})| < \varepsilon'' \qquad \text{for all} \quad \boldsymbol{\theta} \in \Theta.$$

If $\boldsymbol{\theta} \in \Theta - \Theta_{\varepsilon''}$, we therefore get

$$(8.17) \qquad \hat{\eta}(\boldsymbol{\theta}) < \eta(\boldsymbol{\theta}) + \varepsilon'' < \eta(\boldsymbol{\theta}^0) - \varepsilon'' < \hat{\eta}(\boldsymbol{\theta}^0),$$

so in maximizing $\hat{\eta}(\boldsymbol{\theta})$ we need not take such $\boldsymbol{\theta}$ into account. Since $\Theta_{\varepsilon''}$ is closed and $\hat{\eta}(\cdot)$ is continuous, there exists a maximizing value $\boldsymbol{\theta}^* \in \Theta_{\varepsilon''}$.

(iii) *Consistency of* $\boldsymbol{\theta}^*$.   By the definition of $\Theta_{\varepsilon''}$, we have $\Delta(\mathbf{g}(\boldsymbol{\theta}^*)) \leqq 2\varepsilon''$.

Thus $|\boldsymbol{\theta}^* - \boldsymbol{\theta}^0| \leq \delta(2\varepsilon'') \leq \varepsilon'$, and the consistency of $\boldsymbol{\theta}^*$ follows.

The theorem now follows from some general results due to LeCam [51]. $Q.E.D.$

REMARK 9. Note that $\Sigma_{0,0}$ is the most favorable asymptotic covariance matrix we can get by the procedures in Section 6 when $\hat{\boldsymbol{\lambda}}$ is given by (3.7). In this sense, therefore, the method of the present section is at least as good as any of the other general methods we have studied.

9. The choice of a graduating function

9.1. In previous sections, it was presupposed that the applicability of a particular graduating function $\mathbf{g}(\boldsymbol{\theta})$ had been established, and the problem was to estimate $\boldsymbol{\theta}$. In many practical cases, the situation will be different. Instead of a single function $\mathbf{g}$, there is often a finite family $\mathscr{F} = \{\mathbf{g}(s, \boldsymbol{\theta}): s = 1, 2, \cdots, S\}$ of candidates for a graduating function, and one is required to choose one of these on the basis of the data. In the case of human fertility rates, for instance, it is seldom given which function to use, and one may have to select one from among the Pearson family, the Hadwiger function, and the Brass polynomial (1.4), say.

We shall assume that all functions $\mathbf{g}(s, \boldsymbol{\theta})$ have the same parameter space Θ. This need not be the case originally, but it can be achieved by the introduction of dummy parameters if necessary.

9.2. To describe what it means to choose a function from the class $\mathscr{F}$ "on the basis of the data," we shall assume that there is a member $\mathbf{g}(s^0, \cdot)$ of $\mathscr{F}$ which is the "true" graduating function. The choice of a member of $\mathscr{F}$ then amounts to estimating s^0 as an extra parameter. A number of estimation procedures are in use (compare, for example, [45], Section 6.5), but their statistical properties do not seem to have been much investigated, except that one may know something about their consistency as $N \to \infty$. We list some of these procedures.

(i) For choosing among the members of the Pearson family, there exist standard methods [22], [58], [34], [13] based on the first four empirical moments. Keyfitz, [45], p. 160, suggests that this type of criterion can also be used when the Hadwiger function (1.3) is included in $\mathscr{F}$ along with the Pearson type functions.

(ii) In connection with the methods of Section 6, an obvious procedure is to set

$$(9.1) \qquad \hat{Q}_s(\boldsymbol{\theta}) = N(\hat{\boldsymbol{\lambda}} - \mathbf{g}(s, \boldsymbol{\theta}))' \mathbf{M}(\hat{\boldsymbol{\lambda}} - \mathbf{g}(s, \hat{\boldsymbol{\theta}})),$$

let $\hat{\boldsymbol{\theta}}^{(s)}$ be a value of $\boldsymbol{\theta}$ which minimizes $\hat{Q}_s(\boldsymbol{\theta})$ for $s = 1, 2, \cdots, S$, and define $\hat{s}$ as the value of s that subsequently minimizes $\hat{Q}_s(\hat{\boldsymbol{\theta}}^{(s)})$ [27], [73], [80], [81].

(iii) A similar criterion can be used in connection with the method of Section 8. Let

$$(9.2) \qquad \hat{\eta}(s, \boldsymbol{\theta}) = \sum_d \{M(d) \log g_d(s, \hat{\boldsymbol{\theta}}) - L(d) g_d(s, \boldsymbol{\theta})\},$$

let $\boldsymbol{\theta}^{*(s)}$ maximize this quantity, and let $\hat{s}$ be the value of s that subsequently maximizes $\hat{\eta}(s, \boldsymbol{\theta}^{*(s)})$.

(iv) Yntema [28], [81] has suggested calculating

$$(9.3) \qquad \Delta_s = \sum_{d=1}^{D} |\hat{\lambda}_d - g_d(s, \hat{\boldsymbol{\theta}}^{(s)})|$$

and

$$(9.4) \qquad \Delta_s' = \max \{|\hat{\lambda}_d - g_d(s, \hat{\boldsymbol{\theta}}^{(s)})|: d = 1, 2, \cdots, D\},$$

and taking $\hat{s}$ as the s-value that minimizes Δ_s or Δ_s'. Here $\hat{\boldsymbol{\theta}}^{(s)}$ is any suitable estimator for $\boldsymbol{\theta}$ based on $\mathbf{g}(s, \boldsymbol{\theta})$.

9.3. We shall take a look at the consistency properties of $\hat{s}$ as defined in Sections 9.2 (ii) and (iii) above.

By a proper specification of $\mathscr{F}$ and Θ we should be able to get $\mathbf{g}(s', \Theta) \cap \mathbf{g}(s'', \Theta)$ to be empty whenever $s' \neq s''$. (Otherwise, part of the values $\mathbf{g}(s'', \boldsymbol{\theta})$ would be redundant). To prove consistency, however, we need the stronger assumption that

$$(9.5) \qquad |\mathbf{g}(s', \Theta) - \mathbf{g}(s'', \Theta)| > 0 \qquad \text{for} \quad s' \neq s'',$$

where $|A - B| = \inf \{|a - b|: a \in A, b \in B\}$ denotes the Euclidean distance between two subsets A and B of $\mathbf{R}_D$.

If (9.5) holds, if $\hat{\boldsymbol{\lambda}}$ is consistent for $\boldsymbol{\lambda}^0$, and if $\text{plim } \mathbf{M} = \mathbf{M}_0$ *as* $N \to \infty$, *with* $\mathbf{M}_0$ *positive definite, then $\hat{s}$ is consistent in Section 9.2 (ii) above.*

Similarly, by step (i) in the proof of Theorem 9, *$\hat{s}$ is consistent in Section 9.2 (iii) above* when (9.5) holds.

9.4. Let $\{\hat{\boldsymbol{\theta}}^{(s)}\}$ be some estimator which we would use for $\boldsymbol{\theta}$ if it were known that $s^0 = s$, and assume that

$$(9.6) \qquad P_{s^0, \boldsymbol{\theta}^0} \{N^{1/2}(\hat{\boldsymbol{\theta}}^{(s^0)} - \boldsymbol{\theta}^0) \in B\} \to \Phi(B) \text{ as } N \to \infty,$$

where Φ is a limiting probability measure and B is any Φ-continuous measurable set. Let $\hat{s}$ be a consistent estimator for s^0. Then it is easy to show that

$$(9.7) \qquad P_{s^0, \boldsymbol{\theta}^0} \{N^{1/2}(\hat{\boldsymbol{\theta}}^{(\hat{s})} - \boldsymbol{\theta}^0) \in B\} \to \Phi(B) \text{ as } N \to \infty$$

for the same B. Of course, $\hat{\boldsymbol{\theta}}^{(\hat{s})}$ is our estimator for $\boldsymbol{\theta}$ and (9.7) tells us that its limiting distribution is the one we would get if s^0 were known. Similarly, $\mathbf{g}(\hat{s}, \hat{\boldsymbol{\theta}}^{(\hat{s})})$ will be our estimator for $\boldsymbol{\lambda}^0$, and its asymptotic properties follow directly from (9.7).

10. Concluding remarks

10.1. In the models described in Sections 2.1 and 2.2 above, the seniority parameter is continuous. If it is known (or if one assumes) that one of the forces of transition can be represented by a nice and smooth parametric function, say

(10.1) $$\lambda(x) = g(x; \boldsymbol{\theta}),$$

and if one is faced with the problem of estimating λ, using analytic graduation is not necessarily the most obvious line of attack. In fact, it seems more natural to try to construct an estimator $\hat{\boldsymbol{\theta}}$ for $\boldsymbol{\theta}$ directly, without going the way via the $\lambda^{\#}$, as described in Section 3. Grenander [29], pp. 76–91, has shown how this might be done for the force of mortality in the example in 2.2 (i) when the Gompertz-Makeham formula (1.1) (with continuous x) applies. A similar investigation could be carried out for other forces of transition, like the forces of fertility of the example in 2.2 (ii).

If one does not know enough about the function $\lambda(\cdot)$ to specify a parametric $g(\cdot, \boldsymbol{\theta})$ which can represent it, one may turn to nonparametric methods, such as those developed within reliability theory [5], [6]. The force of mortality in Section 2.2 (i) appears there under the name of failure rate or hazard rate, and quite a lot of energy has gone into finding suitable methods of estimating this function.

Although both of these types of approach were initiated by Grenander's paper [29] on mortality measurement, such techniques do not seem to be much in use in demography and related fields. One would be curious to know why this is so. Part of the explanation is, no doubt, that these developments are largely unknown among people working in those fields of application, but there are more valid reasons. We shall suggest some of them.

10.2. The following types of argument seem to be among the ones leading people to base their inferences from the data on the $M(a, d)$ and $L(a, d)$ only, and sometimes on the $\hat{\lambda}_{a, d}$ only. (Note that least squares and moments method estimation procedures of $\boldsymbol{\theta}$ only require knowledge of the $\hat{\lambda}_{a, d}$.)

(i) In Section 3.2 we described how the points $\{\zeta_d : d = 1, 2, \cdots, D\}$ partitioning the seniority interval were selected according to conventional rules. Similarly, it is standard procedure to calculate "occurrence/exposure" rates of the kind developed in Sections 3.4 and 4.3. The use of standard techniques, standard tabulations, and so on, facilitates comparison with other investigations of the same subject matter. This encourages the continued use of techniques which are already widely known and widely applied even when other methods may be known to a few people.

(ii) The reliability of the data which demographers have to work with, can be very weak due to phenomena such as age misreporting, underenumeration, and so on. Also, one frequently does not know more than approximate dates (for example, the calendar year only) of occurrences of the events studied. This calls for the application of rather robust statistical techniques, such as those which we have described. Even though demographic data may be deficient, they may still be reliable enough to permit the use of the aggregated values $M(a, d)$ and $L(a, d)$ or at least the $\hat{\lambda}_{a, d}$.

In many cases, the investigator does not even have access to the original data, but only to standard tabulations made from them. Such tabulations will often permit the use of methods described here, and rule out others.

(iii) A similar argument applies to the reliability of the *models* used. For example, most current models, including those considered in this paper, leave seasonal variation over the calendar year out of account. There is plenty of evidence of the importance of such variation in the occurrence of vital events, but in many cases this is just a nuisance factor which one wants to eliminate. Current methods relying on seniority interval lengths of at least a year seem to effectively do so.

(iv) Even in cases where the data are reliable and sufficiently detailed (and the present author believes that not nearly enough attention has been given to such cases), the information extracted by a statistical procedure should be geared to the needs of the user. It seems that a standard table of rates, like Table I, and certain other tables derived from it, contains just about as much information as can be handled in a substantive study. In fact, the prevalence of summary indices derived from such tables, and the extent to which argumentation is carried out in terms of such indices, suggests that the standard tables contain even too much information. The use of analytic graduation can be seen as another piece of evidence in the same direction, since it enables one to substitute the formula of a function and a (small) set of parameter values for a whole table. (This argument does not rule out the parametric procedure suggested at the beginning of Section 10.1.)

(v) Each of the estimators $\hat{s}$ listed in Section 9.2 is a function of the data via $\hat{\lambda}$ only. This reflects the fact that an investigator faced with the problem of selecting a graduating function from a class $\mathscr{F}$ of candidates is likely to calculate $\hat{\lambda}$, plot the corresponding diagram, and use this to decide which member to choose from $\mathscr{F}$. In fact, this is the way in which certain graduating functions historically have been pinpointed as more suitable than others.

Once $\hat{s}$ has been determined, however, the investigator should not necessarily continue to use $\hat{\lambda}$ in the estimation of θ, but should feel free to choose among all available procedures as far as the quality of his data permits.

10.3.   It is probably appropriate to underline once more (compare Section 1.1) that there exist many types of graduation methods in addition to analytic graduation techniques. Most of them were first developed for use in mortality studies, and in that context they are apparently applied at least as often as analytic methods are. Many of them must have been intended for use in other connections as well, for example in fertility studies. With the exception of graphic methods, however, their application to other types of vital rates than mortality rates seems much less popular. (Compare [54], p. 53.)

I wish to express my great indebtedness to L. LeCam and P. J. Bickel for many helpful discussions on the large sample theory used in this paper, and to M. L. Eaton, who has read the paper in manuscript and has given valuable comments. I am also grateful to W. Simonsen and L. Boza who independently suggested problems which led to the formulations in Section 4 above. The data for the diagrams were furnished by the Central Bureau of Statistics of Norway, and I carried out their graduation with help from the Department of Statistics, University of California, Berkeley.

REFERENCES

[1] H. Ammeter, "Wahrscheinlichkeitstheoretische Kriterien für die Beurteilung der Güte der Ausgleichung einer Sterbetafel," *Mitt. Verein. Schweiz. Versich.-Math.*, Vol. 52 (1952), pp. 19–72.

[2] ———, "Der doppelseitige und die einseitigen $(I\chi^2)$-Tests und ihre Leistungsfähigkeit für die wahrscheinlichkeitstheoretische Überprüfung von Sterbetafeln," *Blätter der Deutschen Gesellschaft für Versicherungsmathematik (Deutscher Aktuarverein) E. V.*, Vol. 1 (1953), pp. 39–60.

[3] T. W. Anderson, *An Introduction to Multivariate Statistical Analysis*, New York, Wiley, 1958.

[4] E. W. Barankin and F. Gurland, "On asymptotically normal, efficient estimators: I," *Univ. California Publ. Statist.*, Vol. 1 (1951), pp. 89–130. Compare [23].

[5] R. E. Barlow, "Some recent developments in reliability theory," *Amsterdam Mathematisch Centrum, Mathematical Centre Tracts*, No. 27 (1968), pp. 49–65.

[6] R. E. Barlow and W. R. van Zwet, "Asymptotic properties of isotonic estimators for the generalized failure rate function. Parts I and II," University of California, Berkeley, Operations Research Center Reports, No. ORC 69–5 and 69–10, 1969.

[7] H. A. R. Barnett, "Graduation tests and experiments," *J. Inst. Actuaries*, Vol. 72 (1951), pp. 15–74.

[8] B. Benjamin, *Health and Vital Statistics*, London, George Allen & Unwin Ltd., 1968.

[9] P. Bickel, "Lecture notes for asymptotic theory, Statistics 217B, Winter 1970," University of California, Berkeley, Department of Statistics (mimeographed).

[10] P. Billingsley, *Convergence of Probability Measures*, New York, Wiley, 1968.

[11] W. Brass, "The graduation of fertility distributions by polynomial functions," *Population Studies*, Vol. 14 (1960), pp. 148–162.

[12] J. M. Callies, "Utilisation de modèles mathématiques pour l'estimation des données démographiques dans les pays en voie de developpement," *Rev. Inst. Internat. Statist.*, Vol. 34 (1966), pp. 341–359.

[13] C. Chandrasekaran and P. P. Talwar, "Forms of age-specific birth rates by orders of birth in an Indian community," *Eugenics Quarterly*, Vol. 15 (1968), pp. 264–272.

[14] C. L. Chiang, "On regular best asymptotically normal estimates," *Ann. Math. Statist.*, Vol. 27 (1956), pp. 336–351.

[15] ———, *Introduction to Stochastic Processes in Biostatistics*, New York, Wiley, 1968.

[16] H. Cramér and H. Wold, "Mortality variations in Sweden. A study in graduation and forecasting," *Skand. Aktuarietidskr.*, Vol. 18 (1935), pp. 161–240.

[17] M. Davies, "Linear approximation using the criterion of least total deviations," *J. Roy. Statist. Soc. Ser. B*, Vol. 29 (1967), pp. 101–109.

[18] J. M. Dickey, "Smoothed estimates for multinomial cell probabilities," *Ann. Math. Statist.*, Vol. 39 (1968), pp. 561–566.

[19] ———, "Smoothing by cheating," *Ann. Math. Statist.*, Vol. 40 (1969), pp. 1477–1482.

[20] H. F. Dorn, "Methods of analysis for follow-up studies," *Human Biology*, Vol. 22 (1950), pp. 238–248.

[21] L. G. Du Pasquier, "Mathematische Theorie der Invaliditätsversicherung," *Mitt. Verein. Schweiz. Versich.-Math.*, Vol. 7 (1912), pp. 1–7, and Vol. 8 (1913), pp. 1–153.

[22] W. P. Elderton and N. L. Johnson, *Systems of Frequency Curves*, London, Cambridge University Press, 1969.

[23] G. U. FENSTAD, "A note on a theorem of Barankin and Gurland," University of Oslo, Institute of Mathematics, Statistical Research Report, No. 2, 1965.

[24] T. S. FERGUSON, "A method of generating best asymptotically normal estimates with application to the estimation of bacterial densities," *Ann. Math. Statist.*, Vol. 29 (1953), pp. 1046–1062.

[25] E. FIX and J. NEYMAN, "A simple stochastic model of recovery, relapse, death and loss of patients," *Human Biology*, Vol. 23 (1951), pp. 205–241.

[26] D. A. S. FRASER, *Nonparametric Methods in Statistics*, New York, Wiley, 1957.

[27] E. GILJE, "Fitting curves to age-specific fertility rates. Some examples," *Statistical Review*, Ser. III, Vol. 7 (1969), pp. 118–134 (Swedish Central Bureau of Statistics, Stockholm).

[28] E. GILJE and L. YNTEMA, "The shifted Hadwiger fertility function," to appear.

[29] U. GRENANDER, "On the theory of mortality measurement," *Skand. Aktuarietidskr.*, Vol. 39 (1956), pp. 70–96 and 125–153.

[30] R. D. GROVE and A. M. HETZEL, *Vital Statistics Rates in the United States, 1940–1960*, Washington, D.C., National Center for Health Statistics, U.S. Department of Health, Education and Welfare, 1968.

[31] H. HADWIGER, "Eine analytische Reproduktions-funktion für biologische Gesamtheiten," *Skand. Aktuarietidskr.*, Vol. 23 (1940), pp. 101–113.

[32] T. E. HARRIS, P. MEIER, and J. W. TUKEY, "Timing of the distribution of events between observations," *Human Biology*, Vol. 22 (1950), pp. 249–270.

[33] L. HENRY, "D'un problème fondamental de l'analyse démographique," *Population*, Vol. 14 (1959), pp. 9–32.

[34] A. B. HOADLEY, "Use of the Pearson densities for approximating a skew density whose left terminal and first three moments are known," *Biometrika*, Vol. 55 (1968), pp. 559–563.

[35] J. M. HOEM, "Markov chain models in life insurance," *Blätter der Deutschen Gesellschaft für Versicherungsmathematik (Deutscher Aktuarverein) E. V.*, Vol. 9 (1969), pp. 91–107.

[36] ———, "Fertility rates and reproduction rates in a probabilistic setting," *Biométr.-Praxim.*, Vol. 10 (1969), pp. 38–66.

[37] ———, "A probabilistic model for primary marital fertility," *Yearbook of Population Research in Finland*, Vol. 11 (1969), pp. 73–86.

[38] ———, "Probabilistic fertility models of the life table type," *Theoretical Population Biology*, Vol. 1 (1970), pp. 12–38.

[39] ———, "Some results concerning the number of jumps in a finite time interval of a Markov chain with a continuous time parameter," University of Oslo, Institute of Economics, Memorandum, February 2, 1970.

[40] ———, "Point estimation of forces of transition in demographic models," *J. Roy. Statist. Soc. Ser. B*, Vol. 32 (1970), to appear.

[41] ———, "A probabilistic approach to nuptiality," *Biométr.-Praxim.*, to appear.

[42] L. HØYLAND, "Estimation in follow-up studies," University of Oslo, Institute of Mathematics, Statistical Research Report, No. 4, 1967.

[43] H. JECKLIN and P. STRICKLER, "Wahrscheinlichkeitstheoretische Begründung mechanischer Ausgleichung und deren praktische Anwendung," *Mitt. Verein. Schweiz. Versich.-Math.*, Vol. 54 (1954), pp. 125–161.

[44] D. A. JONES, "Bayesian statistics," *Trans. Soc. Actuaries*, Vol. 17 (1965), pp. 33–57.

[45] N. KEYFITZ, *Introduction to the Mathematics of Population*, Reading, Addison-Wesley, 1968.

[46] G. S. KIMELDORF and D. A. JONES, "Bayesian graduation," *Trans. Soc. Actuaries*, Vol. 19 (1967), pp. 66–112.

[47] YU. L. KRUOPIS, "On estimates of transition intensities with migration," *Theor. Probability Appl.*, Vol. 14 (1969), pp. 219–228.

[48] I. LAH, "Generalization of Yastremsky's formula for analytical graduation of fertility rates," *J. Roy. Statist. Soc. Ser. A*, Vol. 121 (1958), pp. 100–104.

[49] ———, "Analytische Ausgleichung der aus den Ergebnissen der Volkszählungen berechneten demographischen Tafeln," *International Population Conference, Wien, 1959*, International Union for the Scientific Study of Population, pp. 192–201.

[50] ——, "Analytically graduated fertility of married women in Australia with respect to the duration of marriage," *Contributed Papers, Sydney Conference, 1967*, International Union for the Scientific Study of Population, pp. 266–276.

[51] L. LeCam, "Asymptotic least squares theory," mimeographed notes, University of California, Berkeley, 1955.

[52] S. Lipton and C. A. McGilchrist, "The derivation of methods for fitting exponential regression curves," *Biometrika*, Vol. 51 (1964), pp. 504–507.

[53] A. J. Lotka, *Théorie Analytique des Associations Biologiques. Part II. Analyse Démographique avec Application Particulière à l'Espèce Humaine*, Paris, Hermann & Cie, 1939.

[54] D. P. Mazur, "The graduation of age-specific fertility rates by order of birth of child," *Human Biology*, Vol. 39 (1967), pp. 53–64.

[55] M. D. Miller et al., *Elements of Graduation*, New York, Actuarial Society of America, 1942.

[56] S. Mitra, "The pattern of age-specific fertility rates," *Demography*, Vol. 4 (1967), pp. 894–906.

[57] H. D. Patterson, "A further note on a simple method for fitting an exponential curve," *Biometrika*, Vol. 47 (1960), pp. 177–180.

[58] E. S. Pearson and H. O. Hartley, *Biometrika Tables for Statisticians*, Cambridge University Press, 1966.

[59] W. Saxer, *Versicherungsmathematik. Zweiter Teil*, Berlin, Springer-Verlag, 1958.

[60] B. Schneider, "Die Bestimmung der Parameter im Ertragsgesetz von E. A. Mitscherlich," *Biometr. Z.*, Vol. 5 (1963), pp. 78–95.

[61] H. L. Seal, "Tests of a mortality table graduation," *J. Inst. Actuaries*, Vol. 71 (1941), pp. 5–67.

[62] M. C. Sheps, "Characteristics of a ratio used to estimate failure rates: occurrences per person year of exposure," *Biometrics*, Vol. 22 (1966), pp. 310–321.

[63] ——, "On the person year concept in epidemiology and demography," *Milbank Memorial Fund Q.*, Vol. 44 (1966), pp. 69–91.

[64] M. C. Sheps, J. A. Menken, and A. P. Radick, "Probability models for family building, an analytical review," *Demography*, Vol. 6 (1969), pp. 161–183.

[65] M. C. Sheps and J. A. Menken, "On closed and open birth intervals in a stable population," paper prepared for the Segunda Conferencia Regional de Poblacion, Mexico City, August, 1970.

[66] W. Simonsen, *Forsikringsmatematik Hefte I og II*, Kφbenhavns Universitets Fond til Tilvejebringelse af Laeremidler, 1966–67.

[67] M. Spiegelman, *Introduction to Demography*, Cambridge, Harvard University Press, 1968 (revised ed.).

[68] W. L. Stevens, "Asymptotic regression," *Biometrics*, Vol. 7 (1951), pp. 247–267.

[69] E. Sverdrup, "Basic concepts in life assurance mathematics," *Skand. Aktuarietidskr.*, Vol. 35 (1952), pp. 115–131.

[70] ——, "Estimates and test procedures in connection with stochastic models for deaths, recoveries and transfers between different states of health," *Skand. Aktuarietidskr.*, Vol. 46 (1965), pp. 184–211.

[71] ——, *Laws and Chance Variations*, Vol. 1, Amsterdam, North Holland, 1967.

[72] P. P. Talwar, "Age patterns of fertility," University of North Carolina, Chapel Hill, Institute of Statistics Mimeo Series, No. 656, 1970.

[73] K. Tekse, "On demographic models of age-specific fertility rates," *Statistical Review, Ser. III*, Vol. 5 (1967), pp. 189–207 (Swedish Central Bureau of Statistics, Stockholm).

[74] E. J. Wegman, "Maximum likelihood histograms," University of North Carolina, Chapel Hill, Institute of Statistics Mimeo Series, No. 629, 1969.

[75] ——, "Nonparametric probability density estimation," University of North Carolina, Chapel Hill, Institute of Statistics Mimeo Series, No. 638, 1969.

[76] K. Weichselberger, "Über eine Theorie der gleitende Durchschnitte und verschiedene Anwendungen dieser Theorie," *Metrika*, Vol. 8 (1964), pp. 185–230.

[77] S. D. Wicksell, "Nuptiality, fertility, and reproductivity," *Skand. Aktuarietidskr.*, Vol. 14 (1931), pp. 125–157.

[78] R. A. WIJSMAN, "On the theory of BAN-estimates," *Ann. Math. Statist.*, Vol. 30 (1959), pp. 185–191. (Correction note, pp. 1268–1270.)

[79] H. H. WOLFENDEN, *Fundamental Principles of Mathematical Statistics with Special Reference to the Requirements of Actuaries and Vital Statisticians, and an Outline of a Course in Graduation*, New York, Actuarial Society of America, 1942.

[80] L. YNTEMA, "The graduation of net fertility tables," *Bolm Instit. Actuários Portugueses*, Vol. 8 (1953), pp. 29–43.

[81] ———, "On Hadwiger's fertility function," *Statistical Review*, Ser. III, Vol. 7 (1969), pp. 113–117 (Swedish Central Bureau of Statistics, Stockholm).

[82] S. ZAHL, "A Markov process model for follow-up studies," *Human Biology*, Vol. 27 (1955), pp. 90–120.

[83] E. ZWINGGI, *Versicherungsmathematik*, Basel, Birkhäuser, 1958.

GENERALIZED INVERSE OF A MATRIX AND ITS APPLICATIONS

C. RADHAKRISHNA RAO

and

SUJIT KUMAR MITRA

INDIAN STATISTICAL INSTITUTE

1. Introduction

The concept of an inverse of a singular matrix seems to have been first introduced by Moore [1], [2] in 1920. Extensions of these ideas to general operators have been made by Tseng [3], [4], [5], but no systematic study of the subject was made until 1955 when Penrose [6], [7], unaware of the earlier work, redefined the Moore inverse in a slightly different way. About the same time one of the authors, Rao [8], gave a method of computing what is called a pseudoinverse of a singular matrix, and applied it to solve normal equations with a singular matrix in the least squares theory and to express the variances of estimators. The pseudoinverse defined by Rao did not satisfy all the restrictions imposed by Moore and Penrose. It was therefore different from the Moore–Penrose inverse, but was useful in providing a general theory of least squares estimation without any restriction on the rank of the observational equations. In a later paper, Rao [9] showed that an inverse with a much weaker definition than that of Moore and Penrose is sufficient in dealing with problems of linear equations. Such an inverse was called a generalized inverse (g inverse) and its applications were considered by Rao in [10], [11], [12], [13], and [14].

Some of the principal contributors to the subject since 1955 are Greville [15], Bjerhammer [16], [17], [18], Ben-Israel and Charnes [19], Chipman [20], [21], Chipman and Rao [22], and Scroggs and Odell [23]. Bose [24] mentions the use of g inverse in his lecture notes, "Analysis of Variance" [24]. Bott and Duffin [25] defined what is called a constrained inverse of a square matrix, which is different from a g inverse and is useful in some applications. Chernoff [26] considered an inverse of a singular nonnegative definite (n.n.d.) matrix, which is also not a g inverse but is useful in discussing some estimation problems.

The g inverse satisfying the weaker definition given by Rao [9] is not unique and thus presents an interesting study in matrix algebra. In a publication in 1967 [27], Rao showed how a variety of g inverses could be constructed to suit different purposes and presented a classification of g inverses. The work was later pursued by Mitra [28], [29], who introduced some new classes of g inverses, and Mitra and Bhimasankaram [30], [31]. Further applications of g inverses were considered in a series of papers, Mitra and Rao [32], [33], [34], and Rao [35].

In the present paper we discuss a calculus of g inverses and show how it provides an elegant tool for the discussion of the Gauss-Markov problem of linear estimation, multivariate analysis when the variables have a singular covariance matrix, maximum likelihood estimation when the information matrix is singular, and so forth.

A systematic development of the calculus of generalized inverses and their applications are given in a forthcoming book by the authors, entitled *Generalized Inverse of Matrices and Its Applications* (Wiley, 1971).

2. Generalized inverse of a matrix

If A is an $m \times m$ nonsingular matrix, then there exists an inverse A^{-1} with the property $AA^{-1} = A^{-1}A = I$. If A is an $m \times n$ rectangular matrix with rank $n \leq m$ then $(A*A)^{-1}$ exists, and defining $A_L^{-1} = (A*A)^{-1}A*$ we find that $A_L^{-1}A = I$. In such a case A_L^{-1} is called a left inverse of A. Similarly a right inverse of A exists if its rank is $m \leq n$ with the property $AA_R^{-1} = I$. When A^{-1}, A_L^{-1}, or A_R^{-1} exists we can express a solution of the equation $Ax = y$ in the form $x = A^{-1}y$ or $A_L^{-1}y$, or $A_R^{-1}y$. When such inverses do not exist, can we represent a solution of the consistent equation $Ax = y$ (where A may be rectangular or a square singular) in the form $x = Gy$? If such a G exists, we call it a generalized inverse of A, and represent it by A^-.

We provide three equivalent definitions of a g inverse.

DEFINITION 2.1.   *An $n \times m$ matrix G is said to be a g inverse of an $m \times n$ matrix A if $x = Gy$ is a solution to the equation $Ax = y$ for any y such that the equation $Ax = y$ is consistent.*

DEFINITION 2.2.   *G is a g inverse of A if $AGA = A$.*

DEFINITION 2.3.   *G is a g inverse of A if AG is idempotent and $R(AG) = R(A)$ or GA is idempotent and $R(GA) = R(A)$, where $R(\cdot)$ denotes the rank of the matrix.*

A matrix G satisfying any one of these definitions is denoted by A^- and is called a g inverse. The following theorems establish the existence of A^- and its applications in solving equations, obtaining projections, and so forth. The proofs of some of these theorems are omitted as they are contained in Rao [27], and proofs of other theorems will appear in the forthcoming book by the authors, already cited.

THEOREM 2.1.   *Let A be an $m \times n$ matrix. Then A^- exists. The entire class of g inverses is generated from any given inverse A^- by the formula*

$$(2.1) \qquad\qquad A^- + U - A^- AUA A^-$$

where U is arbitrary, or by the formula

$$(2.2) \qquad\qquad A^- + V(I - AA^-) + (I - A^-A)W$$

where V and W are arbitrary. Further a matrix is uniquely determined by the class (2.1) or (2.2) of its g inverses.

THEOREM 2.2.   *Let $Ax = y$ be a consistent equation and A^- be a g inverse of A.*

(i)   *Then $x = A^-y$ is a solution.*

(ii)   *The class of all solutions is provided by $A^-y + (I - A^-A)z$, z arbitrary.*

(iii)   *Let q be an n vector. Then $q'x$ has a unique value for all solutions x of $Ax = y$ if and only if $q' = q'A^-A$ or $q \in \mathcal{M}(A')$, the vector space generated by the columns of A'.*

THEOREM 2.3.   *Let A be an $m \times n$ matrix and $\mathcal{M}(A) \subset \mathscr{E}^m$. The projection operator P onto $\mathcal{M}(A)$ can be expressed in the form*

$$(2.3) \qquad\qquad P = A(A^*MA)^-A^*M,$$

where the inner product in $\mathscr{E}^m$ is defined as $(y, x) = x^*My$, M being a positive definite matrix and $(A^*MA)^-$ is any g inverse of A^*MA. Further P is unique for any choice of $(A^*MA)^-$.

It would be useful to recognize the situations in which a g inverse behaves like a regular inverse. Theorem 2.4 contains the main result in this direction.

THEOREM 2.4.   *A necessary and sufficient condition that $BA^-A = B$ is that $B = DA$ for some D. Similarly for $B = AA^-B$ to hold, it is necessary and sufficient that $B = AD$.*

The following results are consequences of Theorem 2.4:

(a) $A(A^*A)^-(A^*A) = A$;

(b) $(A^*A)(A^*A)^-A^* = A^*$;

(c) $A(A^*VA)^-(A^*VA) = A$ and $(A^*VA)(A^*VA)^-A^* = A^*$ for any matrix V such that $R(A^*VA) = R(A)$;

(d) $A(A^*VA)^-A^*$ is invariant for any choice of $(A^*VA)^-$ and is of rank equal to the rank of A if $R(A^*VA) = R(A)$. Further, $A(A^*VA)^-A^*$ is hermitian if A^*VA is hermitian.

We provide a decomposition theorem involving g inverses of matrices which has a number of applications.

Let A be a matrix of order $m \times n$, and let A_i, B_i be matrices of order $m \times p_i$, $n \times q_i$, $i = 1, \cdots, k$. Write $A = (A_1 \vdots \cdots \vdots A_k)$ and $B = (B_1 \vdots \cdots \vdots B_k)$. Consider the following statements:

$$(2.4) \qquad\qquad A_i^*\Lambda B_j = 0 \qquad \text{for all} \quad i \neq j.$$

$$(2.5) \qquad\qquad G = \sum_i B_i(A_i^*\Lambda B_i)^-A_i^*$$

is a g inverse of Λ where $(A_i\Lambda B_i)^-$ is any g inverse.

THEOREM 2.5.   (i) *Statement (2.4) implies statement (2.5) if and only if $R(A^*\Lambda B) = R(\Lambda)$.*

(ii) *Statement (2.5) implies statement (2.4) if and only if $\Sigma\, R(A_i^*\Lambda) = \Sigma\, R(\Lambda B_i) = R(\Lambda)$.*

An interesting corollary to Theorem 2.5 is the following.

COROLLARY 2.1.   *Let A_i be an $m \times p_i$ matrix of rank r_i, $i = 1, \cdots, k$ such that $\Sigma\, r_i = m$. Further, let Λ be a positive definite (p.d.) matrix. Then the following two statements are equivalent:*

(2.6) $$A_i^* \Lambda A_j = 0 \qquad \text{for all} \quad i \neq j,$$

(2.7) $$\Lambda^{-1} = \Sigma \, A_i (A_i^* \Lambda A_i)^- A_i^*.$$

The true inverse of a nonsingular square matrix has the property that the inverse of the inverse is equal to the original matrix. This may not hold for any g inverse as defined in this section. We shall however show that a subclass of g inverses possesses an analogous property. We give the following definition.

DEFINITION 2.4. *An $n \times m$ matrix G is said to be a reflexive g inverse of an $m \times n$ matrix A if*

(2.8) $$A G A = A \quad \text{and} \quad G A G = G.$$

We use the notation A_n^- to denote a reflexive g inverse.

THEOREM 2.6. *Any two of the following conditions imply the third*
  (i) $A = AGA$,
  (ii) $G = GAG$,
  (iii) $R(G) = R(A)$.

For a proof of this theorem see Mitra [27]. It is seen that a reflexive g inverse could be equivalently defined by any two of the conditions (i), (ii) and (iii) in the theorem. Frame [36] uses the term *semi-inverse* to denote a matrix G obeying (i) and (iii).

3. Three basic types of g inverses

3.1. *Mathematical preliminaries and notations.* Let $\mathscr{E}^m$ represent an m dimensional vector space furnished with an inner product. The symbol (x, y) is used to denote the inner product between vectors x and y. The norm of a vector x is denoted by $\|x\| = [(x, x)]^{1/2}$.

Let A be an $m \times n$ matrix mapping vectors of $\mathscr{E}^n$ into $\mathscr{E}^m$. The adjoint of A denoted by $A^{\#}$ is defined by

(3.1) $$(Ax, y)_m = (x, A^{\#} y)_n$$

where $(\cdot, \cdot)_m$ and $(\cdot, \cdot)_n$ denote inner products in $\mathscr{E}^m$ and $\mathscr{E}^n$, respectively. If $(y, x)_m = x^* M y$ and $(y, x)_n = x^* N y$ where M and N are positive definite matrices and $*$ denotes the conjugate transpose of a matrix, then relation (3.1) reduces to

(3.2) $$x^* A^* M y = x^* N A^{\#} y \Rightarrow N A^{\#} = A^* M.$$

If A is an $m \times m$ square matrix mapping E^m into E^m, then $M A^{\#} = A^* M$.

We denote P_B the projection operator onto the space $\mathscr{M}(B)$ generated by the columns of B. It is characterized by the conditions:
  (a) it is idempotent $P_B P_B = P_B$;
  (b) it is selfadjoint $P_B = P_B^{\#}$.
If the inner product $(y, x)_m = x^* M y$, then condition (b) is equivalent to $M P_B$ being hermitian.

3.2. *The g inverse for minimum norm solution.* It has been shown that $x = Gy$ is a solution of the consistent equation $Ax = y$ for any g inverse G of A (that is, satisfying the condition $AGA = A$), and the general solution is $x = Gy + (I - GA)z$ where z is arbitrary from, Theorem 2.2 (ii). We raise the question whether there exists a choice of G independently of y such that the solution Gy has a minimum norm in the class of all solutions of $Ax = y$. If such a G exists

$$(3.3) \qquad \|Gy\| \leq \|Gy + (I - GA)z\| \qquad \text{for all } z \text{ and } y \in \mathscr{M}(A),$$

that is,

$$(3.4) \qquad \|GAx\| \leq \|GAx + (I - GA)z\| \qquad \text{for all } z \text{ and } x.$$

This implies $(GAx, (I - GA)z) = 0$ for all z and x which implies in turn

$$(3.5) \qquad (GA)^{\#}(I - GA) = 0 \quad \text{or} \quad (GA)^{\#} = (GA).$$

We now state the conditions for a g inverse G to provide a minimum norm solution of a consistent equation $Ax = y$.

THEOREM 3.1. *Let $Ax = y$ be a consistent equation and G be a g inverse of A such that Gy is a minimum norm solution. Then it is necessary and sufficient that any one of the following equivalent conditions is satisfied:*
  (i) $AGA = A, (GA) = (GA)^{\#}$,
  (ii) $AGA = A, (GA)^*N = N(GA)$ *if* $(y, x)_n = x^*Ny$,
  (iii) $GA = P_{A^{\#}}$.
Condition (i) is already established and the equivalences of conditions (ii) and (iii) with (i) follow from the definitions of adjoint and projection operators.

We denote a g inverse which provides a minimum norm solution of $Ax = y$ by A_m^- or more explicitly $A_{m(N)}^-$, where N defines the inner product as in condition (ii), and refer to it as minimum N norm g inverse. Such an inverse exists; for example, $G = N^{-1}A^*(AN^{-1}A^*)^-$ satisfies the conditions of the theorem for any choice of the g inverse $(AN^{-1}A^*)^-$.

3.3. *The g inverse for a least square solution.* Let $Ax = y$ be an inconsistent equation in which case we seek a least squares solution by minimizing $\|Ax - y\|$. We raise the question whether there exists a matrix G such that $x = Gy$ is a least squares solution. If such a G exists

$$(3.6) \qquad \|AGy - y\| \leq \|Ax - y\| \quad \text{for all } x, y.$$

This implies $(Aw, (AG - I)y) = 0$ for all y, and $w = x - Gy$ implies $A^{\#}(AG - I) = 0$. Thus

$$(3.7) \qquad AG = (AG)^{\#}, \qquad AGA = A.$$

THEOREM 3.2. *Let $Ax = y$ be a possibly inconsistent equation, and let G be a matrix such that Gy is a least squares solution of $Ax = y$. Then each of the following equivalent conditions is necessary and sufficient:*
  (i) $AGA = A, \quad (AG) = (AG)^{\#}$,
  (ii) $AGA = A, \quad (AG)^*M = M(AG)$ *if* $(y, x)_m = x^* My$,

(iii) $AG = P_A$.

Condition (i) is already established, and the equivalences of conditions (ii) and (iii) with (i) follow from the definitions. We denote a g inverse which provides a least squares solution of $Ax = y$ by A_l^- or more explicitly $A_{l(M)}^-$, where M defines the inner product as in condition (3.10), and refer to it as an M least squares g inverse. Such an inverse exists; for example, $G = (A^*MA)^- A^*M$ satisfies the condition of the theorem.

3.4. *The g inverse for minimum norm least squares solution.*   A least squares solution of an inconsistent equation $Ax = y$ may not be unique, in which case we may seek for a matrix G such that Gy has minimum norm in the class of least squares solutions. If such a G exists,

(3.8) $\|Gy\|_n \leq \|\xi\|_n, \{\xi : \|A\xi - y\|_m \leq \|Ax - y\|_m \text{ for all } x\}$ for all y

where $\|\cdot\|_m$ and $\|\cdot\|_n$ denote norms in $\mathscr{E}^m$ and $\mathscr{E}^n$, respectively. The condition (3.8) may be written

(3.9) $\|Gy\| \leq \|\xi\|,$ $\{\xi : A^\# A\xi = A^\# y\}$ for all y.

This implies $A^\# (I - AG) = 0$ and $G^\# (I - GA) = 0$ which in turn imply

(3.10) $AGA = A, (AG) = (AG)^\#,$ $GAG = G,$ $(GA) = (GA)^\#$.

THEOREM 3.3.   *Let $Ax = y$ be a possibly inconsistent equation and $x = Gy$ be a minimum norm least squares solution. Then each of the following equivalent conditions is necessary and sufficient:*

(i) $AGA = A,$ $GAG = G,$ $AG = (AG)^\#,$ $GA = (GA)^\#$;

(ii) $AGA = A,$ $GAG = G,$ $(AG)^*M = MAG,$ $(GA)^*N = NGA$,

when $(y, x)_m = x^* My$ *and* $(y, x)_n = x^* Ny$;

(iii) $AG = P_A,$ $GA = P_G$.

A matrix G satisfying any one of the above conditions is unique.

Conditions (i) are already established and the equivalences of (ii) and (iii) with (i) follow from the definitions. The uniqueness of G follows from the fact that a minimum norm solution of a linear equation is unique.

We denote the g inverse which provides a minimum norm least squares solution of $Ax = y$ by A^+ or more explicitly by A_{MN}^+, where M, N are matrices defining the inner products in $\mathscr{E}^m, \mathscr{E}^n$ as in condition (ii). Such an inverse exists; for example

(3.11) $G = A^\# A(A^\# AA^\# A)^- A^\# = A^\# (A^\# AA^\#)^- A^\# = P_{A^\#} A^- P_A$

satisfies the conditions of Theorem 3.3. We refer to A_{MN}^+ as the minimum N norm M least squares g inverse.

3.5. *Duality relationships between different g inverses.*   An important theorem which establishes a duality relationship between minimum norm and least squares g inverses and which plays a key role in the Gauss-Markov theory of linear estimation is as follows.

THEOREM 3.4.   *Let A be an $m \times n$ matrix and $(y, x)_m = x^* My$. Then*

(3.12) $(A^*)_{m(M)} = [A_{l(M^{-1})}^-]^*$.

PROOF. Let G be a minimum M norm g inverse of A^*. Then using condition (ii) of Theorem 3.1,

$$(3.13) \qquad A^*GA^* = A^*, \qquad (GA^*)^*M = MGA^*.$$

Taking transposes and rewriting, (3.13) becomes

$$(3.14) \qquad AG^*A = A, \qquad (AG^*)^*M^{-1} = M^{-1}(AG^*)$$

which, using condition (ii) of Theorem 3.2, shows that G^* is an M^{-1} least squares g inverse of A. Then equation (3.12) is true. The duality result (3.12), in the special case when $M = I$, is also noted by Sibuya [37].

Another important theorem which has application in linear estimation is as follows.

THEOREM 3.5. Let A be an $m \times n$ matrix and $(y, x)_m = x^*My$ and $(y, x)_n = x^*Ny$. Then,

$$(3.15) \qquad (A^*)^+_{NM} = (A^+_{M^{-1}N^{-1}})^*.$$

The result follows from condition (ii) of Theorem 3.3.

The different types of g inverses considered in Sections 2 and 3 and the properties characterising them are given in Table I.

TABLE I

SOME TYPES OF g INVERSES

P_X denotes projection operator onto $\mathcal{M}(X)$, # denotes adjoint.

Symbol	Equivalent Conditions	Purpose
A^-	$AGA = A$	solving consistent equations
A_r^-	$AGA = A, GAG = G$	solving consistent equations
A_m^-	(i) $AGA = A, (GA)^\# = GA$	minimum norm solution
	(ii) $GA = P_{A^\#}$	
A_l^-	(i) $AGA = A, (AG)^\# = AG$	least squares solution
	(ii) $AG = P_A$	
A^+	(i) $AGA = A, GAG = G$	minimum norm least squares solution
	$(GA)^\# = GA, (AG)^\# = AG$	
	(ii) $AG = P_A, GA = P_G$	

In Theorems 3.1 to 3.3, we used norms defined by p.d. matrices M and N. We can extend the results to cases where M and N are n.n.d. matrices. In such a case we will be minimizing seminorms. Some results in this direction will appear in a forthcoming paper in Sankhyá.

3.6. *Singular value decomposition.* Let A be an $m \times n$ matrix of rank r and M and N are p.d. matrices of order m and n, respectively. Then A can be expressed in the form

$$(3.16) \qquad MAN = a_1\xi_1\eta_1^* + \cdots + a_r\xi_r\eta_r^*,$$

where $a_1^2, \cdots, a_r^2$ are the nonzero eigenvalues of A^*MA with respect to N^{-1}

or of ANA^* with respect to M^{-1}; ξ_i is the eigenvector of ANA^* with respect to M^{-1} corresponding to the eigenvalue a_i^2; and η_i is the eigenvector of A^*MA with respect to N^{-1} corresponding to the eigenvalue a_i^2. The representation (3.16) is called the singular value decomposition of A with respect to M and N. Using such a decomposition we can compute A_{MN}^+ as

$$(3.17) \qquad A_{MN}^+ = a_1^{-1}\eta_1\xi_1^* + \cdots + a_r^{-1}\eta_r\xi_r^*.$$

4. Constrained inverse

Bott and Duffin [25] introduced what is called a constrained inverse of a square matrix, which is different from a g inverse, and considered its application in mechanics and in network theory. In this section we extend the concept of a constrained inverse to a general matrix and give some applications.

Let A be a matrix of order $m \times n$, $\mathscr{V}$ and $\mathscr{U}$ be subspaces in $\mathscr{E}^n$ and $\mathscr{E}^m$, respectively. In what follows we shall impose constraints of two different types to define a constrained inverse G of A.

CONSTRAINTS OF TYPE 1.

c: G maps vectors of $\mathscr{E}^m$ into $\mathscr{V}$;

r: G^* maps vectors of $\mathscr{E}^n$ into $\mathscr{U}$.

CONSTRAINTS OF TYPE 2.

C: GA is an identity in $\mathscr{V}$,

R: $(AG)^*$ is an identity in $\mathscr{U}$.

Inverses obtained by choosing various combinations of these constraints are listed below in Table II along with necessary and sufficient conditions for existence, and explicit forms, where F and E are matrices such that $\mathscr{V} = \mathscr{M}(E)$ and $\mathscr{U} = \mathscr{M}(F^*)$.

TABLE II

CONSTRAINED INVERSES OF VARIOUS TYPES

V and U are arbitrary matrices.

Notation	N.S. Condition for Existence	Algebraic Expression	Reference to Theorem
A_{cC}	$R(AE) = R(E)$	$E(AE)^-$	4.1
A_{rR}	$R(FA) = R(F)$	$(FA)^-F$	4.3
A_{cR}	$R(FAE) = R(F)$	$E(FAE)^-F + E[I - (FAE)^-FAE]U$	4.5
A_{rC}	$R(FAE) = R(E)$	$E(FAE)^-F + V[I - FAE(FAE)^-]F$	4.6
A_{crCR}	$R(FAE) = R(F) = R(E)$	$E(FAE)^-F$	4.7

THEOREM 4.1. A_{cC} exists if and only if $R(AE) = R(E)$. In such a case A_{cC} is of the form $E(AE)^-$.

PROOF. Using constraint c, $G = EX$ for some matrix X. Then constraint C gives

$$(4.1) \qquad E X A E = E.$$

Equation (4.1) is solvable only if $R(AE) = R(E)$ in which case, (4.1) is equivalent to $AEXAE = AE$, or

$$(4.2) \qquad X = (AE)^- \Rightarrow G = E(AE)^-.$$

The "if" part is trivial.

THEOREM 4.2. *A is a g inverse of A_{cC} but not necessarily the other way. A_{cC} is a g inverse of A if and only if $R(AE) = R(A)$.*

PROOF. Theorem 4.2 follows from Theorems 4.1 and 2.4. Theorems 4.3 and 4.4 follow on similar lines.

THEOREM 4.3. *A_{rR} exists if and only if $R(FA) = R(F)$. In such a case A_{rR} is of the form $(FA)^-F$.*

THEOREM 4.4. *A is a g inverse of A_{rR} but not necessarily the other way. A_{rR} is a g inverse of A if and only if $R(FA) = R(A)$.*

THEOREM 4.5. *A_{cR} exists if and only if $R(FAE) = R(F)$. In such a case A_{cR} is of the form*

$$(4.3) \qquad E(FAE)^-F + E[I - (FAE)^-FAE]U,$$

where U is arbitrary.

PROOF. Using constraint c, $G = EX$, for some matrix X. Then constraint R gives

$$(4.4) \qquad FAEX = F.$$

Equation (4.4) is solvable if and only if $R(FAE) = R(F)$, in which case a general solution is given by

$$(4.5) \qquad X = (FAE)^-F + [I - (FAE)^-FAE]U,$$

where U is arbitrary. The "if" part is easy. Thus Theorem 4.5 is established. Theorem 4.6 can be proved on similar lines.

THEOREM 4.6. *A_{rC} exists if and only if $R(FAE) = R(E)$. In such a case A_{rC} is of the form,*

$$(4.6) \qquad E(FAE)^-F + V[I - FAE(FAE)^-]F,$$

where V is arbitrary.

THEOREM 4.7. *A_{crCR} exists if and only if $R(FAE) = R(F) = R(E)$. In such a case A_{crCR} is unique and is given by the expression $E(FAE)^-F$.*

PROOF. The "if" part is trivial. The necessity of the rank condition follows as in Theorems 4.5 and 4.6. The uniqueness follows, since under the condition $R(FAE) = R(F) = R(E)$ both A_{cR} and A_{rC} are uniquely determined by the expression $E(FAE)^-F$. Look for example at the expression (4.3), for A_{cR} and check that when $R(FAE) = R(E)$,

$$(4.7) \qquad FAE[I - (FAE)^-FAE] = 0 \Rightarrow E[I - (FAE)^-FAE] = 0.$$

NOTE 1. Let E_1 and F_1 be matrices such that $\mathcal{M}(E_1) = \mathcal{M}(E)$ and $\mathcal{M}(F_1) = \mathcal{M}(F)$, where F and E are as defined in Theorem 4.1. Then

(4.8) $$R(AE) = R(E) \Rightarrow R(AE_1) = R(E_1),$$

(4.9) $$R(FAE) = R(E) \Rightarrow R(F_1AE_1) = R(E_1),$$

(4.10) $$E_1(F_1AE_1)^- F_1 = E(FAE)^- F,$$

so that A_{crCR} is unique for any choice of the matrices generating the subspaces $\mathscr{V}$ and $\mathscr{U}$.

NOTE 2.   In particular let P and Q be projection operators onto $\mathscr{V}$ and $\mathscr{U}$, respectively. Then

(4.11) $$A_{crCR} = P(QAP)^- Q.$$

NOTE 3. A is g inverse of A_{crCR} but the converse is true only under the additional condition $R(FAE) = R(A)$.

NOTE 4.   When $\mathscr{V} = \mathscr{M}(A^{\#})$ and $\mathscr{U} = \mathscr{M}(A)$, A_{crCR} coincides with A_{MN}^+. It may be of some historical interest to observe that Moore [1], [2] introduced his general reciprocal of a matrix as a constrained inverse of the type we are considering in this section.

Now we consider the special case where A is an $m \times m$ (square) matrix and the subspaces $\mathscr{V}$ and $\mathscr{U}$ are the same and discuss it in some detail. The constrained inverse G in such a case may be defined by the following conditions:

(a) G^* maps vectors of $\mathscr{E}^m$ into the subspace $\mathscr{V} \subset \mathscr{E}^m$;
(b) GA is an identity in $\mathscr{V}$.

This is a special case of A_{rC}, but we shall represent a matrix G satisfying the above two conditions by T, following the notation used by Bott and Duffin. (In condition (i) above Bott and Duffin used G instead of G^*, which does not characterize the matrix T used by them. Their definition leads to an inverse of the type A_{cC} which is not unique and so on.)

THEOREM 4.8.   *Let E be a matrix such that $\mathscr{V} = \mathscr{M}(E)$. Then T exists if and only if $R(E^*AE) = R(E)$ in which case it is unique, and is of the form*

(4.12) $$T = E(E^*AE)^- E^*.$$

Further T is independent of the choice of E.

The proof is on the same lines as in Theorem 4.7.

THEOREM 4.9.   *Let P be the projection operator onto $\mathscr{V}$ and $R(PAP) = R(P)$. Then*

(4.13) $$T = P(PAP)^- P = P(AP + I - P)^{-1}.$$

PROOF.   The first part of equation (4.13) follows from Theorem 4.8 as we can choose E to be P. For the second part, it is easy to see that $(AP + I - P)$ is nonsingular and admits a regular inverse when $R(PAP) = R(P)$. Further

(4.14) $$[P(PAP)^- P - P(AP + I - P)^{-1}](AP + I - P) = 0$$

giving $P(PAP)^- P = P(AP + I - P)^{-1}$ which is the expression used by Bott and Duffin.

THEOREM 4.10. *Let A be an $m \times m$ matrix and T be the constrained inverse as obtained in Theorem 4.8. Then:*

(i) *any arbitrary vector h admits a unique decomposition $h = Au + w$, $u \in \mathscr{V}$ and $w \in \mathscr{V}^{\perp}$;*

(ii) *the quadratic function $Q = (v - e)^*A(v - e) - 2f^*v$, where e and f are given vectors, attains a stationary value for variations of v in $\mathscr{V}$. If A is an n.n.d. matrix, Q attains the minimum.*

PROOF. (i) Let $h = Au + w$. Multiplying by T on both sides $Th = TAu + Tw = u$. Then $w = h - ATh$. It is easily checked that $Th \in \mathscr{V}$ and $h - ATh \in \mathscr{V}^{\perp}$. Further, if $Au_1 + w_1$ is another decomposition, $0 = A(u - u_1) + w - w_1$. Multiplying both sides by T, $u - u_1 = 0$ and hence $w - w_1 = 0$, so that the decomposition is unique.

(ii) Substituting $v = v_0 + \delta$, $v_0 \in \mathscr{V}$, $\delta \in \mathscr{V}$, and retaining only linear terms in δ, the quadratic form becomes

(4.15) $\qquad (v_0 - e)^*A(v_0 - e) - 2f^*v_0 - 2\delta^*(f + Ae - Av_0).$

Then v_0 is a stationary point if $\delta^*(Av_0 - f - Ae) = 0$ or $Av_0 + w = Ae + f = h$, say, where $w \in \mathscr{V}^{\perp}$. Applying result (i) of the theorem, v_0 exists and has the value $v_0 = Th = T(Ae + f)$.

To show that Q attains a minimum at v_0 when A is n.n.d., let us observe that for any $v \in \mathscr{V}$,

(4.16) $\quad (v - e)^*A(v - e) - 2f^*v = (v_0 - e)^*A(v_0 - e) - 2f^*v_0$
$$+ (v - v_0)^*A(v - v_0).$$

This completes the proof of Theorem 4.10.

5. Method of least squares

We show how Theorem 3.4 expressing the duality between minimum norm and least squares inverses provides a simple and an elegant demonstration of the minimum variance property of least squares estimators in the Gauss-Markov model. It also shows how the least squares method comes in a natural way while seeking for minimum variance estimators.

The Gauss-Markov model is characterized by the triplet $(Y, X\beta, \Lambda)$ where Y is $n \times 1$ vector of random variables such that $E(Y) = X\beta$, $D(Y) = \Lambda$ (variance-covariance matrix of Y).

5.1. *Unbiased estimation.* Let $p'\beta$ be a parametric function where $p \in \mathscr{M}(X')$. We wish to find a linear function $L'Y$ of Y such that $E(L'Y) \equiv p'\beta$ and the variance $V(L'Y) = L'\Lambda L$ is a minimum. The condition on expectation gives that $L'X\beta \equiv p'\beta \Leftrightarrow X'L = p$. The equation $X'L = p$ is consistent and what we need is a minimum norm solution, norm being defined as $\|L\|^2 = L'\Lambda L$. The optimum value of L is obviously, using a minimum Λ norm g inverse

(5.1) $\qquad\qquad\qquad L = (X')^{-}_{mm(\Lambda)}p,$

giving the minimum variance linear estimator

$$(5.2) \qquad L'Y = p'[(X')^-_{mm(\Lambda)}]'Y = p'X^-_{l(\Lambda^{-1})}Y = p'\hat{\beta}$$

using the duality Theorem 3.4, where $\hat{\beta}$ is the Λ^{-1} least squares solutions of the equation $Y = X\beta$, that is, which minimizes

$$(5.3) \qquad \|Y - X\beta\|^2 = (Y - X\beta)^*\Lambda^{-1}(Y - X\beta).$$

5.2. *Minimum bias estimation.* If $p \notin \mathcal{M}(X')$, the parametric function $p'\beta$ does not admit an unbiased linear estimator. The magnitude of bias in $L'Y$ is $(X'L - p)'\beta$. The bias may be minimized by choosing L such that

$$(5.4) \qquad \|X'L - p\|^2 = (X'L - p)'N(X'L - p)$$

is a minimum, where N is a specified positive definite matrix. Subject to minimum bias we wish to minimize the variance $L'\Lambda L$. The problem then is that of finding minimum Λ norm N least squares solution of the equation $X'L = p$. Then the optimum value of L is

$$(5.5) \qquad L = (X')^+_{N\Lambda}p,$$

giving the least bias minimum variance linear estimator

$$(5.6) \qquad L'Y = p'[(X')^+_{N\Lambda}]'Y = p'X^+_{\Lambda^{-1}N^{-1}}Y = p'\hat{\beta}$$

using Theorem 3.5, where $\hat{\beta}$ is the N^{-1} norm Λ^{-1} least squares solution of the equation $Y = X\beta$.

6. Maximum likelihood estimation when the information matrix is singular

Let $p(x, \theta)$ be the probability density, where x stands for observed data and θ for n unknown parameters $\theta_1, \cdots, \theta_n$. Then

$$(6.1) \qquad L(\theta, x) = \log p(x, \theta)$$

as a function of θ for given x is known as the log likelihood of parameters. Let $f_i(x, \theta)$ or simply f_i be defined by

$$(6.2) \qquad f_i = \frac{\partial L}{\partial \theta_i}, \qquad\qquad i = 1, \cdots, n,$$

and let the vector $(f_1, \cdots, f_n)'$ be f. The information matrix on θ is defined by

$$(6.3) \qquad H = E(ff')$$

The maximum likelihood m.l. estimate of θ is usually obtained from the equation $f = 0$, and the asymptotic theory of estimation is well known when the matrix H is not singular in the neighborhood of the true value.

If $L(\theta, x)$ depends essentially on $s < n$ independent functions $\phi_1, \cdots, \phi_s$ of θ, then H becomes singular and not all the parameters $\theta_1, \cdots, \theta_n$ are estimable.

Only $\phi_1, \cdots, \phi_s$ and their functions are estimable. In such a case we can define the log likelihood $L(\theta, x)$ in terms of fewer parameters as $L(\phi, x)$ where $\phi' = (\phi_1, \cdots, \phi_s)$ such that J, the information matrix on ϕ, is nonsingular. Then the usual theory would apply. Of course, there is some arbitrariness in the choice of ϕ but this does not cause any trouble. However, the calculus of g inverses enables us to deal with the likelihood as a function of the original parameters and obtain their m.l. estimates and the associated asymptotic variance-covariance matrix. When all the parameters are not estimable, the individual estimates and the variance-covariance matrix so obtained are not meaningful, but they are useful in computing m.l. estimates and standard errors of estimable parametric functions. (We have learned from H. Rubin at the Symposium that he considered such an approach and obtained results similar to ours.)

6.1. *Method of scoring with a singular information matrix.* The m.l. estimates are obtained by solving the equations

$$(6.4) \qquad\qquad f_i(x, \theta) = 0, \qquad\qquad i = 1, \cdots, n.$$

The equations (6.4) are usually complicated in which case one obtains solutions by successive approximations using a technique such as Fishers' method of scoring (see Rao [10], pp. 302–309). Let θ_0 be an approximate solution and $\delta\theta$ the correction. Then neglecting higher order terms in $\delta\theta$

$$(6.5) \qquad\qquad -f(x, \theta_o) = H\delta\theta,$$

where H is computed at θ_0. Since H is singular, there is no unique solution to (6.5) and therefore, the question of choosing a suitable solution arises. A natural choice is a solution with a minimum norm

$$(6.6) \qquad\qquad \delta\theta = -H_m^- f(x, \theta_0)$$

We may terminate the iterative procedure when the correction needed is negligible. Let $\hat\theta$ be the approximate solution thus obtained and H^- any g inverse of H computed at $\hat\theta$. As observed earlier $\hat\theta$ and H^- are not meaningful when H is singular.

A parametric function $\psi(\theta)$ is said to be estimable if ξ_θ, the vector of derivatives of $\psi(\theta)$ with respect to $\theta_1, \cdots, \theta_n$, belongs to $\mathcal{M}(H)$. For such a function $\psi(\theta)$, $\psi(\hat\theta)$ is the unique m.l. estimate for any choice $\hat\theta$ of m.l. estimate of θ and the asymptotic variance of $\psi(\hat\theta)$ is

$$(6.7) \qquad\qquad \xi_\theta' H^- \xi_\theta$$

which is unique for any choice of the g inverse of H.

Chernoff [26] defined in inverse of a singular information matrix, which is not a g inverse in our sense. For instance, when no individual parameter is estimable, Chernoff's inverse does not exist (all the entries become infinite), while H^- exists and can be used as in formula (6.7) to find standard errors of estimable parametric functions.

7. Distribution of quadratic functions in normal variates

In this section we shall study the distribution of a quadratic function $Y'AY + 2b'Y + c$ in normally distributed variables $Y_1, Y_2, \cdots, Y_n$ and obtain conditions under which such a function would have a chi square (χ^2) distribution (central or noncentral). We denote a central χ^2 distribution with k degrees, of freedom by $\chi^2(k)$ and the noncentral distribution with parameter δ by $\chi^2(k, \delta)$. Also we denote a p variate normal distribution by $N_p(\mu, \Sigma)$ where μ is the mean vector and Σ is the dispersion matrix which may be singular (see Rao [10], p. 437).

THEOREM 7.1. *Let* $Y \sim N_n(\mu, I)$. *Then*

$$(7.1) \qquad \Sigma \lambda_i Y_i^2 + 2\Sigma b_i Y_i + c \sim \chi^2(k, \delta)$$

if and only if
   (i)   *each* λ_i *is either* 0 *or* 1,
   (ii) $b_i = 0$ *if* $\lambda_i = 0$, *and*
   (iii) $c = \Sigma b_i^2$,
in which case the number of degrees of freedom is $k = \Sigma \lambda_i$ *and the noncentrality parameter* $\delta = \Sigma \lambda_i (\mu_i + b_i)^2$.

PROOF. The theorem is easy to establish by comparing the characteristic function of $\Sigma_{i=1}^n \lambda_i Y_i^2 + 2\Sigma_{i=1}^n b_i Y_i + c$ and of $\Sigma_{i=1}^k (X_i + v_i)^2$ where the X_i are independent standard univariate normal variables.

THEOREM 7.2. *Let* $Y \sim N_n(\mu, I)$. *Then*

$$(7.2) \qquad Y'AY + 2b'Y + c \sim \chi^2(k, \delta)$$

if and only if
   (i) $A^2 = A$,
   (ii) $b \in \mathcal{M}(A)$, *and*
   (iii) $c = b'b$, *in which case the d.f.,* $k = R(A) = \operatorname{tr} A$ *and*

$$(7.3) \qquad \delta = (b + \mu)'A(b + \mu).$$

PROOF. There exists an orthogonal matrix P such that $A = P'\Delta P$ where Δ is diagonal. Under the transformation $Z = PY$

$$(7.4) \qquad Y'AY + 2b'Y + c = Z'\Delta Z + 2(Pb)'Z + c.$$

Further $Z \sim N_n(P\mu, I)$. Hence by Theorem 7.1

$$(7.5) \qquad Y'AY + 2b'Y + c \sim \chi^2(k, \delta)$$

if and only if
   (i) each diagonal element of Δ is either 0 or 1, that is, $\Delta^2 = \Delta$ or equivalently $A^2 = A$;
   (ii) the ith coordinate of Pb is 0 if the ith diagonal element of Δ is 0, that is, $Pb \in \mathcal{M}(\Delta)$ is equivalent to $b \in \mathcal{M}(A)$; and
   (iii) $c = (Pb)'Pb = b'b$.

Check that $k = \operatorname{tr} \Delta = \operatorname{tr} A = R(A)$ and

$$(7.6) \qquad \delta = (Pb + P\mu)'\Delta(Pb + P\mu) = (b + \mu)'A(b + \mu).$$

THEOREM 7.3. *Let* $Y \sim N_n(\mu, \Sigma)$ *where* Σ *could be singular. Then*

$$(7.7) \qquad Y'AY + 2b'Y + c \sim \chi^2(k, \delta)$$

if and only if
   (i) $\Sigma A \Sigma A \Sigma = \Sigma A \Sigma$ *or equivalently* $(\Sigma A)^3 = (\Sigma A)^2$,
   (ii) $\Sigma(A\mu + b) \in \mathcal{M}(\Sigma A \Sigma)$,
   (iii) $(A\mu + b)'\Sigma(A\mu + b) = \mu'A\mu + 2b'\mu + c$, *in which case* $k = \operatorname{tr} A\Sigma$,
$\delta = (b + A\mu)'\Sigma A \Sigma(b + A\mu)$.
   PROOF. We express $Y = \mu + FZ$, where F is an $n \times r$ matrix of rank r such that $\Sigma = FF'$ and $Z \sim N_r(0, I)$. In terms of Z we have

$$(7.8) \qquad Y'AY + 2b'Y + c = Z'F'AFZ + 2(A\mu + b)'FZ + \mu'A\mu + 2b'\mu + c.$$

Applying Theorem 7.2 to the quadratic function in Z we have therefore the following necessary and sufficient conditions for $\chi^2(k, \delta)$ distribution:
   (i) $(F'AF)^2 = F'AF \Leftrightarrow \Sigma A \Sigma A \Sigma = \Sigma A \Sigma \Leftrightarrow (\Sigma A)^3 = (\Sigma A)^2$,
   (ii) $F'(A\mu + b) \in \mathcal{M}(F'AF) \Leftrightarrow \Sigma(A\mu + b) \in \mathcal{M}(\Sigma A \Sigma)$, and
   (iii) $(A\mu + b)'\Sigma(A\mu + b) = \mu'A\mu + 2b'\mu + c$.
Observe that $k = \operatorname{tr} F'AF = \operatorname{tr} A\Sigma$ and

$$(7.9) \qquad \delta = (A\mu + b)'FF'AFF'(A\mu + b) = (A\mu + b)'\Sigma A \Sigma(A\mu + b)$$

   COROLLARY 7.1. *Let* $Y \sim N_n(\mu, \Sigma)$. *Then* $Y'\Sigma^- Y \sim \chi^2(k, \delta)$ *if and only if* $\mu'(\Sigma^-\Sigma\Sigma^- - \Sigma^-)\mu = 0$ *in which case* $k = R(\Sigma)$ *and* $\delta = \mu'\Sigma^-\mu$.
   The required condition is satisfied for all μ if Σ is a reflexive inverse of Σ and is satisfied for all Σ if and only if $\mu \in \mathcal{M}(\Sigma)$. We note further that if $\mu \in \mathcal{M}(\Sigma)$ then $Y \in \mathcal{M}(\Sigma)$ with probability 1. In such a case with probability 1, $Y'\Sigma^- Y$ is invariant with respect to choice of Σ^-.
   It has come to our notice after the Berkeley Symposium that Bhapkar [38] has obtained the result stated in the Corollary 7.1 to our Theorem 7.3. But the result of Theorem 7.3 is more general and that of the corollary is only a particular case.
   Condition (i), $\Sigma A \Sigma A \Sigma = \Sigma A \Sigma$, of Theorem 7.3 seems to have been found first by Ogasawara and Takahashi [39].

8. Discriminant function in multivariate analysis

   8.1. *Singular multivariate normal distribution.* The book *Linear Statistical Inference and its Applications* [10] develops a density free approach to study the distribution and inference problems associated with a multivariate normal distribution. The approach is more general than the usual one since it includes the study of the normal distribution with a singular covariance matrix which does not admit a density in the usual sense. The elegance of the density free approach was further demonstrated by Mitra [40]. However, in some problems, as in the construction of a discriminant function, it is useful to have an explicit expression for the density. The density function of a multivariate normal distribution, as it is usually written, involves the inverse of the variance-covariance (dispersion) matrix, which necessitates the assumption that the dispersion matrix is non-singular. In this section we demonstrate how the g inverse is useful in defining the density function and in extending some of the results developed for the non-singular case to the singular distribution.

Let Y be a $p \times 1$ vector random variable. In Rao [10], Y is defined to have a p variate normal distribution if $m'Y$ has a univariate normal distribution for every vector $m \in \mathscr{R}^p$. In such a case it is shown that the distribution is characterized by the parameters

$$(8.1) \qquad \mu = E(Y), \ \Sigma = E[(Y - \mu)(Y - \mu)'],$$

called the mean vector and the dispersion matrix of Y, respectively; the symbol $N_p(\mu, \Sigma)$ is used to denote the p variate normal distribution. The distribution is said to be singular if $R(\Sigma) = \rho < p$ in which case ρ is called the rank of the distribution and we may use the symbol $N_p(\mu, \Sigma(\rho))$ to specify the rank in addition to the basic parameters.

Let N be $p \times (p - \rho)$ matrix of rank $p - \rho$ such that $N'N = I$, $N'\Sigma = 0$ and A be a $p \times \rho$ matrix of rank ρ such that $N'A = 0$ and $A'A = I$. By construction $(N : A)$ is an orthogonal matrix. We make the transformation

$$(8.2) \qquad Z_1 = N'Y, \qquad Z_2 = A'Y.$$

Then

$$E(Z_1) = N'\mu, \qquad E(Z_2) = A'\mu,$$
$$(8.3)$$
$$D(Z_1) = N'\Sigma N = 0, \qquad D(Z_2) = A'\Sigma A.$$

It follows that there exists a constant vector ζ such that

$$(8.4) \qquad Z_1 = N'Y = N'\mu = \zeta,$$

with probability 1 and since $A'\Sigma A$ is nonsingular Z_2 has the ρ variate normal density

$$(8.5) \qquad (2\pi)^{-\rho/2}|A'\Sigma A|^{-1/2} \exp\left\{-\tfrac{1}{2}(Z_2 - A'\mu)'(A'\Sigma A)^{-1}(Z_2 - A'\mu)\right\}.$$

We observe that

$$(8.6) \qquad |A'\Sigma A| = \lambda_1 \cdots \lambda_\rho,$$

where $\lambda_1, \cdots, \lambda_\rho$ are the nonzero eigenroots of Σ and

$$(8.7) \qquad (Z_2 - A'\mu)'(A'\Sigma A)^{-1}(Z_2 - A'\mu) = (Y - \mu)'\Sigma^-(Y - \mu)$$

where Σ^- is any g inverse of Σ. Thus the density of Y on the hyperplane $N'(Y - \mu) = 0$ or $N'Y = \zeta$ is defined by

$$(8.8) \qquad \frac{(2\pi)^{-\rho/2}}{(\lambda_1 \cdots \lambda_\rho)^{1/2}} \exp\left\{-\tfrac{1}{2}(Y - \mu)'\Sigma^-(Y - \mu)\right\}$$

which is an explicit function of the vector Y and its associated parameters μ and Σ. The expression (8.8) was considered by Khatri [41] in deriving some distributions in the case of a singular normal distribution.

8.2. *Discriminant function*. The density function derived in (8.8) can be used in determining the discriminant function (ratio of likelihoods) for assigning an individual as a member of one of two populations to which it may belong.

Let Y be a $p \times 1$ vector of observations which has the distribution $N_p[\mu_1, \Sigma_1(\rho_1)]$ in the first population and $N_p(\mu_2, \Sigma_2(\rho_2))$ in the second population. We shall construct the discriminant function applicable to different situations.

Case 1. $\Sigma_1 = \Sigma_2 = \Sigma$, $R(\Sigma) = \rho < p$, $N'\mu_1 \neq N'\mu_2$.

The distribution of Y consists of two parts as shown in (8.4) and (8.5) of which (8.4) is the almost sure part. If $N'\mu_1 \neq N'\mu_2$,

$$(8.9) \qquad N'Y = N'\mu_1 = \zeta_{11} \quad \text{with probability 1}$$

if Y comes from the first population, and

$$(8.10) \qquad N'Y = N'\mu_2 = \zeta_{21} \quad \text{with probability 1}$$

if Y comes from the second population. Then the discriminant function is $N'Y$, and in fact it provides perfect discrimination. No use need be made of the other part (8.5) of the distribution of Y.

Case 2. $\Sigma_1 = \Sigma_2 = \Sigma$, $R(\Sigma) = \rho < p$, $N'\mu_1 = N'\mu_2$.

In this case $N'Y$ does not provide any discrimination and we have to consider the density (8.8). The log densities for the two populations are (apart from a constant)

$$(8.11) \qquad -\tfrac{1}{2} \log (\lambda_1 \cdots \lambda_\rho) - \tfrac{1}{2}(Y - \mu_1)'\Sigma^-(Y - \mu_1),$$

and

$$(8.12) \qquad -\tfrac{1}{2} \log (\lambda_1 \cdots \lambda_\rho) - \tfrac{1}{2}(Y - \mu_2)'\Sigma^-(Y - \mu_2).$$

Taking the difference and retaining only the portion depending on Y we obtain the discriminant function

$$(8.13) \qquad \delta'\Sigma^- Y, \qquad \text{where} \quad \delta = \mu_1 - \mu_2$$

which is of the same form as in the nonsingular case ($\delta'\Sigma^{-1}Y$). Now

$$(8.14) \qquad V(\delta\Sigma^- Y) = \delta'\Sigma^- \delta$$

which is the analogue of Mahalanobis distance $D^2(= \delta'\Sigma^{-1}\delta)$ in the singular case.

Case 3. $\Sigma_1 \neq \Sigma_2$, $\mathscr{M}(\Sigma_1) \neq \mathscr{M}(\Sigma_2)$.

The discrimination is perfect as in Case 1. Let N be a matrix of maximum rank such that $N'\Sigma_1 N = 0 = N'\Sigma_2 N$, and let A be a matrix of maximum rank such that $(N:A)'\Sigma_1 = 0$, and let B be a matrix of maximum rank such that $(N:B)'\Sigma_2 = 0$. Finally let C be such that $(N:A:B:C)$ is a $p \times p$ matrix of rank p. Consider the transformation

$$(8.15) \qquad Z_1 = N'Y, \, Z_2 = A'Y, \, Z_3 = B'Y, \, Z_4 = C'Y$$

The distributions of these variables in the two populations are given in Table III.

TABLE III

DISTRIBUTION OF THE DISCRIMINANT VARIABLES

Case 3.

Population	Z_1	Z_2	Z_3	Z_4
1	$\zeta_{11} = N'\mu_1$ with prob. 1	$\zeta_{12} = A'\mu_1$ with prob. 1	$N(B'\mu_1, B'\Sigma_1 B)$ $\zeta_{22} = B'\mu_2$	$N(C'\mu_1, C'\Sigma_1 C)$
2	$\zeta_{21} = N'\mu_2$ with prob. 1	$N(A'\mu_2, A'\Sigma_2 A)$	with prob. 1	$N(C'\mu_2, C'\Sigma_2 C)$

It is seen that the variables Z_1, Z_2 and Z_3 provide perfect discrimination unless $\zeta_{11} = \zeta_{21}$, $A = 0$ and $B = 0$, which can happen only when $\mathcal{M}(\Sigma_1) = \mathcal{M}(\Sigma_2)$.

Case 4. $\Sigma_1 \neq \Sigma_2$, $\mathcal{M}(\Sigma_1) = \mathcal{M}(\Sigma_2)$.

Let N be as defined in Case 3 and consider $N'Y$ which is a constant for both the populations. If $N'\mu_1 \neq N'\mu_2$, then we have perfect discrimination. If $N'\mu_1 = N'\mu_2$, then we have to consider the densities

$$(8.16) \qquad (\lambda_1 \cdots \lambda_\rho)^{-1/2} \exp \left\{ -\tfrac{1}{2}(Y - \mu_1)' \Sigma_1^-(Y - \mu_1) \right\}$$

and

$$(8.17) \qquad (\lambda_1' \cdots \lambda_\rho')^{-1/2} \exp \left\{ -\tfrac{1}{2}(Y - \mu_2)' \Sigma_2^-(Y - \mu_2) \right\},$$

where $\lambda_1, \cdots, \lambda_\rho$ are the nonzero eigenvalues of Σ_1, $\lambda_1', \cdots, \lambda_\rho'$ are those of Σ_2 and Σ_1^-, Σ_2^- are any g inverses of Σ_1, Σ_2. Taking logarithm of the ratio of densities and retaining only the terms depending on Y we have the quadratic discriminant function

$$(8.18) \qquad (Y - \mu_1)' \Sigma_1^-(Y - \mu_1) - (Y - \mu_2)' \Sigma_2^-(Y - \mu_2)$$

analogous to the expression in the nonsingular case.

REFERENCES

[1] E. H. MOORE, *General Analysis*, Philadelphia, American Philosophical Society, 1935.
[2] ———, "On the reciprocal of the general algebraic matrix" (abstract), *Bull. Amer. Math. Soc.*, Vol. 26 (1920), pp. 394–395.
[3] Y. Y. TSENG, "Generalized inverses of unbounded operators between two unitary spaces," *Dokl. Akad. Nauk. SSSR.*, Vol. 67 (1949), pp. 431–434.
[4] ———, "Properties and classifications of generalized inverses of closed operators," *Dokl. Akad. Nauk. SSSR*, Vol. 67 (1949), pp. 607–610.
[5] ———, "Virtual solutions and general inversions," *Uspehi. Mat. Nauk.*, Vol. 11 (1956), pp. 213–215.
[6] R. PENROSE, "A generalized inverse for matrices," *Proc. Cambridge Philos. Soc.*, Vol. 51 (1955), pp. 406–413.

[7] ———, "On best approximate solutions of linear matrix equations," *Proc. Cambridge Philos. Soc.*, Vol. 52 (1956), pp. 17–19.

[8] C. RADHAKRISHNA RAO, "Analysis of dispersion for multiply classified data with unequal numbers in cells," *Śankhyā*, Vol. 15 (1955), 253–280.

[9] ———, "A note on a generalized inverse of a matrix with applications to problems in mathematical statistics," *J. Roy. Statist. Soc. Ser. B*, Vol. 24 (1962) pp. 152–158.

[10] ———, *Linear Statistical Inference and its Applications*, New York, Wiley, 1965.

[11] ———, "On the theory of least squares when parameters are stochastic and its application to analysis of growth curves," *Biometrika*, Vol. 52 (1965), pp. 447–458.

[12] ———, "A study of large sample test criteria through properties of efficient estimates," *Śankhyā Ser. A*, Vol. 23 (1961), pp. 25–40.

[13] ———, "Generalized inverse for matrices and its applications in mathematical statistics," *Research papers in Statistics, Festschrift for J. Neyman*, New York, Wiley, 1966.

[14] ———, "Least squares theory using an estimated dispersion matrix and its application to measurement of signals," *Proceedings of the Fifth Berkeley Symposium on Statistics and Probability*, Berkeley and Los Angeles, University of California Press, 1967, Vol. 1, pp. 355–372.

[15] T. N. E. GREVILLE, "The pseudo-inverse of a rectangular matrix and its application to the solution of systems of linear equations," *SIAM Rev.*, Vol. 1 (1959), pp. 38–43.

[16] A. BJERHAMMAR, "Rectangular reciprocal matrices with special reference to geodetic calculations," *Bull. Géodésique*, Vol. 52 (1951), pp. 188–220.

[17] ———, "Application of the calculus of matrices to the method of least squares with special reference to geodetic calculations," *Kungl. Tekn. Högsk. Hand. Stockholm*, No. 49 (1951), pp. 1–86.

[18] ———, "A generalized matrix algebra," *Kungl. Tekn. Högsk. Handl. Stockholm*, No. 124 (1958), pp. 1–32.

[19] A. BEN-ISRAEL and A. CHARNES, "Contributions to the theory of generalized inverses," *SIAM J. Appl. Math.*, Vol. 11 (1963), pp. 667–699.

[20] J. S. CHIPMAN, "On least squares with insufficient observations," *J. Amer. Statist. Assoc.*, Vol. 59 (1964), pp. 1078–1111.

[21] ———, "Specification problems in regression analysis," *Theory and Application of Generalized Inverses and Matrices*, Symposium Proceedings, Texas Technological College, Mathematics Series No. 4 (1968), pp. 114–176.

[22] J. S. CHIPMAN and M. M. RAO, "Projections, generalized inverses and quadratic forms," *J. Math. Anal. Appl.*, Vol. 9 (1964), pp. 1–11.

[23] J. E. SCROGGS and P. L. ODELL, "An alternative definition of the pseudo-inverse of a matrix," *SIAM J. Appl. Math.*, Vol. 14 (1966), pp. 796–810.

[24] R. C. BOSE, "Analysis of Variance," unpublished lecture notes, University of North Carolina, 1959.

[25] R. BOTT and R. J. DUFFIN, "On the algebra of networks," *Trans. Amer. Math. Soc.*, Vol. 74 (1953), pp. 99–109.

[26] H. CHERNOFF, "Locally optimal designs for estimating parameters," *Ann. Math. Statist.*, Vol. 24 (1953), pp. 586–602.

[27] C. RADHAKRISHA RAO, "Calculus of generalized inverse of matrices. Part I: General theory," *Śankhyā Ser. A*, Vol. 29 (1967), pp. 317–342.

[28] S. K. MITRA, "On a generalized inverse of a matrix and applications," *Śankhyā Ser. A*, Vol. 30 (1968), pp. 107–114.

[29] ———, "A new class of g-inverse of square matrices," *Śankhyā Ser. A*, Vol. 30 (1968), pp. 323–330.

[30] P. BHIMAŚANKARAM and S. K. MITRA, "On a theorem of Rao on g-inverses of matrices," *Śankhyā Ser. A*, Vol. 31 (1969) pp. 365–368.

[31] S. K. MITRA and P. BHIMAŚANKARAM, "Some results on idempotent matrices and a matrix equation connected with the distribution of quadratic forms," *Śankhyā Ser. A.*, Vol. 32 (1970), pp. 353–356.

[32] S. K. MITRA and C. RADHAKRISHNA RAO, "Simultaneous reduction of a pair of quadratic forms," *Śankhyā Ser. A*, Vol. 30 (1968), pp. 313–322.

[33] ———, "Some results in estimation and tests of hypotheses under the Gauss-Markov model," *Śankhyā Ser. A*, Vol. 30 (1968), pp. 281–290.

[34] ———, "Conditions for optimality and validity of simple least squares theory," *Ann. Math. Statist.*, Vol. 40 (1969), pp. 1617–1624.

[35] C. R. RAO, "A note on a previous lemma in the theory of least squares and some further results," *Śankhyā Ser. A*, Vol. 30 (1968), pp. 245–252.

[36] J. S. FRAME, "Matrix operations and generalized inverses," *IEEE Spectrum*, March (1964), pp. 209–220.

[37] M. SIBUYA, "Subclasses of generalized inverses of matrices," *Ann. Inst. Statist., Math.*, in press.

[38] V. P. BHAPKAR, "The invariance and distribution of the quadratic form $x'\Sigma^- x$," Technical Report, University of Kentucky, Department of Statistics, 1970.

[39] T. OGASAWARA and M. TAKAHASHI, "Independence of quadratic forms in normal system," *Journal of Science of the Hiroshima University*, Vol. 15 (1951), pp. 1–9.

[40] S. K. MITRA, "A density-free approach to matrix variate beta distribution," *Śankhyā Ser. A*, Vol. 32 (1970), pp. 81–88.

[41] C. G. KHATRI, "Some results for the singular multivariate regression models," *Śankhyā, Ser. A*, Vol. 30 (1968), pp. 267–280.

METRIC CONSIDERATIONS IN CLUSTER ANALYSIS

HERMAN CHERNOFF
STANFORD UNIVERSITY

1. Introduction and summary

A variation of the "k means" method of cluster analysis is described which is designed to take into account and profit from the possibility that the separate clusters resemble samples from multivariate normal distributions with substantially different covariance structures. This is preceded by a brief description of a standard version of the method. Indications are given when metric considerations can play an important role and a suitably modified version of the standard method is presented.

While the new method has not yet been applied it is anticipated that its most useful applications will be to situations where the clusters tend to be concentrated in nonparallel hyperplanes of the space of observations. The dimensionality of this space should not be very large. The method should require substantial sample sizes to make the implicit estimates of the covariance matrices useful.

One may expect metric considerations also to be useful in modifying other cluster analysis techniques.

2. The standard k means method

In this section we describe the k means method in the spirit of MacQueen [2]. Suppose that p represents the probability distribution of a random variable (r.v.) Z in an r dimensional Euclidean space and $|y - z|$ represents the distance between points y and z of this space. Let $S = (S_1, S_2, \cdots, S_k)$ be a *decomposition* of the space into k pairwise disjoint measurable subsets (classes) and let $x = (x_1, x_2, \cdots, x_k)$ represent k *reference* points in the space. Then

$$(2.1) \qquad R(x, S) = \sum_{i=1}^{k} \int_{S_i} |z - x_i|^2 \, dp(z)$$

is a measure of the corresponding within class variance. From one point of view of the notion of cluster it would be expected that if the probability measure p corresponds to k *natural clusters*, these clusters would relate in a simple way to an (x, S) which minimizes R.

For given S, $R(x, S)$ can be minimized by selecting the reference points to be the centers of gravity, that is, $x = u(S) = (u_1(S), u_2(S), \cdots, u_k(S))$, where

$$(2.2) \qquad u_i(S) = \int_{S_i} z \, dp(z)/p(S_i), \qquad\qquad i = 1, 2, \cdots, k$$

assuming $p(S_i) > 0$, $i = 1, 2, \cdots, k$. Then we have the measure

$$(2.3) \qquad V(S) = R[u(S), S] = \sum_{i=1}^{k} \int_{S_i} |z - u_i(S)|^2 \, dp(z).$$

Alternatively, for given reference points x, we can minimize R with respect to S by selecting a decomposition $S = T(x) = (T_1(x), \cdots, T_k(x))$ which assigns to T_i those points which are closest to x_i, that is,

$$(2.4) \qquad \text{if} \quad z \in T_i(x) \quad \text{then} \quad |z - x_i| = \min_j |z - x_j|.$$

This gives us the within class variance

$$(2.5) \qquad W(x) = R[x, T(x)] = \sum_{i=1}^{k} \int_{T_i(x)} |z - x_i|^2 \, dp(z).$$

Thus $V(S)$ and $W(x)$ are similar but somewhat different measures of within class variance. They do coincide in the case where $T(x) = S$ and $u(S) = x$. Then $u[T(x)] = x$ and the reference points x are said to be *unbiased*.

The above described minimization properties imply that

$$(2.6) \qquad V(S) = R[u(S), S] \geqq W[u(S)]$$

and

$$(2.7) \qquad W(x) = R[x, T(x)] \geqq V[T(x)].$$

Hence, given an arbitrary x, the iteration $x^1 = x$, $S^1 = T(x^1)$, $\cdots$, $x^n = u[S^{n-1}]$, $S^n = T(x^n)$, $\cdots$, yields the decreasing sequence

$$(2.8) \qquad W(x^1) \geqq V(S^1) \geqq W(x^2) \geqq V(S^2) \geqq \cdots,$$

and hopefully converges to a pair (x, S) with a low within class variance and unbiased x. MacQueen attributes consideration of this procedure to Forgy [1] and Jennrich.

The k means method to be described may be motivated by considerations such as given above and by the aim of reducing computational labor and information storage requirements. Suppose that in place of the probability distribution p one is given a sample of independent observations on the r.v. Z. Described informally, the k means method is an iterative method of generating a sequence of reference points $x^n = (x_1^n, x_2^n, \cdots, x_k^n)$ where x^1 consists of the first k distinct observations $Z_1, Z_2, \cdots, Z_k$ on Z. Afterwards each new observation is assigned to the closest reference point which is then modified to be the average of all observations assigned to it.

More precisely let $x^1 = (Z_1, Z_2, \cdots, Z_k)$ and $w^1 = (1, 1, \cdots, 1)$. If we observe Z after x^n and $w^n = (w_1^n, w_2^n, \cdots, w_k^n)$ are formed, let

$$(2.9) \qquad x_i^{n+1} = x_i^n, \quad w_i^{n+1} = w_i^n, \qquad\qquad \text{if} \quad Z \notin T_i(x^n)$$

and

(2.10)
$$x_i^{n+1} = \frac{w_i^n x_i^n + Z}{w_i^n + 1}, \quad w_i^{n+1} = w_i^n + 1, \quad \text{if} \quad Z \in T_i(x^n).$$

The *weight* w_i^n is the number of observations whose mean is represented by the ith reference point x_i^n. We shall loosely refer to $T_i(x^n)$ as the ith cluster at the nth stage.

Compared with many other techniques of cluster analysis, the iterative procedure of the k means method seems to be rather economical in storage and computational requirements. At the nth stage one needs to store x^n, w^n, and the latest observation Z. The computation consists mainly of evaluating the k distances $|Z - x_i^n|$, $i = 1, 2, \cdots, k$. A sound comparison of computational efficiency would require some insight into the number of iterations required with this technique. Many alternative approaches seem to require the computation and storage of $_mC_2$ distances to study a sample of m points. When m is large this may be excessive. However it is possible that sampling techniques may be applied to reduce these requirements.

MacQueen [3] has proved two theorems listed below which indicate that the $W(x^n)$ of the k means method converge to $W(x)$ for an unbiased x and that the x^n converges to $u(S(x^n))$ in a weak sense. These theorems are proved under the assumptions: (i) p is absolutely continuous with respect to Lebesgue measure, and (ii) $p(R) = 1$ for a closed and bounded convex set R and $p(A) > 0$ for every open set $A \subset R$.

THEOREM 1. *The sequence of random variables* $W(x^1)$, $W(x^2)$, $\cdots$, *converges a.s. and* $W_\infty = \lim_{n \to \infty} W(x^n) = V(T(x))$ a.s. *for some unbiased* $x = (x_1, x_2, \cdots, x_k)$ *for which* $x_i \neq x_j$ *if* $i \neq j$.

THEOREM 2. *Let* $u_i^n = u_i(x^n)$ *and* $p_i^n = p(T_i(x^n))$; *then as* $m \to \infty$

(2.11)
$$m^{-1} \sum_{n=1}^{m} \sum_{i=1}^{k} p_i^n |x_i^n - u_i^n| \to 0 \quad a.s.$$

MacQueen presents examples which show that the k means method *cannot* be counted on to provide minimum within class variance.

The k means method can be modified to increase or decrease the number of clusters under suitable conditions. Typically if a new point Z is too far from each of the reference points it can be made the first reference point of a $(k + 1)$st cluster (*refinement*). If two reference points are too close to each other one can combine their clusters by replacing the two reference points by a suitable weighted average (*coarsening*). The criteria for too far and too close can be set in advance by two parameters R for refinement and C for coarsening.

3. Distance considerations

The preceding section was based on the implicit assumption that Euclidean distance is the appropriate measure of distance. However it is known that in dealing with a random variable Z with a multivariate normal distribution with mean μ and nonsingular covariance matrix A, the Mahalanobis distance measure

$$(3.1) \qquad d(x, y) = (x - y)'A^{-1}(x - y)$$

is highly meaningful. The "ability" to test the hypothesis $H_1 : \mu = \mu_1$ versus the alternative $H_2 : \mu = \mu_2$ is an increasing function of $d(\mu_1, \mu_2)$. It is known that $d(Z, \mu)$ has the χ^2 (chi square) distribution with r degrees of freedom (d.f.) and that $d(Z, x)$ has the noncentral χ^2 distribution with r d.f. and noncentrality parameter $d(x, \mu)$.

It has been suggested that the Euclidean metric $(x - y)'(x - y)$ be replaced by the Mahalanobis metric in measuring the distance used in various cluster analysis techniques. A glance at Figure 1 indicates that this suggestion could lead to undesirable results.

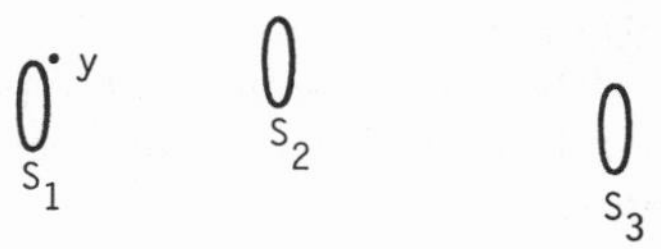

FIGURE 1

Figure 1 represents the diagram of three clusters in two dimensional space. Each cluster is described by a set of points lying in a thin ellipsoid and resembles a sample from a multivariate normal distribution. The three ellipsoids are similar with long vertical axes indicating that horizontal distances are important in determining the cluster to which a point belongs. On the other hand, the Mahalanobis distance using the covariance matrix for the *overall* set of points would tend to give most emphasis to the relatively unimportant vertical distances. The use of this metric might serve to persuade one to assign the point marked y to S_2 rather than to S_1.

It has been suggested that the above undesirable attribute be discounted by using Mahalanobis distance with a covariance matrix corresponding to within cluster variance rather than overall variance. That is, one should use

$$(3.2) \qquad d^*(x, y) = (x - y)'A_w^{-1}(x - y),$$

where A_w is the within cluster covariance matrix

$$(3.3) \qquad A_w = \sum_{i=1}^{k} \sum_{x \in S_i} (x - \bar{x}_i)(x - \bar{x}_i)'$$

with the $S = (S_1, S_2, \cdots, S_k)$ representing a decomposition of the sample into k pairwise disjoint sets (clusters) whose averages are the $\bar{x}_i$.

Elaborations of this proposal have been treated by Friedman and Rubin [2] who consider the characteristic values and vectors of A_w with respect to the overall covariance matrix,

$$(3.4) \qquad A_T = \sum_{x} (x - \bar{x})(x - \bar{x})'.$$

The potential disadvantage of using A_w becomes manifest in Figure 2 where two sharply defined clusters with almost singular covariance matrices combine to yield an A_w which is a multiple of the identity and corresponds to a multiple of the Euclidean metric. It is clear that for cluster 1 vertical distance is most important whereas for cluster 2 horizontal distance is crucial. To decide whether an arbitrary point belongs to S_1 or S_2, it seems most advisable to compare the appropriate metric in each case. Thus the point labeled y is more naturally associated with S_1 though it is closer to the center of S^2.

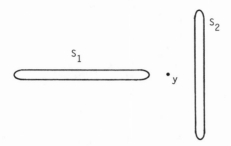

FIGURE 2

It is indicated in Figure 3 that two clusters could conceivably intersect with the result that each cluster effectively divides the other into two disconnected pieces. While this represents an undesirable, if uncommon situation the lack of connection hardly seems as serious a problem as the presence of the common part where points are difficult to classify into one cluster or the other. This more serious problem of difficulty in classification occurs often in less pathological appearing examples.

FIGURE 3

To overcome the difficulties rising from the above approaches we propose to introduce a modification of the k means method where each cluster determines its own metric, the Mahalanobis distance for that cluster. The possibility of using these metrics seems to have been previously considered only by Rohlf [4] and in connection with an hierarchical approach.

4. The modified procedure

Suppose that at the nth stage we have k clusters represented by reference points $x^n = (x_1^n, x_2^n, \cdots, x_k^n)$, weights $w^n = (w_1^n, w_2^n, \cdots, w_k^n)$ and covariances $A^n =$

$(A_1^n, A_2^n, \cdots, A_k^n)$. Corresponding to the reference points x and covariances A, we have a measurable decomposition of the r dimensional Euclidean space into $T(x, A) = \big(T_1(x, A), T_2(x, A), \cdots, T_k(x, A)\big)$, where $z \in T_i(x, A)$ only if

$$(4.1) \qquad (z - x_i)'A_i^{-1}(z - x_i) \leqq (z - x_j)'A_j^{-1}(z - x_j), \qquad 1 \leqq j \leqq k.$$

If a new observation Z is selected at random, we let $x_i^{n+1} = x_i^n$, $w_i^{n+1} = w_i^n$, and $A_i^{n+1} = A_i^n$ if $Z \notin T_i(x^n, A^n)$. If $Z \in T_i(x^n, A^n)$ we let

$$(4.2) \qquad x_i^{n+1} = \frac{x_i^n w_i^n + Z}{w_i^n + 1}, \qquad\qquad w_i^{n+1} = w_i^n + 1$$

and, setting $B_i^n = w_i^n A_i^n$, we let

$$(4.3) \qquad B_i^{n+1} = B_i^n + \frac{w_i^n}{w_i^n + 1}(Z - x_i^n)(Z - x_i^n)',$$

$$B_i^{n+1} = B_i^n + \frac{w_i^n + 1}{w_i^n}(Z - x_i^{n+1})(Z - x_i^{n+1})'.$$

The motivation for this formula derives from the following algebra. Given m observations $y_1, y_2, \cdots, y_m$ with mean u_m let

$$(4.4) \qquad B_m^* = mA_m^* = \sum_{\alpha=1}^m (y_\alpha - u_m)(y_\alpha - u_m)'.$$

Then

$$(4.5) \qquad B_{m+1}^* = B_m^* + y_{m+1}y_{m+1}' + mu_m u_m' - (m+1)u_{m+1}u_{m+1}',$$

where

$$(4.6) \qquad (m+1)u_{m+1} = mu_m + y_m.$$

It is easily seen that

$$(4.7) \qquad B_{m+1}^* - B_m^* = \frac{m}{m+1}(y_{m+1} - u_m)(y_{m+1} - u_m)'$$

$$= \frac{m+1}{m}(y_{m+1} - u_{m+1})(y_{m+1} - u_{m+1})'.$$

Thus, except for a scale factor, the covariance matrix A_m changes by the addition of a matrix of rank 1. This has a desirable aspect, for if C is nonsingular and

$$(4.8) \qquad D = C + hh',$$

then

$$(4.9) \qquad D^{-1} = C^{-1} - \frac{(C^{-1}h)(h'C^{-1})}{1 + h'C^{-1}h}.$$

Therefore the A_i^n can be inverted recursively.

Thus far we have an algorithm for applying the modified method once initial values x^1, w^1, A^1 are obtained. Some suggestion is necessary for initiating the iterative procedure. It is possible to take the first k observations as the components of x^1. To avoid singularity it is desirable to start with the A_i as positive definite symmetric matrices. These can be arbitrary, say the identity matrix. Alternatively some prior insights or previous information could lead to alternative suggestions. The value of w^1 could be $(1, 1, \cdots, 1)'$ or may also be assigned in a more or less arbitrary fashion. One can avoid unnecessarily rapid early fluctuations in A^n by starting w^1 with large components. Alternatively by selecting the B_i^1 to be relatively "large" in magnitude one can reduce the early fluctuations in A^n. An advantage of large B_i^1 with small w^1 over large w^1 is that the initial reference points are not given undue weight.

5. Comments

5.1 *Economic variations*. The computational cost per iteration of the proposed modification is greater than that for the Euclidean metric version by an amount necessary to compute the revised A_i^{-1} and the k distances. This extra cost is of the order of magnitude of kr^2 where r is the dimensionality of the space. If r is large the extra cost may necessitate the use of some short cuts.

One possibility that may be worth exploring is to decompose A_i into principal components. Then one can confine attention to a smaller dimensional space which is spanned by characteristic vectors corresponding to a few of the largest characteristic values from each of the A_i. Thus using three vectors from each of five groups yields a 15 dimensional space which leads to a considerable savings if r is say 50. To use this technique effectively one would have to carry out a substantial number of iterations with a given subspace before recomputing the principal components.

A variation of this approach is to separate the characteristic roots of A_i into two groups and to approximate distances. To illustrate suppose A_1 has characteristic roots $\lambda_1 \geqq \lambda_2 \geqq \cdots \lambda_{r_1} \geqq \lambda_{r_1+1} \geqq \cdots \geqq \lambda_r$ and corresponding vectors $u_1, u_2, \cdots, u_r$. Then an arbitrary vector z may be decomposed so that $z = \Sigma\, z_i u_i$ where the z_i are the projections of z on u_i and the distance $z'A_1^{-1}z = \Sigma\, \lambda_i^{-1}z_i^2$. This may be approximated (from below) by

$$(5.1) \qquad \sum_{i=1}^{r} \lambda_i^{-1}z_i^2 + \lambda_{r_1+1}^{-1} \sum_{i=r_1+1}^{r} z_i^2 = \sum_{i=1}^{r_1} \lambda_i^{-1}z_i^2 + \lambda_{r_1+1}^{-1}\left(\sum_{i=1}^{r} z_i^2 - \sum_{i=1}^{r_1} z_i^2 \right).$$

Thus all that are required are the lengths of the projections of z on the first r_1 characteristic vectors and the Euclidean length of z. With this variation it is also desirable to use the λ_i and u_i for several iterations before recomputing.

One could elaborate on this variation by dividing the λ_i into several groups. It is questionable that this would help much.

It should be remarked that as the dimensionality of the space increases, the sample size necessary to obtain reliable results tends to increase. At this point it would be difficult to anticipate the extent.

5.2. *Singular covariances.*   If some of the clusters lie in proper linear subspaces of the Euclidean space, particularly if they lie in different subspaces, the ability to distinguish the clusters should be very great. In practice this desirable situation means that we deal with covariance matrices which are singular or almost so. Initiating the iteration with nonsingular matrices A_i leads, through equation (4.3) to a sequence of matrices which are nonsingular. However, in the above described situation, these matrices should tend to be successively closer to singular matrices. While no enormous difficulty is anticipated in the inversion through equation (4.9) one should be prepared to recognize the phenomenon as it develops, since its presence points to potentially important properties of the data.

5.3. *Relevance of modified approach.*   The situation in which the procedures described seem most relevant is when the population consists of a set of clusters each of which resembles the ellipsoidal form of a multivariate normal distribution and the covariance structures of these clusters are substantially different and preferably confined to different linear subspaces. Clusters which tend to curve, particularly those cases where some points of one cluster tend to be closer to points of another cluster than to other points of the same cluster (see Figure 4), should not yield much information to our modified approach.

FIGURE 4

5.4. *Coarsening and refining.*   Assuming multivariate normal distributions one would expect that $(Z - x_i^n)'A_i^n(Z - x_i^n)$ would have the chi square distribution if z is in the ith cluster. Thus large or small values of this statistic could be used for coarsening and refining the clusters where the discrimination between large and small should relate in part with the percentiles of the chi square distribution. Such a procedure is recommended here with reservations since nonnormal behavior will alter the distribution of $(Z - x_i^n)'A_i(Z - x_i^n)$ and it seems to be more conservative to keep track of the empirical distribution of means and variances of substantial numbers of recent values of these distances to see what is unusually large or small.

5.5. *What is a cluster?*   The k means method is an approach to cluster analysis based mainly on the "metric" concept of a cluster as a set of points which are closer to one another than to other points. Another general approach which requires more calculation, if one does not use sampling creatively, is to regard a cluster as a set of points each of which has a nearby neighbor of the cluster.

Although in most illustrative examples in two dimensional space the informal application of both approaches yields the same common sense idea of cluster; this is not necessarily the case. It is important to prove some theorems which will establish properties of this distribution p which will imply that both approaches yield the same results. Otherwise one must wonder whether one computationally convenient method will yield unrecognizable clusters from another point of view. Part of the weakness of the highly nontrivial MacQueen conclusions derives in part from requiring too general a domain of applicability. It may be easier to get stronger results for those distributions p for which various approaches coincide.

REFERENCES

[1] E. FORGY, "Cluster analysis of multivariate data: Efficiency versus interpretability of classification abstract," *Biometrics*, Vol. 21 (1965), p. 768.
[2] H. R. FRIEDMAN and J. RUBIN, "On some invariant criteria for grouping data," *J. Amer. Statist. Assoc.*, Vol. 62 (1967), pp. 1159–1178.
[3] J. MACQUEEN, "Some methods for classification and analysis on multivariate observations," *Proceedings of the Fifth Berkeley Symposium on Mathematical Statistics and Probability*, Berkeley and Los Angeles, University of California Press, 1966, Vol. 1, pp. 281–297.
[4] F. J. ROHLF, "Adaptive hierarchical clustering schemes," *Syst. Zool.*, Vol. 19 (1970), pp. 58–83.

MEASUREMENT OF DIVERSITY

F. N. DAVID

UNIVERSITY OF CALIFORNIA, RIVERSIDE

1. Introduction

The several measurements used by ecologists to measure diversity in plant and animal populations have been summarized by Pielou [6]. This present paper is concerned with an extension of the idea of diversity in plant populations and in particular with the description of data produced by a densitometer. Further papers applying the present ideas to actual forests counts where there is more than one observation to a cell will appear elsewhere.

2. The problem

A film is taken by an airplane flying over a natural forest. The film is put through a densitometer which prints out at equal intervals a letter corresponding to its optical density. In the particular experiment which was presented to us there were 120 letters printed out for the scan across the film, the letters being A through G inclusive. The number of letters for the scan down the film is dependent only on the length of the film. The optical density of the film and therefore the letter corresponding to it is supposedly representative of the type of tree. A measure of the clustering of the trees is required.

Essentially the same problem arises if the forest is gridded, the fuel bed computed for each square, and the results of the computations assigned to one or other of ten classes.

It will be recognized that if there are m letters one way and n letters the other the problem reduces to that of a board with $m \times n$ cells on which $m \times n$ letters are arranged, one letter to a cell, the arrangement under the null hypothesis being that of randomness.

3. Notation

Let there be s kinds of letters, with k_t of the tth kind. Denote

$$(3.1) \qquad \sum_{t=1}^{s} k_t^{(r)} = K_r$$

and

$$(3.2) \qquad K_1 = mn = \sum_{t=1}^{s} k_t.$$

With the partial support of the National Institutes of Health

631

Consider the lattice of $(m - 1)(n - 1)$ lines formed by the boundaries of the cells, excluding the border framework. Define a random variable $\alpha_{i,j}$, $i = 1, \cdots, m - 1; j = 1, \cdots, n - 1$, associated with the i,jth node of the lattice.

Let $\alpha_{i,j}$ score 6 if all four letters surrounding the node are the same,

score 3 if three letters are the same and one different,

score 2 if two are the same and two are the same,

score 1 if two are the same and two different,

score 0 if all four are different.

(Other methods of scoring are possible and will be discussed later.)
We propose

$$(3.3) \qquad S = \sum_{i=1}^{m-1} \sum_{j=1}^{n-1} \alpha_{i,j}$$

as a measure of clustering and of diversity.

4. Properties of S

The basic hypothesis is that the $m \times n$ letters are randomly placed in the $m \times n$ squares. The alternate hypothesis, as far as forestry problems of position, of disease, and so on, are concerned is that letters of like kind tend to cluster together. Consequently, a large value of S will indicate possibly that there is clustering, a small value will indicate that like letters are more widely dispersed from each other than might be anticipated. From the point of view of the forestry problems the acceptance or rejection of an hypothesis of randomness is not important. The field ecologist is concerned chiefly in the calculation of an index of diversity, range zero to unity, which will be large when the numbers of species are equal and the species are clumped together, and small when the converse is true. The maximum and minimum values of S are reasonably easy to compute so that one possible index is

$$(4.1) \qquad I_1 = \frac{S - \min S}{\max S - \min S}.$$

From a statistical point of view this is not very satisfactory since the distribution of S may have very long tails in which case I_1 could give a very misleading result. We have tended up to the present to use

$$(4.2) \qquad I_2 = \frac{S - E(S)}{\sigma_S}$$

which, although familiar to the statistician, is not liked by the biologist because it has not got a zero-one range. Undoubtedly in the present writer's opinion the best index would be to compute

$$(4.3) \qquad I_3 = P\{S \leqq S_o\},$$

where S_o is the observed value. But to compute I_3 the distribution of S must be known and this for the present we do not have. The algebraic derivation of the mean and variance of S is given in succeeding sections.

5. Algebraic attack

For the mean and variance of S under any system of scoring we rely heavily on the tables of Augmented Symmetric Functions constructed by David and Kendall [3]. The algebra is elementary and simple in principle. For example, it is clear that we may write

$$(5.1) \qquad K_1^{(4)} = \sum_{\ell=1}^{s} k_\ell^{(4)} + 4 \sum_{\ell \neq h} k_\ell^{(3)} k_n + 3 \sum_{\ell \neq h} k_\ell^{(2)} k_h^{(2)} + 6 \sum_{\ell \neq h \neq t} k_\ell^{(2)} k_h k_t + \sum_{\ell \neq h \neq t \neq r} k_\ell k_h k_t k_r,$$

formally corresponding to the expansion of power sums in terms of the Augmented Monomial Symmetric Functions (David and Kendall [3]), namely,

$$(5.2) \qquad (1)^4 = \sum_{\ell=1}^{s} k_\ell^4 + 4 \sum_{\ell \neq h} k_\ell^3 k_h + 3 \sum_{\ell \neq h} k_\ell^2 k_h^2 + 6 \sum_{\ell \neq h \neq t} k_\ell^2 k_h k_t + \sum_{\ell \neq h \neq t \neq r} k_\ell k_h k_t k_r = [4] + 4[31] + 3[2^2] + 6[21^2] + [1^4].$$

This raises the possibility of expressing what may be called the Augmented Factorial Monomial Symmetric Functions (AFMSF for short) in terms of powers and products of the K and vice versa, with numerical coefficients which are the same as those in the power sum AMSF relationship. The product of the K-functions is easy to compute algebraically but is difficult to express in any succinct fashion. Thus if we write

$$(5.3) \qquad \sum_{\ell \neq h \neq t} k_\ell^{(a)} k_h^{(b)} k_t^{(c)} = [k_a k_b k_c],$$

the first relationship given above is, symbolically,

$$(5.4) \qquad K_1^{(4)} = [k_4] + 4[k_3 k_1] + 3[k_2^2] + 6[k_2 k_1^2] + [k_1^4],$$

but few of the others show such simplicity. Table I gives the formal correspondence between some of the power sums and the K-functions, w being the

TABLE I

K-FUNCTIONS CORRESPONDING TO GIVEN POWER SUMS

Power Sum	K-Functions
$(n)(1)^{w-n}$	$K_n(K_1 - n)^{(w-n)}$
$(n)(2)$	$K_n(K_2 - n^{(2)}) - 2nK_{n+1}$
$(n)(2)^2$	$K_n(K_2 - n^{(2)})(K_2 - n^{(2)} - 2) - 4nK_{n+1}(K_2 - n^{(2)} - 2)$
	$- 4K_n(K_3 - n^{(3)}) + 4n(n+3)K_{n+2} + 16n^{(2)}K_{n+1}$
$(2)^2(1)^{w-4}$	$(K_2(K_2 - 2) - 4K_3)(K_1 - 4)^{(w-4)}$
$(3)^2(1)^{w-6}$	$(K_3(K_3 - 6) - 9K_5 - 18K_4)(K_1 - 6)^{(w-6)}$
$(4)^2(1)^{w-8}$	$(K_4(K_4 - 24) - 16K_7 - 72K_6 - 96K_5)(K_1 - 8)^{(w-8)}$

weight. Generally we will have, corresponding to $(m)(n)$, the K-function

$$(5.5) \qquad K_n(K_m - n^{(m)}) - {}^mC_1{}^nC_1K_{n+m-1} - 2!{}^mC_c{}^nC_2K_{n+m-2}$$
$$- \cdots - {}^mC_{m-1}{}^nC_{m-1}(m-1)!K_{n+1}.$$

These enable Tables AI of the appendix to be written down immediately from the tables already in existence. Further relationships between the K-products and the AFMSF were worked out and tables of higher weights were constructed. These however—and the complete tables—proved to be unnecessary for our particular problem and so we do not reproduce them here.

6. Mean of S (one, two, and three dimensions)

It is clear that the two dimensional problem described earlier is just a special case of scoring over a lattice in any number of dimensions. The method used, involving the use of the AFMSF is applicable for any number of dimensions and any method of scoring.

6.1. *One dimension.* For one dimension the problem is that of the multicolored run which has been completely solved (Barton and David [1]). Define

$$(6.1) \qquad \alpha_i = \begin{cases} 1 & \text{if same color either side of } i\text{th gap} \\ 0 & \text{otherwise.} \end{cases}$$

Then

$$(6.2) \qquad S = \sum_{i=1}^{K_1-1} \alpha_i.$$

We have

$$(6.3) \qquad K_1^{(2)} = \sum_{\ell=1}^{s} k_\ell^{(2)} + \sum_{\ell \neq v} k_\ell k_v,$$

$$(6.4) \qquad E(\alpha_i) = \frac{1}{K_1^{(2)}} \left[1 \sum_{\ell=1}^{s} k_\ell^{(2)} + 0 \sum_{\ell \neq v} k_\ell k_v \right] = \frac{K_2}{K_1^{(2)}}$$

and

$$(6.5) \qquad ES = \frac{K_2}{K_1}.$$

6.2. *Two dimensions.* Corresponding to the (i,j) node of an $mn(=K_1)$ board, we have

$$(6.6) \qquad S = \sum_{i=1}^{m-1} \sum_{j=1}^{n-1} \alpha_{i,j}.$$

Starting from the relationship

$$(6.7) \qquad K_1^{(4)} = \sum_\ell k_\ell^{(4)} + 4 \sum_{\ell \neq v} k_\ell^{(3)} k_v + 3 \sum_{\neq} k_\ell^{(2)} k_v^{(2)} + 6 \sum_{\neq} k_\ell^{(2)} k_v k_h$$
$$+ \sum_{\neq} k_\ell k_v k_h k_r,$$

each permutation of a of one color, b of another, and so on ($a + b + \cdots = 4$) will give rise to the same score. So we may write

$$(6.8) \qquad E(\alpha_{i,j}) = \frac{1}{K_1^{(4)}} \left[s_4 \sum_\ell k_\ell^{(4)} + 4s_3 \sum_{\ell \neq v} k_\ell^{(3)} k_v + 3s_2 \sum_{\neq} k_\ell^{(2)} k_v^{(2)} \right.$$
$$\left. + 6s_1 \sum_{\neq} k_\ell^{(2)} k_v k_n + s_0 \sum_{\neq} k_\ell k_v k_n k_r \right].$$

For the scoring system given previously (Section 3)

$$(6.9) \qquad s_4 = 6, \quad s_3 = 3, \quad s_2 = 2, \quad s_1 = 1, \quad s_0 = 0,$$

and

$$(6.10) \qquad E(\alpha_{i,j}) = \frac{6}{K_1^{(4)}} \left[\sum_\ell k_\ell^{(4)} + 2 \sum_{\ell \neq v} k_\ell^{(3)} k_v + \sum_{\neq} k_\ell^{(2)} k_v^{(2)} + \sum_{\neq} k_\ell^{(2)} k_v k_h \right].$$

It is possible to write each of the AFMSF in terms of sums of products of the K-functions, using the upper half of Table AI in the appendix, but it is easier to note that the expression in the brackets is the expansion of $K_2(K_1 - 2)^{(2)}$. Accordingly

$$(6.11) \qquad E(\alpha_{i,j}) = \frac{6K_2}{K_1^{(2)}}.$$

An alternative method of scoring is not to count diagonal elements. Thus we score

$$\begin{matrix} A \, A \\ \cdot \\ A \, A \end{matrix} \; 4, \quad \begin{matrix} A \, A \\ \cdot \\ A \, B \end{matrix} \; 2, \quad \begin{matrix} A \, B \\ \cdot \\ A \, B \end{matrix} \; 2, \quad \begin{matrix} A \, B \\ \cdot \\ B \, A \end{matrix} \; 0, \quad \begin{matrix} A \, B \\ \cdot \\ A \, C \end{matrix} \; 1,$$

$$\begin{matrix} A \, C \\ \cdot \\ B \, A \end{matrix} \; 0, \quad \begin{matrix} A \, C \\ \cdot \\ B \, D \end{matrix} \; 0.$$

This will mean that not all permutations, for given $a, b, c, \cdots$, and so on, will have the same weight and we will have

$$(6.12) \qquad E(\alpha_{i,j}) = \frac{1}{K_1^{(4)}} \left[4 \sum k_1^{(4)} + 8 \sum_{\ell \neq v} k_\ell^{(3)} k_v + 4 \sum_{\neq} k_\ell^{(2)} k_v^{(2)} + 4 \sum_{\neq} k_\ell^{(2)} k_v k_h \right]$$
$$= \frac{4K_2}{K_1^{(2)}}.$$

These two methods of scoring are the only ones that have been worked out in some detail but clearly any other method of scoring is easily applied.

6.3. *Three dimensions.* Corresponding to the (i, j, ℓ) node of an $m \times n \times t(= K_1)$ rectangular parallelepiped built up from unit cubes we have

$$(6.13) \qquad S = \sum_{i=1}^{m-1} \sum_{j=1}^{n-1} \sum_{\ell=1}^{t-1} \alpha_{i,j,\ell}.$$

The fundamental relationship is

$$(6.14) \qquad K_1^{(8)} = \sum_a k_a^{(8)} + 8 \sum_{\neq} k_a^{(7)} k_b + 28 \sum_{\neq} k_a^{(6)} k_b^{(2)} + \cdots$$

$$+ \sum_{\neq} k_a k_b k_c k_d k_e k_f k_g k_h.$$

If diagonal joins between like letters are allowed then the score for each partition, and therefore for each permutation, is written down immediately and we have

$$(6.15) \qquad E(\alpha_{i,j,\ell}) = \frac{28 K_2}{K_1^{(2)}}.$$

If diagonal joins between like letters are not allowed then the score has to be found for each permutation within a given partition and

$$(6.16) \qquad E(\alpha_{i,j,\ell}) = \frac{12 K_2}{K_1^{(2)}}.$$

It proved more convenient in subsequent manipulations to denote by Z the quantity $K_2(K_1 - 2)^{(2)}$. We have then for mean S

one dimension $\qquad Z(K_1 - 1)/K_1^{(4)};$

two dimensions $\qquad 6Z(m - 1)(n - 1)/K_1^{(4)}, \qquad$ diagonal joins allowed;

$\qquad\qquad\qquad\quad 4Z(m - 1)(n - 1)/K_1^{(4)}, \qquad$ diagonal joins not allowed;

three dimensions $\quad 28Z(m - 1)(n - 1)(t - 1)/K_1^{(4)}, \quad$ diagonal joins allowed;

$\qquad\qquad\qquad\qquad 12Z(m - 1)(n - 1)(t - 1)/K_1^{(4)}, \quad$ diagonal joins not allowed.

7. Free sampling

The algebraic approach for obtaining the expected value of S is conditional on the $\{k_\ell\}$, $\ell = 1, \cdots, s$. Some simplification in the formulae results if it is supposed that the k are obtained as the result of some sampling procedure either, say, free multinomial sampling or a Pólya multiple urn process. Suppose K_1 letters are drawn from an urn in which the proportion of the ℓth kind is kept constant and equal to p_ℓ, and

$$(7.1) \qquad \sum_{\ell=1}^{s} p_\ell = 1.$$

These K_1 letters are put down randomly as before. We have then, if we write

$$(7.2) \qquad \sum_{\ell=1}^{s} p_\ell^r = P_r,$$

that

$$(7.3) \qquad \text{one dimension} \qquad E(S) = (K_1 - 1)P_2;$$

two dimensions

$$(7.4) \qquad E(S) \begin{cases} = 6(m-1)(n-1)P_2, & \text{diagonal joins allowed,} \\ = 4(m-1)(n-1)P_2, & \text{diagonal joins not allowed;} \end{cases}$$

three dimensions

$$(7.5) \qquad E(S) \begin{cases} = 28(m-1)(n-1)(t-1)P_2, & \text{diagonal joins allowed,} \\ = 12(m-1)(n-1)(t=1)P_2, & \text{diagonal joins not allowed.} \end{cases}$$

If we use a Pólya model where for each letter the proportion of letters returned to the urn after drawing is the same and is equal to δ, then, writing

$$(7.6) \qquad P_2 = \sum_{\ell=1}^{s} \frac{p_\ell(p_\ell + \delta)}{1(1 + \delta)},$$

the expectations are formally the same as above.

8. Second moment of S

The second moment of S represents no intrinsic difficulty although the enumeration of the scores for the different permutations within a given partition allows scope for error.

8.1. *One dimension.* Here

$$(8.1) \qquad S^2 = \sum_{i=1}^{K_1-1} \alpha_i^2 + 2 \sum_{i=1}^{K_1-2} \alpha_i\alpha_{i+1} + 2 \sum_{i=1}^{K_1-3} \sum_{j=i+2}^{K_1-1} \alpha_i\alpha_j$$

and if we write $Y = K_3(K_1 - 3)$, $X = K_2(K_2 - 2) - 4K_3$, with Z as before, we have

$$(8.2) \qquad E(\alpha_i^2) = \frac{K_2}{K_1^{(2)}} = \frac{Z}{K_1^{(4)}},$$

$$(8.3) \qquad E(\alpha_i\alpha_{i+1}) = \frac{1}{K_1^{(3)}} \left[1 \sum_{a=1}^{s} k_a^{(3)} + (2)(0) \sum_{a \neq b} k_a^{(2)}k_b + (1)(0) \sum_{a \neq b \neq c} k_a k_b k_c \right]$$

$$= \frac{K_3}{K_1^{(3)}} = \frac{Y}{K_1^{(4)}},$$

$$(8.4) \qquad E(\alpha_i\alpha_j) = \frac{1}{K_1^{(4)}} \left[1 \sum k_a^{(4)} + (4)(0) \sum k_a^{(3)}k_b + (3)\left(\frac{1}{3}\right) \sum k_a^{(2)}k_b^{(2)} \right.$$

$$\left. + (6)(0) \sum k_a^{(2)}k_b k_c + (1)(0) \sum k_a k_b k_c k_d \right],$$

$$= \frac{K_2(K_2 - 2) - 4K_3}{K_1^{(4)}} = \frac{X}{K_1^{(4)}}.$$

Accordingly,

$$(8.5) \qquad E(S^2) = \frac{1}{K_1^{(4)}} \left[(K_1 - 1)Z + 2(K_1 - 2)Y + (K_1 - 2)(K_1 - 3)X \right].$$

It may be noted for purposes of symmetry with the larger dimensions that

$$(8.6) \qquad (K_1 - 2)(K_1 - 3) = (K_1 - 1)^2 - (3K_1 - 5).$$

8.2. *Two dimensions.* Following previous notation

$$(8.7) \qquad S^2 = \left(\sum_{i=1}^{m-1} \sum_{j=1}^{n-1} \alpha_{i,j} \right)^2.$$

The expansion of the square of this double sum may be written out as in Table II.

TABLE II

Number of Terms in Summations of (8.7)

	i^2	$i(i+1)$	$i, h, (\lvert i-h \rvert \geqq 2)$	Totals
j^2	$(m-1)(n-1)$	$2(m-2)(n-1)$	$(m-2)^{(2)}(n-1)$	$(m-1)^2(n-1)$
$j(j+1)$	$2(m-1)(n-2)$	$4(m-2)(n-2)$	$2(m-2)^{(2)}(n-2)$	$2(m-1)^2(n-2)$
$j, r (\lvert j-r \rvert \geqq 2)$	$(m-1)(n-2)^{(2)}$	$2(m-2)(n-2)^{(2)}$	$(m-2)^{(2)}(n-2)^{(2)}$	$(m-1)^2(n-2)^{(2)}$
Totals	$(m-1)(n-1)^2$	$2(m-2)(n-1)^2$	$(m-2)^{(2)}(n-1)^2$	$(m-1)^2(n-1)^2$

There are four types of terms for which it will be necessary to calculate expectations.

(i) $\alpha_{i,j}^2$, $(m-1)(n-1)$ *terms.* From the expansion for $K_1^{(4)}$, allowing diagonal joins, we have

$$(8.8) \qquad E(\alpha_{i,j}^2) = \frac{1}{K_1^{(4)}} \left[36 \sum k^{(4)} + 36 \sum k^{(3)}k + 12 \sum k^{(2)}k^{(2)} + 6 \sum k^{(2)}kk \right]$$

This may be split up as follows

$$
\begin{aligned}
6K_1(K_1 - 2)^{(2)} &= 6[k_4] + 12[k_3k_1] + 6[k_2^2] + 6[k_2k_1^2] \\
(8.9) \qquad 6K_2(K_2 - 2) - 4K_3 &= 6[k_4] + 6[k_2^2] \\
24K_3(K_1 - 3) &= 24[k_4] + 24[k_3k_1]
\end{aligned}
$$

so that

$$(8.10) \qquad E(\alpha_{i,j}^2) = \frac{1}{K_1^{(4)}} [6X + 24Y + 6Z]$$

If diagonal joins are not allowed

$$(8.11) \qquad E(\alpha_{i,j}^2) = \frac{1}{K_1^{(4)}} [4Z + 8Y + 4Z].$$

(ii) $\alpha_{i,j}\alpha_{i,j+1}$, $2(m-1)(n-2)$ *terms.* The expectation will be the same for $\alpha_{i,j}\alpha_{i+1,j}$ although the number of terms will be $2(m-2)(n-1)$. We start from the expansion of $K_1^{(6)}$ in terms of the AFMSF. For each permutation of a given partition it is necessary to calculate the score having regard to the fact that two squares or cells are held in common by each of the other two. Table AII of the appendix gives the total score for each permutation with diagonals allowed and not allowed. Using this table we have for example

$$(8.12) \qquad E(\alpha_{i,j}\alpha_{i,j+1}) = \frac{1}{K_1^{(6)}} \{36[k_6] + 90[k_5k_1] + 112[k_4k_2] + 58[k_3^2]$$
$$+ 73[k_4k_1^2] + 140[k_3k_2k_1] + 22[k_2^3] + 20[k_3k_1^3]$$
$$+ 25[k_2^2k_1^2] + [k_2k_1^4]\}.$$

This may be split up to give Table III, using the formal AMSF correspondence

TABLE III

ILLUSTRATION OF CALCULATION OF THE EXPECTATION OF $\alpha_{i,j}\alpha_{i,j+1}$

	$[k_6]$	$[k_5k_1]$	$[k_4k_2]$	$[k_4k_1^2]$	$[k_3^2]$	$[k_3k_2k_1]$	$[k_3k_1^3]$	$[k_2^3]$	$[k_2^2k_1^2]$,	$[k_2k_1^4]$
$Z(K_1-4)^{(2)}$	1	4	7	6	4	16	4	3	6	1
$19X(K_1-4)^{(2)}$	19	38	57	19	38	76	·	19	19	
$16Y(K_1-4)^{(2)}$	16	48	48	48	16	48	16			
	36	90	112	73	58	140	20	22	25	1

(Table I·1·6 [4]), so that

$$(8.13) \qquad E(\alpha_{i,j}\alpha_{i,j+1}) = \frac{1}{K_1^{(4)}}[19X + 16Y + Z] = E(\alpha_{i,j}\alpha_{i+1,j}).$$

If diagonals are not allowed then

$$(8.14) \qquad E(\alpha_{i,j}\alpha_{i,j+1}) = \frac{1}{K_1^{(4)}}[9X + 6Y + Z] = E(\alpha_{i,j}\alpha_{i+1,j}).$$

The process is essentially the same for the other two expectations required. For $E(\alpha_{i,j}\alpha_{i+1,j+1})$ we start from $K_1^{(7)}$ and calculate the score for each partition for two sets of four squares which have one square in common. For $E(\alpha_{i,j}\alpha_{\ell,h})$ we start from $K_1^{(8)}$ scoring two sets of four squares which do not have a square in common. The results are

$$E(\alpha_{i,j}\alpha_{i+1,j+1}) = \frac{1}{K_1^{(4)}}[27X + 9Y], \quad \frac{1}{K_1^{(4)}}[12X + 4Y],$$

$$(8.15)$$

$$E(\alpha_{i,j}\alpha_{\ell,h}) = \frac{1}{K_1^{(4)}}[36X], \qquad \frac{1}{K_1^{(4)}}[16X],$$

the first expression being where diagonals are allowed and the second where diagonals are not allowed in each case.

While the expressions for the expectations of products of the α may be written comparatively simply, no real simplicity appears possible for the second moment (or the variance) of S. We have

if diagonals are allowed

$$(8.16) \quad E(S^2) = \frac{1}{K_1^{(4)}} \{(m-1)(n-1)[6X + 24Y + 6Z]$$
$$+ [2(m-2)(n-1) + 2(m-1)(n-2)][19X + 16Y + Z]$$
$$+ 4(m-2)(n-2)[27X + 9Y] + [(m-1)^2(n-1)^2$$
$$- (3m-5)(3n-5)][36X]\},$$

and if diagonals are not allowed

$$(8.17) \quad E(S^2) = \frac{1}{K_1^{(4)}} \{(m-1)(n-1)[4X + 8Y + 4Z]$$
$$+ [2(m-2)(n-1) + 2(m-1)(n-2)][9X + 6Y + Z]$$
$$+ 4(m-2)(n-2)[12X + 4Y] + [(m-1)^2(n-1)^2$$
$$- (3m-5)(3n-5)][16X]\}.$$

The simplicity which resulted, in the one dimensional case, from calculating the second factorial moment of S does not follow for two or three dimensions.

8.3. *Three dimensions.*   For three dimensions we have

$$(8.18) \quad E(\alpha_{i,j,h}^2)$$
$$= \frac{1}{K_1^{(4)}} [420X + 336Y + 28Z], \qquad \frac{1}{K_1^{(4)}} [84X + 48Y + 12Z],$$

$$(8.19) \quad E(\alpha_{i,j,h}\alpha_{i,j,h+1})$$
$$= \frac{1}{K_1^{(4)}} [594X + 184Y + 6Z], \qquad \frac{1}{K_1^{(4)}} [112X + 28Y + 4Z],$$

$$(8.20) \quad E(\alpha_{i,j,h}\alpha_{i,j+1,h+1})$$
$$= \frac{1}{K_1^{(4)}} [687X + 96Y + Z], \qquad \frac{1}{K_1^{(4)}} [119X + 24Y + Z],$$

$$(8.21) \quad E(\alpha_{i,j,h}\alpha_{i+1,j+1,h+1})$$
$$= \frac{1}{K_1^{(4)}} [735X + 49Y], \qquad \frac{1}{K_1^{(4)}} [135X + 9Y],$$

$$(8.22) \quad E\alpha_{i,j,h}\alpha_{a,b,c})$$
$$= \frac{1}{K_1^{(4)}} [784X], \qquad \frac{1}{K_1^{(4)}} [144X].$$

The expressions to the left are those in which diagonal counts are allowed and those to the left are those where they are not allowed. The numbers of terms are, respectively,

$(m - 1)(n - 1)(t - 1),$

$2(m - 1)(n - 1)(t - 2) + 2(m - 1)(n - 2)(t - 1) + 2(m - 2)(n - 1)(t - 1),$

$4(m - 1)(n - 2)(t - 2) + 4(m - 2)(n - 1)(t - 2) + 4(m - 2)(n - 2)(t - 1),$

$8(m - 2)(n - 2)(t - 2),$

$(m - 1)^2(n - 1)^2(t - 1)^2 - (3m - 5)(3n - 5)(3t - 5).$

The second moment of S follows by simple multiplication of these expressions but, since no simplification results, I have not formally written it out.

9. Second moment under free sampling

The quantities intervening in the second moment which have to be taken into account in passing from conditional to unconditional expectations are X, Y, Z. The two latter present no difficulty:

(9.1) $$E(Y) = K_1^{(4)} P_3, \qquad E(Z) = K_1^{(4)} P_2,$$

while

(9.2) $$X = K_2^2 - 2K_2 - 4K_3 = \sum_\ell k_{2\ell} k_{2\ell} + \sum_{\neq} k_{2\ell} k_{2h} - 2 \sum k_{2\ell} - 4 \sum k_{3\ell}$$

$$= \sum k_{4\ell} + \sum_{\neq} k_{2\ell} k_{2h}$$

and

(9.3) $$E(X) = \left[P_4 + \sum_{\neq} p_1^2 p_h^2 \right] K_1^{(4)} = P_2^2 K_1^{(4)}.$$

Accordingly, the unconditional expectation of the second moment of S will be obtained by substituting P_2, P_3, and P_2^2 for $Z/K_1^{(4)}$, $Y/K_1^{(4)}$, and $X/K_1^{(4)}$, respectively. The variance of S (2D, diagonals) may be written as

(9.4) $$\text{Var } S = (m - 1)(n - 1)[10P_2 + 124P_3 - 134P_2^2]$$
$$- [(m - 1) + (n - 1)][2P_2 + 68P_3 - 70P_2^2] + 36[P_3 - P_2^2].$$

For the other cases we have

(9.5) $$\text{Var } S \text{ (2D, no diagonals)} = (m - 1)(n - 1)[8P_2 + 48P_3 - 56P_2^2]$$
$$- [(m - 1) + (n - 1)][2P_2 + 28P_3 - 30P_2^2] + 16[P_3 - P_2^2],$$

(9.6) $$\text{Var } S \text{ (3D, diagonals)}$$
$$= (m - 1)(n - 1)(t - 1)[76P_2 + 2984P_3 - 3060P_2^2]$$
$$- [(m - 1)(n - 1) + (n - 1)(t - 1)$$
$$+ (t - 1)(m - 1)][20P_2 + 1528P_3 - 1548P_2^2]$$
$$+ [(m - 1) + (n - 1) + (t - 1)][4P_2 + 776P_3 - 780P_2^2]$$
$$- 392[P_3 - P_2^2],$$

(9.7)   Var S (3D, no diagonals)

$$= (m-1)(n-1)(t-1)[48P_2 + 576P_3 - 624P_2^2]$$
$$- [(m-1)(n-1) + (n-1)(t-1)$$
$$+ (t-1)(m-1)][16P_2 + 320P_3 - 336P_2^2]$$
$$+ [(m-1) + (n-1) + (t-1)][4P_2 + 168P_3 - 172P_2^2]$$
$$- 72[P_3 - P_2^2].$$

10. Special cases

(i) When all the probabilities are the same, that is, when

(10.1) $$p_\ell = \frac{1}{s}$$ $\ell = 1, \cdots, s,$

the last terms of each expression are zero and $(s-1)/s^2$ is a factor. Thus, in order of writing above we have

$$\frac{s-1}{s^2}\{10(m-1)(n-1) - 2[(m-1) + (n-1)]\},$$

$$\frac{s-1}{s^2}\{8(m-1)(n-1) - 2[(m-1) + (n-1]\},$$

$$\frac{s-1}{s^2}\{76(m-1)(n-1)(t-1) - 20[(m-1)(n-1) + (m-1)(t-1)$$
$$+ (n-1)(t-1)] + 4[(m-1) + (n-1) + (t-1)]\},$$

$$\frac{s-1}{s^2}\{48(m-1)(n-1)(t-1) - 16[(m-1)(n-1) + (m-1)(t-1)$$
$$+ (n-1)(t-1)] + 4[(m-1) + (n-1) + (t-1)]\}.$$

(ii) When there are only two kinds of letters

(10.2) $p_1 = p,$ $p_2 = 1 - p_1 = q,$ $p_3 = \cdots = p_s = 0$

and

(10.3) $$P_2 = 1 - 2pq,   P_3 = 1 - 3pq.$$

In order of writing the variances now become

$$pq[(m-1)(n-1)(144 - 536pq) - [(m-1) + (n-1][72 - 280pq]$$
$$+ 36(1 - 4pq)],$$

$$pq[(m-1)(n-1)(64 - 224pq) - [(m-1 + (n-1)][32 - 120pq]$$
$$+ 16(1 - 4pq)],$$

$$pq[(m-1)(n-1)(t-1)[3136 - 12240pq] - [(m-1)(n-1)$$
$$+ (n-1)(t-1) + (t-1)(m-1)][1568 - 6192pq]$$
$$+ [(m-1) + (n-1) + (t-1)][748 - 3120pq] - 392[1 - 4pq],$$

$$pq[(m - 1)(n - 1)(t - 1)[672 - 2469pq] - [(m - 1)(n - 1)$$
$$+ (n - 1)(t - 1) + (t - 1)(m - 1)][352 - 1344pq]$$
$$+ [(m - 1) + (n - 1) + (t - 1)][176 - 688pq] - 72(1 - 4pq)].$$

11. Second moment under a multiple Pólya model

Let

$$(11.1) \qquad P\{k_i\} = \frac{K_i! \displaystyle\prod_{j=1}^{s} \prod_{i=0}^{k_j-1} (p_j + i\delta)}{\displaystyle\prod_{i=1}^{s} k_i! \prod_{h=0}^{K_1-1} (1 + h\delta)}$$

($\delta = 0$ gives the multinomial). Taking for example the two dimensional case, diagonals allowed, we have

$$(11.2) \qquad E(S) = \frac{6(P_2 + \delta)(m - 1)(n - 1)}{1 + \delta}$$

and writing

$$(11.3) \qquad D_4 = 1(1 + \delta)(1 + 2\delta)(1 + 3\delta),$$

we have for the unconditional variance of S

$$(11.4) \ \operatorname{Var} S = \frac{1}{D_4(1 + \delta)} \big[(m - 1)^2(n - 1)^2 \{144\,\delta(P_3 - P_2^2)$$
$$+ 72\,\delta^2(P_2 + 2P_3 - 3P_2^2) + 72\,\delta^3[1 - P_2]\}$$
$$+ (m - 1)(n - 1)\{(10P_2 + 124P_3 - 134P_2^2)$$
$$+ \delta(10 + 164P_2 - 40P_3 - 134P_2^2) + \delta^2(174 - 10P_2 - 164P_3)$$
$$+ \delta^3(164 - 164P_2)\}$$
$$+ ((m - 1) + (n - 1))\{(2P_2 + 68P_3 - 70P_2^2)$$
$$+ \delta(2 + 76P_2 - 8P_3 - 70P_2^2) + \delta^2(72 - 2P_2 - 70P_3)$$
$$+ \delta^3(76 - 76P_2)\}$$
$$+ 36\{(P_3 - P_2^2) - \delta(P_2 - P_3) + \delta^2(1 - P_3) + \delta^3(1 - P_2)\}\big].$$

The results for other methods of scoring and for other dimensions follow similarly.

12. Higher moments of S

The higher moments of S present no intrinsic difficulty although a certain amount of counting is required. It may be noted that the third moment will involve X, Y, Z, and

$$(12.1) \quad \begin{aligned} W &= K_2(K_2 - 2)(K_2 - 4) - 12K_3(K_2 - 2) + 40(K_4 + K_3), \\ V &= K_3(K_2 - 6) - 6K_4, \qquad\qquad\qquad\qquad U = K_4, \end{aligned}$$

but no others. The fourth moment will involve the previous six quantities and

$$(12.2) \quad R = K_2(K_2 - 2)(K_2 - 4)(K_2 - 6) - 24K_3(K_2 - 3)(K_2 - 5)$$
$$+ 112K_3(K_2 - 3) + 160K_4(K_2 - 4) + 48K_3(K_3 - 3) - 672K_5$$
$$- 1616K_4 - 408K_3,$$

$$T = K_3(K_2 - 3)(K_2 - 5), \quad Q = K_4(K_2 - 4), \quad M = K_3(K_3 - 3), \quad N = K_5.$$

13. Examples

A knowledge of the movements of S will enable an approximation to its distribution to be made and therefore a calculation of the index of diversity I suggested earlier. The distribution of S will depend on the size and configuration of the lattice and the letter specification, as will be illustrated below. The difficulty is to find a functional form sufficiently flexible to take account of all these factors, but two approaches are possible.

EXAMPLE 1.   Consider a letter specification (12,4) on a 4 × 4 tableau, with the diagonals counting in the scoring. The distribution of S is shown in Table IV.

TABLE IV

DISTRIBUTION OF S FOR A 4 × 4 TABLEAU
WITH OBJECT SPECIFICATION (12, 4)

S	24	25	26	27	28	29	30	31	32	33	
Frequency	28	20	83	96	140	124	132	216	140	108	
S	34	35	36	37	38	39	40	41	42	43	Total
Frequency	148	180	40	108	128	40	36	32	17	4	1820

The momental constants are

$$\mu_1' = 32.4, \qquad \mu_3 = 13.90735, \quad \beta_1 = 0.03811, \quad \sqrt{\beta_1} = 0.1952,$$
$$\mu_2 = 17.18505, \quad \mu_4 = 687.15027, \quad \beta_2 = 2.32675, \qquad \sigma = 4.1455.$$

(i) Pearson and Hartley [5], p. 206, give the standardized deviates for Pearson curves for given percentage points. Using the (β_1, β_2) of the distribution above, we get Table V. These figures indicate that the *Biometrika* table will be good enough for significance levels if they are ever required.

TABLE V

PERCENTAGE POINTS OF S FROM PEARSON CURVES

Percentage Point	Deviate from Tables	Actual Percentages
5	39.546	4.9
2.5	40.583	2.9
1	41.649	1.15

(ii) Fisher-Cornish [2] give an inverse Edgeworth expansion for the percentage points of a distribution with known momental constants. Using their expansion we have Table VI. The deviates obtained by the Fisher-Cornish expansion do not differ markedly from those obtained from the Pearson system.

TABLE VI

PERCENTAGE POINTS OF S FROM THE EDGEWORTH SERIES

Percentage Point	Deviate	Actual Percentage
5	39.502	4.9
2.5	40.694	2.9
1	41.927	1.15
50	32.2651	46.2

EXAMPLE 2. Table AIII of the Appendix gives the distribution of S for eight letters (different specifications) on a 4×2 board. All these distributions will be, to a certain extent, atypical in that the edge effects will play an undue part. The difference between the Pearson system and the Fisher-Cornish expansion is a little more marked than for Example 1. If we take, for instance, the distribution with object specification (421^2) the (β_1, β_2) point is $(0.755, 3.321)$ and the 5 percent points are 7.306 and 7.198 for Pearson and Fisher-Cornish, respectively. The actual percentage in each case is 4.29.

It is clear that if we obtain formulae for the third and fourth moments of S, we may approximate to its distribution reasonably well as far as the biological problem is concerned. Further, sampling experiments seem to indicate that with a large lattice the distribution of S can be approximated by a constant times χ^2. The question of the configuration of the lattice is one which is of small importance for a large lattice, but may be of importance when the numbers are small. Table AIV and Figure 1 of the Appendix illustrates this point.

14. Conclusion

It is clear from Section 13 that with a knowledge of the third and fourth moments an approximation to the distribution of S is possible although it would be idle to suggest that either of the methods suggested will be that ultimately chosen. The algebraic derivation of the third and fourth moments, as indicated earlier, is simple in principle but tedious in procedure. Some headway has been made with the third moment. If, as appears likely, it is close to the third moment of a constant times χ^2, then little would be gained from the derivation of the fourth moment.

The practical applications of the foregoing theory and the extension of this theory to the case where there is more than one letter per square, are numerous in the ecological field. It is proposed to publish these elsewhere.

The problem (and others allied to it) set out at the beginning of this paper arose from consultations at the Pacific Southwest Forest and Range Experiment Station, Forest Service, U.S. Department of Agriculture.

APPENDIX

TABLE AIa

AUGMENTED FACTORIAL MONOMIAL SYMMETRIC FUNCTIONS
WEIGHT 2

$(w = 2)$	$[k_2]$	$[k_1^2]$
K_2	1	-1
$K_1^{(2)}$	1	**1**

TABLE AIb

AUGMENTED FACTORIAL MONOMIAL SYMMETRIC FUNCTIONS
WEIGHT 3

$(w = 3)$	$[k_3]$	$[k_2 k_1]$	$[k_1^3]$
K_3	1	-1	2
$K_2(K_1 - 2)$	1	**1**	-3
$K_1^{(3)}$	1	3	**1**

TABLE AIc

AUGMENTED FACTORIAL MONOMIAL SYMMETRIC FUNCTIONS
WEIGHT 4

$(w = 4)$	$[k_4]$	$[k_3 k_1]$	$[k_2^2]$	$[k_2 k_1^2]$	$[k_1^4]$
K_4	1	-1	-1	2	-6
$K_3(K_1 - 3)$	1	**1**	·	-2	8
$K_2(K_2 - 2) - 4K_3$	1	·	**1**	-1	3
$K_2(K_1 - 2)^{(2)}$	1	2	1	**1**	-6
$K_1^{(4)}$	1	4	3	6	**1**

TABLE AId

AUGMENTED FACTORIAL MONOMIAL SYMMETRIC FUNCTIONS
WEIGHT 5

$(w = 5)$	$[k_5]$	$[k_4 k_1]$	$[k_3 k_2]$	$[k_3 k_1^2]$	$[k_2^2 k_1]$	$[k_2 k_1^3]$	$[k_1^5]$
K_5	1	-1	-1	2	2	-6	24
$K_4(K_1 - 4)$	1	**1**	·	-2	-1	6	-30
$K_3(K_2 - 6) - 6K_4$	1	·	**1**	-1	-2	5	-20
$K_3(K_1 - 3)^{(2)}$	1	2	1	**1**	·	-3	20
$[K_2(K_2 - 2) - 4K_3][K_1 - 4]$	1	1	2	·	**1**	-3	15
$K_2(K_1 - 2)^{(3)}$	1	3	4	3	3	**1**	-10
$K_1^{(5)}$	1	5	10	10	15	10	**1**

To express factorial power sums in terms of the augmented factorial symmetric functions read from left to right up to and including the heavy type diagonal. To express the augmented factorial symmetric functions in terms of the factorial power sums read vertically downwards as far as and including the heavy type diagonal.

NOTATION. For example

$$\left[k_3 k_2\right] = \sum_{\neq} k_\ell^{(3)} \cdot k_h^{(2)}, \qquad K_r = \sum_{i=1}^{s} k_i^{(r)},$$

(A.1)

$$\left[k_2^2 k_1\right] = \sum_{\neq \neq} k_\ell^{(2)} k_h^{(2)} k_t,$$

TABLE AII

EIGHT SQUARES: TWO SQUARES IN COMMON

Number of Permutations	Partition	Total score for partition Diagonals (Yes)	Diagonals (No)
1	(6)	36	16
6	(51)	90	40
15	(42)	112	52
10	(3^2)	58	28
15	(41^2)	73	33
60	(321)	140	70
15	(2^3)	22	12
20	(31^3)	20	10
45	$(2^2 1^2)$	25	15
15	(21^4)	1	1
1	(1^6)	0	0

TABLE AIII

EXACT DISTRIBUTION OF S
$K_1 = 8, \quad m = 4, \quad n = 2$

S	(8)	(71)	(62)	(61^2)	(53)	(521)	(51^3)	(4^2)	(42^2)	(421^2)	(41^4)	(21^6)
18	1											
17												
16												
15		4										
14			2					1				
13				2								
12		4	4	4	4							
11		8		8		4						
10		2	8			20	4	1	3			
9		8	8			4	4	8	4	4		
8		4	2		24	28		8	2	14	1	
7			4		20	44	24	8	32	24	2	
6						36		9	19	48	8	
5						32	20		96	90	8	
4							4		18	132	16	
3									36	92	26	
2										16	9	2
1												14
0												12
Totals	1	8	28	28	56	168	56	35	210	420	70	28

TABLE AIV

DISTRIBUTION OF S ACCORDING TO
LATTICE CONFIGURATION IN FIGURE 1
OBJECT SPECIFICATION (42^2)

S	A	B
10	3	1
9	4	2
8	2	10
7	32	28
6	19	32
5	96	64
4	18	45
3	36	28
Total	210	210
μ'_1	5.143	5.167
μ_2	2.199	2.106
β_1	0.443	0.201
β_2	3.811	2.851

A   B

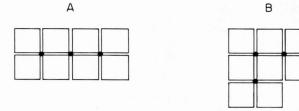

FIGURE AI

Two versions of small lattice configuration

REFERENCES

[1] D. E. BARTON and F. N. DAVID, "Multiple runs," *Biometrika*, Vol. 44 (1957), pp. 168–178.
[2] E. A. CORNISH and R. A. FISHER, "Moments and cumulants in the specifications of distributions," *Rev. Inst. Internat. Statist.*, Vol. 4 (1937) pp. 307–320.
[3] F. N. DAVID and M. G. KENDALL, "Tables of symmetric functions—part I," *Biometrika*, Vol. 36 (1949), pp. 431–449.
[4] F. N. DAVID, M. G. KENDALL, and D. E. BARTON, *Symmetric Function and Allied Tables*, Cambridge, Cambridge University Press, 1966.
[5] E. S. PEARSON and H. O. HARTLEY, *Biometrika Tables for Statisticians*, London, Cambridge University Press, 1958.
[6] E. C. PIELOU, *An Introduction to Mathematical Ecology*, New York, Wiley, 1970.

SOME MULTIPLICATIVE MODELS FOR THE ANALYSIS OF CROSS CLASSIFIED DATA

LEO A. GOODMAN
UNIVERSITY OF CHICAGO

1. Introduction and summary

In the present article, we shall present some multiplicative models for the analysis of $R \times C$ contingency tables (that is, contingency tables with R rows and C columns), and shall apply these models to cross classified data in ways that will lead to a more complete analysis of these data than has heretofore been possible.

For the $R \times C$ contingency table, the usual model of "independence" in the table (that is, independence between the row classification and column classification of the table) is a simple example of a multiplicative model. For short, I shall call this model (that is, the model of independence between the row classification and column classification of the table) the I model. The model of "quasi-independence" in the $R \times C$ table, which was introduced and developed in my earlier work, and which I shall comment upon again later, is another example of a multiplicative model (see, for example, Goodman [9], [10], [12], [13], Caussinus [4], Bishop and Fienberg [2]). For short, I shall call this model the Q model. The various multiplicative models which I shall present here can be viewed as modifications or generalizations of the I model and/or the Q model.

To illustrate the application of these models, we shall analyze a 5×5 contingency table (Table I) in which there is a one to one correspondence between the five classes of the row classification and the five classes of the column classification, and in which the classes of the row (and column) classification can be ordered (from high to low). Although some of the particular models, which we shall consider herein (see Section 2), are particularly well suited to square contingency tables of this kind (in which there is this one to one correspondence and in which the classes of the row (and column) classification can be ordered), we wish to draw the reader's attention to the fact that the general class of multiplicative models presented in this article (see Sections 3 and 4) also includes a variety of models that can be applied more generally to rectangular contingency tables (as well as to square tables), where there may or may not be some kind of correspondence between the classes of the row and

This research was supported in part by Research Contract No. NSF GS 1905 from the Division of the Social Sciences of the National Science Foundation.

649

TABLE I

CROSS CLASSIFICATION OF BRITISH MALE SAMPLE ACCORDING TO EACH SUBJECT'S STATUS
CATEGORY AND HIS FATHER'S STATUS CATEGORY, USING FIVE STATUS CATEGORIES

| | | Subject's Status | | | | |
		1	2	3	4	5
	1	50	45	8	18	3
	2	28	174	84	154	55
Father's Status	3	11	78	110	223	96
	4	14	150	185	714	447
	5	0	42	72	320	411

column classifications and where the classes of the row (and/or column) classifications may or may not be ordered.

Although the models presented here are described in different terms from the models in Haberman's fundamental work [20], it can be shown that the general theory and methods developed by Haberman are applicable to the various models considered herein. In most respects, the class of models considered by Haberman is broader than the class considered here; but by adopting a somewhat different perspective here and by confining our attention to a more limited class of models, we shall obtain some new results.

We shall formulate the models for the case where a random sample of n observations is drawn from the population cross classification table (the case where the sample size n is fixed), but the methods that we shall present here can also be applied to the case where a sample of n_i. observations is drawn from the ith row, $i = 1, 2, \cdots, R$, of the population table (the case where the row marginals are fixed), or where a sample of $n._j$ observations is drawn from the jth column, $j = 1, 2, \cdots, C$, of the population table (the case where the column marginals are fixed). The analysis in the case where the row marginals or the column marginals are fixed is similar, in most respects, to the analysis in the case where only the sample size n is fixed; but there are differences in the way some of the parameters of interest are defined in these cases, and there also are differences in the way these parameters are estimated. These differences will be discussed later (see Section 5).

The table that we shall use for illustrative purposes (Table I) presents data on intergenerational social mobility in Britain, which were collected by Glass and his coworkers [6]. The data in this table were obtained by a kind of stratified random sampling, but for our present exposition we view this table as a contingency table; that is, as if simple random sampling had been used. Table I was used earlier by Svalastoga [25], Levine [22], Mosteller [24], and Goodman [14] for purposes of comparison with a comparable 5 × 5 table (Table II) describing social mobility in Denmark, and it is a condensation of a 7 × 7 British table (Table III) that used a more detailed set of classes (status categories). Table I was formed from Table III by combining status category 2 with 3 and status category 6 with 7 in Table III, in order to make the status

TABLE II

CROSS CLASSIFICATION OF DANISH MALE SAMPLE ACCORDING TO EACH SUBJECT'S STATUS
CATEGORY AND HIS FATHER'S STATUS CATEGORY, USING FIVE STATUS CATEGORIES

		Subject's Status				
		1	2	3	4	5
Father's Status	1	18	17	16	4	2
	2	24	105	109	59	21
	3	23	84	289	217	95
	4	8	49	175	348	198
	5	6	8	69	201	246

TABLE III

CROSS CLASSIFICATION OF BRITISH MALE SAMPLE ACCORDING TO EACH SUBJECT'S STATUS
CATEGORY AND HIS FATHER'S STATUS CATEGORY, USING SEVEN STATUS CATEGORIES

		Subject's Status						
		1	2	3	4	5	6	7
Father's Status	1	50	19	26	8	18	6	2
	2	16	40	34	18	31	8	3
	3	12	35	65	66	123	23	21
	4	11	20	58	110	223	64	32
	5	14	36	114	185	714	258	189
	6	0	6	19	40	179	143	71
	7	0	3	14	32	141	91	106

categories more comparable to the corresponding categories in the Danish
5×5 table. For expository purposes, when illustrating the application of the
models and methods presented here, we shall focus our attention on the
analysis of Table I; but for the sake of completeness, we shall present corresponding results for Tables II and III as well, and shall also comment briefly
upon these results (see Sections 6 and 7); which will further enrich our understanding of the data.

The main part of this paper will be concerned with the development of models
and methods for the analysis of a given $R \times C$ table (or a given set of R or C
multinomial populations). We shall also comment briefly in the final section
(Section 8) on the extension and application of the models and methods presented to the analysis and comparison of two (or more) $R \times C$ cross classification tables, and to the analysis of multidimensional cross classification tables.

2. Some examples of multiplicative models

For expository purposes, we shall begin by considering first the usual model
of "independence" between the row classification and the column classification
in an $R \times C$ cross classification table (the I model), then the model of "quasi-
independence" in the $R \times C$ table (the Q model), and then various modifications
or generalizations of these models.

2.1. *The model of independence* (*the I model*).   We shall now define the usual model of "independence" between the row classification and the column classification in an $R \times C$ population cross classification table. This model can be defined in various ways, and for our present development of the subject we shall proceed as follows. Let $\pi_{i,j}$ denote the probability that an individual in the $R \times C$ population table will fall in cell (i, j) of the table (that is, in the ith row and jth column), for $i = 1, 2, \cdots, R$, and $j = 1, 2, \cdots, C$. Then the row and column classifications are defined as "independent" if the probability $\pi_{i,j}$ can be written as

$$(2.1.1) \qquad \pi_{i,j} = \alpha_i \beta_j \qquad \text{for} \quad i = 1, 2, \cdots, R; \qquad j = 1, 2, \cdots, C,$$

for a set of positive constants α_i and β_j, for $i = 1, 2, \cdots, R$, and $j = 1, 2, \cdots, C$. When (2.1.1) is satisfied, α_i can be interpreted as the probability that an individual will fall in the ith row of the population table (when the α_i have been scaled so that $\Sigma_i \alpha_i = 1$), and a similar interpretation can be given to the β_j. If some of the rows or columns are empty, we consider the table consisting of the nonempty rows and columns.

For a sample of n individuals, let $f_{i,j}$ denote the number of individuals that fall in cell (i, j) of the table, and let $\hat{f}_{i,j}$ denote the maximum likelihood estimate of the expected number that would fall in cell (i, j) under a given model. Under the usual model of independence between the row classification and the column classification of the table, the $\hat{f}_{i,j}$ can be written as

$$(2.1.2) \qquad \hat{f}_{i,j} = \frac{f_i^\alpha f_j^\beta}{n},$$

where f_i^α and f_j^β denotes the ith row marginal and jth column marginal, respectively, in the table of the $f_{i,j}$. From (2.1.2) we see that the $\hat{f}_{i,j}$ can be written in the form

$$(2.1.3) \qquad \hat{f}_{i,j} = a_i b_j,$$

where the a_i and b_j are such that

$$(2.1.4) \qquad \sum_j \hat{f}_{i,j} = f_i^\alpha, \qquad \sum_i \hat{f}_{i,j} = f_j^\beta.$$

Conditions (2.1.4) can be rewritten as

$$(2.1.5) \qquad \hat{f}_i^\alpha = f_i^\alpha, \qquad \hat{f}_j^\beta = f_j^\beta,$$

where the $\hat{f}_i^\alpha$ and $\hat{f}_j^\beta$ denote the ith row marginal and jth column marginal, respectively, in the table of the $\hat{f}_{i,j}$. Thus, the row and column marginals of the $\hat{f}$ fit the corresponding observed quantities.

Although the number of the a_i and the b_j in (2.1.3) is $R + C$, we can ignore one of these quantities (say, a_1) since $\hat{f}_{i,j}$ is unaffected by scaling the a_i and the b_j so that $a_1 = 1$; that is, by replacing a_i and b_j by $\tilde{a}_i = a_i/a_1$ and $\tilde{b}_j = a_1 b_j$, respectively. Similarly, although the number of restrictions described by (2.1.5) is $R + C$, we can ignore one of these restrictions (say, the first restriction), since if $\hat{f}_j^\beta = f_j^\beta$ for $j = 1, 2, \cdots, C$, then $\Sigma_{i,j} \hat{f}_{i,j} = n$; and thus, if also $\hat{f}_i^\alpha = f_i^\alpha$ for $i = 2, 3, \cdots, R$, then $\hat{f}_1^\alpha = f_1^\alpha$. Thus, to calculate the degrees of freedom for testing the model of independence, we subtract $R + C - 1$ from $R \times C$, obtaining thereby the usual quantity, namely, $(R - 1)(C - 1)$.

2.2. *The model of quasi-independence (the Q model).* The model (2.1.1) in the preceding section applies to all $R \times C$ cells (i, j) for $i = 1, 2, \cdots, R$, $j = 1, 2, \cdots, C$. Now we shall consider a subset S of these cells; for example, the subset consisting of the cells that are not on the main diagonal of the table (that is, the cells (i, j) with $i \neq j$). For a given subset S, the row and column classification is defined as "quasi-independent" (with respect to S) if the probability $\pi_{i,j}$ can be written as

$$(2.2.1) \qquad \pi_{i,j} = \alpha_i \beta_j \qquad \text{for cells } (i, j) \text{ in } S,$$

for a set of positive constants α_i and β_j.

Since we are not concerned here with the cells that are not in S, we can assign zero probability to those cells. Thus, (2.2.1) can be rewritten as

$$(2.2.2) \qquad \pi_{i,j} = \delta_{i,j}^S \alpha_i \beta_j \qquad \text{for} \quad i = 1, 2, \cdots, R; \quad j = 1, 2, \cdots, C,$$

where

$$(2.2.3) \qquad \delta_{i,j}^S = \begin{cases} 1 & \text{for cells } (i, j) \text{ in } S, \\ 0 & \text{otherwise.} \end{cases}$$

Letting $f_{i,j}$ denote the observed number of individuals that fall in cell (i, j), the maximum likelihood estimate $\hat{f}_{i,j}$ of the corresponding expected number (under model (2.2.1)) can be written as

$$(2.2.4) \qquad \hat{f}_{i,j} = \delta_{i,j}^S a_i b_j,$$

where the a_i and b_j are such that

$$(2.2.5) \qquad \hat{f}_i^\alpha = f_i^\alpha, \qquad \hat{f}_j^\beta = f_j^\beta,$$

and where now f_i^α and f_j^β are defined as

$$(2.2.6) \qquad f_i^\alpha = \sum_j \delta_{i,j}^S f_{i,j}, \qquad f_j^\beta = \sum_i \delta_{i,j}^S f_{i,j},$$

and the $\hat{f}_i^\alpha$ and $\hat{f}_j^\beta$ are defined similarly (with the $f_{i,j}$ in (2.2.6) replaced by the corresponding $\hat{f}_{i,j}$). Methods for calculating the $\hat{f}_{i,j}$ were discussed in, for example, Goodman [10], [13], and we shall return to them later herein.

To facilitate our understanding of matters that will be discussed later, we now introduce some new terminology. For each cell (i, j), let $\Lambda_{i,j}$ denote the set of parameters that appear in the formula for $\pi_{i,j}$ under a given model (see, for example, (2.2.1)). Thus, for the Q model, $\Lambda_{i,j}$ contains α_i and β_j if cell (i, j) is in S, and $\Lambda_{i,j}$ is empty if (i, j) is not in S. With this terminology, we can rewrite (2.2.6) as

$$(2.2.7) \qquad f_i^\alpha = \sum_{g,h} \delta_{g,h}^{\alpha_i} f_{g,h}, \qquad f_j^\beta = \sum_{g,h} \delta_{g,h}^{\beta_j} f_{g,h},$$

where
$$\delta_{g,h}^{\alpha_i} = \begin{cases} 1 & \text{if } \alpha_i \in \Lambda_{g,h}, \\ 0 & \text{otherwise}, \end{cases}$$

(2.2.8)

$$\delta_{g,h}^{\beta_j} = \begin{cases} 1 & \text{if } \beta_j \in \Lambda_{g,h}, \\ 0 & \text{otherwise}. \end{cases}$$

Note that (2.2.7) states that f_i^α is the sum of the $f_{g,h}$ for cells (g, h) for which $\alpha_i \in \Lambda_{g,h}$; and that f_j^β is the corresponding sum for cells (g, h) for which $\beta_j \in \Lambda_{g,h}$. A similar statement applies to the $\hat{f}_i^\alpha$ and $\hat{f}_j^\beta$ (with the $f_{g,h}$ in (2.2.7) replaced by the corresponding $\hat{f}_{g,h}$).

To calculate the degrees of freedom for testing the model in this section, we consider first the case where the set S includes at least one cell from each row and column of the table, and where S is "inseparable" in the sense that it cannot be partitioned into two mutually exclusive (and exhaustive) subsets S_1 and S_2 that have no rows and no columns in common (see, for example, Goodman [13], Caussinus [4]). Note that S would be separable if the subsets S_1 and S_2 had no parameters in common; in other words, if the set of parameters that are contained in S_1 (that is, in $\Lambda_{i,j}$ for one or more of the cells (i, j) in S_1) and the set of parameters that are contained in S_2 (that is, in $\Lambda_{i,j}$ for one or more of the cells (i, j) in S_2) were mutually exclusive. For the case where S is inseparable, the remarks in the final paragraph of Section 2.1 can be directly applied; and in the present case, the degrees of freedom are obtained by subtracting $R + C - 1$ from $R \times C - V$, where V is the number of cells in the $R \times C$ table that are not included in S. Thus, as in the earlier literature on "quasi-independence," we find that there are $(R - 1)(C - 1) - V$ degrees of freedom.

In cases where entire rows and/or entire columns are not included in S, the above formula for the degrees of freedom can still be applied, except that "R" and "C" in that formula should be taken as the number of rows and columns, respectively, that contain at least one cell from S, and similarly the quantity V should be calculated for this (modified) "R" $\times$ "C" table. For cases in which S is separable, the results presented above can be applied separately to each subset that is itself not separable. A separable set S can always be partitioned into such subsets.

2.3. *The QO model, the QP model, the QN model, and the QPN model.* The results in the preceding section pertain to the case where S is any given subset of the cells in the $R \times C$ table. Consider now the case where the cross classification table is square (that is, $R = C$), where there is a one to one correspondence between the ith class of the row classification and the ith class of the column classification for $i = 1, 2, \cdots, R$, and where S is the set of cells that are not on the main diagonal (that is, the cells (i, j) with $i \neq j$). Since S consists of the off diagonal cells, we shall call the model of quasi-independence (with respect to this set S) the QO model. The remarks in the preceding section can be applied directly to the QO model. Note, for example, that the number of degrees of freedom for testing this model will be $(R - 1)(R - 1) - R = R^2 - 3R + 1$.

Consider now the case where the classes of the row (and column) classification in the $R \times R$ table can also be ordered from 1 to R, and where S is the set of cells (i, j) with $i > j$. For each cell in S, the difference $i - j$ is positive, and so we shall call this particular model of quasi-independence the QP model. From the remarks in the preceding section, we see that the number of degrees of freedom for testing the QP model will be $(R - 2)(R - 2) - [(R - 1)(R - 2)/2] = (R - 2)(R - 3)/2$. Note that the degrees of freedom are zero for $R = 3$; we shall be concerned here mainly with models for cases where $R > 3$.

Consider now the case where S is the set of cells (i, j) with $i < j$. For each cell in S, the difference $i - j$ is negative, and so we shall call this particular model of quasi-independence the QN model. As in the preceding paragraph, the number of degrees of freedom for testing the QN model will be $(R - 2)(R - 3)/2$.

Consider now the model in which the probability $\pi_{i,j}$ can be written as

$$(2.3.1) \qquad \pi_{i,j} = \begin{cases} \alpha_i \beta_j & \text{for} \quad i > j, \\ \alpha_i' \beta_j' & \text{for} \quad i < j. \end{cases}$$

This model states that both the QP model and the QN model are true, and so we shall call it the QPN model. Note that this model is not the same as the QO model, although both models pertain to the cells (i, j) with $i \neq j$. The QO model is a special case of the QPN model in which

$$(2.3.2) \qquad \begin{aligned} \alpha_i' &= \Delta \alpha_i & \text{for} \quad i = 2, 3, \cdots, R - 1, \\ \beta_j' &= \Delta^* \beta_j & \text{for} \quad j = 2, 3, \cdots, R - 1, \end{aligned}$$

and in which

$$(2.3.3) \qquad \Delta^* = \frac{1}{\Delta}.$$

The QPN model is not, strictly speaking, an example of a model of quasi-independence as this term was defined in the preceding section. Nevertheless, the methods developed earlier for the quasi-independence model (see, for example, Goodman [13]) can be applied to the QPN model, by analyzing separately the subsets S_1 and S_2, where S_1 is the set of cells (i, j) with $i > j$, and S_2 is the set of cells (i, j) with $i < j$; and applying the corresponding models (the QP and QN models) to those sets. Note that, although the sets S_1 and S_2 do have rows and columns in common, these sets are separable for the QPN model, since the set of parameters in S_1 (the α_i and β_j) and the set of parameters in S_2 (the α_i' and β_j') are mutually exclusive.

The number of degrees of freedom for testing the QPN model is the sum of the degrees of freedom for testing the QP model in S_1 and the QN model in S_2. Thus, there are $(R - 2)(R - 3)$ degrees of freedom for testing the QPN model. Note that the difference between the degrees of freedom for the QO model and the QPN model is $2R - 5$, which corresponds to the sum of the degrees of freedom associated with testing condition (2.3.2) and condition (2.3.3) (namely, $2(R - 3)$ degrees of freedom for (2.3.2) and one degree of freedom for (2.3.3)).

We now extend the concept of "quasi-independence" to include models of the kind described by (2.3.1). Let $S_1, S_2, \cdots, S_K$ denote mutually exclusive subsets of the cells (i, j) in an $R \times C$ cross classification table. The model of "quasi-independence" (with respect to the subsets S_k, for $k = 1, 2, \cdots, K$) is defined by the condition that the probability $\pi_{i,j}$ can be written as

$$(2.3.4) \qquad \pi_{i,j} = \alpha_i^{(k)} \beta_j^{(k)} \qquad \text{for cells } (i, j) \text{ in } S_k,$$

for $k = 1, 2, \cdots, K$. The sets $S_1, S_2, \cdots, S_K$ are separable for the model defined by (2.3.4), since the set of parameters in S_k (namely, $\alpha_i^{(k)}$ and $\beta_j^{(k)}$) and the set of parameters in $S_{k'}$ (namely, $\alpha_i^{(k')}$ and $\beta_j^{(k')}$) are mutually exclusive for $k \neq k'$. To analyze this more general model of "quasi-independence," we can apply the methods developed earlier for the more usual quasi-independence model, by analyzing separately the sets $S_1, S_2, \cdots, S_K$, applying the corresponding model of quasi-independence to each set.

2.4. *The triangles parameter model (the T model).* We return again to the square contingency table ($R = C$) in which there is a one to one correspondence between the classes of the row and column classification, and in which the classes of the row (and column) classification are ordered from 1 to R. Consider now the special case of the QPN model in which condition (2.3.2) is satisfied (but condition (2.3.3) may or may not be satisfied). This special case of the QPN model is equivalent to the model in which the probability $\pi_{i,j}$ can be written as

$$(2.4.1) \qquad \pi_{i,j} = \alpha_i \beta_j \tau_k \qquad \text{for cells } (i, j) \text{ in } S_k,$$

for $k = 1, 2$, where S_1 is the set of cells (i, j) with $i > j$, and S_2 is the set of cells (i, j) with $i < j$. For cell (i, j) in S_k, under model (2.4.1) the set $\Lambda_{i,j}$ contains α_i, β_j, and τ_k; and $\Lambda_{i,j}$ is empty if (i, j) is not in S_1 or S_2. The model (2.4.1) differs from the QO model in that it introduces an additional set of parameters τ_k that pertains differentially to the triangular subsets S_k, for $k = 1$ and 2, and so we call this model the triangles parameter model (the T model).

Note that the sets S_1 and S_2 are not separable for the T model defined by (2.4.1) since the set of parameters in S_1 (namely, α_i, β_j, and τ_1) and the set of parameters in S_2 (namely, α_i, β_j, and τ_2) are not mutually exclusive—the parameters α_i for $i = 2, 3, \cdots, R - 1$ and β_j for $j = 2, 3, \cdots, R - 1$ are included in both sets. Although the T model is not an example of a quasi-independence model as defined in Section 2.2 (nor of the more general model of "quasi-independence" defined by (2.3.4)), the remarks in Section 2.2 can be directly extended to the T model. By direct extension of (2.2.4) to (2.2.8), we find that the estimate $\hat{f}_{i,j}$ under the T model can be written as

$$(2.4.2) \qquad \hat{f}_{i,j} = a_i b_j t_k \qquad \text{for cells } (i, j) \text{ in } S_k,$$

for $k = 1$ and 2, where the a_i, b_j, and t_k are such that

$$(2.4.3) \qquad \hat{f}_i^{\alpha} = f_i^{\alpha}, \qquad \hat{f}_j^{\beta} = f_j^{\beta}, \qquad \hat{f}_k^{\tau} = f_k^{\tau},$$

where the f_i^{α} and f_j^{β} are defined by (2.2.7), and f_k^{τ} is defined by

(2.4.4)
$$f_k^\tau = \sum_{g,h} \delta_{g,h}^{\tau_k} f_{g,h},$$

where

(2.4.5)
$$\delta_{g,h}^{\tau_k} = \begin{cases} 1 & \text{if } \tau_k \in \Lambda_{g,h}, \\ 0 & \text{otherwise}; \end{cases}$$

and with the $\hat{f}_i^\alpha$, $\hat{f}_j^\beta$, $\hat{f}_k^\tau$ defined similarly (with the $f_{g,h}$ in (2.4.4) replaced by the corresponding $\hat{f}_{g,h}$).

Although there are two t_k in (2.4.2) (namely, t_1 and t_2), we can ignore one of them (say, t_1) since $\hat{f}_{i,j}$ is unaffected by scaling the t_k (and the a_i) so that $t_1 = 1$; that is, by replacing t_k and a_i by $\tilde{t}_k = t_k/t_1$ and $\tilde{a}_i = a_i t_1$, respectively. Similarly, although there are two restrictions described by the third condition of (2.4.3) for $k = 1$ and 2, we can ignore one of these restrictions (say, the first restriction) since if $\hat{f}_i^\alpha = f_i^\alpha$ for $i = 1, 2, \cdots, R$, then $\Sigma \hat{f}_{i,j} = n$ (where the $\hat{f}$ are summed over the off diagonal cells and n is similarly calculated for the f); and thus, if $\hat{f}_2^\tau = f_2^\tau$, then $\hat{f}_1^\tau = f_1^\tau$. Since there is one more parameter in the T model than in the QO model, the number of degrees of freedom for testing the T model will be $R^2 - 3R = R(R - 3)$, for $R \geqq 3$.

2.5. *The diagonals parameter model (the D model) and related models.* Consider now the model in which the probability $\pi_{i,j}$ can be written as

(2.5.1)
$$\pi_{i,j} = \alpha_i \beta_j \delta_k \qquad \text{for cells } (i, j) \text{ in } S_k',$$

where S_k' is the set of cells (i, j) with $i - j = k$, for $k = \pm 1, \pm 2, \cdots, \pm(R - 1)$. The model (2.5.1) differs from the QO model in that it introduces an additional set of parameters (the δ_k) that pertains differentially to the minor diagonals S_k for $k = \pm 1, \pm 2, \cdots, \pm(R - 1)$, and so we call this model the diagonals parameter model (the D model).

The T model (see (2.4.1)) is a special case of the D model in which the following condition is satisfied:

(2.5.2)
$$\delta_k = \begin{cases} \delta^* & \text{for } k = 1, 2, \cdots, R - 1, \\ \delta^{**} & \text{for } k = -1, -2, \cdots, -(R - 1). \end{cases}$$

The remarks about the analysis of the T model in the preceding section can be extended directly to the D model. For example, under the D model, the $\hat{f}_{i,j}$ can be written as

(2.5.3)
$$\hat{f}_{i,j} = a_i b_j d_k \qquad \text{for cells } (i, j) \text{ in } S_k',$$

for $k = \pm 1, \pm 2, \cdots, \pm(R - 1)$, where the a_i, b_j and d_k are such that

(2.5.4)
$$\hat{f}_i^\alpha = f_i^\alpha, \qquad \hat{f}_j^\beta = f_j^\beta, \qquad \hat{f}_k^\delta = f_k^\delta,$$

where the f_i^α and f_j^β are defined by (2.2.7) and f_k^δ is defined similarly to (2.4.4) and (2.4.5), with δ replacing τ in those formulae. (The corresponding $\hat{f}$ are defined similarly.)

Although there are $2(R-1)$ statistics d_k in (2.5.3) (namely, d_k for $k = \pm 1$, $\pm 2, \cdots, \pm(R-1)$), we can ignore two of them (say, d_k for $k = \pm 1$) since $\hat{f}_{i,j}$ is unaffected by transforming the d_k (and the a_i and the b_j) so that $d_k = 1$ for $k = \pm 1$; that is, by replacing d_k by

$$(2.5.5) \qquad \tilde{d}_k = \frac{d_k d_*^{k-1}}{d_1} \qquad \text{for} \quad k = \pm 1, \pm 2, \cdots, \pm(R-1),$$

with

$$(2.5.6) \qquad\qquad\qquad d_* = \left(\frac{d_{-1}}{d_1}\right)^{1/2},$$

and replacing a_i and b_j by

$$(2.5.7) \qquad\qquad\qquad \tilde{a}_i = \frac{a_i d_1}{d_*^{i-1}}, \qquad \tilde{b}_j = b_j d_*^j,$$

respectively. Thus since there are $2(R-2)$ more parameters in the D model than in the QO model, the number of degrees of freedom for testing the D model will be $R^2 - 5R + 5$, for $R \geq 4$. There will be zero degrees of freedom for testing the D model in the case where $R = 3$.

Consider now the special case of the D model in which the following condition is satisfied:

$$(2.5.8) \qquad \delta_k = \delta_{k*} \qquad \text{for} \quad k* = -k, \text{with } k = 1, 2, \cdots, R-1.$$

In this case, the parameter δ_k pertains to the pair of minor diagonals S'_k and S'_{k*} with $k* = -k$; that is, to the cells (i, j) for which the absolute value $|i - j|$ is equal to k. We shall call this case the DA model. The earlier remarks about the analysis of the D model can be directly extended to the DA model, where now δ_k pertains to the paired minor diagonals, for $k = 1, 2, \cdots, R-1$. The DA model has $R - 2$ more parameters than the QO model for $R \geq 4$, and so the number of degrees of freedom for testing the DA model will be $R^2 - 4R + 3 = (R-1)(R-3)$, for $R \geq 4$. For $R = 3$, the DA model is equivalent to the QO model, so that in this special case there will be one degree of freedom for testing the model. Note that the difference between the degrees of freedom for the DA model and the D model is $R - 2$ for $R \geq 4$, which corresponds to the degrees of freedom associated with testing condition (2.5.8), for $k = 2, 3, \cdots, R - 1$ (with $\delta_k = 1$, for $k = \pm 1$).

Consider now the special case of the D model in which the following condition is satisfied:

$$(2.5.9) \qquad\qquad \delta_k = 1, \qquad \text{for} \quad k = -1, -2, \cdots, -(R-1).$$

In this case, the parameter δ_k pertains only to the minor diagonals S'_k for which k is positive; that is, to the cells (i, j) for which $i - j = k$ is positive. We shall call this case the DP model. The condition (2.5.9) is actually equivalent to the following condition, which might appear (at first sight) to be more general;

namely,

(2.5.10) $\qquad \delta_k = \delta^k \qquad$ for $\quad k = -1, -2, \cdots, -(R-1),$

for some positive constant δ. Conditions (2.5.9) and (2.5.10) are equivalent because the $\pi_{i,j}$ of (2.5.1) is unaffected by replacing δ_k, α_i, and β_j by

(2.5.11) $\qquad \tilde{\delta}_k = \dfrac{\delta_k}{\delta^k}, \qquad \tilde{\alpha}_i = \alpha_i \delta^{i-1}, \qquad \tilde{\beta}_j = \dfrac{\beta_j}{\delta^{j-1}},$

respectively. When the parameters are transformed so that $\delta_k = 1$ for $k = \pm 1$, the condition (2.5.10) is replaced by the following condition:

(2.5.12) $\qquad \delta_k = (\delta')^{k+1} \qquad$ for $\quad k = -1, -2, \cdots, -(R-1),$

for a positive constant δ'.

The earlier remarks about the analysis of the D model can also be directly extended to the DP model, where now δ_k pertains to the minor diagonals S'_k for which k is positive for $k = 1, 2, \cdots, R-1$. The DP model has $R-1$ more parameters than the QO model (namely, the δ_k for $k = 2, 3, \cdots, R-1$; and δ' from (2.5.12)), and so the number of degrees of freedom for testing the DP model will be $R^2 - 4R + 2$, for $R \geq 4$. There be zero degrees of freedom for testing the DP model in the case where $R = 3$.

Consider now the special case of the D model in which the following condition is satisfied:

(2.5.13) $\qquad \delta_k = 1 \qquad$ for $\quad k = 1, 2, \cdots, R-1.$

We shall call this case the DN model. By modifying in an obvious manner the remarks (pertaining to the DP model) in the preceding two paragraphs, they can be applied to the DN model.

2.6. *The crossings parameter model* (*the C model*). We return again to the QPN model, and consider the special case where the following condition is satisfied:

(2.6.1) $\qquad \alpha_i \beta_i = \alpha'_i \beta'_i \qquad$ for $\quad i = 2, 3, \cdots, R-1.$

This special case of the QPN model is equivalent to the model in which the probability $\pi_{i,j}$ can be written as

(2.6.2) $\qquad \pi_{i,j} = \alpha_i \beta_j \gamma'_{i,j} \qquad$ for $\quad i \neq j,$

where

(2.6.3) $\qquad \gamma'_{i,j} = \begin{cases} \displaystyle\prod_{u=j}^{i-1} \gamma_u & \text{for } i > j, \\ \displaystyle\prod_{u=i}^{j-1} \gamma_u & \text{for } i < j. \end{cases}$

The model (2.6.2)–(2.6.3) differs from the QO model in that it introduces an additional set of parameters, namely, the γ_u, for $u = 1, 2, \cdots, R-1$, that pertains differentially (and multiplicatively) to each crossing between adjacent classes (from class u to $u + 1$ or from class $u + 1$ to u). In this model, the parameter γ_u pertaining to the crossing from class u to $u + 1$ is equal to the parameter pertaining to the crossing from $u + 1$ to u. We shall call this model the crossings parameter model (the C model). Note that the symbol C will be used to refer to this model and also, as earlier, to the number of columns in a rectangular contingency table; the meaning of the symbol will be clear in either case.

The remarks in the preceding section about the analysis of the models considered there can be directly extended to the C model. Under the C model, the $\hat{f}_{i,j}$ can be written as

$$(2.6.4) \qquad \hat{f}_{i,j} = a_i b_j c'_{i,j},$$

where

$$(2.6.5) \qquad c'_{i,j} = \begin{cases} \displaystyle\prod_{u=j}^{i-1} c_u & \text{for} \quad i > j, \\ \displaystyle\prod_{u=i}^{j-1} c_u & \text{for} \quad i < j, \end{cases}$$

and where the a_i, b_j, and c_u are such that

$$(2.6.6) \qquad \hat{f}_i^\alpha = f_i^\alpha, \qquad \hat{f}_j^\beta = f_j^\beta, \qquad \hat{f}_u^\gamma = f_u^\gamma,$$

where f_i^α and f_j^β are defined by (2.2.7), and f_u^γ is defined similarly to (2.4.4) and (2.4.5) with τ replaced by γ in those formulae. The corresponding $\hat{f}$ are defined similarly.

Although there are $(R - 1)$ statistics c_u in (2.6.4) and (2.6.5) (namely, c_u, for $u = 1, 2, \cdots, R - 1$), we can ignore two of them (c_1 and c_{R-1}) since $\hat{f}_{i,j}$ is unaffected by setting $c_1 = c_{R-1} = 1$; that is, by replacing these two c_u by $\tilde{c}_1 = \tilde{c}_{R-1} = 1$ and by replacing a_1, a_R, b_1, b_R by $\tilde{a}_1 = a_1 c_1$, $\tilde{a}_R = a_R c_{R-1}$, $\tilde{b}_1 = b_1 c_1$, $\tilde{b}_R = b_R c_{R-1}$. Since there are $R - 3$ more parameters in the C model than in the QO model, the number of degrees of freedom for testing the C model is $R^2 - 4R + 4 = (R - 2)^2$. Note that the difference between the degrees of freedom for the C model and the QPN model is $R - 2$, which corresponds to the degrees of freedom associated with testing condition (2.6.1).

From (2.6.2) and (2.6.3) we see that the factor $\gamma'_{i,j}$ associated with a change from the ith class (with respect to the row classification) to the jth class (with respect to the column classification) was a product of the γ_u parameters pertaining to successive one step changes (crossings) from class i to j. Thus, the crossings parameter model (the C model) could also have been called the "one step Markov" model. (This terminology is appropriate, in a certain sense, since the parameter γ_u pertaining to an individual's one step change from class u to $u + 1$ depends only upon the class u and not upon the earlier history of changes that may have led to the individual's presence in class u. On the other hand, the terminology is not quite appropriate since the direction of change (from class u to $u+1$ or from class u to $u-1$) does depend upon this earlier history.) Although the models in Haberman [20] are described there in different terms from the models presented here, it can be shown that the C model (see (2.6.4) and (2.6.5)) is equivalent, in most respect, to the "variable distance" model applied by Haberman. With the present formulation of the C model, some of the parameters and their estimates will be defined and calculated differently from the corresponding definitions and calculations that were applied to the "variable distance" model (see, for example, the related comments at the end of Section 7).

Let us suppose now that we wished to generalize the C model by replacing (2.6.3) by

$$(2.6.7) \qquad \gamma'_{i,j} = \begin{cases} \prod_{u=j}^{i-1} \gamma^*_u & \text{for} \quad i > j, \\ \prod_{u=i}^{j-1} \gamma^{**}_u & \text{for} \quad i < j. \end{cases}$$

This model distinguishes (at first sight) between the parameter γ^*_u pertaining to the crossing from the uth row class to the $(u+1)$th column class and the parameter γ^{**}_u pertaining to the crossing from the $(u+1)$th row class to the uth column class. Actually, this model is equivalent to the C model (with $\gamma^*_u = \gamma^{**}_u$) since the $\pi_{i,j}$ defined by (2.6.2) and (2.6.7) are unaffected by replacing the γ^*_u and γ^{**}_u by $\tilde{\gamma}_u = (\gamma^*_u \gamma^{**}_u)^{1/2}$, and by replacing α_i and β_j by $\tilde{\alpha}_i = \alpha_i \gamma'_{i-1}$ and $\tilde{\beta}_j = \beta_j / \gamma'_{j-1}$, with

$$(2.6.8) \qquad \gamma'_i = \begin{cases} 1 & \text{for} \quad i = 1, \\ (\gamma^*_i / \gamma^{**}_i)^{1/2} & \text{for} \quad i = 2, 3, \cdots, R-1. \end{cases}$$

Note, in particular, that the special case of the model defined by (2.6.2) and (2.6.7), in which $\gamma^{**}_u = 1$ for $u = 1, 2, \cdots, R-1$, is equivalent to the C model as defined by (2.6.2) and (2.6.3). A similar remark applies for the special case of the model defined by (2.6.2) and (2.6.7) in which $\gamma^*_u = 1$ for $u = 1, 2, \cdots, R-1$.

2.7. *The diagonals crossings parameter model (the DC model) and other combined models.* Consider now the model in which the probability $\pi_{i,j}$ can be written as

$$(2.7.1) \qquad \pi_{i,j} = \alpha_i \beta_j \gamma'_{i,j} \delta_k \qquad \text{for cells } (i,j) \text{ in } S'_k,$$

where S'_k is defined as in (2.5.1), and $\gamma'_{i,j}$ is defined as in (2.6.3). The remarks in Sections 2.5 and 2.6 can be directly extended to apply to this model (the DC model).

The DC model differs from the D model in that it includes an additional set of parameters (namely, the γ_u, for $u = 1, 2, \cdots, R-1$). We noted earlier that γ_1 and γ_{R-1} could be ignored in the C model, and now for the DC model we also find this to be the case and that, in addition, this model is unaffected by a change in scale for the γ_u (with corresponding changes made in the other parameters). Thus, there are $R-4$ more parameters in the DC model than in the D model, and so the number of degrees of freedom for testing the DC model will be $R^2 - 6R + 9 = (R-3)^2$.

The C model can be combined in a similar way with the other models in Section 2.5, obtaining thereby the DAC model, the DPC model, and the DNC model. In these cases too the number of degrees of freedom for the models are obtained by subtracting $R - 4$ from the number for the corresponding model in Section 2.5. Thus, for the DAC model, we obtain $R^2 - 5R + 7$ degrees of freedom; and for the DPC and DNC models, we obtain $R^2 - 5R + 6 = (R - 2)(R - 3)$ degrees of freedom. For R = 3, the DAC model is equivalent to the DA model and the QO model. In addition, the T model of Section 2.3 can be combined with the DA model, the C model, and the DAC model. The number of degrees of freedom for these combined models are obtained by subtracting one from the number for the corresponding (uncombined) model. Thus, the degrees of freedom for the DAT model, the CT model, and the DACT model are $R^2 - 4R + 2$, $R^2 - 4R + 3 = (R - 1)(R - 3)$, and $R^2 - 5R + 6 = (R - 2)(R - 3)$, respectively. The latter two formulas apply for $R \geq 3$, while the first formula applies for $R \geq 4$. For $R = 3$, the DAT model actually has zero degrees of freedom, as do the other models (discussed in the present section and in Section 2.4) that include the triangles parameter τ_k.

All of the models considered in the present section and in Sections 2.4 to 2.6 are special cases of the DC model. For models that do not include the crossings parameters, we have

$$(2.7.2) \qquad \gamma_u = 1 \qquad \text{for} \quad u = 2, 3, \cdots, R - 2,$$

in (2.7.1) (see (2.6.3)). For models that do not include the diagonals parameters, we have

$$(2.7.3) \qquad \delta_k = 1 \qquad \text{for} \quad k = \pm 1, \pm 2, \cdots, \pm(R - 1),$$

in (2.7.1). For models that include the triangles parameter (but not the parameter pertaining to the paired minor diagonals), the δ_k in (2.7.1) will satisfy condition (2.5.2). Similarly, the δ_k in (2.7.1) will satisfy condition (2.5.8) for models that include parameters pertaining to the paired minor diagonals (but not the triangles parameter), and they will satisfy the following condition for models that include both the parameters pertaining to the paired minor diagonals and the triangles parameters:

$$(2.7.4) \qquad \delta_k = \tau' \delta_{k*} \qquad \text{for} \quad k^* = -k, \text{with } k = 1, 2, \cdots, R - 1.$$

Furthermore, the δ_k in (2.7.1) will satisfy condition (2.5.9) for models that do not include parameters pertaining to the "negative" diagonals, and they will satisfy (2.5.13) for models that do not include parameters pertaining to the "positive" diagonals.

2.8. *The DC Model for the full table (the DCF model) and related models.* The models in Sections 2.3 to 2.7 were concerned with the analysis of the off diagonal cells, and were extensions of the QO model. Now we shall present models for the analysis of the full table—models that are extensions of the *I* model.

Consider the model in which the probability $\pi_{i,j}$ is given by (2.7.1), except that now we also apply (2.7.1) to the set S_0' (that is, the cells (i, j) with $i - j = 0$), as well as to the sets S_k' as defined in (2.5.1) for $k = \pm 1, \pm 2, \cdots, \pm(R - 1)$; and we set $\gamma'_{i,j} = 1$ for $i = j$. This model (the DCF model) differs from the DC model in that the R cells on the main diagonal are now included in the analysis, an additional diagonals parameter δ_0 is included, and two additional crossings parameters γ_1 and γ_{R-1} are also included. Although the probability $\pi_{i,j}$ for the DC model is unaffected by ignoring the parameters γ_1 and γ_{R-1} (that is, by setting $\gamma_1 = \gamma_{R-1} = 1$ and making corresponding changes in $\alpha_1, \alpha_R, \beta_1, \beta_R$), this is not so for the DCF model; the $\pi_{i,j}$ for the DCF model would be affected by ignoring γ_1 and γ_{R-1}. Therefore, the number of degrees of freedom for the DCF model can be obtained by adding $R - 3$ to the number for the DC model, thus obtaining $R^2 - 5R + 6 = (R - 2)(R - 3)$.

The relationship between the DC model and the DCF model, which we noted in the preceding paragraph, can be extended to other models. To each of the models introduced in Sections 2.4 to 2.7 for the analysis of the off diagonal cells, there is a corresponding model defined for the full table. The number of degrees of freedom for the model for the full table can be obtained by adding $R - 3$ to the number for the corresponding model (for the analysis of the off diagonal cells) if that model includes both diagonals and crossings parameters; by adding $R - 1$ to the number for the corresponding model if that model includes diagonals parameters but not crossings parameters, by adding $R - 2$ to the number for the corresponding model if that model includes crossings parameters but not diagonals parameters. (In the special case where $R = 3$, the above calculation is modified slightly if the corresponding model is one of those models for which the particular formula given earlier herein for the degrees of freedom does not apply when $R = 3$.) A similar calculation can be made when the model also includes the triangles parameter. Alternatively, the degrees of freedom for the models for the full table could be calculated directly using the same methods that were applied in Sections 2.4 to 2.7 (but without first calculating the results for the models for the off diagonal cells and then adding the appropriate quantities). The number of degrees of freedom for testing each of the models considered herein is given in Table IV.

3. The general case

We return now to the general $R \times C$ table. Let $\pi_{i,j}$ denote the probability that an observation will fall in cell (i, j), let $f_{i,j}$ denote the observed frequency in cell (i, j) for a sample of n observations, and let $\hat{f}_{i,j}$ denote the maximum likelihood estimate of the expected frequency under a given model. We shall now denote the parameters in the model as $\lambda_1, \lambda_2, \cdots, \lambda_W$ (with $\lambda_w > 0$ for $w = 1, 2, \cdots, W$), and we let Λ denote the full set of parameters; that is, $\Lambda = \{\lambda_1, \lambda_2, \cdots, \lambda_W\}$. For each cell (i, j) in that table, let $\Lambda_{i,j}$ denote a given subset of Λ. Consider now the model in which the probability $\pi_{i,j}$ can be

TABLE IV

THE DEGREES OF FREEDOM FOR TESTING VARIOUS MODELS
APPLIED TO THE RECTANGULAR $R \times C$ TABLE AND TO THE
SQUARE $R \times R$ TABLE (FOR $R \geqq 3$)

*The asterisk indicates that the formula does not apply for $R = 3$.
For this special case, see comments in article.
**For the case of quasi-independence, see further details in the article.

Model	Degrees of Freedom
Independence	$(R - 1)(C - 1)$
Quasi-Independence	$[(R - 1)(C - 1) - V]$**
QO	$R^2 - 3R + 1$
QP (or QN)	$(R - 2)(R - 3)/2$
QPN	$(R - 2)(R - 3)$
T	$R(R - 3)$
D	$(R^2 - 5R + 5)$*
DA	$(R - 1)(R - 3)$*
DP (or DN)	$(R^2 - 4R + 2)$*
C	$(R - 2)^2$
DC	$(R - 3)^2$
DAC	$R^2 - 5R + 7$
DPC (or DNC)	$(R - 2)(R - 3)$
DAT	$(R^2 - 4R + 2)$*
CT	$(R - 1)(R - 3)$
DACT	$(R - 2)(R - 3)$
TF	$R^2 - 2R - 1$
DF	$(R - 2)^2$
DAF	$(R - 1)(R - 2)$
DPF (or DNF)	$R^2 - 3R + 1$
CF	$(R - 1)(R - 2)$
DCF	$(R - 2)(R - 3)$
DACF	$(R - 2)^2$
DPCF (or DNCF)	$(R - 1)(R - 3)$
DATF	$R^2 - 3R + 1$
CTF	$R(R - 3)$
DACTF	$(R - 1)(R - 3)$

written as

$$(3.1) \qquad \pi_{i,j} = \prod_{\Lambda_{i,j}} \lambda_w,$$

where $\prod_{\Lambda_{i,j}}$ denotes a product over the indices w for which $\lambda_w \in \Lambda_{i,j}$ with the product defined as zero when $\Lambda_{i,j}$ is empty.

Let f_w^λ and $\hat{f}_w^\lambda$ be defined by

$$(3.2) \qquad f_w^\lambda = \sum_{g,h} \delta_{g,h}^{\lambda,w} f_{g,h}, \qquad \hat{f}_w^\lambda = \sum_{g,h} \delta_{g,h}^{\lambda,w} \hat{f}_{g,h},$$

where

$$(3.3) \qquad \delta_{g,h}^{\lambda,w} = \begin{cases} 1 & \text{if } \lambda_w \in \Lambda_{g,h}, \\ 0 & \text{otherwise.} \end{cases}$$

Then for the model (3.1), it is easy to show that $\hat{f}_{i,j}$ can be written as

$$(3.4) \qquad\qquad \hat{f}_{i,j} = \prod_{\Lambda_{i,j}} \ell_w,$$

where the ℓ_w are such that

$$(3.5) \qquad\qquad \hat{f}_w^\lambda = f_w^\lambda \qquad \text{for} \quad w = 1, 2, \cdots, W.$$

To calculate the $\hat{f}_{i,j}$ defined by (3.4) we can proceed by the following iterative scaling method. At the initial step we define

$$(3.6) \qquad\qquad \hat{f}_{i,j}(0) = \begin{cases} 0 & \text{if} \quad \Lambda_{i,j} \text{ is empty,} \\ 1 & \text{otherwise.} \end{cases}$$

Then we set $v = w$ in the following formula, and we use (3.7) to calculate $\hat{f}_{i,j}(v)$ for $w = 1, 2, \cdots, W$:

$$(3.7) \qquad \hat{f}_{i,j}(v) = \begin{cases} \hat{f}_{i,j}(v-1) f_w^\lambda / [\hat{f}(v-1)]_w^\lambda & \text{if} \quad \lambda_w \in \Lambda_{i,j}, \\ \hat{f}_{i,j}(v-1) & \text{otherwise,} \end{cases}$$

where

$$(3.8) \qquad\qquad [\hat{f}(v-1)]_w^\lambda = \sum_{g,h} \delta_{g,h}^{\lambda,w} \hat{f}_{g,h}(v-1).$$

This is the first cycle of iterations. For the second cycle, we set $v = W + w$ in (3.7), and we use (3.7) to calculate $\hat{f}_{i,j}(v)$ considering again $w = 1, 2, \cdots, W$. For the third cycle, we set $v = 2W + w$ in (3.7), and we proceed as in the preceding cycle of iterations. The cycles of iterations are continued until the $\hat{f}_{i,j}(v)$ satisfy condition (3.5).

The above method is a generalization of a corresponding procedure that was used earlier to calculate the maximum likelihood estimate $\hat{f}_{i,j}$ under the quasi-independence model and under other related models (see, for example, Caussinus [4], Bishop and Fienberg [2], Goodman [13], [17]). This method (as described above) does not provide estimates of the parameters λ_w. To calculate the ℓ_w in (3.4) we can proceed by any of the following three methods.

(1) After calculating the $\hat{f}_{i,j}$ by the iterative procedure given above, equation (3.4) can be solved for the ℓ_w. Explicit expressions for the ℓ_w as functions of the $\hat{f}_{i,j}$ can be obtained for the models of Sections 2.2 to 2.8. See the Appendix where such expressions are given.

(2) Instead of calculating the $\hat{f}_{i,j}$ by the iterative method (3.6) to (3.8), the ℓ_w in (3.4) can be calculated by a direct extension of the iterative procedure that was used earlier by Goodman [10], [13] to estimate the parameters in the quasi-independence model. To do this, we first note from (3.4) and (3.5) that

$$(3.9) \qquad\qquad \hat{f}_w^\lambda = \ell_w \sum_{g,h} \delta_{g,h}^{\lambda,w} \prod_{\Lambda_{g,h,w}} \ell_u,$$

where $\Lambda_{g,h,w}$ is the set consisting of all those λ in $\Lambda_{g,h}$ except λ_w, and where $\Pi_{\Lambda_{g,h,w}}$

denotes a product over the indices u for which $\lambda_u \in \Lambda_{g, h, w}$. We can rewrite (3.5) as

$$(3.10) \qquad \ell_w = \frac{f_w^\lambda}{\displaystyle\sum_{g, h} \delta_{g, h}^{\lambda, w} \prod_{\Lambda_{g, h, w}} \ell_u}.$$

To start the iterative procedure for calculating ℓ_w, we define

$$(3.11) \qquad \ell_w(0) = 1 \qquad \text{for} \quad w = 1, 2, \cdots, W.$$

For the first cycle of iterations, we use the following formula to calculate $\ell_w(v)$ for $v = 1, 2, \cdots, W$:

$$(3.12) \qquad \ell_w(v) = \begin{cases} f_w^\lambda / \displaystyle\sum_{g, h} \delta_{g, h}^{\lambda, w} \prod_{\Lambda_{g, h, w}} \ell_u(v - 1) & \text{if} \quad v = w, \\ \ell_w(v - 1) & \text{otherwise.} \end{cases}$$

For the second cycle of iterations, we replace the condition that $w = v$, which appears on the right side of (3.12), by the condition that $w = v - W$, and then apply (3.12) for $v = W + 1, W + 2, \cdots, 2W$. For the third cycle, we replace the condition that $w = v$ by the condition that $w = v - 2W$, and then apply (3.12) for $v = 2W + 1, 2W + 2, \cdots, 3W$, and so on.

(3) Instead of calculating the ℓ_w in (3.4) by the methods described under (1) or (2) above, they can be determined by the following formula, making use of the terms $f_w^\lambda / [\hat{f}(v - 1)]_w^\lambda$ calculated in the iterative scaling method described by (3.6) to (3.8):

$$(3.13) \qquad \ell_w'(T) = \prod_{t=0}^{T-1} \left\{ \frac{f_w^\lambda}{[\hat{f}(tW + w - 1)]_w^\lambda} \right\}$$

where $\prod_{t=0}^{T}$ denotes the product of the term in braces for $t = 0, 1, \cdots, T$; with $\prod_{t=0}^{0}$ denoting the term in braces for $t = 0$. If the iterative scaling method is completed when T cycles of iterations have been carried out, then $\ell_w'(T)$ can be used as the ℓ_w in (3.4).

The following formula describes the relationship between the $\ell_w'(T)$ defined by (3.13) and the $\ell_w(v)$ defined by (3.12):

$$(3.14) \qquad \ell_w'(T) = \ell_w(v) \qquad \text{for} \quad v = (T - 1)W + w.$$

Formula (3.14) can be proved by mathematical induction on T. (For some related (but different) results on the relationship between the iterative scaling method described by (3.6) to (3.8) and the iterative method described by (3.11) and (3.12), see Goodman's article [13] on the model of quasi-independence, and Haberman's work [20] on more general models.)

It should be noted that, for a particular model, the ℓ_w in (3.4), which we can calculate by any of the methods described under (1), (2), or (3) above, may still need to be scaled (transformed) in order to obtain maximum likelihood estimates of the corresponding scaled (transformed) parameters. (The λ_w in (3.1) and the ℓ_w in (3.4) may not be uniquely defined until they have been scaled (transformed) in a suitable manner.) Thus, for a particular model, if the $\hat{f}_{i, j}$ are unaffected by setting, say, $\ell_1, \ell_2, \cdots, \ell_U$ equal to one (where $U < W$) and by transforming $\ell_{U+1}, \ell_{U+2}, \cdots, \ell_W$ accordingly so that the transformed ℓ are uniquely defined, then these transformed quantities are the maximum likelihood estimates of the corresponding transformed parameters. (Actually, since (3.1) pertains to the probabilities $\pi_{i,j}$ whereas (3.4) pertains to the expected frequencies $\hat{f}_{i,j}$, the particular transformations that are used will determine whether a given transformed ℓ is an estimate of the corresponding transformed λ, or whether it is an estimate of the corresponding transformed λ multiplied by the sample size n.) For further comments on the transformed estimates, see Sections 4 and 7, and the Appendix.

When $f_{i,j} > 0$ for all cells (i, j) for which $\Lambda_{i,j}$ is not empty, then the results of Haberman [20] can be applied to show that the iterative procedures described here will converge to quantities that can be used to obtain the maximum likelihood estimates (that is, the $\hat{f}_{i,j}$ and the transformed ℓ). For cases where $f_{i,j} = 0$ for one or more cells (i, j) for which $\Lambda_{i,j}$ is not empty, the iterative procedures can still be used to obtain the maximum likelihood estimates, so long as the iterative procedures converge to a solution with $\hat{f}_{i,j} \neq 0$ when $\Lambda_{i,j}$ is not empty. The modified Newton-Raphson method, which was applied by Haberman [20] in his numerical example, has a more rapid convergence rate than do the iterative procedures described here (see Haberman [20]); but for data analysis in which a number of different models (of the kind presented here) are applied, the procedures described here have the advantage of being easier to program for a computer.

Before closing this section, we shall give an example of a multiplicative model that is not included within the class of models defined by (3.1), and that cannot be analyzed using the methods presented above. Consider again the QPN model in the special case where

$$(3.15) \qquad \begin{aligned} \alpha_i' &= \Delta\alpha_i\phi^i & \text{for} \quad i = 2, 3, \cdots, R - 1, \\ \beta_j' &= \frac{\Delta^*\beta_j}{\phi^j} & \text{for} \quad j = 2, 3, \cdots, R - 1. \end{aligned}$$

This special case of the QPN model is equivalent to the special case of the DAT model in which the additional condition

$$(3.16) \qquad \delta_k = \delta^k,$$

where $\delta = 1/\sqrt{\phi}$, is imposed. Although the DAT model, as defined earlier, can be analyzed by the methods presented in the present section, different methods are required for the special case in which condition (3.16) is imposed. The modified Newton-Raphson method presented in Haberman [20] can be applied to this special case. (The "fixed distance" model, which was applied in Haberman [20], is equivalent, in most respects, to the special case of the DA model in which condition (3.16) is imposed. In the present paragraph (see (3.15)) we have been considering the more general DAT model, rather than the DA model, in the special case where condition (3.16) is imposed.)

4. Unrestricted and restricted multiplicative models

The model defined by (3.1) was "intrinsically unrestricted" in the sense that no restrictions (or conditions) were imposed upon the parameters λ_w for $w = 1, 2, \cdots, W$, except for the fact that these parameters were such that

(4.1) $$\sum_{i,j} \pi_{i,j} = 1.$$

(Of course, since the parameters λ_w were used in (3.1) to form the probabilities $\pi_{i,j}$ for the cells (i,j) where $\pi_{i,j} > 0$, we also assumed that $\lambda_w > 0$ for $w = 1, 2, \cdots, W$.) Formula (3.1) can also be written as

(4.2) $$\pi_{i,j} = \Big[\prod_{\Lambda_{i,j}} \lambda_w \Big] \Big(\sum \Big)^{-1}$$

where $\sum$ denotes the summation of the $\prod_{\Lambda_{i,j}} \lambda_w$ over all cells (i,j) in the $R \times C$ table. Expressing the model in the form (4.2), we see that the λ_w in (4.2) are even unrestricted in the sense that the condition (4.1) does not impose any restrictions upon them.

For the various models introduced in Sections 2.1 to 2.8, we noted that the $\pi_{i,j}$ were unaffected by scaling (transforming) some of the λ parameters in certain ways. The parameters as described in formulas of the form (3.1) or (4.2) (see, for example, (2.1.1), (2.2.1), (2.3.1), and so on) were not uniquely defined until certain kinds of restrictions were imposed upon them (for example, the restriction that $\alpha_1 = 1$ in formula (2.1.1)). For each of the models in Sections 2.1 to 2.8, in order to calculate the degrees of freedom for testing the model, we described restrictions that could be imposed upon the λ parameters that (a) would uniquely define these parameters, and (b) would not affect the $\pi_{i,j}$. In some of the earlier sections, these restrictions were imposed upon the corresponding estimates of the λ parameters, but they could as well have been imposed upon the λ parameters. Since these particular kinds of restrictions did not affect the $\pi_{i,j}$, the models obtained when the restrictions were imposed were equivalent to the unrestricted models. This was the case for each of the models considered in Sections 2.1 to 2.8.

In addition to introducing restrictions upon the λ parameters that did not affect the $\pi_{i,j}$ (in order to calculate the degrees of freedom for testing the model or to uniquely define the parameters), we also introduced certain kinds of restrictions upon the parameters that did affect the $\pi_{i,j}$. For example, conditions (2.5.8), (2.5.9), and (2.5.10) were restrictions imposed upon the parameters that changed the D model into the DA model, the DP model, and the DN model, respectively. Despite the fact that these particular restrictions did affect the $\pi_{i,j}$, the models obtained when these restrictions were imposed could also be expressed in the general form (3.1) or (4.2), and so these models were also equivalent to unrestricted models. On the other hand, we noted at the end of Section 3, that if condition (3.16) were imposed upon the parameters of the DAT model (or the DA model, the DAC model, or the DACT model), the model obtained thereby would not be within the class of models defined by (3.1).

We shall call a model "intrinsically unrestricted" if it can be expressed in the general form (3.1) or (4.2). (Thus, the DA model, the DP model and the DN model are intrinsically unrestricted in this sense, despite conditions (2.5.8), (2.5.9), (2.5.10); whereas, the modification of the DAT model, which is obtained when condition (3.16) is imposed, is not unrestricted.) We shall now describe some of the kinds of restrictions that can be imposed upon a model of the general form (3.1) which are such that the modified models obtained thereby would also be intrinsically unrestricted.

Let H denote a given model of the form (3.1). The model H can be described by the set of parameters $\Lambda = \{\lambda_1, \lambda_2, \cdots, \lambda_W\}$, and by the subsets $\Lambda_{i,j}$ that are defined for each cell (i, j) in the table (see Section 3). Let Λ' be a given subset of the set Λ, and let H' denote the modification of model H that is obtained by imposing the condition

$$(4.3) \qquad \qquad \lambda_u = 1 \qquad \text{for} \quad \lambda_u \in \Lambda'.$$

Despite the fact that this modification (that is, the model H') is a special case of model H that satisfied condition (4.3), it is also intrinsically unrestricted. Model H' can be expressed in the form (3.1) simply by deleting from Λ and from each set $\Lambda_{i,j}$ any λ_u that are included in Λ'. (For some examples of the modification H', note that the QO model and each of the models in Sections 2.4 to 2.7, except the models that include the paired minor diagonals parameters and/or the triangles parameter (for example, the DA model, the DAT model, and so forth), can be formed from the DC model by this type of modification.)

Now let $\Lambda_1^*, \Lambda_2^*, \cdots, \Lambda_K^*$ denote mutually exclusive subsets of the parameters in Λ; and for each parameter λ_w in Λ, let S_w^λ denote the set of cells (i, j) for which $\lambda_w \in \Lambda_{i,j}$. Consider the modification of model H that is obtained by imposing the conditions

$$(4.4) \qquad \qquad \lambda_u = \lambda_{u'} \qquad \text{for all } \lambda_u \text{ and } \lambda_{u'} \in \Lambda_k^*,$$

for $k = 1, 2, \cdots, K$. This modification will be called a H'' type of modification if each Λ_k^* is such that the sets S_u and $S_{u'}$ are mutually exclusive for all λ_u and $\lambda_{u'}$ in Λ_k^*, for $k = 1, 2, \cdots, K$ (that is, if all λ_u and $\lambda_{u'}$ in Λ_k^* are "separable"). For some examples of the H'' type of modification, note that the QO model can be formed from the QPN model by this type of modification, and the models that include the paired minor diagonals parameters or the triangles parameter can also be formed from the corresponding models that include the diagonals parameters, by this type of modification. The model obtained by the H'' type of modification is also intrinsically unrestricted. It can be expressed in the form (3.1) by removing from Λ and from each $\Lambda_{i,j}$ any λ that are included in Λ_k^*, and by replacing each of these λ by a single parameter, say λ_k^* for $k = 1, 2, \cdots, K$.

The modification described by (2.7.4) might appear (at first sight) to differ from a H'' type of modification, but the change from the DC model to the DACT (or from the D model to the DAT model), which is described by the condition (2.7.4), can also be expressed in the following equivalent way. Without affecting the probabilities $\pi_{i,j}$ in the DC model (or the D model), the triangles parameters could be included in the expression (2.7.1) (or the expression (2.5.1)) for the $\pi_{i,j}$ (as a multiplicative factor of the kind appearing in (2.4.1)); and with the inclusion of these parameters in the model, the change from the DC model to the DACT model (or from the D model to the DAT model) can be expressed by condition (2.5.8) (rather than (2.7.4)), which is obviously a condition of the H'' type. A similar kind of remark can be made about the change from the QPN model to the T model, the change from the QPN to the C model, and the change from the QPN model to the CT model.

For a given model H of the form (3.1), we noted in Section 3 that the estimate $\hat{f}_{i,j}$ of the expected frequency satisfied condition (3.5). For the modification H' of H obtained by imposing condition (4.3), the estimate $\hat{f}_{i,j}$ of the expected frequency under model H' will satisfy the following modification of (3.5):

(4.5) $$\hat{f}_w^\lambda = f_w^\lambda \quad \text{if} \quad \lambda_w \notin \Lambda'.$$

For the modification H'' of H obtained by imposing condition (4.4), the estimate $\hat{f}_{i,j}$ of the expected frequency under model H'' will satisfy the following modification of (3.5):

(4.6)
$$\hat{f}_w^\lambda = f_w^\lambda \quad \text{if} \quad \lambda_w \notin \Lambda_k^* \qquad \text{for} \quad k = 1, 2, \cdots, K,$$
$$\sum_{\Lambda_k^*} \hat{f}_u^\lambda = \sum_{\Lambda_k^*} f_u^\lambda \qquad \text{for} \quad k = 1, 2, \cdots, K,$$

where $\Sigma_{\Lambda_k^*}$ denotes summation over the indices u for which $\lambda_u \in \Lambda_k^*$.

We noted earlier that all of the models in Section 2.4 to 2.7 can be obtained from the DC model by using modification H' and/or H''. We shall now show how the models of Section 2.8 can be obtained from the DC model by using these kinds of modifications.

The DC model (as defined by (2.7.1)) was concerned with the analysis of the off diagonal cells, and so for simplicity we could set $\pi_{i,i} = 0$ in (2.7.1). On the other hand, we need not have imposed this restriction on the $\pi_{i,i}$, and could instead have defined the DC model by writing the probability $\pi_{i,j}$ as

(4.7) $$\pi_{i,j} = \begin{cases} \alpha_i \beta_j \gamma'_{i,j} \delta_k & \text{for cells } (i,j) \text{ in } S'_k(k = \pm 1, \pm 2, \cdots, \pm(R-1)), \\ \pi_{i,i} & \text{for cells } (i,i) \text{ in } S'_0. \end{cases}$$

The $\pi_{i,i}$ on the right side of (4.7) can be viewed as "intrinsically unrestricted" λ parameters, as can the parameters α_i, β_j, γ_u, and δ_k. If the condition

(4.8) $$\pi_{i,i} = \alpha_i \beta_i \delta_0$$

is imposed upon the $\pi_{i,i}$ in (4.7), the DC model will be changed to the DCF model. To express condition (4.8) in the form (4.4) (that is, as a modification of the H'' type), we first note that the $\pi_{i,i}$ in (4.6) can be written as $\alpha'_i \beta'_i \delta_0$ (where the $\alpha'_i, \beta'_i, \delta_0$ are intrinsically unrestricted λ parameters) without affecting the probabilities in the DC model (see 4.7); and we then impose the condition

(4.9) $\qquad \alpha_i = \alpha_i', \qquad \beta_j = \beta_j' \qquad$ for $\quad i = 1, 2, \cdots, R; j = 1, 2, \cdots, R,$

to change the DC model to the DCF model.

All of the models in Section 2.8 can be obtained from the DCF model by using modifications H' and/or H'' in the same ways that these modifications would be used to obtain the corresponding models of Sections 2.4 to 2.7 from the DC model. Since we noted in the preceding paragraph that the DCF model can be obtained from the DC model using a modification of the H'' type, we now see that all of the models in Section 2.8 can also be obtained from the DC model by using the modifications H' and/or H''. Furthermore, each of the models in Section 2.8 can be obtained from the corresponding model in Sections 2.4 to 2.7 by using a modification of the H'' type in the same way that it was used above to obtain the DCF model from the DC model.

As we noted earlier, the present article is limited to multiplicative models of the general form (3.1), that is, models that are intrinsically unrestricted. These include all the models of Sections 2.1 to 2.8 and many others as well, but they do not include "restricted" models of the kind referred to at the end of Section 3.

5. The ratio index and the relative difference index for the cells on the main diagonal

Consider again the DC model for the off diagonal cells, and the corresponding model for the full table (that is, the DCF model). For the DCF model (as defined in Section 2.8), the probability $\pi_{i,j}$ can be written as

(5.1) $\qquad \mu_{i,j} = \alpha_i \beta_j \gamma_{i,j}' \delta_k \qquad$ for cells (i, j) in S_k',

for $k = 0, \pm 1, \pm 2, \cdots, \pm(R-1)$, with $\gamma_{i,j}' = 1$ for $i = j$. Thus, for both the DCF model and the DC model, the probability $\pi_{i,j}$ can be written as

(5.2) $\qquad \pi_{i,j} = \alpha_i \beta_j \gamma_{i,j}' \delta_k \mu_{i,j} \qquad$ for cells (i, j) in S_k',

for $k = 0, \pm 1, \pm 2, \cdots, \pm(R-1)$, where the $\mu_{i,j}$ satisfy condition

(5.3) $\qquad \mu_{i,j} = 1 \qquad$ for all cells (i, j),

for the DCF model, and condition

(5.4) $\qquad \mu_{i,j} = \begin{cases} 1 & \text{for } i \neq j, \\ \pi_{i,i}/(\alpha_i \beta_i \delta_0) & \text{for } i = j \end{cases}$

for the DC model. Note that the $\pi_{i,j}$ defined by (5.2) and (5.4) are equivalent to the $\pi_{i,j}$ defined by (4.7).

In view of (5.4), the quantity

(5.5) $$\mu_i = \frac{\pi_{i,i}}{\alpha_i \beta_i \delta_0}$$

is of interest. We shall call μ_i the ratio index for the ith cell on the main diagonal.

(This index was defined earlier in Goodman [14] for the model of quasi-independence, and it was called there a "new index of immobility.") For models that do not include diagonals parameters (or the triangles parameter), we set $\delta_0 = 1$; for the other models of Section 2.4 to 2.7, the parameter δ_0 cannot be estimated from the data without introducing some additional assumptions (which we shall discuss in Section 7); for the models of Section 2.8, we set $\mu_i = 1$ for $i = 1, 2, \cdots, R$, and the parameter δ_0 (and the other δ_k) in (5.1) can be estimated from the data.

In models that include crossings parameters for the analysis of the off diagonal cells (for example, the C model or DC model), we noted earlier (see Sections 2.6 and 2.7) that the probability $\pi_{i,j}$ is unaffected by setting $\gamma_1 = \gamma_{R-1} = 1$, and by making corresponding changes in α_1, α_R, β_1, β_R. On the other hand, instead of setting $\gamma_1 = \gamma_{R-1} = 1$, the $\pi_{i,j}$ would also be unaffected by making some other assumptions about γ_1 and γ_{R-1} (with corresponding changes made in α_1, α_R, β_1, β_R); and with these assumptions (which we shall discuss in Section 7), the data could be used to estimate γ_1 and γ_{R-1}. These assumptions about γ_1 and γ_{R-1} will affect $\alpha_1\beta_1$ and $\alpha_R\beta_R$; and they thereby will affect μ_1 and μ_R.

For the DC model for analyzing the off diagonal cells, the remarks in the preceding two paragraphs indicate that there is an element of arbitrariness about μ_1, μ_R and δ_0 (that is, the additional assumptions, to which we referred in those paragraphs, will affect these quantities). This was not the case for the DCF model (see Section 2.8). Note that the difference between the degrees of freedom for the DCF model and the DC model is $R - 3$, which corresponds to the degrees of freedom associated with testing the hypothesis

$$(5.6) \qquad \mu_i = \mu \qquad \text{for} \quad i = 2, 3, \cdots, R - 1,$$

where μ is unspecified. A similar kind of remark can be applied to the difference between the degrees of freedom for each of the models in Section 2.8 and the corresponding model from Sections 2.4 to 2.7. For the models in Section 2.8 that do not include crossings parameters, replace (5.6) by the same condition applied for $i = 1, 2, \cdots, R$ (rather than for $i = 2, 3, \cdots, R - 1$); and for models that do not include diagonals parameters (or the triangles parameter), replace the unspecified quantity μ in (5.6) by one.

Let us now denote $\alpha_i\beta_i\delta_0$ by $\pi^*_{i,i}$, and let $\pi^*_{i,j} = \pi_{i,j}$ for $i \neq j$, with the $\pi_{i,j}$ satisfying condition (4.7) for the DC model. With this notation, the index μ_i defined by (5.5) can be written as

$$(5.7) \qquad \mu_i = \frac{\pi_{i,i}}{\pi^*_{i,i}}.$$

Instead of this ratio index, consider now the quantity

$$(5.8) \qquad \mu^*_i = \frac{(\pi_{i,i} - \pi^*_{i,i})}{\pi_{i\cdot}},$$

where $\pi_{i\cdot} = \Sigma^R_{j=1} \pi_{i,j}$, which we shall call the relative difference index for the ith cell on the main diagonal. This index was defined earlier in Goodman [16] for the model of quasi-independence, and it was called there the "index of persistence."

Let $\tilde{\pi}_{i,j}$ and $\tilde{\pi}_{i,j}^*$ denote $\pi_{i,j}/\pi_{i\cdot}$ and $\pi_{i,j}^*/\pi_{i\cdot}^*$, respectively, where $\pi_{i\cdot}^* = \sum_{j=1}^{R} \pi_{i,j}^*$. From (4.7) and (5.8) we find that $\tilde{\pi}_{i,j}$ can be written as

$$(5.9) \qquad \tilde{\pi}_{i,j} = \begin{cases} (1 - \mu_i^*)\tilde{\pi}_{i,j}^* & \text{for } i \neq j, \\ \mu_i^* + [(1 - \mu_i^*)\tilde{\pi}_{i,j}^*] & \text{for } i = j. \end{cases}$$

Because of (5.9), the index μ_i^* can be interpreted as the proportion of "stayers" among those individuals who are in the ith row in the population. This interpretation would apply when $\mu_i^* > 0$.

For the case where $\mu_i^* < 0$, we can rewrite (5.9) as

$$(5.10) \qquad \tilde{\pi}_{i,j} = \begin{cases} \tilde{\pi}_{i,j}^*(1 + v_i) & \text{for } i \neq j, \\ \tilde{\pi}_{i,i}^* - v_i(1 - \tilde{\pi}_{i,i}^*) & \text{for } i = j, \end{cases}$$

where $v_i = -\mu_i^*$. Because of (5.10), the index v_i (that is, the index $-\mu_i^*$) can be interpreted (when $\mu_i^* < 0$) as the proportion of individuals who have a "second chance" to move out of the ith row class among those individuals who are in the ith row in the population. This interpretation can be applied when $\tilde{\pi}_{i,i}^* > v_i$. The interpretations presented in the present paragraph, and in the preceding one, are extensions of the interpretations introduced in Goodman [16] for the model of quasi-independence.

From (5.7) and (5.8), we see that μ_i and μ_i^* are related as follows:

$$(5.11) \qquad \begin{aligned} \mu_i &= \frac{\tilde{\pi}_{i,i}}{\tilde{\pi}_{i,i} - \mu_i^*} \\ \mu_i^* &= \frac{\tilde{\pi}_{i,i}(\mu_i - 1)}{\mu_i}. \end{aligned}$$

In addition, the index μ_i^* can be rewritten as

$$(5.12) \qquad \mu_i^* = \frac{\tilde{\pi}_{i,i} - \tilde{\pi}_{i,i}^*}{1 - \tilde{\pi}_{i,i}^*}.$$

Thus, the index μ_i^* measures the difference $(\tilde{\pi}_{i,i} - \tilde{\pi}_{i,i}^*)$ relative to the difference $(1 - \tilde{\pi}_{i,i}^*)$.

From (5.12) we see that the index μ_i^* can be expressed as a function of $\tilde{\pi}_{i,i}$ and $\tilde{\pi}_{i,i}^*$ (rather than $\pi_{i,i}$ and $\pi_{i,i}^*$), whereas the corresponding ratio index μ_i was a function of $\pi_{i,i}$ and $\pi_{i,i}^*$. (Compare (5.12) with (5.7).) Note that $\tilde{\pi}_{i,j}$ is the conditional probability that an individual will fall in the jth column class of the population table, given that he is in the ith row class; and $\tilde{\pi}_{i,j}^*$ is the corresponding conditional probability obtained by replacing $\pi_{i,i}$ by $\alpha_i\beta_i\delta_0$. The conditional probabilities $\tilde{\pi}_{i,j}$ and $\tilde{\pi}_{i,j}^*$ are particularly relevant when the data to be analyzed have been obtained from a sample of n_i individuals drawn from the ith row class for $i = 1, 2, \cdots, R$ (that is, when the row marginals are fixed), rather than from a sample of n individuals drawn from the population cross classification table. It is in this context that the index μ_i^* (rather than the index μ_i) is particularly relevant.

When a sample of n individuals is drawn from the population cross classification table, the index μ_i may be viewed as one of the parameters of the model, since the $\pi_{i,j}$ can be expressed by (5.2) with the parameter $\mu_{i,j}$ in (5.2) set equal to μ_i for $i = j$ (see (5.4) and (5.5)). This remark can be applied to any of the models of Sections 2.4 to 2.7; and for the models of Section 2.8 the parameter μ_i is set equal to one. Similarly, when a sample of n_i individuals is drawn from the ith row class for $i = 1, 2, \cdots, R$, the index μ_i^* may be viewed as one of the parameters of the model, since the $\tilde{\pi}_{i,j}$ can be expressed by (5.9).

Note that the index μ_i is "symmetric" in the sense that it is invariant when the row classes are interchanged with the column classes. Of course, this will not be the case for the index μ_i^*. In addition to the index μ_i^* defined herein for the case where the row marginals are fixed, there is the relative difference index that would be defined in a directly analogous way for the case where the column marginals are fixed.

Applying the general methods of estimation described in Section 3 to the DC model, we can calculate the estimate

$$(5.13) \qquad \hat{f}_{i,j} = \begin{cases} a_i b_j c'_{i,j} d_k & \text{for} \quad i \neq j, \\ f_{i,i} & \text{for} \quad i = j, \end{cases}$$

of the expected frequency under the model. (Compare (5.13) with (4.7).) Note that d_0 does not appear in (5.13), and the corresponding δ_0 did not appear in (4.7) or (2.7.1). For models that do not include diagonals parameters (or the triangles parameter), we set $d_0 = 1$; for models that do include these parameters, the value of d_0 will be estimated from the data after we introduce some additional assumptions (which we shall discuss in Section 7). Denoting $a_i b_i d_0$ by $f_{i,i}^*$, and letting $f_{i,j}^* = \hat{f}_{i,j}$ for $i \neq j$ (with $\hat{f}_{i,j}$ defined by (5.13)), we can estimate the index μ_i, which was defined by (5.7),

$$(5.14) \qquad m_i = \frac{f_{i,i}}{f_{i,i}^*}.$$

Similarly, the index μ_i^*, which was defined by (5.8), can be estimated as

$$(5.15) \qquad m_i^* = \frac{f_{i,i} - f_{i,i}^*}{f_{i \cdot}}$$

where $f_{i \cdot} = \Sigma_{j=1}^R f_{i,j}$. Formulae that are directly analogous to (5.11) and (5.12) can be obtained for m_i and m_i^*, by replacing μ_i, μ_i^*, $\tilde{\pi}_{i,i}$, and $\tilde{\pi}_{i,i}^*$ in (5.11) and (5.12) by m_i, m_i^*, $f_{i,i}/f_{i \cdot}$, and $f_{i,i}^*/f_{i \cdot}^*$, respectively, where $f_{i \cdot}^* = \Sigma_{j=1}^R f_{i,j}^*$. The estimates m_i and m_i^* can be calculated for the QO model and for each of the models in Sections 2.4 to 2.7.

6. The comparison of observed frequencies and expected frequencies under the various models

The usual chi square goodness of fit statistic for comparing the observed frequency $f_{i,j}$ with the corresponding estimate $\hat{f}_{i,j}$ of the expected frequency under a given model, can be written as

$$(6.1) \qquad \sum \frac{(f_{i,j} - \hat{f}_{i,j})^2}{\hat{f}_{i,j}},$$

and the corresponding chi square statistic based upon the likelihood ratio criterion can be written as

$$(6.2) \qquad 2\sum f_{i,j} \log\left(\frac{f_{i,j}}{\hat{f}_{i,j}}\right),$$

where the summation in (6.1) and (6.2) is taken over the off diagonal cells for the QO model and for the models of Sections 2.4 to 2.7, over all the cells in the table for the model of "independence" and for the models of Section 2.8, and over the set (or sets) of cells that are to be analyzed for a given model of quasi-independence of the kind described in Section 2.2 and 2.3. Both the statistic (6.1) and the statistic (6.2) have an asymptotic chi square distribution under the given model, with the degrees of freedom equal to the number of parameters in the model (calculating this number after the parameters have been uniquely defined). In certain contexts, the statistic (6.2) has some advantages over (6.1) (see, for example, Bahadur [1], Good [7], Goodman [13], [15], [17], [19], and Hoeffding [21]). We shall give in Table V the numerical values of both (6.1) and (6.2), for various models applied to the British 5×5 table (Table I), the Danish 5×5 table (Table II), and the British 7×7 table (Table III); but, for the sake of simplicity, when we discuss Table V later in the present section, we shall confine our attention to the numerical values of the statistic (6.2).

Since the estimate $\hat{f}_{i,j}$ of the expected frequency under each of the models considered here can be expressed in the general form (3.4), we see that the statistic (6.2) can also be written as

$$(6.3) \qquad 2\left[\sum f_{i,j} \log f_{i,j} - \sum_{w=1}^{W} f_w^{\lambda} \log \ell_w\right],$$

where the first summation sign $\sum$ in (6.3) has the same meaning that it did in (6.1) and (6.2); namely, the summation over all cells (i, j) for which $\Lambda_{i,j}$ is not empty. Note that formula (6.3) provides a method for calculating the statistic (6.2) using the W terms ℓ_w for $w = 1, 2, \cdots, W$ without calculating the $\hat{f}_{i,j}$ terms.

Now let H denote a given model of the form (3.1), and let H^+ denote the modified model that is obtained by applying to H a given modification of the H' and/or H'' type (see Section 4). As we noted in Section 4, the model H^+ is also "intrinsically unrestricted." Let $[H^+|H]$ denote the hypothesis that H^+

TABLE V

COMPARISON OF THE OBSERVED FREQUENCIES AND THE EXPECTED FREQUENCIES
UNDER VARIOUS MODELS APPLIED TO THE BRITISH AND DANISH 5 × 5 TABLES, AND THE
BRITISH 7 × 7 TABLE

		British and Danish 5 × 5 Tables			
		British Sample		Danish Sample	
Model	Degrees of Freedom	Goodness of Fit Chi Square	Likelihood Ratio Chi Square	Goodness of Fit Chi Square	Likelihood Ratio Chi Square
Ind	16	1199.4	811.0	754.1	654.2
QO	11	328.7	249.4	270.3	248.7
QP	3	8.5	12.6	6.9	7.4
QN	3	1.3	1.4	2.4	2.5
QPN	6	9.9	14.0	9.4	9.9
T	10	313.1	242.3	269.3	248.5
C	9	11.9	15.4	12.2	12.8
DA	8	15.9	19.1	6.7	6.9
CT	8	10.2	14.1	12.1	12.7
DAT	7	14.4	17.8	6.6	6.8
DP	7	10.3	10.6	6.8	7.0
DN	7	18.6	23.8	10.3	10.9
DAC	7	8.6	11.1	6.4	6.6
DACT	6	7.2	10.0	6.4	6.5
DPC	6	2.2	2.2	6.6	6.9
DNC	6	9.5	13.4	10.0	10.5
D	5	9.0	9.5	4.7	4.8
DC	4	1.5	1.6	4.4	4.5
DCF	6	6.7	6.9	6.3	6.3
DPCF	8	7.6	7.7	8.2	8.3
DACF	9	13.8	16.7	8.1	8.4
DF	9	52.1	50.4	10.1	10.2
DAF	12	59.5	60.6	12.2	12.4

is true assuming that H is true. Since H^+ is a modification of H in which some given conditions of the type (4.3) and/or (4.4) are imposed, the hypothesis $[H^+|H]$ states that the given conditions are true, assuming the H is true. Let $\chi^2(H)$ and $\chi^2(H^+)$ denote the statistic (6.2) with $\hat{f}_{i,j}$ calculated under H and H^+, respectively. The statistics $\chi^2(H)$ and $\chi^2(H^+)$ can be used to test the models H and H^+, respectively, and the following statistic can be used to test the hypothesis $[H^+|H]$:

$$(6.4) \quad \chi^2(H^+|H) = \chi^2(H^+) - X^2(H)$$

$$= 2 \sum f \log\left[\frac{\hat{f}}{\hat{f}^+}\right] = 2 \sum \hat{f} \log\left[\frac{\hat{f}}{\hat{f}^+}\right],$$

where $\hat{f}$ and $\hat{f}^+$ are the estimated expected frequencies under H and H^+, respectively. The final equality in (6.4) holds because the $\hat{f}$ satisfy (3.5) and

TABLE V (Continued)

Model	Degrees of Freedom	British 7 × 7 Table Goodness of Fit Chi Square	Likelihood Ratio Chi Square
Ind	36	1361.7	897.5
QO	29	523.0	408.4
QP	10	9.4	13.4
QN	10	7.4	7.5
QPN	20	16.7	20.9
T	28	517.8	404.1
C	25	20.6	24.6
DA	24	20.1	22.1
CT	24	19.9	24.1
DAT	23	19.5	21.6
DP	23	21.3	22.3
DN	23	19.3	23.6
DAC	21	15.1	17.1
DACT	20	14.6	16.6
DPC	20	14.6	15.8
DNC	20	13.8	18.0
D	19	13.6	14.6
DC	16	8.4	9.4
DCF	20	25.0	26.3
DPCF	24	38.8	39.7
DACF	25	31.7	33.8
DF	25	54.2	54.8
DAF	30	59.8	61.9

the $\hat{f}^+$ satisfy conditions of the form (4.5) and/or (4.6). The statistic (6.4) is the chi square statistic based upon the likelihood ratio criterion for testing the hypothesis $[H^+|H]$. This statistic has an asymptotic chi square distribution under the hypothesis $[H^+|H]$, with the degrees of freedom equal to the difference between the corresponding number of degrees of freedom for testing H^+ and H, respectively.

Let us now examine the numerical values of $\chi^2(H)$, which are given in Table V for each of the models in Sections 2.3 to 2.7, and for some of the models in Section 2.8, applied to Table I. For the DC model, we see that $\chi^2(DC) = 1.6$ with 4 degrees of freedom, which indicates that this model fits the data very well. Comparing the D model with the DC model using the statistic (6.4), we obtain $\chi^2(D|DC) = 7.9$ with one degree of freedom, which indicates that the crossings parameter makes a statistically significant contribution. In other words, assuming that the DC model is true, a test of the null hypothesis (2.7.3) would lead to rejection of the hypothesis. Comparing the C model with the DC model, we obtain $\chi^2(C|DC) = 13.8$ with 5 degrees of freedom, which indicates that the diagonals parameters make a statistically significant contribution. In other words, assuming that the DC model is true, a test of the null hypothesis (2.7.2) would lead to the rejection of the hypothesis.

Comparing the CT model, the DAC model, and the DNC model with the DC model, we obtain $\chi^2(CT|DC) = 12.5$, $\chi^2(DAC|DC) = 9.5$, and $\chi^2(DNC|DC) = 11.8$, with 4, 3, and 2 degrees of freedom, respectively. Thus, assuming that the DC model is true, a test of each of the null hypotheses (2.5.2), (2.5.8), and (2.5.13) would lead to their rejection. A similar result is obtained comparing the DACT model with the DC model. Comparing the DCF model and the DPC model with the DC model, we obtain $\chi^2(DCF|DC) = 5.3$ and $\chi^2(DPC|DC) = 0.6$ each with 2 degrees of freedom. Thus, assuming that the DC model is true, a test of the null hypothesis (5.6) would lead to rejection of the hypothesis at the 10 per cent level of significance, and a test of the null hypothesis (2.5.9) would lead to acceptance of that hypothesis. Indeed, the DPC model fits the data very well.

Having noted that the DPC model fits the data well, we now compare various models with the DPC model. Comparing the DP and the C models with the DPC model, we obtain $\chi^2(DP|DPC) = 8.4$ and $\chi^2(C|DPC) = 13.2$, with 1 and 3 degrees of freedom, respectively, which indicates that both the crossings parameters and the parameters pertaining to the "positive" diagonals in the DPC model make a statistically significant contribution. Comparing the DPCF model with the DPC model, we obtain $\chi^2(DPCF|DPC) = 5.5$ with 2 degrees of freedom. Thus, assuming that the DPC model is true, a test of the null hypothesis (5.6) would lead to rejection of the hypothesis at the 10 per cent level of significance.

The preceding comments pertained to the results given in Table V for the analysis of Table I. The corresponding results, which are given in Table V for the analysis of Tables II and III, do not lead to conclusions that are as clear cut as those obtained for Table I. We shall now comment briefly on the results for Tables II and III without presenting a full analysis of them.

As was the case for Table I, the DC model fits the data very well for Table III; and it fits the data rather well for Table II, but not as well as for Tables I and III. Among the models that fit the data well (or rather well), we find the DA model for Table II and the DAC model for Table III. Comparing the models for the analysis of the full tables with the corresponding models for the analysis of the off diagonal cells, we find that a test of the null hypothesis (5.6) would lead to rejection of the hypothesis for Table III and acceptance of the hypothesis for Table II. Comparison of the QP and QN models indicates that the QN model fits the data better than the QP model for Tables II and III (and also for Table I).

For each chi square statistic, the corresponding number of degrees of freedom can be obtained from Tables IV and V. These tables give the degrees of freedom of the corresponding asymptotic distribution under the null hypothesis. This is the appropriate number of degrees of freedom to use in testing the hypothesis if the hypothesis were decided upon before the data were studied. On the other hand, if a set of hypotheses were tested simultaneously (or if the particular hypothesis that was tested was contained within a larger set of hypotheses that were studied), the degrees of freedom could be adjusted in a similar way to the adjustment made in calculating simultaneous confidence intervals and simultaneous tests in the present context (see Goodman [11], [14], [19]). This adjustment will limit the risks of rejecting hypotheses that are true, even when the hypotheses are suggested by the data. Of course, the risks of accepting false hypotheses are also affected if the hypotheses are suggested by the data.

The various hypotheses and models that we have tested and compared here could also have been assessed by adapting to the present context some of the concepts that arise in stepwise regression (for example, some of the concepts of backward regression and/or forward regression). For a discussion of this kind of adaptation, and for an example of its application, the reader is referred to Goodman [18].

7. The estimated parameters and indices

In this section we shall comment briefly upon the estimates of some of the parameters, and of the ratio and relative difference indices, which are obtained when some of the models of Section 2 are applied to Tables I, II, and III. These estimates will be presented in Tables VI to X later in this section.

For the models that included the triangles parameters τ_k for $k = 1$ and 2 (see, for example, (2.4.1)), we noted earlier that these parameters were not uniquely defined until one restriction was imposed upon them. This restriction could be expressed in several different ways; for example, as (a) the condition that $\tau_1 = 1$ (as we did in Section 2.4), or as (b) the condition that $\tau_2 = 1$, or as (c) the condition

$$(7.1) \qquad \tau_1 \tau_2 = 1.$$

For our present purposes, it is convenient to impose condition (7.1), and to take τ_1 as the uniquely defined triangles parameter τ. From (7.1), we see that τ can also be expressed as

$$(7.2) \qquad \tau = \left(\frac{\tau_1}{\tau_2}\right)^{1/2}$$

(In contexts where condition (7.1) is replaced by one of the other conditions given above, the quantity defined by (7.2) would also be replaced as the triangles parameter.)

Table VI gives the maximum likelihood estimate t of τ obtained under the four models that include the triangles parameter; namely, the T model, the CT model, the DAT model, and the DACT model. Note that the estimate t is less than one for each of the cases considered in Table VI, which indicates that, aside from the effects of the other parameters in the model, the estimated triangles parameter will diminish the estimated expected frequencies for the cells in set S_1 (where $i - j > 0$) relative to the estimated expected frequencies for the cells in set S_2 (where $i - j < 0$). For each of the three mobility tables

TABLE VI

THE ESTIMATE OF THE TRIANGLES PARAMETER UNDER VARIOUS MODELS
APPLIED TO THE BRITISH AND DANISH SAMPLES

	Models			
	T	CT	DAT	DACT
British 5 × 5 Table	0.855	0.935	0.935	0.937
Danish 5 × 5 Table	0.970	0.981	0.980	0.983
British 7 × 7 Table	0.904	0.966	0.966	0.968

(Tables I, II, and III), this would indicate that the estimated τ parameter has the direct effect of introducing "downward mobility" (as expressed in the fact that $(\tau_2/\tau_1)^{1/2}$ is estimated as being larger than one), over and above the indirect effects (with respect to upward or downward mobility) that are due to the other parameters. The effect of t appears to be more pronounced for the T model (particularly for Table I), and it becomes less pronounced as more parameters (for example, the crossings parameters and/or the parameters pertaining to the paired minor diagonals) are included in the model. To test whether $\tau = 1$ in, say, the DACT model applied to Tables I, II, and III, we compare the DAC model with the DACT model using the results given in Table V, and we find that the τ parameter does not have a statistically significant effect when the other parameters are included in the model.

With respect to the diagonals parameters (Table VII), we note that d_k decreases as $|k|$ increases in all cases, except for d_4 for Table II under the DPC, D, and DC models. This would indicate that aside from the effects of the other parameters in the model, generally speaking the estimated diagonals parameters diminish the estimated expected frequencies in a progressively more pronounced way for cells that are on minor diagonals that are further away from the main diagonal. In other words, the estimates d_k have the direct effect of introducing "status inertia" in the mobility table. With respect to Table II, it is worth noting (see Table V) that the modifications of the DA model and the DAC model that are obtained by distinguishing diagonals parameters on the "positive" diagonals from those on the "negative" diagonals (for example, the D model, the DC model, and the DPC model) did not improve the fit markedly. Note also that the difference in Table VII between d_k and the corresponding d_{-k} is, generally speaking, smaller for Table II than for the other tables. When comparing the DA model with the DAC model (or the D model with the DC model), we find that the effect of d_k is somewhat more pronounced in the former model than in the latter one (which included the crossings parameters as well). To facilitate the comparison of models in Table VII, condition (2.5.12) was used, rather than the equivalent (2.5.9), in the DPC model; and condition (7.3) below was used in the DAC, DPC, and DC models.

With respect to the crossings parameters (Table VIII), for models that include both the γ_u and δ_k parameters, we noted earlier that the γ_u, for $u =$

TABLE VII

THE ESTIMATE OF THE DIAGONALS PARAMETERS UNDER VARIOUS MODELS
APPLIED TO THE BRITISH AND DANISH SAMPLES

	Models							
	DA	DAC	DPC		D		DC	
British 5 × 5 Table	$d_k = d_{-k}$	$d_k = d_{-k}$	d_k	d_{-k}	d_k	d_{-k}	d_k	d_{-k}
$k = 1$	1.00	1.00	1.00	1.00	1.00	1.00	1.00	1.00
$k = 2$	0.59	0.64	0.74	0.54	0.68	0.52	0.73	0.56
$k = 3$	0.26	0.32	0.39	0.29	0.32	0.22	0.39	0.28
$k = 4$	0.08	0.13	0.00	0.16	0.00	0.11	0.00	0.17
Danish 5 × 5 Table								
$k = 1$	1.00	1.00	1.00	1.00	1.00	1.00	1.00	1.00
$k = 2$	0.48	0.51	0.52	0.45	0.49	0.47	0.52	0.50
$k = 3$	0.15	0.16	0.14	0.20	0.13	0.16	0.14	0.17
$k = 4$	0.11	0.12	0.17	0.09	0.15	0.06	0.17	0.06
British 7 × 7 Table								
$k = 1$	1.00	1.00	1.00	1.00	1.00	1.00	1.00	1.00
$k = 2$	0.65	0.70	0.69	0.61	0.64	0.65	0.69	0.70
$k = 3$	0.34	0.40	0.51	0.38	0.41	0.28	0.49	0.34
$k = 4$	0.20	0.26	0.30	0.23	0.21	0.18	0.28	0.25
$k = 5$	0.08	0.11	0.08	0.14	0.05	0.09	0.08	0.13
$k = 6$	0.03	0.04	0.00	0.09	0.00	0.04	0.00	0.06

TABLE VIII

THE ESTIMATE OF THE CROSSINGS PARAMETERS UNDER VARIOUS MODELS
APPLIED TO THE BRITISH AND DANISH SAMPLES

	Models			
	C	DAC	DPC	DC
British 5 × 5 Table				
c_2	0.40	0.46	0.45	0.46
c_3	0.60	0.64	0.63	0.64
Danish 5 × 5 Table				
c_2	0.46	0.51	0.48	0.51
c_3	0.43	0.47	0.46	0.48
British 7 × 7 Table				
c_2	0.54	0.60	0.55	0.60
c_3	0.52	0.58	0.53	0.58
c_4	0.60	0.67	0.61	0.66
c_5	0.64	0.70	0.65	0.70

$2, 3, \cdots, R - 2$, were not uniquely defined until one restriction was imposed upon them—a restriction that has the effect of fixing the scale of the γ_u. This restriction could be expressed in several different ways; for example, as (a) the condition that

$$(7.3) \qquad \max_k \gamma_k = 1 \qquad \text{for} \quad 2 \leqq k \leqq R - 2,$$

which is equivalent to replacing the γ_u by $\tilde{\gamma}_u = \gamma_u / \gamma^*$, where $\gamma^* = \max_k \gamma_k$ for $2 \leqq k \leqq R - 2$; or (b) the condition that

$$(7.4) \qquad \delta_2 \delta_{-2} = 1,$$

which will also have the effect of fixing the scale of the γ. For models that included the δ_k parameters, we noted earlier that the restriction that $\delta_1 = \delta_{-1} = 1$ would uniquely define the δ_k; this restriction together with condition (7.4) would uniquely define the δ_k and γ_u for $u = 2, 3, \cdots, R - 2$, in models that included both sets of parameters. In order to facilitate comparison between the crossings parameters for the C model (for which the γ_u for $u = 2, 3, \cdots, R - 2$, are uniquely defined without imposing any restrictions upon them) and the corresponding parameters in models that also include the δ_k parameters, the restriction (7.4) was used in calculating the results presented in Table VIII. To make the c_u in Table VIII consistent with the corresponding d_k in Table VII for a model that includes both sets of parameters (for example, the DAC, the DPC, and the DC models), it is only necessary to divide each of the c_u in Table VIII by $\max_k c_k$ for $2 \leq k \leq R - 2$, for the given model (see condition (7.3)). The estimates c_u in Table VIII have the direct effect of introducing "status barriers" in the mobility tables. Comparing the results in Table VIII for the 5×5 tables (that is, Tables I and II), we note that the effect of c_2 is more pronounced than that of c_3 for Table I, and the reverse is true for Table II. For Table III, the effect of c_u becomes more pronounced as u decreases, except for $u = 2$. From Table VIII we see that the relative difference between the c is less for Table II than for Table I. From Table V we also take note of the fact that the modification of the various models (for example, the DA, the DP, the D models) that is obtained by including the crossing parameters (thus obtaining the DAC, the DPC, and the DC models) did not improve the fit markedly for Table II, but it did for Table I. Finally, we note that the corresponding c_u in Table VIII are very similar for the DAC model and the DC model.

With respect to the ratio index and the relative difference index (Tables IX and X), we first note that the numerical values obtained for the QO model (which did not fit the mobility tables well) are grossly misleading. For models that fit the data well, the numerical values obtained for these indices can differ greatly from the values obtained with models that do not fit the data. Note also that the corresponding values for the DA model and the D model are very similar, and so are the corresponding values of the DAC and the DC models. With the introduction of the additional parameters (for example, the δ_k and/or the γ_u) into the QO model, the effect of the m_i diminishes (see Table IX).

TABLE IX

THE ESTIMATE OF THE RATIO INDEX FOR THE CELLS ON THE MAIN DIAGONAL
OF THE $R \times R$ CROSS CLASSIFICATION TABLE, UNDER VARIOUS MODELS APPLIED TO THE
BRITISH AND DANISH SAMPLES

| | Models | | | | | |
	QO	DA	DAC	DPC	D	DC
British 5 × 5 Table						
m_1	34.5	3.8	4.6	4.5	3.8	4.5
m_2	4.0	0.9	0.6	0.6	0.9	0.6
m_3	1.7	1.0	1.0	1.0	1.0	1.0
m_4	1.0	0.7	0.8	0.7	0.8	0.8
m_5	2.9	0.9	1.1	1.1	0.9	1.1
Danish 5 × 5 Table						
m_1	13.8	1.1	1.2	1.2	1.1	1.2
m_2	4.8	0.9	1.0	1.0	0.9	1.0
m_3	1.8	0.8	0.8	0.8	0.8	0.8
m_4	1.2	0.7	0.7	0.7	0.7	0.7
m_5	3.4	0.7	0.8	0.7	0.7	0.8
British 7 × 7 Table						
m_1	35.0	2.4	2.6	2.4	2.4	2.6
m_2	8.6	1.5	1.3	1.2	1.5	1.3
m_3	2.2	0.7	0.6	0.5	0.7	0.6
m_4	1.7	1.0	1.0	1.0	1.0	1.0
m_5	1.2	0.8	0.8	0.7	0.8	0.8
m_6	2.3	1.3	1.4	1.3	1.3	1.4
m_7	2.9	0.7	0.8	0.7	0.7	0.8

TABLE X

THE ESTIMATE OF THE RELATIVE DIFFERENCE INDEX FOR THE CELLS ON THE
MAIN DIAGONAL OF THE $R \times R$ CROSS CLASSIFICATION TABLE
UNDER VARIOUS MODELS APPLIED TO THE BRITISH AND DANISH SAMPLES

| | Models | | | | | |
	QO	DA	DAC	DPC	D	DC
British 5 × 5 Table						
m_1^*	0.38	0.29	0.30	0.30	0.29	0.30
m_2^*	0.26	−0.05	−0.23	−0.24	−0.04	−0.22
m_3^*	0.09	−0.01	0.00	0.00	−0.01	0.00
m_4^*	−0.01	−0.17	−0.16	−0.16	−0.16	−0.14
m_5^*	0.32	−0.04	0.05	0.04	−0.04	0.05
Danish 5 × 5 Table						
m_1^*	0.29	0.02	0.06	0.04	0.02	0.06
m_2^*	0.26	−0.02	0.01	−0.01	−0.02	0.01
m_3^*	0.18	−0.09	−0.09	−0.13	−0.09	−0.09
m_4^*	0.08	−0.16	−0.16	−0.19	−0.16	−0.16
m_5^*	0.33	−0.22	−0.15	−0.20	−0.21	−0.14
British 7 × 7 Table						
m_1^*	0.38	0.23	0.24	0.23	0.23	0.24
m_2^*	0.24	0.09	0.06	0.05	0.09	0.06
m_3^*	0.10	−0.07	−0.11	−0.17	−0.07	−0.12
m_4^*	0.09	0.01	0.01	−0.01	0.00	0.01
m_5^*	0.08	−0.14	−0.12	−0.20	−0.14	−0.12
m_6^*	0.17	0.06	0.08	0.08	0.06	0.08
m_7^*	0.18	−0.14	−0.06	−0.09	−0.14	−0.06

To test hypotheses about the μ_i, we return to our earlier discussion of (5.6). As we noted there, for a given model that does not include crossings parameters (but does include diagonals parameters or the triangles parameter), the hypothesis that the relative differences among the μ_i are nil for $i = 1, 2, \cdots, R$, can be tested by comparing the corresponding model for the analysis of the full table with the model for the analysis of the off diagonal cells (for example, comparing the DF model with the D model, or the DAF model with the DA model). Also, for a given model that does include crossings parameters (and also diagonals parameters and/or the triangles parameter), the hypothesis that the relative differences among the μ_i are nil for $i = 2, 3, \cdots, R - 2$, excluding $i = 1$ and $i = R$, can be tested by a similar kind of comparison (for example, the comparison of the DCF model with the DC model, or the DPCF model with the DPC model, or the DACF model with the DAC model). Examination of the corresponding relative magnitudes in Table IX shed further light on the results obtained by comparing the corresponding chi squares given in Table V. Recall, for example, that the comparison of the DCF model with the DC model indicated that the relative differences among the μ_i were statistically significant for Table III; they were not as statistically significant for Table I (but they were significant at the 10 percent level); and they were not statistically significant for Table II.

We remarked in Section 5 that, in calculating the m_i and m_i^* for models that include diagonals parameters and/or the triangles parameter, we must first make some assumptions about δ_0. For the calculations in Tables IX and X, we have assumed that

$$(7.5) \qquad \frac{\delta_0}{\delta_1''} = \frac{\delta_1''}{\delta_2''},$$

where δ_k'' is defined as

$$(7.6) \qquad \delta_k'' = (\delta_k \delta_{-k})^{1/2}.$$

Denoting δ_0 by δ_0'', condition (7.5) states that $\delta_k''/\delta_{k+1}''$ is constant, for $k = 0$ and 1. Although there is an element of arbitrariness in this way of defining δ_0, there are good reasons for using (7.5) here. (An alternative procedure would be to set $\delta_0 = 1$, which would yield the same result as obtained with (7.5) in cases where the model does not include diagonals parameters (and/or the triangles parameter), and also in cases where the model includes both the crossings parameters and diagonals parameters (and/or the triangles parameter) when the parameters are uniquely defined by condition (7.4) together with the condition that $\delta_1 = \delta_{-1} = 1$. In addition, for models that include both γ_u and δ_k (and/or τ_k), the m_i for $i = 2, 3, \cdots, R - 1$, would be the same when $\delta_0 = 1$ as the corresponding quantities obtained when the parameters are uniquely defined by condition (7.3) together with the condition that $\delta_1 = \delta_{-1} = 1$.) Since we set $\delta_1 = \delta_{-1} = 1$ in all models that include diagonals parameters, condition (7.5) can be simplified to

(7.7)
$$\delta_0 = \frac{1}{\delta_2''}.$$

In order to calculate m_1 and m_R (and also m_1^* and m_R^*) in Tables IX and X, we have set $\gamma_1 = \gamma_{R-1} = \gamma^*$, where $\gamma^* = \max_k \gamma_k$ for $2 \leq k \leq R - 2$. This procedure provides a way of calculating m_1 and m_R (and also m_1^* and m_R^*) that is conservative in the sense that, while it acknowledges the possible existence of the crossings parameters γ_1 and γ_{R-1}, it estimates their effects as being equal to the least pronounced estimated effects among the γ_u for $u = 2, 3, \cdots, R - 2$. (In his analysis of the "variable distance" model, Haberman [20] made the assumption that $\mu_1 = \mu_R = 1$ (that is, that $\log \mu_1 = \log \mu_R = 0$), and under this assumption he then estimated parameters corresponding to γ_1 and γ_{R-1} (namely, $\log \gamma_1$ and $\log \gamma_{R-1}$). In contrast to this, in the present article the μ_1 and μ_R (and also the μ_1^* and μ_R^*) are estimated from the data as indicated in the first sentence of this paragraph. Also, in contrast to the earlier analysis of the "variable distance" model, the parameters μ_i^* for $i = 1, 2, \cdots, R$, introduced here provide an additional index of interest (see Section 5).)

8. Extensions and applications to the analysis of two (or more) cross classification tables, and to the analysis of multidimensional tables

The results presented in Sections 3 and 4 for the general case of the $R \times C$ table can be directly extended to the analysis of two $R \times C$ tables (that is, to the analysis of the $R \times C \times 2$ table), to the analysis of G such $R \times C$ tables (that is, to the analysis of the $R \times C \times G$ table), and more generally to the analysis of multidimensional contingency tables. Indeed, the results presented in those sections can be applied directly to the multidimensional table simply by replacing each reference to the cells (i, j) of the $R \times C$ table throughout those sections by a corresponding reference to the cells of the multidimensional table.

I shall give here only a few examples (although there are many) of the possible extensions of the particular models introduced earlier to the analysis of three way tables (or to the analysis of contingency tables of higher dimensions). Suppose we applied, say, the DC model to G different $R \times R$ tables and found that the model fit the data for each of these tables. For the gth table, $g = 1, 2, \cdots$, the parameters $\alpha_i^{(g)}$, $\beta_j^{(g)}$, $\delta_k^{(g)}$, $\gamma_u^{(g)}$ could be estimated by the methods described earlier, and we could consider various hypotheses of the following kind:

(8.1)
$$\delta_k^{(g)} = \delta_k \quad \text{for} \quad g = 1, 2, \cdots, G,$$

(8.2)
$$\gamma_u^{(g)} = \gamma_u \quad \text{for} \quad g = 1, 2, \cdots, G,$$

(8.3)
$$\delta_k^{(g)} = \delta_k \quad \text{and} \quad \gamma_u^{(g)} = \gamma_u \quad \text{for} \quad g = 1, 2, \cdots, G.$$

Each of these hypotheses can be expressed as an "intrinsically unrestricted" model for the $R \times C \times G$ table, since the restrictions (8.1), (8.2), and (8.3) are of the H'' type (as defined in Section 4). Thus, the method presented earlier could also be used to analyze the model H'' obtained by modifying the DC model for each of the G tables by imposing conditions of the kind described above (that is, (8.1) to (8.3)).

Let H_g denote a given model (for example, the DC model) that will be applied to the gth $R \times R$ table for $g = 1, 2, \cdots, G$, and let H denote the hypothesis (model) that states that the H_g are true for $g = 1, 2, \cdots, G$. Let H'' denote the model obtained by modifying H by imposing a given condition (for example, a condition of the kind described under (8.1) to (8.3)) that still permits H'' to be expressed as an "intrinsically unrestricted" model for the $R \times C \times G$ table. Let $\chi^2(H_g)$, $\chi^2(H)$, and $\chi^2(H'')$ denote the chi square statistic based upon the likelihood ratio criterion for testing H_g, H, and H'', respectively. Each of these statistics can be calculated by the methods presented earlier applied to the G different $R \times R$ tables (for calculating $\chi^2(H_g)$) and to the $R \times R \times G$ table (for calculating $\chi^2(H)$ and $\chi^2(H'')$). Note that

$$(8.4) \qquad \chi^2(H) = \sum_{g=1}^{G} \chi^2(H_g).$$

Letting $[H''|H]$ denote the hypothesis that H'' is true assuming that H is true, we see that $[H''|H]$ states that the given condition, which was used to modify H to form H'', is true assuming that H is true. The following statistic is the chi square statistic based upon the likelihood ratio criterion for testing $[H''|H]$:

$$(8.5) \qquad \chi^2(H''|H) = \chi^2(H'') - \chi^2(H)$$
$$= 2 \sum f \log \left[\frac{\hat{f}}{\hat{f}''} \right] = 2 \sum \hat{f} \log \left[\frac{\hat{f}}{\hat{f}''} \right],$$

where $\hat{f}$ and $\hat{f}''$ denote the estimated expected frequencies in the $R \times C \times G$ table under H and H'', respectively. Note that the $\hat{f}$ pertaining to the gth $R \times R$ table can be calculated separately for each of the G tables, but the $\hat{f}''$ need to be calculated from the $R \times R \times G$ table (if H'' is a hypothesis of the kind described by (8.1) to (8.3)).

The statistic (8.5) has an asymptotic chi square distribution under the hypothesis $[H''|H]$, with the degrees of freedom equal to the difference between the corresponding number of degrees of freedom for testing H'' and H, respectively. The degrees of freedom of $\chi^2(H''|H)$ can also be calculated from the number of restrictions on the parameters (calculating this number after the parameters have been uniquely defined). Thus, for example, for testing hypotheses (8.1), 8.2), and (8.3), the corresponding number of degrees of freedom will be $G - 1$ multiplied by $2(R - 2) - 1 = 2R - 5$, $(R - 3) - 1 = R - 4$, and $2(R - 2) + (R - 4) = 3R - 8$, for hypotheses (8.1), (8.2), and (8.3), respectively.

We shall now comment briefly on the relationship between hypotheses of the kind considered above for the three way table (or for tables of higher dimension) and the usual hypothesis H_0 of zero three factor interaction in the three way table (or the corresponding kind of hypotheses for the table of higher dimension) (see, for example, Goodman [15], [17]). The usual hypothesis H_0 in the three way $R \times C \times G$ table states that the probability $\pi_{i,j,g}$ can be written as

$$(8.6) \qquad \pi_{i,j,g} = \alpha_i^{(g)} \beta_j^{(g)} \theta_{i,j}$$

$$\text{for } i = 1, 2, \cdots, R, \quad j = 1, 2, \cdots, C, \quad g = 1, 2, \cdots, G.$$

For the $R \times R \times G$ table, consider now the hypothesis H that states that the DCF model holds true for the gth $R \times R$ table, $g = 1, 2, \cdots, G$, and let H'' denote the modification of H obtained by imposing condition (8.3) on model H. The model H'' can be expressed as

$$(8.7) \qquad \pi_{i,j,g} = \alpha_i^{(g)} \beta_j^{(g)} \gamma_{i,j}' \delta_k \quad \text{for cells } (i, j, g) \text{ for which } (i, j) \text{ is in } S_k',$$

for $k = 0, \pm 1, \pm 2, \cdots, \pm(R-1)$; and with $\gamma_{i,j}'$ defined by (2.6.3), with $\gamma_{i,j}' = 1$ for $i = j$. To test the hypotheses H_0, H, and H'' in the $R \times R \times G$ table, the corresponding number of degrees of freedom will be $(R-1)^2(G-1)$, $G(R-2)(R-3)$, and $G(R-1)^2 - (3R-5)$, respectively. Since there are $(R-2)(R-3)$ degrees of freedom for testing the DCF model (see Table IV), the corresponding number of degrees of freedom for the H model will be $G(R-2)(R-3)$. Note also that the degrees of freedom for the H'' model can be calculated by subtracting from GR^2 the number of estimated parameters under the model (namely, $G(2R-1) + (3R-5)$) or by adding $(G-1)$ $(3R-5)$ to $G(R-2)(R-3)$, since condition (8.3) actually imposes $(G-1)$ $(3R-5)$ restrictions upon the parameters of H. The hypothesis $[H''|H]$ states that condition (8.3) is true, assuming that H is true; and the hypothesis $[H''|H_0]$ states that the following condition is true, assuming that H_0 is true:

$$(8.8) \qquad \theta_{i,j} = \gamma_{i,j}' \delta_k \quad \text{for } (i, j) \text{ in } S_k',$$

for $k = 0, \pm 1, \pm 2, \cdots, \pm(R-1)$, where the parameters in (8.8) are defined by the expressions (8.6) and (8.7). To test the hypotheses $[H''|H]$ and $[H''|H_0]$, the corresponding number of degrees of freedom will be $(G-1)(3R-5)$ and

$$(8.9) \qquad (R-1)^2 - (3R-5) = R^2 - 5R + 6 = (R-2)(R-3),$$

respectively.

Note that the same number of degrees of freedom are obtained for the hypothesis $[H''|H_0]$ as for the DCF model in the $R \times R$ table. The hypothesis $[H''|H_0]$ states that the parameters $\theta_{i,j}$ in (8.6) will satisfy condition (8.8); and the DCF model states that the probabilities $\pi_{i,j}$ can be written as

$$(8.10) \qquad \pi_{i,j} = \alpha_i \beta_j \theta_{i,j}',$$

where the $\theta_{i,j}'$ are of the form

$$(8.11) \qquad \theta_{i,j}' = \gamma_{i,j}' \delta_k \quad \text{for cells } (i, j) \text{ in } S_k',$$

for $k = 0, \pm 1, \pm 2, \cdots, \pm(R-1)$. Similarly, each of the hypotheses considered in Sections 2.1 to 2.8 states that the probabilities $\pi_{i,j}$ can be written as

(8.10) where the $\theta'_{i,j}$ are subject to certain specified conditions. For example, the DC model states that the $\theta'_{i,j}$ are of the form

$$
(8.12) \qquad \theta'_{i,j} = \begin{cases} \gamma'_{i,j}\delta_k & \text{for cells } (i,j) \text{ in } S'_k \text{ for } k = \pm 1, \pm 2, \cdots, \pm(R-1), \\ \mu'_i & \text{for cells } (i,i) \text{ in } S'_0, \end{cases}
$$

where μ'_i is an "intrinsically unrestricted" parameter.

For the usual hypothesis H_0 of zero three factor interaction in the $R \times C \times G$ table (see (8.6)), the probabilities $\pi_{i,j,g}$ can be rewritten as

$$
(8.13) \qquad\qquad\qquad \pi_{i,j,g} = \alpha_{i,g}\beta_{j,g}\theta_{i,j}.
$$

In the present section, we have considered hypotheses about the form of the $\theta_{i,j}$ in the $R \times R \times G$ table. It should also be noted that the general methods presented in Sections 3 and 4 can be directly applied to any given hypothesis about the form of the $\alpha_{i,g}$, $\beta_{j,g}$, and/or $\theta_{i,j}$ in the $R \times C \times G$ table, as long as the corresponding model is "intrinsically unrestricted" in the three way table.

The methods presented here can be directly applied not only to models for the three way table that are formed by imposing conditions on the parameters in (8.13), but also more generally to any hypothesis about the probabilities $\pi_{i,j,g}$ in the three way table (or in a table of higher dimension), as long as the corresponding model is intrinsically unrestricted. As an example of such models (that is, intrinsically unrestricted models) that are not formed by imposing conditions on the parameters in (8.13), see Goodman [13], [14] where the hypothesis of zero three factor interaction is extended to the case where a given subset of the cells in the three way table are deleted. These models for the three way table can be further extended in the same ways as we have here extended the model of quasi-independence in the two way table. For example, in the same way that the T model was formed by introducing the triangles parameters into the QO model, we could also introduce the three dimensional analogues of the triangles parameters (considering now parameters pertaining to certain triangles and tetrahedra in the three way table) into the models for the three way table considered in Goodman [13] (that is, into models of zero three factor interaction applied to a given subset of the cells of the table). In the same way that the triangles parameters are introduced in order to describe a particular kind of two factor interaction (between the row and column classifications) in the two way table, the three dimensional analogues of the triangles parameters can be introduced in order to describe a particular kind of three factor interaction in the three way table. For three way tables that do not conform to the usual hypothesis H_0 of zero three factor interaction, we can now provide a wide variety of multiplicative models (that is, the three dimensional analogues of the models in Sections 2.2 to 2.8) that can be used to analyze the data. In addition, for three way tables that do conform to the usual hypothesis H_0, we noted earlier in the present section that the general methods presented here can be used to test given hypotheses about the form of the parameters in the H_0 model (that is, hypotheses about the form of the two factor interactions in the three way table).

The preceding remarks can be directly extended to the multidimensional contingency table. We noted earlier that the usual hypothesis H_0 of zero three factor interaction in the three way table could be expressed as an intrinsically unrestricted multiplicative model, and similarly each of the hierarchical hypotheses described in Goodman [17] for the multidimensional table can also be expressed as an intrinsically unrestricted multiplicative model. The various extensions and modifications of H_0, which we described earlier in the present section for the analysis of the three way table, can also be directly extended in order to provide further extensions and modifications of the hierarchical hypotheses that were considered in the earlier literature on multidimensional contingency tables. Since these extensions and modifications are directly analogous to those already presented in the present article, we need not discuss this further here.

Before closing this section, we return for a moment to the models of Sections 2.3 to 2.8. Note that (2.3.4) can be viewed as a model for an $R \times C \times K$ table (R rows, C columns, K layers) that describes conditional quasi-independence between the row and column classifications, given the kth layer classification, $k = 1, 2, \cdots, K$, when certain cells (i, j, k) have been deleted, namely, the cells (i, j, k) for which the (i, j) is not in S_k (see, for example, Goodman [13], [17]). In particular, (2.3.1) is a model that describes conditional quasi-independence in the $R \times R \times 2$ table, where one layer pertains to positive $(i - j)$ and the other layer pertains to negative $(i - j)$. Similarly, (2.4.1) is a model for quasi-mutual independence in the $R \times R \times 2$ table (that is, a model of "complete independence" among the three variables in the three way table when certain cells have been deleted); and (2.5.1) is a model of quasi-mutual independence in an $R \times R \times [2(R - 1)]$ table. The other models in Sections 2.3 to 2.8 can also be viewed as models in three way or multi-way tables.

APPENDIX

Explicit Formulae for the Estimates of the Parameters in the Models

For any given model of the kind described in Sections 2.2 to 2.8, we shall now show how the maximum likelihood estimates of the parameters in the model can be expressed explicitly as functions of the estimates $\hat{f}_{i,j}$ of the expected frequencies under the model. Other ways to calculate the maximum likelihood estimates of the parameters were described in Section 3. The results, which we shall now present, will provide (a) further insight into the meaning of the parameters (expressed explicitly as functions of the probabilities $\pi_{i,j}$), and (b) a method for calculating the maximum likelihood estimates of the parameters (after the estimates $\hat{f}_{i,j}$ have been calculated as described in Section 3) which some readers may find easier to apply than the other methods that were described in Section 3 for estimating the parameters.

For simplicity, let us first consider the QO model (see Section 2.3). For this model, the maximum likelihood estimate $\hat{f}_{i,j}$ can be written as follows (see (2.2.4)):

(A.1) $\hat{f}_{i,j} = a_i b_j$ for $i \neq j$.

As we noted earlier (see Sections 2.1 to 2.3), the $\hat{f}_{i,j}$ in (A.1) will be unchanged if a_1 is set equal to one (and the other a_i and b_j in (A.1) are changed accordingly). With $a_1 = 1$, we see from (A.1) that the b_j can be written as

(A.2) $b_j = \begin{cases} \hat{f}_{1,j} & \text{for } j = 2, 3, \cdots, R, \\ \hat{f}_{3,1}\,\hat{f}_{1,2}/\hat{f}_{3,2} & \text{for } j = 1. \end{cases}$

Similarly, having calculated b_1 from (A.2), we obtain the following formula for the a_i

(A.3) $a_i = \begin{cases} 1 & \text{for } i = 1, \\ \hat{f}_{i,1}/b_1 & \text{for } i = 2, 3, \cdots, R. \end{cases}$

If we are interested in the a_i after they have been scaled so that $\Sigma_i\, a_i = 1$, then a_i of (A.3) can be replaced by $\tilde{a}_i = a_i/\Sigma_{h=1}^R\, a_h$. The $\tilde{a}_i$ are the maximum likelihood estimates of the scaled parameters $\tilde{\alpha}_i = \alpha_i/\Sigma_{h=1}^R\, \alpha_h$, which can be interpreted (for the QO model) as the hypothetical proportion of individuals in the ith row class in the hypothetical population in which none of cells on the main diagonal needs to be deleted, and there is independence between the row and column classifications in the table (see Goodman [13], [14]). A similar comment applies to b_j and to the corresponding $\tilde{b}_j$ and $\tilde{\beta}_j$. For the QO model, we find that

(A.4) $\tilde{\beta}_j = \tilde{\pi}_{i,j}^*$ for $i = 1, 2, \cdots, R,\quad j = 1, 2, \cdots, R,$

where $\tilde{\pi}_{i,j}^* = \pi_{i,j}^*/\pi_i.$ is the hypothetical conditional probability defined in Section 5. Thus, by applying the mover-stayer interpretation described by (5.9) to the QO model, we see from (A.4) that $\tilde{\beta}_j$ can be interpreted as the hypothetical probability that a "mover" will be in the jth column class (see, for example, Goodman [16]). By interchanging the row and column classifications, and then applying the mover-stayer interpretation described by (5.9), we see that the $\tilde{\alpha}_i$ can be interpreted in a similar way to the above interpretation of the $\tilde{\beta}_j$. (Analogous kinds of interpretations of the $\tilde{\beta}_j$ and $\tilde{\alpha}_i$ can be obtained when (5.10) is applicable, rather than (5.9).)

Consider now the QP and QN models (see Section 2.3). For each of these models, the parameters can be estimated by formulae similar to (A.2) and (A.3), based upon the fact that the b_j are proportional to the $\hat{f}_{R,j}$ for $j = 1, 2, \cdots, R - 1$, and the a_i are proportional to the $\hat{f}_{i,1}$ for $i = 2, 3, \cdots, R$, for the QP model, and upon the fact that the b_j are proportional to the $\hat{f}_{1,j}$ for $j = 2, 3, \cdots, R$ and the a_i are proportional to the $\hat{f}_{i,R}$ for $i = 1, 2, \cdots, R - 1$, for the QN model. More generally, for any model of quasi-independence for an inseparable set of cells (or for the inseparable subsets of a separable set of cells), the parameters can also be estimated by formulae similar to (A.2) and (A.3).

Now consider the DC model (see Section 2.7). For this model, the maximum likelihood estimate $\hat{f}_{i,j}$ can be written as follows (see (2.7.1)):

(A.5) $\qquad \hat{f}_{i,j} = a_i b_j c'_{i,j} d_k \qquad$ for cells (i, j) in S'_k,

for $k = \pm 1, \pm 2, \cdots, \pm(R - 1)$, where

$$
\text{(A.6)} \qquad c'_{i,j} = \begin{cases} \displaystyle\prod_{u=j}^{i-1} c_u & \text{for} \quad i > j, \\ \displaystyle\prod_{u=i}^{j-1} c_u & \text{for} \quad i < j. \end{cases}
$$

We shall let v_i, w_i, y_i, and z_i denote the quantities

$$
\text{(A.7)} \qquad v_i = \frac{(\hat{f}_{i,i+2}\hat{f}_{i+1,i-1})}{(\hat{f}_{i,i-1}\hat{f}_{i+1,i+2})} \qquad \text{for} \quad i = 2, 3, \cdots, R - 2,
$$

$$
\text{(A.8)} \qquad w_i = \frac{(\hat{f}_{i,1}\hat{f}_{2,i+1})}{(\hat{f}_{2,1}\hat{f}_{i,i+1})} \qquad \text{for} \quad i = 3, 4, \cdots, R - 1,
$$

$$
\text{(A.9)} \qquad y_i = \left[\prod_{j=1}^{i} \left(\frac{\hat{f}_{j,j+1}}{\hat{f}_{j+1,j}} \right) \right] \left[\left(\frac{\hat{f}_{i+1,1}}{\hat{f}_{1,i+1}} \right) \right] \qquad \text{for} \quad i = 2, 3, \cdots, R - 1,
$$

$$
\text{(A.10)} \qquad z_i = \begin{cases} \dfrac{(\hat{f}_{R,1}\hat{f}_{R-1,2})}{(\hat{f}_{R-1,1}\hat{f}_{R,2})} & \text{for} \quad i = 1, \\[3mm] \dfrac{(\hat{f}_{1,R}\hat{f}_{2,R-1})}{(\hat{f}_{1,R-1}\hat{f}_{2,R})} & \text{for} \quad i = -1. \end{cases}
$$

(For simplicity, we shall first assume that $\hat{f}_{i,j} > 0$ for $i \neq j$.) From (A.5) to (A.10), we find that

$$
\text{(A.11)} \qquad v_i = \frac{c_i^2 d'_2}{d'_1} \qquad \text{for} \quad i = 2, 3, \cdots, R - 2,
$$

$$
\text{(A.12)} \qquad w_i = \frac{d'_{i-1}\left(\displaystyle\prod_{j=2}^{i-1} c_j^2 \right)}{d'_1} \qquad \text{for} \quad i = 3, 4, \cdots, R - 1,
$$

$$
\text{(A.13)} \qquad y_i = \left(\frac{d_{-1}}{d_1} \right)^i \left(\frac{d_i}{d_{-i}} \right) \qquad \text{for} \quad i = 2, 3, \cdots, R - 1,
$$

$$
\text{(A.14)} \qquad z_i = \begin{cases} \dfrac{d_{R-1}d_{R-3}}{d_{R-2}^2} & \text{for} \quad i = 1, \\[3mm] \dfrac{d_{-(R-1)}d_{-(R-3)}}{d_{-(R-2)}^2} & \text{for} \quad i = -1, \end{cases}
$$

where d'_i is defined by

$$
\text{(A.15)} \qquad d'_i = d_i d_{-i}.
$$

As we noted earlier (see Section 7), the c_u in (A.5) and (A.6) can be defined uniquely for $u = 2, 3, \cdots, R - 2$, either by introducing a restriction to be imposed directly upon them (for example, that the maximum c_u be set equal to one) or by introducing an additional restriction to be imposed upon the d_k (for example, that $d_2' = 1$, in addition to the restriction that $d_1 = d_{-1} = 1$). The former kind of restriction would be appropriate when the DC model is viewed as an extension of the D model, and the latter kind of restriction would be appropriate when the DC model is viewed as an extension of the C model. If the maximum c_u is equal to one for $u = 2, 3, \cdots, R - 2$, then from (A.11) we obtain the following formula for the c_u,

$$(A.16) \qquad c_u = \left(\frac{v_u}{v^*}\right)^{1/2} \qquad \text{for} \quad u = 2, 3, \cdots, R - 2,$$

where

$$(A.17) \qquad v^* = \max_i v_i \qquad \text{for} \quad 2 \leqq i \leqq R - 2.$$

If instead of the above restriction on the maximum c_u, we set $d_2' = 1$ (in addition to setting $d_1 = d_{-1} = 1$), we see from (A.11) that c_u would be calculated by a modified form of (A.16) in which v^* is replaced by one. Since the $\hat{f}_{i,j}$ in (A.5) are unaffected if d_1 and d_{-1} are set equal to one (and the a_i, b_j, and other d_k are changed accordingly), we see from (A.12) that the d_i' can be calculated by

$$(A.18) \qquad d_i' = \frac{w_{i+1}}{\prod\limits_{j=2}^{i} c_j^2} \qquad \text{for} \quad i = 2, 3, \cdots, R - 2,$$

where the c_j are calculated from (A.16). From (A.13) and (A.15), we see that the d_i can be calculated as

$$(A.19) \qquad d_i = \begin{cases} (d_i' y_i)^{1/2} & \text{for} \quad i = 2, 3, \cdots, R - 2, \\ (d_i'/y_i)^{1/2} & \text{for} \quad i = -2, -3, \cdots, -(R - 2), \end{cases}$$

where d_i' is calculated from (A.18). From (A.14) we obtain the following formulae for d_{R-1} and $d_{-(R-1)}$:

$$(A.20) \qquad d_i = \begin{cases} z_1 d_{R-2}^2/d_{R-3} & \text{for} \quad i = R - 1, \\ z_{-1} d_{-(R-2)}^2/d_{-(R-3)} & \text{for} \quad i = -(R - 1). \end{cases}$$

We next consider

$$(A.21) \qquad \hat{f}_{i,j}' = \frac{\hat{f}_{i,j}}{d_k c_{i,j}'} \qquad \text{for cells } (i, j) \text{ in } S_k',$$

for $k = \pm 1, \pm 2, \cdots, \pm(R - 1)$, where the d_k are calculated from (A.19) and (A.20) and the $c_{i,j}'$ are calculated from (A.6), with c_u calculated from (A.16). From (A.5) and (A.21), we see that

(A.22) $$\hat{f}'_{i,j} = a_i b_j \qquad \text{for} \quad i \neq j.$$

Since the $\hat{f}'_{i,j}$ are of the same form as described by (A.1), the a_i and b_j can be calculated from (A.2) and (A.3) by replacing $\hat{f}_{i,j}$ by $\hat{f}'_{i,j}$ in these formulae. Thus, by applying the methods described in the present and preceding paragraphs, all of the parameters in the DC model can be estimated.

In the preceding discussion of the DC model, we assumed that $\hat{f}_{i,j} > 0$ for $i \neq j$. In cases where this assumption is not true, the above methods require some modification. For example, if $f_{R,1} = 0$ (as is the case for Tables I and III), then $\hat{f}_{R,1} = 0$ and $d_{R-1} = 0$ when the DC model is applied to these tables, and $\hat{f}'_{R,1}$ will be undefined. In this case, the $\hat{f}'_{i,j}$ that are well defined will have the same form as (A.22), and the methods described earlier for estimating the parameters in the model of quasi-independence can be applied to this set of $\hat{f}'_{i,j}$.

The methods described in this section can be applied, either directly or indirectly, to estimate the parameters in any of the models of Sections 2.2 to 2.8. For example, with the DPC model, by a direct application of the above formulae, we would obtain, among other things, the estimate d_k of the δ_k that satisfy (2.5.12). By an indirect application, we would first set $d_k = 1$ for $k = -1, -2, \cdots, -(R-1)$, in accordance with condition (2.5.9), and then from (A.11), (A.12), and (A.15), we would obtain

(A.23) $$d_1 = \frac{v_2}{y_2 c_2^2},$$

with c_2 calculated from (A.16). (For the DPC model, if the d_k are set at one for $k = -1, -2, \cdots, -(R-1)$, then d_1 will not be set at one.) To calculate d_k for $k = 2, 3, \cdots, R-1$, for this model, we then obtain, from (A.12) and (A.15),

(A,24) $$d_i = d_1 d'_i, \qquad \text{for} \quad i = 2, 3, \cdots, R-2,$$

with d_1 and d'_i calculated from (A.23) and (A.18), respectively. The value of d_{R-1} can be calculated from (A.20), using the d_i, for $i = R-2$ and $R-3$, calculated from (A.24).

In cases where the given model includes the triangles parameter (for example, in the DACT model), the parameter τ can be expressed as

(A.25) $$\tau = \left(\frac{\delta_k}{\delta_{-k}}\right)^{1/2} \qquad \text{for} \quad k = 1, 2, \cdots, R-1,$$

where $\tau = \sqrt{\tau'}$, for τ' defined by (2.7.4). From (A.13) and (A.25) we would obtain the following formula for the maximum likelihood estimate t of τ:

(A.26) $$t = \frac{1}{\sqrt{y_2}}.$$

If the model includes both the triangles parameter and the parameters pertaining to the paired minor diagonals, then after introducing the τ parameter explicitly into the model, the model will be unaffected by setting $\delta_k = \delta_{-k}$ for $k = 1, 2, \cdots, R - 1$. With the parameter τ estimated by (A.26), we can estimate $\delta_k = \delta_{-k}$ by

$$(A.27) \qquad d_k = \sqrt{d'_k} \qquad \text{for} \quad k = 2, 3, \cdots, R - 2,$$

where d'_k is calculated from (A.18). We can set $d_1 = d_{-1} = 1$ in this model, and d_{R-1} can be calculated from (A.20), using d_i for $i = R - 2$ and $R - 3$ calculated from (A.27).

For the QO model or for any of the models described in Sections 2.4 to 2.7, the ratio index μ_i and the relative difference index μ_i^*, which we defined in Section 5, can be estimated by the corresponding quantities m_i and m_i^* (see (5.14) and (5.15)), with the a_i and b_j calculated as described in the present section, and with d_0 as described in Section 7. For the models described in Section 2.8, we set $\mu_i = 1$, and therefore

$$(A.28) \qquad \hat{f}_{i,i} = a_i b_i d_0 \qquad \text{for} \quad i = 1, 2, \cdots, R.$$

With the $\hat{f}_{i,j}$ calculated by the iterative scaling method (for all cells (i, j) in the $R \times R$ table under the models of Section 2.8), we can calculate d_0 from

$$(A.29) \qquad d_0 = \frac{\hat{f}_{i,i}}{a_i b_i},$$

with the a_i and b_i calculated as described earlier in the present section.

For any given model of the kind described in Section 2.8 (for example, the DCF model), an alternative method for calculating d_0 can be based upon the fact that

$$(A.30) \qquad d_0^2 = d'_1 c_i^2 u_i,$$

where

$$(A.31) \qquad u_i = \frac{(\hat{f}_{i,i} \hat{f}_{i+1,i+1})}{(\hat{f}_{i,i+1} \hat{f}_{i+1,i})} \qquad \text{for} \quad i = 1, 2, \cdots, R - 1,$$

with the c_i calculated as earlier for $i = 2, 3, \cdots, R - 2$. Formula (A.30) can be used to calculate d_0 (applying the formula for any given value of $i = 2, 3, \cdots, R - 2$), and then c_1 and c_{R-1} can be calculated by rewriting (A.30)

$$(A.32) \qquad c_i = \frac{d_0}{(d'_1 u_i)^{1/2}}.$$

Now for all cells (i, j) in the table, we can consider

$$(A.33) \qquad \hat{f}'_{i,j} = \frac{\hat{f}_{i,j}}{d_k c'_{i,j}} \qquad \text{for cells } (i, j) \text{ in } S'_k,$$

for $k = 0, \pm 1, \pm 2, \cdots, \pm(R - 1)$. (Compare (A.33) with (A.21).) The $\hat{f}'_{i,j}$ will be of the form

$$(A.34) \qquad \hat{f}'_{i,j} = a_i b_j \qquad \text{for all cells } (i, j)$$

(see (A.22)), and so the a_i and b_j can be calculated here by the same methods used to estimate the parameters in the usual model of independence between the row and column classifications. For the models of Section 2.8, the method just described provides an alternative to the method described following (A.22) for calculating a_i and b_j. It should also be noted, as we did earlier, that in cases where the assumption that $\hat{f}_{i,j} > 0$ is not met for all cells in the table (under a given model of Section 2.8), the above methods require modifications of the kind which we described in the paragraph following (A.22).

The results of the present section can be applied to any of the models in Section 2.2 to 2.8 whenever the maximum likelihood estimate $\hat{f}_{i,j}$ exists. For comments concerning the existence of the $\hat{f}_{i,j}$, see Section 3 herein and Haberman [20].

For helpful comments, the author is indebted to R. Fay, S. Fienberg, S. Haberman, and T. Pullum.

REFERENCES

[1] R. R. BAHADUR, "Rates of convergence of estimates and test statistics," *Ann. Math. Statist.*, Vol. 38 (1967), pp. 303–324.

[2] Y. M. M. BISHOP and S. E. FIENBERG, "Incomplete two-dimensional contingency tables," *Biometrics*, Vol. 25 (1969), pp. 118–128.

[3] I. BLUMEN, M. KOGAN, and P. J. McCARTHY, *The Industrial Mobility of Labor as a Probability Process*, Cornell Studies in Industrial and Labor Relations, Vol. 4, Ithaca, Cornell University, 1955.

[4] H. CAUSSINUS, "Contribution à l'analyse statistique des tableaux de corrélation," *Annales de la Faculté des Sciences de l'Université de Toulouse*, Vol. 29 (1965), pp. 77–182.

[5] S. E. FIENBERG, "Quasi-independence and maximum-likelihood estimation in incomplete contingency tables," *J. Amer. Statist. Assoc.*, Vol. 65 (1970) pp. 1610–1616.

[6] D. V. GLASS (editor), *Social Mobility in Britain*, Glencoe, Free Press, 1954.

[7] I. J. GOOD, "Saddle-point methods for the multinomial distribution," *Ann. Math. Statist.*, Vol. 28 (1957), pp. 861–881.

[8] L. A. GOODMAN, "Statistical methods for the mover-stayer model," *J. Amer. Statist. Assoc.*, Vol. 56 (1961), pp. 841–868.

[9] ———, "Statistical methods for the preliminary analysis of transaction flows," *Econometrica*, Vol. 31 (1963), pp. 197–208.

[10] ———, "A short computer program for the analysis of transaction flows," *Behav. Sci.*, Vol. 9 (1964), pp. 176–186.

[11] ———, "Simultaneous confidence limits for cross-product ratios in contingency tables," *J. Roy. Statist. Soc., Ser. B*, Vol. 26 (1964), pp. 86–102.

[12] ———, "On the statistical analysis of mobility tables," *Am. J. Sociology*, Vol. 70 (1965), pp. 564–585.

[13] ———, "The analysis of cross-classified data: Independence, quasi-independence, and interactions in contingency tables with or without missing entries," *J. Amer. Statist. Assoc.*, Vol. 63 (1968), pp. 1091–1131.

[14] ———, "How to ransack social mobility tables and other kinds of cross-classification tables," *Am. J. Sociology*, Vol. 75 (1969), pp. 1–40.

[15] ——, "On partitioning χ^2 and detecting partial association in three-way contingency tables," *J. Roy. Statist. Soc.*, Ser. B, Vol. 31 (1969), pp. 486–498.

[16] ——, "On the measurement of social mobility: An index of status persistence," *American Sociological Review*, Vol. 34 (1969), pp. 832–850.

[17] ——, "The multivariate analysis of qualitative data: Interactions among multiple classifications," *J. Amer. Statist. Assoc.*, Vol. 65 (1970), pp. 226–256.

[18] ——, "The analysis of multidimensional contingency tables: Stepwise procedures and direct estimation methods for building models for multiple classifications," *Technomet.*, Vol. 13 (1971), pp. 33–61.

[19] ——, "A simple simultaneous test procedure for quasi-independence in contingency tables," *J. Roy. Statist. Soc.*, Ser. C, Vol. 20 (1971).

[20] S. HABERMAN, "The general log-linear model," Ph.D. thesis, University of Chicago, 1970.

[21] W. HOEFFDING, "Asymptotically optimal tests for multinomial distributions," *Ann. Math. Statist.*, Vol. 36 (1965), pp. 369–408.

[22] J. LEVINE, "Measurement in the study of intergenerational status mobility," Ph.D. thesis, Harvard University, 1967.

[23] N. MANTEL, "Incomplete contingency tables," *Biometrics*, Vol. 26 (1970), pp. 291–304.

[24] F. MOSTELLER, "Association and estimation in contingency tables," *J. Amer. Statist. Assoc.*, Vol. 63 (1968), pp. 1–28.

[25] K. SVALASTOGA, *Prestige, Class and Mobility*, London, William Heinemann, 1959.

A MISSING INFORMATION PRINCIPLE: THEORY AND APPLICATIONS

TERENCE ORCHARD
and
MAX A. WOODBURY
DUKE UNIVERSITY MEDICAL CENTER

1. Introduction

The problem that a relatively simple analysis is changed into a complex one just because some of the information is missing, is one which faces most practicing statisticians at some point in their career. Obviously the best way to treat missing information problems is not to have them. Unfortunately circumstances arise in which information is missing and nothing can be done to replace it for one reason or another. In analogy to other accidents—we don't plan on accidents, nevertheless they do occur and safety measures must be aimed at palliating consequences as well as at prevention. Consequently, a great volume of literature has been produced, dealing with a number of specific situations. An indication of the content of many of these papers is given in the Appendix. In this paper we propose to try to present a general philosophy for dealing with the problem of missing information, and to give a method which will lead quite easily to maximum likelihood estimates of the parameters obtained from the incomplete data using as nearly as possible the same techniques as if the data were all present.

Our first simple use of the missing information principle resulted from a conversation in 1946 between Max A. Woodbury and C. W. Cotterman resulting from the latter's interest (Cotterman [20]) in estimating gene frequencies from phenotypic frequency data. The observation was made that if one has the genotypic *frequencies NAA, NAB, NBA, NBB, NAO, NBO, NOB,* and *NOO* of red blood cell genotypes, indicated by the second and third letters of each symbol, then the gene frequencies are easily computed. If N is the total of the above frequencies then the estimates would be

$$\hat{p}_A = \frac{1}{2N}(2NAA + NAB + NBA + NAO + NOA),$$

(1.1) $$\hat{p}_B = \frac{1}{2N}(2NBB + NBA + NAB + NBO + NOB),$$

$$\hat{p}_o = \frac{1}{2N}(2NOO + NOA + NAO + NOB + NBO).$$

Research conducted under PHS-NIH Grant GM 16725-02 from the National Institute of General Medical Sciences.

However, only the phenotypic frequencies

$$
\begin{aligned}
MA &= NAA + NAO + NOA, \\
MB &= NBB + NBO + NOB, \\
MO &= NOO, \\
MAB &= NAB + NBA,
\end{aligned}
$$
(1.2)

are available.

If, however, one makes use of Bayes' theorem and the gene frequencies one can obtain estimates of the genotypic frequencies from the phenotypic frequencies

$$
N\hat{A}A = MA \frac{p_A^2}{(p_A^2 + 2p_A p_O)} = MA \frac{p_A}{(p_A + 2p_O)},
$$

$$
N\hat{A}O + N\hat{O}A = 2MA \frac{p_A}{(p_A + 2p_O)},
$$

(1.3)
$$
N\hat{B}B = MB \frac{p_B}{(p_B + 2p_O)},
$$

$$
N\hat{B}O + N\hat{O}B = 2MB \frac{p_O}{(p_B + 2p_O)},
$$

$$
N\hat{A}B + N\hat{B}A = MAB,
$$

$$
N\hat{O}O = MO.
$$

If one solves (1.1) and (1.3) simultaneously by equating the genotypic frequencies in (1.1) to their estimates in (1.3), one can obtain estimates $\tilde{p}_A$, $\tilde{p}_B$, and $\tilde{p}_O$, which of course are not as good as those obtainable from the true genotype frequencies but which are as efficient as the maximum likelihood estimates based only on the phenotypic frequencies.

The problem with estimating p_A, p_B, and p_O by this method is the difficulty of finding rapid and accurate solutions of these equations and estimates of their error variances. These difficulties are shared with the method of Maximum Likelihood (ML). This is not too surprising since in fact the two methods are equivalent. One way in which the two problems of slow convergence and loss of information may be handled is by the method of scoring which can be modified to work in the presence of missing information. The solution of the problem of estimating gene frequencies in more general circumstances is provided by Ceppellini, Siniscialco, and Smith [15], who demonstrated that the procedure implied by the principle indicated above is in fact ML in all cases. These authors also considered the increased variance of the estimates due to the loss of genotypic information under the heading "hidden variance."

This missing information principle has been applied to missing observations in a linear model, and in a multivariate normal, and to mixture problems. A few examples will be presented later in order to demonstrate the relative facility with which the principle may be applied.

2. Theory

The method proposed is to regard the values of the missing data as random variables within the framework of a model of the data. Thus, estimates which are well defined when all the data are present become random variables (being functions of the missing data). This variation of the estimates is in addition to the usual sampling variation so that the error variances of the estimates are increased. The consequences of the data's being missing and some insight to the approach used here are obtained if one considers replacing the missing data by sample values from the appropriate distribution function. The question is, from which distribution function should we sample?

In the independent, identically distributed case where the vector x_i has the distribution function $f(x_i|\theta)$ and $x_i = (Y_i, z_i)$ where the vector z_i contains the missing components, then $f(x_i|\theta) = f_1(y_i|\theta) \cdot f_2(z_i|\theta, y_i)$ is the factorization of the distribution function into the marginal distribution for y_i and the conditional distribution of z_i given y_i. The proper distribution to sample for the missing data then is the conditional distribution $f(z_i|\theta, y_i)$, but θ is unknown so that some estimated value $\hat{\theta}$, must be used. One could draw many samples from the distribution f and from these completed data samples obtain the distribution of the parameter estimates due to the missing data. Call this distribution $\text{MID}(\hat{\theta}, Z)$. If this distribution is asymptotically normal, then the mean will be the obvious statistic to use to provide an estimate in the presence of missing data. If this mean value should be $\hat{\theta}$, then the estimate has not been affected by the assumed missing data distribution. That is the missing data tells you nothing. This interpretation of the principle is due to Jacquez. The remaining part of the missing information principle is to equate the mean of the $\text{MID}(\hat{\theta}, Z)$ to $\hat{\theta}$, or take some action equivalent to this. The effect of the variance of $\text{MID}(\hat{\theta}, Z)$ ("the hidden variances") on the error variance is best understood in another context.

Before continuing with the problem of missing information, it will be of value to review the method of maximum likelihood estimation. The likelihood function of a multivariate data matrix X is denoted by $L(X|\theta)$, and is defined to be $L(X|\theta) = \prod_{n=1}^{N} f(X_n|\theta)$, where $f(X_n|\theta)$ is the density function of X_n and θ is an $(s \times 1)$ vector of parameters. The Score (Sco) for the parameter θ is then defined to be

$$(2.1) \qquad \text{Sco}(\theta_j|X) = \frac{\partial}{\partial \theta_j} \log L(X|\theta) = \frac{1}{L} \frac{\partial}{\partial \theta_j} L(X|\theta).$$

The maximum likelihood estimates for the parameters are obtained by solving the set of equations. It may be readily deduced that the mean value of the score is zero at the true parameter point; that is,

$$(2.2) \qquad \text{Sco}(\theta_j|X) = 0 \qquad \text{for} \quad j = 1, \cdots, s,$$

$$(2.3) \qquad E[\text{Sco}(\hat{\theta}_j|X)] = 0.$$

The information matrix for θ is defined to be the matrix with (j, k)th element

$$
\begin{aligned}
(2.4) \qquad J(\theta_j, \theta_k | X) &= -E\left[\frac{\partial}{\partial \theta_k} \text{Sco}\,(\theta_j | X)\right] \\
&= -E\left[\frac{\partial^2}{\partial \theta_j \partial \theta_k} \log L(X | \theta)\right] \\
&= E[\text{Sco}\,(\theta_j | X)\,\text{Sco}\,(\theta_k | X)] \\
&= \text{Cov}\,[\text{Sco}\,(\theta_j | X), \text{Sco}\,(\theta_k | X)]
\end{aligned}
$$

Under certain general regularity conditions (see Rao [36], p. 295), for example, the existence of second and third derivatives of the log likelihood function, it may be shown that the joint distribution of the maximum likelihood estimators is asymptotically multivariate normal with a covariance matrix given by the inverse of the information matrix. For a detailed discussion of this subject, the paper by Wald [41] can be consulted. A condition for the existence of maximum likelihood estimates is that the information matrix is positive definite. However, in some instances, such as in the case of the multinomial distribution, the information matrix has rank less than s and is therefore singular. If the information matrix is of rank $s - t$, then we are required to impose t restrictions, $h_1(\theta) = 0, \cdots, h_t(\theta) = 0$, on the parameters in order to achieve identifiability.

The problem of maximum likelihood estimation of parameters subject to constraints has been studied by Aitchison and Silvey [5], [6], and Silvey [38]. These authors also obtained a test statistic for the hypothesis $h(\theta) = 0$. However, the situation of interest here is when $h(\theta)$ is a $t \times 1$ vector of constraints necessary to produce an identifiable parameter set. In this case the constrained likelihood function may be written as

$$
(2.5) \qquad L^* = \log L(X | \theta) - \lambda^T h(\theta),
$$

where λ is a $t \times 1$ vector of Lagrangian multipliers. If we define the constrained score

$$
(2.6) \qquad \text{Sco}^*(\theta | X) = \frac{\partial L^*}{\partial \theta}
$$

and

$$
(2.7) \qquad \underset{(s \times t)}{H} = \left(\frac{\partial h_i(\theta)}{\partial \theta_j}\right),
$$

then we may deduce that the expected value of the constrained score is zero if the true parameter satisfies the constraints.

Considering the various definitions of the information matrix listed in (2.4), we may determine that a nonsingular matrix, denoted by J^*, will be obtained by taking the negative of the expected value of the derivative of the constrained scores. However, if we take the covariance of the constrained scores we will call the singular information matrix J. This is related to the required nonsingular matrix by the equation

(2.8)
$$J^* = J + NHH^T,$$

where N is the number of observations. This form will lead to the asymptotic covariance matrix of the parameter estimates given by

(2.9)
$$V = (J^*)^{-1} = (J^{-1} - J^{-1}H(H^T J^{-1} H)^{-1} H^T J^{-1}).$$

It may also be remarked that a nonsingular information matrix will not result simply from reparametrizing so that the new parameters satisfy the constraints, unless some are eliminated.

Let us now return to the situation where we have missing information. Suppose that we cannot observe the random variable X, but can instead observe some image Y, of it. The likelihood function for Y may be obtained by integrating $L(X|\theta)$ over the appropriate range, and thus we may write

(2.10)
$$L(X|\theta) = L_1(X|Y, \theta) L_2(Y|\theta)$$

giving

(2.11)
$$\text{Sco } (\theta_j|X) = \frac{1}{L_1} \frac{\partial L_1}{\partial \theta_j} + \text{Sco } (\theta_j|Y).$$

The item $(1/L_1)(\partial L_1/\partial \theta_j)$ is called the conditional score of θ_j from X, given Y, and this is denoted by $\text{Sco } (\theta_j, X|Y)$. It may be noted that

(2.12)
$$E[\text{Sco } (\theta_j, X|Y)|Y] = 0,$$

the truth of this following from the same reasoning as was used to establish (2.3).

Finally we have

(2.13)
$$\text{Sco } (\theta_j|Y) = E[\text{Sco } (\theta_j|X)|Y]$$

and

(2.14)
$$E[\text{Sco } (\theta_j|X) \text{ Sco } (\theta_k|X)] = \text{Cov } \{E[\text{Sco } (\theta_j|X)|Y], E[\text{Sco } (\theta_k|X)|Y]\}$$
$$+ E\{\text{Cov } [\text{Sco }(\theta_j|X), \text{Sco }(\theta_k|X)|Y]\}$$

which leads to

(2.15)
$$E[\text{Sco } (\theta_j|X) \text{ Sco } (\theta_k|X)] = E[\text{Sco } (\theta_j|Y) \text{ Sco } (\theta_k|Y)]$$
$$+ E\{\text{Cov } [\text{Sco } (\theta_j|X), \text{Sco } (\theta_k|X)|Y]\}.$$

For the information matrix this gives

(2.16)
$$J(\theta, \theta|X) = J(\theta, \theta|Y) + J(\theta, \theta; Y|X).$$

The last quantity on the right $J(\theta, \theta; Y|X)$ is what is termed the lost information. For brevity, equation (2.16) may be written $J_X = J_Y + J_{X/Y}$, where $J_{X/Y} = J_X - J_Y =$ the lost information.

We may now easily obtain a relationship between the lost information and the increase in variance of the parameter estimates (the "hidden variance" of Ceppellini, Siniscialco, and Smith [15]). This relationship derives from

$$(2.17) \qquad J_X(J_Y^{-1} - J_X^{-1}) = (J_X - J_Y)J_Y^{-1}$$

which we may write as

$$(2.18) \qquad J_Y^{-1} - J_X^{-1} = J_X^{-1}(J_X - J_Y)J_Y^{-1},$$

$$(2.19) \qquad J_X - J_Y = J_X(J_Y^{-1} - J_X^{-1})J_Y.$$

If, for simplicity, we write $A \geq B$ when the matrix difference, $C = A - B$, is positive semidefinite, then we have

$$(2.20) \qquad J_Y^{-1} \geq J_X^{-1},$$

$$(2.21) \qquad J_X \geq J_Y.$$

We may now obtain bounds on the hidden variance in terms of the lost information, and vice versa. These are

$$(2.22) \qquad J_X^{-1}J_{X|Y}J_X^{-1} \leq (J_Y^{-1} - J_X^{-1}) \leq J_Y^{-1}J_{X|Y}J_Y^{-1},$$

$$(2.23) \qquad J_Y(J_Y^{-1} - J_X^{-1})J_Y \leq J_{X|Y} \leq J_X(J_Y^{-1} - J_X^{-1})J_X.$$

These may be of value in practical situations where some of the quantities are easier to obtain than others. The widths of these limits depend on the amount of missing information and are "tight" when this is small.

The usefulness of the above theory, and in particular result (2.13), in estimating parameters in the presence of missing information is that it is often quite easy to obtain the right side even in those cases when it is extremely difficult to obtain the left side.

3. Examples

3.1. *Example 1.*   Consider first the case of missing observations in a linear model. Suppose that Y is a set of independent, normally distributed, random variables having a common variance of σ^2, and a mean of $X\theta$, where X is an $n \times k$ design matrix and θ is a $k \times 1$ vector of unknown parameters. Then we have

$$(3.1.1) \qquad \mathrm{Sco}\,(\theta|Y) = \frac{X^T(Y - X\theta)}{\sigma^2},$$

$$(3.1.2) \qquad \mathrm{Sco}\,(\sigma^2|Y) = \frac{n}{2\sigma^2} - \frac{(Y - X\theta)^T(Y - X\theta)}{2\sigma^4}.$$

Equating these to zero gives

$$(3.1.3) \qquad \hat{\theta} = (X^TX)^{-1}X^TY,$$

$$(3.1.4) \qquad \hat{\sigma}^2 = \frac{1}{n}(Y - X\theta)^T(Y - X\theta)$$

This estimator for σ^2 is, of course, well known to be biased, the unbiased estimator being

$$(3.1.5) \qquad \hat{\sigma}^2 = \frac{(Y - X\theta)^T(Y - X\theta)}{n - k},$$

where k is the rank of the design matrix X.

Suppose that there are n potential observations but that there are m missing, leading to an image of Y being the vector Y_0 of observed values. Due to the assumed independence of the observations the conditional expectation of the scores may be easily computed. Also the observed values are unaltered whilst functions of the missing values are replaced by their expected values. Hence

$$(3.1.6) \qquad \text{Sco}\,(\theta | Y_0) = \frac{1}{\sigma^2} X^T(Y_0 - X_m\theta - X\theta)$$

and

$$(3.1.7) \qquad \text{Sco}\,(\sigma^2 | Y_0) = \frac{n}{2\sigma^2} - \frac{(Y_0 - X_0\theta)^T(Y_0 - X_0\theta)}{2\sigma^2} - \frac{m}{2\sigma^2};$$

here X_m and X_0 are $(n \times k)$ matrices such that $X = X_m + X_0$, and Y_m and Y_0 are $(n \times 1)$ vectors such that $Y = Y_m + Y_0$. The following estimators are thus obtained:

$$(3.1.8) \qquad \hat{\theta} = (X^TX)^{-1}X^T\hat{Y},$$

$$(3.1.9) \qquad \hat{Y} = Y_0 + X_m\hat{\theta},$$

$$(3.1.10) \qquad \hat{\sigma}^2 = (Y_0 - X_0\hat{\theta})^T \frac{(Y_0 - X_0\hat{\theta})}{n - m - k}$$

$$= (\hat{Y} - X\hat{\theta})^T \frac{(\hat{Y} - X\hat{\theta})}{n - m - k}.$$

It may be noted that $\hat{\theta}$ can be eliminated between (3.1.8) and (3.1.9) to obtain

$$(3.1.11) \qquad \hat{Y}_m = [I - X_m(X^TX)^{-1}X_m^T]^{-1}X_m(X^TX)^{-1}X_0^TY_0.$$

This equation provides a form of estimating missing data which is easy to compute for a single value. It is proposed to use this simple form to obtain an initial set of estimates for the missing values, and then to cycle iteratively through equations (3.1.8) and (3.1.9) until the parameter estimates stabilize. This is the modification of Yates' [50] approach to the problem, as proposed by Tocher [39].

The lost information for estimating $\hat{\theta}$ may be shown to be $(X_m^TX_m)\sigma^{-2}$, thus the covariance matrix for $\hat{\theta}$ may be written

$$(3.1.12) \qquad \text{Cov}\,(\hat{\theta}, \hat{\theta}) = (X^TX - X_m^TX_m)^{-1}\sigma^2$$

$$= \{(X^TX)^{-1} + (X^TX)^{-1}X_m^T[I - X_m(X^TX)^{-1}X_m^T]^{-1}X_m$$
$$(X^TX)^{-1}\}\sigma^2.$$

Therefore the quantity to be added to $\mathrm{Cov}\,(\hat{\theta}, \hat{\theta})$ so as to correct for "bias" is

$$(3.1.13) \qquad B = \{(X^TX)^{-1}X_m^T[I - X_m(X^TX)^{-1}X_m^T]^{-1}X_m(X^TX)^{-1}\}\sigma^2.$$

Since we recover none of the lost information, in general the main reason for using the procedure proposed here is that the design matrix X for the complete data, where we have a balanced design, may be chosen such that X^TX is diagonal, and hence much easier to invert than the general matrix $X_0^TX_0$ which would result from using the available data only. However, it should be noted that in order to correct for the bias in $\mathrm{Var}\,(\theta)$, it is necessary to invert the matrix $[I - X_m^T(X^TX)^{-1}X_m]$. This is of order m and generally has a regular pattern, but it may not be diagonal. It is therefore felt that if the number of missing values m equals the rank of the design matrix then there is less to be gained from using the procedure outlined here.

3.2. *Example 2.* A second example, which has received considerable attention, is that of missing components in a multivariate normal distribution. This has been considered by Woodbury and Hasselblad [47] but it is being included here due to its great interest. The log of the likelihood function, for a sample of size N, may be written

$$(3.2.1) \qquad \log L(X \,|\, \mu, \Sigma) = C + \tfrac{1}{2}N|\Sigma^{-1}| - \tfrac{1}{2}N\,\mathrm{tr}\,\Sigma^{-1}S,$$

where $S = \Sigma_{n=1}^N (X_n - \mu)(X_n - \mu)^T/N$.

The parameter scores are easily obtained and are

$$(3.2.2) \qquad \mathrm{Sco}\,(\mu\,|\,X) = N\Sigma^{-1}(\bar{X} - \mu)$$

and

$$(3.2.3) \qquad \mathrm{Sco}\,(\Sigma\,|\,X) = -\tfrac{1}{2}N(\Sigma^{-1} - \Sigma^{-1}S\Sigma^{-1}).$$

By equating these to zero we can obtain the parameter estimates

$$(3.2.4) \qquad\qquad\qquad\qquad \hat{\Sigma} = S$$

and

$$(3.2.5) \qquad\qquad\qquad\qquad \hat{\mu} = \bar{X} = \sum_{n=1}^N \frac{X_n}{N}.$$

Equation (3.2.3) is obtained by observing that the derivative of the logarithm of a determinant of a matrix with respect to an element of that matrix is simply the corresponding element of the inverse, and that $\partial\Sigma^{-1} = -\Sigma^{-1}(\partial\Sigma)\Sigma^{-1}$. We also used such properties as $\mathrm{tr}\,(AB) = \mathrm{tr}\,(BA)$, and $\mathrm{tr}\,(a) = a$ if a is a scalar. Additionally we note that $\sigma^{i,j} = \sigma^{j,i}$, although this fact is not used in obtaining $\mathrm{Sco}\,(\sigma_{i,j})$. The solution of the normal equations will not be affected since we obviously have $\mathrm{Sco}\,(\sigma_{i,j}) = \mathrm{Sco}\,(\sigma_{j,i})$. The reason for forming the score of the covariance matrix instead of the score of the inverse, as is more normal, is that this will greatly simplify the computation of the information for Σ, as will be developed later.

The image Y of X consists of the observed components of the data matrix X. If we assume that there are p components then an individual observation is a $p \times 1$ vector which may be written in the form

$$(3.2.6) \qquad \hat{Y}_k = Y_{k,0} + \hat{Y}_{k,m},$$

where $Y_{k,0}$ is the observed portion, with zero in each position corresponding to a missing component, and $\hat{Y}_{k,m}$ is the estimated missing portion, with again zero in the positions corresponding to an observed component. It should of course be remarked that if there are no missing components then $Y_{k,0}$ constitutes the entire vector. To obtain the conditional expectation of the scores for the mean we have to solve the regression of the missing data $Y_{k,m}$ on the observed data $Y_{k,0}$. Similarly the conditional expectation of the scores for the covariance matrix requires the conditional covariance matrix of the missing data given the observed data. The following estimators are obtained:

$$(3.2.7) \qquad \hat{\mu} = \frac{1}{N} \sum_{n=1}^{N} Y_n,$$

$$(3.2.8) \qquad \hat{\Sigma} = \frac{1}{N} \sum_{n=1}^{N} \left[(\hat{Y}_n - \hat{\mu})(\hat{Y}_n - \hat{\mu})^T + V_n \right],$$

$$(3.2.9) \qquad \hat{Y}_{k,m} = \hat{\mu}_m + \Sigma_{m,0} \Sigma_{0,0}^{-1} (Y_{k,0} - \hat{\mu}_0)$$
$$= \hat{\mu}_m + (\Sigma^{m,m})^{-1} \Sigma^{m,0}(Y_{k,0} - \hat{\mu}_0),$$

where V_n is a $p \times p$ matrix, for the nth observation, with $\Sigma_{m,m} - \Sigma_{m,0}\Sigma_{0,0}^{-1}\Sigma_{0,m}$, $\left(= (\Sigma^{m,m})^{-1}\right)$, in the positions corresponding to the missing components and zero elsewhere. It should be noted that partitions such as $(Y_{k,m}, Y_{k,0})$ vary from observation to observation depending on which components, if any, are missing.

The lost information for the mean due to a single observation may be shown to be

$$(3.2.10) \qquad LI(\mu) = \begin{bmatrix} \Sigma^{m,m} & \Sigma^{m,0} \\ \Sigma^{0,m} & \Sigma^{0,0} - \Sigma_{0,0}^{-1} \end{bmatrix}$$

This may be deduced intuitively since we would lose Σ^{-1} by discarding a complete observation, whereas the procedure described here recovers the amount $\Sigma_{0,0}^{-1}$ contained in the observed portion. Once again the components of $\Sigma_{0,0}$ vary from observation to observation and will be the entire matrix if no component is missing.

Since there are only $q = \frac{1}{2}p(p + 1)$ distinct elements in the covariance matrix, $J(\Sigma, \Sigma)$ will be a $q \times q$ matrix. However, it is proposed to regard it as being composed of p^2 distinct elements and then to gather the terms in the $p^2 \times p^2$ information matrix $J^*(\Sigma, \Sigma)$ to give the required $J(\Sigma, \Sigma)$. Thus we have

$$(3.2.11) \quad J^*(\Sigma, \Sigma) = \frac{N^2}{4} E[\Sigma^{-1} \otimes \Sigma^{-1} - \Sigma^{-1} \otimes \Sigma^{-1} S \Sigma^{-1}$$

$$- \Sigma^{-1} S \Sigma^{-1} \otimes \Sigma^{-1} + \Sigma^{-1} S \Sigma^{-1} \otimes \Sigma^{-1} S \Sigma^{-1}]$$

$$= \frac{N^2}{4} E[(\Sigma^{-1} \otimes \Sigma^{-1})(S \otimes S)(\Sigma^{-1} \otimes \Sigma^{-1}) - \Sigma^{-1} \otimes \Sigma^{-1}]$$

A typical element in $E(S \otimes S)$ may be shown to be

$$(3.2.12) \quad E(s_{i,j} s_{u,v}) = \frac{1}{N} [\sigma_{i,j}\sigma_{u,v} + \sigma_{i,u}\sigma_{j,v} + \sigma_{i,v}\sigma_{j,u}]$$

Hence (3.2.11) reduces to

$$(3.2.13) \quad J^*(\Sigma, \Sigma) = \frac{N}{4} (\Sigma^{-1} \otimes \Sigma^{-1})(\sigma_{i,u}\sigma_{v,j} + \sigma_{i,v}\sigma_{u,j})(\Sigma^{-1} \otimes \Sigma^{-1})$$

$$= \frac{N}{4} (\sigma^{i,u}\sigma^{v,j} + \sigma^{i,v}\sigma^{u,j}).$$

To obtain these equations it is necessary to use one of the properties of the Kronecker product, namely, $(A \otimes B)(C \otimes D) = AC \otimes BD$. To obtain $J(\Sigma, \Sigma)$ we simply collapse this matrix noting that $J(\sigma_{i,i}, \sigma_{u,u})$ consists of one element like that shown on the right side of (3.2.13), $J(\sigma_{i,i}, \sigma_{u,v})$ consists of the sum of two such elements, and $J(\sigma_{i,j}, \sigma_{u,v})$ consists of the sum of four such elements. The point of obtaining $J^*(\Sigma, \Sigma)$ is that it provides a means of computing the information matrix in the presence of missing data since J^* is the sum of the corresponding matrices for each observation vector, which may differ from observation to observation in the case of missing data. If we write each observation in the form $\hat{Y}_k = Y_{k,0} + \hat{Y}_{k,m}$ then $\Sigma_{0,0}$ is the submatrix of Σ corresponding to $Y_{k,0}$. The retained information is then obtained by accumulating those elements of (3.2.13) that correspond to $\Sigma_{0,0}^{-1}$. Once this has been done then J can be obtained by combining terms and from this the lost information may be computed, as can the covariance matrix of the estimated covariance matrix. covariance matrix of the estimated covariance matrix.

Computationally, the procedure followed is to group the observations into classes of identical patterns of missing and observed components. Initial estimates of the mean and covariance matrix are obtained using (3.2.4) and (3.2.5) on the complete vectors, if there are any. Then (3.2.9) is used to get initial estimates of the missing values and the completed data used in (3.2.4) and (3.2.5) to get new estimates of the parameters. Finally the covariance matrix is corrected for bias by adding quantities like $(\Sigma^{m,m})^{-1}$ as indicated by (3.2.8). If there are no complete vectors, it is proposed to use some good initial guess of the missing data and then to start the cycle with (3.2.4) and (3.2.5) as before. It should be noted that the theory of partitioned matrices is quite useful in reducing the amount of computation since the inverse of the quite large matrix $\Sigma_{0,0}$ can be easily expressed in terms of the submatrices of Σ^{-1}, as can $\Sigma_{m,0}\Sigma_{0,0}^{-1}$. (See Woodbury, M.A., "Inverting modified matrices," *Stat. Res. Group. Princeton, N.J. Memo.*, Vol. 42, 1950.)

The convergence is, in certain cases, quite slow and hence methods of speeding it must be used. It must also be remembered that the correct procedure to follow for any analysis, such as multiple regression, is to use the corrected covariance matrix and the mean as data, since using only the values predicted by (3.2.9) will give rise to a biased covariance matrix. The necessary theory is quite easy to work out and computer programs have been written. The available information for estimating the mean is most easily obtained by accumulating the portion $\Sigma_{0,0}^{-1}$ contained in the observed components.

3.3. *Example 3.* Mixture problems may be regarded as missing information problems by noting that the indicator variable $Z_{n,k}$ (which is 1 if the nth observation is in the kth class and zero otherwise) is missing. If this was available then the constrained log likelihood would be

$$(3.3.1) \qquad L = \sum_{n=1}^{N} \sum_{k=1}^{K} Z_{n,k} [\log p_k + f_k(x_n | \theta_k)] - N \left(\sum_{k=1}^{K} p_k - 1 \right);$$

from this we may obtain the score for p_k as

$$(3.3.2) \qquad \text{Sco}\,(p_k | x_n, Z_n) = \text{Sco}\,(p_k | Z_n) = \sum_{n=1}^{N} \frac{Z_{n,k}}{p_k} - N,$$

whilst the score for θ_k is

$$(3.3.3) \qquad \text{Sco}\,(\theta_k | x_n, Z_n) = \sum_{n=1}^{N} Z_{n,k} \frac{1}{f_k} \frac{\partial}{\partial \theta_k} f_k(x_n | \theta_k).$$

If there is no missing data these equations may be separated into K classes, one set for each class.

If we do not observe the $Z_{n,k}$, then the image of $(x_n, Z_{n,1}, \cdots, Z_{n,k})$ is just x_n and hence we must find $E(Z_{n,k} | x_n)$ which is the posterior probability

$$(3.3.4) \qquad P[k | x_n] = \frac{p_k f_k(x_n | \theta_k)}{f(x_n | \theta)},$$

where

$$(3.3.5) \qquad f(x_n | \theta) = \sum_{k=1}^{K} p_k f_k(x_n | \theta_k).$$

The image scores are

$$(3.3.6) \qquad \text{Sco}\,(p_k | X) = \sum_{n=1}^{N} \frac{p[k | x_n]}{p_k} - N$$

and

$$(3.3.7) \qquad \text{Sco}\,(\theta_k | X) = \sum_{n=1}^{N} P[k | x_n] \frac{1}{f_k} \frac{\partial}{\partial \theta_k} f_k(x_n | \theta_k)$$

$$= \sum_{n=1}^{N} P[k | x_n] \, \text{Sco}_k\,(\theta_k | x_n).$$

The following estimating equations are obtained

$$(3.3.8) \qquad \hat{N}_k = \sum_{n=1}^{N} P[k|x_n]$$

and

$$(3.3.9) \qquad \hat{p}_k = \frac{\hat{N}_k}{N}.$$

The information computations require expressions for the expected values of the second derivative of the likelihood functions. Although these expressions do not have a form which can be easily evaluated for the mixture of normals, or any other standard distribution, they are being recorded here for the sake of completeness:

$$(3.3.10) \qquad -\frac{\partial^2 L}{\partial p_i \partial p_j} = \frac{1}{p_i p_j} \sum_{n=1}^{N} P[i|x_n]P[j|x_n],$$

$$(3.3.11) \qquad J^*(p_i p_j) = N \int \left[\frac{f_i(x|\theta_i)f_j(x|\theta_j)}{f(x|\theta)} \right] dx = N J^*_{i,j}(\theta),$$

where $J^*_{i,j}(\theta)$ is the above integral,

$$(3.3.12) \qquad -\frac{\partial^2 L}{\partial p_i \partial \theta_j} = \frac{1}{p_i} P[i|x_n]P[j|x_n] \operatorname{Sco}(\theta_j|x_n),$$

$$(3.3.13) \qquad J^*(p_i, \theta_j) = N p_i \frac{\partial J_{i,j}(\theta)}{\partial \theta_j}$$

$$(3.3.14) \qquad -\frac{\partial^2 L}{\partial \theta_i \partial \theta_j} = \sum_{n=1}^{N} P[i|x_n]P[j|x_n] \operatorname{Sco}_i(\theta_j|x_n) \operatorname{Sco}_j(\theta_j|x_n),$$

$$(3.3.15) \qquad J^*(\theta_i, \theta_j) = N p_i p_j \frac{\partial^2 J^*_{i,j}(\theta)}{\partial \theta_i \partial \theta_j}.$$

The overall nonsingular (unconstrained) information matrix is

$$(3.3.16) \qquad J^* = \begin{bmatrix} J^*(p, p) & J^*(p, \theta) \\ J^*(\theta, p) & J^*(\theta, \theta) \end{bmatrix}$$

and its inverse is V^*. We note that

$$(3.3.17) \qquad J^* \begin{bmatrix} p \\ 0 \end{bmatrix} = \begin{bmatrix} e \\ 0 \end{bmatrix},$$

so that

$$(3.3.18) \qquad V^* \begin{bmatrix} e \\ 0 \end{bmatrix} = \begin{bmatrix} p \\ 0 \end{bmatrix}$$

and

$$(3.3.19) \qquad \begin{bmatrix} e & 0 \end{bmatrix} V^* \begin{bmatrix} e \\ 0 \end{bmatrix} = 1$$

where e is a vector of ones. Thus the final covariance matrix of the estimator is

$$(3.3.20) \qquad V^* - \begin{bmatrix} p \\ 0 \end{bmatrix} \begin{bmatrix} p^T & 0 \end{bmatrix} = \frac{1}{N} \begin{bmatrix} V^*(p, p) - pp^T & V^*(p, \theta) \\ V^*(\theta, p) & V^*(\theta, \theta) \end{bmatrix}.$$

The properties of J^* and V^* discussed in the papers of Aitchison and Silvey [5], [6] on constrained maximum likelihood estimation are also shared by the approximating sums of partial derivatives.

3.4. *Example 4.* Consider now the problem of estimating the parameters of a mixture of multivariate normal populations assumed to have equal covariance matrices but different means. Suppose that there are K populations and that we sample N observations. Thus, the data matrix X would have consisted of K submatrices, of N_k observations from population k for $k = 1, \cdots, K$. Instead, we have the image Y which consists of N observations from the mixture. If we had considered Y to consist of a single population then we would have obtained

$$(3.4.1) \qquad \text{Sco}\, (\bar{\mu} | Y) = \Sigma^{-1} \sum_{n=1}^{N} (Y_n - \bar{\mu})$$

leading to

$$(3.4.2) \qquad \bar{\mu} = \frac{1}{N} \sum_{n=1}^{N} Y_n$$

and

$$(3.4.3) \qquad \text{Sco}\, (\Sigma^{-1} | Y) = -\tfrac{1}{2} N \Sigma + \tfrac{1}{2} \sum_{n=1}^{N} (Y_n - \bar{Y})(Y_n - \bar{Y})^T$$

leading to

$$(3.4.4) \qquad \Sigma = \frac{1}{N} \sum_{n=1}^{N} (Y_n - \bar{Y})(Y_n - \bar{Y})^T.$$

However, regarding Y as a mixture, we may write the likelihood function as

$$(3.4.5) \qquad L(Y | \mu, \Sigma) = \sum_{n=1}^{N} \left[\sum_{k=1}^{k} p_k f_k(X_n) \right]$$

$$(3.4.6) \qquad \log L = \sum_{n=1}^{N} \log \left[\sum_{k=1}^{K} p_k (2\pi)^{-p/2} |\Sigma|^{-1/2} \right.$$

$$\left. \exp \left\{ -\tfrac{1}{2} (X_n - \mu_k)^T \Sigma^{-1} (X_n - \mu_k) \right\} \right]$$

$$= -\frac{Np}{2} \log 2\pi + \frac{N}{2} \log |\Sigma^{-1}|$$

$$+ \sum_{n=1}^{N} \log \left[\sum_{k=1}^{k} p_k \exp \left\{ -\tfrac{1}{2} (X_n - \mu_k)^T \Sigma^{-1} (X_n - \mu_n) \right\} \right].$$

Due to the imposed restriction, that $\Sigma_{k=1}^{K} p_k = 1$, we are required to maximize the function

$$(3.4.7) \qquad L^* = \log L - \lambda \left(\sum_{k=1}^{k} p_k - 1 \right),$$

where λ is a Lagrangian multiplier.

Defining Sco $(\theta | X) = (\partial L^* / \partial X) \log L^*$ we obtain

$$(3.4.8) \qquad \text{Sco} (p_k | Y) = \sum_{n=1}^{N} \frac{f_k(X_n)}{f(X_n)} - \lambda,$$

where

$$(3.4.9) \qquad f(X_n) = \sum_{k=1}^{K} p_k f_k(X_n).$$

Multiplying by p_k and summing over k gives $\lambda = N$, hence

$$(3.4.10) \qquad \sum_{n=1}^{N} \frac{f_k(X_n)}{f(X_n)} = N \qquad \text{for} \quad k = 1, \cdots, K.$$

If we introduce the posterior probability

$$(3.4.11) \qquad P[k | X_n] = \frac{p_k f_k(X_n)}{f(X_n)}$$

and define

$$(3.4.12) \qquad \hat{N}_k = \sum_{n=1}^{N} P[k | X_n],$$

we obtain

$$(3.4.13) \qquad p_k = \frac{\hat{N}_k}{N}$$

and

$$(3.4.14) \qquad \hat{\mu}_k = \sum_{n=1}^{N} P[k | X_n] \frac{X_n}{\hat{N}_k}.$$

Finally

$$(3.4.15) \qquad \hat{\sigma}_{i,j} = \sum_{n=1}^{N} \sum_{k=1}^{K} \frac{1}{N} P[k | X_n] (X_{n,i} - \hat{\mu}_{k,i}) (X_{n,j} - \hat{\mu}_{k,j}).$$

If we call this estimate of Σ the within class covariance matrix and denote it by $\Sigma^{(w)}$, then we may write

$$(3.4.16) \qquad N \Sigma^{(w)} = \sum_{n=1}^{N} \sum_{k=1}^{K} P[k | X_n] (X_n - \bar{\mu} + \bar{\mu} - \mu_k) (X_n - \bar{\mu} + \bar{\mu} - \mu_k)^T$$

$$= \sum_{n=1}^{N} \sum_{k=1}^{K} P[k|X_n] [(X_n - \bar{\mu})(X_n - \bar{\mu})^T + (X_n - \bar{\mu})(\bar{\mu} - \mu_k)^T$$
$$+ (\bar{\mu} - \mu_k)(X_n - \bar{\mu})^T + (\bar{\mu} - \mu_k)(\bar{\mu} - \mu_k)^T]$$
$$= \sum_{n=1}^{N} (X_n - \bar{\mu})(X_n - \bar{\mu})^T$$
$$- \sum_{n=1}^{N} \sum_{k=1}^{K} P[k|X_n](\bar{\mu} - \mu_k)(\hat{\mu} - \mu_k)^T.$$

Thus we may write

(3.4.17) $$\Sigma^{(w)} = \Sigma^{(T)} - \Sigma^{(B)},$$

where $\Sigma^{(T)}$ is the total variance and $\Sigma^{(B)}$ is the between population variance.

The computational procedure proposed is to take some good guess as an initial set of estimates for the parameters p_k, μ_k, Σ. Then to cycle iteratively through (3.4.14), (3.4.15), and (3.4.13) until the parameter estimates stabilize. At each stage we use the best parameter estimates available. As is usual with such iterative procedures the convergence may be slow and will require speeding in practice.

APPENDIX

Brief Review of the Literature and Historical Development

Although considered earlier by Allan and Wishart [7], the first general approach to the problem of missing data, in field experiments, was that of Yates [50], who provided formulae enabling the least squares estimate of a single missing datum to be computed from the row and column sums. He also provided a similar formula to correct for the bias introduced into the sums of squares, and suggested that the estimation formula could be used iteratively if there was more than one missing datum. However, no general correction for the bias in the sums of squares was given. The basic ideas behind the approach of Yates have been used by many authors since that time to cover most common linear models (see code U1). It has also been used for a factor analytic model (see Woodbury, Clelland, and Hickey [46], Woodbury and Siler [48]), but it is restricted to the univariate case with independent observations. A second approach to the problem was the "covariance method" (code U2) of Bartlett [11], (see also Coons [17]) which results in the same estimates as Yates but which also gives unbiased sums of squares. Tocher [39] described a method (code U3) which appears to be a combination of the approaches of Yates and Bartlett. This method gives biased sums of squares but a general correction is given. A final method (code U4) for univariate random variables is the iterative maximum likelihood method described by Hartley [28]. This involves replacing the missing values by their expected values, given the model and the parameters. Examples were given, by Hartley, for a number of discrete distributions having sufficient statistics and for which the maximum likelihood parameter scores were linear in the observations.

The multivariate normal has been extensively studied; the first approach being the direct application of maximum likelihood (code M1). The first step was made by Wilks [45] who considered the general bivariate case. Special trivariate cases were considered by Lord [34] and Edgett [23]. Their approaches were greatly simplified by Anderson [9] who considered the general "nested" case for which the likelihood function could be factored. Trawinski and Bargmann [40] used the notation of Roy's general linear model to write the likelihood equations, and gave an example of a trivariate case for which each type of incomplete vector was observed an equal number of times. Another approach to the problem of testing hypotheses in such situations, where the components are missing by design rather than accident, is given by Kleinbaum [31]. Hocking and Smith [29] approached the problem by sequentially combining covariance estimators by adding one group at a time, starting with the complete observations. This method is statistically efficient and is the same as Anderson's maximum likelihood solution in the nested case. The above methods all appear to be valid approaches to the multivariate case. A method very similar to that of Hartley [28] was described by Federspiel, Monroe and Greenberg [24] (although used earlier by Greenberg). This is an iterative method which involves replacing the missing components by their conditional expectation, given the observed components. Although computationally simple, it gives rise to biased estimates of the covariance matrix (and, hence, of the mean), since the score contains quadratic functions of the observations. Buck [14] used a similar approach and also corrected for the bias in the covariance matrix, in the case of a single missing component. However, he failed to give the correct extension to more than a single missing component. Another approach (code M3), which could give rise to a nonpositive definite covariance matrix, is the use of all the available data to estimate each component of the mean and covariance matrix separately. This has been described by Glasser [25] and Haitovsky [27]. An extensive literative review was given by Afifi and Elashoff [2], who also considered some of the methods discussed here, as well as many of the methods in common use. One can determine from their extensive analysis that most approximations are best "quick and dirty" and at worst misleading.

Missing information problems, as distinct from missing data problems, are many and varied, some being recognized as such, some not. Examples of the lack of identifiability in mixture problems have been presented by Behboodian [12] for a mixture of univariate normal populations, by Wolfe [49] for a mixture of multivariate normal populations, and by Cohen [16] for the negative binomial, while the iterative methods used by Hartley [28] can deal, in addition, with the problem of lost information due to grouping, censoring and truncating. The problem, (code I2), of obtaining genotype frequencies from phenotype frequencies has been dealt with by Ceppellini, Siniscialco, and Smith [15].

REFERENCES

[1] A. A. AFIFI and R. M. ELASHOFF, "Missing observations in multivariate statistics I: Review of the literature," *J. Amer. Statist. Assoc.*, Vol. 61 (1966), pp. 595–604. (M2)

[2] ———, "Missing observations in multivariate statistics II: Point estimation in simple linear regression," *J. Amer. Statist. Assoc.*, Vol. 62 (1967), pp. 11–29. (M1)

[3] ——— , "Missing observations in multivariate statistics III," *J. Amer. Statist. Assoc.*, Vol. 64 (1969), pp. 337–358.

[4] ———, "Missing observations in multivariate statistics IV," *J. Amer. Statist. Assoc.*, Vol. 64 (1969), pp. 358–365.

[5] J. AITCHISON and S. D. SILVEY, "Maximum likelihood estimates of parameters subject to restraints," *Ann. Math. Statist.*, Vol. 29 (1958), pp. 813–828.

[6] ———, "Maximum likelihood estimation procedures and associated tests of significance," *Ann. Math. Statist.*, Vol. 31 (1960), pp. 154–171.

[7] F. E. ALLAN and J. WISHART, "A method of estimating the yield of a missing plot in field experimental work," *J. Agric. Sci.: Camb.*, Vol. 20 (1930), pp. 399–406.

[8] R. L. ANDERSON, "Missing plot techniques," *Biometrics*, Vol. 2 (1946), pp. 41–47. (U1)

[9] T. W. ANDERSON, "Maximum likelihood estimates for a multivariate normal distribution when some observations are missing," *J. Amer. Statist. Assoc.*, Vol. 52 (1957), pp. 200–203. (M1)

[10] H. R. BAIRD and C. Y. KRAMER, "Analysis of variance of a balanced incomplete block design with missing observations," *Appl. Statist.*, Vol. 10 (1961), pp. 189–198. (U1)

[11] M. S. BARTLETT, "Some examples of statistical methods of research in agriculture," *J. Roy. Statist. Soc.*, Ser. B, Vol. 4 (1937), pp. 137–183. (U2)

[12] J. BEHBOODIAN, "On a mixture of normal distributions," *Biometrika*, Vol. 57 (1970), pp. 215–217.

[13] J. D. BIGGERS, "The estimation of missing and mixed-up observations in several experimental designs," *Biometrika*, Vol. 46 (1959), pp. 91–105. (U1)

[14] S. F. BUCK, "A method of estimation of missing values in multivariate data, suitable for use with an electronic computer," *J. Roy. Statist. Soc.*, Ser. B, Vol. 22 (1960), pp. 302–306. (M2)

[15] R. CEPPELLINI, M. SINISCIALCO, and C. A. B. SMITH, "The estimation of gene frequencies in a random mating population," *Ann. Hum. Genet.*, Vol. 20 (1955), pp. 97–115. (I2).

[16] A. C. COHEN, "A note on certain discrete mixed distributions," *Biometrics*, Vol. 22 (1966), pp. 566–571. (I1)

[17] I. COONS, "The analysis of covariance as a missing plot technique," *Biometrics*, Vol. 13 (1957), pp. 387–402. (U2)

[18] E. A. CORNISH, "The estimation of missing values in incomplete randomized block experiments," *Ann. Eugenics*, Vol. 10 (1940), pp. 112–118. (U1)

[19] ———, "The estimation of missing values in quasi-factorial designs," *Ann. Eugenics*, Vol. 10 (1940), pp. 137–143. (U1)

[20] C. W. COTTERMAN, "A weighting system for the evaluation of gene frequencies from family records," *Contributions from the Laboratory of Vertebrate Biology*, Ann Arbor, University of Michigan Press, 1947, pp. 1–21. (I2)

[21] D. B. DELURY, "The analysis of latin squares when some observations are missing," *J. Amer. Statist. Assoc.*, Vol. 41 (1946), pp. 370–389 (U1)

[22] N. R. DRAPER and D. M. STONEMAN, "Estimating missing values in unreplicated 2-level factorial and fractional factorial designs," *Biometrics*, Vol. 20 (1964), pp. 443–458. (U1)

[23] G. L. EDGETT, "Multiple regression with missing observations among the independent variables," *J. Amer. Statist. Assoc.*, Vol. 51 (1956), pp. 122–131. (M1)

[24] C. F. FEDERSPIEL, R. J. MONROE, and B. G. GREENBERG, "An investigation of some multiple regression methods for incomplete samples," *U. N. C. Inst. Statist. Memo*, No. 236, 1959. (M1), (M2), (M4)

[25] M. GLASSER, "Linear regression with missing observations among the independent variables," *J. Amer. Statist. Assoc.*, Vol. 59 (1964), pp. 834–844. (M3)

[26] W. A. GLENN and C. Y. KRAMER, "Analysis of variance of a randomized block design with missing observations," *Appl. Statist.*, Vol. 7 (1958), pp. 173–185. (U1)

[27] Y. HAITOVSKY, "Missing data in regression analysis," *J. Roy. Statist. Soc.*, Ser. B, Vol. 30 (1968), pp. 67–82. (M3)

[28] H. O. HARTLEY, "Maximum likelihood estimation from incomplete data," *Biometrics*, Vol. 14 (1958), pp. 174–194. (U4)

[29] R. M. HOCKING and W. B. SMITH, "Estimation of parameters in the multivariate normal distribution with missing observations," *J. Amer. Statist. Assoc.*, Vol. 63 (1968), pp. 159–173. (M4)

[30] E. C. JACKSON, "Missing values in linear multiple discriminant analysis," *Biometrics*, Vol. 24 (1968), pp. 835–844. (M2)

[31] D. G. KLEINBAUM, "A general method for obtaining test criteria for multivariate linear models with more than one design matrix and/or incomplete response variates," *U.N.C.* Inst. Statist. Memo, No. 614, 1969. (M4)

[32] C. Y. KRAMER and S. GLASS, "Analysis of variance of a latin square design with missing observations," *Appl. Statist.*, Vol. 9 (1963), pp. 43–50. (U1)

[33] W. KRUSKAL, "The coordinate-free approach to Gauss-Markov estimation, and its application to missing and extra observations," *Proceedings of the Fourth Berkeley Symposium, Mathematical Statistics and Probability*, Berkeley and Los Angeles, University of California Press, 1961, Vol. 1, pp. 435–451. (U1)

[34] F. M. LORD, "Estimation of parameters from incomplete data," *J. Amer. Statist. Assoc.*, Vol. 50 (1955), pp. 870–876. (M1)

[35] G. E. NICOLSON, "Estimation of parameters from incomplete multivariate samples," *J. Amer. Statist. Assoc.*, Vol. 52 (1957), pp. 523–526. (M1)

[36] C. R. RAO, *Linear statistical inference and its applications*, New York, Wiley, 1965.

[37] ———, "Analysis of dispersion with incomplete observations on one of the characters," *J. Roy Statist. Soc.*, Ser. B, Vol. 18 (1956), pp. 259–264. (M4)

[38] S. D. SILVEY, "The Lagrangian multiplier test," *Ann. Math. Statist.*, Vol. 30 (1959), pp. 389–407.

[39] K. D. TOCHER, "The design and analysis of block experiments," *J. Roy. Statist., Soc. Ser. B*, Vol. 14 (1952), pp. 45–100. (U3)

[40] I. M. TRAWINSKI and R. E. BARGMAN, "Maximum likelihood estimates with incomplete multivariate data," *Ann: Math. Statist.*, Vol. 35 (1964), pp. 647–657. (M4)

[41] A. WALD, "Tests of statistical hypotheses when the number of observations is large," *Trans. Amer. Math. Soc.*, Vol. 34 (1943), pp. 426–482.

[42] G. N. WILKINSON, "The analysis of covariance with missing data," *Biometrics*, Vol. 13 (1957), pp. 363–372. (U1)

[43] ———, "The estimation of missing values for the analysis of incomplete data," *Biometrics*, Vol. 14 (1958), pp. 257–286.. (U1)

[44] ———, "The analysis of variance and derivation of standard errors for incomplete data," *Biometrics*, Vol. 14 (1958), pp. 360–384. (U1)

[45] S. S. WILKS, "Moments and distribution of estimates of population parameters from fragmentary samples," *Ann. Math. Statist.*, Vol. 3 (1930), pp. 163–195. (M1)

[46] M. A. WOODBURY, R. C. CLELLAND, and B. J. HICKEY, "Applications of a factor analytic model in the prediction of biological data," *Behavioral Sci.*, Vol. 8 (1963), pp. 347–354. (M4)

[47] M. A. WOODBURY and V. HASSELBLAD, "Maximum likelihood estimates of the variance-covariance matrix from the multivariate normal," *SHARE National Meeting*, Denver, Colorado, March, 1970.

[48] M. A. WOODBURY and W. SILER, "Factor analysis with missing data," *Ann. New York, Acad. Sci.*, Vol. 128 (1966), pp. 746–754. (M4)

[49] J. H. WOLFE, "NORMIX—computational methods for estimating the parameters of multi-variate normal mixtures of distributions," Technical Report, U.S. Naval Personnel Research Activity, San Diego, California, 1967, pp. 1–31.

[50] F. YATES, "The analysis of replicated experiments when field results are incomplete," *Emp. J. Expt. Agric.*, Vol. 1 (1933), pp. 129–142. (U1)

[51] F. YATES and R. W. HALE, "The analysis of latin squares where two or more rows, columns or treatments are missing," *J. Roy. Statist. Soc., Ser. B*, Vol. 6 (1939), pp. 67–79. (U1)

RECENT RESULTS ON USING THE PLAY THE WINNER SAMPLING RULE WITH BINOMIAL SELECTION PROBLEMS

MILTON SOBEL[1]
UNIVERSITY OF MINNESOTA
and
GEORGE H. WEISS
NATIONAL INSTITUTES OF HEALTH

1. Introduction

The theory of clinical trials has been studied from many different points of view in recent years. Perhaps the feature of principal interest that distinguishes the clinical trial from analogous problems that arise in industrial statistics is the ethical factor. The doctor treating a patient in a clinical trial is not only obliged to derive information relevant for the treatment of a larger statistical population, but is also obliged to treat each patient in the best way that he is able. These two requirements are contradictory to a certain extent, and lend urgency to the design of clinical trials that goes at least part of the way towards incorporating both requirements in some rational fashion. Armitage's monograph [2] and a subsequent review by Anscombe [1] did much to frame the general problem and bring it to the attention of statisticians. Although Armitage's original thinking had envisaged a fairly straightforward application of sequential analysis to the choice of the better of two treatments, Colton, working at his suggestion [4] developed a different formulation that has attracted some interest.

In brief, Colton's model assumes that the total patient horizon N is known. Of these, a total of $2n$ patients are to be used to derive information about the relative worth of the two treatments, and the remaining $N - 2n$ patients are given the treatment designated as better in the testing phase. Under various assumptions about the underlying distributions, Colton has derived optimal fixed and sequential rules for calculating the optimal value of n. Zelen [16] considered a more specialized version of the Colton model, in which the response was assumed to be dichotomous rather than continuous. The new and interesting feature of Zelen's work was the suggestion that the sampling technique could be adapted to reduce the number of patients on the poorer treatment. Zelen applied

[1] Research supported by NSF Grant GP-11021, Department of Statistics, University of Minnesota and NSF Grant GP-17172, Department of Statistics, Stanford University.

the "Play the Winner" rule (to be abbreviated PW rule) to the clinical trial problem; it prescribes that a success with a given treatment generates a further trial on the same treatment while a failure generates a trial on the alternative treatment. This particular assignment procedure was first suggested by Robbins [10] in a discussion of the two-arm bandit problem and was subsequently elaborated by Isbell [7] and by Pyke and Smith [9]. Although the PW rule is not optimal in the context of either the two-arm bandit problem or the clinical trial problem, it is a simple and easily implemented sampling rule that does introduce a bias in favor of testing the better treatment. Zelen analyzed the PW rule both with respect to indefinitely extended trials and with respect to the finite patient horizon model of Colton. More recently, Cornfield, Halperin, and Greenhouse [5] have analyzed a Bayesian generalization of the Colton model. The model is generalized in three ways: the first being in the assumption of a prior distribution, the second being in the distribution of patients in the data gathering phase to the two treatments in the ratio $\theta : (1 - \theta)$ rather than $1 : 2$, and the third being in the repeated application of the procedure for $k > 2$ stages. Their results still depend on some estimate of the patient horizon, but the dependence is much less sensitive to variations in N than the original Colton model. The possible improvement that might be afforded through application of the PW rule to this formulation has not been discussed to date.

In the present paper we will summarize some of our own recent research on the formulation of clinical trial problems. So far we have considered only the case of dichotomous response, where the effect of treatment is immediately available. These are considerable restrictions and leave open many interesting and useful problems for further investigation, but the general problem area even in its simplest formulation leads to difficult mathematical problems. Since we restrict ourselves to a version of the two-arm bandit problem, it is well to motivate later mathematical developments. A rigorous solution to the two-arm bandit problem would lead to an indefinitely extended series of trials in which the probability of choosing the better treatment at trial N trends to 1 as $N \to \infty$. Economic factors at the very least would imply a desire to terminate testing in some finite number of trials either with a decision that one or the other treatment is better or, in some applications, with a third decision that the two treatments are equally good. Furthermore, the Colton model seems somewhat artificial because some knowledge, Bayesian or otherwise, is implied about the patient horizon N, which is usually not available. If these premises are granted, it would seem desirable to develop a test method that allows one to reach a decision with a finite number of tests without any suppositions about the patient horizon.

We have recently discussed various aspects of the problem of choosing the best of $k \geq 2$ binomial populations [13], [14], [15] using a formulation explored in some detail in a monograph by Bechhofer, Kiefer, and Sobel [3]. Let us consider two treatments A and B with success probabilities p and p' respectively, where $p > p'$. A correct selection corresponds to identifying A as the better treatment after a series of tests. In the formulation of [13], one assumes that two constants P^* and Δ^* are given with $\frac{1}{2} < P^* < 1$ and $\Delta^* > 0$ and the termination rule is set up so that the probability of a correct selection, or $P\{CS\}$, $\geqq P^*$ whenever $p - p' \geqq \Delta^*$. Most of the work to be described below requires that one of the drugs be declared the better even though the condition $p - p' \geqq \Delta^*$ may not hold; some results have also been obtained for the three decision problem in which one is allowed to conclude that $p - p' < \Delta^*$.

In all of the models to be described, a decision is made after a finite number of tests, with probability one, and no assumptions are made about a patient horizon. The first question of interest relates to the effect of different sampling rules on the number of patients put on the poorer treatment. It is clear that in an infinitely extended series of trials, the PW rule is superior to the rule that prescribes alternate sampling $ABABA \ldots$ (this rule will be designated as the "Vector at a Time" rule or VT rule). For finite sets of trials, a comparison is of interest because it is not clear whether a decision can be reached more quickly by testing the same number of patients on both treatments, or whether the bias introduced by the PW rule does not extend the trial by so much that ultimately more patients are given the poorer drug than if the VT rule had been used.

The two sampling rules must, of course, be supplemented by stopping rules. Here again, a wide choice of rules can be considered, but we focus mainly on two stopping rules partly because of likelihood ratio considerations and partly because they lead to analytically tractable problems. The first rule depends on keeping track of the number of successes for both treatments at each stage and computing $|S_i - S_j| = \Delta S$ where S_i and S_j are the numbers of successes on the two treatments. The clinical trial is terminated when $\Delta S = r$, where r is a critical integer determined from P^* and Δ^*. The second stopping rule corresponds to inverse sampling in which the trial terminates when either S_i or S_j reaches a critical value r'. Although the first stopping rule is usually better since it incorporates more information, the inverse sampling procedure warrants analysis because it is easily generalized to choosing the best of $k > 2$ populations and because it can be used as a basis for comparisons.

In the next section we compare the performance of the VT rule with the PW rule when only two decisions are allowed, that is, that one or the other treatment is better, and the trials are potentially of unlimited duration. It should be noted, however, that with probability one the trials continue for only a finite number of tests and the expected number of trials to termination is finite for any fixed pair (p, p') with $0 < p' \leqq p < 1$. This will be followed by some results on selecting the best of $k > 2$ populations. Section 3 presents results obtained so far on truncated testing in which a maximum number of tests is prescribed. In Section 4 we describe some sequential procedures for $k \geqq 3$ and compare the PW and VT sampling rules by Monte Carlo Studies. In Section 5 some alternative and allied lines of current research are described. Finally, in the last section, we discuss some of the many open questions in this area of research.

2. Comparison of procedures R_{PW} and R_{VT}

Several results will be quoted here from different papers and some new tables will be appended to confirm that these results hold not only for $P^* \to 1$ and/or $\Delta^* \to 0$ but also for fixed smaller values of P^* and fixed larger values of Δ^*.

The basic termination rule is to stop sampling when $\Delta S = r$. For any pair $p = 1 - q$ and $p' = 1 - q'$ the integer $r = r_{\mathrm{PW}}$ needed under PW sampling to satisfy the basic requirement

(2.1) $$P\{\mathrm{CS}\} \geqq P^* \quad \text{whenever} \quad p - p' \geqq \Delta^*$$

is found in [13] to be the smallest integer at least as large as the root in x of

(2.2) $$\lambda^x = \frac{1}{2qP^*}\{\bar{q} - [\bar{q}^2 - 4qq'P^*(1 - P^*)]^{1/2}\},$$

where $\lambda = p'/p < 1$ and $\bar{q} = \frac{1}{2}(q + q')$. In the least favorable (LF) configuration we set $p' = p - \Delta^*$ (or $q' = q + \Delta^*$) and minimize the $P\{\mathrm{CS}\}$ or maximize the solution of (2.2) in x as a function of p for $\Delta^* \leqq p \leqq 1$. This minimum occurs at a p value close to 1 and has been obtained exactly for the calculation in Tables IIa through IId. Denote the resulting precedure by R_{PW}.

For the same termination rule we also consider the VT sampling rule and denote the resulting procedure by R_{VT}. Then, corresponding to (2.2), the integer $s = r_{\mathrm{VT}}$ needed under VT sampling to satisfy (2.1) was found in [3] and in [12] to be the smallest integer at least as large as the root in x of

(2.3) $$P^*(1 + \delta^x) = 1,$$

where $\delta = p'q/pq' < 1$. Under the LF configuration we set $p = \frac{1}{2}(1 + \Delta^*)$ and $p' = \frac{1}{2}(1 - \Delta^*)$ so that x is given explicitly by

(2.4) $$x = \frac{\log\left[(1 - P^*)/P^*\right]}{2\log\left(\dfrac{1 - \Delta^*}{1 + \Delta^*}\right)}.$$

To compare the two sampling rules for the same pair (Δ^*, P^*) and the same termination rule we use two different criteria which are calculated exactly in [13] after the values of r_{PW} and r_{VT} are obtained. The first is the expected loss or risk $E\{L\}$ (which we write as $\bar{L}$ below) defined by

(2.5) $$\mathrm{E}\{L\} = (p - p')\mathrm{E}\{N_B\} = \bar{L},$$

say, where $E\{N_B\}$ denotes the number of patients put on the poorer drug. This has considerable interest in medical applications since it represents the difference in the expected number of successes between a conceptual set of trials in which the better treatment is always used and the actual set of trials; thus we have $E\{L \,|\, R_{\mathrm{PW}}\}$ for PW sampling and $E\{L \,|\, R_{\mathrm{VT}}\}$ for VT sampling.

The second criterion is the expected total number of trials needed for termination which can be written as

(2.6) $$E\{N\} = E\{N_A\} + E\{N_B\}.$$

This is the standard criterion used in [3] for comparing procedures that satisfy the same (Δ^*, P^*) requirement (2.1); thus we have $E\{N \mid R_{PW}\}$ for PW sampling and $E\{N \mid R_{VT}\}$ for VT sampling.

It is found in [13] by an asymptotic $(P^* \to 1)$ analysis that the ratio $E\{L \mid R_{PW}\}/E\{L \mid R_{VT}\} < 1$ when

(2.7) $$p > \frac{3}{4} - \frac{\Delta^*}{8} + O(\Delta^{*3}).$$

Since Δ^* is usually small we can disregard the term of order (Δ^{*3}) and say that PW sampling is preferred (L) when $p > \frac{1}{8}(6 - \Delta^*)$ and that VT sampling is preferred when the reverse inequality holds.

Similarly, according to the $E\{N\}$ criterion it is found in [13] that the ratio $E\{N \mid R_{PW}\}/E\{N \mid R_{VT}\} < 1$ when

(2.8) $$\bar{p} > \frac{3}{4} - \frac{\Delta^*}{8} + O(\Delta^{*3})$$

where $\bar{p} = \frac{1}{2}(p + p')$ and $\bar{q} = 1 - \bar{p}$.

These results are corroborated in this paper by Tables IIa through IId and shown to hold for the four (Δ^*, P^*) pairs: $(0.50, 0.75)$, $(0.20, 0.75)$, $(0.05, 0.95)$, and $(0.20, 0.95)$. The exact integer values of r_{PW} and r_{VT} are obtained and exact values of

(2.9) $$E\{L \mid R_{PW}\} = \frac{(p + 2qr)(1 - \lambda^r)(q' - q\lambda^r)}{2(q' - q\lambda^{2r})},$$

$$E\{N \mid R_{PW}\} = \frac{(1 - \lambda^r)(q' - q\lambda^r)(\bar{p} + 2r\bar{q})}{(1 - \lambda)(q' - q\lambda^{2r})p},$$

(2.10) $$E\{L \mid R_{VT}\} = \frac{s(1 - \delta^s)}{1 + \delta^s}, \qquad E\{N \mid R_{VT}\} = \frac{2s(1 - \delta^s)}{(p - p')(1 + \delta^s)}$$

are computed for the generalized least favorable (GLF) configuration in which $p' = p - \Delta^*$ and $\Delta^* \leqq p \leqq 1$. (A misprint in $E\{N \mid R_{PW}\}$ in (2.18) of [13] is corrected in (2.9) above.)

In addition, we have tabulated $E\{N_B\}$ for the equal parameter (EM) configuration (that is, for $p = p'$) as a function of p. Note that this function approaches infinity as $p \to 0$ or $p \to 1$ under VT sampling but this occurs only as $p \to 0$ under PW sampling.

These tables show a definite crossover pattern in which PW sampling is better for large p values and poorer for small p values. The results for VT sampling are more constant and tend to be symmetrical about $\frac{1}{2}$. The need for an adaptive procedure which switches from VT sampling to PW sampling when p (or when $\bar{p}$) appears to be larger than $\frac{1}{8}(6 - \Delta^*)$ is clearly demonstrated, but such a procedure has not yet been studied.

From (2.9) and (2.10) we can find exact expressions for $E\{N\}$ for $p = p'$ (or $\lambda = 1$) when Δ^* and P^* are fixed; our principal interest is in the case where the common p tends to 1 or 0. From (2.9) we have for $p = p'$

$$(2.11) \qquad E\{N \mid R_{\mathrm{PW}}\} = r + \frac{r^2 q}{p}$$

and from (2.10) we find that

$$(2.12) \qquad E\{N \mid R_{\mathrm{VT}}\} = \frac{2s(1 - \delta^s)}{(p - p')(1 + \delta^s)} \sim \frac{s^2}{pq},$$

where the last expression is the limiting value as $p' \to p$.

If we now let $\Delta^* \to 0$ then we obtain from (2.2) and (2.4) for the respective LF configurations

$$(2.13) \qquad r \sim \frac{\log \big((1 - P^*)/P^*\big)}{\log (1 - \Delta^*)}; \quad s \sim \frac{\log \big((1 - P^*)/P^*\big)}{2 \log (1 - \Delta^*)},$$

so that r is twice s for small Δ^*. For Δ^* sufficiently small we can disregard the r term in (2.11) and, using the fact that $r = 2s$, we note from (2.11) and (2.12) that for $p > \frac{1}{2}$ (and hence for $p \to 1$) $E\{N \mid R_{\mathrm{PW}}\}$ is smaller than $E\{N \mid R_{\mathrm{VT}}\}$. On the other hand for $p \leqq \frac{1}{2}$ (and hence for $p \to 0$) it is clear that $E\{N \mid R_{\mathrm{VT}}\}$ is smaller. In fact, $E\{N \mid R_{\mathrm{VT}}\}$ is smaller for $p < 0.70$ in most of our calculations. Hence there is no uniform result in this comparison.

In [14] and [15] the termination rule employed is the so-called inverse sampling where we stop when any one population has attained r' ($= r$) successes; for convenience we drop the prime on r. New values r'_{PW} and r'_{VT} are obtained to satisfy the same basic requirement (2.1), let the resulting procedures be denoted by R'_{PW} and R'_{VT}, respectively. For this purpose we use the exact expression for any integer $r \geqq 1$

$$(2.14) \qquad P\{\mathrm{CS} \mid R'_{\mathrm{PW}}\} = \tfrac{1}{2} E_r\{I_{q'}(X, r) + I_{q'}(X + 1, r)\},$$

where $I_p(x, y)$ is the usual incomplete beta function $I_q(0, r) = 1 = 1 - I_p(r, 0)$ for $r \geqq 1$, and the E_r denotes expectation of the random variable X which has the discrete negative binomial probability law

$$(2.15) \qquad f(x) = p^r \binom{x + r - 1}{x} q^x, \qquad x = 0, 1, \cdots.$$

For the Vector at a Time procedure R'_{VT} we obtain exactly the same result (2.14) so that the LF configurations must be identical for R'_{PW} and R'_{VT} and hence $r'_{\mathrm{PW}} = r'_{\mathrm{VT}}$ for any pair (Δ^*, P^*). (Similar results were found for $k > 2$, and E. Nebenzahl has noted the same result for a fixed sample size problem, see (2.26) and (5.2) below). This does not imply any equality in $E\{L\}$ or in $E\{N\}$ for R'_{PW} and R'_{VT} and we need these to compare these procedures. In [14] we obtain for $k = 2$ the exact expressions

$$(2.16) \quad E\{N \mid R'_{\mathrm{PW}}\} = \left(\frac{1}{q} + \frac{1}{q'}\right)\left[\frac{rq}{p} + \frac{r}{p'}E_r\{I_{p'}(r+1, X)\} - \frac{r}{p}E_{r+1}\{I_{p'}(r, X)\}\right]$$

$$+ \frac{1}{2q'} - \frac{1}{2q'}E_r\{I_{p'}(r, X+1)\} + \frac{1}{2q}E_r\{I_{p'}(r, X)\},$$

$$(2.17) \quad E\{N \mid R'_{\mathrm{VT}}\} = \frac{2r}{p} + \frac{2r}{p'} - \frac{r}{p}E_{r+1}\{I_{p'}(r, X) + I_{p'}(r, X+1)\}$$

$$- \frac{r}{p'}E_r\{I_{q'}(X, r+1) + I_{q'}(X+1, r+1)\}.$$

In this form all the expectations in (2.16) and (2.17) tend to zero as $r \to \infty$. (This can occur because $\Delta^* \to 0$ or because $P^* \to 1$). Hence, it easily follows that for large r we have

$$(2.18) \quad E\{N \mid R'_{\mathrm{PW}}\} < E\{N \mid R'_{\mathrm{VT}}\}.$$

Again we note for the case of a common p that $\mathrm{EN} \to \infty$ as $p \to 0$ or as $p \to 1$ under VT sampling but this happens only as $p \to 0$ under PW sampling. In another formulation David Hoel includes failures of the opponent and defines the score for A and B, respectively, as

$$(2.19) \quad \begin{aligned} R_A &= S_A + F_B, \\ R_B &= S_B + F_A, \end{aligned}$$

and then uses inverse sampling with these R values. The results are similar to those given here but one important difference is that $E\{N\}$ is bounded in his case.

For the $E\{L\}$ criterion we obtain from (2.19) in [14]

$$(2.20) \quad E\{L \mid R'_{\mathrm{PW}}\} = \frac{\Delta}{q'}\left[\frac{rq}{p} + \frac{r}{p'}E_r\{I_{p'}(r+1, X)\} - \frac{r}{p}E_{r+1}\{r, X\}\right.$$

$$\left. + \frac{1}{2} - \frac{1}{2}E_r\{I_{p'}(r, X+1)\}\right],$$

and using the result in (2.17) above

$$(2.21) \quad E\{L \mid R'_{\mathrm{VT}}\} = \frac{\Delta}{2}E\{N \mid R'_{\mathrm{VT}}\}$$

Here again the expectations approach 0 as $r \to \infty$ and we obtain for large r

$$(2.22) \quad E\{L \mid R'_{\mathrm{PW}}\} < E\{L \mid R'_{\mathrm{VT}}\}.$$

In the common LF configuration, we found in [14] the interesting feature that the minimum of the $P\{\mathrm{CS}\}$ (subject to $p - p' \geqq \Delta^*$) occurs when p and p' are centered at $\frac{2}{3}$ with difference Δ^*; this was obtained by disregarding terms of order $(\Delta^*)^2$. Then the appropriate r value for both R'_{PW} and R'_{VT} as a function of Δ^* and P^* is

$$(2.23) \quad r = \frac{8}{27}\left(\frac{\lambda}{\Delta^*}\right)^2$$

where $\lambda = \lambda(P^*)$ is the P^* percentage point of the standard normal distribution. This is obtained by assuming r is large and approximating the difference of two independent negative binomial chance variables by the appropriate normal approximation. It is estimated that (2.22) will hold when $r > p/2\Delta$ and the same estimate is obtained from (2.18) by dropping all the expectations in (2.16) and (2.17).

The question of whether waiting for a fixed number of failures is better than waiting for a fixed number of successes is also considered and it is found that for $p > \frac{1}{2}$ the latter is preferable but for $p < \frac{1}{2}$ the former is preferable.

In [15] we consider the inverse sampling procedure with $k \geqq 3$ populations and again compare the two procedures R'_{PW} and R'_{VT}. Here we order the populations (say, A, B, C for $k = 3$) and consider a cyclic variation PWC of the PW sampling rule. Observe A until it produces a failure. Switch to B until it produces a failure; then to C until it produces a failure. Then return to A and repeat the cycle. The results obtained are quite similar to those for $k = 2$; procedure R'_{PW} is uniformly better than R'_{VT} for large values of r whether the criterion $E\{N\}$ or

$$(2.24) \qquad E\{L\} = \sum_{i=1}^{k} (p_1 - p_i)E\{N_i\}, \qquad\qquad p_1 = \max_i p_i,$$

is used. The exact formulas and the correction term to the normal approximation are of interest. For R'_{PWC} we obtain

$$(2.25) \qquad P\{CS \,|\, R'_{PWC}\} = \frac{1}{k} E_r \left\{ \prod_{j=2}^{k} I_{q_j}(X, r) \right.$$
$$\left. + \sum_{\alpha=2}^{k} \left[\prod_{j=2}^{\alpha-1} I_{q_j}(X, r) \right] \left[\prod_{j=\alpha}^{k} I_{q_j}(X + 1, r) \right] \right\},$$

where X has the negative binomial probability law (2.15) with index $r \geqq 1$, success parameter p_1 and mean rq_1/p_1. For R'_{VT} we obtain, after the first step of minimization, exactly the same PCS in the form

$$(2.26) \qquad P\{CS \,|\, R'_{VT}\} = \frac{1}{k} E_r \left\{ \frac{I_{q_2}^k(X, r) - I_{q_2}^k(X + 1, r)}{I_{q_2}(X, r) - I_{q_2}(X + 1, r)} \right\} = P(CS \,|\, R'_{PWC}).$$

Hence, the LF configurations are the same and $r'_{PW} = r'_{VT}$. The value of r needed for both R'_{PW} and R'_{VT} and used in the formulas below is again approximately given by (2.23), except that $\lambda = \lambda(P^*, \rho, k)$ is the value of H that satisfies

$$(2.27) \qquad \int_{-\infty}^{\infty} \Phi^{k-1}\left(\frac{x\sqrt{\rho} + H}{\sqrt{1 - \rho}} \right) d\Phi(x) = P^*$$

where $\rho = \frac{1}{2} - \frac{3}{2}\Delta^* + O\{\Delta^{*2}\}$ (which we approximate by $\frac{1}{2}$), and $\Phi(x)$ (resp., $\varphi(x)$) is the standard normal distribution (resp., density) function. Because $\rho \neq \frac{1}{2}$, it is desirable to introduce a correction term to the normal ranking integral (2.27). This is accomplished by proving a lemma about the left member of (2.27), which we call $A_{k-1}(\rho, H)$.

For $k \geq 3$, any fixed H and positive $\rho < 1$

$$(2.28) \qquad \frac{d}{d\rho} A_{k-1}(\rho, h) = \frac{(k-1)(k-2)\varphi(H)\varphi\left[H\left(\frac{1-\rho}{1+\rho}\right)^{1/2}\right]}{2(1+\rho)^{1/2}}$$

$$\cdot A_{k-3}\left[\frac{\rho}{1+2\rho}, H\left(\frac{1-\rho}{(1+\rho)(1+2\rho)}\right)^{1/2}\right].$$

Then the $P\{CS|R'_{PWC}\}$ in the LF configuration is approximated by

$$(2.29) \qquad A_{k-1}\left(\frac{1}{2}, H\right) - \frac{3\Delta^*}{2} \frac{(k-1)(k-2)}{\sqrt{6}} \varphi(H)\varphi\left(\frac{H}{\sqrt{3}}\right) A_{k-3}\left(\frac{1}{4}, \frac{H}{\sqrt{6}}\right),$$

where $H = \Delta^*(27r/8)^{1/2}$. For example if $k = 3$, $P^* = 0.90$ and $\Delta^* = 0.10$, then the first term of (2.29) gives $H = 1.58$ and $r = 74$, but if we try $H = 1.59$ and 1.60 in (2.29) we find that the latter is closer and this leads to $r = 78$ and a minimum PCS $= 0.9008$.

For $E\{N\}$ and $E\{L\}$ the exact expressions are of less interest but the normal approximations to these are for large r with $p_1 = \max_i p_i$

$$(2.30) \qquad E\{N|R'_{PWC}\} \sim \frac{rq_1}{p_1}\left(\sum_{i=1}^{k} \frac{1}{q_i}\right),$$

$$(2.31) \qquad E\{L|R'_{PWC}\} \sim \frac{rq_1}{p_1} \sum_{i=1}^{k} \left(\frac{p_1 - p_i}{q_i}\right).$$

A Monte Carlo simulation by D. G. Hoel shows that these approximations are very close to those obtained in 1000 Monte Carlo experiments for $k = 3$, $\Delta^* = 0.2$ and $P^* = 0.95$. Some typical excerpts from his table are given in Table I.

TABLE I

| | $E\{L|R'_{PWC}\}$ | | $E\{N|R'_{PWC}\}$ | |
$p_1 = \max p_i$	Observed (Monte Carlo)	Approximate (2.24)	Observed (Monte Carlo)	Approximate (2.23)
0.2	45.5	44.8	369	364
0.4	21.3	21.0	177	175
0.6	12.6	12.4	110	109
0.8	7.4	7.0	72	70
1.0	1.1	0.0	33	28

For the case of equal parameters (EM configuration) we obtain in [15] for $q < 1$

$$(2.32) \qquad E\{N|R'_{PWC}\} = \frac{1}{q} \sum_{\alpha=0}^{\infty} I_q(\alpha + 1, r) \left[\frac{I_q^k(\alpha, r) - I_q^k(\alpha + 1, r)}{I_q(\alpha, r) - I_q(\alpha + 1, r)}\right],$$

which converges for $0 \leq q < 1$; the case $q = 0$ is obtained by continuity. This is then approximated by

$$(2.33) \qquad E\{N \,|\, R'_{\text{PWC}}\} \sim \frac{kr - \lambda_1(k - p)(r/q)^{1/2}}{p} - \left(\frac{k - p}{2q}\right),$$

where $\lambda_1 = \lambda_1(k)$ is the $(100k/k + 1)$st percentile of the standard normal distribution. The corresponding result in [15] for the VT rule is

$$(2.34) \qquad E\{N \,|\, R'_{\text{VT}}\} = \frac{kr}{p} E_{r+1} \left\{\frac{I_q^k(X, r) - I_q^k(X + 1, r)}{I_q(X, r) - I_q(X + 1, r)}\right\},$$

where E_{r+1} is defined as in (2.15) with r increased to $r + 1$. This is then approximated by

$$(2.35) \qquad E\{N \,|\, R'_{\text{VT}}\} \sim \frac{k}{p} \left[r - \lambda_1(rq)^{1/2}\right] - \frac{k}{2},$$

where λ_1 is defined after (2.33).

For $k = 3$, $\Delta^* = 0.2$ and $P^* = 0.95$ we need an r value equal to 29 for both procedures R'_{PW} and R'_{VT}. Using $r = 29$ the exact value of $E\{N \,|\, R_{\text{PW}}\}$ for $p_1 = p_2 = p_3 = 0.9$ from (2.32) is 59.8 and the approximate value from (2.33) is 59.4. For smaller values of the common p, the approximation is even more accurate.

For any common p, $0 < p < 1$, and any r we find that $E\{L\} = 0$ for both procedures.

In summary, the procedure R'_{PWC} is asymptotically ($r \to \infty$) superior to R'_{VT} throughout the parameter space regardless of whether we use the criterion $E\{N\}$ or the criterion $E\{L\}$.

3. A truncated sequential procedure for k = 2 that uses vector-at-a-time

Another truncated version has been discussed by Kiefer and Weiss, [8], for the VT Rule. For this case a decision is made at or before test N (where each test is assumed to contain both treatments). The termination rule again requires $|S_A - S_B| = s$. Let us first calculate the probability of a correct selection for this procedure. If we let $U_k(n)$ be the probability of a correct selection on or before test pair n (for $p > p'$) given $S_A - S_B + s = k$ then the $U_k(n)$ satisfy

$$(3.1) \qquad U_k(n + 1) = \alpha U_{k+1}(n) + \beta U_k(n) + \gamma U_{k-1}(n),$$

where

$$(3.2) \qquad \alpha = pq', \qquad \beta = pp' + qq', \qquad \gamma = p'q.$$

Equation (3.1) is to be solved subject to the boundary conditions $U_0(n) = 0$, $U_{2s}(n) = 1$, and the initial conditions $U_k(0) = 0$, $k \neq 2s$, $U_{2s}(0) = 1$. Equation (3.1) is more conveniently handled by defining a new set of variables

$$(3.3) \qquad V_k(n) = U_k(\infty) - U_k(n).$$

TABLE IIa

COMPARISON OF EXACT RESULTS FOR VT AND PW SAMPLING RULES
USING THE TERMINATION RULE BASED ON $|S_A - S_B| \geqq r$
IN THE BINOMIAL TWO ARM SELECTION PROBLEM

For $\Delta^* = 0.05$ and $P^* = 0.75$, we use $r_{\text{PW}} = 17$ and $r_{\text{VT}} = 6$, without randomization.

Note. Under the generalized least favorable (GLF) configuration, $p = p_A$ is the larger of the two probabilities of success and $p - \Delta^* = p_B$ is the smaller of the two. The GLF comparisons are made only for pairs $(p, p - \Delta^*)$ with $\Delta^* \leqq p \leqq 1$. Under the equal parameter (EM) configuration, p denotes the *common* probability of success on a single trial. $E\{N_B\}$ denotes the expected number of observations on the "poorer drug" and, for the EM configuration, $L = 0$ and $E\{N\} = 2E\{N_B\}$.

$p = p_A$	GLF configuration				EM configuration	
	$E\{L\|R_{\text{PW}}\}$	$E\{L\|R_{\text{VT}}\}$	$E\{N\|R_{\text{PW}}\}$	$E\{N\|R_{\text{VT}}\}$	$E\{N_B\|R_{\text{PW}}\}$	$E\{N_B\|R_{\text{VT}}\}$
0	—	—	—	—	∞	∞
0.05	16.2	6.0	663	240	2754	379
0.10	15.3	5.9	630	235	1309	200
0.20	13.5	4.7	556	187	587	113
0.30	11.0	3.8	457	153	346	86
0.40	8.5	3.4	354	136	225	75
0.50	6.3	3.2	266	129	153	72
0.60	4.6	3.3	194	131	105	75
0.70	3.2	3.6	137	143	70	86
0.75	2.6	3.8	112	153	57	96
0.80	2.0	4.2	90	167	45	113
0.85	1.5	4.7	71	187	34	141
0.90	1.1	5.3	53	212	25	200
0.95	0.7	5.9	36	235	16	378
1.00	0.3	6.0	21	240	9	∞

TABLE IIb

COMPARISON OF EXACT RESULTS FOR VT AND PW SAMPLING RULES
USING THE TERMINATION RULE BASED ON $|S_A - S_B| \geqq r$
IN THE BINOMIAL TWO ARM SELECTION PROBLEM

For $\Delta^* = 0.20$ and $P^* = 0.75$, we use $r_{\text{PW}} = 4$ and $r_{\text{VT}} = 2$, without randomization.

$p = p_A$	GLF configuration				EM configuration	
	$E\{L\|R_{\text{PW}}\}$	$E\{L\|R_{\text{VT}}\}$	$E\{N\|R_{\text{PW}}\}$	$E\{N\|R_{\text{VT}}\}$	$E\{N_B\|R_{\text{PW}}\}$	$E\{N_B\|R_{\text{VT}}\}$
0	—	—	—	—	∞	∞
0.20	3.3	2.0	37	20	34	13
0.30	2.9	1.7	32	17	21	10
0.40	2.3	1.5	26	15	14	8
0.50	1.8	1.4	21	14	10	8
0.60	1.4	1.3	16	13	7	8
0.70	1.0	1.4	12	14	5	10
0.75	0.9	1.4	11	14	5	11
0.80	0.7	1.5	9	15	4	13
0.85	0.6	1.6	8	16	3	16
0.90	0.5	1.7	7	17	3	22
0.95	0.4	1.9	6	19	2	42
1.00	0.3	2.0	5	20	2	∞

TABLE IIc

COMPARISON OF EXACT RESULTS FOR VT AND PW SAMPLING RULES
USING THE TERMINATION RULE BASED ON $|S_A - S_B| \geq r$
IN THE BINOMIAL TWO ARM SELECTION PROBLEM

For $\Delta^* = 0.05$ and $P^* = 0.95$, we use $r_{PW} = 50$ and $r_{VT} = 15$, without randomization.

$p = p_A$	GLF configuration				EM configuration	
	$E\{L\|R_{PW}\}$	$E\{L\|R_{VT}\}$	$E\{N\|R_{PW}\}$	$E\{N\|R_{VT}\}$	$E\{N_B\|R_{PW}\}$	$E\{N_B\|R_{VT}\}$
0	—	—	—	—	∞	∞
0.05	47.5	15.0	1951	600	23,775	2368
0.10	45.1	15.0	1852	600	11,275	1250
0.20	40.1	14.8	1653	594	5,025	703
0.30	35.1	14.3	1455	573	2,941	536
0.40	30.1	13.8	1254	553	1,900	469
0.50	25.0	13.6	1049	544	1,275	450
0.60	19.8	13.7	841	547	858	469
0.70	14.7	14.1	634	562	561	536
0.75	12.1	14.3	533	573	442	600
0.80	9.7	14.6	433	584	338	703
0.85	7.3	14.8	337	594	246	882
0.90	5.0	15.0	243	599	164	1250
0.95	2.7	15.0	152	600	91	2368
1.00	0.5	15.0	64	600	25	∞

TABLE IId

COMPARISON OF EXACT RESULTS FOR VT AND PW SAMPLING RULES
USING THE TERMINATION RULE BASED ON $|S_A - S_B| \geq r$
IN THE BINOMIAL TWO ARM SELECTION PROBLEM

For $\Delta^* = 0.20$ and $P^* = 0.95$, we use $r_{PW} = 11$ and $r_{VT} = 4$, without randomization.

$p = p_A$	GLF configuration				EM configuration	
	$E\{L\|R_{PW}\}$	$E\{L\|R_{VT}\}$	$E\{N\|R_{PW}\}$	$E\{N\|R_{VT}\}$	$E\{N_B\|R_{PW}\}$	$E\{N_B\|R_{VT}\}$
0	—	—	—	—	∞	∞
0.20	8.9	4.0	100	40	248	50
0.30	7.8	4.0	89	40	147	38
0.40	6.8	3.8	78	38	96	33
0.50	5.7	3.7	68	37	66	32
0.60	4.6	3.7	56	37	46	33
0.70	3.5	3.7	45	37	31	38
0.75	3.0	3.8	40	38	26	43
0.80	2.4	3.8	34	38	21	50
0.85	1.9	3.9	29	39	16	63
0.90	1.4	4.0	24	40	12	89
0.95	0.9	4.0	19	40	9	168
1.00	0.5	4.0	14	40	6	∞

which satisfy equation (3.1), but have the boundary and initial conditions

$$(3.4) \qquad V_0(n) = V_{2s}(n) = 0, \qquad V_k(0) = U_k(\infty), \qquad k \neq 2s, \qquad V_{2s}(0) = 0.$$

If we define

$$(3.5) \qquad \mathbf{V}(n) = \begin{pmatrix} V_1(n) \\ V_2(n) \\ \vdots \\ V_{2s-1}(n) \end{pmatrix}, \qquad \mathbf{A} = \begin{bmatrix} \beta & \alpha & 0 & 0 & \cdots & 0 \\ \gamma & \beta & \alpha & 0 & \cdots & 0 \\ 0 & \gamma & \beta & \alpha & \cdots & 0 \\ & & \vdots & & & \\ 0 & 0 & 0 & \cdots & \gamma & \beta \end{bmatrix},$$

then $\mathbf{V}(n + 1) = \mathbf{A}\mathbf{V}(n)$ or

$$(3.6) \qquad \mathbf{V}(n) = \mathbf{A}^n \mathbf{V}(0).$$

For tridiagonal matrices of the form of $\mathbf{A}$ one can calculate $\mathbf{A}^n$, and therefore $\mathbf{V}(n)$, by using a spectral decomposition. First we note that $\mathbf{A}$ is similar to a symmetric matrix $\mathbf{B}$ through the transformation $\mathbf{B} = \mathbf{T}^{-1}\mathbf{A}\mathbf{T}$, where $\mathbf{T}$ is a diagonal matrix with elements $T_{j,j} = (\gamma^{j-1}\alpha^{2s-1-j})^{1/2}$. The transformed matrix $\mathbf{B}$ is

$$(3.7) \qquad \mathbf{B} = \begin{bmatrix} \beta & \zeta & 0 & 0 & 0 & \cdots & 0 \\ \zeta & \beta & \zeta & 0 & 0 & \cdots & 0 \\ 0 & \zeta & \beta & \zeta & 0 & \cdots & 0 \\ & & & \vdots & & & \\ 0 & 0 & 0 & 0 & \cdots & \zeta & \beta \end{bmatrix},$$

where $\zeta = (\alpha\gamma)^{1/2} = (pp'qq')^{1/2}$. The spectral properties of $\mathbf{B}$ were studied by Rutherford [11]. Denoting the jth eigenvector of $\mathbf{B}$ by $\mathbf{u}_j$ and the corresponding eigenvalue by λ_j we have

$$(3.8) \qquad \begin{aligned} (\mathbf{u}_j)_r &= \frac{1}{\sqrt{s}} \sin\left(\frac{\pi j r}{2s}\right), \qquad r = 1, 2, \cdots, 2s - 1 \\ \lambda_j &= \beta + 2\zeta \cos\left(\frac{\pi j}{2s}\right). \end{aligned}$$

If $\mathbf{V}(n)$ and $\mathbf{V}(0)$ are expanded in terms of the eigenvectors $\mathbf{u}_j$ then a straightforward argument suffices to show that the elements of interest (the probability of a correct selection) are given by

$$(3.9) \qquad V_s(n) = \frac{1}{s}\left(\frac{\alpha}{\gamma}\right)^{(s+1)/2} \sum_{r=1}^{2s-1} \frac{(-1)^{r+1} \lambda_r^n \sin\left(\frac{\pi r}{2}\right) \sin\left(\frac{\pi r}{2s}\right)}{1 + \frac{\alpha}{\gamma} - 2\left(\frac{\alpha}{\gamma}\right)^{1/2} \cos\left(\frac{\pi r}{2s}\right)}.$$

A similar set of calculations can be used to show that the probability that no decision will be made (that is, $|S_A - S_B| < s$ in all n trials) is

$$(3.10) \qquad W(n) = \frac{1}{s}\left(\frac{\gamma}{\alpha}\right)^{(s-1)/2} \sum_{r=1}^{2s-1} \frac{\left[1 + (-1)^r\left(\frac{\alpha}{\gamma}\right)^s\right] \lambda_r^n \sin\left(\frac{\pi r}{2}\right) \sin\left(\frac{\pi r}{2s}\right)}{1 + \frac{\alpha}{\gamma} - 2\left(\frac{\alpha}{\gamma}\right)^{1/2} \cos\left(\frac{\pi r}{2s}\right)}.$$

TABLE III

VALUES OF s AND P_{max} FOR FIXED Δ^* AND N

N	$\Delta^* = 0.1$		$\Delta^* = 0.2$		$\Delta^* = 0.3$		$\Delta^* = 0.4$	
	s	P_{max}	s	P_{max}	s	P_{max}	s	P_{max}
25	2	0.678	3	0.837	3	0.949	3	0.989
50	3	0.751	4	0.934	5	0.991	5	0.9996
75	4	0.804	5	0.970	6	0.999		
100	5	0.844	6	0.986	8	0.9997		
150	7	0.895	6	0.992				
200	8	0.931						
300	10	0.968						
400	12	0.985						

TABLE IV

VALUES OF s AND N CORRESPONDING TO FIXED P^* AND Δ^*

Δ^*	$P^* = 0.75$		$P^* = 0.90$		$P^* = 0.95$		$P^* = 0.99$	
	s	N	s	N	s	N	s	N
0.1	3	50	7	155	9	241	14	453
0.2	2	13	3	39	4	61	7	112
0.3	1	5	2	16	3	26	5	50
0.4	1	3	2	10	2	14	3	26

We have not succeeded in deriving an exact expression for the expected number of trials to reach a decision or the expected number of patients on the poorer treatment, but good approximations are available that have been verified by Monte Carlo calculations.

Having obtained the general results of the last two paragraphs, we can now discuss the uses to which they can be put in designing a clinical trial. Two methods of using the exact information were discussed by Kiefer and Weiss [8]. The first assumed that Δ^* and N were fixed and that the probability of a correct selection was to be maximized, and the second that P^* and Δ^* were given and that the smallest N consistent with these requirements was to be chosen. Both of these problems were studied for the least favorable configuration which was shown to be $p = \frac{1}{2}(1 + \Delta^*)$ and $p' = \frac{1}{2}(1 - \Delta^*)$ as in the untruncated case. Table III gives values of s and P_{max}, where P_{max} is the maximum (over s) of the probability of a correct selection for fixed Δ^* and N. Table IV contains values of s and N for fixed P^* and Δ^*. The expected trial lengths have been calculated and are not appreciably shorter than those for unrestricted testing. Hence the most significant feature of the truncated tests is that an absolute upper bound can be placed on the number of tests, with the trial design retaining the same discriminatory ability as in the unrestricted design. A similar analysis can be made for PW sampling but no detailed calculations have been made so far to compare PW and VT sampling rules in their truncated versions.

4. Comparisons of PW and VT sampling rules using sequential likelihood rules

Several sequential procedures have been investigated but here we shall only report on one or two that do not include early elimination of noncontenders. The VT rule for k binomial populations based on likelihoods is the procedure described on pages 9, 270, and 324 of [3]; we refer to it as R_{BKS}. A remarkable feature of this procedure is that both $E\{N\}$ and $E\{L\}$ remain essentially constant for the GLF configurations with $p_{[1]} = p_{[2]} = \cdots = p_{[k-1]}$, $p_{[k]} = p_{[k-1]} + \Delta^*$ and Δ^* fixed, so that $p_{[k]}$ is the only variable. For example, for $k = 3$, $\Delta^* = 0.2$ and $P^* = 0.95$ a Monte Carlo study based on 1000 trials gave the results 78 ± 3.5 for $E\{N\}$ and 10.4 ± 0.5 for $E\{L\}$ for the GLF configurations $p = 0.2(0.05)1.0$ and $p_{[1]} = p_{[2]} = p - \Delta^*$.

Another procedure based on likelihoods and without elimination is R_{LPWC} (or likelihood exact) which considers the most likely of the three possible assignments of the observed data to the populations with ordered S values and then stops when the infimum over all configurations with $p_{[k]} - p_{[k-1]} \geqq \Delta^*$, of this maximum likelihood is at least P^*. To write down this stopping rule for $k = 3$ let $S_1 \leqq S_2 \leqq S_3$ denote the ordered S value (that is, the current numbers of successes) and let F_i denote the current number of failures associated with S_i, $i = 1, 2, 3$. After some simplification, the explicit rule is to stop when

$$(4.1) \quad \sup_{\Delta^* \leqq p \leqq 1} \left\{ \left(1 - \frac{\Delta^*}{p}\right)^{S_3 - S_2} \left(\frac{q}{q + \Delta^*}\right)^{F_2 - F_3} \right.$$
$$\left. + \left(1 - \frac{\Delta^*}{p}\right)^{S_3 - S_1} \left(\frac{q}{q + \Delta^*}\right)^{F_1 - F_3} \right\} \leqq \frac{1 - P^*}{P^*}.$$

A modification of R_{LPWC} is R_{LCPWC} (or likelihood conservative) which replaces both factors $q/(q + \Delta^*)$ in (4.1) by their upper bound 1. For $P^* > \frac{1}{2}$ this procedure stops if $F_3 \leqq \min (F_1, F_2)$ and

$$(4.2) \quad (1 - \Delta^*)^{S_3 - S_2} + (1 - \Delta^*)^{S_3 - S_1} \leqq \frac{1 - P^*}{P^*}.$$

In both procedures we do not stop if $S_3 > \max (S_1, S_2)$ and $F_3 > \min (F_1, F_2)$. If $S_3 = S_2$ (or $S_3 = S_2 = S_1$) then we assign the subscript 3 to the population with fewer (fewest) failures. If $S_3 = S_2$ and $F_3 = F_2$ then the left side of (4.1) is larger than 1 and, for $P^* > \frac{1}{2}$, the inequality (4.1) will not hold. Under the PWC sampling rule, the differences $F_i - F_3$, $i = 1, 2$, can only take on the values 0 or 1.

Monte Carlo results for $k = 3$ show that R_{LPWC} and R_{LCPWC} procedures give similar results with R_{LPWC} showing a reduction of about 20 per cent in $E\{N\}$ over R_{LCPWC}. In comparing R_{LPWC} with the "constant" R_{BKS} we find, as for $k = 2$ in (2.7) and (2.8) above, that there is the same crossover pattern in which the PWC sampling rule is better for $p_{[3]} > 0.75$ and the VT sampling rule is better for $p_{[3]} < 0.75$ (approximately). These results have been confirmed by Hoel. Table V shows the Monte Carlo results obtained for $k = 3$, $\Delta^* = 0.2$ and $P^* = 0.95$.

TABLE V

SELECTING THE BEST ONE OF $k = 3$ BINOMIAL POPULATIONS (THREE ARMED BANDIT)

EXPECTED TOTAL NUMBER OF OBSERVATIONS $E\{N\}$ UNDER VARIOUS PROCEDURES FOR $k = 3$, $\Delta^* = 0.2$, $P^* = 0.95$ AND GLF CONFIGURATIONS $p_{[1]} = p_{[2]} = p - \Delta^*$

$p = \max p_i$ $i = 1, 2, 3$	Inverse Sampling Stopping Rules (asympt. ($\Delta^* \to 0$) normal approx.)		Sequential Stopping Rules Based on Likelihood (Monte Carlo results based on 1000 observations per entry)		
	R'_{VT}	R'_{PW}	R_{BKS}	Likelihood conservative R_{LCPWC}	Likelihood exact R_{LPWC}
0.20		364.0	74.6	223.3	184.8
0.25		288.8	77.1	210.3	171.2
0.30	$= E\{N \mid R'_{\text{VT}}\}$	238.5	77.2	196.6	164.1
0.35		202.4	77.6	188.2	155.5
0.40		175.0	76.9	178.8	145.9
0.45		153.5	77.9	162.5	134.5
0.50	$= \dfrac{3r}{p} > \dfrac{r}{p}\left(1 + \dfrac{2q}{q + \Delta^*}\right)$	136.0	78.0	153.5	126.1
0.55		121.4	76.5	141.8	115.1
0.60		108.9	76.7	126.8	103.0
0.65		97.9	77.4	116.3	94.5
0.70	$= E\{N \mid R'_{\text{PW}}\}$	88.0	78.3	102.9	83.0
0.75		78.8	81.3	90.3	74.7
0.80		70.0	80.2	75.9	61.4
0.85		61.2	81.6	62.4	51.0
0.90		51.9	80.9	49.6	39.3
0.95		41.3	80.4	36.5	30.2
1.00		28.0	80.1	23.7	20.2
	see (2.17) see [15]	based on (2.16)	see [3], pp. 259, 270, 324	see (4.2)	see (4.1)

5. Related problems under investigation

Some other variations of the PW, VT and allied sampling schemes are under investigation. E. Nebenzahl while at the University of Minnesota was studying the PW sampling rule for a fixed sample size procedure $R_{\text{PW}}^{(N)}$ with $k = 2$; that is, he wants to determine a total sample size N_{PW} (from both populations) such that

$$(5.1) \qquad P\{CS \mid R_{\text{PW}}^{(N)}\} = P^* \qquad \text{when} \quad p - p' = \Delta^*,$$

where $\Delta^* > 0$ and $P^* < 1$ are preassigned. He then compares N_{PW} with the N_{VT} required for the corresponding procedure $R_{\text{VT}}^{(N)}$ based on VT sampling. He has found that for any fixed even N

$$(5.2) \qquad P\{CS \mid R_{\text{PW}}^{(N)}\} = P\{CS \mid R_{\text{VT}}^{(N)}\}$$

and hence $N_{\text{PW}} = N_{\text{VT}}$ for any pair (Δ^*, P^*).

In another three decision problem with $k = 2$ populations he allows the three decisions:

D_1: population 1 is better, that is, $p_1 > p_2$;
D_2: population 1 and 2 have approximately equal values of p;
D_3: population 2 is better, that is, $p_2 > p_1$.

Based on a total of N observations and the PW sampling rule, he uses the statistic

$$(5.3) \qquad W = \frac{S_1 - S_2}{N},$$

where S_i is the number of successes from population i, $i = 1, 2$. We make the decision

$$(5.4) \qquad \begin{array}{lll} D_1 & \text{if} & W > d, \\ D_2 & \text{if} & -d \geqq W \geqq d, \\ D_3 & \text{if} & W < -d, \end{array}$$

where $d > 0$ and N are to be determined. Let A and p_A correspond to the better population so that $p_A \geqq p_B$. If $p_1 = p_2$ and the common value is 0 or 1 then the power of the procedure based on PW sampling is clearly very poor. However, for specified P_1^*, P_2^*, Δ^* and γ^*, we can choose a pair (N, d) satisfying simultaneously the conditions

$$(5.5) \qquad \begin{array}{ll} \text{(a)} & P(\text{selecting } A) \geqq P_1^* \quad \text{when} \quad p_A - p_B \geqq \Delta^*, \\ \text{(b)} & P(\text{deciding } D_2) \geqq P_2^* \quad \text{when} \quad p_A = p_B \leqq \gamma^*, \end{array}$$

where $P_i^* < 1$, $i = 1, 2$, $\Delta^* > 0$ and $0 < \gamma^* < 1$; for convenience it is also assumed that $\gamma^* \geqq \frac{1}{2}(1 + \Delta^*)$. The corresponding problem for the VT sampling rule is also considered (the condition $p_A = p_B \leqq \gamma^*$ in (b) is now replaced by $p_A = p_B$). A comparison of the results shows that the number of observations $N_{\text{PW}}(N_{\text{VT}})$ required by the PW sampling rule (VT sampling rule) asymptotically $(\Delta^* \to 0)$ satisfies the inequality

$$(5.6) \qquad N_{\text{PW}} \geqq N_{\text{VT}}.$$

In general for a fixed number of observations the VT sampling rule is preferable, but for a sequential problem neither sampling rule is uniformly preferable.

In another avenue of investigation, Nebenzahl showed that if we consider a class of inverse sampling procedures in which the ith procedure switches only after i successive failures, $i = 1, 2, \cdots$, then the best ranking and selection results are obtained by taking $i = 1$. More explicitly, he finds an r_i for procedure R_i' based on inverse sampling and switching after i successive failures such that (2.1) holds. He then computes $E\{N \mid R_i'\}$ and finds that: (1) for all i the supremum occurs when $p_A = p_B$, and (2) the smallest supremum is obtained by taking $i = 1$.

Y. S. Lin, at the University of Minnesota, is applying the PW sampling scheme to the problem of finding the "fairest" of two or more coins, that is, the one with p closest to 1/2. A trial consists of three tosses of a coin and if we observe 1 or 2 heads it is a success, otherwise it is a failure. The probability of success on a single trial, $3p^2q + 3pq^2 = 3pq$, is a maximum for $p = \frac{1}{2}$. Hence the coin with p closest to $\frac{1}{2}$ should give the most successes. In this way the problem is again brought into the framework of selecting the population with the highest probability of success on a single trial. The PW scheme is then applied to the trials and a comparison of the PW and the VT sampling scheme is made. It will also be compared with a straightforward (ranking and selection) approach to this problem of finding the fairest coin which was solved for $k = 2$ (the case $k > 2$ is still incomplete) by Sobel and Starr [12].

Along different lines D. Feldman with one of the authors has studied a modification of the PW sampling rule which is called the "Follow the Leader" (or FL) sampling rule. Here we again stick to any population as long as it provides successes. When it gives a failure and the current numbers of failures (F_1, F_2) are not equal then we switch as before. If $F_1 = F_2$ and $S_1 = S_2$ then we randomize, that is, perform an independent experiment with equal probability for each.

Exact formulas for the PCS, $E\{L\}$, and $E\{N\}$ were obtained for the procedure R_{FL} based on the FL sampling rule and termination rule $\Delta S = r$. Our first result was that for any fixed r

(5.7) $P\{CS|R_{FL}\} < P\{CS|R_{PW}\}.$

It follows that the r value r_{PW} for R_{PW} is not greater than r_{FL} for R_{FL}. For any P^*, if Δ^* is *not* too small then $r_{FL} = r_{PW}$. When this happens the procedure R_{FL} becomes a serious competitor to R_{PW} and is superior (in the sense of a smaller $E\{L\}$ and $E\{N\}$) for most of the GLF configurations $(p = p' + \Delta^*)$ with p varying from Δ^* to 1. On the other hand if $\Delta^* \to 0$ then R_{PW} is uniformly preferable to R_{FL}.

The corresponding comparisons for the fixed sample size problem and for the inverse sampling problem have not yet been investigated.

6. Open questions

Any enumeration of the practical difficulties associated with clinical trials is a good source of problems for future research. The principal shortcomings of the trial designs just described are the assumptions of dichotomous and instantaneous response to treatment. The first of these can be handled by some of the techniques given by Bechhofer, Kiefer, and Sobel [3] which principally use the VT rule and also more complicated termination rules than those described so far. No analogue of the PW rule has been proposed for the case in which the response is continuous—a very important consideration in the testing of anti-cancer treatments for which a natural measure of effectiveness is lifetime.

The example of testing anticancer treatments also suggests the second difficulty of a response at some random time after administration of the treatment. Zelen [16] has proposed an analogue to the PW rule for dichotomous response with random reporting intervals. His idea is to bias future assignments to the two treatments according to present trial results. Thus, every success with A would generate a future test with A and every failure with A generates a future test with B. Whenever a new patient arrives, the treatment assignment is made by randomly choosing one of the possibilities generated by past trials, or the choice is made with probability $\frac{1}{2}$ if all past results have been accounted for. The analysis of this sampling scheme in the context of our present formulation of terminating rules has not been made nor has it been compared to other alternatives.

Flehinger, Miller, and Louis [6] have discussed a fixed sample analogue of the PW rule for testing differences in mean survival time in two populations, each of which has an underlying negative exponential distribution of lifetimes. The method is a fairly complicated one in which the information at any stage is summarized in a six dimensional vector including the number of patients on each treatment who have died, those who are alive, and the total time lived by patients on the two treatments. Because of mathematical difficulties, results could only be obtained by Monte Carlo methods, and no comparison with other methods was made.

Much remains to be done on the problem of choosing the best of $k \geq 2$ dichotomous populations. We have encountered great difficulties in trying to deal analytically with any termination rule other than the one prescribed by inverse sampling. Inverse sampling is probably inefficient for small Δ^*, as we have demonstrated in the case $k = 2$. It would certainly be of interest to consider the ranking problem for $k > 2$, keeping in mind the objective of doing so with as small a number of patients as possible on the poorer treatments. Another problem that has not been touched is that of multiple patient entries. It would seem of some interest to investigate multistage trials to alleviate the possibly overpessimistic designs generated by using the least favorable configuration.

Many further problems suggest themselves based both on practical difficulties in the expanding use of clinical trials and on the basic theory developed so far. There is clearly room for much further investigation in the area of clinical trials based on ethical and technical considerations.

The authors are indebted to Dr. David G. Hoel presently of the National Institute of Environmental Health for the computation in Tables I and V. We also wish to thank Leo May of the University of Minnesota for his help with Tables IIa through IId.

REFERENCES

[1] F. J. Anscombe, "Sequential medical trials," *J. Amer. Statist. Assoc.*, Vol. 58 (1963), pp. 365–383.

[2] P. Armitage, *Sequential Medical Trials*, Springfield, Thomas, 1960.

[3] R. E. Bechhofer, J. Kiefer, and M. Sobel, *Sequential Identification and Ranking Problems*, Chicago, University of Chicago Press, 1968.

[4] T. Colton, "A model for selecting one of two medical treatments," *J. Amer. Statist. Assoc.*, Vol. 58 (1963), pp. 338–400.

[5] J. Cornfield, M. Halperin, and S. W. Greenhouse, "An adaptive procedure for sequential clinical trials," *J. Amer. Statist. Assoc.*, Vol. 64 (1969), pp. 759–770.

[6] B. J. Flehinger, J. M. Miller, and T. A. Louis, "Clinical trials to test for difference in mean survival time: Data dependent assignment of patients," IBM Research Report RC 2737, 1969, pp. 1–46.

[7] J. R. Isbell, "On a problem of Robbins," *Ann. Math. Statist.*, Vol. 30 (1959), pp. 606–610.

[8] J. Kiefer, and G. H. Weiss, "A truncated test for choosing the better of two binomial populations," *J. Amer. Statist. Assoc.*, to appear.

[9] R. Pyke, and C. V. Smith, "The Robbins-Isbell two-armed bandit problem with finite memory," *Ann. Math. Statist.*, Vol. 36 (1965), pp. 1375–1386.

[10] H. Robbins, "A sequential procedure with a finite memory," *Proc. Nat. Acad. Sci., U.S.A.*, Vol. 42 (1956), pp. 920–923.

[11] D. E. Rutherford, "Some continuant determinants arising in chemistry and physics," *Proc. Roy. Soc. Edinburgh Sect.*, Vol. 63A (1951), pp. 232–241.

[12] M. Sobel, and N. Starr, "Selecting the *t* fairest coins," University of Minnesota, Department of Statistics, Technical Report, No. 99, 1968, pp. 1–8.

[13] M. Sobel, and G. H. Weiss, "Play-the-winner sampling for selecting the better of two binomial populations," *Biometrika*, Vol. 57 (1970), pp. 357–365.

[14] ———, "Play-the-winner rule and inverse sampling in selecting the better of two binomial populations," *J. Amer. Statist. Assoc.*, to appear.

[15] ———, "Play-the-winner rule and inverse sampling for selecting the best of $k \geq 3$ binomial populations," University of Minnesota, Department of Statistics, Technical Report, No. 126, 1969, pp. 1–25.

[16] M. Zelen, "Play the winner rule and the controlled clinical trial," *J. Amer. Statist. Assoc.*, Vol. 64 (1969), pp. 131–146.

EXACT SIGNIFICANCE TESTS FOR CONTINGENCY TABLES EMBEDDED IN A 2^n CLASSIFICATION

MARVIN ZELEN
State University of New York at Buffalo

1. Introduction

This paper considers the analysis of multidimensional contingency tables when the contingency table can be regarded as being embedded in a 2^n factorial classification. The model assumes that the response variable is binary and is observed over n factors each at two levels. This gives rise to 2^{n-1} 2×2 contingency tables. The theoretical development is in the spirit of the Fisher-Irwin treatment of the 2×2 table. The work reported here can be regarded as a generalization and extension of their work.

The new techniques for analyzing contingency tables derived here are based on conditional reference sets. This allows derivation of exact tests of significance for testing interactions arising in a contingency table context. These tests are conditional tests and have the property that they are uniformly most powerful unbiased tests.

Although this paper only discusses binary response random variables embedded in a 2^n classification, the methods are readily extended to multinomial response embedded in an arbitrary cross classification structure. In a later paper, analyses for more general contingency tables will be developed.

The classical method for analyzing the interactions associated with a complex classification is based on chi square goodness of fit tests. More recently Kullback and his associates [5], [6] have used the ideas of information theory to analyze multidimensional contingency tables. These techniques are equivalent to likelihood ratio tests. However, both the chi square and likelihood ratio techniques are based on asymptotic distributions. The methods of analysis which use a logit model or a multiplicative model for the probability of a response also are based on asymptotic theory. It is interesting that recent reviews of the analysis of contingency tables do not refer to any exact tests for testing interactions (see Lewis [10], Goodman [4], and Plackett [11]).

This work was supported by Public Health Service Research Grant No. CA-10810 from the National Cancer Institute.

2. Preliminary results

The analysis of contingency tables will be made using the analogue of linear regression for the logistic model. These results have served as the basis of several papers by Cox [2], [3] dealing with the analysis of quantal response data and are implicit in his work. In this section the results are summarized for completeness. Let $\{Y_i\}$ be a sequence of independent random variables such that $\theta_i = P\{Y_i = 1\}$ and $1 - \theta_i = P\{Y_i = 0\}$. Let $\mathbf{x}_i' = (x_{i,1}, x_{i,2}, \cdots, x_{i,p})$ be a vector of known constants and $\boldsymbol{\beta}' = (\beta_1, \beta_2, \cdots, \beta_p)$ be a vector of unknown parameters. The θ_i will be assumed to have the form

$$(2.1) \qquad\qquad \theta_i = \frac{\exp\{\boldsymbol{\beta}'\mathbf{x}_i\}}{1 + \exp\{\boldsymbol{\beta}'\mathbf{x}_i\}} \qquad \text{for} \quad i = 1, 2, \cdots, n.$$

If λ_i is defined by $\lambda_i = \log\{\theta_i/(1 - \theta_i)\}$, then we have the model

$$(2.2) \qquad\qquad \lambda_i = \boldsymbol{\beta}'\mathbf{x}_i = \sum_{j=1}^{p} \beta_j x_{i,j} \qquad \text{for} \quad i = 1, 2, \cdots, n.$$

The joint frequency function of $\{Y_i\}$ is

$$(2.3) \qquad f(y_1, y_2, \cdots, y_n) = \prod_{i=1}^{n} \theta_i^{y_i}(1 - \theta_i)^{1-y_i} = \frac{\exp \sum_{j=1}^{p} \beta_j t^{(j)}}{\prod_{i=1}^{n} (1 + \exp\{\boldsymbol{\beta}'\mathbf{x}_i\})}$$

where $t^{(j)} = \sum_{i=1}^{n} x_{i,j} y_i$. Hence $\mathbf{t}' = (t^{(1)}, t^{(2)}, \cdots, t^{(p)})$ are jointly sufficient statistics for $\boldsymbol{\beta}' = (\beta_1, \beta_2, \cdots, \beta_p)$. The frequency function of the sufficient statistics is

$$(2.4) \quad f(\mathbf{t}) = P\{T^{(1)} = t^{(1)}, T^{(2)} = t^{(2)}, \cdots, T^{(p)} = t^{(p)}\} = \frac{C(\mathbf{t}) \exp \sum_{j=1}^{p} t^{(j)} \beta_j}{\prod_{i=1}^{n} (1 + \exp\{\mathbf{x}_i'\boldsymbol{\beta}\})}$$

where $C(\mathbf{t}) = C(t^{(1)}, t^{(2)}, \cdots, t^{(p)})$ is the number of ways of permuting $y_1, y_2, \cdots, y_n$ such that $T^{(1)} = t^{(1)}, T^{(2)} = t^{(2)}, \cdots, T^{(p)} = t^{(p)}$. The combinatorial coefficient $C(\mathbf{t})$ may be found as the coefficient of

$$(2.5) \qquad\qquad \zeta_1^{t^{(1)}} \zeta_2^{t^{(2)}} \cdots \zeta_p^{t^{(p)}}$$

in the generating function

$$(2.6) \qquad\qquad \varphi(\zeta) = \prod_{i=1}^{n} (1 + \zeta_1^{x_{i,1}} \zeta_2^{x_{i,2}} \cdots \zeta_p^{x_{i,p}}).$$

If one is interested in making an inference about only a single β_i, say β_p, one can use the distribution of $T^{(p)}$ conditional on $T^{(j)} = t^{(j)}$ for $j = 1, 2, \cdots, p - 1$. This results in

$$(2.7) \qquad f(t^{(p)} \,|\, t^{(1)}, \cdots, t^{(p-1)}) = \frac{C(t^{(1)}, t^{(2)}, \cdots, t^{(p)}) \exp\{t^{(p)}\beta_p\}}{\sum_z C(t^{(1)}, t^{(2)}, \cdots, t^{(p-1)}, z) \exp\{z\beta_p\}},$$

where the summation in the denominator is taken over the range of $t^{(p)}$. Note that (2.4) is of the exponential form. Hence use of the conditional distribution given by (2.7) results in uniformly most powerful unbiased tests for testing the null hypothesis $H_0 : \beta_p = 0$ versus one sided or two sided alternatives (see Lehmann [9]). Under $H_0 : \beta_p = 0$, the conditional distribution of $T^{(p)}$ takes the form

$$(2.8) \qquad f_0\big(t^{(p)}\big|t^{(1)}, \cdots, t^{(p-1)}\big) = \frac{C\big(t^{(1)}, t^{(2)}, \cdots, t^{(p)}\big)}{\sum_z C\big(t^{(1)}, t^{(2)}, \cdots, t^{(p-1)}, z\big)}.$$

3. Testing for interaction in two 2×2 contingency tables

3.1. *Choice of model.* Suppose we have a 2^2 factorial experiment where the observations are "successes" or "failures." The two factors will be denoted by A and B and the factor levels by 0 and 1. The observation of the kth measurement made on the ith level of factor A and the jth level of B will be denoted by $Y_{i,j,k}$, where $i, j = 0, 1; k = 1, 2, \cdots, n_{i,j}$. Define the quantities $\theta_{i,j,k} = P\{Y_{i,j,k} = 1\}$ and $\lambda_{i,j,k} = \log \{\theta_{i,j,k}/(1 - \theta_{i,j,k})\}$. The model which we shall use for the $\lambda_{i,j,k}$ is

$$(3.1) \qquad \lambda_{i,j,k} = \mu + i\alpha + j\beta + ij(\alpha\beta), \qquad i, j = 0, 1.$$

Note that this is not the usual model associated with a 2^2 experiment. If the parameters μ, α, β, and $(\alpha\beta)$ are written in terms of $\lambda_{i,j}$ (the subscript k has been dropped because $\lambda_{i,j,k}$ is constant for all k), we have

$$(3.2) \qquad \begin{aligned} \mu = \lambda_{0,0}, \qquad \alpha = \lambda_{1,0} - \lambda_{0,0}, \qquad \beta = \lambda_{0,1} - \lambda_{0,0}, \\ (\alpha\beta) = (\lambda_{1,1} - \lambda_{0,1}) - (\lambda_{1,0} - \lambda_{0,0}). \end{aligned}$$

An interpretation of these parameters can be made in terms of relative risks or odds ratios. For this purpose define

$$(3.3) \qquad \begin{aligned} \psi_A(j) = \exp \{\lambda_{1,j} - \lambda_{0,j}\} = \frac{\theta_{1,j}/(1 - \theta_{1,j})}{\theta_{0,j}/(1 - \theta_{0,j})}, \qquad j = 0, 1, \\ \psi_B(i) = \exp \{\lambda_{i,1} - \lambda_{i,0}\} = \frac{\theta_{i,1}/(1 - \theta_{i,1})}{\theta_{i,0}/(1 - \theta_{i,0})}, \qquad i = 0, 1, \end{aligned}$$

where $\theta_{i,j} = \theta_{i,j,k}$ for all k. Then we have

$$(3.4) \qquad \begin{aligned} \exp \{\alpha\beta\} = \frac{\psi_A(1)}{\psi_A(0)} = \frac{\psi_B(1)}{\psi_B(0)}, \\ \exp \{\alpha\} = \psi_A(0), \qquad \exp \{\beta\} = \psi_B(0). \end{aligned}$$

The quantity $\exp \{\alpha\beta\}$ is simply Bartlett's definition of a second order interaction for a $2 \times 2 \times 2$ table [1]. The parameter $\psi_A(j)$ is the relative risk (odds ratio) comparing the odds ratio of success for factor A at level 1 versus 0, holding factor B at level j. An analogous interpretation holds for $\psi_B(i)$. Thus $\exp\{\alpha\beta\}$ is the ratio of two relative risks (or odds ratios).

A model for $\lambda_{i,j}$ in terms of the usual main effect and interaction terms associated with a 2^2 experiment would be

$$(3.5) \qquad \lambda_{i,j} = \mu + (2i-1)(A) + (2j-1)(B) + (2i-1)(2j-1)(AB),$$
$$i, j = 0, 1,$$

where (A) and (B) are main effect terms and (AB) is the interaction term. Clearly the relations among the parameters in the two models are

$$(3.6) \qquad \begin{aligned} 2^2(A) &= -\lambda_{0,0} + \lambda_{1,0} - \lambda_{0,1} + \lambda_{1,1} = 2\alpha + (\alpha\beta), \\ 2^2(B) &= -\lambda_{0,0} - \lambda_{1,0} + \lambda_{0,1} + \lambda_{1,1} = 2\beta + (\alpha\beta), \\ 2^2(AB) &= \lambda_{0,0} - \lambda_{1,0} - \lambda_{0,1} + \lambda_{1,1} = (\alpha\beta). \end{aligned}$$

Therefore a test for $H_0\colon (AB) = 0$ is equivalent to $H_0\colon (\alpha\beta) = 0$. However tests on the main effects $H_0\colon (A) = 0$ or $H_0\colon (B) = 0$ *do not* correspond to $H_0\colon \alpha = 0$ or $H_0\colon \beta = 0$. On the other hand, tests on the null hypotheses $H_0\colon \alpha = 0$ and $H_0\colon \beta = 0$ may be regarded as tests on main effects *conditional on* $(\alpha\beta) = 0$. Hence, the parameters α and β will be referred to as *conditional main effects*. The statistical analyses are carried out by investigating the hypothesis $H_0\colon (\alpha\beta) = 0$. If the answer is in the affirmative, then one may carry out tests on the conditional main effects.

3.2. *Exact test for* $H_0\colon (\alpha\beta) = 0$. The data from the 2^2 experiment may be summarized in the two 2×2 tables depicted below. In order to avoid triple subscript notation we use the notation $s_j = \Sigma_k Y_{1,j,k}$, $r_j = \Sigma_k Y_{0,j,k}$, $t_j = r_j + s_j$, $N_j = m_j + n_j$,

	B_0				B_1		
	S	F			S	F	
A_0	r_0	$m_0 - r_0$	m_0		r_1	$m_1 - r_1$	m_1
A_1	s_0	$n_0 - s_0$	n_0		s_1	$n_1 - s_1$	n_1
	t_0	$N_0 - t_0$	N_0		t_1	$N_1 - t_1$	N_1

(3.7)

We shall utilize the results of Section 2 to obtain a uniformly most powerful unbiased test for $H_0\colon (\alpha\beta) = 0$ versus a one sided or two sided alternative. Define $\mathbf{1}_{i,j}$ as a column vector of identity elements of length $n_{i,j}$; also let $\lambda' = (\lambda_{0,0,1}, \cdots, \lambda_{1,1,n_1})$ be the vector of the logits. The model described by equation (3.1) can then be written in matrix notation as

$$(3.8) \qquad \lambda = \begin{bmatrix} \mathbf{1}_{0,0} & \mathbf{0} & \mathbf{0} & \mathbf{0} \\ \mathbf{1}_{1,0} & \mathbf{1}_{1,0} & \mathbf{0} & \mathbf{0} \\ \mathbf{1}_{0,1} & \mathbf{0} & \mathbf{1}_{0,1} & \mathbf{0} \\ \mathbf{1}_{1,1} & \mathbf{1}_{1,1} & \mathbf{1}_{1,1} & \mathbf{1}_{1,1} \end{bmatrix} \begin{bmatrix} \mu \\ \alpha \\ \beta \\ (\alpha\beta) \end{bmatrix}$$

Let $\mathbf{Y}_{i,j}$ correspond to the $n_{i,j} \times 1$ column vector of observations $\{Y_{i,j,k}\}$ made at condition (A_i, B_j). Then the vector of sufficient statistics is

$$
(3.9) \quad
\begin{bmatrix} t^{(1)} \\ t^{(2)} \\ t^{(3)} \\ t^{(4)} \end{bmatrix}
=
\begin{bmatrix}
\mathbf{1}'_{0,0} & \mathbf{1}'_{1,0} & \mathbf{1}'_{0,1} & \mathbf{1}'_{1,1} \\
\mathbf{0} & \mathbf{1}'_{1,0} & \mathbf{0} & \mathbf{1}'_{1,1} \\
\mathbf{0} & \mathbf{0} & \mathbf{1}'_{0,1} & \mathbf{1}'_{1,1} \\
\mathbf{0} & \mathbf{0} & \mathbf{0} & \mathbf{1}'_{1,1}
\end{bmatrix}
\begin{bmatrix} \mathbf{Y}_{0,0} \\ \mathbf{Y}_{1,0} \\ \mathbf{Y}_{0,1} \\ \mathbf{Y}_{1,1} \end{bmatrix}
$$

$$
=
\begin{bmatrix}
\sum_{i,j} \mathbf{1}'_{i,j} \mathbf{Y}_{i,j} \\
\sum_{j=0} \mathbf{1}'_{1,j} \mathbf{Y}_{1,j} \\
\sum_{i=0} \mathbf{1}'_{i,1} \mathbf{Y}_{i,1} \\
\mathbf{1}'_{1,1} \mathbf{Y}_{1,1}
\end{bmatrix}
=
\begin{bmatrix}
t_0 + t_1 = t \\
s_0 + s_1 = s \\
t_1 \\
s_1
\end{bmatrix}
$$

Therefore to test the hypothesis $H_0 : (\alpha\beta) = 0$ we require the distribution of $P\{S_1 = s_1 \,|\, T = t, S = s, T_1 = t_1\}$ which will be denoted by $f(s_1 \,|\, t, s, t_1)$. This distribution is given by equation (3.1); that is,

$$
(3.10) \qquad f(s_1 \,|\, t, s, t_1) = \frac{C(t, s, t_1, s_1) \exp\{s_1(\alpha\beta)\}}{\sum_z C(t, s, t_1, z) \exp\{z(\alpha\beta)\}}.
$$

The coefficient $C(t, s, t_1, s_1)$ can be found from the coefficient of $\zeta_\mu^t \zeta_\alpha^s \zeta_\beta^{t_1} \zeta_{(\alpha\beta)}^{s_1}$ in the generating function

$$
(3.11) \qquad \varphi(\zeta) = (1 + \zeta_\mu)^{m_0} (1 + \zeta_\mu \zeta_\alpha)^{n_0} (1 + \zeta_\mu \zeta_\beta)^{m_1} (1 + \zeta_\mu \zeta_\alpha \zeta_\beta \zeta_{(\alpha\beta)})^{n_1}.
$$

Expanding $\varphi(\zeta)$ results in

$$
(3.12) \qquad \varphi(\zeta) = \sum \cdots \sum_{i,j,k,\ell} \binom{m_0}{i}\binom{n_0}{j}\binom{m_1}{k}\binom{n_1}{\ell} \zeta_\mu^{i+j+k+\ell} \zeta_\alpha^{j+\ell} \zeta_\beta^{k+\ell} \zeta_{(\alpha\beta)}^{\ell},
$$

which after making the transformations

$$
(3.13) \qquad t = i + j + k + \ell, \quad s = j + \ell, \quad t_1 = k + \ell, \quad s_1 = \ell,
$$

results in

$$
(3.14) \qquad C(t, s, t_1, s_1) = \binom{m_0}{t_0 - s_0}\binom{n_0}{s_0}\binom{m_1}{t_1 - s_1}\binom{n_1}{s_1}
$$

$$
= \binom{m_0}{t_0 - s + s_1}\binom{n_0}{s - s_1}\binom{n_0}{t_1 - s_1}\binom{n_1}{s_1}.
$$

Note that the joint conditions $(T = t, S = s, T_1 = t_1)$ are equivalent to $T_0 = t_0$, $T_1 = t_1, S = s$ by virtue of $T = T_1 + T_1$. Hence we shall write $f(s_1 \,|\, t, s, t_1)$ as $f(s_1 \,|\, t_0, t_1, s)$. Furthermore in what follows it will be convenient to define

$$
(3.15) \qquad C(s_j, t_j) = \binom{m_j}{t_j - s_j}\binom{n_j}{s_j}, \qquad\qquad j = 0, 1.
$$

Therefore

(3.16) $$C(t, s, t_1, s_1) = C(s - s_1, t_0)C(s_1, t_1)$$

and we shall write (3.10) as

(3.17) $$f(s_1|t_0, t_1, s) = \frac{C(s - s_1, t_0)C(s_1, t)\exp\{s_1(\alpha\beta)\}}{\sum_z C(s - z, t_0)C(z, t_1)\exp\{z(\alpha\beta)\}}.$$

The conditional distribution of S_1 when $(\alpha\beta) = 0$ is thus

(3.18) $$f_0(s_1|t_0, t_1, s) = \frac{C(s - s_1, t_0)C(s_1, t_1)}{\sum_z C(s - z, t_0)C(z, t_1)}.$$

Consequently the test of significance for $H_0: (\alpha\beta) = 0$ against $H_1: (\alpha\beta) > 0$ employs the tail probability

(3.19) $$P\{S_1 \geqq s_1|T_0 = t_0, T_1 = t_1, S = s\} = \sum_{w \geqq s_1} f_0(w|t_0, t_1, s).$$

The test of significance against the two sided alternative $H_1: (\alpha\beta) \neq 0$ is calculated by defining the set $W = \{w: f_0(w|t_0, t_1, s) \leqq f_0(s_1|t_0, t_1, s)\}$ and evaluating the tail probability $P = \Sigma_{w \in W} f_0(w|t_0, t_1, s)$.

3.3. *Test for main effects and decomposition of probabilities.* Analogous to the decomposition of the sums of squares associated with the general linear hypothesis for continuous type data is the decomposition of the frequency function of the observations. To see this, we note that the joint probability function is

(3.20) $$f(t, s, t_1, s_1|\mu, \alpha, \beta, (\alpha\beta))$$
$$= \frac{C(t, s, t_1, s_1)\exp\{\mu t + \alpha s + \beta t_1 + (\alpha\beta)s_1\}}{\sum_i \sum_j \sum_k \sum_\ell C(i, j, k, \ell)\exp\{\mu i + \alpha j + \beta k + (\alpha\beta)\ell\}}.$$

This probability function can be further decomposed into

(3.21) $$f(t, s, t_1, s_1|\mu, \alpha, \beta, (\alpha\beta))$$
$$= f(s_1|t_0, t_1, s, (\alpha\beta)) f(s, t_1|t, \alpha, \beta, (\alpha\beta)) f(t|\mu, \alpha, \beta, (\alpha\beta)),$$

where $f(s_1|t_0, t_1, s, (\alpha\beta))$ is given by (3.17) and

(3.22) $$f(s, t_1|t, \alpha, \beta, (\alpha\beta)) = \frac{\sum_\ell C(t, s_1, t_1, \ell)\exp\{\alpha s + \beta t_1 + (\alpha\beta)\ell\}}{\sum_j \sum_k \sum_\ell C(t, j, k, \ell)\exp\{\alpha j + \beta k + (\alpha\beta)\ell\}},$$

(3.23) $$f(t|\mu, \alpha, \beta, (\alpha\beta)) = \frac{\sum_j \sum_k \sum_\ell C(t, j, k, \ell)\exp\{\mu t + \alpha j + \beta k + (\alpha\beta)\ell\}}{\sum_i \sum_j \sum_k \sum_\ell C(i, j, k, \ell)\exp\{\mu i + \alpha j + \beta k + (\alpha\beta)\ell\}}.$$

Note that $f(s, t_1 | t, \alpha, \beta, (\alpha\beta))$ is the joint distribution of (s, t_1) conditional on t and $f(t | \mu, \alpha, \beta, (\alpha\beta))$ is the marginal distribution of t. Further we can decompose

$$(3.24) \quad f(s, t_1 | t, \alpha, \beta, (\alpha\beta)) = f(s | t_0, t_1, \alpha, (\alpha\beta)) f(t_1 | t, \alpha, \beta, (\alpha\beta))$$
$$= f(t_1 | s, t, \beta, (\alpha\beta)) f(s | t, \alpha, \beta, (\alpha\beta)),$$

where

$$(3.25) \quad f(s | t_0, t_1, \alpha, (\alpha\beta)) = \frac{\sum\limits_{s_1} C(s - s_1, t_0) C(s_1, t_1) \exp\{\alpha s + (\alpha\beta)s_1\}}{\sum\limits_{s} \sum\limits_{s_1} C(s - s_1, t_0) C(s_1, t_1) \exp\{\alpha s + (\alpha\beta)s_1\}},$$

$$(3.26) \quad f(t_1 | s, t, \beta, (\alpha\beta)) = \frac{\sum\limits_{s_1} C(s - s_1, t - t_1) C(s_1, t_1) \exp\{\beta t_1 + (\alpha\beta)s_1\}}{\sum\limits_{t_1} \sum\limits_{s_1} C(s - s_1, t - t_1) C(s_1, t_1) \exp\{\beta t_1 + (\alpha\beta)s_1\}},$$

$$(3.27) \quad f(t_1 | t, \alpha, \beta, (\alpha\beta))$$
$$= \frac{\sum\limits_{s_1} \sum\limits_{s} C(s - s_1, t - t_1) C(s_1, t_1) \exp\{\alpha s + \beta t_1 + (\alpha\beta)s_1\}}{\sum\limits_{t_1} \sum\limits_{s_1} \sum\limits_{s} C(s - s_1, t - t_1) C(s_1, t_1) \exp\{\alpha s + \beta t_1 + (\alpha\beta)s_1\}},$$

and

$$(3.28) \quad f(s | t, \alpha, \beta, (\alpha\beta))$$
$$= \frac{\sum\limits_{s_1} \sum\limits_{t_1} C(s - s_1, t - t_1) C(s_1, t_1) \exp\{\alpha s + \beta t_1 + (\alpha\beta)s_1\}}{\sum\limits_{s} \sum\limits_{s_1} \sum\limits_{t_1} C(s - s_1, t - t_1) C(s_1, t_1) \exp\{\alpha s + \beta t_1 + (\alpha\beta)s_1\}}.$$

If the hypothesis $H_0: (\alpha\beta) = 0$ is correct then the probability function $f(s | t_0, t_1, \alpha, (\alpha\beta) = 0) = f(s | t_0, t_1, \alpha)$ can be used to make an inference about the null hypothesis $H_0: \alpha = 0$. That is, the appropriate tail probability for the alternate hypothesis $H_1: \alpha > 0$ is

$$(3.29) \quad \sum_{z \geq s} f(z | t_0, t_1, \alpha = 0) = \frac{\sum\limits_{z \geq s} \sum\limits_{s_1} C(z - s_1, t_0) C(s_1, t_1)}{\sum\limits_{z} \sum\limits_{s_1} C(z - s_1, t_0) C(s_1, t_1)}.$$

The tail probability associated with the two sided alternative $H_1: \alpha \neq 0$ is

$$(3.30) \quad P = \sum_{w \in W} f(s | t_0, t_1, \alpha = 0),$$

where $W = \{w: f(w | t_0, t_1, \alpha = 0) \leq f(s | t_0, t_1, \alpha = 0)\}$.

3.4. *Normal approximation to tests of significance.* In this section, normal approximations will be obtained for tests of significance associated with the test on the interaction and the conditional main effects. The probability of $S_j = s_j$ conditional on $t_j, j = 0, 1$, associated with a single 2×2 table is

$$(3.31) \quad p(s_j | t_j) = \frac{C(s_j, t_j) \psi_A(j)^{s_j}}{\sum\limits_{z_j} C(z_j, t_j) \psi_A(j)^{z_j}}, \qquad j = 0, 1,$$

where $\psi_A(j)$ is the relative risk for the 2×2 table with factor B held at level j. Under the null hypothesis $H_0: \psi_A(j) = 1$, the distribution of S_j follows the hypergeometric distribution; that is,

$$(3.32) \qquad p_0(s_j | t_j) = \frac{C(s_j, t_j)}{\binom{N_j}{s_j}}$$

having mean and variance

$$(3.33) \qquad \begin{aligned} \mu_j &= E\{S_j | T_j = t_j\} = \frac{t_j n_j}{N_j}, \\ \sigma_j^2 &= \mathrm{Var}\,\{S_j | T_j = t_j\} = \frac{t_j m_j n_j (N_j - t_j)}{N_j^2 (N_j - 1)}. \end{aligned}$$

Note that if $(\alpha\beta) = 0$, the distribution of S_0 conditional on $\{T_j = t_j, j = 0, 1; S = s\}$ given by (3.17) is simply

$$(3.34) \qquad f(s_1 | t_0, t_1, s) = \frac{p_0(s - s_1 | t_0) p_0(s_1 | t_1)}{\sum_z p_0(s - z | t_0) p_0(z | t_1)}.$$

For notational simplicity we shall define the random variables W_0 and W_1 by

$$P\{W_0 = s_0\} = P\{S_0 = s_0 | T_0 = t_0\}, \quad P\{W_1 = s_1\} = P\{S_1 = s_1 | T_1 = t_1\}.$$

When N_j is relatively large, the distribution of W_j tends to an independent normal distribution with mean u_j and variance σ_j^2. Therefore, when the normal approximation to the hypergeometric distribution holds we have

$$(3.35) \quad P\{S_1 = s_1 | T_j = t_j, j = 0, 1; S_0 + S_1 = s\} = P\{W_1 = s_1 | W_0 + W_1 = s\}$$

$$\cong \frac{\varphi_0(s - w_1)\varphi_1(w_1)}{\varphi(s)},$$

where $\varphi_j(x)$ is the p.d.f. of the normal distribution with parameters (μ_j, σ_j^2), and $\varphi(s)$ is the normal p.d.f. with mean $(\mu_0 + \mu_1)$ and variance $(\sigma_0^2 + \sigma_1^2)$. Substituting the appropriate p.d.f., an easy calculation shows that the approximate distribution of W_1 conditional on $W_0 + W_1 = s$ is normal with mean and variance given by

$$(3.36) \qquad \begin{aligned} E\{W_1 | W_0 + W_1 = s\} &= \mu_{01} = \mu_1 + \frac{\rho_{01}\sigma_1}{\sigma_0}(s - \mu_0 - \mu_1), \\ \mathrm{Var}\,\{W_1 | W_0 + W_1 = s\} &= \sigma_{01}^2 = \rho_{01}^2(\sigma_0^2 + \sigma_1^2), \end{aligned}$$

where

$$(3.37) \qquad \rho_{01} = \frac{\sigma_0 \sigma_1}{(\sigma_0^2 + \sigma_1^2)}.$$

Hence since the conditional distribution of W_1 is approximately $N(\mu_{01}, \sigma_{01}^2)$, a test of significance may be conducted by taking $(W_1 - \mu_{01})/\sigma_{01}$ to be a $N(0, 1)$ random variable.

If one accepts the inference that the interaction is nonexistent, then one could test the null hypotheses $H_0: \alpha = 0$ and $H_0: \beta = 0$ which refer to the conditional main effects. The appropriate conditional distributions for these hypotheses are given by (3.25) to (3.28). Under H_0, these distributions do not depend on any parameters. We shall illustrate the normal approximation for $H_0: \alpha = 0$.

Since the test of the hypothesis $H_0: \alpha = 0$ depends on the distribution of $S = S_0 + S_1$ conditional on $T_j = t_j, j = 0, 1$, we have

(3.38)
$$E\{S | T_0 = t_0, T_1 = t_1\} = \mu_0 + \mu_1,$$
$$\text{Var}\,\{S | T_0 = t_0, T_1 = t_1\} = \sigma_0^2 + \sigma_1^2.$$

Consequently the large sample approximation to S, if $\alpha = 0$, is to take S to be normal with mean $\mu_0 + \mu_1$ and variance $\sigma_0^2 + \sigma_1^2$.

3.5. *Binomial approximation to nonnull distribution.* The expression for the frequency function $f(s_1 | t_0, t_1, s)$, equation (3.17), associated with the test for interaction may be approximated by using the binomial approximation to the hypergeometric distribution. The hypergeometric distribution can often be approximated quite accurately by the corresponding terms of the binomial distribution which has the same mean, and as closely as possible, the same variance [5]. This approximation is

(3.39)
$$\frac{C(s_j, t_j)}{\binom{N_j}{t_j}} = \frac{\binom{m_j}{t_j - s_j}\binom{n_j}{s_j}}{\binom{N_j}{t_j}} \sim \binom{n_j^*}{s_j} p_j^{* s_j} q_j^{* n_j^* - s_j},$$

where n_j^* is the nearest integer to $t_j n_j / [N_j - m_j(N_j - t_j)/(N_j - 1)]$ and $p_j^* = t_j n_j / N_j n_j$.

Using this approximation in (3.17) yields

(3.40)
$$f(s_1 | t_0, t_1, s) \sim \frac{\binom{n_0^*}{s - s_1}\binom{n_1^*}{s_1}\{\psi^* e^{(\alpha\beta)}\}^{s_1}}{\sum\limits_{z=z_1}^{z_2} \binom{n_0^*}{s - z}\binom{n_1^*}{z}\{\psi^* e^{(\alpha\beta)}\}^z},$$

where

(3.41)
$$\psi^* = \frac{p_1^*/q_1^*}{p_0^*/q_0^*}, \qquad z_1 = \max\,(0, s - n_0^*), \qquad z_2 = \min\,(s, n_1^*).$$

Note that (3.40) is exactly the same expression as the frequency function associated with the nonnull conditional distribution of a single 2×2 contingency table. Again, employing the binomial approximation to the hypergeometric distribution; that is,

$$(3.42) \qquad \frac{\binom{n_0^*}{s - s_1}\binom{n_1^*}{s_1}}{\binom{N^*}{s}} \sim \binom{\eta}{s_1} \pi^{s_1}(1 - \pi)^{\eta - s_1},$$

where $N^* = n_0^* + n_1^*$, $\eta = $ nearest integer to $sn_1^*/[N^* - n_0^*(N^* - s)/(N^* - 1)]$, and $\pi = sn_1^*/N^*\eta$, results in

$$(3.43) \qquad f(s_1|t_0, t_1, s) \sim \frac{\binom{\eta}{s_1}\{\pi\psi^* e^{(\alpha\beta)}\}^{s_1}\{1 - \pi\}^{\eta - s_1}}{[\pi\psi^* e^{(\alpha\beta)} + 1 - \pi]^\eta} = \binom{\eta}{s_1} P^{s_1} Q^{\eta - s_1},$$

where

$$(3.44) \qquad P = \frac{\pi\psi^* e^{(\alpha\beta)}}{[\pi\psi^* e^{(\alpha\beta)} + 1 - \pi]}, \qquad Q = 1 - P.$$

Consequently, an approximation to $f(s_1|t_0, t_1, s)$ is to take the random variable S_1 to be approximately distributed as a binomial random variable with sample size η and success probability P. Hence a confidence interval on P can, by suitable transformation, be made into a confidence interval on $e^{(\alpha\beta)}$. That is, if (P_1, P_2) are $100(1 - 2\alpha)$ per cent confidence limits on P, then

$$(3.45) \qquad \frac{(1 - \pi)/(1 - P_i)}{\psi^*(\pi/P_i)}, \qquad\qquad i = 1, 2,$$

are approximate confidence limits for $e^{(\alpha\beta)}$. Furthermore, another approximation of the significance test for the null hypothesis $H_0: (\alpha\beta) = 0$ versus $H_1: (\alpha\beta) > 0$ is to compute the tail area probabilities

$$(3.46) \qquad P\{S_1 \geqq s_1|(\alpha\beta) = 0\} = \sum_{k=s_1}^{\eta} \binom{\eta}{k} P_0^k Q_0^{\eta - k},$$

where $P_0 = \pi\psi^*/[\pi\psi^* + 1 - \pi]$.

4. The general case of n factors

4.1. *Preliminaries and notation.* In this section we extend the treatment of the analysis of 2×2 contingency tables for the situation where the number of tables is a power of two. It is convenient to use the notation associated with factorial experiments. Let $A_1, A_2, \cdots, A_n$ represent n factors each at two levels. There will then be 2^n factorial combinations. If we fix on one factor, say A_1, this situation may be regarded as having 2^{n-1} 2×2 contingency tables involving the two levels of A_1 where each contingency table represents a fixed combination of the remaining $(n - 1)$ factors.

The general development is eased if one adopts an operational calculus suited to factorial experiments (see Kurkjian and Zelen [8]). Let a factorial combination be denoted by the n tuple binary number $x = (x_1, x_2, \cdots, x_n)$ where $x_i = 0$ or 1, $i = 1, 2, \cdots, n$, depending on whether the ith factor is at the "low" level or "high" level, respectively. Throughout this section, we will have need for ordering the 2^n n digit binary numbers in a standard order. For this purpose we use the operation of the symbolic direct product which is designated by $\otimes$ (see [8]). Define the vector δ by $\delta' = (0, 1)$. Then the standard order for $n = 2$ (four treatment combinations) is given by the rows of $\delta \otimes \delta$ which is

$$
(4.1) \qquad \delta \otimes \delta = \begin{pmatrix} 0 \\ 1 \end{pmatrix} \otimes \begin{pmatrix} 0 \\ 1 \end{pmatrix} = \begin{pmatrix} 0 & 0 \\ 0 & 1 \\ 1 & 0 \\ 1 & 1 \end{pmatrix}
$$

The standard order for $n = 3$ is given by the rows of

$$
(4.2) \qquad \delta \otimes \delta \otimes \delta = \begin{pmatrix} 0 \\ 1 \end{pmatrix} \otimes \begin{pmatrix} 0 \\ 1 \end{pmatrix} \otimes \begin{pmatrix} 0 \\ 1 \end{pmatrix} = \begin{bmatrix} 0 & 0 & 0 \\ 0 & 0 & 1 \\ 0 & 1 & 0 \\ 0 & 1 & 1 \\ 1 & 0 & 0 \\ 1 & 0 & 1 \\ 1 & 1 & 0 \\ 1 & 1 & 1 \end{bmatrix}.
$$

The generalization to arbitrary n is clear.

Let $Y(x)$ denote a binary random variable representing the outcome of the xth treatment combination. Also define

$$
(4.3) \qquad \begin{aligned} \theta(x) &= P\{Y(x) = 1\}, \qquad 1 - \theta(x) = P\{Y(x) = 0\}, \\ \lambda(x) &= \log\{\theta(x)/(1 - \theta(x))\}. \end{aligned}
$$

The generalized interaction among the factors $A_{i_1}, A_{i_2}, \cdots, A_{i_p}$ will be denoted by $a_{i_1 i_2 \cdots i_p}$. Another way of designating this generalized interaction is to define

$$
z_i = \begin{cases} 1 & \text{if factor } A_i \text{ is included in the generalized interaction,} \\ 0 & \text{otherwise,} \end{cases}
$$

$$
(4.4) \qquad z = (z_1, z_2, \cdots, z_n),
$$

and let $a(z) = a_{i_1 i_2 \cdots i_p}$. That is, the generalized interaction can also be designated by an n digit binary number. The binary number $(0, 0, \cdots, 0)$ will refer to a constant term; that is, $a(0, 0, \cdots, 0) = \mu$.

For the purpose of writing $\lambda(x)$ as a function of the generalized interactions, define

$$
(4.5) \qquad \delta^{z_i} = \begin{cases} (0, 1)' & \text{if } z_i = 1, \\ (1, 1)' & \text{if } z_i = 0, \end{cases}
$$

and

$$(4.6) \qquad \delta^z = (\delta_1^{z_1} \times \delta_2^{z_2} \times \cdots \times \delta_n^{z_n}),$$

where $\times$ denotes the Kronecker product. Then if $\boldsymbol{\lambda}$ is the vector of $\{\lambda(x)\}$ arranged in the standard order we have

$$(4.7) \qquad \boldsymbol{\lambda} = \sum_z \delta^z a(z),$$

where the summation is over all n digit binary numbers. The $\lambda(x)$ corresponding to a particular treatment combination can then be written as

$$
\begin{aligned}
(4.8) \qquad \lambda(x) &= \left[\binom{1 - x_1}{x_1} \times \binom{1 - x_2}{x_2} \times \cdots \times \binom{1 - x_n}{x_n} \right]' \boldsymbol{\lambda} \\
&= \sum_z \left[\binom{1 - x_1}{x_1}' \delta^{z_1} \times \binom{1 - x_2}{x_2}' \delta^{z_2} \times \cdots \times \binom{1 - x_n}{x_n}' \delta^{z_n} \right] a(z),
\end{aligned}
$$

where $\times$ denotes Kronecker product multiplication. It is easy to verify that

$$(4.9) \qquad \binom{1 - x_i}{x_i}' \delta^{z_i} = 1 - z_i(1 - x_i), \qquad i = 1, 2, \cdots, n.$$

Hence we have

$$(4.10) \qquad \lambda(x) = \sum_z \prod_{i=1}^n [1 - z_i(1 - x_i)] a(z) = \sum_z \varphi(x, z) a(z),$$

where

$$\varphi(x, z) = \prod_{j=1}^n [1 - z_i(1 - x_i)].$$

4.2 *The logistic model.*   At condition x let

$$(4.11) \qquad \theta(x) = P\{Y(x) = 1\} = \frac{\exp \{\lambda(x)\}}{1 + \exp \{\lambda(x)\}}.$$

Then if $Y_1(x), Y_2(x), \cdots, Y_{m(x)}(x)$ denote $m(x)$ independent binary random variables made at treatment x, we have

$$(4.12)$$
$$
\begin{aligned}
P\{\mathbf{Y}(x) = \mathbf{y}(x)\} &= P\{Y_1(x) = y_1(x), Y_2(x) = y_2(x), \cdots, Y_{m(x)}(x) = y_{m(x)}(x)\} \\
&= \frac{\exp \{s(x)\lambda(x)\}}{[1 + \exp \{\lambda(x)\}]^{m(x)}},
\end{aligned}
$$

where $s(x) = \sum_{j=1}^{m(x)} y_j(x)$ is the total number of one's at condition x. Hence if $\mathbf{Y}$ denotes the vector of $\{\mathbf{Y}(x)\}$ over all treatment combinations and $\mathbf{y}$ is the vector of outcomes we have

$$(4.13) \qquad P\{\mathbf{Y} = \mathbf{y}\} = \prod_x \frac{\exp \{s(x)\lambda(x)\}}{[1 + \exp \{s(x)\lambda(x)\}]^{m(x)}} = \frac{\exp \sum_x s(x)\lambda(x)}{\prod_x [1 + \exp \{\lambda(x)\}]^{m(x)}},$$

where all the products and sums are taken over the n digit binary numbers x.

Since $\sum_x s(x)\lambda(x) = \sum_x s(x) \sum_z \varphi(x, z)a(z) = \sum_z t(z)a(z)$ where

$$(4.14) \qquad t(z) = \sum_x \varphi(x, z)s(x),$$

we have

$$(4.15) \qquad P\{\mathbf{Y} = \mathbf{y}\} = \frac{\exp \sum_z t(z)a(z)}{\prod_x [1 + \exp \sum_z \varphi(x, z)a(x)]^{m(x)}}.$$

Thus we have shown that $\{t(z)\}$ is jointly sufficient for the 2^n parameters $\{a(z)\}$.

Since a set of sufficient statistics exists which is actually a minimal set, we can reduce the distribution of $\mathbf{Y}$ given by (4.13) to the joint distribution of the sufficient statistics $\{t(x)\}$. Consequently, if $\mathbf{t}$ is the vector of sufficient statistics, we have

$$(4.16) \qquad p(\mathbf{t}) = P\{\mathbf{T} = \mathbf{t}\} = \frac{C(\mathbf{t}) \exp \sum_z t(z)a(z)}{\prod_x [1 + \exp \sum_z \varphi(x, z)a(z)]^{m(x)}},$$

where $C(\mathbf{t})$ is a quantity dependent on $\mathbf{t}$ and *not* on $\{a(z)\}$ which makes $\Sigma_t\, p(\mathbf{t}) = 1$.

In order to find $C(\mathbf{t})$, note that

$$(4.17) \qquad \sum_t C(\mathbf{t}) \exp \sum_z t(z)a(z) = \prod_x [1 + \exp \sum_z \varphi(x, z)a(z)]^{m(x)}.$$

Letting $\xi(z) = \exp\{a(z)\}$ results in (4.17) being written as

$$(4.18) \qquad \sum_t C(\mathbf{t}) \prod_z \xi(z)^{t(z)} = \prod_x [1 + \prod_z \xi(z)^{\phi(x, z)}]^{m(x)}.$$

Thus

$$(4.19) \qquad \Phi(\xi) = \prod_x [1 + \prod_z \xi(z)^{\phi(x, z)}]^{m(x)}$$

is the generating function which enables the coefficients $C(\mathbf{t})$ to be found.

Expanding $\Phi(\xi)$ gives

$$(4.20) \qquad \Phi(\xi) = \prod_x \sum_{r(x)=0}^{m(x)} \binom{m(x)}{r(x)} \prod_z \xi(z)^{\Sigma_x \phi(x, z)r(x)}$$

and therefore setting

$$(4.21) \qquad t(z) = \sum_x \varphi(x, z)r(x),$$

we have

$$(4.22) \qquad C(\mathbf{t}) = \prod_x \binom{m(x)}{r(x)},$$

where the $\{r(x)\}$ in (4.22) are replaced by solving the 2^n simultaneous equations (4.21) for $\{r(x)\}$.

We shall now solve the system of linear equations (4.21) for $\{r(x)\}$. Writing $t(z)$ as

$$(4.23) \qquad t(z) = \sum_x \varphi(x, z) r(x) = \sum_x \prod_z \left(\frac{1 - x_i}{x_i} \right)' \delta^z r(x)$$

$$= [\delta^z]' \sum_x \left[\left(\frac{1 - x_1}{x_1} \right) \times \left(\frac{1 - x_2}{x_2} \right) \times \cdots \times \left(\frac{1 - x_n}{x_n} \right) \right] r(x)$$

and substituting

$$(4.24) \qquad r(x) = \left[\left(\frac{1 - x_1}{x_1} \right) \times \left(\frac{1 - x_2}{x_2} \right) \times \cdots \times \left(\frac{1 - x_n}{x_n} \right) \right]' \mathbf{r}$$

where $\mathbf{r}$ is the column vector of $\{r(x)\}$ arranged in standard order, results in

$$(4.25) \quad t(z) = [\delta^z]' \sum_x \left[\left(\frac{1 - x_1}{x_1} \right) (1 - x_1, x_1) \right.$$

$$\left. \times \left(\frac{1 - x_2}{x_2} \right) (1 - x_2, x_2) \times \cdots \times \left(\frac{1 - x_n}{x_n} \right) (1 - x_n, x_n) \right] \mathbf{r}.$$

Since

$$(4.26) \qquad \left(\frac{1 - x_i}{x_i} \right) (1 - x_i, x_i) = \begin{cases} \begin{pmatrix} 1 & 0 \\ 0 & 0 \end{pmatrix} & \text{if } x_i = 0, \\ \begin{pmatrix} 0 & 0 \\ 0 & 1 \end{pmatrix} & \text{if } x_i = 1, \end{cases}$$

the matrix of order 2^n

$$(4.27) \qquad \left[\left(\frac{1 - x_1}{x_1} \right) (1 - x_1, x_1) \right.$$

$$\left. \times \left(\frac{1 - x_2}{x_2} \right) (1 - x_2, x_2) \times \cdots \times \left(\frac{1 - x_n}{x_n} \right) (1 - x_n, x_n) \right]$$

consists of all zeros except for a single entry of unity on the main diagonal which is in the same position as $x = (x_1, x_2, \cdots, x_n)$ is in the standard order. Therefore

$$(4.28) \qquad \sum_x \left[\left(\frac{1 - x_1}{x_1} \right) (1 - x_1, x_1) \right.$$

$$\left. \times \left(\frac{1 - x_2}{x_2} \right) (1 - x_2, x_2) \times \cdots \times \left(\frac{1 - x_n}{x_n} \right) \right] = I$$

and thus

$$(4.29) \qquad\qquad t(z) = [\delta^z]' \mathbf{r}.$$

Now if the matrix M is defined by

(4.30) $$M = \begin{pmatrix} 1 & -1 \\ 0 & 1 \end{pmatrix},$$

the solution of (4.29) is

(4.31) $$\mathbf{r} = [M \times M \times \cdots \times M]\mathbf{t}.$$

The proof is immediate by substituting (4.31) in (4.29); that is,

(4.32) $$t(z) = [\delta^z]'[M \times M \times \cdots \times M]\mathbf{t}$$

$$= \left[\begin{pmatrix} 1 - z_1 \\ z_1 \end{pmatrix} \times \begin{pmatrix} 1 - z_2 \\ z_2 \end{pmatrix} \times \cdots \times \begin{pmatrix} 1 - z_n \\ z_n \end{pmatrix} \right]' \mathbf{t} = t(z)$$

as $(\delta^{z_i})'M = (1 - z_i, z_i)$. Furthermore, since

(4.33) $$r(x) = \left[\begin{pmatrix} 1 - x_1 \\ x_1 \end{pmatrix} \times \begin{pmatrix} 1 - x_2 \\ x_2 \end{pmatrix} \times \cdots \times \begin{pmatrix} 1 - x_n \\ x_n \end{pmatrix} \right]' \mathbf{r},$$

we have

(4.34) $$r(x) = \left[\begin{pmatrix} 1 - x_1 \\ 2x_1 - 1 \end{pmatrix} \times \begin{pmatrix} 1 - x_2 \\ 2x_2 - 1 \end{pmatrix} \times \cdots \times \begin{pmatrix} 1 - x_n \\ 2x_n - 1 \end{pmatrix} \right]' \mathbf{t}$$

by virtue of

(4.35) $$(1 - x_i, x_i)M = (1 - x_i, 2x_i - 1).$$

Thus, we have shown that the coefficient $C(\mathbf{t})$ in (4.16) is given explicitly by $C(\mathbf{t}) = \Pi_x\binom{m(x)}{r(x)}$ where $r(x)$ is found from (4.34).

Another way of viewing this analysis problem is to consider the data arranged in 2^{n-1} 2×2 contingency tables where each table records the number of successes and failures for factor A_1 at its two levels. The 2^{n-1} different tables correspond to all possible combinations of the remaining factors $A_2, A_3, \cdots, A_n$. An $(n - 1)$ digit binary number, say $y = (y_2, y_3, \cdots, y_n)$, can be used to denote a particular combination of the $(n - 1)$ factors. It may also be convenient to number the 2^{n-1} contingency tables in the base ten number system. For this purpose let the ith table, corresponding to the combination $y = (y_2, y_3, \cdots, y_n)$, be

(4.36) $$i = y_2 2^{n-2} + y_3 2^{n-3} + \cdots + y_n,$$

where $i = 0, 1, \cdots, N$ ($N = 2^{n-1} - 1$). Then from (4.14) we have

(4.37)
$$t(0, y) = \sum_x \varphi(x, 0, y)s(x)$$
$$= \sum_x \varphi(x, 0, y)[s(0, x_2, \cdots, x_n) + s(1, x_2, \cdots, x_n)],$$
$$t(1, y) = \sum_x \varphi(x, 1, y)s(x) = \sum_x \varphi(x, 1, y)[s(1, x_2, \cdots, x_n)].$$

Thus we have that an equivalent set of sufficient statistics are $\{s(0, y) + s(1, y)\}$ and $\{s(1, y)\}$ where y ranges over all $(n - 1)$ digit binary numbers. It will be convenient to let

$$t_i = s(0, y) + s(1, y) = \text{total number of successes in the } i\text{th contingency table,}$$

$$s_i = s(1, y) = \text{total successes for level one of factor } A_1 \text{ in the } i\text{th contingency table,}$$

(4.38)

$$m_i = m(0, y) = \text{total number of trials in the } i\text{th contingency table for level zero of factor } A_1,$$

$$n_i = m(1, y) = \text{total number of trials in the } i\text{th contingency table for level one of factor } A_1.$$

Using the above change of notation, $C(\mathbf{t})$ may be written as

$$(4.39) \qquad C(\mathbf{t}) = \prod_{i=0}^{N} C(s_i, t_i), \qquad\qquad N = 2^{n-1} - 1,$$

where

$$(4.40) \qquad C(s_i, t_i) = \binom{m_i}{t_i - s_i}\binom{n_i}{s_i}.$$

Note that $t(1, y)$ (4.37) can be written as

$$(4.41) \qquad t(1, y) = [\delta_2^{y_2} \times \cdots \times \delta_n^{y_n}]' \mathbf{S}_1,$$

where $\mathbf{S}_1' = (s_0, s_1, \cdots, s_N)$. However (4.41) is the same form as (4.34). Consequently, the solution of the $\mathbf{S}_1$ vector is

$$(4.42) \qquad \mathbf{S}_1 = [M \times \cdots \times M]\mathbf{t}_1,$$

where $\mathbf{t}_1$ is the vector of $\{t(1, y)\}$. Also the analogue of (4.34) is

$$(4.43) \qquad s_i = s(1, y) = \left[\binom{1 - y_2}{2y_2 - 1} \times \binom{1 - y_3}{2y_3 - 1} \times \cdots \times \binom{1 - y_n}{2y_n - 1}\right]' \mathbf{t}_1.$$

Similarly $t(0, y)$ (4.37) may be written as

$$(4.44) \qquad \mathbf{t}_0 = [\delta_2^{y_2} \times \delta_2^{y_3} \times \cdots \times \delta_n^{y_n}]' \mathbf{t},$$

where $\mathbf{t}_0$ is the vector of $\{t(0, y)\}$ and $\mathbf{t}' = (t_0, t_1, \cdots, t_N)$. Hence $\mathbf{t} = [M \times M \times \cdots \times M]\mathbf{t}_0$.

4.3. *Conditional test of significance for interactions.* The analysis of the set of the 2^{n-1} contingency tables proceeds by testing the highest order interaction. If the inference is made that this interaction exists, the 2^{n-1} contingency tables are partitioned into two sets, each of 2^{n-2} tables each. An independent analysis is done on each set, first testing for the highest order interaction. On the other hand, if the inference is made that the n factor interaction is zero, the next step in the analysis is to make inferences on the n interactions involving $(n - 1)$

factors, assuming the n factor interaction is zero. The analysis proceeds in this way, partitioning the set of tables whenever the highest order interaction is real and testing the next lowest order interactions if the highest order interaction is negligible. In this section we exhibit the appropriate tests of significance for carrying out the necessary significance tests. The significance tests are all based on conditional reference sets and are parameter free in the same sense as the Fisher-Irwin analysis of the 2×2 contingency table.

Let x be a given n digit binary number and let $\bar{t}(x)$ denote the $2^n - 1$ vector of the $\{t(z)\}$ excluding $t(x)$. Then we can write $\mathbf{t}$ as $\mathbf{t} = (\bar{t}(x), t(x))$. Using (4.16), we have

$$(4.45) \qquad p(t(x)|\bar{t}(x)) = P\{T(x) = t(x)|\bar{T}(x) = \bar{t}(x)\}$$

$$= \frac{C(\bar{t}(x), t(x)) \exp \{t(x)a(x)\}}{\sum_{t(x)} C(\bar{t}(x), t(x)) \exp \{t(x)a(x)\}},$$

where the summation in the denominator is over the range of $t(x)$. When the hypothesis $H_0: a(x) = 0$ is true, then (4.45) becomes

$$(4.46) \qquad p_0(t(x)|\bar{t}(x)) = \frac{C(\bar{t}(x), t(x))}{\sum_{t(x)} C(\bar{t}(x), t(x))}.$$

Thus $p_0(t(x)|\bar{t}(x))$ can be used to carry out a test of significance for the null hypothesis $H_0: a(x) = 0$. The test may be one sided or two sided depending on the alternative hypothesis.

The test for the highest order interaction corresponds to taking $x = (1, 1, \cdots, 1)$. Using this value in (4.14) we find that

$$(4.47) \qquad t(1, 1, \cdots, 1) = s(1, 1, \cdots, 1) = \text{total number of ones (or successes) at factorial combination where all factors are at the upper level.}$$

Hence, the appropriate distributions for carrying out inferences on $a(1, 1, \cdots, 1)$ are (4.45) and (4.46).

If one concludes that the n factor interaction is zero, then the next step in the analysis is to make an inference on the interactions involving $(n-1)$ factors conditional on the n factor interaction being zero. Let $\mathbf{1}$ be a row vector having n elements. Then the marginal distribution of $\bar{T}(\mathbf{1})$ may be written

$$(4.48) \qquad P\{\bar{T}(\mathbf{1}) = \bar{t}(\mathbf{1})\} = \frac{\sum_{t(\mathbf{1})} C(\bar{t}(\mathbf{1}), t(\mathbf{1})) \exp \sum_z t(z)a(z)}{\prod_x [1 + \exp \sum_z \varphi(x, z)a(z)]^{m(x)}},$$

and the conditional distribution of $T(x)$ (corresponding to $a(x)$) is

$$(4.49) \qquad P\{T(x) = t(x)|\bar{T}(x) = \bar{t}(x) \text{ excluding } T(\mathbf{1}) = t(\mathbf{1})\}$$

$$= \frac{\{\sum_{t(\mathbf{1})} C(\bar{t}(\mathbf{1}), t(\mathbf{1}))\} \exp \{t(x)a(x)\}}{\sum_{t(x)} \{\sum_{t(\mathbf{1})} C(\bar{t}(\mathbf{1}), t(\mathbf{1}))\} \exp \{t(x)a(x)\}}.$$

The distribution under the null hypothesis $H_0: a(x) = 0$ is obtained from (4.49) by setting $a(x) = 0$.

The general procedure for making inferences on lower order interactions is clear. A test on an interaction among p factors is only carried out if all higher order interactions involving the p factors are assumed to be zero. That is, define:

$$(4.50) \quad \mathscr{A}_p = \left\{ x : a(x) = 0, \sum_{i=1}^{n} x_i > p \right\} = \text{subset of all } n \text{ digit binary numbers}$$

associated with interactions involving at least $(p + 1)$ factors;

$$(4.51) \quad \bar{\mathscr{A}}_p = \text{complement of } \mathscr{A}_p;$$

$$(4.52) \quad \mathscr{B}(a(x)) = \{ x_i : x_i = 1, a(x) \} = \text{set of } x_i \text{ elements for which } x_i = 1 \text{ in } a(x);$$

$$(4.53) \quad \mathscr{C}_p = \bigcap_{x \in \mathscr{A}} \mathscr{B}(a(x)) = \text{subset of } x_i \text{ elements which are equal to unity in } \mathscr{A}_p.$$

Then if $a(x)$ corresponds to an interaction involving p factors such that all $x_i = 1$ in $a(x)$ belong to $\mathscr{C}_p$ we have

$$(4.54) \quad P\{T(x) = t(x) | T(y) = t(y), y \neq x \text{ and all } y \in \bar{\mathscr{A}}_p\}$$

$$= \frac{\{\sum \cdots \sum_{t(z)} C(\mathbf{t})\} \exp t(x)a(x)}{\sum_{t(x)} \{\sum \cdots \sum_{t(z)} C(\mathbf{t})\} \exp t(x)a(x)},$$

where the summation in brackets in both numerator and denominator is over the range of $t(z)$ for all $z \in \mathscr{A}_p$.

4.4 *The analysis for four 2×2 tables $(n = 3)$.* In this section we illustrate how special cases can easily be obtained from the general results of the preceeding sections. We shall take the case $n = 3$ corresponding to four 2×2 tables. The first step in the analysis is to obtain the $C(\mathbf{t})$ coefficients using (4.40) and (4.41). The identification between the binary and base ten notation for two digit numbers is given in Table I.

TABLE I

IDENTIFICATION BETWEEN BINARY AND BASE TEN NOTATION

Binary (y)	Base Ten (i)
$(0, 0)$	0
$(0, 1)$	1
$(1, 0)$	2
$(1, 1)$	3

Let the four tables be identified with the above indices. Then if s_i is the total success for level one of A_1 and t_i is the total number of successes for table i, $i = 0, 1, 2, 3$. We easily calculate from (4.41) and (4.43),

$$(4.55) \qquad t_1(0) = t(1, 0, 0) = \sum_{i=0}^{3} s_i, \; t_1(2) = t(1, 1, 0) = s_2 + s_3,$$

$$t_1(1) = t(1, 0, 1) = s_1 + s_3, \; t_1(3) = t(1, 1, 1) = s_3,$$

and

$$
\begin{aligned}
s_0 &= t_0(1) - t_1(1) - t_1(2) + t_1(3), \\
s_1 &= t_1(1) - t_1(3), \\
s_2 &= t_2(1) - t_1(3), \\
s_3 &= t_1(3).
\end{aligned}
$$

(4.56)

Therefore, from (4.39) and (4.40) we have

$$(4.57) \qquad\qquad C(\mathbf{t}) = \prod_{i=0}^{3} C(s_i, t_i),$$

where (4.56) gives the values of s_i in terms of $t_1(i)$. Thus, the conditional probability distribution associated with the test of the highest order interaction is

$$(4.58) \quad P\Big\{ T(1, 1, 1) = s_3 \Big| T_i = t_i (i = 0, 1, 2, 3), S_2 + S_3 = t_1(2),$$

$$S_1 + S_3 = t_1(1), \sum_{i=0}^{3} S_i = t_1(0) \Big\}$$

$$= \frac{\displaystyle\prod_{i=0}^{3} C(s_i, t_i) \exp\{s_3 a(1, 1, 1)\}}{\displaystyle\sum_{s_3} \prod_{i=0}^{3} C(s_i, t_i) \exp\{s_3 a(1, 1, 1)\}}.$$

Tail area probabilities for $H_0 : a(1, 1, 1) = 0$ may be calculated from (4.58) by setting $a(1, 1, 1) = 0$.

If one concludes that the $a(1, 1, 1)$ interaction is zero, the next step in the analysis is to make an inference on the conditional interactions involving two factors; that is, $a(1, 1, 0)$, $a(1, 0, 1)$, and $a(0, 1, 1)$. We shall illustrate the appropriate conditional test for the $a(1, 1, 0)$ and $a(0, 1, 1)$ interactions. Using (4.49) with $x = (1, 1, 0)$, we have

$$(4.59) \quad P\Big\{ T_1(2) = t_1(2) \Big| T_i = t_i (i = 0, 1, 2, 3), S_1 + S_3 = t_1(1), \sum_{i=0}^{3} S_i = t_1(0) \Big\}$$

$$= \frac{\Big\{ \displaystyle\sum_{s_3} \prod_{i=0}^{3} C(s_i, t_i) \Big\} \exp\{t_1(2) a(1, 1, 0)\}}{\displaystyle\sum_{t_1(2)} \Big\{ \sum_{s_3} \prod_{i=0}^{3} C(s_i, t_i) \Big\} \exp\{t_1(2) a(1, 1, 0)\}}.$$

Note that both the inference on $a(1, 1, 1)$ and $a(1, 1, 0)$ were made conditional on $T_i = t_i$, $i = 0, 1, 2, 3$. On the other hand, the inference for $a(0, 1, 1)$ is conditional on the $S_i = s_i$, $i = 0, 1, 2, 3$. The appropriate conditional test for $H_0: a(0, 1, 1) = 0$ is obtained by finding the probability of $T(0, 1, 1) = t(0, 1, 1)$ conditional on $S_i = s_i, i = 0, 1, 2, 3$, $T(0, y) = t(0, y)$ for $y = (0, 0), (0, 1), (1, 0)$.

From (4.37) we have

$$t_0(0) = t(0, 0, 0) = \sum_{i=0}^{3} t_i, \qquad t_0(2) = t(0, 1, 0) = t_2 + t_3$$

(4.60)

$$t_0(1) = t(0, 0, 1) = t_1 + t_3, \qquad t_0(3) = t(0, 1, 1) = t_3.$$

Also, solving for the $\{t_i\}$ terms of $\{t_0(i)\}$ results in

(4.61)
$$
\begin{aligned}
t_0 &= t_0(0) - t_0(1) - t_0(2) + t_0(3), \\
t_1 &= t_0(1) - t_0(3), \\
t_2 &= t_0(2) - t_0(3), \\
t_3 &= t_0(3).
\end{aligned}
$$

The conditional distribution of T_3 is

$$(4.62) \quad P\left\{ T_3 = t_3 \,\middle|\, S_i = s_i(i = 0, 1, 2), \sum_{i=0}^{3} T_i = t_0(0),\right.$$

$$\left. T_1 + T_3 = t_0(1), T_2 + T_3 = t_0(2)\right\}$$

$$= \frac{\sum\limits_{s_3} \prod\limits_{i=0}^{3} C(s_i, t_i) \exp\{t_3 a(0, 1, 1)\}}{\sum\limits_{t_3} \sum\limits_{s_3} \prod\limits_{i=0}^{3} C(s_i, t_i) \exp\{t_3 a(0, 1, 1)\}},$$

where the t_i is replaced in the $\{C(s_i, t_i)\}$ by (4.61). The test of significance is obtained by setting $a(0, 1, 1) = 0$ in (4.62) and calculating the appropriate tail probability.

$$\diamond \qquad \diamond \qquad \diamond \qquad \diamond \qquad \diamond$$

Added in proof. The recently published book by D. R. Cox, *Analysis of Binary Data*, London, Methuen, 1970, summarizes much of the results of [2] and [3] as well as several generalizations.

REFERENCES

[1] M. S. BARTLETT, "Contingency table interactions," *J. Roy. Statist. Soc. Supplemnt*, Vol. 2 (1935), pp. 248–252.

[2] D. R. COX, "The regression analysis of binary sequences," *J. Roy. Statist. Soc. Ser. B*, Vol. 20 (1958), pp. 215–242.

[3] ———, "Some procedures connected with the logistic qualitative response curve," *Research Papers in Statistics* (edited by F. N. David), New York, Wiley, 1966.

[4] L. A. GOODMAN, "The multivariate analysis of qualitative data: interactions among multiple classifications," *J. Amer. Statist. Assoc.*, Vol. 65 (1970), pp. 226–256.

[5] HARVARD UNIVERSITY, *Tables of the Cumulative Binomial Probability Distribution*, Cambridge, Harvard University Press, 1955.

[6] H. H. KU and S. KULLBACK, "Interaction in multidimensional contingency tables: an information theoretic approach," *J. Res. Nat. Bur. Standards Section B*, Vol. 72 (1968), pp. 159–199.

[7] S. KULLBACK, M. KUPPERMAN, and H. H. KU, "An application of information theory to the analysis of contingency tables, with a table of $2n \log n$, $n = 1(1)10,000$," *J. Res. Nat. Bur. Standards Section B*, Vol. 66 (1962), pp. 217–243.

[8] B. KURKJIAN and M. ZELEN, "A calculus for factorial arrangements," *Ann. Math. Statist.*, Vol. 33 (1962), pp. 600–619.

[9] E. L. LEHMANN, *Testing Statistical Hypotheses*, New York, Wiley, 1959, Chapter 4.

[10] B. N. LEWIS, "On the analysis of interaction in multidimensional contingency tables," *J. Roy. Statist. Soc. Ser. A*, Vol. 125 (1962), pp. 88–117.

[11] R. L. PLACKETT, "Multidimensional contingency tables: a survey of models and methods," *Proceedings of the 37th Session, Bulletin of the I.S.I.*, Vol. 43, Book 1 (1969), pp. 133–142.

AUTHOR REFERENCE INDEX

This index includes authors from Volumes I, II, III only. A reference index for Volumes IV, V, and VI is contained in those Volumes.